AUTOMOTIVE MECHANICS

EIGHTH EDITION

William H. Crouse

Gregg Division
McGraw-Hill Book Company

New York
St. Louis
Dallas
San Francisco
Auckland
Bogotá
Düsseldorf
Johannesburg
London
Madrid
Mexico
Montreal
New Delhi
Panama
Paris
São Paulo
Singapore
Sydney
Tokyo
Toronto

ABOUT THE AUTHOR

William H. Crouse

Behind William H. Crouse's clear technical writing is a background of sound mechanical engineering training as well as a variety of practical industrial experience. He has worked in General Motors plants and as Director of Field Education in the Delco-Remy Division of General Motors Corporation.

He has written a number of technical manuals for the Armed Forces and served as editor of technical books for the McGraw-Hill Book Company. He has contributed numerous articles to automotive and engineering magazines and has written many books about science and technology.

William H. Crouse's outstanding work in the automotive field has earned for him membership in the Society of Automotive Engineers and the American Society of Engineering Education.

Library of Congress Cataloging in Publication Data

Crouse, William Harry, (date)
Automotive mechanics.

Includes index.
1. Automobiles. 2. Automobiles—Maintenance and repair. 3. Automobiles—Testing. I. Title.
TL205.C86 1980 629.2'3 79-12845
ISBN 0-07-014820-1

AUTOMOTIVE MECHANICS, eighth edition

34567890 DODO 78654321

The editors for this book were Eugene Gilmore and Myrna Breskin, the interior designer was Lisa Delgado, the cover designer was Nancy Axelrod, the art supervisor was George T. Resch, and the production supervisor was Kathleen Morrissey. The cover photographs were by Belott/Wolfson Photography, Inc., and the technical art was by Vantage Art, Inc. It was set in Trade Gothic Light by York Graphic Services, Inc.
Printed and bound by R.R. Donnelley & Sons Company.

Contents

PART EIGHT AUTOMOTIVE POWER TRAINS 453

PART NINE AUTOMOTIVE CHASSIS 519

PART TEN AUTOMOTIVE SAFETY DEVICES AND SAFETY INSPECTION 609

Preface

This is the eighth edition of *Automotive Mechanics.* In the five years since the publication of the last edition, there have been many important developments in automotive vehicles. New federal legislation on automotive safety, emission controls, and fuel economy have required automotive engineers to develop numerous new devices and systems to meet legal requirements. The automotive industry has met the challenges imposed by these new regulations.

This eighth edition of *Automotive Mechanics* is designed to help automotive mechanics teachers to also meet the challenges of these regulations. It covers the new developments in engine and emission controls, front-drive vehicles, automotive diesel engines, and suspension systems. Other important additions to the book are: new material on safety in the shop; new carburetors including the latest variable-venturi units; variable-displacement engines; the new General Motors V-8 diesel engine (with a chapter on service); electronic ignition systems; electronic spark control such as the Chrysler Lean Burn system; the Ford Electronic Control (FEC) system; the General Motors Microprocessing Sensing Automatic Regulation (MISAR) system; stratified charge; electronic control of air-fuel ratios such as the General Motors Electronic Fuel Control (EFC) system; the Ford Feedback Carburetor Electronic Engine Control system; three-way catalytic converters; gasoline and diesel fuel-injection systems; MacPherson front suspension; new tire markings; and many other innovations the student must understand.

This edition continues the feature of including the metric equivalents of all United States Customary measurements. The previous edition of *Automotive Mechanics* was the first completely metricated book in the field.

In addition to the new developments covered in this edition, textual material has been largely rewritten to simplify explanations, shorten sentences, and improve readability. Objectives have been added at the beginning of each chapter so that the student and teacher both know what the learning outcome should be.

Several ancillary materials have been developed for use with *Automotive Mechanics*. These include a study guide, a shop workbook, a testbook, and an instructor's planning guide. The book is also correlated with the eight McGraw-Hill Automotive Transparencies books, eight sets of automotive room charts (nearly 200 charts in all), and a set of trouble-shooting cards.

All these materials, with the book, provide the instructor with a flexible teaching package that should fit any type of teaching situation. The instructor's planning guide explains how the various materials can be used, either alone or with others, to meet any teaching requirement.

The *Automotive Mechanics* program is correlated with the latest recommendations of the Motor Vehicle Manufacturers Association—American Vocational Association Industry Planning Council. The program also incorporates new guidelines for Automotive Mechanics certification, state plans for vocational education, and recommendations for automotive trade preapprenticeship and apprenticeship training. The program is flexible; it will fit classroom instruction, shop activities, individual instructions, or "how-to" courses for hobbyists and consumers.

This edition of *Automotive Mechanics* meets, with few exceptions, the standards set by the Motor Vehicle Manufacturers Association for an associate degree in automotive servicing and in service management. These standards are described in their booklet "Community College Guide for Associate Degree Programs in Auto and Truck Service and Management." *Automotive Mechanics 8/e* also covers the subjects recommended by the American National Standards Institute in their detailed standard D18.1-1972 "American National Standards for Training of Automotive Mechanics for Passenger Cars and Light Trucks."

In addition, *Automotive Mechanics* covers in depth the subjects tested by the National Institute for Automotive Service Excellence (NIASE). These tests are used for certifying general automotive mechanics and other automotive service technicians working in specific areas of specialization under the NIASE voluntary mechanic testing and certification program.

During the planning and preparation of the eighth edition of *Automotive Mechanics,* the author and publisher had the advice and assistance of many, many people—educators, researchers, artists, editors, and automotive industry service specialists. It would take pages to acknowledge them all individually but special thanks must go to the following: Donald W. Patten, Lou Salvadore, and John Steck who took time from their busy schedules to attend the master planning sessions for the new edition; and to Frank C. Derato and George Whitehouse, who spent many hours reviewing the manuscript and offering suggestions for improving it. The author gratefully acknowledges his indebtedness and offers his sincere thanks to those many people. Above all, the author acknowledges with deep gratitude the devoted assistance of Ruth, his wife, who through the production of more than 50 books served efficiently and untiringly as associate producer.

William H. Crouse

To the Student

This book—*Automotive Mechanics*—was designed with you, the student, in mind. It was put together to help you learn, in the quickest and easiest way possible, all about automobiles: how they are constructed, how they operate, and how to service and repair them. *Automotive Mechanics* covers all the fundamentals. By the time you have finished studying this book, and have done the related shop work, you will be ready to enter the world of automotive servicing as a wage earner. This means that now, as you do your job of studying and working in the school shop, you are getting ready for a job in the automotive service business. Various special materials have been developed to make learning about the automobile easier. These special materials include a study guide, a shop workbook, a testbook, automotive room charts, and a set of troubleshooting review cards. You may not have the chance to use all of these because some schools do not have them available. But whatever is available, remember that it was prepared with one thought in mind—to help you learn all about automobiles. Whenever you use one of these special materials, you will be told how to use it and how to get the most out of using it.

Your major job now is to study this book and to do your shop work. Studying is usually hard for everyone. It takes will power to sit down and read this technical material. But if you follow the suggestions listed below, you will find studying much easier. You will also be making good progress in your automotive mechanics course.

1. The first thing to do when you pick up your textbook to study your assignment is to turn the pages one by one. Look at the pictures. Read the numbered section headings. Study each section heading carefully. This will give you an idea of what the assignment is about.
2. If you are starting a new chapter in the textbook, read the objective that tells you what you should be able to do after you have studied the chapter. This emphasizes what you should learn from the chapter. Also read the introductory paragraph at the start of the chapter. It tells you more about what you are going to study in the chapter.
3. Read the first section in your assignment. Then read the first section again slowly and carefully so that you make sure you understand it.
4. Continue studying the pages assigned to you. Read each section carefully.
5. If you come to a sentence that you don't understand, read it aloud. Think about it. If this does not help, write the sentence on a piece of paper. When you have a chance, ask your instructor to explain the sentence to you.
6. Don't hesitate to admit that something puzzles you. Everybody gets stuck once in a while. Your instructor is there to help you understand.
7. Don't worry about not getting everything the first time you read it. Most good students read and reread their lessons several times. That is the way to make it stick! Reread it!
8. If you feel yourself getting sleepy, or if your attention begins to wander—wake up! Get up. Stretch. Put cold water on your face. Have a cup of coffee or a soft drink. Then get back to work!

1

Automobiles and Automotive Service

Part One of *Automotive Mechanics* is your introduction to the automobile and to automotive service. The automobile is a marvelous machine. But like all machines, it requires periodic service to keep going. And this is where you, as an automotive mechanic or technician, come in. The business of servicing automobiles is huge and varied. Once you have learned the fundamentals of servicing automobiles, you will have many opportunities for a good future in the automotive service business. We describe these opportunities in this part, and discuss safety in the shop. There are three chapters in Part One of *Automotive Mechanics.*

Chapter 1 Introduction to the Automobile

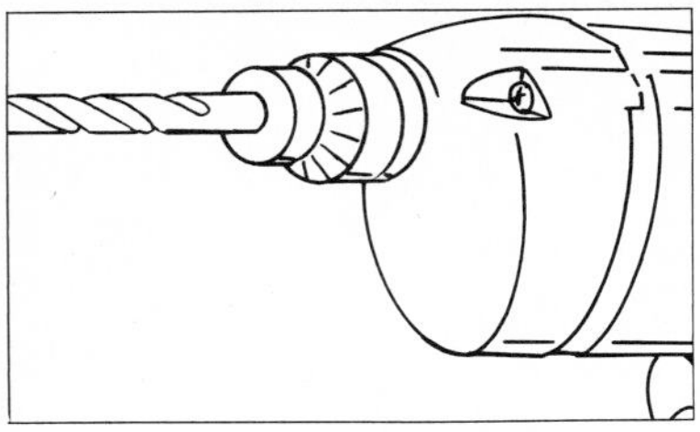

Chapter 2 Shop Work and Shop Manuals

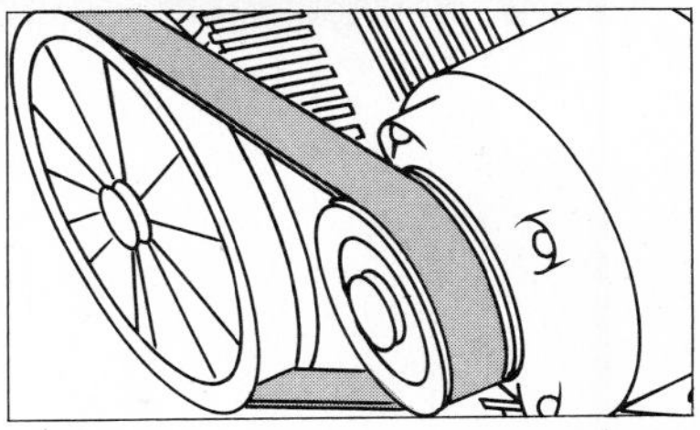

Chapter 3 Safety in the Shop

CHAPTER 1
Introduction to the Automobile

After reading this chapter, you should be able to:

- **Define interchangeability and mass production.**
- **List eight opportunities in the automotive service business.**
- **List the various types of automotive service business enterprises.**
- **List the four basic components in the automobile.**

This chapter discusses the history of gasoline automobiles and the development of the automotive service business.

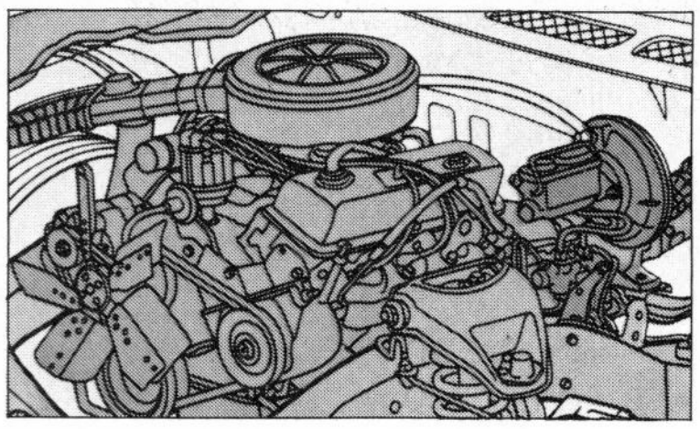

1-1 The First Gasoline Automobiles

Automobiles have been around only a hundred years or so. The first automobiles were very crude affairs. They were little more than old-fashioned buggies with engines. Naturally, they were called "gas buggies." The early engines had one cylinder that could wheeze out only a horsepower or two. (A horsepower is the power of one horse.)

Around 1885, Daimler and Benz were building their gas buggy in Germany. Later, French firms were building automobiles. In 1893, the Duryea brothers built the first American car. By 1895, Henry Ford, Ransom Olds, and others were building automobiles. They were very crude compared to today's cars, but they ran—most of the time.

By 1900, there were several factories, in Detroit and elsewhere, making automobiles. The engineers who set up these factories used two basic ideas—interchangeability and mass production.

1. Interchangeability The idea of interchangeability was almost unknown in 1900. Before then, many of the parts for machines were handmade. Each was different from the others. It could fit only the machine it was made for. So each part had to be hand-fitted to its machine. This was a long and costly process. When a part wore out—a gear, for example—a new one had to be made by hand.

But with interchangeability, parts are made to be exactly alike. For example, all connecting rods for one model engine are made the same. Then any of these connecting rods will fit any engine of that model. Likewise the valves and other engine parts are made to be the same. Also, automatic machines make bolts, nuts, washers, and other small parts so that all are alike. All this requires machinery that can turn out thousands of the same part. Then, when an engine is put together, everything fits with very little extra work.

2. Mass production Along with interchangeability came the idea of mass production. If all similar parts are interchangeable, you do not have to hand-fit each part to the machine. The job of assembling an engine, for example, is simplified. The engine block is put on a moving belt which is in the center of the assembly line. As the belt moves along, workers and automatic machines add pistons, rods, crankshafts, and other parts. At the end of the line, the engines are complete, ready to be tested and installed in cars.

1-2 The Automotive Service Business

Today, more than 133 million automobiles, trucks, and buses operate on the streets and highways of the United States. There are millions more off-the-road vehicles, such as farm machines, power mowers, construction equipment, mining equipment, dune buggies, snowmobiles, and motorboats. More than a million men and women work to keep all this equipment going. The automotive service business includes:

- Dealerships where cars are sold and serviced.
- Service stations, where the cars get gas and oil and related products and services.
- Tire and battery dealers.
- Independent garages.
- Specialty shops, where front alignment, transmissions, body work, carburetors, engine rebuilding, and ignition work are handled.
- Fleet garages, where truck and bus lines have their own service shops.
- Parts dealers, where automotive parts are sold.

● 1-3 Opportunities in the Automotive Service Business

Automotive service offers many excellent opportunities for men and women who know automobiles. You are making a fine start toward learning about automobiles by studying this book. The better jobs require a knowledge of how the automobile works and how to service and maintain it.

What are some of these job opportunities?

The efficient automotive mechanic is a highly respected, well-paid worker who earns as much as men and women in other skilled professions. The mechanic's services are always in demand. And there is a shortage of good automotive mechanics.

The job of automotive mechanic can be a steppingstone to greater things. The automotive mechanic can become an automotive technician, service manager, or parts manager of an automotive shop or new-car agency. The automotive mechanic might some day want to open his or her own business. Most independent garages, specialty shops, and service stations are owned by automotive mechanics who have moved up the ladder.

Many automotive salespeople, manufacturers' representatives, and even top executives in the factories were once automotive mechanics. The opportunities are almost unlimited. Men and women who know automobiles and have the right personality will achieve success.

All successful men and women have one thing in common, the ability to study and work hard. Success for you lies in the same direction, and you can succeed in the same way. Get everything you can from this book and your courses at school.

This book was written to help you achieve success. It was prepared with the assistance of experts in the automotive service business. The book has one major purpose, to prepare you for an interesting and well-paid job in the automotive business. This is your book to study, to help you become a success in life. Now, let us take a closer look at the opportunities in the various types of automotive service business we listed previously.

● 1-4 The Service Station

There are more than 180,000 service stations in the United States, varying from the small two- or three-pump station adjacent to the country store to the huge complexes you see along the interstate highways. The services they offer are just as varied. Some are simply "filling" stations, where you get gas and oil. Others are equipped to do a complete tune-up job. Some have mechanics on the job who can take care of many of the services a car might need—balance wheels, align front ends, replace brake linings, make minor engine repairs, and the like.

Many service stations carry a line of automotive items such as shock absorbers, seat covers, batteries, ignition-system parts, car polish, horns, and so on. In addition, most service stations sell soft drinks and food snacks. Also, some have a sizable novelty store (Fig. 1-1) which sells commonly used drugs (like aspirin and milk of magnesia), hand creams, toothbrushes and paste, paper products, laxatives, toys, trinkets, paperback books, and other items that the traveling public might buy.

You can see, therefore, that the jobs in service stations vary greatly. In some, your main job would be to meet and greet the public as cars drive in, and to see that they get the gasoline, oil, water, and air they need. Your job is to be pleasant and to see that the driver's needs are satisfied. If your service station has things to sell, you should be alert to possible sales opportunities. Are the tires worn? Is the battery run down? How about the windshield wipers? The shock absorbers? Is it time for an oil and oil-filter change?

Your work in a service station can be quite varied, as you can see. And this can make it very interesting, with

Fig. 1-1 Some service stations have extra facilities in which they sell snacks, gifts, toiletries, and novelties. Service stations on well-traveled highways often find that a good income can be derived from such merchandise. (*Continental Oil Company*)

no two days the same. One hour you might work the service island, pumping gas, checking under the hood, cleaning windshields. The next you might be replacing shock absorbers or setting ignition timing. Or you might be waiting on customers in the novelty shop.

Working in a service station can lead to bigger, more important, and better-paying jobs which could be anything from a trained automotive service technician to a service-station manager or owner.

1-5 Automotive Dealers

There are about 24,000 new-car dealers in the United States. These are franchised dealers. By this, we mean that the dealer has a contract—a written agreement—with the vehicle manufacturer, spelling out the dealer-manufacturer agreement. The manufacturer supplies the dealer with vehicles at wholesale prices. The manufacturer also provides many promotional aids and services. For example:

- Advertising in newspapers, magazines, over the radio, on television.
- Sales promotional aids for the showroom.
- Aid to new dealers in setting up their salesrooms, offices, service and parts departments.
- Technical assistance, service manuals, and schools to train service personnel on new equipment and service procedures.
- A parts distribution network so service parts are readily available when needed.

These are but a few of the many things the manufacturer does for the vehicle dealer.

The dealer, in turn, is obligated to keep all customers satisfied with the vehicles sold and with the service work done in the service shop. The dealer must make sure that all new cars are properly prepared for delivery to customers. Service work on cars driven into the service department by customers must be good so that customers are satisfied. At the same time, the work must make a profit for the department. If trouble develops in a new car, it is the dealer's responsibility to see that it is corrected under the warranty agreement. This agreement is a guarantee that new cars will be trouble-free for a given mileage or time. If trouble does develop, then the dealer must fix it at little or no cost to the car owner. Under the warranty agreement, the cost of parts and labor is paid by the manufacturer.

As you can see, the vehicle dealer operates a complicated business. But it offers many opportunities for the intelligent man or woman who wants to get ahead in the automotive service business. The service mechanic can rise to become a service technician in engine, brake, electrical, suspension and steering, or some other specialty. Another step up the ladder would be service or sales manager and possibly, at last, an automotive dealer. Many of the most prosperous dealers today started out in the service shop.

1-6 Independent Garages

There are an estimated 90,000 independent garages of any size in the United States. These vary from the small shops with only a few workers to the big, all-purpose garage with a large staff of automotive mechanics, parts people, and related personnel. The fact that there are so many independent garages means that many car owners do not go back to the dealer for service. These owners, for one reason or another, prefer to go to an independent garage. Also, many cars are second- or third-hand, bought from a used-car dealer not connected with a new-car dealership. These cars are most apt to end up at an independent garage when service is required.

The independent garage, if large, offers a complete line of automotive service, from tune-ups, alignments, and body work, to complete overhaul of engines, transmissions, differentials, and other major car components. As a result, you can see that the automotive mechanics in these garages work on a great variety of jobs. Here again, the best mechanics rise to the top. Mechanics can become department managers, service managers, and garage managers or owners.

1-7 Specialty Shops

A great variety of specialty shops provide backup service for the automotive service industry. They include shops that do major machine work such as brake-drum or disk turning, crankshaft and camshaft grinding, brake-shoe relining, and the like. Some shops specialize in engine rebuilding, or remanufacturing, as it is also called. Also, there are such places as the AAMCO franchised automotive-transmission shops, which deal basically with one component. Another example is the Midas Muffler shops specializing in replacing mufflers. Some service shops specialize in repairing or rebuilding small components such as alternators, carburetors, and distributors. Some specialty shops are set up to handle body repair and repainting or interior upholstery work.

There are many opportunities in these shops for the person who likes to stay in one place and do one kind of job, rather than hop from one job to another as in a service station. Here again, the intelligent man or woman who knows the job can rise to a better job—shop supervisor, manager, owner.

1-8 Fleet Garages

Small fleet operators with only a few trucks, buses, or other vehicles usually have their service work done at an independent garage equipped to handle their heavy vehicles. However, the larger fleet owner usually sets up a garage to handle service and maintenance of the vehicles in the fleet. Often, the work is done on a preset schedule that calls for periodic checks and servicing of various components. Some items are checked every time the truck or bus goes out on the road. Other items are checked less frequently. The procedure is called *preventive maintenance.* The equipment is *maintained* in good working condition to *prevent* failures. You see how important this is for the trucker hauling frozen food. A breakdown on the highway could result in a loss of thousands of dollars if the food thawed and spoiled. Likewise, a bus breaking down in some remote place could result in many angry passengers and much lost business.

Here again, there are many opportunities for the man or woman who knows the vehicles and can handle the service jobs. Because it is very important to the fleet owner to keep the vehicles in top condition, the pay is unusually high. Top performance means top pay.

1-9 Parts Dealer

The parts dealer operates much like a retail store. The dealer keeps thousands of different automotive service parts on hand. Also, the dealer has contact with a network of distributors and suppliers who can get needed parts quickly. Service stations and garages needing parts order them and the parts dealer delivers them. Often, the garage sends someone in to the parts dealer to pick up parts needed in a hurry.

Working in a parts store is similar to working in any other retail store. You must know the merchandise. This makes being an automotive mechanic very helpful, because the merchandise is all automotive parts. You must find the proper parts numbers in the parts books, know where the parts are located, keep an inventory of what is in stock, and order more parts from the suppliers when the stock gets low. For example, a customer might want a set of main bearings for a 1980 Chevrolet. First you find the part number in the parts catalog (Fig. 1-2). Then you locate the bearings and bring them to the counter. Parts dealers have parts books or catalogs that list and sometimes picture the various parts. These parts books are very large, sometimes 2 feet [0.61 meter] thick.

Today, many parts dealers have replaced the parts books with a microfiche system. Each fiche is a small sheet of microfilm that may have on it hundreds of pages reduced to almost microscopic size (Fig. 1-3). To find a part number on a microfiche, you pick out the correct fiche from the file and put the fiche into the microfiche reader. Then, by operating a control, the fiche is shifted until the reader shows the enlarged page on a screen.

Fig. 1-2 Automotive service parts books can be very large.

Fig. 1-3 A microfiche can contain many pages with thousands of parts listed, reduced to almost microscopic size.

1-10 Department, Accessory, and Automotive-Supply Stores

Many department stores, chains such as Sears Roebuck and K-Mart, and specialty stores offer automotive service. They do a big business in selling all sorts of accessories and necessities from tires and batteries to complete rebuilt engines. And many have complete servicing facilities. There are many job opportunities in these stores, ranging from actual service work to selling automotive parts and accessories. The range of service offered by these stores is large. Thus, there is a great variety of jobs, with many opportunities to advance to better and more responsible positions.

1-11 Components of the Automobile

We have described briefly the various types of business associated with the automotive service business. All of these which we have discussed play some part in keeping the more than 133 million automotive vehicles in the United States rolling. Now let us take an introductory look at the various components in the automobile that may require attention from the automotive mechanic. Later in the book we will describe these components in detail, and find out what makes them work and what to do when they do not work as they should.

There are four basic components of the automobile:

1. The *engine,* or source of power that moves the car.
2. The *framework,* or support for the engine and wheels. This includes the frame and the steering and braking system. The assembly is often called the chassis (Figs. 1-4 and 1-5).

3. The *power train*, which carries the engine power to the car wheels (Fig. 1-6).
4. The *body*, which encloses and protects the passengers (Fig. 1-7).

To these we could add the heater, air conditioner, radio, CB transceivers, lights, and other devices. These accessories add to the safety and comfort of the driver and passengers.

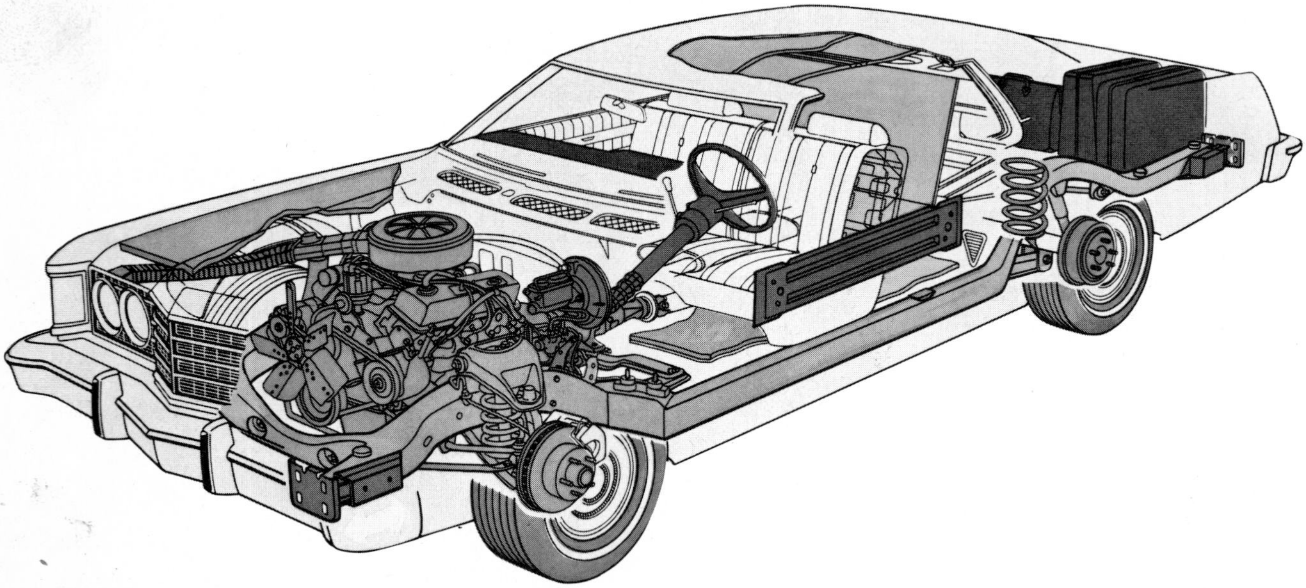

Fig. 1-4 Chassis of a passenger car with the body outlined around it. The chassis contains the source of power, or engine; the frame, which supports the engine, wheels, and body; the power train, which carries the engine power to the wheels; and the steering and braking systems. (*Ford Motor Company*)

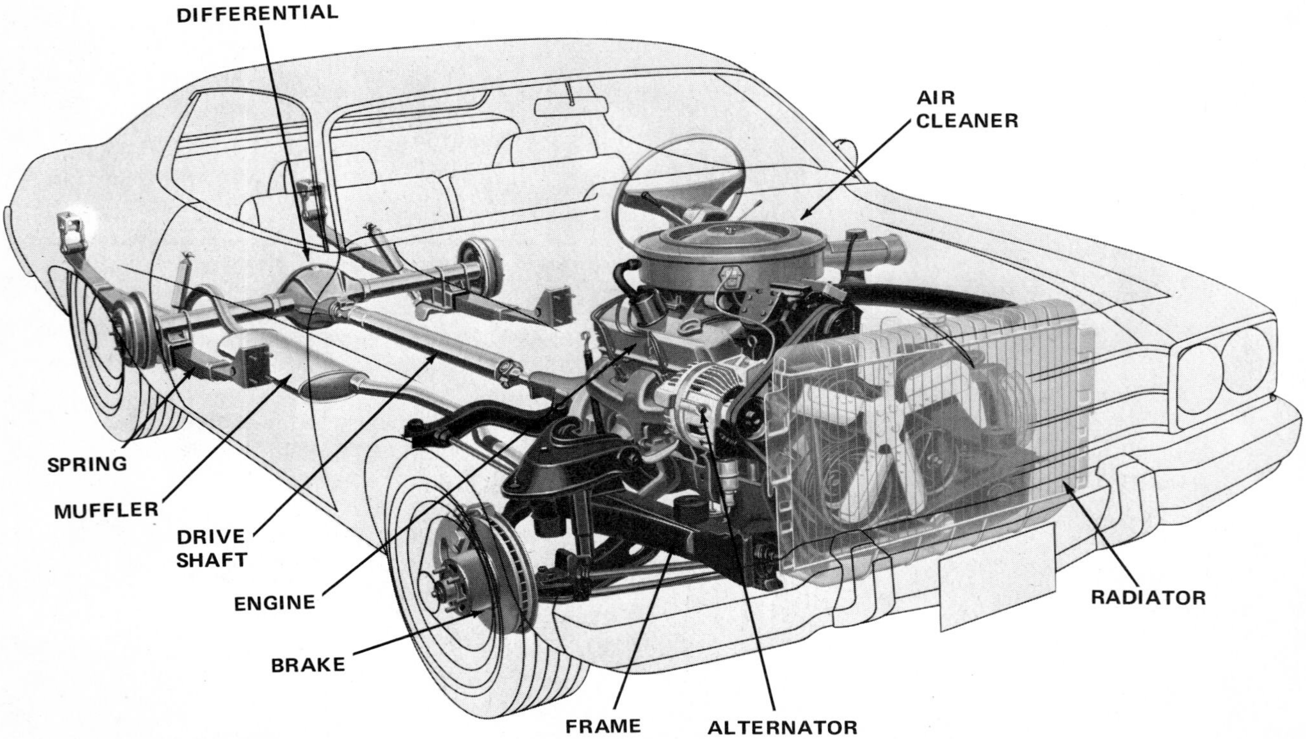

Fig. 1-5 Location of the major components in the automobile.

1-12 The Engine

The engine is the source of power that makes the car move. "Engine" is often confused with "motor." A "motor" is an electrical device that runs on electricity. The automobile engine runs on gasoline. The confusion arises from the fact that some people call automobiles "motor cars," and refer to driving as "motoring." Actually, we should call them "engine cars," and driving as "engining." But then, it is hard to change people's habits. So just remember that the power plant in the car is an engine and the electrical devices in refrigerators and electric fans are motors.

Two major types of engine are used in automobiles today. The first is the piston engine, in which pistons move up and down, or reciprocate. The second is the rotary engine, in which a rotor spins (Fig. 1-8). The piston

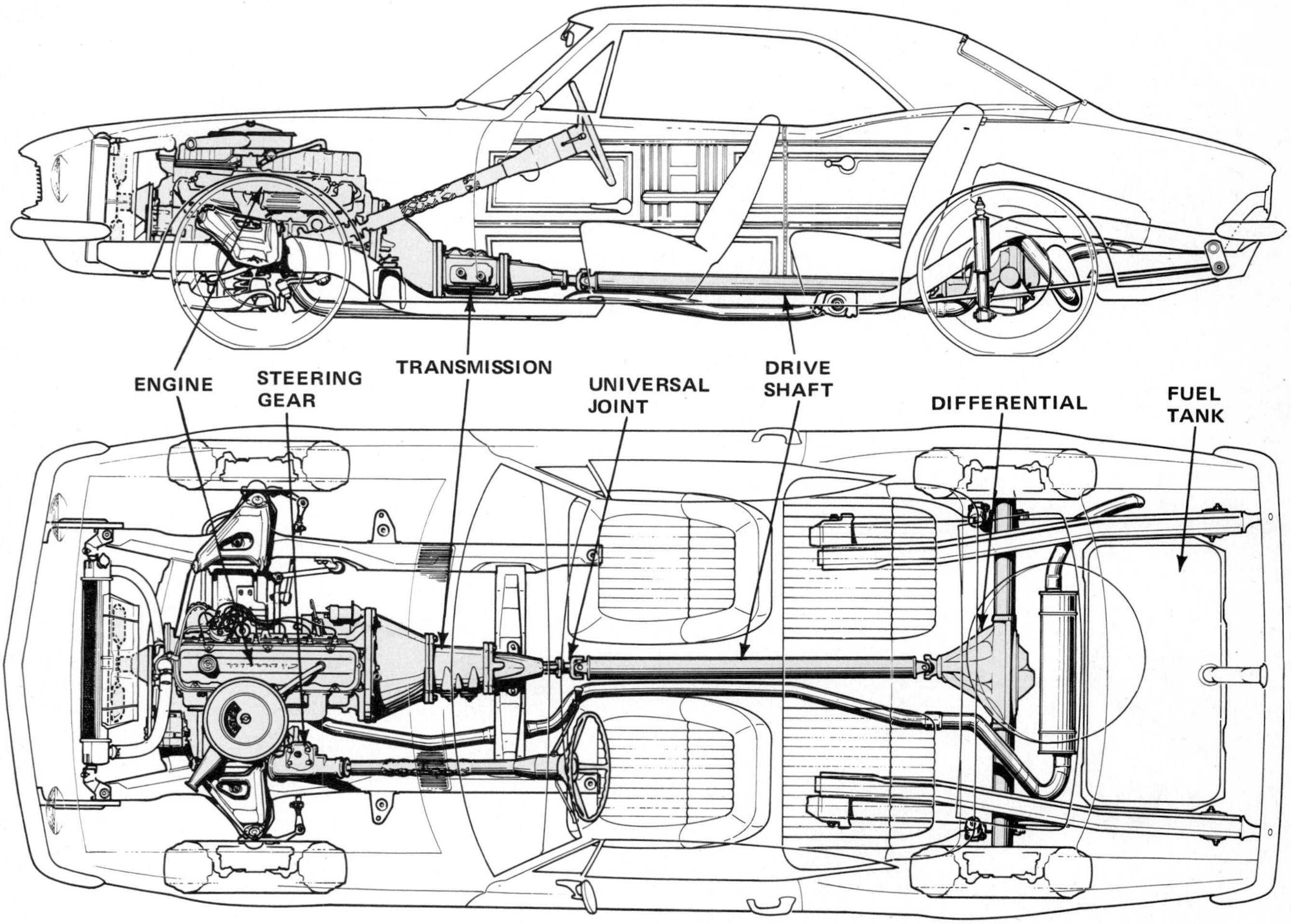

Fig. 1-6 Power train of an automobile.

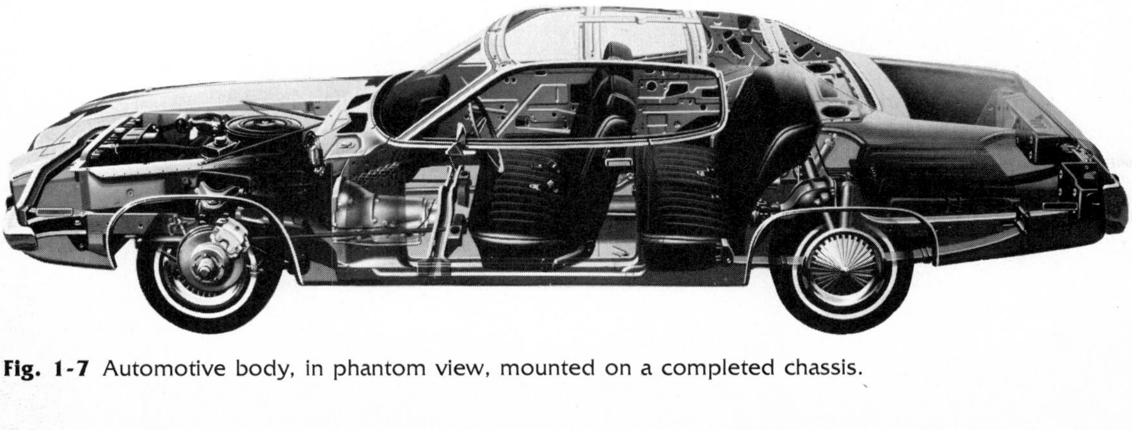

Fig. 1-7 Automotive body, in phantom view, mounted on a completed chassis.

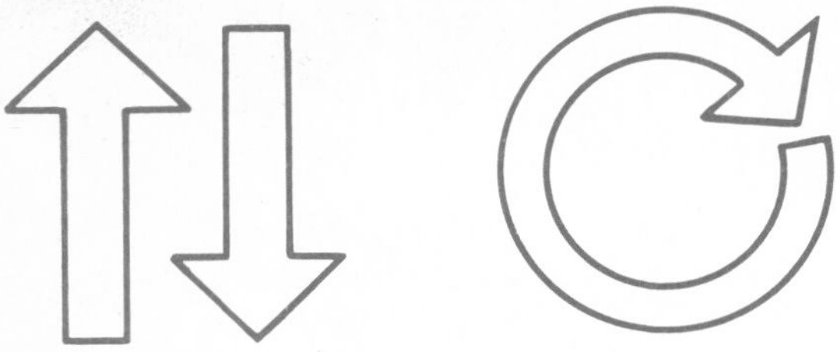

Fig. 1-8 Reciprocating motion is up and down, or back and forth, as contrasted with rotary (rotating) motion.

engine is by far the most common. It is the kind of engine used in Chevrolets, Fords, Plymouths, and other cars. Figure 1-9 shows a piston engine partly cut away so you can see the parts inside. Later in the book we will see how all the parts work together.

The rotary engine has rotors, or parts that spin. The two kinds of rotary engines are the Wankel and the turbine. There are no turbines in automobiles today. However, the Wankel is operating in hundreds of thousands of cars. The most widely known is the Mazda, made by Toyo Kogyo of Japan. Figure 1-10 shows a Mazda Wankel engine.

All these engines must have four basic systems:

- Fuel system
- Ignition system (not in diesel engines)
- Lubricating system
- Cooling system

The fuel system (Fig. 1-11) mixes gasoline with air to make a mixture that will burn. When this mixture burns in the engine, it produces high pressure. The high pressure forces the pistons to move. The movement turns shafts that turn the wheels so the car moves.

Fig. 1-9 Cutaway view of a V-8 piston engine. (*Ford Motor Company*)

The ignition system (Fig. 1-12) supplies a steady stream of sparks to the engine cylinders. These sparks set fire to, or *ignite,* the mixture of air and fuel. The mixture burns to produce the power.

The lubricating system (Fig. 1-13) keeps all moving parts coated with oil so they will move easily.

The cooling system (Fig. 1-14) circulates a mixture of water and antifreeze between the engine and a radiator. This mixture is called the *coolant.* The coolant carries heat away from the engine to prevent it from getting too hot.

All these systems are described in more detail later in the book.

1-13 Power Train

The power train (Fig. 1-6) includes the parts that carry the engine power to the car wheels. These are:

1. The transmission.
2. The clutch (on some cars).
3. The propeller shaft or drive line.
4. The differential.
5. The wheel axles.

Let's take a quick look at these now. In later chapters we describe them in detail.

1-14 Transmission

There are two types of transmission, manual or hand shifted, and automatic. Both do the same job. They allow the engine to run fast when the car first starts to move. This sends more power to the wheels, so the car can accelerate, or increase in speed. Once the car is moving, the transmission gears are changed, or shifted. This reduces the difference between engine speed and wheel speed. Less gasoline is used, and the engine does not work so hard. To start, the engine shaft might rotate 12 times to turn the car wheels once. This is called a "12-to-1 gear ratio" (12:1). Then the transmission might shift so that 8 engine shaft rotations turn the car wheels once. This is an 8-to-1 gear ratio (8:1). In high, or direct, gear the shaft would rotate 4 times to turn the wheels once.

How this is done and why are covered in later chapters. If the car has a manual transmission, the driver moves the shift lever to shift gears. Automatic transmissions do the job automatically, without any effort by the driver.

1-15 Clutch

In a car using a manual transmission, a clutch is needed. The clutch allows the driver to temporarily disconnect the engine from the power train while the gears are shifted. This is necessary because it is very difficult to shift from one gear to another if power is flowing through the gears.

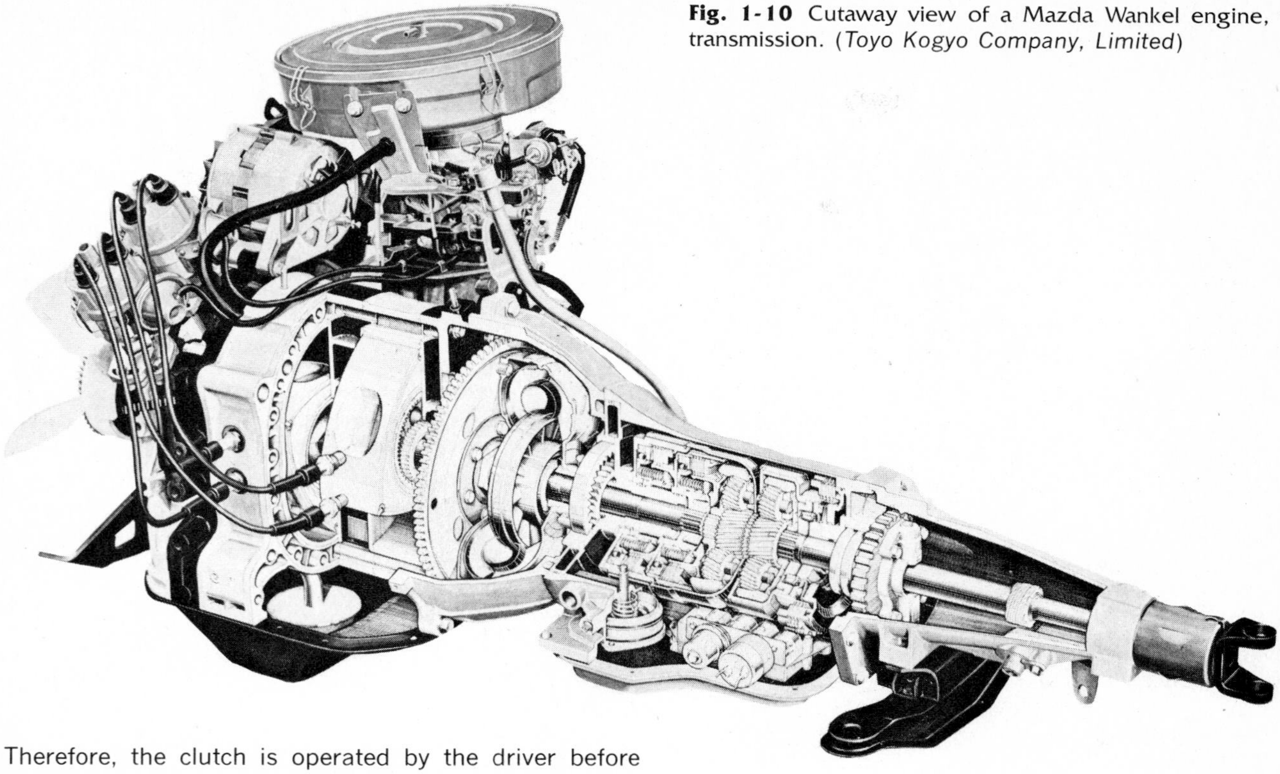

Fig. 1-10 Cutaway view of a Mazda Wankel engine, with transmission. (*Toyo Kogyo Company, Limited*)

Therefore, the clutch is operated by the driver before shifting so that no power flows from the engine through the gears. The gears can therefore be shifted easily. To operate the clutch, the driver pushes down on the clutch pedal. This disconnects the engine from the gears. Then, after the gears are shifted, the driver releases the clutch pedal to reengage the clutch. Now power can flow through the gears.

● 1-16 Drive Line

The drive line, or propeller shaft, carries power from the transmission toward the car wheels. It is just a hollow shaft with two types of flexible joints. The joints permit the shaft to drive the wheels when the wheels are moving up and down.

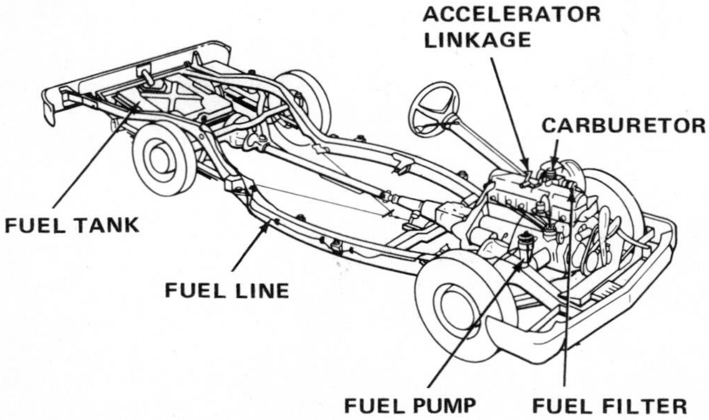

Fig. 1-11 Chassis of an automobile, with the fuel-system parts named.

● 1-17 Differential and Rear Axles

The differential receives the power from the propeller shaft. It splits the power and sends it through the two rear-wheel axles to the wheels. The differential allows the rear wheels to turn different amounts when the car goes around a curve. In going around a curve, the outer wheel

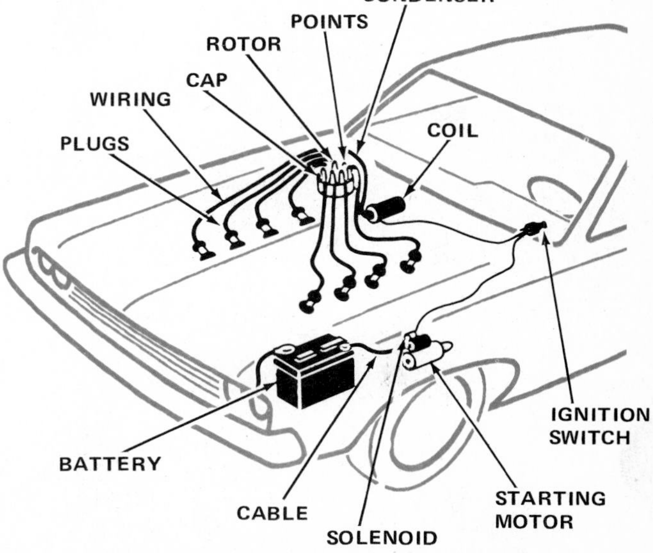

Fig. 1-12 Ignition system on a V-8 engine.

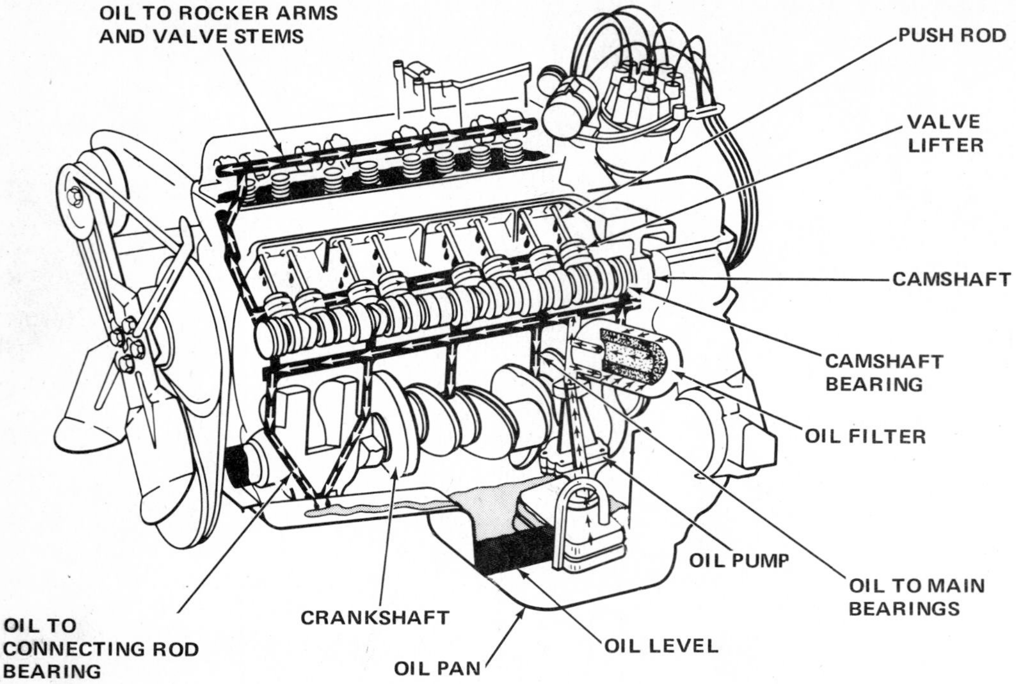

Fig. 1-13 Lubricating system of an engine, showing oil flow to moving parts.

travels farther than the inner wheel. The differential allows this and still delivers power to both wheels.

More and more cars today have front wheel drive. That is, the front wheels, rather than the rear wheels, take the power from the engine and drive the car. We go into detail on front-drive and rear-drive cars later in the book.

1-18 Frame and Chassis

The frame supports the car body, engine, and power train. The frame is supported by springs placed between the frame and the wheel axles. The assembly is called the chassis.

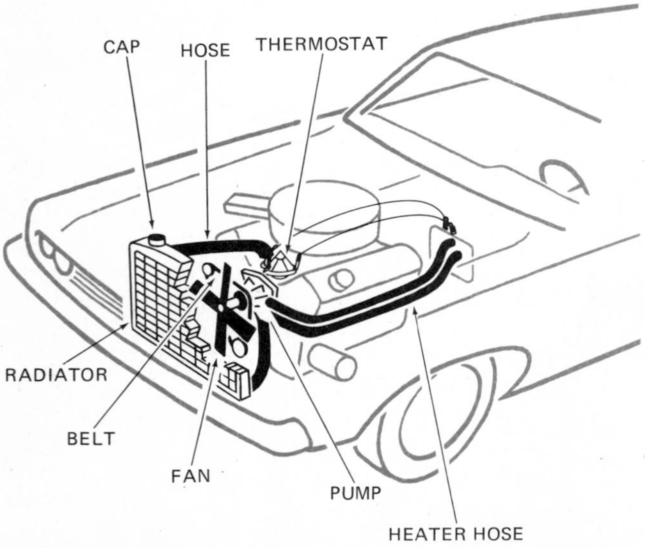

Fig. 1-14 Engine cooling system.

1-19 Spring and Shock Absorbers

The wheels are attached to the frame by springs (Figs. 1-4 and 1-5). The springs absorb, or "soak up," up-and-down motions of the wheels due to holes and bumps in the road. Thus, these up-and-down motions do not reach the frame or passengers in the car.

Springs alone do not give a good ride. They need a control to keep them from too much bouncing. That is the job of the shock absorbers. Shock absorbers are connected between the car frame and the wheel axles. They

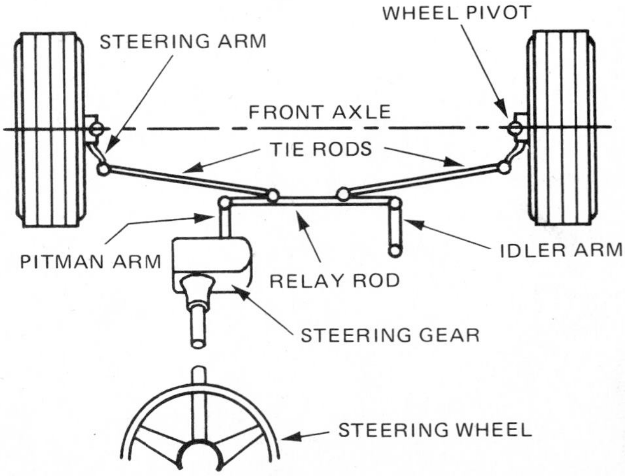

Fig. 1-15 Simplified drawing of a steering system.

restrain, or hold back, the springs. This prevents too much movement, or bounce, after a hole or bump has been passed. Springs and shock absorbers are discussed in detail later in the book.

1-20 Steering Systems

The steering system (Fig. 1-15) allows the front wheels to be pivoted, or turned, to the right or left. This steers the car. The steering wheel is mounted on a steering shaft that extends down into the steering gear. When the wheel is turned, the steering gear swings an arm to the right or left. As the arm swings, it pushes or pulls on rods connected to arms on the front axles. This movement pivots, or turns, the wheels for steering.

1-21 Brakes

Brakes are necessary to slow or stop the car. Practically all cars use brakes that operate by putting pressure on a fluid. These are called *hydraulic* brakes. The system contains a fluid-filled cylinder, called the *master cylinder.* When the brake pedal is pushed down by the driver, fluid is forced out of the cylinder. The fluid flows through pipes and hoses to the brake mechanisms at the wheels. There the fluid forces the brakes to work. This slows or stops the wheels so the car slows or stops. You can see the brake mechanisms at the wheels in Figs. 1-4 and 1-5.

There are two different kinds of wheel-brake mechanisms. They are the drum-and-shoe type and the disk type. In the drum-and-shoe type, curved plates, called *shoes,* are pressed against round drums on the wheels to produce braking. In the disk type, flat shoes are pressed against flat round disks on the wheels to produce braking. Both types are discussed in detail in later chapters.

1-22 Automotive Service

Automotive service is big business, as we have seen. There are two general kinds of automotive service, preventive maintenance and repair of defects.

Preventive maintenance means taking care of the car to prevent troubles. Checking the battery, tires, front alignment, cooling system, brakes, and steering at regular times are parts of preventive maintenance. Tuning the engine is another part. Changing the engine oil, lubricating the chassis, and draining, flushing, and refilling the cooling system are part of the preventive-maintenance job. In other words, the job means keeping everything in good condition. It also means preventing troubles that could come from lack of maintenance.

Repair work is required when something wears, gets out of adjustment, breaks, or has some other problem. A well-cared-for car usually gives long, trouble-free mileage. Repair work is usually called automotive service. It includes working on every component of the automobile. That's what this book is all about: To describe preventive maintenance and all of automotive service.

Keeping a Notebook

Keeping a notebook is a valuable part of your training to become an expert automotive mechanic. Start it now, at the beginning of your study of this textbook. Your notebook will help you in many ways. It will be a record of your progress in your studies. It will become a storehouse of valuable information you will refer to time after time. It will help you learn. And it will help you organize your training program to do you the most good. Do not overlook this valuable way of becoming the automotive expert you want to be. Keep a notebook!

How to Keep Your Notebook

Here's how to keep your notebook. Get yourself a large $8\frac{1}{2}$ by 11 inch [215.9 by 279.4 mm] ring binder and a set of notebook dividers or index tabs. Organize your notebook into ten parts, just as *Automotive Mechanics* is divided into ten parts. These parts are:

1. Automobiles and Automotive Service
2. Automotive Engines
3. Automotive-Engine Systems
4. Automotive Electrical and Electronic Equipment
5. Automotive Emission Controls
6. Automotive-Engine Service
7. Engine Trouble Diagnosis and Tune-up
8. Automotive Power Trains
9. Automotive Chassis
10. Automotive Safety Devices and Safety Inspection

When you study a lesson in the textbook, have your notebook open in front of you. Start with a fresh notebook page at the beginning of the lesson. Write the lesson number or textbook page numbers and date at the top of the page. As you read your lesson, jot down the highlights.

You may not want to carry your big notebook around with you while you work on a car. You can put a small scratch pad or a few 3 by 5 inch [76.2 by 127 mm] cards in your pocket.[1] (You will probably need this paper anyway to write down part numbers, measurements, and so on.) And be sure you always have a pencil with you in the shop. Keep your big notebook in a drawer or on the bench. After completing a job, jot down on the pad or cards the points covered or special problems. These notes are merely reminders. Redo the notes on separate pages in your big notebook in the evening.

You can also make sketches in your notebook. These could be wiring or hose diagrams, fuel circuits, sketches of parts, or anything that is more easily drawn than described in words. You can use different colored pencils to draw different circuits in the same system. This will help when you use your notes for review or as a reference.

At the end of each chapter in this textbook, you will find a section called "Self-Projects." Here you will find suggestions for projects you can do on your own. These

[1] Note that in this text, metric or United States Customary equivalents are shown in brackets [ ].

projects will help you to understand the automotive service business. Whatever the project, write a paragraph or two in your notebook about what you did and learned.

Then, too, you can save articles and illustrations you find in the technical and hot-rod magazines. For instance, you might find a description of a new fuel-injection system, with pictures. You could clip this, punch the pages to fit your notebook, and insert it in the fuel section.

Other important material you should save in your notebook includes the instructions you find in service parts. For example, a set of piston rings will include an instruction sheet explaining how to install the rings. Save this sheet. Cement or scotch-tape it on a sheet of paper and file it in your notebook.

Can you see it now? Your notebook will become one of your most valuable possessions—a permanent record of how you made your way to the top, to becoming an expert automotive mechanic!

So, *keep a notebook!*

REVIEW QUESTIONS

At the end of each chapter in this book is a series of review questions. These questions will help you to review what you have just studied. Try to answer them all. If you are not sure about the answer to any question, reread the section that gives you the answer.

1. What is interchangeability?
2. Why is interchangeability necessary for mass production?
3. How many automotive vehicles are there in the United States?
4. Name six kinds of automotive service businesses.
5. Name six jobs in the automotive service business.
6. What are the four basic components of the automobile?
7. What are the two basic types of engines used in automobiles?
8. What are the two basic types of rotary engines?
9. What are the four basic systems that every engine must have?
10. What is the purpose of the lubricating system?
11. What is the basic purpose of the cooling system?
12. What are the five basic parts of the power train?
13. What is the purpose of the differential?
14. Why are shock absorbers needed?
15. What are the two kinds of brake mechanisms at the car wheels?

SELF PROJECTS

Self projects are projects you do yourself, on your own. Their purpose is to help you understand more about the automotive service business. They will help you learn about the many kinds of jobs there are in this business. In addition, they will help you develop more self-reliance. When you start and carry through a self project it will increase your self-confidence. Some self projects are easy, others are not so easy. But don't worry, you can handle them all.

1. Look in the yellow pages of your local telephone directory under "Automobiles." Make a list of the various main headings. Probably the first heading will be "Automobile Accessories." You may be surprised at the number and variety of firms in the automotive business.
2. Look under "Service Stations" in the yellow pages and list the different oil companies, such as American, Gulf, and Texaco. Count the number of service stations under each oil-company name.
3. Now figure out how many automotive vehicles, on the average, each service station serves. For example, suppose your city has a population of 200,000. That means there should be about 121,000 automotive vehicles in your city. In the United States, there are about 220,000,000 people and 133,000,000 automotive vehicles (or more than 1 vehicle for every 2 persons). So if a city has 100,000 automotive vehicles and 200 service stations, each service station serves about 500 vehicles. Now figure it out for your own home town.

CHAPTER 2
Shop Work and Shop Manuals

After reading this chapter, you should be able to:

- **List the six steps in automotive service and explain each.**
- **List the various sources of automotive servicing information.**

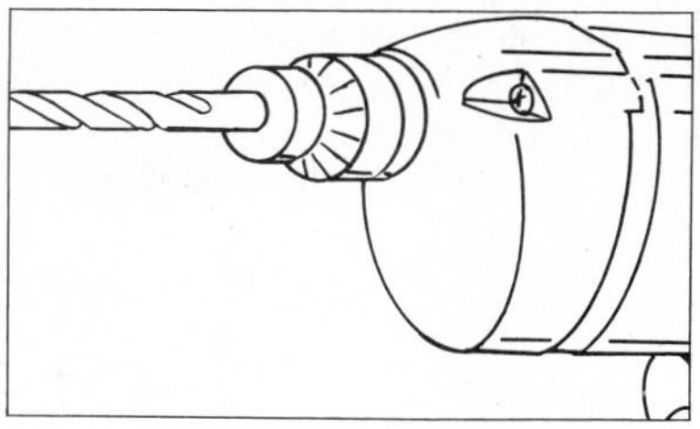

This chapter describes the six steps in doing an automotive service job. It also explains where automotive mechanics get the information they need. The servicing procedures are a part of this information. So are the specifications, that is, the measurements of parts when they are adjusted and fit properly. The chapter also discusses the paperwork required in the automotive service business.

2-1 The Six Steps in Automotive Service

Servicing jobs vary from simple to difficult, but no job requires more than six steps. These are measuring, disassembling, machining, installing new parts, reassembling, and adjusting. Some jobs require fewer steps. Let's look at the six basic steps.

2-2 Measuring

Before you can work on a car, you must find out what is wrong with it. You begin by measuring.

Linear measurements are the most common kind of measurement. They are measurements you take in a straight line. For example, you might measure an opening or a diameter. Using the familiar United States Customary system (USCS), you take measurements in inches or fractions of an inch. Using the metric system, you take measurements in millimeters, centimeters, or meters. All foreign cars are measured with the metric system. The United States is moving toward the metric system, so you should know about it.

There are other ways to measure. Sometimes the measuring is done by listening—as when you listen to a running engine. When you check the oil in an engine, you measure its level in the crankcase. You use test instruments to measure battery conditions. When you check engine vacuum or compression, you measure engine performance. The results of your measurements tell you what sort of job you have to do.

The metric system is often referred to as the SI system (for International System of Units). We will, however, use the term "metric" in this book.

2-3 Disassembly

Sometimes the measurements show that there is trouble. You then have to disassemble, or "take apart," the component to get at the trouble. Suppose your measurements show that the valves are not doing their job. You then have to take some parts of the engine off, to get to the valves and fix them.

Disassembly is also called *teardown*. For example, you disassemble, or tear down, an engine. Of course you do it carefully, part by part.

2-4 Machining

Sometimes you have to remove metal from a part. Using a machine to remove metal is called *machining.* Suppose you find valve trouble. This could require machining, or "grinding," the valves and valve seats. Or you might find that the engine cylinders require machining. Special machines, called *power tools,* are required to do these jobs.

2-5 Installing New Parts

You might find that some parts are so worn that they must be thrown away. Then new parts must be installed in their place. For example, engine bearings sometimes wear out, and new ones must be installed in place of the old ones. Even new parts may require machining to make them fit.

2-6 Reassembly

After you fix a trouble, you may have to put some parts back together. This is called *reassembly.* That is, you put the parts back together to make a complete assembly.

2-7 Adjustments

As an automobile is operated, parts naturally wear. This requires adjustments from time to time. Also, adjustments may be required after a service job. For example, after grinding the valves, you put everything back together. Then you measure the valve action. If it is not right, you have to make some adjustments.

2-8 Measuring Tools

In the automotive shop, you will use these measuring tools:

1. Steel rule
2. Feeler gauge
3. Caliper
4. Micrometer
5. Depth and small hole gauges
6. Dial indicators

2-9 Hand Tools

In the automotive shop, you will use these hand tools:

1. Screwdrivers
2. Hammers
3. Wrenches—open-end and box
4. Sockets
5. Torque wrenches
6. Pliers
7. Pullers
8. Bench vise
9. Special tools for brake, ignition, steering, and other service work

2-10 Shop Cutting Tools

In the automotive shop, you will use these cutting tools:

1. Chisels
2. Hacksaws
3. Files
4. Punches
5. Drills
6. Taps and dies
7. Stud extractors

2-11 Shop Power Tools and Equipment

1. Air tools for cutting, driving, wrenching
2. Car lifts
3. Jacks operated by hand or hydraulically
4. Lubricating equipment
5. Electric tools—grinders, drills, lathes, valve-seat and valve grinders, engine-cylinder hones, etc.
6. Cleaning equipment—solvent and hot tanks, steam cleaners, sandblast and glass-bead cleaners

2-12 Automotive Fasteners

Various fasteners are used to hold the engine and automobile together. You must have a good working knowledge of all these fasteners when you work in the automotive shop.

2-13 Specifications

You will hear the word "specification," or "spec," quite often in your shop work. The specs give you the right measurements for the cars you work on. The car manufacturer sets the specs. You find the specs in the manufacturer's shop manual. These specs include valve setting, ignition timing, piston clearances, piston-ring clearances, and hundreds of other measurements. We explain these terms in later chapters.

2-14 Manufacturers' Service Manuals

Each year, every car manufacturer issues a service manual covering the cars manufactured that year. Each manual covers all service procedures, provides the specs, names the tools needed, and explains how to do all service jobs on the car models produced that year. You should take a careful look at any manufacturer's service manual you get your hands on. Note how it is arranged in groups or sections. Figure 2-1 shows the section index for a recent Chevrolet service manual. Each section is further divided into descriptions of specific components. These explain how to find and fix troubles with those components, and list the special tools required to service them. Get acquainted with car manufacturers' service manuals. You will be using them when you work in the shop. Practice using these manuals. Look up "Engine," "Transmission," and so on in a manual. Note how each section is divided into specific service jobs.

2-15 Flat Rate

Two ways that automotive mechanics are paid are by fixed income and by flat rate. With a fixed income, the mechanic is paid so much per hour or per week. When a mechanic is paid a flat rate, he gets so much per job. "Flat rate" tells the number of hours and minutes it normally takes to do a service job. Car makers and independent companies print flat-rate manuals. They list every service job and the amount of time it takes to do the job. For example, one flat-rate manual says that it takes 10 hours to install new piston rings in a certain engine. If the labor charge is $15.00 per hour, the total labor charge for the job is $150.00.

Many shops split the labor charge with the mechanic. With the right tools, the mechanic can often "beat" flat rate, that is, do the job in less time. Suppose the mechanic does a 10-hour job in 8 hours. The mechanic would still get paid for 10 hours even though the job took less time.

SECTION	NAME
0A	GENERAL INFORMATION
0B	LUBRICATION
1A	HEATER
1B	MANUAL AIR CONDITIONING
1C	AUTOMATIC AIR CONDITIONING
2A	FRAME
2B	BUMPERS
2C	SHEET METAL
2D	BODY
3A	FRONT ALIGNMENT
3B	STEERING
3C	FRONT SUSPENSION
3D	REAR SUSPENSION
3E	WHEELS AND TIRES
4A	PROPSHAFT
4B	DRIVE AXLE
5	BRAKES
6A	ENGINE MECHANICAL
6B	ENGINE COOLING
6C	FUEL SYSTEM
6D	ENGINE ELECTRICAL
6E	EMISSION CONTROLS
6F	ENGINE EXHAUST
7A	AUTOMATIC TRANSMISSION
7B	MANUAL TRANSMISSION
7C	CLUTCH
8	CHASSIS ELECTRICAL
9	ACCESSORIES

Fig. 2-1 Section index for a recent Chevrolet service manual.

2-16 Other Useful Publications

Many automotive service magazines are published. They contain information on servicing cars and articles on specific service jobs. They often have tips on how to make hard jobs a little easier.

Testing-equipment manufacturers, parts makers, and tool and service-equipment manufacturers publish manuals on how to install their parts or use their equipment. These publications can be very helpful.

2-17 Paperwork

You cannot run any kind of business without paperwork, and the automotive service business is no exception. From the time that a car drives in for service, until the time that the bill for service is paid, there is paperwork to be done. Actually, paperwork begins well before the car arrives. Let's see how it works.

1. The parts department The parts department has to have on hand all fast-moving parts that might be needed to service cars. These parts include spark plugs, fan belts, antifreeze, engine oil, brake fluid, distributor contact points, ignition coils, gaskets, screws and bolts, and so on. All these parts and supplies have to be ordered by the parts manager. That is, the parts manager wrote out orders for the things needed to stock the department. The parts manager also has to keep an inventory of stock. That is, a record of what stock is on hand. That way, more parts or supplies can be ordered when stock runs low. All this record keeping and ordering is paperwork.

2. The service department The phone rings and the service manager or an assistant answers. Someone has car trouble. The caller wants to bring a car in for service. The service manager consults the work schedule (often called the "dispatch sheet"), and tells the caller that the department can handle the car at ten the next morning. The service manager notes this down on the schedule.

When the car arrives the next morning, it is met by the service manager or a service-order writer, who is usually called a service writer or a service salesperson. The driver explains the problem. The service writer writes the customer complaint on the repair order, together with labor instructions and notes to the mechanic. The repair order accompanies the car as it is "dispatched" or sent to the mechanic assigned to do the service job. The mechanic follows the labor instructions on the repair order, doing whatever is necessary to repair the car. If new parts are needed, the mechanic gets them from the parts department. To get parts from some parts departments, the mechanic fills out a requisition, or "rec" as it is called. The parts counter attendant issues the parts listed on the rec and writes the price of the parts on the repair order. When the job is completed, the service writer or service manager writes, on the repair order, the flat rate time or the time taken by the mechanic to do the job. To pick up the car, the customer pays the price of the parts, plus the price of the mechanic's labor charge, and any taxes required by law.

3. Billing Everything is added up on the service order to get the total cost of the service job. If it is a cash job, the car owner pays the bill when taking delivery of the car. If it is a credit job, the car owner is sent a bill which should be paid after so many days. Even here the paperwork does not end. When the payment comes in, it has to be credited against the car owner's account. And, if the car owner is slow to pay, letters have to go out asking for payment. If the car owner refuses to pay, it may be necessary to take legal action. All this is paperwork.

4. Other paperwork The owner or operator of an automotive service business also keeps books of various kinds—periodic P&L (profit and loss) statements, employee records, general inventories, local, state, and federal tax data, and so on.

You now have a brief review of the paperwork involved in the automotive service business. Whatever type of job you take in the business will involve some paperwork. You will learn exactly how to handle your paperwork when you go out into the shop.

REVIEW QUESTIONS

Here is another group of questions to help you review what you have just been studying. See if you can answer all the questions. If you are not sure about any of them, reread the section that gives you the answer.

1. What are the six basic steps in any automotive service job?
2. Name four measurements you might make on an engine.
3. In automotive terms, what does "teardown" mean?
4. Name three matching jobs you might do on an engine.
5. What does the word "spec" mean?
6. Where do you find specs?
7. Name three things that a car manufacturer's service manual gives you.
8. Name two useful automotive-service publications besides a manufacturer's service manual.

SELF PROJECTS

Self projects are designed to do two things: help you build confidence in yourself, and help you learn more about automotive servicing. Here are some self projects that will interest you.

1. Look at a car manufacturer's service manual. At the front, on the first or second page, there is a quick reference index. Find out how to use it. Notice that you bend the manual back to expose the black spots on the first page of each section. These correspond to the black spots and titles on the quick reference index.
2. Now make a list of the sections listed in the quick reference index.
3. Next, turn to the first page of the section on engines. Notice that this page has a table of contents, listing the various subsections on engines. Look through the engine section to see how it is arranged.
4. Now find the section on engine specifications. Make a list of some of the important specs for the engine. You may not understand some of the terms, but we shall explain them all later in the book. Meanwhile, this will give you an idea of what the service manual contains.

CHAPTER 3
Safety in the Shop

After reading this chapter, you should be able to:

- **Discuss shop layouts.**
- **Discuss shop safety.**
- **List shop hazards due to various causes such as faulty working habits, working conditions, equipment defects, and incorrect use of hand tools.**
- **Discuss fire prevention in the shop, including various types of fire extinguishers and how to use them.**
- **List the safety rules.**

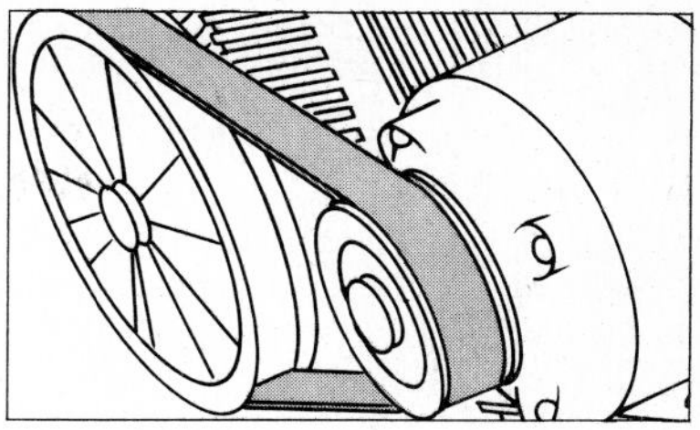

Shopwork is varied and interesting. The shop is where you will learn how to do all sorts of automotive jobs, including (1) adjusting front alignment, (2) checking charging systems, (3) pulling an engine from a car and tearing it down, (4) grinding valves, and (5) adjusting differentials. These and the other basic automotive service jobs are discussed in this book.

Before you work in the shop, you should know about safety. Safety in the shop means protecting yourself and your fellow workers from possible danger and injury. This chapter describes the rules you should follow in the shop to protect yourself from harm. Remember this: When everybody obeys the rules, the shop is a much safer place to work than your home! Many more people are hurt in the home than in the shop.

● 3-1 Safety is Your Job

Yes, safety *is* your job. In the shop, you are "safe" when you protect your eyes, your fingers, your hands—all of yourself—from danger. And, just as important, when you look out for the safety of those around you.

The rules of safety are listed and discussed in the next few pages. Follow the rules for your protection, and for the protection of your fellow workers.

● 3-2 Shop Layouts

The term "shop layout" means the locations of workbenches, car lifts, machine tools, and so on. Shop layouts vary. Figure 3-1 shows the layout of a typical general automotive repair shop. Figure 3-2 shows the layout of a

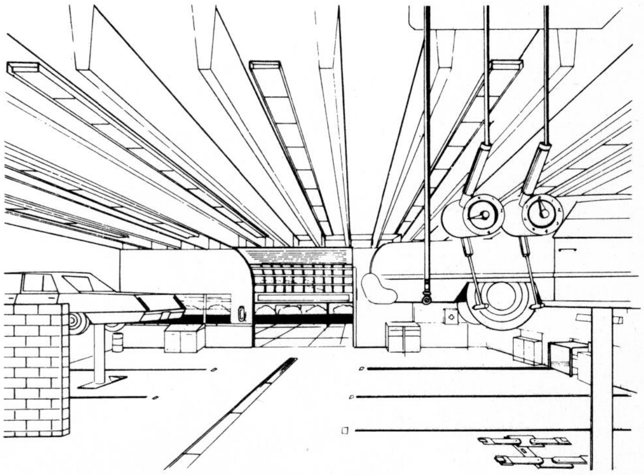

Fig. 3-1 Typical shop layout. (*Motor Vehicle Manufacturers Association*)

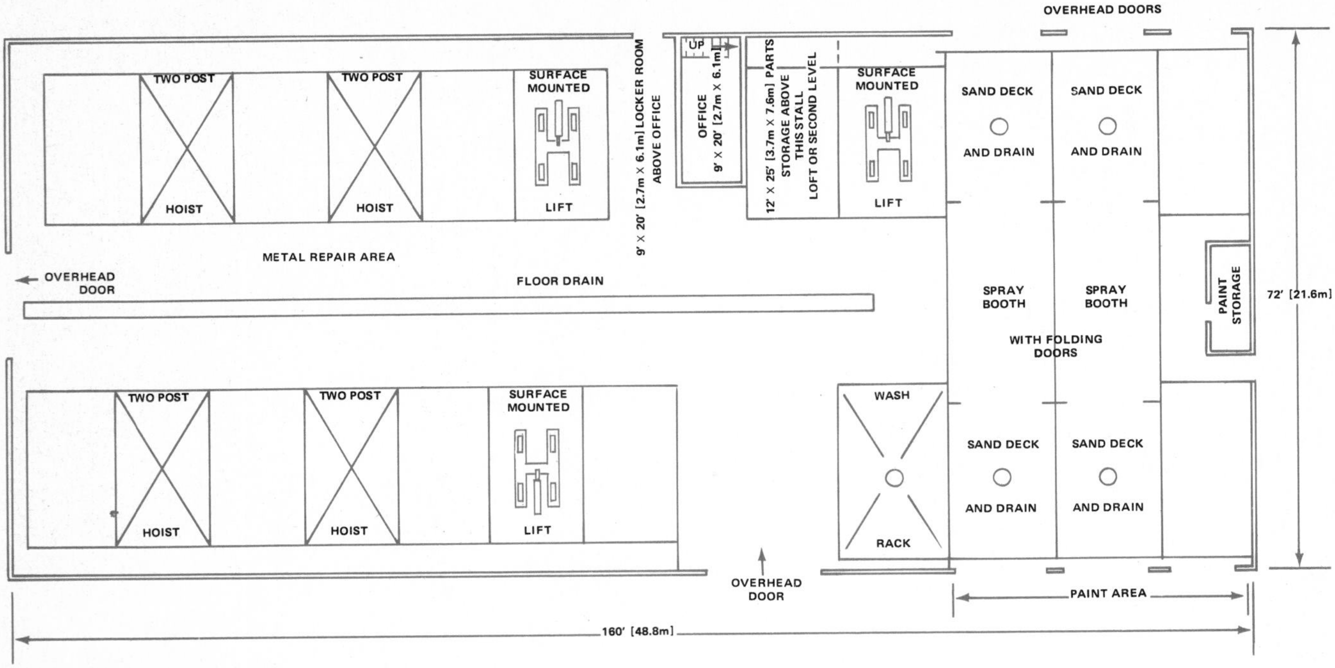

Fig. 3-2 Floor plan for a body shop with a dozen body and paint technicians. (*Fisher Body Division of General Motors Corporation*)

body shop. These are only two of many possible variations. So the first thing you should do in a shop is find out where everything is located. This includes the different machine tools and the workbenches, car lifts, and work areas. Many shops have painted lines on the floor to mark off work areas. These lines guide customers and workers away from danger zones where machines are being operated. The lines also remind workers to keep their tools and equipment inside work area lines.

Many shops have warning signs posted around machinery. Some of these signs may have the name OSHA, which stands for Occupational Safety and Health Administration. This is the federal agency with the responsibility of studying and correcting conditions and equipment that present hazards to workers (see ● 3-3). The signs are posted to remind everyone about safety, and about how to use machines safely. Follow the posted instructions at all times. The most common cause of accidents in the shop is failure to follow instructions.

● 3-3 Shop Hazards to Watch Out For

Federal laws have been enacted which are designed to assure healthful and safe working conditions for working men and women. The federally established National Institute for Occupational Safety and Health (NIOSH) makes studies of shop working conditions and reports on potential hazards that should be corrected. Further, the law requires that the shops with such hazards must eliminate them. Hazards found are sometimes the fault of management, and sometimes the fault of the workers. In the following three sections, we discuss hazards that might be considered as due to working habits, hazards due to faulty or improperly used shop equipment, and hazards due to faulty or improperly used hand tools.

● 3-4 Hazards Due to Faulty Working Habits or Conditions

Here are some of the major hazards that might be due to working habits of the employees or to the general working conditions.

1. Smoking while handling dangerous materials such as gasoline or solvents. See Fig. 3-3. This can result in a major fire or explosion.
2. Careless or incorrect handling of paint, thinners, solvents, or other flammable substance. Figure 3-4 shows the correct arrangement for pumping a flammable substance from a large container into a small one. Note the bond and ground wires. Without these, a spark might jump from the nozzle to the small container, causing a disastrous explosion and fire.
3. Blocking exits (Fig. 3-5). Areas around exit doors and passageways leading to exits must be kept free of all obstructions. If you wanted to get out in an emergency—as, for instance, when a fire or explosion occurred—a blocked exit could mean serious injury or even death.
4. Failure to wear the proper respirator (Fig. 3-6) when handling thinners, paints, and similar substances, when working with body-filler lead, or when sanding the car body or body filler. The vapor or dust from these substances, when breathed for a long time, can

Fig. 3-3 Do not smoke or have open flames around combustibles such as gasoline or solvents.

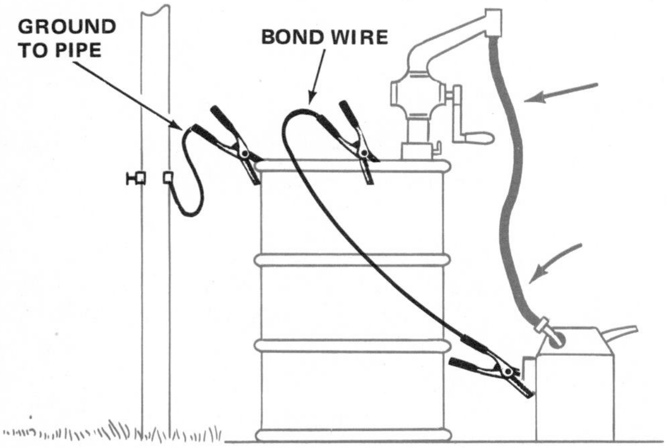

Fig. 3-4 Setup for pumping a flammable liquid from a large to a small container. Note that if the hose is nonconducting, a ground wire should be connected, as shown.

Fig. 3-5 Exits should be neat.

Fig. 3-6 A respirator shoud be worn when painting, sanding, or doing similar jobs.

cause liver and lung damage. Proper ventilation of the work area and wearing of a respirator reduces this hazard to a minimum.

5. Here are some other hazards caused by working habits and conditions in the body shop: Washing paint from the hands with thinner. Thinner can cause skin rashes and may be absorbed through the skin to cause liver damage. Plastic body filler, in the raw or uncured state, can also irritate the skin and should be washed off at once with soap and water. Using rubber gloves to handle these materials is a good working habit.

3-5 Hazards Due to Equipment Defects or Misuse

Here are the most common hazards in the shop due to equipment that is faulty or which is improperly used:

1. Incorrect safety guarding of moving machinery. For example, fans should have adequate guards, as shown in Fig. 3-7. Air compressors should have proper guards over the belt and pulley (Fig. 3-8).
2. Misuse of flexible electric cords, cords that are worn or frayed, or cords that have been improperly spliced. Flexible cord should not be run through holes in the wall, nor tacked onto the wall. See Fig. 3-9. Any of these could cause a fire or could electrocute someone.
3. Compressed-gas cylinders improperly stored or misused. These cylinders should never be stored near radiators or other sources of heat. They should never be kept in unventilated enclosures such as lockers or closets. There should be at least 20 feet [6.1 meters] between stored oxygen and stored acetylene cylinders. Cylinders should not stand free

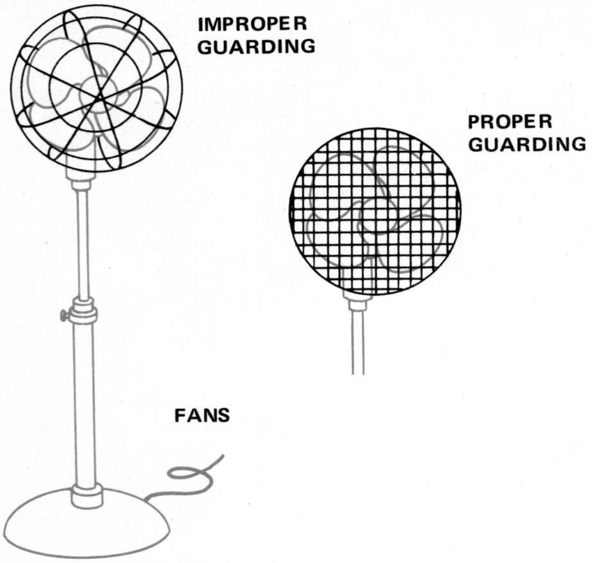

Fig. 3-7 A fan, improperly and properly guarded.

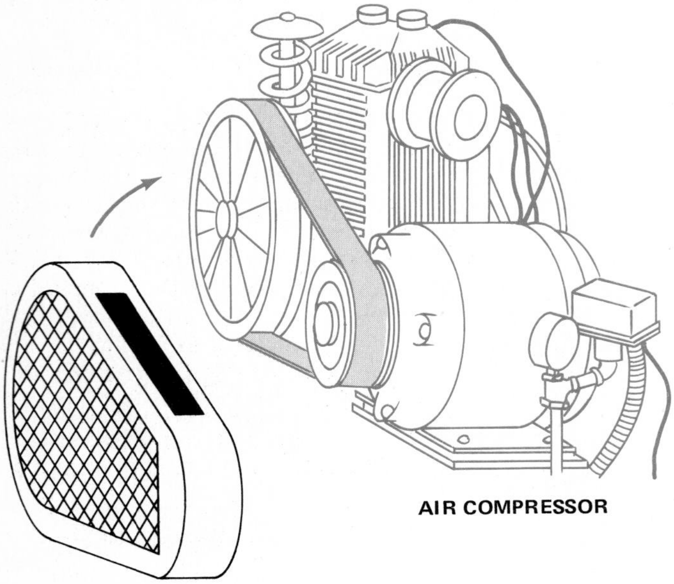

Fig. 3-8 Belts and pulleys should always be protected with guards.

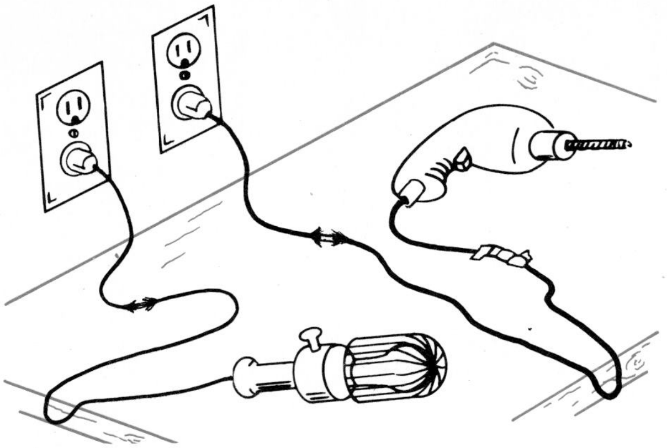

Fig. 3-9 All electrical appliances should be used carefully. Plugs should be grounded and appliances stored neatly. What safety rules are violated here?

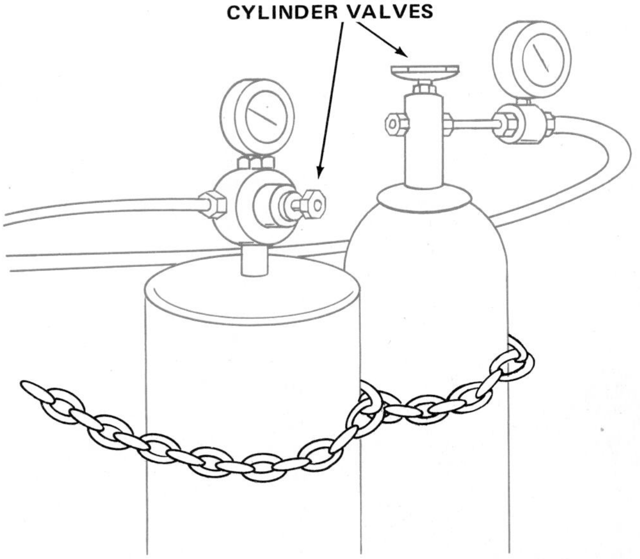

Fig. 3-10 Gas cylinders should not be allowed to stand freely but should always be secured with chains or lashing.

but should be secured with a chain or lashing (Fig. 3-10). Cylinders should never be used as supports or as rollers to move an object. Such treatment could cause the cylinder to explode with terrible effect on anyone standing nearby.

4. Hand-held electric tools not properly grounded. All such tools must have a separate ground lead (Fig. 3-11), or be doubly insulated to guard against shock.
5. Hydraulic car lifts improperly used. Passengers should not remain in the car when it is lifted. All doors, hood, and trunk lid should be closed. Otherwise, they could be damaged as the car is lifted. If the lift has a mechanical locking device, it should be engaged before you go under the lift. Do not use a lift that is not working properly and jumps or jerks when raised, settles slowly when it should not, or rises or settles too slowly, blows oil out of the exhaust line, or leaks oil at the packing gland.
6. Using a wheel-and-tire balancer which does not have the hood in place. OSHA regulations call for all wheel balancers of the spinner type to have hoods (Fig. 3-12). These hoods protect the workers in case a stone or other object is thrown from the spinning tire tread.
7. Letting a tester lead fall into the engine fan when the engine is running. This, at the least, can ruin the tester. At worst, the fan can throw out leads or even the tester and injure the worker.
8. Leaving a running machine unattended. Whenever you are using a power tool, and have to leave it for a moment, turn it off! If you leave it running, someone else might come along and, not realizing it is running, get a hand in the way of moving parts and be injured.

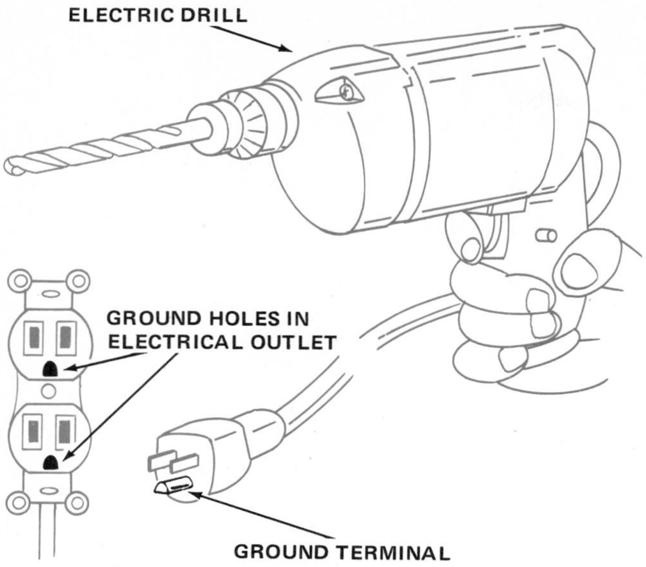

Fig. 3-11 Electric drill with three-wire cord. The third wire and terminal are to ground the drill.

Fig. 3-12 An electric wheel balancer used to check tire and wheel balance. Note the hood to protect the operator from stones or other objects that might be thrown from the tire tread. The hood pivots down to a horizontal position before the wheel is spun. (*Ford Motor Company*)

9. Playing with fire extinguishers. There have been cases where the workers thought it was fun to play with the fire extinguishers, but someone got hurt from slipping and falling down, or injured their eyes from the liquid or foam. Also, this leaves the extinguishers empty so they are useless if a fire should break out.
10. Using electric-arc- or gas-welding equipment which is faulty, or using it incorrectly. Gas-welding hose must be in good condition and not have leaks, burns, or worn places (Fig. 3-13). Gas welding should be done in an area protected with screens, and there should be no combustibles nearby. Oil and grease must be kept away from gas-welding equipment. If high-pressure oxygen hits oil or grease, it will burst into flame with explosive violence. Electric-arc welding must also be done in a protected place, and the worker must wear a welding helmet or face shield. Arc welders should wear fire-resistant clothes and gloves. Fire-extinguisher equipment should be kept handy. The welding cable should never be coiled and must never be coiled around the body of the welder. Hot material should be marked to prevent someone from attempting to handle it.

● 3-6 Hand-Tool Hazards to Watch Out For

Hand tools should be kept clean and in good condition. Greasy and oily tools are hard to hold and use. Always wipe them before trying to use them. Do not use a hardened hammer or punch on a hardened surface. Hardened steel is brittle and can shatter from heavy blows. Slivers may fly out and enter the hand, or worse, the eye. Hammers with broken or cracked handles, chisels and punches with mushroomed heads, or broken or bent wrenches are other tool hazards that should be avoided.

Never use a tool that is in poor condition or that does not fit the job. Avoid hand-tool hazards so you won't get hurt.

● 3-7 Fire Prevention

Gasoline is used so much in the shop that people forget it is very dangerous if not handled properly. A spark or

Fig. 3-13 Hoses used in gas-welding equipment should be in good condition. Frayed or worn spots are potentially dangerous and could let gas leak out, with disastrous results.

lighted match in a closed place filled with gasoline vapor can cause an explosion. Even the spark from a light switch can set off an explosion. So you must always be careful with gasoline. Here are some hints.

There will be gasoline vapors around, if gasoline is spilled or a fuel line is leaking. You should keep the shop doors open or keep the ventilating system going. Wipe up the spilled gasoline at once, and put the rags outside to dry. Never smoke or light cigarettes around gasoline. When you work on a leaky fuel line, carburetor, or fuel pump, catch the leaking gasoline in a container or with rags. Put the soaked rags outside to dry. Fix the leak as quickly as possible. And don't make sparks around the car, for instance by connecting a trouble light to the battery.

Fig. 3-14 Store gasoline and all flammable liquids in approved safety containers.

Fig. 3-15 Never store gasoline or other flammable liquids in glass bottles, jars, or jugs. If the container should break, a disastrous explosion and fire could result.

Fig. 3-16 Safety can for the storage of oily rags.

Store gasoline in an approved safety container (Fig. 3-14). Never, *never* store gasoline in a glass jug. The jug could break and could cause a terrible explosion and fire (Fig. 3-15).

Oily rags can also be a source of fire. They can catch fire without a spark or flame. Oily rags and waste should be put into a special closed container where they can do no harm (Fig. 3-16).

● 3-8 Fire Extinguishers

Note the location of the fire extinguishers in the shop. Make sure you know how to use them. Figure 3-17 is a chart showing different types of fires and the types of fire extinguisher to use for each type. Remember that the quicker you get at a fire, the easier it is to control. But you have to use the right kind of fire extinguisher, and use it correctly. The chart explains this. Talk it over with your instructor if you have any questions.

● 3-9 The Safety Rules

Some people say, "Accidents will happen!" But safety experts do not agree. They say, "Accidents are caused; they are caused by careless actions, by inattention to the job at hand, by using damaged or incorrect tools. And sometimes accidents are caused by just plain stupidity!"

To keep accidents from happening, follow these simple rules.

1. Work quietly and give the job your full attention.
2. Keep your tools and equipment under control (Fig. 3-18).
3. Keep jack handles out of the way. Stand creeper against the wall when it is not in use (Fig. 3-19). Or push it under the car being worked on (Fig. 3-20).
4. Never indulge in horseplay or other foolish activities. You could cause someone to get seriously hurt.
5. Don't put sharp objects, such as screwdrivers, in your pocket. You could cut yourself or get stabbed. Or you could ruin the upholstery in a car.
6. Make sure your clothes are right for the job. Dangling sleeves or ties can get caught in machinery and cause serious injuries. Do not wear sandals or open-toe shoes. Wear full leather shoes with nonskid rubber heels and soles. Steel-toe safety shoes are best for shop work. Keep long hair out of machinery by wearing a cap.
7. Wipe excess oil and grease off your hands and tools so you can get a good grip on tools or parts.
8. If you spill oil, grease, or any liquid on the floor, clean it up so that no one will slip and fall.
9. Never use compressed air to blow dirt from your

clothes. Never point a compressed-air hose at another person. Flying particles could put out an eye.

10. Always wear goggles or a face shield when there are particles flying about. Always wear an eye protector when using a grinding wheel (Fig. 3-21).

11. Watch out for sparks flying from a grinding wheel or welding equipment. The sparks can set your clothes on fire.

12. To protect your eyes, wear goggles when using chemicals, such as solvents. If you get a chemical in your eyes, wash them with water at once (Fig. 3-22). Then see the school nurse or a doctor as soon as possible.

13. When using a car jack, make sure it is centered so that it won't slip. And never, *never* jack up a car while someone is working under it! People have been killed when the jack slipped and the car fell on them! Always use car stands or supports, properly placed, when going under a car (Fig. 3-23).

14. Always use the right tool for the job. The wrong tool could damage the part being worked on and could cause you to get hurt.

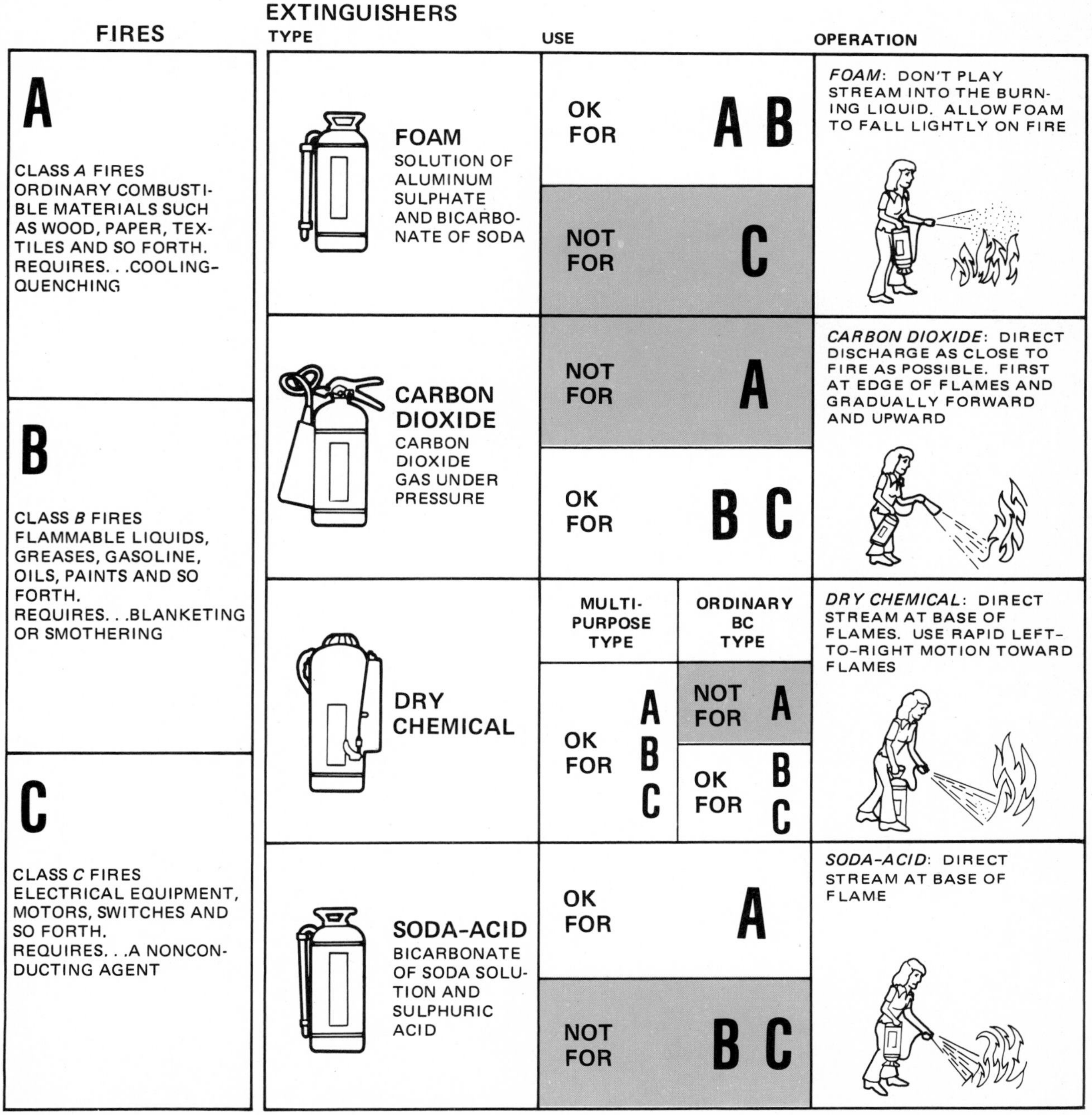

Fig. 3-17 Chart showing types of fire extinguishers and the classification of fires. (*Ford Motor Company*)

Fig. 3-18 Keep your tools within easy reach, neatly arranged. Do not scatter them around.

Fig. 3-19 When jacking up a car, always finish with the handle pointing up so no one can trip over the handle. When creepers are not in use, stand them up against the wall, wheels out, where no one can stumble over them.

Fig. 3-20 After you use a creeper but you are not finished with the job, shove it under the car temporarily.

15. Keep hands away from the engine fan and belt when the engine is running. You could be badly cut or even lose fingers if your hand got caught in the fan or fan belt.

16. Do not stand directly in line with the engine fan when the engine is running. Some fans, especially the flex fans, have been known to throw off a blade when spinning. Anyone standing in line with the fan could get seriously hurt if hit by the blade.

CAUTION: Never run an engine in a closed garage that does not have a ventilating system. The exhaust gases

Fig. 3-21 Always wear goggles or a face mask when using a machine that can throw chips or sparks.

Fig. 3-22 If solvent or some other chemical splashes in your eye, immediately wash your eye with water.

contain carbon monoxide. Carbon monoxide is a colorless, odorless, tasteless, poisonous gas that can kill you! In a closed one-car garage, enough carbon monoxide to kill you can collect in only three minutes.

3-10 Using Power-Driven Equipment

A lot of power-driven equipment is used in the automobile shop. The instructions for using any equipment should be studied carefully before the equipment is operated. Hands and clothes should be kept away from moving machinery. Keep hands out of the way when using any cutting device, such as a drum lathe. Do not attempt to feel the finish while the machine is in operation. There may be slivers of metal that will cut your hands badly. When using grinding equipment, keep hands away from rotating parts. Do not try to feel the finish with the machine in operation. Sometimes you will work on a device with compressed springs, such as a clutch or valves. Use great care to prevent the springs from slipping and jumping loose. If this happens, the spring may take off at high speed and hurt someone.

Never attempt to adjust or oil moving machinery unless the instructions tell you that this should be done.

3-11 What To Do in Emergencies

If there is an accident and someone gets hurt, notify your instructor at once. The instructor will know what to do—give first aid, phone for the school nurse, a doctor, or an ambulance. Be very careful in giving first aid. You must know what you are doing. Trying first aid on an injured person can do more harm than good if it is done wrong. For example, a serious back injury could be made worse if the injured person is moved improperly. On the other hand, quick mouth-to-mouth resuscitation may save the life of a person who has suffered an electric shock. Talk to your instructor if you have any questions about this.

Fig. 3-23 Stands should be properly placed to support the car before you work under the car.

And remember this about fires: As we said before, the quicker you get at a fire, the easier it is to control. But you have to use the right kind of fire extinguisher and use it correctly. See Fig. 3-17.

3-12 Driving Cars in the Shop

Cars have to be moved in the shop. They must be brought in for service, and may have to be moved from one work area to another. When the job is finished, they have to be moved out of the work area. You must be extremely careful when you drive a car in the shop. Make sure the way is clear. Make sure no one is under a nearby car. Someone might suddenly stick out an arm or a leg. Make sure there are no tools on the ground.

When you take a car out for a road test, *fasten your seat belt,* even though you're going only a short distance.

CAUTION: Always fasten your safety belt in a moving car, whether you are the driver or a passenger. Seat belts save lives; your seat belt could save yours. Buckle up for safety!

3-13 Tow-Truck Operation

Driving a tow truck and doing emergency work at the scene of a wreck takes experience. Each wreck is a special problem. But here are some general comments that can be made about tow trucks:

1. Make sure the fire extinguisher is properly serviced, in good working condition, and mounted securely on the truck.
2. Do not exceed the unit's maximum hoisting capacity.
3. Make sure the truck floodlights are in good working condition.
4. The control mechanism for the hoist should be inspected periodically to make sure it is in good operating condition, and that the cable, hooks, drum, and other parts are okay.

REVIEW QUESTIONS

Here are some more review questions to try on yourself. If any of them baffle you, reread the section that gives you the answer.

1. What is the first caution to observe when you are using a grinding wheel?
2. What is wrong with using gasoline to clean the floor or workbench with the garage doors closed?
3. Why are oily rags dangerous?
4. What is wrong with operating an engine in a garage with the garage doors closed?
5. What is the purpose of the lines painted on the floor of the shop?
6. How many fire extinguishers are there in your shop, and where are they located?

7. What is mouth-to-mouth resuscitation?
8. Why should you never store gasoline in a glass jug?
9. Why must you never point the compressed-air hose at another person?
10. What is the purpose of safety belts, and how do they work?
11. What is OSHA?
12. Name four hazards due to faulty working habits or conditions.
13. Name six hazards due to equipment defects or misuse.
14. Why should you never stand in line with an engine fan when the engine is running?
15. Name the four types of fire extinguishers and explain which class of fire each is to be used on. Also, explain how to use each.

SELF PROJECTS

Here are some self projects that will help you understand the automotive service business.

1. Make a floor plan of the school shop, showing the various working areas and the location of the car lifts and other power equipment.
2. Go to a service station and make a floor plan of the station and gasoline pump areas, showing where the service station equipment is located. First, ask the station manager's permission, explaining that you are making the floor plan as a school project. Then, put the floor plan in your notebook.
3. Go to a large new-car agency and ask the service manager or shop foreman for a tour of the parts and service departments. Explain that this is for a school project. If they are too busy, ask if you could walk around. Look for the safety lanes in the shop area. Notice the signs cautioning customers to stay away from dangerous areas in the shop. See if you can locate the fire extinguishers and first-aid kits. Ask the shop foreman or service manager if they ever had a fire or accident that injured someone and what their procedure is for handling such emergencies. *Do not get in anyone's way and do not bother the mechanics while they are working!*

Automotive Engines

2

This part of *Automotive Mechanics* describes the construction and operation of automobile engines. Part Three discusses the engine systems that keep the engine going: the fuel system, lubricating system, and cooling system. In this part of the book—Part Two—there are seven chapters.

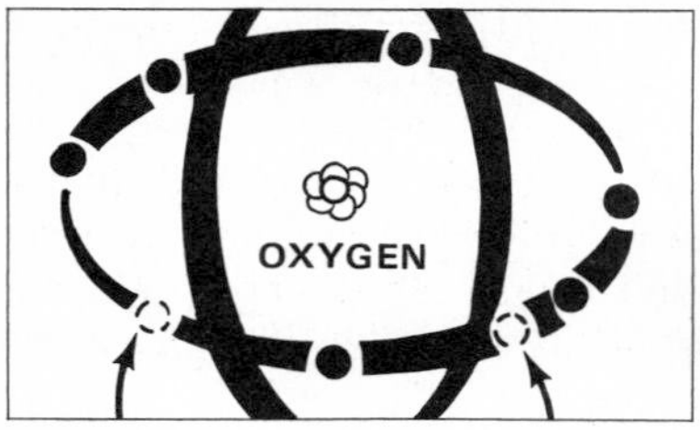

Chapter 4 Engine Fundamentals

Chapter 5 Piston-Engine Operation

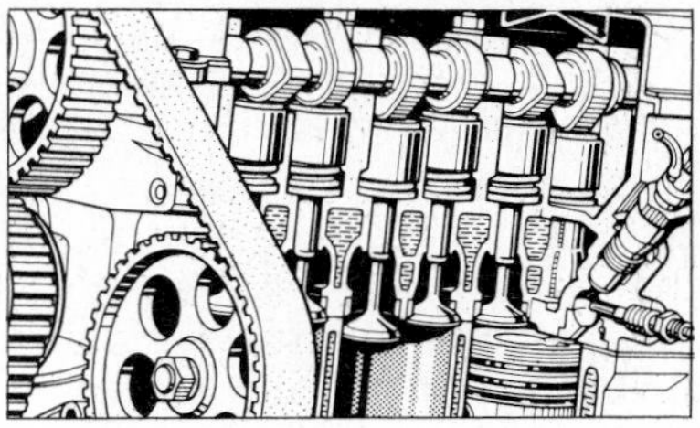

Chapter 6 Engine Types

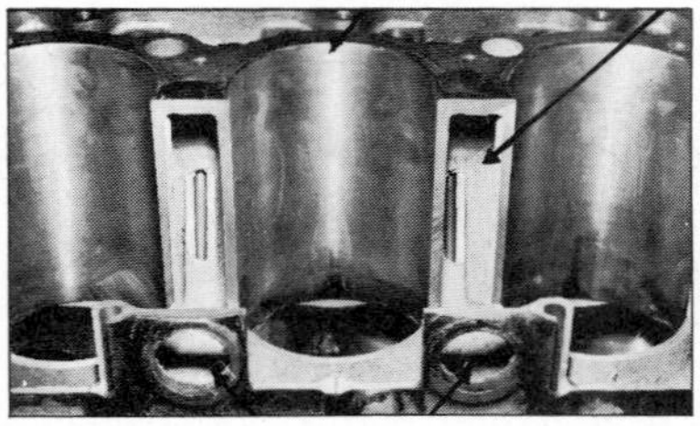

Chapter 7 Engine Construction: Cylinder Block, Cylinder Head, Crankshaft, and Bearings

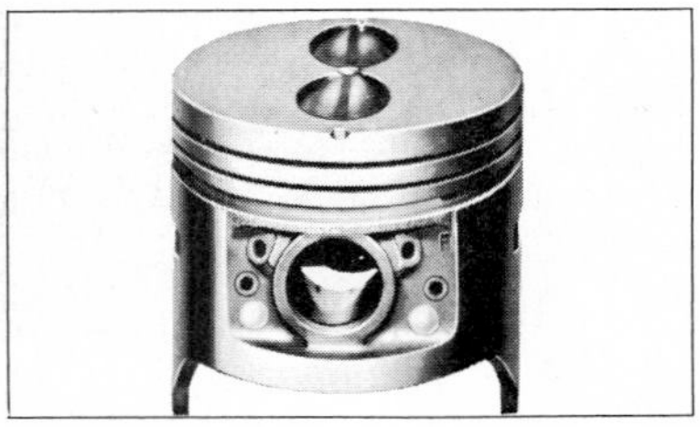

Chapter 8 Engine Construction: Pistons, Rings, and Rods

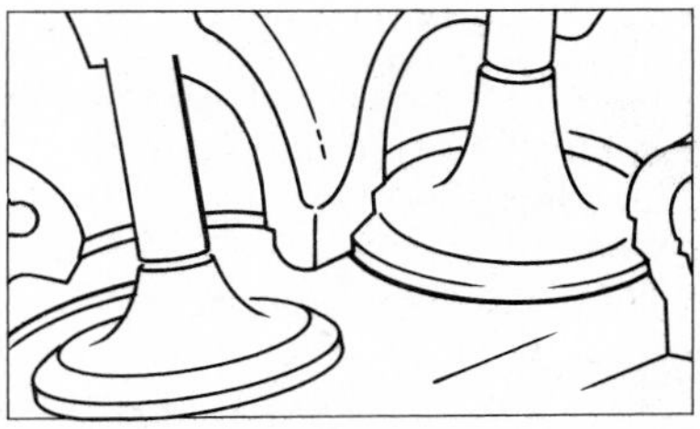

Chapter 9 Engine Construction: Valves and Valve Trains

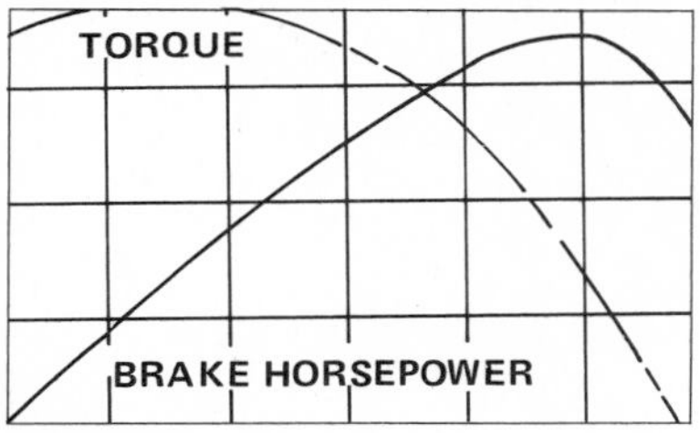

Chapter 10 Engine Measurements and Performance

CHAPTER 4
Engine Fundamentals

After reading this chapter, you should be able to:

- **Discuss chemical reactions such as combustion and how it produces heat.**
- **Discuss the effects of heat on various materials such as ice, iron, and gases.**
- **Explain atmospheric pressure and vacuum.**

This chapter explains what makes engines run. Later chapters discuss the different kinds of automotive engines in detail.

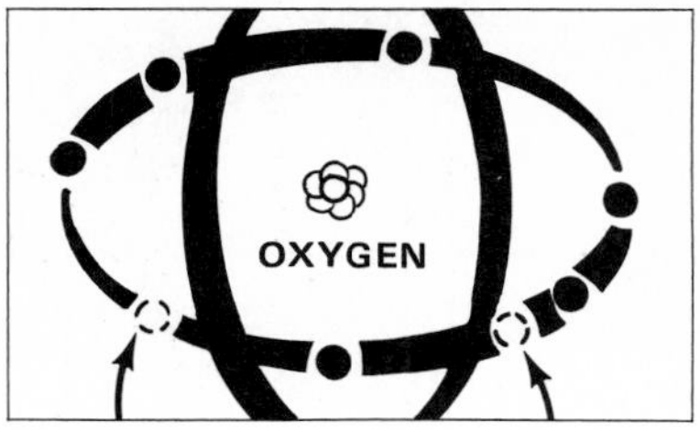

4-1 Atoms

You might think it strange to start this chapter on engines by talking about atoms. But an engine will not run until atoms start getting together inside the engine. So let's see what atoms are.

Take a look at all the different things around you—this book, your chair, the window, trees, buildings, and so on. You're looking at atoms. All these things are made of metals, wood, paper, plastics, clay, water, air, and thousands of other materials. But all these different materials are made up of only a few basic "building blocks" we call atoms. There are only about 100 different kinds of atoms. But they can be put together in millions of ways to form millions of different substances. You can compare this with the 26 letters of our alphabet. These letters can be put together in many different ways to make the several hundred thousand words in our language.

Now, about those 100 or so kinds of atoms. We have a special name for each kind: copper, iron, carbon, oxygen, silver, gold, uranium, aluminum, and so on. The copper in a penny is made up of trillions of one kind of atom. The oxygen in the air you breathe is made up of a great number of another kind of atom. Any substance made up of only one kind of atom is called an *element.* Copper is an element. So is oxygen. So are hydrogen and all the others listed in the table of elements in Fig. 4-1.

4-2 Size of Atoms

Atoms are very small. There are more than 100 billion billion atoms in a single drop of water! That's about 30 billion atoms for every person on earth. Suppose you tried to count your share—your 30 billion. It would take you 1,000 years if you counted one atom every second, day and night. You must admit that atoms are very small!

4-3 Inside the Atom

If you were able to look inside an atom, you would find that it is mostly empty space. Take the hydrogen atom, the simplest atom of all, as an example. In the center, or "nucleus," you would find a single particle. Whirling around this particle, in an almost circular path, or "orbit," you would find a second particle. And that's all. The center particle in hydrogen is called a *proton.* The outer particle is called an *electron.* We shall say much more about electrons in the chapters on electricity, because electrons in motion make up an *electric current.*

Now back to our atom of hydrogen. How close are the two particles? Well, suppose the proton were the size of a marble. If you laid this marble under one basket of a basketball court (84 feet long) [25.603 m], the electron would be at the other basket. And in between—nothing! Nothing but empty space.

4-4 The Proton and Electron

The proton has a tiny charge of positive electricity. Positive electricity is indicated by a plus (+) sign. The electron has a tiny charge of negative electricity. Negative electricity is indicated by a negative (−) sign.

Opposites attract. Positive attracts negative. Negative attracts positive. Thus, the negative electron is attracted toward the positive proton. But this inward-pulling force is balanced by the outward pull of centrifugal force. This is somewhat like the balancing of forces you get when you whirl a ball on a rubber band around your hand

TABLE OF ELEMENTS

NAME	SYMBOL	ATOMIC NUMBER	APPROXIMATE ATOMIC WEIGHT	ELECTRON ARRANGEMENT
Aluminum	Al	13	27	.2)8)3
Calcium	Ca	20	40	.2)8)8)2
Carbon	C	6	12	.2)4
Chlorine	Cl	17	35.5	.2)8)7
Copper	Cu	29	63.6	.2)8)18)1
Hydrogen	H	1	1	.1
Iron	Fe	26	56	.2)8)14)2
Magnesium	Mg	12	24	.2)8)2
Mercury	Hg	80	200	.2)8)18)32)18)2
Nitrogen	N	7	14	.2)5
Oxygen	O	8	16	.2)6
Phosphorus	P	15	31	.2)8)5
Potassium	K	19	39	.2)8)8)1
Silver	Ag	47	108	.2)8)18)18)1
Sodium	Na	11	23	.2)8)1
Sulfur	S	16	32	.2)8)6
Zinc	Zn	30	65	.2)8)18)2

Fig. 4-1 A list of a few common elements.

(Fig. 4-2). The rubber band pulls the ball toward your hand. But centrifugal force pushes the ball away. As a result, the ball moves in an orbit, a circle, around your hand.

● 4-5 Helium

The simplest atom is hydrogen, a gas. It has one proton and one electron. Next, as we go from the simplest to the more complex atoms, is helium, another gas. The helium atom has two protons (two + charges) in its nucleus, or center. It has two electrons (two − charges) circling the nucleus (Fig. 4-3). The nucleus also contains two particles called *neutrons*. They have no electric charge. The neutrons seem to serve as a nuclear "glue" to hold the protons together. For, just as unlike charged particles attract, so do like charged particles repel each other. Without neutrons, the protons in the nucleus would fly apart, and we would not have any atoms except hydrogen atoms.

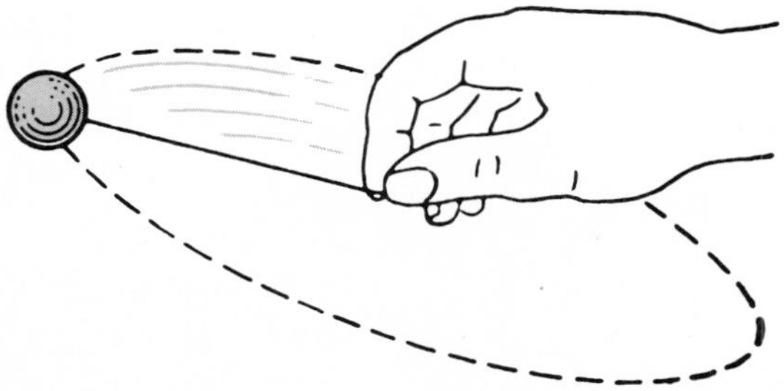

Fig. 4-2 The electron in a hydrogen atom circles the proton. This is like a ball on a rubber band swung in a circle around your hand.

● 4-6 More Complex Atoms

The next more complex element after helium is lithium, a very light metal. The lithium atom has three protons and four neutrons in its nucleus. Three electrons, one to balance each proton, orbit the nucleus.

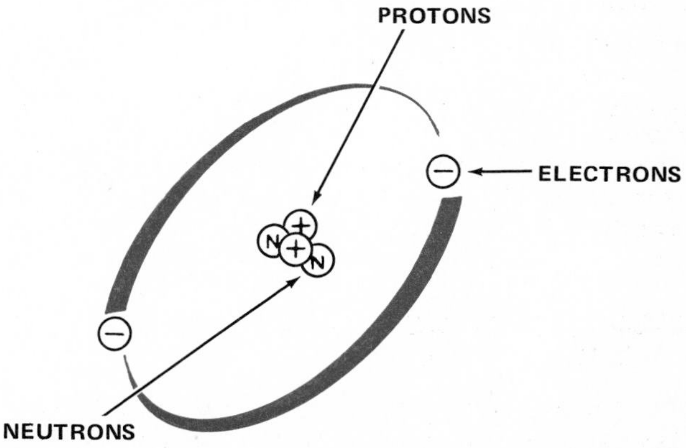

Fig. 4-3 The helium atom has two electrons circling the nucleus. The nucleus contains two protons and two neutrons.

Next comes beryllium, another metal, with four protons, five neutrons, and five electrons. Then carbon with six, six, and six; then nitrogen with seven, seven, and seven; then oxygen with eight, eight, and eight; and so on. Note that each atom normally has the same number of electrons as protons. This makes the atom electrically neutral, since negative charges balance positive charges. However, many kinds of atoms do not hold their electrons very well. Electrons are forever wandering off, leaving electrically unbalanced atoms behind (with + charges). These wandering electrons give us electricity, as we shall see later.

● 4-7 Chemical Reactions

Atoms of different elements can get together, and actually link up. The process is called a *chemical reaction.* Chemical reactions go on around us all the time. The burning of gasoline in the automobile engine is a chemical reaction. Whenever atoms react, or link up, they form molecules. For example, atoms of hydrogen and oxygen can react to form a molecule of water (Fig. 4-4). Water has the chemical formula H_2O. This means that each water molecule has two atoms of hydrogen and one atom of oxygen.

Another chemical reaction occurs when one atom of carbon unites with two atoms of oxygen, as shown in Fig. 4-5. The result is a molecule of carbon dioxide, or CO_2.

During a chemical reaction, electrons move from an atom of one element to an atom of another element. This exchange of electrons produces an unbalance of electric charge. That is, the atoms that gain electrons take on negative electric charges. The atoms that lose electrons take on positive electric charges. This difference in electric charges holds the atoms together in a molecule.

● 4-8 Combustion

Combustion, or fire, is a common chemical reaction. In combustion, oxygen in the air combines with other elements, such as hydrogen or carbon. This is what happens in the automotive engine. A mixture of air and gasoline vapor is compressed and then ignited, or set on fire. The air is about 20 percent oxygen. Gasoline is mostly hydrogen and carbon (and is thus called a *hydrocarbon*).

In the engine, oxygen atoms unite with hydrogen atoms in the gasoline to form H_2O as water (Fig. 4-4). Oxygen atoms also unite with carbon atoms to form CO_2 or carbon dioxide (Fig. 4-5). During combustion, the temperature may go as high as 6000°F [3316°C]. This high temperature produces the pressure which makes the engine run. More about that later.

With ideal (perfect) combustion, all the hydrogen and carbon in the gasoline would be made into harmless H_2O (water) and CO_2 (carbon dioxide). However, in the engine we do not get ideal combustion. Some hydro-carbons are left over. Also, some carbon monoxide (CO) is produced

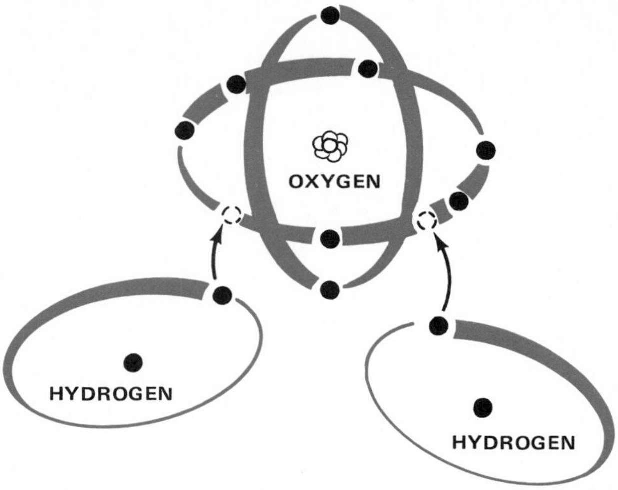

Fig. 4-4 One atom of oxygen uniting with two atoms of hydrogen to form one molecule of water (H_2O).

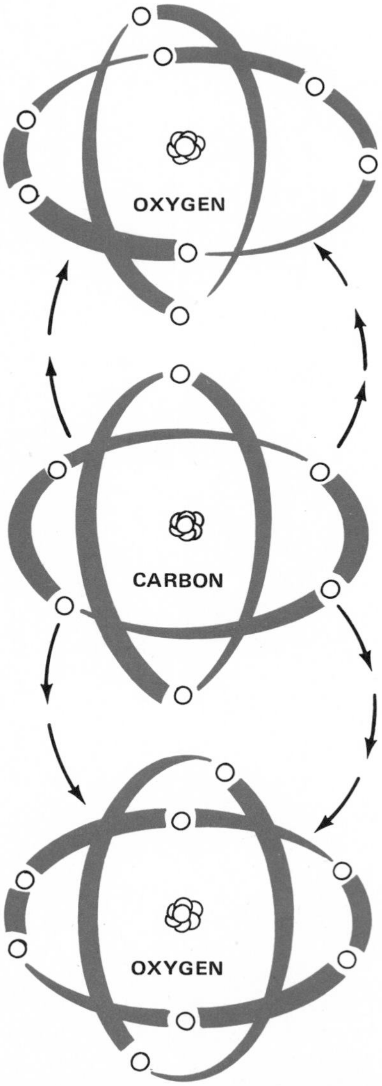

Fig. 4-5 Two atoms of oxygen uniting with one atom of carbon to form a molecule of carbon dioxide (CO_2).

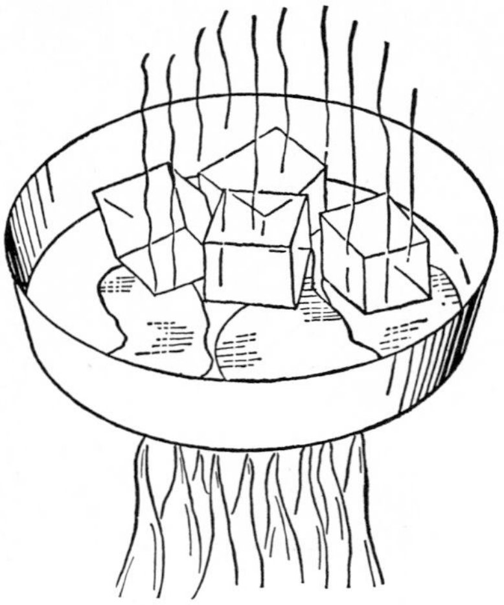

Fig. 4-6 The ice cubes in a pan over a fire will first melt, or turn to water. Then the water will evaporate, or vaporize. Each of these changes, from a solid to a liquid to a vapor, is called a *change of state*.

instead of some CO_2. These products contribute to the pollution problem. They are the reason for the emission controls on modern cars. These controls and how they work are described in later chapters.

● 4-9 Heat

We mentioned that combustion produces high temperatures. But what do we mean by "high temperatures" and "heat"? Actually, heat is the rapid motion of atoms or molecules in a substance.

That may be a little hard to understand. First, you must realize that the atoms and molecules of any substance are in rapid motion. Even though a piece of iron appears solid and motionless, its atoms and molecules are moving rapidly. The atoms in a piece of hot iron move faster than the atoms in a piece of cold iron. It's that simple.

● 4-10 Change of State

Atoms and molecules in motion may be a little hard to understand, as we said. But let's look at "change of state" to see if we can't make it a little clearer. Suppose we put a pan of ice cubes over a fire (Fig. 4-6). First, the ice cubes melt, or turn to water. Then the water boils, or vaporizes. Each of these changes is a change of state. There are three states: solid, liquid, and gas or vapor.

A change of state is caused by changes in the speed of the atoms and molecules in a substance. In the ice, for example, the molecules of water move slowly and in restricted paths. That is, they don't wander very far. But as the temperature increases, the molecules move faster. Soon, the molecules are moving so fast that they break out of their restricted paths. That is when the ice turns to water (at 32°F) [0°C]. As the temperature increases still

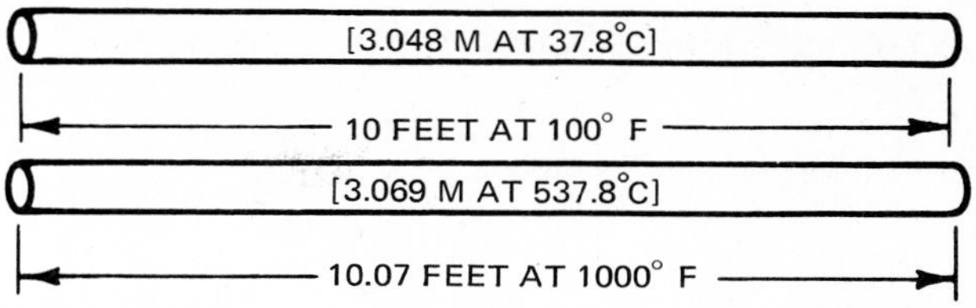

Fig. 4-7 A steel rod that measures 10 feet [3.048 m] at 100°F [37.8°C] will measure 10.07 feet [3.069 m] at 1000°F [537.8°C].

more, the molecules reach speeds that allow them to fly clear out of the water. That is, the water boils, or turns to vapor. This happens when the water reaches 212°F [100°C] (at sea-level pressure).

● 4-11 Expansion of Solids with Heat

When a piece of iron is heated, it expands. Suppose a steel rod measures 10 feet [3.048 m] in length at 100°F [37.8°C]. It will measure 10.07 feet [3.069 m] in length at 1000°F [537.8°C] (Fig. 4-7). Here is the reason: As the rod is heated, the molecules in it move faster and faster. They need more room for this, and so they push neighboring molecules away. The rod gets longer and wider.

Expansion of solids with heat is important in the engine. As the engine goes from cold to hot, the engine parts expand. This expansion must be taken into account when the engine is designed. It is also important in engine servicing.

● 4-12 Expansion of Liquids and Gases

Liquids and gases also expand when heated. One cubic foot [0.0283 m^3] of water at 39°F [3.89°C] becomes 1.01 cubic feet [0.0286 m^3] when heated to 100°F [37.8°C]. One cubic foot of air at 32°F [0°C] becomes 1.14 cubic feet [0.0323 m^3] when heated to 100°F [37.8°C] without a change of pressure. These expansions are also the result of molecules moving faster. The faster-moving molecules spread out more and take up more room.

● 4-13 Increase of Pressure

Suppose we don't let the molecules spread out when we heat them up. We can do this by holding the volume constant and heating a cubic foot of air from 32° to 100°F [0° to 37.8°C]. If we start with a pressure of 15 psi (pounds per square inch) [103.4 kPa], we find that the pressure increases to about 17 psi [117.2 kPa].

The higher pressure is also due to the higher speed of the moving molecules. The molecules keep hitting the inside surfaces of the container (Fig. 4-8) as they move about. With higher temperatures, the molecules move faster. They hit the surfaces harder and more often. The result is an increase in pressure.

Another way to increase the pressure in a container of gas is to compress it into a smaller space. This is what

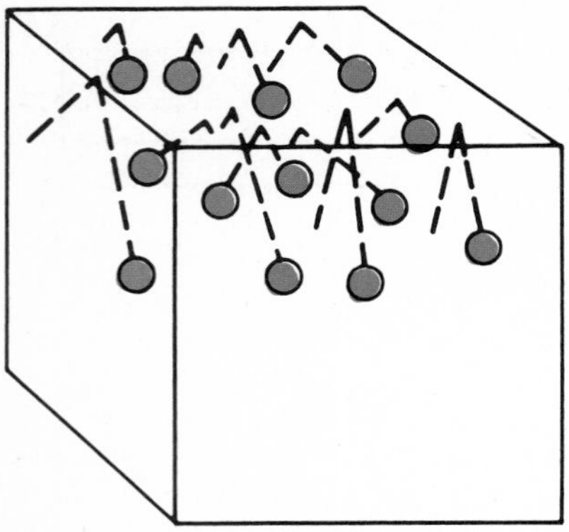

Fig. 4-8 Gas pressure in a container is the result of the ceaseless bombardment of the inner sides of the container by the fast-moving molecules of gas. The bombardment is shown on only one side of the container, for simplicity. Actually, the bombardment is taking place on all inside surfaces. And there are billions upon billions of molecules, and not just the few shown here.

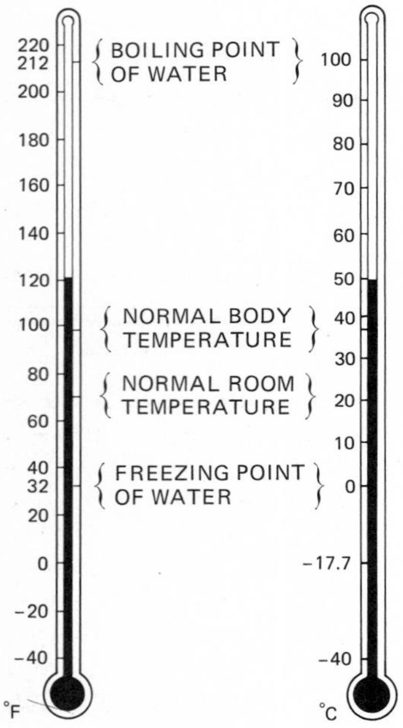

Fig. 4-9 Fahrenheit and Celsius thermometers.

happens in the engine cylinders. The mixture of air and gasoline vapor is squeezed into one-eighth or one-ninth of its original volume. This puts the molecules closer together. They hit the pistonhead far more often. That is, the pressure goes up.

The pressure gets even higher when the compressed air-fuel mixture is ignited. As we have explained, this makes the molecules move much faster. They bombard the pistonhead still harder. These billions upon billions of molecules, moving at great speeds, hit the pistonhead so hard and so often that they add up to a total push of a ton or more. This push, or pressure, is due only to the pounding of the fast-moving molecules.

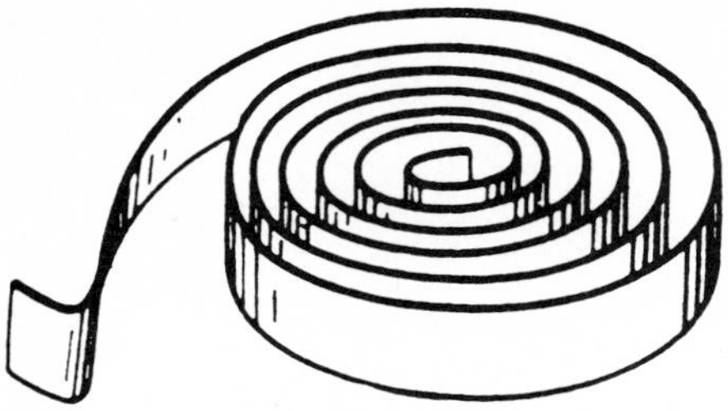

Fig. 4-10 Coil-type thermostat. The coil winds up or unwinds as the temperature changes. This motion can be used to operate a control.

4-14 Increase of Temperature

When a gas is compressed, its temperature also increases. Having the molecules closer together causes them to bump into one another more often. That makes them move faster. In other words, the temperature of the gas goes up. In the diesel engine, air is compressed to one-sixteenth of its original volume. This sends the temperature of the air as high as 1000°F [537.8°C].

Of course, the heat produced by compressing a gas does not stay in the gas. It soon escapes from the compressed gas to the container and then to the outside air. Any hot object loses heat until its temperature is the same as the temperature around it.

4-15 The Thermometer

The thermometer (Fig. 4-9) is a familiar example of expansion with heat. It is just a hollow glass tube partly filled with a liquid, such as mercury. As the temperature increases, the mercury expands. Part of it is forced up the tube. The higher the temperature, the more the mercury expands, and the farther up the tube the mercury rises. The tube is marked off in degrees to indicate the temperature.

4-16 The Thermostat

Different metals expand different amounts when they are heated. For example, aluminum expands about twice as much as iron. Such differences in expansion are used in many thermostats. One type is shown in Fig. 4-10. It is a coil made of two strips of different metals, such as brass and steel, welded together. When the coil is heated, one metal expands more than the other. This causes the coil to wind up or unwind. The motion of the coil can be used to control liquids or electricity. It is used in this manner in several places in the automobile.

4-17 Gravity

Gravity is the attractive force between the earth and all other objects. When we release a stone from our hand, it falls to earth. When a car is driven up a hill, part of the

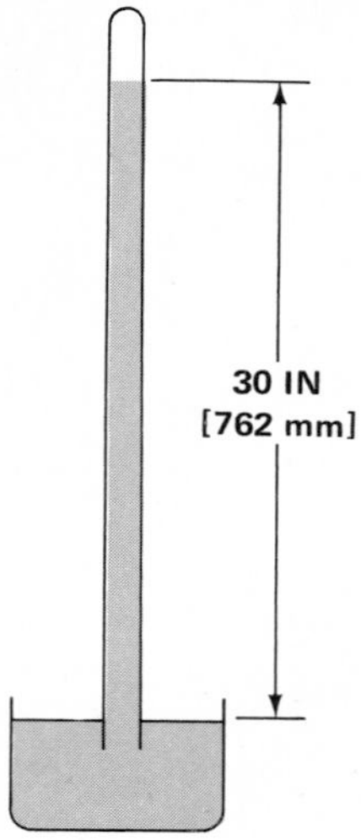

Fig. 4-11 Barometer. The mercury will stand in the tube about 30 inches [762 mm] above the surface of the mercury in the dish when atmospheric pressure is 15 psi [103.42 kPa].

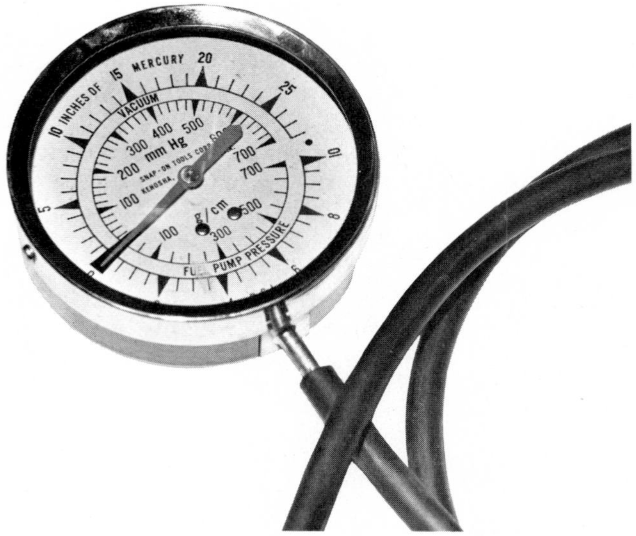

Fig. 4-12 Vacuum gauge. (*Snap-On Tools Corporation*)

engine power is used to raise the car against gravity. Likewise, a car coasts down a hill with the engine turned off, because gravity pulls downward on the car.

Gravitational attraction is usually measured in terms of weight. We put an object on a scale and note that it weighs 10 pounds [4.536 kg]. What we mean is that the earth registers that much pull on it. Gravitational attraction gives any object its weight. Astronauts out in space, where the earth's gravitational attraction is almost zero, are weightless. They float freely around in the interior of their spaceships.

● 4-18 Atmospheric Pressure

The air is also pulled toward the earth by gravity. At sea level and average temperature, a cubic foot of air weighs about 0.08 pound, or about 1.25 ounces (0.035 kilogram). This seems like very little. But the blanket of air surrounding the earth—our atmosphere—is many miles thick. It is like thousands of cubic feet of air piled one on top of another, all adding their weight. The total weight, or downward push, is about 15 pounds for every square inch [103.4 kPa]. Since the human body has a total surface area of several square feet, the weight of the air pressing on your skin amounts to several tons!

Why doesn't all that pressure crush you? It can't because pressures inside your body balance this atmospheric pressure.

Atmospheric pressure also affects the amount of power that an engine can produce. At sea level, an engine gets more air and produces more power. In Denver, Colorado, which is a mile above sea level, an engine gets less air and produces less power.

● 4-19 Vacuum

A vacuum is the absence of air or any other matter. Astronauts on their way to the moon soon pass through our atmosphere and into a vast region of empty space. This is a vacuum.

● 4-20 Producing a Vacuum

There are many ways to produce a vacuum here on earth, in a small volume. The automobile engine, as it operates, produces a partial vacuum in the engine cylinders. Atmospheric pressure then pushes the air-fuel mixture into the cylinders. This is how the engine gets the air-fuel mixture it needs to run.

1. Barometer The barometer is another device that works by vacuum. You can make a barometer by filling a long tube with mercury. (See Fig. 4-11.) Then close the end with your finger. Next, turn the tube upside down, and put the end into a dish of mercury. Now open the end. Some of the mercury will run out of the tube. This will leave the upper end of the tube empty (a vacuum). The barometer is used to measure atmospheric pressure. Changes in atmospheric pressure tell what the weather is going to be. If the atmospheric pressure drops, the mercury "falls," and a storm is probably on the way.

2. Vacuum gauge Vacuum gauges are used in automotive service. It is really a pressure gauge. It measures the pressure in a closed space and compares it with atmospheric pressure. If the pressure in the closed space is lower, the vacuum gauge records so many "inches of mercury." Figure 4-12 shows a vacuum gauge used to check automotive engines. When the vacuum gauge is connected, it measures the amount of vacuum in the engine. If the measurement is low or unsteady, then there is engine trouble.

The vacuum gauge contains a bellows, or diaphragm, linked to a needle on the dial face. When vacuum is applied, the bellows or diaphragm moves.This causes the needle to move, to show the amount of vacuum.

● 4-21 Humidity

Almost all air has some water vapor (evaporated water) in it. When the air is carrying a good deal of water vapor, the air is said to be *humid*. That is, it has high humidity. You find humid air around bodies of water. Air over deserts has low humidity. It has very little water vapor. Humidity is measured in terms of percentages. Zero percent humidity means the air has no water vapor. Hundred percent humidity means the air is holding all the water vapor it can hold. A reading of 50 percent humidity means that the air is holding half as much water vapor as it could.

Humidity affects engines. An engine puts out less power in hot, dry conditions than in cool, moist conditions. The cool air is denser (molecules are closer together) so that more air enters the engine and this means more power. At the same time, the moisture seems to make the engine run more smoothly.

● 4-22 Atmospheric Factors Affecting Combustion

Changes in temperature, atmospheric pressure, and humidity affect combustion in the engine. They affect the way the fuel burns and the power output of the engine. Accurate testing of engines requires that all readings be corrected to account for temperature, atmospheric pressure, and humidity.

REVIEW QUESTIONS

See if you can answer the following questions about engine fundamentals.

1. What two particles make up an atom of hydrogen?
2. How many particles are there in an atom of helium? In an atom of oxygen?
3. When atoms link up during a chemical reaction, what is formed?
4. With ideal combustion of gasoline, what two substances would be formed?
5. In terms of atoms and molecules, what is heat?
6. What is a change of state? How many states are there?
7. What happens to the temperature when a gas is compressed?
8. What is a coil thermostat? How does it work?
9. What is atmospheric pressure? What causes it?

SELF PROJECTS

For complete automotive servicing, several thermometers are needed to measure various temperatures. For example, you may need to know the temperature of the electrolyte (liquid in the battery) when checking the battery state of charge. You may need to measure the temperature of the coolant (liquid) in the engine cooling system. Other temperatures to be measured include the cooling-system thermostat operating temperature, temperature of the air coming from the air-conditioning ducts in the car, operating temperature of the thermostat in the carburetor air cleaner, and so on. As you study the later chapters in the book, you will find explanations of the various temperature readings and how to take them. Obviously, you cannot use any one thermometer to take all these different measurements. Therefore, if you are going to be an all-round mechanic, you would need several thermometers. Collect them as you go along and keep them in a safe place in your toolbox.

CHAPTER 5
Piston-Engine Operation

After reading this chapter, you should be able to:

- **Explain how the reciprocating motion of the piston is changed to rotary motion of the crankshaft.**
- **List the parts in the valve train and explain how the valve train works.**
- **Describe the four piston strokes and explain what happens during each.**

There are two basic types of engines, external-combustion and internal-combustion. The external-combustion engine has its fuel burned outside the engine. Steam engines are external-combustion engines. The steam engine runs on steam produced by boiling water outside the engine. The internal-combustion engine burns fuel inside the engine. The engines in automobiles are all internal-combustion engines.

● 5-1 Internal-Combustion Engines

Internal-combustion (or IC) engines can be of two types, reciprocating and rotary (Fig. 5-1). The type used in almost all automobiles is the reciprocating engine. In this engine, pistons move up and down, or reciprocate (Fig. 5-1*A*). The other kind of IC engine is the rotary type, in which rotors turn or rotate (Fig. 5-1*B*). We describe the reciprocating engine in this chapter. In a following chapter we describe rotary engines, the Wankel and turbine. The Wankel is used in a number of automobiles, but the reciprocating engine is by far the more common.

● 5-2 The Engine Cylinder

Most automotive engines have four, six, or eight cylinders. Since the same actions take place in all cylinders, let us study one cylinder to learn about piston-engine operation. Figure 5-2 shows an end view of an engine, cut away so the essential parts can be seen. These parts include the piston, cylinder, combustion chamber, and valves. We shall come back to these later.

First, however, we want to explain what happens to the piston in the cylinder. Essentially, the cylinder is only a long, round air pocket, like a tin can with one end cut out. The piston is a movable metal plug that fits snugly into the cylinder. Although the fit is snug, the piston is loose enough to slide up and down in the cylinder. Figure 5-3 shows what happens.

In Fig. 5-3*A*, the piston is below the cylinder. (The cylinder is drawn as though it were transparent, so you can see what goes on inside.) In Fig. 5-3*B*, the piston has been pushed up into the cylinder. This upward movement of the piston traps air above it. The air has no place to go, so it is compressed, or squeezed, into a smaller volume. Now, suppose that the air had some gasoline vapor in it. If we could ignite this compressed-air–gasoline-vapor mixture, we would have an explosion. As shown in Fig. 5-3*C*, the explosion would blow the piston out of the cylinder.

In the actual engine, other parts keep the piston from blowing out of the cylinder. These other parts also push the piston back up to the top of the cylinder again. That is, the piston is kept moving up and down, or reciprocating, in the cylinder.

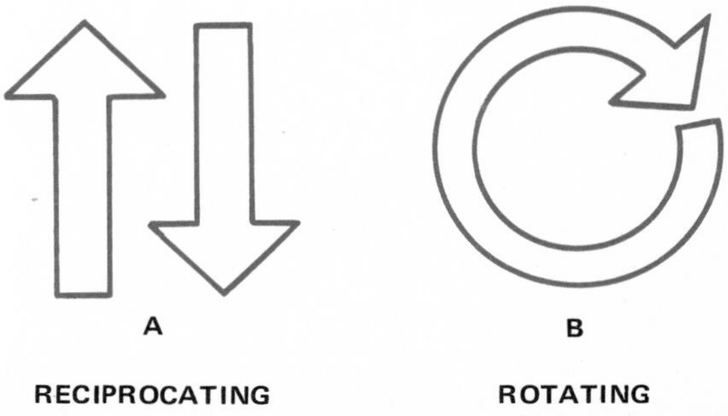

Fig. 5-1 Reciprocating motion is up-and-down or back-and-forth motion as contrasted with rotary (rotating) motion.

5-3 Pistons and Piston Rings

Figure 5-4 shows an actual piston for a modern automotive engine. It is made of aluminum alloy (a mixture of aluminum and other metals). It is strong enough to take the heavy pressure of the burning air-fuel mixture in the cylinder. The typical piston is about 4 inches [101.6 cm] in diameter and weighs about 1 pound [0.454 kg].

The piston must be fairly tight in the cylinder, but not too tight. If it were too loose, part of the air-fuel mixture would leak past the piston and be wasted. If the piston were too tight, it could stick and ruin the engine.

Piston rings are used to provide a good sliding seal between piston and cylinder. Figure 5-5 shows a piston

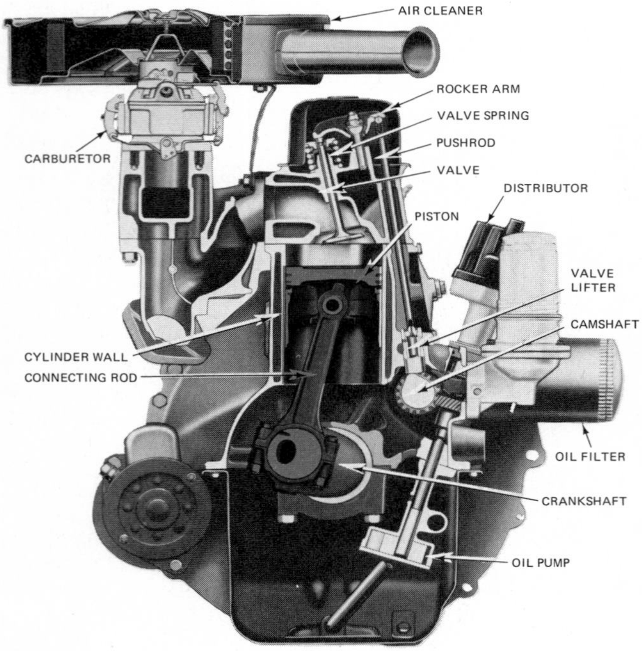

Fig. 5-2 Sectional view of an engine from the end, showing the piston in one of the cylinders. (*Ford Motor Company*)

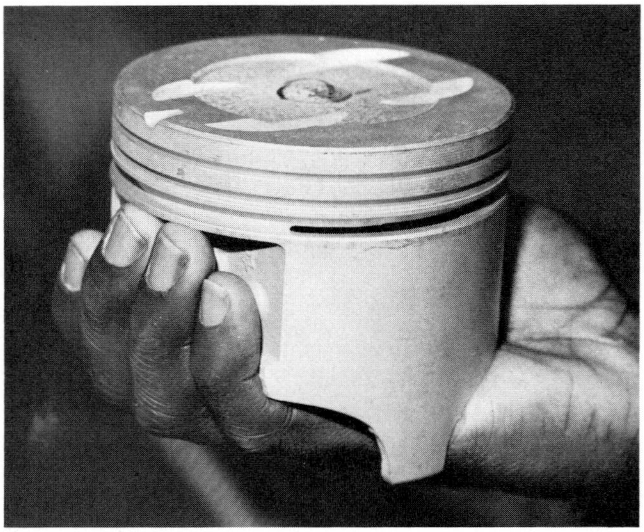

Fig. 5-4 Piston from a piston engine. (*Piedmont Virginia Community College*)

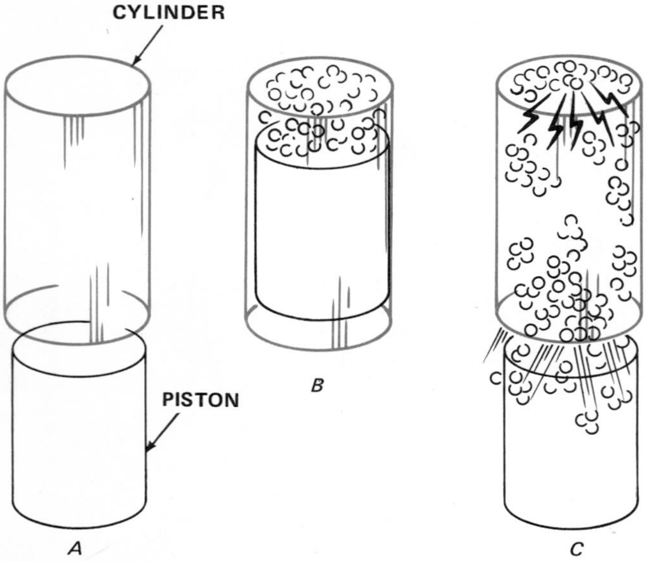

Fig. 5-3 Three views showing the actions in an engine cylinder. (*a*) The piston is a metal plug that fits snugly into the cylinder. (*b*) When the piston is pushed up into the cylinder, air is trapped and compressed. The cylinder is drawn as though it were transparent, so the action can be seen. (*c*) The increase in pressure, as the mixture of air and gasoline ignites, pushes the piston out of the cylinder.

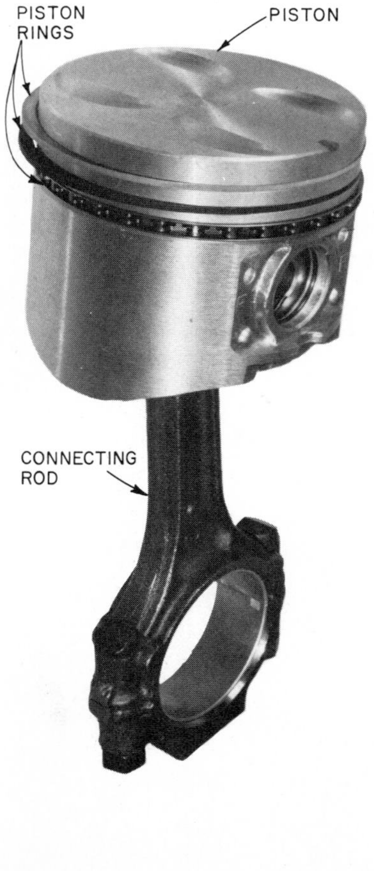

Fig. 5-5 Piston with piston rings. (*Chrysler Corporation*)

with piston rings ready to be put in a cylinder. Each ring is split so it can be opened up and slipped over the piston. The rings then are slipped into grooves in the piston. When the piston is placed in the cylinder, the rings are compressed into the grooves. The split ends of the ring almost come together. The rings fit tightly against the cylinder wall and against the sides of the ring grooves in the piston. Thus, they form a good seal between the piston and cylinder wall. However, the piston and piston rings can slide up and down easily in the cylinder.

● 5-4 Reciprocating to Rotary Motion

The piston moves up and down in the cylinder. It reciprocates. This reciprocating, or straight-line, motion must be changed to turning, or rotary, motion so the car wheels will turn. A crank and connecting rod change the reciprocating motion to rotary motion.

The crank is part of the crankshaft. Figure 5-6 shows a crankshaft. There is one crank for each cylinder in an engine that has all its cylinders in a line. The crank is an offset section of the crankshaft. As the crankshaft rotates, the crank swings in a circle around the shaft (Fig. 5-7).

The connecting rod connects the piston to the crank (Figs. 5-8 and 5-9). In Fig. 5-8, the essential parts are shown separated, in a disassembled view. The connecting-rod assembly has two parts, the rod and the rod cap. They are fastened to each other by rod-cap bolts (Fig. 5-9). This is how the connecting rod is attached to the crankpin on the crank: The connecting rod is brought down into place over the crankpin. Then the rod cap is fastened to the rod.

There is a bearing between the rod and rod cap and the crankpin. The bearing lets the crankpin rotate easily in the rod assembly. We shall discuss bearings later.

The upper end of the connecting rod is attached to the piston by a piston pin (Fig. 5-9). The piston has two holes in it; the upper end of the connecting rod has one hole. The piston pin is pushed through these holes so the piston and upper end of the rod move together.

The crank end of the connecting rod is often called the rod " big end." The piston end is called the rod "small end." The piston pin is also called the "wrist pin."

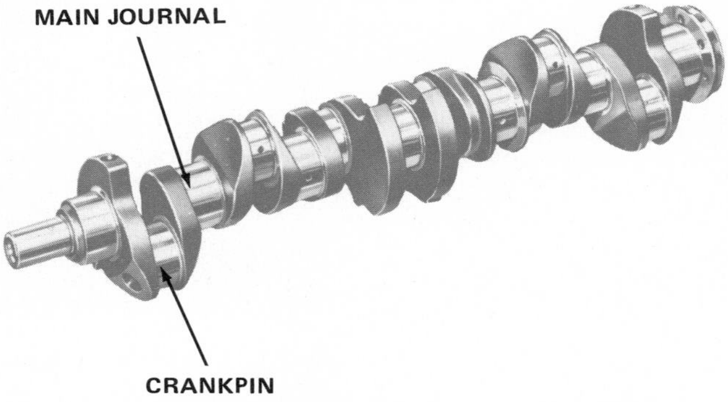

Fig. 5-6 Engine crankshaft. (*Ford Motor Company*)

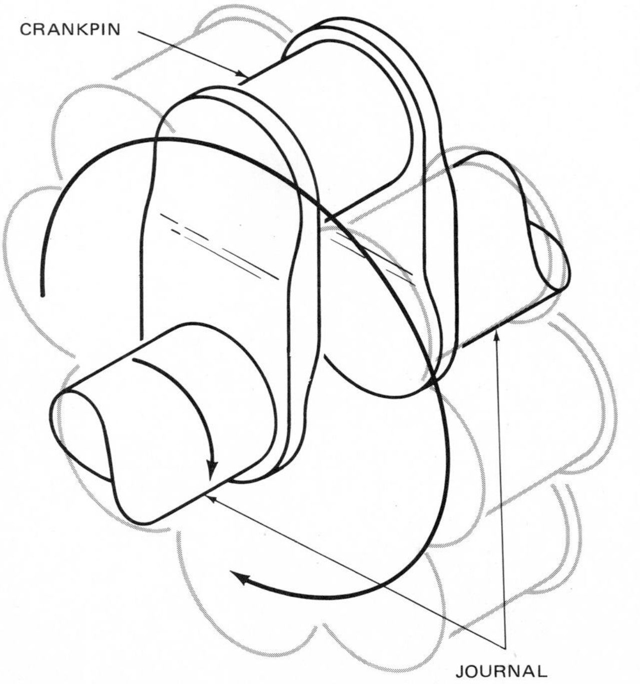

Fig. 5-7 As the crankshaft rotates, the crankpin swings in a circle around the shaft.

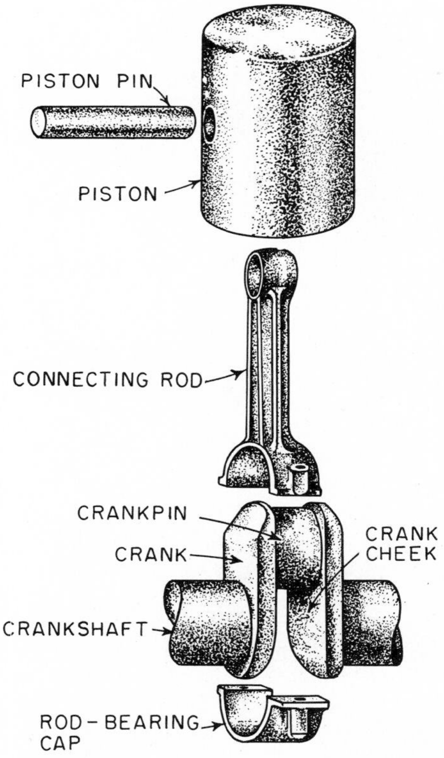

Fig. 5-8 Piston, connecting rod, piston pin, and crank of a crankshaft in disassembled view. The piston rings are not shown.

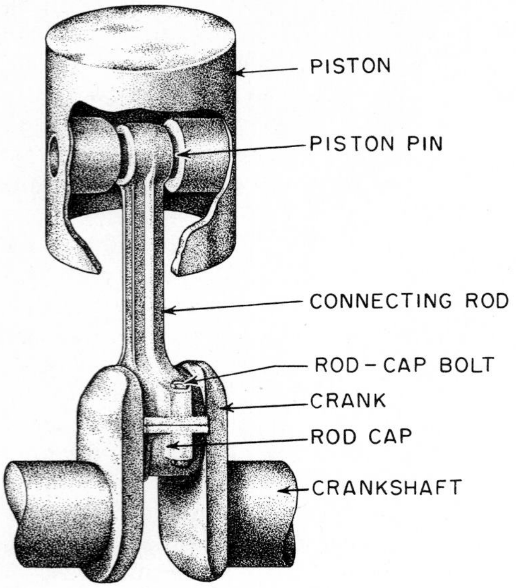

Fig. 5-9 Piston and connecting-rod assembly attached to a crankpin on the crankshaft. The piston rings are not shown. The piston is partly cut away so you can see how it is attached to the connecting rod.

Now, let us see what happens as the piston moves up and down in the cylinder (Fig. 5-10). As the piston starts down (at 2 in Fig. 5-10), the connecting rod tilts to one side. This lets the lower end of the rod follow the crankpin. The crankpin is moving in a circle around the crankshaft. Now follow the actions shown in Fig. 5-10 (from 1 through 8 and back to 1 again). It shows how the connecting rod tilts first to one side, and then to the other. The lower end of the rod moves in a circle along with the crankpin. To sum up, the crankpin moves in a circle while the piston moves up and down (Fig. 5-11).

5-5 The Valves

Combustion makes the piston move up and down in the cylinder. For that, we need some means of getting air-fuel mixture into the cylinder. And, after combustion, we need a means of getting the burned mixture out of the cylinder. Two round openings at the top of the cylinder, called the *valve ports,* permit these actions.

Valves are fitted into these two valve ports. Valves (Fig. 5-12) are carefully machined metal plugs on long stems. They fit tightly into the round openings, the valve ports. Figure 5-13 shows a valve and valve port in a

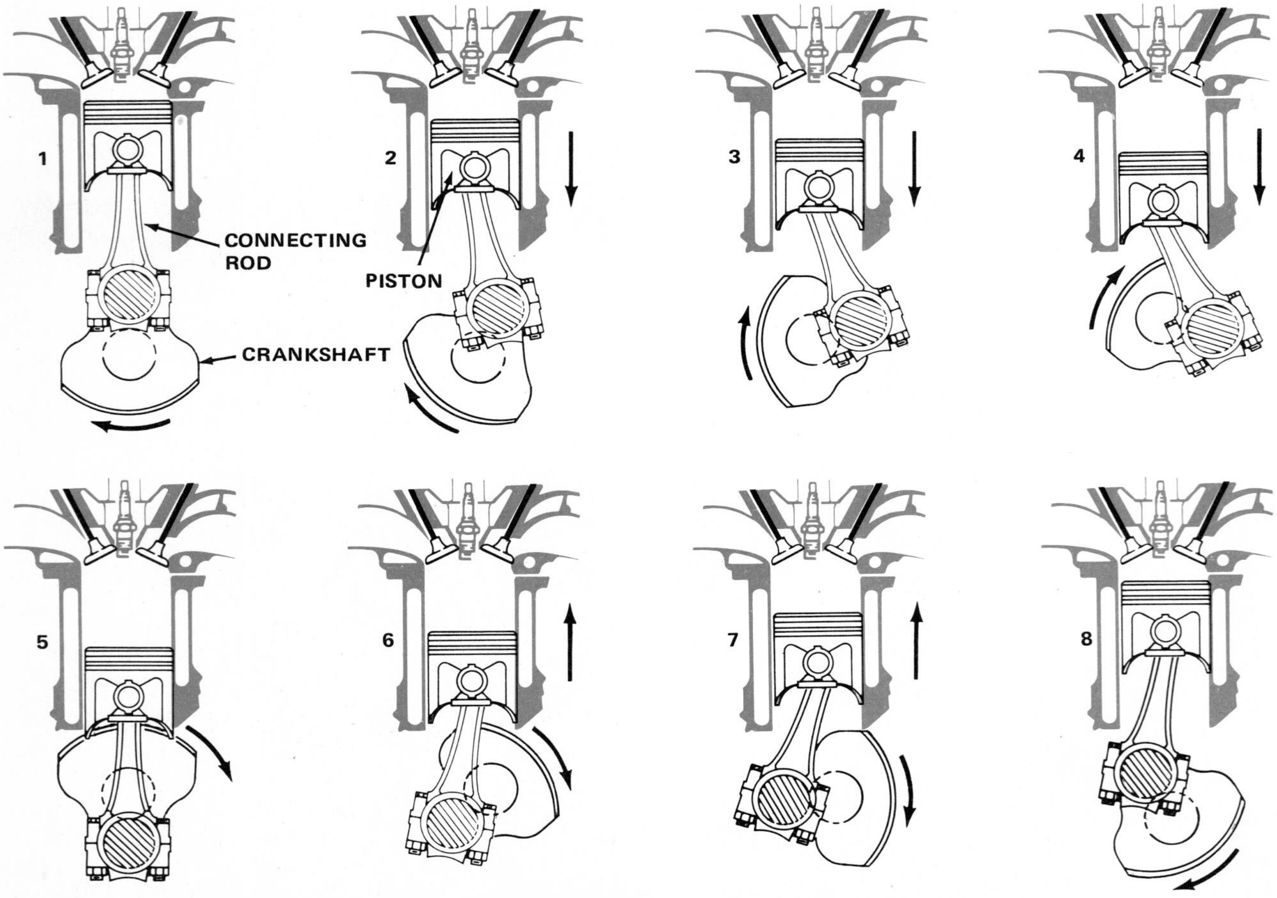

Fig. 5-10 Sequence of actions as the crankshaft completes one revolution, and the piston moves from top to bottom to top again.

cylinder head. In Fig. 5-13, the valve is in the opened position. In this position, air or burned gas can flow around the valve. At other times, the valve moves up into the closed, or "seated," position. That is, the valve completely plugs the opening, so no air or gas can get by. In the seated position, the edge of the valve (the valve *face*) fits tightly against a matching seat in the valve port.

One of the valves is called the "intake" valve. The other is the "exhaust" valve. These terms are explained later.

5-6 The Valve Train

A series of parts, called the "valve train," moves the valves up and down (closes and opens them) at the proper time. Figure 5-14 shows valve trains for several cylinders in an automotive engine. This is a little complicated, so Fig. 5-15 shows the main parts of the valve train for only one cylinder. The valve action starts at a *cam lobe* on the *camshaft*. Figure 5-16 shows a camshaft with its related parts. These are the bearings in which the camshaft rotates, and the *chain* and *sprocket* that drive the camshaft. The camshaft is driven from the crankshaft. When the crankshaft turns, the camshaft turns. This brings the cam lobes into the action. Figure 5-17 shows a cam. The lobe is a bump, or high spot, on the cam. Riding on the cam is a round cylinder called the

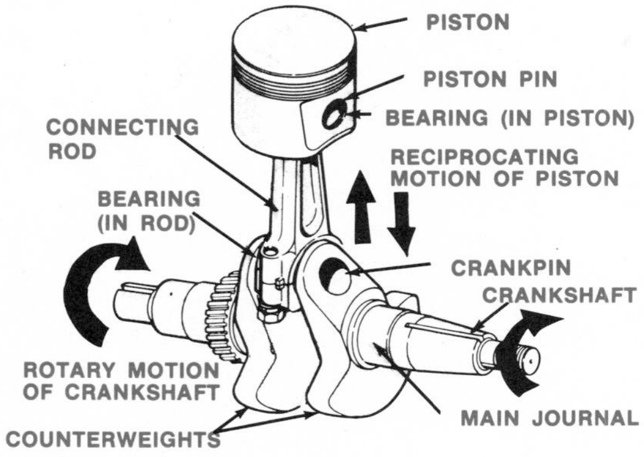

Fig. 5-11 The crankpin moves in a circle around the crankshaft while the piston moves up and down.

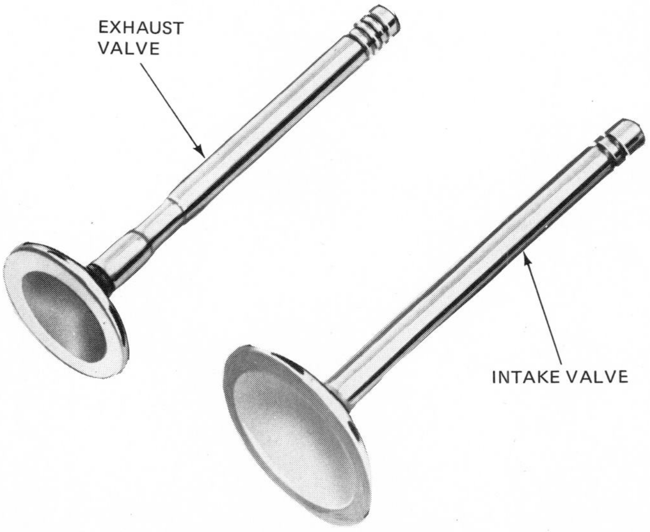

Fig. 5-12 Two valves for an engine cylinder.

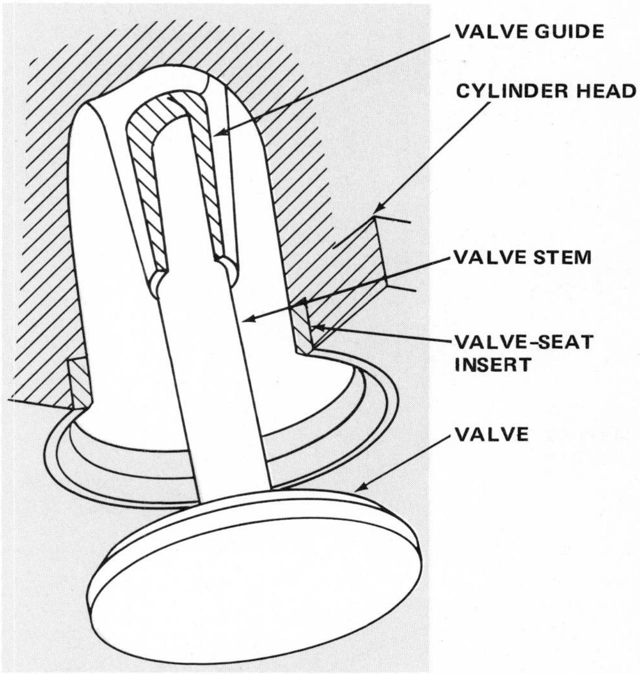

Fig. 5-13 Valve and valve seat in a cylinder head. The cylinder head and valve guide have been cut away so the valve stem can be seen. Some engines have valve-seat inserts, as shown.

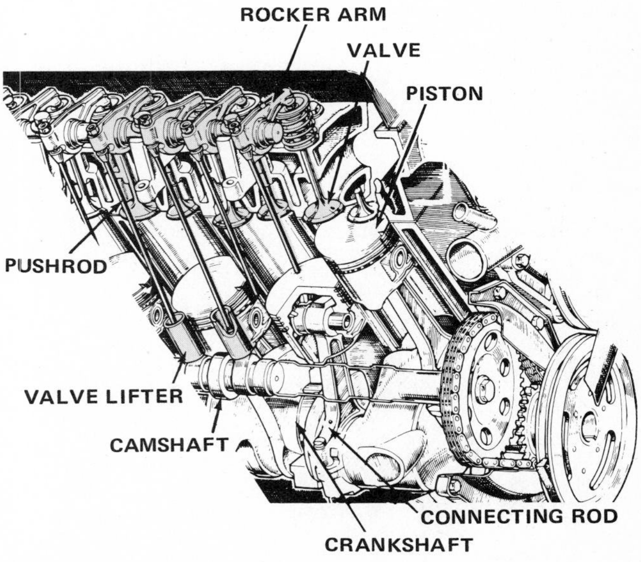

Fig. 5-14 Valve-train parts in an engine. (*Chrysler Corporation*)

valve lifter. When the lobe comes up under the lifter, it causes the lifter to move up.

Now look at Fig. 5-15 again. Note that the upward movement of the valve lifter causes the push rod to move up. The push rod pushes up on the *rocker arm.* The rocker arm is attached to a shaft at its center. When one end of the rocker arm moves up, the other end moves down, like a seesaw. As the end of the rocker arm moves down, it pushes down on the valve stem. This pushes the valve down off its seat. That is, the valve opens.

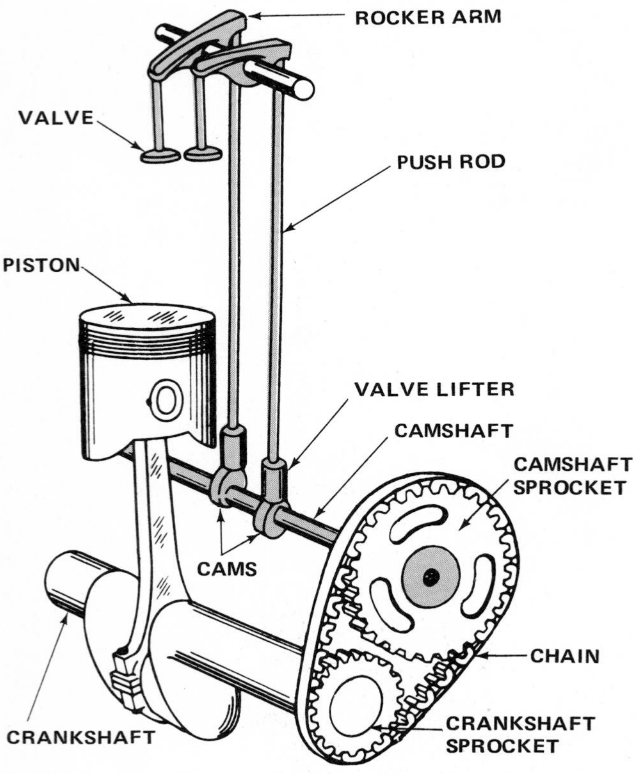

Fig. 5-15 Valve-train parts for a single cylinder of an engine. Only the essential moving parts are shown.

At other times, the valve is held closed by a spring. The lower end of the valve spring rests against the cylinder head (Fig. 5-14). The upper end of the spring rests against a washer which is locked to the end of the valve stem. When the valve is pushed open, this spring is compressed. It stays compressed until the cam lobe moves away from under the valve lifter. Then the valve spring pulls the valve back up onto its seat and closes the port.

Figures 5-18 and 5-19 show the actions in a cylinder of an engine that uses a pair of gears to drive the camshaft from the crankshaft.

Engines with overhead valves and push rods, as in Figures 5-14 and 5-15, are often called "push-rod engines." Most automotive engines are push-rod engines.

5-7 Actions in the Cylinder

Now, let's put it all together and see just what happens in an engine cylinder. The actions in a cylinder can be divided into four stages, or "strokes." A stroke is the movement of the piston from the top position to the bottom position, or from the bottom position to the top position. The top position, or upper limit of piston movement, is called top dead center (TDC). The bottom position, or lower limit of piston movement, is called bottom dead center (BDC). A piston stroke is piston movement from TDC to BDC, or from BDC to TDC.

The four strokes are called the *intake* stroke, the *compression* stroke, the *power* stroke, and the *exhaust* stroke. These four strokes make up the complete cycle of events in the cylinder. An engine that runs on this four-stroke-cycle principle is called a *four-stroke-cycle* engine. This is often shortened to "four-cycle" engine. The term "Otto cycle" is also applied to this type of engine (after Friedrich Otto, a German scientist of the nineteenth cen-

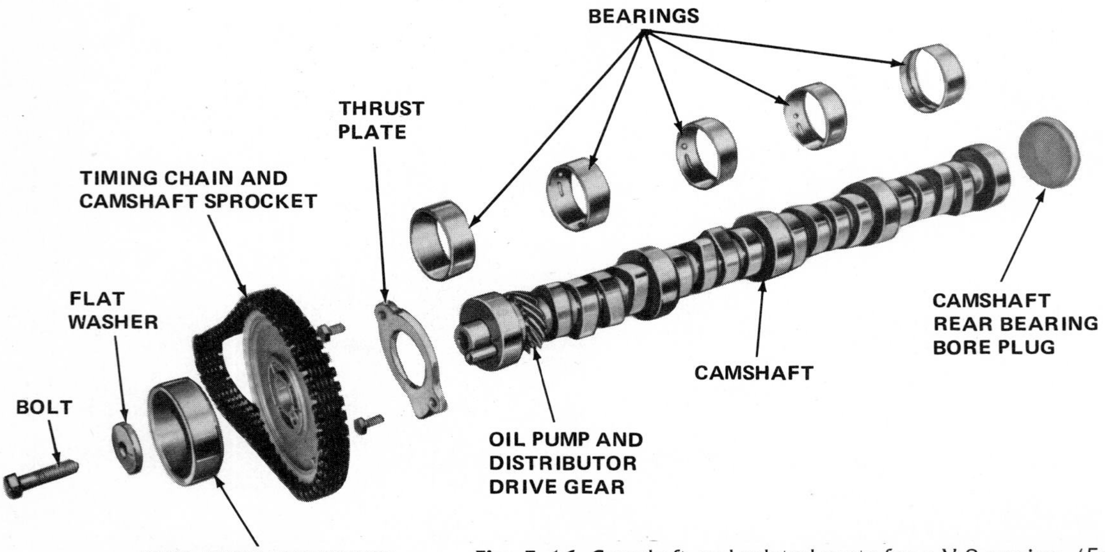

Fig. 5-16 Camshaft and related parts for a V-8 engine. (*Ford Motor Company*)

tury). Some engines are two-stroke-cycle, or two-cycle, engines. These are described later in the book.

The term "cycle" is sometimes misunderstood. A cycle is simply a series of events that repeat themselves. The cycle of the seasons (spring, summer, fall, and winter) is one example. The cycle of the four piston strokes in the four-stroke-cycle engine is another.

To simplify the following explanation, we shall assume that the valves open and close at TDC and BDC. Actually, they are not timed to open exactly at these points, as we shall explain later.

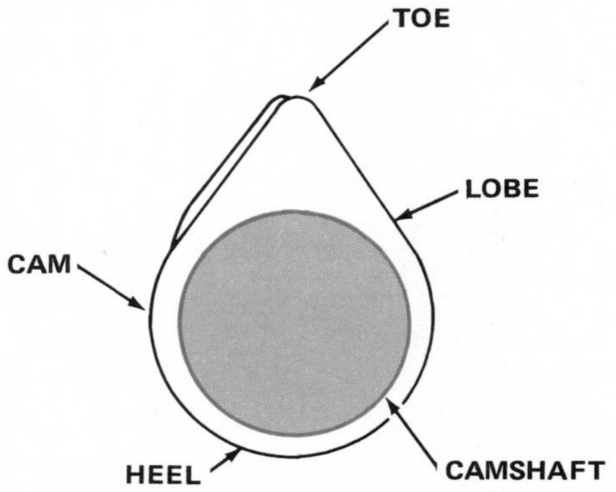

Fig. 5-17 Cam on a camshaft. Note positions of the toe and heel.

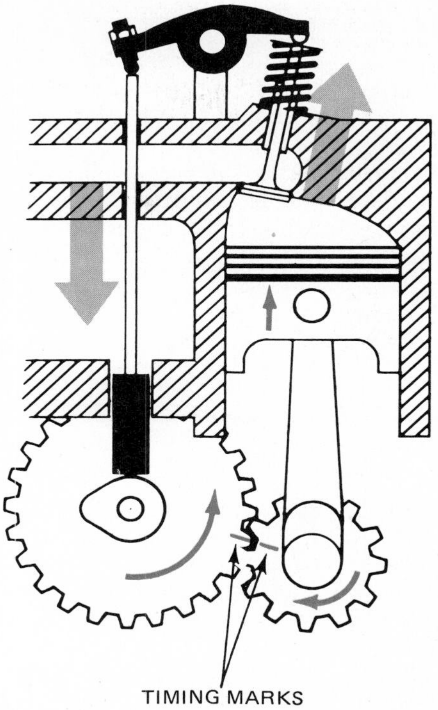

Fig. 5-19 Valve train of previous picture, showing how the valve closes when the cam lobe moves out from under the valve lifter.

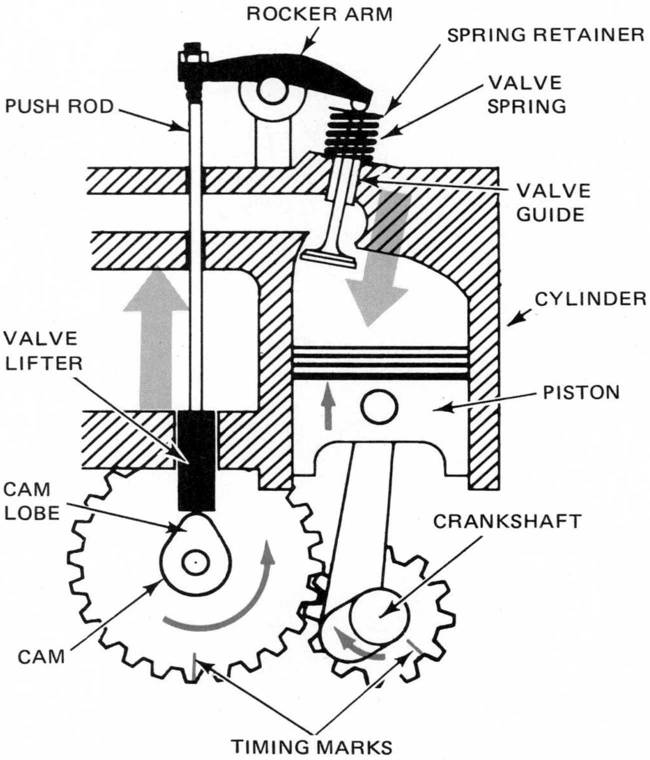

Fig. 5-18 Valve train on an engine using a pair of gears to drive the camshaft. The cam lobe has pushed the push rod up so the valve is opened.

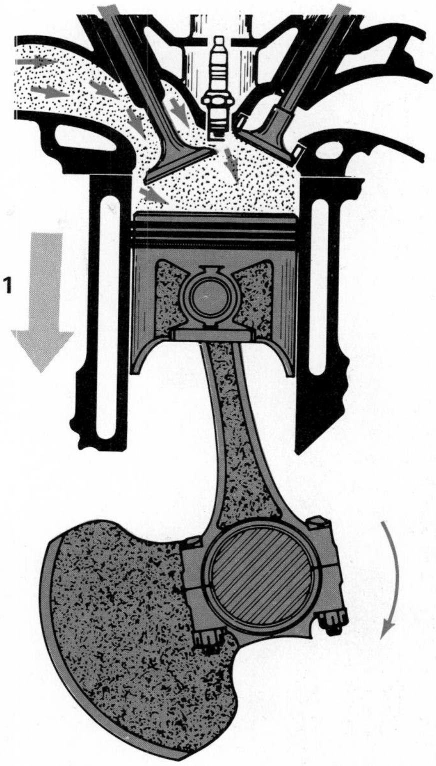

Fig. 5-20 Intake stroke. The intake valve (at left) has opened. The piston is moving downward, drawing in a mixture of air and gasoline vapor into the cylinder.

5-8 Intake Stroke

On the intake stroke, the piston is moving down (Fig. 5-20). The intake valve is open, and an air-fuel mixture is entering the cylinder, as shown by the arrows. The mixture of air and gasoline vapor is supplied by the fuel system (discussed in detail in later chapters).

The reason the air-fuel mixture enters the cylinder is this: When the piston moves down, a partial vacuum is produced in the cylinder. Atmospheric pressure (the pressure of the air), pushes the air-fuel mixture past the opened intake valve and into the engine cylinder.

5-9 Compression Stroke

When the piston nears BDC on the intake stroke, the intake valve closes. That is, the cam lobe has moved out from under the valve lifter. This allows the valve spring to close the intake valve. Next, the piston starts to move up on the compression stroke (Fig. 5-21). Both valves are closed, so that the air-fuel mixture has no place to go. It is compressed. In a modern engine, it is compressed to one-eighth or one-ninth of its original volume. That is like taking a quart of air and compressing it into less than half a cup. The amount that the air-fuel mixture is compressed is called the *compression ratio*. If the mixture is compressed to one-eighth of its original volume, then the compression ratio is 8 to 1 (written 8:1). There is more on compression ratios in a later chapter.

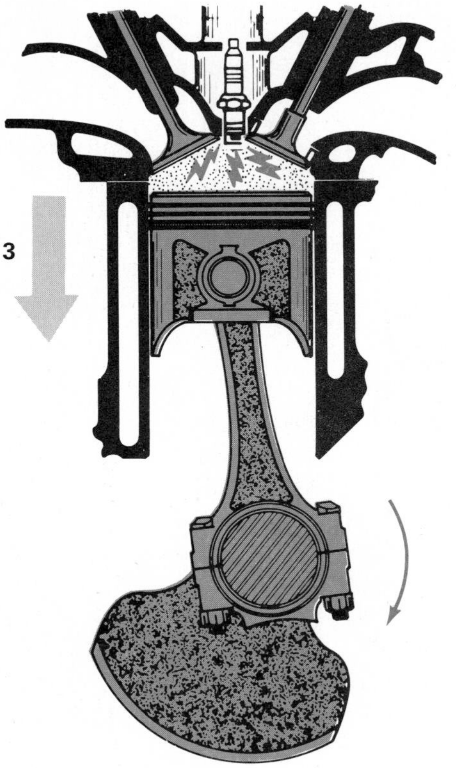

Fig. 5-22 Power stroke. The ignition system has delivered a spark to the spark plug that ignites the compressed mixture. As the mixture burns, high pressure is created, pushing the piston down.

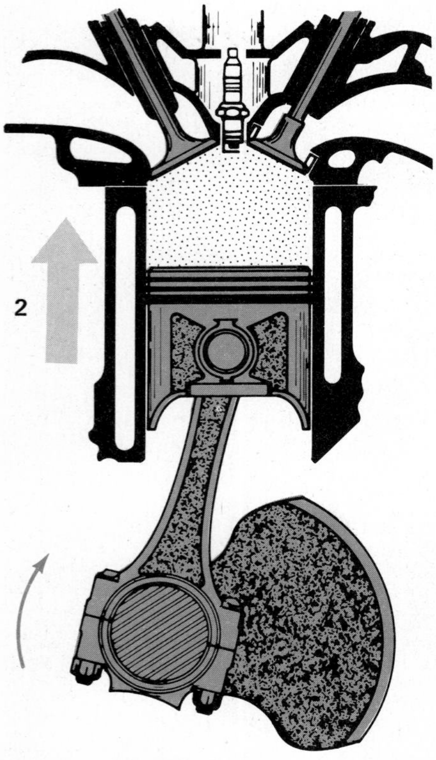

Fig. 5-21 Compression stroke. The intake valve has closed. The piston is moving up, compressing the mixture.

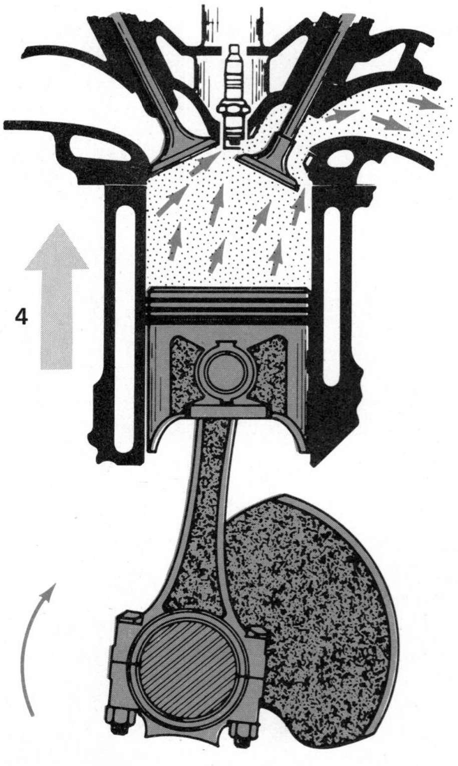

Fig. 5-23 Exhaust stroke. The exhaust valve has opened. The piston is moving upward, forcing burned gases out.

5-10 Power Stroke

The piston moves up during the compression stroke. As it gets near TDC, an electric spark is produced at the spark plug in the cylinder. This spark comes from the ignition system. The spark plug contains two wire *electrodes*, electrically insulated from each other. The spark jumps between these two electrodes. The spark ignites, or sets fire to, the compressed air-fuel mixture. The temperature and pressure go up. The temperature of the burning mixture can reach several thousand degrees. The pressure goes up to as much as 600 psi (pounds per square inch) [42.18 kg/cm^2 or 4,136.9 kPa]. This means that the pressure pushing down on the pistonhead can reach two tons (4,000 pounds) [1814.36 kg].

This pressure pushes the piston down on the power stroke (Fig. 5-22). The pressure is carried through the connecting rod to the crank on the crankshaft. It forces the crankshaft to turn. This rotary motion is carried through shafts and gears to the car wheels. The wheels turn, and the car moves.

The power to move the car comes from the molecules of fuel and air heated to high temperatures. In Chap. 4 we described heat and pressure as the movement of molecules. When molecules move faster, the temperature goes up. The pressure also goes up, because more molecules hit the sides of the container, and they hit harder. In the cylinder, during combustion, the molecules are set into very rapid motion. This not only raises the temperature several thousand degrees, but also produces the push of up to two tons on the piston.

5-11 Exhaust Stroke

The exhaust stroke follows the power stroke. As the piston approaches BDC on the power stroke, the exhaust valve opens. Now the piston passes BDC and starts to move up on the exhaust stroke. As it moves up, it pushes

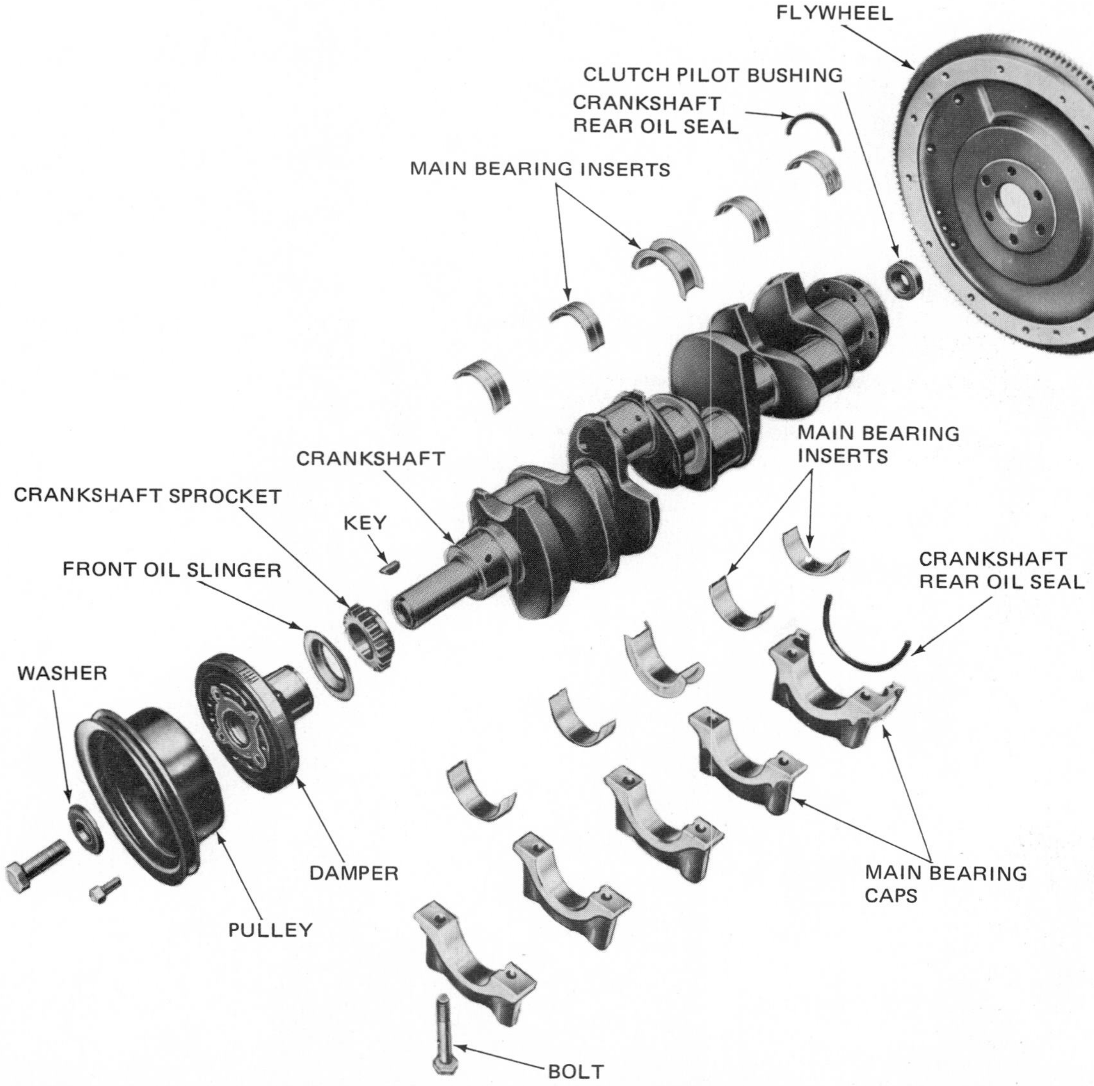

Fig. 5-24 Crankshaft and related parts for a V-8 engine. (*Ford Motor Company*)

the burned gases out of the cylinder (Fig. 5-23). Finally, the piston reaches TDC at the end of the exhaust stroke. The exhaust valve closes, and the intake valve opens. Now the piston starts down again, on the intake stroke. A fresh charge of air-fuel mixture enters the cylinder through the intake-valve port. The complete cycle of intake, compression, power, and exhaust is repeated over and over again.

● 5-12 Multiple-Cylinder Engines

A one-cylinder engine has only one power impulse, or power stroke, for every two crankshaft revolutions. That is, the engine delivers power only one-fourth of its running time. During the intake, compression, and exhaust strokes, it is not delivering power. For a more even flow of power, modern automotive engines have four, five, six, or eight cylinders. The power impulses are arranged so they follow one another in order. On six- and eight-cylinder engines, the power impulses overlap.

● 5-13 Flywheel

The power impulses in a multicylinder engine follow each other or overlap, to provide a fairly even flow of power. To smooth it even more, engines have flywheels. Figure 5-24 shows an engine crankshaft with the related parts, including the flywheel. The flywheel is a large wheel bolted to one end of the crankshaft. The flywheel smooths out the power flow by resisting any change in its speed of rotation. When a power impulse starts, the flywheel resists any sudden increase in speed. When the power impulse ends, the flywheel resists any slowing down.

In general, the fewer cylinders an engine has, the more important the flywheel is in smoothing out the power impulses. Therefore, a four-cylinder engine usually has a much heavier flywheel than an eight-cylinder engine.

The flywheel also serves as part of the clutch on cars with manual transmissions. In addition, the flywheel has teeth on its outer rim. A small drive pinion on the starting motor meshes with these teeth when the starting motor is turned on. This allows the starting motor to spin the flywheel and the crankshaft, so that the engine starts.

REVIEW QUESTIONS

See if you can answer these questions. If any question baffles you, reread the section that gives the answer.

1. What are the two basic types of engines?
2. What are the two types of internal-combustion engines?
3. What is the purpose of the piston rings?
4. What two parts change the reciprocating motion of the piston to the rotary motion of the crankshaft?
5. How many cranks are there on the crankshaft of a six-cylinder in-line engine?
6. What connects the piston to the crank? How is this part fastened to the piston? How is it fastened to the crank?
7. What drives the camshaft?
8. What are the basic moving parts in the valve train?
9. What are the four piston strokes in the four-stroke cycle?
10. Explain what goes on during each of the four piston strokes.
11. What three jobs does the flywheel do on a car with a manual transmission?

SELF PROJECTS

To understand valve action, you should have the movements of the valve-train parts well in mind. It will help if you will draw two pictures. One picture should show the valve-train parts—cam, lifter, push rod, rocker arm, and valve—with the valve closed. Draw the second picture showing the valve open. Study the pictures in the chapter so you will show the parts in the right relationship. You may not get everything just right the first time. Draw them again. When you are satisfied with your pictures, file them in your notebook.

Now draw four pictures showing the four piston strokes. Model your pictures on the illustrations in the chapter. File these in your notebook.

If you are able to examine and handle engine parts—camshafts, lifters, push rods, rocker arms, valves, pistons, and connecting rods—your pictures will be more real.

CHAPTER 6
Engine Types

After reading this chapter, you should be able to:

- **List the various ways in which engines are classified and explain what each classification means.**
- **Discuss dual- and multiple-displacement engines.**
- **Explain what firing order means and why it is important in the design and operation of an engine.**
- **Discuss the difference between gasoline and diesel engines.**

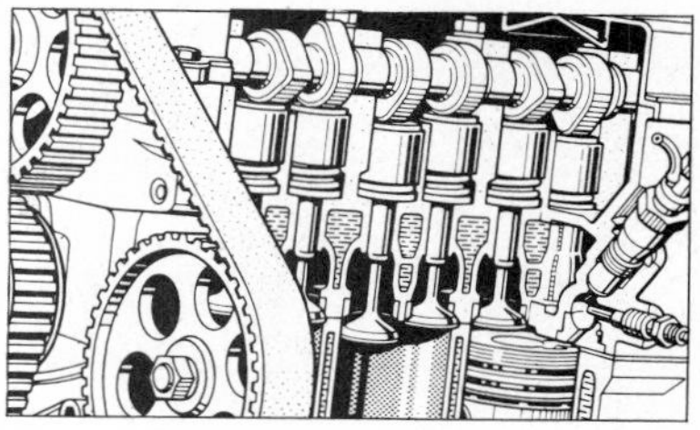

This chapter discusses the ways in which we classify engines. All automobile engines are of the internal-combustion type (except for some experimental external-combustion, or steam, engines). Internal-combustion engines can be classified according to:

1. Number of cylinders.
2. Arrangement of cylinders.
3. Arrangement of valves.
4. Type of cooling.
5. Number of strokes per cycle (two or four).
6. Type of fuel system.
7. Type of cycle (Otto or diesel).
8. Reciprocating or rotary (Wankel and turbine).

● 6-1 Number and Arrangement of Cylinders

American passenger-car engines have four, six, or eight cylinders. Foreign-made cars offer a greater variety, including three, four, five, six, eight, and twelve cylinders. Engines with four, five, six, and eight cylinders are described and illustrated in the next few sections. Cylinders can be arranged in several ways:

1. In a row (in line)
2. In two rows or banks set at an angle (V type)
3. In two rows opposing each other (flat, or pancake)
4. Like spokes on a wheel (radial airplane type)

Figure 6-1 shows various cylinder arrangements.

● 6-2 Four-Cylinder Engines

The cylinders of a four-cylinder engine can be arranged in any of three ways: in-line, V, or opposed. In the V type, the cylinders are in two banks, or rows, of two cylinders each. The two rows are set at an angle to each other. In the opposed type, the cylinders are in two banks of two cylinders each, set opposite each other.

The four-cylinder engine has become increasingly popular in recent years. A basic reason is the trend toward small, lightweight, fuel-efficient cars. This trend has been caused by the oil shortage and stringent government regulations regarding car size and gasoline-mileage requirements. These regulations step up the miles-per-gallon requirements, year by year. For example, the requirements for 1980 are that all cars produced by a manufacturer must average 20 miles per gallon. By 1985, the average must be 27.5 miles per gallon.

The only way these requirements can be met is for the automotive industry to go to the smaller engines, to "down-size" their big cars, and to bring out a whole new generation of small cars. A four-cylinder engine in a small car can give 40 miles per gallon or more on the highway. The large cars, with eight-cylinder engines, can often do no better than 12 to 15 miles per gallon. Another factor to be considered is the pollution problem. Emission

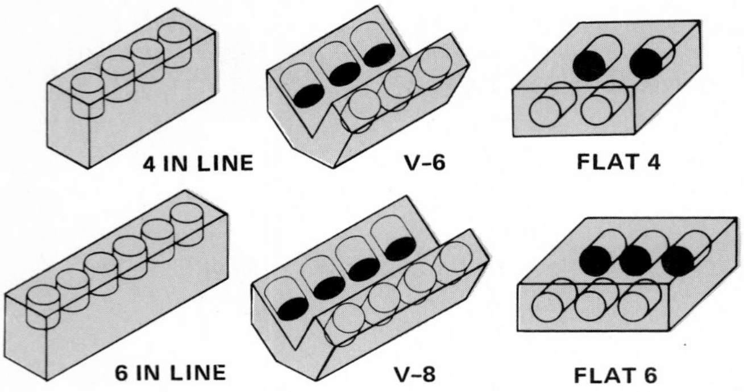

Fig. 6-1 Several cylinder arrangements.

standards set by the federal government and some states (discussed in a later chapter) require additional controls on the car that tend to make gasoline mileage worse. Therefore, the small, lightweight car with a small engine appears to be the car of the future.

The four-cylinder engine does not give the pickup and operating performance of cars with larger engines. A partial solution, which we discuss later, is a six-cylinder or eight-cylinder engine that deactivates, or turns off, some of the cylinders when they are not needed. Thus, there is one type of six-cylinder engine that cuts out three of the cylinders when the car is cruising and power demands are light. One eight-cylinder engine has electronic controls that can progressively cut out up to four of the cylinders when power demands are low. The controls deactivate the cylinders by immobilizing the valves, that is, prevent them from opening and closing. Another solution is to turbocharge the engine. Turbocharging puts more air-fuel mixture into the engine cylinders so that more power is produced. We discuss all this later in detail. First, we look at the various types of four-cylinder engines.

1. In-line engines Figure 6-2 is a cutaway view of a four-cylinder in-line engine. The cylinders are arranged in one row, or line.

Another four-cylinder in-line engine, with overhead camshaft (OHC, camshaft on cylinder head) and attached clutch, transmission, and differential, is shown in Fig. 6-3. This engine mounts at the front of the car and drives the front wheels. You can see one of the takeoff shafts for one front wheel to the lower left in Fig. 6-3.

A relatively new engine with an overhead camshaft has been brought out by Renault (Fig. 6-4). This engine is designed to mount transversely (crosswise) in the car and drive the front wheels. Note the ignition distributor which is directly coupled to the camshaft and the toothed belt which drives the camshaft.

2. V-4 engines The V-4 engine has two rows of two cylinders each, set at an angle, or a V, to each other. The crankshaft has only two cranks. Connecting rods from opposing cylinders in the two rows are attached to the same crankpin. Each crankpin thus has two connecting rods attached to it. Figure 6-5 is a "phantom" view of a

Fig. 6-2 Partial cutaway view of a four-cylinder, in-line, overhead-valve engine. *(Chevrolet Motor Division of General Motors Corporation)*

V-4 engine showing the internal moving parts. This type of engine is rather difficult to balance with counterweights on the crankshaft. The engine illustrated is balanced by a balance shaft that turns in a direction opposite to the crankshaft.

3. Flat four-cylinder engine One of the early flat-fours is the Volkswagen engine, used in their "Beetle," and shown in sectional view in Fig. 6-6. The four cylinders are arranged in two opposing rows of two cylinders each. This engine is mounted at the rear of the car; the car is a rear-wheel-drive car. The engine is air-cooled. Note the cylinders and the flat metal rings, or fins, surrounding them. These fins provide large surfaces from which heat can radiate so the engine will not overheat.

Another more recent flat-four is shown in cutaway view in Fig. 6-7. This engine was developed by the Japanese firm of Subaru, Fuji Heavy Industries, Ltd. This engine is liquid-cooled. It is mounted at the front of the car and drives the front wheels. The transmission and differential are assembled directly to the engine, as shown.

We describe the two types of cooling system—air cooling and liquid cooling—in a later chapter.

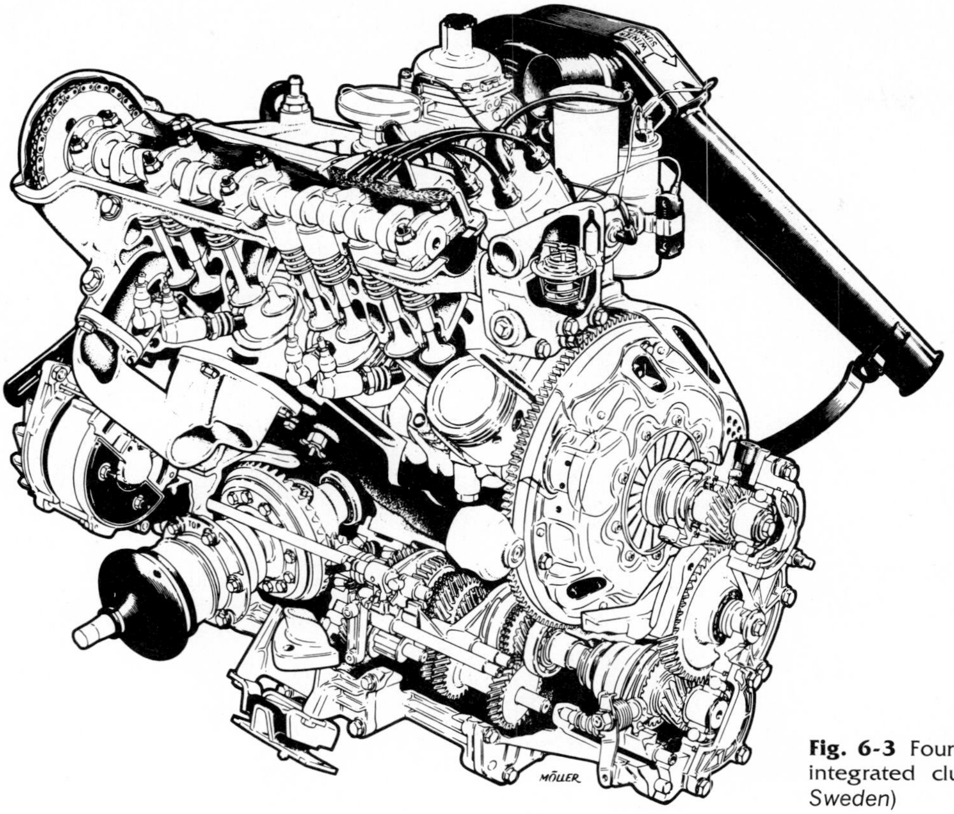

Fig. 6-3 Four-cylinder in-line OHC engine with integrated clutch and transmission. (*Saab of Sweden*)

Fig. 6-4 Partial cutaway view of a four-cylinder in-line OHC engine. Note toothed belt driving camshaft and direct-coupled distributor to camshaft. (*Renault*)

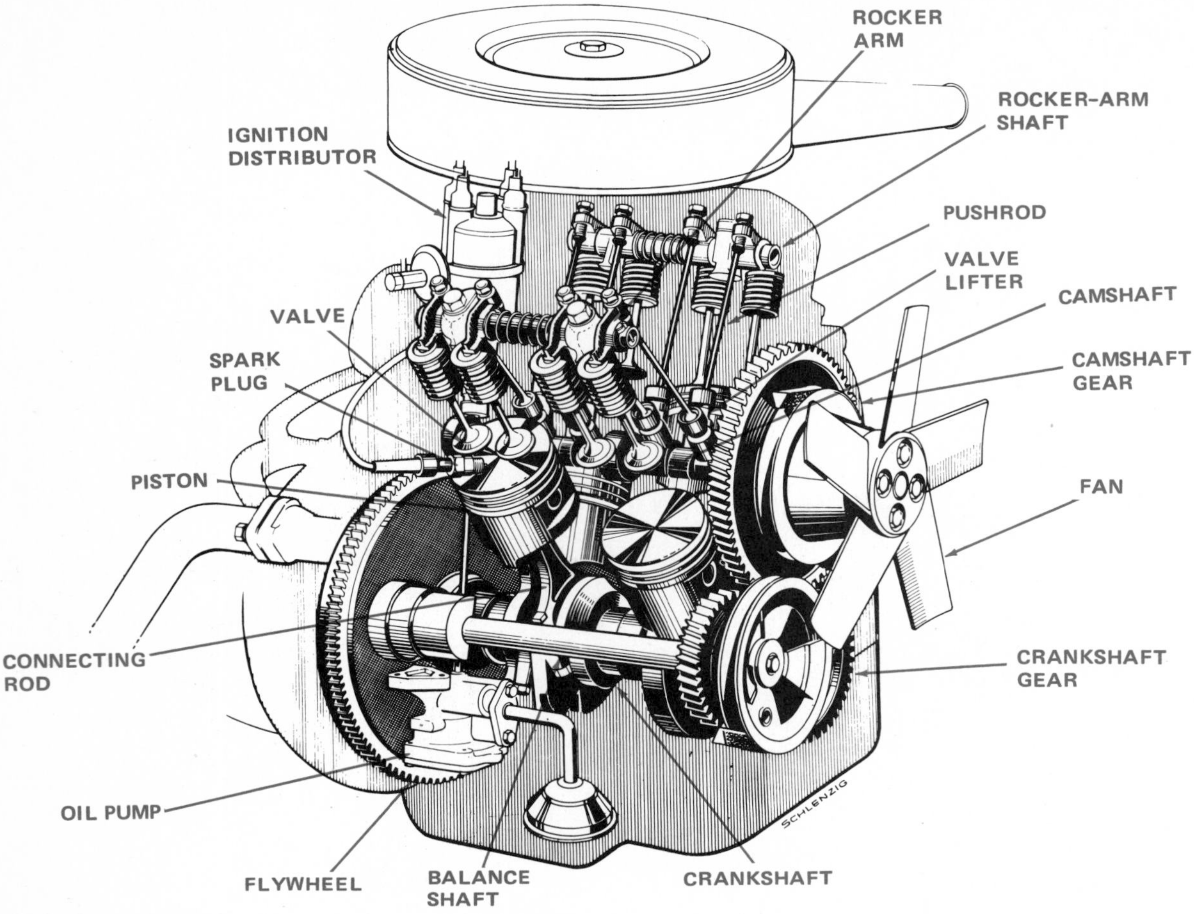

Fig. 6-5 Phantom view of V-4 engine showing major moving parts in engine. (*Ford Motor Company of Germany*)

● 6-3 Five-Cylinder Engines

There are a few five-cylinder automotive engines. Mercedes-Benz produces a five-cylinder diesel engine. Diesel engines are discussed later in the chapter. Audi NSU has a five-cylinder overhead-camshaft gasoline engine (Fig. 6-8). The five-cylinder engine is somewhat of a compromise between a four and a six. It produces more power than a four but is not as long or as heavy as a six. The Audi engine is for a front-drive car. You can see, in Fig. 6-8, the transmission and one front axle coming out of the transmission at right angles. Note that this arrangement has the engine up ahead of the axles. This arrangement puts more weight on the front wheels for good traction.

● 6-4 Six-Cylinder In-Line Engines (Straight Six)

Most six-cylinder engines are in-line, although there are V-6 and flat-six engines. This is similar to four-cylinder engines, which can also be in-line, V-type, or flat.

Figure 6-9 shows a six-cylinder in-line engine partly cut away to show the internal construction. The valves are overhead. This is an "I-head" engine. In this engine, the crankshaft is supported by seven main bearings. Thus, there is a bearing on each side of every crank, for support and strength.

Figure 6-10 shows a slant-six overhead-valve engine. This engine is like other six-cylinder in-line engines, except that the cylinders slant to one side. This is so that the hood line can be lowered. This engine is made with either a cast-iron or a die-cast-aluminum cylinder block.

Fig. 6-6 Partial sectional view from end of flat four-cylinder, air-cooled engine. The engine has two banks of two cylinders each, opposing each other. (*Volkswagen*)

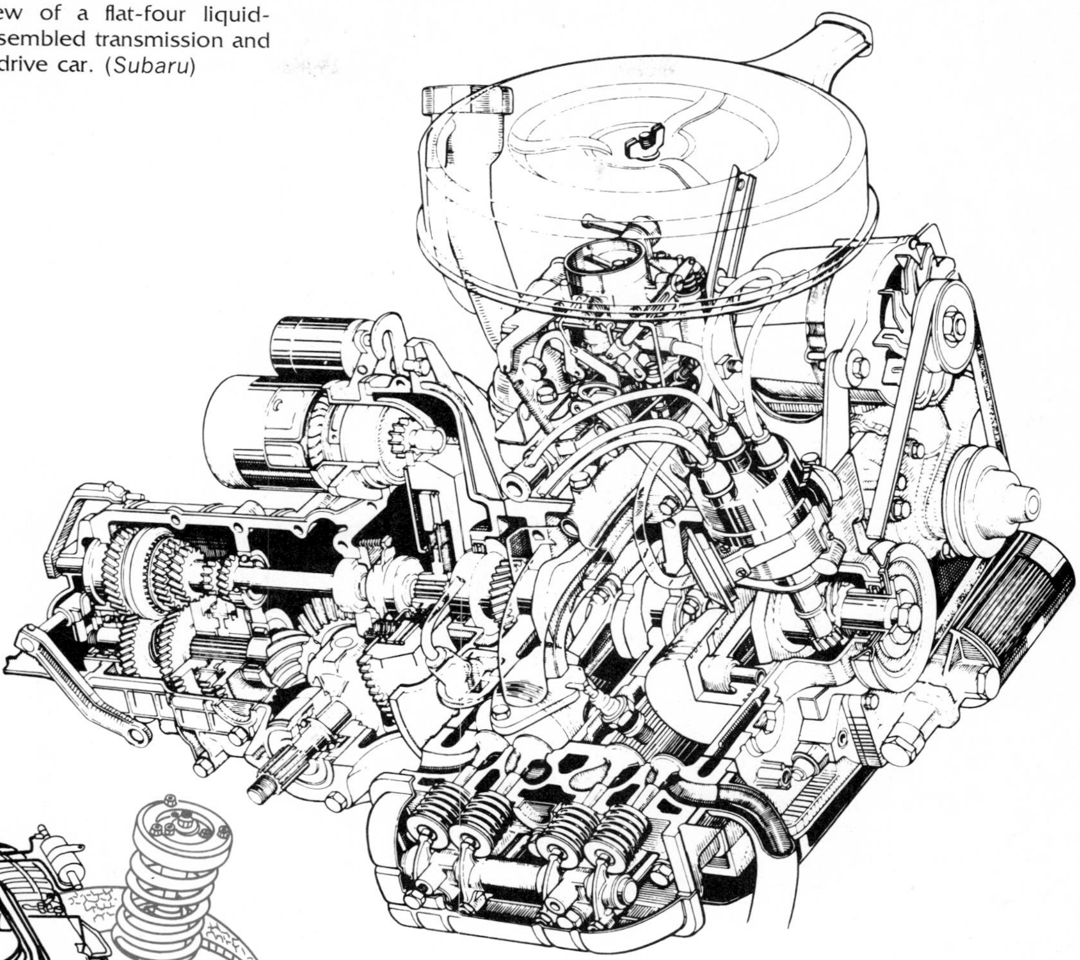

Fig. 6-7 Cutaway view of a flat-four liquid-cooled engine with assembled transmission and differential for a front-drive car. (*Subaru*)

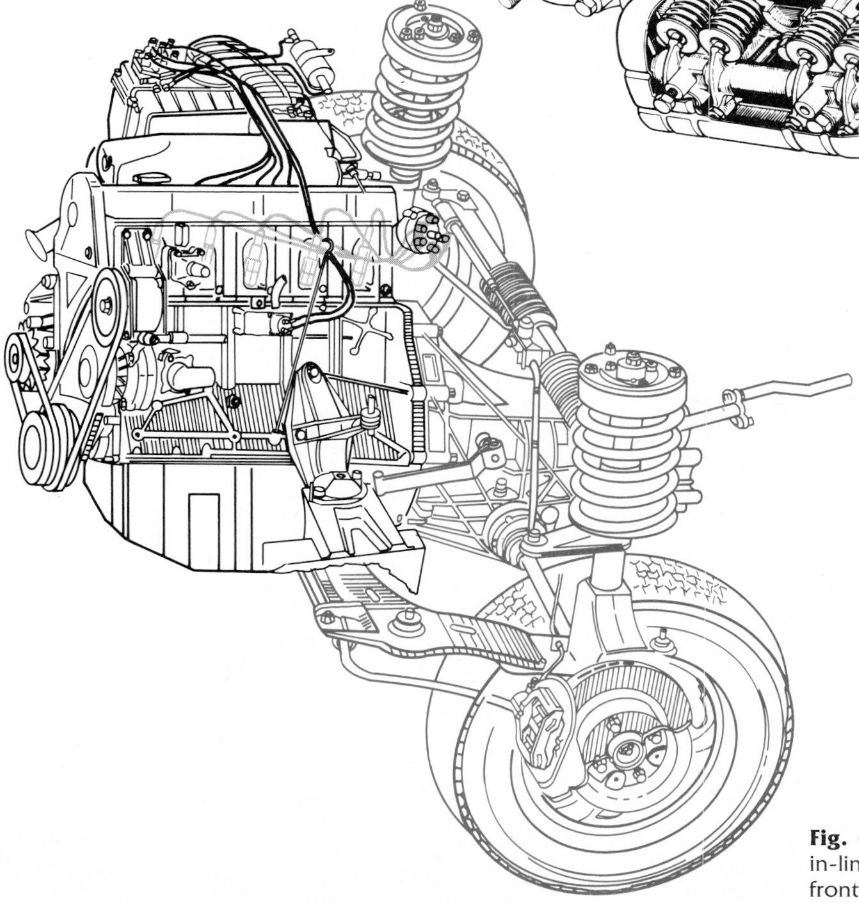

Fig. 6-8 Partial cutaway view of a five-cylinder in-line engine with overhead camshaft for a front-drive car. (*Audi*)

This feature is discussed later, where we compare cast-iron blocks with aluminum blocks.

Many automotive engineers are predicting that the six-in-line engine will be phased out over the next several years. The V-6 (● 6-6) will take its place, they say. This is history repeating itself. For many years, the eight-in-line engine was the "big" engine in some manufacturer's stable of cars. Then the V-8 came along and it superseded the eight-in-line. Now, the V-6 will supersede the six-in-line, according to many engineers.

● 6-5 Dual-Displacement Six-In-Line Engine

This is an experimental engine under development, until recently, at Ford Motor Company. The project has been put aside for the present, but it is an interesting idea you should know about. It is a standard six-in-line, except that under certain driving conditions three of the cylinders can be turned off. This is done electronically by a computer (Fig. 6-11). That is, the engine is either a full-displacement (six-cylinder) engine, or a half-displacement (three-cylinder) engine. Thus, it has, in effect, two displacements. It is therefore a "dual-displacement" engine.

The reasoning behind this design is that an engine is rarely called on to deliver full power, especially on the highway. When cruising, for example, three cylinders can handle the load about as well as six cylinders.

The controlling device is a computer. The computer is an electronic device. Sensors located at various places in the vehicle feed information to the computer. This information includes intake-manifold vacuum, engine speed, throttle position, coolant temperature, and transmission gear. All this information "tells" the computer the operating mode of the engine, and when three of the cylinders should be deactivated. When this time arrives, the computer activates three solenoids located above the valves in three of the cylinders. The solenoids are electromagnetic devices. They are activated by being connected to the battery. When this happens, the solenoids operate mechanisms at the three sets of valves (six in all) at three

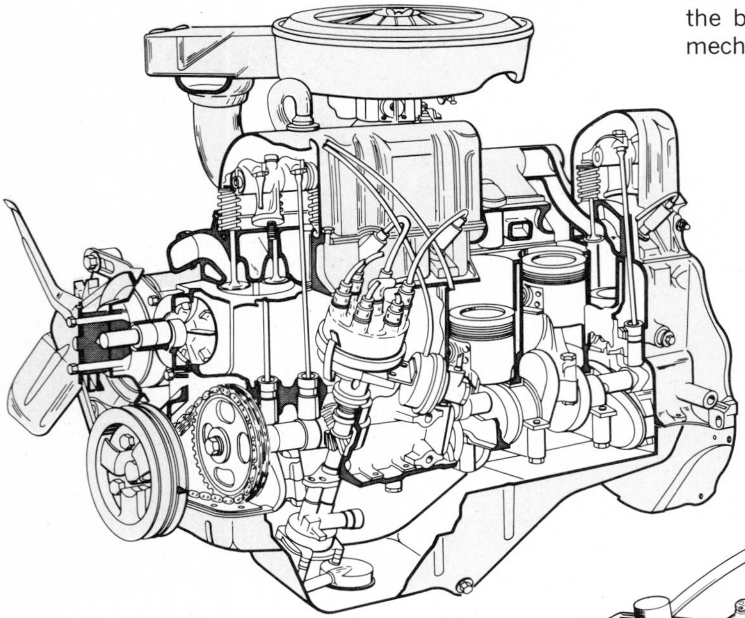

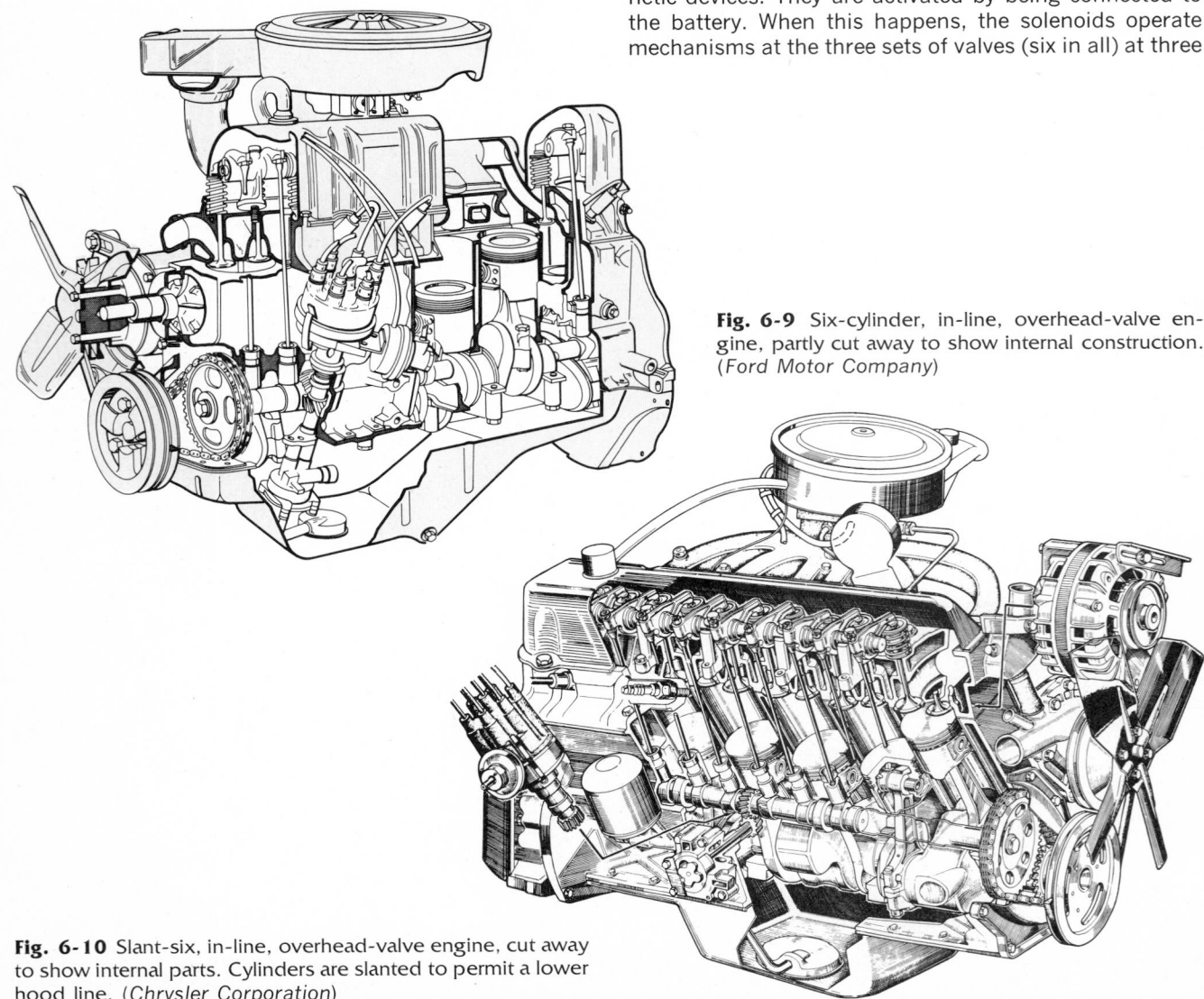

Fig. 6-9 Six-cylinder, in-line, overhead-valve engine, partly cut away to show internal construction. (*Ford Motor Company*)

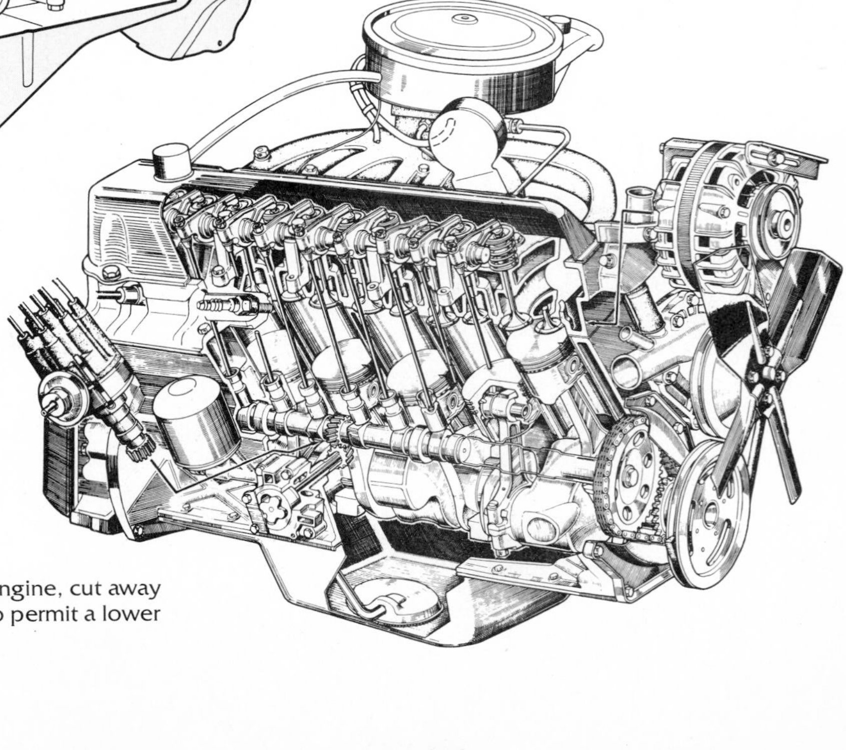

Fig. 6-10 Slant-six, in-line, overhead-valve engine, cut away to show internal parts. Cylinders are slanted to permit a lower hood line. (*Chrysler Corporation*)

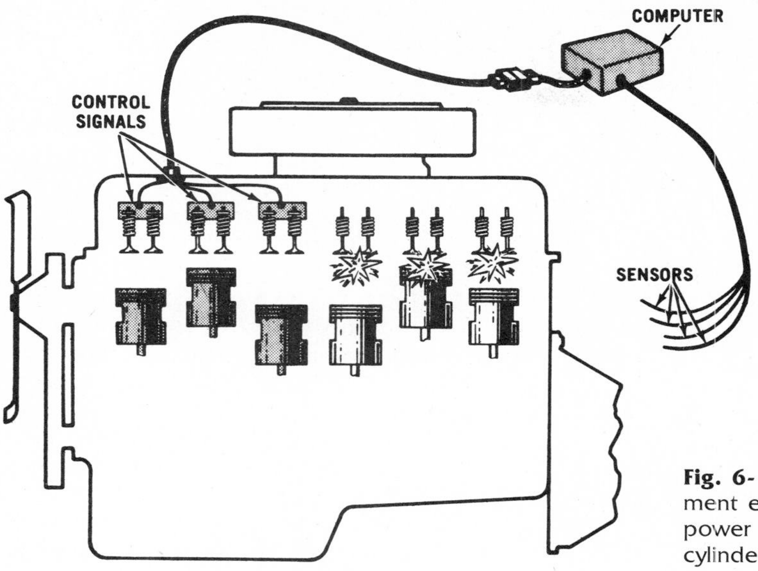

Fig. 6-11 Schematic drawing of the six-in-line dual-displacement engine. Under some operating conditions, when full power is not needed, the computer cuts out three of the cylinders. (*Ford Motor Company*)

of the cylinders. This prevents the valves (intake and exhaust) from opening. As a result, no air-fuel mixture can enter and no exhaust gas can escape. These three cylinders are inactive and the engine runs on the other three cylinders.

The arrangement provides a substantial saving in gasoline while at the same time providing adequate power for steady-speed cruising. When conditions change so more power is required, as for instance when speeding up to pass another car, the computer deactivates the solenoids and the valves become free to operate. They open and close normally so the engine reverts to standard six-cylinder operation.

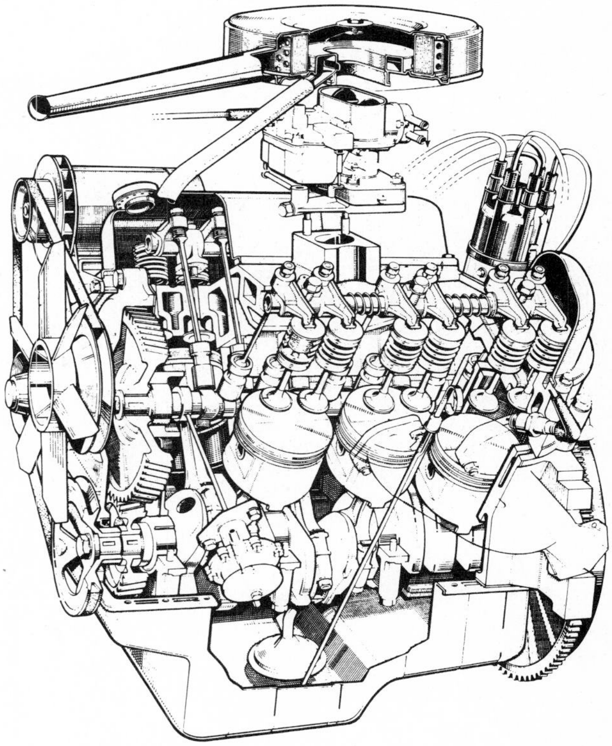

Fig. 6-12 Cutaway view of a V-6, overhead-valve engine. (*Ford Motor Company of Germany*)

● 6-6 V-6 Engines

A number of companies have built V-6 engines. These have two rows of three cylinders each, set at an angle to form a V. Figure 6-12 is a cutaway view of a V-6 engine. You can see many of the parts and the interior construction of the engine in this picture. Figure 6-13 is a partial cutaway view of a V-6 engine that General Motors has developed for some model cars that their divisions build.

As mentioned previously, many automotive engineers wonder if the V-6 will displace the six-in-line. They remember how the V-8 completely wiped out the eight-in-line engine and they think it is possible that the V-6 will do somewhat the same thing to the six-in-lines. The V-6 is a compact, well-engineered engine that fits the shorter cars of today and the future. General Motors is using it in several car models produced by some of their divisions.

● 6-7 V-6 Turbocharged Engine

The turbocharger is a rotary air pump that sends more air-fuel mixture into the engine cylinders. This can substantially increase engine power. Figure 6-14 shows a V-6 turbocharged engine. The exhaust gas spins a rotor which is on the same shaft as a turbine. This spins the turbine. When the turbine spins, it sends additional air, under pressure, into the carburetor. This feeds additional air-fuel mixture to the engine cylinders. More power develops. On many turbocharged engines, the power output

Fig. 6-13 Partial cutaway view of a V-6 overhead-valve engine. (*General Motors Corporation*)

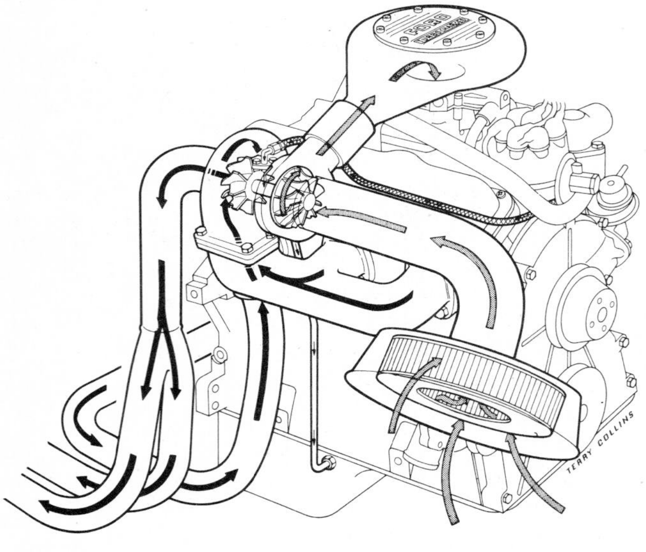

Fig. 6-14 Turbocharged version of the Ford 182 cubic inch [3L] V-6. (*Ford Motor Company*)

is boosted 30 percent or more. Thus, a 1978 Ford 101-hp (horsepower) engine is said to have its power output boosted to 147 hp or more by a turbocharger. Turbochargers are covered in detail in a later chapter.

Turbochargers can be used on almost any internal-combustion engine, regardless of size or number of cylinders. They have been around for a long time. But the recent emphasis on light weight and fuel economy has sparked renewed interest in them. With a turbocharger, you can get more power out of a given engine at a cost of only a little more weight (the turbocharger) and some added complexity. Many automotive engineers predict that we will see many turbocharged engines (fours and sixes) in the cars of the future.

6-8 Flat-Six Engine

The flat-six engine is very similar to the flat-four except that two more cylinders have been added, one to each bank. A flat-six was used in the Chevrolet Corvair. This engine is air-cooled and is mounted at the rear of the car, where it drives the rear wheels.

6-9 V-8 Engines

At one time, the eight-cylinder in-line engine was widely used. Now, though, it has been replaced by the V-8 engine (Fig. 6-15). In the V-8, the cylinders are arranged in two rows, or banks, of four cylinders each. The two rows are set at an angle to each other. This engine is like two four-cylinder in-line engines mounted on the same crankcase with one crankshaft. The crankshaft in the V-8 has four cranks. Connecting rods from opposing cylinders in the two rows are attached to a single crankpin. Two rods are attached to each crankpin, and two pistons work to each crankpin. The crankshaft is usually supported on five bearings.

The V-8 engine in Fig. 6-15 has overhead valves. All the valve lifters, push rods, and rocker arms are worked by one camshaft located between the two cylinder banks. Some high-performance engines have overhead camshafts located in the cylinder heads. One type has a single overhead camshaft in each cylinder head. Another has two overhead camshafts in each cylinder head. One is for the intake valves; the other is for the exhaust valves.

This is a total of four camshafts in the engine. An engine of this type is shown in Fig. 6-16. Advantages of the overhead camshaft are discussed later.

● 6-10 Multiple-Displacement V-8 Engine

This engine (Fig. 6-17) is an eight-cylinder version of the dual-displacement engine described in ● 6-5. However, instead of cutting out half of the cylinders all at once, this engine has electronic controls that selectively cut out one, two, three, or four cylinders at a time. The number of cylinders the controls cut out depends on the power requirements. When the engine is idling, or cruising at a steady speed on a level highway, the controls cut out four of the cylinders. Only four cylinders are needed to provide sufficient power for these operating conditions.

However, when the car encounters a hill, additional power is needed and the electronic controls put additional cylinders to work. The same thing happens when the driver "steps on the gas" to increase car speed, as for example passing another car. Additional cylinders go to work. The number of cylinders that go back to work depends on the amount of additional power that is needed. This arrangement, as with the dual-displacement six, saves gasoline without sacrifice of performance.

● 6-11 Twelve- and Sixteen-Cylinder Engines

Twelve- and sixteen-cylinder engines have been used in passenger cars, buses, trucks, and industrial plants. The cylinders are mostly in two banks (V type or pancake type). Sometimes they are in three banks (W type) or four banks (X type). The pancake engine is similar to a V

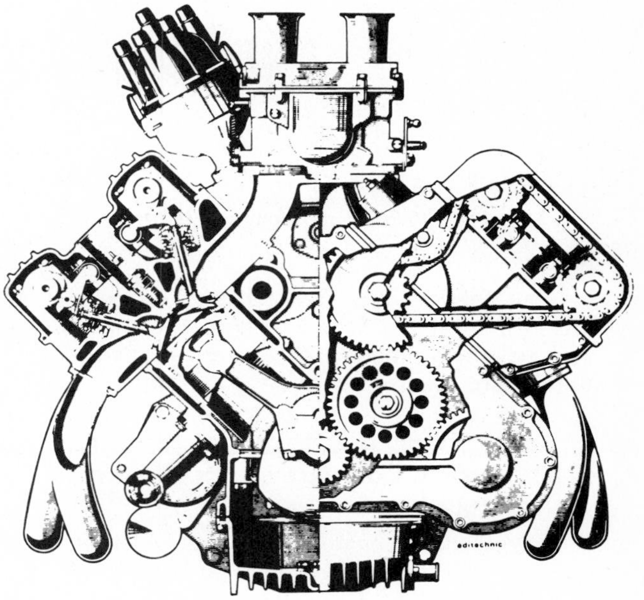

Fig. 6-16 Sectional view from the end of a V-8 engine with four overhead camshafts. Right bank has been cut away to show the camshaft drive arrangement. Left bank has been cut away to show the internal construction of the engine and location of the valves and other components. (*Renault*)

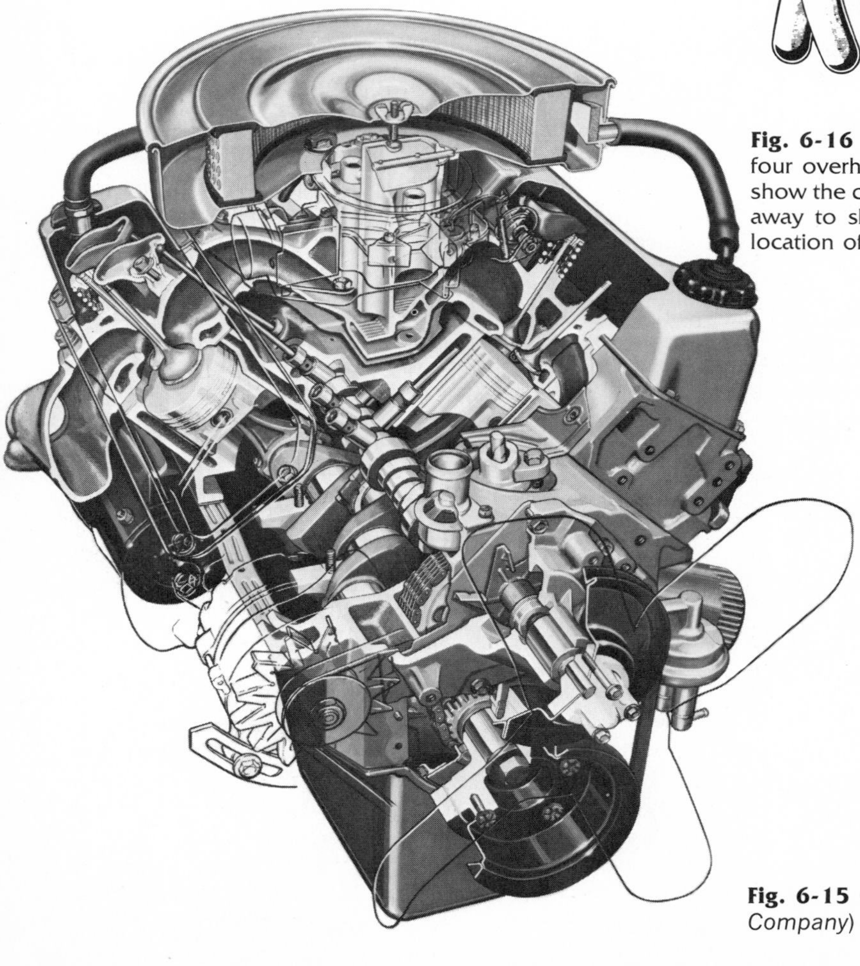

Fig. 6-15 Cutaway view of 365-hp V-8 engine. (*Ford Motor Company*)

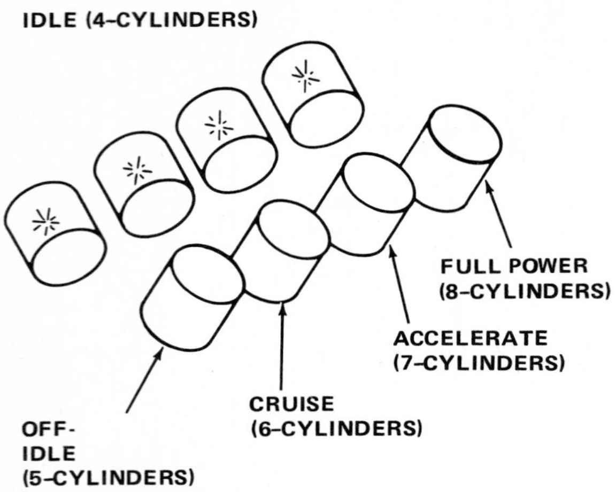

Fig. 6-17 Multiple-displacement engine. (*Ford Motor Company*)

engine, but the two rows are flat and opposing. The cylinders work to the same crankshaft. The only passenger cars now being made with a twelve-cylinder engine are the Ferrari, the Jaguar, and the Maserati.

● 6-12 V-Type Compared with In-Line Engines

In-line eight-cylinder engines were once widely used in automobiles, but now V-8 engines are used. The V-8 engine is shorter, lighter, and more rigid. In a V-8, it is easier to get enough air-fuel mixture to all cylinders (since the cylinders are closer together). Sometimes, in an in-line engine, the center cylinders could be fuel-starved while the end cylinders would be getting enough fuel. Some eight-cylinder in-line engines used two carburetors to help this situation.

The more rigid V-8 engine permits higher running speeds and higher combustion pressures (higher power outputs). There is less difficulty from flexing, or bending, of the cylinder block and crankshaft. Flexing throws the engine out of line, increases frictional losses and wear, and may also set up internal vibrations.

The shorter engine makes possible more passenger space on the same size car, or a shorter car. Also, the V-8 permits a lower hood line and thus a lower car. This is because the carburetor and other parts can be placed between the two rows of cylinders. Then they do not take up headroom above the cylinders.

The same argument is now being applied to the V-6, as we mentioned. Many engineers predict that the V-6 will take over in place of the six-in-line. See comments at the end of ● 6-4.

● 6-13 Cylinder Numbering and Firing Order

The firing order, or order in which the cylinders deliver their power strokes, is selected as part of the engine design. The best design provides a well-distributed pattern along the crankshaft. A design that permits two cylinders at the end of the crankshaft to fire one after the other is avoided as far as possible. In-line engines are numbered from front to back. Several different cylinder-numbering arrangements are used for V-6 and V-8 engines. The General Motors V-6 engine has the cylinders numbered:

Front of Car

Left Bank	Right Bank
①	②
③	④
⑤	⑥

Ford V-8 engine cylinders are numbered:

Front of Car

Left Bank	Right Bank
⑤	①
⑥	②
⑦	③
⑧	④

General Motors cars, except for some Buicks and Cadillacs, have the cylinders of their V-8 engines numbered in this way:

Front of Car

Left Bank	Right Bank
①	②
③	④
⑤	⑥
⑦	⑧

Chrysler Corporation numbers their V-8 engines the same way as the General Motors V-8 engines, above.

The Volkswagen flat-four engine has the cylinders numbered:

Front of Car

Left Bank	Right Bank
③	①
④	②

The Corvair flat-six engine has the cylinders numbered:

Front of Car

Left Bank	Right Bank
⑥	⑤
④	③
②	①

As you can see, these different cylinder-numbering methods can be confusing. Therefore, you should always check the shop manual or tune-up chart when you want to know how the cylinders of an engine you are working on are numbered.

The firing order is the order in which the cylinders deliver their power strokes. This is a built-in part of the engine design. The strokes are scattered along the crankshaft so that a well-distributed pattern results. This minimizes the strain on the crankshaft. In other words, two cylinders firing at one end of the crankshaft one after the other is usually avoided, as we said. In in-line engines, the cylinders are numbered from front to back, as already noted. The two firing orders used in four-cylinder engines are 1–3–4–2 and 1–2–4–3.

Two firing orders are possible in six-cylinder, in-line engines, 1–5–3–6–2–4 or 1–4–2–6–3–5. All modern six-cylinder, in-line engines use 1–5–3–6–2–4.

General Motors V-8 engines use different firing orders. For example, most of the V-8 engines use 1–8–4–3–6–5–7–2. Late-model Cadillac engines use this firing order, 1–5–6–3–4–2–7–8. Chrysler Corporation cars with V-8 engines use 1–8–4–3–6–5–7–2. Ford V-8 engines use two firing orders, 1–5–4–2–6–3–7–8 and 1–3–7–2–6–5–4–8.

So you see, you must look up the firing order in the shop manual when you work on an engine. You must know the proper firing order, especially when you are doing ignition work.

Note: On many engines, the firing order is cast into the intake manifold.

6-14 Valve Arrangements

The intake and exhaust valves can be arranged in various ways in the cylinder head or block (Fig. 6-18). The I-head design is the most common.

1. L-head engine Many years ago, most automobile engines used the L-head arrangement. In the L-head engine, the combustion chamber and cylinder form an inverted L. The intake and exhaust valves are located side by side in the cylinder block, and to one side of the cylinder. Although this is a relatively simple and dependable arrangement, it does not allow a high-compression-ratio design. The overhead-valve engine does allow higher compression ratios. Therefore, to achieve these desirable higher compression ratios, automotive manufacturers switched to the overhead-valve engine. There was another plus in doing this. The large surface area of the L-head combustion chamber makes the L-head engine a relatively heavy polluter of the atmosphere. The layers of air-fuel mixture next to the surfaces do not burn. So they escape into the atmosphere along with the exhaust gas. This is much less of a problem with the overhead-valve engines. So switching to these engines reduced this type of pollution. You will still see many small L-head engines; they are used extensively in power lawn mowers and other such applications where light weight and simplicity are major considerations.

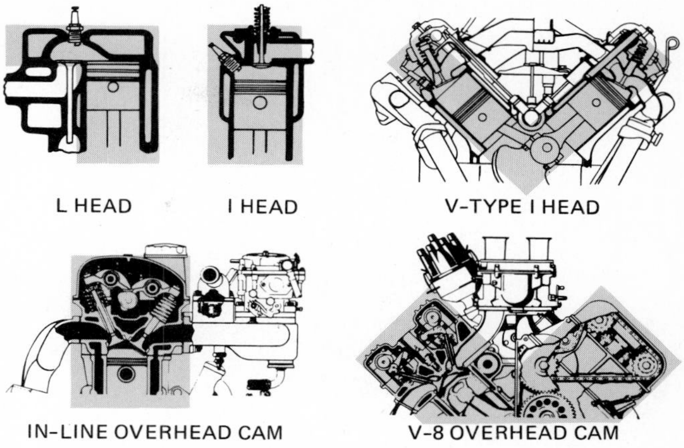

Fig. 6-18 Valve arrangements. Compare these line drawings with the sectional and cutaway views of the engines shown elsewhere.

2. I-head engine In the I-head or overhead-valve engine, the valves are in the cylinder head (Fig. 6-18). In in-line engines, the valves are usually in a single row, as in Figs. 6-2 and 6-10. In V-8 engines, the valves may be in a single row in each bank or in a double row in each bank. Either way, a single camshaft operates all valves. The valve lifters, push rods, and rocker arms carry the motion from the cams to the valves.

Overhead-valve engines are used in all modern automobiles, because they have higher compression ratios. In an engine with overhead valves, the clearance volume can be made smaller than in an L-head engine. The illustrations of the various I-head engines show that having the valves directly above the piston permits a smaller clearance volume. In some I-head engines, there are pockets in the piston heads. The valves can move into these pockets when the valves are open with the piston at TDC. In some engines, the clearances between the pistons and valves are only a few thousandths of an inch.

3. Overhead camshafts (Figs. 6-18 and 6-19) The I-head engine uses push rods and rocker arms to operate the valves. This design is often called a *push-rod* engine. The push rods and rocker arms impose some inertia, however, that affects valve action. That is, the push rods and rocker arms flex or bend slightly before they open the valve. This slows the valve action somewhat. At lower speeds, this does not matter. As speed increases, the flexing also increases. This causes an increasing lag in valve action which tends to limit top engine speed. However, with the overhead camshaft, the cams work directly on the rocker arms or valve lifters. This gives quicker valve response so that higher engine speeds are possible. The single-overhead-camshaft engine (one camshaft in each cylinder head) is called an SOHC engine. The double-overhead-camshaft engine is a DOHC engine.

6-15 Classification by Cooling

Engines can be liquid-cooled or air-cooled. Most present-day American automobile engines are liquid-cooled. The

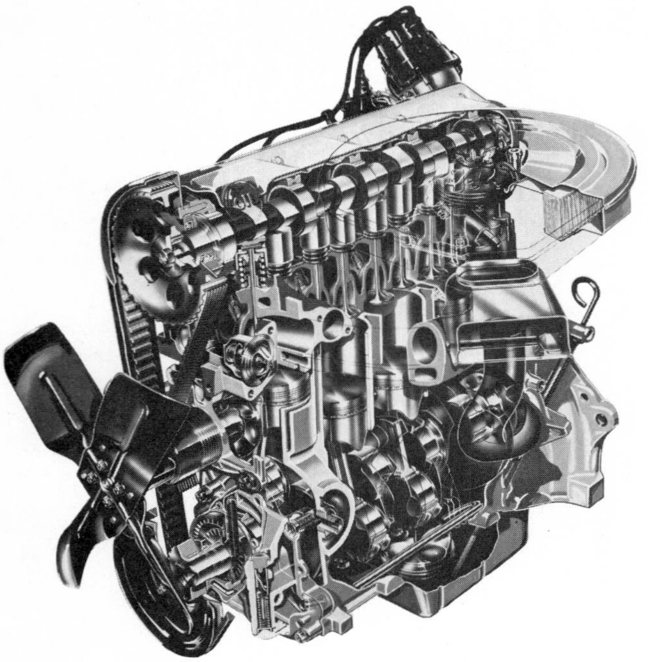

Fig. 6-19 Cutaway view of the Vega four-cylinder overhead-camshaft engine. (*Chevrolet Motor Division of General Motors Corporation*)

Corvair, Volkswagen, and some other automobile engines are air-cooled. (See Fig. 6-6.) Also, the small one- and two-cylinder engines on power mowers and other garden equipment are air-cooled. In air-cooled engines, the cylinders are usually separate. They have metal fins which give a large *radiating* surface. This permits engine heat to radiate away from the cylinders. Air-cooled automotive engines have metal shrouds which direct the airflow around the cylinders for cooling.

Liquid-cooled engines use a liquid to take heat from the engine. The liquid, called the coolant, is water mixed with antifreeze. These engines have water jackets surrounding the cylinders and combustion chambers (Fig. 6-20). A later chapter describes the two basic types of cooling, air cooling and liquid cooling.

● 6-16 Classification by Cycles

Engines can be either two-stroke-cycle or four-stroke-cycle. The four-stroke-cycle engine (usually called a four-cycle engine) was discussed in ● 5-7 to 5-11. In it, the complete cycle requires four piston strokes (intake, compression, power, and exhaust). In the two-stroke-cycle, or two-cycle, engine, the intake and compression strokes, and the power and exhaust strokes, are combined. This permits the engine to produce a power stroke every two piston strokes, or every crankshaft rotation.

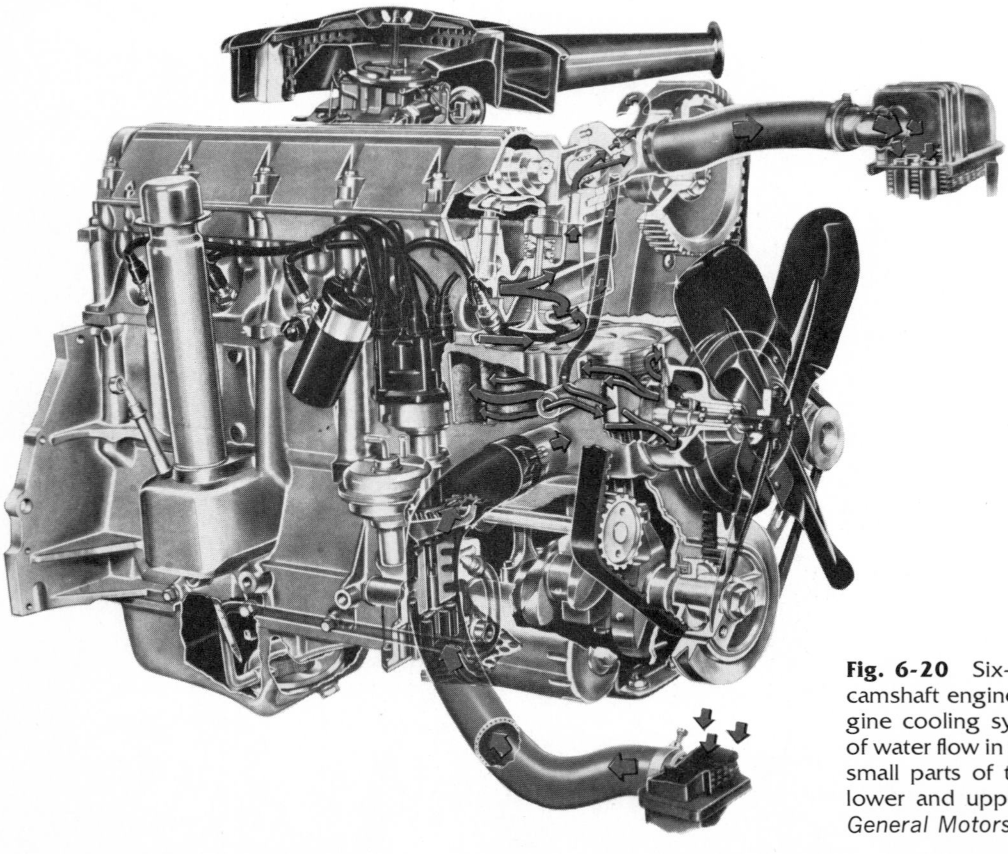

Fig. 6-20 Six-cylinder, in-line, overhead-camshaft engine partly cut away to show engine cooling system. Arrows show direction of water flow in the engine water jackets. Only small parts of the radiator are shown at the lower and upper right. (*Pontiac Division of General Motors Corporation*)

6-17 Classification by Fuel

Internal-combustion engines can be classified according to the type of fuel they use. Automotive engines, in general, use gasoline. Some bus and truck engines use LPG (liquefied petroleum gas). These are really gasoline engines adapted for LPG. Diesel engines use diesel fuel oil. Chapter 11 describes these fuels in detail.

6-18 Diesel Engines

In the diesel engine, air alone enters the cylinder on the intake stroke. So air alone is compressed on the compression stroke. At the end of the compression stroke, the fuel is injected, or sprayed, into the combustion chamber. In diesel engines the air is compressed to as little as a twentieth of its original volume. This compression pro-

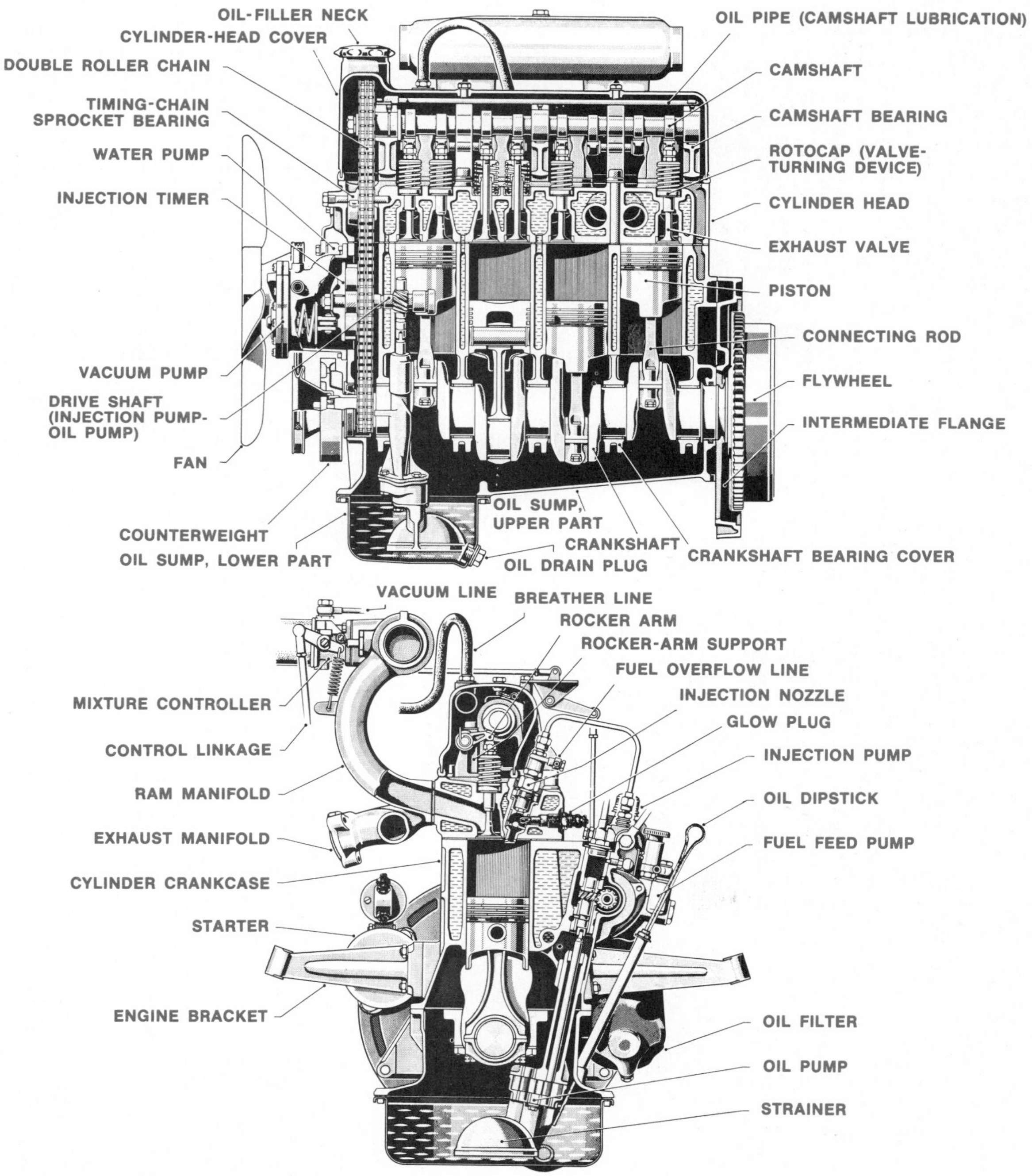

Fig. 6-21 Sectional views of a four-cylinder diesel engine for passenger cars. (*Mercedes-Benz*)

duces pressures of about 500 psi (pounds per square inch) [35.15 kg/cm^2] at the end of the compression stroke. When air is rapidly compressed to this pressure, it is heated to a temperature of about 1000°F [537.8°C]. This temperature is high enough to ignite fuel oil sprayed into the heated air. Thus, no separate ignition system is required. The oil is ignited by the heat resulting from the compression of the air.

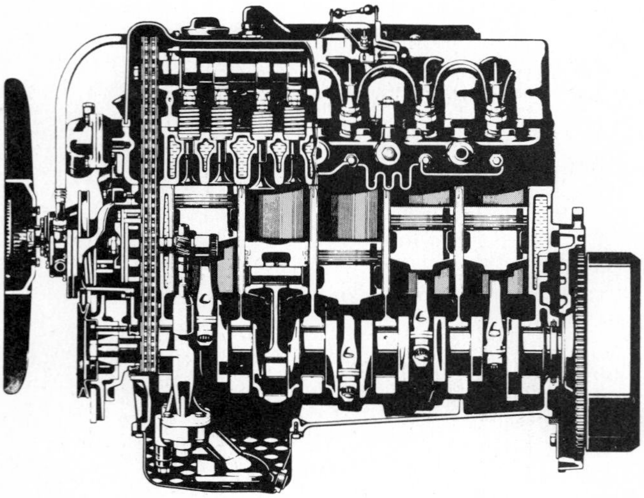

Fig. 6-22 Sectional view of a five-cylinder diesel engine for passenger cars. (*Mercedes-Benz*)

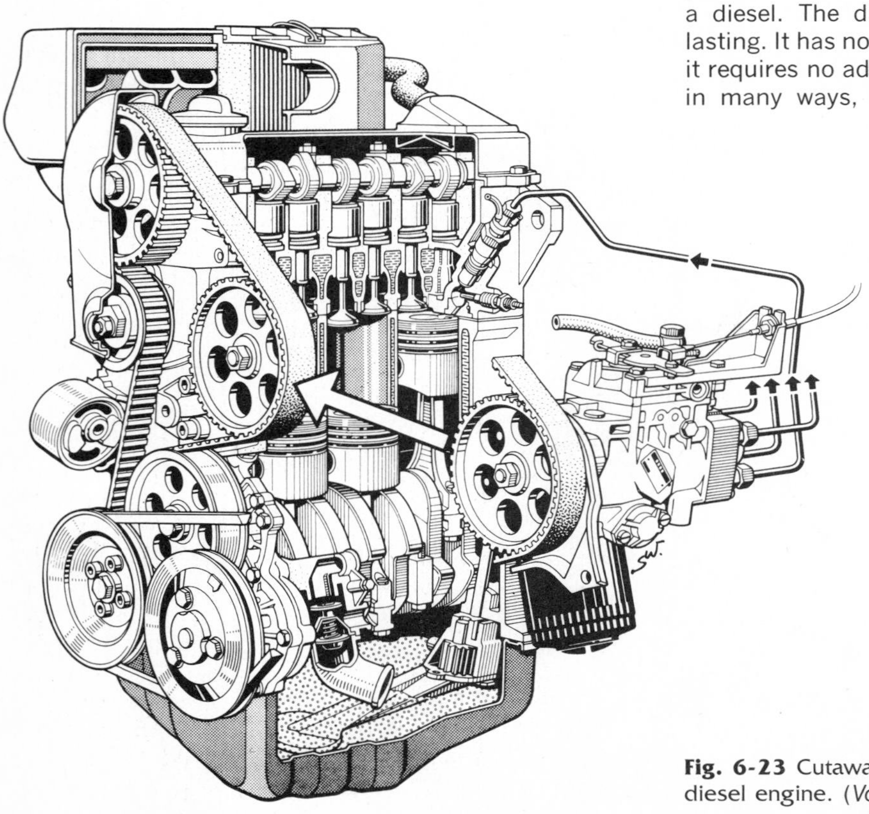

Fig. 6-23 Cutaway view of four-cylinder, overhead-camshaft diesel engine. (*Volkswagen of America, Inc.*)

● 6-19 Diesel-Engine Applications

Diesel engines have been made in many sizes and outputs, from a few horsepower to 5,000 horsepower [3,730 kW]. They are used in passenger cars, trucks, buses, farm and construction machinery, ships, electric power plants (up to about 5,000 kilowatts), and other mobile and stationary applications.

Diesel engines have been used in passenger cars for many years, but until recently no automobile manufacturer in the United States was producing diesels. The picture has changed in recent years, however. For many years, the major user of diesel engines in passenger cars was Mercedes-Benz, which has produced more than half a million diesel-powered cars since 1950. Many of these have been imported into the United States, so perhaps you have seen them. Figure 6-21 is a sectional view of their four-cylinder, in-line diesel engine for passenger cars. Mercedes-Benz recently introduced a five-cylinder diesel (Fig. 6-22). Peugeot is another European automobile manufacturer who has been producing diesel-powered cars.

More recently, Volkswagen and General Motors began to produce diesel-powered cars. Volkswagen has a four-cylinder in-line diesel engine (Fig. 6-23). General Motors recently brought out a V-8 diesel (Fig. 6-24). It is expected that other automotive companies will be looking into the possibility of producing diesel engines. There are several reasons for this increased emphasis on diesels. The diesel is a relatively fuel-efficient engine, especially under part load. That is, you can get more miles per gallon with a diesel. The diesel engine is very reliable and long-lasting. It has no ignition system. For the present at least, it requires no add-on devices to control emissions. Thus, in many ways, it is a simpler mechanism. There are

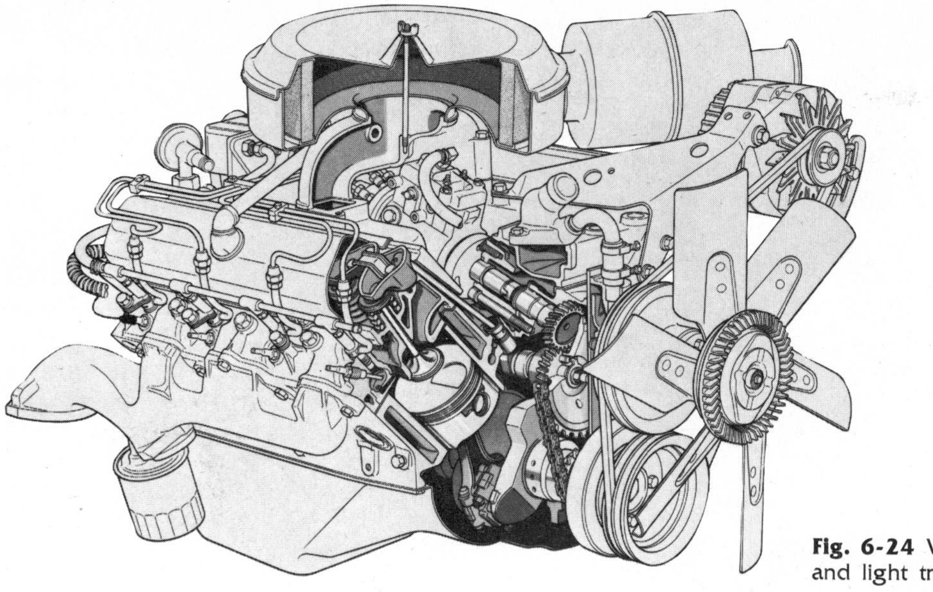

Fig. 6-24 V-8 diesel engine for passenger cars and light trucks. (*General Motors Corporation*)

disadvantages. It requires more cranking power to get it started and thus needs a heavier starting motor and battery. Also, it lacks the acceleration or "pep" of a well-tuned gasoline engine. It is a heavier engine because of the high internal pressures that develop in the diesel-engine combustion chambers. That is, the diesel engine requires more pounds per horsepower than the gasoline engine.

Nevertheless, many authorities believe the diesel has a future in the automotive field. It has already made a name for itself in heavy trucks and buses and in off-the-road construction equipment. General Motors has been making diesel engines for these vehicles for many years. Cummins is another large manufacturer of diesel engines. So it is not a new field. General Motors did not, for example, have to start from scratch to develop their V-8. As a matter of fact, their V-8 diesel engine is based on one of their larger V-8 gasoline engines. We cover diesel engines in detail in a later chapter.

6-20 Rotary Engines

Until now, we have been talking about reciprocating engines. These engines have pistons that move up and down, or reciprocate, in cylinders. Another type of engine has no pistons. Instead, it has a rotor that is spun by the burning of the fuel in the engine. Two general types of rotary engines are the gas *turbine* and the *Wankel*. We look at the gas turbine first.

6-21 Gas Turbines

Gas turbines are used in some buses and trucks, but many engineers believe we will never see them as the main power plant for cars. However, you may be interested in this kind of engine and might like to find out how it works. Figure 6-25 is a simplified sectional view of a gas turbine. Figure 6-26 is a cutaway of an actual unit. There are two sections in the gas turbine: the gasifier section, where the fuel is burned, and the power section, where the power from the burning fuel is produced. The turbine can use gasoline, kerosene, or oil for fuel.

The compressor in the gasifier section has an air-intake rotor with a series of blades on it. When the air-intake rotor spins, it acts as an air pump and supplies the burner with high-pressure air, as shown by the arrows at the left in Fig. 6-25. In the burner, fuel is sprayed into the compressed air. It burns, and the burned gases then flow, at still higher pressures (because of the combustion process), through the blades of the gasifier section. This spins the turbine rotor. The gasifier turbine rotor is mounted on the same shaft as the air-intake rotor. Thus when the turbine rotor spins, the air-intake rotor spins, supplying the gasifier section with compressed air.

The power of the spinning turbine rotor is carried through shafts and gears to the vehicle's wheels.

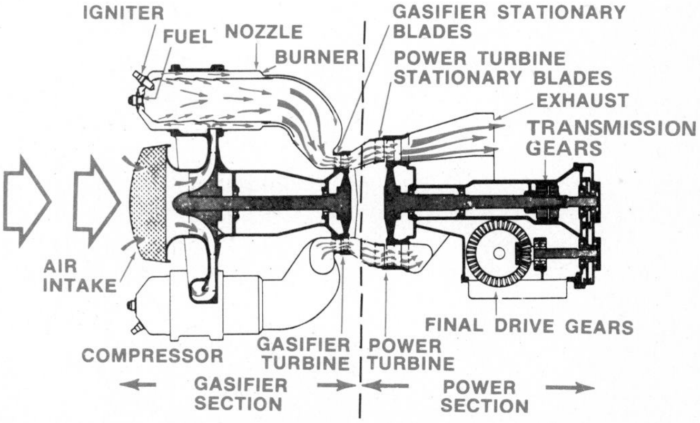

Fig. 6-25 Schematic view of a turbine engine, showing locations of major components. (*Ford Motor Company*)

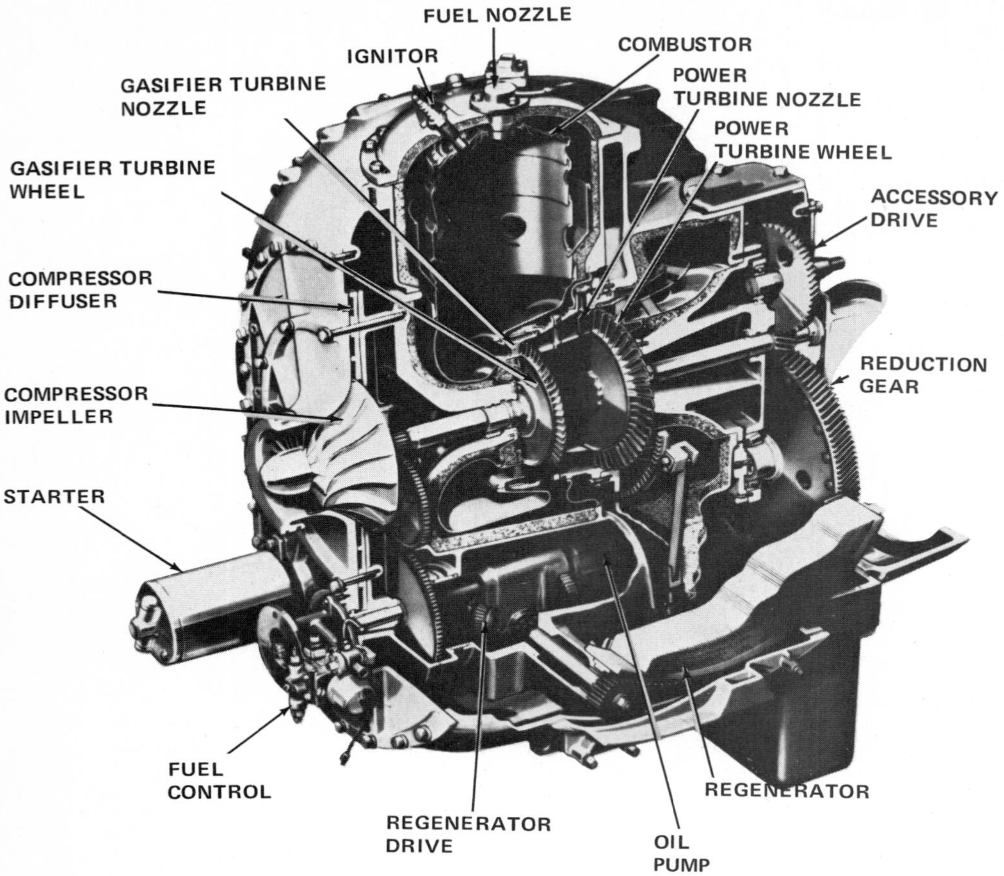

Fig. 6-26 Cutaway view of a turbine engine. *(Ford Motor Company)*

6-22 Wankel Engine

The Wankel engine (Fig. 6-27) is new on the automotive scene. It has a rotor that spins in an oval chamber shaped like a fat figure 8. Several automotive manufacturers have invested a great deal of time and money in researching and developing the Wankel. The Japanese firm of Toyo Kogyo has produced several hundred thousand automobiles powered by Wankel engines. Other manufacturers have also produced Wankel-powered cars. General Motors spent many millions of dollars on the Wankel engine, but then decided not to use it in any of their automobiles. However, you may run across the Wankel and you may want to know how it works.

The Wankel is also called a rotary-combustion, or RC, engine because the combustion chambers rotate, or move in a somewhat circular pattern. The engine uses a three-lobe rotor (Fig. 6-28) that rotates eccentrically in an oval housing (Fig. 6-29). The three lobes are always in contact with the oval housing, and they form a tight seal. This seal compares with the seal formed by the piston rings against the cylinder wall in the reciprocating engine. The rotor is positioned on the crankshaft by external and internal gears. The four actions—intake, compression, power, and exhaust—are going on at the same time around the rotor when the engine is running. Figure 6-30 gives you an idea of how the engine works. The rotor lobes *A*, *B*, and *C* seal tightly against the side of the oval housing. The rotor has recesses in its three faces between the lobes. The dashed lines on the rotors in Fig. 6-30 show the locations of the recesses and how deep they are. It is in these recesses that combustion actually starts. The spaces between the rotor lobes are where intake, compression, power, and exhaust take place.

Let us follow the rotor around as it goes through a complete cycle—intake, compression, power, and exhaust. At I (upper left), lobe *A* has passed the intake port, and the air-fuel mixture is starting to enter. This is shown by the circled 1. As the rotor moves around, at II (upper right), the space between lobes *A* and *C* increases, as shown by the circled 2. This motion produces a vacuum, which causes the air-fuel mixture to enter. This action compares with the intake stroke of the piston in the reciprocating engine. At III (lower right), the air-fuel mixture continues to enter as the space between lobes *A* and *C* continues to increase, as shown by the circled 3. Then lobe *C* starts to move past the intake port, as shown in IV (lower left). Further rotor movement carries lobe *C* past the intake port, so the air-fuel mixture is sealed between lobes *A* and *C*, as shown by the circled 4.

Fig. 6-27 Cutaway view of a two-rotor Wankel engine with attached torque converter and transmission. (*NSU of Germany*)

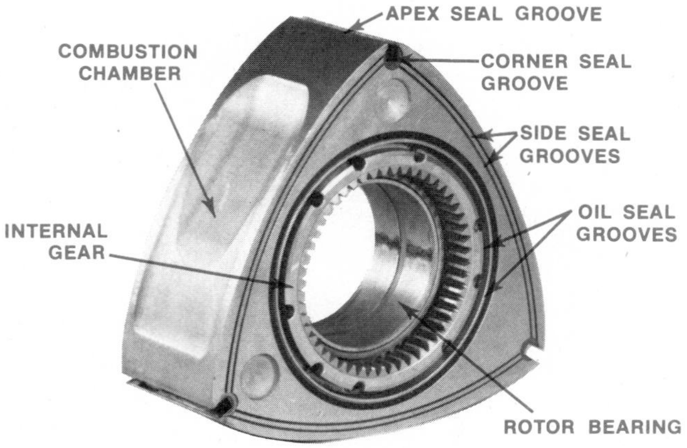

Fig. 6-28 Rotor for a Wankel engine. (*Toyo Kogyo Company, Limited*)

To see what happens to the air-fuel mixture, let us go back to I (upper left) again. Here the air-fuel mixture has been trapped between lobes *A* and *B*, as shown by the circled 5. Further rotation of the rotor decreases the space between lobes *A* and *B*. By the time the rotor reaches the position shown in III, the space (circled 7) is at a minimum. This action is the same as the piston reaching TDC on the compression stroke in the reciprocating engine. Now the spark plug fires and ignites the compressed mixture. Pressure is exerted on the side of the rotor and this forces the rotor to move around. See IV (lower left). This action is the same as the power stroke in the reciprocating engine.

At IV, the high pressure of the burned air-fuel mixture (circled 8) forces the rotor around to position I again. Continued expansion of the burned gases continues to rotate the rotor until the leading lobe passes the exhaust port. Then the burned gases begin to exhaust from between the lobes, as shown by circled 11 and 12 in III and IV. As the rotor continues to rotate, the space between the lobes decreases and the gases are exhausted. This action is the same as the exhaust stroke of the piston in the reciprocating engine.

Following the exhaust stroke, the leading lobe passes the intake port, and the whole cycle is repeated. Note that there are three lobes and three spaces between the lobes. That means that there are three complete cycles of in-

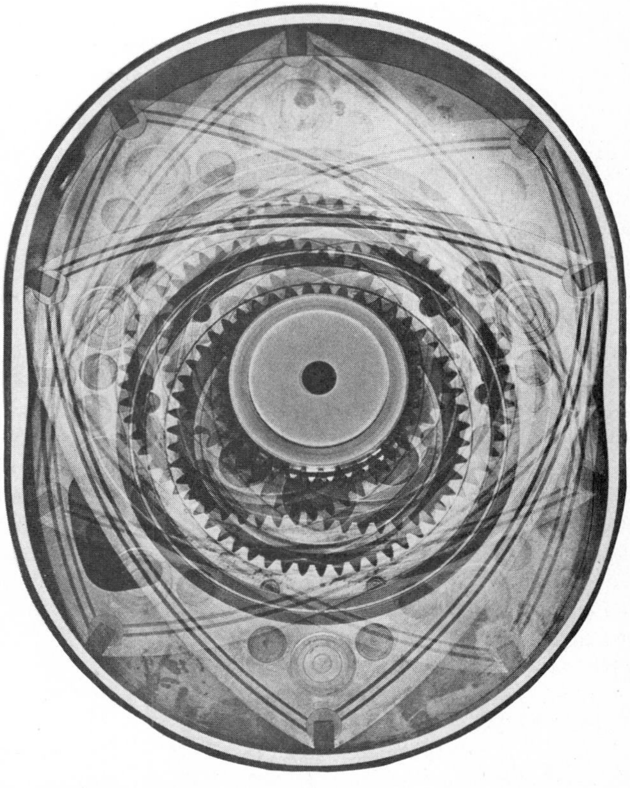

Fig. 6-29 The rotor rotates eccentrically, so the three apexes are always in sliding contact with the inner face of the rotor housing. (*Toyo Kogyo Company, Limited*)

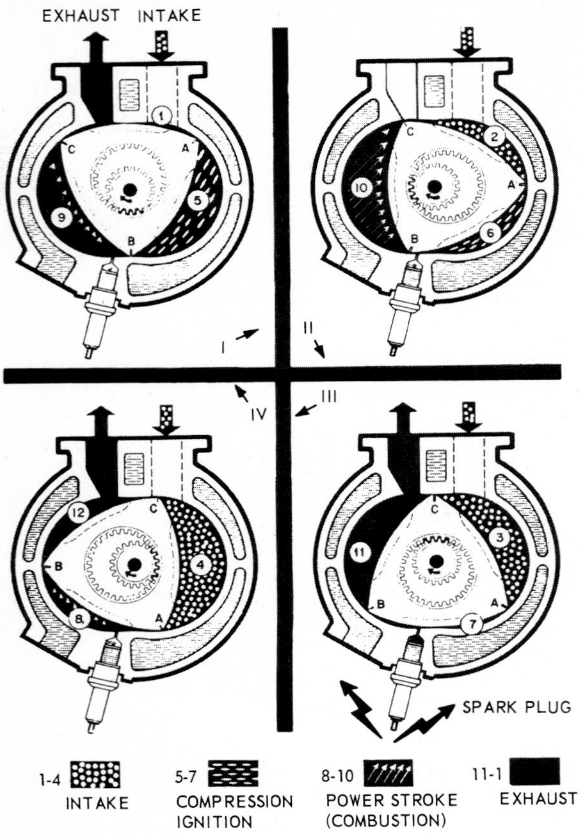

Fig. 6-30 Actions in a Wankel engine during one complete revolution of the rotor.

take, compression, power, and exhaust going on at the same time. The engine is delivering power almost continuously. In a way, the engine is equivalent to a three-cylinder piston engine. A two-rotor Wankel engine would, then, be equivalent to a six-cylinder piston engine.

● 6-23 Stirling Engine

The pressure in a container of gas goes up when the gas is heated and goes down when the gas is cooled. The Stirling engine makes use of that fact. A certain amount of gas is sealed in the engine, and it is first heated and then cooled. When it is heated, its increasing pressure pushes a power piston down. When it is cooled, its lower pressure, in effect, pulls the power piston up.

Figure 6-31 is a simplified drawing of a Stirling engine. The upper, or displacer, piston does not produce power. It simply causes the air to move between the heater and the cooler sections of the engine. The power piston moves up and down between the power chamber and the buffer space. The two pistons are linked by a rhombic drive to a pair of gears, which drive the output shaft. The name "rhombic" comes from the fact that the four links form a rhombus, which is a geometric figure with four parallel sides.

Let us follow the engine through a complete cycle of events and see how it operates. We start with the pistons in the position shown in Fig. 6-31. The working gas has been heated, and therefore its pressure has gone up. This increase of pressure, applied to the head of the power piston, forces the piston to move down. Note that the working gas expands during this action.

The power piston, in moving down, pushes downward through its two connecting links against the two gears. The gears rotate and turn the engine output shaft. In the meantime, the rotation of the gears causes the two links to the displacer piston to pull the displacer piston down. This increases the hot space above the displacer piston. Some of the working gas flows from the power chamber toward this space. As it moves through the cooler on the way to this upper space, the gas is cooled, and its pressure drops.

Note that there is a buffer space below the power piston. This buffer space contains a certain amount of gas. When the power piston moves down, this gas is compressed, and its pressure goes up. Now, after the power piston has reached BDC, the pressure of this buffer gas becomes higher than the pressure of the working gas. (Remember that the working gas has been cooled, so that its pressure has dropped.) Now, with the buffer gas at a higher pressure than the working gas, the power piston is pushed up. This motion is transmitted through the two links to the gears, rotating them to the position shown in Fig. 6-31.

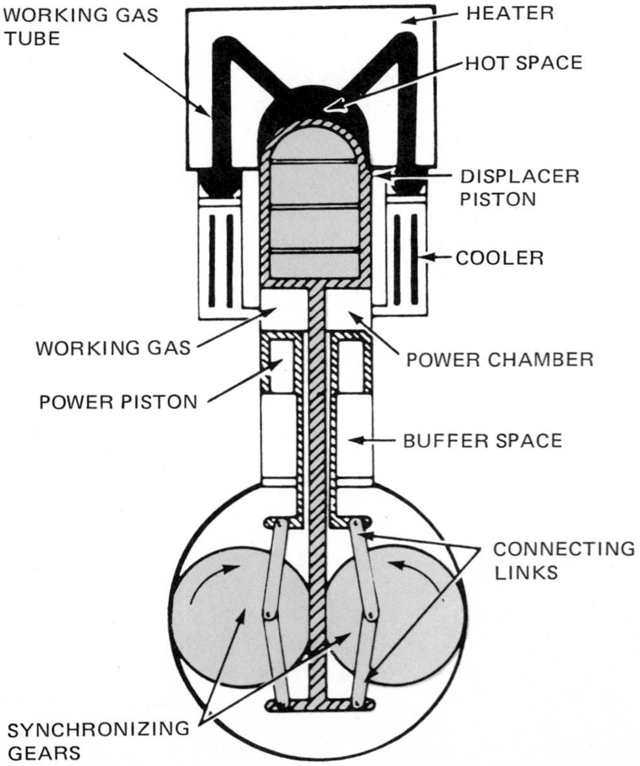

Fig. 6-31 Schematic drawing of a Stirling motor.

In the meantime, the working gas that has flowed into the hot space becomes heated. The heating results from the continuous burning of a fuel such as kerosene or oil. Heating the working gas causes its pressure to increase. Once again, the power piston is forced down by the increasing pressure.

The action, therefore, results from repeated heating and cooling of the working gas. When the working gas is heated, it drives the power piston down. When it cools, the buffer-gas pressure forces the piston up. You might think this heating and cooling would take a lot of time, and the engine would be awkward, slow, and inefficient. However, experimental engines, using helium as the working gas, at an average pressure of 1,500 psi [105.46 kg/cm^2], have operated at 3,000 rpm. Their efficiency is about 30 percent, which is as good as, or better than, most automotive engines.

Research and development of Stirling engines for passenger cars continues. Stirling engines have great appeal to engineers and people who want to protect our world from pollution. They are almost silent in operation, can be operated from any heat source (including the sun), and even when burning fuel, have very little polluting exhaust. Some engineers have proposed the Stirling engine for such small-engine applications as lawn mowers. They are easy to start, quiet, and simple. Other engineers believe that they will appear first for large stationary applications, as, for instance, for electric power plants or pumping stations. One possibility is that they will be used to power space satellites and space stations. In space, no source of heat other than the sun would be needed. With a large reflector to gather heat from the sun and a large radiator to radiate the heat away, a Stirling engine would operate without any fuel at all.

● 6-24 Other Possible Power Plants

In recent years, other power plants to run automobiles have been suggested. These include steam, electric batteries, and spinning flywheels.

1. Steam engines The steam engine is an external-combustion engine. That is, the combustion takes place outside the engine. In a steam engine, water or other liquid is heated in a boiler where it turns into steam, or vapor. Figure 6-32 shows the design of one experimental car using a steam engine. The engine burns kerosene and the heat evaporates the liquid. The liquid used is Freon, the same liquid used in refrigerators. It evaporates

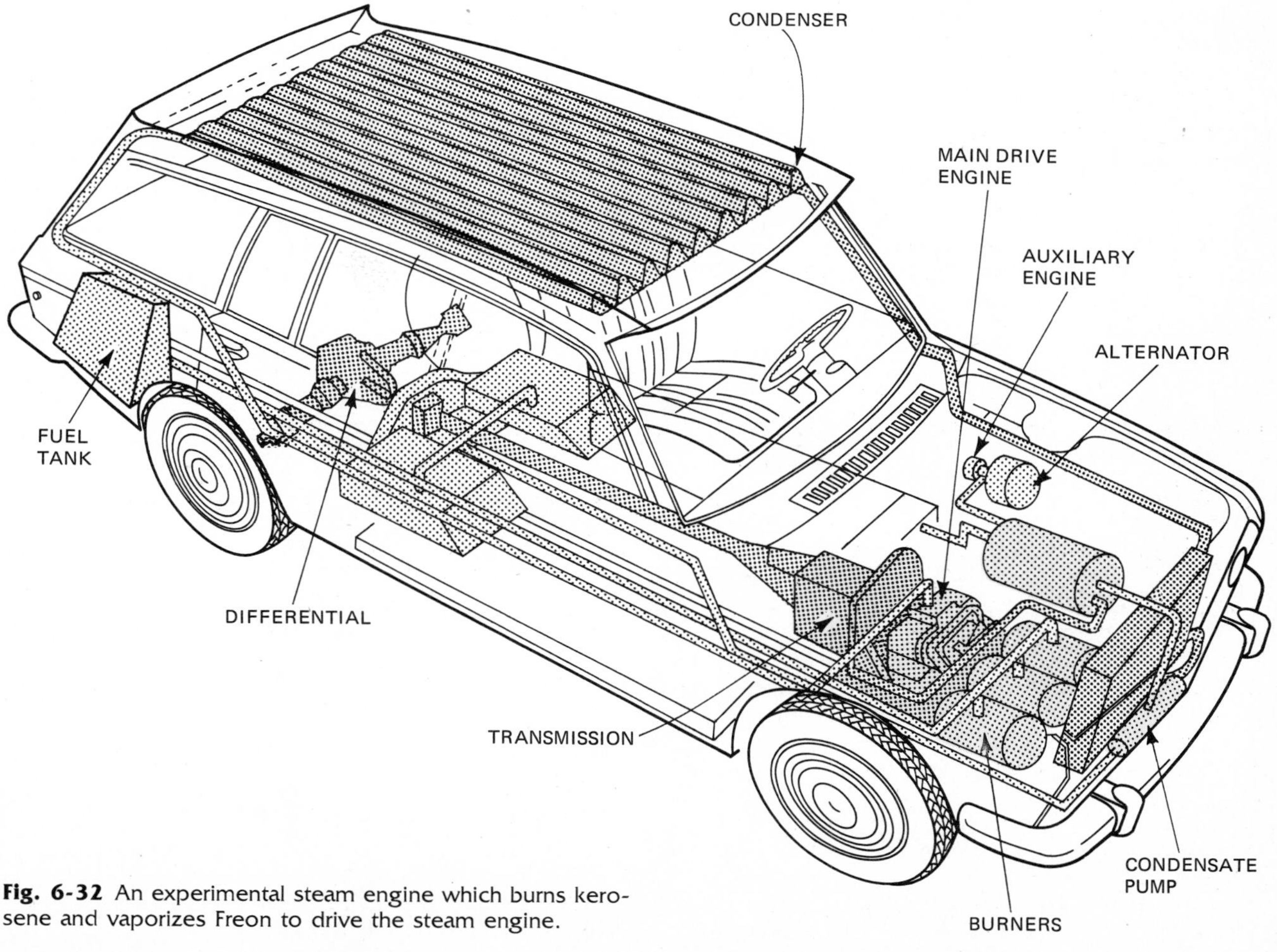

Fig. 6-32 An experimental steam engine which burns kerosene and vaporizes Freon to drive the steam engine.

at a much lower temperature than water. The Freon vapor, at high pressure and temperature, enters the main drive engine where it spins rotors. The rotary motion is then carried through the transmission and differential to the car wheels. As the Freon vapor leaves the engine, it is carried up to the condenser which covers the top of the car. There, the Freon gives up heat and returns to a liquid. It then flows back down to the burners where it is turned into high-pressure vapor again.

2. Electric cars Electric cars use batteries to hold the energy needed to run the car. The lead-acid storage battery used in automobiles to start the engine is used in most electric cars. The trouble with these batteries is that they are very heavy. And they will not provide enough electricity to run the car more than a few miles. That is, after fifty or so miles, the batteries need to be recharged. Despite these drawbacks, a few electric vehicles are still around (in addition to electric golf carts). For example, there are an estimated 50,000 electric vans in London, England, mostly in short mileage delivery service. General Motors developed an experimental car (Fig. 6-33) which uses silver-zinc batteries. These batteries hold much more energy per pound than the heavy lead-acid batteries. Even though this experimental car shows better mileage per battery charge, it still has serious drawbacks. The silver-zinc batteries are very expensive. Also, any electric car requires electric energy to charge the batteries. And the electric-power resources in the United States are already strained to the limit. A vast number of new electric power plants would have to be built if electric cars became common.

3. Hybrid car Figure 6-34 shows a hybrid vehicle developed by Volkswagen. It was designed as a taxi and uses two banks of storage batteries and a small gasoline engine. That is the reason for the name "hybrid." It is neither completely an electric nor completely a gasoline engine. When the vehicle first starts out, the batteries carry the complete load, supplying electricity to the electric motor that drives the rear wheels. Then, as the throttle is opened, the gasoline engine begins to deliver power. Speed can be maintained with the aid of the gasoline engine or with the electric motor alone. If the gasoline engine produces more power than needed, the excess is used to charge the batteries while the vehicle is being driven. With the proper ratio of acceleration to steady speed, the gasoline engine can keep the batteries charged.

4. Flywheel car This is a rather "far out" idea. The car has a heavy flywheel, perhaps located in the back, as in Fig. 6-35. The flywheel would spin at high speed by using an electric motor at a charging station. Then the energy of the flywheel could be used to run the car. One study shows that an ordinary car equipped with a driving flywheel could travel 100 miles between charges. It could accelerate to 60 miles per hour in 15 seconds and reach a top speed of 70 miles per hour.

The idea was actually tried in a bus in Europe. The bus was driven by a 3,300-pound flywheel. It worked, but was

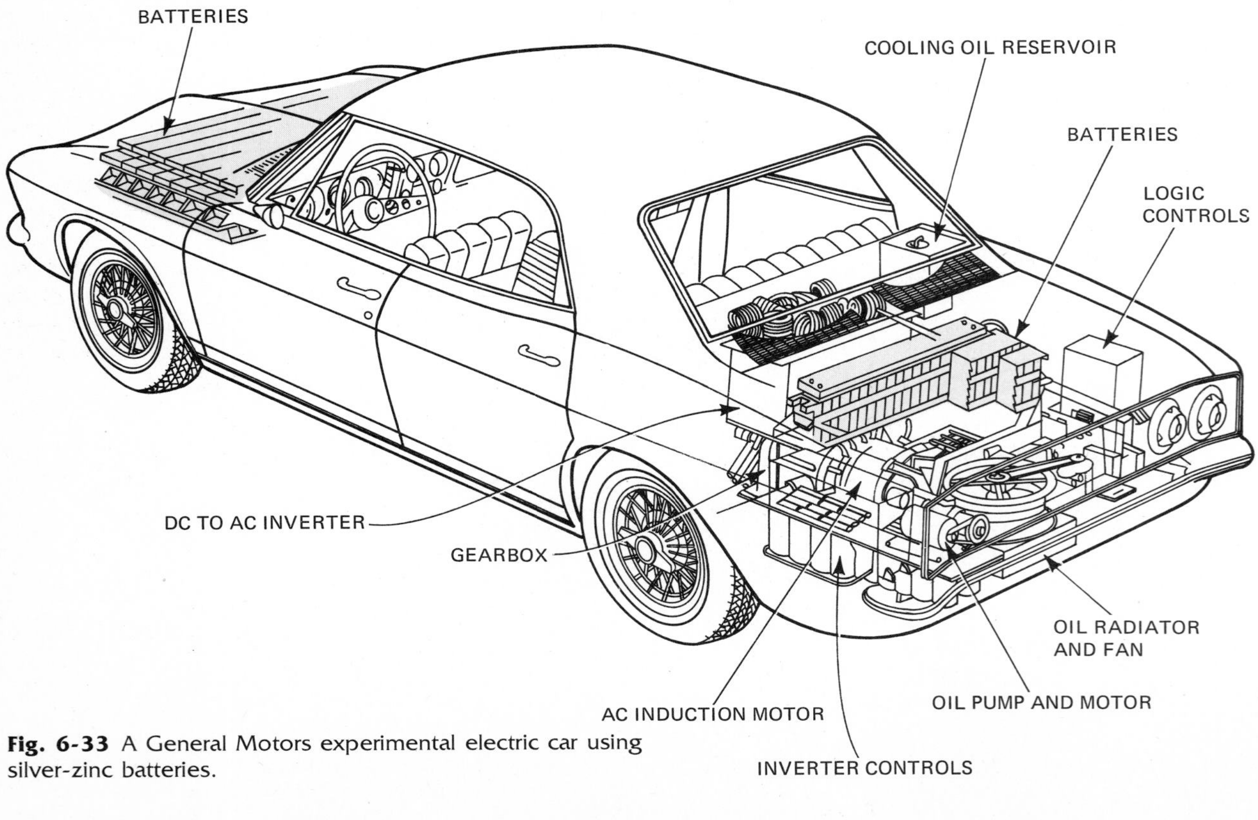

Fig. 6-33 A General Motors experimental electric car using silver-zinc batteries.

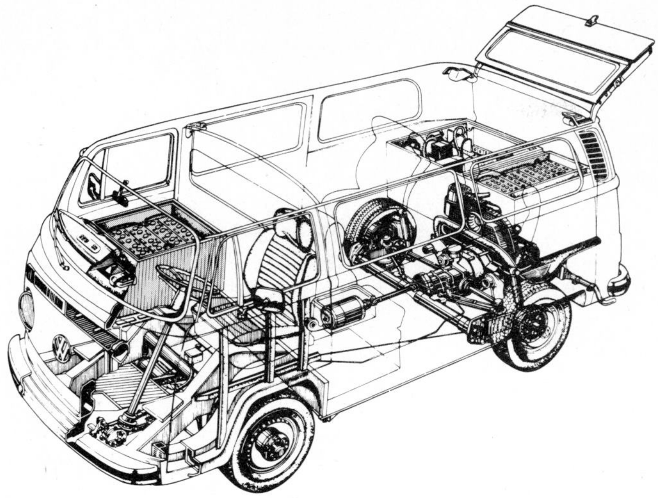

Fig. 6-34 Hybrid vehicle using a small gasoline engine and storage batteries for motive power. (*Volkswagen*)

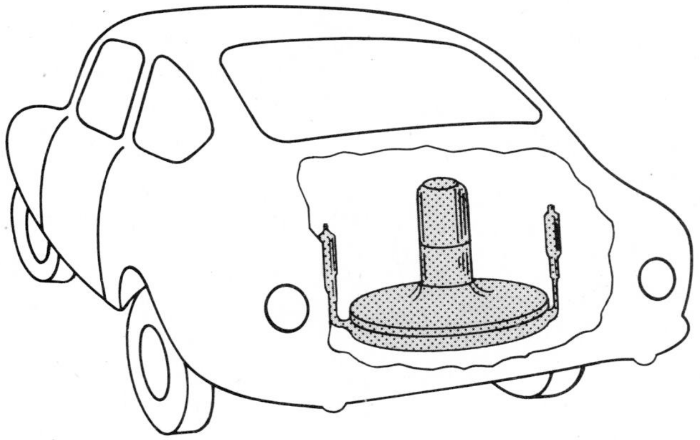

Fig. 6-35 A flywheel car. The large flywheel is shown here located at the back of the car.

not considered practical because the flywheel had to be charged every half mile or so. Nonetheless, experimental and design work continues on this idea. Recent design improvements on flywheels have considerably increased the energy that could be stored per pound of flywheel.

REVIEW QUESTIONS

See if you can answer these questions. They will help you find out how well you remember what you have just studied about engine types.

1. Name seven ways that internal-combustion engines can be classified. (Eight are listed in the book.)
2. Name four different cylinder arrangements.
3. Compare the in-line with the V-type engine.
4. What are three valve arrangements?
5. What do SOHC and DOHC mean?
6. What are the advantages of the overhead-camshaft engine?
7. Some engineers are predicting that the V-6 will supplant the six-in-line. Why?
8. What is the dual-displacement engine and how does it work?
9. What are the advantages of turbocharging and how does it work?
10. What is the multiple-displacement engine and how does it work?
11. What is the firing order?
12. What are the advantages of the diesel engine. Disadvantages?

SELF PROJECTS

Do a little research on the different types of engines. Check with your nearest museum to see if it has a turbine, an electric car, or a steam car. If a friend has a car with a Wankel engine, ask if you can look at it. Or go to a dealer who handles cars with Wankel engines and look the engine over carefully. The dealer may have a display that shows how the engine works. Study the display. Check encyclopedias in your school or public library for more information on Stirling, steam, and gas turbines.

On separate sheets of paper, write the names of the different engines that could be used to power automobiles. Then, write down the most important facts about each. Paste in clippings from newspapers and magazines about these engines. Keep up-to-date on the engines by clipping and filing other articles that appear in papers and magazines. That way, you will know about other engines adopted by automotive manufacturers.

CHAPTER 7
Engine Construction: Cylinder Block, Cylinder Head, Crankshaft, and Bearings

After reading this chapter, you should be able to:

- **Discuss cylinder blocks and how they are made.**
- **Discuss cylinder heads and how they are made.**
- **Describe the three types of cylinder heads for gasoline engines.**
- **Explain the purpose and function of intake and exhaust manifolds, vibration dampers, and flywheels.**
- **Discuss crankshafts and how they are made.**
- **Describe engine bearings, how they work and how they are lubricated.**
- **Identify and name engine parts of disassembled engines.**

This and the next two chapters describe the construction of automotive piston engines. These chapters supply the information you need to go on to the engine troubleshooting and servicing sections of the book.

● 7-1 The Engine

In earlier chapters, we looked at the basics of engine operation and engine types. We saw how the mixture of air and fuel is delivered by the fuel system to the engine cylinders. There, it is compressed, ignited, and burned. This burning, or combustion, produces the high pressure that pushes the piston down so the crankshaft turns and the engine produces power. Now, we shall look at the various parts of the engine in detail.

● 7-2 Cylinder Block

The cylinder block of a liquid-cooled engine is the foundation of the engine. Everything else is put inside the block, or fastened onto it. (In air-cooled engines, the cylinders are often separate parts, as shown in Fig. 7-11.) Figure 7-1 shows the block for a four-cylinder, in-line engine. Figure 7-2 shows three views of a cylinder block for a V-8 engine. The cylinder block is cast in one piece from gray iron or iron mixed with some other metal, such as nickel or chromium. These added metals give the cylinder block greater resistance against wear as well as greater strength. Some cylinder blocks are cast from aluminum. The engine shown in Fig. 6-10, for example, has been supplied with either a die-cast aluminum or cast-iron cylinder block. The cylinder block shown Fig. 7-1

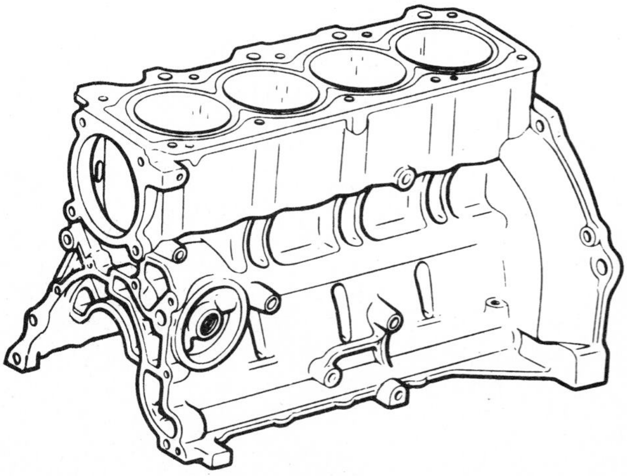

Fig. 7-1 Aluminum cylinder block for a four-cylinder, in-line engine. (*Pontiac Motor Division of General Motors Corporation*)

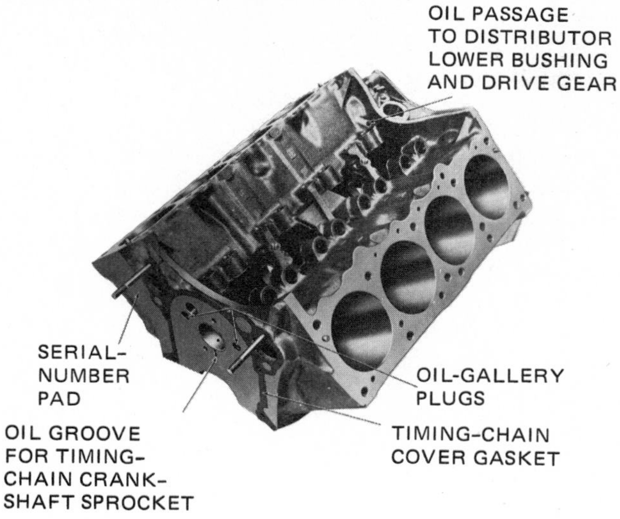

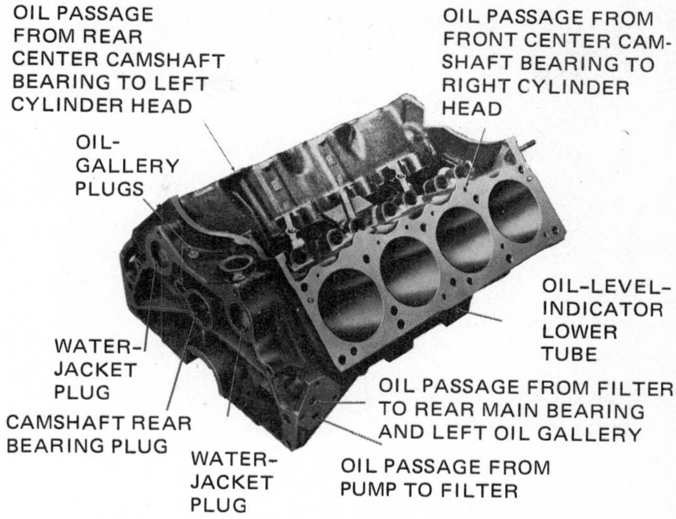

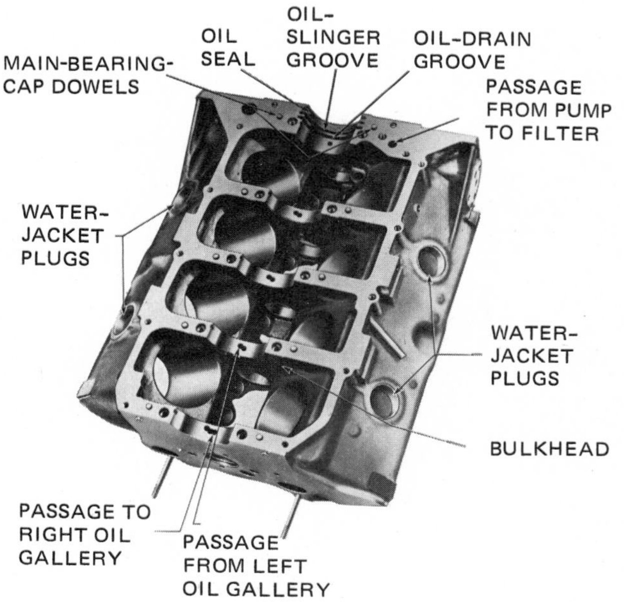

Fig. 7-2 Three views of a cylinder block for a V-8, overhead-valve, liquid-cooled engine, showing locations of water and oil passages and plugs. (*Pontiac Motor Division of General Motors Corporation*)

is cast from aluminum. We describe aluminum cylinder blocks in detail in ● 7-6.

The iron cylinder block is cast by pouring molten iron into a mold. The mold is usually made of sand. The openings in the casting are formed by putting sand cores into the mold. Figure 7-3 shows the lower half of a mold for two six-cylinder engine blocks with the cores in place. After the upper half is put into place, molten iron is poured into the mold. The molten iron fills all the spaces between the mold and the cores. After the casting has cooled and hardened, it is removed from the mold. The cores are broken up and shaken out of the casting. If you examine Fig. 7-2 carefully, you will find several arrows pointing to "water-jacket plugs." These plugs are in the openings through which the cores are removed, to leave the water jackets. The water jackets are the spaces around the cylinders through which coolant (water and antifreeze) flows.

The water jackets are formed by the inner shells of the cylinders and the outer shell of the cylinder block. You can see the water jackets in one bank of a V-6 engine in Fig. 7-4.

A gasoline-engine cylinder block and a diesel-engine cylinder block look much alike. However, the diesel-engine cylinder block is heavier and stronger. It must be stronger because the compression ratio and internal pressures in the combustion chambers are higher for the diesel engine.

● 7-3 Machining the Block

After the cores are removed, the block is machined. Examine the three views of the block in Fig. 7-2 again. Note the many finished surfaces, passages, bores, and tapped holes in the block. The machining operation includes:

1. Drilling holes for attachment of various parts
2. Machining the cylinders
3. Boring the camshaft-bearing holes
4. Smoothing the surfaces to which parts are attached
5. Drilling oil passages
6. Boring the valve-lifter bores
7. Cleaning out water passages

● 7-4 Parts Attached to Block

Figure 7-5 shows the parts that are attached to and mounted on the cylinder block for a V-8 engine. The cylinder heads are mounted at the top, along with the cylinder-head covers, the intake and exhaust manifolds,

Fig. 7-3 Cores for two six-cylinder engine blocks in place in lower half of mold. (*Ford Motor Company*)

Fig. 7-4 One bank of a V-6 engine block partly cut away so the internal construction can be seen. (*General Motors Corporation*)

the carburetor, and other related parts. The oil pan is attached to the bottom of the cylinder block. In addition, the water pump and timing-chain cover (not shown in Fig. 7-5) are attached at the front of the block. Figure 6-15 shows how all these parts go together in the assembled engine. The purpose of all these parts will be explained later.

Figure 7-6 shows parts that are installed inside the cylinder block of a six-cylinder engine. Only one piston and connecting rod are shown, although the engine actually uses six of each. Figure 6-9 shows how all the parts go together in the assembled engine. The crankshaft is hung from the bottom of the cylinder block by bearings and caps. Figure 7-7 is another picture of a crankshaft and related parts for a six-cylinder engine.

The crankshaft bearings, called the main bearings, or "mains," are of the split type. The upper halves fit into half-round sections in the block. The lower halves fit into bearing caps which are attached to the cylinder block by cap screws.

Here is the procedure followed during original engine assembly: The caps are bolted into place. Then the cylinder block and cap are bored to take the bearings. This means that each cap fits in one particular spot on the block. Therefore, when caps are removed for any service job, they must be put back onto the block in the same place and in the same position.

The flywheel is attached to the rear of the crankshaft. The crankshaft gear, or sprocket and pulley with damper, is attached to the front of the crankshaft. We shall discuss all these parts later.

After the crankshaft is installed, the pistons, with their piston rings and connecting rods, are installed in the block. The connecting rods are fastened to the cranks on the crankshaft by bearings and caps with rod bolts and nuts.

The parts shown in Fig. 7-6, when assembled into the cylinder block, make an assembly called a "short block." This is a service item that can be purchased. Sometimes a service problem can be solved by buying a short block. The rest of the old engine, including head and manifolds, is then installed on the new short block. More will be said about this in the chapters on engine service.

● 7-5 Oil Pan

The oil pan (Fig. 7-8) is attached to the bottom of the cylinder block. The bottom of the cylinder block, plus the oil pan, form the *crankcase*. They enclose, or *encase*, the cranks on the crankshaft. The oil pan holds from about 4 to 9 quarts [3.79 to 8.52 L] of oil, depending on the engine design. Note that gaskets (explained later) are used to seal the joint between the cylinder block and the oil pan. The gaskets prevent loss of oil from the oil pan.

When the engine is running, the oil pump sends oil from the oil pan up to the moving engine parts. The engine lubricating system is described in Chap. 19.

● 7-6 Aluminum Cylinder Block

Some engines have aluminum cylinder blocks. Aluminum is a light metal, weighing much less than cast iron. Aluminum also conducts heat more rapidly than cast iron. Lightness and heat conductivity are the two main reasons that aluminum has been used for cylinder blocks. However, aluminum is too soft to use as cylinder-wall material. It would wear very rapidly. Therefore, aluminum cylinder blocks must have cast-iron cylinder liners (with one exception, the General Motors 140 cubic inch [2.2 L] engine). Cylinder liners are sleeves that are either cast into the cylinder block or installed later.

In the cast-in type, the cylinder liners are installed in the mold, and the aluminum is poured around them. They thus become a permanent part of the cylinder block.

Two kinds of liner, dry and wet, can be installed later. The dry types of cylinder liner are pressed into the cylinder block with high pressure. They touch the cylinder bore along their full length. The wet types of liner touch the cylinder block only at the top and bottom. The rest of the liner touches only the coolant. Figure 7-9 is a sectional view of an engine using wet cylinder liners. These liners are removable and can be replaced if they become worn or damaged.

7-7 General Motors 140 Cubic Inch [2.2 L] Engine

This engine has a cast-aluminum cylinder block which does not use cylinder liners. Instead, the aluminum is loaded with silicon particles. Silicon is an extremely hard material. After the cylinder block is cast, the cylinders are honed. They are then treated with a chemical that etches (or eats away) the surface aluminum. This leaves only the silicon particles exposed, so the pistons and rings slide on the silicon (Fig. 7-10). This technique is being developed for other engines, so we may see it on several engines in the future.

7-8 Air-Cooled Engine

Figure 7-11 shows a cylinder for an air-cooled engine, along with the piston and other related parts. Note that,

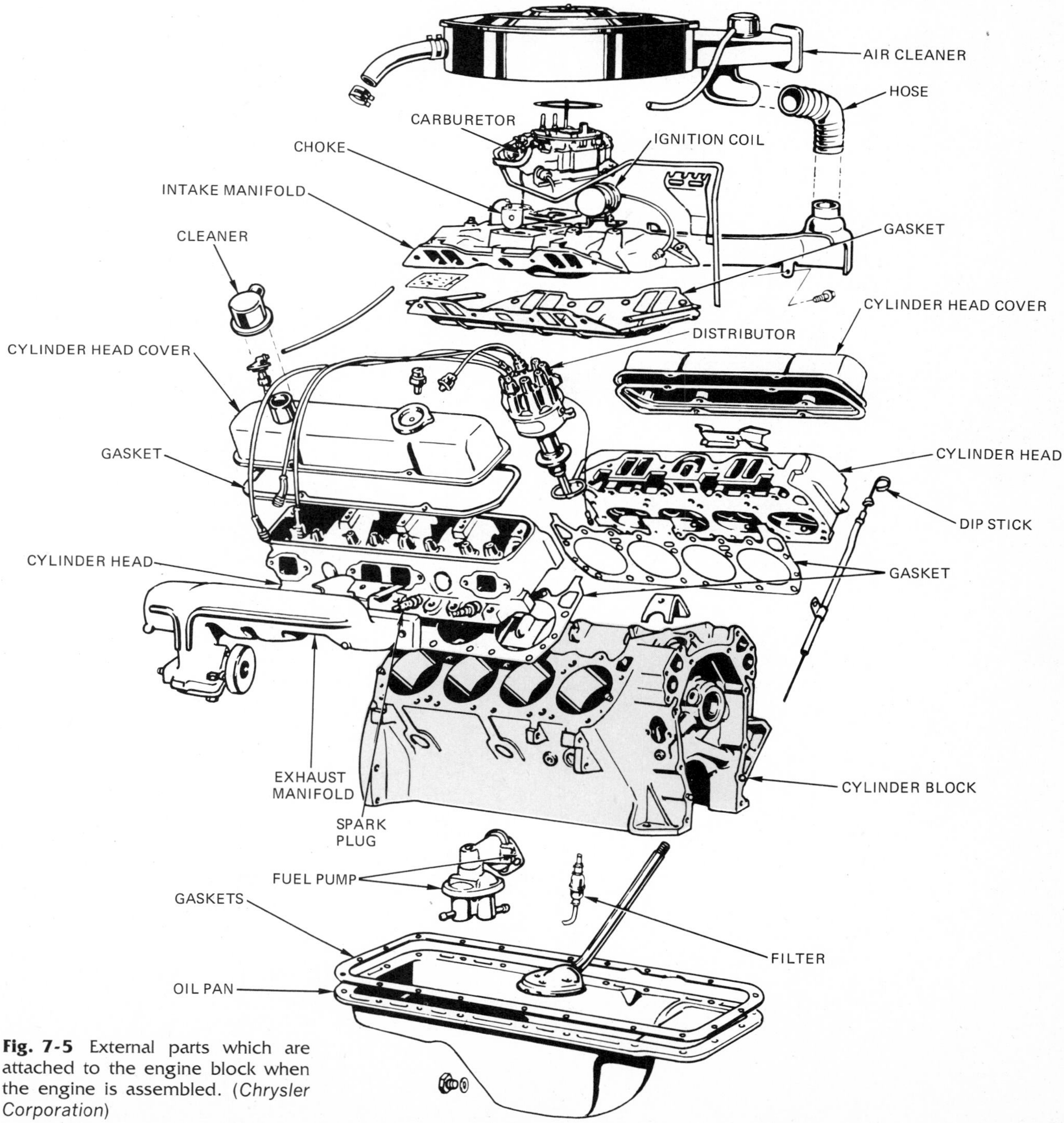

Fig. 7-5 External parts which are attached to the engine block when the engine is assembled. (*Chrysler Corporation*)

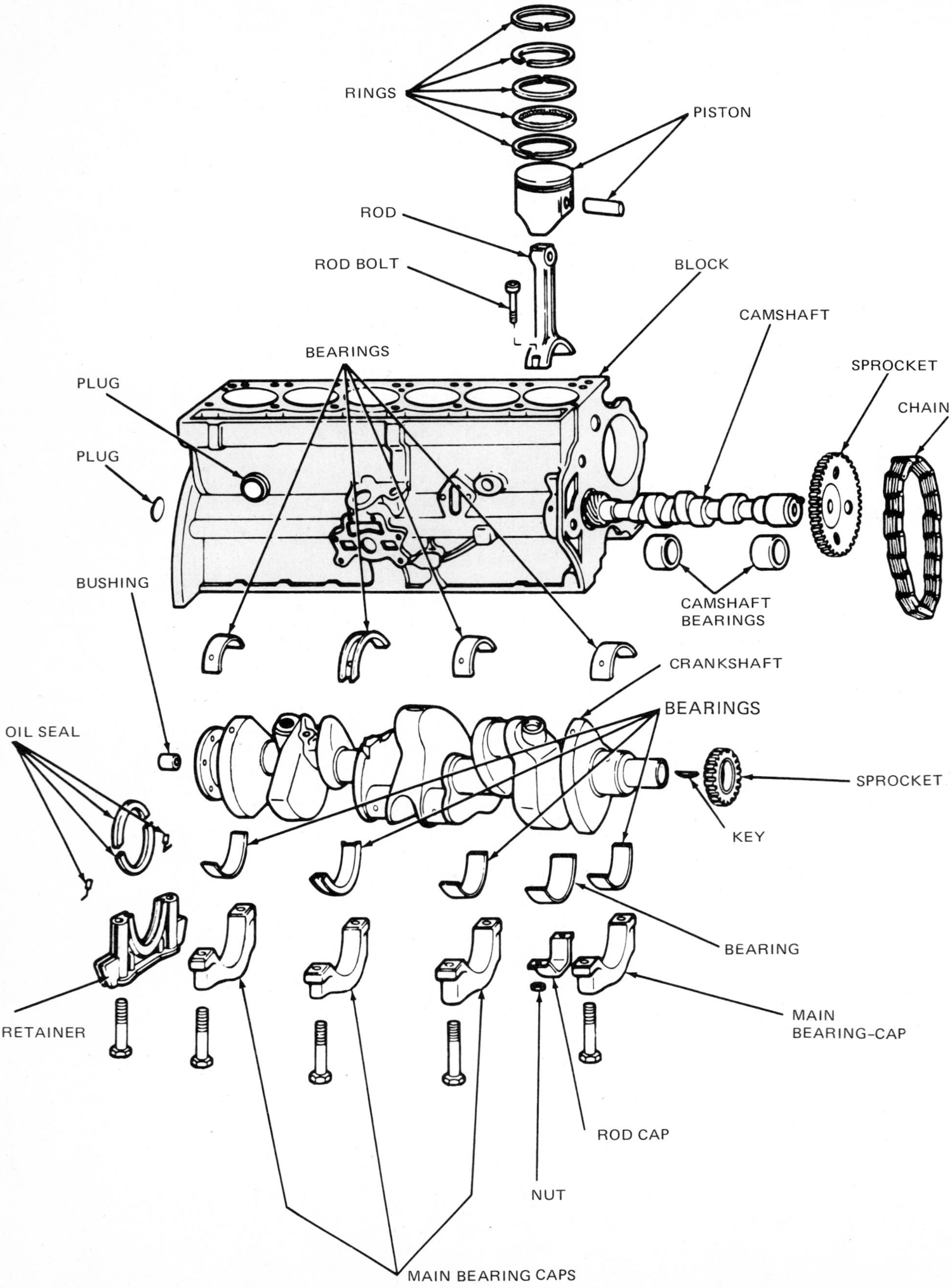

Fig. 7-6 Internal parts from the cylinder block and head of a six-cylinder engine. (*Chrysler Corporation*)

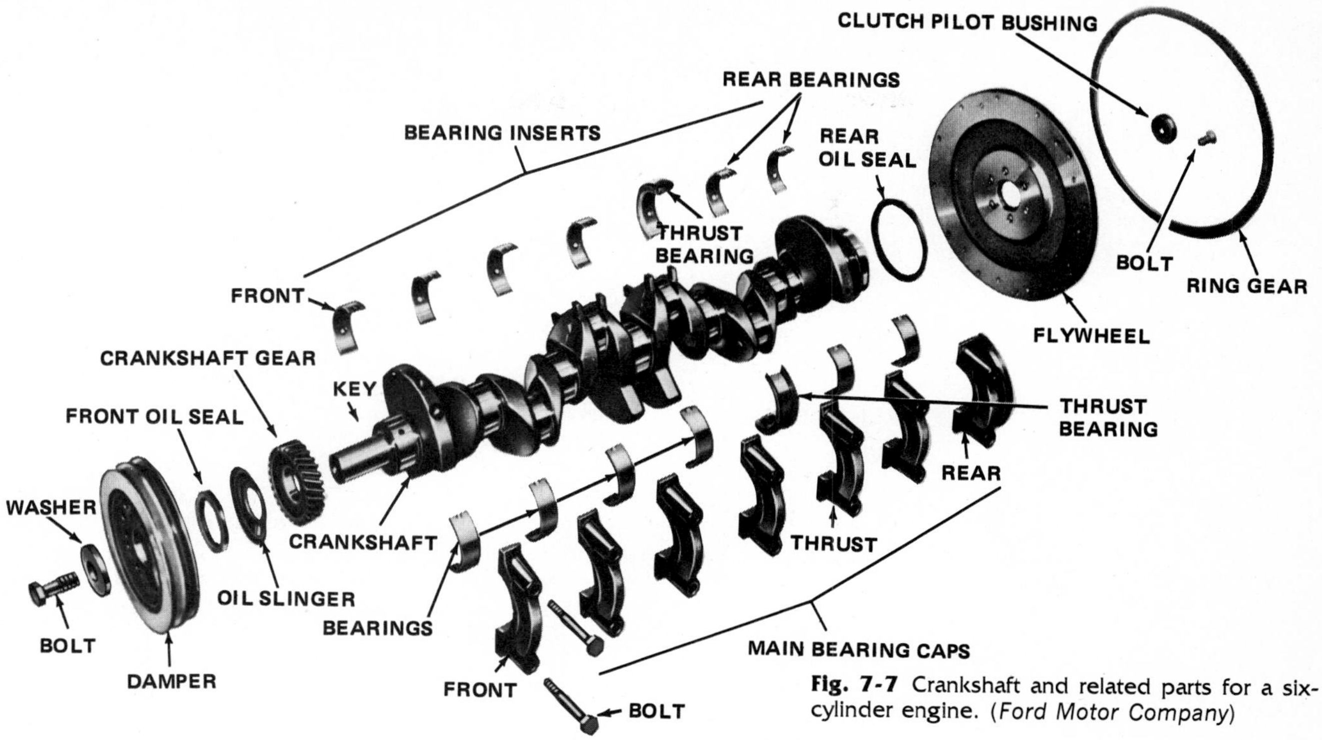

Fig. 7-7 Crankshaft and related parts for a six-cylinder engine. (*Ford Motor Company*)

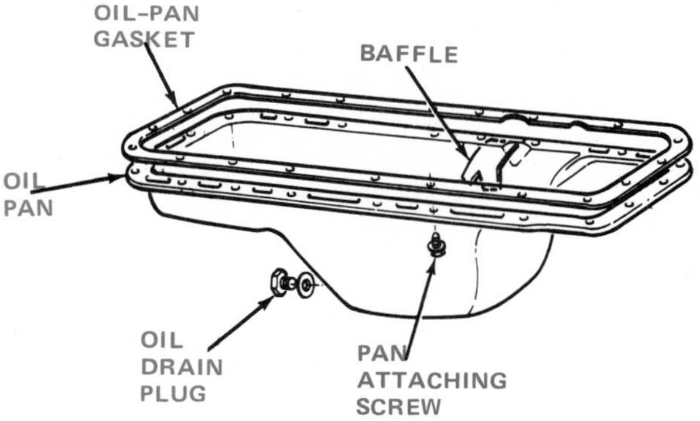

Fig. 7-8 Oil pan with gaskets in place, ready for pan replacement. (*Chrysler Corporation*)

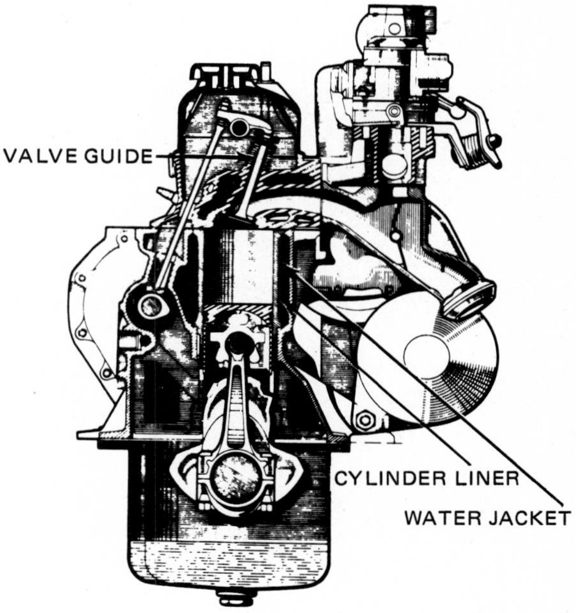

Fig. 7-9 Sectional view of an engine that uses wet cylinder liners. (*Renault*)

in the air-cooled engine, the cylinders are separated and have fins. The fins are metal rings which aid in radiating heat from the cylinder. In the completed air-cooled engine, the cylinders are installed on the crankcase. Then the cylinder head is installed on top of the cylinders. Figure 6-6 shows an air-cooled engine.

● 7-9 Cylinder Head

The cylinder head (or heads for a V-type engine) encloses the engine cylinders and forms the upper end of the combustion chambers. Figure 7-12 shows two views of a cylinder head for a V-8 engine. Two of these are required, one for each cylinder bank in the V-8 engine. Figure 7-5 shows how they fit on the cylinder block.

The cylinder head is cast in one piece from iron, from iron mixed (alloyed) with other metals, or from aluminum alloy. The majority of cylinder heads are made of cast iron. Some engines use aluminum cylinder heads. Aluminum is considerably lighter and transmits heat more rapidly. That is, the aluminum carries the heat away from the hot areas more readily and this is an advantage where high performance is wanted.

The cylinder head includes water jackets and passages from the valve ports to the openings in the manifolds. The water jackets are formed by the upper and lower shells of the cylinder head. The casting procedure is similar to that used for cylinder blocks, described in ● 7-2. Many machining operations are required to convert the rough casting into a finished cylinder head, including:

1. Drilling and tapping holes for the spark plugs (gasoline engines) or fuel injectors (diesel engines)
2. Drilling and tapping holes for bolts and studs to attach the manifolds and other parts
3. Finishing water-passage and oil-passage holes
4. Finishing mounting surfaces
5. Drilling valve-guide holes
6. Finishing valve seats
7. Finishing the supports for overhead camshaft (in OHC or DOHC engines)

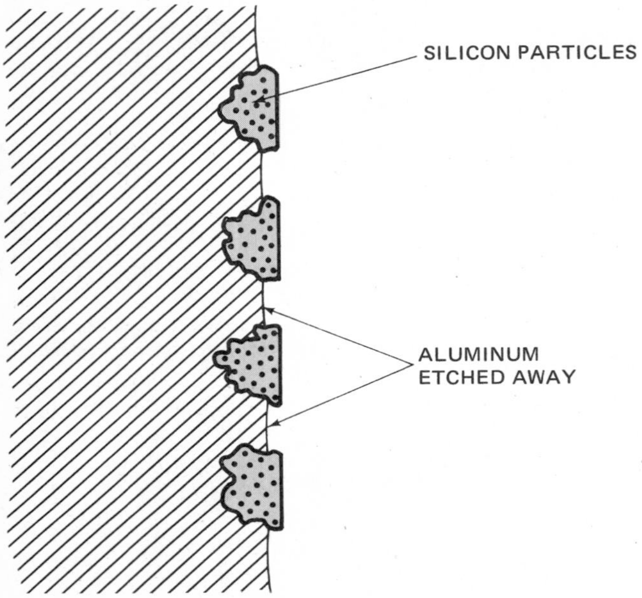

Fig. 7-10 In the finishing operation, the aluminum is etched away, leaving only the hard particles of silicon.

There are three general types of cylinder heads for gasoline engines, L-head, I-head, and overhead camshaft. L-heads are described in ● 6-14. About the only place you will see L-heads is in antique cars and in some small engines, like those used in power lawn mowers. The I-head engine, with the valves in the head, is the design used in modern vehicles. Many of these I-head engines also have the camshaft on the cylinder head (OHC and DOHC).

There are many pictures of I-head engines in the book (Figs. 6-2 to 6-15 show several). Figure 6-16 shows an overhead-camshaft engine. Figure 7-12 shows a cylinder head for an I-head engine. Note the arrangement to accommodate the valves. Figure 7-13 shows the important parts installed in an I-head cylinder head. The valves for only one cylinder are shown, along with one rocker arm and one push rod. To show more would clutter up the picture too much. Note that the valves are shown above the cylinder head. In the assembly, they are installed from the underside of the cylinder head. Chapter 9 describes valve trains in detail.

The cylinder head for an overhead-camshaft engine is shown in Fig. 7-14. Note the bearings to support the camshaft. The camshaft is driven by sprockets and a chain or toothed belt, as we explain in Chap. 9.

The cylinder head for an air-cooled flat six-cylinder engine is shown in Fig. 7-15. A similar cylinder head for a flat-four air-cooled engine is shown assembled in Fig. 7-16.

Some cylinder heads have additional features. For example, a cylinder head developed by Toyota has a turbulence-generating pot (TGP) as shown in Fig. 7-17.

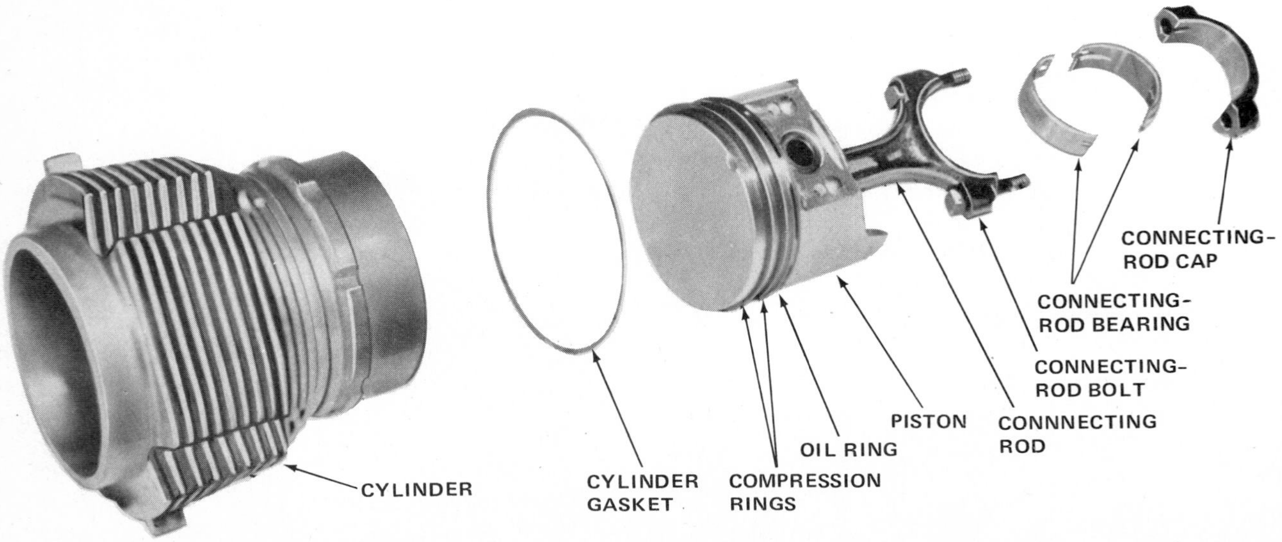

Fig. 7-11 Piston, connecting rod, cylinder, and related parts from a six-cylinder, air-cooled, pancake engine. (*Chevrolet Motor Division of General Motors Corporation*)

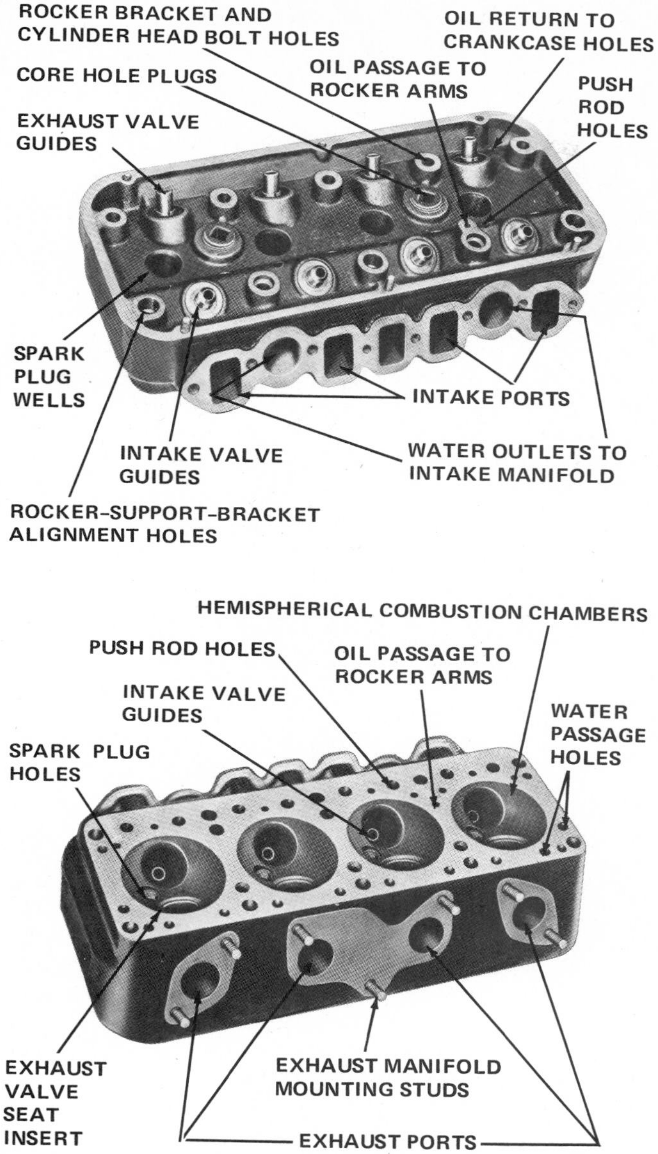

Fig. 7-12 Top and bottom views of cylinder head from a V-8 overhead-valve engine. (*Chrysler Corporation*)

The purpose of this pot is to produce high turbulence, or swirl, of the air-fuel mixture during combustion. During compression, part of the air-fuel mixture is forced into the pot. On ignition, the mixture in the pot starts burning first and streams out at high velocity. This helps spread the flame rapidly so that better burning takes place. We discuss this feature in more detail in a later chapter.

Some Honda engines use the same basic principle but with an added feature. The "pot" is called a precombustion chamber and it includes a third valve—an added intake valve (Fig. 7-18). When the two intake valves open (the large normal one and the small one in the precombustion chamber) air-fuel mixture enters. A lean mixture enters the main combustion chamber. A very rich mixture enters the precombustion chamber. Ignition takes place in the precombustion chamber. The burning air-fuel mixture streams out into the main combustion chamber. This produces good turbulence and rapid burning of the mixture. We cover this design in detail in a later chapter.

7-10 Gaskets

The joint between the cylinder block and cylinder head must be tight. It must withstand the pressure and heat developed in the combustion chambers. The block and head cannot be machined flat enough to provide the necessary seal. Therefore, gaskets are used to make a good seal (Fig. 7-19). Head gaskets are made of thin sheets of soft metal, or of asbestos and metal. All cylinder, water, valve, and head-bolt openings are cut out. When the gasket is installed, tightening the head bolts squeezes the soft metal, so the joint is sealed. Gaskets are also used to seal joints between other parts, such as between the oil pan, manifolds, or water pump and the head or block.

7-11 Exhaust Manifold

The exhaust manifold is a set of tubes that carry the exhaust gases away from the cylinder head and toward the exhaust system. Figure 7-20 shows the exhaust manifold for a six-cylinder in-line engine. The heat-control valve provides quick heating of the air-fuel mixture for good cold-engine operation. It is described in Chap. 13.

Figure 7-21 shows an exhaust system designed for use with a V-8 engine. It is a dual-exhaust system, so-called

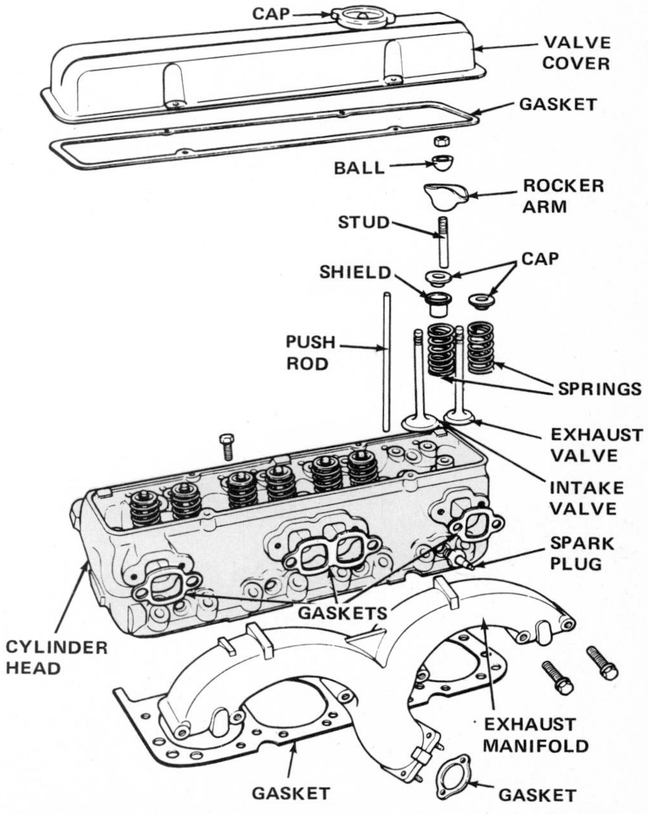

Fig. 7-13 Parts attached to the cylinder head of a V-8 engine. (*Chevrolet Motor Division of General Motors Corporation*)

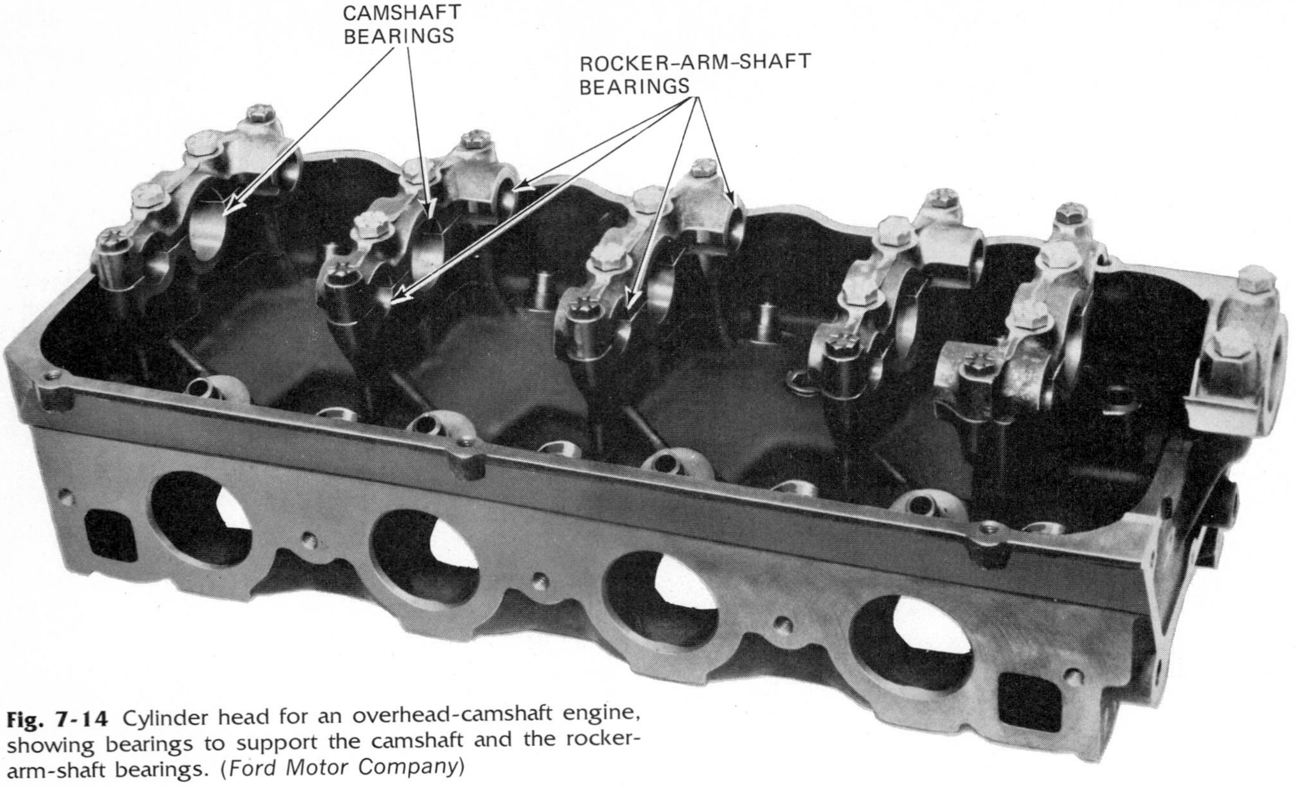

Fig. 7-14 Cylinder head for an overhead-camshaft engine, showing bearings to support the camshaft and the rocker-arm-shaft bearings. (*Ford Motor Company*)

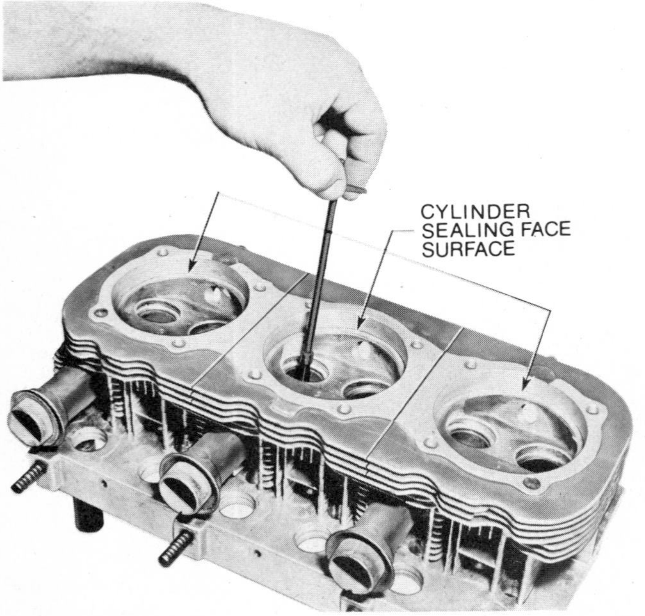

Fig. 7-15 One of the two cylinder heads for six-cylinder, air-cooled, pancake engine. The tool is being used to clean the valve-guide bores. (*Chevrolet Motor Division of General Motors Corporation*)

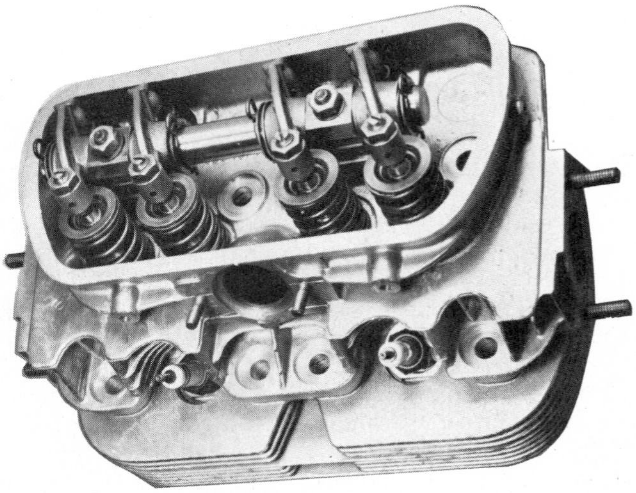

Fig. 7-16 One of the two cylinder-head assemblies for a four-cylinder, air-cooled, pancake engine. (*Volkswagen*)

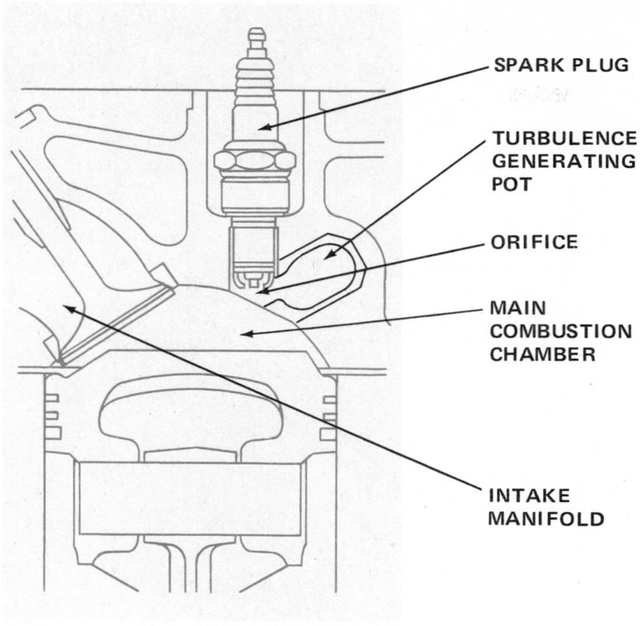

Fig. 7-17 Sectional view of cylinder head showing the location of the turbulence-generating pot. (*Toyota*)

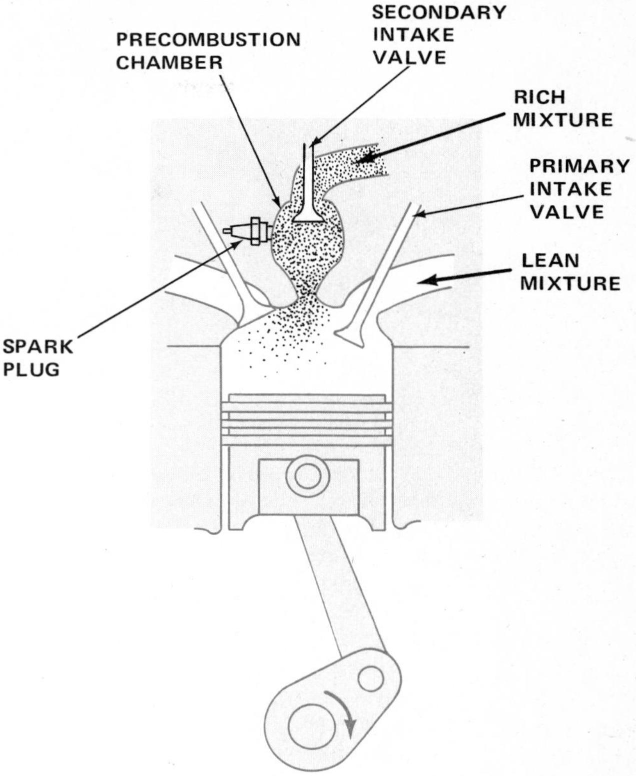

Fig. 7-18 Precombustion chamber in cylinder head with added intake valve. Spark plug is located in the precombustion chamber so ignition starts there. (*Honda*)

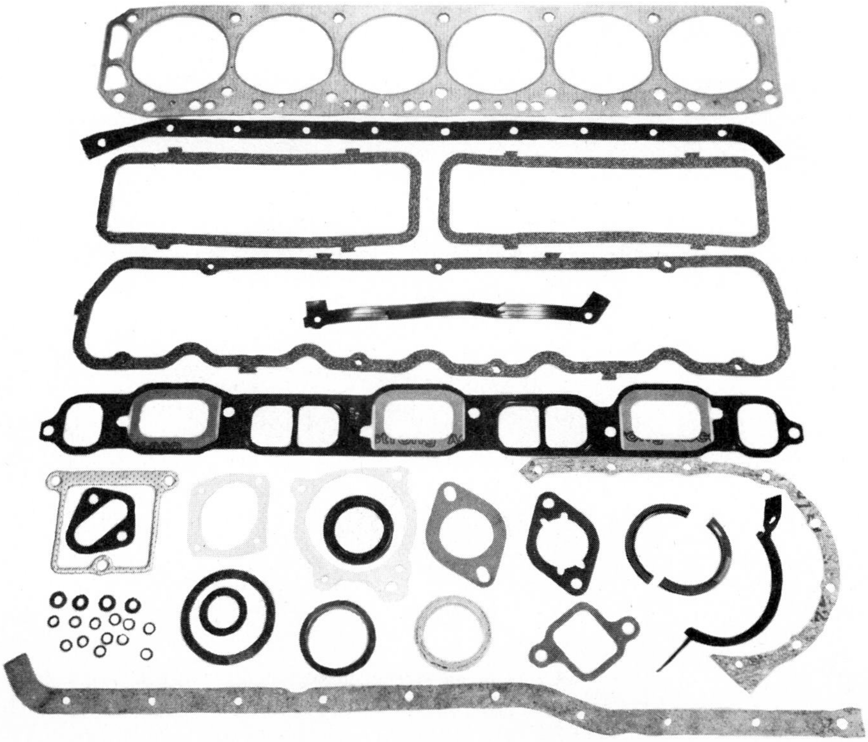

Fig. 7-19 Engine-overhaul gasket set for a six-cylinder engine, showing gaskets and seals used in the engine. (*McCord Replacement Products Division of McCord Corporation*)

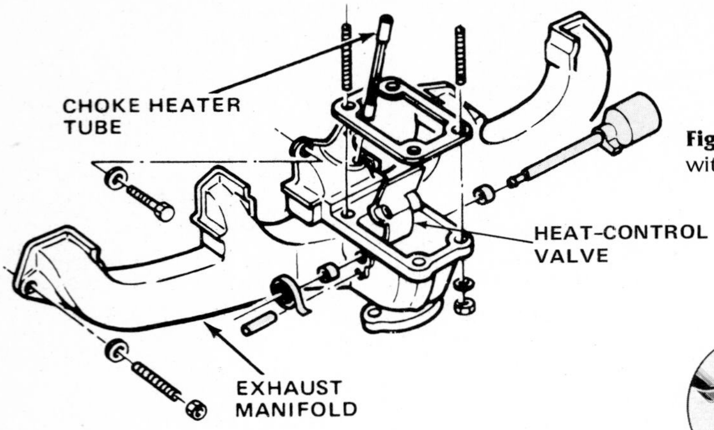

Fig. 7-20 Exhaust manifold for a six-cylinder in-line engine with heat-control valve and parts disassembled.

because there are really two separate exhaust systems, one for each cylinder bank. Each exhaust line has a catalytic converter and muffler to reduce, or muffle, the engine noise. Modern exhaust systems have catalytic converters which reduce the amount of pollutants coming out the tail pipe. Figure 7-22 shows an exhaust system of modern design for some V-8 engines. Note that the two exhaust pipes are connected into a single exhaust line which includes a catalytic converter and a muffler. The exhaust end of the system then branches out after the converter to include two mufflers and tail pipes. Catalytic converters and mufflers are covered in later chapters. Some V-8 exhaust systems have only one muffler and tail pipe (Fig. 35-29).

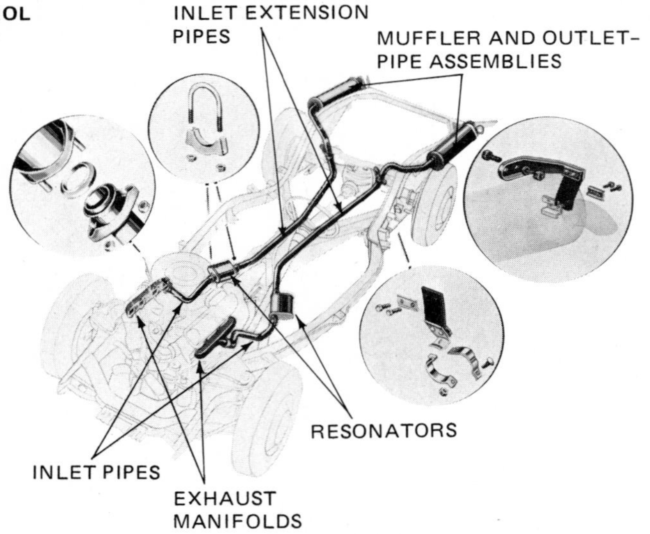

Fig. 7-21 Dual-exhaust system for a V-8 engine. Each bank of cylinders has its own complete exhaust system. The circles show details of assembly and attachment. (*Ford Motor Company*)

7-12 Intake Manifold

The intake manifold (Figs. 7-23 and 7-24) is a set of tubes that carry the air-fuel mixture from the carburetor to the cylinder heads. The carburetor is mounted on the

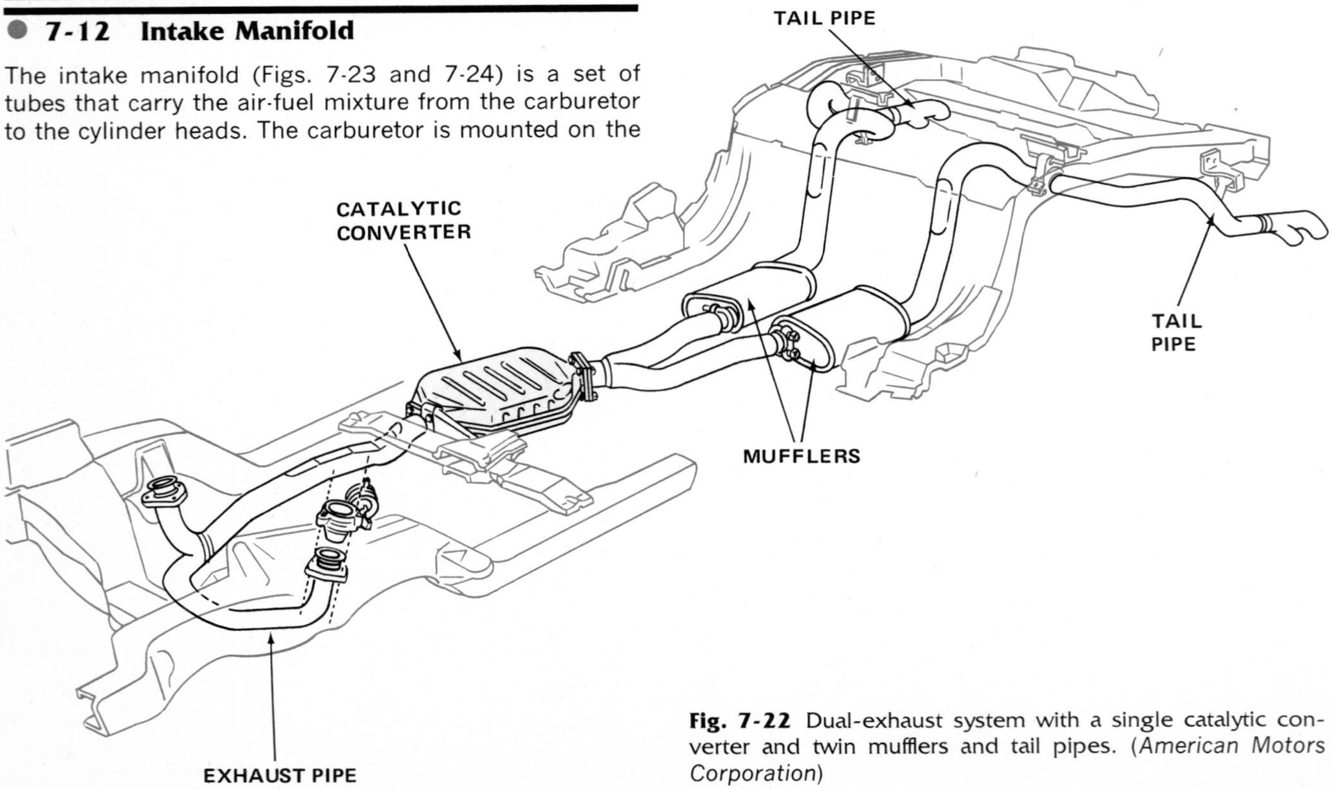

Fig. 7-22 Dual-exhaust system with a single catalytic converter and twin mufflers and tail pipes. (*American Motors Corporation*)

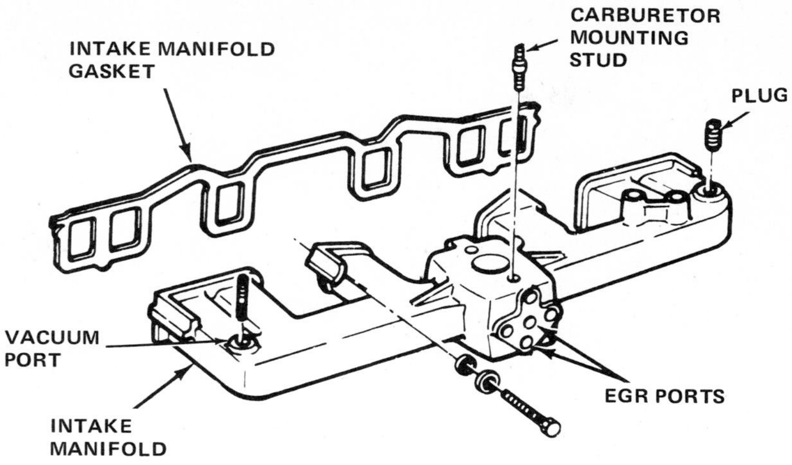

Fig. 7-23 Intake manifold for an in-line six-cylinder engine.

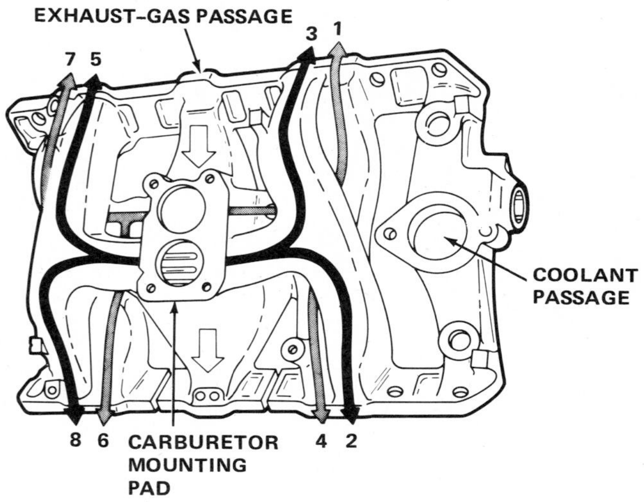

Fig. 7-24 Intake manifold for an I-head, V-8 engine. The arrows show the flow of air-fuel mixture from the two barrels of the carburetor to the eight cylinders in the engine. The central passage connects the two exhaust manifolds. Exhaust gas flows through this passage during engine warm-up. (*Pontiac Motor Division of General Motors Corporation*)

intake manifold. On in-line engines, the intake manifold is mounted on the same side of the cylinder head as the exhaust manifold. On V-4, V-6, and V-8 engines, the intake manifold is mounted between the two cylinder banks. You can see this arrangement in Fig. 7-5, as well as in other pictures of V-type engines in the book. There is more on intake manifolds in a later chapter.

Note that the intake manifold for diesel engines carries only air from the air intake to the cylinders. Recall that in these engines, the fuel is injected directly into the cylinders by the diesel-fuel-injection system.

● 7-13 Crankshaft

The crankshaft is a strong one-piece casting or forging of heat-treated alloy steel (Figs. 7-7 and 7-25). The crankshaft must be strong enough to take the downward push of the pistons during the power strokes without excessive twisting. In addition, the crankshaft must be carefully balanced to eliminate undue vibration resulting from the weight of the offset cranks. To provide balance, crankshafts have counterweights opposite the cranks. Crankshafts have drilled oil passages (Fig. 7-25) through which oil flows from the main to the connecting-rod bearings.

A recent development in crankshafts is the splayed crankpin (connecting-rod journal) in V-6 engines. The word "splay" means to spread out. As applied to the crankpin, it means they are split into two parts. Before we explain the purpose of this, look at the usual V-engine arrangement. Each crankpin has two connecting rods attached to it (Fig. 7-26). The rods are connected to pistons in opposing cylinders in the two cylinder banks. This works fine for V-8 engines, but in a V-6 the arrangement can produce a serious out-of-balance condition. One solution to this is to provide additional counterweighting to the crankshaft. Another is to splay, or spread apart, the crankpins. Figure 7-27 shows how this is done. The lower picture shows a standard arrangement with integral crankpins which will take two connecting rods side by side. The upper picture shows how the crankpins have been split, in effect, and moved apart. Each rod has its own crankpin. Moving the crankpins apart this way reduces the out-of-balance condition.

The new Renault V-6 achieves balance with a special camshaft which has a two-degree difference in valve timing between the two banks. There is also a special ignition system which is really two ignition systems in one. Each fires one bank of cylinders. Thus, a two-degree difference in firing is achieved. Throwing the two banks of cylinders out of synchronism in this way brings the engine into balance.

● 7-14 Flywheel

The flow of power from the engine cylinders is not smooth. The power impulses overlap (on six- and eight-cylinder engines). But there are still times when more

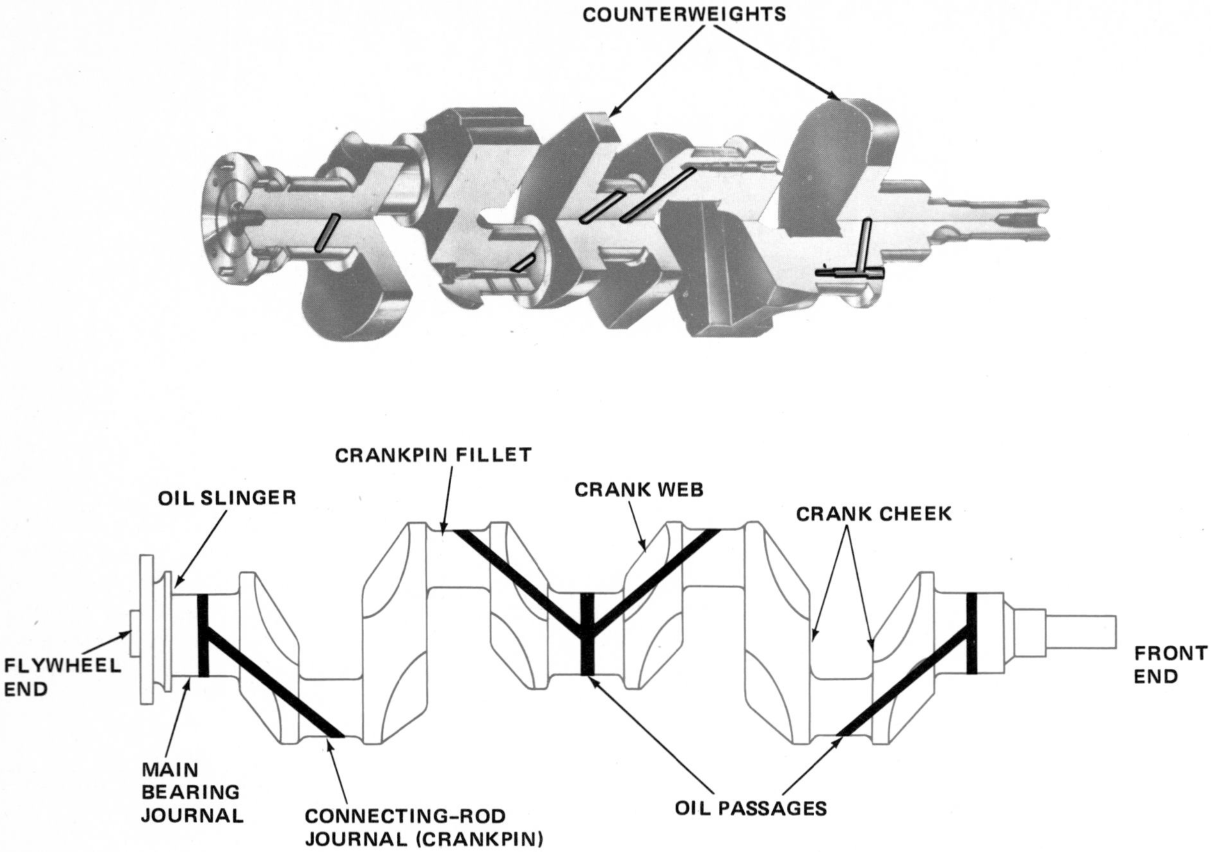

Fig. 7-25 Top, cutaway crankshaft for a V-8 engine, showing the oil passages drilled to the crankpins for lubricating the rod bearings. Bottom, line drawing of a similar crankshaft showing oil flow to the crankpins. (*Ford Motor Company and Johnson Bronze Company*)

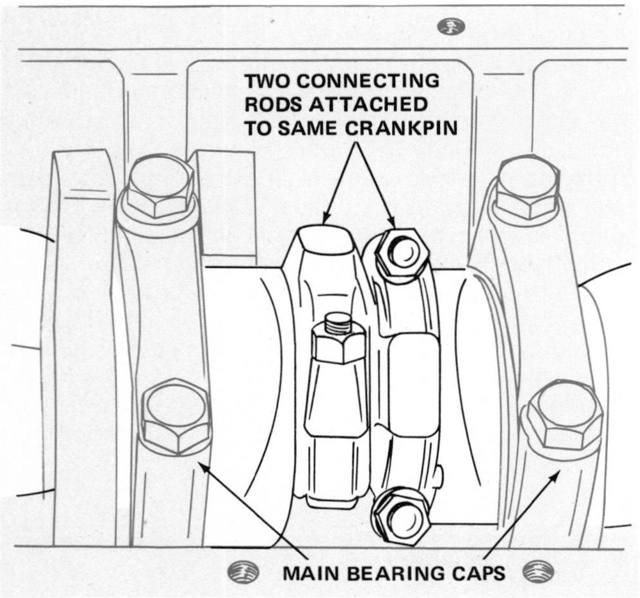

Fig. 7-26 In most V-type engines, connecting rods from cylinders in the two banks are attached to the same crankpin.

power is being delivered, and times when less power is delivered (Fig. 7-28). This tends to make the crankshaft speed up and then slow down. However, the flywheel helps keep the crankshaft turning at the same speed. The flywheel is a heavy wheel bolted to the rear end of the crankshaft (Fig. 7-7). Because it is so heavy, the flywheel tends to turn at constant speed. The flywheel absorbs energy as the crankshaft tries to speed up, and gives back energy as the crankshaft tries to slow down.

The flywheel also has gear teeth around its outer rim (a ring gear). They mesh with the starting motor gear for cranking the engine. The rear face of the flywheel also serves as the driving member of the clutch (on manual-transmission engines).

The front end of the crankshaft carries three devices, the gear that drives the camshaft, the vibration damper (● 7-15), and the drive-belt pulleys. These pulleys drive the engine fan, water pump, power steering pump, air pump, air-conditioner compressor, and alternator.

● 7-15 Vibration Damper

The power impulses tend to set up a twisting vibration in the crankshaft. When a piston moves down on its power

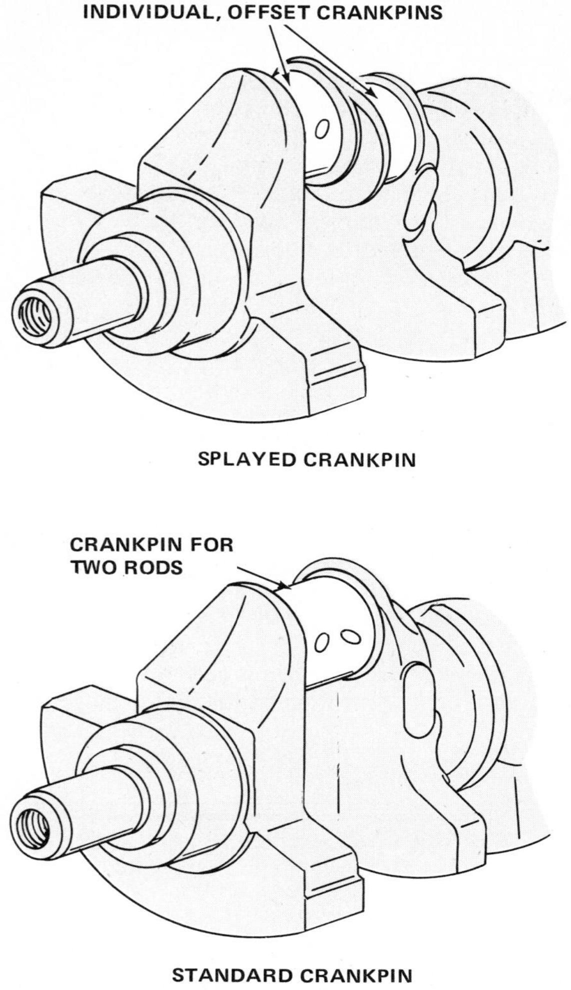

Fig. 7-27 Standard crankshaft for a V-6 engine (bottom) compared with a V-6 crankshaft with splayed crankpins (top). (*Buick Motor Division of General Motors Corporation*)

stroke, it thrusts, through the connecting rod, against a crankpin with a force that may exceed 2 tons. This force tends to twist the crank ahead of the rest of the crankshaft. Then, a moment later, the force against the crank is gone. The crank then tends to untwist, or move back into its original relationship with the rest of the crankshaft. This twist-untwist, repeated with every power impulse, tends to set up a back and forth, or oscillating, motion in the crankshaft. This is called *torsional vibration*. If it were not controlled, it could cause the crankshaft to break at certain speeds. To control torsional vibration, devices which are called *vibration dampers*, or *harmonic balancers*, are used. These dampers are usually mounted on the front end of the crankshaft (Fig. 7-7) and the drive-belt pulleys are incorporated into them.

A typical damper is made in two parts, a small inertia ring or damper flywheel and the pulley. They are bonded to each other by a rubber insert about $\frac{1}{4}$ inch thick (Fig. 7-29). The pulley is mounted to the front end of the crankshaft. As the crankshaft speeds up or slows down, the damper flywheel has a dragging effect. This effect, which slightly flexes the rubber insert, tends to hold the pulley and crankshaft to a constant speed. The action tends to check the twist-untwist, or torsional vibration, of the crankshaft.

7-16 Engine Bearings

Bearings are placed in the engine wherever there is rotary motion between engine parts (Fig. 7-30). These engine bearings are called *sleeve bearings* because they are shaped like sleeves that fit around the rotating shaft. The part of the shaft that rotates in the bearing is called a "journal." Connecting-rod and crankshaft (also called *main*) bearings are of the split, or half, type (Figs. 7-31 and 7-32). The upper half of a main bearing is installed in the counterbore in the cylinder block. The lower half is

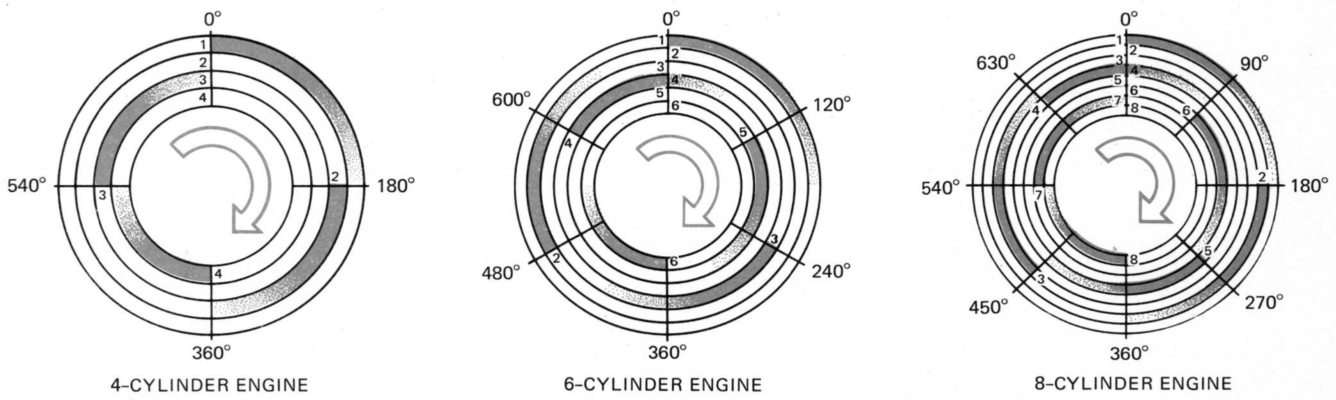

Fig. 7-28 Power impulses in four-, six-, and eight-cylinder engines during two crankshaft revolutions. The complete circle represents two crankshaft revolutions, or 720 degrees. Less power is delivered toward the end of the power stroke. This is indicated by the lightening of the shaded areas that show power impulses. Note the overlap on six- and eight-cylinder engines.

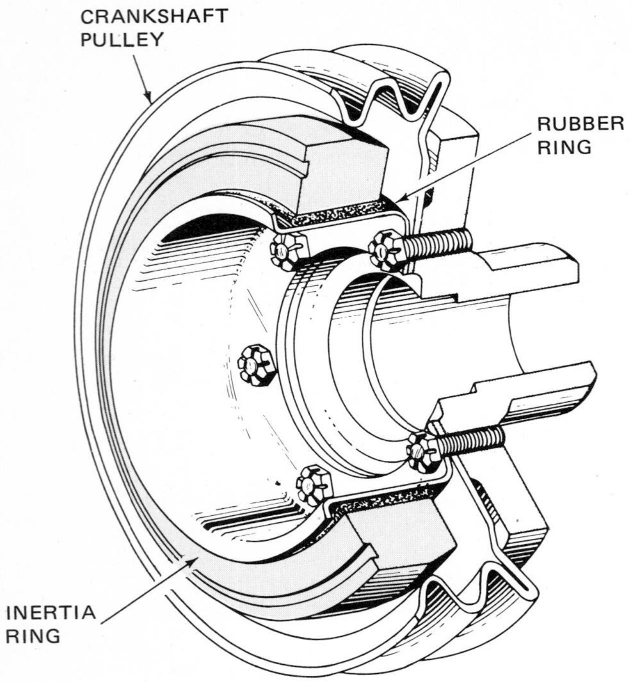

Fig. 7-29 Partial cutaway view of a torsional vibration damper. (*Pontiac Motor Division of General Motors Corporation*)

held in place by the bearing cap (Fig. 7-7). The upper half of a connecting-rod big-end (or crankpin) bearing is installed in the rod. The lower half is placed in the rod cap (Fig. 8-1).

The main bearings in most engines do not have the oil grooves shown in Fig. 7-32. They may or may not have the annular grooves or oil channels; many sleeve bearings do not have these grooves. On some engines, only the upper halves of the main bearings have them. On still other engines, both the upper and lower main-bearing halves have the annular grooves. Connecting-rod big-end bearings usually do not have oil grooves.

The typical bearing half is made up of a steel or bronze back, with one to three linings of bearing material (Fig. 7-33). The bearing material is soft. Thus, it will be the bearing that wears, and not the more expensive engine part. Then, the bearing, and not the engine part, needs to be replaced when it has worn too much.

7-17 Thrust Bearing

The crankshaft has to be kept from moving back and forth in the block. To prevent back-and-forth movement, one of the main bearings is a thrust, or end-thrust, bearing. This bearing has flanges on its two sides, as shown in Fig. 7-34. In the engine, these flanges fit close to the sides of one of the crankpins. If the crankshaft tends to shift forward or backward, the sides of the crankpin come up against the flange. This prevents endwise movement. You can see a thrust bearing in Fig. 7-7.

7-18 Bearing Lubrication

As we have already seen, the bearings have oil holes. Oil from the engine oil pump moves through these holes and flows onto the bearing surfaces. Thus, the rotating shaft journals are supported on layers of oil. The journal must be smaller than the bearing (Fig. 7-35) so that there is a clearance (called *oil clearance*) between the two. In the engine, oil moves through this clearance.

The lubricating system feeds oil to the bearing. It enters through the oilhole (Fig. 7-32), and the rotating journal carries it around to all parts of the bearing. The oil works its way to the outer edges of the bearing. From there, it is thrown off and drops back into the oil pan. The oil thrown off helps lubricate other engine parts, such as the cylinder walls, pistons, and piston rings.

As the oil moves across the faces of the bearings, it also helps to cool them. The oil is relatively cool as it leaves the oil pan. It picks up heat in its passage through the bearings. This heat is carried down to the oil pan and released to the air around the oil pan. The oil also flushes and cleans the bearings. It flushes out particles of grit and dirt from the bearings. The particles are carried back to the oil pan by the oil. They then drop to the bottom of the oil pan, or are removed from the oil by the oil screen or filter.

7-19 Bearing Oil Clearances

The greater the oil clearance (Fig. 7-35), the faster oil flows through the bearing. Proper clearance varies with different engines, but 0.0015 inch [0.037 mm] is a typical clearance. As the clearance becomes greater (owing to bearing wear, for instance), the amount of oil flowing through and being thrown off increases. With a 0.003-inch [0.076-mm] clearance (only twice 0.0015 inch) [0.037 mm], the oil throwoff increases as much as five times. A 0.006-inch [0.152-mm] clearance allows twenty-five times as much oil to flow through and be thrown off.

Thus, as bearings wear, more and more oil is thrown onto the cylinder walls. The piston rings cannot handle so much oil. Part of it works up into the combustion chambers, where it burns and forms carbon. Carbon deposits in the combustion chambers reduce engine power and cause other engine troubles (see Chap. 42).

Excessive oil clearances can also cause some bearings to fail from oil starvation. Here's the reason: The oil pump can deliver only so much oil. If the oil clearances are excessive, most of the oil will pass through the nearest bearings. There won't be enough for the more distant bearings, and these will probably fail from lack of oil. An engine with too-large bearing oil clearances usually has low oil pressure: The oil pump cannot build up normal pressure because of the large oil clearances in the bearings.

On the other hand, if oil clearances are too small, there will be metal-to-metal contact between the bearing and

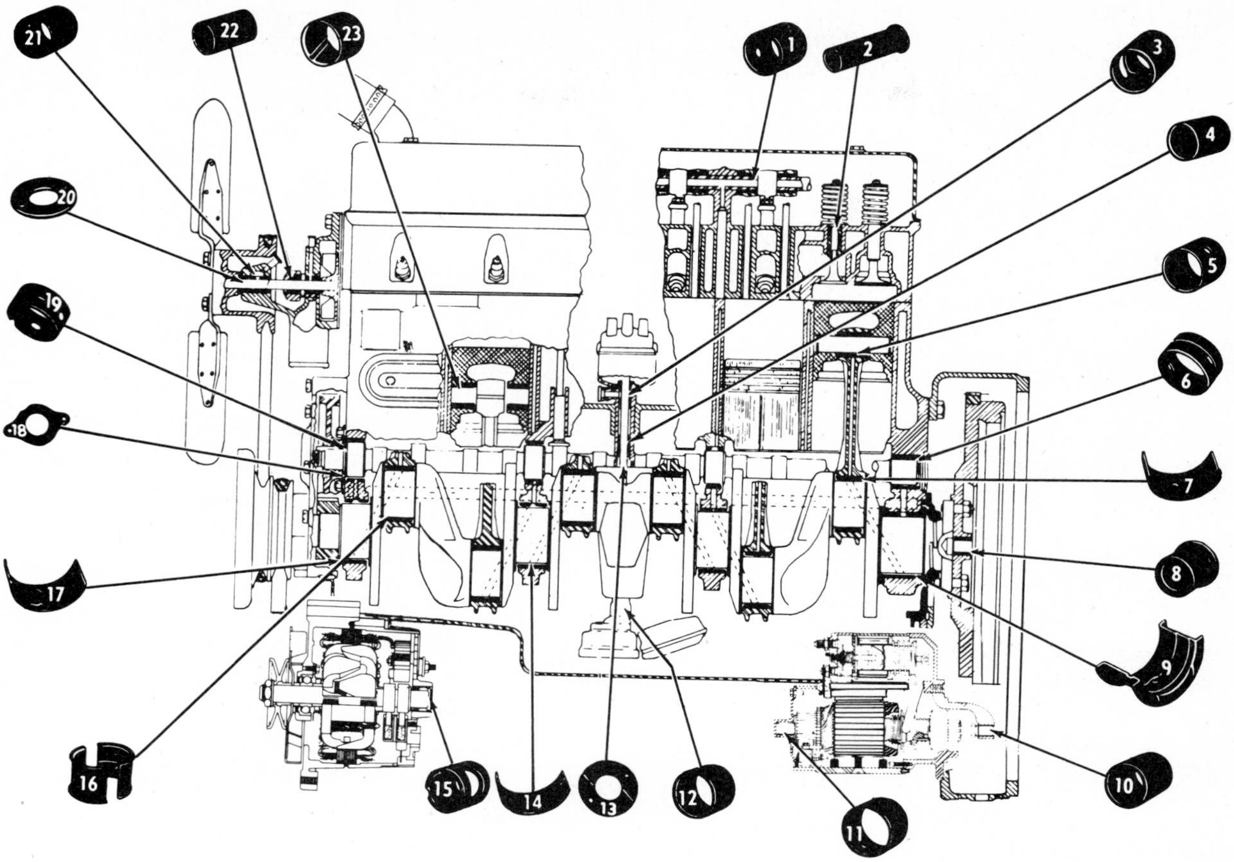

1. ROCKER-ARM BUSHING
2. VALVE-GUIDE BUSHING
3. DISTRIBUTOR BUSHING, UPPER
4. DISTRIBUTOR BUSHING, LOWER
5. PISTON-PIN BUSHING
6. CAMSHAFT BUSHINGS
7. CONNECTING-ROD BEARING
8. CLUTCH PILOT BUSHING
9. CRANKSHAFT THRUST BEARING
10. STARTING-MOTOR BUSHING, DRIVE END
11. STARTING-MOTOR BUSHING, COMMUTATOR END
12. OIL-PUMP BUSHING
13. DISTRIBUTOR THRUST PLATE
14. INTERMEDIATE MAIN BEARING
15. ALTERNATOR BEARING
16. CONNECTING-ROD BEARING, FLOATING TYPE
17. FRONT MAIN BEARING
18. CAMSHAFT THRUST PLATE
19. CAMSHAFT BUSHING
20. FAN THRUST PLATE
21. WATER-PUMP BUSHING, FRONT
22. WATER-PUMP BUSHING, REAR
23. PISTON-PIN BUSHING

Fig. 7-30 Bearings and bushings used in a typical engine. (*Johnson Bronze Company*)

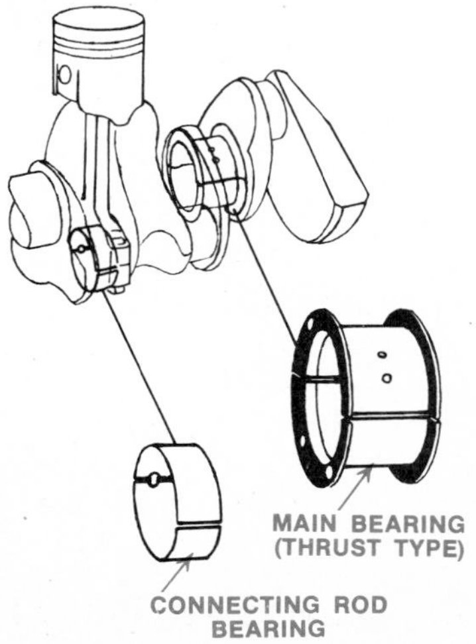

Fig. 7-31 Main bearing of the thrust type and a connecting-rod bearing, showing their positions on the crankshaft.

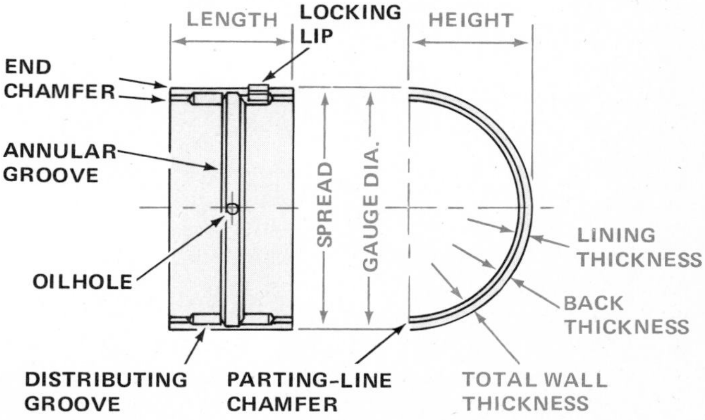

Fig. 7-32 Typical sleeve-type bearing half, with parts named. Many bearings do not have annular and distributing grooves. (*Federal-Mogul Corporation*)

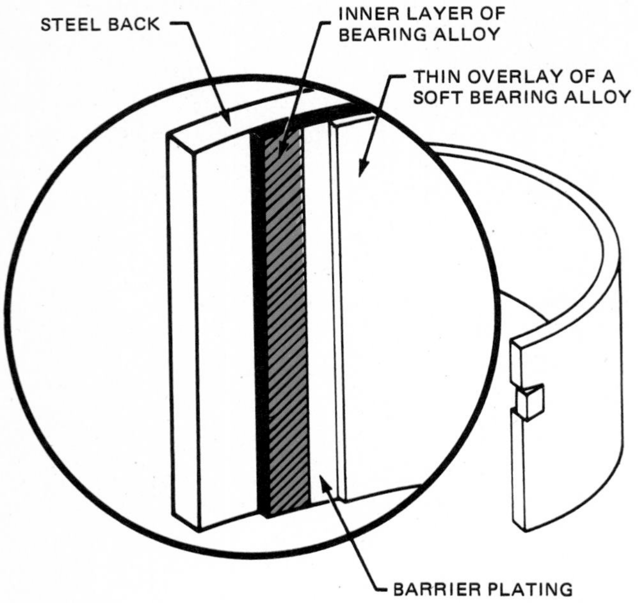

Fig. 7-33 Construction of a three-layer bearings. Some bearings have three layers, as shown. Others have two layers. *(Federal-Mogul Corporation)*

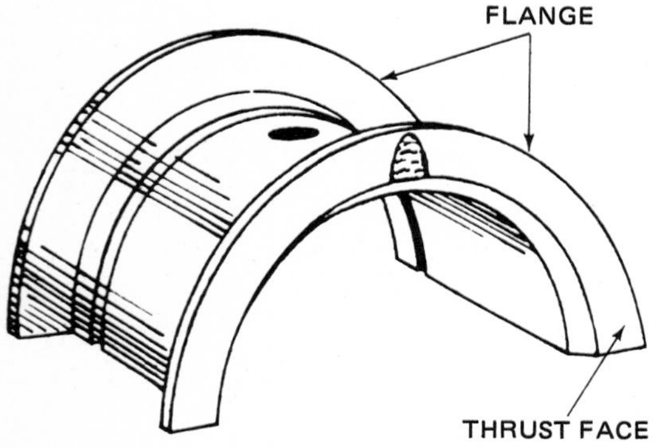

Fig. 7-34 Crankshaft thrust bearing.

the shaft journal. Very rapid wear and quick failure will result. Also, there will not be enough oil throwoff to lubricate cylinder walls, pistons, and rings.

● 7-20 Engine Bearing Types

Early engines and some late-model heavy-duty engines use poured bearings. Such a bearing is prepared by fitting a journal-sized jig, or mold, into the counterbore. Then molten bearing material is poured into the opening. The resulting bearing is scraped and smoothed to provide the final fit and clearance.

In modern automotive engines, *precision-insert* bearings are used. These bearings are so precisely made that they can be installed without any machining or fitting. In many engines, it is possible to replace the main bearings without removing the crankshaft, as explained in a later chapter.

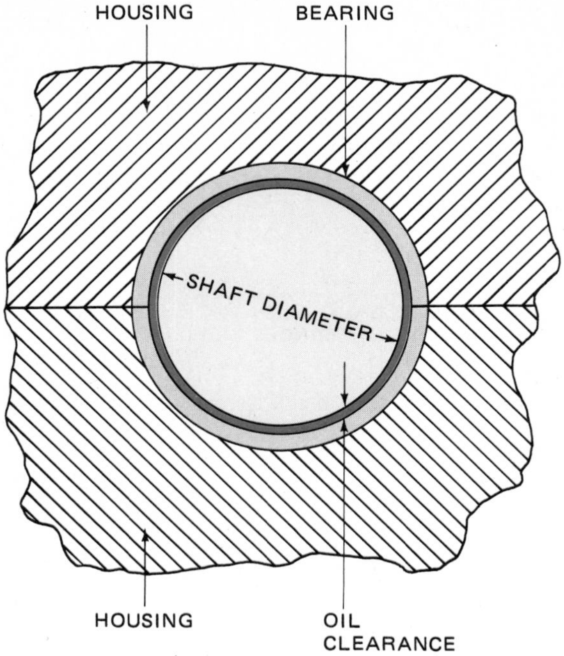

Fig. 7-35 Oil clearance between bearing and shaft journal.

● 7-21 Bearing Requirements

Bearings must be able to do other things besides carry loads. Some of these are listed below.

1. Load-carrying capacity Modern engines are lighter and more powerful. They have higher compression ratios and thus impose greater bearing loads. Only a few years ago, bearing loads were around 1,600 to 1,800 psi (pounds per square inch) [112.48 to 126.54 kg/cm^2 or 11,032 to 12,411 kPa]. Today, connecting-rod bearings carry loads of 6,000 psi [421.82 kg/cm^2 or 41,369 kPa].

2. Fatigue resistance When a piece of metal is bent back and forth, over and over, it hardens and finally breaks. This is called *fatigue failure*. You have probably done this with a piece of wire or sheet metal. Bearings are subject to such loads and must withstand them without failing from fatigue.

3. Embeddability This term refers to the ability of a bearing to permit foreign particles to embed in it. Dirt and dust particles enter the engine despite the air cleaner and oil filter. Some of them work onto the bearings and are not flushed away by the oil. A bearing protects itself by letting such particles sink into, or embed in, the bearing lining material. If the bearing were too hard to allow this, the particles would lie on the surface. They would scratch the shaft journal and probably gouge out the bearing. This would cause overheating and rapid bearing failure. Thus, the bearing material must be soft enough for adequate embeddability.

4. Conformability This is associated with embeddability. It is the ability of the bearing material to conform to

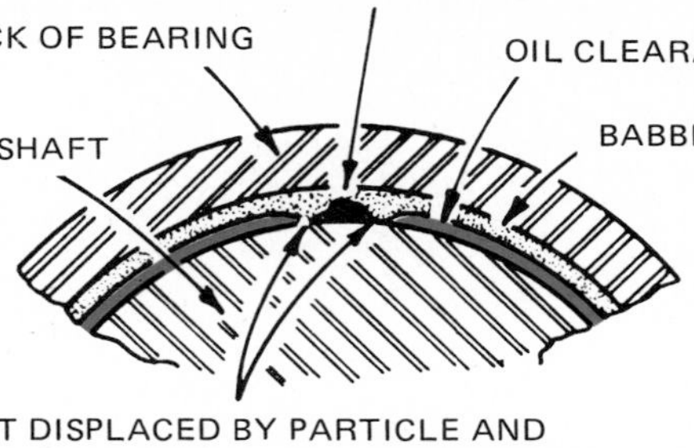

Fig. 7-36 Effect of a metallic particle that is embedded in the bearing material (the babbitt lining). (*Federal-Mogul Corporation*)

variations in shaft alignment and journal shape. For example, suppose that a shaft journal is slightly tapered. The bearing under the larger diameter will be more heavily loaded. If the bearing material has high conformability, it will "flow" slightly, from the heavily loaded areas to the lightly loaded areas. This slight flow evens the load on the bearing. A similar action takes place when foreign particles embed in the bearing. As they embed, they displace bearing material, thus producing local high spots (Fig. 7-36). However, with high conformability, the material flows away from the high spots. This prevents local heavy loading that could cause bearing failure.

5. Corrosion resistance The by-products of combustion may form corrosive substances, harmful to some metals. Thus, the bearing materials must be resistant to corrosion. Unleaded gasoline, required on cars using catalytic converters, changes the chemistry of the engine oil. Catalytic converters, discussed in later chapters, are installed in the exhaust systems to reduce the pollutants coming out the tail pipe. The unleaded gasoline, in changing the chemistry of the oil, tends to increase bearing corrosion. Thus, the composition of engine bearings has been changed. For example, instead of the copper-lead bearings used for years, some engines now have aluminum-lead bearings. These appear to withstand corrosion better.

6. Wear rate The bearing material must be so hard and tough that it will not wear too fast. At the same time, it must be soft enough to permit good embeddability and conformability.

● 7-22 Bearing Materials

The bearing back is usually made of steel. The lining material is a combination of several metals, mixed, or alloyed, to provide the right combination of properties. Such metals as copper, lead, tin, mercury, antimony, cadmium, aluminum, and silver are used. Many combinations are possible. Each ingredient, or metal, supplies certain properties. The designer selects the combination that best suits a particular engine.

REVIEW QUESTIONS

Here are some questions to answer. They will help you check how well you remember what you have just been reading about the cylinder block, head, and crankshaft.

1. Explain how the cylinder block is made.
2. List six machining operations required on the cylinder block.
3. What forms the water jackets in an engine?
4. Name five parts attached to the cylinder block.
5. Explain how the crankshaft is attached to the cylinder block.
6. What forms the crankcase?
7. What is a cylinder liner? What are the two types?
8. Explain why the GM 140-cubic inch [2.2-L] engine does not need cylinder liners.
9. List five machining operations required on the cylinder head.
10. What is the purpose of gaskets?
11. What are head gaskets made of?
12. Does a V-8 engine have one or two exhaust manifolds?
13. Does a V-8 engine have one or two intake manifolds?
14. How many cranks are there on a crankshaft for a six-cylinder in-line engine?
15. What is the purpose of the vibration damper?
16. Describe the construction of a main bearing.
17. What is the bearing oil clearance?
18. What is a thrust bearing? What is its purpose?
19. What is the embeddability of an engine bearing?
20. What is the conformability of an engine bearing?
21. What are splayed crankpins and why are they splayed?

SELF PROJECT

Locate a cylinder block with the head removed. Take a pencil and a sheet of paper and sketch the outline of the top of the block. Then draw big circles to represent the cylinders. Next, examine the holes in the blocks for threads. Threaded holes are for the cylinder-head bolts. Note how four or five bolt holes are spaced around each cylinder so the gasket can be evenly compressed and thus seal the combustion chamber when the bolts are tightened. Then locate and mark on your drawing the positions of dowels that are used to align the head during installation. See if you can find oil in some smaller holes. These would be oil drain holes from the head. Then look at the rest of the holes. These other holes are water passages between the block and head. Mark and label the oil and water holes on your drawing.

Go through this same procedure with several different engines and add the pages to your notebook. Be sure to add the make, model number, and year the engine was made to the drawings. Follow this procedure and soon you will be able to tell instantly what each hole is in any engine block or head.

CHAPTER 8
Engine Construction: Pistons, Rings, and Rods

After reading this chapter, you should be able to:

- **Discuss the construction, purpose, operation, and lubrication of the connecting rods.**
- **Discuss the construction, purpose, operation, and lubrication of pistons and piston rings.**
- **Explain the purpose of piston-expansion control and how it is achieved.**

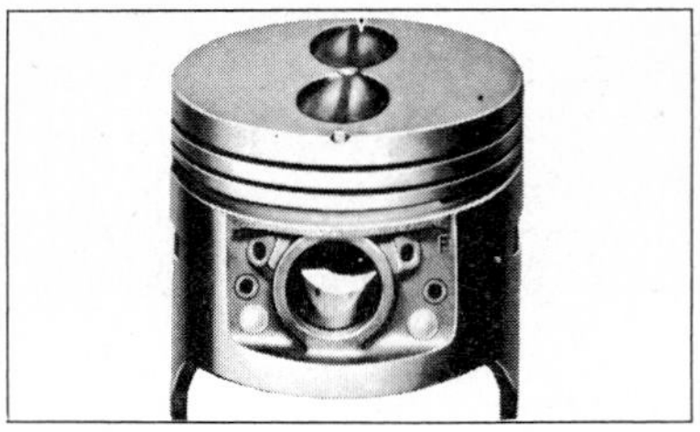

8-1 Connecting Rod

The connecting rod (Fig. 8-1) is attached at one end to a crankpin on the crankshaft. It is attached at the other end to a piston, through a piston pin or wrist pin. The connecting rod must be very strong and rigid, and as light as possible. The connecting rod carries the power thrusts from the piston to the crankpin. At the same time, the rod is in eccentric, or off-center, motion (Fig. 5-10). To minimize vibration and bearing loads, the rod must be light in weight.

The crankpin end of the rod (the big end) is attached to the crankpin by the rod cap and bolts (Fig. 8-1). A split-style bearing is installed between the crankpin and the rod and rod cap.

The piston end of the rod (the small end) is attached to the piston by means of a piston pin. The pin passes through bearing surfaces in both the piston and the connecting rod. There are five ways to connect the rod and piston with the piston pin (Fig. 8-2). The most common method is to press-fit the pin to the connecting rod (Fig. 8-2*D*). The press fit is tight enough to keep the pin from moving out of position. When an aluminum piston is used, as shown in Fig. 8-2*D*, the piston pin rests directly on the aluminum of the piston, with no intervening bushing. Another method is to provide the rod with a bushing so the pin can float freely in both the rod bushing and the bearing surfaces in the piston (Fig. 8-2*A*). The pin is kept from moving out and scoring the cylinder wall by a pair of lock rings. The rings fit into undercuts or grooves in the piston (Fig. 8-3). The other three arrangements, shown in Fig. 8-2 (*B*, *C*, and *E*) which use lock bolts to lock the pin to either the rod or piston, are seldom used today.

To provide connecting-rod big-end bearing lubrication, oil travels the following path: The oil pump sends oil to oil lines or galleries in the cylinder block. From these oil lines, the oil moves to the main bearings. From the main bearings, oil travels through oil passages drilled in the crankshaft (see Fig. 7-25). Finally, the oil moves through the connecting-rod-bearing oil holes onto the bearing surfaces.

On many engines, the piston pins are lubricated by oil scraped from the cylinder walls. As we shall see later in this chapter, the piston rings scrape excess oil from the cylinder walls. Some of this oil flows through slots or holes in the piston to lubricate the piston pin.

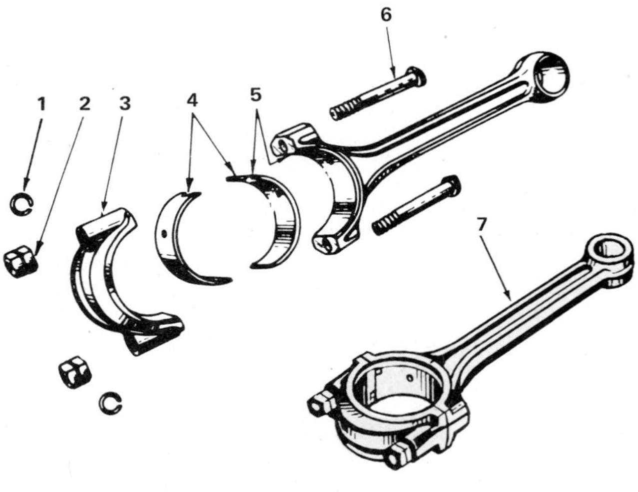

1. CAP-BOLT-NUT LOCK WASHER
2. CAP-BOLT NUT
3. CAP
4. ROD BEARINGS
5. TONGUE AND GROOVE
6. CAP BOLT
7. ASSEMBLED ROD

Fig. 8-1 Connecting rod disassembled (top) and assembled (bottom).

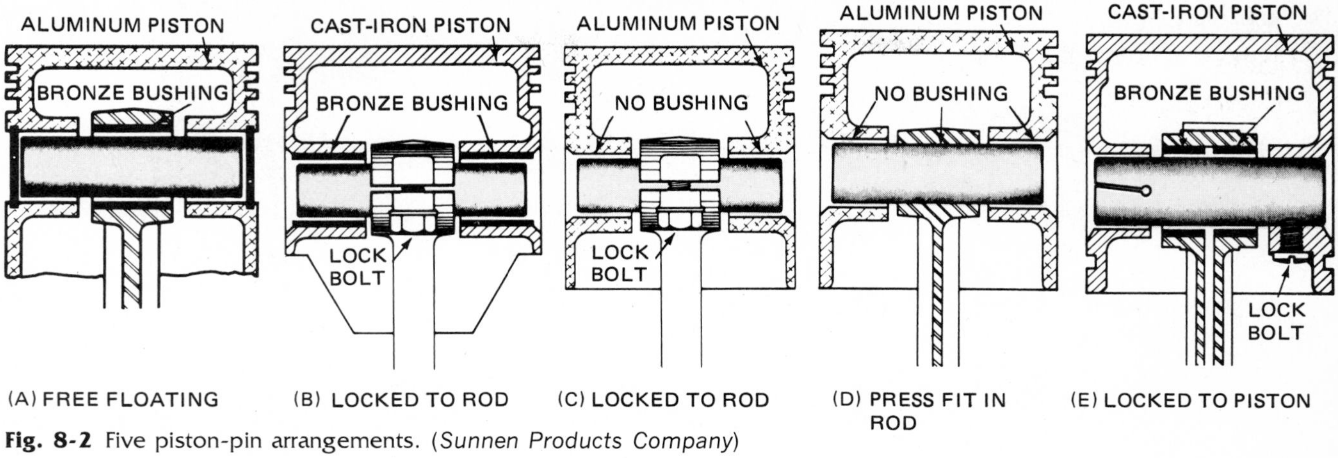

Fig. 8-2 Five piston-pin arrangements. (*Sunnen Products Company*)

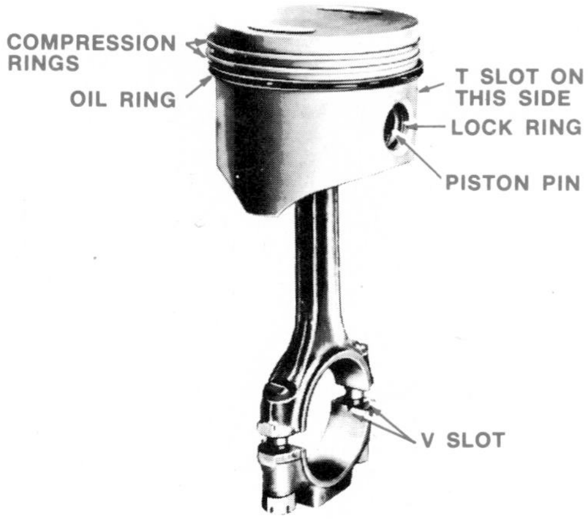

Fig. 8-3 Piston and connecting-rod assembly. This type has lock rings to hold the piston pin in position in the piston and connecting rod. (*Chrysler Corporation*)

Some connecting rods have oil passages drilled from the connecting-rod big end to the piston-pin bushing. The oil gets to the piston-pin bushing in the connecting rod by the following path: From the oil pump, oil travels through oil lines in the block, and through oil passages in the crankshaft, to the connecting-rod big-end bearings. It then moves through the oil passage in the connecting rod to the piston-pin bushing. This arrangement has been used in engines that have free-floating pins in the connecting-rod bushings.

On many V-8 engines, cylinder walls and piston pins are lubricated by oil jets from opposing connecting rods. That is, each rod has a groove or hole that lines up with an oil-passage hole in the crank journal every crankshaft revolution. When this happens, a jet of oil spurts into the opposing cylinder in the other cylinder bank.

Rod caps and rods are carefully matched during original assembly. After a cap is installed on a rod, the bearing bore is machined. That means the cap will fit that rod and only that rod. Caps must not be interchanged between rods during service jobs. If a cap goes on the wrong rod, the assembly will not fit the crank on the crankshaft. If a cap is lost or damaged, then a new or reconditioned cap-rod assembly is required.

● 8-2 Pistons and Piston Rings

As already noted (● 5-2), the piston is a cylindrical plug that moves up and down in the engine cylinder. It has piston rings to provide a good seal between the cylinder wall and piston. Although the piston appears to be a simple part, it is actually quite a complex design. But before we discuss pistons, let us examine piston rings.

● 8-3 Piston Rings

A good seal must be maintained between the piston and cylinder wall to prevent blow-by. "Blow-by" is the name that describes the escape of unburned air-fuel mixture and burned gases from the combustion chamber, past the pistons and into the crankcase. In other words, these gases "blow by" the pistons. It is not practical to fit the piston to the cylinder closely enough to prevent blow-by. Piston rings must be used to provide the necessary seal.

The rings are installed in grooves in the piston, as shown in Fig. 8-3. Actually, there are two types of rings, compression rings and oil-control rings. The compression rings seal in the air-fuel mixture as it is compressed. They also seal in the combustion pressures as the mixture burns. The oil-control rings scrape excessive oil from the cylinder wall and return it to the oil pan. Figure 8-4 shows typical compression and oil-control rings. The rings have joints (they are split) so they can be expanded and slipped over the pistonhead and into grooves in the piston. Rings for automotive engines usually have butt joints, but in some heavy-duty engines, the joints may be angled, lapped, or of the sealed type.

The rings are somewhat larger in diameter than they will be when in the cylinder. Then, when they are installed, they are compressed so that the joints are nearly closed. Compressing the rings puts them in tension. They press tightly against the cylinder wall.

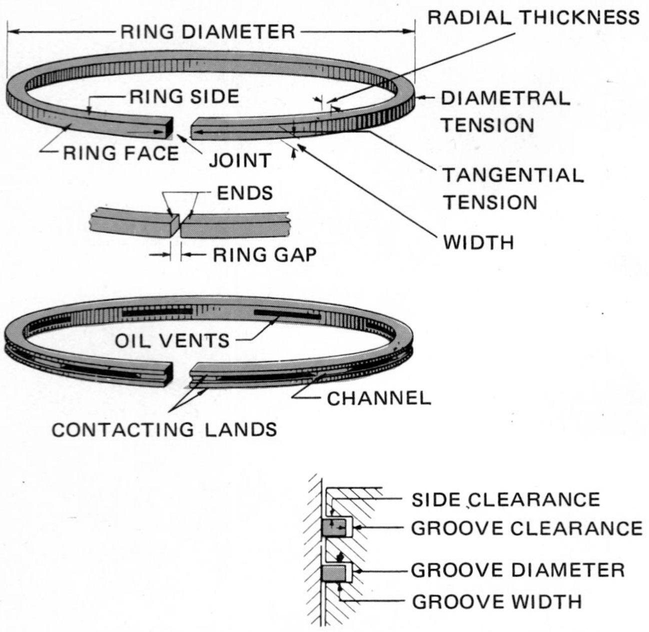

Fig. 8-4 Compression ring (top) and one-piece oil-control ring (bottom) with parts named. (*Sealed Power Company*)

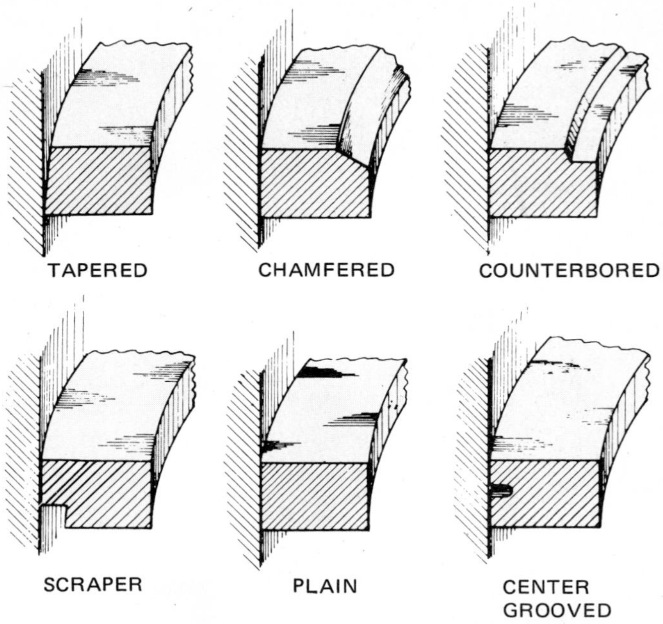

Fig. 8-5 Compression-ring shapes. (*Muskegon Piston Ring Company*)

● 8-4 Compression Rings

Compression rings are made of cast iron. Typical compression rings are shown in Fig. 8-5.

1. Counterbored and scraper rings Counterbored and scraper compression rings are used in many engines for top and second compression rings. Figure 8-6 shows the action of these rings during the intake stroke. Internal forces, produced by cutting away a corner of the rings, cause them to twist slightly, as shown. Thus, as they move down on the intake stroke, they produce a scraping action. They scrape off oil that might have been left on the cylinder wall by the oil-control rings. Then, on the exhaust and compression strokes, when the rings are moving upward, they tend to "skate" over the film of oil on the cylinder wall. As a result, less oil is carried up into the combustion chamber, and wear is minimized.

On the power stroke, combustion pressures press down on top of and back of the rings. This overcomes the internal tension and causes them to untwist. They thus have full-face contact with the cylinder walls for effective sealing (Fig. 8-7).

2. Headland rings This ring has a modified L-shaped cross section (Figs. 8-8 and 8-9). Its name comes from the fact that the ring covers, or shields, the headland area of the piston. This is the area of the piston between the top ring groove and the head, or top, of the piston. Note that the piston is a special type; it is chamfered at the top. This combination of ring and piston largely eliminates the space between the piston and cylinder wall above the top ring. This space, in other piston-and-ring combinations, holds a certain amount of air-fuel mixture that does not burn. The cylinder wall and piston cool this air-fuel mixture below the combustion point. The result is that the unburned gasoline vapor passes out of the engine and can produce smog. (There is more on smog and emission controls in Part Five.) However, with the headland rings and special pistons, the amount of unburned gasoline exhausted from the cylinders is reduced.

A second advantage claimed for the headland ring is that it reacts swiftly to the combustion pressure. As combustion starts, the increasing pressure quickly pushes out on the upper lip of the ring. This forces the lip out and into good sealing contact with the cylinder wall. This reduces blow-by and thus simplifies the requirements of the closed-crankcase ventilating system.

By giving more complete combustion of air-fuel mixture and reducing blow-by, the new ring is said to increase available horsepower up to 10 percent.

3. Ring coatings Coatings of various substances have been applied to compression rings to help wear-in and prevent rapid wear of rings and cylinder walls. By "wear-in" we mean this: When new, the rings and cylinder wall have slight irregularities so the fit is not perfect. However, after a time, these irregularities are worn away so the fit is improved. Relatively soft materials such as phosphate, graphite, and iron oxide, which wear rapidly, help the wear-in. Also, they absorb some oil so that improved lubrication results.

Another factor to consider in selecting compression-ring coatings is the type of wear the rings will meet in service. At one time, the most common type of ring and cylinder-wall wear was caused by abrasives that worked their way through the air filter and got in the engine oil. These abrasives—fine dust particles—would deposit on

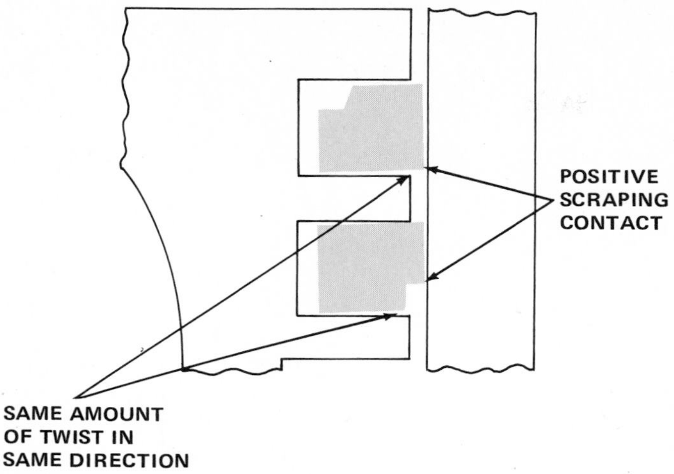

Fig. 8-6 Action of counterbored and scraper compression rings during the intake stroke. Internal forces in rings tend to twist them so that a positive scraping contact is established between rings and cylinder wall. This helps to remove any excessive oil that has worked past the oil-control ring. (*Perfect Circle Company*)

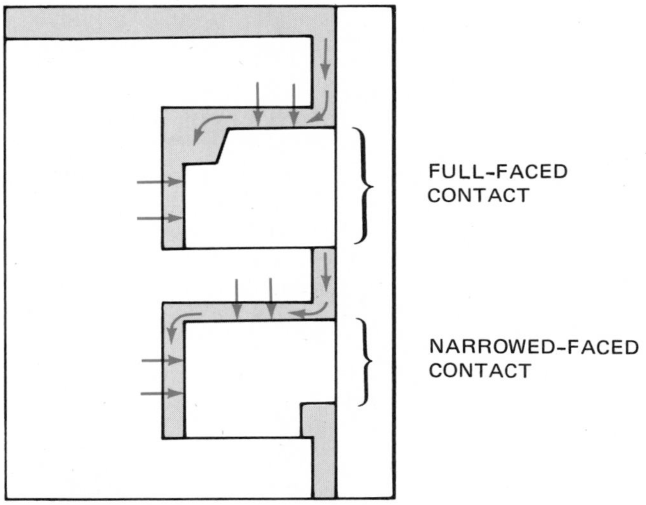

Fig. 8-7 Action of counterbored and scraper compression rings during a power stroke. The combustion pressure presses the rings against the cylinder wall with full-face contact, thus forming a good seal. (*Perfect Circle Company*)

the cylinder walls and cause abrasive wear of the walls and rings. Thus, where abrasive wear is a problem, compression rings are chromium-plated. Chromium (or "chrome" as it is called) is a very hard metal. It is finished to a very smooth surface so the rings cause very little cylinder-wall wear.

Today, however, abrasive wear is less important because of greatly improved air-filtering systems. A relatively new type of wear has cropped up, however. This is *scuff wear.* In the modern high-performance engine, cylinder walls, rings, and pistons operate at higher temperatures. These high temperatures increase the possibility that the rubbing metal surfaces will reach the melting

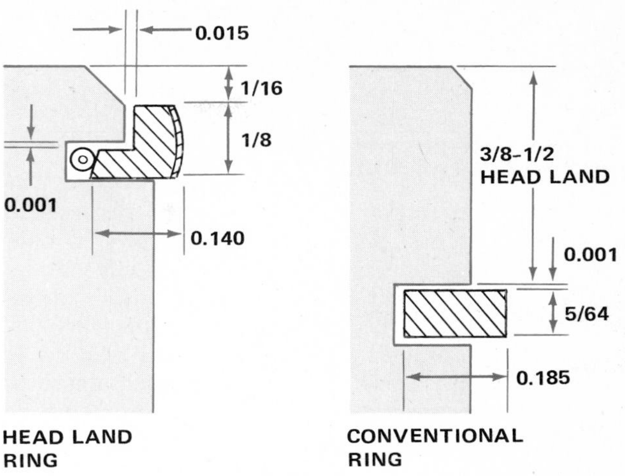

Fig. 8-8 Headland piston-ring installation compared with conventional ring. Note that the headland ring is located nearer the top of the piston. This eliminates the space left between the piston and cylinder wall in the conventional design. (*Sealed Power Corporation*)

point in spots. That is, local hot spots may develop that are so hot the metals will momentarily melt. Small-area welds can then develop. These are only momentary welds which are most apt to occur only when the piston is momentarily at rest at TDC. Then movement of the piston and rings breaks the welds, leaving rough scuffed spots and surfaces. Rings, pistons, and cylinder walls will be scratched and failure may soon occur.

To combat scuff wear, high-temperature ring coatings such as molybdenum are used. Iron, from which the compression rings are made, melts at about 2250°F [1233°C]. Chromium melts at about 3450°F [1898°C]. But molybdenum has to be heated to 4800°F [2449°C] before it melts. Thus, compression rings coated with molybdenum can be worked at higher temperatures without danger of scuff wear.

Still other high-temperature coatings are in the works. Recent developments in plasma thermal plating have made possible the use of such scuff-resistant coatings as carbides of tungsten, titanium, and tantalum. The proc-

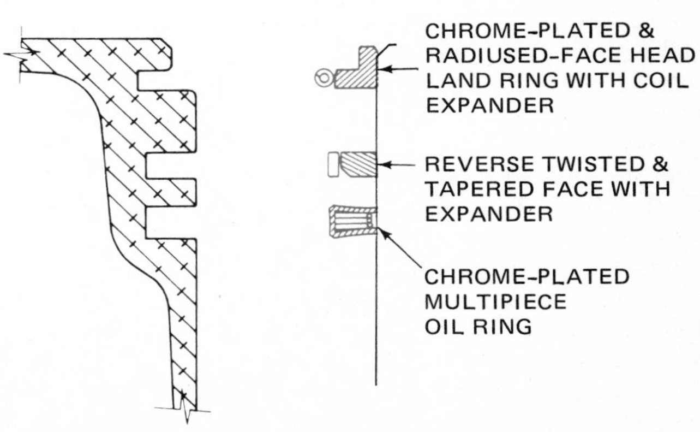

Fig. 8-9 Sectional view of piston and rings showing, at top, the headland compression ring. (*Sealed Power Corporation*)

ess requires a very high vacuum and temperature in which these coating metals are turned into vapor which then condenses on the ring surfaces.

8-5 Why Two Rings?

Two compression rings are used so as to reduce the pressure on each one. At the start of the power stroke, pressures in the combustion chamber may go as high as 1,000 psi [70.31 kg/cm^2 or 6,895 kPa]. Pressure in the crankcase is about atmospheric. Thus, the pressure difference could be around 1,000 psi. It would be difficult for a single compression ring to hold this much pressure. However, with the second ring, the pressure is, in effect, divided between the two. This does more than reduce blow-by, or loss of pressure past the upper ring. It also reduces the load on the upper ring, so that it does not press quite so hard on the cylinder wall. Ring friction and cylinder and ring wear are thus reduced.

8-6 Oil-Control Rings

The oil-control ring or rings prevent oil from working up into the combustion chamber. As already mentioned, oil throwoff from the main and rod bearings lubricates the cylinder walls, pistons, and rings. Some connecting rods have an oil-spit hole which spits oil onto the cylinder wall every time it lines up with the oil hole in the crankpin. Most of the time, more oil is thrown onto the cylinder walls than is needed. Most of it must be scraped off and returned to the oil pan. If too much oil is left on the cylinder walls, it works its way up into the combustion chamber, where it burns. This increases oil consumption, so that the engine requires extra oil quite often. Also, the burned oil fouls the spark plugs, hampers the action of the compression rings, and increases the possibility of engine knock. (Knock is described in Part Seven in the chapter diagnosing engine troubles.)

The oil that is scraped off the cylinder walls does several things. It carries away particles of carbon, dust, and dirt. These particles are then removed by the oil screen or filter. The oil also provides some cooling effect. In addition, the oil on the rings helps provide a seal between the rings and the cylinder wall. Thus, as it circulates, the oil lubricates, cleans, cools, and seals.

There are three types of oil-control rings: the one-piece slotted cast-iron type, the one-piece pressed-steel type, and the three-piece steel rail type with expander (Figs. 8-10 to 8-14).

Some one-piece rings are installed with expander springs (one is shown under the oil-control ring—No. 3—in Fig. 8-15). The expander spring increases the pressure of the ring on the cylinder wall and thus improves the oil-scraping effect.

Recently, the one-piece cast-iron oil-control ring has given way to the one-piece pressed-steel and the dual-rail and expander-spacer types (Figs. 8-11 to 8-13). In the latter type, the expander spacer (or spring) forces the rails outward into contact with the cylinder wall, as usual.

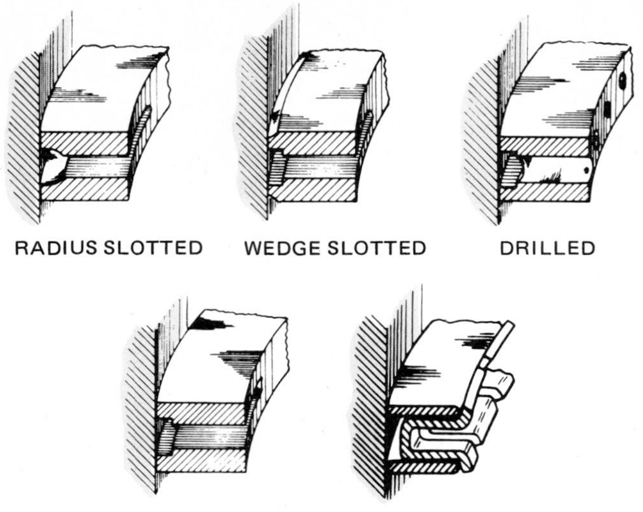

Fig. 8-10 One-piece oil-control rings compared with a three-piece ring. (*Muskegon Piston Ring Company*)

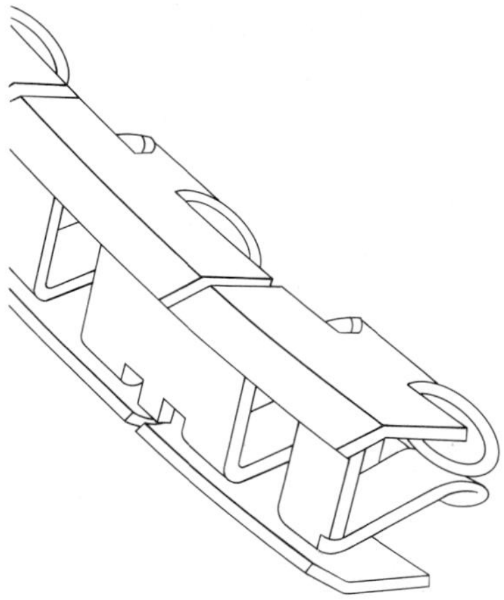

Fig. 8-11 One-piece, pressed-steel, oil-control ring. The segmental construction of this ring, with its three-way spring action, provides pressure against the upper and lower sides of the ring groove as well as against the cylinder wall. (*Muskegon Piston Ring Company*)

It also forces the rails upward and downward against the upper and lower sides of the ring groove in the piston. This provides a better seal at these three vital points, for effective oil control. In contrast, the one-piece cast-iron ring can seal against only one side of the ring groove in the piston at a time. It thus leaves open a path through which oil can pass upward toward the combustion chamber. The one-piece pressed-steel type of oil-control ring is used in engines with worn cylinder walls (see ● 8-9).

8-7 Why Only One Ring?

On earlier passenger-car engines, relatively long pistons were used, with four piston rings installed on them. Pistons of this type are shown in Figs. 8-15 and 8-18. The lower two rings were oil-control rings. However, the trend

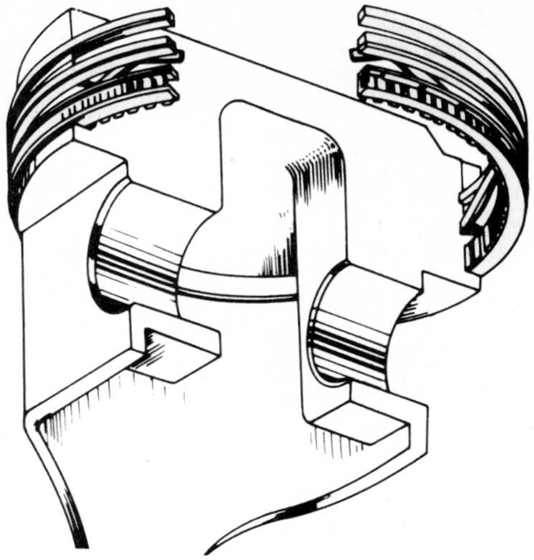

Fig. 8-12 Cutaway views of rings and piston to show construction. The second compression ring has an inner tension ring. The oil ring consists of an expander spacer and two rails. (*Thompson, Ramo, Wooldridge, Inc.*)

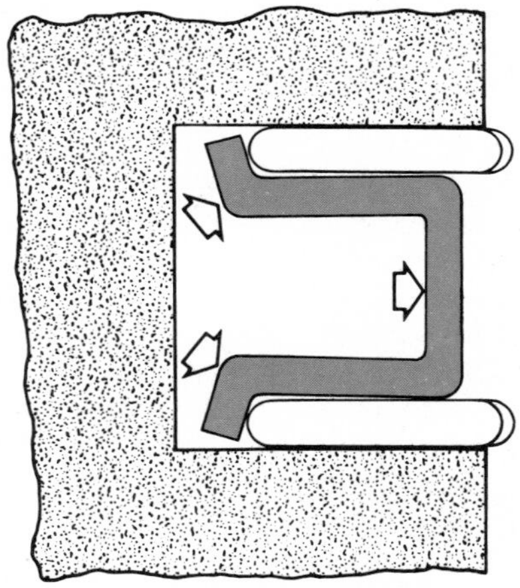

Fig. 8-13 Action of the expander spacer. As shown by the arrows, it forces the rails out against the cylinder wall and up and down against the sides of the ring groove. (*Perfect Circle Corporation*)

is toward lower hood lines and thus squatter engines. This makes it necessary to shorten the piston and reduce the number of rings to three, only one of which is an oil-control ring. This is possible because of many engineering and manufacturing improvements and the more effective action of the modern oil-control ring, such as the three-piece type (Fig. 8-14).

● 8-8 Effect of Speed on Oil Control

As engine speed increases, the oil-control rings have a harder time controlling the oil and preventing excessive amounts from passing them. There are several reasons for this. The engine and engine oil are hotter. Hot oil is thinner and can pass the rings more easily. More oil is pumped at high speed, so that more oil is thrown onto

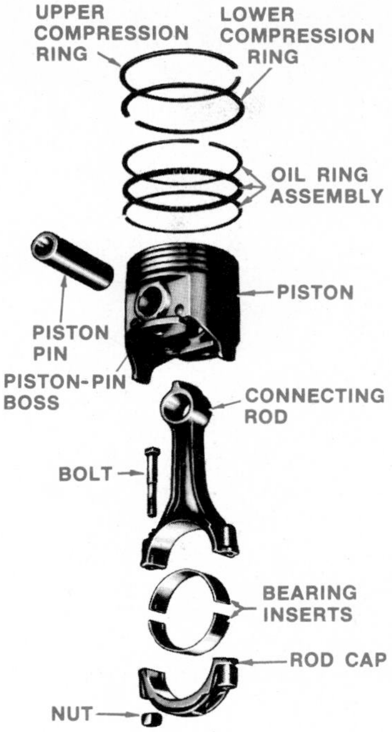

Fig. 8-14 Piston and connecting-rod assembly, disassembled so the various parts can be seen. The oil-control ring is of the three-piece type. The piston is of the slipper type, with the skirt partly cut away. (*Ford Motor Company*)

the cylinder walls. This means the oil-control rings have a harder job to do. And they have less time to do it. Thus, at high speed, more oil gets past the rings and is burned in the combustion chamber. This increases oil consumption considerably. An engine may use two or three times as much oil at high speed as at low speed. Much, but not all, of this is due to the reduced effectiveness of the oil-control rings at high speeds.

● 8-9 Effect of Engine Wear on Oil Control

As engine miles pile up, parts wear. The cylinder walls wear unevenly. They wear more at the top, where the combustion pressures are the greatest. Remember that the higher the combustion pressure, the harder the rings are pressed against the cylinder walls (see Fig. 8-7). Uneven wear makes it harder for the rings to maintain good contact with the wall. The rings have to change in size, expanding as they move up into the worn area. The result is that the rings do a poorer job of scraping off oil. More oil works up into the combustion chamber, where it is burned.

● 8-10 Oil Consumption

The amount of oil that an engine uses depends on engine speed and engine wear. All engines burn some oil. A new engine operated at moderate speed would probably not require any extra oil between oil changes. An old engine operated at high speed could require a quart of oil every few hundred miles.

● 8-11 Replacement Rings

As cylinder walls wear, power is lost, and oil is burned in the combustion chambers. There comes a time when the engine is losing so much power and is burning so much oil that repair is required. When the engine is torn down, the first step is to decide whether the cylinder is so badly worn that it must be honed or bored, or whether new rings will be enough. Honing and boring are explained later. If the wear is not too great, satisfactory repair can often be made by installing a set of special high-tension rings (Fig. 8-15).

The lower of the two compression rings (2 in Fig. 8-15) has a *ring expander* behind it. The ring expander is a steel spring in the shape of a wavy or humped ring. It adds tension, or cylinder-wall pressure, to the compression or oil-control ring. With the ring expander, the ring is made thinner (from back to front) so that it is more flexible. The ring expander more than makes up for any loss of tension from the reduced thickness. The combination (piston ring plus ring expander) offers high flexibility with high tension. In a tapered or out-of-round cylinder, the ring must expand and contract—it must change shape—as it moves up and down. The combination gives the ring a better chance to follow the changing shape of the cylinder as the ring moves up and down.

If the cylinder is only slightly worn, or if it has been serviced by honing or boring, a set of standard rings may be all that is required. The special, high-tension rings are used only if the cylinder wall is worn so much that standard rings cannot maintain contact with the cylinder wall throughout the complete piston stroke.

Some engines use expanders under the oil-control rings. These engines use only one oil-control ring per piston. The oil-control ring is of the one-piece slotted cast-iron type, and the expander helps control the oil.

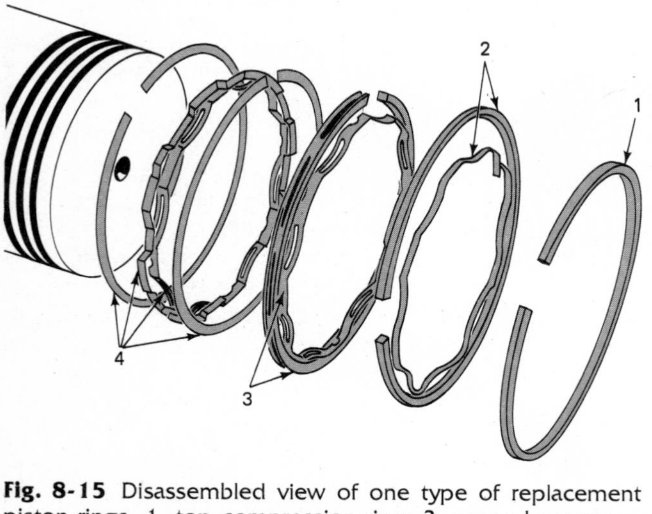

Fig. 8-15 Disassembled view of one type of replacement piston rings: 1, top compression ring; 2, second compression ring, which includes an expander spring; 3, upper oil-control ring, which has an expander ring; 4, lower oil-control ring. The last has a three-part construction; it consists of upper and lower rails with an expander spring. (*Grant Piston Rings*)

The oil-control rings in the replacement set shown in Fig. 8-15 are of two types. The upper ring (3) is of the slotted cast-iron type. It uses a ring expander. The lower one (4) is of the dual-rail and expander spacer type (see also Fig. 8-13).

● 8-12 Pistons

Pistons have been made in a number of shapes and designs. Pistons for the older-style, long-stroke, small-bore engine were generally of the long, or full-skirt, type (Fig. 8-16). Then, as lower hood lines and short-stroke engines became popular, the semislipper and full-slipper pistons came into use. On these pistons, the number of piston rings was reduced to three, two compression and one oil-control (● 8-7). One reason for the slipper piston is that, on the short-stroke engine, the piston skirt had to be cut away to make room for the counterweights on the crankshaft (Figs. 8-14 and 8-17). Also, the slipper piston, being shorter and having part of its skirt cut away, is lighter. This reduces the inertia load on the engine bearings and makes for a more responsive engine.

Inertia is a property of all material objects. Any object in motion resists any effort to change its speed or direction. The piston is continuously accelerating, decelerating, and changing direction as it moves up and down in the cylinder. Overcoming the inertia of the piston places a load on the rod bearings. The lighter the piston, the less the bearing load, and the longer the bearings will last.

Another way to lighten the piston is to make it of light metal. Thus, most automotive-engine pistons today are made of aluminum, which is less than half as heavy as iron. Iron pistons were common in the earlier engines. However, aluminum expands more rapidly than iron with increasing temperature. Since the cylinder block is made of iron, special care must be taken to maintain proper piston clearance at operating temperatures.

● 8-13 Piston Clearance

Piston clearance is the distance between the piston and the cylinder wall. Proper clearance varies with different engines, but it is generally in the neighborhood of 0.001 to 0.004 inch [0.025 to 0.102 mm]. In operation, this clearance is filled with oil so that the piston and rings move on films of oil.

If the clearance is too small, there will be loss of power from excessive friction, severe wear, and possible seizure of the piston in the cylinder. Seizure would cause complete engine failure. On the other hand, if clearances are too large, *piston slap* will result. Piston slap is caused by the sudden tilting of the piston in the cylinder as the piston starts down on the power stroke. The piston shifts from one side of the cylinder to the other, with enough force to produce a noise. Usually, piston slap is a problem only in older engines with worn cylinder walls and worn or collapsed piston skirts. Any of these can produce excessive clearance.

Pistons run many degrees hotter than cylinder walls and therefore expand more. This expansion must be controlled in order to avoid loss of piston clearance. Such loss could lead to serious engine trouble. There is even more of a problem with aluminum pistons, because aluminum expands more rapidly than iron with increasing temperature.

8-14 Expansion Control in Pistons

There are several ways of controlling the expansion of piston skirts. One method is to keep heat away from the lower part of the piston as much as possible. One way of doing this, in the full-skirt piston, is to cut horizontal slots in the piston just below the lower oil-control ring groove (Fig. 8-18). These slots keep some of the cylinder-head heat from reaching the skirt. Thus, the skirt does not become so hot and does not expand so much. Some full-skirt pistons had vertical slots cut in the skirt (Fig. 8-18). These would allow the metal in the skirt to expand without increasing the piston diameter very much.

Another method of reducing heat travel to the piston skirt makes use of a *heat dam* (Fig. 8-19). The dam is a groove cut near the top of the piston (like an extra piston-ring groove). This reduces the size of the path the heat can travel from the pistonhead to the skirt. The skirt runs cooler and does not expand so much.

Most pistons today are made so that they have a slightly oval shape when cold. These pistons are called *cam-ground* pistons (Fig. 8-20). They are finish-ground on a machine that uses a cam to move the piston toward and away from the grinding wheel as the piston is revolved. When a cam-ground piston warms up, it becomes round. Its area of contact with the cylinder wall therefore increases (Fig. 8-21). "Contact" here does not mean actual metal-to-metal contact. There must be clearance between the piston and cylinder wall, as previously noted. What is meant is that, when cold, the oval shape of the piston permits normal clearance in only a small area (with excessive clearance elsewhere). But as the piston warms up, this area of normal clearance increases. Here is what happens: The head of the piston expands uniformly in all directions. But the stiff piston-pin bosses (the round sections that support the piston pins) are

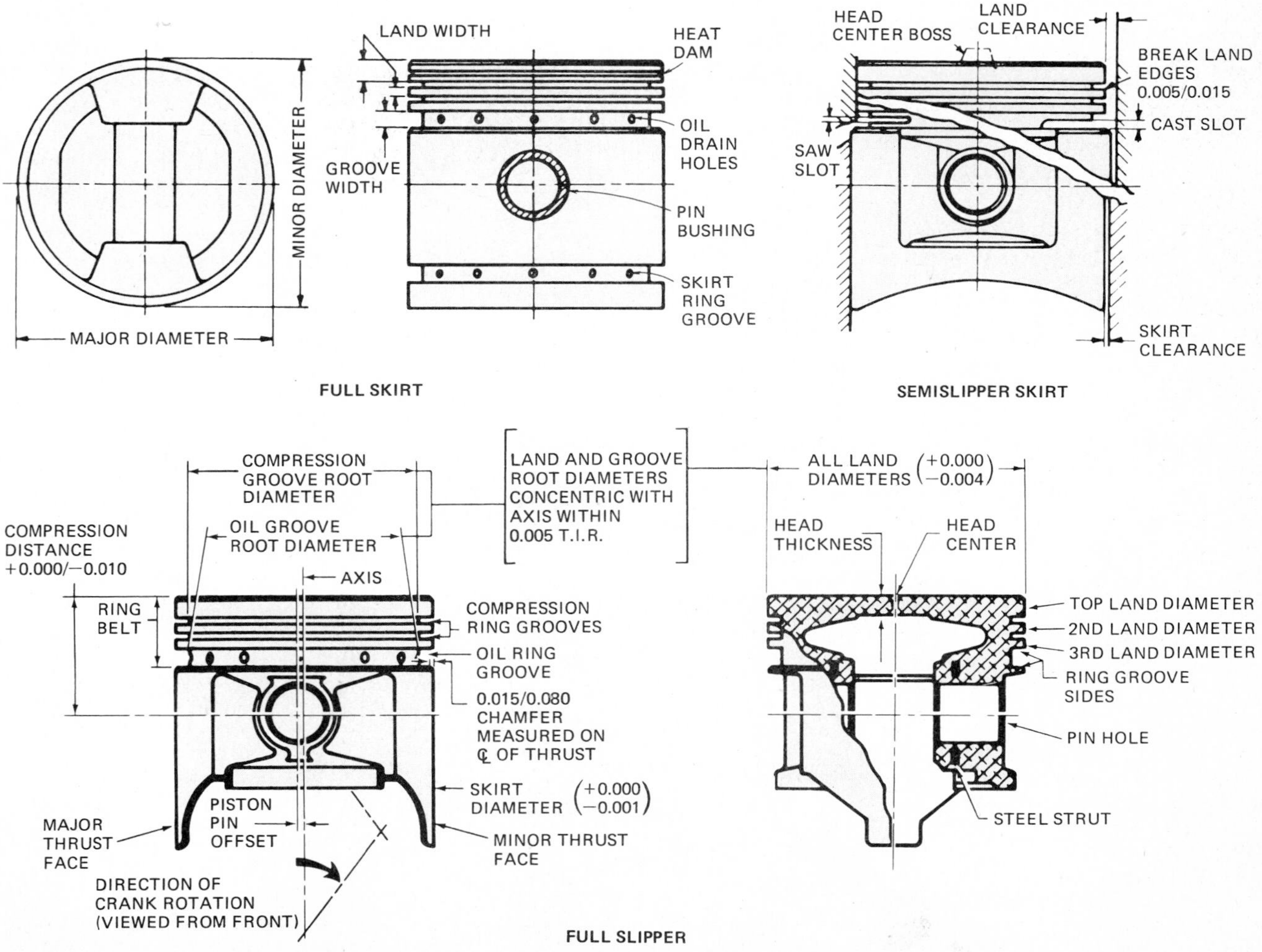

Fig. 8-16 Full-skirt, semislipper, and full-slipper pistons with parts named.

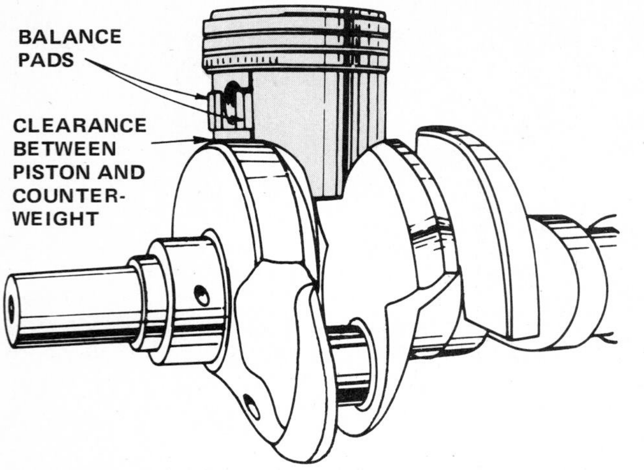

Fig. 8-17 Modern slipper piston and connecting rod assembled to crankshaft. Note the small clearance between the piston and counterweights on the crankshaft. (*Chevrolet Motor Division of General Motors Corporation*)

Fig. 8-18 Full-skirt piston with horizontal and vertical slots in the skirt. The horizontal slot reduces the path for heat travel. The vertical slot allows for expansion without an increase in piston diameter.

more effective in transmitting this outward thrust. Thus, these bosses move outward, causing the piston to assume a round shape.

Another method of controlling piston expansion is to use bands or belts cast into the piston (Fig. 8-22). These cause the outward thrust of the expanding pistonhead to be carried more toward the piston-pin bosses than toward the thrust faces. The effect is similar to that in cam-ground pistons.

8-15 Pistonhead Shape

The simplest pistonhead is the flat head (Fig. 8-19). However, increasing compression ratios have made it necessary to reduce the clearance volume (the volume above the piston at TDC). But the valves must have room to open without striking the pistonhead. Notches or hollows in the pistonhead can be used to provide adequate valve

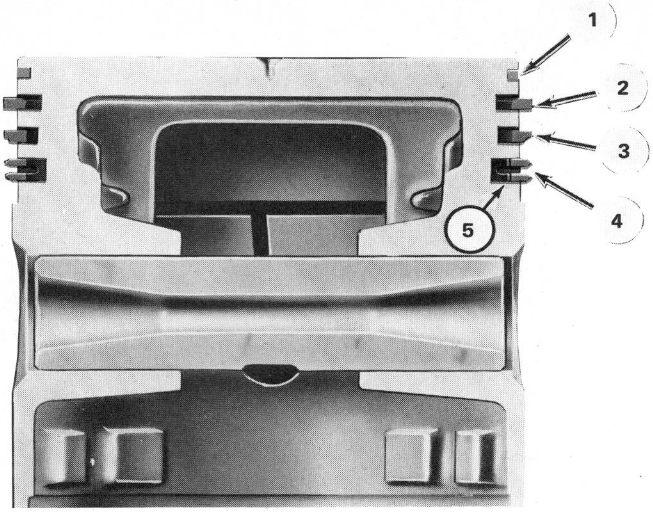

Fig. 8-19 Piston with rings in sectional view, showing heat dam and ring shapes; 1, heat dam; 2, upper compression ring; 3, lower compression ring; 4, oil ring; 5, ring expander.

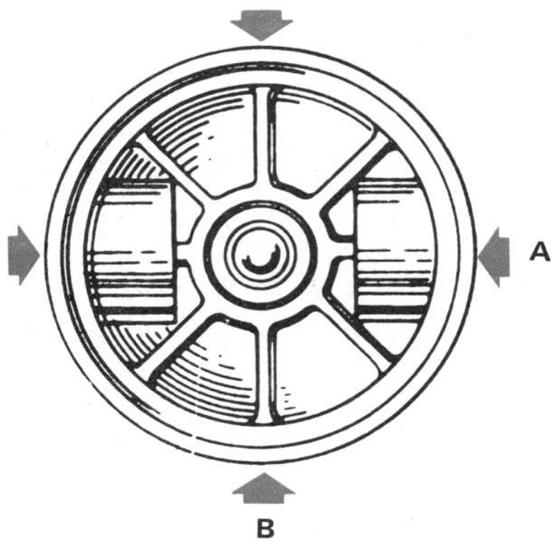

Fig. 8-20 Cam-ground piston viewed from bottom. When the piston is cold, its diameter at *A* (the piston-pin holes) may be 0.002 to 0.003 inch [0.051 to 0.076 mm] smaller than at *B*. (*Chrysler Corporation*)

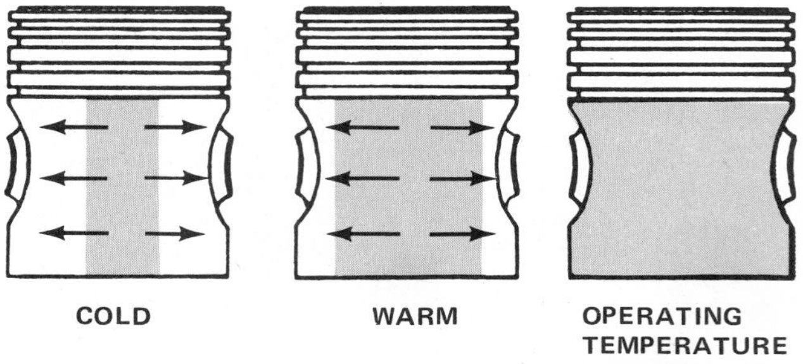

Fig. 8-21 As the cam-ground piston warms up, the expansion of the skirt distorts the piston from an elliptical to a round shape. This increases the area of normal clearance between the piston and the cylinder wall.

clearance when the piston is at TDC. Also, some pistonheads have a trough, or are dished, to improve the turbulence, or swirling, of the air-fuel mixture (Fig. 8-23). Such turbulence improves combustion.

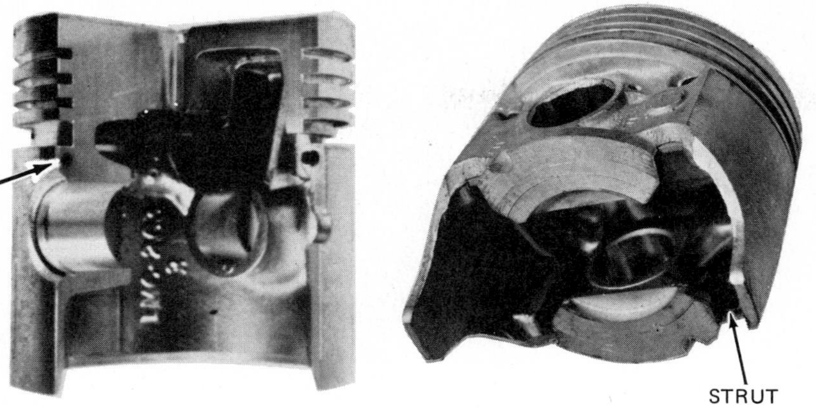

Fig. 8-22 Pistons with cast-in belt and cast-in strut to provide expansion control. (*Thompson, Ramo, Wooldridge, Inc.*)

● 8-16 Piston-Pin Offset

In some engines, the piston pin is offset from the centerline of the piston toward the major thrust face. This is the face that bears most heavily against the cylinder wall during the power stroke (Fig. 8-24). If the pin is centered, the minor thrust face remains in contact with the cylinder wall until the end of the compression stroke. Then, as the power stroke starts, the rod angle changes from left to right (in Fig. 8-24). This causes a sudden shift of the side thrust on the piston, from the minor thrust face to the major thrust face. If there is too much clearance, piston slap will result.

However, if the piston pin is offset (Fig. 8-25), the combustion pressure causes the piston to tilt as the piston nears TDC, as shown. The lower end of the major thrust face makes first contact with the cylinder wall. Then, after the piston passes TDC and the reversal of side thrust occurs, full major-thrust-face contact is made. There is less chance for piston slap to occur.

The tilting action occurs because there is more combustion pressure on the right-hand part of the piston (which measure R + O, or piston radius plus offset) than on the left-hand side of the piston (which measures R − O).

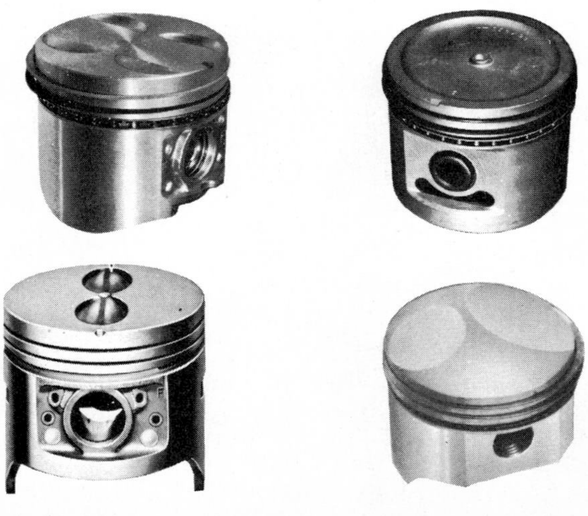

Fig. 8-23 Pistons for modern internal-combustion engines.

● 8-17 Ring-Groove Fortification

The compression rings constantly move up and down in the ring grooves. At the beginning of the power stroke, as the piston goes up to TDC and starts down again, the top compression ring is forced down hard against the lower side of the ring groove. The increasing combustion pressures produce this action. Then, during the intake stroke, the vacuum in the cylinder causes the top compression ring to move up and into contact with the upper side of the ring groove. Thus, the ring repeatedly strikes the upper and lower sides of the ring groove.

In high-performance engines, these repeated impacts can cause rapid ring-groove wear. The top ring groove is most critical, because it receives the greater part of the combustion pressures. To combat this wear, pistons for some high-performance engines have top-ring groove fortification. The fortification consists of a ring of cast iron or nickel-iron alloy which is cast into the piston (Fig. 8-26). For cast pistons in medium duty, the inserts are stamped from steel sheets. In forged pistons, inserts cannot be used. Instead, if ring fortification is required,

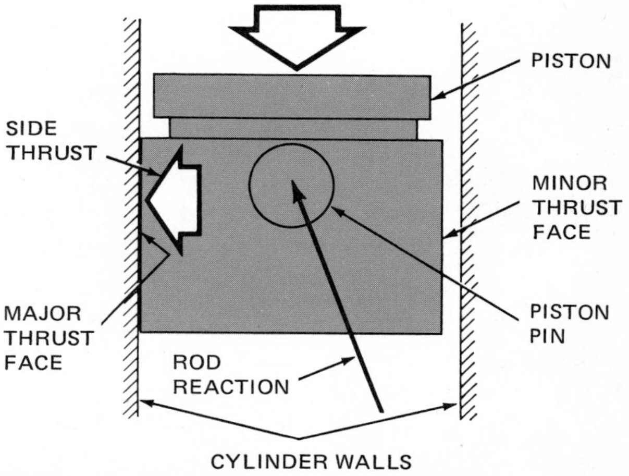

Fig. 8-24 As combustion pressure is applied to the pistonhead, and the connecting-rod angle changes from left to right, side thrust on the piston causes it to shift abruptly toward the major thrust face.

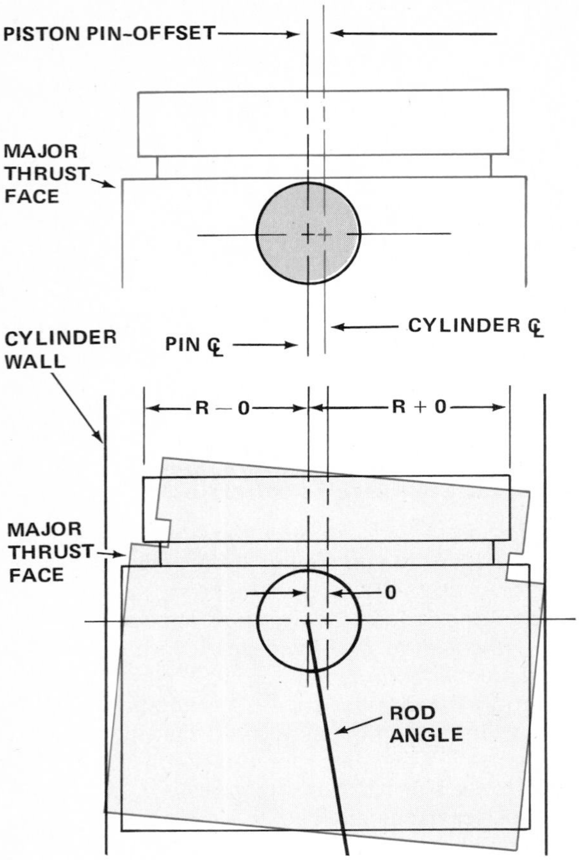

Fig. 8-25 If the piston pin is offset toward the major thrust face, combustion pressure will cause the piston to tilt to the right, as shown, to reduce piston slap. R is the radius of the piston; O is the offset of the piston pin. (*Bohn Aluminum and Brass Company, Division of Universal American Corporation*)

the groove area is sprayed with molten metal having the proper wear resistance. Then, the groove is machined to the proper dimensions.

● 8-18 High-Performance Pistons

Aluminum pistons can be either cast or forged. Cast pistons are made by pouring molten aluminum into molds. Forged pistons are "hammered out" from slugs of aluminum alloy. The alloy, when subjected to high forging pressure, flows, or extrudes, into dies to form pistons. Both must be heat-treated.

The forged piston is denser and forms a better heat path, to allow the heat to get away from the pistonhead. The flags in Fig. 8-27 indicate the temperatures at comparable points when both pistons operate under the same conditions.

REVIEW QUESTIONS

Check up on yourself by answering these questions on pistons, rings, and connecting rods. If any question baffles you, reread the section that gives the answer.

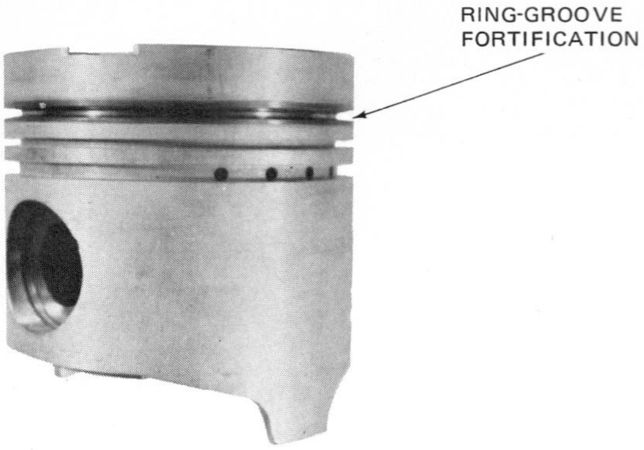

Fig. 8-26 Piston with top-ring-groove cast-in fortification. (*Thompson, Ramo, Wooldridge, Inc.*)

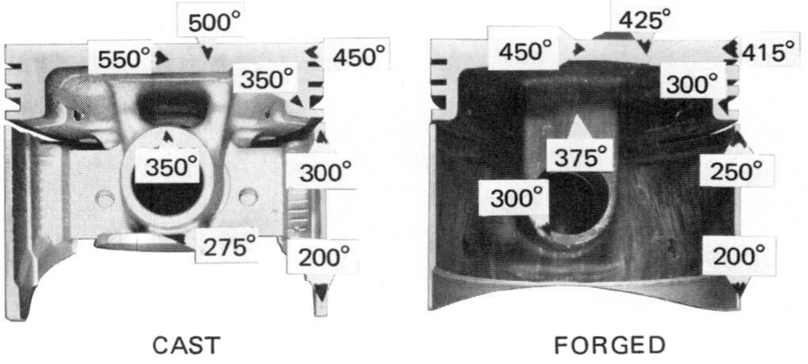

Fig. 8-27 Cast and forged pistons, cut in half to show the operating temperatures at comparable points in the two pistons. (*TRW, Inc.*)

1. What are three ways in which the rod is attached to the piston?
2. Which is the most common method of connecting piston and rod?
3. Describe the path the oil follows to lubricate the rod big-end bearings.
4. Describe the path the oil follows to lubricate the piston pin on the type of rod with an oil passage drilled in it.
5. What are the two basic types of piston rings?
6. What is blow-by?
7. Describe the action of a counterbored piston ring during intake. During the power stroke.
8. Why are two compression rings needed?
9. What is the purpose of the oil-control ring?
10. What are the effects of engine speed and engine wear on oil control?

SELF PROJECT

Engine parts have to fit. There must be proper clearances between parts. While clearances may vary somewhat, certain clearances are pretty much the same in all engines. On a sheet of paper, write "Clearance" and then the subheadings of this chapter: "Pistons, Rings, Connecting Rods." Below this, write "Piston Clearance." Skip four lines and write "Ring End Gap." Skip four more lines and write "Rod Bearing Clearance." Now go to shop manuals and put these specifications under each heading.

CHAPTER 9
Engine Construction: Valves and Valve Trains

After reading this chapter, you should be able to:

- **Discuss the purpose, construction, and operation of engine valves.**
- **Describe the advantages of, and the construction and operation of overhead-camshaft engines.**
- **Explain what valve timing means and why valves are timed to open early and close late.**
- **Explain the purpose of hydraulic valve lifters and how they work.**
- **Discuss the purpose of valve rotators and how they work.**
- **Identify and name the parts in valve trains and explain what they are for.**

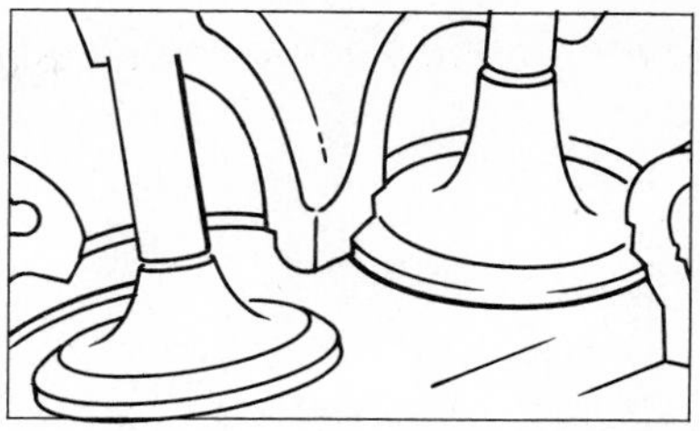

● 9-1 Types of Valve Trains

Three types of valve trains have been mentioned: the L-head, the I-head, and the overhead camshaft. About the only places you will see L-head engines are in antique cars and in such small machines as power lawn mowers. In this chapter, we describe the I-head and the overhead-camshaft arrangements.

● 9-2 Cams and Camshafts

We introduced the I-head engine valve train briefly, in ● 5-6. Now we shall look at this valve train in more detail, starting with the cams and camshaft. Figure 5-14 shows the valve trains for several cylinders in an in-line engine. Recall that there is a cam for each valve, or two cams per cylinder. A six-cylinder engine would therefore have a camshaft with twelve cams on it. Figure 5-16 shows a camshaft with its related parts.

The camshaft is driven by sprockets and chain, or by gears. As it rotates, the cam lobes move up under the valve lifters, raising them. When a valve lifter is pushed up, it pushes up on the push rod. The push rod causes one end of the rocker arm to move up. The other end of the rocker arm pushes down on the valve stem, causing the valve to move down, or open. Then, when the cam lobe passes out from under the valve lifter, the valve spring takes over and moves the valve up to the closed position. This movement causes the rocker arm to rock back. It pushes down on the push rod and valve lifter so they return to their closed-valve positions. To summarize the actions, we can say that the valve is opened by the positive mechanical action of the cam lobe. It is closed by valve-spring pressure.

In many engines, the camshaft also has an eccentric to operate the fuel pump and a gear to drive the ignition distributor and oil pump. (See Fig. 9-1.)

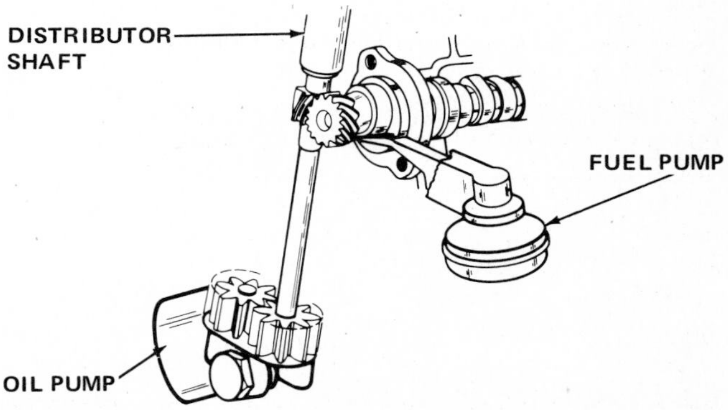

Fig. 9-1 Oil-pump, distributor, and fuel-pump drives. The gear on the camshaft drives the oil pump and distributor. The eccentric on the camshaft drives the fuel pump. (*Buick Motor Division of General Motors Corporation*)

Fig. 9-2 Crankshaft and camshaft sprockets with chain drive for a six-cylinder engine, showing timing marks on the sprockets. The larger of the two sprockets is on the camshaft so that it turns at one-half crankshaft speed. (*Chrysler Corporation*)

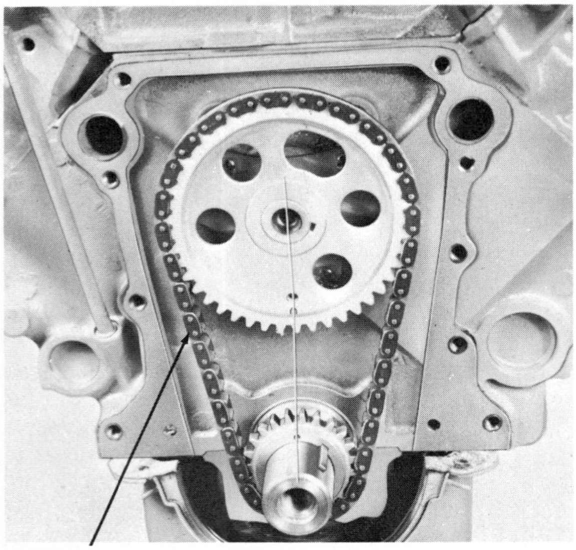

Fig. 9-3 Crankshaft and camshaft sprockets with chain drive for a V-8 engine. (*Chrysler Corporation*)

9-3 Driving the Camshaft

The camshaft is driven by gears or by sprockets and chain. Figure 9-2 shows the chain drive for a six-cylinder engine. Figure 9-3 shows the chain drive for a V-8 engine. Figure 9-4 shows a gear drive for a six-cylinder engine. The crankshaft must turn two times to turn the camshaft once. There are four piston strokes to a complete cycle of actions in the engine. This requires two crankshaft revolutions. Each valve must open only once during a complete cycle. Since each valve opens every camshaft revolution, the camshaft must rotate only once while the crankshaft rotates twice. This 1:2 gear ratio is achieved by making the camshaft gear or sprocket twice as large as the crankshaft gear or sprocket.

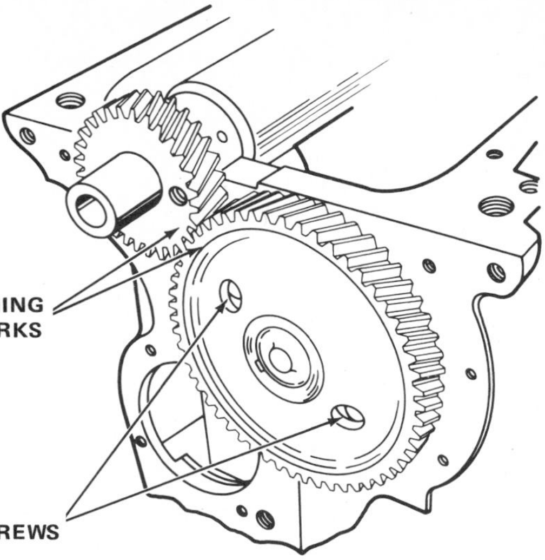

Fig. 9-4 Crankshaft and camshaft gears for a six-cylinder engine. Note timing marks on gears. (*Buick Motor Division of General Motors Corporation*)

The gears are called the "timing gears." The chain is called the "timing chain." The reason for this is that the gears, or chain and sprockets, "time" the valve action. One difference between driving the camshaft by a chain and driving it by gears is this. When a chain is used, the camshaft turns in the same direction as the crankshaft. When gears are used, the camshaft rotates in the opposite direction from the crankshaft.

The camshaft is mounted in bearings, or bushings, in the lower part of the cylinder block, in most engines. In engines with overhead camshafts, the camshaft is mounted in bearings on top of the cylinder head.

9-4 Valves

Each cylinder has two valves, an intake valve and an exhaust valve. Some high-performance engines have four valves per cylinder—two intake valves and two exhaust valves. The intake valve is larger than the exhaust valve. The reason is this: When the intake valve is opened, the only force moving air-fuel mixture into the cylinder is atmospheric pressure. When the exhaust valve is opened, the piston is moving up, and there is a high pressure driving the exhaust gases out. Thus, the intake port must be larger to allow enough air-fuel mixture to enter.

Various types of valves have been used in the past; among them are the sliding-sleeve and rotary types. But the valve in general use today is the mushroom, or poppet, valve (Fig. 9-5). When the valve is closed, it is held on the valve seat by the valve spring.

Modern engines using low-lead or no-lead gasoline work the valves harder. The reason for removing the lead from gasoline is that lead can foul the emission controls. Also, excessive amounts of lead can pollute the atmosphere. We shall discuss this in detail in later chapters. Meantime, the fact that there is little or no lead in gaso-

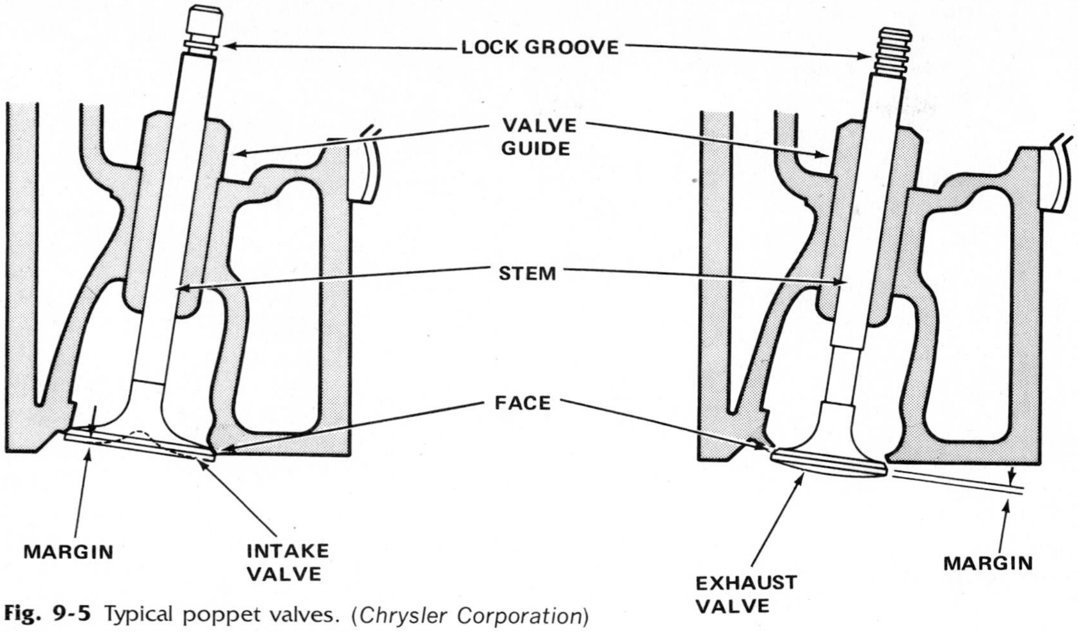

Fig. 9-5 Typical poppet valves. (*Chrysler Corporation*)

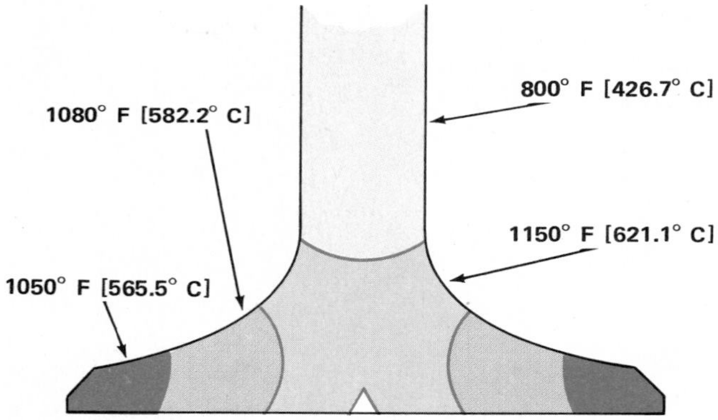

Fig. 9-6 Temperatures in an exhaust valve. The valve is shown in sectional view. (*Eaton Corporation*)

line creates a wear problem for valves and valve seats. Lead in gasoline forms a fine coating on the valve faces and seats. This coating acts as a lubricant. Without the lead, the valve faces and seats lack lubricant and can wear rather rapidly. For this reason, many engines have valves with special coatings on their faces. These coatings reduce wear significantly. For severe service, the valve faces may be made of stellite, a very hard metal (Fig. 9-8).

In addition, many valves and valve seats are ground to give what is called an *interference angle*. This is described in ● 9-7.

● 9-5 Valve Cooling

The intake valve runs relatively cool, since it passes only the air-fuel mixture. But the exhaust valve must pass the very hot exhaust gases. The exhaust valve may actually become red hot in operation, at temperatures well above 1000°F [537.8°C]. Figure 9-6 shows a typical temperature pattern for an exhaust valve. Note that the valve stem is coolest, and the part nearest the valve face is next coolest. The valve stem passes heat to the valve guide, which helps to keep the valve stem cool. Likewise, the valve face passes heat to the valve seat. This helps to keep the valve face cool. The valve seat and valve guide are cooled by the engine cooling system. The cylinder head is carefully designed to permit good coolant circulation through the water jackets around the seat and guide. Some I-head engines have water nozzles that force good circulation around the valve seats. In some late-model engines, deflectors in the head improve coolant circulation around the valve seats.

● 9-6 Sodium-Cooled Valve

To aid valve cooling and thus increase valve life, some heavy-duty engines use sodium-cooled valves. Such a valve has a hollow stem partly filled with metallic sodium. Sodium melts at 208°F [97.8°C]. Thus, at operating temperatures, the sodium is liquid. As the valve moves up and down, the sodium is thrown upward into the hotter part of the valve. It absorbs heat, which it then gives up to the cooler stem as it falls down into the stem again.

CAUTION: Sodium is a highly reactive element. If a piece of sodium is dropped into water, it will burst into flame with explosive violence. If it gets on your skin, it will cause deep and serious burns. Never cut into a sodium-cooled valve stem! The sodium could explode and burn you.

● 9-7 Valve Seat

The exhaust-valve seat is heated to extremely high temperatures by the burned gases. The cylinder-head material is adequate to take these temperatures under normal operating conditions. But where the engine is put to severe service, as for instance in trucking, special provisions must be made. One is to harden the valve seats by a special electric-induction process (Fig. 9-8). The other is to install seat inserts. These are special heat-resistant steel-alloy insert rings (Fig. 9-7). These rings can withstand the higher temperatures without undue wear. If they do wear, they can be replaced.

The use of an interference angle in valves and valve seats has become quite common. As shown at the top of Fig. 9-8, the interference angle is attained by grinding the valve at an angle $\frac{1}{4}$ to 1 degree flatter than the seat angle. This produces greater pressure at the outer edge of the valve seat. The valve-seat edge thus tends to cut through any deposits that have formed. This produces a good seal. As shown at the bottom of Fig. 9-8, an interference angle is not recommended where the valve is faced with stellite and the seat is induction hardened.

The difference in angles between the valve face and the valve seat gradually disappears as the valve face and seat wear. That is, the contact between the two changes from line contact (interference angle) to area contact.

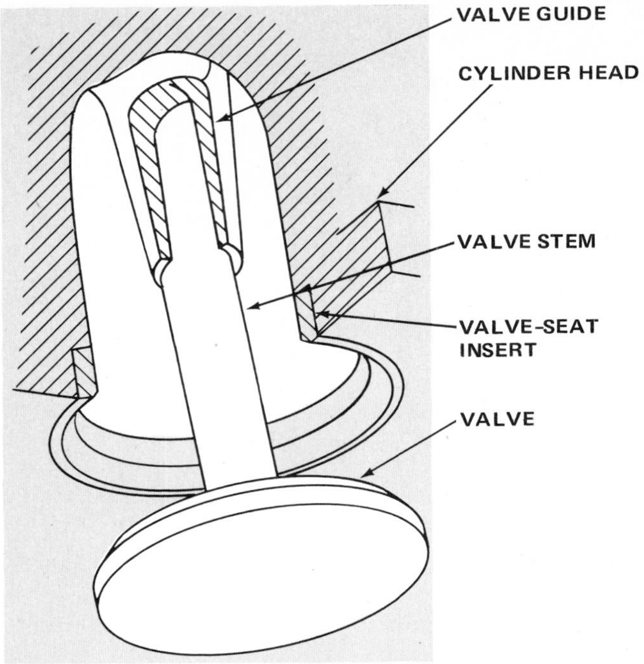

Fig. 9-7 Valve and valve seat in a cylinder head. Note that the seat is formed by a valve-seat insert.

9-8 Valve-Spring Attachment

In I-head engines, one end of the valve spring presses against the cylinder head. The other end is attached to the end of the valve stem with a spring retainer and a retainer lock or keeper (Fig. 9-9). When installed, the spring is compressed with the retainer above it. Then the retainer lock is installed in the groove in the valve stem. When the spring is released, the retainer presses against the lock, holding it in place in the valve stem. The tension spring puts a heavy pull on the valve stem. This keeps the valve seated tightly on the valve seat—except while the valve train keeps the valve open.

9-9 Varieties of I-Head Valve Trains

In the I-head, or overhead-valve, engine, a push rod and rocker arm are required (Fig. 5-14). There are several types of rocker arm. In one type (Fig. 9-10), the rocker arm has an adjustment screw for adjusting valve-train clearance. The adjusting procedure is described in a later chapter.

A different type of rocker arm and support is shown in Fig. 9-11. This rocker arm is a heavy steel stamping. Each T-shaped rocker-arm support supports two rocker arms, as shown. The rocker-arm supports are positioned on the cylinder head by the rocker-arm retainer, as shown. Each retainer positions two rocker-arm supports. Clips on the rocker-arm retainer hold the rocker arms on their supports and prevent them from slipping.

A still different design of rocker arm is shown in Fig. 9-12. This rocker arm is a heavy steel stamping, shaped as shown. The push-rod end is formed into a socket, in which the end of the push rod rides. The rocker

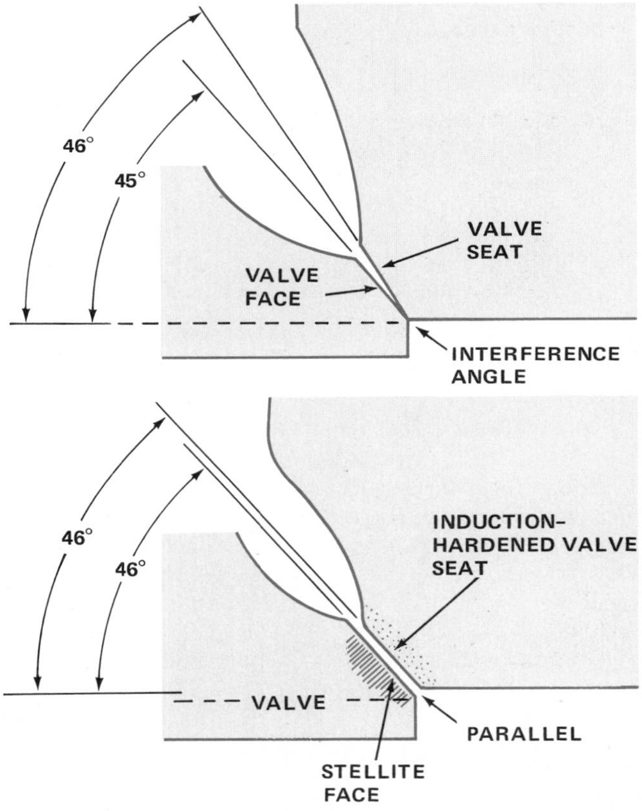

Fig. 9-8 Valve and valve-seat angles. Top, interference angle. Bottom, an induction-hardened valve seat. The valve here is stellite-faced. Stellite is resistant to heat and wear. Note that the faces of the valve and valve seat (bottom) are parallel. *(Chevrolet Motor Division of General Motors Corporation)*

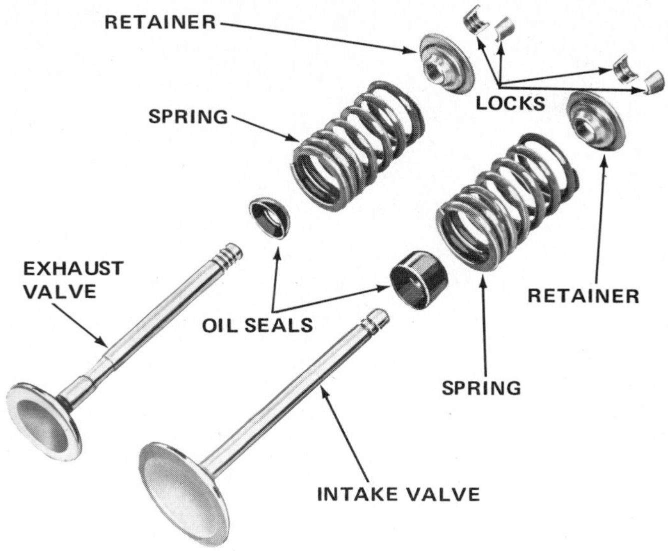

Fig. 9-9 Exhaust and intake valves and related parts in proper assembly relationship. (*Chrysler Corporation*)

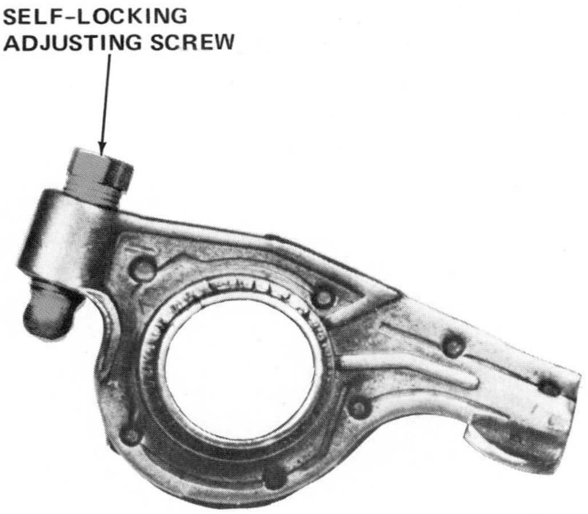

Fig. 9-10 One type of rocker arm used in overhead-valve, push-rod engines. (*Chrysler Corporation*)

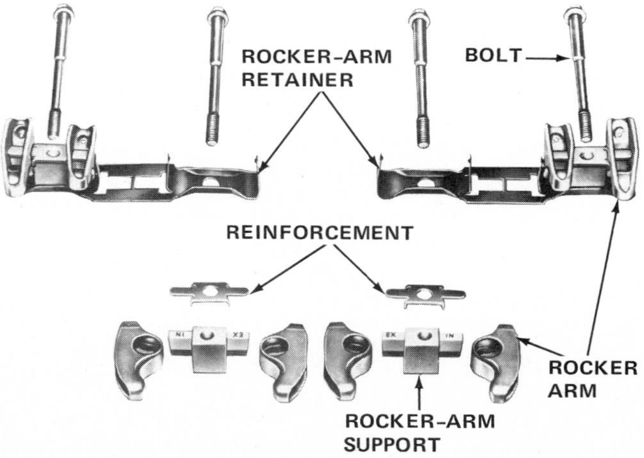

Fig. 9-11 Rocker arms, supports, and retainers. (*Cadillac Motor Car Division of General Motors Corporation*)

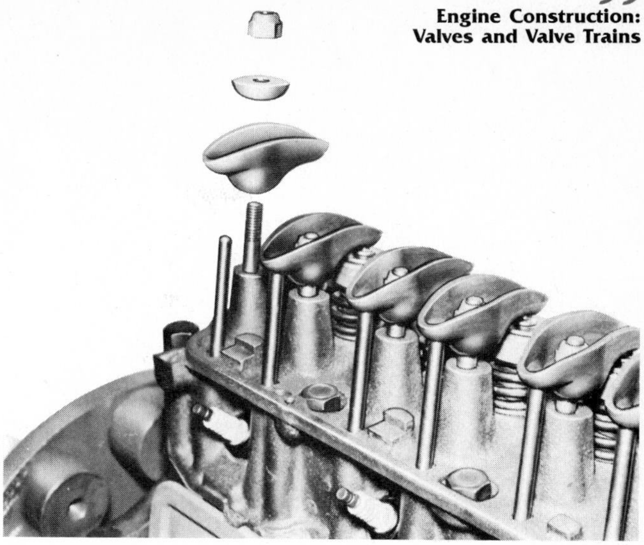

Fig. 9-12 Rocker arm using a ball pivot with related parts. (*Chevrolet Motor Division of General Motors Corporation*)

arm is supported by a ball pivot mounted on a stud. The stud is hollow and opens into an oil gallery in the head. Lubricating oil feeds through the stud to the ball pivot. The push rod is also hollow. It feeds oil from the valve lifter to the contact area between the push rod and the rocker arm. Adequate lubrication of the moving parts of the valve train is thus assured. The valve clearance on this design is measured between the rocker arm and valve stem, as with the other design. But adjustment is made by turning the adjustment nut above the ball pivot up or down on the mounting stud. This raises or lowers the rocker arm to increase or decrease valve clearance.

In some engines using hydraulic valve lifters (● 9-15), there is no provision for adjusting valve lash. The lifters hydraulically take up any clearance when the valve is closed. Some manufacturers specify that, if valves and valve seats are ground, then valve stems should be shortened by grinding off tips. Others supply longer or shorter push rods to correct for changes in stem height. Both restore proper linkage dimensions. Other engines with hydraulic valve lifters do have adjustment provisions.

In many I-head engines, the valves move in replaceable valve guides. In others, the guides are part of the cylinder head. That is, they are holes bored in the cylinder head, and are not separate parts. If such guides become worn, they can be reamed to a larger size. Then valves with oversize stems can be installed.

Modern I-head engines have special provisions to prevent oil seepage past the valve stems into the combustion chamber. Figure 9-13 shows one type of seal and shield, or "shedder," as it is also called. The seal is a rubberlike ring. It fits between the spring retainer skirt and an undercut in the valve stem. The shield covers the top two turns of the spring. The seal prevents oil from seeping down the valve stem past the locks and retainer. The shield prevents large amounts of oil from being thrown through the spring onto the valve stem. Another design is shown in Fig. 9-14. Here, the shield is on the inside of the valve springs, but the purpose is the same: to shield the valve stem from too much oil. See also Fig. 9-9.

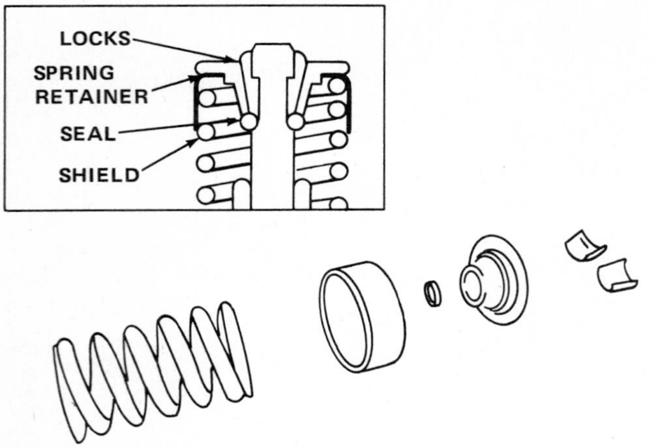

Fig. 9-13 Sectional assembled and disassembled views of valve-and-spring assembly with oil seal and shield. (*Chevrolet Motor Division of General Motors Corporation*)

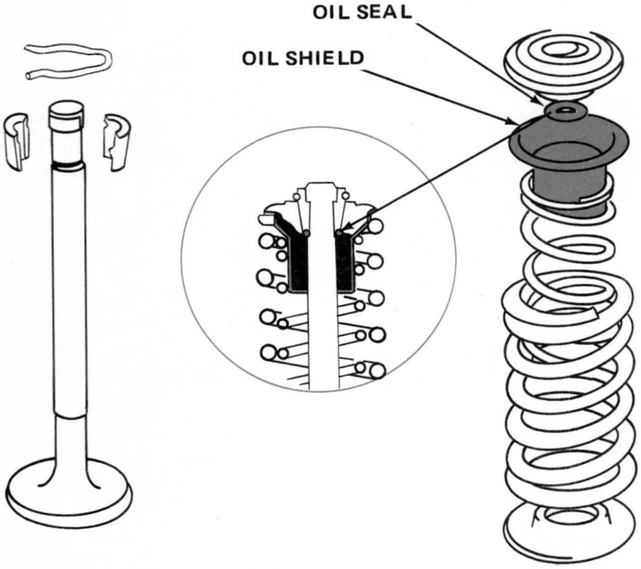

Fig. 9-14 Disassembled and sectional assembled views of valve-and-spring assembly with oil seal and shield. (*British Leyland Motors, Incorporated*)

● 9-10 Overhead-Camshaft Engines

The overhead-camshaft (OHC) engine may have one camshaft per cylinder head (single overhead camshaft, or SOHC) or two camshafts per cylinder head (double overhead camshaft, or DOHC). Figure 6-3 shows a four-cylinder SOHC engine. Figure 7-14 shows a cylinder head for a SOHC engine. Figure 6-16 shows a V-8 DOHC engine, with two camshafts in each cylinder head. The camshaft or camshafts may be driven by sprockets and chain (Figs. 6-3 and 6-16) or by sprockets and a neoprene belt (Figs. 6-19 and 9-20). The belt is reinforced with fiber-glass cord and has a facing of woven nylon fabric on the toothed side. The teeth in the drive belt fit the sprocket teeth.

Fig. 9-15 Sectional view of the Chevrolet Vega overhead-camshaft engine. (*Chevrolet Motor Division of General Motors Corporation*)

Several arrangements are used to carry the cam action to the valve stems. In some OHC engines, cam action is carried directly to the valve stem through a cap on the valve stem called the valve tappet (Figs. 6-3 and 9-15). In other OHC engines, the cam action is carried through a rocker arm (Figs. 9-16, 9-18, and 9-19).

In the Vega four-cylinder SOHC engine, the cam action is carried directly to the valve stem through the valve tappet (see Figs. 6-19 and 9-15). Because the action is direct and not through a push rod and rocker arm, the action is quicker and more accurate. This means higher engine speeds are possible because the valves react more quickly to allow cylinder charging and exhaust.

The Opel four-cylinder SOHC engine uses rocker arms (Fig. 9-16). This rocker arm is like the one shown in Fig. 9-12. That is, it is the ball-pivot type. The rocker arm pivots at its center on a ball stud installed in the cylinder head. This engine has a timing chain (Fig. 9-17). Note the chain tensioner and damper block which take up all slack in the chain. They prevent looseness of the chain which could cause noise.

Another valve-train arrangement is shown in Fig. 9-18. The rocker arm is not pivoted at the center. Instead, its two ends are held in position by a clip (Fig. 9-19). The center of the rocker arm rests on the cam. When the cam lobe comes around, it pushes down on the rocker arm. The arm pivots on the valve-lash adjuster (valve lifter) and the other end of the arm pushes the valve stem down. Note that in this arrangement (Figs. 9-18 and 9-19), as well as the arrangement shown in Figs. 9-16 and 9-17, the valve lifter (or valve-lash adjuster) is special. It is hydraulic and automatically takes up any

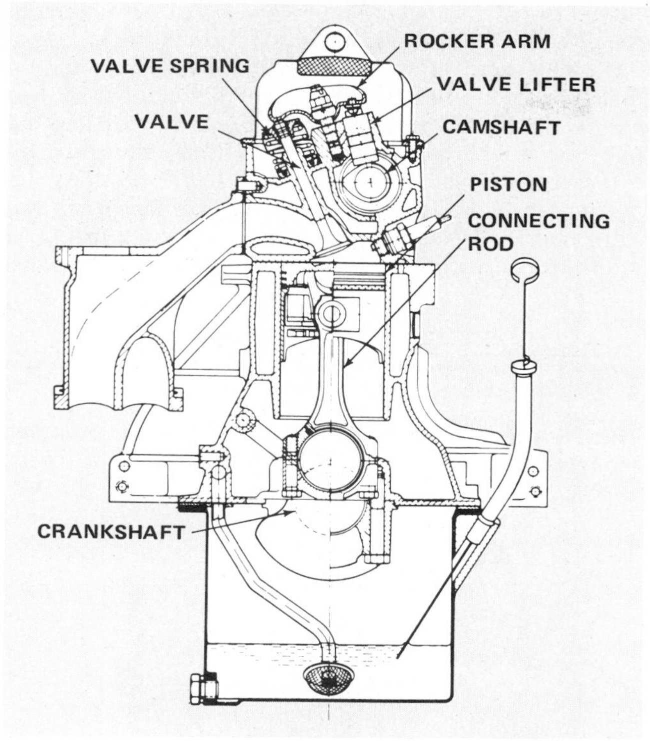

Fig. 9-16 Sectional view from the end of the Opel overhead-camshaft engine. (*Buick Motor Division of General Motors Corporation*)

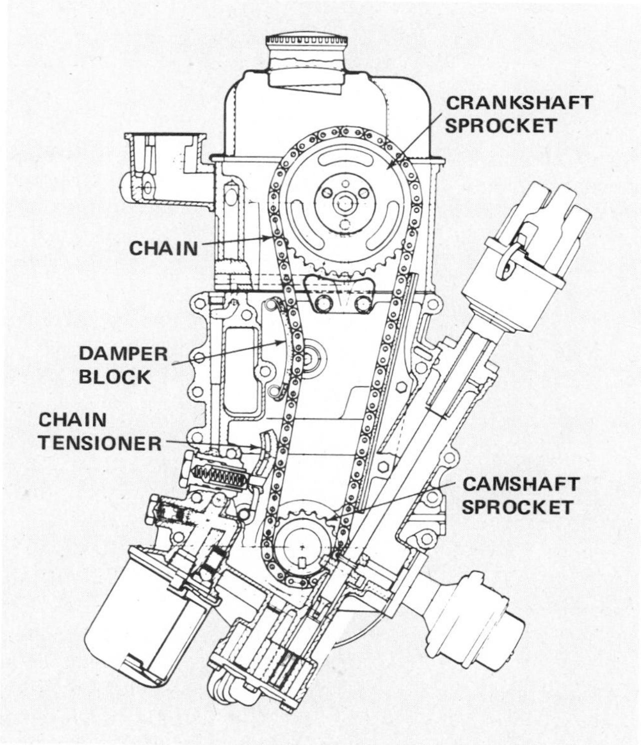

Fig. 9-17 Camshaft drive arrangement for the Opel overhead-camshaft engine. (*Buick Motor Division of General Motors Corporation*)

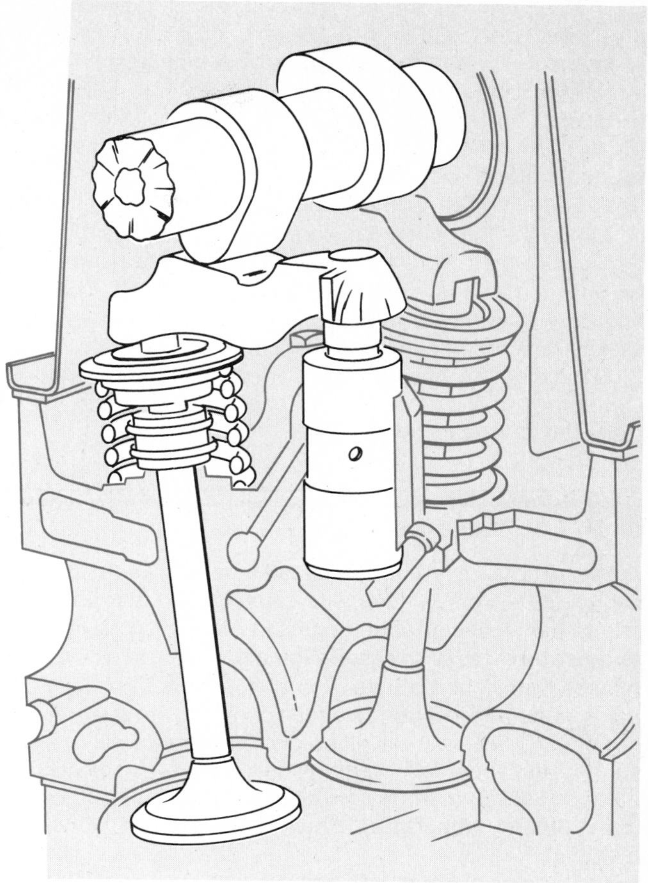

Fig. 9-18 Close-up view of valve train for the Ford 2,300 cubic centimeter overhead-camshaft engine. (*Ford Motor Company*)

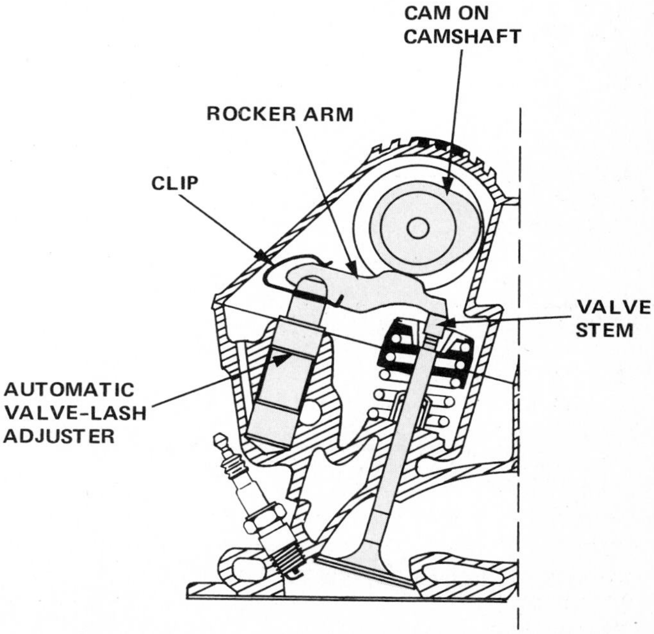

Fig. 9-19 Sectional view of the valve mechanism in a SOHC engine using a toothed belt to drive the camshaft. (*Pontiac Motor Division of General Motors Corporation*)

clearance, or lash, in the valve train. Hydraulic valve lifters are described in ● 9-15.

Figure 9-20 is a cutaway view of a high-performance engine having four valves per cylinder (two intake, two exhaust), and two camshafts in the cylinder head (DOHC). Note that the cams operate directly on the valve tappets on top of the valve stems. This is similar to the arrangement used in the Vega engine (Fig. 9-15).

Although the overhead-camshaft engines are usually given their full names—single-overhead-camshaft (SOHC) engine or double-overhead-camshaft (DOHC) engine—"shaft" is often dropped. That is, they are called simply "overhead-cam engines."

Note that the I-head engine which uses push rods is sometimes called a "push-rod engine" to distinguish it from the overhead-cam engine.

● 9-11 Valve Rotation

If the exhaust valve rotates as it opens, there is less chance of valve-stem deposits causing the valve to stick. In addition, valve rotation results in more even valve-head temperature, for this reason: Some parts of the valve seat may be hotter than others. Actual hot spots may develop. If the same part of the valve face continues to seat on the hot spot, a hot spot develops on the valve face. The hot spot on the valve face wears or burns away faster. But if the exhaust valve rotates, no one part is always subjected to the higher temperature. Thus, longer exhaust-valve life results.

In the typical engine design, the rocker arm is slightly offset from the centerline of the valve (Fig. 9-21). Every time the valve is opened, there is an off-center push on the valve stem. This tends to rotate it. There are also special valve-rotating mechanisms that are part of the valve train. These are of two types, free and positive.

● 9-12 Free-Type Valve Rotator

Figure 9-22 shows one type of free valve rotator. The rocker arm pushes against the tip cup instead of the valve stem. The tip cup pushes against the valve retainer, thereby taking up the spring pressure. Further tip-cup movement brings it into contact with the end of the valve stem so the valve is opened. Note, however, that there is a moment when the valve is free. Engine vibration causes the valve to rotate.

● 9-13 Positive-Type Valve Rotator

The positive-type valve rotator turns the valve, not by engine vibration, but by a positive push. Figure 9-23 shows how the positive valve rotator fits into the valve train. It takes the place of the spring retainer. The valve rotator applies a rotating force on the valve stem each time the valve opens. A seating collar is spun over the outer lip of the spring retainer. The valve spring rests on the seating collar. The collar encloses a flexible washer placed below a series of spring-loaded balls. The tops of the grooves (races) are inclined. When the lifter is raised, the rocker arm lifts the valve and applies increasing pressure on the seating collar. This flattens the flexible washer so that the washer applies the spring load on the balls. As the balls receive this load, they roll up the inclined races. This causes the retainer to turn a few degrees and thus turn the valve a few degrees. When the valve closes, the spring pressure is reduced. The balls return to their original positions, ready for the next valve motion.

● 9-14 Valve Lifters

There are two types of valve lifters (also called *valve tappets*), the solid or mechanical lifter and the hydraulic

Fig. 9-20 Phantom view of the Cosworth Vega overhead-camshaft engine. (*Chevrolet Motor Division of General Motors Corporation*)

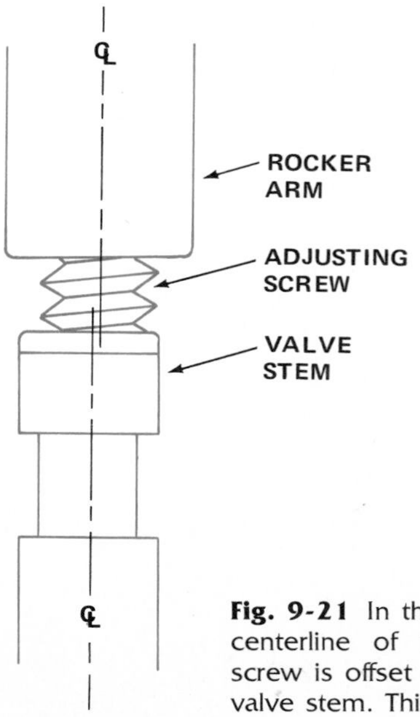

Fig. 9-21 In the Volkswagen engine, the centerline of the rocker-arm adjusting screw is offset from the centerline of the valve stem. This promotes valve rotation.

lifter (see ● 9-15). The solid lifter is just a cylinder placed between the cam (on the camshaft) and the push rod. The valve lifter is rotated in much the same way the valve is rotated by the rocker arm. That is, it is offset slightly from the center of the cam. This rotation of the valve lifter prevents sludge formation from accumulating in the lifter bore in the cylinder block. At the same time, it rotates the push rod, keeping the push-rod bearing surfaces with the lifter and rocker arm clean. A rough or worn cam lobe can often be pinpointed by noting which push rod is not turning, or not turning at the same speed as the other push rods. This rotation distributes the wear from the cam over the face of the lifter. The hydraulic valve lifter has an internal construction that reduces noise and valve-train clearance.

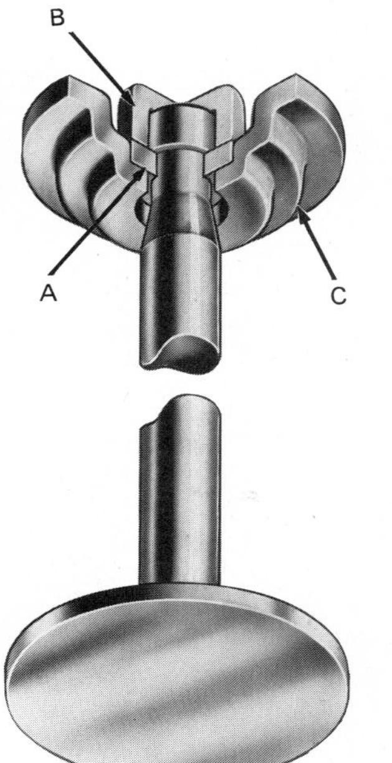

Fig. 9-22 Construction of a "free-type" valve rotator; (*A*) spring-retainer lock; (*B*) tip cup; (*C*) spring retainer. (*TRW Inc.*)

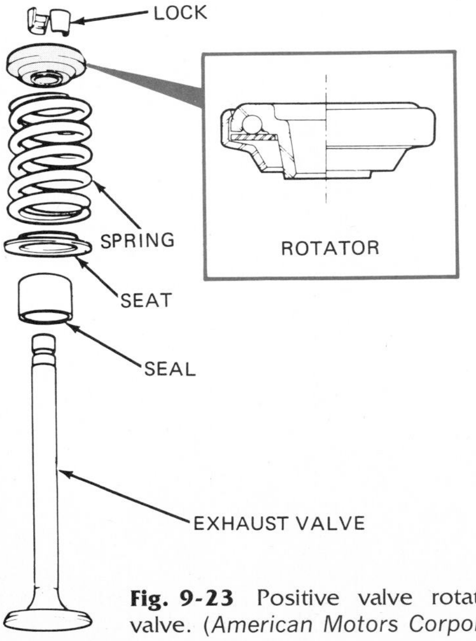

Fig. 9-23 Positive valve rotator on exhaust valve. (*American Motors Corporation*)

● 9-15 Hydraulic Valve Lifter

This type of lifter is used in many engines. It is very quiet because it assures zero tappet clearance (or valve lash). Also, it usually requires no adjustment in normal service. Variations due to temperature changes or to wear are taken care of hydraulically.

Figure 9-24 shows the details of a hydraulic valve lifter used in a V-8 I-head engine. Oil is fed into the valve lifter from the oil pump, through an oil gallery that runs the length of the engine. When the valve is closed, oil from the pump is forced into the valve lifter through oil holes in the lifter body and plunger. The oil forces the ball-check valve in the plunger to open. Oil then passes the ball-check valve and enters the space under the plunger. The plunger is forced upward until it touches the valve push rod (or valve stem in L-head engines). This takes up any clearance in the system.

Now, when the cam lobe moves around under the lifter body, the lifter is raised. Since there is no clearance, there is no tappet noise. The raising of the lifter and the opening of the valve suddenly increases the pressure in the body chamber under the plunger. This causes the ball-check valve to close. Oil is therefore trapped in the chamber. Because liquids such as oil are not compressible, the lifter acts like a simple one-piece lifter. It moves up as an assembly and causes the valve to open. Then, when the valve closes, the lifter moves down, and the pressure on the plunger is relieved. If any oil has been lost from the chamber under the plunger, oil from the engine oil pump causes the ball-check valve to open. Engine oil can then refill the chamber, as noted above.

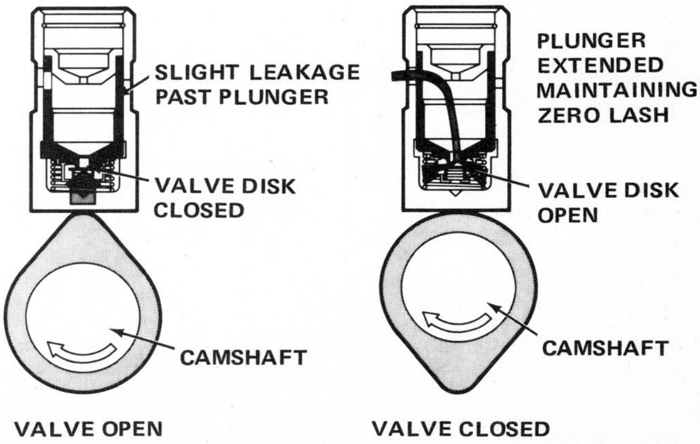

Fig. 9-24 Hydraulic valve lifter with valve open and valve closed. (*Ford Motor Company*)

● 9-16 Valve Timing

Some people assume that the intake and exhaust valves open and close at TDC and BDC. They are not timed in this manner. Figure 9-25 shows a typical valve-timing diagram. Note that, in this diagram, the exhaust valve starts to open at 47 degrees before BDC on the power stroke. It stays open until 21 degrees after TDC on the intake stroke. This gives more time for the exhaust gases to leave the cylinder. By the time the piston reaches 47 degrees before BDC on the power stroke, the combustion pressures have dropped considerably. Little power is lost by giving the exhaust gases this extra time to leave the cylinder.

In a similar manner, the intake valve remains open for 56 degrees past BDC after the intake stroke. This gives additional time for air-fuel mixture to flow into the cylinder. The delivery of adequate amounts of air-fuel mixture to the engine cylinders is a critical item in engine operation. Actually, the cylinders are never quite "filled up" when the intake valve closes.

Timing of the valves is controlled by the shape of the lobe on the cam and the relationship between the gears or sprockets and chain on the camshaft and crankshaft. Changing the relationship between the driving and driven gears or sprockets changes the timing at which the valves open and close. For example, suppose the timing chain is worn and this allows the chain to "jump time." That is, it slips a tooth and this causes the camshaft to fall behind that one tooth. The valves would then open and close earlier. Let us say the valve action has been moved ahead 15 degrees. Then the exhaust valve would open at 62 degrees before BDC on the power stroke. It would close at only 6 degrees before TDC on the exhaust stroke (in the example shown in Fig. 9-25). The intake-valve actions would likewise be moved ahead. These valve-action advances would seriously reduce engine performance and would cause engine overheating. The gears or sprockets are marked so that they can be properly aligned on the original assembly (see Figs. 9-2 to 9-4).

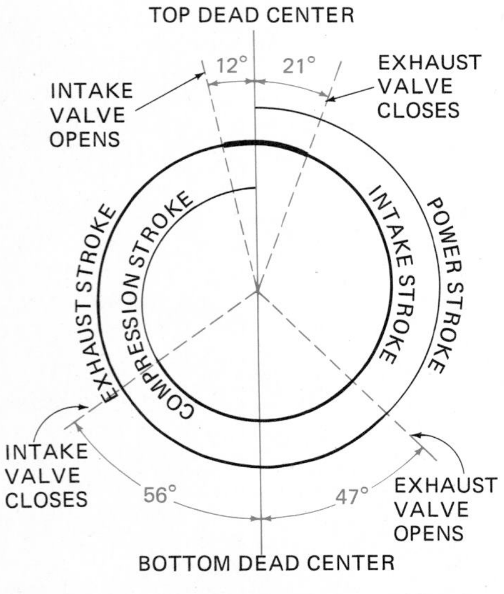

Fig. 9-25 Intake- and exhaust-valve timing. The complete cycle of events is shown as a 720-degree spiral, which represents two complete crankshaft revolutions. Timing of valves differs for different engines.

REVIEW QUESTIONS

1. What are the three types of valve trains?
2. How many cams are there on the camshaft for a four-cylinder engine?
3. What are the two types of camshaft drives?
4. How many times does the crankshaft turn while the camshaft turns once?
5. Which is larger, the intake valve or the exhaust valve? Why?
6. What is the interference angle in the valve and valve face? What is its purpose?
7. Which valve runs hotter, the intake valve or the exhaust valve?
8. What is a sodium-cooled valve? How does it work?
9. What are valve-seat inserts? What are the problems involved in using them?
10. How is the ball-pivot rocker arm adjusted?
11. How is the spring attached to the end of the valve stem?
12. Explain the advantages of the overhead-camshaft engine.
13. What do SOHC and DOHC mean?
14. Why is valve rotation important?
15. What has been done in recent years with valves and valve seats to make up for the loss of lead in gasoline?
16. What are the two types of valve lifters?
17. Explain how the hydraulic valve lifter works.

SELF PROJECT

The valve train is only as good as the camshaft that operates it. So examine camshafts. Count the number of lobes. If there is one more than the engine has valves, check the contours until you locate the eccentric that operates the fuel pump. Take a sheet of notepaper and, starting at the top, write down the make and model of engine from which the camshaft came. Then start at the front of the camshaft and identify each cam. Most cams will start with cylinder number 1 exhaust valve. So write down "1E" to start the column under the line where you wrote in the engine make and model.

You may be in for a surprise or two. For example, if you are checking the camshaft from a Chevrolet 327 V-8, you will find that the next cam is for the exhaust valve for cylinder number 2. (Remember what we said about cylinder numbering in ● 6-13.) The cams interlace like the fingers on your two hands when you interlock your fingers. Continue your examination until you have identified all the cams and have completed the column. Prepare additional cam sheets until you can look at an engine and camshaft and identify all the cams. File the cam sheets in your notebook.

CHAPTER 10
Engine Measurements and Performance

After reading this chapter, you should be able to:

- **Discuss work, energy, power, and torque, and explain how they are measured and how they are related to engines.**
- **Discuss friction and name and explain the three kinds of friction.**
- **Explain what such engine measurements as bore, stroke, piston displacement, compression ratio, and volumetric efficiency mean.**

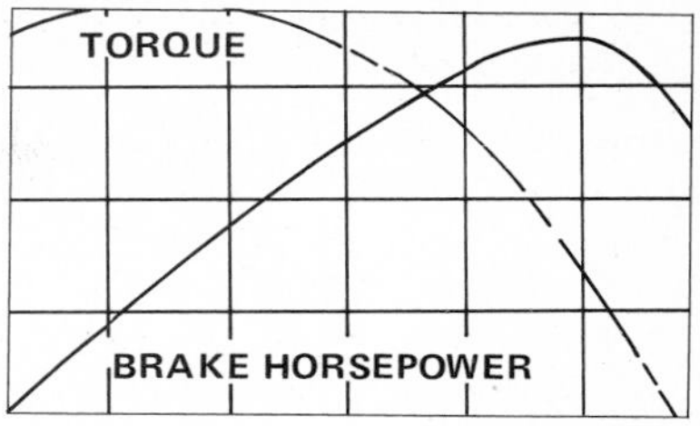

This chapter describes the ways in which engines and engine performance are measured. These include physical measurements such as cylinder diameter and length of piston stroke, and performance measurements such as torque and horsepower.

10-1 Work

Work is the moving of an object against an opposing force. The object is moved by a push, a pull, or a lift. For example, when a weight is lifted, it is moved upward against the pull of gravity. Work is done on the weight. Also, when a coil spring is compressed, work is done on the spring (Fig. 10-1).

Work is measured in terms of distance and force. If a 5-pound weight is lifted off the ground 5 feet, the work done on the weight is 25 foot-pounds (ft-lb), or 5 feet times 5 pounds (Fig. 10-2). **Distance times force equals work.**

In the metric system, work is measured in meter-kilograms (mkg). Thus, one example would be: Lifting a 5-kilogram (kg) weight [11 lb] 1 meter [3.28 ft] requires 5 mkg of work [36.08 ft-lb]. If the 5-kg weight is lifted 2 m, the work done is 10 mkg [72.16 ft-lb].

10-2 Energy

Energy is the ability, or capacity, to do work. When work is done on an object, energy is stored in that object. Lift a 20-pound [9.072 kg] weight 4 feet [1.219 m] and you have stored up energy in the weight. The weight can do 80 ft-lb [11.056 mkg] of work. If a spring is compressed, energy is stored in it, and it can do work (Fig. 10-3).

10-3 Power

Work can be done slowly, or it can be done rapidly. The rate at which work is done is measured in terms of power. A machine that can do a great deal of work in a short time is called a *high-powered* machine: **Power is the rate, or speed, at which work is done.**

10-4 Torque

Torque is twisting, or turning, effort. You apply torque to the top of a screw-top jar when you loosen it (Fig. 10-4). You apply torque to the steering wheel when you take a car around a turn. The engine applies torque to the wheels to make them rotate.

Torque, however, must not be confused with power. Torque is turning effort which *may or may not result in motion*. Power is something else again. It is the rate at

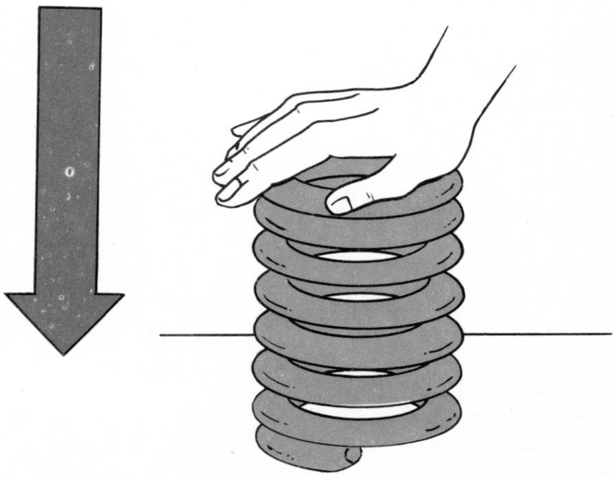

Fig. 10-1 When a spring is compressed, work is done on the spring and energy is stored in it.

which work is being done, and this means that something must be moving.

Torque is measured in pound-feet (or lb-ft, not to be confused with ft-lb of work). For example, suppose you push on a crank with a 20-pound push, and the crank is $1\frac{1}{2}$ feet long. You would be applying 30 pound-feet of torque to the crank (Fig. 10-5). You would be applying this torque regardless of whether or not the crank was turning. The torque is there as long as you continue to apply the 20-pound push to the crank handle.

Note that in the metric system, torque is measured in kilogram-meters (kgm), and not pound-feet (lb-ft).

● 10-5 Horsepower

A horsepower (hp) is the power of one horse, or a measure of the rate at which a horse can work. A 10-hp engine, for example, can do the work of 10 horses.

A horsepower is 33,000 ft-lb of work per minute. In Fig. 10-6, the horse walks 165 feet in 1 minute, lifting the 200-pound weight. The amount of work involved is 33,000 ft-lb (165 feet × 200 pounds). The time is 1 minute. If the horse did this work in 2 minutes, then it would be only "half" working; it would be putting out only $\frac{1}{2}$ hp. One formula for horsepower is

$$\text{hp} = \frac{\text{ft-lb per minute}}{33{,}000} = \frac{L \times W}{33{,}000 \times t}$$

where hp = horsepower
L = length, in feet, through which W is exerted
W = force, in pounds, exerted through distance L
t = time, in minutes, required to move W through L

In the metric system, power output from an engine is often measured in kilowatts (kW). That is, the power output is the amount of electricity the engine could produce if it were used to drive an electric generator. 1.34 hp is equal to 1 kW, and 1 hp is equal to 0.746 kW. Thus, a 200-hp engine is equal to a 149-kW engine.

A second formula for horsepower, used more often today, is:

$$\text{hp} = \frac{\text{torque} \times \text{rpm}}{5{,}252.1}$$

It is more commonly used because modern dynamometers (described in ● 10-13) measure engine performance in rpm (revolutions per minute), torque, and horsepower. Thus, the second formula is easier to work with. Torque is defined in ● 10-4.

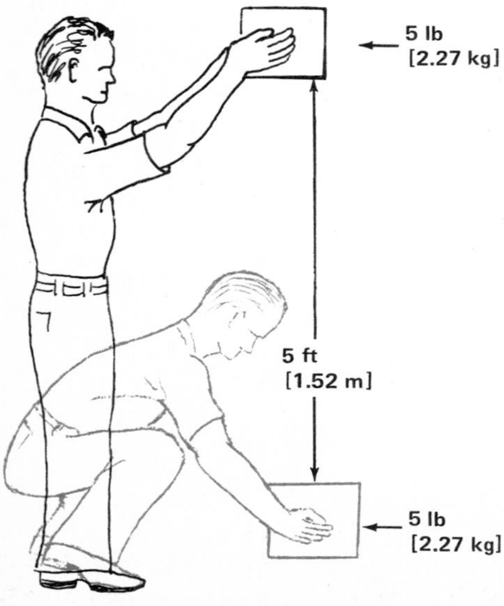

Fig. 10-2 When a weight is lifted, work has been done on it.

● 10-6 Inertia

Inertia is a property of all material objects. It causes them to resist any change of speed or direction of travel. A

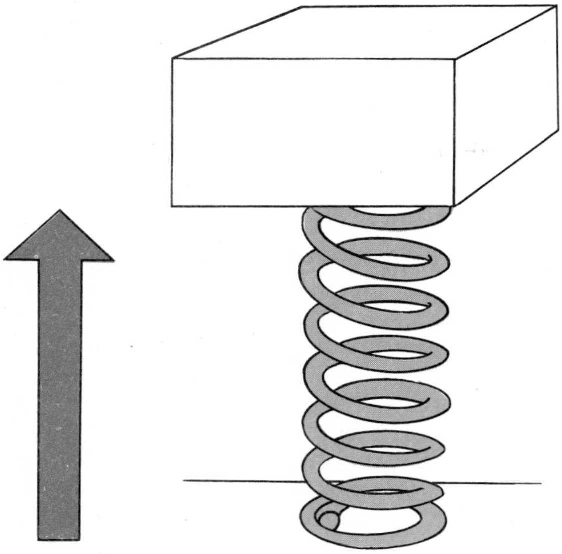

Fig. 10-3 When the spring is released, it can do work on another body. For example, it can lift a weight against gravity.

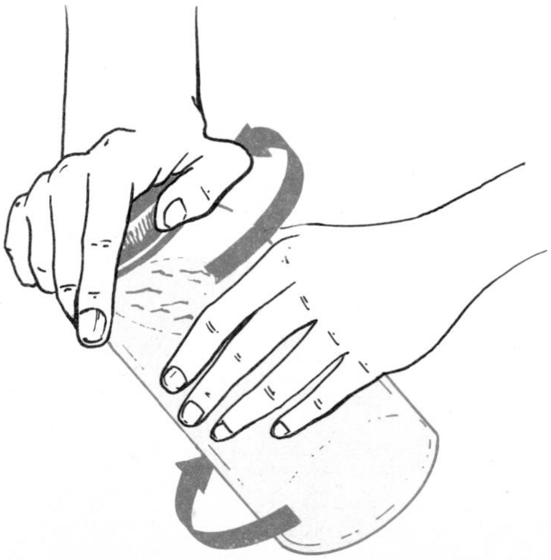

Fig. 10-4 Torque, or twisting effort, must be applied to loosen and remove the top from a screw-top jar.

motionless object tends to remain motionless. A moving object tends to keep moving at the same speed and in the same direction.

Consider the automobile. When it is standing still, its inertia must be overcome by applying power to make it move. To increase its speed, more power must be applied. To decrease its speed, the brakes must be applied. The brakes must overcome the car's inertia to slow it down. Also, when the car goes around a curve, its inertia tends to keep it moving in a straight line. The tires on the road must overcome this tendency, or else the inertia of the car will send it into a skid.

10-7 Friction

Friction is resistance to motion between two objects in contact with each other. If you put a book on a table and then pushed the book, you would find that it took a certain amount of push (Fig. 10-7). If you put a second book on top of the book, you would find that you had to push harder to move the two books on the table top (Fig. 10-8). Thus, friction, or resistance to motion, increases with the load. The higher the load, the greater the friction. There are three kinds of friction: dry, greasy, and viscous.

1. Dry friction This is the resistance to motion between two dry objects, for instance, a board being dragged across a floor.

2. Greasy friction This is the friction between two objects thinly coated with oil or grease. In an automobile engine, greasy friction may occur in an engine on first starting. Most of the lubricating oil may have drained away from the bearing surfaces and from the cylinder walls and piston rings. When the engine is started, only the small amount of oil remaining on these surfaces protects them from undue wear. Of course, the lubricating system quickly supplies additional oil. But before this happens, greasy friction exists on the moving surfaces. The lubrication between the surfaces where greasy friction exists is not enough to prevent wear. This is why automotive engineers say that initial starting and warm-up of the engine is hardest on the engine and wears it the most.

3. Viscous friction "Viscosity" is a term that refers to the tendency of liquids, such as oil, to resist flowing. A heavy oil is more viscous than a light oil and flows more slowly. (It has a higher viscosity, or higher resistance to flowing.) Viscous friction is the friction, or resistance to motion, between layers of liquid. In an oiled engine bearing, layers of oil adhere to the bearing and shaft surfaces. Layers of oil clinging to the shaft are carried around by the rotating shaft. They wedge between the shaft and the bearing (Fig. 10-9). The wedging action lifts the shaft so that the oil supports the weight, or load. Now, since the shaft is supported ("floats") on layers of oil, there is no metal-to-metal contact. However, the layers of oil must move over each other. Some energy is needed to make them do so. The resistance to motion between these oil layers is *viscous friction.*

4. Bushings and bearings In the engine, as in almost all machinery, the moving parts are lubricated with oil. The surfaces that move against each other are thus protected

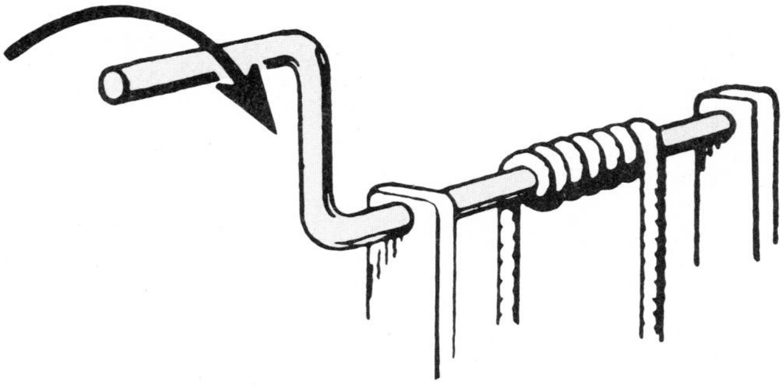

Fig. 10-5 Torque is measured in pound-feet (lb-ft) [kilogram-meters or kgm]. It is calculated by multiplying the push by the crank offset, or the distance of the push from the rotating shaft.

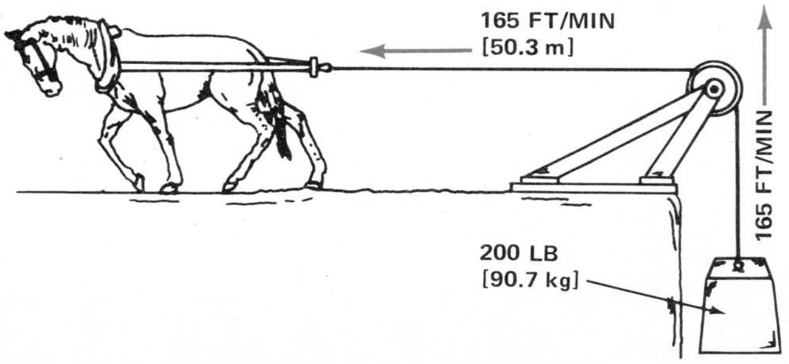

Fig. 10-6 One horse can do 33,000 foot-pounds of work a minute.

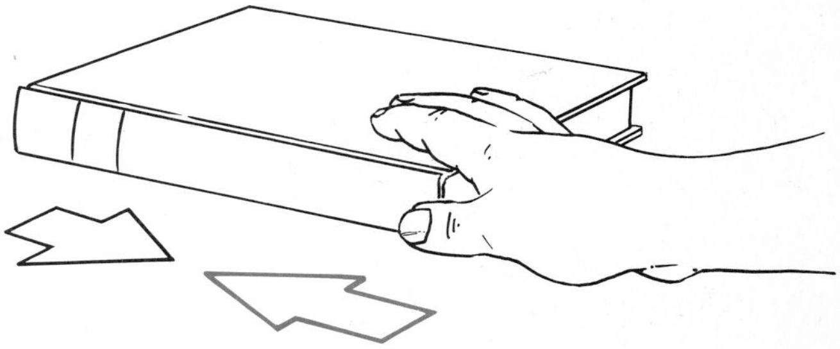

Fig. 10-7 Friction resists the push on the book.

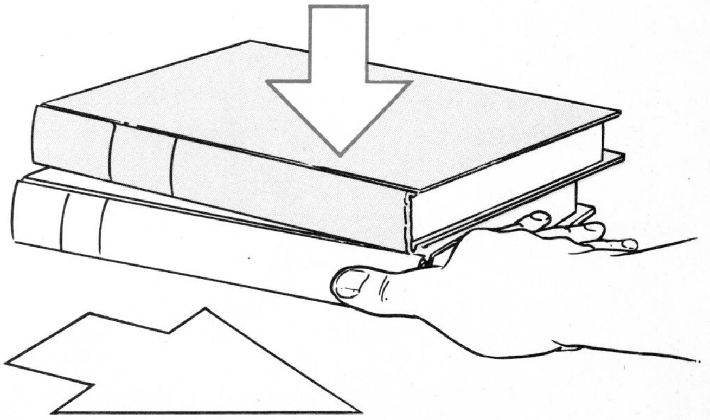

Fig. 10-8 Increasing the weight, or load, increases the friction.

against dry friction. These surfaces are of special materials, specially prepared. The cylinder walls, for example, against which the pistons and piston rings slide, are of smooth gray iron or other metal with good wearing qualities. The cylinder walls in some small engines are chrome-plated to improve their resistance to wear. The piston rings are also made of material that gives long life. Shafts are supported by bushings or bearings. Three types of bearing surfaces found in engines are shown in Fig. 10-10.

● 10-8 Bore and Stroke

The size of an engine cylinder is given by its bore and stroke. The *bore* is the diameter of the cylinder. The *stroke* is the distance the piston travels from BDC (bottom dead center) to TDC (top dead center). (See Fig. 10-11.) The bore is always mentioned first. For example, in a 4- by $3\frac{1}{2}$-inch cylinder, the diameter, or bore, is 4 inches, and the stroke is $3\frac{1}{2}$ inches. These measurements are used to figure the piston displacement.

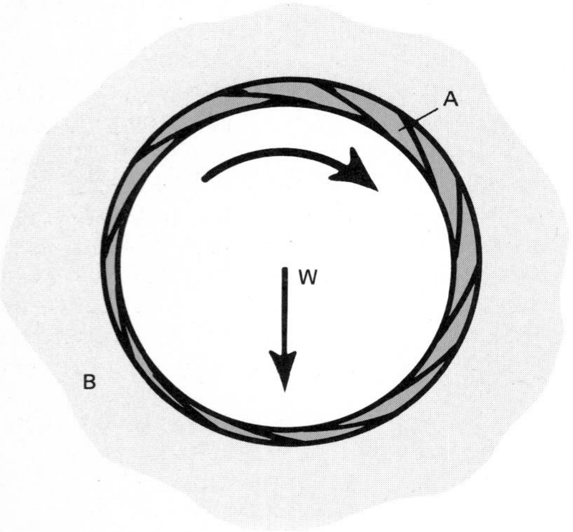

Fig. 10-9 Shaft rotation causes layers of clinging oil to be dragged around with it. The oil moves from the wide clearance *A* and is wedged into the narrow clearance *B*, thereby supporting the shaft weight *W* on an oil film. The clearances are greatly exaggerated in the illustration.

Before about 1955, most engines were built with a long stroke and smaller bore, like, for instance, a 3 by 4 engine. More recently, engines have been designed with a shorter stroke and large bore. For example, one recent 350-hp Chevrolet engine has a 4-inch bore and a $3\frac{1}{4}$-inch stroke. Such engines are called "oversquare." A "square" engine has a bore and stroke of equal lengths.

There are several reasons for the swing to the oversquare engine. With the shorter piston stroke, there is less friction loss (● 10-15) and shorter piston-ring travel (which means less wear). Also, the shorter stroke reduces the loads on the engine bearings. In addition, the shorter stroke permits a reduction of engine height and, thus, a lower hood line.

In the metric system, bore and stroke are given in millimeters (mm). Thus, a 4- by 3.5-inch cylinder would be a 101.6- by 88.9-mm cylinder.

● 10-9 Piston Displacement

Piston displacement is the volume that the piston displaces, or "sweeps out," as it moves from BDC to TDC. The piston displacement of a 4- by $3\frac{1}{2}$-inch cylinder, for example, is the volume of a cylinder 4 inches in diameter and $3\frac{1}{2}$ inches long. That is,

$$\frac{\pi \times D^2 \times L}{4} = \frac{3.1416 \times 4^2 \times 3\frac{1}{2}}{4}$$

$$= \frac{3.1416 \times 16 \times 3\frac{1}{2}}{4} = 43.98 \text{ in.}^3$$

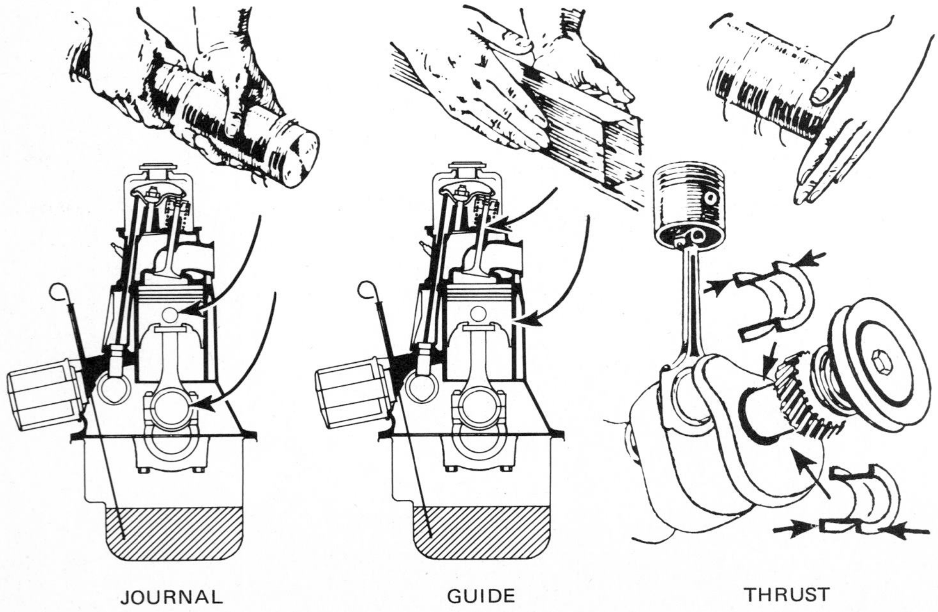

Fig. 10-10 Three types of friction-bearing surfaces in an automobile engine.

If the engine has eight cylinders, the total displacement is 43.98 times 8, or 351.84 cubic inches.

In the metric system, displacement is given in cubic centimeters (cc). Thus, a 200-cubic-inch displacement would be 3,280 cc in metric measurements. And, since 1,000 cc equals 1 liter, 3,280 cc is 3.28 liter. One cubic inch (in^3) equals 16.39 cubic centimeters (cm^3).

In competitive racing, displacement limitations are set. Thus, at the Indianapolis 500—the "Indy 500"—the maximum allowable displacement for a recent race was set at 305.1 cubic inches for nonsupercharged engines. In many races, the displacement is given in terms of liters. Thus, the Indy-500 specification (305.1 cubic inches) is 5 liters. (1 liter is 61.02 cubic inches, so 305.1 divided by 61.02 is 5.)

The Wankel engine does not have pistons, so you cannot figure piston displacement on the Wankel. But you can figure the displacement the rotor produces as the volume in the combustion chamber goes from maximum to minimum (see Fig. 10-12). For example, suppose the volume is reduced 490 cc as it goes from maximum to minimum (Fig. 10-12). This is the displacement in one of the three chambers of the rotor. Instead of using the term "piston displacement," this figure is called "single-chamber capacity."

10-10 Compression Ratio

The compression ratio of an engine is a measure of how much the air-fuel mixture is compressed in an engine cylinder. It is calculated by dividing the air volume in one cylinder with the piston at BDC by the air volume with the piston at TDC (Fig. 10-13).

The air volume with the piston at TDC is called the *clearance volume.* It is the clearance that remains above the piston at TDC.

For example, the engine of one car has a cylinder volume of 42.35 cubic inches [694 cm^3] at BDC (*A* in Fig. 10-13). It has a clearance volume of 4.45 cubic inches [73 cm^3] (*B* in Fig. 10-13). The compression ratio, therefore, is 42.35 divided by 4.45 [694 ÷ 73], or 9.5/1 (that is, 9.5:1). In other words, during the compression stroke, the air-fuel mixture is compressed from a volume of 42.35 cubic inches [694 cm^3] to 4.45 cubic inches [73 cm^3], or to 1/9.5 of its original volume.

10-11 Increasing Compression Ratio

Up until recently, the compression ratios of automotive engines were gradually being increased, year by year. This increase offers several advantages. The power and economy of an engine increase as the compression ratio goes up (within limits). This does not require an increase in engine size or weight. An engine with a higher compression ratio "squeezes" the air-fuel mixture harder (compresses it more). This causes the air-fuel mixture to produce more power on the power stroke. Here is the reason: A higher compression ratio means a higher pres-

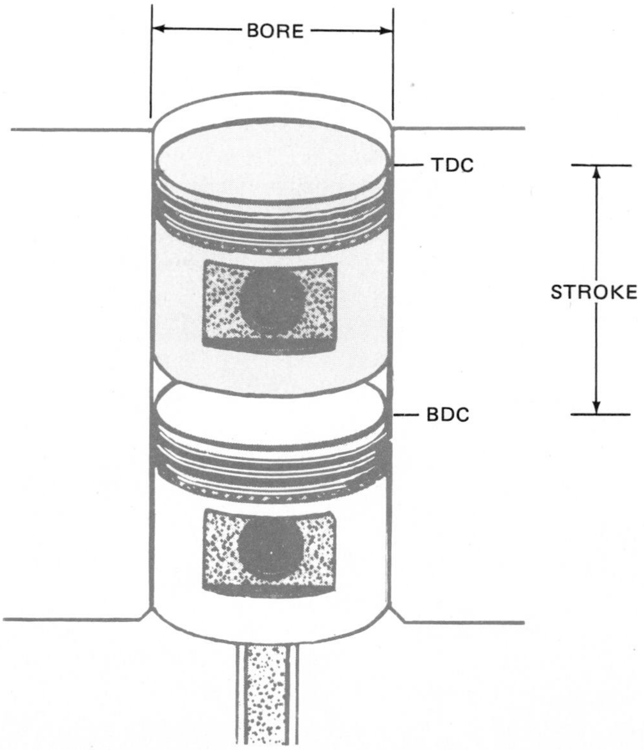

Fig. 10-11 Bore and stroke of an engine cylinder.

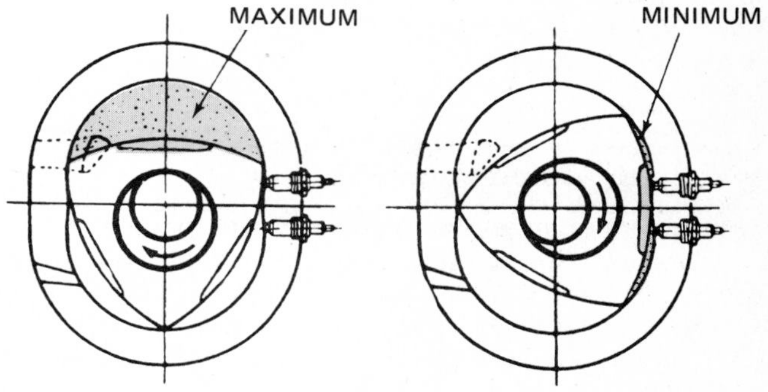

Fig. 10-12 Single-chamber capacity of a Wankel engine.

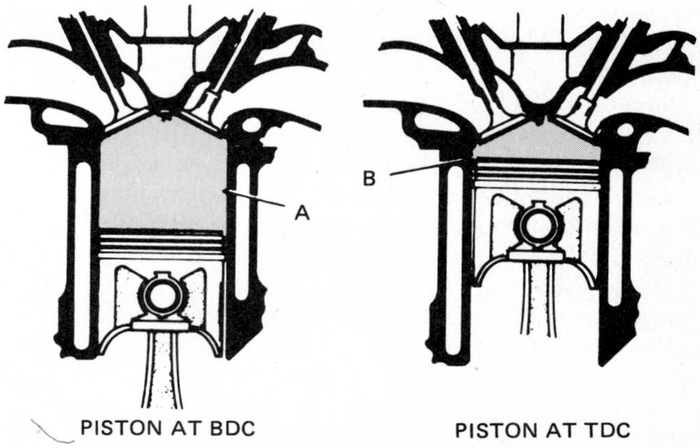

Fig. 10-13 Compression ratio is the volume in a cylinder with the piston at BDC divided by its volume with the piston at TDC, or *A* divided by *B*.

sure at the end of the compression stroke. This means higher combustion pressures during the power stroke; the piston is pushed harder. The burning gases also expand to a greater volume. It all adds up to this: There is more push on the piston for a larger part of the power stroke. More power is obtained from each power stroke.

Increasing the compression ratio does, however, make special problems. As the compression ratio goes up, detonation, or "knocking," becomes more of a problem. There is a detailed discussion of knocking in Chap. 11. In addition, higher compression ratios produce more pollutants in the exhaust gases. There is a detailed discussion of automotive emission controls in Chaps. 33 to 36.

Compression ratios have been reduced somewhat in late-model engines. The reason is this. The higher the compression, the higher the combustion temperatures. These high temperatures cause the formation of an atmospheric pollutant (NO_x, or nitrogen oxide). Federal law sets a limit on the amount of NO_x that is permitted in the exhaust gas. To reduce NO_x to this limit, compression ratios had to come down. Here is another reason the compression ratios had to come down. In recent years, all cars have been equipped with catalytic converters. These are devices which convert certain pollutants in the exhaust gases into harmless gases. (They are discussed in detail in a later chapter.) Catalytic converters are damaged by lead in gasoline. The lead reduces the tendency for the engine to knock. But the converters require lead-free gasoline. Removing the lead increases the tendency for the engine to knock. Therefore, the compression ratios have been reduced to combat this tendency. There is more on this in later chapters.

● 10-12 Volumetric Efficiency

The amount of air-fuel mixture taken into the cylinder on the intake stroke is a measure of the engine's volumetric efficiency. If the mixture were drawn into the cylinder very slowly, a full measure could get in. But the mixture must pass rapidly through narrow openings and bends in the carburetor and intake manifold. In addition, the mixture is heated (from engine heat); it therefore expands. The rapid movement and heating reduce the amount of mixture that can get into the cylinder. A full charge of air-fuel mixture cannot enter, because the time is too short and because the air becomes heated.

Volumetric efficiency is the ratio of the amount of air-fuel mixture that actually enters the cylinder to the amount that could possibly enter. For example, a certain cylinder has an air volume (*A* in Fig. 10-13) of 47 cubic inches [770 cm^3]. If the cylinder were allowed to completely "fill up," it would take in 0.034 ounce [0.964 g] of air. However, suppose that the engine is running at a high speed, so that only 0.027 ounce [0.765 g] of air can enter during each intake stroke. This means that the volumetric efficiency is only about 80 percent (0.027 is 80 percent of 0.034). Actually, 80 percent is a good volumetric efficiency for an engine running at fairly high speed. The volumetric efficiency of some engines may drop to as low as 50 percent at high speeds. This is another way of saying that the cylinders are only "half-filled" at high speeds.

This is one reason why engine speed and output cannot increase without limit. At higher speed, the engine has a harder time "breathing," or drawing in air. It is "starved" for air and cannot produce any further increase in power output.

To improve volumetric efficiency, intake valves can be made larger. In addition, the number of valves per cylinder can be increased. Also, valve lift can be increased. That is, the cam lobes on the cams can be made larger so the valve opens wider. However, when this is done, there is danger of the pistonhead striking the valve head. Unless the piston design takes this into account, serious engine damage could result (see ● 8-15).

Volumetric efficiency can also be increased by making the intake-manifold passages wider, and as straight and short as possible. Also, the smoothness of the inside surfaces of the intake manifolds is important. Rough surfaces slow down the flow of air-fuel mixture. Another way to improve volumetric efficiency is to use carburetors with extra circuits, or air passages (called "barrels"), which open at high speed to improve engine breathing. Chapter 13 describes carburetors in detail. All these changes help produce more power at higher speeds.

● 10-13 Brake Horsepower

The horsepower output of engines is measured in terms of *brake horsepower* (bhp). The name comes from the braking device that is used to hold engine speed down while horsepower is measured. When an engine is rated at 300 horsepower [223.8 kW], for example, it is really brake horsepower that is meant. This is the amount of power the engine can produce at a certain speed at wide-open throttle.

The usual way to rate an engine is with a *dynamometer* (Fig. 10-14). This device has a mechanism (an electric

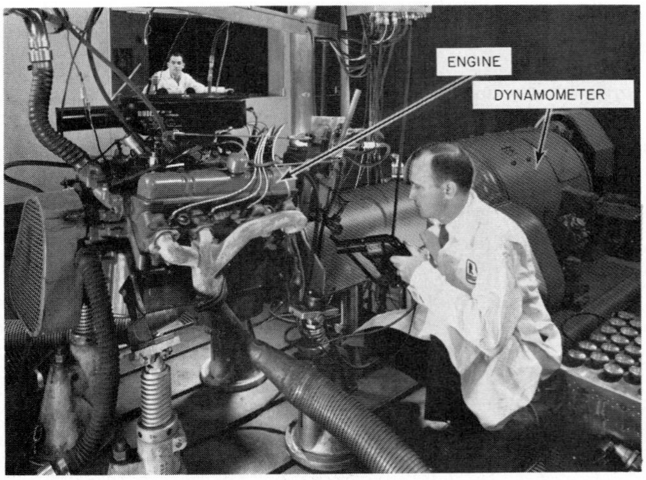

Fig. 10-14 Dynamometer used for testing engines and measuring their power output. (*General Motors Corporation*)

generator or a water brake) which can put different loads on the engine. Thus, the dynamometer can measure the amount of horsepower the engine can develop under various operating conditions.

Some dynamometers are used to test engines that have been removed from cars (Fig. 10-14). However, the dynamometer used in the service shop checks the engine *in* the car. This type of unit is called a *chassis dynamometer* (Figs. 10-15 and 10-16). On these, the rear wheels of the car are placed on rollers (front wheels on front-drive cars). Then, the engine drives the wheels, and the wheels drive the rollers. The rollers can be loaded varying amounts so that engine output can be measured. The use of the chassis dynamometer is becoming more common in the automotive servicing field. It can give a very quick report on engine conditions (by measuring output at various speeds and loads). This type of dynamometer is also used to test and adjust automatic transmissions right in the shop; no road testing is necessary.

10-14 Indicated Horsepower

Indicated horsepower (ihp) is the power that the engine develops inside the combustion chambers during the combustion process. A special device is required to measure ihp. It measures the pressures in the engine cylinders (Fig. 10-17). The four small drawings show the four piston strokes, and the curve shows the pressures in the cylinder during these four strokes. These pressures are used to figure ihp. Of course, ihp is well above bhp, because some of the power developed in the engine cylinders is used up to overcome friction.

10-15 Friction Horsepower

Friction horsepower (fhp) is the power required to overcome the friction of the moving parts in the engine. One of the major causes of friction loss (or fhp) is piston-ring friction. Under some conditions, the fricton of the rings moving on the cylinder walls accounts for 75 percent of all friction losses in the engine. This points up one advantage of the short-stroke, oversquare engine. With a short stroke, the piston rings do not have as far to move, and thus ring friction is lower. Figure 10-18 shows a curve of friction horsepower for one engine operating under certain specified conditions.

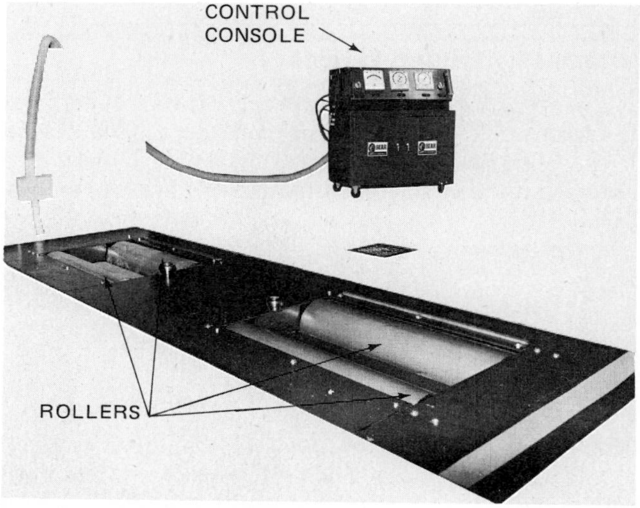

Fig. 10-15 Chassis dynamometer of the flush-floor type. The rollers are set at floor level. (*Charlottesville-Albemarle Technical Education Center*)

10-16 Relating bhp, ihp, and fhp

Brake horsepower is the power delivered, ihp is the power developed in the engine, and fhp is the power lost due to friction. The relationship among the three is:

$$\text{bhp} = \text{ihp} - \text{fhp}$$

That is, the horsepower delivered by the engine (bhp) is equal to the horsepower developed (ihp) minus the power lost due to friction (fhp).

10-17 Engine Torque

Torque is turning effort. When the piston is moving down on the power stroke, it applies torque to the engine crankshaft (through the connecting rod). The harder the push on the piston, the greater the torque applied. Thus, the higher the combustion pressures, the greater the amount of torque.

The dynamometer is normally used to check engine torque. Torque can be measured at the same time as horsepower on the dynamometer.

Fig. 10-16 Automobile in place on a chassis dynamometer. The rear wheels drive the dynamometer rollers, which are flush with the floor. Instruments on the test panel measure car speed, engine power output, engine vacuum, air-fuel ratio, and so on. (*Clayton Manufacturing Company*)

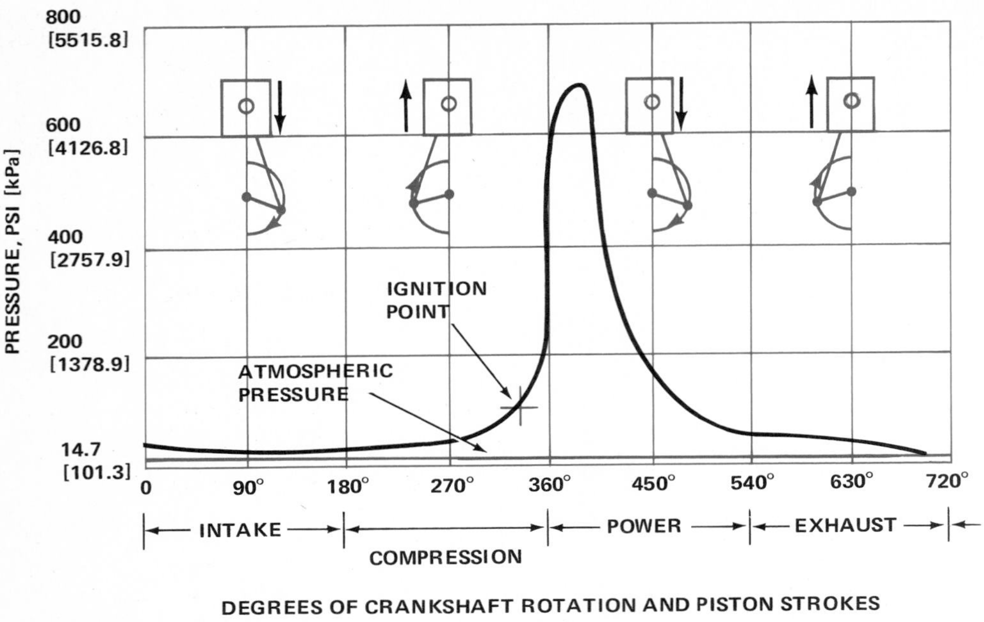

Fig. 10-17 Pressures in an engine cylinder during the four piston strokes. The four strokes require two crankshaft revolutions (360 degrees each), a total of 720 degrees of rotation. This curve is for a particular engine operating at one definite speed and throttle opening. Changing the speed and throttle opening would change the power curve for this engine.

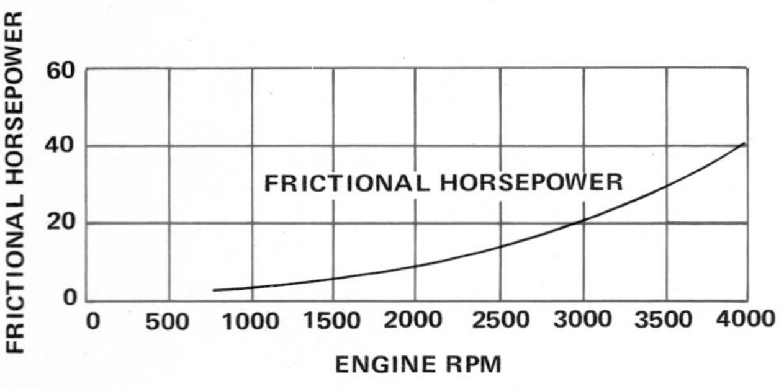

Fig. 10-18 Friction-horsepower curve, showing the relationship between fhp and engine speed.

● 10-18 Brake Horsepower Versus Torque

The torque that an engine can develop changes with gine speed (see Fig. 10-19). During intermediate speeds, volumetric efficiency is high (there is sufficient time for the cylinders to become fairly well "filled up"). This means that with a fairly full charge of air-fuel mixture, higher combustion pressures will develop. With higher combustion pressures, the engine torque is higher.

But, at higher speed, volumetric efficiency drops off (there is not enough time for the cylinders to become filled up with air-fuel mixture). Since there is less air-fuel mixture to burn, the combustion pressures do not go so high. There is less push on the pistons, and thus engine torque is lower. Note, in the graph (Fig. 10-19), how the torque drops off as engine speed increases.

The bhp curve of an engine is different from the torque curve. Figure 10-20 is the bhp of the same engine for which the torque curve is shown in Fig. 10-19. It starts low at low speed and increases until a high engine speed is reached. Then, at still higher engine speeds, bhp drops off.

The drop-off of bhp is due to reduced torque at higher speed and to increased fhp at the higher speed. Figure 10-21 compares the curves of torque, bhp, and fhp for an engine.

Note that the curves (Figs. 10-18 to 10-21) are for one particular engine only. Different engines have different torque, bhp, and fhp curves. Peaks may be at higher or lower speeds, and the relationships may not be as shown in the curves.

● 10-19 Engine Efficiency

The term "efficiency" compares the effort exerted and the results obtained. For engines, efficiency is the relation between the power delivered and the power that could be obtained if the engine operated without any power loss. Engine efficiency can be computed in two ways, as *mechanical* efficiency and as *thermal* efficiency.

1. Mechanical efficiency This is the relationship between bhp and ihp. It is

$$\text{Mechanical efficiency} = \frac{\text{bhp}}{\text{ihp}}$$

EXAMPLE At a certain speed, the bhp of an engine is 116, and its ihp is 135. Mechanical efficiency is thus bhp/ihp = 116/135 = 0.86, or 86 percent. That is, 86 percent of the power developed in the cylinders is delivered by the engine. The remaining 14 percent, or 19 hp [14.17 kW], is consumed as fhp.

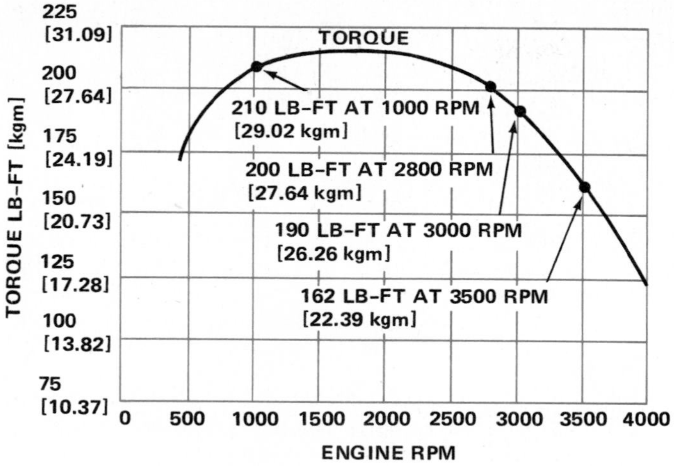

Fig. 10-19 Torque curve of an engine, showing the relationship between torque and speed.

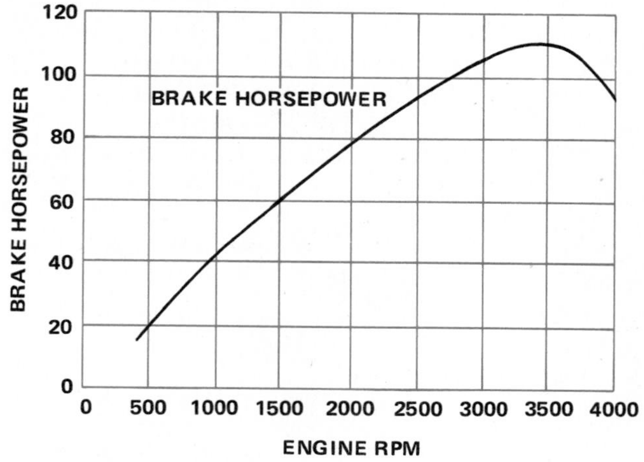

Fig. 10-20 Curve showing the relationship between bhp and engine speed.

2. Thermal efficiency "Thermal" means "of or related to heat." The thermal efficiency of an engine is the relation between the power output and the energy in the fuel burned to produce this output.

Some of the heat produced by combustion is carried away by the engine cooling system. Some of it is lost in the exhaust gases, which are hot when they leave the cylinder. These are heat (thermal) losses that reduce the thermal efficiency of the engine. They do not add to the power output of the engine. The remainder of the heat is used by the engine to develop power. Because a great deal of heat is lost during engine operation, thermal efficiencies may be as low as 20 percent. They are seldom higher than 25 percent.

10-20 Overall Efficiency

The gasoline enters the engine with a certain energy content, a certain ability to do work. At every step in the process, from the burning of the gasoline in the cylinders to the rotation of the car wheels, energy is lost. Figure 10-22 illustrates these losses for one engine and car during one test run. Note that as little as 15 percent of the energy in the gasoline remains to actually propel the car. This energy is used to overcome rolling resistance, air resistance, and power-train resistance; and to accelerate the car.

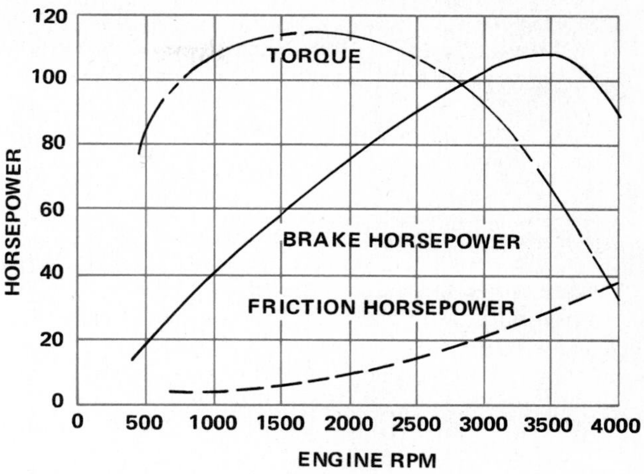

Fig. 10-21 Torque-bhp-fhp curves for an engine.

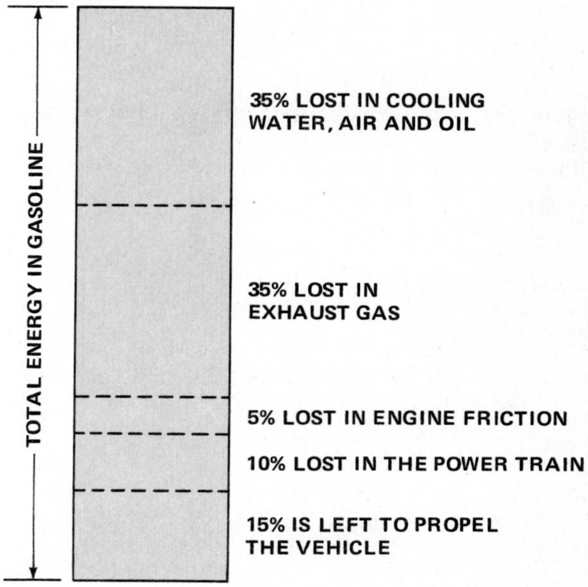

Fig. 10-22 Energy loss from cylinders to wheels.

1. Rolling resistance This results from irregularities in the road over which the wheels ride. It is also a result of the flexing of the tires as they turn under the car.

Rolling resistance is less with radial tires. See Chap. 57.

2. Air resistance Air resistance is the resistance of the air to the passage of the car body through it. As car speed increases, so does the air resistance. At 90 mph (miles per hour) [144.8 km/h], tests show that as much as 75 percent of engine power is used up in overcoming air resistance. Streamlining the car body reduces power loss from air resistance.

3. Acceleration Power is required to increase car speed. The power applied to accelerate the car overcomes the inertia of the car. Energy in the form of speed is stored in the car.

REVIEW QUESTIONS

1. Define work. In what terms is it measured?
2. Define energy.
3. Define power.
4. Define a horsepower. In what terms is horsepower measured in the United States Customary system? In the metric system?
5. An engine develops 225 lb-ft of torque at 2,000 rpm. What is the horsepower?
6. What are the three kinds of friction? Define each.
7. What are the bore and stroke of an engine? How are bore and stroke given in the metric system?
8. What is piston displacement? In what terms is the piston displacement given in the metric system? How many cubic inches displacement does a 6-liter engine have?
9. What does the term "single-chamber displacement" mean?
10. Define volumetric efficiency.
11. Explain what can be done to the valves and intake manifold to improve volumetric efficiency.
12. Will a two-barrel or a four-barrel carburetor give better volumetric efficiency? Why?
13. What is brake horsepower? How is it measured in the shop?
14. What is compression ratio? How can it be increased in an engine?
15. What is ihp?
16. What is fhp? What is the major cause of fhp in an engine?
17. What is the relationship among bhp, ihp, and fhp?
18. What is mechanical efficiency?
19. What is thermal efficiency?
20. What is inertia?

SELF PROJECT

At the top of a sheet of notebook paper write down "Finding Dynamometer Horsepower." Then write down the second formula for horsepower given in ● 10-5. This formula calls for only torque and rpm. Now look up these two figures in manufacturers' shop manuals. Then work out the horsepower ratings, using the formula. These are the ratings you would find if you tested the engines on the dynamometer and the engines were in good condition.

Note that the horsepower ratings based on torque and rpm are different from the horsepower ratings the engine manufacturers advertise. And the rpms are different, too.

Automotive-Engine Systems

3

Part Three of *Automotive Mechanics* describes the construction, operation, and servicing of automotive-engine systems. These include fuel systems (both carbureted and fuel-injected), lubrication systems, and cooling systems. Ignition systems are covered in Part Four, Automotive Electrical and Electronic Equipment. There are twelve chapters in Part Three.

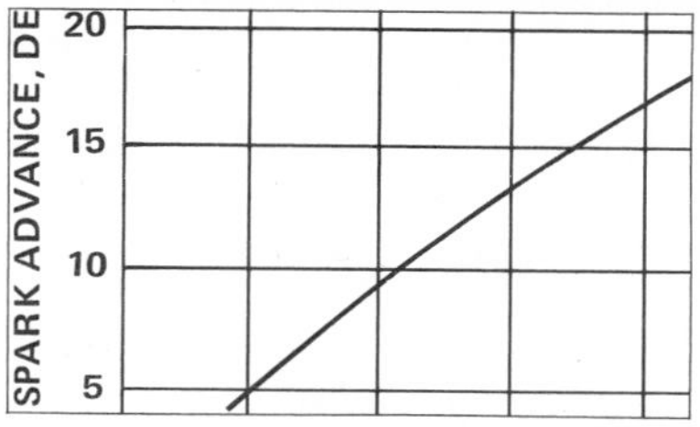

Chapter 11 Automotive-Engine Fuels

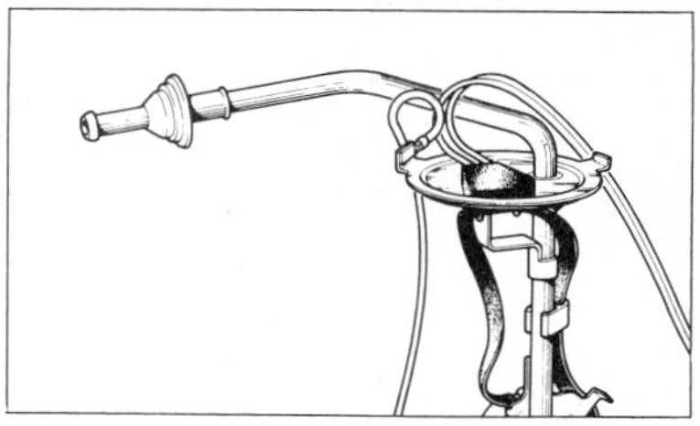

Chapter 12 Automotive Carbureted Fuel Systems

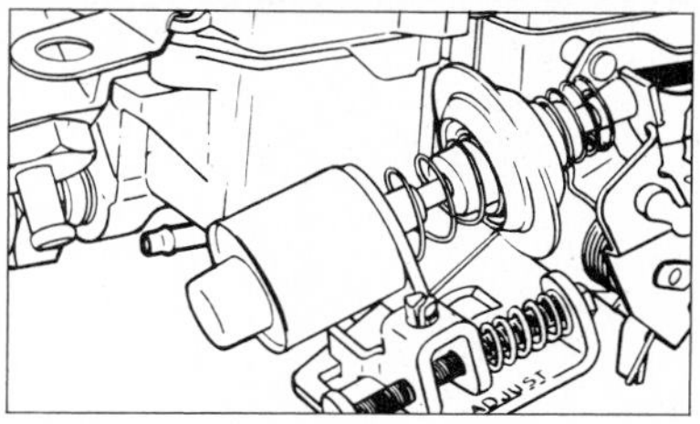

Chapter 13 Automotive Carburetors

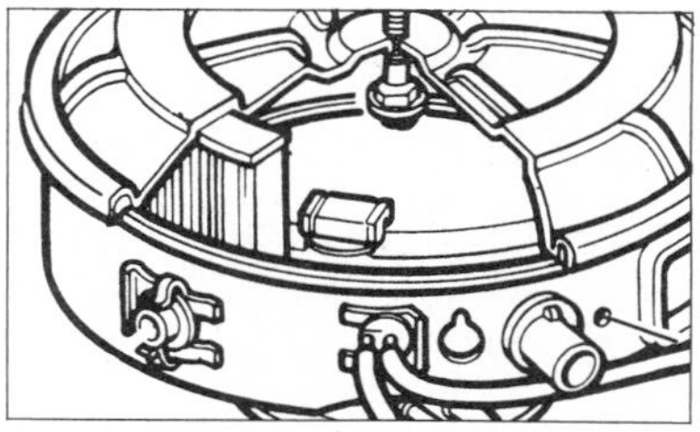

Chapter 14 Carbureted Fuel-System Service

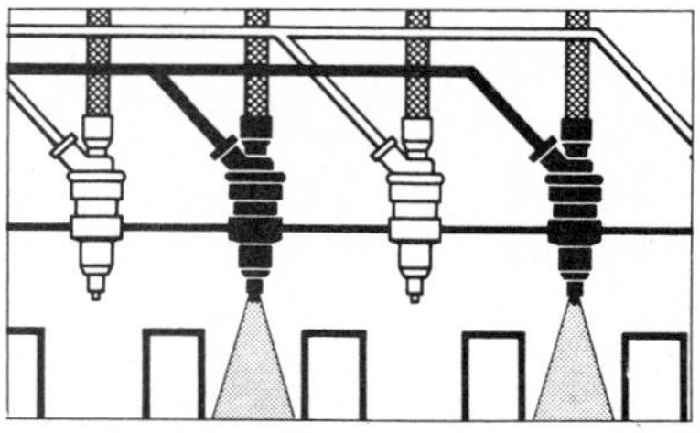

Chapter 15 Fuel-Injection System for Gasoline Engines

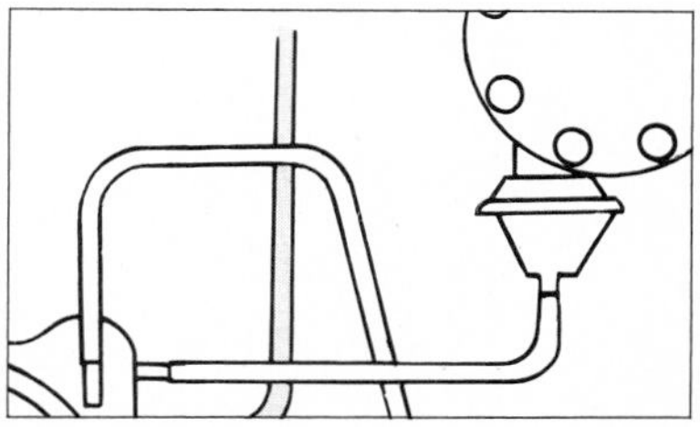

Chapter 16 Servicing Gasoline-Engine Fuel-Injection Systems

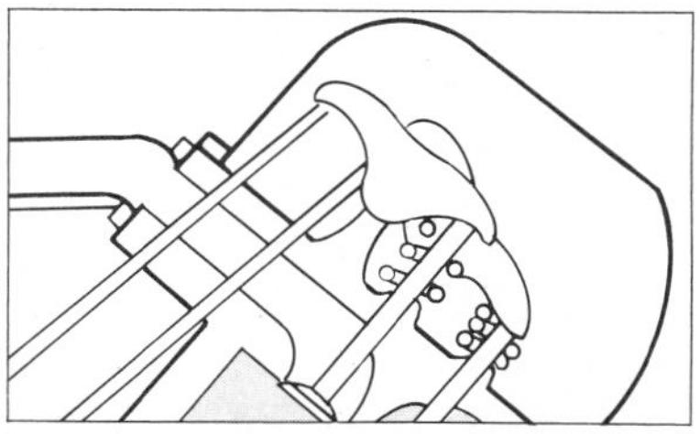

Chapter 17 Diesel-Engine Fuel-Injection Systems

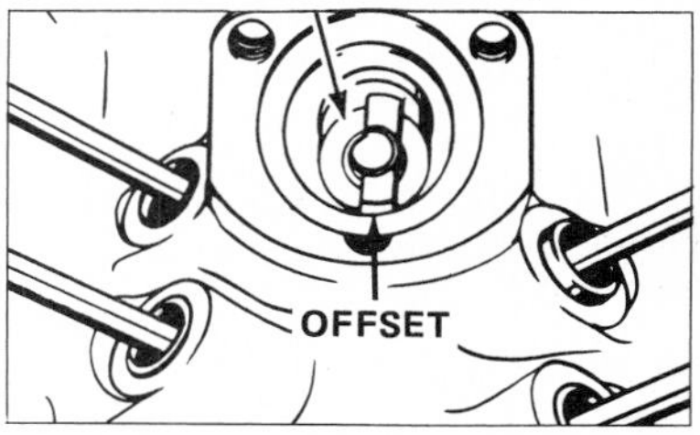

Chapter 18 Diesel-Engine Fuel-Injection System Service

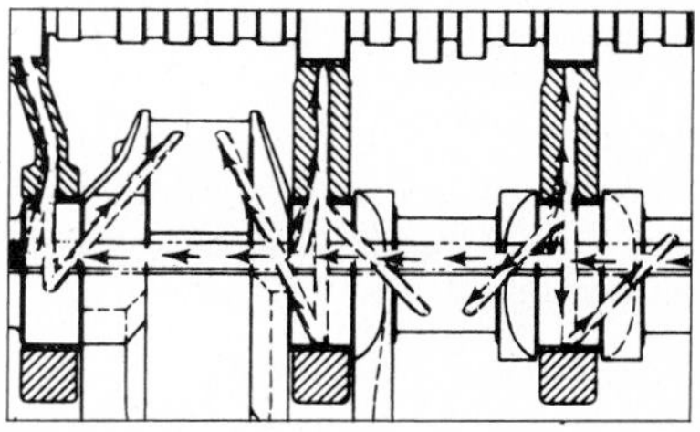

Chapter 19 Engine Lubricating System

Chapter 20 Lubricating-System Service

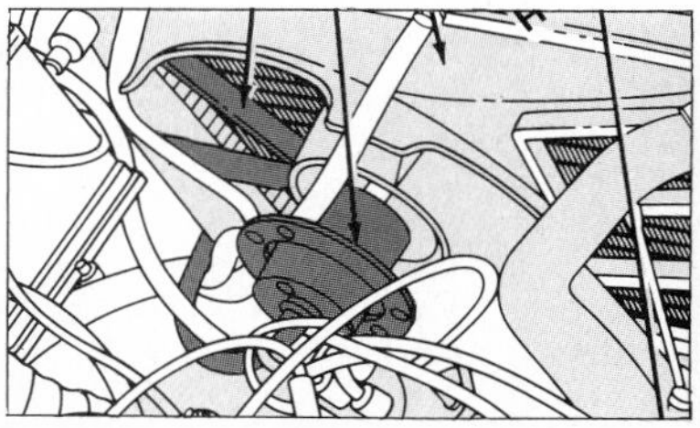

Chapter 21 Engine Cooling System

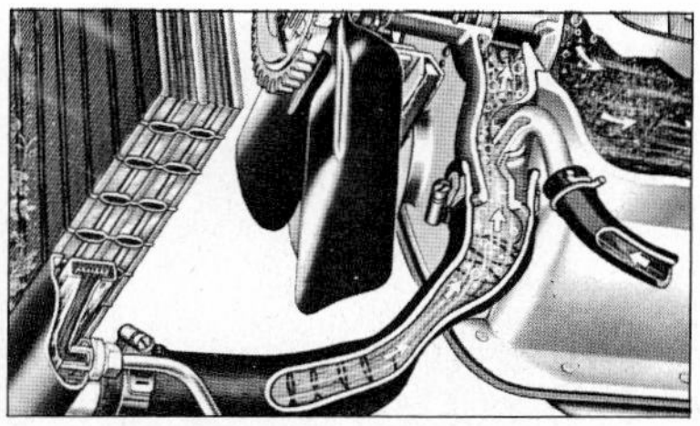

Chapter 22 Cooling-System Service

CHAPTER 11
Automotive-Engine Fuels

After reading this chapter, you should be able to:

- **Discuss the composition of and purpose of the additives used in gasoline.**
- **Explain volatility and octane ratings and why they are important in gasoline.**
- **Discuss knocking, detonation, and preignition, and explain what causes them and how they can be prevented.**
- **Discuss the composition and special characteristics of diesel-engine fuel.**

This chapter discusses the various kinds of fuel used in automotive-type engines: gasoline, LPG (liquefied petroleum gas), and diesel-engine fuel oil.

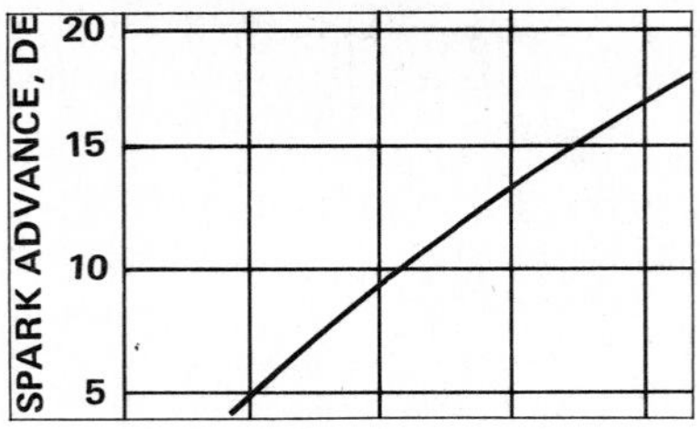

● 11-1 Gasoline

Gasoline is a hydrocarbon (abbreviated HC) made up largely of hydrogen and carbon compounds. These two elements unite readily with oxygen, a common element which makes up about 20 percent of our air. When hydrogen unites with oxygen, water (H_2O) is formed. When carbon unites with oxygen, carbon monoxide (CO) and carbon dioxide (CO_2) are formed. If all the gasoline burned completely in the engine, all that would come out would be water (H_2O) and carbon dioxide (CO_2). However, perfect combustion is not achieved in the engine. So some CO and HC are present in the exhaust gases. These two compounds, plus a third compound (nitrogen oxides or NO_x), are the pollutants emitted from automobiles. (There are several nitrogen oxides, NO, NO_2, NO_3, etc. The term "NO_x" includes all of these.) Part Five of the book explains in detail how control of these emissions is achieved with automotive emission controls.

CAUTION: Gasoline can be a very dangerous material if it is not handled properly and treated with respect. Vapor rising from an open pan of gasoline can become ignited from a spark, a flame, or a lighted cigarette. This could result in a drastic explosion and fire. A leaky fuel pump can become a killer if not treated properly. That is, a spark from turning on a trouble lamp, or from connecting a light across the battery, could ignite the vapor and cause a terrible fire. Reread ● 3-7, which discusses hazards of misusing gasoline.

Gasoline is often referred to as "gas," which can cause some confusion. The sort of gas you burn in a gas stove or use to heat a house is actually a *gas* that is delivered through gas lines or pipes. So there is gas that is a *gas*, and a "gas" that is gasoline.

● 11-2 Source of Gasoline

Gasoline is made from crude oil, or petroleum. No one knows exactly how crude oil was originally formed. It is found in "pools" or reservoirs in the ground. A well is drilled down into a reservoir. Underground pressure (or pressure artificially applied from above ground) forces the oil up and out of the well. The crude oil is then put through a refining process. From this process come gasoline, lubricating oil, greases, fuel oil, and many other products.

In the refining process, several compounds, called *additives*, are put into the gasoline. These additives give the gasoline the properties a good fuel should have. They are described in ● 11-17.

● 11-3 Volatility of Gasoline

Actually, gasoline is not a simple substance. It is a mixture of a number of different hydrocarbons. Each has its own characteristics. Aside from its combustibility, one of the important properties of gasoline is *volatility*.

Volatility refers to the ease with which a liquid vaporizes. The volatility of a simple compound like water is found by increasing its temperature until it boils, or vaporizes. A liquid that vaporizes at a low temperature has a high volatility; it is highly volatile. If its boiling point is high, its volatility is low. A heavy oil with a boiling point of 600°F [315°C] has a low volatility. Water has a high volatility. It boils at 212°F [100°C] at atmospheric pressure.

Gasoline is blended from different hydrocarbon compounds. Each has a different volatility, or boiling point which can vary from 122°F [50°C] to 437°F [225°C]. The proportions of high-volatility and low-volatility hydrocarbons in gasoline must be correct for engine operating requirements. These include:

1. Easy starting For easy starting with a cold engine, gasoline must vaporize readily at a low temperature. Thus, a percentage of the gasoline must be highly volatile. The percentage must be higher for the colder northern states than for the warmer south.

2. Freedom from vapor lock If the gasoline is too volatile, engine heat will cause it to vaporize in the fuel pump. This can cause vapor lock. Vapor lock prevents normal fuel delivery to the carburetor and would probably stall the engine. Thus, the percentage of highly volatile gasoline must be kept low to prevent vapor lock. The use of a vapor-return line to return vaporized fuel from the fuel pump to the fuel tank is discussed in ● 12-8.

3. Quick warm-up The speed with which the engine warms up depends in part on how much gasoline vaporizes immediately after the engine starts. Volatility for this purpose does not have to be quite so high as for easy starting. But, all the same, it must be fairly high.

4. Smooth acceleration When the throttle is opened for acceleration, there is a sudden increase in the amount of air passing through the throttle valve. At the same time, the accelerator pump delivers extra gasoline. If this gasoline does not vaporize quickly, there are a few seconds during which the air-fuel mixture is too lean. This causes the engine to hesitate, or stumble. Immediately after, as the gasoline begins to evaporate, the mixture becomes too rich. Here again, there is poor combustion and a tendency for the engine to hesitate. Enough of the gasoline must be highly volatile to assure adequate vaporization for smooth acceleration.

5. Good economy For good fuel economy, or maximum miles per gallon, the fuel must have high heat content, or energy, and low volatility. High overall volatility tends to reduce economy. It may produce an overrich mixture under many operating conditions. On the other hand, lower-volatility fuels tend to burn more efficiently. They have a higher heat content. However, the lower-volatility fuels increase starting difficulty. They reduce speed of warm-up, and do not give quite as good acceleration. Thus, only a limited percentage of the gasoline can be of low volatility.

6. Freedom from crankcase dilution Crankcase dilution results when part of the gasoline is not vaporized when it enters the engine cylinders. The liquid gasoline does not burn. It runs down the cylinder walls and enters the oil pan, where it dilutes the oil. This washes lubricating oil from the cylinder walls (thus increasing the wear of walls, rings, and pistons). Also, diluted oil is less able to lubricate other engine parts, such as the bearings. To avoid damage from crankcase dilution, the gasoline must be volatile enough so that little enters the cylinders in liquid form.

7. The volatility blend As you can see, no one volatility satisfies all engine operating requirements. The fuel must be of high volatility for easy starting and good acceleration. But it must also be of low volatility to give good fuel economy and combat vapor lock. Thus, gasoline is blended from different hydrocarbons having different volatilities. Such a blend satisfies the various operating requirements.

● 11-4 Antiknock Value

During normal combustion in the engine cylinder, the pressure increases evenly. Under some conditions, the last part of the compressed air-fuel mixture explodes, or detonates. This produces a sudden and sharp pressure increase that may cause a rapping or knocking noise. It sounds almost as though the pistonhead had been struck a hard hammer blow. Actually, the sudden pressure increase does put a sudden heavy load on the piston, almost like a hammer blow. This can be very damaging to the engine. It can wear moving parts rapidly and even cause parts to break. Also, some of the energy in the gasoline is wasted. The sudden pressure increase does not permit best utilization of the fuel energy.

Some types of gasoline produce much more knock than others. Because knocking is such an undesirable characteristic, gasoline producers try to reduce knocking tendencies. Certain chemicals have been found to reduce knocking when added to gasoline. The actual rating of the antiknock tendencies of a gasoline is given in terms of *octane number,* or ONR (octane number rating).

● 11-5 Heat of Compression

To understand why knocking occurs, remember what happens to air or any other gas when it is compressed. We noted (● 6-18) that the diesel engine compresses air to about one-twentieth of its original volume. This increases the air temperature to about 1000°F [537.8°C]. The temperature rise is called *heat of compression.* Let us see how heat of compression affects knocking.

● 11-6 Cause of Knocking or Detonation

Normally, the spark at the spark plug starts the fuel burning in the combustion chamber. A wall of flame spreads out in all directions from the spark (moving outward almost like a rubber balloon being blown up). The flame travels rapidly through the compressed mixture, until all the charge is burned. The speed with which the flame travels is called the *rate of flame propagation.* The movement of the flame wall during normal combus-

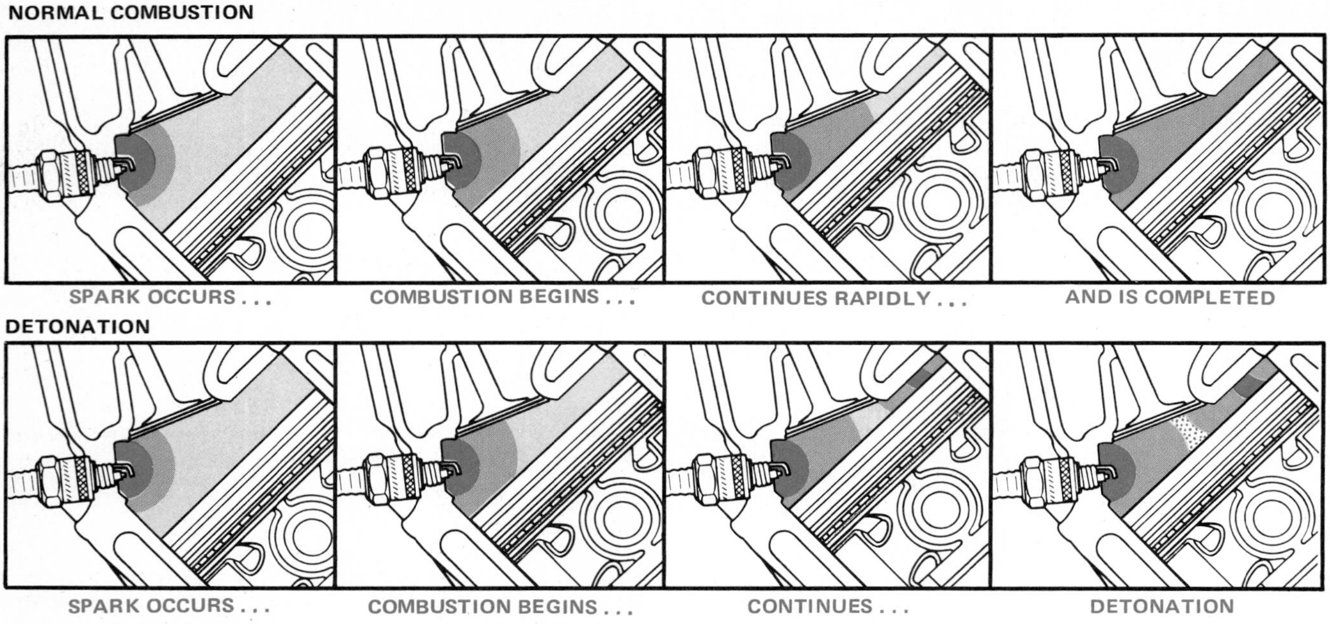
NORMAL COMBUSTION
SPARK OCCURS . . .
COMBUSTION BEGINS . . .
CONTINUES RAPIDLY . . .
AND IS COMPLETED
DETONATION
SPARK OCCURS . . .
COMBUSTION BEGINS . . .
CONTINUES . . .
DETONATION

Fig. 11-1 Normal combustion without detonation is shown in the top row. The fuel charge burns smoothly from beginning to end, providing an even, powerful thrust to the piston. Detonation is shown in the bottom row. The last part of the fuel explodes, or burns, almost at once, to produce detonation, or spark knock. (*Champion Spark Plugs*)

tion is shown in the row of pictures at the top in Fig. 11-1. During combustion, the pressure increases to several hundred psi (pounds per square inch). It may exceed 1,000 psi [70.3 kg/cm^2] in modern high-compression engines.

Under certain conditions, the last part of the compressed air-fuel mixture, or the end gas, explodes before the flame front reaches it (Fig. 11-1, at the bottom). Remember that the unburned mixture is subject to increasing pressure as the flame moves through the combustion chamber. This increases the temperature of the end gas (because of heat of compression and heat from the combustion process). If the temperature gets high enough, this end gas explodes before the flame front arrives. The effect is almost the same as if the pistonhead had been struck a heavy hammer blow. In fact, it sounds as though this had happened. The sudden shock load due to detonation of the end gas increases wear on bearings. It may actually break engine parts if the knocking is severe enough.

11-7 Compression Ratio Versus Knocking

As compression ratios of engines have gone up, so has the tendency for engines to knock. Here is the reason. With a higher compression ratio, the mixture, at TDC (top dead center), is more highly compressed. *It is thus at a higher initial temperature.* With higher initial pressure and temperature, the temperature at which detonation occurs is reached sooner. Thus, high-compression engines have a greater tendency to knock. However, special fuels which burn slower have been developed for higher-compression engines. These fuels have a greater resistance to being set off suddenly by heat of compression. They are less apt to explode suddenly. They depend, for ignition, only on the wall of flame traveling through the air-fuel mixture.

11-8 Measuring Antiknock Values

There are several methods of measuring the antiknock value of fuels. The rating is made in terms of octane number. As noted above, this rating is called the ONR (octane number rating). A high-octane gasoline is highly resistant to knock. A low-octane fuel knocks rather easily. The octane-testing procedure establishes the knock resistance of the fuel being tested at various car speeds (Fig. 11-2). Note that the curve shows the amount of spark advance the fuel will tolerate at each speed without knocking. As we explain in a later chapter on ignition systems, a spark advance is necessary to utilize the energy in the gasoline. The spark at the spark plug should occur before the piston reaches TDC on the compression stroke. This allows the combustion to start and pressure to build up before the piston passes TDC and starts down on the power stroke. If the spark is late, the piston will be moving down before the pressure has a chance to build up. The power stroke will be weak.

On the other hand, if the spark is advanced, or moved ahead too much, the pressure will reach a peak before the piston reaches TDC. This can cause a severe and possibly damaging knock. The ignition distributor is designed to provide the best spark advance for each engine speed. That is, the spark occurs at the spark plug at the proper time for each engine speed. At higher speeds, the spark is more advanced because things are happening faster in the engine. The piston is moving faster. So the spark must occur earlier to give the compressed mixture time to burn and build up pressure. Note that the pressure should approach its peak at the moment the piston reaches TDC. If it peaks too early, knock occurs. If it peaks too late, power is lost.

● 11-9 Detonation Versus Preignition

Detonation is a secondary explosion that occurs after the spark at the spark-plug gap. *Preignition* is ignition of the air-fuel mixture prior to the occurrence of the spark at the spark-plug gap. (See Fig. 11-3.)

Thus far, we have discussed the type of knocking that results from detonation, or sudden explosion, of the last part of the fuel charge. This type of knocking is usually regular in character. It is most noticeable when the engine is accelerated or is under heavy load, as when climbing a hill. Under these conditions, the throttle valve is nearly or fully wide-open. The engine is taking in a full air-fuel charge on every intake stroke. This means the compression pressures reached are maximal. Detonation

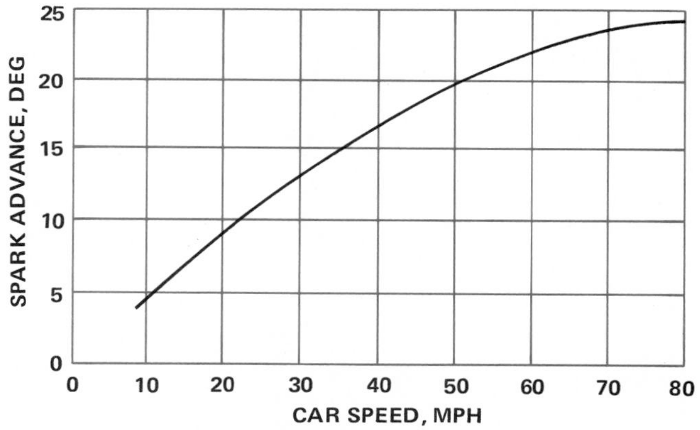

Fig. 11-2 Borderline knock curve. The fuel being tested will knock if the ignition spark is advanced to any value above the curve at any speed.

pressures are more likely to be reached after the mixture is ignited.

There are other types of abnormal combustion, however, including surface ignition, preignition, and rumble. Surface ignition can start at hot spots in the combustion chamber, that is, from a hot exhaust valve or spark plug, or from combustion-chamber deposits. In some cases the deposits break loose, so that particles float free and become hot enough to produce ignition. Surface ignition can occur before (preignition) or after the spark occurs at the spark plug. It can cause engine rumble and rough operation, or mild to severe knocking. In some cases, the hot spots act as substitutes for the spark plugs. Then the engine continues to run even after the ignition switch is turned off. This can cause serious engine damage.

Preignition, surface ignition, and rumble are usually service problems. They result from inadequate servicing of the engine and from the installation of the wrong spark plugs (which run too hot). They may also result from the use of incorrect fuels and lubricating oils. (See Fig. 11-3.)

With incorrect fuel or oil, engine deposits may occur, and lead to surface ignition and rumble. Engine deposits also increase the compression ratio, so the engine tends to knock more.

● 11-10 Chemical Control of Knocking

Several chemicals, when added to gasoline, tend to prevent detonation of the end gas during combustion. One theory is that the chemical increases the reaction time of the fuel. That is, it increases the time required for the end gas to explode. This increased time gives the flame front time to reach the end gas. The result is that it enters into the normal combustion process instead of exploding. One of the compounds most successful in preventing knocking is tetraethyl lead, commonly called *ethyl* or *tel*. A small amount added to gasoline raises the ONR of the gasoline. See ● 11-12.

Special scavengers are also added. These prevent the combustion products of the lead from depositing in the combustion chambers (on plugs, valves, walls, and pistons). These compounds (ethylene dibromide and ethylene dichloride, for example) change the lead into compounds which vaporize and exit with the exhaust gases.

PREIGNITION

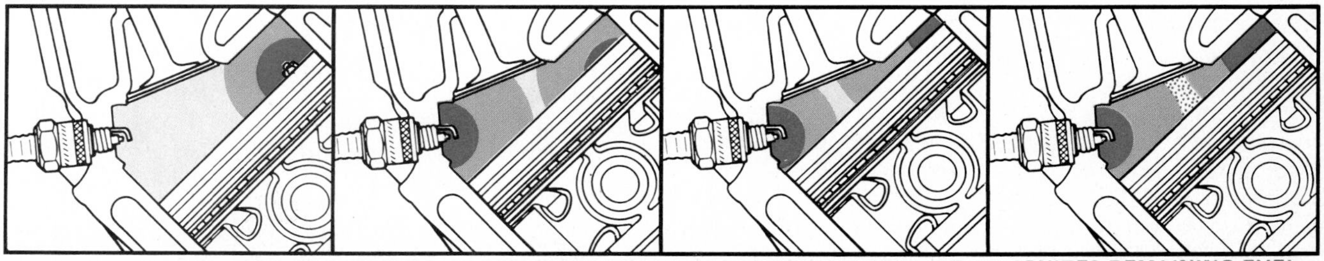

Fig. 11-3 One cause of preignition. The hot spot ignites the compressed mixture before the spark occurs at the spark plug gap. (*Champion Spark Plugs*)

● 11-11 Octane Ratings

Antiknock values of gasoline, or the octane number ratings (ONR), are measured in different ways. Research octane is a measure of the antiknock properties under relatively mild operating conditions. A second measure is engine octane (also called motor octane), which is made under more severe operating conditions. The third octane rating now in widespread use is the Environmental Protection Agency (EPA) octane rating. This is the rating you find on the white circular stickers posted on gasoline pumps at service stations. The EPA octane rating is actually the average between the research-octane and the motor-octane ratings. Generally, the EPA octane ratings you find posted on gasoline pumps are 87 for unleaded regular gasoline, 90 for leaded regular gasoline, and 95 for leaded premiums. Note, however, that these figures may change, so the figures you see on the pump may be different.

● 11-12 Tetraethyl Lead

Tetraethyl lead raises the ONR of gasoline and thus reduces knocking tendencies. It also improves valve and valve-seat life. The lead coats the valve and valve seat and thus provides lubrication. However, lead in gasoline has a bad effect on the catalysts used in catalytic converters. These converters are connected into the exhaust system to convert certain pollutants in the exhaust gases into harmless compounds. (See ● 11-13.) Lead from the gasoline deposits on the catalysts and stops them from doing their job. That is why gasoline for late-model cars has no lead in it. Without lead, the antiknock value (ONR) of the gasoline drops. Therefore, to prevent knock, the compression ratios of the engines must be lowered. This causes some reduction of power.

Taking the lead out of gasoline can also have a harmful effect on valves and valve seats. Without the lubrication provided by the lead, the valve and valve seat can wear more rapidly. To prevent this, modern automotive-engine valves have special coatings, or facings, on their faces. Also, many engines use valve-seat inserts or have valve seats that are induction-hardened (● 9-4 and 9-7).

● 11-13 Two Kinds of Gasoline

One of the emission controls used on late-model cars is the catalytic converter. The exhaust gases from the engine flow through this device. Its purpose is to encourage any residue of unburned gasoline (HC) or partly burned gasoline (CO) to convert into harmless water (H_2O) and carbon dioxide (CO_2). However, if the engine uses leaded gasoline, the lead will coat the active materials in the converter.The converter will then stop working.

This is the reason you see, in most service stations, pumps labeled "No Lead." These pumps dispense gasoline without tetraethyl lead for those cars having catalytic converters. These cars have special small tank-filler necks marked "No Lead." The filler neck requires a special nozzle on the dispensing hose (see Fig. 12-7). This combination is designed to prevent a service-station attendant from putting leaded gasoline in a "no lead" car.

● 11-14 Mechanical Factors Affecting Knocking

The shape of the combustion chamber has a great effect on engine knock. The top of the combustion chamber of a I-head engine is actually the cylinder head, intake and exhaust valves, and spark plug. The bottom is the piston-head and top compression ring (Fig. 11-4). There are two general shapes, wedge and hemispheric (Fig. 11-5). The shape determines turbulence, squish, and quench, and these three factors affect knock.

1. Turbulence When you stir coffee, you swirl or impart turbulence to it so that the cream and sugar mix with the coffee. In the same way, turbulence in the air-fuel mixture entering the combustion chamber assures more even mixing. This makes the combustion more even. Turbulence also reduces the time required for the flame front to sweep through the compressed mixture.

2. Squish In some combustion chambers, the piston squishes, or squeezes, a part of the air-fuel mixture at the end of the compression stroke. Figure 11-5, left, shows the squish area in a combustion chamber. As the piston nears TDC, the mixture is squished, or pushed, out of the squish area. As it squirts out, it promotes turbulence and further mixing of the air-fuel mixture.

3. Quench We mentioned that knocking results when the end-gas temperature goes too high. The end gas explodes before the flame front reaches it. However, if some heat is taken from the end gas, then its temperature will not reach the detonation point. In the cylinder shown to the left in Fig. 11-5, the squish area is also a quench area. The cylinder head is close to the piston, and these metallic surfaces are cooler than the end gas. This causes heat to be removed from the end gas. Thus, the end gas becomes too cool to burn. The tendency for detonation to occur is quenched. However, this causes a problem with exhaust-gas emissions, as explained in a later chapter. The end gas, which contains gasoline vapor (HC), exits with the exhaust gas; this increases the HC in the exhaust gas.

4. The hemispheric combustion chamber With the hemispheric combustion chamber, the spark plug can be located near the center of the dome (Fig. 11-5, right). Then, when combustion starts, the flame front has a shorter distance to travel. There are no distant pockets of end gas to detonate. The chamber has no squish or quench areas. However, there is relatively little turbulence.

5. The wedge combustion chamber With the wedge combustion chamber, the spark plug is located to one

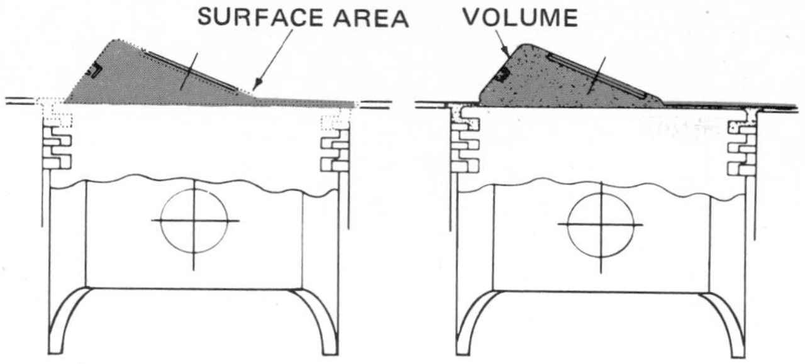

Fig. 11-4 Combustion chamber. The surface area is shown by the dotted lines.

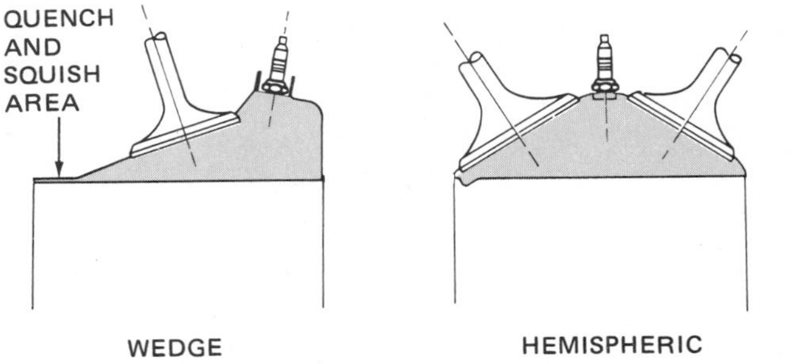

Fig. 11-5 Wedge and hemispheric combustion chambers. *(General Motors Corporation)*

side. The flame front must travel a greater distance to reach the end of the wedge (Fig. 11-5, left). The end of the wedge has a squish and quench area which cools the end gas to prevent detonation. At the same time it imparts turbulence to the mixture.

6. Smog The shape of the combustion chamber also affects the amount of contaminants that appear in the exhaust gases. The cooler metal surfaces of the cylinder head and piston top slow combustion. Therefore, the layers of air-fuel mixture next to these metal surfaces do not burn completely. The wedge combustion chamber has a larger surface area. It thus produces a greater percentage of contaminants than the hemispheric combustion chamber.

● 11-15 Reducing Compression Ratios

The addition of catalytic converters to cars has resulted in a reduction of the compression ratios for the engines. As we have explained, catalytic converters require gasoline without lead. Gasoline without lead has a lower octane rating. To use this lower-octane gasoline, the engine compression ratio must be lowered. This reduces engine performance and efficiency to some extent. But that is the price we must pay for cleaner air.

● 11-16 Other Factors Affecting Knocking

Many operating conditions in an engine affect knocking. For example, higher air temperatures increase the tendency to knock. Higher humidity (or damper air) and higher altitudes (or lower-density air) reduce the tendency to knock. Engine deposits (carbon in the combustion chamber) increase knock tendency. Advancing the spark increases the tendency to knock. Less fuel in the air-fuel mixture increases the tendency of the engine to knock.

All these factors show the need for good maintenance of the modern high-compression engine. Buildups of scale in the cooling system reduce cooling efficiency. Clogged fuel lines or nozzles in the carburetor lean out the mixture. These increase knocking in the engine, as do improper ignition timing and engine deposits.

● 11-17 Other Gasoline Additives

Antiknock compounds and lead-compound-vaporizing substances are put into gasoline to raise its octane rating (● 11-10). Other additives are also used. Major types include:

1. Oxidation inhibitors to help prevent the formation of gum while the gasoline is in storage
2. Metal deactivators to protect the gasoline from the harmful effects of certain metals. Such metals can be picked up in the refining process or in the vehicle fuel system
3. Antirust agents to protect the vehicle fuel system
4. Anti-icers to combat carburetor icing and fuel-line freeze
5. Detergents to keep the carburetor clean
6. Phosphorus compounds to combat surface ignition and spark-plug fouling
7. Dye for identification

In addition, the refining process is carefully controlled to keep to a minimum sulfur compounds and gum-forming substances. Sulfur compounds, in excess, form sulfur acids which seriously damage metal parts and bearings. They also contribute to air pollution as they burn with the fuel. Gum-forming substances form deposits in carburetor circuits and intake manifolds, and on valves, pistons, and rings. Proper refining minimizes the amount of these harmful substances in gasoline.

● 11-18 Chemistry of Combustion

We have already noted (in ● 11-1) that gasoline, in burning, forms water (H_2O) and carbon dioxide (CO_2). *This occurs when enough oxygen is present to combine with all the hydrogen and carbon atoms.* However, in the gasoline engine, sufficient amounts of oxygen are not usually present. As a result, the carbon does not attain complete combustion. Some atoms of carbon unite with only one atom of oxygen (instead of two). This produces carbon monoxide, or CO. Carbon monoxide is a dangerously poisonous gas. It has no color, is tasteless, and has practically no odor. But 15 parts of carbon monoxide in 10,000 parts of air makes the air dangerous to breathe.

Larger amounts may cause quick paralysis and death. An engine should never be operated in a closed space without some means of exhausting the gas to the outside air. Remember this fact: Enough carbon monoxide is produced in 3 minutes by an automobile engine running in a closed one-car garage to cause paralysis and death. *Never operate an engine with the garage doors closed!*

When there is incomplete combustion, the exhaust gases contain unburned hydrocarbons. These contribute to smog, which is a health hazard in populous places.

Automobile companies are working in several ways to reduce these unwanted pollutants in the exhaust gas—HC and CO. For one thing, electronic control devices have been developed that exercise much greater control over the fuel and ignition system. For gasoline engines, fuel-injection systems have been developed which are electronically controlled. This type of system delivers the fuel much more accurately and uniformly to the engine cylinders. As a result, combustion is more complete and there is less HC or CO left over. We have already mentioned catalytic converters which take care of these pollutants. Other devices and systems have been developed to take care of the problem of atmospheric pollution from the automobile. We discuss all of these devices in Part Five, Automotive Emission Controls.

● 11-19 Diesel-Engine Fuel

Diesel engines use diesel fuel oil. The fuel oil is sprayed, or injected, into the engine cylinders toward the end of the compression stroke (Fig. 11-6). Heat of compression ignites the fuel oil, and the power stroke follows.

Diesel fuel is made from crude oil with the same basic refining process by which gasoline is made. It is a light oil having a boiling point not higher than 700°F [371°C]. It is blended from several base materials so it will have the proper characteristics. These characteristics include volatility, ignition quality, viscosity, gravity, sulfur content, and so on.

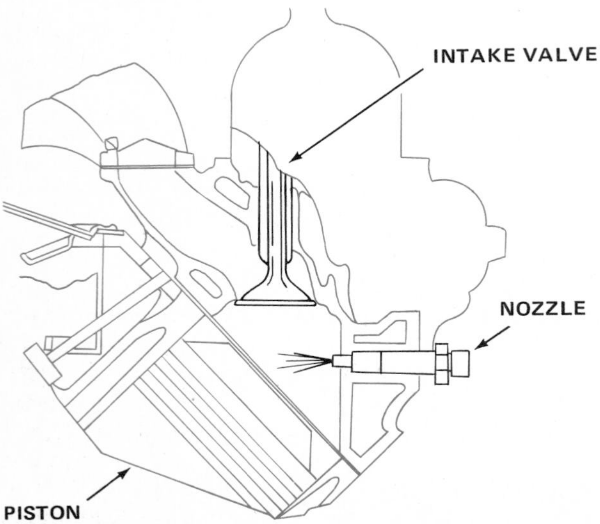

Fig. 11-6 Simplified view of fuel injection into the combustion chamber in a diesel engine.

Diesel fuels are classified according to fuel types as follows:

Type C-B Diesel fuel oil for automobiles, city buses, etc.
Type T-T Fuels for diesel engines in trucks, tractors, etc.
Type R-R Fuels for railroad diesel service.
Type S-M Heavy and residual fuels for large stationary and marine diesel engines.

Type C-B is the lightest oil with the lowest viscosity. Other characteristics of diesel oil are cetane number, pour point and cloud point, residue left after combustion, and flash point. We discuss these characteristics in following sections.

● 11-20 Diesel-Fuel Volatility

We explained, in ● 11-3, that volatility refers to the ease with which a liquid vaporizes. A liquid with high volatility (very volatile) vaporizes at a relatively low temperature. So far as diesel fuel is concerned, engine power and economy are not directly related to fuel volatility. However, the less volatile fuels do have a higher heating value. On the other hand, starting and warm-up of the engine are easier with higher-volatility fuels.

● 11-21 Diesel-Fuel Viscosity

Viscosity refers to the tendency of a liquid to resist flowing. Water has a low viscosity; it flows easily. A light oil is more viscous than water, but it still flows easily. It too has a low viscosity. Heavy oil has a high viscosity; it flows slowly. The fuel oil used in diesel engines must have low viscosity so that it flows easily through the fuel-pumping system. But it must have sufficient viscosity to lubricate the moving parts in the pumping system. However, if the viscosity is too high, the fuel will not spray, or atomize, easily and thus will not burn well.

● 11-22 Cetane Number of Diesel Fuel

The cetane number of diesel fuel refers to the ease with which the fuel ignites. With a high cetane number, the fuel ignites easily (or at a relatively low temperature). The lower the cetane number, the higher the temperature needed to ignite the fuel. The lower the cetane number, the more likely the fuel is to knock. The fuel being sprayed into the cylinder will not ignite quickly, so it tends to accumulate. Then, when ignition does take place, there is a combustion knock as the fuel suddenly burns. On the other hand, if the cetane number is high enough, the fuel will ignite and burn as soon as the injection spray starts. There will thus be an even pressure rise and no knock.

● 11-23 Pour-Point and Cloud Point

These are important in frigid areas. If the oil becomes too thick to pour freely, it will not go through the fuel-injection system easily and poor engine performance will result. The cloud point is the temperature at which some component of the oil starts to turn to a solid. It thus causes the oil to become cloudy. Cloudy oil can clog the oil filter and cause poor engine performance. In very cold areas, the oil must be specifically refined so it will flow freely and will not cloud at the lowest temperatures encountered.

● 11-24 Residue After Combustion

There are two general kinds of residue that may be left after the oil is burned; carbon and ash. The carbon can clog piston rings and exhaust valves. Ash comes from suspended solids or soluble materials; they can cause wear of injector, fuel pump, and piston rings. The oil-refining process normally removes these substances.

● 11-25 Flash Point

The flash point is the lowest fuel temperature at which a flame held above the oil will ignite the vapor rising from the oil. The flash point is not important so far as engine operation is concerned. However, it is important in connection with legal requirements and safety precautions involved in fuel handling and storage. An abnormally low flash point usually means the oil is contaminated with some lighter substance such as gasoline.

● 11-26 Other Diesel-Oil Characteristics

There are several other characteristics that oil refiners must consider when processing crude into gasoline, diesel fuel, and other products. These include sulfur content, water or sediment contamination, corrosion or rust protection, oxidation stability, and heat content.

Sulfur content must be low in diesel fuel to meet legal requirements. Also, high sulfur can cause corrosion, engine wear, and deposits in the engine.

Water and sediment can get into the oil during fuel delivery from the refiner to the service station to the vehicle fuel tank. Water and sediment can lead to engine corrosion and plugging of filters or fuel injectors. Therefore, everyone handling diesel oil must be careful to avoid such contamination.

Oxidation stability is important when diesel oil is stored for any length of time. Oxidation in the fuel can cause deposits which will clog the filter, fuel pump, and injectors.

Heat content is the amount of heat the fuel oil will produce when it is burned. This varies with the composition of the oil and the refining process. The less volatile oils (with higher viscosity) have a higher heating content. As a rule, oils with a higher heat content are favored but these oils must not have too low viscosity or volatility.

● 11-27 Diesel-Fuel Additives

Several substances, or additives, are put into diesel fuel to improve its desirable properties. These include ignition-quality improvers, oxidation inhibitors, biocides, rust preventives, metal deactivators, pour-point depressors, smoke suppressants, and detergent dispersants.

The purpose of these various additives is obvious, with the possible exception of the biocides. These are chemicals that inhibit or prevent the growth of bacteria or fungi in the oil. If such life forms do develop in the oil, they can produce serious contamination. This can degrade the oil to a point where it cannot be used in diesel engines.

● 11-28 Liquefied Petroleum Gas (LPG)

This fuel requires a special fuel system (● 12-19). There are actually two types of LPG that have been used for automotive-engine fuel, *propane* and *butane*. Of these, propane is the more widely used. Sometimes, small amounts of butane are added. Propane boils at −44°F [−42.2°C] (at atmospheric pressure). Thus, it can be used in any climate where temperatures below this are not reached. Butane cannot be used in temperatures below 32°F [0°C], since it is liquid below that temperature. If it remains liquid, it will not vaporize in the fuel system and will never reach the engine.

REVIEW QUESTIONS

1. With perfect combustion, what two compounds would be formed when gasoline burns?
2. Name three pollutants emitted from automobiles.
3. What is volatility? Why is it important in gasoline?
4. What does the term "antiknock value" mean?
5. What is heat of compression?
6. Explain the cause of knocking produced by heat of compression.
7. What effect does increasing the compression ratio have on knocking? Why?
8. Describe one method of measuring the antiknock value of a gasoline.
9. What does ONR mean?
10. What is the difference between detonation and preignition?
11. What effect does lead have on valves and valve seats?
12. Why has lead been removed from gasoline?
13. What is quench?
14. What is squish?
15. What are the two basic combustion-chamber shapes?
16. Name six gasoline additives.
17. Why is CO dangerous?

18. Can you tell by the odor whether or not CO is present in a garage?
19. What does cetane number mean?
20. Name and describe six characteristics that a good diesel-engine fuel must have.
21. What is the purpose of the biocide added to diesel-engine fuel?
22. What does LPG mean?

SELF PROJECTS

Find out how gasoline is made. Go to your local library and look in an encyclopedia for information on oil and gasoline. Make notes on how engineers prospect (look for) oil in the earth. Prospectors drill many wells into the earth in their search for oil. Most of these are dry wells—they contain no oil. But when they do find oil, a whole series of events sets in. The oil must be controlled as it comes from the well. It must be transported to refineries. The refineries put the crude oil through a series of processes. Out of this come many different products from grease and other lubricants to fuel oil and gasoline.

As you read articles, make notes on how it all happens. File your notes in your notebook. Knowledge of this sort will help you understand what the making of gasoline and diesel-fuel oil is all about. When you pump gasoline or fuel oil into a customer's tank, or drive a car down the highway, you will appreciate more than ever how important the products of crude oil are to all of us.

CHAPTER 12
Automotive Carbureted Fuel Systems

After you read this chapter, you should be able to:

- **List and describe the purpose, construction, and operation of all the components in the carbureted fuel system except the carburetor. Identify and find the locations of these components on cars.**
- **Explain the basic difference between the carbureted fuel system and the fuel-injection system.**

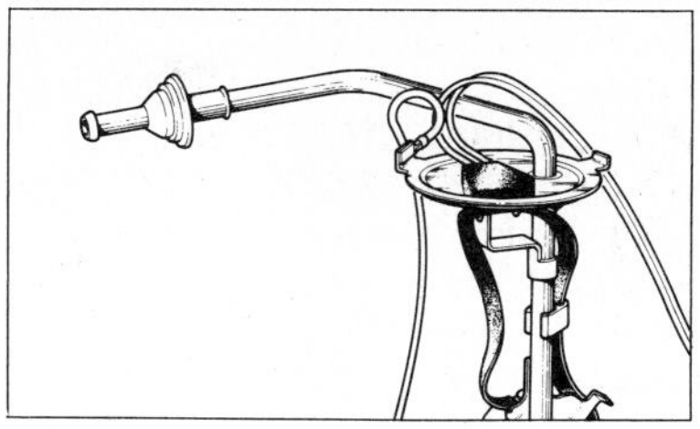

This chapter covers all the components in the fuel system using a carburetor except the carburetor itself, which is covered in the following chapter. The chapter discusses the fuel tank, fuel lines, fuel pump, and the air cleaner.

12-1 Two Fuel Systems for Gasoline Engines

For many years, the gasoline engines used in automobiles have had carburetors in their fuel systems. The carburetor is a mixing device which mixes air and gasoline vapor in the proper proportions to produce a combustible mixture. Figure 12-1 is a simplified version of the carburetor action. The combustible mixture flows from the carburetor to the engine cylinders, as shown.

In recent years, some automobiles have come out with a different kind of fuel system. This is a fuel-injection system. Figure 12-2 is a simplified drawing of one design. In this system, the carburetor is replaced by a throttle body which has only one job. That is, it controls the amount of air flowing into the intake manifold. The intake manifold has a series of injectors assembled to it. The injector is a spraying device that, at the proper moment, sprays a metered amount of gasoline into the intake manifold, opposite the intake valve (Fig. 12-3). When the intake valve opens, this sprayed gasoline and air enters

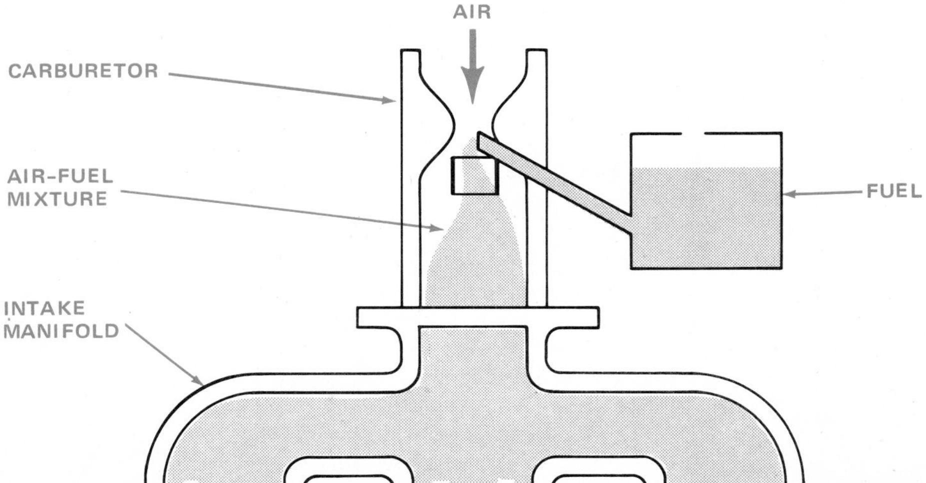

Fig. 12-1 Simplified view of a carbureted fuel system. Arrows show flow of fuel and air through the system.

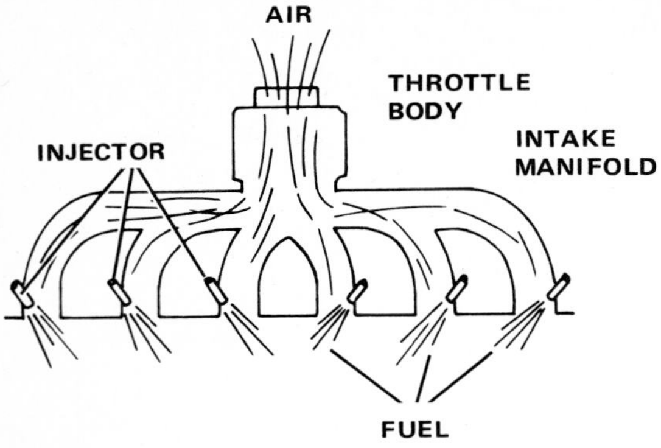

Fig. 12-2 Simplified view of a fuel-injection system.

the cylinder. There is an injector for every cylinder. Each injector is placed opposite an intake valve.

Some fuel-injection systems do not use the pulsed, or timed, injection system. Instead, the fuel injectors spray continuously, varying the amount of fuel sprayed to suit operating requirements. We describe these systems in detail later.

The basic difference between the carbureted system and the fuel-injection system is the point at which the gasoline enters the air going to the cylinders. In the carburetor, it enters the air passing through the carburetor. In the fuel-injection system, the gasoline is sprayed into the air in the intake manifold. We cover the fuel-injection system in a later chapter (Chap. 15).

You will notice we have been using the term "carbureted." This is a word used in the automotive business when referring to the fuel systems using carburetors. These are carbureted systems. An engine using this system would be called a carbureted engine.

● 12-2 Purpose of the Fuel System

Both the carbureted fuel system and the fuel-injection system have the same job—supplying a combustible mixture of air and fuel to the engine. The fuel system must change the proportions of air and fuel for different operating conditions. When the engine is cold, for example, the mixture must be rich (have a high proportion of fuel). This is because the fuel does not vaporize readily at low temperatures. Extra fuel must be added to the mixture so that enough will vaporize to form a combustible mixture.

● 12-3 Carbureted-Fuel-System Components

The carbureted fuel system consists of the fuel tank, fuel pump, fuel filter, carburetor, intake manifold, and fuel lines. The fuel lines are tubes connecting the tank, fuel pump, and carburetor (Fig. 12-4). Most of these components are the same in both the carbureted and the fuel-injection systems. We now look at each component in detail.

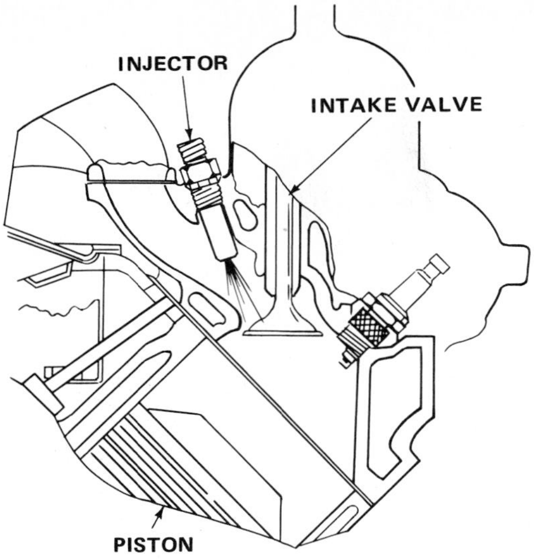

Fig. 12-3 Simplified view showing the method of injecting fuel into the intake manifold in back of the intake valve.

● 12-4 Fuel Tank

The fuel tank (Fig. 12-5) is normally located at the rear of the vehicle. It is usually made of sheet metal and is attached to the frame. The filler opening of the tank is closed by a cap. The fuel line is attached at or near the bottom of the tank. In some tanks, there is a filter at the fuel-line connection. The tank also contains the sending unit of the fuel gauge. Fuel tanks in older cars also have a vent pipe to allow air to escape when the tank is being filled.

Vaporized gasoline escaping from the fuel tank through the vent pipe contributes to the formation of smog. Thus, cars manufactured since 1970 have been equipped with a fuel-vapor emission-control system. In this system, the fuel-tank vent pipe is connected to a charcoal canister, which holds the vapor and prevents its escape into the air (● 12-9).

Cars with a fuel-vapor emission-control system have a special fuel-tank filler cap (Fig. 12-6). The cap has a two-way relief valve, operating on both pressure and vacuum. When gasoline is withdrawn from the tank, a slight vacuum develops. The vacuum valve opens to admit air. If the pressure builds up excessively, the pressure valve opens to relieve the pressure. Excessive pressure would not normally develop because any pressure is released through the fuel-vapor emission-control system. Note that the filler cap, shown in Fig. 12-6, also has a rollover check valve which closes if the car is in an accident that rolls it over. When the car, and cap, are upside down, the check valve closes to prevent leakage of gasoline from the tank.

All automotive vehicles manufactured after 1976 have catalytic converters (discussed in Part Five). These cars must use nonleaded gasoline, as explained in ● 11-13. To guard against a service-station attendant putting leaded gasoline in the tank, these cars have a restrictor (Fig.

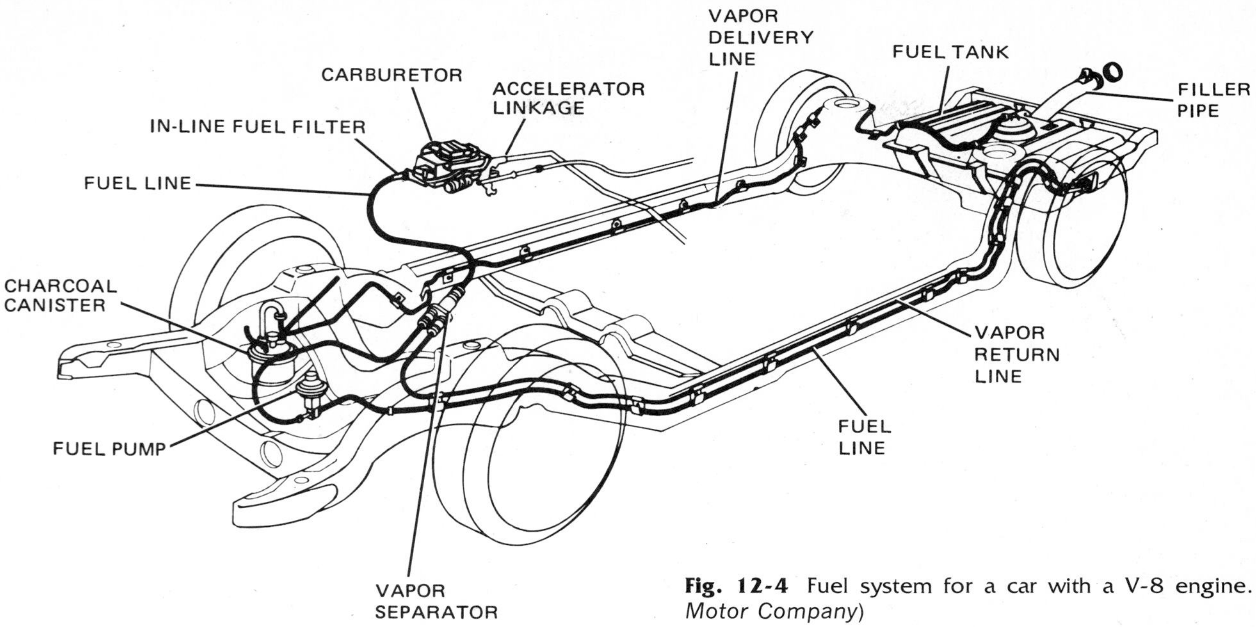

Fig. 12-4 Fuel system for a car with a V-8 engine. (*Ford Motor Company*)

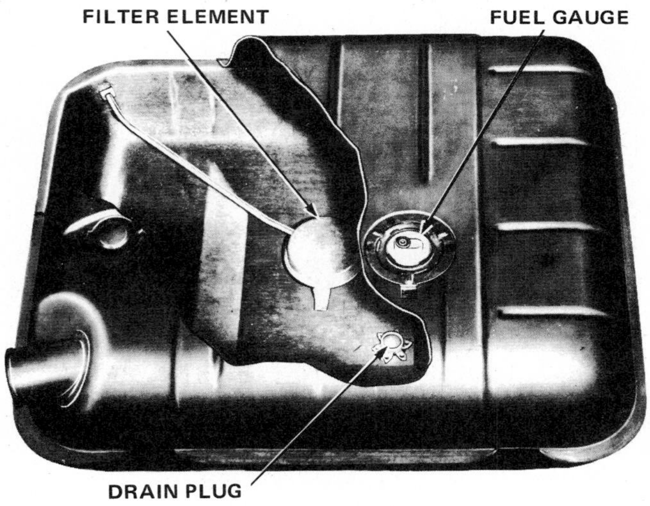

Fig. 12-5 Fuel tank, partly cut away to show the filter element and the drain plug. (*Chrysler Corporation*)

12-7). This restrictor is smaller in diameter and has a trap door, as shown. The pump nozzles that dispense leaded gasoline are too large to fit. Only the smaller pump nozzles that dispense no-lead gasoline can enter the neck and open the trap door so gasoline can flow into the tank.

12-5 Fuel Filters and Screens

Fuel systems have filters and screens to prevent dirt in the fuel from entering the fuel pump or carburetor. Dirt could prevent normal operation of these units and cause poor engine performance. One type of filter is part of the fuel pump. It may also be a separate unit connected into

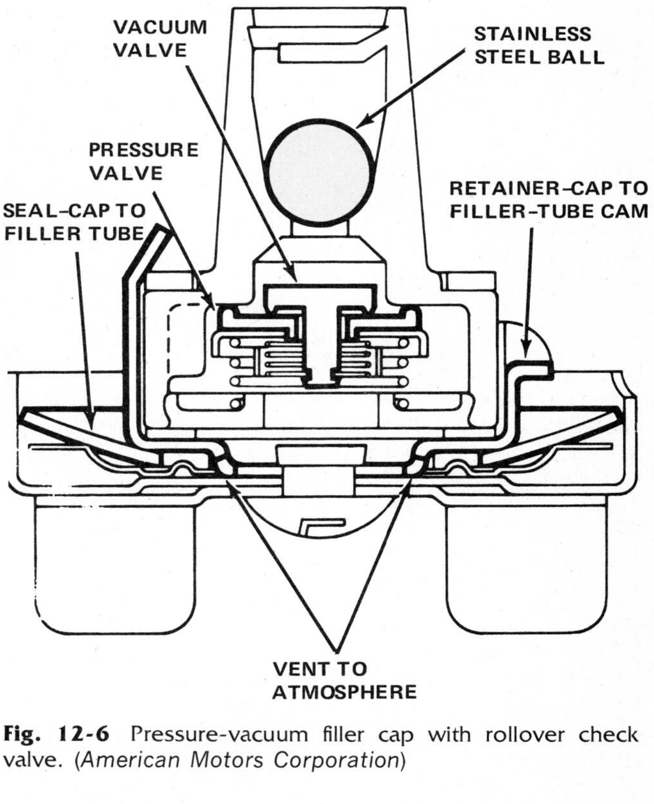

Fig. 12-6 Pressure-vacuum filler cap with rollover check valve. (*American Motors Corporation*)

the fuel line between the tank and fuel pump. Or, it may be located between the fuel pump and carburetor. It may be in or on the carburetor itself. Figure 12-8 shows the type that is outside the carburetor but mounted on it. The screw threads enter a tapped hole in the carburetor.The fuel line fits on the opposite end of the filter. Figure 12-9 shows the type that has a magnet to pick up metal particles in the fuel.

12-6 Fuel Gauges

There are two types of fuel gauge, *balancing coil* and *thermostatic.* Each of these gauges has a tank unit and a dash unit.

1. Balancing coil (Fig. 12-10) The tank unit in this fuel gauge contains a sliding contact. The contact slides back and forth on a resistor as the float moves up and down in the fuel tank. This changes the amount of electric resistance the tank unit offers. Thus, as the tank empties, the float drops and the sliding contact moves to reduce the resistance. The instrument panel unit contains two coils, as shown in Fig. 12-10. When the ignition switch is turned on, current from the battery flows through the two coils. This produces a magnetic pattern that acts on the armature, to which the pointer is attached. When the resistance of the tank unit is high (tank filled and float up), the current through the E (empty) coil also flows through the F (full) coil. Thus the armature is pulled to the right, so that the pointer is on the F (full) side of the dial. But when the tank begins to empty, the resistance of the tank unit drops. Thus, more of the current flowing through the empty coil passes through the tank unit. Since less is flowing through the full coil, its magnetic pull is weaker. As a result, the empty coil pulls the armature toward it. The pointer swings around toward the E (empty) side of the dial.

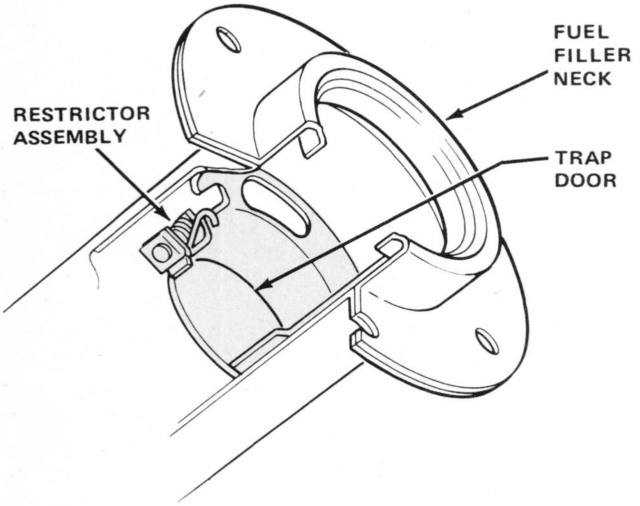

Fig. 12-7 One type of filler-neck restrictor for cars requiring no-lead gasoline. (*American Motors Corporation*)

2. Thermostatic Figure 12-11 is the wiring circuit of a thermostatic fuel gauge. It has a fuel-tank unit much like the balancing-coil system. That is, it has a float and a sliding contact that moves on a resistor. Current flows from the battery through a heater wire in the fuel-gauge instrument-panel unit, and through the resistance in the tank unit. When the fuel is low in the tank, most of the resistance is in the circuit. Very little current can flow. When the tank is filled, the float moves up, and the sliding contact cuts most of the resistance out of the circuit. Now more current flows. As it flows through the heater coil in the dash unit, the current heats the thermostat. The thermostat blade bends because of the heat. This moves the needle to the right, toward the F or full mark.

Note that the system in Fig. 12-11 has an instrument voltage regulator. This device is thermostatic; its purpose is to keep the voltage to the fuel-gauge system low.

3. Low-level-fuel indicator The system, shown in Fig. 12-11, also has a low-fuel-level indicator. It includes a thermistor assembly in the fuel tank, a warning light, and a warning relay. A *thermistor* is a special sort of resistor that loses resistance as it gets hot. As long as there are more than a few gallons of fuel in the tank, the thermistor is submerged and is kept cool. However, when the fuel level is low, the thermistor is exposed to air. It

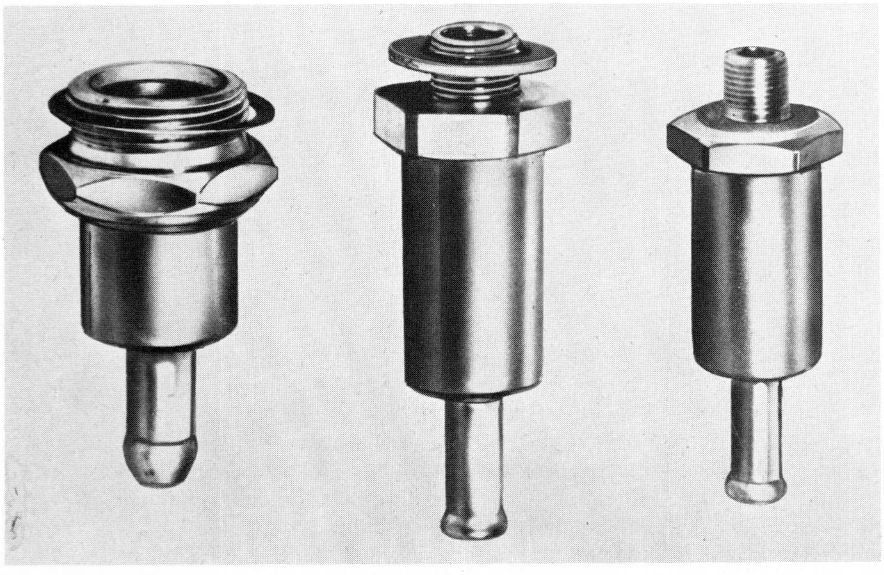

Fig. 12-8 In-line fuel filters. (*Ford Motor Company*)

gets hotter. Its resistance decreases, and more current flows. The increased current flow is sufficient to operate the warning relay. It connects the warning light to the battery. The light comes on to warn the driver that the fuel is getting low.

● 12-7 Fuel Pumps

The fuel system uses a fuel pump to deliver fuel from the tank to the carburetor. There are two types of fuel pump, mechanical and electric. Electric fuel pumps are discussed in ● 12-10. Mechanical fuel pumps are operated by an eccentric (an off-center section) on the engine camshaft (see Fig. 9-1), as explained below. The mechanical fuel pump is mounted on the side of the cylinder block in in-line engines. In some V-8 engines, the pump is mounted between the two cylinder banks. Most modern V-8 engines have the fuel pump on the side of the cylinder block, at the front of the engine.

The mechanical fuel pump has a rocker arm whose end rests on the camshaft eccentric. Many V-8 engines

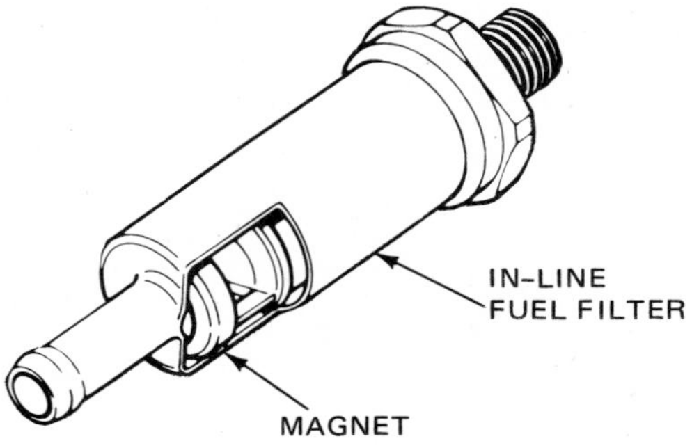

Fig. 12-9 Fuel filter located in the carburetor. (*Ford Motor Company*)

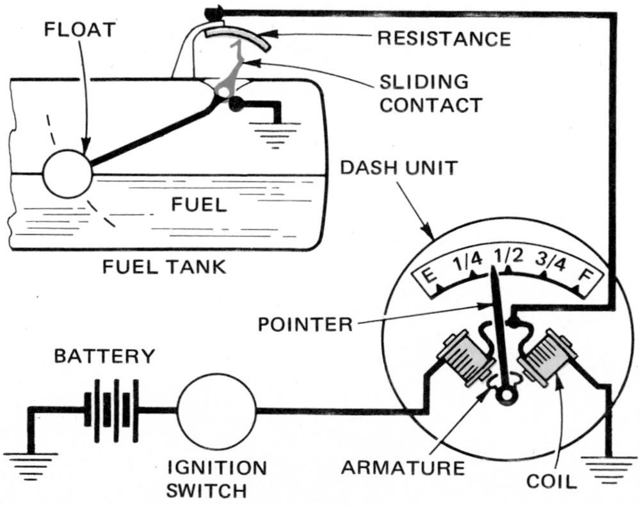

Fig. 12-10 Schematic wiring diagram for a balancing-coil fuel-gauge indicating system.

also use a push rod from the eccentric to the rocker arm.

As the camshaft rotates, the eccentric rocks the rocker arm back and forth. The inner end of the rocker arm is linked to a flexible diaphragm. The diaphragm is clamped between the upper and lower pump housings (Fig. 12-12). There is a spring over the diaphragm that keeps tension on it. As the rocker arm rocks, it pulls the diaphragm up and then releases it. The spring then forces the diaphragm down. Thus, the diaphragm moves up and down as the rocker arm rocks.

This diaphragm movement produces partial vacuums and pressures in the space above the diaphragm. When the diaphragm moves up, a partial vacuum is produced. Then, atmospheric pressure, acting on the fuel in the tank, forces fuel through the fuel line and into the pump. The inlet valve in the pump opens to admit fuel, as shown by the arrows in Fig. 12-12.

When the diaphragm is released by the rocker arm,the spring forces the diaphragm downward. This produces pressure in the space under the diaphragm. The pressure closes the inlet valve and opens the outlet valve. Now fuel is forced from the fuel pump through the fuel line to the carburetor, as shown by the arrows.

The actions in the pump as the eccentric rotates are shown in Figs. 12-12 and 12-13. The fuel from the fuel pump enters the carburetor past a needle valve in the float bowl. If the bowl is full, the needle valve closes so that no fuel can enter. When this happens, the fuel pump cannot deliver fuel to the carburetor. In this case, the rocker arm continues to rock. However, the diaphragm remains at or near its upper limit of travel. Its spring cannot force the diaphragm downward so long as the carburetor float bowl will not accept fuel. However, as the carburetor uses up fuel, the needle valve opens to admit fuel to the float bowl. Now the diaphragm can move down (on the rocker-arm return stroke) to force fuel into the carburetor float bowl.

● 12-8 Vapor-Return Line

Some cars have a vapor-return line running from the fuel pump or the fuel filter to the fuel tank. Figure 12-12 shows the connection at the fuel pump (to lower left) for the vapor-return line. The purpose of this line is to return to the fuel tank any vapor that forms in the fuel pump. The fuel pump can handle liquid only. Pumping stops if vapor forms in the fuel pump. With air conditioning, under-the-hood temperatures are likely to be higher. The air-conditioning condenser delivers more heat under the hood. Also, during idle, the engine cooling system is not very efficient. This allows under-the-hood temperatures to increase. The higher temperatures tend to cause vapor to form in the fuel pump.

To understand why vapor can form in the fuel pump, note first that the pump alternately produces vacuum and pressure. During the vacuum phase, the boiling, or vaporizing, temperature of the fuel goes down. The lower the pressure, the lower the temperature at which any liquid vaporizes. For example, water boils at 212°F

[100°C] at sea level—atmospheric pressure 14.7 psi (pounds per square inch) [1.078 kg/cm^2]—see ● 4-18. But at 16,000 feet [4876.8 m] above sea level, where the pressure is around 7 psi [0.492 kg/cm^2], water boils at 185°F [85°C].

The combination of increased temperature and partial vacuum in the fuel pump can cause fuel to vaporize. This produces vapor lock, a condition that prevents normal delivery of fuel to the carburetor. The engine stalls.

The vapor-return line is connected to a special outlet in the fuel pump. It allows the vapor to return to the fuel tank. The vapor-return line also permits excess fuel being pumped by the fuel pump to return to the fuel tank. This excess fuel, in constant circulation, helps keep the fuel pump cool. It thus prevents vapor from forming.

Some cars have a vapor separator connected between the fuel pump and the carburetor (Fig. 12-14). It consists of a sealed can, a filter screen, an inlet and outlet fitting,

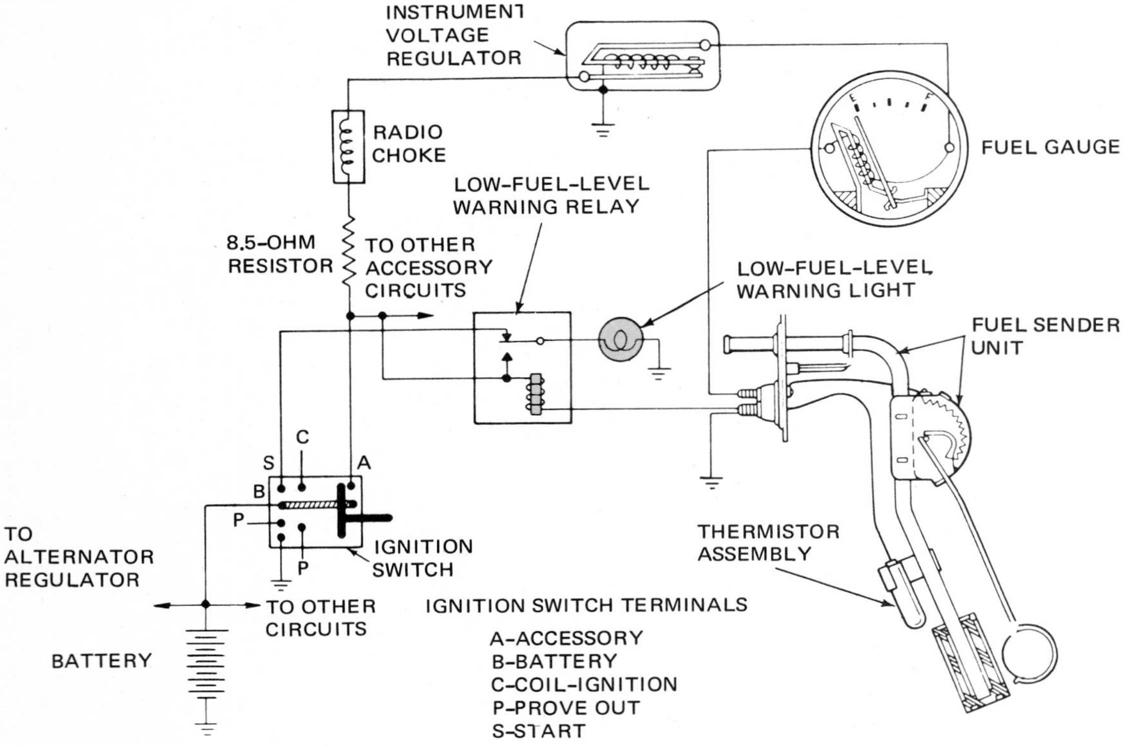

Fig. 12-11 Schematic wiring diagram for a thermostatic fuel-gauge indicating system. This system uses a variable-resistance tank unit and a thermostatic dash unit. (*Ford Motor Company*)

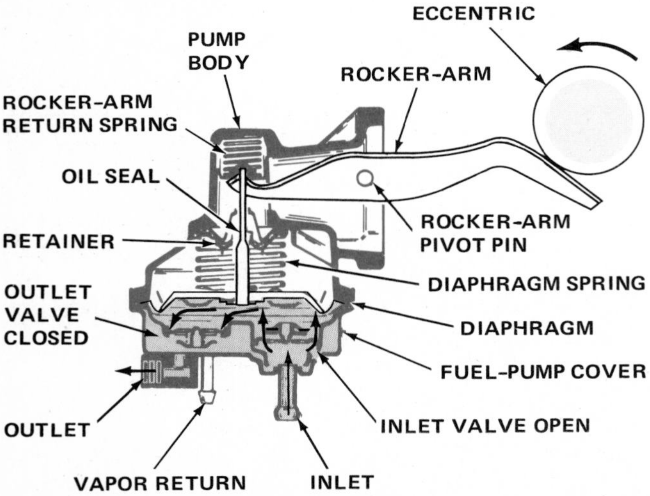

Fig. 12-12 When the eccentric rotates so as to push the rocker arm down, the arm pulls the diaphragm up. The inlet valve opens to admit fuel into the space under the diaphragm.

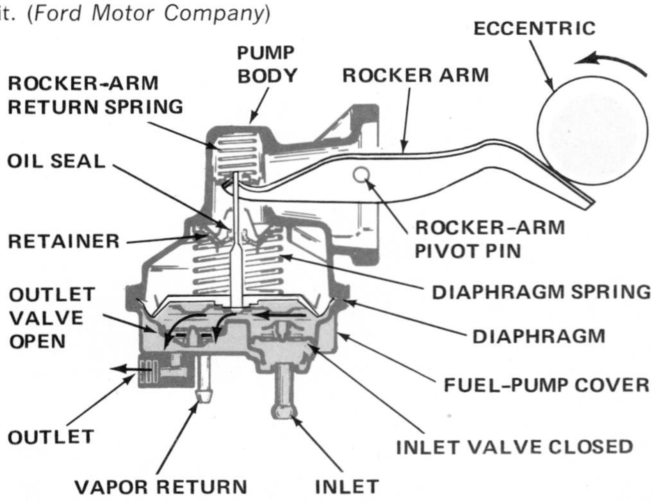

Fig. 12-13 When the eccentric rotates, the rocker arm moves up under it. The diaphragm is then released so it can move down, producing pressure under it. This pressure closes the inlet valve and opens the outlet valve, so fuel flows to the carburetor.

and a metered orifice, or outlet, for the return line to the fuel tank. Any fuel vapor that the fuel pump produces enters the vapor separator (as bubbles) along with fuel. These bubbles of vapor rise to the top of the vapor separator. The vapor then is forced, by fuel-pump pressure, to pass through the fuel-return line and back to the fuel tank. In the tank, it condenses back into liquid fuel.

Some vapor-return lines have an in-line check valve (Fig. 12-15). The purpose of this check valve is to prevent fuel from feeding back to the carburetor from the fuel tank through the vapor-return line. If fuel does attempt to feed back in this way, the pressure of the fuel forces the steel ball to seat, thus blocking the line. In normal operation, the pressure of the fuel vapor from the fuel pump unseats the ball and allows the fuel vapor to flow to the fuel tank.

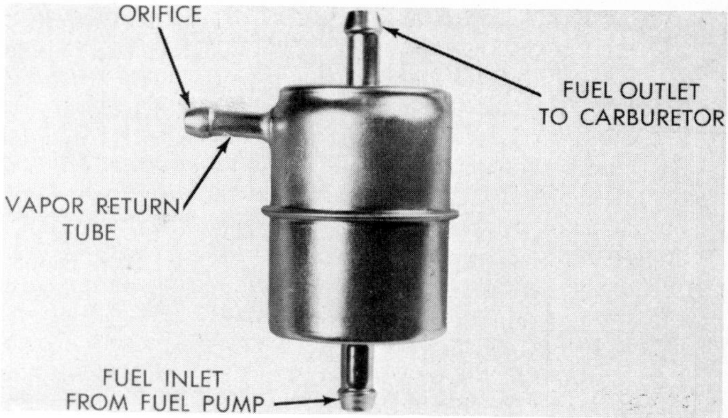

Fig. 12-14 A combination fuel filter and vapor separator. It filters the fuel and allows fuel vapor to return to the fuel tank through the vapor-return line. (*Chrysler Corporation*)

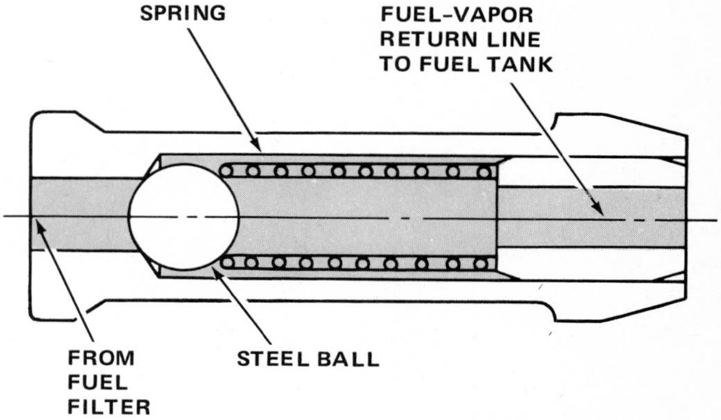

Fig. 12-15 In-line check valve which is located between the fuel filter and the fuel tank. (*American Motors Corporation*)

12-9 Fuel-Vapor Emission-Control System

Gasoline vapor can escape from uncontrolled fuel tanks and carburetors. This happens when the car is sitting idle and the engine is not running. At other times, when the engine is running, gasoline is being pumped from the fuel tank and carburetor, so gasoline vapor is not likely to escape. Large quantities of gasoline vapor (HC) can escape from parked cars without vapor controls and can add to atmospheric pollution. To prevent this loss of gasoline vapor, modern automobiles are equipped with fuel-vapor emission-control systems. Figure 12-4 shows one system. The fuel tank is sealed. Escaping gasoline vapor has to flow through the vapor recovery line to the charcoal canister. Gasoline vapor from the carburetor float bowl also flows to the charcoal canister. There, the charcoal particles pick up the gasoline vapor and hold it. Then, when the engine is started and runs, air flows through the charcoal canister on the way to the carburetor. This air picks up the gasoline vapor trapped in the canister and carries it to the carburetor. There, it mixes with the air-fuel mixture and enters the engine. It is thus burned, instead of being allowed to enter the atmosphere as HC (unburned gasoline vapor).

Some vapor-emission control lines have a rollover check valve (Fig. 12-16). This valve blocks the fuel-vapor recovery line in case of a rollover accident. That is, if the car is upside down, the steel ball will force the plunger to seat against the guide plate. This blocks the line and prevents gasoline from flowing out of the tank through the line to the charcoal canister or carburetor. On some cars, the rollover check valve is part of the carburetor assembly.

Part Five of the book, on emission controls, describes fuel-vapor emission control in detail.

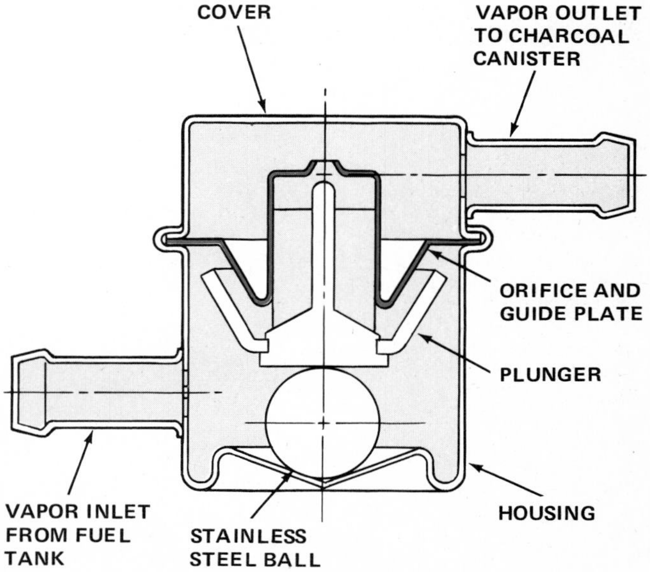

Fig. 12-16 Rollover check valve in vapor line between the fuel tank and charcoal canister. (*American Motors Corporation*)

12-10 Electric Fuel Pumps

Electric fuel pumps have certain advantages over mechanical fuel pumps. Fuel is at the carburetor as soon as the ignition is turned on. The pump can deliver more fuel than the engine will require even under maximum operating conditions. Thus, the engine will never be fuel-starved. They are therefore used in many high-performance and heavy-duty applications.

There are various types of electric fuel pumps. One of the latest types is mounted in the fuel tank (Fig. 12-17). It contains an impeller driven by an electric motor (Fig. 12-18). This pushes fuel through the fuel line to the carburetor. Other types of electric fuel pumps are mounted in the engine compartment. One design is shown in Fig. 12-19. It contains a flexible metal bellows that is operated by an electromagnet. The electromagnet is connected to the battery by turning on the ignition switch. Then it pulls down the armature and extends the bellows. This produces a vacuum in the bellows. Fuel from the fuel tank enters the bellows through the inlet valve. Then, as the armature reaches its lower limit of travel, it opens a set of contact points. This disconnects the electromagnet from the battery. The return spring therefore pushes the armature up and collapses the bellows. This forces fuel from the bellows through the outlet valve and to the carburetor. As the armature reaches the upper limit of its travel, it closes the contacts. The electromagnet is again energized and again pulls the armature down. These actions are repeated as long as the ignition is on.

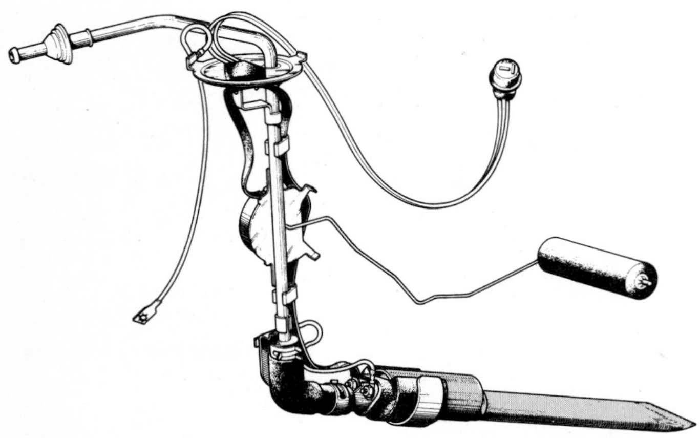

Fig. 12-17 The tank-type electric fuel pump is mounted on the same support as the fuel-gauge tank unit. (*Buick Motor Division of General Motors Corporation*)

12-11 Air Cleaners

As already noted, the fuel system mixes air and fuel to produce a combustible mixture. A great deal of air passes through the carburetor and engine—as much as 100,000 cubic feet [2831.7 m^3] of air every 1,000 car miles [1,609.3 km]. This is a lot of air, and it probably contains a lot of floating dust and grit. The grit and dust could cause serious damage if they entered the engine. Therefore, an air cleaner is mounted on the air horn, or air entrance, of the carburetor, to keep out the dirt (Fig. 12-20).

All air entering the engine through the carburetor must first pass through the air cleaner. The upper part of the air cleaner contains a ring of filter material (fine-mesh metal threads or ribbons, special paper, cellulose fiber, or polyurethane). The air must pass through this ring. This material provides a fine maze that traps most of the dust particles. Some air cleaners have an oil bath. This is a reservoir of oil which the incoming air flows past. The moving air picks up particles of oil and carries them up into the filter. There the oil washes any dust back down into the oil reservoir. The oiliness of the filter material also improves the filtering action.

The air cleaner also muffles the noise of the intake of air through the carburetor, manifold, and valve ports. This noise would be quite noticeable if it were not for the air cleaner. In addition, the air cleaner acts as a flame arrester in case the engine backfires through the carburetor. Backfiring may occur if the air-fuel mixture is ignited in the cylinder before the intake valve closes. When

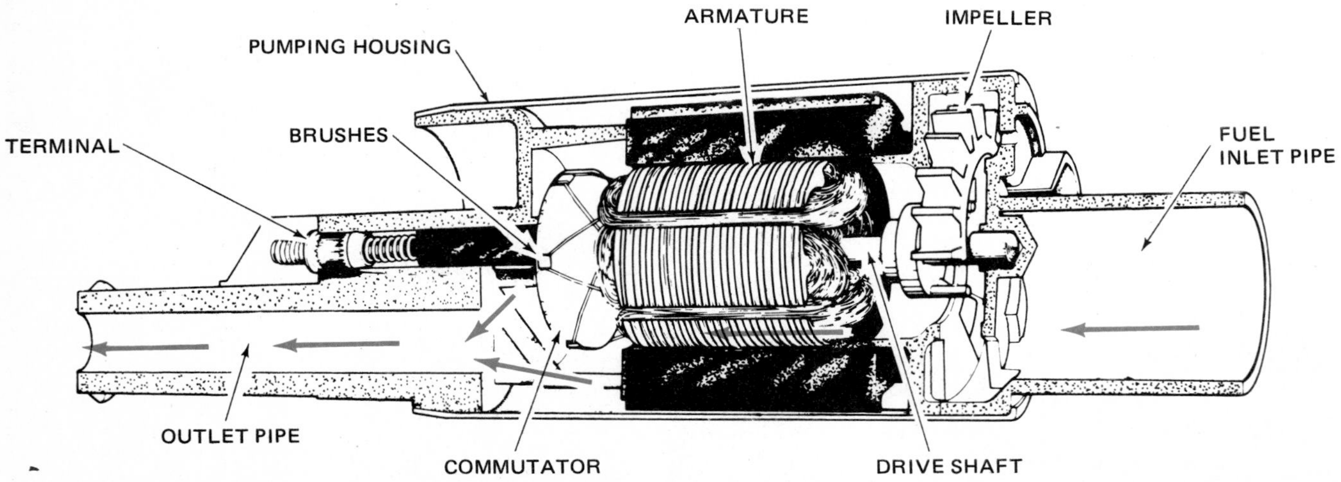

Fig. 12-18 Cutaway view of a tank-mounted electric fuel pump. (*Buick Motor Division of General Motors Corporation*)

this happens, there is a momentary flashback through the carburetor. The air cleaner keeps this flame from leaving the carburetor and igniting gasoline fumes outside the engine.

12-12 Heated-Air System with Thermostatic Air Cleaner

The heated-air system which has a thermostatic air cleaner is part of the controlled combustion system used on late-model cars. It is one component of the emission-control equipment discussed in detail in Part Five. In order to reduce engine emissions, carburetors are adjusted to give leaner mixtures at idle and part-throttle. That is, the amount of gasoline entering into the air-fuel mixture is reduced. These leaner mixtures assure a more complete burning of the gasoline. This means there is less HC coming out the tail pipe.

However, these leaner mixtures can reduce engine performance when the engine is cold. To correct this, a thermostatically controlled air cleaner is used. This is also called the heated air system (HAS) by General Motors. It sends heated air to the carburetor during cold weather, when the engine is cold (Fig. 12-21). This improves engine performance after a cold start and during engine warm-up. Thus leaner mixtures can be used to reduce smog, without affecting cold-engine performance.

One air cleaner of this type is shown in Fig. 12-22. It contains a sensing spring which reacts to the temperature of the air entering the carburetor through the air cleaner. This spring controls an air-bleed valve (see Fig. 12-23). When the entering air is cold, the sensing spring

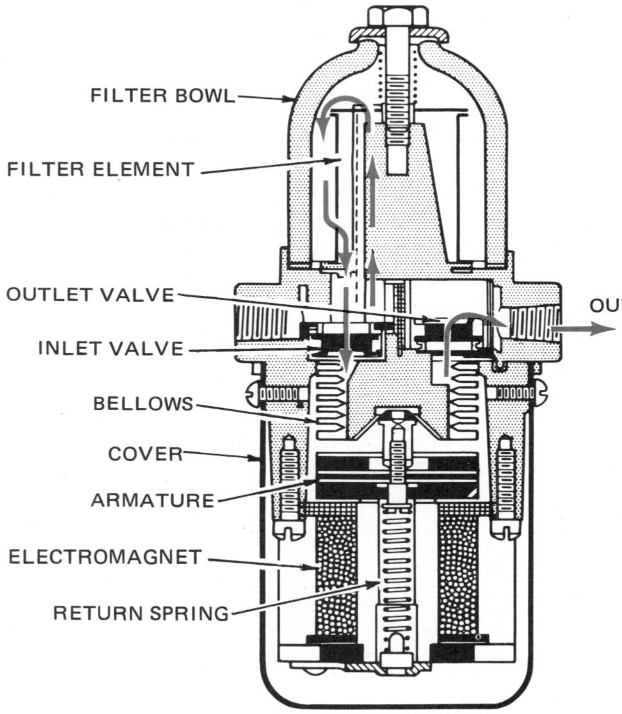

Fig. 12-19 Sectional view of an electric fuel pump.

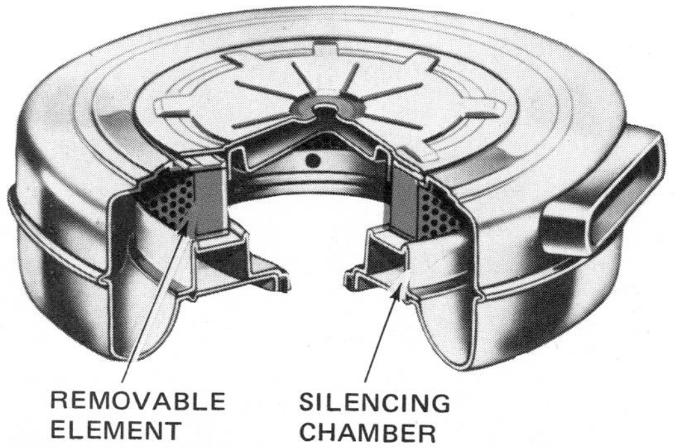

Fig. 12-20 Typical air cleaner, partly cut away to show the filter element. (*Ford Motor Company*)

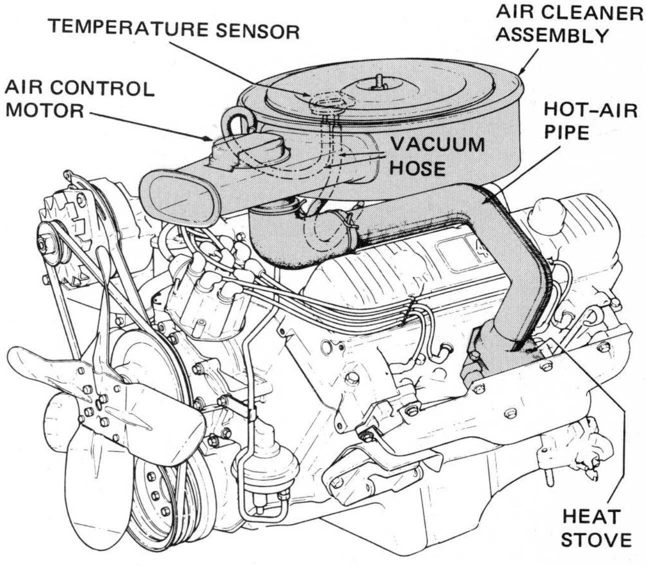

Fig. 12-21 Heated-air system on a V-8 engine. (*Buick Motor Division of General Motors Corporation*)

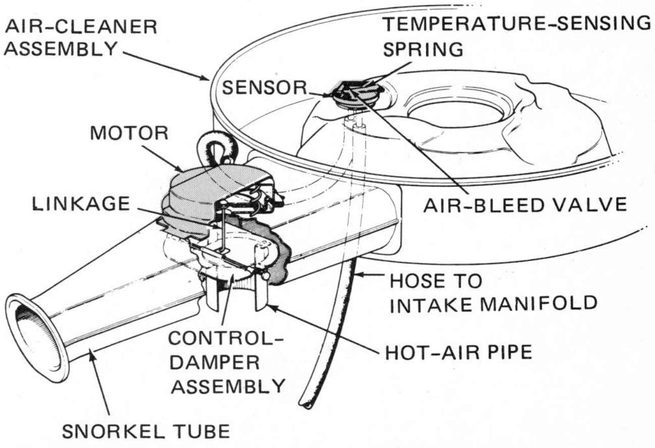

Fig. 12-22 Air cleaner with thermostatic control. (*Chevrolet Motor Division of General Motors Corporation*)

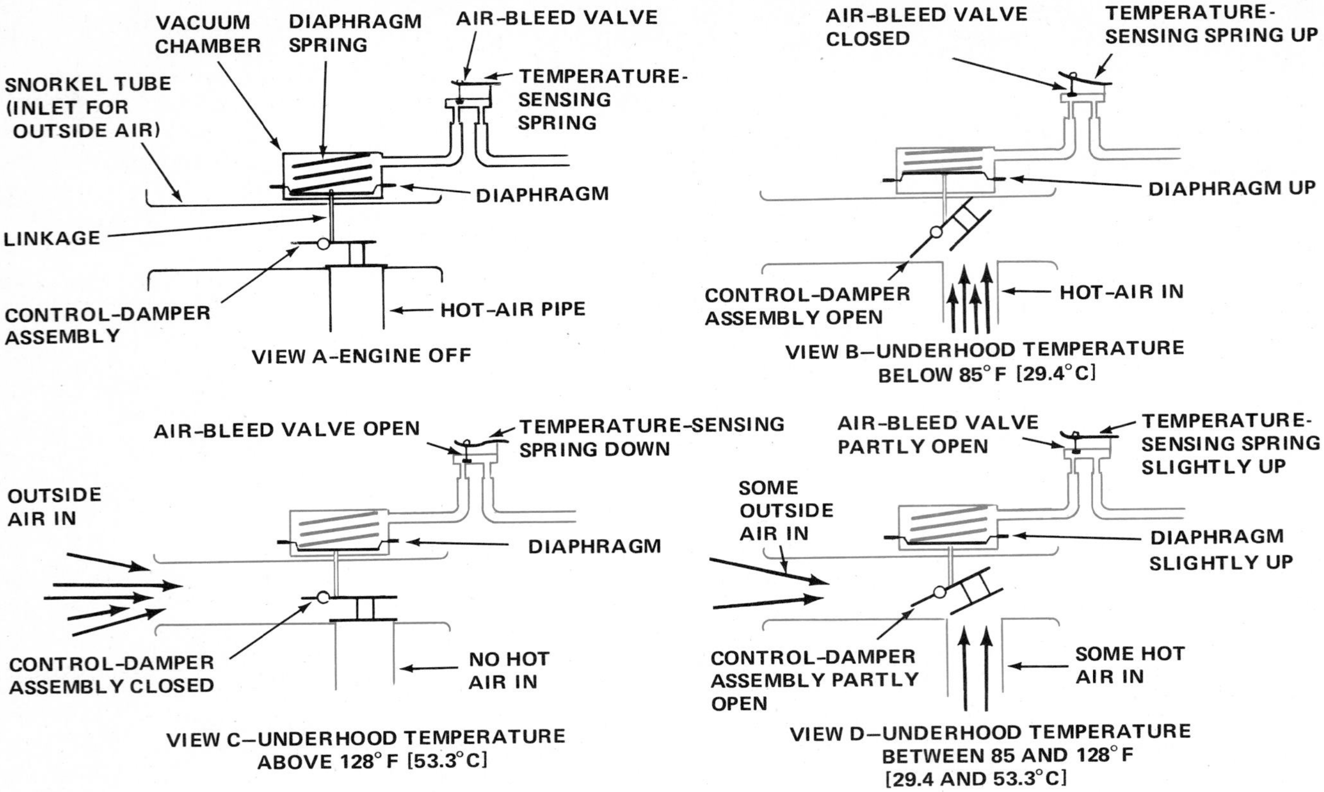

Fig. 12-23 Operational modes for air cleaner with thermostatic control. (*Chevrolet Motor Division of General Motors Corporation*)

holds the bleed valve closed. Now, intake-manifold vacuum is applied to the vacuum chamber. The diaphragm is pushed upward by atmospheric pressure, and the diaphragm spring is compressed. In this position, linkage from the diaphragm raises the control-damper assembly. The snorkel tube is blocked off. All air now has to enter from the hot-air pipe (view *B* in Fig. 12-23). This pipe is connected to the heat stove on the exhaust manifold. Therefore, as soon as the engine starts and the exhaust manifold begins to warm up, hot air is delivered to the carburetor and engine. This improves cold and warm-up operation.

As the engine begins to warm up, the under-the-hood temperature increases. If the under-the-hood temperature goes above 128°F [53.3°C] (in the application shown), the conditions are as shown in view *C* in Fig. 12-23. That is, the temperature-sensing spring has bent enough to open the air-bleed valve. This reduces the vacuum above the diaphragm so that the diaphragm spring pushes the control damper all the way down. Now, all air entering the carburetor comes from under the hood, and none comes from the hot-air pipe.

If the temperature under the hood stays somewhere between 85 and 128°F [29.4 and 53.3°C], conditions are as shown in view *D* in Fig. 12-23. That is, the temperature-sensing spring holds the air-bleed valve partly open. Some vacuum therefore gets to the vacuum chamber above the diaphragm. This vacuum holds the control damper partly open. In this position, some air enters from

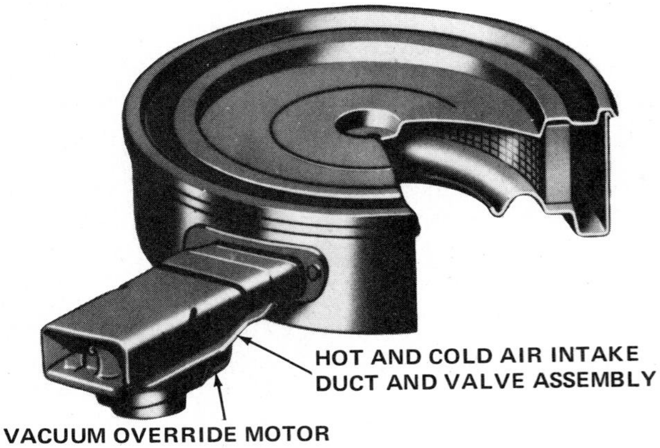

Fig. 12-24 Air cleaner with thermostatic control. (*Ford Motor Company*)

the hood; and some air comes up, through the hot-air pipe, from the heat stove around the exhaust manifold.

A similar thermostatically controlled air cleaner is shown in Figs. 12-24 and 12-25. This design has a thermostatic bulb that acts directly on the valve plate. When the engine is cold, the thermostatic bulb positions the valve plate as shown in Fig. 12-25. Now, all ingoing air must come from the hot-air duct, which is connected to a shroud around the exhaust manifold. As the engine warms up, the hotter air from the shroud causes the thermostatic bulb to move the valve plate. Thus, some air

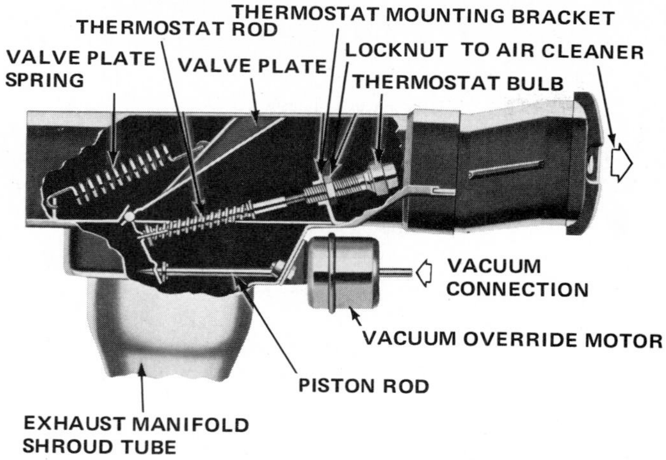

Fig. 12-25 Cutaway view of hot- and cold-air intake ducts and valve assembly. (*Ford Motor Company*)

begins to enter from the engine compartment. With further increases of temperature, the valve plate moves farther so that more compartment air enters. When the engine compartment becomes hot, then most or all ingoing air comes from the engine compartment.

The design shown in Figs. 12-24 and 12-25 includes a vacuum override motor. This motor operates on intake-manifold vacuum. During cold-engine acceleration, when additional air is needed, the motor overrides the thermostatic control. This opens the system to both engine-compartment and heated air, so that adequate air is delivered to the carburetor.

Some air cleaners have a vacuum motor that opens and closes an auxiliary hole in the air-cleaner housing. This type of air cleaner is shown in Fig. 12-26. The vacuum motor operates if a partial vacuum develops in the air cleaner. During cold-weather acceleration, not enough air may get through the heat stove to satisfy engine requirements. When this happens, the partial vacuum in the air cleaner operates the vacuum motor. It then opens the auxiliary air-inlet passage. Now, extra air can enter the air cleaner so engine requirements are satisfied.

● 12-13 Exhaust-Gas Recirculation

The higher the combustion temperature, the more nitrogen oxide forms. Nitrogen oxides (NO_x) contribute to smog, so changes have been made in engines and fuel systems to reduce NO_x. One method is to use an exhaust-gas recirculation (EGR) system. The system sends part of the exhaust gas back through the engine. The exhaust gas reduces combustion temperatures and thus the amount of NO_x coming out the tail pipe. Figure 12-27 shows one system for sending some of the exhaust gas back through the engine. The system has an exhaust-gas-recirculation valve that is controlled by engine vacuum. During part-throttle operation, the vacuum causes the valve to raise its diaphragm and open the port. This allows some of the exhaust gas to enter the intake manifold.

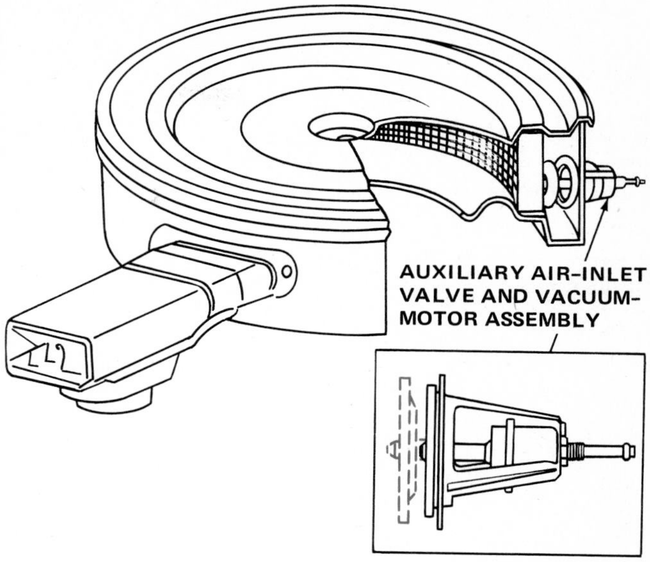

Fig. 12-26 Air cleaner with auxiliary air-inlet valve and vacuum motor. (*Ford Motor Company*)

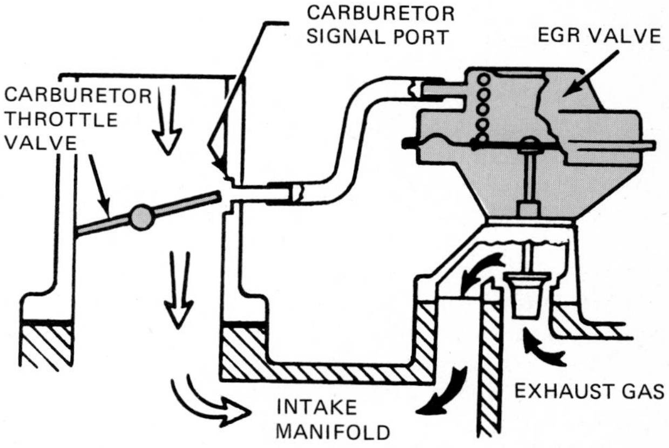

Fig. 12-27 Schematic drawing of an exhaust-gas-recirculating (EGR) system. (*Chevrolet Motor Division of General Motors Corporation*)

Another system uses high valve overlap. That is, the camshaft cams allow the exhaust and intake valves to stay open longer at the same time. This allows more of the exhaust gases to stay behind in the cylinders and mix with the incoming air-fuel mixture. Part Five describes these systems in detail.

● 12-14 Crankcase Ventilation

Crankcases must be ventilated. Some blow-by always gets past the piston rings and enters the crankcase. In addition, water and liquid fuel appear in the crankcase during cold-engine operation. These must be cleared from the crankcase before they cause trouble. In earlier engines, the crankcase was ventilated by an opening at the front of the engine and a vent tube at the back. The forward motion of the car, plus rotation of the camshaft, caused air to flow through and remove the blow-by, water, and fuel. Today, however, closed-crankcase ventilating systems are used on all cars. Air circulates through the crankcase as before. But the air then enters the carburetor and intake manifold. Any blow-by, water vapor, or unburned fuel is therefore sent back through the engine, instead of being emitted into the air. Part Five covers closed-crankcase ventilating systems.

● 12-15 Exhaust System

The exhaust system includes the exhaust manifold, exhaust pipe, muffler, and tail pipe (Figs. 7-21 and 7-22). Some V-8 engines have a crossover pipe to connect their two exhaust manifolds. Other cars with V-8 engines use two separate exhaust systems (dual system), one for each cylinder bank (Fig. 7-21). This improves the "breathing" ability of the engine, allowing it to exhaust more freely and thus increase power output to some extent.

Exhaust systems on most modern cars include a catalytic converter, mentioned in Chap. 17 and discussed in detail in Part Five.

Exhaust manifolds have heat control valves. These valves close when the engine is cold, and this directs heat to the intake manifold. The heat helps vaporize the ingoing gasoline and thus improves cold-engine operation. Manifold heat control valves are covered in detail in ● 13-26.

Some engines have exhaust manifolds equipped with air-injection systems. The system includes an air pump and a series of injection tubes in the exhaust manifold. In operation, the air pump sends a flow of air into the exhaust manifold opposite the exhaust valves. This extra air helps to burn any HC or CO still left in the exhaust gases. The air-injection system is covered in Part Five.

● 12-16 Muffler

The muffler (Fig. 12-28) contains a series of holes, passages, and resonance chambers to absorb and damp out the high-pressure surges introduced into the exhaust system as the exhaust valves open. This quiets the exhaust. Some new systems do not use a muffler. Instead, the exhaust pipe has a series of scientifically shaped restrictions that damp out the exhaust noises without unduly restricting the flow of exhaust gases.

To further reduce exhaust noises, many exhaust pipes are made from a three-ply laminate. The laminate consists of a plastic film sandwiched between two metal skins. The laminate has very good sound-deadening properties and thus holds in the exhaust noises.

● 12-17 Catalytic Converters

Late-model cars have catalytic converters in the exhaust system. These are muffler-like containers filled with catalysts. The catalyts convert the pollutants coming from the engine into harmless gases (see Part Five).

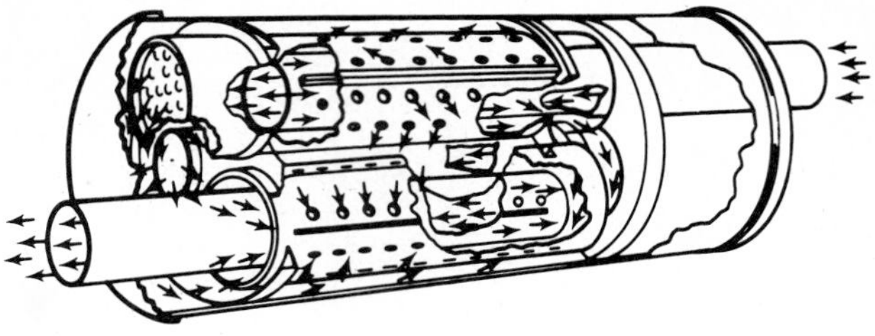

Fig. 12-28 Exhaust muffler in cutaway view. The arrows show the path of exhaust-gas flow through the muffler. (*Chevrolet Motor Division of General Motors Corporation*)

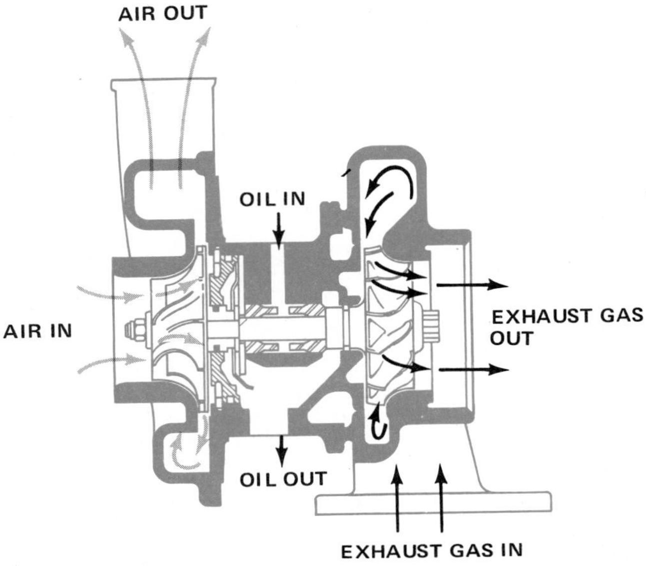

Fig. 12-29 Operation of a turbocharger. (*Schwitzer Division, Wallace-Murray Corporation*)

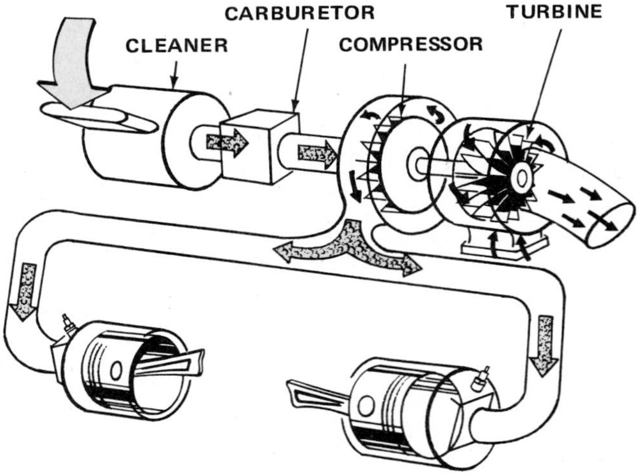

Fig. 12-30 Schematic layout of the turbocharger on a flat-six-cylinder engine. (*Chevrolet Motor Division of General Motors Corporation*)

● 12-18 Turbocharger

To get more air-fuel mixture into the engine, a *supercharger*, or *turbocharger*, can be used. This is a rotary air pump that boosts engine power, sometimes dramatically. It has been called a supercharger because it delivers a "super" charge of air-fuel mixture to the engine. Some early superchargers were driven by mechanical means (gears or chain) from the engine. The supercharger contains a compressor rotor, which is a wheel with blades. The wheel is much like the impeller in a water pump. When it spins, it moves air by centrifugal force. The air between the rotor blades is pushed outward and through the outlet port of the supercharger. The air exits at relatively high pressure.

The mechanical drive gave trouble and used considerable engine power. To avoid these problems, the turbosupercharger, or turbocharger, was developed. This assembly has a turbine that drives the compressor rotor (Fig. 12-29). The turbine itself is driven by the pressure of the exhaust gases from the engine. Figure 12-30 shows schematically the installation of a turbocharger on a flat-six engine. In the installation shown, the compressor is compressing the air-fuel mixture after it leaves the carburetor. In most systems, the turbocharger compresses air alone (Figs. 12-31 and 12-32).

A somewhat different turbocharger installation is shown in Fig. 12-31. This shows the turbocharger mounted on a V-8 engine. Note that the crossover pipe brings the exhaust gases from the opposing bank of cylinders to the turbine of the turbocharger. There it unites with the exhaust gases from the cylinder bank on which the turbocharger mounts. Note also that the compressor is compressing air alone and sending it to the carburetor.

Figure 12-32 shows a turbocharger installation on a four-cylinder, fuel-injected engine. Air alone is compressed and is delivered to the intake manifold. Fuel is injected into the intake manifold by the fuel-injection system. The mixture enters the cylinders when the intake valves open. Fuel injection is discussed in Chap. 15. On this engine, the power output is boosted 30 percent, or to 170 hp [126.82 kW (kilowatt)]. More dramatic power boosts have been made with turbochargers. For example, engines designed specifically for competition racing put out as much as 200 hp [149.2 kW] *per cylinder*. The Offenhauser (or "Offy") four-cylinder engine used in the Indianapolis 500 has a 159 cubic inch [2,605 cc] piston displacement (● 10-9). In other words, it is a relatively small engine. Yet with the turbocharging, this engine produces in the range of 600 to 800 hp [448 to 597 kW]. This is an extreme example, however, because the engine is designed from the ground up for competitive racing using a turbocharger.

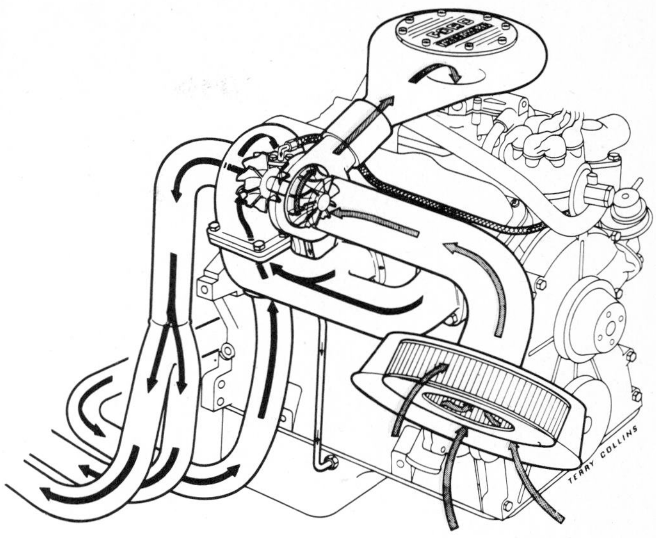

Fig. 12-31 Turbocharged version of the Ford 182 cubic inch [3 L] V-6 engine. *(Ford Motor Company)*

Fig. 12-32 Turbocharger on a 121 cubic inch [1,982 cc] engine. *(BMW Corporation)*

● 12-19 LPG Fuel System

Liquefied petroleum gas (LPG) is a fuel that is liquid only under pressure (see ● 11-28). When the pressure is reduced, the fuel vaporizes. Thus, the system must have a pressure-tight fuel tank to store the fuel at high pressures. A typical LPG fuel system is shown in Fig. 12-33. Pressure forces fuel through the filter, high-pressure regulator, and vaporizer. The high-pressure regulator reduces the pressure so that the fuel starts to turn to vapor. This vaporizing process is completed in the vaporizer. The vaporizer has an inner tank surrounded by a water jacket through which cooling-system water passes. The water adds heat to the fuel so that it is well vaporized. It then passes through the low-pressure regulator, where the pressure is further reduced. It then enters the carburetor. The carburetor is simply a mixing valve that mixes the vaporized fuel and air. The low-pressure regulator reduces pressure to slightly *below* atmospheric pressure. This prevents it from flowing into the carburetor when the engine is off. Fuel will flow only when the engine is running and there is a vacuum in the carburetor venturi (or air horn).

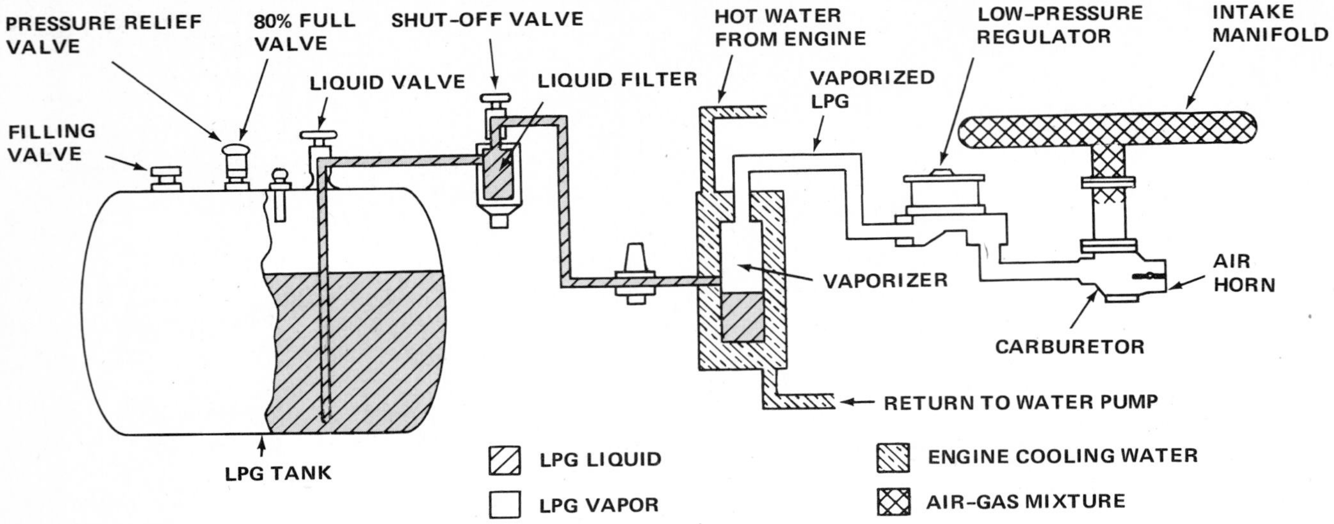

Fig. 12-33 Schematic view of an LPG fuel system.

LPG fuel systems have been used on some cars, trucks, buses, and fork-lift and platform trucks. This system is well suited for such applications since the fuel burns clean. The exhaust gases have few contaminants.

REVIEW QUESTIONS

1. Name four components of the carbureted fuel system.
2. What are the two types of fuel gauges?
3. Explain how a fuel pump works.
4. What is the purpose of the vapor-return line?
5. What is the purpose of the vapor recovery system?
6. Describe how two types of electric fuel pump work.
7. What is the purpose of the air cleaner? How does it work?
8. What is the purpose of the heated air system? How does it work?
9. What is the purpose of the EGR system? How does it work?
10. Why does high valve overlap reduce NO_x?
11. What is the purpose of crankcase ventilation? How does it work?
12. What is the purpose of the catalytic converter?
13. What is the purpose of the muffler? How does it do its job?
14. What is the purpose of the turbocharger? How does it work?
15. Explain how the LPG system works.

SELF PROJECTS

"Hands-on" experience will help you understand how the components work. For example, when you read about fuel filters, fuel pumps, charcoal canisters, and so on, try to handle these parts. If you cannot locate the separate assemblies, locate them on cars. Do this:

When you raise the hood of a car, note the locations of the fuel pump, the fuel lines, the charcoal canister, the air cleaner on top of the carburetor, the exhaust manifold. Make a drawing locating each of these parts. At the top of the sheet, write the model year and name of the car. Compare these drawings. As you build up a file of drawings, you will begin to see similarities and differences in the locations of the fuel-system components. File these drawings in your notebook.

CHAPTER 13
Automotive Carburetors

After studying this chapter, you should be able to:

- **Discuss carburetors and explain how a simple fixed-venturi carburetor works.**
- **Explain the difference between a fixed-venturi and a variable-venturi carburetor.**
- **List and describe the operation of the six systems in a fixed-venturi carburetor. Locate the components of these systems in typical carburetors.**
- **Explain how a variable-venturi carburetor works.**

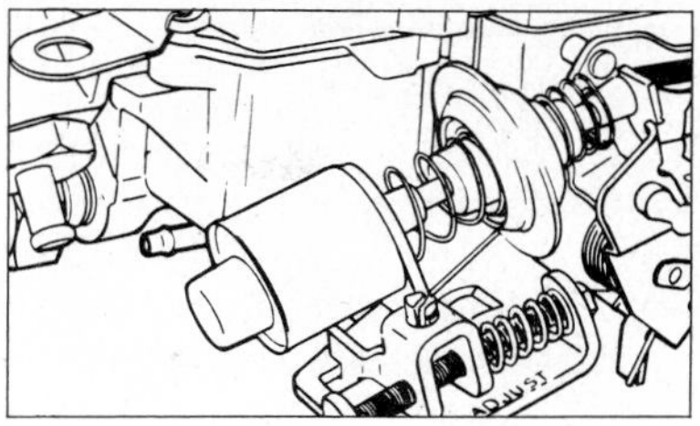

In previous chapters, we discussed automotive fuels and carbureted fuel systems. Now, in this chapter, we look at automotive carburetors and find out about the various systems in carburetors and how they work.

13-1 Carburetor Types

There are two basic types of carburetor, fixed-venturi and variable-venturi (VV). The venturi is the restricted place in the carburetor air passage through which the air must flow. As we explain later, this restriction produces a partial vacuum. The vacuum causes a fuel nozzle to discharge gasoline into the air passing through. The gasoline mixes with the air to produce the combustible mixture the engine needs to run.

Most carburetors installed in cars made in the United States are of the fixed-venturi type, and we discuss these first. Many foreign cars, and an increasing number of cars made in the United States are using the variable-venturi type of carburetor. We discuss this type of carburetor later in the chapter.

13-2 Carburetion

Carburetion is the mixing of the gasoline fuel with air so that a combustible mixture is obtained. The carburetor performs this job, supplying a combustible mixture of varying degrees of richness to suit engine operating conditions. The mixture must be rich (have a higher percentage of fuel) for starting, acceleration, and high-speed operation. A less rich (leaner) mixture is desirable at intermediate speed with a warm engine. The carburetor has several systems through which air-fuel mixture flows during different operating conditions. These systems produce the varying richness of the air-fuel mixture required for the varying operating conditions. All this is explained in sections that follow.

13-3 Vaporization

When a liquid changes to a vapor (undergoes a change of state), it is said to evaporate. Water placed in an open pan will evaporate: it changes from a liquid to a vapor. Clothes hung on a line dry: the water in the clothes turns to vapor. When the clothes are well spread out, they dry more rapidly than when they are bunched together. This illustrates an important fact about evaporation. The greater the surface exposed, the more rapidly evaporation takes place. A pint of water in a tall glass takes quite a while to evaporate. But a pint of water in a shallow pan evaporates much more quickly (Fig. 13-1).

Fig. 13-1 Water evaporates from the shallow pan faster than from the glass. The greater the area exposed to air, the faster the evaporation.

13-4 Atomization

In order to produce very quick vaporization of the liquid gasoline, it is sprayed into the air passing through the carburetor. Spraying the liquid turns it into many fine droplets. This effect is called *atomization* because the liquid is broken up into small droplets (but not actually into atoms, as the name implies). Each droplet is exposed to air on all sides so that it vaporizes very quickly. Thus, during normal running of the engine, the gasoline sprayed into the air passing through the carburetor turns to vapor, or vaporizes, almost instantly.

13-5 Carburetor Fundamentals (Fixed Venturi)

A simple fixed-venturi carburetor could be made from a round cylinder with a constricted section, a fuel nozzle, and a round disk, or valve (Fig. 13-2). The round cylinder is called the *air horn,* the constricted section the *venturi,* and the valve the *throttle valve.* The throttle valve can be tilted more or less to open or close the air horn (Fig. 13-3). In the horizontal position, it shuts off, or *throttles,* the air flow through the air horn. When the throttle is turned away from this position, air can flow through the air horn.

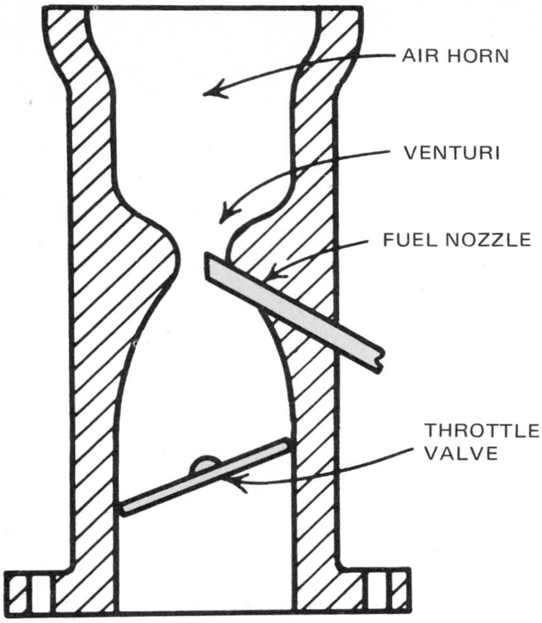

Fig. 13-2 Simple carburetor consisting of a round air horn, a fuel nozzle, and a throttle valve.

Fig. 13-3 Throttle valve in the air horn of a carburetor. When the throttle is closed, as shown, little air can pass through. But when the throttle is opened, as shown dashed, there is little throttling effect.

13-6 Venturi Effect

As air flows through the constriction, or venturi, a partial vacuum is produced at the venturi. This vacuum causes the fuel nozzle to deliver a spray of gasoline into the passing air stream. The venturi effect (of producing a vacuum) can be illustrated with the setup shown in Fig. 13-4. Here, three dishes of mercury (a very heavy metallic liquid) are connected by tubes to an air horn with a venturi. The greater the vacuum, the higher the mercury is pushed up in the tube by atmospheric pressure. Note that the greatest vacuum is right at the venturi. Also, it should be remembered that the faster the air flows through, the greater the vacuum.

Why is there a vacuum at the venturi? A simple explanation might be as follows: The air is made up of countless molecules. As air moves into the top of the air horn, all the air molecules move at the same speed. But if all are to get through the venturi, they must speed up and move through faster. For instance, let us look at two molecules, one behind the other. As the first molecule enters the venturi, it speeds up, tending to leave the second molecule behind. The second molecule also speeds up as it enters the venturi. But the first molecule has, in effect, a head start. Thus, the two molecules are farther apart in the venturi than they were before they entered it. Now imagine a great number of particles going through the same action. As they pass through the venturi, they are farther apart than before they entered. This is just another way of saying that a partial vacuum exists in the venturi. A partial vacuum is a thinning out of the air, a more-than-normal distance between air molecules.

13-7 Fuel-Nozzle Action

The partial vacuum occurs in the venturi, just where the end of the fuel nozzle is located. The other end of the fuel

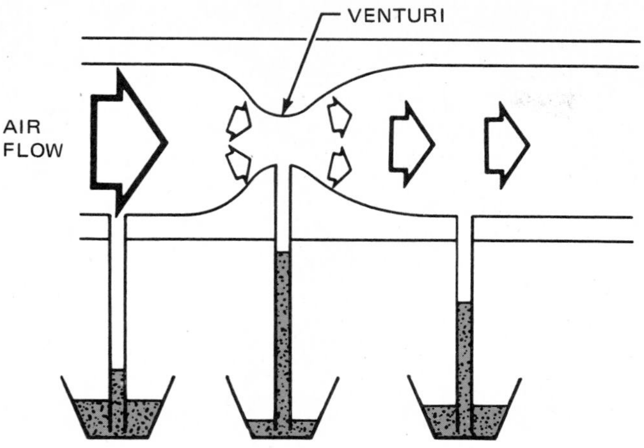

Fig. 13-4 Three dishes of mercury and tubes connected to an air horn. They show differences in vacuum by the distances the mercury rises in the tubes. The venturi has the highest vacuum.

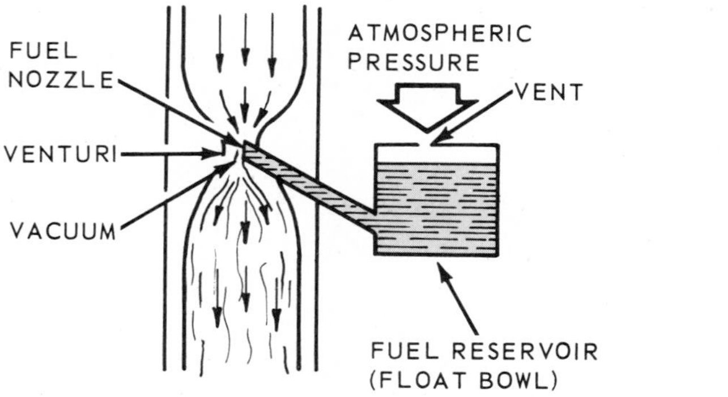

Fig. 13-5 The venturi, or constriction, causes a vacuum to develop in the air stream just below the constriction. Then atmospheric pressure pushes fuel up and out the fuel nozzle.

nozzle is in a fuel reservoir (the float bowl), as shown in Fig. 13-5. Atmospheric pressure pushes on the fuel through a vent in the float-bowl cover. With the vacuum at the upper end of the nozzle, fuel is pushed up through the nozzle. The fuel enters the passing air stream as a fine spray. It quickly turns to vapor as the droplets of fuel evaporate. The more air that flows through the air horn, the greater the vacuum in the venturi. The greater the vacuum, the more fuel is delivered.

13-8 Throttle-Valve Action

The throttle valve can be tilted in the air horn to allow more or less air to flow through (Fig. 13-3). When it is tilted to allow more air to flow, larger amounts of air-fuel mixture are delivered to the engine. The engine develops more power and tends to run faster. But if the throttle valve is tilted to throttle off most of the air, then only small amounts of air-fuel mixture are delivered. The engine produces less power and tends to slow down. The throttle valve is linked to an accelerator pedal in the driver's compartment. The linkage may consist of a series

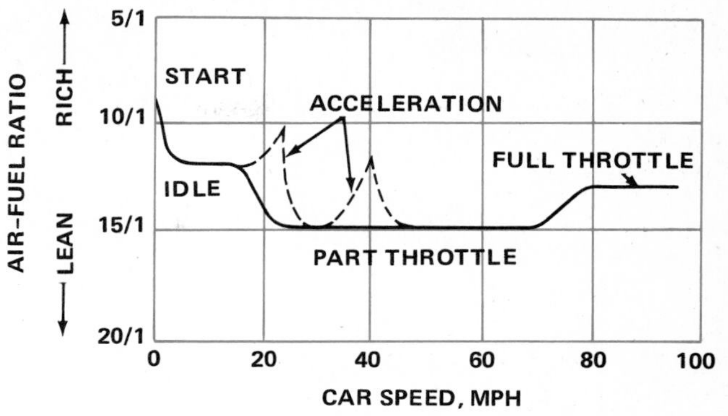

Fig. 13-6 Graph of air-fuel ratios for different car speeds. The graph is typical. Car speeds at which the various ratios are obtained may vary with different cars and engines. Also, there may be some variations in the ratios.

of interconnected levers and rods or levers and a cable. The cable consists of a flexible outer covering and an inner sliding wire. With either arrangement, movement of the accelerator pedal causes the throttle valve to change its position in the carburetor air horn. Thus, the driver can position the throttle valve to suit operating requirements. Whenever you look under the hood of a car, check the throttle-valve linkage and become acquainted with the various arrangements. Note that the throttle linkage not only controls the throttle valve, but performs other functions such as producing a downshift of the automatic transmission under the right operating conditions.

13-9 Air-Fuel-Ratio Requirements

As already noted, the fuel system must vary the air-fuel ratio to suit different operating requirements. The mixture must be rich (have a high proportion of fuel) for starting. It must be leaner (have a lower proportion of fuel) for part-throttle medium-speed operation. Figure 13-6 is a graph showing typical air-fuel ratios as related to various car speeds. Ratios, and the speeds at which they are obtained, vary with different cars. In the example shown, a rich mixture of about 9:1 (9 pounds [4.082 kg] of air for each pound [0.454 kg] of fuel) is supplied for starting. Then, during idle, the mixture leans out to about 12:1. At medium speeds, the mixture further leans out to about 15:1. But at higher speeds, with a wide-open throttle, the mixture is enriched to about 13:1. Opening the throttle for acceleration at any speed causes a momentary enrichment of the mixture. This results from special carburetor systems which we will study later. Two examples are shown in Fig. 13-6 (at about 23 mph [37.01 km/h] and at 40 mph [64.36 km/h]).

You might think that the engine itself demands varying air-fuel ratios for different operating conditions. This is not quite true. For example, the mixture must be very rich for starting because fuel vaporizes very poorly under starting conditions. The engine and carburetor are cold, the air speed is low, and much of the fuel does not vaporize. Thus, an extra amount of fuel must be delivered by the carburetor so that enough will vaporize for starting. Likewise, sudden opening of the throttle for acceleration allows a sudden inrush of air. Extra fuel must enter at the same time (that is, the mixture must be enriched). This is because only part of the fuel vaporizes and mixes with the ingoing air to provide the proper proportions of air and fuel in the engine.

The following sections describe the various systems in carburetors that supply the air-fuel mixture required for different operating conditions.

In some late-model cars, the air-fuel ratio, or mixture richness, is controlled electronically. The electronic control system is explained in a later chapter.

● 13-10 Carburetor Systems

The systems (or circuits as they are sometimes called) in the carburetor are:

1. Float system
2. Idle system
3. Main-metering system
4. Power system
5. Accelerator-pump system
6. Choke system

These systems are discussed in detail in ● 13-11 to 13-25.

● 13-11 Float System

The float system includes the float bowl and a float and needle-valve arrangement. The float and the needle valve maintain a constant level of fuel in the float bowl. If the level is too high, then too much fuel will feed from the fuel nozzle. If it is too low, too little fuel will feed. In either event, poor engine performance will result. Figure 13-7 is a simplified drawing of the float system. If fuel enters the float bowl faster than it is withdrawn, the fuel level rises. This causes the float to move up and push the needle valve into the valve seat. This, in turn, shuts off the fuel inlet so that no fuel can enter. Then, if the fuel level drops, the float moves down and releases the needle so that the fuel inlet is opened. Now fuel can enter. In actual operation, the fuel is kept at an almost constant level. The float tends to hold the needle valve partly closed so that the incoming fuel just balances the fuel being withdrawn.

Figure 13-8 shows an actual carburetor with a dual float assembly partly cut away so that the two floats can be seen. The carburetor has a float bowl that partly surrounds the carburetor air horn. The two floats are attached by a U-shaped lever and operate a single needle valve. Some carburetors have an auxiliary fuel valve and inlet, as shown in Fig. 13-9. During heavy-load or high-speed operation, fuel may be withdrawn from the float bowl faster than it can enter through the main fuel inlet. If this happens, the fuel level drops. The end of the float lever presses against the auxiliary valve, pushing it up-

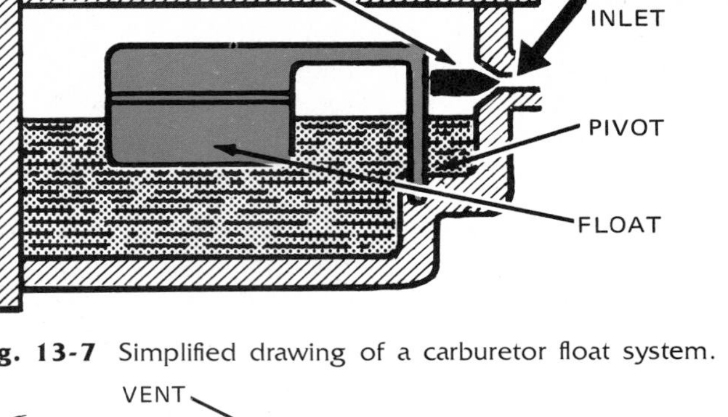

Fig. 13-7 Simplified drawing of a carburetor float system.

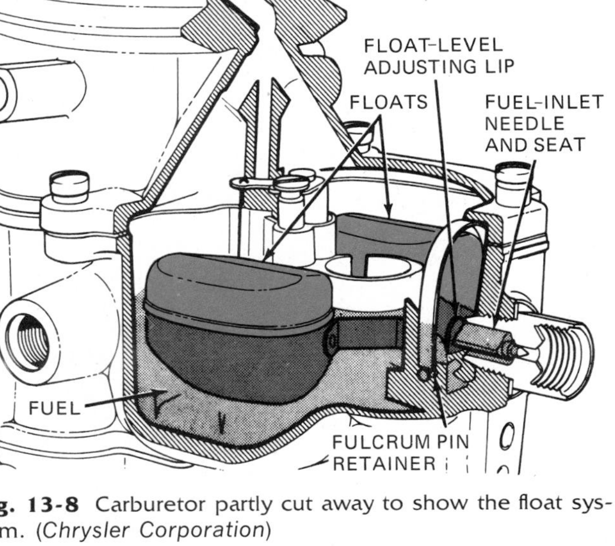

Fig. 13-8 Carburetor partly cut away to show the float system. (*Chrysler Corporation*)

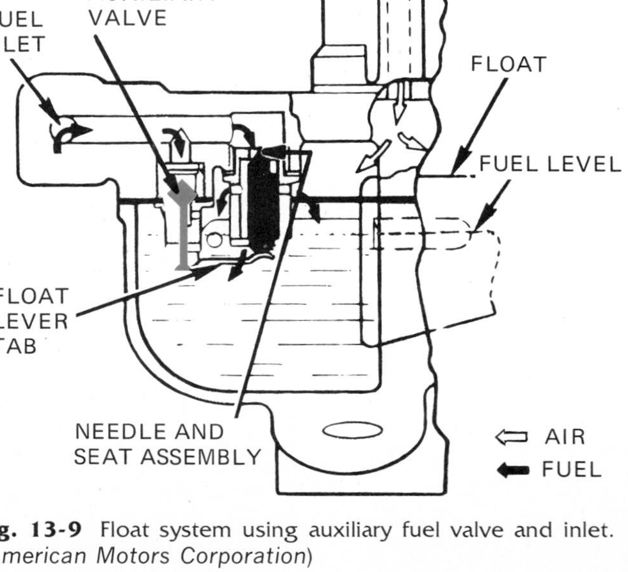

Fig. 13-9 Float system using auxiliary fuel valve and inlet. (*American Motors Corporation*)

ward. This opens the auxiliary fuel inlet so additional fuel can enter.

A number of years ago, some four-barrel carburetors had two sets of floats. The four-barrel carburetor is, in effect, two two-barrel carburetors. As will be explained later, the primary barrels supply the engine during most operating conditions. But the secondary barrels come into operation during acceleration and high speed, for improved performance. The purpose of using two sets of floats was to provide, in effect, separate float systems for each pair of barrels. However, more recent four-barrel carburetors have a single centrally located float. The two-float system requires a large float bowl and has other disadvantages that the single or dual floats do not. For example, the single or dual floats are more centrally located so they respond more accurately to fuel needs. Also, the float bowl is smaller, so there is less of a problem with fuel evaporation and the possibility of atmospheric pollution from escaping HC.

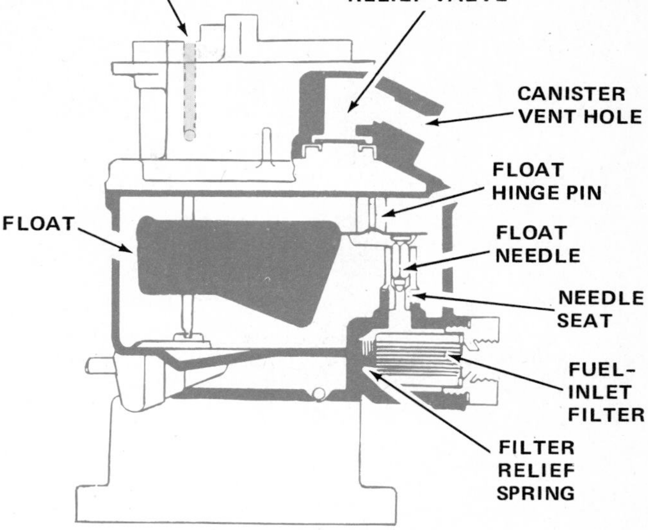

Fig. 13-10 Float system showing the two vents, one internal and the other to the charcoal canister. (*Chevrolet Motor Division of General Motors Corporation*)

● 13-12 Float-Bowl Vents

The float bowls of carburetors are vented into the carburetor air horn at a point above the choke valve (see top of Fig. 13-8; upper left in Fig. 13-10). The purpose of the vent is to equalize the effects of a clogged air cleaner. For example, suppose the air cleaner has become clogged with dirt. The passage of air through it is then restricted. As a result, a partial vacuum develops in the carburetor air horn. Therefore, a somewhat greater vacuum is applied to the fuel nozzle (since this vacuum is added to the venturi vacuum). However, the partial vacuum resulting from the clogged air cleaner is also applied to the float bowl (through the vent). Therefore, the only driving force that pushes fuel from the fuel nozzle is the air pressure in the air cleaner. This is less than atmospheric pressure. Thus, the vent makes up for the effect of a clogged air cleaner. If the float bowl were vented to the atmosphere, then atmospheric pressure would be the driving force. This would produce a greater fuel flow from the fuel nozzle, and the mixture would be too rich.

The carburetor which has the bowl vented into the carburetor air horn is called a *balanced carburetor.* Almost all carburetors used today are of this type.

The float bowl has another vent, shown to the upper right in Fig. 13-10. This vent is connected by a tube to the charcoal canister which is part of the fuel-vapor recovery system (● 12-9). In the carburetor in Fig. 13-10, the float bowl has a pressure-relief valve. The valve opens when vapor pressure increases in the float bowl. This allows the fuel vapor to flow to the charcoal canister. In other carburetors the vent to the charcoal canister has a valve operated by the accelerator-pump lever. The valve is opened when the engine is idling or when it has been turned off.

Another valve-opening arrangement is shown in Fig. 13-11. In this carburetor, the pressure-relief valve is controlled by a solenoid. When the ignition is turned on, the solenoid is connected to the battery. It pulls the solenoid plunger down to position *E*. This blocks off the vent to the charcoal canister (*G* and *H*). Now, the float bowl is vented to the top of the air horn. When the ignition is turned off, the solenoid is disconnected from the battery. Now, the solenoid spring pushes the plunger up into position *E*. This blocks the vent to the upper part of the air horn. At the same time, the vent to the charcoal canister (*G* and *H*) is open.

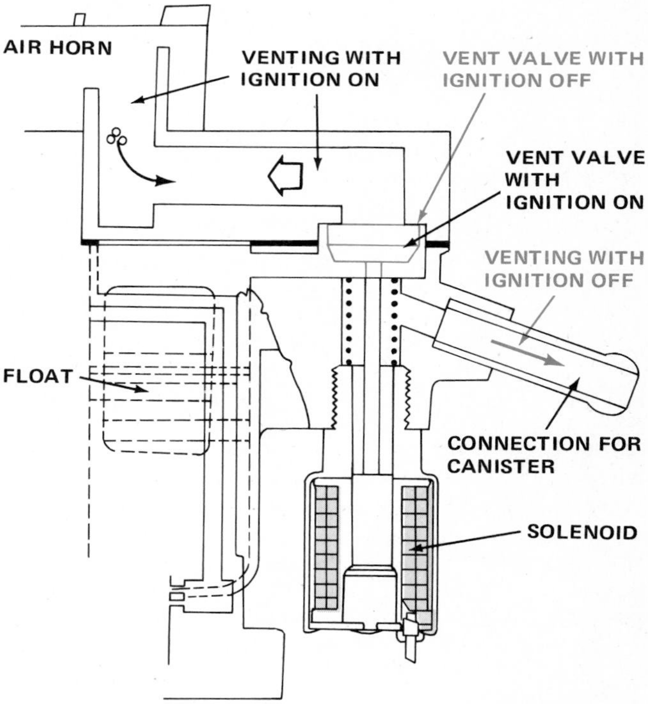

Fig. 13-11 Sectional view of carburetor showing solenoid-controlled vent to charcoal canister. (*Ford Motor Company*)

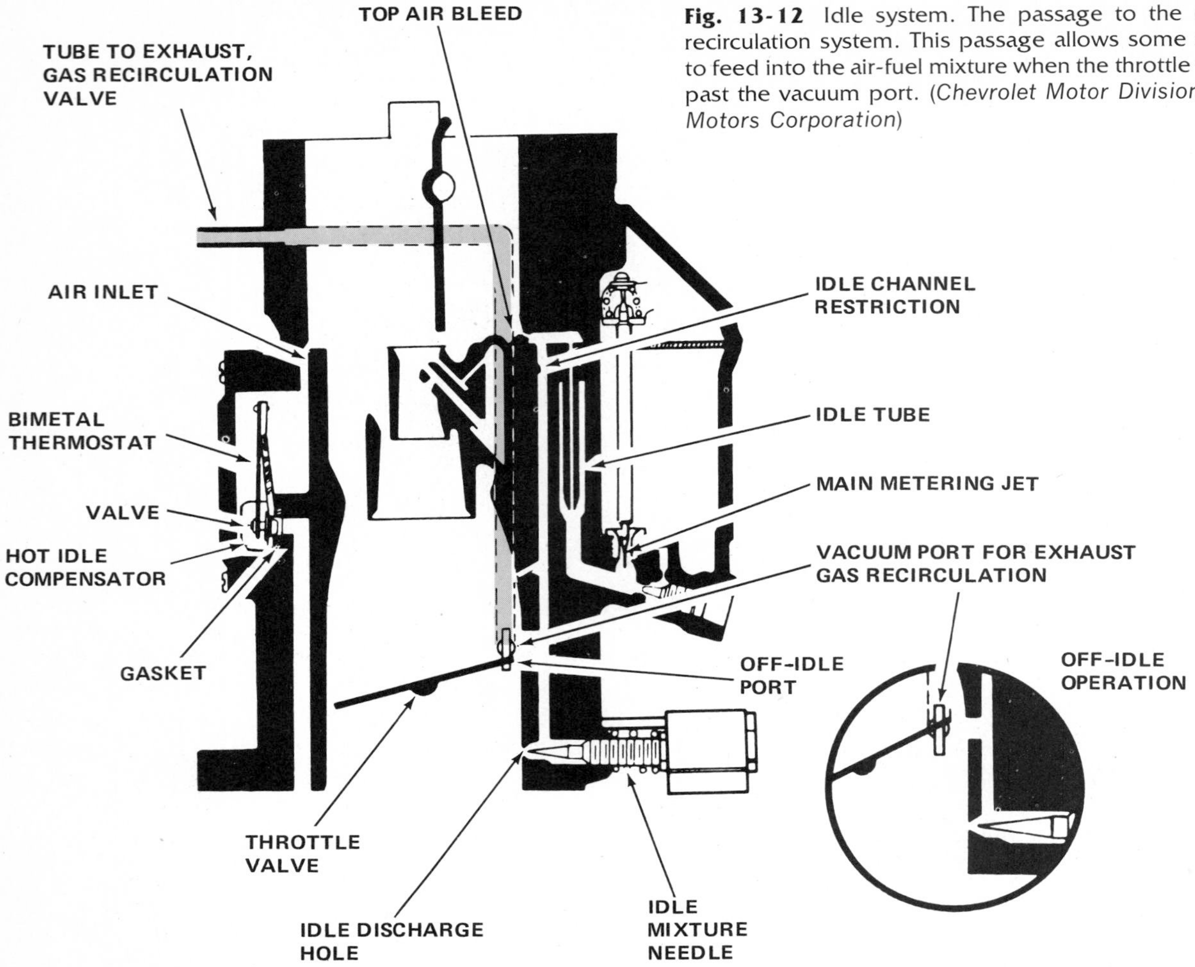

Fig. 13-12 Idle system. The passage to the exhaust-gas recirculation system. This passage allows some exhaust gas to feed into the air-fuel mixture when the throttle valve opens past the vacuum port. (*Chevrolet Motor Division of General Motors Corporation*)

13-13 Hot-Idle Compensator Valve

The internal vent could be a problem during idling or low-speed operation, especially during hot weather. Gasoline vapor from the float bowl can pass through the internal vent in sufficient amounts to upset the air-fuel ratio. That is, the gasoline vapor adds to the normal air-fuel mixture, and the mixture becomes too rich. To take care of this, some carburetors have a hot-idle compensator valve, as shown to the left in Fig. 13-12. This valve is operated by a thermostatic blade. When temperatures reach a preset valve, the blade bends enough to open the valve port. Now additional air can flow through the auxiliary air passage. This additional air bypasses the idle system. It leans out the mixture enough to make up for the added gasoline vapor coming from the float bowl.

13-14 Idle System

When the throttle is closed or only slightly opened, only a small amount of air can pass through the air horn. The air speed is low, and very little vacuum develops in the venturi. This means that the fuel nozzle does not feed fuel. Thus, the carburetor must have another system to supply fuel when the throttle is closed or slightly opened.

This system, called the *idle system,* is shown in operation in Fig. 13-13. It includes passages through which air and fuel can flow. The air passage is called the *air bleed.* With the throttle closed as shown, there is a high vacuum below the throttle valve from the intake manifold. Atmospheric pressure pushes air and fuel through the passages as shown. They mix and flow past the tapered point of the idle air-fuel-mixture adjustment screw. The mixture has a high proportion of fuel (is very rich). It leans out somewhat as it mixes with the small amount of air that gets past the closed throttle valve. But the final mixture is still rich enough (see Fig. 13-6) for good idling. The richness can be adjusted by turning the idle air-fuel-mixture adjustment screw in or out. This permits less or more air-fuel mixture to flow past the screw.

CAUTION: In late-model cars, the idle air-fuel-mixture adjustment screw is fixed or has a locking cap on it. The reason for this is that it is illegal to adjust the idle mixture beyond specific limits. It has been set according to federal standards and must not be tampered with.

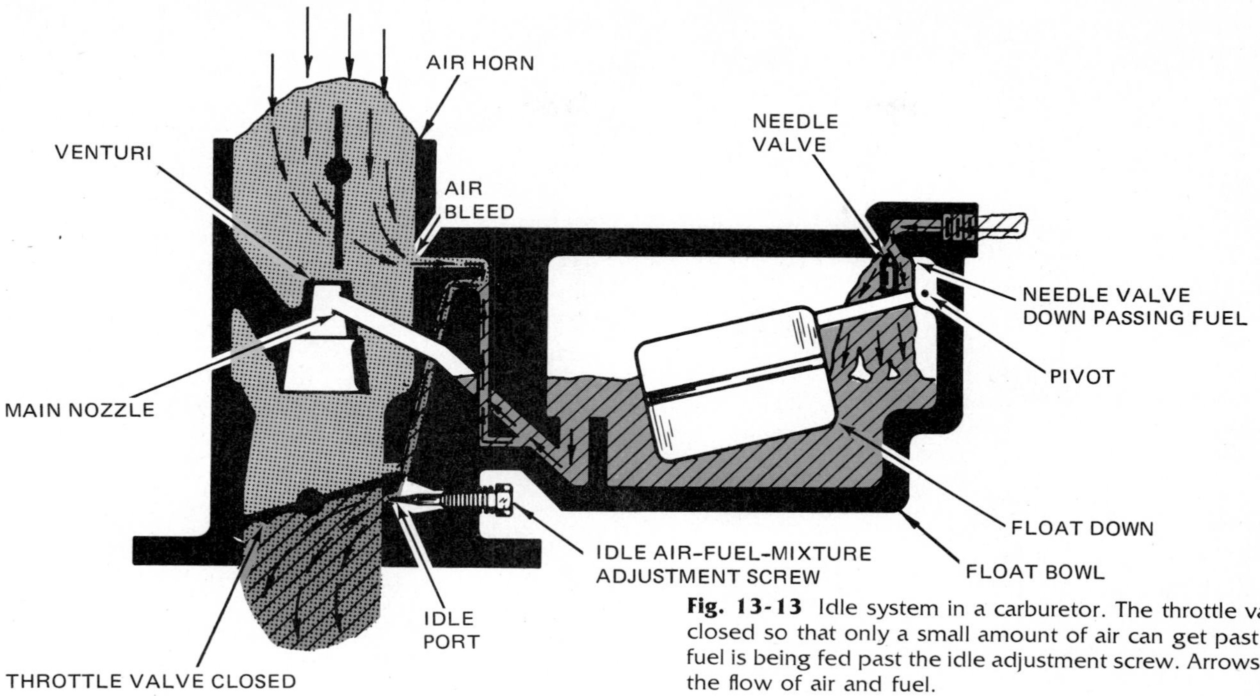

Fig. 13-13 Idle system in a carburetor. The throttle valve is closed so that only a small amount of air can get past it. All fuel is being fed past the idle adjustment screw. Arrows show the flow of air and fuel.

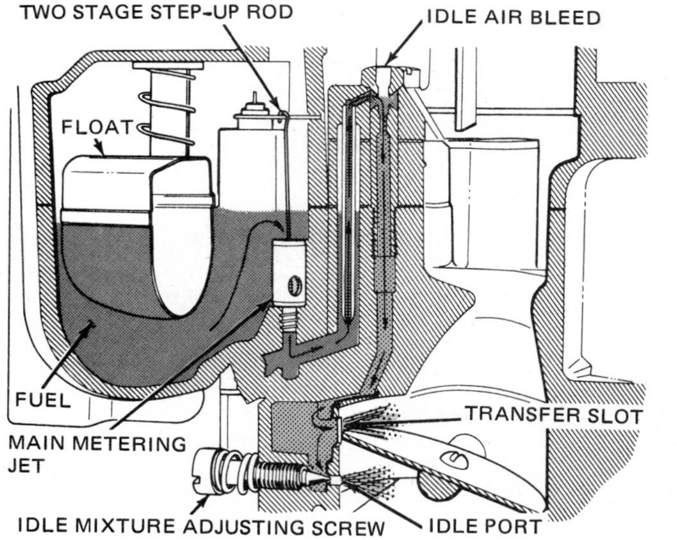

Fig. 13-14 Idle system in a carburetor. Carburetor cut away to show the internal arrangement. (*Chrysler Corporation*)

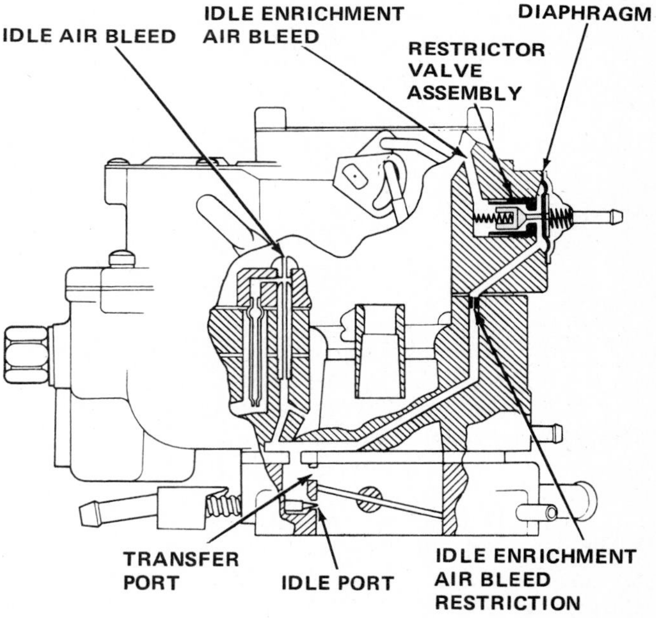

Fig. 13-15 Location of the restrictor valve in the cold idle enrichment system. (*Chrysler Corporation*)

Figure 13-14 shows a cutaway view of a carburetor with the idle system in operation.

Some carburetors have an extra feature in the idle system to provide some enrichment of the idle mixture during cold start-up. This enrichment is in addition to the choke action (● 13-24). The arrangement includes a restrictor valve in the air bleed to the idle system (Fig. 13-15). This restrictor valve is controlled by a vacuum diaphragm. When vacuum is applied to the vacuum diaphragm, the restrictor valve partly closes off the air flow. The air loss in the idle system then causes additional fuel to feed from the idle system. This improves cold idle and combats any tendency for the engine to stall. The vacuum diaphragm gets its vacuum through a thermal switch that is mounted in the cooling system. When the coolant is cold, the thermal switch opens to allow intake-manifold vacuum to operate the vacuum diaphragm. Therefore, the idle air-fuel mixture is enriched. When the engine warms up, the thermal switch closes to shut off the vacuum to the vacuum diaphragm. Now, normal hot-engine idle results, with normal hot-engine air-fuel mixture feeding from the idle system.

● 13-15 Low-Speed Operation

When the throttle is opened slightly, as shown in Fig. 13-16, the edge of the throttle valve moves past the low-speed port in the side of the air horn (also called the transfer slot—see Fig. 13-14). This port is a vertical slot or a series of small holes, one above the other. Additional fuel is thus fed into the intake manifold through the low-speed port. This fuel mixes with the additional air moving past the slightly opened throttle valve. It provides sufficient mixture richness for part-throttle low-speed operation.

Some air bleeds around the throttle plate through the low-speed port when the edge of the throttle is only partway past this port. This air improves the atomization of the fuel coming from the low-speed port.

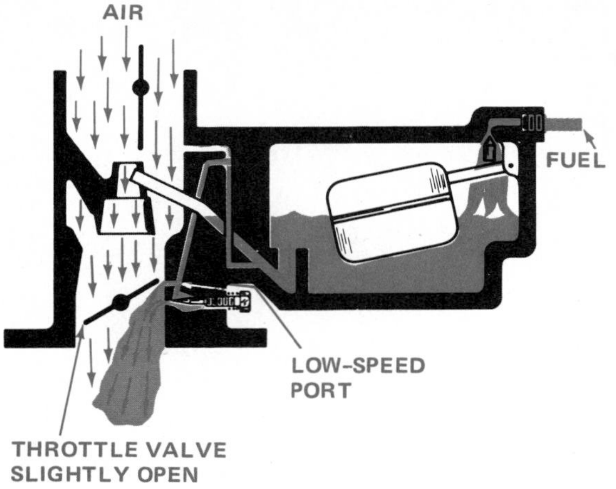

Fig. 13-16 Low-speed operation. The throttle valve is slightly open, and fuel is being fed through the low-speed port and through the idle port. The dark color is fuel; the light color is air.

● 13-16 Other Idle Systems

There are many varieties of idle systems in addition to that shown in Figs. 13-14 to 13-16. In two-barrel carburetors, each barrel has its own idle system. In many four-barrel carburetors, only the primary barrels have idle systems (● 13-32).

● 13-17 Main Metering System

Suppose the throttle valve is opened enough so that its edge moves well past the low-speed port. Now there is little difference in vacuum between the upper and lower parts of the air horn. Thus, little air-fuel mixture discharges from the low-speed port. However, under this condition, enough air moves through the air horn to produce a vacuum in the venturi. As a result, the fuel nozzle centered in the venturi (called the *main nozzle* or the *high-speed nozzle*) begins to discharge fuel (as explained in ● 13-7). The main nozzle supplies the fuel during operation with the throttle partly to fully opened. Figure 13-17 shows this action. The circuit from the float bowl to the main nozzle is called the *main metering system*.

The wider the throttle is opened and the faster the air flows through the air horn, the greater the vacuum in the venturi. This means that additional fuel will be discharged from the main nozzle (because of the greater vacuum). As a result, a nearly constant air-fuel ratio is maintained by the main metering system from part- to wide-open throttle.

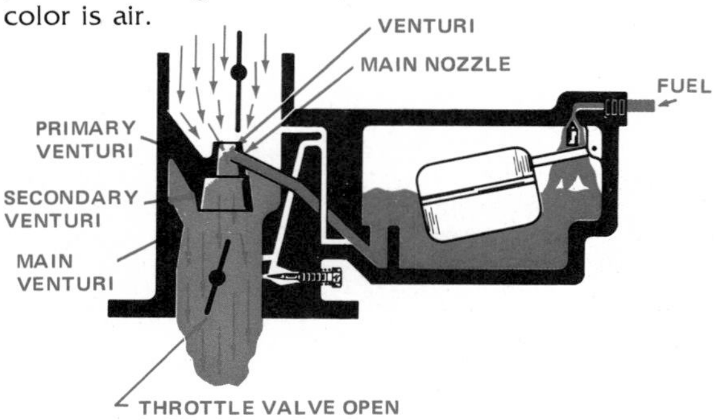

Fig. 13-17 Main metering system in the carburetor. The throttle is open, and fuel is being fed through the high-speed, or main, nozzle. The dark color is fuel; the light color is air.

● 13-18 Power System

For high-speed full-power wide-open-throttle operation, the air-fuel mixture must be enriched (see Fig. 13-6). Additional devices are incorporated in the carburetor to provide this enriched mixture during high-speed full-power operation. They are operated mechanically or by intake-manifold vacuum.

● 13-19 Mechanically Operated Power System

This system includes a metering-rod jet (a carefully calibrated orifice or opening) and metering rod with two or more steps of different diameters (Fig. 13-18). The metering rod is attached to the throttle linkage (Fig. 13-19). When the throttle is opened, the metering rod is lifted. But when the throttle is partly closed, the larger diameter of the metering rod is in the metering-rod jet. This partly restricts fuel flow to the main nozzle. However, enough fuel does flow for normal part-throttle operation. When the throttle is opened wide, the rod is lifted enough to cause the smaller diameter of the rod to move up into the metering-rod jet. Now, the jet is less restricted, and more fuel can flow. The main nozzle is therefore supplied with more fuel, and the resulting air-fuel mixture is richer.

● 13-20 Vacuum-Operated Power System

This system is operated by intake-manifold vacuum. It includes a vacuum piston or diaphragm linked to a valve or a metering rod similar to the one shown in Fig. 13-18.

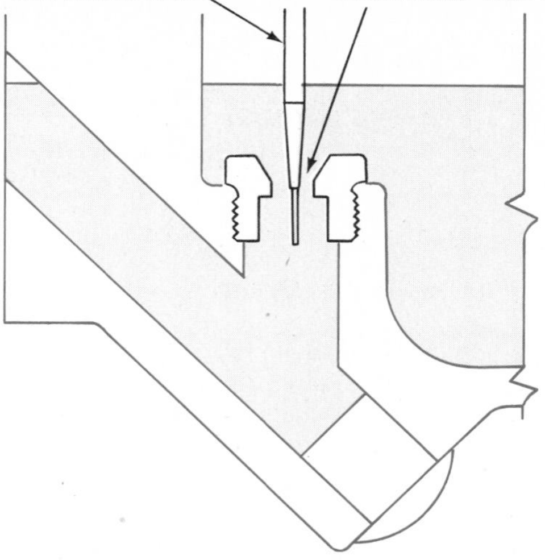

Fig. 13-18 Metering rod and metering-rod jet for better performance at full throttle.

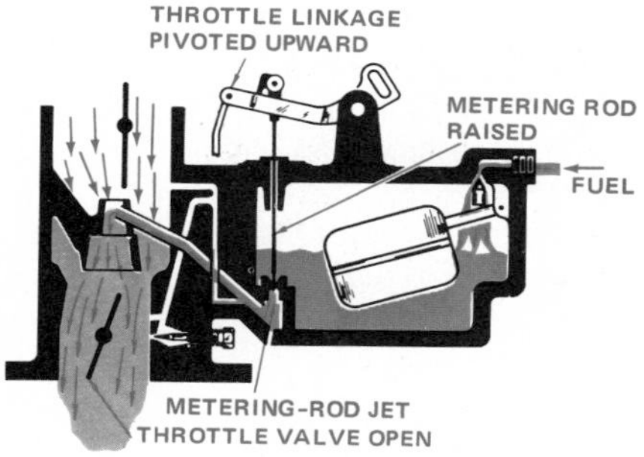

Fig. 13-19 Mechanically operated power system. When the throttle is open, as shown, the metering rod is raised so the smaller diameter of the rod clears the jet. This allows additional fuel to flow.

One design is shown in Fig. 13-20. During part-throttle operation, the piston is held in the lower position by intake-manifold vacuum. However, when the throttle is opened wide, manifold vacuum is reduced. This allows the spring under the vacuum piston to push the piston upward. This motion raises the metering rod so that the smaller diameter of the rod clears the jet. Now, more fuel can flow to handle the full-power requirements of the engine.

A carburetor using a spring-loaded diaphragm to control the position of the metering rod is shown in Fig. 13-21. The action is similar to that of the carburetor using a spring-loaded piston. When the throttle is opened so that intake-manifold vacuum is reduced, the spring raises the diaphragm. This allows the metering rod to be lifted so that its smaller diameter clears the jet, allowing more fuel to flow.

● 13-21 Combination Power Systems

In some carburetors, a combination full-power system is used. It is operated both mechanically and by vacuum from the intake manifold. In one such carburetor, a metering rod is linked to a vacuum diaphragm as well as to the throttle linkage (Fig. 13-21). Movement of the throttle to "full open" lifts the metering rod to enrich the mixture. Also, loss of intake-manifold vacuum (as during a hard pull up a hill or during acceleration) causes the vacuum-diaphragm spring to raise the metering rod for an enriched mixture.

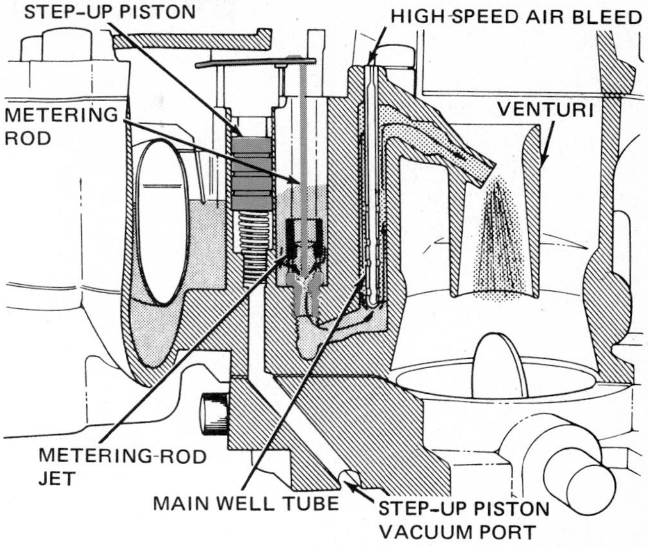

Fig. 13-20 Sectional view of a carburetor using a power or step-up piston, actuated by intake-manifold vacuum, to control the position of the metering rod. (*Chrysler Corporation*)

● 13-22 Air-Fuel Ratios with Different Systems

Figure 13-22 shows the air-fuel ratios with the different carburetor systems in operation. This is a typical curve only. Actual air-fuel ratios may be different for different carburetors and different operating conditions. Note that the idle system supplies a very rich mixture to start with but that as engine speed increases, the mixture leans out. From about 25 to 40 mph [40.24 to 64.37 km/h], the throttle is only partly opened and both the idle and the main metering system are supplying air-fuel mixture. Then, in the curve shown, the main metering system takes over at about 40 mph [64.37 km/h] and continues by itself to about 60 mph [96.56 km/h]. Note that the air-fuel ratio increases somewhat as speed increases (that is, the mixture becomes leaner). Somewhere around 60 mph [96.56 km/h], the power system comes into operation (earlier, of course, if the throttle is opened wide at lower speeds). Now, the mixture richness goes up with the higher speeds.

● 13-23 Accelerator-Pump System

For acceleration, the carburetor must deliver additional fuel (see ● 13-9). Rapid opening of the throttle allows a sudden inrush of air. Thus there is a sudden demand for

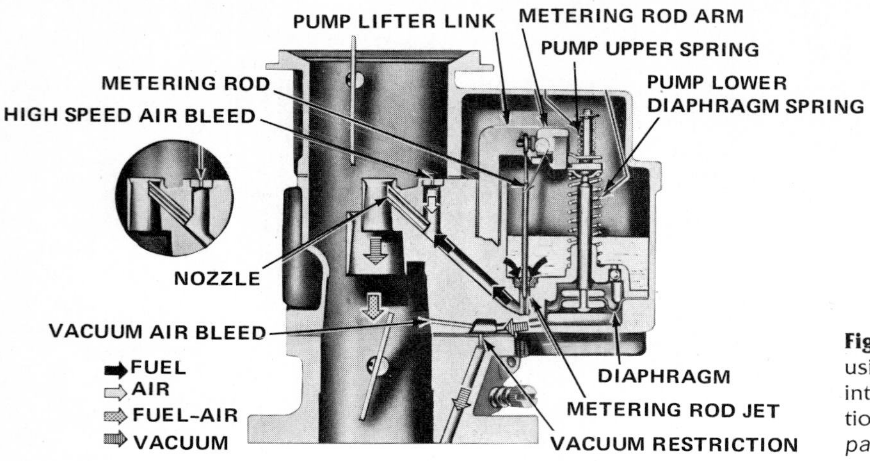

Fig. 13-21 Sectional view of a carburetor using a spring-loaded diaphragm, actuated by intake-manifold vacuum to control the position of the metering rod. (*Ford Motor Company*)

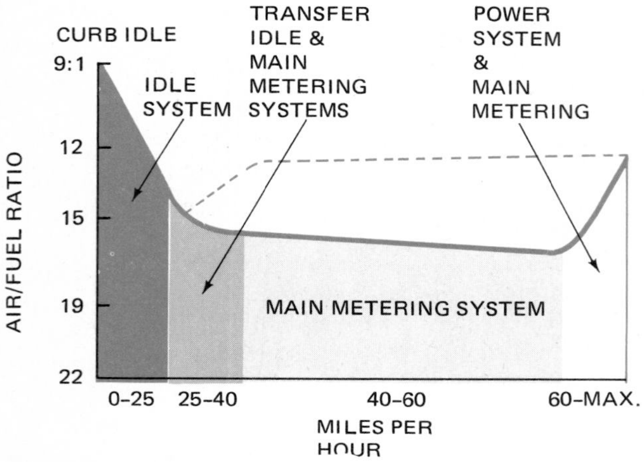

Fig. 13-22 Air-fuel ratios with different carburetor systems operating at different speeds. (*Chevrolet Motor Division of General Motors Corporation*)

additional fuel. Carburetors have accelerator-pump systems to provide this extra fuel. Figure 13-23 shows one type. It includes a pump plunger which is forced downward by a pump lever that is linked to the throttle. When the throttle is opened, the pump lever pushes the pump plunger down. This forces fuel to flow through the accelerator-pump system and out the pump jet (Fig. 13-23). This fuel enters the air passing through the carburetor to supply the additional fuel needed.

The action carries through a plunger spring (Fig. 13-24). When the throttle is opened, the spring is compressed. This applies a pressure against the plunger which pushes it down. The spring maintains this pressure all the time that the throttle is held open until the pump plunger is all the way down, as shown in Fig. 13-23. This arrangement allows the accelerator-pump system to discharge fuel for several seconds, or until the full-power system can take over. The result is smooth acceleration.

One type of accelerator-pump circuit that uses a diaphragm instead of a plunger is shown in Fig. 13-25. When the throttle is opened, the pump lower diaphragm spring lifts the diaphragm. This forces additional fuel from the chamber above the diaphragm through the accelerator-pump circuit and out the pump jet.

13-24 Choke System

When the engine is being started, the carburetor must deliver a very rich mixture to the intake manifold. With the engine and carburetor cold, only part of the fuel vaporizes. Thus, extra fuel must be delivered. In this way, enough evaporates to make a combustible mixture that permits the engine to start.

During cranking, air speed through the carburetor air horn is very low. Vacuum from the venturi action and vacuum below the throttle would be insufficient to produce adequate fuel flow for starting. Thus, to produce enough fuel flow during cranking, the carburetor has a choke (Fig. 13-26). The choke is a round valve, shaped like the throttle valve, located in the top of the air horn. It is controlled mechanically or by an automatic device. When the choke is closed, it is almost horizontal, as shown in Fig. 13-26. Only a small amount of air can get past it. The valve has "choked off" the air flow. Then, when the engine is cranked, a fairly high vacuum develops in the air horn. This vacuum causes the main nozzle to discharge a heavy stream of fuel. The quantity delivered is sufficient to produce the correct air-fuel mixture needed for starting the engine.

As soon as the engine starts, the speed increases from a cranking speed of around 250 to 300 rpm (revolutions per minute) to over 600 rpm. Now more air and a somewhat leaner mixture are required. One method of getting more air into the engine as soon as it starts is to mount the choke valve off center on its shaft in the air horn. Then, by adding a spring to the choke linkage, we have an arrangement that works as follows. The vacuum produced by the running engine causes the valve to partly open against the spring pressure. More air can then flow through. Another arrangement includes a small spring-loaded section in the valve. This section opens to admit the additional air.

13-25 Automatic Chokes

Mechanically controlled chokes are operated by a pull rod on the dash. The pull rod is linked to the choke valve.

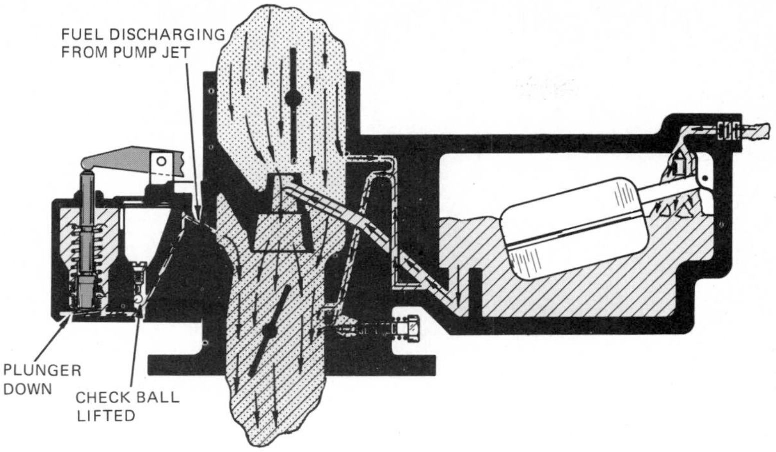

Fig. 13-23 Accelerator-pump system of type using a pump plunger. When the throttle is opened, the pump lever pushes the pump plunger down. This forces fuel to flow through the accelerator-pump system and out the jet.

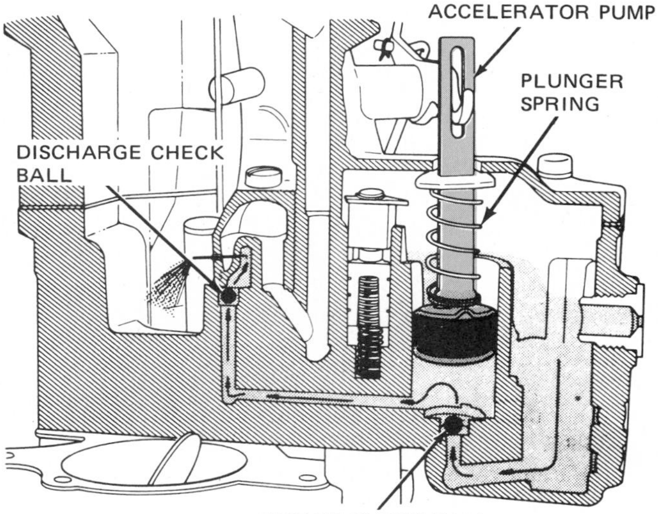

Fig. 13-24 Accelerator-pump system using a pump of the piston type. *(Chrysler Corporation)*

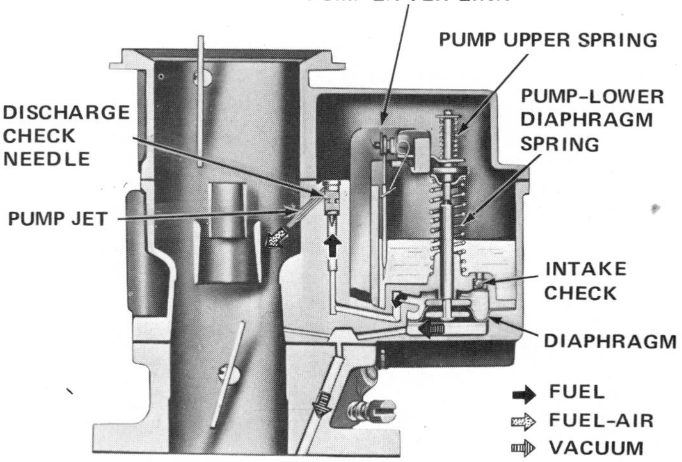

Fig. 13-25 Accelerator-pump system of the type using a spring-loaded diaphragm. Opening of the throttle allows the lower diaphragm spring to lift the diaphragm. This forces fuel through the accelerator-pump system and out through the jet. *(Ford Motor Company)*

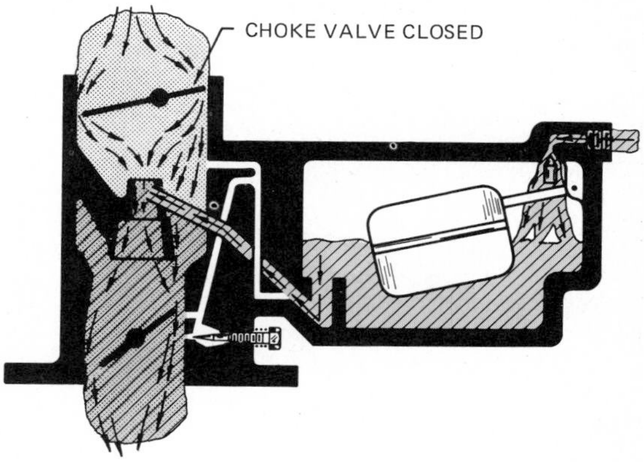

Fig. 13-26 With the choke closed, intake-manifold vacuum is introduced into the carburetor air horn. This causes the main nozzle to discharge fuel.

When it is pulled out, the choke valve is closed. The driver must remember to push the control rod into the de-choked position as soon as the engine begins to warm up. If not, the carburetor will continue to supply a very rich mixture to the engine. This excessive richness will cause poor engine performance, high levels of exhaust emissions, fouled spark plugs, and poor fuel economy, as well as many other problems.

To prevent such troubles, most cars now have an automatic choke. Most automatic chokes operate on exhaust-manifold temperature and intake-manifold vacuum. Figure 13-27 shows an automatic choke on a carburetor. It includes a thermostatic spring and a vacuum piston, both linked to the choke valve. The thermostatic spring is made up of two different metal strips welded together and formed into a spiral. Owing to differences in the expansion rates of the two metals, the thermostatic spring winds up or unwinds with changing temperature. When the engine is cold, the spring is wound up enough to close the choke valve and spring-load it in the closed position. When the engine is cranked, a rich mixture is delivered to the intake manifold. As the engine starts, air movement through the air horn causes the choke valve to open slightly (working against the thermostatic-spring tension). In addition, the vacuum piston is pulled out-

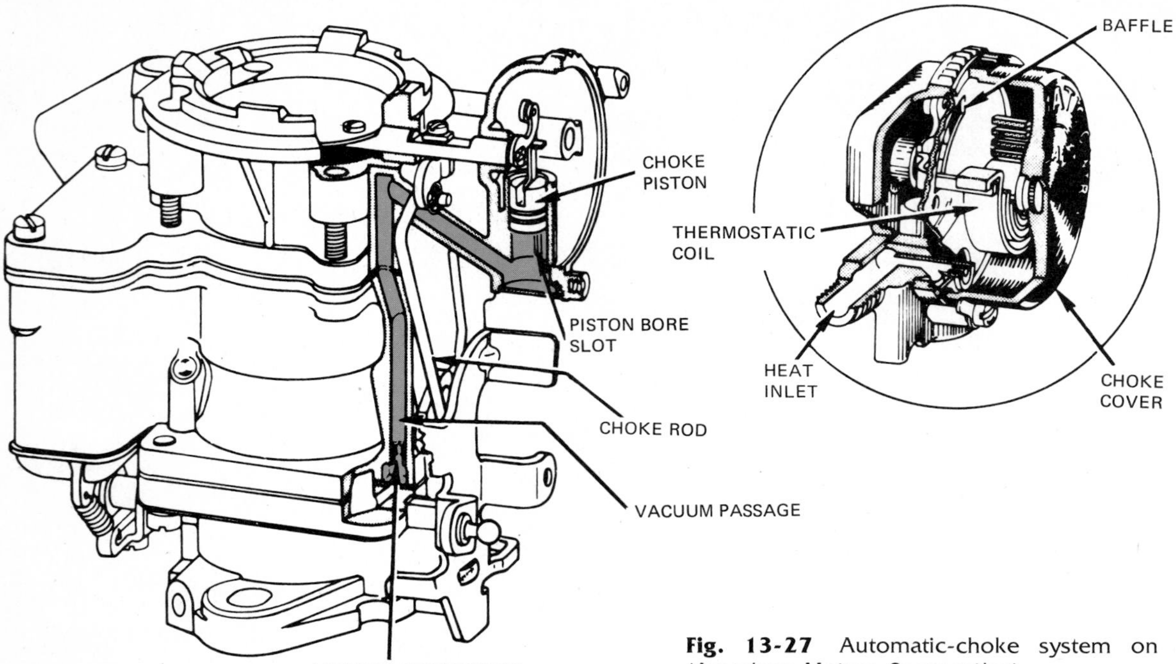

Fig. 13-27 Automatic-choke system on a carburetor. *(American Motors Corporation)*

ward by intake-manifold vacuum. This produces some further opening of the choke valve. The choke valve is now properly positioned to cause the carburetor to supply the proper rich mixture for cold-engine idling.

When the throttle is opened, the mixture must be enriched. The accelerator pump provides some extra fuel, but still more fuel is needed when the engine is cold. This additional fuel is secured by the action of the vacuum piston. When the throttle is opened, intake-manifold vacuum is lost. The vacuum piston releases and is pulled inward by the thermostatic-spring tension. The choke valve therefore moves toward the "closed" position and causes the mixture to be enriched. During the first few moments of operation, the choke valve is controlled by the vacuum piston.

However, the thermostatic spring begins to take over as the engine warms up. The thermostatic spring is in a housing that is connected to the exhaust manifold through a small tube. Heat passes through this tube and enters the thermostatic-spring housing. Soon, the thermostat begins to warm up. As it warms up, the spring unwinds. This causes the choke valve to move toward the "opened" position. When operating temperature is reached, the thermostatic spring has unwound enough to fully open the choke valve. No further choking takes place.

When the engine is stopped and cools, the thermostatic spring again winds up. This closes the choke valve and spring-loads it in the "closed" position.

Figure 13-27 shows a carburetor partly cut away so that the construction of the automatic choke can be seen. The vacuum passage to the vacuum piston is shown, but the heat tube to the exhaust manifold is not. The heat tube sends heat from the exhaust manifold to the thermostatic-spring housing.

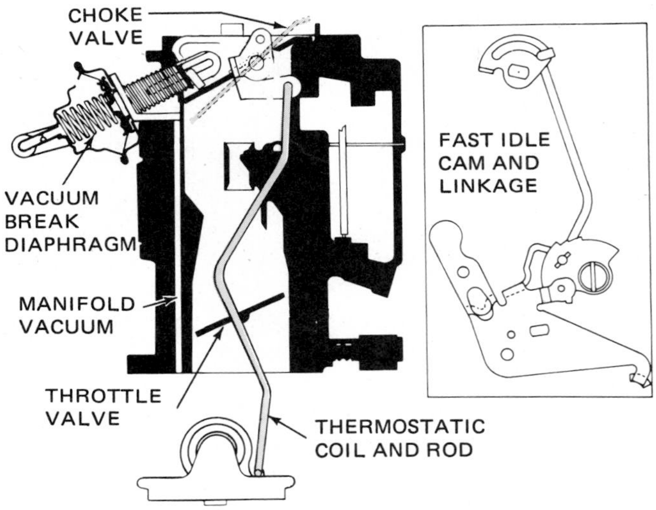

Fig. 13-28 Choke system with the thermostat located in a well in intake manifold. Note vacuum-break diaphragm. *(Chevrolet Motor Division of General Motors Corporation)*

In many engines, the thermostat is located in a well in the intake manifold. There it can quickly react to the manifold heat as the engine starts (see Fig. 13-28). The thermostat is connected by a link to the carburetor. Some carburetors using this arrangement have vacuum pistons. Others have vacuum diaphragms. Both work with the thermostat as previously noted to control the choke-valve position during warm-up.

Some carburetors use heat from the engine cooling water to operate the thermostat. That is, the thermostat housing has a passage through which the cooling water flows. The action is similar to other automatic chokes previously discussed.

Instead of a vacuum piston, many automatic chokes now use a vacuum-operated diaphragm (called a vacuum-break or vacuum-kick diaphragm—see Fig. 13-28). The operation is quite similar. However, the diaphragm provides more force to break the choke valve loose if it gets stuck. The linkage from the diaphragm to the choke-valve lever rides freely in a slot in the lever. During certain phases of warm-up operation, the changing vacuum causes the linkage to ride to the end of the slot in the choke lever and move the choke valve. For example, when first starting, vacuum develops in the intake manifold. This vacuum acts on the vacuum-break diaphragm. The linkage rides to the end of the slot with enough force to "break" the choke valve away from the full choke position. This partly opens the choke so that an overrich mixture is prevented. The vacuum-break diaphragm exerts a more positive and stronger force than a choke piston, such as shown in Fig. 13-27.

Many late-model cars have electric automatic chokes. This type of choke includes an electric heating element (Fig. 13-29). The purpose of this heater is to assure faster choke opening. This helps reduce emissions from the engine. Emissions (HC and CO) are relatively high during the early stages of engine warm-up. At low temperatures, the electric heater adds to the heat coming from the intake manifold. This reduces choke-opening time to as short as $1\frac{1}{2}$ minutes. Figure 13-30 shows the arrangement for a choke mounted in a well in the intake manifold.

Some automatic chokes do not rely on heat from the engine to actuate the thermostatic coil. Instead, they operate only on electricity and intake-manifold vacuum. The electric heating element is carefully calibrated to produce as quick a choke opening as possible without unduly penalizing cold-engine operation. This quick choke opening, as previously noted, reduces the time that a rich mixture is going to the engine cylinders and thus the time that there will be excessive HC and CO in the exhaust gas.

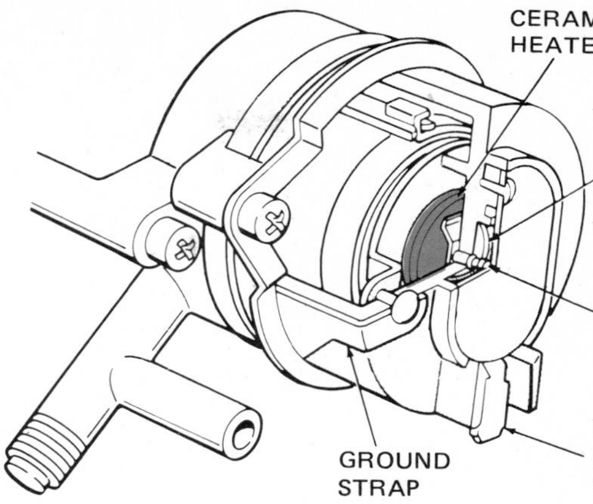

Fig. 13-29 Cutaway view of an electric-assist choke. At low temperature, the ceramic heater turns on, adding heat to the choke so it opens more quickly. (*Ford Motor Company*)

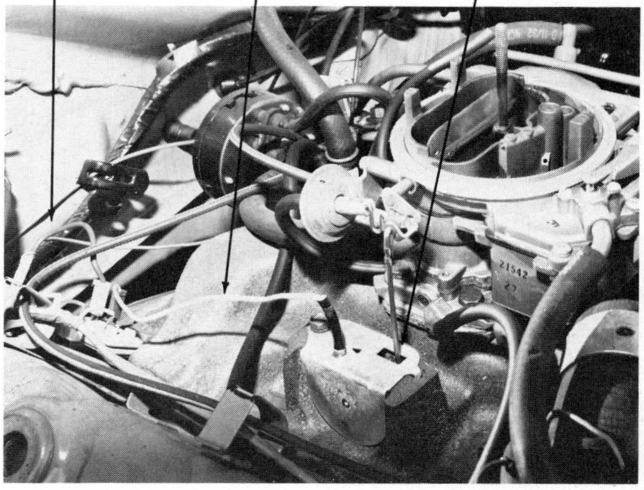

Fig. 13-30 Arrangement of an electric-assist choke mounted in a well in the intake manifold. (*Chrysler Corporation*)

● 13-26 Manifold Heat Control

During initial warm-up of the engine, just after starting, vaporization of the fuel is poor. Gasoline vaporizes more slowly when it is cold (see ● 11-3). To improve fuel vaporization and therefore cold-engine operation, a device is provided to heat the intake manifold when it is cold. This device, called the *manifold heat-control valve,* is built into the exhaust and intake manifolds. Two arrangements are used, one for in-line engines and another for V-8 engines.

1. In-line engines In these engines, the exhaust manifold is located under the intake manifold. At a central point, there is an opening from the exhaust manifold into a chamber, or oven, surrounding the intake manifold (Fig. 13-31). A butterfly valve is placed in this opening (see Fig. 7-20). When the valve is turned one way, the opening is closed off. The position of the valve is controlled by a thermostat. When the engine is cold, the thermostatic spring winds up and moves the valve to the "closed" position (top in Fig. 13-31). Now, when the engine is started, the hot exhaust gases pass through the opening and circulate through the oven around the intake manifold. Heat from the exhaust gas quickly warms the intake manifold and helps the fuel to vaporize. Thus, cold-engine operation is improved. As the engine warms up, the thermostatic spring unwinds and the valve moves to the "opened" position (bottom in Fig. 13-31). Now, the exhaust gases pass directly into the exhaust pipe. The gases no longer circulate in the oven around the intake manifold.

2. V-8 engines In V-8 engines, the intake manifold is placed between the two banks of cylinders. It has a special passage (Figs. 7-24 and 13-32) through which exhaust gases can pass. One of the exhaust manifolds has a thermostatically controlled valve that closes when the

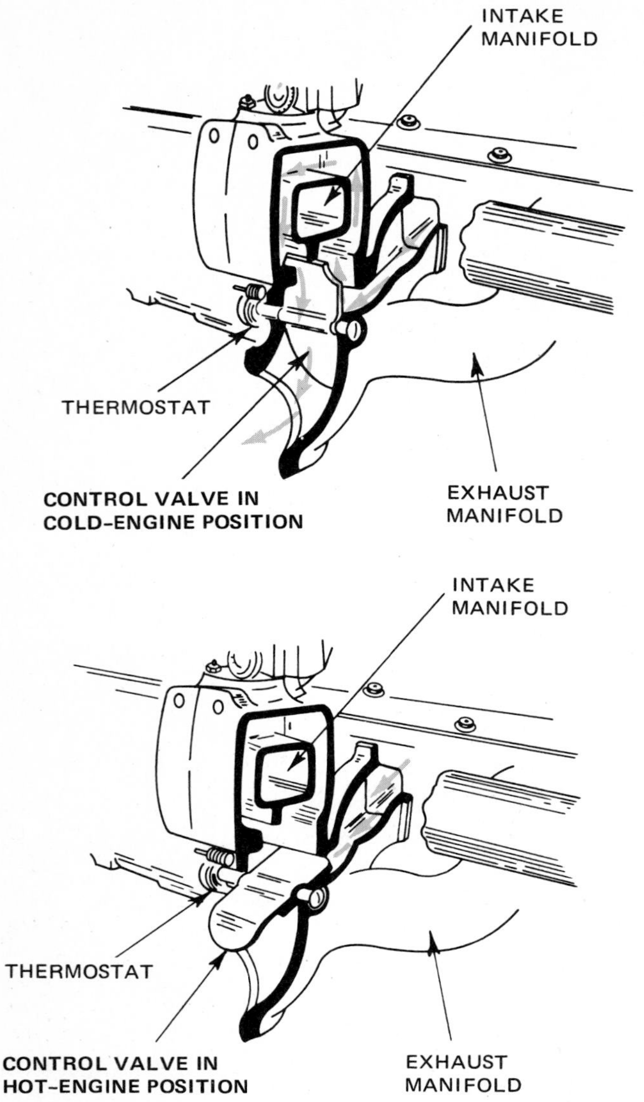

Fig. 13-31 Intake and exhaust manifolds of a six-cylinder in-line engine, cut away to show the location and action of the manifold heat control. At top, the heat-control valve is in the "heat on" position. It is directing hot exhaust gases up and around the intake manifold, as shown by the arrows. At bottom, the valve is in the hot-engine position. (*Ford Motor Company*)

engine is cold. This causes exhaust gases to pass from that exhaust manifold through the special passage in the intake manifold. The exhaust gases then enter the other exhaust manifold. Heat from the exhaust gases thus heats the air-fuel mixture in the intake manifold for improved cold-engine operation. As the engine warms up, the thermostatically controlled valve opens. Then, the exhaust gases from both exhaust manifolds pass directly into the exhaust pipes.

Since the introduction of thermostatic air cleaners (heated air systems ● 12-12), some engines no longer use a heat-control valve. To do so might add too much heat to the incoming air-fuel mixture. This would expand the mixture (heat expands gas) so much that the amount of the mixture entering the engine would be lowered, and the engine power would be reduced.

● 13-27 Anti-Icing

When fuel is sprayed into the air passing through the air horn, it evaporates, or turns to vapor. During evaporation, the fuel takes on heat. That is, it takes heat from the surrounding air and metal parts. This is the same effect you get when you pour alcohol on your hand. Your hand feels cold. If you blow on your hand, thus causing the alcohol to evaporate faster, your hand will feel still colder. The faster the evaporation takes heat from your hand, the cooler your hand feels.

Now, let us see how this affects the carburetor. Spraying and evaporation of the fuel "rob" the surrounding air and carburetor of heat. Under certain conditions, the surrounding metal parts are so cooled that moisture in the air condenses and actually freezes on the metal parts. The ice can build up sufficiently, if conditions are right, to cause the engine to stall. This is most apt to occur during the warm-up period following the first start-up of the day. It happens more with air temperatures in the range of 40 to 60°F [4.4 to 15.6°C] and fairly humid air.

To prevent such icing, many carburetors have a special anti-icing system. One arrangement for a V-8 engine is shown in Fig. 13-33. During the warm-up period, the manifold heat-control valve sends hot exhaust gases from one exhaust manifold to the other (see ● 13-26). Part of this hot exhaust gas circulates around the carburetor idle ports and near the throttle-valve shaft. This adds enough heat to guard against ice formation. Another system has water passages in the carburetor. The water (coolant) comes from the engine cooling system. A small amount of the coolant passes through a special water manifold in the carburetor throttle body. This adds enough heat to the carburetor to prevent icing.

● 13-28 Fast Idle

When the engine is cold, some throttle opening must be maintained. This is done so that the engine will idle faster than it would when warm. Otherwise, the slow idle and cold engine might cause the engine to stall. With fast idle, enough air-fuel mixture gets through, and air speeds are great enough, to produce adequate vaporization and a sufficiently rich mixture. Fast idle is obtained by a fast-idle cam linked to the choke valve (Fig. 13-34). When the engine is cold, the automatic choke holds the choke valve closed. In this position, the linkage has revolved the fast-idle cam so that the adjusting screw rests on the high point of the cam. The adjusting screw prevents the throttle valve from moving to the "fully closed" position. The throttle valve is held partly open for fast idle. As the engine warms up, the choke valve opens. This rotates the fast-idle cam so that the high point moves from under the adjusting screw. The throttle valve closes for normal hot-engine slow idle.

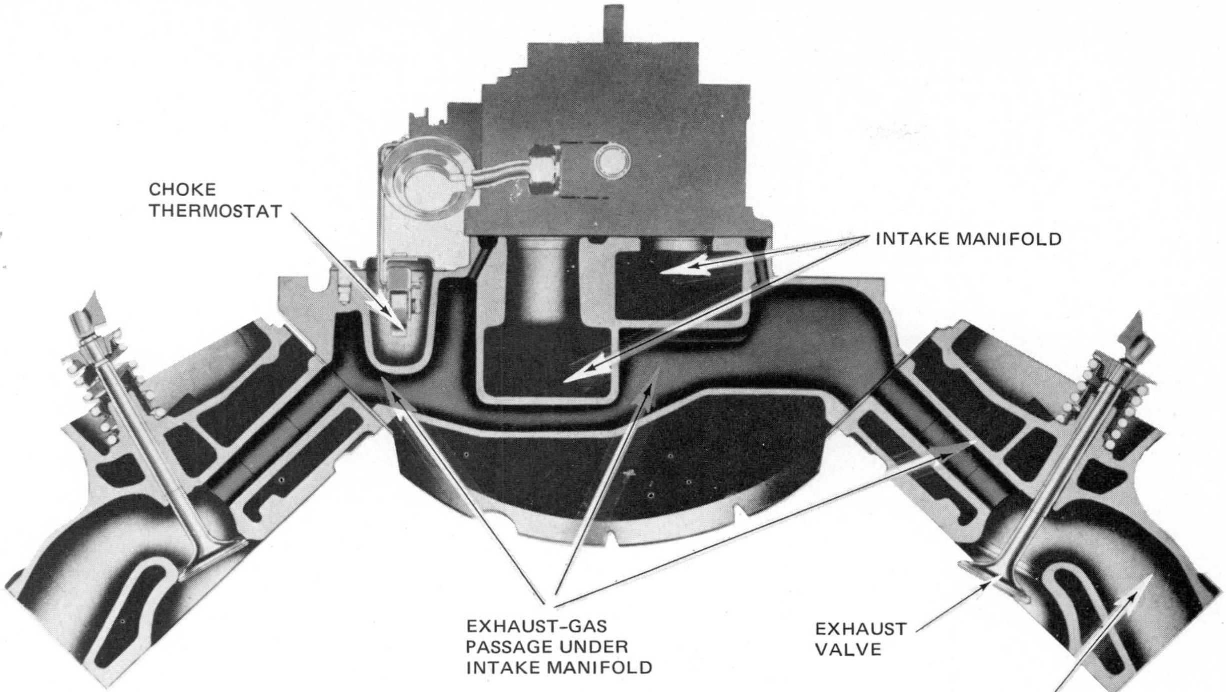

Fig. 13-32 Exhaust-gas passage under the intake manifold in a V-8 engine. Note the well in which the carburetor choke thermostat is located. (*Buick Motor Division of General Motors Corporation*)

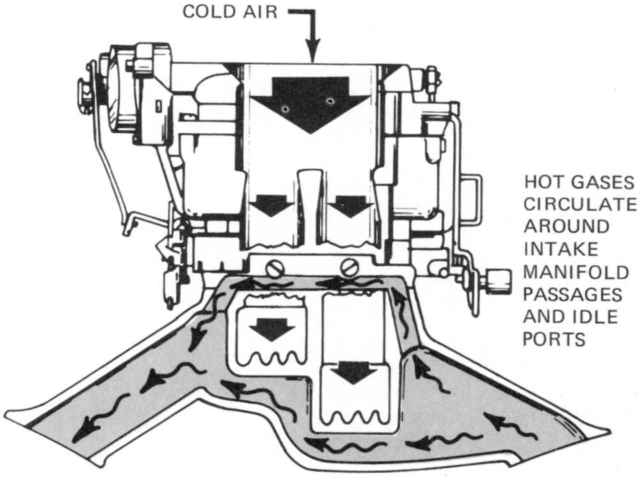

Fig. 13-33 Intake manifold and carburetor-idle-port heating passages. Hot exhaust gases heat these areas as soon as the engine starts. (*Cadillac Motor Car Division of General Motors Corporation*)

13-29 Air-Bleed and Antisiphon Passages

In all the systems of the carburetor except the accelerator-pump system, there are small openings to permit air to enter, or *bleed* into, the system (Fig. 13-20). This produces some premixing of the air and fuel, so that better atomization and vaporization result. It also helps maintain a more uniform air-fuel ratio. At higher speeds, a larger amount of fuel tends to discharge from the main nozzle. But at the same time the faster fuel movement through the high-speed circuit causes more air to bleed into the circuit. Thus, the air-bleed holes tend to equalize the air-fuel ratio.

Air-bleed passages are also sometimes called *antisiphon* passages. They act as air vents to prevent the siphoning of fuel from the float bowl at intermediate engine speeds.

If air-bleed passages become plugged, they may cause the float bowl to be emptied after the engine shuts off. When the engine is shut off, the intake manifold cools down and a slight vacuum forms as a result. With open air bleeds, air can move through the bleeds to satisfy the vacuum. If air bleeds are plugged, then the vacuum will cause the float bowl to empty through the idle circuit.

13-30 Special Carburetor Devices

Carburetors for late-model cars have other devices to improve driveability and fuel economy and reduce air pollution. We describe some of these in detail in later chapters on emission controls (in Part Five). Here we list them and discuss them briefly.

1. Antidieseling solenoid Some modern engines have a tendency to continue to run after the ignition switch is turned off. This is called "run-on" or dieseling (because the engine runs like a diesel engine, without electric ignition). Dieseling can be caused by high engine-cylinder temperatures, particles of carbon that are hot, or hot spark-plug electrodes. Enough air-fuel mixture can seep around the throttle valve and through the idle system to

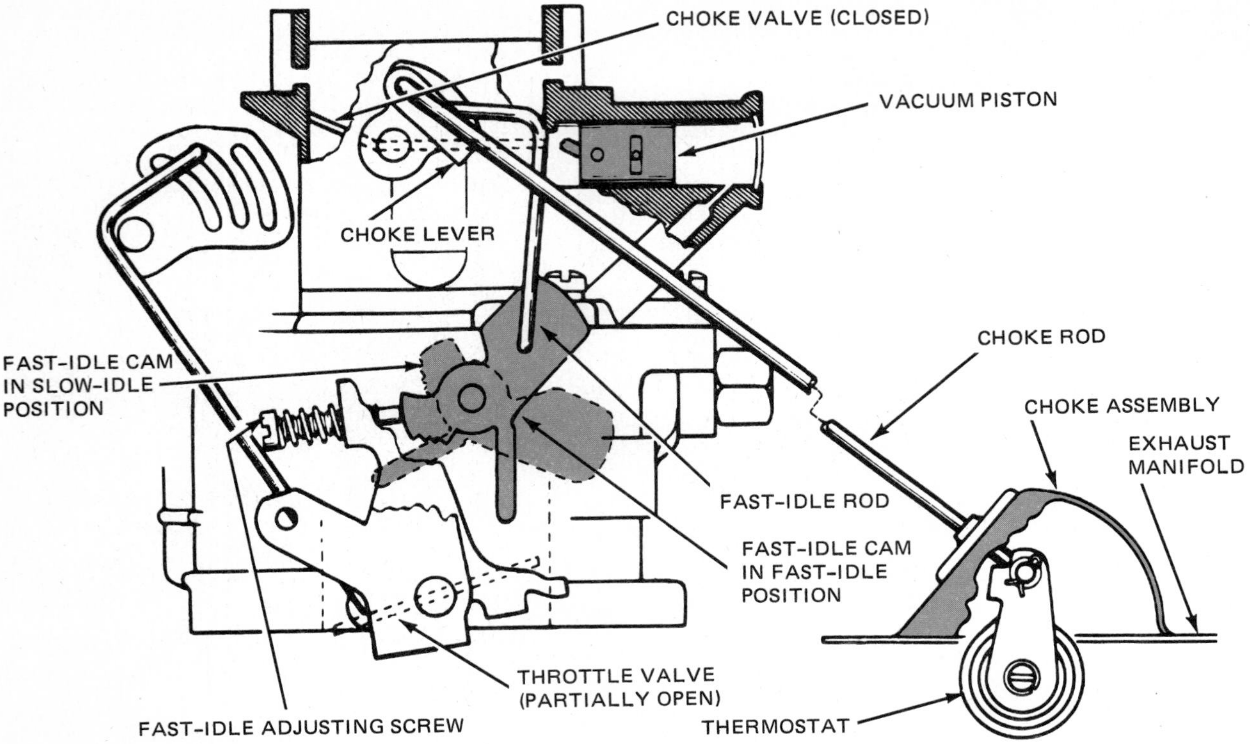

Fig. 13-34 Vacuum and thermostatically operated choke with the thermostat located in the exhaust manifold. Note two positions of the fast-idle cam. (*Chrysler Corporation*)

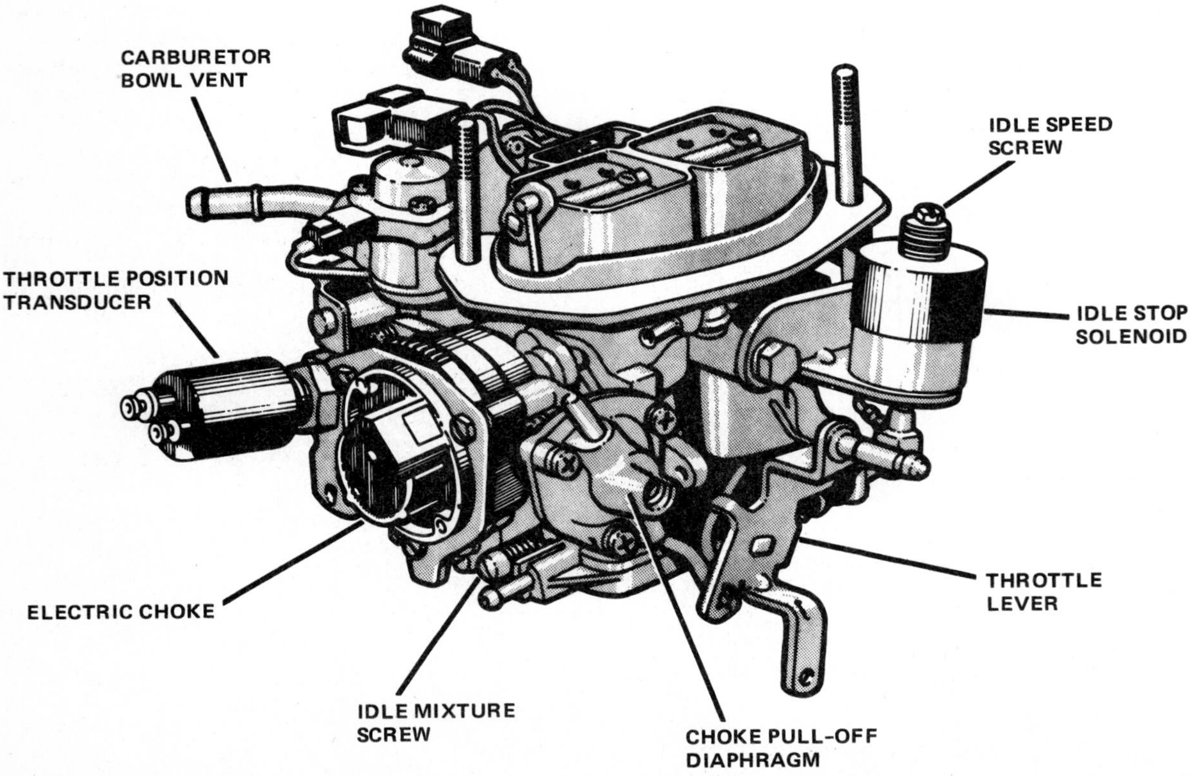

Fig. 13-35 Carburetor with an idle-stop, or antidieseling, solenoid. (*Buick Motor Division of General Motors Corporation*)

maintain the dieseling. To prevent this, many carburetors are equipped with an antidieseling solenoid. Figure 13-35 shows one arrangement which controls the closing of the throttle valve. When the engine is running, the solenoid is connected to the battery. It extends its plunger. The plunger serves as the idle stop and prevents complete closing of the throttle. Therefore, normal hot-idle operation results when the driver releases the accelerator pedal. However, when the engine is turned off, the solenoid is disconnected from the battery. It pulls in its plunger. Now, the throttle closes almost completely, shutting off all air flow. The engine stops running.

A second arrangement uses the antidieseling solenoid to shut off the fuel flow in the idle system when the engine is turned off. That is, during normal operation, the solenoid pulls in its plunger so normal fuel flow can continue through the idle system. However, when the engine is shut off, the solenoid releases its plunger and this blocks the idle system so no fuel can flow through it. The engine therefore stops.

2. Throttle return check If the throttle closes too fast after the driver releases the accelerator pedal, the air-fuel mixture can be momentarily excessively enriched. This is because the fuel nozzle continues to dribble fuel for a moment even though the air flow is largely shut off. The idle system will also momentarily feed a rich mixture. This is due to the high vacuum that results when the engine is running fairly fast with the throttle closed. This very rich mixture can cause the engine to stumble, or hesitate, because it will not burn properly. Also, the high level of HC and CO in the exhaust, because of poor combustion, may damage the catalytic converter. To prevent this, many carburetors are equipped with a throttle-return check (Fig. 13-36). It slows down the closing of the throttle enough so that the momentary excessive richness is prevented.

3. Electric kickdown switches on some cars equipped with automatic transmissions These switches provide an electrical means to downshift the transmission into a lower gear when the throttle is opened wide.

4. Governors to control or limit top engine speed The use of governors is largely confined to heavy-duty vehicles. They prevent overspeeding and rapid wear of the engine. One type directly controls the throttle valve. It tends to close the valve as rated engine speed is reached. Another type has a throttle plate between the carburetor throttle valve and the intake manifold. The throttle plate moves toward the closed position as rated speed is reached. This prevents delivery of additional amounts of air-fuel mixture and any further increase in engine speed.

5. Vacuum vents There is a vacuum in the carburetor air horn when the engine is running. This vacuum is used by various other devices on the engine and elsewhere. We list these uses here and cover them in detail later in the book.
a. Ignition-distributor vacuum-advance mechanism. This advances the spark during part-throttle operation. In electronically-controlled ignition systems such as MISAR, the vacuum is carried to an electronic control unit and this unit "reads" the vacuum and decides when and how much to advance the spark.
b. Positive crankcase-ventilating system. This is a system for ventilating the crankcase without polluting the atmosphere.
c. Fuel-vapor emission-control system. This system traps gasoline vapor from the fuel tank and carburetor float bowl.
d. Heated-air system. This system provides rapid heating of the air entering the carburetor when the engine is cold. This improves cold-engine operation.
e. Exhaust-gas recirculation system. The exhaust-gas recirculation (EGR) system introduces some exhaust gas into the air-fuel mixture going into the engine cylinders. This reduces the formation of one of the atmospheric pollutants during the combustion process. (See ● 13-31.)
f. Vacuum motors for air conditioners. Many air conditioners have vacuum motors for operating the air-conditioner doors.
g. Vacuum for power brakes.

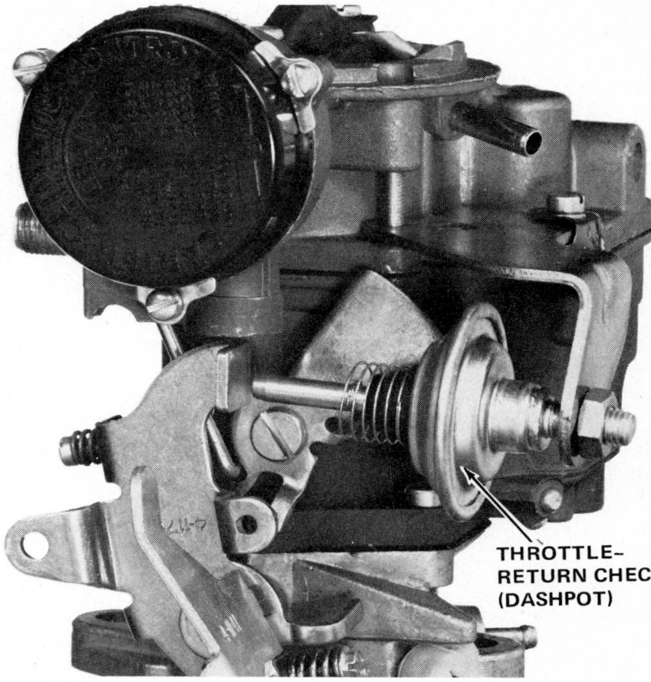

Fig. 13-36 Carburetor with a throttle-return check, or dashpot. (*Carber Carburetor Division of ACF Industries*)

As mentioned previously, we cover these various devices and systems in later chapters.

Diesel engines do not develop any appreciable vacuum in the intake manifold. Therefore, to operate the various vacuum devices on a diesel vehicle, a separate vacuum pump must be installed.

● 13-31 Exhaust-Gas Recirculation (EGR)

In many late-model cars, some exhaust gas is sent into the ingoing air-fuel mixture before it reaches the engine

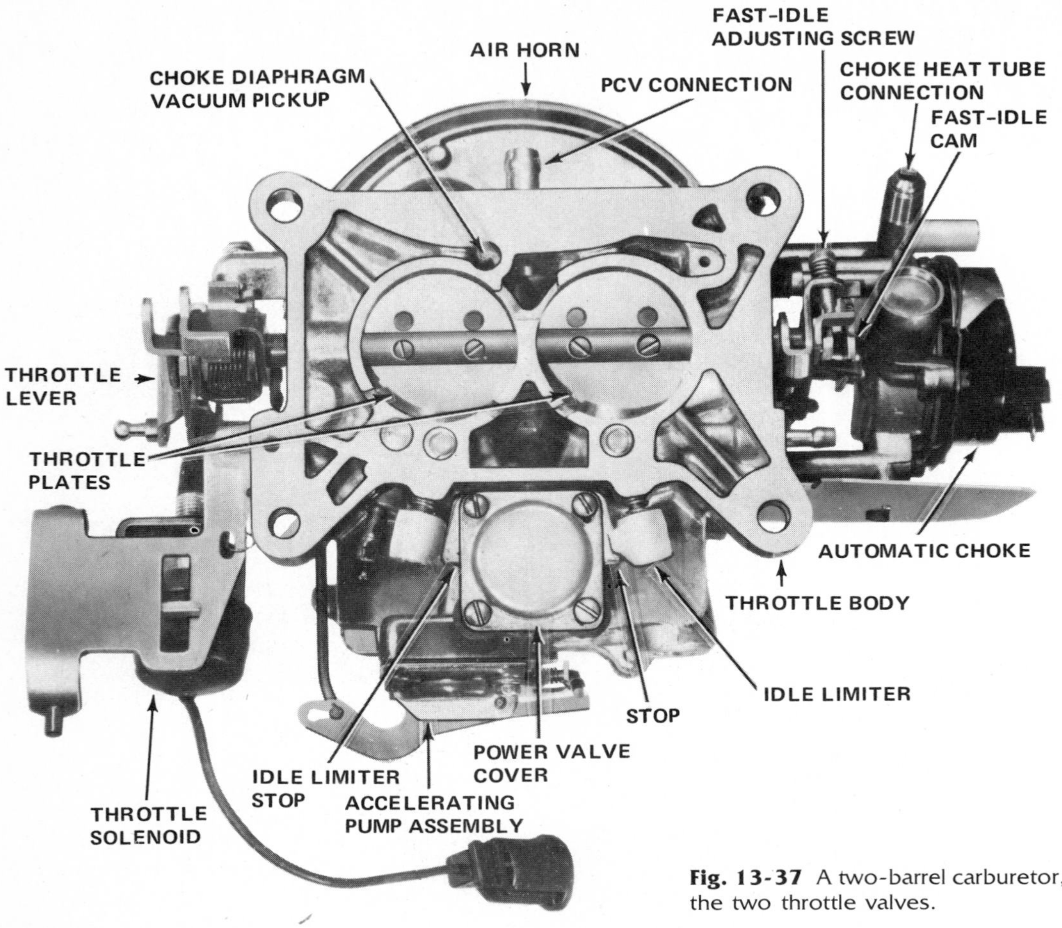

Fig. 13-37 A two-barrel carburetor, showing the locations of the two throttle valves.

cylinders. This exhaust gas lowers the temperature of the burning gas and reduces the formation of nitrogen oxide (NO_x). This is an undesirable atmospheric pollutant. The EGR system is mentioned in ● 12-13 and is covered in detail in Part Five.

● 13-32 Two-Barrel and Four-Barrel Carburetors

Carburetors with more than a single barrel are used on many engines. Thus, many carburetors have two barrels (dual carburetors) and others have four barrels (quad carburetors). The purpose of the additional barrels is to improve engine "breathing," particularly at high speeds. That is, the extra barrels permit more air and fuel to enter the engine. Of course, if air were the only consideration, then a single large-diameter barrel could be used. But with only a single large barrel, venturi action would be poor. Proper air-fuel ratios would be hard to achieve under varying operating conditions.

1. Two-barrel carburetors The two-barrel carburetor is essentially two single-barrel carburetors in a single assembly (Fig. 13-37). The second barrel is used in two different ways, according to the carburetor design. In one arrangement, each barrel handles the air-fuel requirements of half the engine cylinders. For example, Fig. 7-24 shows the air-fuel delivery pattern in a V-8 engine. One carburetor barrel supplies cylinders 2, 3, 5, and 8. The other barrel supplies cylinders 1, 4, 6, and 7. The arrows indicate the pattern in Fig. 7-24. Each barrel has a complete set of systems. The throttle valves are fastened to a single throttle shaft so that both valves open and close together.

The second design uses the second barrel in a different way, and this is an arrangement most commonly used on four-cylinder in-line engines. In this arrangement, the secondary barrel comes into operation only after the primary throttle valve has opened some 45 degrees (more or less depending on the design). When the primary throttle valve moves past the 45-degree opening, linkage to the secondary throttle valve starts to open it. The secondary barrel then comes into operation and starts to supply additional air-fuel mixture to the four cylinders. This action increases the supply of air-fuel mixture to the engine for improved medium- to high-speed operation.

2. Four-barrel carburetor The four-barrel carburetor (Fig. 13-38 and 13-39) consists essentially of two two-barrel carburetors combined in a single assembly. The carburetor has four barrels and four main nozzles and thus is often called a *quadrajet, or quad, carburetor.* One pair of barrels makes up the primary side, the other pair the secondary side (Fig. 13-38). Under most operating conditions, the primary side alone takes care of engine requirements. However, when the throttle is moved toward the "wide-open" position for acceleration or full-power operation, the secondary side comes into operation. It supplies additional amounts of air-fuel mixture. Thus, engine breathing improves. That is, the engine receives more air-fuel mixture. Volumetric efficiency (● 10-12) is higher, and the engine produces more horsepower.

There are two ways of controlling the secondary-barrel action—by mechanical linkage from the primary-throttle shaft, or by a vacuum device. Figure 13-39 is a sectional view of a carburetor using mechanical linkage. During part throttle, only the primary throttle valves are open. However, whenever the throttle is opened wide for additional power, the linkage between the primary and secondary throttle valve causes the secondary throttle valves to open. The secondary barrels now supply air-fuel mixture for full-power engine operation.

In the vacuum-controlled system, there is a vacuum-operated diaphragm. The vacuum is picked up from one of the primary-barrel venturis. As air speed through the primary barrels increases, so does this vacuum. When the vacuum reaches a predetermined amount—indicating a rather high engine rpm—the vacuum actuates the diaphragm. This opens the secondary throttle valves so the secondary barrels begin to supply air-fuel mixture.

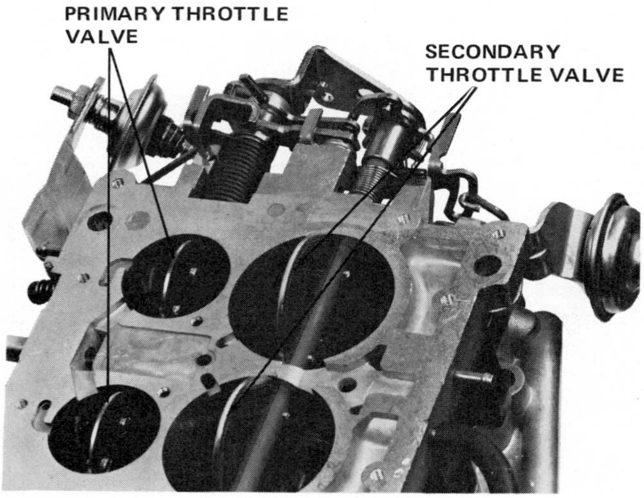

Fig. 13-38 A four-barrel or quad carburetor, showing the locations of the four throttle valves. The small throttle valves are on the primary side.

● 13-33 Factory-Adjusted Part Throttle (APT)

In order to provide a more accurate factory adjustment of the fuel flow during part-throttle operation, some late-model carburetors have an additional metering rod and a fixed metering-rod jet. These are shown to the left in Fig. 13-39. The metering rod is adjusted at the factory by turning the adjustment screw. If it is turned to lift the metering rod, more fuel can flow to the main metering-rod jets and rods. If it is turned to lower the metering rod, less fuel can flow. The purpose of this additional APT metering rod and jet is to fine-tune the carburetor during its manufacture. Adjusting the metering rod at the factory compensates for any slight differences in the power system resulting from manufacturing tolerances.

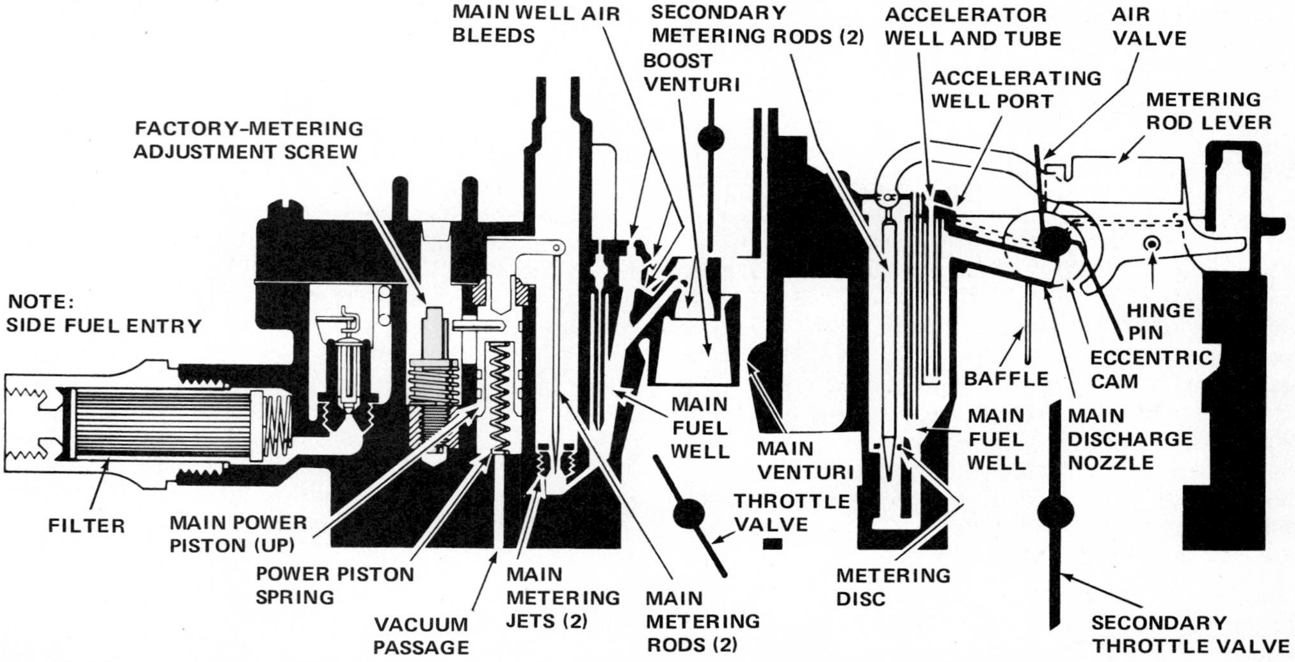

Fig. 13-39 Sectional view of a four-barrel carburetor which has the extra system for making an adjustment of the metering rod at the factory. (*Chevrolet Motor Division of General Motors Corporation*)

Some engineers believe that the APT system is another step in the direction of a no-adjustment carburetor, that is, a carburetor which cannot be adjusted in the field. They believe that if the part-throttle system can be fine-tuned, so can the idle mixture.

13-34 Altitude Compensation

Some carburetors with APT also have an automatic altitude adjustment. This is in the form of an *aneroid* surrounding the APT metering rod (Fig. 13-40). An aneroid is a sealed bellows which is sensitive to changes in atmospheric pressure. When pressure is increased, the aneroid is squeezed so it shortens. When pressure is reduced, the aneroid expands. These actions lower or raise the APT metering rod as atmospheric pressure changes.

For example, when a car is driven up a mountain, it encounters reduced atmospheric pressure. Without any compensating device, the air-fuel mixture would become enriched. This happens because less air enters the carburetor (air pressure is lower). To compensate for this, the aneroid expands as a result of the reduced pressure. This action lowers the APT metering rod so that less fuel can flow. As a result, a more even air-fuel ratio is maintained.

13-35 Multiple Carburetors

To achieve still better engine "breathing," some high-performance engines are equipped with more than one carburetor. The additional carburetors supply more air and fuel and thus improve high-speed, full-power engine performance. Two carburetors mounted on an engine are called "dual carburetors." Three carburetors mounted on an engine are called "triple carburetors" or sometimes "tri-power." The ultimate in this is to equip each engine cylinder with its own carburetor. Many racing and drag-strip or hot-rod cars are equipped in this manner; an eight-cylinder engine would have eight carburetors.

One step beyond a carburetor for each cylinder is fuel injection (Chap. 15). That is, the fuel-injection system feeds fuel directly to each cylinder by means of fuel injectors in the intake manifold, just opposite the intake valves. The fuel-injection system provides positive hydraulic pressure to force the fuel through the injectors. In the carburetor system, atmospheric pressure (a considerably lower pressure) is the driving force.

13-36 Variable-Venturi (VV) Carburetors

All the carburetors described so far are fixed-venturi units. The size and shape of the venturi does not change. There is another design of carburetor, however, in which the size of the venturi can change. These are called variable-venturi or VV carburetors. Many foreign cars, and an increasing number of cars made in the United States, are using variable-venturi carburetors. The size of the venturi varies as operating conditions change.

We describe two types here, the round-piston type and the rectangular venturi-valve type. The round-piston type has been used on some foreign cars and motorcycles for years. The rectangular venturi-valve type is a relatively new development from the Ford Motor Company. Note that both types have float systems similar to those used in the fixed-venturi carburetors.

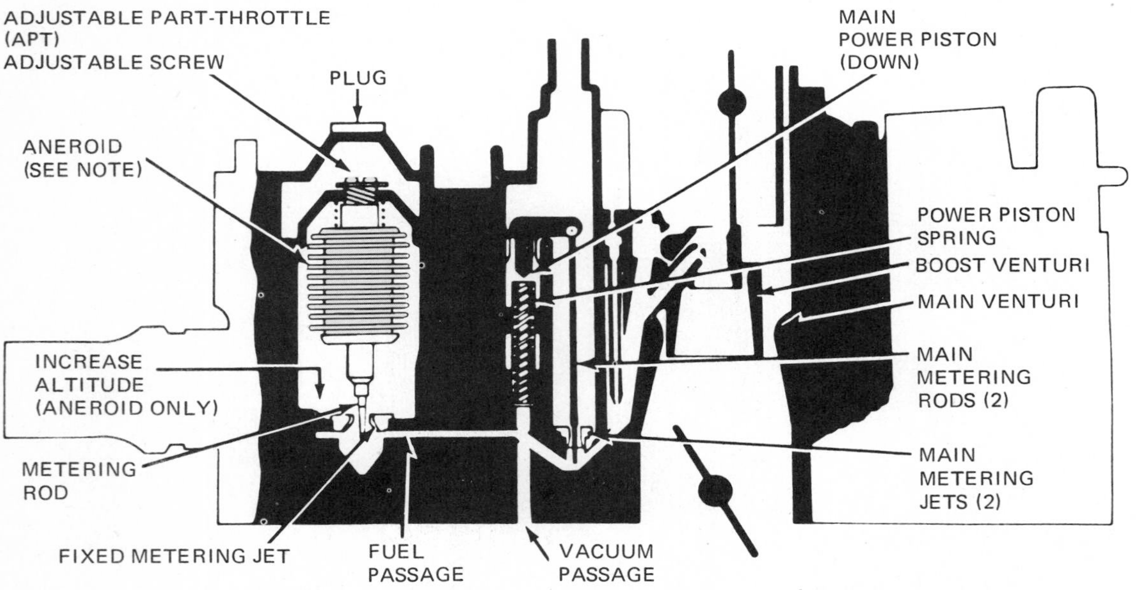

Fig. 13-40 Sectional view of a carburetor showing the main metering system and the factory-adjustable part-throttle metering rod with aneroid. (*Chevrolet Motor Division of General Motors Corporation*)

13-37 Round-Piston VV Carburetor

Figures 13-41 and 13-42 show exterior and sectional views of one model round-piston VV carburetor. The carburetor is relatively simple in construction, as can be seen from the disassembled view in Fig. 13-43.

Here is how the carburetor works. Refer to Figs. 13-42 and 13-43 as you read the following so that you can see the shapes of the parts we discuss. The piston is an assembly of two basic parts: the outer two-diameter piston and the inner oil-damper reservoir. The piston moves up and down in the piston chamber in response to the amount of vacuum between the piston and throttle valve. When the piston moves down, it reduces the size of the venturi. The venturi is formed by the end of the piston and the throttle body. Figure 13-44 shows how downward movement of the piston reduces the size of the venturi.

Movement of the piston also moves the tapered needle. The needle is fastened to the bottom of the piston. When the piston moves downward, the needle moves down into the fuel jet. This reduces the operating size of the jet opening and thus the amount of fuel that can flow. At the same time, the downward movement of the piston reduces the size of the venturi and thus the amount of air that can flow through. You thus have a balanced condition. As the size of the venturi changes, the size of the jet opening also changes. Thus, the proper proportions of air and fuel are maintained. The air-fuel ratio stays approximately constant.

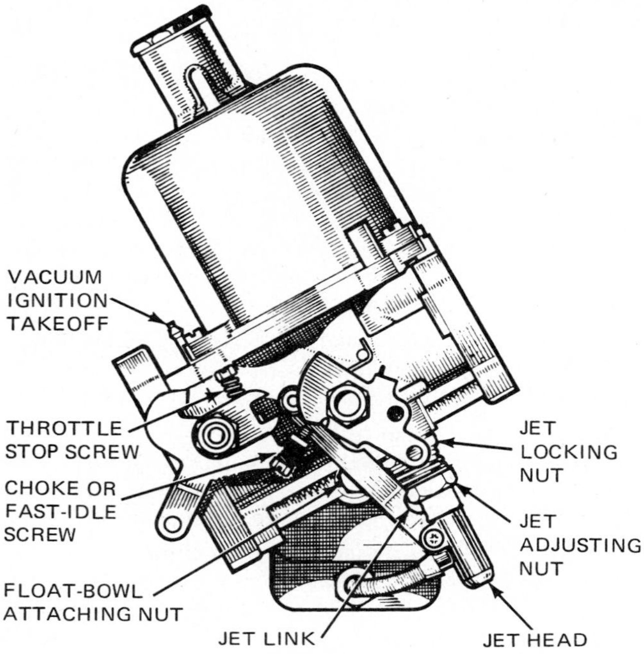

Fig. 13-41 Variable-venturi (VV) carburetor. (*British Leyland Motors Incorporated*)

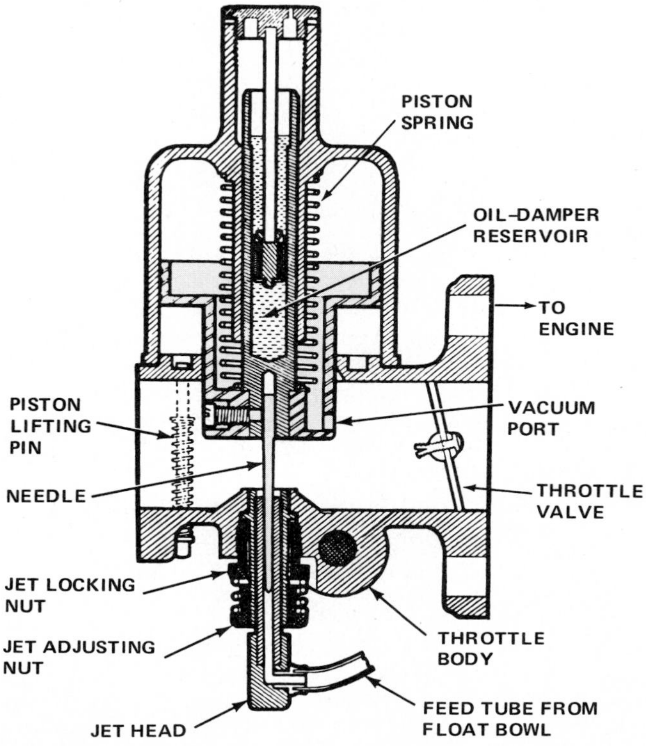

Fig. 13-42 Sectional view of a variable-venturi carburetor. (*British Leyland Motors Incorporated*)

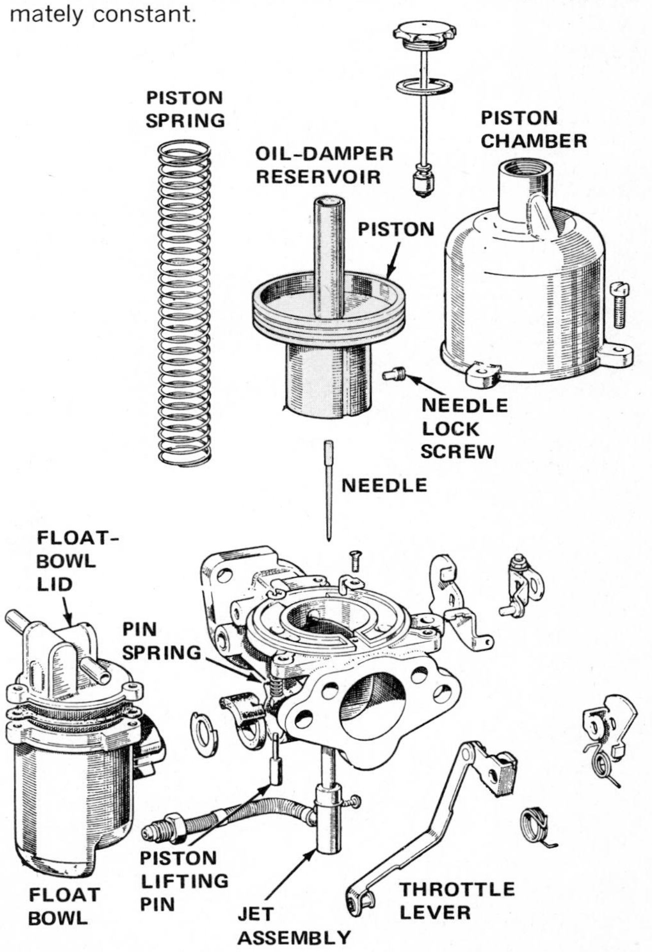

Fig. 13-43 Disassembled view of a variable-venturi carburetor. (*British Leyland Motors Incorporated*)

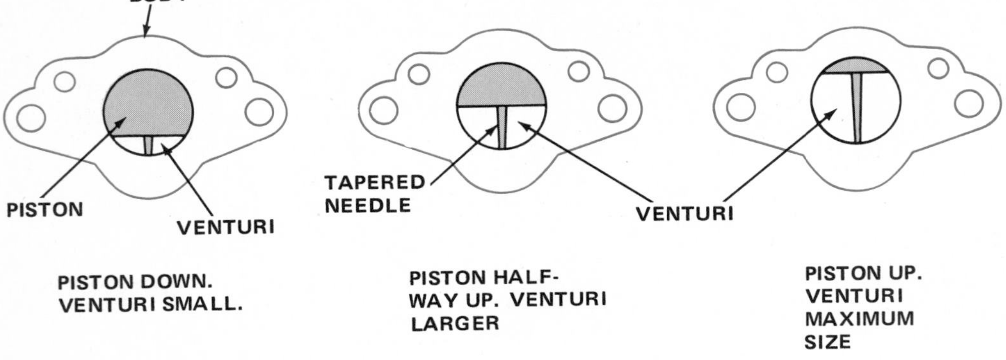

Fig. 13-44 Looking into the throttle body to see how the up and down movement of the piston changes the size of the venturi.

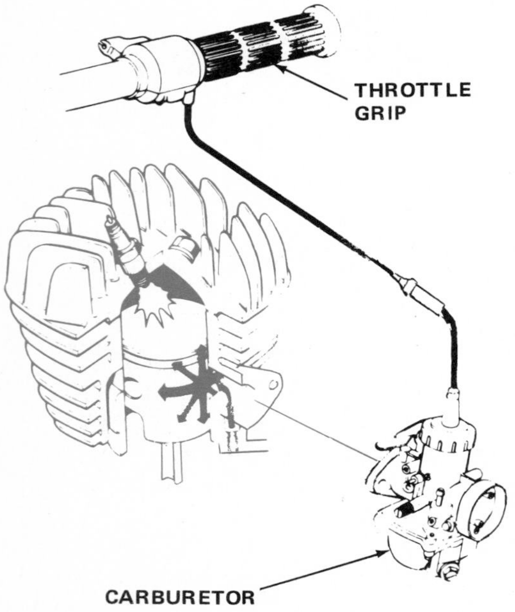

Fig. 13-45 In many motorcycles, the piston position is varied by twisting the throttle grip. This raises or lowers the piston in the throttle body. (*Suzuki Motor Company, Ltd.*)

The piston is raised or lowered in response to the movement of the throttle valve (Fig. 13-42). When the throttle valve is closed to the idling position, intake-manifold vacuum is cut off from the throttle body. The piston spring pushes the piston down to its lowest position. A small amount of air flows around the throttle valve and through the venturi. It produces just enough vacuum at the venturi to cause the fuel jet to deliver enough fuel for idling.

When the throttle is opened, intake-manifold vacuum enters the throttle body. This vacuum draws air from the space above the piston, acting through the vacuum port in the lower part of the piston. The piston is raised by the vacuum, partly compressing the piston spring. As the piston moves up, more air can pass through the venturi. At the same time, the needle moves up in the jet, thus increasing the effective size of the jet. More fuel flows and mixes with the air passing through. As you can see, the piston and tapered needle work together to provide a properly proportioned mixture of air and fuel for the operating condition. The taper on the needle is of critical importance in balancing the air-fuel ratio so that it will be uniform through the entire range of throttle opening. Actually, the pin taper permits additional fuel to flow at full throttle so the mixture is somewhat enriched for acceleration and full-power operation.

The oil-damper reservoir that is part of the piston acts like a tiny shock absorber, preventing excessive movements of the piston as the throttle valve is moved and vacuum conditions change. Without this shock-absorber action, the piston could bounce up and down, causing erratic carburetor action and poor engine performance.

A variation of the VV carburetor used on many motorcycles has a positive means of varying the position of the piston. This consists of a flexible cable connected to the throttle grip on the motorcycle handlebar (Fig. 13-45). As the throttle grip is twisted, the cable pushes or pulls on the piston, thus lowering or raising the piston. This reduces or increases the amount of air-fuel mixture being fed to the engine and thus changes engine speed.

13-38 Ford Rectangular VV Carburetor

This carburetor (Fig. 13-46) is Ford's version of the VV carburetor, which they introduced on some 1978 cars. Externally, this carburetor is quite similar in appearance to the fixed-venturi carburetor (Fig. 13-47). Internally, however, it is quite different. It has two air horns, which Ford calls "throats." In each throat there is a rectangular-shaped piston, or venturi valve which slides back and forth across the throat (Fig. 13-48). This changes the size of the opening (the venturi) above the throttle valve. The two venturi valves are connected together. Their positions are controlled by intake-manifold vacuum and throttle position. Each is connected to a tapered needle or metering rod which is positioned in a fuel jet. Thus, you can see that this carburetor has all the main elements of the round-piston VV carburetor we discussed in the previous section. The major difference is in the shapes and locations of the elements. We will now explain how the venturi valves are controlled, and how the various special systems in the carburetor work. These special systems include the float system, main metering system, cold-

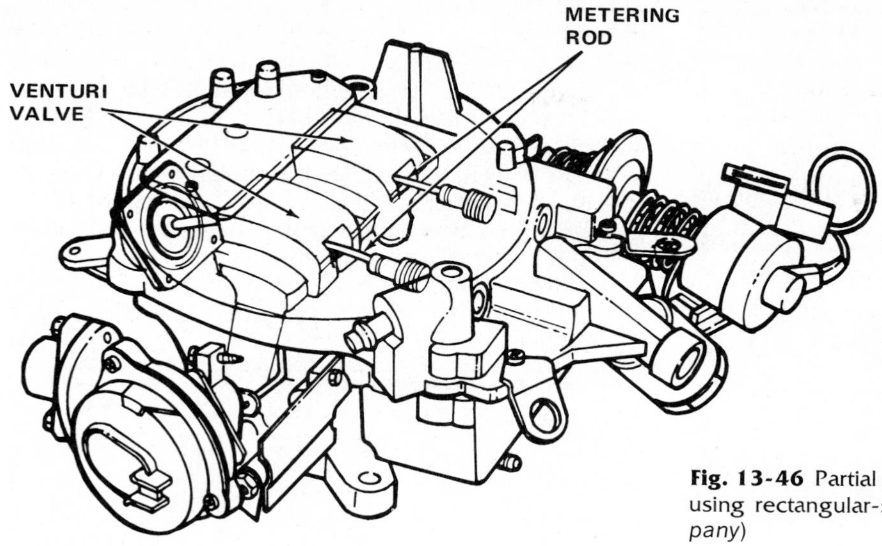

Fig. 13-46 Partial cutaway view of a two-barrel VV carburetor using rectangular-shaped venturi valves. (*Ford Motor Company*)

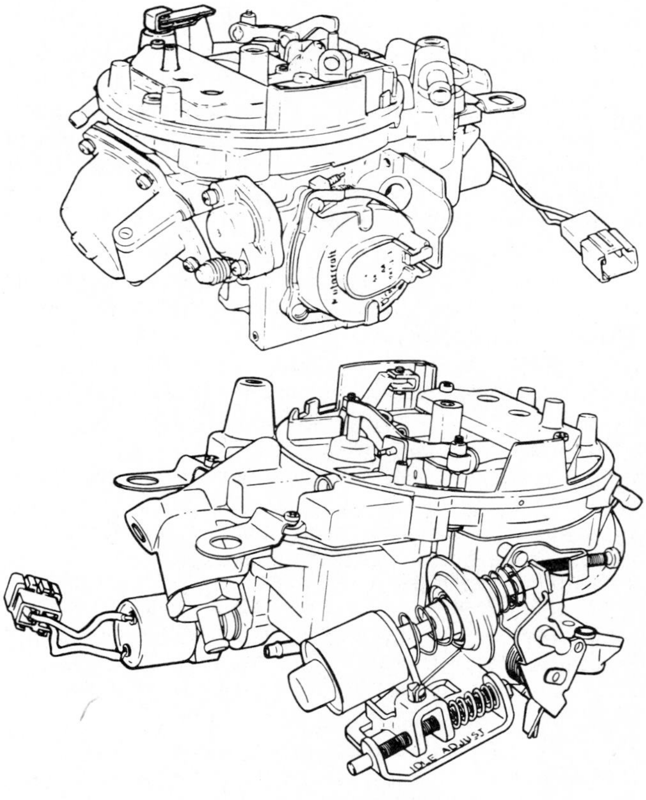

Fig. 13-47 Front and rear views of the Ford VV carburetor. (*Ford Motor Company*)

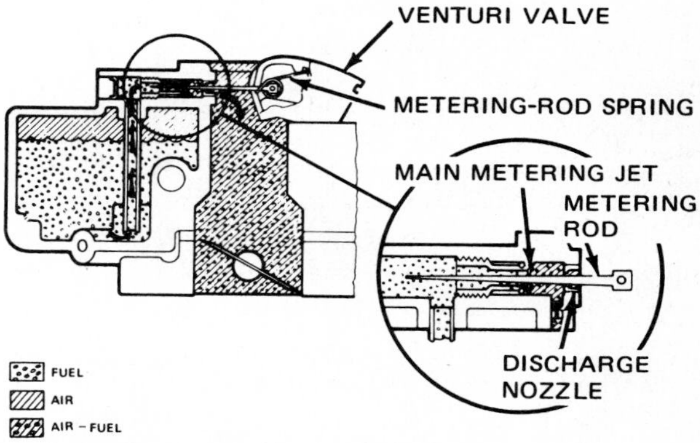

Fig. 13-48 Main metering system of the Ford VV carburetor. (*Ford Motor Company*)

cranking enrichment system, cold-running enrichment system, the accelerator-pump system, and others. You can see that this carburetor has all the special controls built into it that the fixed-venturi carburetors have. However, this VV carburetor does not need a separate idle system, choke valve, and enrichment-valve system such as the fixed-venturi carburetors require.

1. Vacuum control The position of the venturi valves and thus the tapered metering rods is controlled by vacuum. The vacuum control includes a spring-loaded vacuum diaphragm (Fig. 13-49) which is connected by a rod to the venturi valve (Fig. 13-50). When the throttle is opened, intake-manifold vacuum can work on the vacuum diaphragm. The vacuum causes the venturi valve to move back to increase the venturi opening. More air can flow through. At the same time, the tapered metering rod is raised in the fuel jet. Therefore, more fuel can flow through. This additional fuel matches the additional air flowing through so that the proper air-fuel ratio is maintained.

2. Float system This system, shown in Fig. 13-51, is similar to the float system used in fixed-venturi carburetors and works in the same way. Note the fuel filter to the left.

3. Venturi-valve limiter At wide-open throttle, under some conditions the control vacuum will not be strong enough to open the venturi valve fully. In such case, the valve is opened by the venturi-valve limiter lever on the throttle shaft (Fig. 13-52).

4. Cold-cranking enrichment system This system (Fig. 13-53) provides extra fuel when a cold engine is being

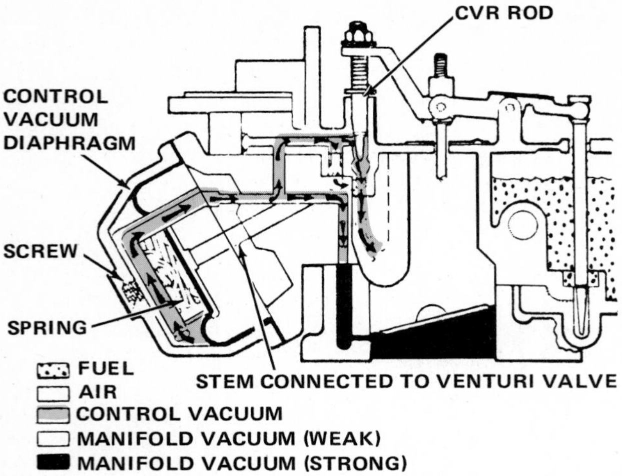

Fig. 13-49 Control vacuum circuit. (*Ford Motor Company*)

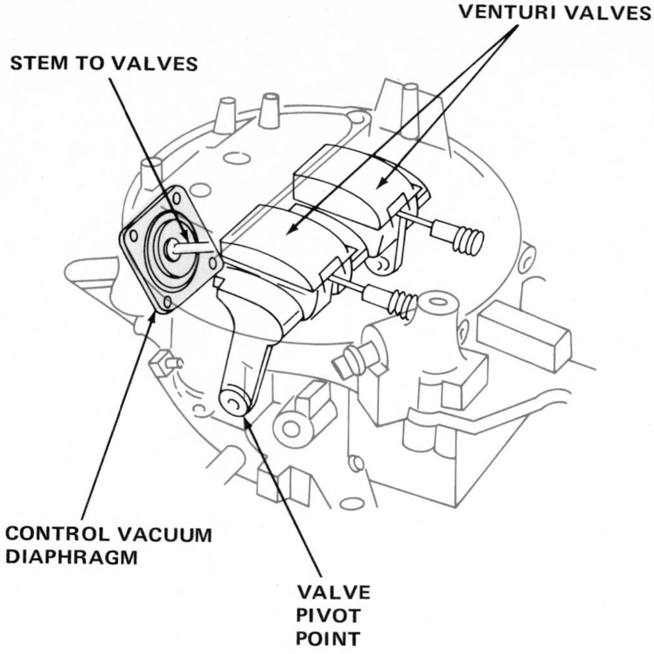

Fig. 13-50 The control vacuum diaphragm is connected by a stem to the venturi valves. As vacuum conditions in the intake manifold change, the vacuum diaphragm changes the position of the venturi valves, thereby changing the size of the venturis. (*Ford Motor Company*)

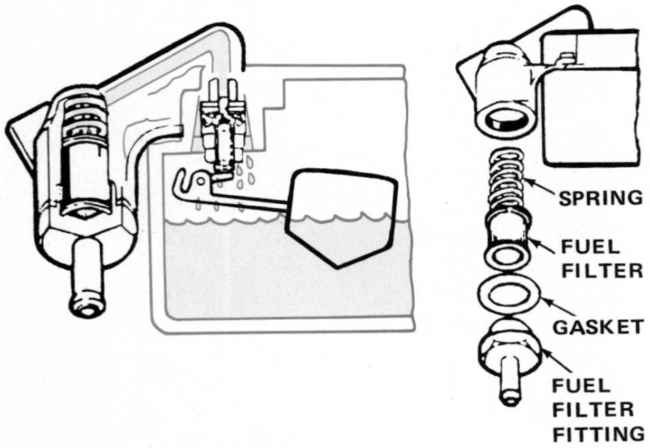

Fig. 13-51 Fuel inlet and float system. (*Ford Motor Company*)

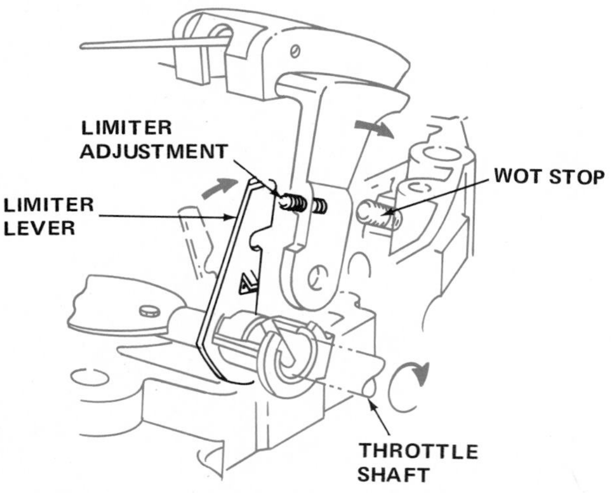

Fig. 13-52 Venturi-valve limiter. (*Ford Motor Company*)

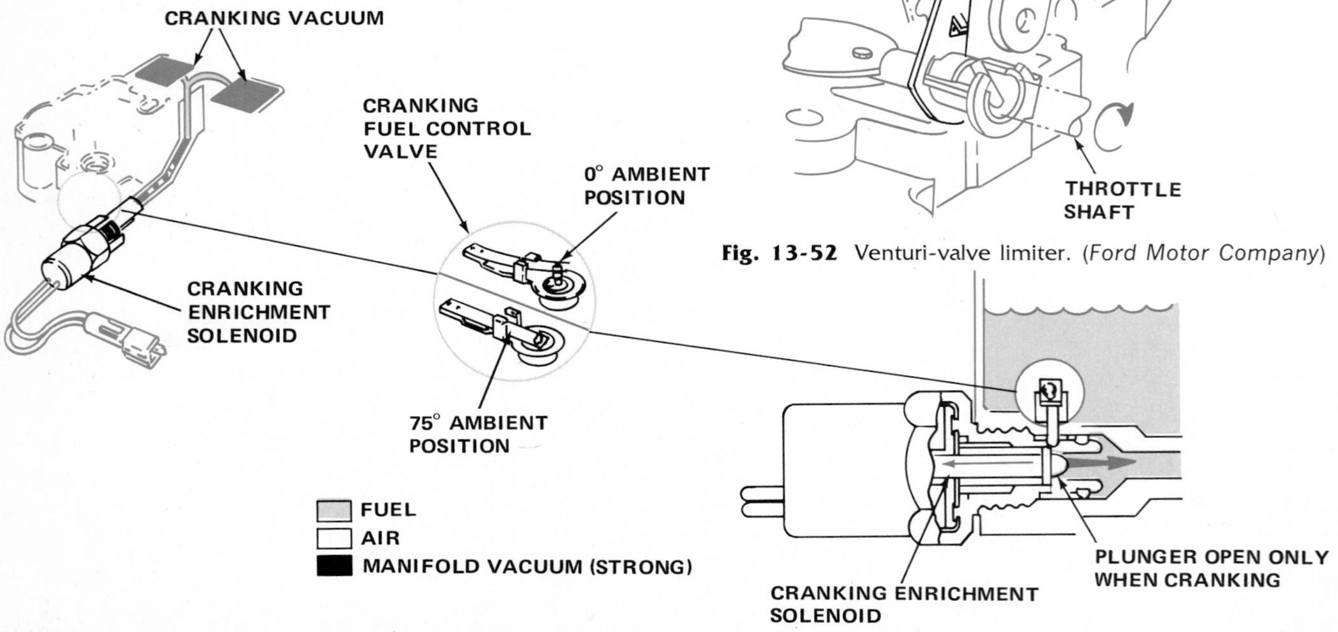

Fig. 13-53 Cold-cranking enrichment system. (*Ford Motor Company*)

cranked. It includes a solenoid valve that is normally closed. Fuel to this valve is controlled by a thermostatic-blade valve that is closed above 75°F [24°C]. When the ignition key is turned to START, the engine is cranked. At the same time, the solenoid valve is opened. Now, fuel can flow if the thermostatic-blade valve is also opened. This provides extra fuel for starting.

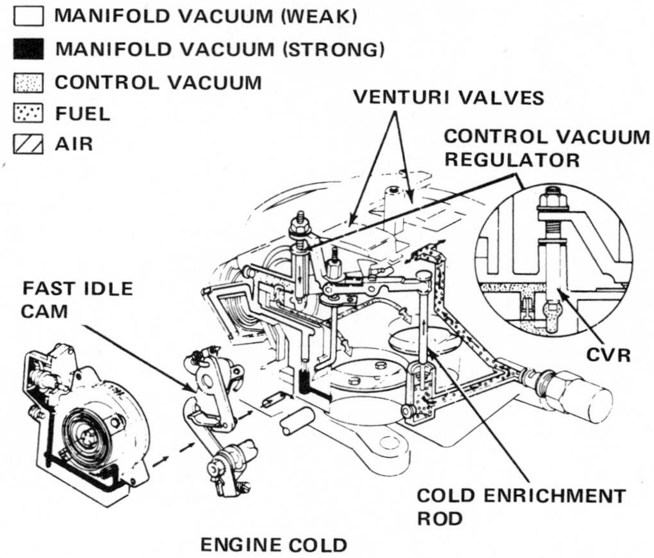

Fig. 13-54 Cold-running enrichment system. (*Ford Motor Company*)

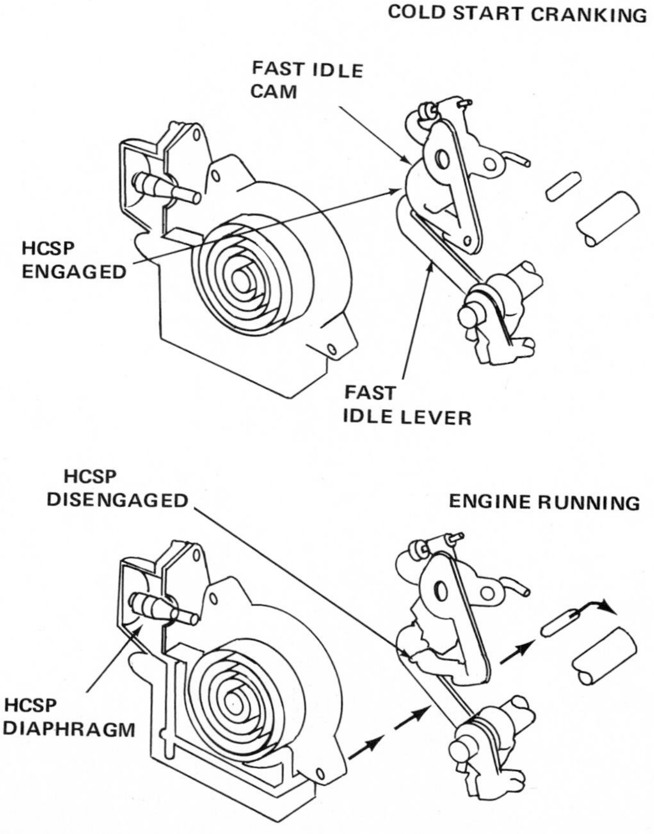

Fig. 13-55 Fast-idle cam and high-cam-speed positioner (HCSP) operation. (*Ford Motor Company*)

5. Cold-running enrichment system After a cold engine starts, it must be supplied with a rich mixture for a short time until it warms up. In the fixed-venturi carburetor, this is taken care of by the choke system. In the Ford VV carburetor, it is taken care of by a cold-running enrichment system that includes a bimetal spring heated from the exhaust (Fig. 13-54). This bimetal spring controls a vacuum regulator and a cold-running enrichment rod. After a cold start, the bimetal spring has raised the cold-running enrichment rod to allow additional fuel to flow. At the same time, the vacuum regulator cuts off part of the vacuum to the vacuum-control unit (Fig. 13-49). Now, intake-manifold vacuum overrides the control and determines the position of the venturi valve. The combination produces an extra-rich mixture for cold-engine operation. As the engine warms up, the bimetal spring relaxes and allows the cold-running enrichment rod to seat, thus cutting off the flow of additional fuel. At the same time, the vacuum regulator releases the vacuum to the vacuum-control unit. Then, normal hot-engine operation results.

6. Fast-idle cam The fast-idle cam is tied in with the cold-running enrichment system, as shown in Fig. 13-54. The position of the cam is controlled by the bimetal spring, just as in other carburetors. When the engine is cold, the cam is positioned to provide a fast idle. There is one unique feature of the system, however. During cold starts, a special lever is inserted between the fast-idle cam and fast-idle lever. This provides additional throttle-plate opening for starting. When the engine starts, manifold vacuum is applied to a separate vacuum diaphragm (called the high-cam-speed positioner or HCSP). This causes the lever to be withdrawn after the first throttle movement. The action lowers the fast-idle speed for all cold-driving modes. Figure 13-55 shows the two positions of the HCSP lever.

7. Other special systems The Ford VV carburetor also has other systems to meet varying operating conditions. The accelerator-pump system (Fig. 13-56) is similar to those used in fixed-venturi carburetors. Also included is a hot-idle compensator (Fig. 13-57), similar to those used

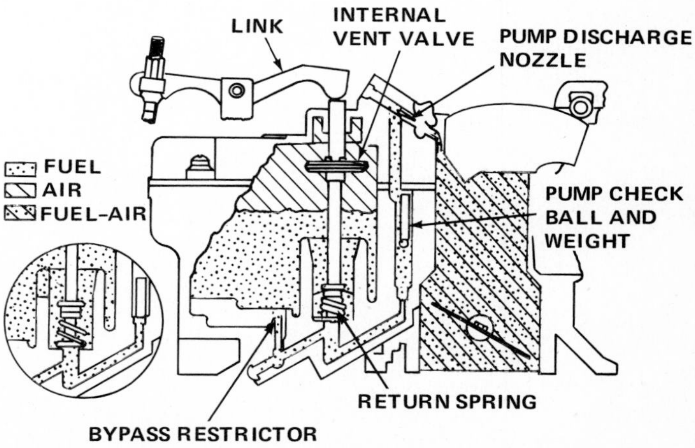

Fig. 13-56 Accelerator-pump system. (*Ford Motor Company*)

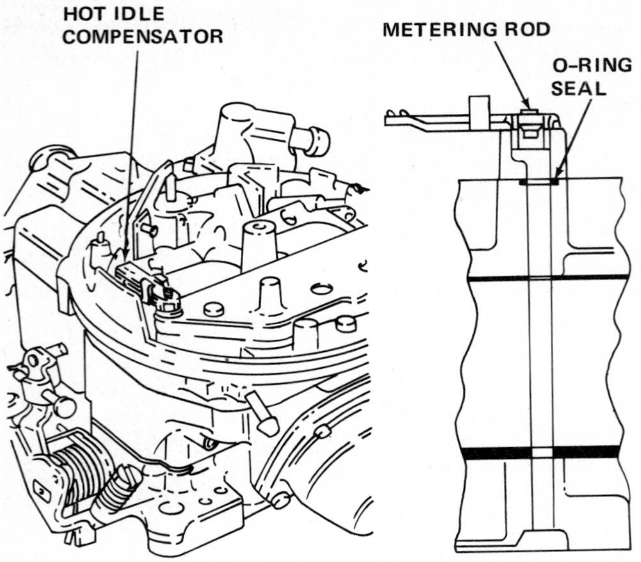

Fig. 13-57 Hot-idle compensator. (*Ford Motor Company*)

in fixed-venturi carburetors. Some of these VV carburetors also include automatic altitude compensation, similar in construction and operation to those used in some fixed-venturi carburetors and described in ● 13-34.

REVIEW QUESTIONS

1. What is carburetion?
2. What is vaporization?
3. What is atomization?
4. What are the three basic parts of the simple carburetor?
5. Describe the venturi effect.
6. What drives the fuel up out of the fuel nozzle and into the air passing through the carburetor air horn?
7. Describe the operation of the carburetor float system.
8. Describe the operation of the carburetor idle system.
9. Describe the operation of the carburetor main metering system.
10. Describe the operation of the power system that is mechanically operated. Describe the operation of the power system that is vacuum operated.
11. Why is the float bowl vented into the carburetor air horn just above the choke?
12. What is the purpose of the float bowl vent that is connected to the charcoal canister?
13. What is the purpose of the compensator valve in the carburetor?
14. Why is it that the idle adjustment screw is fixed or has a locking cap on it?
15. Why is an accelerator-pump system required?
16. How does the accelerator-pump system work?
17. What is the purpose of the choke? How does the automatic choke work?
18. What is the purpose of the electric choke? How does it work?
19. What is the purpose of the manifold heat-control valve? Why do some engines no longer have this valve?
20. What is the purpose of the antidieseling solenoid and how does it work?
21. What is the purpose of the altitude-compensation system in some carburetors and how does it work?
22. What is the purpose of using two-barrel and four-barrel carburetors?
23. Explain how the round-piston VV carburetor works.
24. Explain how the Ford VV carburetor works.

SELF PROJECTS

Examine as many different carburetors as you can. When you examine a carburetor, take a separate sheet of paper to write down the facts about it. At the top of the sheet write the make and model of car from which the carburetor came, and the model and type of carburetor.

Note the location of the venturi, main nozzle or nozzles, throttle plate or plates, and so on. Identify all the openings in the barrel or air horn. If you are able to do so, disassemble a carburetor and identify the internal parts such as the float, needle valve, accelerator-pump system, and so on. Make notes about all the carburetor parts. If possible, make sketches showing the locations of the essential parts and passages in the carburetor. As you examine carburetor parts, and make notes and sketches, you are really getting to know carburetors. After you have gone through this routine with several carburetors, you will be well on your way to being a carburetor expert.

Read the cautions about fuel-system work (● 14-16) if you handle carburetors that have gasoline in them, or if you clean carburetor parts in solvent.

CHAPTER 14
Carbureted Fuel-System Service

After studying this chapter, you should be able to:

- **List and describe the various troubles that might occur in fuel systems using carburetors and explain what could cause each trouble.**
- **Discuss the cautions to observe in fuel-system work.**
- **Explain how to service air cleaners and then perform this service on a car.**
- **Explain how to remove and replace a fuel pump and then perform this service on a car.**
- **Explain how to remove, service, and replace a carburetor, and then perform this job.**

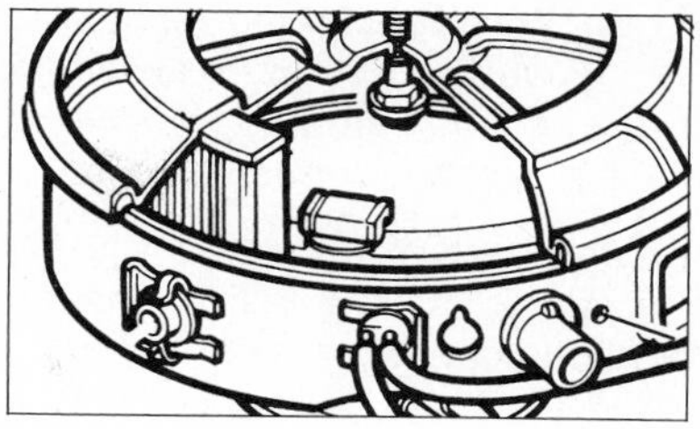

This chapter discusses various troubles that can develop in a fuel system using a carburetor, and how to correct them. This is only a part of the complete engine troubleshooting procedure which is discussed in detail in Chap. 42, Engine Trouble Diagnosis.

● 14-1 Defining the Terms

Before we get into fuel-system troubles, let us define the terms we will use to describe various engine troubles. These troubles can be caused by conditions in the fuel system. They can also be caused by conditions in the other engine systems, or the engine itself. We will be using these terms throughout the rest of the book dealing with the engine and engine systems. Therefore, you should familiarize yourself with these terms and understand what they mean.

1. Detonation This is a secondary explosion that occurs after the spark at the spark plug. It produces excessively rapid burning of the air-fuel mixture and a pinging or knocking sound. Usually worse under acceleration. Can damage the engine if severe.

2. Dieseling The engine continues to run after the ignition is turned off. Runs unevenly and may knock. Ignition is caused by hot spots in the combustion chambers—carbon, exhaust valve, spark plugs, etc. Supported by air and fuel leaks past the throttle valve and through the idle system.

3. Hesitation Momentary lack of response as the throttle is opened. Can occur at all car speeds. Most severe when first starting out. May cause engine to stall.

4. Knock See Detonation.

5. Miss Failure of a cylinder to fire. Causes steady pulsation or jerking. Usually more noticeable when the load increases. Not normally noticeable at higher engine speeds. Exhaust can have a steady spitting sound at low speeds.

6. Roll Engine speed varies under steady throttle or cruise. Feels as if the car speeds up and slows down with no change in throttle.

7. Preignition Ignition before the spark occurs, due to hot spots in the combustion chamber. Can cause a rumbling sound and rough engine operation.

8. Rough idle Engine runs roughly at idle, sometimes badly enough to cause the car to shake.

9. Sag See Hesitation.

10. Sluggish Engine power limited under load or when climbing hills. Slow on acceleration. Loses too much speed going up hills.

11. Spongy Less than anticipated response from engine when throttle is opened. Little or no speed increase when throttle is opened a small amount. Further throttle opening will finally get some response.

12. Stalls Engine quits running. This may occur when idling or when car is moving, as for instance when the throttle is suddenly opened. See Stumble.

13. Stumble Opening of the throttle causes the engine power to drop off. Can even cause engine to stall. Like hesitation but more marked. Engine will recover if throttle is closed and then patted a couple of times.

As we said previously, these troubles can be caused by conditions in the fuel system, and also by conditions in the other engine systems or even in the engine itself.

● 14-2 Analyzing Carbureted Fuel-System Troubles

Fuel-system troubles usually show up as faulty engine operation, or what is called car "driveability." A car that runs well is "driveable." A car with such problems as poor acceleration, hard starting, missing, loss of power, stumble, hesitation, stalling, and so on, has poor "driveability." All engine or driveability problems caused by the fuel system, ignition system, cooling system, or by troubles in the engine itself, are discussed in detail in Chap. 42, Engine Trouble Diagnosis.

In following sections, we discuss briefly the driveability problems that could be due to troubles in the fuel system. Remember that the conditions we discuss here could be caused by many conditions *outside* the fuel system. However, in this chapter, we are concentrating on driveability problems that might arise from faulty conditions in the carburetor fuel system.

● 14-3 Engine Cranks Normally but Will Not Start

This could be due to ignition problems, lack of fuel, underchoking or overchoking, or to the failure of the carburetor to deliver air-fuel mixture normally. Make sure that there is gasoline in the fuel tank and in the carburetor. The engine may be flooded due to overchoking, to a high float-level adjustment, or to a leaky float-bowl needle valve and seat. Open the throttle wide and try cranking. This clears the engine of excess fuel and will allow the engine to start if the trouble is engine flooding. If this doesn't work, perhaps the choke is not closed so the trouble is underchoking. Remove the air cleaner to check the choke position with the engine cold.

Modern chokes have an electric heater element. If you wait too long between turning on the ignition key and trying to start, the choke valve may open. Then you have underchoking. The remedy here is to turn the ignition off and wait long enough for the choke to cool so the choke valve closes. Then turn on the ignition and start without any delay.

CAUTION: Do not crank the engine with the air cleaner off. The engine could backfire up through the open carburetor and cause a fire, or you could get your face badly burned.

You can check for fuel in the carburetor by pushing the throttle to the floor several times. This should cause the accelerator pump to squirt fuel into the carburetor air horn. Watch for this with the air cleaner off (engine not running). If fuel squirts out, you can assume there is fuel in the float bowl. Reinstall the air cleaner. Try starting now, after having primed the engine with the accelerator pump. If the engine starts and runs briefly, the carburetor is probably at fault. If the engine does not start, then there may be other trouble, such as a clogged fuel-tank strainer, a defective fuel pump, a stopped-up fuel-tank vent or cap, or the wrong tank cap.

● 14-4 Engine Starts, Speeds Up Momentarily, Then Dies

This could be due to the choke vacuum-break (vacuum-kick) setting being too wide. That is, when the vacuum drops, the vacuum break kicks the choke too far open. It could also be due to a defect in the exhaust-gas-recirculation (EGR) system. That is, the EGR system is allowing exhaust gas to reenter the engine cylinders when the engine is cold. This can upset the combustion process so the engine dies. The trouble could be a stuck EGR valve, or some abnormal condition in the EGR control system. Other conditions that could cause this trouble include a fast idle set too low, a defective fuel pump not delivering enough fuel, or a low fuel level in the float bowl.

● 14-5 Engine Runs but Misses

This can be caused by many things—defective ignition, spark plugs, valve action, piston rings, leaking intake-manifold or gasket, or a vacuum line disconnected or split. In the fuel system, it could be caused by too lean or too rich a mixture. This could mean a problem in the carburetor, fuel pump, flex line, or fuel tank.

● 14-6 Engine Lacks Power

This complaint is best checked with a dynamometer and oscilloscope (Chap. 41). If the problem is poor acceleration, check to see if the throttle valve opens fully. On four-barrel carburetors make sure that the secondary throttle valves are opening properly. Also make sure the accelerator-pump system is working (● 14-2). If the complaint is lack of power cold, it could be due to a defective choke or the manifold heat control or EGR valve stuck open. Lack of power hot could mean that the choke or manifold heat-control valve is stuck closed. A defective fuel pump or clogged fuel filter could starve the engine so it does not get enough gasoline. The result is loss of power under almost any condition, particularly when full-power performance is called for. Many other conditions outside of the fuel system can cause lack of power, including automatic-transmission problems, as explained in Engine Trouble Diagnosis (Chap. 42).

● 14-7 Engine Stalls Cold or As It Warms Up

This could be due to a choke opening too quickly or to a manifold heat-control valve that is stuck open. It could also mean that not enough fuel is getting to or through the carburetor because of fuel-pump or carburetor troubles. There are several other possibilities in other engine components (Chap. 42).

● 14-8 Cold-Engine Driveability Problems

Now let us look at the special problems that might occur when driving a car after first starting a cold engine. Modern cars are somewhat more susceptible to these problems because of antipollution controls and close carburetor settings. No one worried in years gone by, if an engine ran rich during and after warm-up. Today, however, such richness is not allowed. Lean carburetor settings and fast-acting chokes prevent it. The lean settings and fast-acting chokes can cause cold-running problems if everything is not just right. Here are some possible troubles and their causes:

1. Engine stalls when transmission is shifted into gear. This could be due to improper choke-vacuum-break setting, fast-idle setting, or incorrect ignition timing.
2. Engine stalls, hesitates, or sags during first mile when throttle is opened. This could be due to improper choke-vacuum-break setting, to electric choke kicking off too fast, to EGR system releasing exhaust gas to the intake manifold, to low float-level adjustment, defective accelerator pump, or no distributor vacuum advance.
3. Engine stalls, hesitates, or sags after first warm-up mile. This could be due to everything mentioned in the previous paragraph. It could also be due to a defective fuel pump that is not delivering enough fuel for sustained intermediate speeds.

● 14-9 Warm-Engine Driveability Problems

Many warm-engine driveability problems cause loss of engine power as discussed in ● 14-5. Other problems include hesitation, stumble, or sag. These can result from incorrect ignition timing or defective spark-advance actions, from a defective accelerator pump, or from the heated-air-system inlet door in the air-cleaner snorkel being stuck. Also, consider the conditions in ● 14-5.

● 14-10 Surge

Another driveability problem that sometimes occurs is surge. That is, while you are driving along at a steady speed, the engine suddenly rolls—increasing and then decreasing its power output. This can result from a vacuum leak (one of the hoses to the carburetor loose, or carburetor mounting loose), incorrect spark advance or faulty mechanism, a defective PCV valve (stuck in high-flow position), lack of sufficient fuel to engine (low float level, defective fuel pump, clogged fuel filter), or from the heated-air-system inlet door being stuck.

● 14-11 Dieseling or Run-On

This is a condition where the engine continues to run after the ignition is turned off. It could be due to a high idle-speed adjustment or to the idle-stop solenoid not allowing the throttle to close completely when the ignition is turned off. As a result, enough air-fuel mixture gets through the carburetor to allow the engine to continue to run. Ignition is by hot spots in the combustion chamber. (See ● 37-14.)

● 14-12 Excessive Fuel Consumption

This can be due to many causes—driving habits, high speed, short runs, choke partly closed, high carburetor float level, worn float-bowl needle valve or seat, worn jets in the carburetor, internal carburetor leaks, external gasoline leaks, or stuck metering rod or accelerator pump. Engine troubles such as defective rings or valve action, excessive friction in the drive line, low tire pressure, improperly operating automatic transmission, and so on, can also increase fuel consumption.

● 14-13 Air-Fuel-Mixture Test

A rough test of mixture richness that does not require any testing instruments can be made by inspecting the spark plugs after a short run. Start by installing a set of new or cleaned spark plugs of the correct heat range for the engine. Operate the car for about 15 or 20 minutes. Stop the car. Let the engine cool enough to work on, and then remove and examine the spark plugs. If they are coated with a black carbon deposit, the mixture is probably too rich. Black exhaust smoke also indicates an excessively rich mixture. The mixture is too rich to burn fully, so the exhaust gas contains "soot," or unburned gasoline.

● 14-14 Carburetor Troubles

In previous sections, we took a look at various engine troubles and mentioned possible troubles in the fuel system and elsewhere that might cause the troubles. Now, let us sum up the possible troubles that might be caused by conditions in the carburetor and related components. Remember that engine troubles can come from many other things besides problems in the fuel system and carburetor. In the chapter on engine trouble diagnosis (Chap. 42) we list and describe all the engine troubles and possible trouble causes. Here, we focus on the carburetor to see what troubles can be caused by conditions inside the carburetor itself.

1. Excessive fuel consumption can result from:
   a. A high float level or a leaky float
   b. A sticking or dirty float needle valve
   c. Worn jets or nozzles
   d. A stuck metering rod or full-power piston
   e. Idle too rich or too fast
   f. A stuck accelerator-pump check valve
   g. A leaky carburetor
   h. Dirty air cleaner
2. Lack of engine power, acceleration, or high-speed performance can result from:
   a. The power step-up on the metering rod not clearing the jet
   b. Dirt or gum clogging the fuel nozzle or jets
   c. A stuck power piston or valve
   d. A low float level
   e. A dirty air filter
   f. The choke stuck or not operating
   g. Air leaks into the manifold
   h. The throttle valve not fully opening
   i. A rich mixture, due to causes listed under item 1, above
3. Poor idle can result from a leaky vacuum hose, stuck PCV valve, or retarded timing. Also, it could be due to an incorrectly adjusted idle mixture or speed, a clogged idle system, or any of the causes listed under item 2, above.
4. Failure of the engine to start unless primed could be due to no gasoline in the fuel tank or carburetor, the wrong tank cap (1970 cars and later), or a stopped-up tank or cap vent. The latter causes a vacuum to develop in the tank, which prevents delivery of fuel to the carburetor. Holes in the fuel-pump flex line will allow air leakage which prevents fuel delivery. In addition, consider carburetor jets or lines clogged, a defective choke, a clogged fuel filter, or air leaks into the manifold.
5. Hard starting with the engine warm could be due to a defective choke, a closed choke valve, or improperly adjusted throttle-cracker linkage.
6. Slow engine warm-up could be due to a defectively operating choke or manifold heat-control valve.
7. A smoky, black exhaust is due to a very rich mixture. Carburetor conditions that could cause this are listed in item 1 above.
8. If the engine stalls as it warms up, this could be due to a defective choke or a closed choke valve.
9. If the engine stalls after a period of high-speed driving, this could be due to a malfunctioning antipercolator.
10. If the engine backfires, this could be due to an excessively rich or lean mixture. If the backfire is in the exhaust system, it is usually caused by an excessively rich mixture in the exhaust. This results from a defective air-injection-system anti-backfire valve. Lean mixtures usually cause a pop-back in the carburetor.
11. If the engine runs but misses, the most likely cause is that a vacuum hose has come off an intake-manifold fitting (causing nearest cylinders to miss), and a leaky intake manifold gasket. In addition, it could be that the proper amount and ratio of air-fuel mixture are not reaching the engine, and this might be due to clogged or worn carburetor jets or to an incorrect fuel level in the float bowl.

Some of the conditions noted above can be corrected by carburetor adjustment. Others require removal of the carburetor from the engine so that it can be disassembled, repaired, and reassembled. Sections at the end of the chapter describe briefly carburetor adjustment and servicing procedures.

● 14-15 Quick Carburetor Checks

A number of quick checks can be made that will give you a rough idea of how the carburetor is working. More accurate analysis requires test instruments such as an exhaust-gas analyzer and an intake-manifold vacuum gauge, as explained in Chap. 41.

1. Float-level adjustment With the engine warmed up and running at idle speed (so there is no danger of backfiring), remove the air cleaner. Note the condition of the high-speed nozzle. If the nozzle tip is wet or is dripping gasoline, chances are that the float level is too high. This could cause a continuous discharge of gasoline from the nozzle, even on idle.

2. Idle system If the engine does not idle smoothly after it is warmed up, the idle system could be at fault. Slowly open the throttle to about a 25-mph [40.23 km/h] engine speed. If the speed does not increase evenly and the engine runs roughly through this speed range, chances are the idle or main metering system is out of order.

3. Accelerator-pump system With the air cleaner off and the *engine not running,* open the throttle suddenly. See if the accelerator-pump system discharges a squirt of gasoline into the air horn. The flow should continue for a few seconds after the throttle reaches the open position.

4. Main metering system With the engine warmed up and running at approximately 25 mph [40.23 km/h], slowly cover part of the air horn with a piece of stiff cardboard. The engine should speed up slightly, since this causes a normally operating main metering system to discharge more gasoline.

CAUTION: Do not use your hand to cover the air horn. The vacuum could pull your fingers into the air horn and give you a painful injury.

● 14-16 Cautions in Fuel-System Work

The following cautions should be carefully observed in fuel-system work:

1. Remember that even a trace of dirt in a carburetor or fuel pump can cause fuel-system and engine trouble. Be very careful about dirt when repairing these units. Your hands, the workbench, and the tools should be clean.
2. Gasoline vapor is very explosive. Wipe up spilled gasoline at once, and put the cloths outside to dry. Never smoke or bring an open flame near gasoline! Reread the CAUTION at the end of ● 11-11.
3. When using the solvent tank, be careful not to splash solvent in your eyes. Do not add gasoline to the solvent tank; this increases the danger of fire. Dump any gasoline from the carburetor or fuel pump in a container before cleaning them in the solvent.
4. Remember that the gasoline may be under some pressure in the fuel lines. Whenever you disconnect a fuel line, especially between the fuel pump and carburetor, have a cloth handy to soak up any gasoline that spurts or leaks out. Wipe up all gasoline and put the cloth outside in a safe place to dry. The procedure of loosening fuel-line connections is described in ● 14-21.
5. When drying parts with the air hose, handle the hose with care, as noted in Chap. 3. Also observe the other safety cautions listed in that chapter.

● 14-17 Air-Cleaner Service

Air-cleaner-service recommendations vary with different automotive manufacturers. Always check the manufacturer's shop manual for specifications. Here are sample recommendations and service procedures.

1. Chrysler The paper filter element (Fig. 14-1) in the air cleaner should be replaced every 30,000 miles [48,000 km]. It should be replaced more often if the car is driven in dusty or sandy areas.

Disconnect the air-cleaner hose at the air cleaner. Remove the cleaner from the carburetor and take the filter element out of the cleaner. Examine it for oil. If the element appears saturated for more than half its circumference, check the rest of the crankcase ventilating system. The system is carrying oil up to the air cleaner.

Several years ago, Chrysler was recommending that you could clean the filter and reuse it if it was in good condition. The cleaning process required an air hose which blew compressed air through the filter from the inside, as shown in Fig. 14-2. This procedure is no longer recommended.

When replacing the filter and air cleaner, be sure the plastic rings on both sides of the element are smooth and seal properly at the top and bottom.

2. Ford Figure 14-3 shows a typical air cleaner and duct system for a Ford V-8 installation. Recommendation is to replace the air cleaner every 30,000 miles [48,000 km] or 30 months, whichever occurs first. More often, of course, if the car is driven in dusty areas.

3. Chevrolet The recommendations for Chevrolet apply, in general, to all General Motors automobiles. They are similar to those for Chrysler and Ford. That is, replace the filter every 30,000 [48,000 km], or more often for dusty driving conditions. Chevrolet also recommends inspection of the element for dust leaks after the first 15,000 miles [24,000 km] of operation. Look for the presence of dust inside the cleaner housing filter element. Dust could enter through holes in the filter element, and under or over the element, past the end seals. If everything looks okay, rotate the element 180° and replace it.

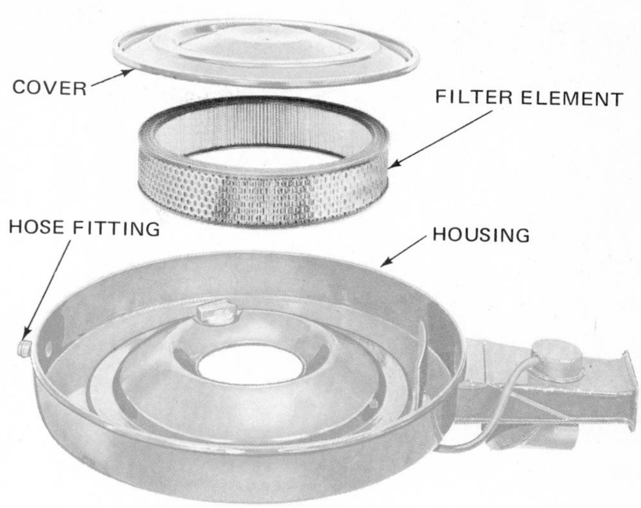

Fig. 14-1 Carburetor air cleaner of the type using a paper element. *(Chrysler Corporation)*

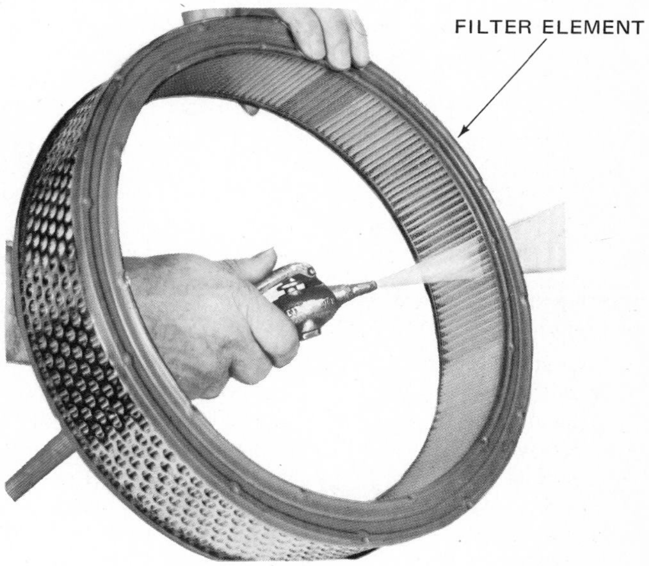

Fig. 14-2 Cleaning the paper element with compressed air. Note that air is being blown from inside the element. *(Chrysler Corporation)*

CAUTION: Do not operate the engine with the air cleaner off unless it is absolutely necessary when making tests. If the engine backfires with the air cleaner off, it could cause a fire in the engine compartment.

Air filters which are not installed directly on the carburetor are connected to it by a flexible hose. This hose must be connected airtight to both the filter and the

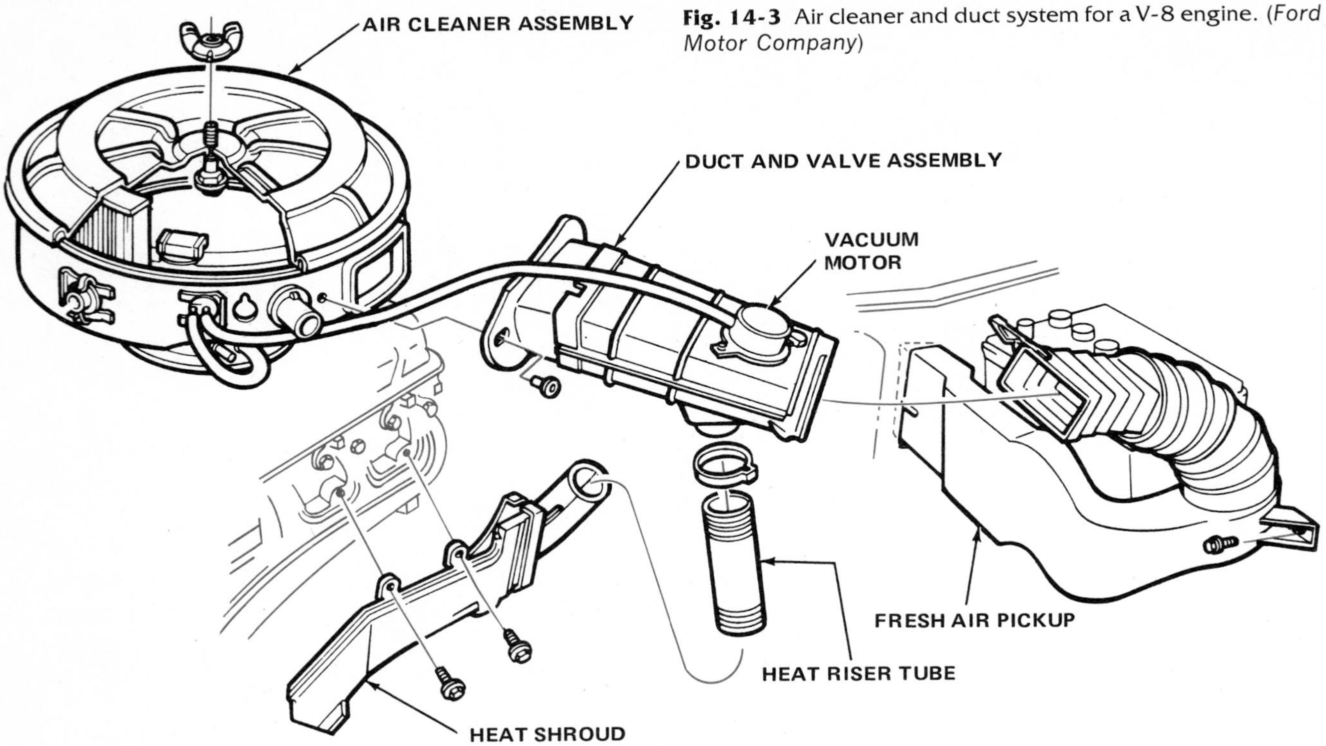

Fig. 14-3 Air cleaner and duct system for a V-8 engine. (*Ford Motor Company*)

carburetor. The hose must have no tears or punctures that could leak unfiltered air into the carburetor.

● 14-18 Thermostatically Controlled Air Cleaner

To check the system, first make sure the hoses and heat pipe are tightly connected. See that there are no leaks in the system. The system can be checked with a temperature gauge. Remember that failure of the thermostatic system usually results in the damper door staying open. This means that the driver would probably not notice anything wrong in warm weather. But, in cold weather, the driver will notice hesitation, surge, and stalling. A typical checking procedure follows.

Remove the air-cleaner cover. Install the temperature gauge as close to the sensor as possible. Allow the engine to cool below 85°F [29.4°C] if it is hot. Replace the air-cleaner cover without the wing nut.

Start and idle the engine. When the damper begins to open, remove the air-cleaner cover and note the temperature reading. It should be between 85 and 115°F [29.4 and 46.1°C]. If it is difficult to see, use a mirror.

If the damper does not open at the correct temperature, check the vacuum motor and sensor.

With the engine off, the control damper should be in the compartment or cold-air-delivery position (see Fig. 12-24). To determine if the vacuum motor is operating, apply at least 9 inches [228.6 mm] of vacuum to the fitting on the vacuum motor. The vacuum can be from the engine, from a distributor tester, or from a hand vacuum pump. With vacuum applied, the damper should move to the hot-air-delivery position (Fig. 12-24).

If the vacuum motor does not work satisfactorily, it should be replaced. This can be done by drilling out the spot welds, and unhooking the linkage. The new motor can be installed with a retaining strap and sheet-metal screws.

If the vacuum motor does not work well, the sensor should be replaced. This is done by prying up the tabs on the retaining clip. The new sensor is then installed, and the tabs bent down again.

● 14-19 Automatic Chokes

As a rule, automatic chokes require no service once they are adjusted for the operating conditions and engine. Some chokes, used years ago, had an adjustable cover that could be turned one way or the other to enrich or lean out the mixture (Fig. 14-4). Other types of choke are adjusted by bending a linkage rod.

A thermometer is needed to check the operation of the electric automatic choke. Specifications vary, so look up the testing procedure in the shop manual for the car you are checking. A typical procedure for a carburetor-mounted choke is to tape a bulb thermometer to the choke housing, and then start the engine. Note the temperature at which the choke opens and the amount of time required for the choke to open.

● 14-20 Fuel Gauges

Fuel gauges require very little in the way of service. Defects in either the dash unit or the tank unit usually require replacement of the unit. However, on the type of gauge with vibrating thermostatic blades, dirty contact

Fig. 14-4 An automatic choke which has an adjustable cover. (*Ford Motor Company*)

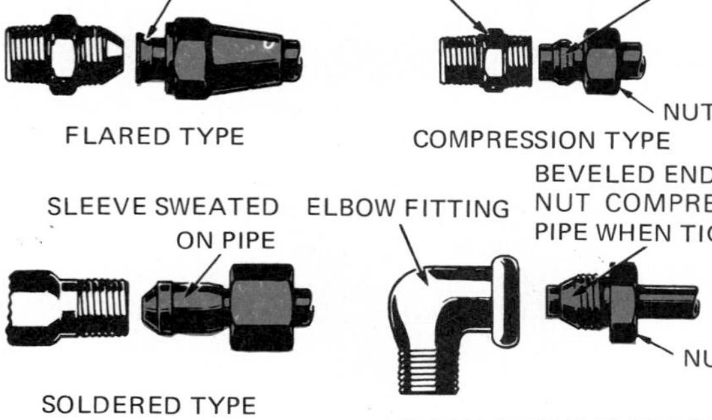

Fig. 14-5 Types of fuel-line couplings, or connectors.

points may cause fluctuations of the needle. Points can be cleaned by pulling a strip of clean bond paper between them.

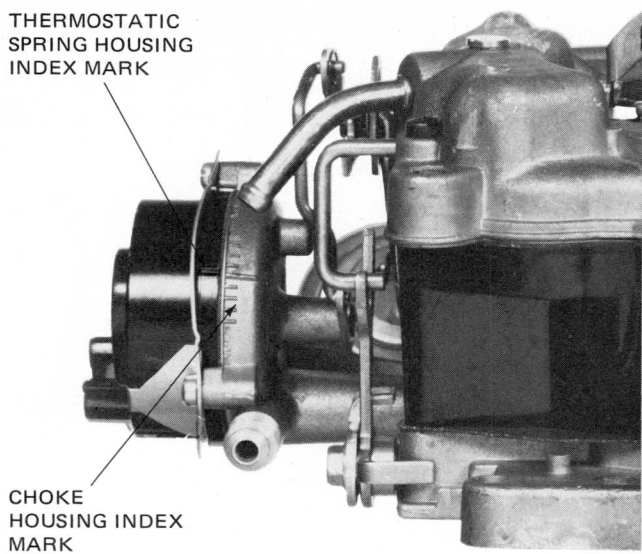

If you run into a fuel-gauge problem, substitute a good tank unit for the one on the car. That is, connect the good tank unit between the dash unit and ground. Move the float arm up and down to see if the movement registers on the dash unit. Remember that the thermostatic fuel gauge (thermostatic blade in dash unit) takes a few moments to register. If the dash unit works normally, the tank unit on the car is defective. If the dash unit does not work, it is defective.

Another check is to momentarily ground the wire between the tank unit and dash unit. If the dash unit registers, the trouble is probably in the tank unit or in the circuit between the two units.

● 14-21 Fuel-Line Connections

When you remove and install fuel pumps and carburetors, you must disconnect and reconnect fuel lines. Fuel-line connections, or couplings, take various forms (Fig. 14-5). When loosening a coupling of the type having two nuts, use two wrenches, as shown in Fig. 14-6, to avoid damaging the line. If available, a broad-jaw, flare-nut wrench should be used. When using two wrenches, hold one steady while you turn the other, as shown in Fig. 14-6.

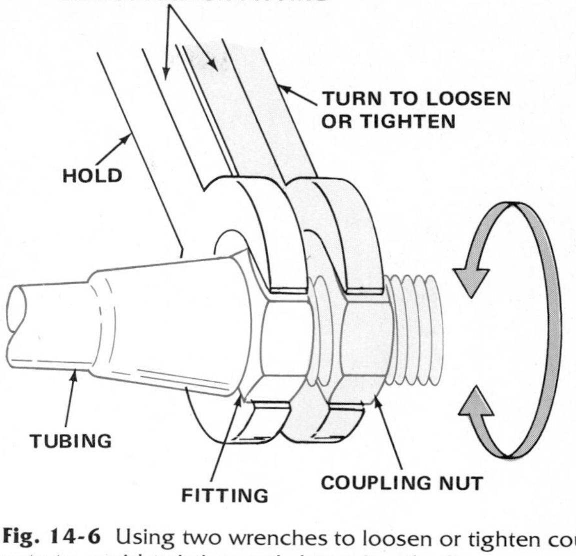

Fig. 14-6 Using two wrenches to loosen or tighten coupling nuts to avoid twisting and damaging the line.

CAUTION: Note caution 4 in ● 14-16 regarding the gasoline being under pressure. Be ready with a cloth to soak up gasoline that spurts or leaks out. Put cloth outside to dry.

● 14-22 Fuel-Pump and Filter Service

Fuel-pump pressure and capacity can be checked with special testers. Low pump pressure causes fuel starvation and poor engine performance. High pressure will cause an overrich mixture, excessive fuel consumption, and such troubles as fouled spark plugs, rings, and valves (from excessive carbon deposits). Fuel-pump testers are connected into the fuel line from the pump. They measure either the pressure that the pump can develop or the amount of fuel the pump can deliver during a timed interval. The vacuum that the fuel pump can develop should also be checked. This is done by connecting the tester to the suction side of the pump.

Fuel filters require no service except periodic checks to make sure that they are not clogged, and replacement of the filter element or cleaning of the filter, according to type. On many models, the filter is part of the fuel pump (Fig. 14-7) and can be removed so that the element can be replaced. Another type is the in-line filter (Figs. 12-8 and 12-9). In the type shown in Fig. 12-8, the filter is replaced by unclamping and detaching the fuel hose from the filter. Then, the filter can be unscrewed from the carburetor and replaced. In the type shown in Fig. 14-8, the fuel line is detached. Then the nut is removed so that

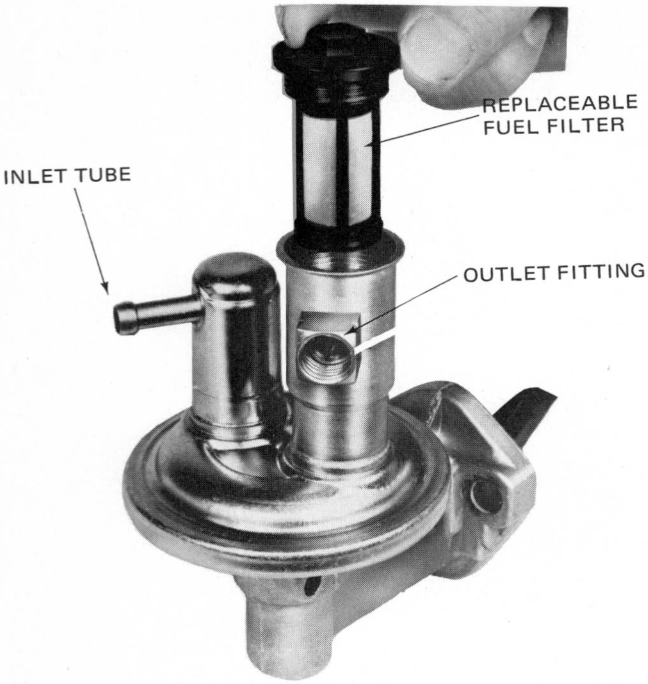

Fig. 14-7 Removing fuel filter from fuel pump. (*Chrysler Corporation*)

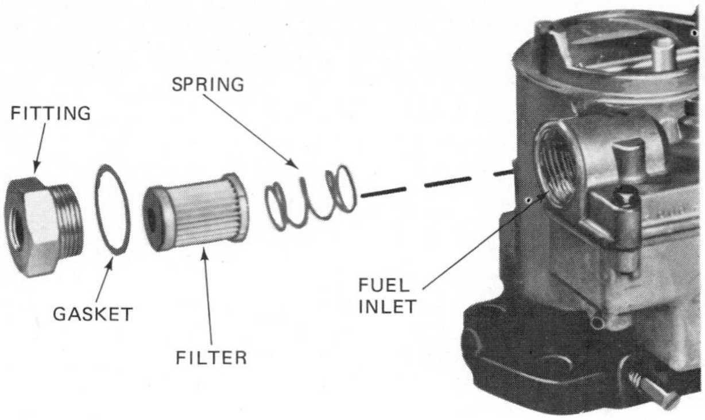

Fig. 14-8 Removing fuel filter from carburetor. (*Chevrolet Motor Division of General Motors Corporation*)

the old filter element can be slipped out and replaced (see Fig. 14-8).

● 14-23 Fuel-Pump Troubles

Fuel-system troubles that might be caused by the fuel pump are discussed below.

1. Insufficient fuel delivery This could result from low pump pressure, which in turn could be due to any of the following:

a. Broken, worn-out, or cracked diaphragm
b. Improperly operating fuel-pump valves
c. Broken or damaged rocker arm
d. Clogged pump-filter screen or filter
e. Leakage of air into sediment bowl because of loose bowl or worn gasket

These are all causes of insufficient fuel delivery due to conditions within the pump. In addition, there are conditions outside the pump that could prevent delivery of normal amounts of fuel. These include such things as a clogged fuel-tank-cap vent, clogged fuel line or filter, air leaks into the fuel line, and vapor lock. Of course, in the carburetor, an incorrect float level, a clogged inlet screen, or a malfunctioning inlet needle valve would prevent delivery of adequate amounts of fuel to the carburetor.

2. Excessive pump pressure High pump pressure causes delivery of too much fuel to the carburetor. The excessive pressure tends to lift the needle valve off its seat. As a result, the fuel level in the float bowl is too high. This would be a very unusual condition. If a fuel pump has been operating satisfactorily, its pressure should not increase enough to cause trouble.

3. Fuel-pump leaks If the fuel pump leaks, gasoline can accumulate in the oil pan faster than the crankcase ventilator can remove it. Most modern fuel pumps are crimped on assembly (Fig. 14-5). If the crimp is not properly made, the diaphragm may leak around the edges. The diaphragm itself could be damaged. In either case, the fuel pump should be replaced. Note, also, that leaks may occur at fuel-line connections which are loose or improperly coupled.

4. Fuel-pump noises A noisy pump is usually the result of worn or broken parts within the pump. These include a weak or broken rocker-arm spring, a worn or broken rocker-arm pin or rocker-arm, or a broken diaphragm spring. In addition, a loose fuel pump or a scored rocker arm or cam on the camshaft may cause noise. Fuel-pump noise may sound something like engine-valve tappet noise. Its frequency is the same as camshaft speed. If the noise is bad enough, it can actually be "felt" by gripping the fuel pump firmly by hand.

CAUTION: Don't do this if reaching into the fuel pump puts your hand too close to the moving fan or belts. Careful listening will usually tell you that the noise is coming from the fuel pump. Tappet noise is usually heard all along the engine or in the engine valve compartment.

● 14-24 Fuel-Pump Removal

Before removing the fuel pump, wipe off any dirt or grease so it will not get into the engine. Then take off the heat shield (where present), and disconnect the fuel lines. Remove the attaching nuts or bolts, and lift off the pump. If it sticks, work it gently from side to side or pry lightly under its mounting flange or attaching studs. On an engine using a push rod to operate the fuel pump, remove the rod and examine it for wear or sticking.

● 14-25 Fuel-Pump Service

Most late-model fuel pumps are assembled by crimping and cannot be disassembled. If defective, they must be replaced with a new assembly.

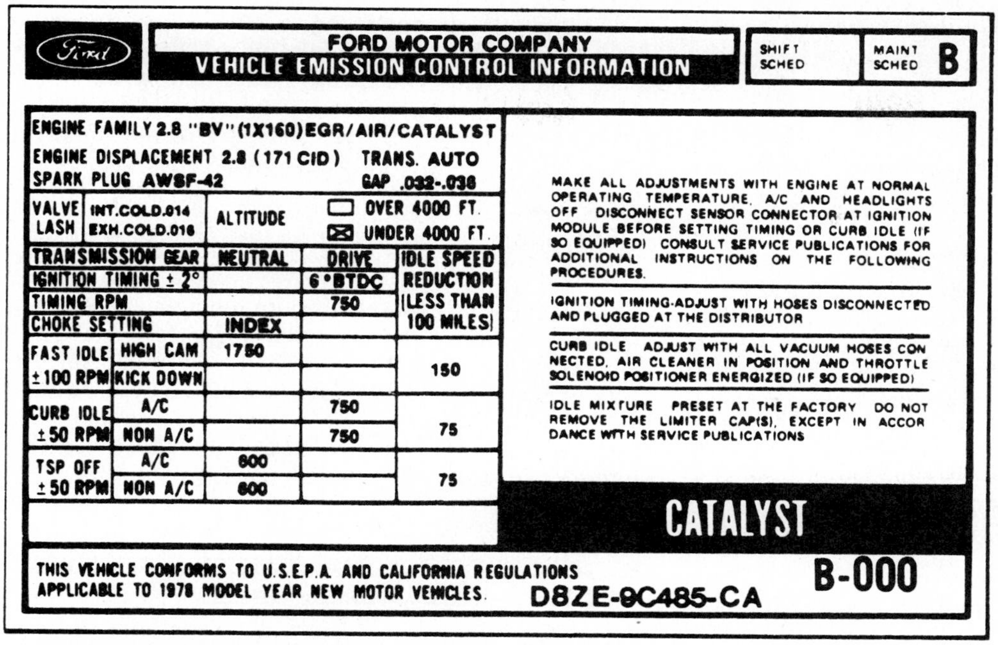
Ford
FORD MOTOR COMPANY
VEHICLE EMISSION CONTROL INFORMATION
SHIFT SCHED
MAINT SCHED B
ENGINE FAMILY 2.8 "BV" (1X160) EGR/AIR/CATALYST
ENGINE DISPLACEMENT 2.8 (171 CID) TRANS. AUTO
SPARK PLUG AWSF-42 GAP .032-.036
VALVE LASH INT.COLD.014 EXH.COLD.016
ALTITUDE OVER 4000 FT. UNDER 4000 FT.
TRANSMISSION GEAR NEUTRAL DRIVE
IDLE SPEED REDUCTION (LESS THAN 100 MILES)
IGNITION TIMING ± 2° 6°BTDC
TIMING RPM 750
CHOKE SETTING INDEX
FAST IDLE ±100 RPM HIGH CAM 1750 KICK DOWN 150
CURB IDLE ±50 RPM A/C 750 NON A/C 750 75
TSP OFF ±50 RPM A/C 600 NON A/C 600 75
MAKE ALL ADJUSTMENTS WITH ENGINE AT NORMAL OPERATING TEMPERATURE, A/C AND HEADLIGHTS OFF. DISCONNECT SENSOR CONNECTOR AT IGNITION MODULE BEFORE SETTING TIMING OR CURB IDLE (IF SO EQUIPPED). CONSULT SERVICE PUBLICATIONS FOR ADDITIONAL INSTRUCTIONS ON THE FOLLOWING PROCEDURES.
IGNITION TIMING-ADJUST WITH HOSES DISCONNECTED AND PLUGGED AT THE DISTRIBUTOR
CURB IDLE ADJUST WITH ALL VACUUM HOSES CONNECTED, AIR CLEANER IN POSITION AND THROTTLE SOLENOID POSITIONER ENERGIZED (IF SO EQUIPPED)
IDLE MIXTURE PRESET AT THE FACTORY. DO NOT REMOVE THE LIMITER CAP(S), EXCEPT IN ACCORDANCE WITH SERVICE PUBLICATIONS
CATALYST
THIS VEHICLE CONFORMS TO U.S.E.P.A. AND CALIFORNIA REGULATIONS APPLICABLE TO 1978 MODEL YEAR NEW MOTOR VEHICLES.
D8ZE-9C485-CA
B-000

Fig. 14-9 Tune-up decal typical of the decals found in the engine compartment. (*Ford Motor Company*)

● 14-26 Fuel-Pump Installation

Make sure fuel-line connections are clean and in good condition. Connect the fuel lines to the pump before attaching the pump to the engine (see ● 14-21). Then place a new gasket on the studs of the fuel-pump mounting or over the opening in the crankcase. The mounting surface of the engine should be clean. Insert the rocker arm of the fuel pump into the opening. Make sure that the arm goes on the proper side of the camshaft (or that it is centered over the push rod). If it is hard to get the holes in the fuel-pump flange to align with the holes in the block or timing cover, do this: Turn the engine over until the low side of the camshaft eccentric is under the fuel-pump rocker arm. Now the pump can be installed without forcing or prying it into place. Attach it with bolts or nuts. Check the pump operation, as explained in ● 14-22).

● 14-27 Carburetor Adjustments

At one time, there were several adjustments that could be made on carburetors. In recent years, however, the increasingly rigid requirements of the Clean Air Act regarding automotive emissions have eliminated most adjustments. That is, the only adjustment recommended for late-model cars during tune-up is to adjust idle speed. The idle mixture is preset at the factory and a limiter cap is installed to prevent tampering. The adjustment procedure is spelled out in a special tune-up decal in the engine compartment. The decal (Fig. 14-9) lists the specific procedure that *must* be followed. However, if the carburetor requires disassembly for servicing, the limiter caps may be removed. Then the idle mixture must be readjusted and new limiter caps installed.

In 1977, the procedure of setting the idle mixture was changed for some cars. The new procedure requires the use of a cartridge of propane. It is described in the following section (● 14-28).

First, we describe the procedure as outlined in the decal (Fig. 14-9). This procedure is typical for recent-model cars, except for those now requiring the propane procedure.

Setting idle speed Here is the procedure as given in Fig. 14-9.

1. Disconnect the fuel-tank hose from the vapor canister.
2. Disconnect the vacuum hose to the distributor. Plug the hose leading to the carburetor.
3. Make sure distributor contact-point dwell and ignition timing are correct (covered in Chap. 31).
4. Adjust idle speed by the means provided. In earlier models, this was a screw in the throttle linkage at the carburetor. In later models equipped with idle-speed solenoids, the adjustment screw is in the solenoid. Regardless of the location of the adjustment screw, use a tachometer to measure engine speed. Make the adjustment to get the specified idle speed.
5. Reconnect the distributor vacuum hose and the fuel-tank hose.

Setting idle mixture This adjustment is permissible only if the carburetor has required major service. A typical procedure is:

1. With limiter caps off, turn mixture screws in until they lightly touch the seats. Then back them off two full turns.
2. Adjust idle speed as already noted.
3. Connect a CO meter to the exhaust system. Adjust the idle-mixture screws to get a satisfactory idle at the

specified rpm and a CO reading at or below the specified allowable maximum. Engine should be running at normal idle with automatic transmission in D (drive) or manual transmission in neutral.

4. After setting the idle mixture, recheck the idle speed as previously. If everything checks, install new limiter caps on the idle-mixture screws.

● 14-28 Adjusting Idle Mixture with Propane

The catalytic converter cleans up the exhaust gas by converting the HC and CO into harmless gases. It also makes it practically impossible to adjust the idle mixture by reading the CO content of the exhaust gas. Therefore, a new idle-mixture adjustment procedure was developed, using propane to artificially enrich the mixture during the adjustment. The engine rpm is used to determine the proper setting. Figure 14-10 shows the setup for doing the job.

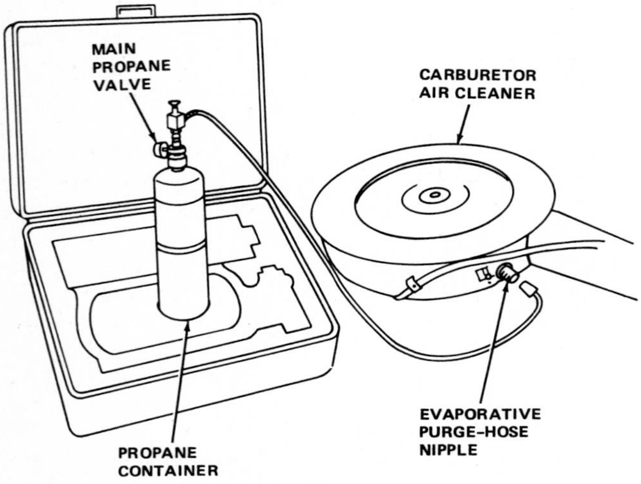

Fig. 14-10 Setup for adjusting idle mixture with propane. *(Ford Motor Company)*

Cars requiring the propane adjustment procedure have an Emission Control Information label under the hood (Fig. 14-11). The label explains how to disconnect and plug hoses before the adjustment is made. Factory shop manuals also carry detailed explanations of how to use propane to make the adjustment.

Note: Making any adjustment contrary to the detailed official procedure may violate federal, state, or municipal laws.

Essentially, what the procedure requires is the use of propane to temporarily enrich the idle mixture so that the idle speed and mixture screws can be adjusted as necessary to achieve the proper engine rpm. The procedure is called the *best lean idle* adjustment. The propane is fed into the air going into the air cleaner and thus the carburetor, as shown in Fig. 14-10.

● 14-29 Carburetor Removal

To remove a carburetor, first disconnect the battery ground terminal and wires from any electrical devices on the carburetor. This guards against sparks that might ignite any spilled gasoline. Then disconnect the air and vacuum lines and take off the air cleaner. Then disconnect the throttle and choke linkages. Disconnect the hot-air tube to the choke (if present). Disconnect the fuel line and the distributor vacuum-advance line from the carburetor. Use two wrenches, as necessary, to avoid damage to the lines or couplings.

CAUTION: Note Caution 4 in ● 14-16 about the gasoline being under pressure in the line between the fuel pump and carburetor.

Take off the carburetor attaching nuts or bolts, and lift off the carburetor. Try to avoid jarring the carburetor. It may have accumulations of dirt in the float bowl. Rough treatment may stir up this dirt and cause it to get into carburetor jets or circuits.

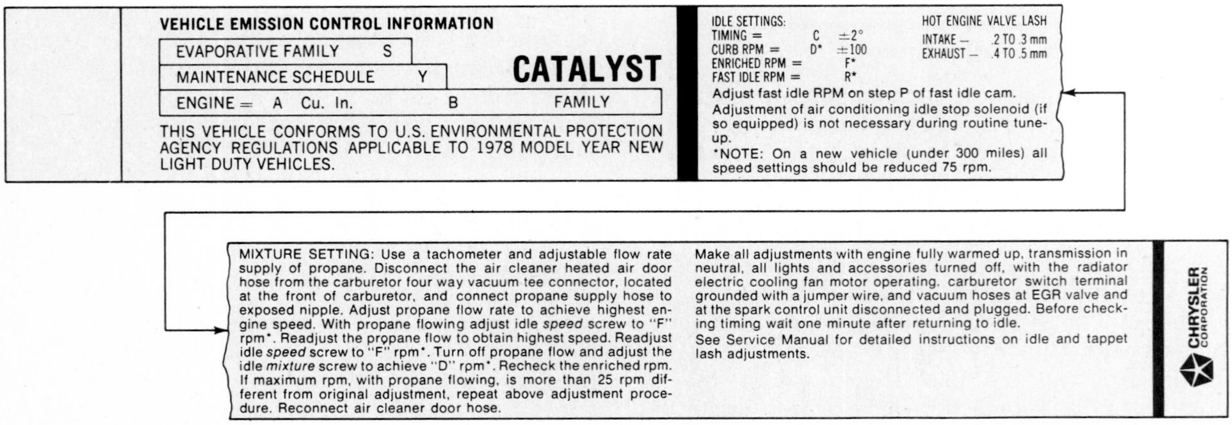

VEHICLE EMISSION CONTROL INFORMATION

EVAPORATIVE FAMILY S

MAINTENANCE SCHEDULE Y

CATALYST

ENGINE = A Cu. In. B FAMILY

THIS VEHICLE CONFORMS TO U.S. ENVIRONMENTAL PROTECTION AGENCY REGULATIONS APPLICABLE TO 1978 MODEL YEAR NEW LIGHT DUTY VEHICLES.

IDLE SETTINGS:
TIMING = C ±2°
CURB RPM = D* ±100
ENRICHED RPM = F*
FAST IDLE RPM = R*

HOT ENGINE VALVE LASH
INTAKE — .2 TO .3 mm
EXHAUST — .4 TO .5 mm

Adjust fast idle RPM on step P of fast idle cam.
Adjustment of air conditioning idle stop solenoid (if so equipped) is not necessary during routine tune-up.
*NOTE: On a new vehicle (under 300 miles) all speed settings should be reduced 75 rpm.

MIXTURE SETTING: Use a tachometer and adjustable flow rate supply of propane. Disconnect the air cleaner heated air door hose from the carburetor four way vacuum tee connector, located at the front of carburetor, and connect propane supply hose to exposed nipple. Adjust propane flow rate to achieve highest engine speed. With propane flowing adjust idle *speed* screw to "F" rpm*. Readjust the propane flow to obtain highest speed. Readjust idle *speed* screw to "F" rpm*. Turn off propane flow and adjust the idle *mixture* screw to achieve "D" rpm*. Recheck the enriched rpm. If maximum rpm, with propane flowing, is more than 25 rpm different from original adjustment, repeat above adjustment procedure. Reconnect air cleaner door hose.

Make all adjustments with engine fully warmed up, transmission in neutral, all lights and accessories turned off, with the radiator electric cooling fan motor operating, carburetor switch terminal grounded with a jumper wire, and vacuum hoses at EGR valve and at the spark control unit disconnected and plugged. Before checking timing wait one minute after returning to idle.
See Service Manual for detailed instructions on idle and tappet lash adjustments.

CHRYSLER
CORPORATION

Fig. 14-11 Information decal that is under the hood or on the engine of new cars. It includes the procedure for adjusting the idle mixture with propane. *(Chrysler Corporation)*

After the carburetor is off, it should be put in a clean place where dirt and dust cannot get into openings.

If the carburetor is to be off the engine for any length of time, cover the exposed manifold holes with masking tape (Fig. 14-12). Do not use shop cloths because threads and lint can drop off and get into the manifold. Protecting the manifold holes in this way will prevent engine damage from loose parts dropped into the manifold. Parts dropped in the manifold could end up in the engine combustion chambers where they could cause serious damage.

● 14-30 Carburetor Overhaul Procedures

Disassembly and reassembly procedures on carburetors vary according to design; the manufacturer's recommendations should be carefully followed. The time required to overhaul a carburetor varies from approximately $\frac{3}{4}$ to 2 hours, according to type. A few special carburetor tools may be required. Gauges needed to measure float clearance, float centering, float height, choke clearance, and so on are usually included in the carburetor overhaul kit.

Complete carburetor overhaul kits are supplied for many carburetors. These kits contain instructions and all parts (jets, gaskets, and so on) required to overhaul the carburetor and restore it to performing condition.

CAUTION: When removing and handling a carburetor, be extremely careful to avoid spilling gasoline. Remember that the carburetor float bowl will have gasoline in it, so keep the carburetor upright. Gasoline is extremely inflammable. Any spilled gasoline should be wiped up immediately. Put gasoline-soaked towels outside to dry.

Disassembly and assembly procedures on carburetors vary widely according to design. When overhauling a carburetor, get the overhaul kit supplied for that carburetor. Follow the instruction sheet included in the kit. Do not use drills or wires to clean out fuel passages or bleed holes in the carburetor or throttle body.This can enlarge the openings and upset the carburetor calibration. Instead, clean out the openings with a chemical cleaner and blow them out with compressed air. Heed the following:

1. Do not splash cleaner in your eyes. It can seriously harm them. Wear goggles to protect your eyes. Wash your eyes thoroughly if you get chemicals in them (Fig. 3-22). See a doctor without delay.
2. Use the air hose with care. Wear goggles. See ● 3-9 on the dangers of misusing compressed air.
3. Always be aware of the fact that gasoline and other solvents are highly flammable. Never smoke or have open flames around these materials.

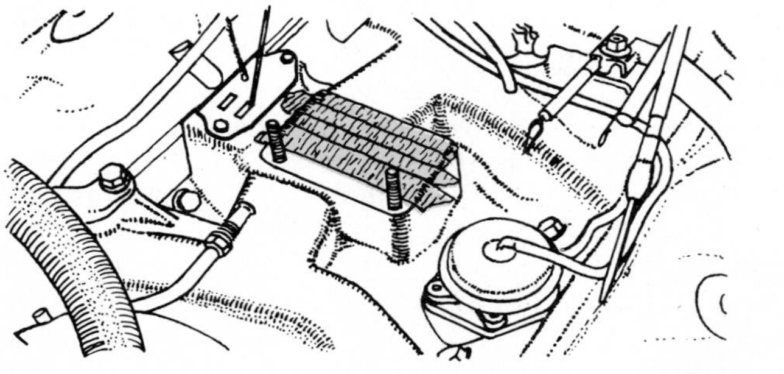

Fig. 14-12 If the carburetor is to be off the engine during service, cover the intake-manifold holes with masking tape to keep parts and dirt from falling into the manifold.

● 14-31 Carburetor Installation

Use a new gasket to assure a good seal between the carburetor and the mounting pad. Put the carburetor into position on the intake manifold and attach it with nuts or bolts. Connect the fuel line and the distributor vacuum-advance line to the carburetor, using two wrenches if necessary to avoid damage to the lines or couplings. Connect wires to switches and other electric controls (where present). Make idle-speed, idle-mixture, and other adjustments. Install the air cleaner.

REVIEW QUESTIONS

1. Explain what "driveability" means.
2. List the 13 terms used to describe engine troubles and explain each.
3. Sections 14-3 to 14-13 describe various engine troubles that could be caused by the fuel system. List these troubles and explain what, in the fuel system, could cause each.
4. If the trouble is overchoking so the engine is flooded, what can you do to start the engine?
5. Why should you not crank the engine with the air cleaner off?
6. Explain how to remove and replace a fuel pump. What precaution should you observe as you loosen the fuel-line connections?
7. Explain how to use new or cleaned spark plugs of the correct heat range to check mixture richness.
8. Why is it not possible to service most fuel pumps by overhaul?
9. Describe the procedure for making an idle-speed adjustment, not using propane.
10. Explain the best lean-idle adjustment procedure, using propane.

SELF PROJECTS

There are instruction sheets in carburetor overhaul kits. These kits tell you how to overhaul carburetors. They have pictures showing procedure. You should be able to get several of these from local shops or service stations that handle carburetor work. Also, your school shop may have copies of these sheets. Make a collection of them. Scotch-tape them in your notebook. In the shop-work part of your automotive-mechanics course, you will be working on carburetors. Studying the sheets will give you a head start on becoming a carburetor expert.

CHAPTER 15
Fuel-Injection System for Gasoline Engines

After reading this chapter, you should be able to:

- **Explain the purpose and advantages of fuel injection for gasoline engines.**
- **Describe the operation of a gasoline fuel-injection system.**
- **Locate the components of a gasoline fuel-injection system on a car, and explain how each works.**
- **Discuss the different designs of gasoline fuel-injection systems and explain how each works.**

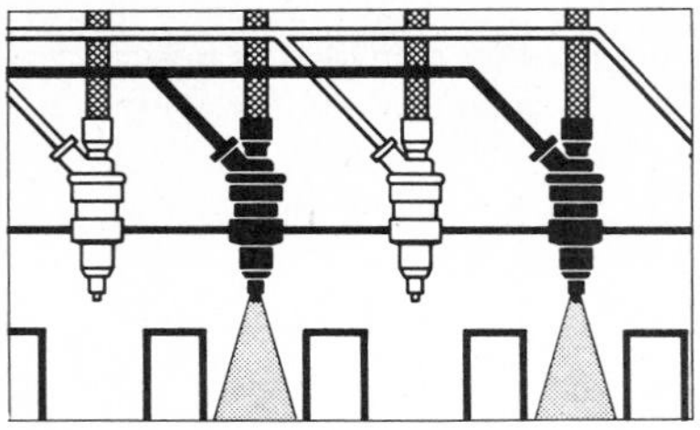

In previous chapters, we discussed fuel systems using carburetors. These are by far the most widely used fuel systems. However, fuel injection is gaining popularity with both the driving public and the automotive manufacturers. There are several reasons for this which we discuss in this chapter. The next chapter describes servicing procedures for these systems. Later chapters discuss fuel-injection systems for automotive diesel engines.

● 15-1 What Is Fuel Injection?

The engine must have a continuous supply of combustible air-fuel mixture to run. In most engines, the carburetor supplies this mixture (Fig. 15-1). The carburetor is mounted on the intake manifold. The manifold has passages which connect the carburetor with the combustion chambers in the engine. As you learned in previous chapters, the carburetor mixes air and fuel to produce the combustible mixture. The combustible mixture then flows through the intake manifold to the combustion chambers as the intake valves open.

In the gasoline fuel-injection system, air only enters the intake manifold. As the air approaches the intake valves, fuel injectors spray gasoline into the air. See Fig. 15-2. The basic difference, therefore, is the point at which the gasoline enters the ingoing air stream. In the carbureted system the fuel enters the air in the carburetor. In the fuel-injection system, the fuel enters the air inside the intake manifold. There are certain advantages to the fuel-injection system that make it the preferred system, aside from basic costs. We look at this further in ● 15-3.

● 15-2 Types of Fuel Injection

There are two basic ways to classify fuel-injection systems. One way is according to where, in the engine, the fuel is injected. The other way is according to whether the injection of fuel is continuous or timed.

Let's look at the first classification—the point of injection. Fuel can be injected directly into the combustion chambers (Fig. 15-3), or into the intake manifold (Figs. 15-2 and 15-4). The direct injection method is used in diesel engines. During the compression stroke in the diesel engine, the heat of compression raises the air temperature to 1000°F [537°C] or above. The fuel is injected (sprayed) directly into this superheated air. The high temperature ignites the fuel as it enters.

Fuel can also be injected into the intake manifold, and this is the system used for gasoline engines. The fuel is injected into the air approaching the intake port. The system is known as *port* injection.

The other classification is the injection procedure—whether the fuel is injected in pulses—timed injection—or continuously. In the diesel engine, timed injection

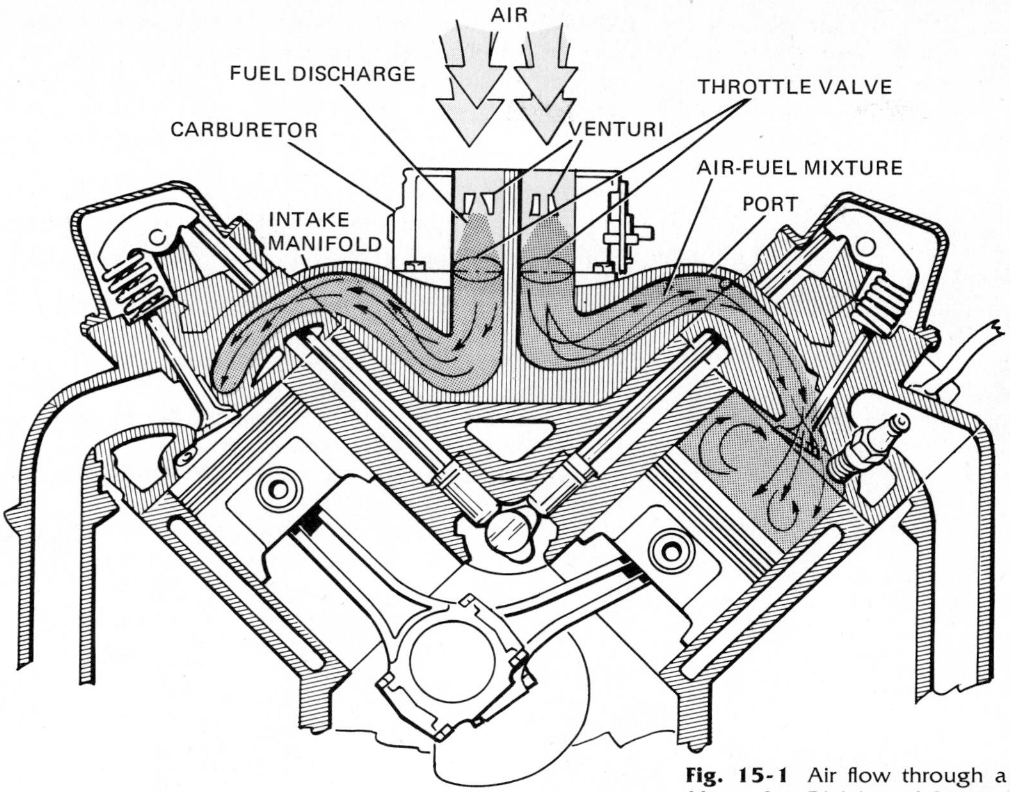

Fig. 15-1 Air flow through a carbureted engine. (*Cadillac Motor Car Division of General Motors Corporation*)

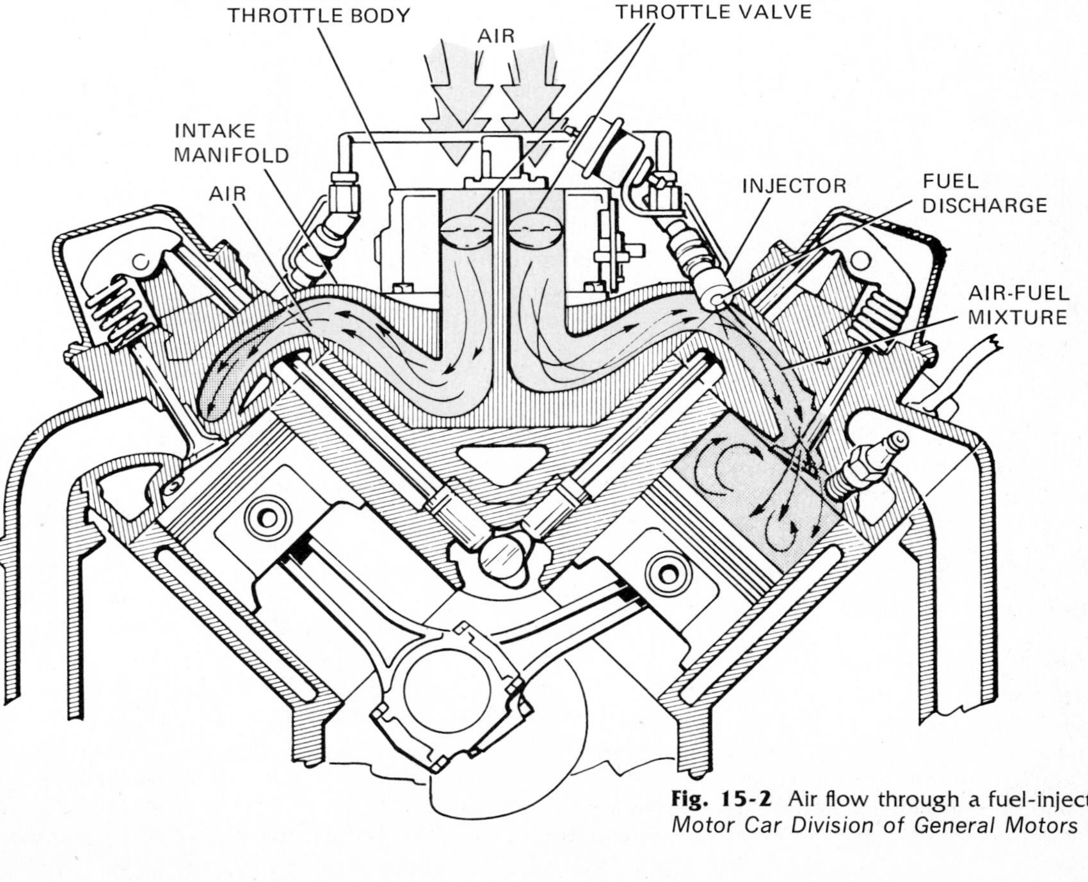

Fig. 15-2 Air flow through a fuel-injected engine. (*Cadillac Motor Car Division of General Motors Corporation*)

must be used because the fuel must be sprayed into the combustion chambers at the exactly correct instant. This instant must change as engine speed changes. That is, the injection must take place earlier at high engine speeds. This compares with the ignition-system action for gasoline engines. The spark must occur earlier at higher engine speed. In both cases, the advance gives the fuel enough time to burn and deliver its power to the piston.

Continuous injection systems (CIS) are used in some gasoline engines. In these, the gasoline is sprayed continuously into the intake manifold. The amount of fuel sprayed varies with the engine speed and power demands. The continuous injection system is simpler in many ways and thus less expensive. That is, it does not require a control system that changes the instant of injection as operating conditions change. This control system can be rather complex and thus more expensive, as we shall see later in the chapter.

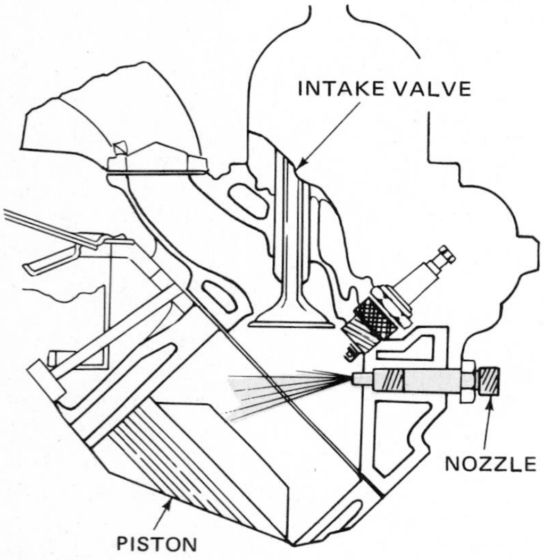

Fig. 15-3 Simplified view showing the method of injecting fuel directly into the combustion chamber of an engine.

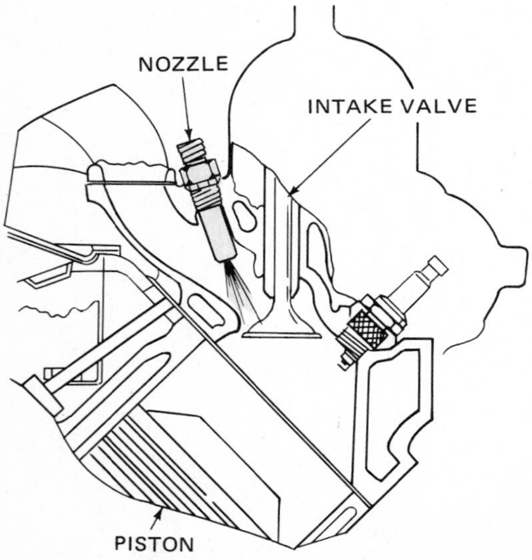

Fig. 15-4 Simplified view showing the method of injecting fuel into the intake port just ahead of the intake valve.

● 15-3 Advantages of Gasoline Injection

Regardless of whether the injection system is pulsed (timed) or continuous, this system eliminates many intake-manifold distribution and carburetion problems. One of the most difficult problems with a carbureted system is to get the same air-fuel mixture into each cylinder. An intake manifold is a casting with passages and ports of different widths and lengths. Because of this, it is difficult to design the system so that all cylinders will receive the same amount and richness of air-fuel mixture. The air flows readily around corners and through the variously shaped passages. However, the gasoline, because it is heavier, is unable to travel as easily through the bends and around the corners in the intake manifold. As a result, some of the gasoline particles continue to move to the end of the manifold, accumulating or puddling there (Fig. 15-5). This enriches the mixture going to the end cylinder. In the example, shown in Fig. 15-5, the center cylinder, closest to the carburetor, gets the leanest mixture.

The air-fuel mixture leaving the carburetor must be rich enough to take care of this difference and supply the center cylinders with a sufficiently rich mixture. Otherwise, these cylinders will not fire. But this means that the end cylinders will get a richer mixture than necessary to fire and will thus tend to produce excessive amounts of HC and CO in the exhaust.

Figure 15-6 shows how a fuel-injection system solves the intake-manifold distribution problem. A calibrated nozzle, or fuel injector, is located near the intake valve of each cylinder. At the right instant, fuel under pressure is sprayed out of the injector (timed injection system). Or, with the continuous injection system, the fuel sprays continuously. In either case, each injector sprays the same amount of fuel into the same amount of air entering every cylinder. As a result, every cylinder gets the same amount of air-fuel mixture having the same mixture richness. The engine can operate on a leaner overall air-fuel ratio. This reduces HC and CO in the exhaust gas.

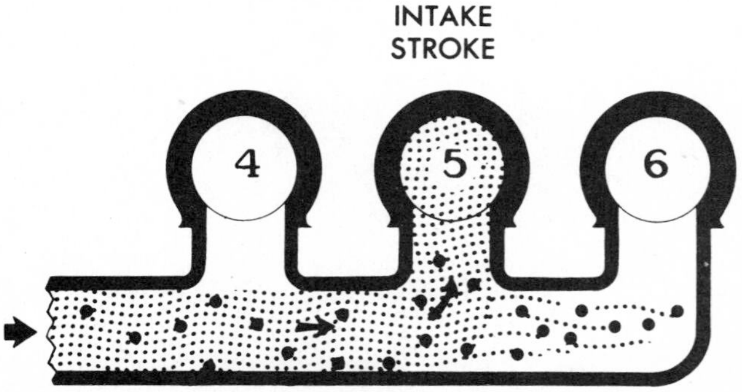

Fig. 15-5 Distribution pattern in an intake manifold. The gasoline particles tend to continue to the end of the manifold. This enriches the mixture going into the end cylinders. *(Chevrolet Motor Division of General Motors Corporation)*

Another advantage of the fuel-injection system is that it allows a more efficient design of the intake manifold. That is, the intake manifold can be designed for the most efficient flow of air only. It does not have to handle fuel, too. Also, because only a throttle body is needed, instead of a full carburetor, and because of the redesigned intake manifold, the hood height of the car can be reduced. Then too, with fuel injection, no extra heat to the fuel mixture is required during warm-up. That is, no manifold heat-control valve or heated-air system is required. Throttle response is faster because the fuel is under pressure at the injector at all times. All that is required is to open the injector; it sprays out instantly. Exhaust emissions during deceleration can be practically eliminated. The system can be designed to have a positive, or complete, fuel shutoff when the throttle is released and the car decelerates. That eliminates exhaust emissions during deceleration, and saves gasoline and improves fuel economy.

With all these advantages, you may wonder why everyone has not gone to fuel injection by now. The answer is twofold. First, there is the cost. Modern pulsed or timed fuel-injection systems rely on rather complex sensors and electronic controls. These are needed to control the timing of injection and the amount of fuel to be injected. They can be relatively expensive, particularly the electronic components. As we will note later, the continuous injection system does not require a complex timing arrangement, is less expensive, and is being viewed with favor by some car companies.

The second part of the answer is that it is only recently that electronics has been brought into the picture. That is, electronics is a relatively new science and it has only lately produced the controlling devices at a price reasonable enough to be acceptable.

Prior to the electronic age, some completely mechanical fuel-injection systems were developed but they did not work well and were soon abandoned.

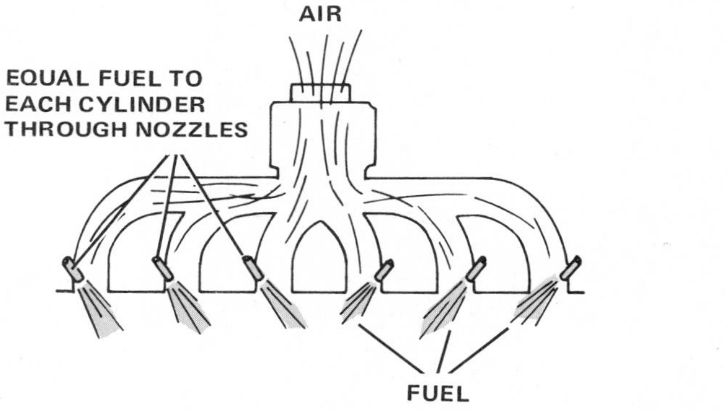

Fig. 15-6 In a fuel-injected engine, the same amount of fuel reaches each cylinder. *(Cadillac Motor Car Division of General Motors Corporation)*

15-4 Early Mechanical Fuel-Injection Systems

In the late 1950s and early 1960s, Chevrolet and Pontiac offered a continuous-flow fuel-injection system (CIS) as a high-priced option. It was built by the Rochester Products Division of General Motors Corporation. The system uses vacuum signals from the engine to control the air-fuel mixture. Fuel sprayed continuously from nozzles into the intake manifold at points opposite the intake valves. The nozzle was simply a calibrated tube, with no valve of any sort.

There is always pressure behind the fuel in the nozzle. This comes from the fuel-injection pump. However, on the discharge end of the nozzle, the vacuum can vary from about 22 inches [558.8 mm] with the throttle closed and the engine idling, to almost zero at wide-open throttle. This widely varying vacuum affects the amount of fuel spraying from the nozzles. For this reason, it becomes more difficult to control fuel flow, particularly at idle and low speeds.

Figure 15-7 is a schematic view of the Rochester system. It is no longer in use. It consists essentially of a special intake manifold, an air meter, and a fuel meter.

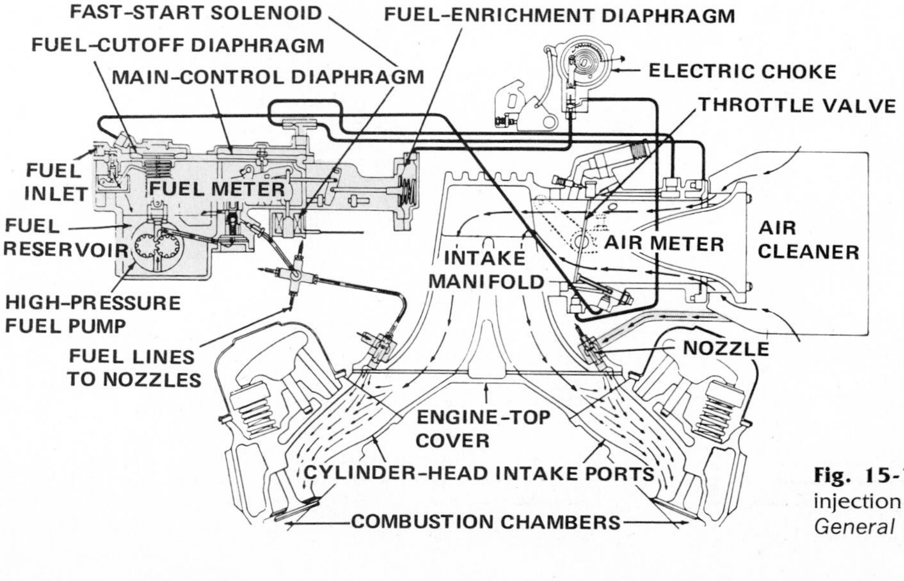

Fig. 15-7 Sectional view of a mechanical fuel-injection system. *(Chevrolet Motor Division of General Motors Corporation)*

The air meter controls the flow of air through the intake manifold. The fuel meter controls the flow of fuel to the fuel nozzles in the intake manifold. The system is basically simple. Linkage from the accelerator pedal actuates a throttle valve in the air meter. More air is admitted when more engine power is desired. The fuel meter provides varying amounts of fuel. It supplies more fuel as more air is admitted. Other mechanical mechanisms enrich the mixture for acceleration, warm-up, hill climbing, and so on.

● 15-5 Electronic Fuel Injection

In the late 1950s the Chrysler Corporation built a few cars with an early type of electronic fuel-injection system. This system, known as the Bendix Electrojector, was developed and patented by the Bendix Corporation. The electronic part of this system was made with vacuum tubes. It was not widely used.

With the development of solid-state electronics, such as transistors and diodes, a new type of electronic fuel injection appeared. In 1968, Volkswagen began installing a solid-state electronic fuel-injection system built by Bosch on some car models imported into the United States. We discuss the Volkswagen system again in ● 15-10. Now, let's examine the basic Bosch electronic fuel-injection system for gasoline engines.

Figure 15-8 shows the system schematically. The electric fuel-supply pump maintains a high pressure in the fuel line to the injection valves in the intake manifold (Fig. 15-9).

At the proper instant, trigger contact points in the ignition distributor close. Figure 15-10 shows the distributor, cut away so that the trigger contacts can be seen. They are opened and closed by a cam on the distributor shaft. This cam is very different from the cam that opens and closes the ignition contact points.

When the trigger contacts close, they send an electric signal to the electronic control unit (ECU). The ECU then connects half the solenoid injection valves to the battery. (In a four-cylinder engine, this would be two valves; in a six-cylinder engine, three valves; and in an eight-cylinder engine, four valves.) The important point is that the solenoid injection valves are not individually actuated. Half of them are actuated at a time. Figure 15-11 shows three of

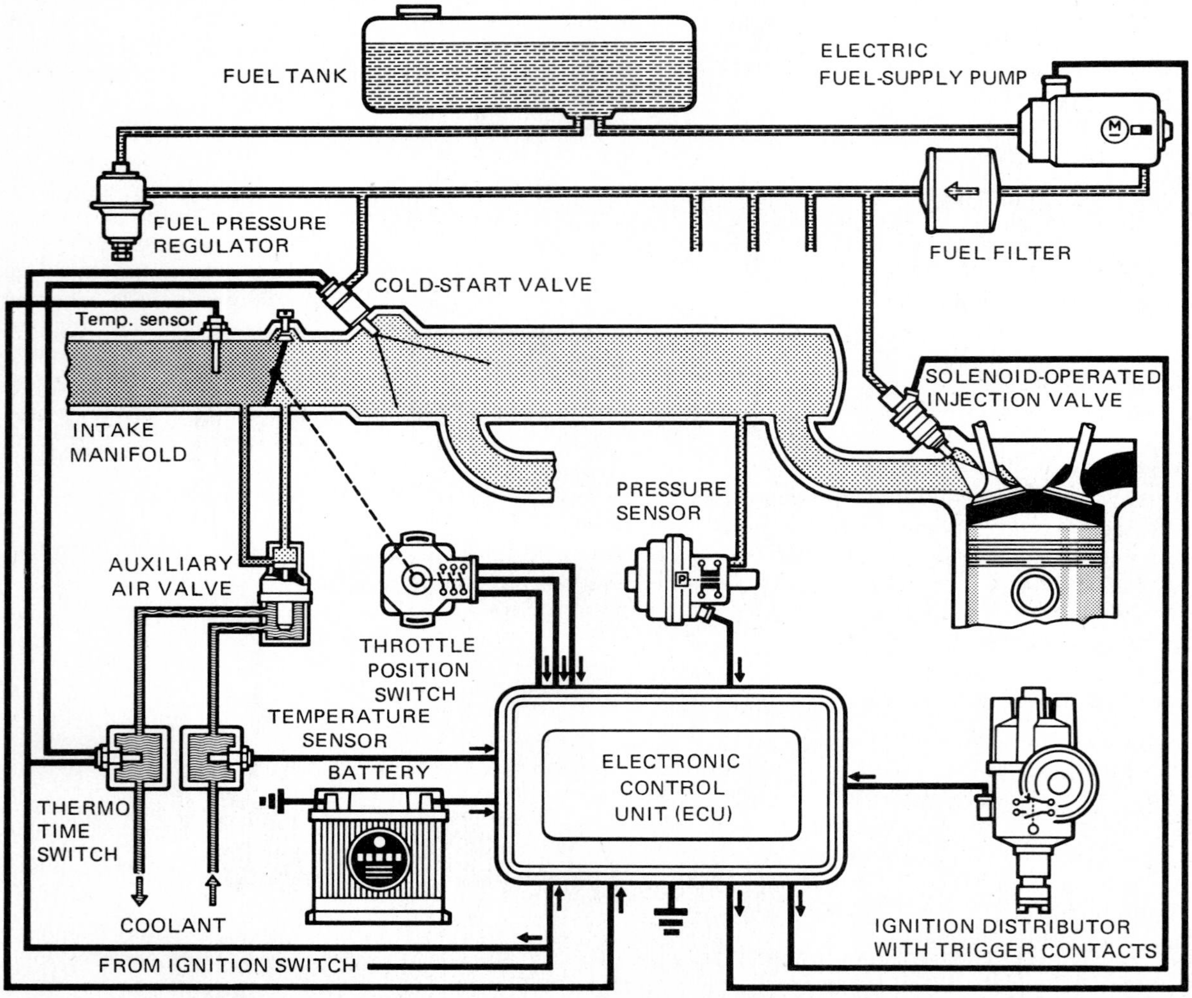

Fig. 15-8 Schematic diagram of an electronic gasoline-injection system. (*Robert Bosch GmbH*)

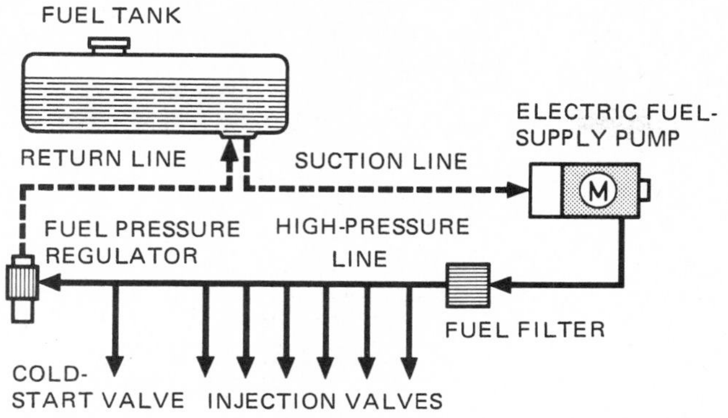

Fig. 15-9 Schematic diagram of the fuel-supply system. *(Robert Bosch GmbH)*

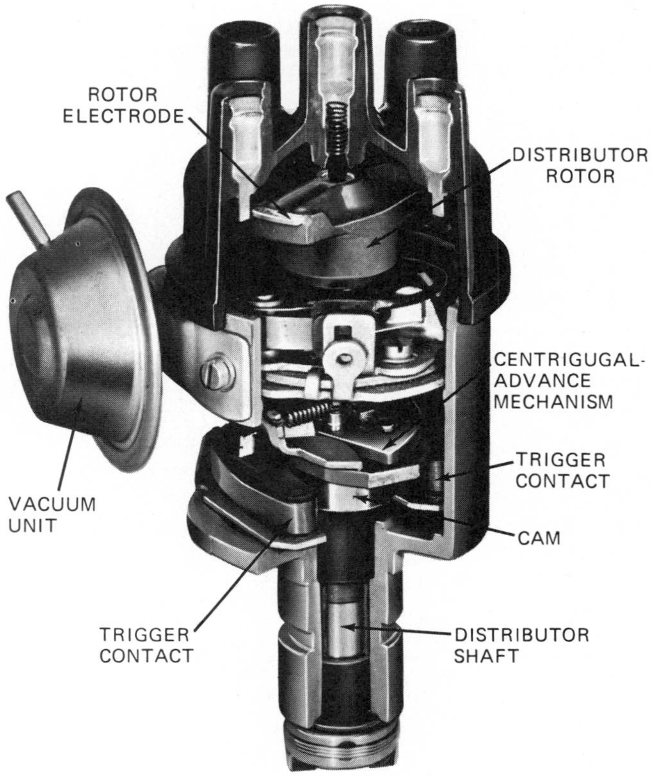

Fig. 15-10 Cutaway view of the distributor, showing the trigger contacts which activate the electronic control. *(Robert Bosch GmbH)*

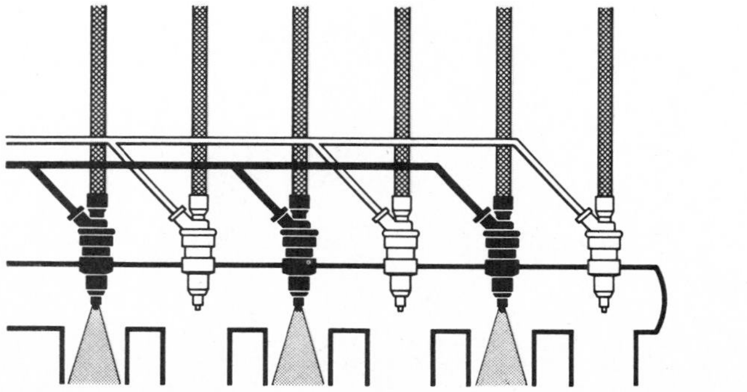

Fig. 15-11 Injection-valve grouping. *(Robert Bosch GmbH)*

the valves in operation, spraying fuel into the intake manifold.

The fuel enters just opposite the intake valves, as shown in Fig. 15-8. Figure 15-12 is the injection timing chart for a six-cylinder engine. Note that the individual intake valves open at varying times (crankshaft degrees) after injection. For example, look at the top line, which is for No. 1 cylinder. Injection takes place at 300 degrees of crankshaft rotation. Almost 60 degrees later (near 360 degrees), the No. 1 intake valve opens and the intake stroke starts. Cylinder No. 5 is next in the firing order. Its intake valve opens near 480 degrees, or about 180 degrees after injection. The intake valve for No. 3 cylinder opens about 300 degrees of crankshaft rotation after injection. During these varying intervals between fuel injection and intake-valve opening, the fuel is "stored" in the intake manifold, opposite the intake valves.

Having only two groups of injection valves simplifies the system. No appreciable loss of engine performance results from this storage of the fuel. Remember, the whole action takes place in a small fraction of a second. At highway speed, for example, the time between injection and opening of the intake valve averages only about one-hundredth of a second.

The fuel pump is shown in sectional view in Fig. 15-13. It is an electric motor of the "wet-pump" type. This means that the fuel flows through the pump and motor,

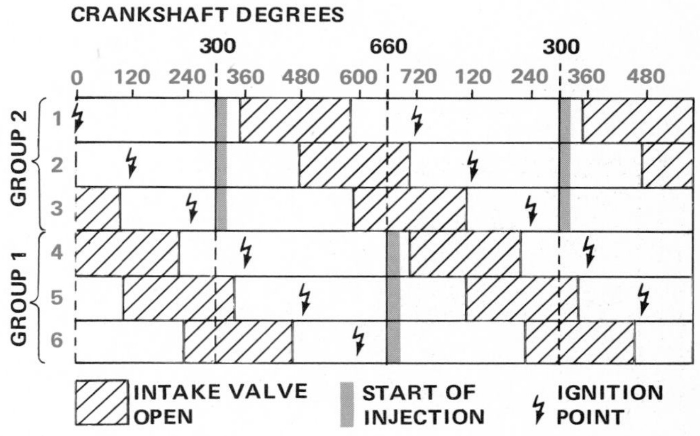

Fig. 15-12 Injection timing chart for a six-cylinder engine. *(Robert Bosch GmbH)*

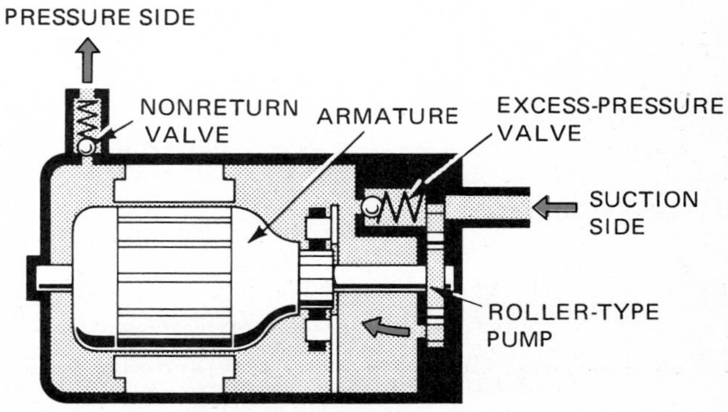

Fig. 15-13 Sectional view of the electric fuel-supply pump. *(Robert Bosch GmbH)*

as shown by the arrows. The pump drives an off-center rotor with a series of notches in which rollers are located (Fig. 15-14). When the pump armature rotates, the rollers are forced out by centrifugal force. They trap fuel between the rotor and the inner face of the pump. This fuel is forced out as the distance between the rotor and inner face decreases, on the outlet or pressure side.

The fuel pressure is controlled by a pressure regulator. The regulator is shown to the upper left in Fig. 15-8. It regulates by dumping some of the fuel back into the fuel tank if the pressure gets too high. Figure 15-15 is a section view of the pressure regulator. If the pressure exceeds a preset value, the diaphragm is pushed back against the spring. This opens the valve, allowing some of the fuel to flow out through the return line to the fuel tank. It is very important to maintain a constant pressure. The amount of fuel injected must depend entirely on how long the solenoid injection valve is open, and not on the fuel pressure.

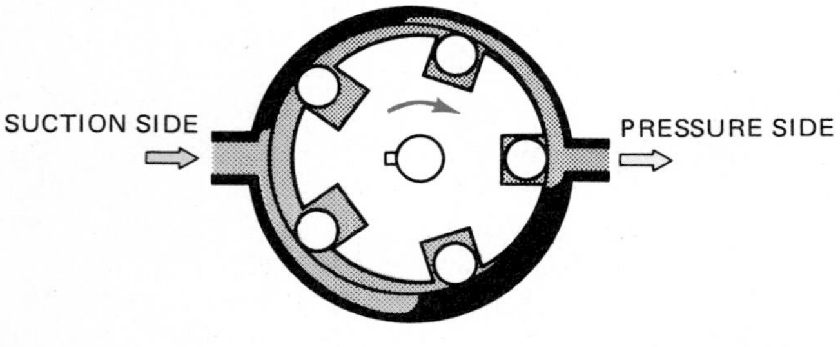

Fig. 15-14 Pumping action of the fuel-supply pump. (*Robert Bosch GmbH*)

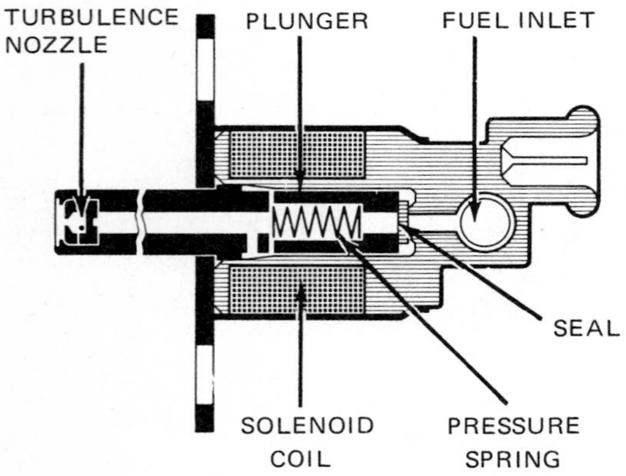

Fig. 15-15 Sectional view of the pressure regulator. (*Robert Bosch GmbH*)

The solenoid injection valve is shown in sectional view in Fig. 15-16. It simply pulls the plunger and needle away from the nozzle jet when it is connected to the battery. When it is connected, the solenoid produces a magnetic field. This pulls the plunger in toward the solenoid, and the needle is lifted off the nozzle. The fuel can then spray through the nozzle into the intake manifold. The longer the needle is off the nozzle, the more fuel is sprayed.

● 15-6 Electronic Fuel-Injection Controls

Several factors determine how long an injection valve delivers fuel to the intake manifold. They include the throttle position, intake-manifold vacuum, ingoing-air temperature, and coolant (engine) temperature. See Fig. 15-8. Sensing devices continuously monitor these factors and "report" (electronically) to the ECU. The ECU puts the varying signals together and determines the length of the injection cycle.

For example, consider the intake-air temperature sensor (Fig. 15-17). It constantly measures the temperature of the air entering the intake manifold. The intake-air temperature sensor contains an element that passes varying amounts of electric current as the temperature changes. At low temperatures, for example, it passes more current; it sends a stronger electric signal to the ECU. The ECU then increases the time during which the fuel-injection valves are open. More fuel is delivered to compensate for the colder and denser air.

Similarly, the coolant temperature sensor sends varying amounts of current to the ECU, depending on the temperature of the engine coolant (and thus the engine temperature). When the engine is cold, the engine must receive more fuel, so that the mixture will be rich enough. The ECU therefore increases the injection time when the coolant temperature sensor reports that the engine is cold. As the engine warms up, the ECU decreases the injection time.

The intake-manifold pressure sensor (Fig. 15-18) measures intake-manifold pressure and compares it with atmospheric pressure. It contains a pair of *aneroids*. These are flat, hollow disks. In many models, both aneroids are evacuated. That is, they hold a vacuum. As the outside pressure changes, the sides of the aneroid bulge out or in, depending on whether the outside pressure is relatively low or high.

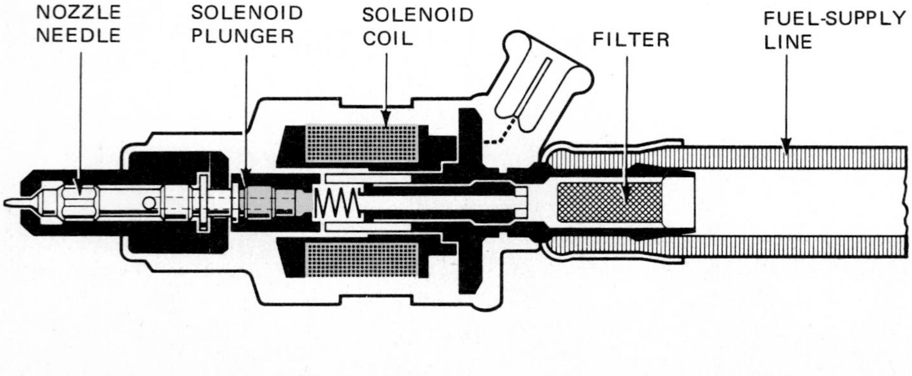

Fig. 15-16 Sectional view of the solenoid-operated injection valve. (*Robert Bosch GmbH*)

Intake-manifold vacuum is introduced into the end of the pressure sensor. This vacuum acts on the aneroids. For example, if the throttle valve is open and there is little vacuum in the intake manifold, the aneroids are collapsed (Fig. 15-18). But if intake-manifold vacuum is high, the aneroids bulge out (Fig. 15-19). This repositions the plunger in the coils. Changing the position of the plunger in the coils changes the magnetic activity of the coil. What this means is that the coils send a changed electric signal to the ECU. The ECU then changes the injection time so that the correct amount of fuel is injected to meet intake-manifold vacuum conditions. For example, the manifold vacuum is high when the throttle valve is closed or nearly closed. This means that only a little air is getting through to the cylinders. Therefore, only a little fuel should be injected. As a result, the ECU shortens the injection time. This is the same as the carburetor feeding fuel to the air through the idle system.

When the throttle is opened, the vacuum in the intake manifold is reduced. The aneroids collapse somewhat (see Fig. 15-18) and pull the plunger into the coils. The electric signal from the coils then changes. This change causes the ECU to increase the injection time. More fuel is injected so that the air-fuel ratio remains constant.

Remember, however, that other factors also affect the injection time. If the intake air is cold, the injection time is increased, as noted above. Likewise, if the engine (coolant) is cold, the injection time is increased.

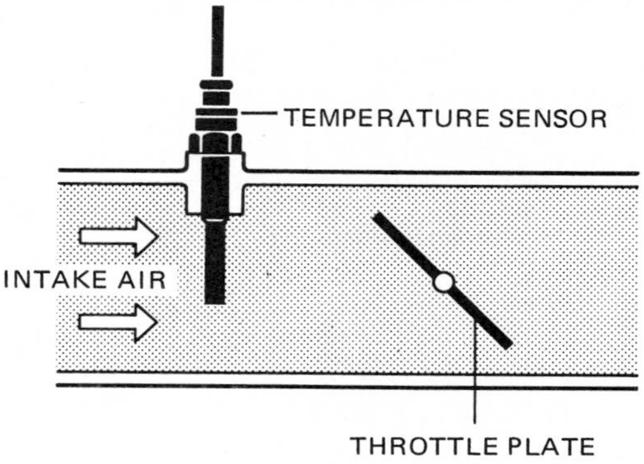

Fig. 15-17 Temperature sensor in the intake manifold. (*Robert Bosch GmbH*)

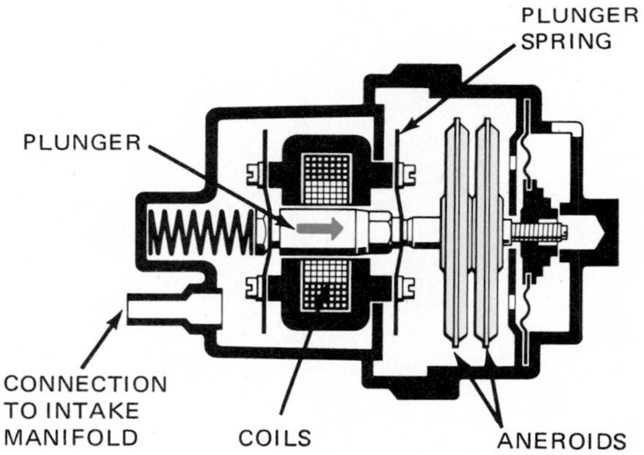

Fig. 15-18 Sectional view of the intake-manifold pressure sensor with the throttle valve open. The aneroid disks are compressed. (*Robert Bosch GmbH*)

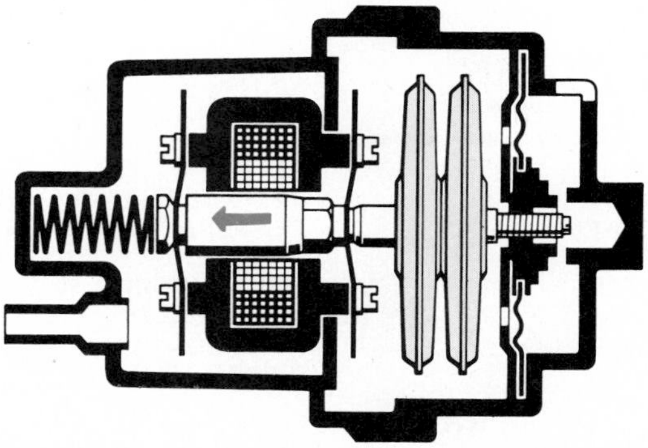

Fig. 15-19 Sectional view of the intake-manifold pressure sensor with the throttle closed. Increased vacuum allows the aneroid disks to expand. (*Robert Bosch GmbH*)

15-7 Cold Starts

In the carburetor fuel system, the choke increases the amount of fuel delivered when the engine is being started cold. In the electronic fuel-injection system, a cold-start valve increases the amount of fuel delivered when the engine is being cold-started. Note the location of the valve in Fig. 15-8. A sectional view is shown in Fig. 15-20. The cold-start valve is triggered by the ECU. The ECU receives information from the engine temperature sensor, the air-intake temperature sensor, and the pressure sensor. If the information indicates that the engine is cold, the ECU connects the cold-start valve to the battery. The solenoid then moves the plunger to allow fuel to spray into the air entering the intake manifold. This results in a sufficiently rich starting mixture.

15-8 Throttle-Position Switch

For more exact control of the injection time, the throttle has a position switch with several contact strips and a sliding contact. As the throttle is opened, the sliding contact connects to the contact strips, one after another. As each connection is made, a different signal is sent to the ECU. This causes the ECU to modify the injection

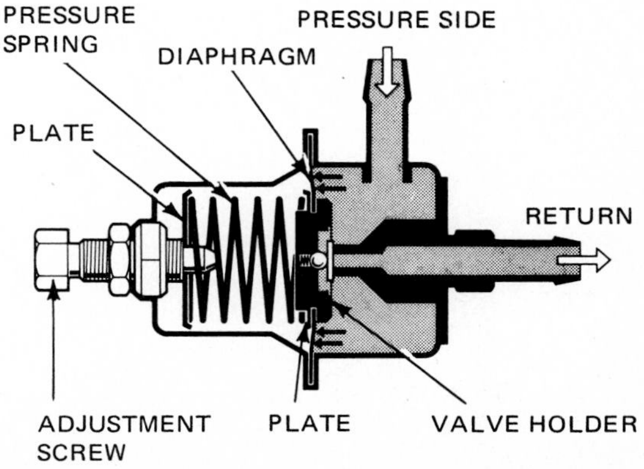

Fig. 15-20 Sectional view of the cold-start valve. (*Robert Bosch GmbH*)

time according to throttle position. For instance, on later models, there is a full-load-enrichment contact strip. When the throttle is opened wide, the sliding contact connects to this contact strip. As a result, the ECU increases the injection time so that additional fuel is injected. This meets the need for a richer mixture at full power.

● 15-9 ECU System Operation

Now let us review the operation of the system. When it is first started, the engine is cold. The air going into the intake manifold is cold. The vacuum in the intake manifold is fairly high because the throttle is nearly closed. The ECU gets signals from the air-intake temperature sensor, the coolant temperature sensor, the pressure sensor, and the throttle-position switch. It puts all of these together and "decides" just how long the injection valves should stay open. The ECU also "decides" whether to open the cold-start valve. When the trigger contacts close in the ignition distributor, the ECU opens half the injection valves. A rich mixture is delivered to the intake manifold before the intake valves open.

Then, when the engine starts, the changing intake-air temperature, coolant temperature, throttle position, and intake-manifold vacuum all modify the signals going from the ECU to the injection valves.

The auxiliary air valve, not previously mentioned, senses the temperature of the engine coolant. Note its location in Fig. 15-8, and the sectional view in Fig. 15-21. When the engine is cold, the auxiliary air valve opens to allow some air to flow around the closed throttle valve. This provides the extra air needed when the cold-start valve is open and discharging fuel. However, as the engine (and the coolant) begins to warm up, the auxiliary valve closes, shutting off the flow of extra air.

● 15-10 Volkswagen Electronic Fuel-Injection System

The Volkswagen electronic fuel-injection system is similar to the one we just described. It is illustrated in Figs. 15-22 to 15-24. This system was especially designed for the Volkswagen flat-four air-cooled engine. One major reason was to improve the combustion process and reduce the amount of smog-producing substances in the exhaust. (Part Five of the book discusses smog.) The fuel is injected into the intake manifolds behind the intake valves. The injection is timed to coincide with valve opening by triggering contacts in the ignition distributor. The amount of fuel injected is controlled by the length of time the fuel injectors are open. This, in turn, is determined by a number of sensors, which send electric signals to the transistorized control unit (Fig. 15-22). Figure 15-23 illustrates the air-supply system and its controls. Figure 15-24 illustrates the fuel-supply system.

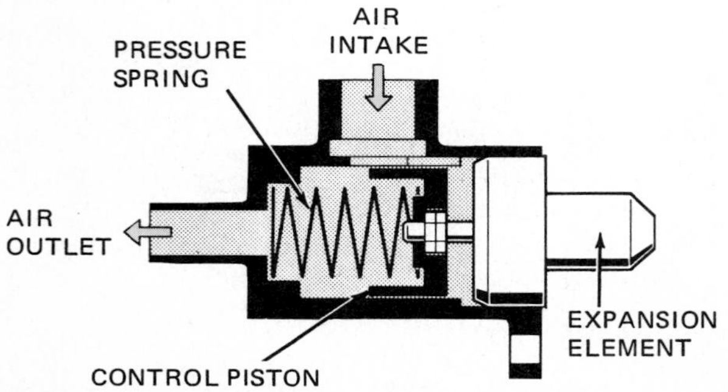

Fig. 15-21 Sectional view of the auxiliary air valve. (*Robert Bosch GmbH*)

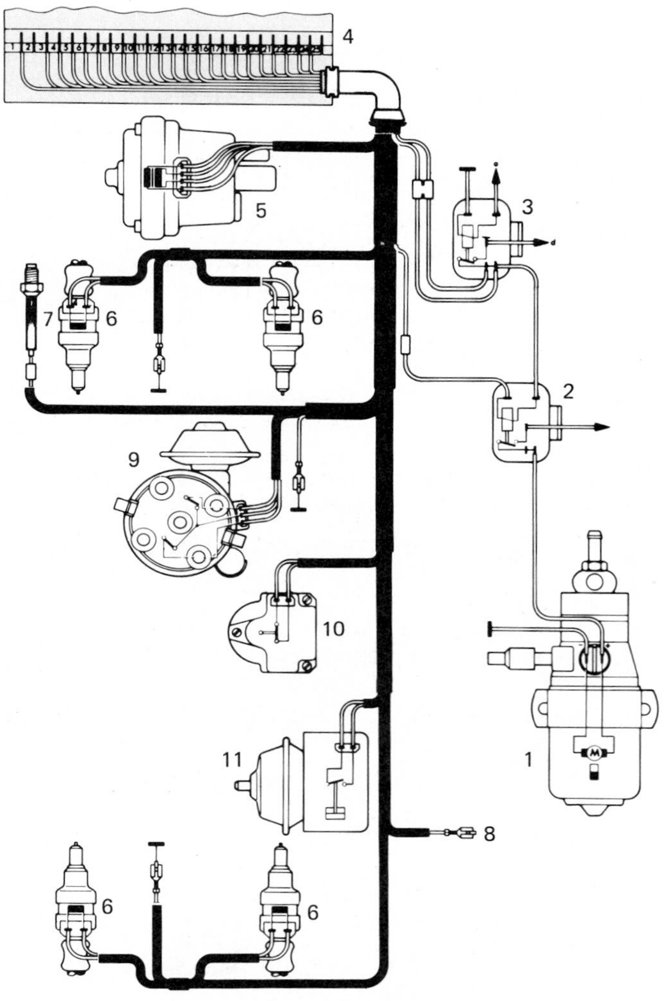

1. FUEL PUMP
2. PUMP RELAY
3. MAIN RELAY
4. CONTROL UNIT
5. INTAKE-MANIFOLD PRESSURE SENSOR
6. INJECTOR
7. CYLINDER-HEAD TEMPERATURE SENSOR
8. CRANKCASE SENSOR
9. IGNITION DISTRIBUTOR
10. THROTTLE SWITCH
11. FULL-LOAD PRESSURE SWITCH

Fig. 15-22 Schematic layout of the control system for the Volkswagen fuel-injection system. The electronic control unit (4) receives signals from various sensors and integrates them to determine the timing and amount of fuel to be injected. (*Volkswagen of America, Incorporated*)

● 15-11 Cadillac Electronic Fuel-Injection System

The Cadillac electronic fuel-injection system (Fig. 15-25) is very similar to the Volkswagen system. Cadillac uses a separate fuel-injector valve for each cylinder (Fig. 15-26). The eight injectors on the Cadillac V-8 engine are connected to a fuel rail; they are divided into two groups of four injectors each. Each group of injectors is alternately turned on and off by the electronic control unit. The injectors are turned on once for each two revolutions of the crankshaft. Figure 15-27 is a sectional view of the Cadillac V-8 engine with electronic fuel injection.

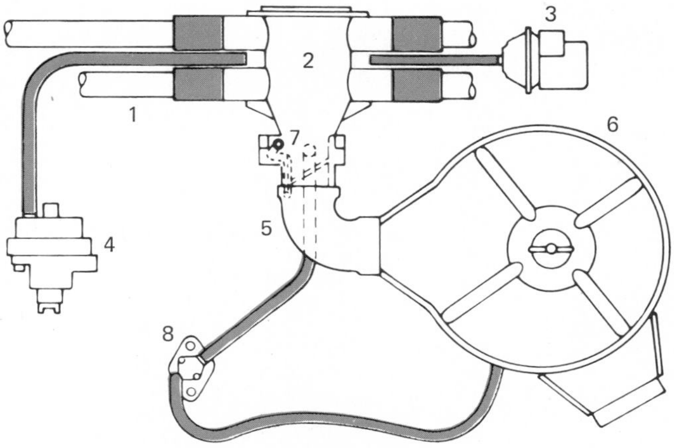

1. Air pipes to cylinders
2. Air distributor
3. Pressure switch
4. Pressure sensor
5. Idling circuit
6. Air cleaner
7. Adjusting screw
8. Auxiliary air regulator (rotary valve)

Fig. 15-23 Air-supply control for the Volkswagen electronic fuel-injection system. (*Volkswagen of America, Incorporated*)

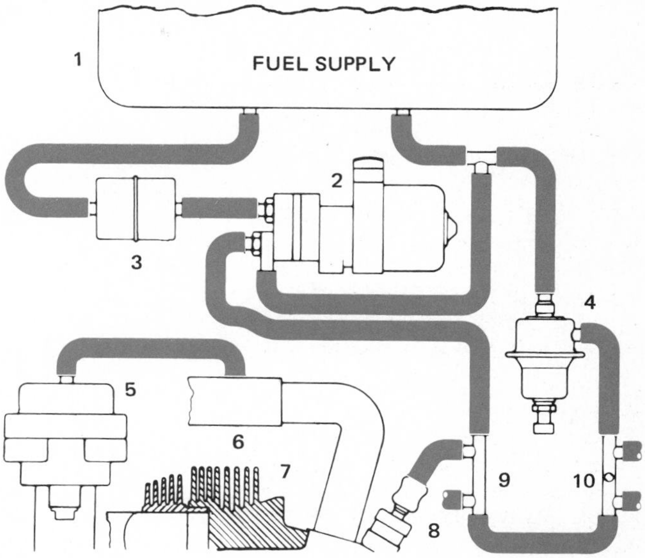

1. Fuel tank
2. Electric fuel pump
3. Filter
4. Pressure regulator
5. Intake-manifold pressure sensor
6. Air pipe
7. Cylinder head
8. Fuel injector
9,10. Distributor pipes to injector

Fig. 15-24 Fuel-supply system for the Volkswagen electronic fuel-injection system. (*Volkswagen of America, Incorporated*)

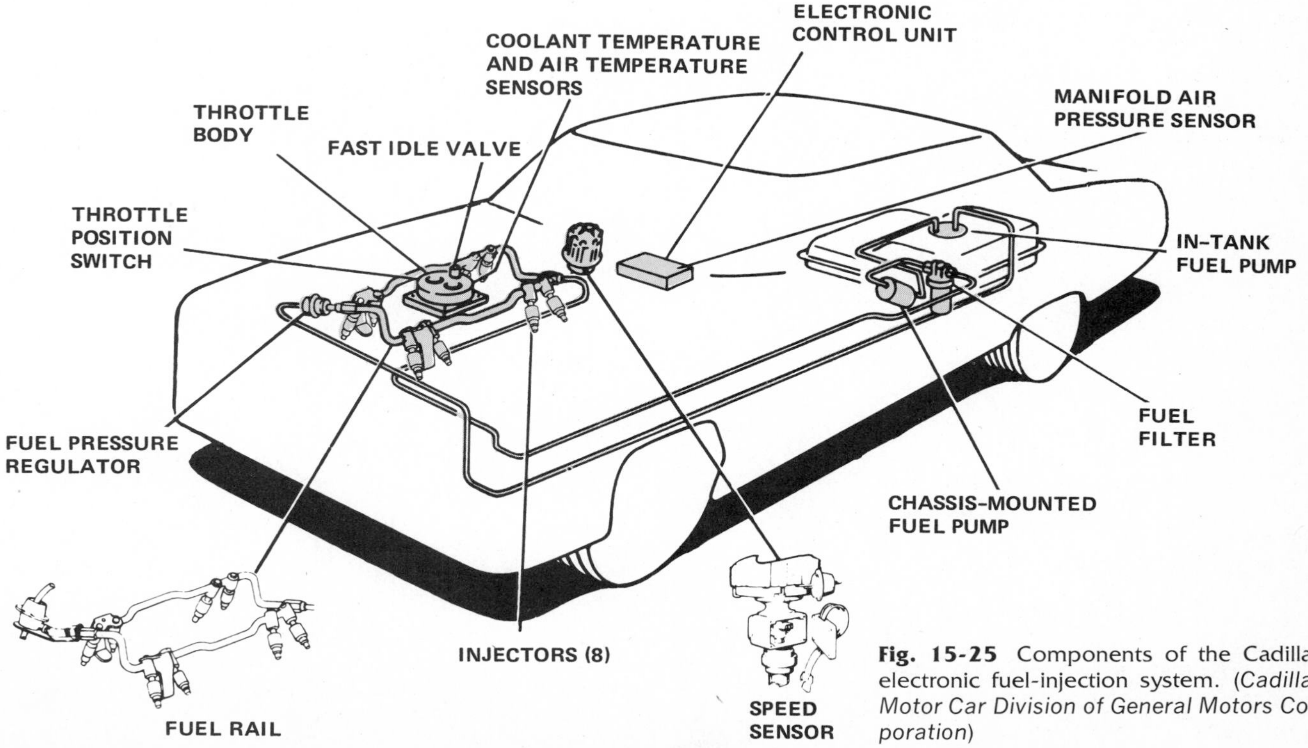

Fig. 15-25 Components of the Cadillac electronic fuel-injection system. (*Cadillac Motor Car Division of General Motors Corporation*)

Figure 15-28 is a block diagram showing (on the left) the sensors that send information to the ECU. With this and other information (such as engine displacement and volumetric efficiency), the ECU computes the amount of fuel the engine needs and sends signals to the injectors and other parts of the system (on the right in Fig. 15-28).

The ECU is a preprogrammed computer installed above the glove box within the passenger compartment. It converts the input information from the sensors into an electric signal which opens the injectors for the proper duration at the proper time. The ECU cannot be adjusted or serviced. When a malfunction is traced to the ECU, it is

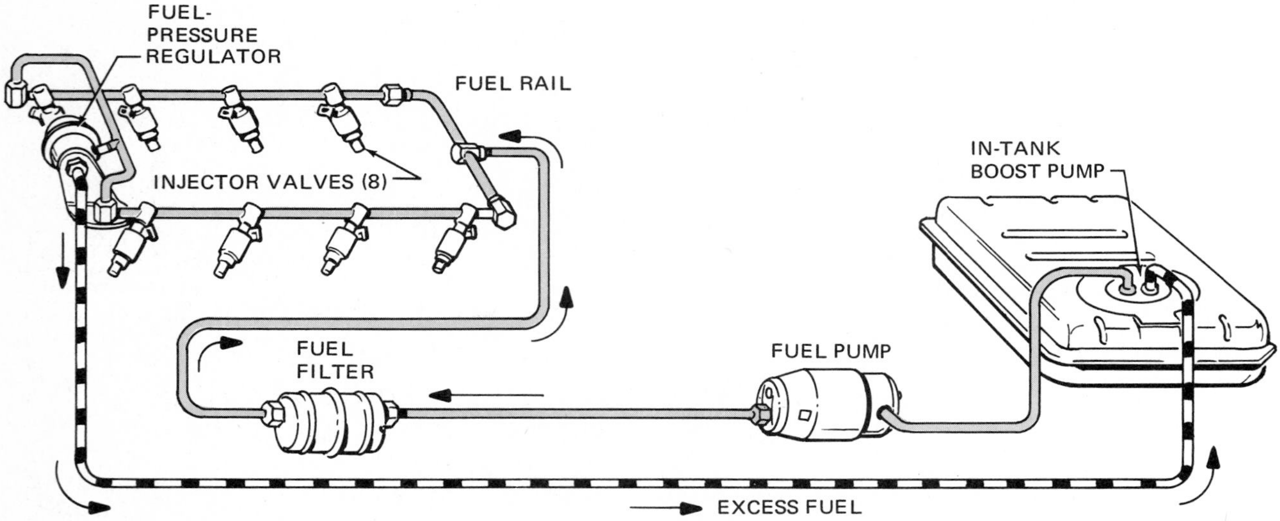

Fig. 15-26 Schematic view of the fuel system used on the Cadillac fuel-injected engine. (*Cadillac Motor Car Division of General Motors Corporation*)

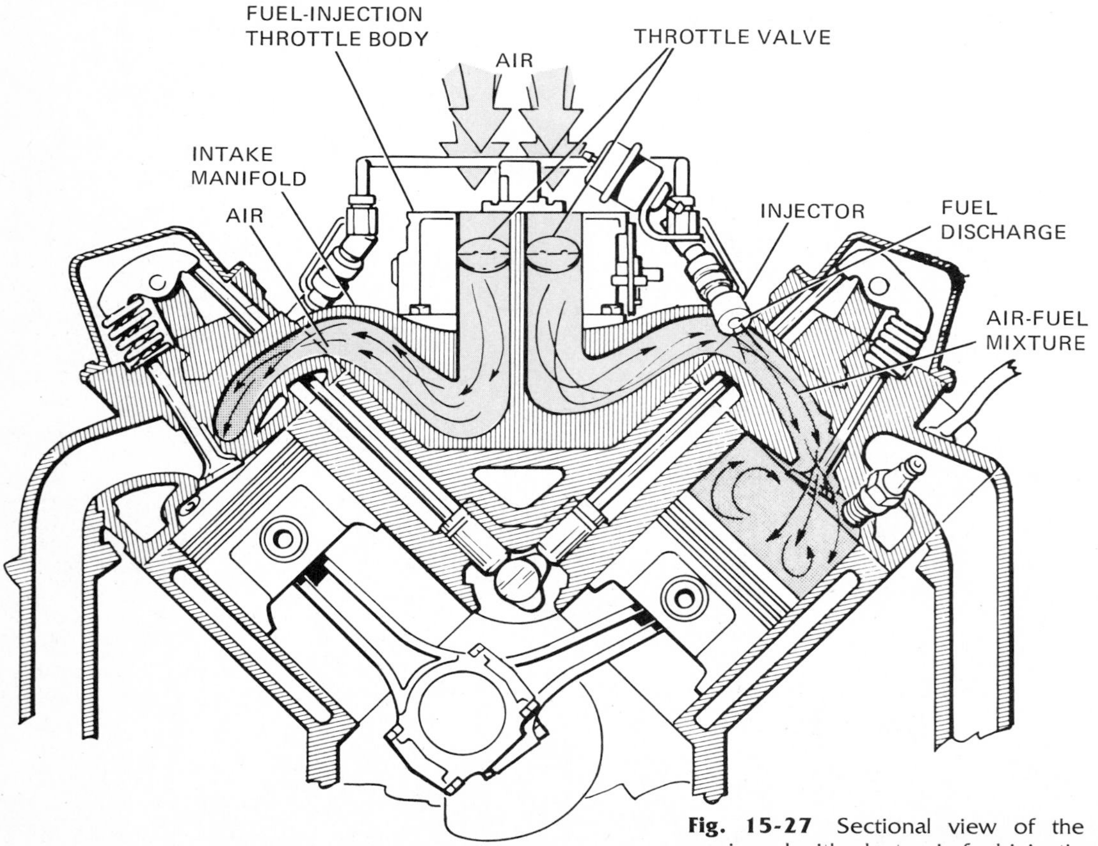

Fig. 15-27 Sectional view of the Cadillac V-8 engine equipped with electronic fuel injection. (*Cadillac Motor Car Division of General Motors Corporation*)

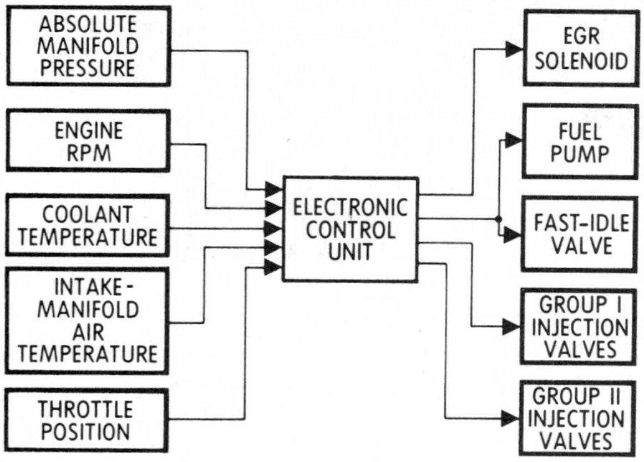

Fig. 15-28 Block diagram showing the sensors (left) that provide information to the electronic control unit. (*Cadillac Motor Car Division of General Motors Corporation*)

removed from the car and a new one is installed. Accurate diagnosis of ECU operation requires a special tester.

● 15-12 Electromechanical Fuel-Injection System for Gasoline Engines

Figure 15-29 is a schematic view of a fuel-injection system that is essentially mechanical in nature. Several relays are used to control the system. The system has a two-plunger pump. Each plunger feeds three cylinders through a metering unit. Fuel is injected into the intake manifold, as in the systems described above. The metering units are controlled by a linkage to the accelerator pedal. The amount of fuel delivered by the pump is controlled by a centrifugal governor. Other controls are included to increase the richness of the air-fuel mixture for starting, cold operation, and high-speed, full-power running. There are also pressure cells built into the diaphragm of the injection pump. They alter the amount of fuel delivered, in accordance with the altitude and the density of the air. At higher altitudes, the air is less dense. Therefore, less fuel is required to achieve the normal air-fuel ratio. The pressure cells take care of this automatically.

● 15-13 Late-Model Mechanical Continuous Fuel Injection

The electronic fuel-injection system is more expensive than the simpler mechanical fuel-injection system. Also, its complicated electronic components cannot be diagnosed except with special testers. To overcome these disadvantages, Bosch has introduced a new continuous injection system that operates primarily by mechanical means (Fig. 15-30). This system uses an electric fuel pump to pressurize the fuel. To control the amount of fuel injected, the volume of intake air is measured by an airflow sensor plate. The sensor plate is located in an air funnel through which all ingoing air must pass.

As the airflow through the funnel increases, it lifts the sensor plate higher in the funnel. This causes a lever to lift the control plunger in the fuel distributor. The action increases the amount of fuel that can flow to the injectors or nozzles which are located in the intake manifold. A properly balanced air-fuel ratio results. The fuel is sprayed continuously from the injection valves as long as the engine is running.

There are special subsystems to provide extra fuel for cold starts and during warm-up operation. One of these is a special start valve located in the common intake manifold. It includes a solenoid which is connected to the battery through a thermo-time switch. This switch is mounted on the side of the engine and senses the coolant temperature. When the engine is cold, the switch is closed. Now, when the engine is cranked, the start-valve solenoid is actuated and the start valve sprays extra gasoline into the air going to the cylinders. The mixture is enriched for starting. This system serves the same purpose as the choke in the carbureted system. As the engine warms up, the thermo-time switch opens, shutting off the start valve.

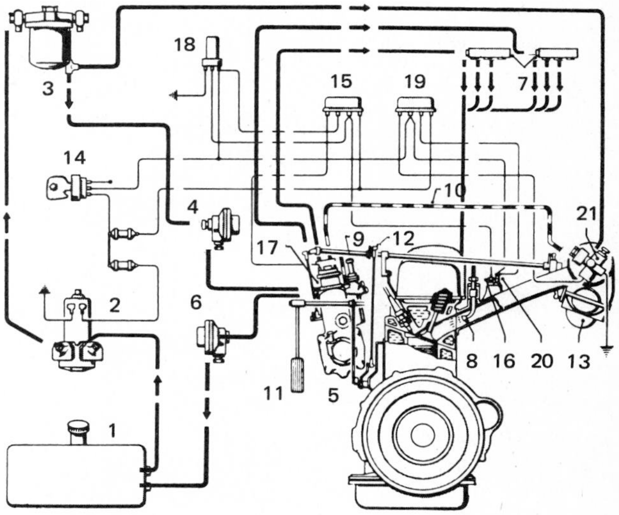

1. Fuel tank
2. Fuel-feed pump
3. Fuel filter
4. Damper container (inlet)
5. Injection pump
6. Damper container (outlet)
7. Fuel-metering units
8. Injection valves
9. Cooling-water thermostat
10. Additional air duct
11. Accelerator
12. Control linkage
13. Throttle connector
14. Ignition-starter switch
15. Relay
16. Thermo switch in cooling-water circuit
17. Magnetic switch for mixture control
18. Time switch
19. Relay
20. Thermo time switch in cooling-water circuit
21. Electromagnetic starter valve with atomizing jet

Fig. 15-29 Schematic layout of a fuel-injected six-cylinder engine. (*Mercedes-Benz of North America, Incorporated*)

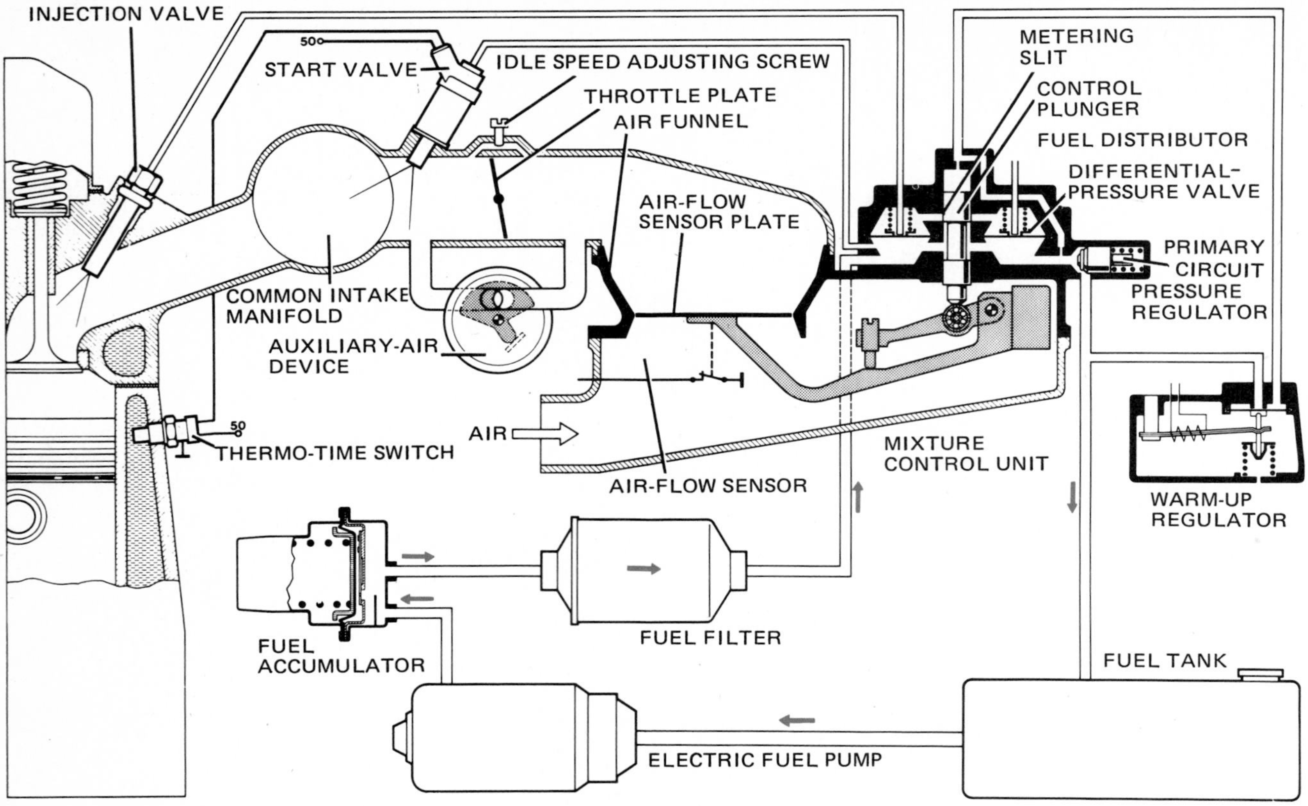

Fig. 15-30 Late-model mechanical continuous fuel-injection system used on some imported cars. (*Robert Bosch GmbH*)

This system is used on some imported cars and is being considered, with some changes, for some domestic cars.

REVIEW QUESTIONS

1. What does ECU stand for? What does it do?
2. Explain the difference between a diesel-engine fuel-injection system and the fuel-injection system for gasoline engines.
3. Explain the basic difference between the CIS and the timed injection system.
4. Explain the advantages of the gasoline fuel-injection system over the carbureted fuel system.
5. Are the injectors in the timed system timed individually or in groups?
6. Name and describe the operation of the various sensors that feed information into the ECU which determine the timing and duration of the injection.

SELF PROJECTS

A great deal is being published in the trade magazines and newspapers about fuel-injection systems. Whenever you run across one of these articles, clip it, paste it on sheets of paper, and file it in your notebook. Study the ads in the trade magazines; some of them offer pamphlets on their equipment or testers. Often, you can write in for this material. File it in your notebook.

If you can get manufacturer's service manuals that discuss the operation, trouble diagnosis, and servicing of fuel-injection systems, read this material carefully. Make notes of the important points for filing in your notebook. If possible, talk to an automotive technician who has been trained in fuel-injection service. See if you can get some tips on how to handle this service. Write down the important points that the technician brings up and file them in your notebook.

CHAPTER 16
Servicing Gasoline-Engine Fuel-Injection Systems

After reading this chapter, you should be able to:

- **Discuss the special procedures required to service gasoline fuel-injection systems.**
- **Discuss typical service problems in electronic fuel-injection systems and their possible causes.**
- **Under the direction of the instructor, make the checks and corrections listed in the chapter.**

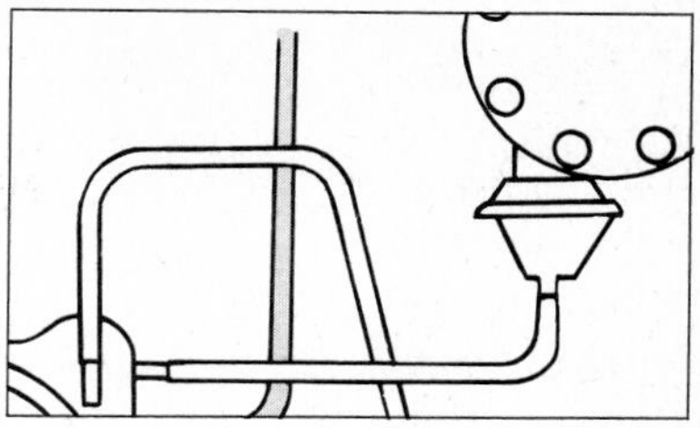

This chapter discusses servicing of gasoline fuel-injection systems, possible complaints and their corrections. A variety of these systems are in use; each has its own unique set of possible troubles and corrections. When you are faced with troubleshooting and servicing a fuel-injection system, have the service manual that applies, at hand. Refer to it for the specific step-by-step procedures. As one example of what you will find in the specific service manuals, we include here a condensation of the procedure outlined in the Cadillac service manual.

16-1 Servicing the Cadillac Electronic Fuel-Injection System

Before checking for defects in the electronic fuel-injection system, make sure the engine and especially the ignition system are not causing the complaint. If they are not at fault, Cadillac recommends the following visual inspection of the fuel-injection-system components.

1. Visually check all wiring-harness connections for:
   a. Loose or detached connectors
   b. Broken or detached wires
   c. Terminals not completely seated in connector housings
   d. Partially broken or frayed wires at terminal connections
   e. Excessive corrosion

2. Start the engine. Plug the idle bypass passage on top of the throttle body with a clean shop towel, to make it easier to hear any vacuum leaks. Then visually check all vacuum lines:
   a. To ensure all vacuum lines are securely connected to their proper fittings (Fig. 16-1)
   b. For broken, pinched, or cracked lines

3. Visually check the fuel lines for:
   a. Leakage
   b. Kinks

CAUTION: Do not loosen any fittings in the fuel system until after you have relieved the pressure in the system! Fuel in the system may be pressurized to 41 psi [2.81 kg/cm^2] or more. When loosening a fitting, observe the proper precautions to prevent fuel from spraying out and causing a fire or injuring you. Cover the fitting to be removed with a shop towel, so the fuel will be absorbed. Then place the gasoline-soaked towel in a safety container, or put it in a safe place outside to dry.

16-2 Preliminary Diagnosis

If the cause of the problem is not found during the visual inspection, then a preliminary diagnosis is in order. Cad-

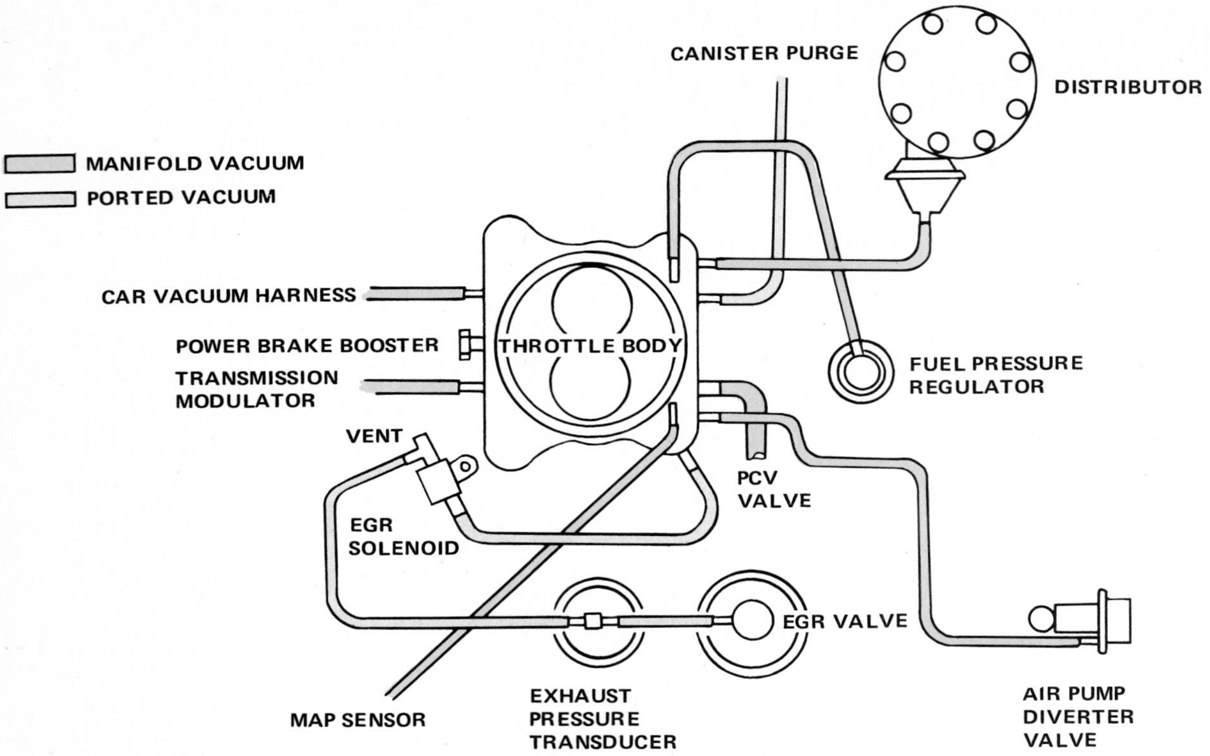

Fig. 16-1 Vacuum routings for one Cadillac model with electronic fuel injection. *(Cadillac Motor Car Division of General Motors Corporation)*

Carburetor	Electronic Fuel-Ignition
1. Accelerator pump	1. Throttle position switch
2. Fast-idle cam	2. Electric fast idle valve
3. Float	3. Fuel-pressure regulator
4. Power valve/metering rods	4. Manifold absolute pressure sensor
5. Metering jets and idle fuel system	5. Injection valves, electronic control unit (ECU)

Fig. 16-2 Chart showing the relationship between carburetor components and electronic-fuel-injection components. *(Cadillac Motor Car Division of General Motors Corporation)*

illac lists nine different types of problems that might occur in cars with electronic fuel injection. If the engine and all other systems are operating properly, these problems may be traced to the electronic fuel-injection system. The problems are:

1. Engine cranks but will not start
2. Hard starting
3. Poor fuel economy
4. Engine stalls after start
5. Rough idle
6. Prolonged fast idle
7. No fast idle
8. Engine hesitates or stumbles on acceleration
9. Lack of high-speed performance

After the visual inspection, a special tester, or fuel-injection-system analyzer, must be used to locate the cause of any of these problems.

During the preliminary diagnosis, you should keep in mind certain relationships between parts in the carburetor systems and components in the electronic fuel-injection system (Fig. 16-2). These can help you sort out, identify, and pinpoint a defective component.

16-3 Trouble-Diagnosis Chart

The chart that follows lists various complaints that can be traced to electronic fuel-injection systems, their possible causes, and the checks or corrections to be made. The information in the chart will shorten the time you need to correct a trouble. If you follow a logical procedure, you can usually find the cause of a trouble quickly. On the other hand, haphazard guesswork wastes time and effort.

Note that the troubles and possible causes are not listed according to how often they occur. That is, item 1 (or item a under "Possible Cause") does not necessarily occur more often than item 2 (or item b).

For further information on the testing of electronic fuel-injection systems and the use of the special tester, refer to the automobile manufacturer's service manual.

ELECTRONIC-FUEL-INJECTION TROUBLE-DIAGNOSIS CHART

Complaint	Possible Cause	Check or Correction
1. Engine will not start; fuel pump does not run	a. Fuse to pump relay terminal blown; cables to pump or pump relay or cables on fuel-pump relay defective	Renew fuse; check whether pump relay energizes (switch ignition on and off, and listen for audible action of relay); if necessary, test with voltmeter; check plug on fuel pump for good connection
	b. No voltage at terminal of pump relay (12 V), because main relay is not operating or cable is defective	Eliminate open circuit
	c. Voltage present at terminal of pump relay, but no ground connection	Fuel pump operates for 1 to 2 seconds after ignition is switched on; check with meter (ground is made by control unit); replace ECU
	d. Open circuit in cable from pump relay to pump plug	Check plug connection; eliminate open circuit
2. Engine will not start; fuel pump runs	a. Connection from cable harness to starter terminal defective	Check with special tester
	b. Pressure-sensor cable not connected; open circuit	Push on pressure-sensor cable; repair
	c. Open circuit in cable connection at temperature sensor (coolant)	Check cables; if necessary, replace temperature sensor
	d. No fuel pressure building up (pipe compressed or pressure regulator defective)	Check pressure with gauge; if necessary, replace pressure regulator
3. Engine starts cold but stalls	a. Cable connector for triggering contacts not pushed on at ignition distributor; open circuit in cable	If necessary, connect special tester and locate the fault; replace trigger contacts or cable harness
	b. Trigger contacts defective	Replace
	c. Pressure sensor defective	Replace
	d. See also item 1, "Engine will not start"	
4. Engine cuts out when driving (usually preceded by misfiring)	a. Trigger contacts have excessive contact resistance or are dirty	Replace trigger contacts
	b. Plug loose	Check
	c. No fuel pressure	Check pressure; determine cause
5. Engine runs irregularly, one cylinder not firing; exhaust white	a. One injector sticking	Replace injector
	b. Connection to injector or injector coil defective	Check connections; replace injector; test system with special tester
6. Engine misfiring, not caused by ignition system	a. Loose connections; main ground cable has poor connection to car body	Check connections; tighten ground connection
7. Engine not reaching full power	a. Fuel pressure low	Check pressure regulator
	b. Pressure sensor defective	Replace
	c. Throttle valve does not open sufficiently	Check throttle valve
8. Fuel consumption too high	a. Sensors of ECU not functioning correctly; electrical connections have too high resistance	Test system with special tester
	b. Throttle-valve switch incorrectly adjusted	Adjust, using special tester
	c. Fuel pressure incorrect	Check pressure regulator; if necessary, replace it

ELECTRONIC-FUEL-INJECTION TROUBLE-DIAGNOSIS CHART

Complaint	Possible Cause	Check or Correction
9. Engine hunts excessively at idle (between 1,000 and 1,800 rpm)	a. Hose between auxiliary air regulator and induction manifold detached or defective	Push hose into position or replace
	b. Throttle-valve stop incorrectly adjusted (open too wide)	Readjust throttle-valve stop
	c. Idling speed set too high	Adjust idling speed
10. Engine misfires when accelerating	Temporary enrichment device in throttle-valve switch not functioning; plug incorrectly connected	Check throttle-valve switch with special tester
11. Too high idling speed; idling speed cannot be adjusted	a. Idling air system leaks	Check idling air system
	b. Rubber sealing ring under the injector	Replace rubber sealing ring
	c. Throttle-valve adjustment incorrect	Readjust throttle valve

REVIEW QUESTIONS

1. What caution should you observe when loosening any fittings in the fuel system?
2. What two components should be visually inspected before actual testing of a fuel-injection system starts?
3. The trouble-diagnosis chart for the Cadillac fuel-injection system lists 11 possible troubles. Make a list of these and study them until you can recite them from memory.
4. As you mention each complaint in the chart, explain what might cause it and how the trouble can be fixed.

SELF PROJECTS

Whenever you get your hands on a service manual covering a fuel-injection system, study the trouble-diagnosis sections. Make lists of the possible troubles in the system you are studying, along with possible causes of the troubles. File these in your notebook. In your discussions with technicians who handle fuel-injection-system servicing, learn what testing equipment they use and how it is used. If possible, borrow the instruction manuals that are supplied with the special testers. Study them and make notes of the procedures required. File this in your notebook.

CHAPTER 17
Diesel-Engine Fuel-Injection Systems

After reading this chapter, you should be able to:

- **Discuss the difference in the operation of diesel and gasoline engines.**
- **Explain the operating requirements of the diesel-engine fuel-injection system.**
- **Explain the operation of the rotary fuel-injection pump.**
- **Locate the components of a diesel-engine fuel-injection system on a car and explain how each works.**

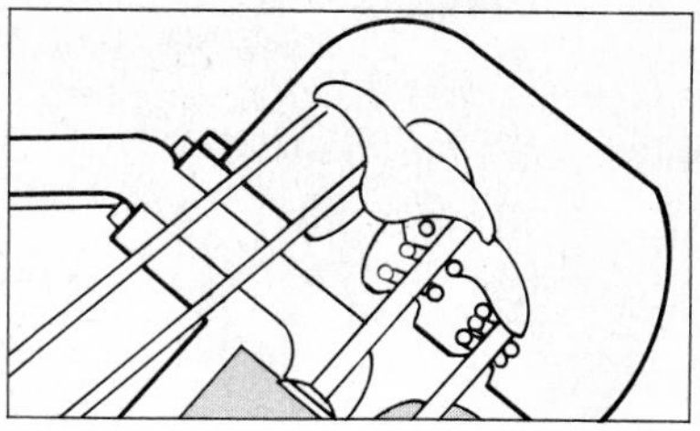

In this chapter, we discuss fuel-injection systems used with automotive diesel engines. Diesel engines have been used for many years for trucks, buses, and off-the-road equipment. The diesel engine has made a name for itself as a reliable, heavy-duty power plant. In recent years, several manufacturers have begun to install four-cycle diesel engines in some car models. We discussed this, and the reason for the swing toward diesels, in ● 6-19.

● 17-1 Differences between Gasoline and Diesel Engines

We have previously mentioned that the diesel engine must be more heavily constructed because of the higher pressure in the combustion chambers and in the actions during the four strokes. Let us compare the strokes:

1. Diesel-engine intake stroke The intake valve is open and the piston is moving down, producing a partial vacuum in the cylinder (Fig. 17-1). Atmospheric pressure pushes air through the air filter and intake manifold, past the open intake valve, and into the cylinder. Notice that there is no throttle valve or carburetor venturi to impede the movement of the air. On the intake stroke the cylinder is filled completely with air. No fuel is present.

2. Diesel-engine compression stroke During the compression stroke, both valves are closed and the upward-moving piston compresses the air (Fig. 17-2). The compression ratio of the diesel engine is much higher than that of the gasoline engine. It may be as high as 22:1. This contrasts with gasoline-engine compression ratios, which are in the neighborhood of 9:1. The reason the diesel engine can have such high compression ratios is that air alone is compressed. Compressing air makes it

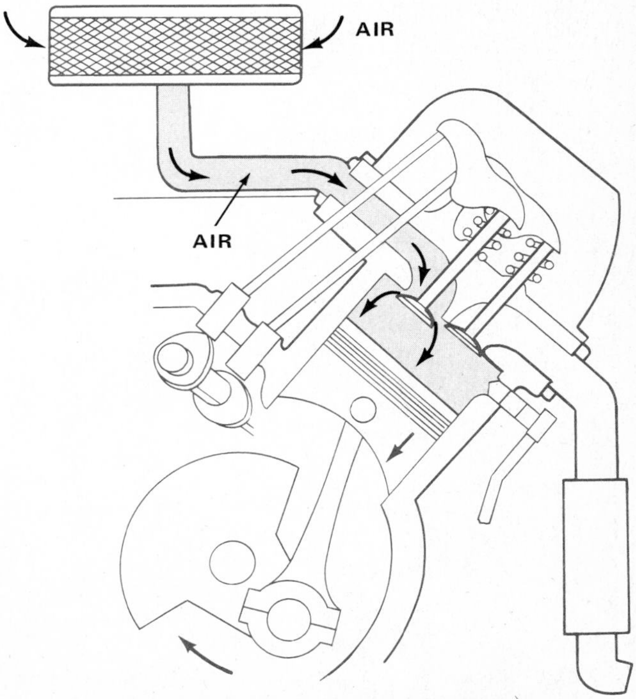

Fig. 17-1 Diesel-engine intake stroke. (*Oldsmobile Division of General Motors Corporation*)

hot. Compressing air to a twentieth of its original volume (compression ratio 20:1) increases the temperature to above 1000°F [537.8°C]. This temperature is high enough to ignite any fuel. That is the reason why we cannot have such high compression ratios with gasoline engines. The air-fuel mixture would ignite before the piston reached TDC.

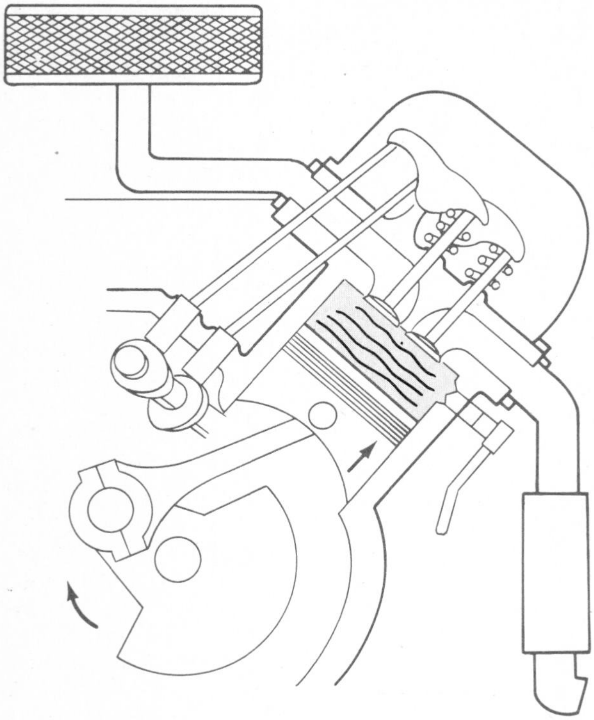

Fig. 17-2 Diesel-engine compression stroke. (*Oldsmobile Division of General Motors Corporation*)

3. Power stroke As the piston approaches TDC on the compression stroke, the diesel-engine fuel-injection system starts to inject a spray of fuel into the cylinder (Fig. 17-3). The high temperature of the compressed air ignites the fuel and the pressure rapidly rises. The power stroke then takes place.

4. Exhaust stroke The exhaust stroke is similar to that of the gasoline engine. The piston moves up and pushes out the exhaust gases past the opened exhaust valve (Fig. 17-4).

To sum up, the diesel engine:

1. Has no throttle valve to restrict air flow into the engine.
2. Compresses only air on the compression stroke.
3. Has a much higher compression ratio.
4. Does not have an ignition system. Instead, heat of compression ignites the fuel as it is sprayed into the cylinders.
5. Engine power and speed are controlled only by the amount of fuel sprayed into the cylinders. For more power, more fuel is injected. For less power, less fuel is injected.
6. Many diesel engines have glow plugs which make it easier to start a cold engine (see ● 17-9).

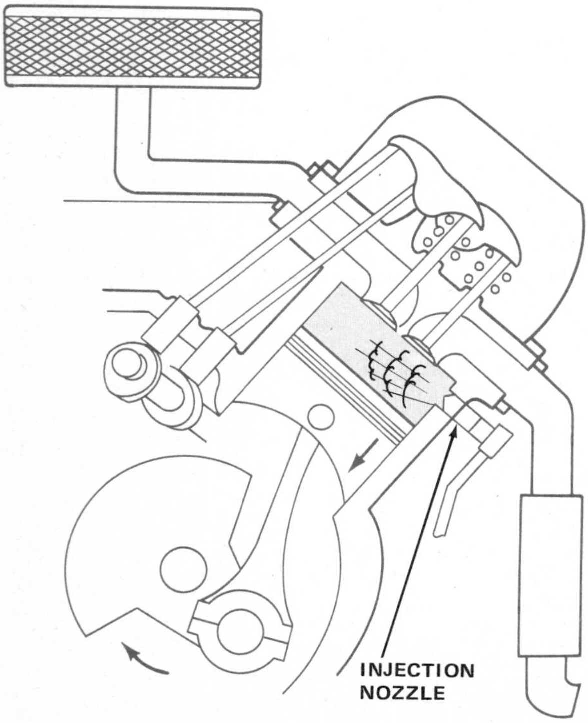

Fig. 17-3 Diesel-engine power stroke. (*Oldsmobile Division of General Motors Corporation*)

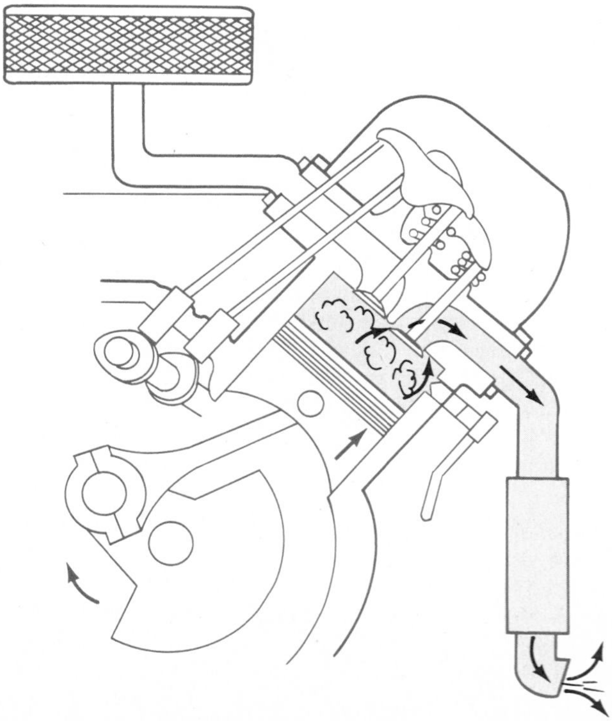

Fig. 17-4 Diesel-engine exhaust stroke. (*Oldsmobile Division of General Motors Corporation*)

17-2 Diesel Fuel-System Requirements

The diesel fuel system must:

1. Deliver the exactly right amount of fuel to meet the operating requirements.
2. Change the timing of fuel delivery as engine speed changes. That is, as engine speed increases, the fuel delivery must start earlier. This compares with the advance of the spark in the gasoline engine as engine speed increases. The purpose is the same, to get the ignition of the fuel started before the piston reaches TDC. Without an advance, the piston would be over TDC and starting down on the power stroke before ignition was well started. The piston movement would keep ahead of the pressure rise so that most of the power in the fuel would be wasted.
3. Deliver the fuel to the cylinders under high pressure. The pressure in the cylinder at the end of the compression stroke is more than 500 psi (pounds per square inch) [35.15 kg/cm^2 or 3,447.4 kPa]. The fuel must be under pressure much higher than this in order for it to be sprayed into the compressed air.

17-3 Two Systems

There are, basically, two types of systems used to feed the fuel to the engine. In one, a centrally located pump pressurizes the fuel, meters it, times it, and delivers it to the cylinders through tubes.

In the other system, the fuel is sent to the injectors under a relatively low pressure. The injectors have cam-operated plungers (like the valves) which are adjustable. At the proper instant the cams operate the plungers and they force the fuel at high pressure into the cylinders.

The centrally located pump system, used on a majority of diesel engines, can be further divided into two types, the cam-operated in-line plunger type, and the rotary-distributor type. The rotary type is the most commonly used for passenger-car diesel engines and we will concentrate on that in the remainder of the chapter. First, however, we describe the cam-operated, in-line plunger type.

17-4 Cam-Operated In-Line Plunger Pump

This pump has a cylinder with plunger for each engine cylinder. Figure 17-5 is a partial cutaway view of a fuel-injection pump for a six-cylinder engine. It has six plungers working in six barrels, one for each engine cylinder. The camshaft is driven from the engine and it has a cam for each plunger. When the lobe of a cam comes up under a plunger, the plunger is raised and fuel is sent at high pressure through a high-pressure tube to an injector nozzle in an engine cylinder. Figure 17-6 is a schematic view of a fuel-injection system using this type of pump.

The injection pump has speed-advance and metering systems which time the moment of injection and determine the amount of fuel to be injected.

17-5 Rotary-Distributor Pump

This is the pump used on a majority of automotive-type diesel engines. Figure 17-7 is a simplified schematic view of the system for a V-6 engine using this pump. Figure 17-8 is a partial cutaway view of a V-8 engine using the rotary fuel-injection pump. Note its position and the method of drive through bevel gears from the camshaft. The pump sits between the cylinder banks and high-pressure tubes connect the pump to the injectors in each cylinder. Note that the injectors are connected by a second set of tubes to the fuel tank. These are called the fuel leak-off return lines. The injectors leak some oil and these lines carry this excess fuel back to the fuel tank. The fuel-injection pump includes a fuel-supply pump which delivers the fuel to the distributor part of the pump at high pressure. The distributor then sends the fuel to the engine cylinders in the proper firing order. This compares with the ignition system, on gasoline engines, sending high-voltage surges to the spark plugs in the proper firing order.

Each nozzle of the fuel injector has a spring-loaded check valve that is closed except when high pressure is applied to the fuel. When this happens, the check valve opens, allowing fuel to pass through. The fuel exits from the nozzle tip through small holes. The holes are located so as to send the fuel into the center of the compressed air. It ignites the instant it hits this hot air. When the fuel pressure drops, the check valve closes so the flow of fuel through the nozzle stops.

17-6 Rotary Injection Pump Operation

Figure 17-9 is a cutaway view of the complete rotary injection pump. Figure 17-10 is the external view of the pump. Note the eight high-pressure connectors, one for each of the eight cylinders in the V-8 engine. Although the pump looks complex, its operation is basically simple. The drive shaft operates at one-half crankshaft speed. As you can see from Fig. 17-8, it is driven from the camshaft by a pair of bevel gears. The bevel gears are necessary because of the slight angle of drive from the camshaft to the pump drive shaft.

The rotating members inside the pump include a transfer pump which builds up fuel pressure, and the heart of the system, the injection pump rotor itself (Fig. 17-11). The injection pump rotor, as it rotates, causes the two plungers to move in and out. The pump rotor rotates in a semistationary internal cam ring. This ring has cam lobes on its inner surface. You can see the ring in Fig. 17-15. As the rotor rotates, the rollers ride up and down on these cam lobes. When they move in this way, they cause the two plungers to move up and down in their holes. This causes the size of the chamber between the inner ends of the plungers to increase and decrease in size. See Fig. 17-12. When the chamber increases in size, fuel from the transfer pump flows into the chamber. When the plungers move toward each other, the chamber decreases in size, forcing fuel to flow out of the chamber. The fuel flows through ports to the high-pressure line connected to the nozzle in the cylinder ready for ignition. That is, to the cylinder in which the piston is nearing TDC on the compression stroke.

This action is shown in Fig. 17-13. The action compares to the rotor in the ignition distributor. The rotor sends the high-voltage surge from the ignition coil through the high-voltage cable to the spark plug in the cylinder that is ready to fire. In a similar manner, the rotor in the injection pump aligns holes that allow the fuel under high pressure between the two plungers to flow to the cylinder that is ready for ignition.

● 17-7 Governor

The governor feeds the proper amount of fuel through the system to allow the engine to operate at the proper speed. The governor has a pair of weights that are fastened to a rotating shaft (Fig. 17-14). As engine speed increases, they move out due to centrifugal force. This motion moves a thrust sleeve which, in turn, moves the governor arm. As the governor arm pivots, it actuates the fuel-metering valve. This valve controls the amount of fuel that is fed to the pump rotor and thus the amount of fuel the cylinders receive. For example, when the car starts down a hill, the load on the engine decreases. The engine starts to speed up. But the governor operates to prevent this by cutting back on the amount of fuel going to the engine. On the other hand, suppose the car meets a hill.

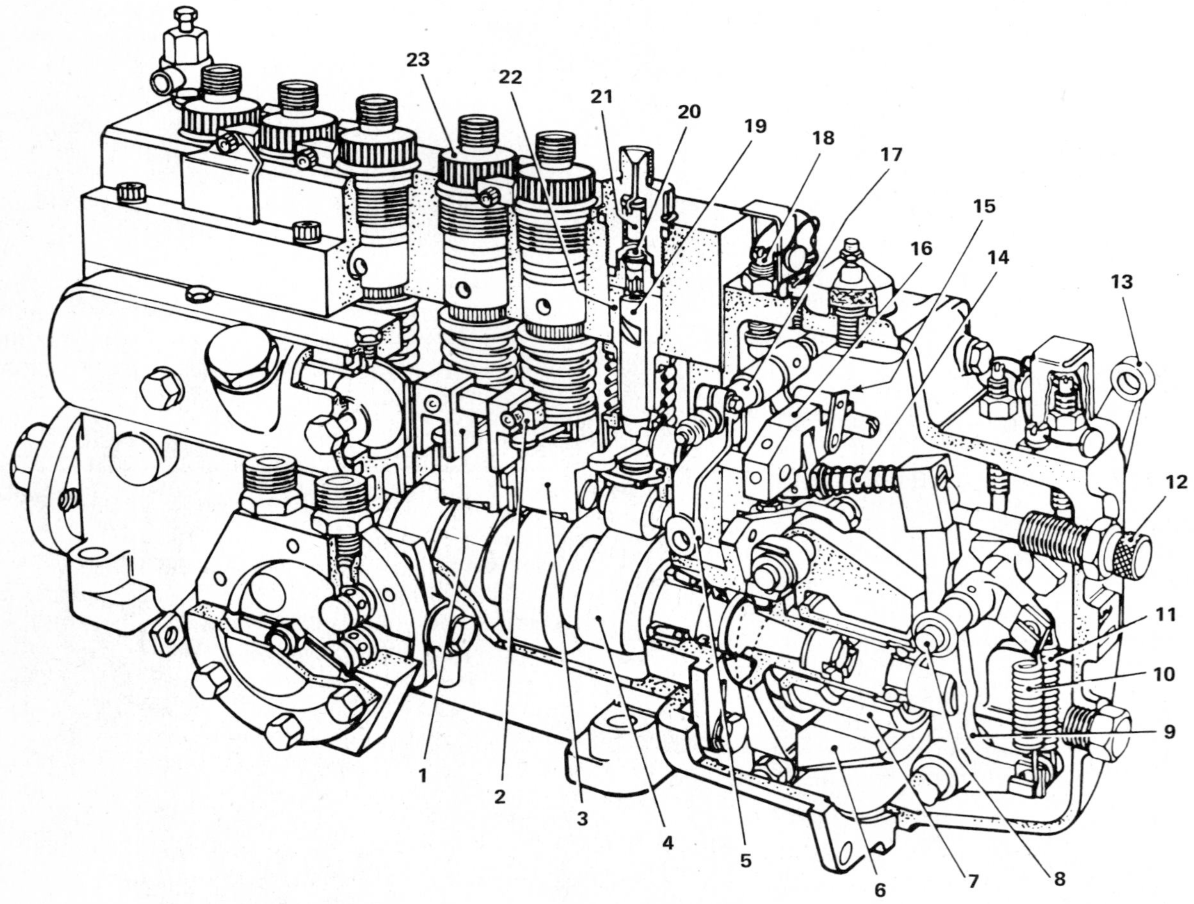

1 CONTROL FORK
2 CONTROL ROD
3 TAPPET ASSEMBLY
4 CAMSHAFT
5 STOP CONTROL LEVER
6 GOVERNOR FLYWEIGHT
7 GOVERNOR SLEEVE
8 SPEED LEVER SHAFT
9 CRANK LEVER
10 GOVERNOR MAIN SPRING
11 GOVERNOR IDLING SPRING
12 DAMPER
13 SPEED CONTROL LEVER
14 TELESCOPIC LINK
15 TRIP LEVER
16 BRIDGE LINK
17 EXCESS FUEL DEVICE
18 MAXIMUM FUEL STOP SCREW
19 PLUNGER
20 DELIVERY VALVE
21 VOLUME REDUCER
22 BARREL
23 DELIVERY VALVE HOLDER

Fig. 17-5 In-line plunger pump for six-cylinder diesel engine. (*CAV Ltd.*)

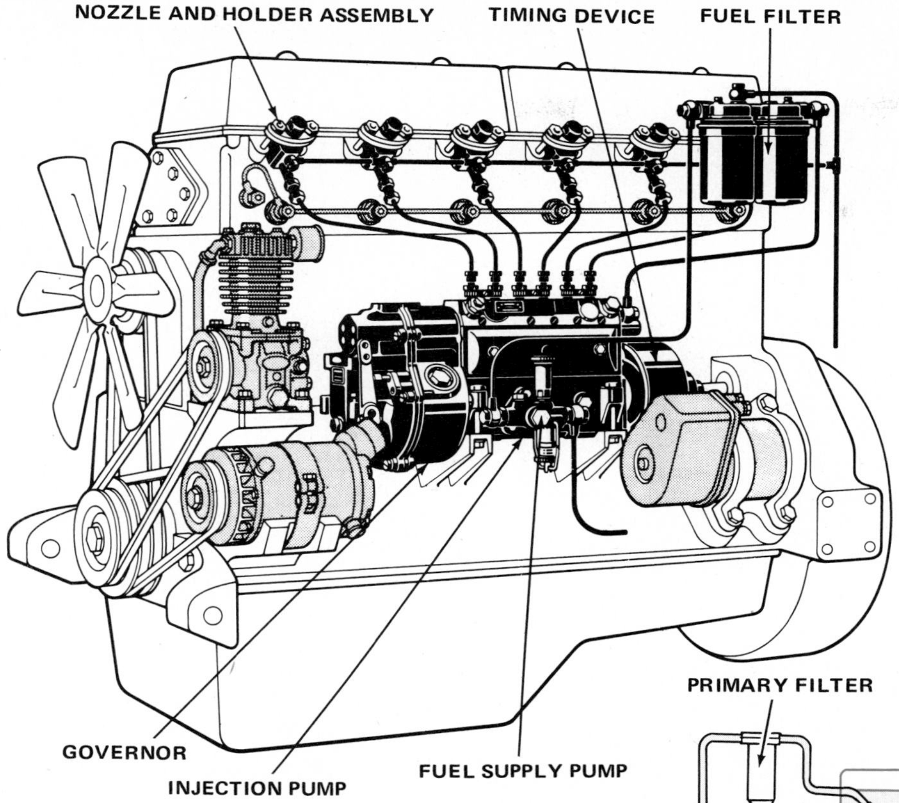

Fig. 17-6 A fuel-injection system on a six-cylinder diesel engine. (*Robert Bosch GmbH*)

The engine tends to slow down. The governor therefore allows more fuel to flow so the engine produces more power. This maintains engine speed. Note that the throttle position enters into the action. Any particular throttle opening, in effect, presets the amount of fuel and engine speed. Any variation from this speed is countered by the governor as it changes the fuel flow so as to maintain the preset speed. The throttle movement puts more or less tension on the governor spring. With more tension (greater throttle opening), the engine and governor speed must go higher before the governor acts to cut back on fuel flow. With less tension (lighter throttle), the governor can cut back on fuel flow at lower engine and governor speed.

● 17-8 Automatic Advance

The ignition system for gasoline engines has spark-advance mechanisms that move the spark ahead as engine speed increases. This allows the combustion to start earlier so it will be well advanced by the time the piston reaches TDC and starts down on the power stroke. The diesel-engine fuel-injection system also includes a speed advance. It is built into the injection pump, as shown in Fig. 17-15. The advance mechanism uses the hydraulic pressure from the transfer pump to control the position of the cam ring. As engine and transfer-pump speed increase, the transfer-pump hydraulic pressure also increases. This increasing pressure, acting on a piston located below the cam, forces the advance pin to move (to the left in Fig. 17-15). This rotates the cam as shown

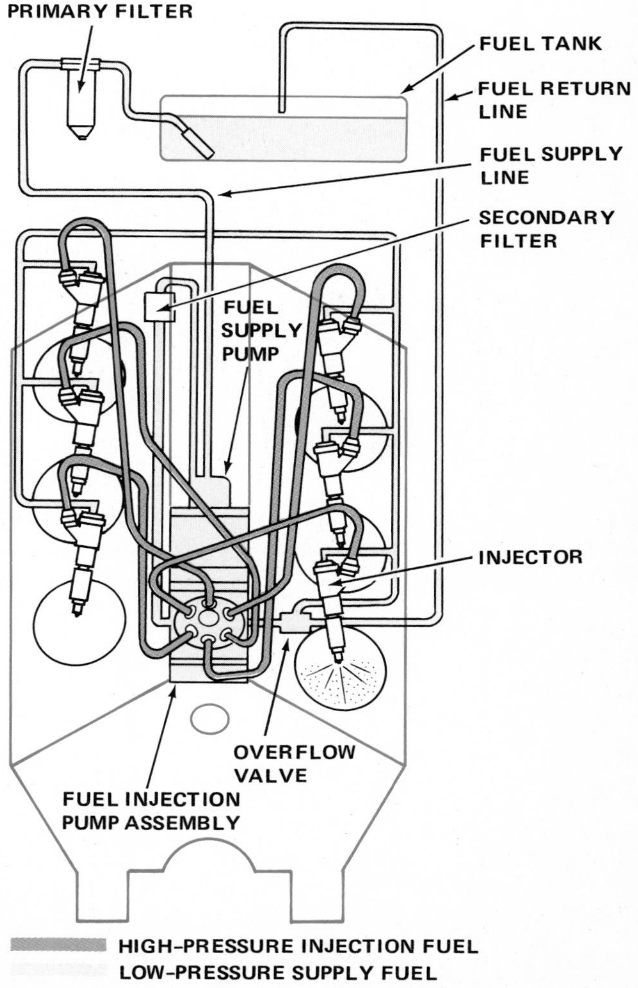

Fig. 17-7 Rotary distributor pump system for a V-6 diesel engine. (*General Motors Corporation*)

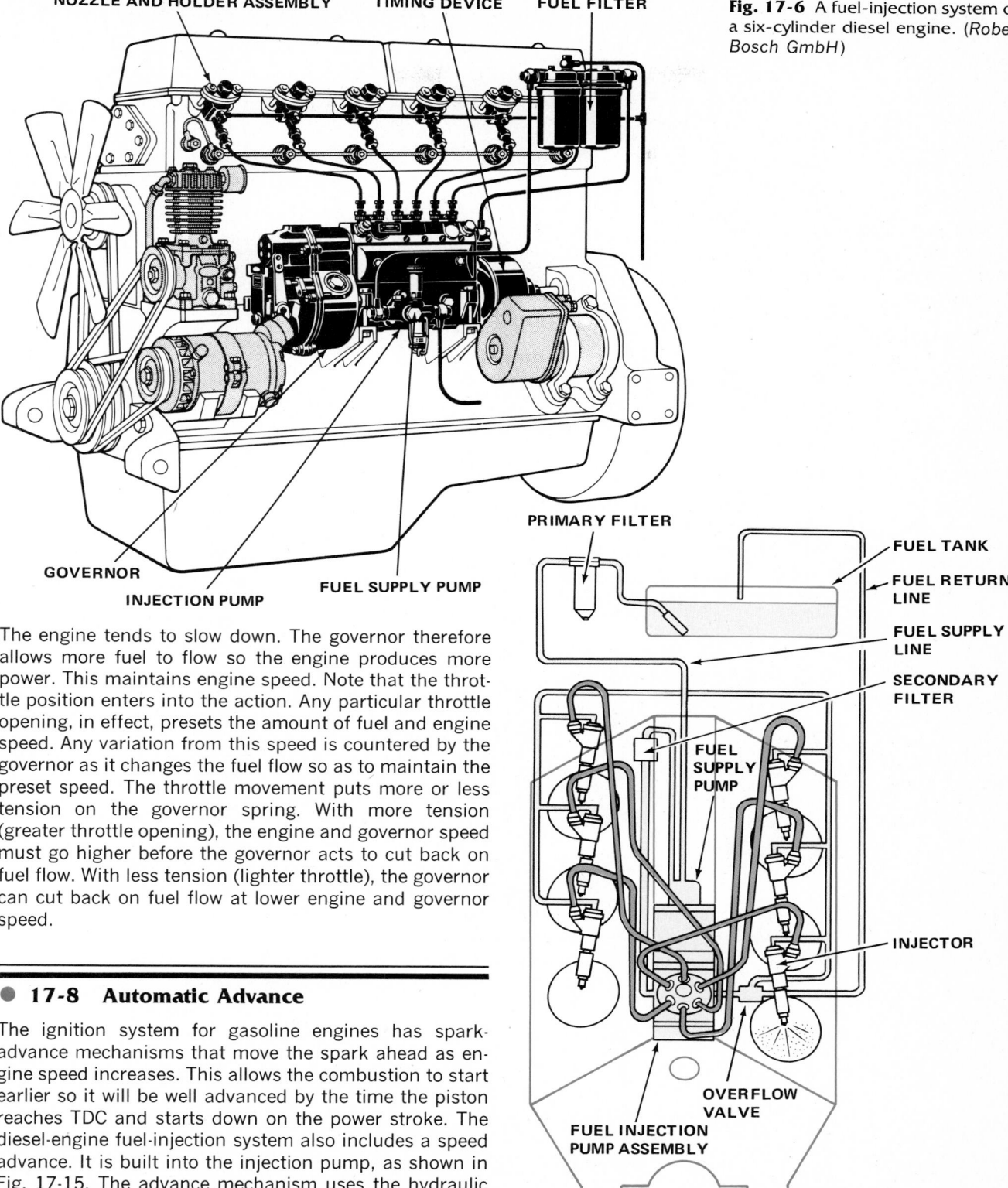

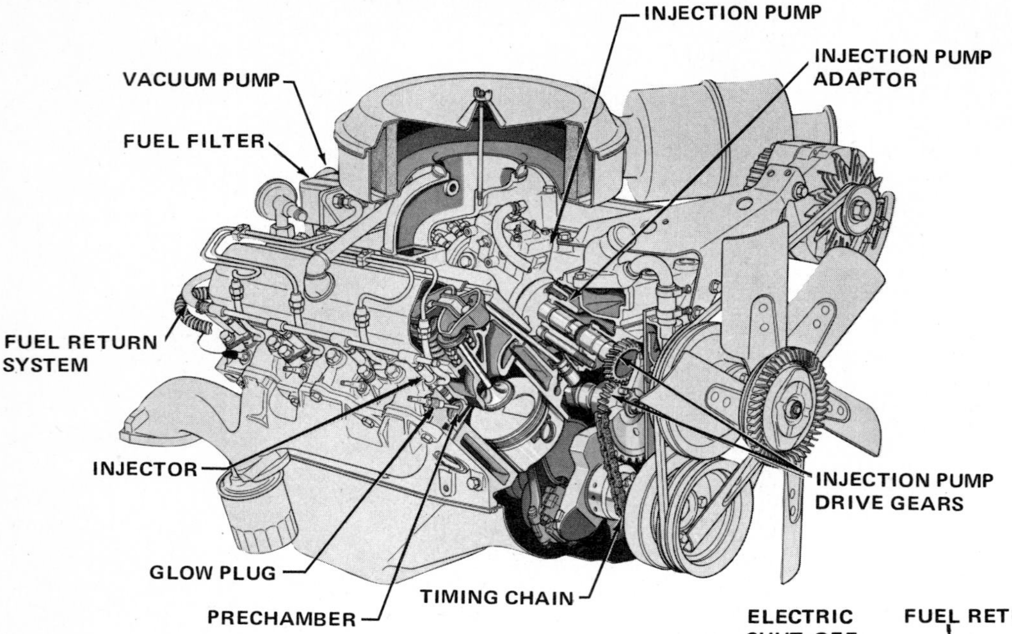

Fig. 17-8 V-8 diesel engine using rotary fuel-injection-distributor pump, partly cut away to show pump drive arrangement. ((*Oldsmobile Division of General Motors Corporation*)

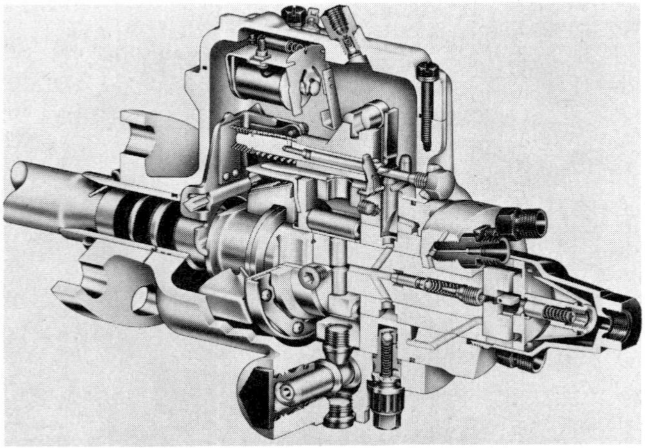

Fig. 17-9 Rotary fuel-injection pump, partly cut away to show internal construction. (*Oldsmobile Division of General Motors Corporation*)

by the heavy arrows. As the cam moves, the rotor rollers meet the lobes earlier and cause the plungers to be pushed together earlier. The fuel therefore is sent to the cylinder injection nozzles earlier. This produces the fuel-injection advance required as engine speed increases. The advance is proportional to the transfer-pump pressure, which is proportional to engine speed. The advance, therefore, is proportional to engine speed. The higher the speed, the more the advance.

17-9 Glow Plugs

For easy starting, especially in cold weather, many diesel engines use glow plugs. The glow plugs have electric heating elements that become very hot when connected

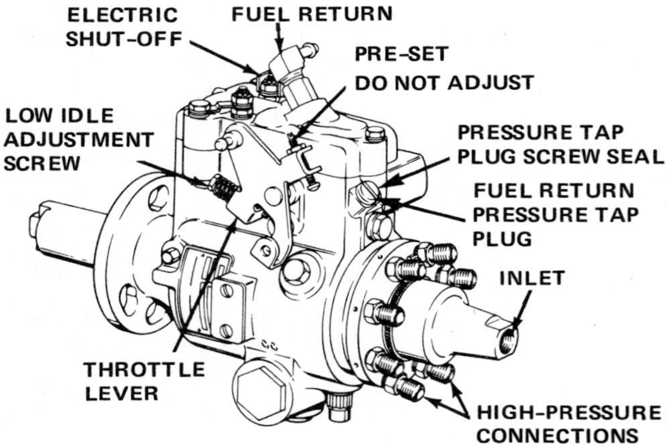

Fig. 17-10 External view of rotary fuel-injection pump. (*Chevrolet Motor Division of General Motors Corporation*)

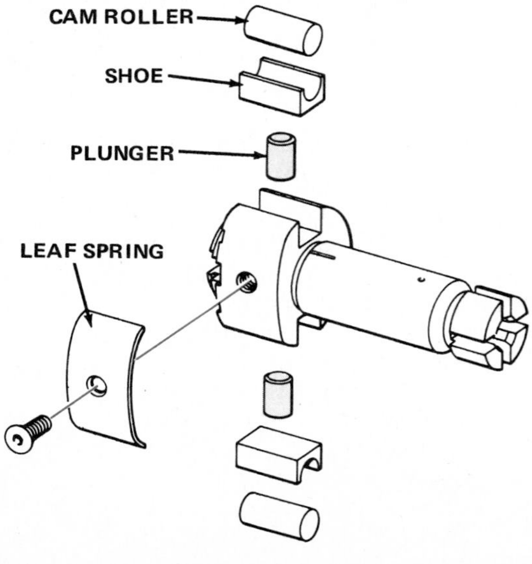

Fig. 17-11 Parts in the injection pump rotor. (*Chevrolet Motor Division of General Motors Corporation*)

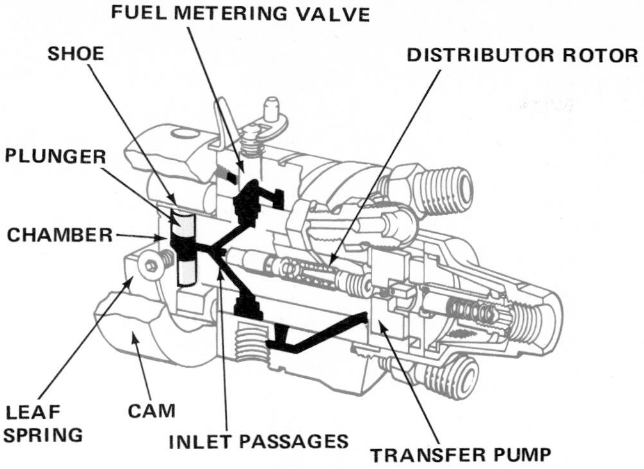

Fig. 17-12 Charging cycle during which the two plungers move apart and provide space for fuel to enter the chamber. (*Oldsmobile Division of General Motors Corporation*)

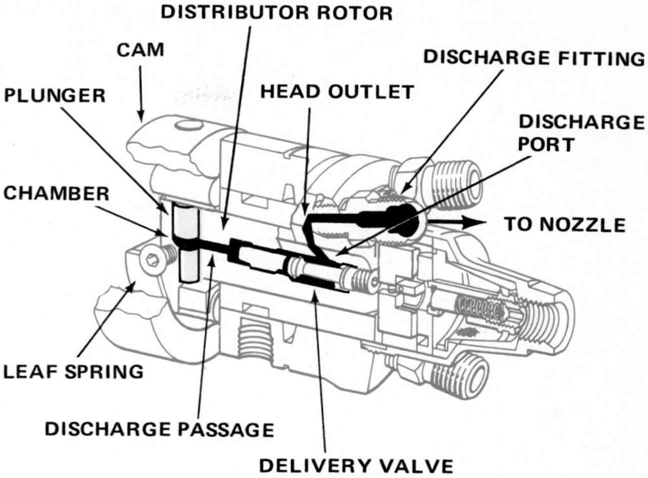

Fig. 17-13 Discharge cycle. The plungers are moving together, forcing the fuel from the chamber, past the delivery valve and through the discharge port. Note that the rotor has turned far enough so the port in the rotor has registered with the discharge port that is connected through a fuel line to the injector in the cylinder that is ready to fire. That is, to the cylinder in which the piston is approaching TDC on the compression stroke. (*Oldsmobile Division of General Motors Corporation*)

to the battery. Figure 17-16 shows the location of a plug in a cylinder. Note that it is in a precombustion chamber and is close to the fuel-injection nozzle. The precombustion chamber is where the fuel is injected and where the combustion starts. After combustion begins, the burning air-fuel mixture streams out of the precombustion chamber and into the main combustion chamber. There, it mixes with the combustion-chamber air and combustion is completed. Note that there is always an excess of air so that combustion of fuel can be relatively complete.

When the engine is cold, and the air temperature is low, the glow plugs are turned on to put some heat into the precombustion chambers. This greatly improves starting because the fuel is sprayed into air that has been preheated by the glow plugs.

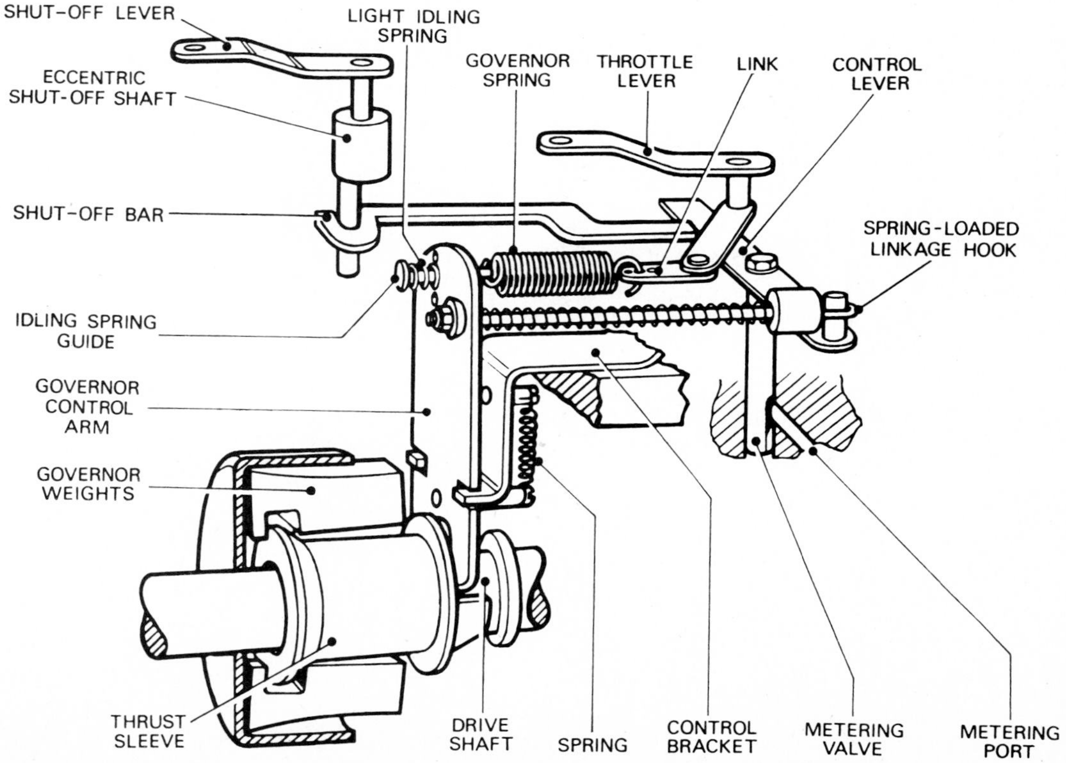

Fig. 17-14 Simplified layout of a governor. It works with the throttle to allow the correct amount of fuel to flow for the operating condition. (*CAV Ltd.*)

On some applications, the glow plugs can be turned on manually if the operator feels the engine needs them. On others, such as the General Motors V-8 diesel used on several vehicles (including some Oldsmobiles), the system is semiautomatic. The instrument panel has two special lights, WAIT and START (Fig. 17-17). The starting procedure is as follows:

1. Put the transmission lever in PARK or N. PARK is preferred.
2. Turn the ignition switch (Fig. 17-18) to RUN not START. Do not turn it to START just yet. When you turn the switch to RUN, an amber WAIT light comes on (if engine is cold). This tells you that the glow plugs are on, heating the engine precombustion chambers.

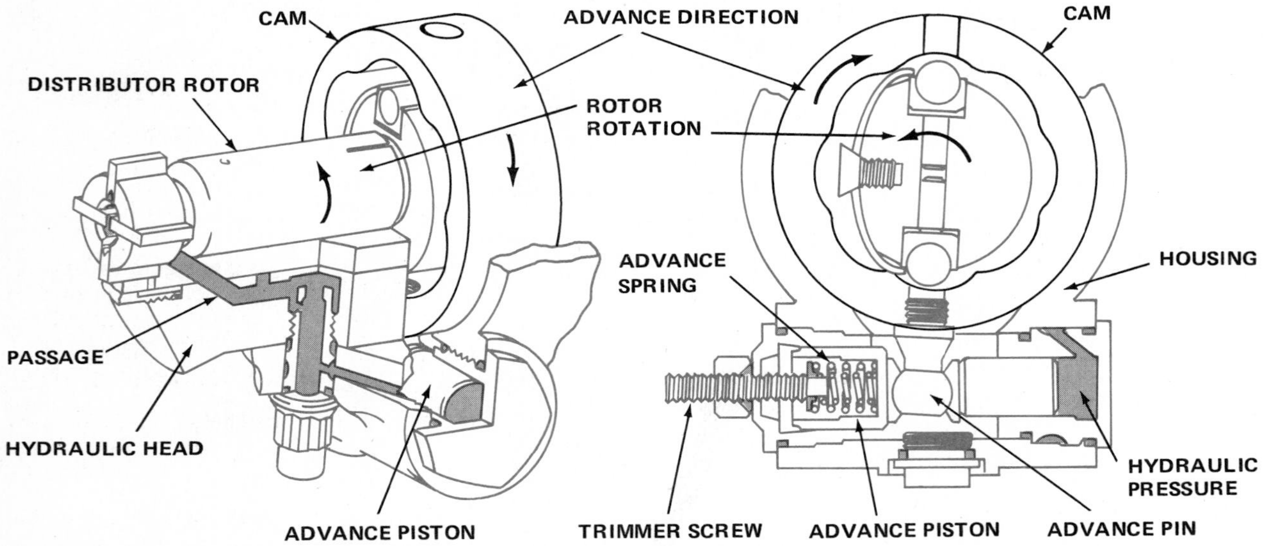

Fig. 17-15 Two cutaway views of the automatic advance system. Hydraulic pressure, which increases with speed, moves the cam ahead. (*Oldsmobile Division of General Motors Corporation*)

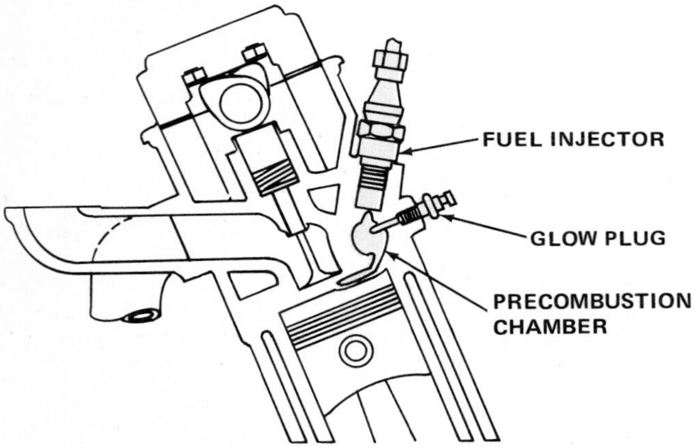

Fig. 17-16 Simplified drawing of precombustion chamber showing locations of glow plug and fuel injector.

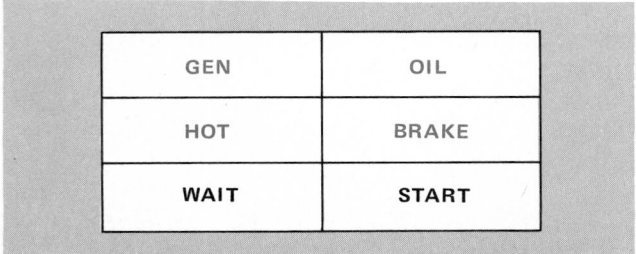

GEN	OIL
HOT	BRAKE
WAIT	**START**

Fig. 17-17 The instrument panel has two special lights, WAIT and START. (*Oldsmobile Division of General Motors Corporation*)

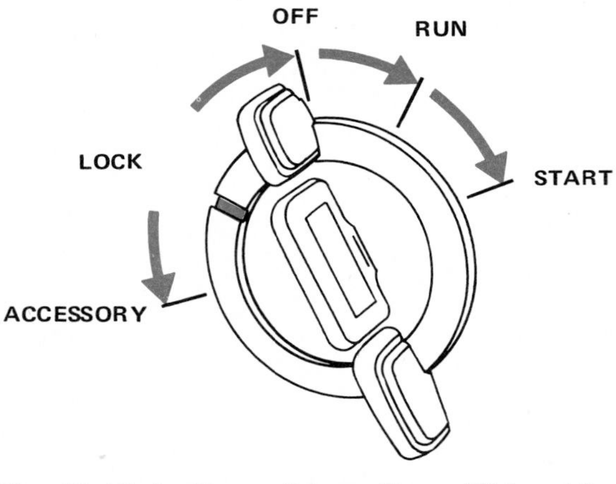

Fig. 17-18 Ignition-switch positions. (*Oldsmobile Division of General Motors Corporation*)

After the precombustion chambers have been sufficiently heated (usually only a few seconds depending on the temperature), the WAIT light will go out and the START light will come on. Then, push the accelerator pedal halfway down to the floor and hold it there. Turn the ignition switch to START. Normally, the engine will start in only a few seconds. If it does not start in 15 seconds, release the ignition switch. If the WAIT light comes on again, leave the ignition switch in the RUN position until the WAIT light goes off and the START light comes on. Now, try the starting procedure again.

Pumping the accelerator before or during cranking will not aid in starting. The system has no accelerator pump system to force fuel into the air passing through the carburetor on its way to the combustion chambers, as does the gasoline-engine fuel system. In the diesel-fuel system, the accelerator works with the governor to provide the amount of fuel needed to meet the operating conditions, as previously explained.

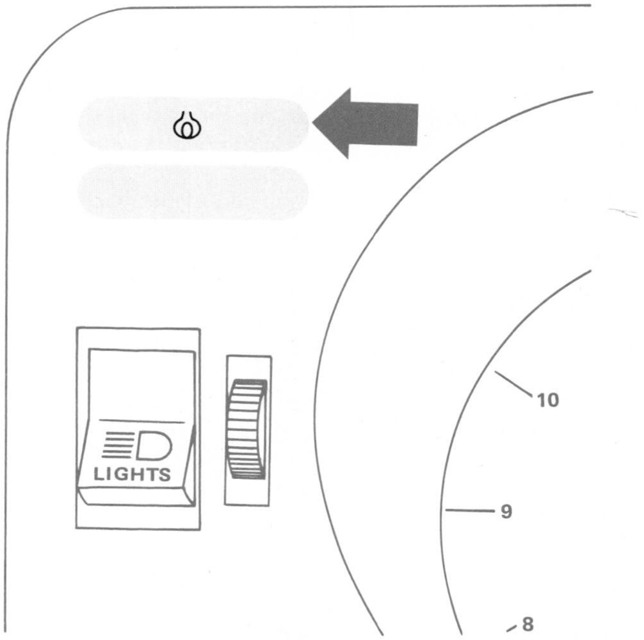

Fig. 17-19 The diesel Volkswagen Rabbit uses a single light to indicate glow-plug action. When the plugs are on, the light comes on. (*Volkswagen*)

● 17-10 Other Starting Instructions

The starting instructions above are for the V-8 General Motors diesel engine used in some Oldsmobiles and other vehicles. Other manufacturers have slightly different instructions and different indicating devices. The Volkswagen Rabbit with a diesel engine, for example, uses a single light to indicate glow-plug action (Fig. 17-19). Also, the ignition switch has only three positions—OFF, ON, and START (Fig. 17-20). Here is the recommended starting procedure:

1. Temperatures above 32°F [0°C], engine cold—turn ignition switch ON. The glow-plug light (Fig. 17-19) will come on and remain on as long as the glow plugs are heating. As soon as the light goes off, the plugs have heated enough and you are ready to start. Turn the ignition switch to START. Do not depress the throttle pedal.
2. Temperatures below 32°F [0°C], engine cold—turn ignition switch to ON. After the glow-plug light goes off, push the throttle all the way open. Pull out the cold-start knob under the instrument panel (Fig. 17-21). This provides extra fuel to the cylinders. Turn the switch to START. Two minutes after starting, push the cold-start knob in.
3. Starting a warm engine—do not depress the throttle. Do not use the glow plugs. That is, turn the ignition switch past ON to START immediately.

Do not accelerate the engine excessively immediately after starting. Wait for the oil pressure in the lubricating system to build. Do not operate the starting motor longer than 30 seconds. If the engine does not start, turn ignition switch to OFF. Wait for about 30 seconds. Then

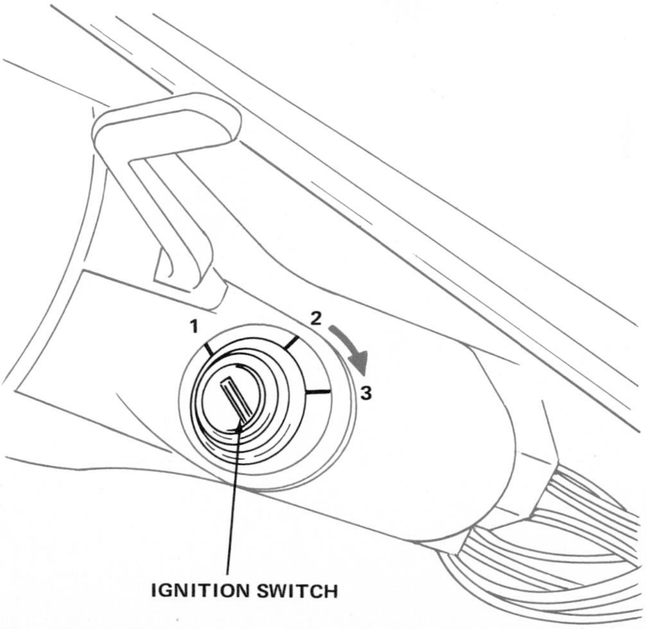

Fig. 17-20 The ignition switch on the Rabbit has three positions, OFF, ON, and START. (*Volkswagen*)

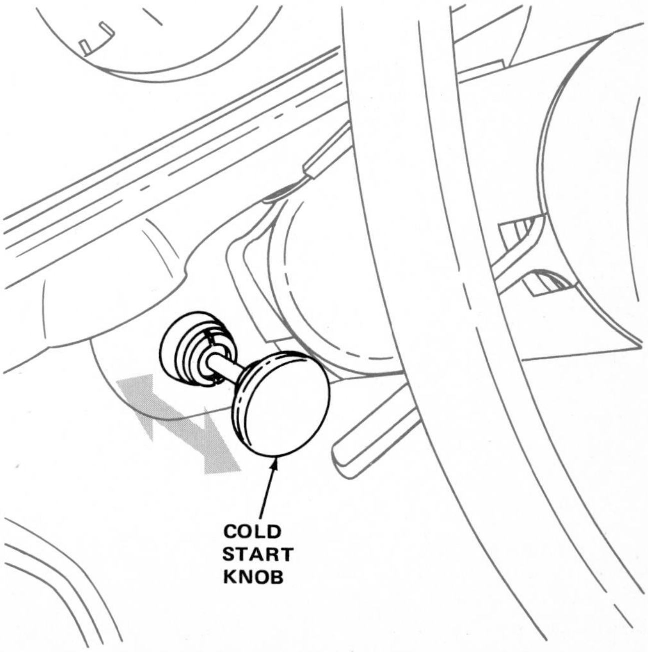

Fig. 17-21 Location of the cold-start knob. (*Volkswagen*)

pre-glow again and after the glow-plug light goes off, try another start.

● 17-11 Block Heater

For very cold weather operation, where temperatures get down to zero or below [−18°C or under], block heaters are often used to assist in starting. This type of heater has an electrical element that works off house current (115 volts). It is located in the engine block and has a special electric cord that is plugged into a regular electric outlet.

CAUTION: Make sure to plug into a three-prong outlet. The third prong is the ground and is essential to protect you from electric shock. If the electrical cord connected to the heater is not long enough, do not use an ordinary extension cord with only two prongs. This is not heavy enough and does not have the ground wire in it.

The length of time that the block heater should be plugged in depends on the type of oil used and the temperature. See the chart (Fig. 17-22). The notation OIL NOT RECOMMENDED means that you should not use the oil indicated for the temperature shown.

Do not use starting "aids" such as ether, gasoline, or similar materials, in the air intake. These so-called aids can actually delay starting and they can also damage the engine.

● 17-12 Vacuum Pump

Because there is no throttle valve or venturi in the air stream, there is no vacuum source that can be tapped in the diesel-engine fuel system. Therefore, the diesel engine requires a vacuum pump to provide the vacuum to operate various devices used on the modern automobile, such as power brakes and air-conditioner vacuum doors. On the General Motors diesel engine we have been discussing, the vacuum pump is located at the back of the engine (Fig. 17-8).

ENGINE BLOCK HEATER USAGE*			
OIL	32° to 0° F (0° to −18° C)	0° to −10° F (−18° to −23° C)	Below −10° F (Below −23° C)
30W	Two Hours Minimum	Eight Hours or Overnight	Oil Not Recommended
15W−40	Not Required	Two Hours Minimum	Eight Hours or Overnight
10W−30	Oil Not Recommended	Not Required	Eight Hours or Overnight

Fig. 17-22 Chart showing proper usage of the engine block heater. (*Oldsmobile Division of General Motors Corporation*)

REVIEW QUESTIONS

1. Compare the four strokes of the diesel and the gasoline engine and explain the differences between them.
2. What is the purpose of glow plugs?
3. Explain the basic differences between the gasoline fuel-injection systems and the diesel-engine fuel-injection system.
4. Why must the diesel-engine fuel-injection system inject the fuel at high pressure?
5. Name and describe the two basic types of centrally located fuel-injection pumps.
6. Explain how the rotary distributor pump works.
7. How do the governor and the advance mechanisms work?
8. Describe the starting instructions for an automobile using a diesel engine.
9. What is a block heater and how does it work?
10. Why must a car with a diesel engine have a separate vacuum pump?

SELF PROJECTS

Whenever you can raise the hood on a diesel-equipped car, examine the engine carefully. Note the locations of the various components of the fuel system. Make sketches showing these locations, to file in your notebook. Study the shop manuals issued by car manufacturers covering their diesel engines. Write down how the fuel systems work. Make sketches showing the locations of the various components for filing in your notebook.

CHAPTER 18
Diesel-Engine Fuel-Injection-System Service

After you have read this chapter, you should be able to:

- **Explain how to remove and replace the components of the diesel-engine fuel-injection systems used in automobiles.**
- **Explain how to remove and replace a fuel-injection pump.**
- **Following instructor's directions, remove and replace components of a diesel fuel-injection system such as hydraulic lines, pump, and nozzles.**

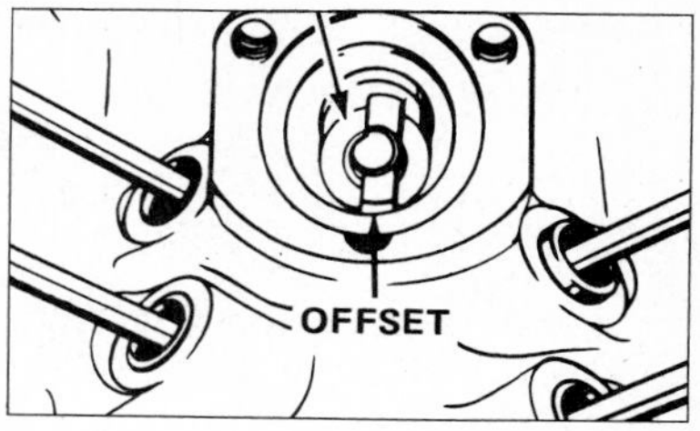

The fuel-injection system is so closely bound up with the diesel engine that it is difficult to separate the servicing of the fuel system from the servicing of the engine. For that reason, we confine ourselves in this chapter to the servicing of the fuel-injector nozzles and hydraulic lines and fittings. Servicing the fuel-injection pump is a specialty that requires special tools and training, and for this reason we do not cover it in this book. Instead, we recommend unit replacement of a pump that has developed trouble. As a rule, the old pump can be traded in on a new or rebuilt unit.

● 18-1. Hydraulic Lines and Fittings

The hydraulic lines are special. They must withstand pressures of several thousand pounds per square inch and must be noncorrosive. If a hydraulic line requires replacement, it should be replaced with the appropriate service part as recommended by the engine manufacturer.

Figure 14-5 shows various types of fittings, or couplings. When disconnecting a line of the type shown in Fig. 14-6, use two wrenches. Hold the fitting with one wrench and turn the coupling nut. If you try to loosen the nut without using the backup wrench, you can twist the line and damage it. It is not necessary to use the backup wrench when disconnecting the lines from the fuel-injection pump.

Whenever lines are disconnected, the lines, nozzles, and pump fittings must be capped. This will prevent dirt from entering the fuel system. Cleanliness is of utmost importance when working on the fuel system. A dirt particle so tiny as to be almost invisible to the naked eye can clog an injector.

The hydraulic lines supplied as service items are preformed, and care should be used in installing them to avoid twisting or bending them out of shape. If the line that is to be replaced is under other lines, you may have to remove these upper lines.

● 18-2 Fuel Filter

It is of utmost importance for the fuel to be clean and free of contaminants. Tiny, almost invisible specks of dirt can clog the fuel nozzles or damage the injector-pump parts. The fuel filter is, therefore, an essential part of the system. Figure 18-1 shows its location on one engine. It works like the filter used in engine lubricating systems. That is, it contains a cartridge of filtering material (special pleated paper or fiber mat) through which the fuel must pass. The filter traps any particles and keeps them from entering the fuel system. The filter should be replaced at the specified intervals. The interval recommended by the manufacturer varies from one company to another. Always follow the recommendations.

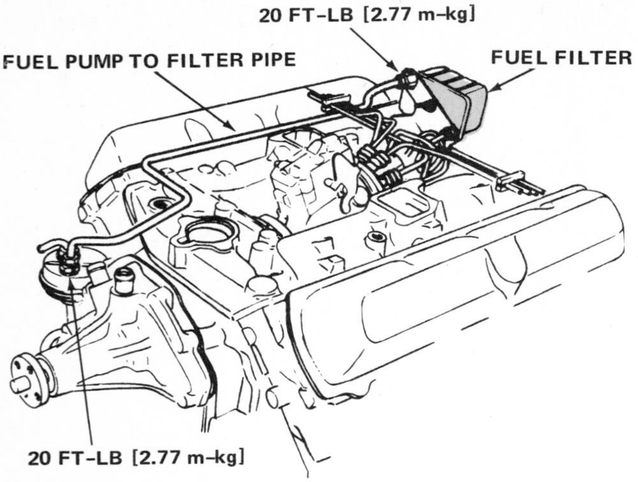

Fig. 18-1 Location of fuel filter on a V-8 engine. (*Chevrolet Motor Division of General Motors Corporation*)

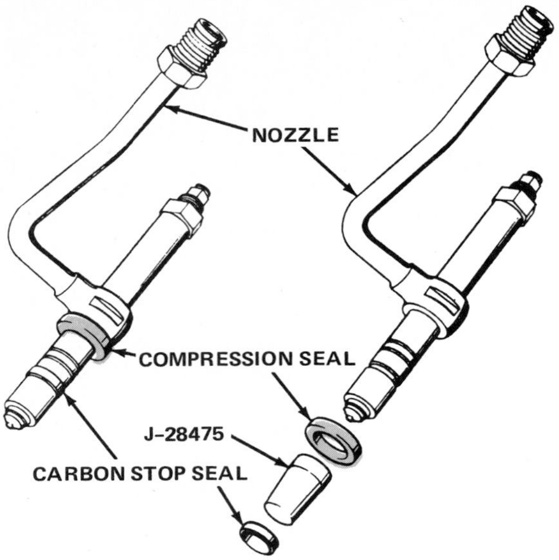

Fig. 18-2 Injector nozzles with seals. (*Chevrolet Motor Division of General Motors Corporation*)

18-3 Injector Nozzle

Injector nozzles should not be removed unless there is evidence that they require servicing or replacement. Usual indications of trouble include:

One or more cylinders knocking
Loss of power
Smoky black exhaust
Engine overheating
Excessive fuel consumption

One way to check for a faulty fuel injector is to run the engine at fast idle. Then loosen the connector at each nozzle in turn, one at a time. Wrap a cloth around the connection before you loosen it to keep fuel from spurting out. If loosening a connector causes the engine speed to drop off, the injector is probably working okay. If the engine speed remains the same, then the injector is probably not performing properly. It could be clogged so that no fuel flows through. Or the holes could be partly clogged so that the spray is inadequate or does not have the required pattern.

A variety of nozzles are used, but all have a check valve. The check valve opens when spray pressure is applied so the fuel can flow through. When the pressure drops, the check valve closes to shut off the flow rapidly and completely. To remove an injection nozzle, first remove the fuel return line clamps and return line. Then remove the nozzle hold-down clamp and spacer or other connector arrangement. Remove the nozzle. Cap the nozzle inlet line and the tip of the nozzle. Figure 18-2 shows one installation arrangement which includes a compression seal and a carbon stop seal.

Some manufacturers recommend a spray test of the detached nozzle. This requires a special hydraulic pump which has a pressure gauge. You attach the nozzle to the pump and work the pump. The fuel should spray when the pump pressure reaches the specified valve. When the pressure is released, the spray should stop abruptly and the nozzle should not drip.

CAUTION: Direct the spray from the nozzle into a suitable container. Do not allow the spray to hit your skin. The pressure is high enough to force fuel oil through the skin. This can cause serious trouble. You can be seriously injured because the oil could cause infections or could get into the blood stream and produce a general infection.

If the nozzle does not work properly, it can be disassembled and cleaned. Some manufacturers recommend unit replacement. If you do disassemble a nozzle, work carefully to avoid damaging the tip or enlarging the holes.

18-4 Injector-Pump Service

We now cover the removal and replacement of the injector pump used on the General Motors V-8 diesel engine (Fig. 17-8). This is one of several types of injection pump and the procedure that follows cannot be said to be completely typical. That is, there will be variations in the procedure from one model diesel engine to another. Always follow the special instructions supplied by the engine manufacturer.

The servicing of the pump is a specialized activity, requiring special training and tools. For this reason, the service manuals supplied by General Motors do not carry any disassembly-assembly information on the pump.

To remove the pump, first remove the following:

1. Air cleaner, filters, and pipes from valve covers and air crossover (Fig. 18-3). Cap intake manifold with special screened covers.
2. Disconnect the throttle rod and return spring (Fig. 18-4).
3. Remove bell crank and throttle cable (from intake-manifold brackets), and push the cable away from the engine.

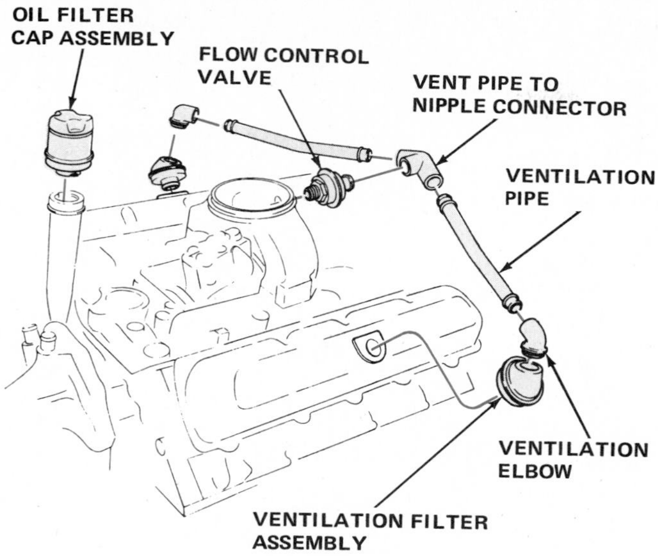

Fig. 18-3 Ventilation system for crankcase. (*Chevrolet Motor Division of General Motors Corporation*)

4. Remove lines to fuel filter and then remove filter and bracket (Fig. 18-1). Disconnect the fuel line at the fuel pump. Disconnect the fuel-return line from the injection pump.
5. Use two wrenches (Fig. 14-8) and disconnect hydraulic lines from the fuel nozzles. Remove the three nuts attaching the injection pump. This takes a special tool. Now, remove the pump and cap all open lines and nozzles.

To replace the pump, remove the protective caps. Line up the offset tang on the pump drive shaft with the pump-driven gear and install the pump (Fig. 18-5). Attach it with three nuts and lock washers, lightly run down on the studs. Connect the hydraulic lines to the nozzles, using two wrenches (Fig. 14-8).

Align the mark on the injector pump with the line on the adapter and tighten the attaching nuts. Use a $\frac{3}{4}$-inch end wrench on the boss at the front of the injection pump to aid in rotating the pump to align the marks.

Install the fuel line from the fuel pump to the fuel filter. Adjust the throttle rod. Install bell crank and throttle return spring.

Start the engine and check for fuel leaks. Tighten connections as necessary. Then install the air crossover (Fig. 18-3) and air cleaner.

REVIEW QUESTIONS

1. Should you rebuild the old pump or trade it in on a new one?
2. Can you bend the hydraulic lines to make them fit?
3. What special precaution should you observe when testing a fuel injector?

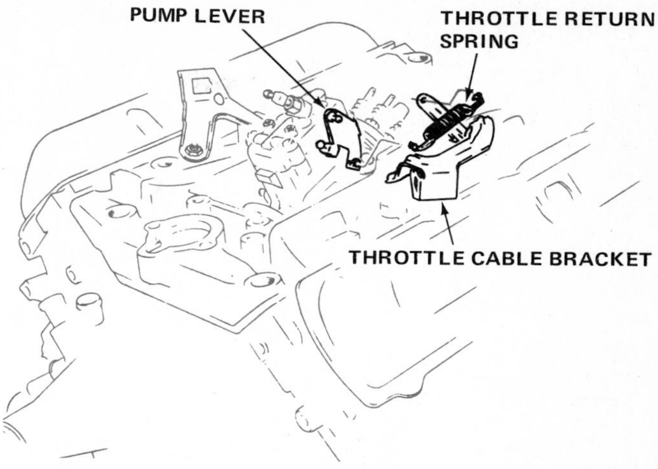

Fig. 18-4 Location of throttle return spring. (*Chevrolet Motor Division of General Motors Corporation*)

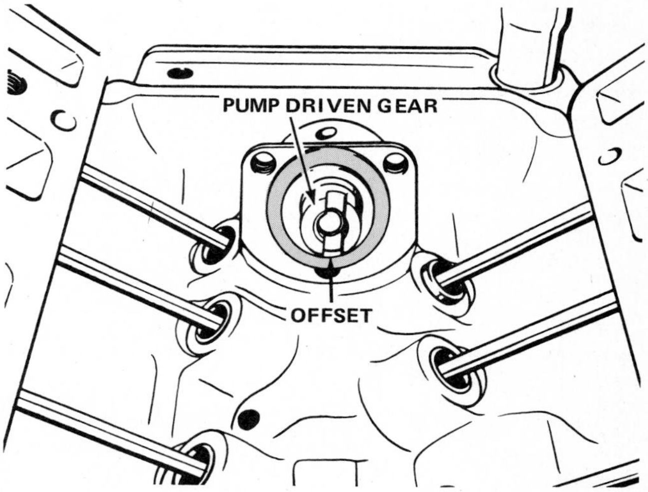

Fig. 18-5 Location of offset tang on the pump-driven gear. (*Chevrolet Motor Division of General Motors Corporation*)

4. What is the purpose of the check valve in the nozzle?
5. Explain how to remove and replace the rotary distributor pump.

SELF PROJECTS

Study as many different shop manuals as you can which cover the servicing of diesel-engine fuel-injection systems. As you study each manual, make notes of the important points explaining how the components work and how they are tested and serviced. File these notes in your notebook. Many magazines carry explanatory stories on the automotive diesel engines and their fuel-injection systems. Clip these articles and file them in your notebook.

CHAPTER 19
Engine Lubricating System

After you have read this chapter, you should be able to:

- **Explain the six jobs the lubricating oil does in the engine.**
- **Discuss the properties of a good lubricating oil and the additives it should have.**
- **Explain the service ratings of lubricating oil.**
- **Explain how a typical lubricating system works.**

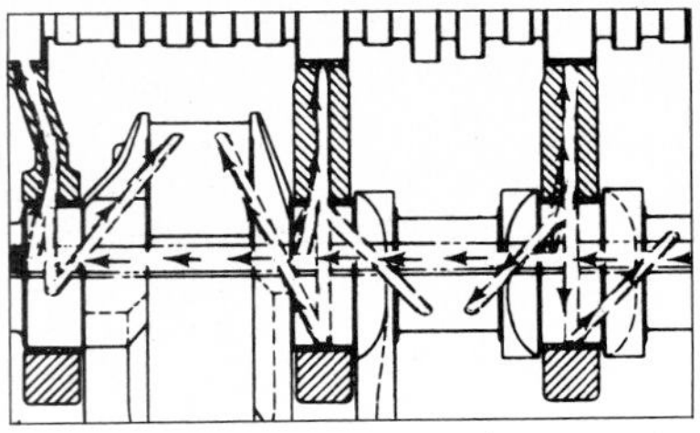

This chapter describes the purpose and properties of engine lubricating oils and the jobs they do in the engine lubricating system. Also, typical lubricating systems are described. We have already discussed, in earlier chapters, engine friction, engine bearings, and the action of the piston rings in controlling oil on the cylinder walls. Now, we apply this information in our discussions of lubricating oil and engine lubricating systems.

● 19-1 Purpose of Lubricating System

Lubricating oil reduces wear and friction between moving surfaces. However, the lubricating oil circulating through the engine performs other jobs. The lubricating oil must:

1. Lubricate moving parts to minimize wear
2. Lubricate moving parts to minimize power loss from friction
3. Remove heat from engine parts by acting as a cooling agent
4. Absorb shocks between bearings and other engine parts, reducing engine noise and extending engine life
5. Form a good seal between piston rings and cylinder walls
6. Act as a cleaning agent

1 and 2. Minimizing wear and power loss from friction Friction has been discussed in some detail (● 10-7). Friction in the engine is normally viscous friction, that is, the friction between moving layers of oil. If the lubricating system does not function properly, sufficient oil will not be supplied to moving parts. Greasy or even dry friction will result between moving surfaces. This would cause considerable power loss, since power must be used to overcome these types of friction. At worst, major damage would occur to engine parts as greasy or dry friction developed. Bearings would wear with extreme rapidity. The heat resulting from dry or greasy friction would cause bearing failure, so that connecting rods and other parts would be broken. Insufficient lubrication of cylinder walls would cause rapid wear and scoring of walls, rings, and pistons. A properly operating engine lubricating system supplies moving parts with enough oil so friction is of the viscous type only.

3. Removing heat from engine parts Engine oil circulates rapidly through the engine lubrication system. All bearings and moving parts are bathed in streams of oil. In addition to lubricating, the oil absorbs heat from engine parts and carries it back into the oil pan. The oil pan absorbs heat from the oil, transferring it to the surrounding air. The oil thus acts as a cooling agent.

4. Absorbing shocks between bearings and other engine parts As the piston approaches the end of the compression stroke, the mixture in the cylinder is ignited. Pressure in the cylinder suddenly increases many times. A load of as much as 2 tons [1,814 kg] is suddenly placed on the top of a 3-inch [76.2-mm] piston. This sudden increase in pressure causes the piston to thrust down hard through the piston-pin bearing, connecting rod, and connecting-rod bearing. There is always some space, or clearance, between bearings and journals; this space is filled with oil. When the load suddenly increases as described above, the layers of oil between bearings and journals must act as cushions. They must resist penetration or "squeezing out." A film of oil must remain between metal surfaces. In thus absorbing and cushioning the hammerlike effect of the sudden loads, the oil quiets the engine and reduces wear of parts.

5. Forming a seal between piston rings and cylinder walls Piston rings must form a gastight seal with the cylinder walls. The lubricating oil that is delivered to the cylinder walls helps the piston rings to accomplish this. The oil film on the cylinder walls makes up for microscopic unevenness in the fit between the rings and walls. The film fills in any gaps through which gas might escape. The oil film also lubricates the rings, so that they move easily in the ring grooves and on the cylinder walls.

6. Acting as a cleaning agent The oil, as it circulates, tends to wash off and carry away dirt, carbon, and other foreign matter. The oil picks up this material and carries it back to the crankcase. There, larger particles drop to the bottom of the oil pan. Smaller particles are removed from the oil by the oil filter.

● 19-2 Properties of Oil

A satisfactory engine lubricating oil must have certain characteristics, or properties. It must have proper viscosity (body and fluidity). It must resist oxidation, carbon formation, corrosion, rust, extreme pressures, and foaming. Also, it must act as a good cleaning agent, must pour at low temperatures, and must have good viscosity at very high and low temperatures.

No mineral oil, by itself, has all these properties. Lubricating-oil manufacturers therefore put a number of additives into the oil during the manufacturing process. An oil for severe service may have many additives:

1. Usually a viscosity-index improver
2. Pour-point depressants
3. Oxidation inhibitors
4. Corrosion inhibitors
5. Rust inhibitors
6. Foam inhibitors
7. Detergent-dispersants
8. Extreme-pressure agents

Here are some of the properties of oil and some oil additives:

1. Viscosity (body and fluidity) Viscosity is the most important property of lubricating oil. Viscosity refers to the tendency of oil to resist flowing. In a bearing and journal, layers of oil adhere to the bearing and journal surfaces. These layers must move, or slip, with respect to each other. The viscosity of the oil determines the ease with which this slipping can take place.

Temperature influences viscosity. Increasing temperature reduces viscosity. That is, it makes the oil thin out. Decreasing temperature causes oil viscosity to increase. The oil thickens.

2. Viscosity numbers Viscosity of oil is determined by use of a *viscosimeter.* This device determines the length of time required for a definite amount of oil to flow through an opening of a definite size. Temperature is taken into consideration during this test. High temperature decreased viscosity. Low temperature increases viscosity. In referring to viscosity, oils with lower numbers are of lower viscosity (thinner). The Society of Automotive Engineers (SAE) rates oil viscosity in two different ways, for winter and for other than winter. Winter-grade oils are tested at 0°F [−18°C]. There are three grades, SAE5W, SAE10W, and SAE20W. The "W" indicates winter grade. For other than winter use oils, tested at 210°F [99°C], the grades are SAE20, SAE30, SAE40, and SAE50, all without the "W."

3. Viscosity index (VI) Viscosity index is a number indicating how much the viscosity of an oil changes with its temperature. When oil is cold, it is thicker and runs more slowly than when it is hot. It is more viscous when it is cold. The engine is harder to start when it is cold because the oil is more viscous. In recent years, oil chemists have developed viscosity-index improvers. These compounds tend to hold oil viscosity constant, whether cold or hot. Thus, the oil makes cold starting easier and yet does not thin out too much. Oils with these characteristics are called multiple-viscosity oils. For example, an oil may be rated SAE10W-30. This means that the oil is the same as SAE10W when cold, and SAE30 when hot.

4. Pour-point depressant This is an additive that depresses, or lowers, the temperature at which the oil becomes too thick to flow. This additive keeps the oil fluid at low temperatures for cold-weather starts.

5. Resistance to carbon formation Cylinder walls, pistons, and rings operate at temperatures of several hundred degrees. This temperature is high enough to cause the oil to break down and form carbon. The less carbon in the engine cylinders, the better. So oil chemists regulate the refining process to make sure that lubricating oil has good resistance to forming carbon.

6. Oil oxidation When oil is heated and then stirred up—as happens in the crankcase—oxygen in the air tends to combine with, or oxidize, the oil. As oil oxidizes, various harmful substances can form, including some that are like tar, and others like varnish. To prevent this, additives are put into the oil to inhibit oxidation.

7. Corrosion and rust inhibitors At high temperatures, especially if there is excessive blow-by, acids may form in the oil which can corrode engine bearings and other parts. Corrosion inhibitors are added to the oil to inhibit this corrosion. Also, rust inhibitors are added. These displace water from metal surfaces so that oil coats them. They also neutralize acids.

8. Foaming resistance The churning action in the engine crankcase also causes engine oil to foam. This is like an egg beater causing an egg white to form a frothy foam. As the oil foams up, it tends to overflow, or to be lost through the crankcase ventilator (● 12-14). In addition, foaming oil does not provide normal lubrication of bearings and other moving parts. Foaming oil in hydraulic valve lifters causes them to function poorly, work noisily, wear rapidly, and possibly break. To prevent foaming, antifoaming additives are mixed with the oil.

9. Detergent-dispersants Despite the filters and screens at the carburetor and crankcase ventilator, dirt does get into the engine. In addition, as the engine runs, the combustion processes leave deposits of carbon on piston rings, valves, and other parts. Also, some oil oxidation may take place, resulting in still other deposits. Then, too, metal wear in the engine puts particles of metal into the oil. As a result, deposits tend to build up on and in engine parts. The deposits reduce the performance of the engine and speed up wear of parts. To prevent or slow down the formation of these deposits, some engine oils contain a detergent additive.

The detergent acts much like ordinary hand soap. When you wash your hands with soap, the soap surrounds the particles of dirt on your hands. They become detached from your hands so that the water can rinse them away. In a similar manner, the detergent in the oil loosens and detaches the deposits of carbon, gum, and dirt. The oil then carries the loosened material away. The larger particles drop to the bottom of the crankcase. Smaller particles remain suspended in the oil. These impurities, or contaminants, are flushed out when the oil is changed.

To prevent the particles from clotting, and to keep them in a finely divided state, a dispersant is added to the oil. Without the dispersant, the particles would tend to collect and form larger particles. These larger particles might then block the oil filter and reduce its effectiveness. They could also build up in oil passages and plug them, thus depriving bearings and other engine parts of oil. The dispersant prevents this. It thus greatly increases the amount of contaminants the oil can carry and still function effectively.

Lubricating-oil manufacturers now place more emphasis on the dispersant qualities of the additive than on its detergent qualities. If the contaminants can be kept suspended as small particles, they will not deposit on engine parts. There is thus less need of detergent action.

10. Extreme-pressure resistance The modern automotive engine subjects the lubricating oil to very high pressures in the bearings and in the valve train. Modern valve trains have heavy valve springs and high-lift cams. This means that the valves must move farther against heavier spring loads. To prevent the oil from squeezing out, extreme-pressure additives are put into the oil. They react chemically with metal surfaces to form very strong, slippery films which may be only a molecule or so thick. Thus, they supplement the oil by providing protection during extreme pressure.

● 19-3 Sludge Formation

Sludge is a thick, creamy, black substance that often forms in the crankcase. It clogs oil screens and oil lines, preventing normal circulation of lubricating oil to engine parts. This can result in engine failure from oil starvation.

1. How sludge forms Water collects in the crankcase in two ways. First, water is formed as a product of combustion. Second, the crankcase ventilating system (described in ● 12-14) carries air, with moisture in it, through the crankcase. If the engine parts are cold, the water condenses and drops into the crankcase. There, it is churned up with the lubricating oil by the action of the crankshaft. The crankshaft acts much like a giant egg beater. It whips the oil and water into the thick, black, mayonnaiselike "goo" known as *water sludge*. The black color comes from dirt and carbon.

2. Why sludge forms If a car is driven for long distances each time it is started, water in the crankcase quickly evaporates. The crankcase ventilating system then removes the water vapor. Thus, no sludge will form. However, if the engine is operated when cold most of the time, then sludge will form. For example, the home-to-shop-to-home sort of driving is sludge-forming. When a car is used for short-trip start-and-stop driving, the engine never has a chance to warm up enough to get rid of the water. The water remains and forms sludge.

3. Preventing sludge To prevent sludge, the car must be driven long enough for the engine to heat up and get rid of the water in the crankcase. This means trips of 12 or more miles in winter (but fewer miles in summer). If trips of this length are impractical, then the oil must be changed frequently. Naturally, during cold weather, it takes longer for the engine to warm up. Thus, in cold weather, the trips must be longer, or oil must be changed more often, to prevent sludge formation.

● 19-4 Service Ratings of Oil

We have already mentioned that lubricating oil is rated by viscosity number (SAE 10W-30, for example). Lubricating oil is also rated in another way, by what is called its *service designation*. That is, it is rated according to the type of service for which it is best suited. There are five service ratings for gasoline-engine lubricating oils: SA, SB, SC, SD, and SE. There are four service ratings for diesel-engine lubricating oils: CA, CB, CC, and CD. The oils differ in their properties and their additives.

1. SA oil This oil is for utility gasoline and diesel engines operating under mild conditions so that protection by additives is not required. This oil may have pour-point and foam depressants.
2. SB oil This oil is for service in gasoline engines operated under such mild conditions that only minimum protection by additives is required. Oils designed for this service have been used since the 1930s. They provide only antiscuff capability and resistance to oil oxidation and bearing corrosion.
3. SC oil This oil is for service typical of gasoline engines in the 1964–1967 models of passenger cars and trucks. It is intended primarily for use in passenger cars. This oil

provides control of high- and low-temperature engine deposits, wear, rust, and corrosion.

4. SD oil This oil is for service typical of gasoline engines in passenger cars and trucks beginning with 1968 models. This oil provides more protection from high- and low-temperature engine deposits, wear, rust, and corrosion than the SC oils.

5. SE oil This oil is for service typical of gasoline engines in passenger cars and some trucks beginning with 1972 (and some 1971) models. This oil provides more protection against oil oxidation, high-temperature engine deposits, rust, and corrosion than do oils with the SC and SD ratings.

Diesel-engine oils must have different properties than oils for gasoline engines. The CA, CB, CC, and CD ratings indicate oils for increasingly severe diesel-engine operation. For example, CA oil is for light-duty service. CD oil is for severe-duty service typical of high-speed, high-output diesel engines.

Modern high-speed automotive diesel engines require a special "combination" type of lubricating oil. For example, General Motors specifies the oil for their V-8 passenger-car diesel engine should be "first line" oil which has the service designation SE/CD marked on the can. They specifically state you *must not* use oil labeled only SE or CD. This oil will not do the job and will probably cause engine damage. They also state that a single-viscosity grade oil such as SAE 20W or SAE 30 is better for their V-8 automotive diesel than multiviscosity oils for sustained high-speed driving.

All car manufacturers recommend the use of a high-detergent engine oil. A high-detergent oil is designated by HD on the can. Here, HD means "high detergency" as well as "heavy duty."

Do not confuse viscosity and service ratings of oil. Some people think that a high-viscosity oil is a "heavy-duty" oil. This is not necessarily so. Viscosity ratings refer to the thickness of the oil; thickness is not a measure of heavy-duty quality. Remember that there are two ratings, viscosity and service. Thus, an SAE10 oil can be an SC, SD, or SE oil. And oil of any other viscosity rating can have any one of the service ratings.

● 19-5 Oil Changes

From the day that fresh oil is put into the crankcase, it begins to lose its effectiveness as an engine lubricant. This gradual loss of effectiveness is largely due to contamination. For instance, water sludge may collect, as already noted (● 19-3). In addition, during engine operation, carbon tends to form in the combustion chamber. Some of this carbon gets into the oil. Gum, acids, and certain lacquerlike substances may also be left by the combustion of the fuel. Or they may be produced in the oil itself by the high engine temperatures. In addition, the air that enters the engine (in the air-fuel mixture) carries a certain amount of dust. Even when the air filter is operating efficiently, it will not remove all dust. Then, too, the engine releases fine metal particles as it wears.

All these substances tend to circulate with the oil. As the mileage piles up, the oil collects more and more of these contaminants. Even though the engine has an oil filter, some of these contaminants remain in the oil. Finally, after many miles of operation, the oil is so loaded with contaminants that it is not safe to use. Unless it is drained and clean oil is put in, engine wear increases.

Modern engine oils are made to fight contamination. They contain certain chemicals (called *additives*) which fight corrosion and foaming. They help keep the engine clean by detergent action. Yet they cannot keep the oil in good condition forever. After many miles of service the oil is bound to become contaminated, and it should be changed. The additives wear out or are used up.

Different automotive manufacturers have different recommendations on how often engine oil should be changed. For example, Chrysler Corporation recommends, for their 1977–1978 line of cars, an oil change every 6 months or 7,500 miles [12,067 km] whichever comes first. For severe service (extended periods of idling and short-trip operation, dusty driving conditions, towing trailers, and the like), the oil should be changed at least twice as often (3 months or 3,750 miles [6,033 km]).

Ford has, for their 1978 cars, the same recommendations for the intervals between oil changes as Chrysler.

General Motors has gone to a more extended period for many of their cars. Their recommendation for many of their 1978 automobiles is to change oil every 12 months or 7,500 miles [12,067 km], whichever occurs first. The exception to this is the new V-8 diesel engine being installed in some Oldsmobiles, Cadillacs, and other cars. On these, the recommendation is to change oil every 3,000 miles [4,827 km].

The oil filter should be changed at the first oil change on a new engine, and *every other* oil change after that. However, for severe service, when the oil is being changed more often, the filter should be changed *every* oil change.

● 19-6 Oil Consumption

Oil is lost from the engine in three ways: by burning in the combustion chambers, by leakage in liquid form, and by passing out of the crankcase in the form of a mist. (In the closed-crankcase ventilating system, this mist will also be burned in the combustion chambers. See ● 12-14.) Two main factors affect oil consumption, *engine speed* and *the amount that engine parts have worn.*

High speed produces high temperature. This, in turn, lowers the viscosity of the oil. Now it can more readily work past the piston rings into the combustion chamber, where it is burned. In addition, the high speed exerts a centrifugal effect on oil feeding through the crankshaft to the connecting-rod journals. Thus, more oil is fed to the bearings and thrown on the cylinder walls. Also, high speeds cause "ring shimmy." With this condition, the oil-control rings cannot function effectively. Crankcase ventilation (● 12-14) causes more air to pass through the crankcase quickly. This causes oil to be lost as mist.

As engine parts wear, oil consumption increases. Worn bearings tend to throw more oil onto the cylinder walls. Tapered and worn cylinder walls prevent normal oil-control-ring action. The rings cannot change shape rapidly enough to conform with the worn cylinder walls as the rings move up and down. More oil thus gets into the combustion chamber, where it burns and fouls spark plugs, valves, rings, and pistons. Carbon formation worsens the condition, since it further reduces the effectiveness of the oil-control rings. Where cylinder-wall wear is not excessive, special oil-control rings can be used to reduce oil consumption. They improve the wiping action so that less oil can move past the rings. After cylinder walls have worn beyond a certain point, the cylinders must be machined and new rings installed to reduce oil consumption.

Worn intake-valve guides also increase oil consumption. Oil leaks past the valve stems and is pulled into the combustion chamber along with the air-fuel mixture every time the intake valves open. Worn exhaust-valve guides can also cause high oil consumption. In this case, the oil is burned as the hot exhaust gases hit it when the exhaust valve opens. Installation of new valve guides, reaming of guides and installation of valves with oversize stems, or installation of valve-stem seals will reduce oil consumption from these causes.

19-7 Types of Lubricating Systems

Two types of lubricating systems have been used on four-cycle engines. They are the splash and pressure-feed systems. Two-cycle engines require a different kind of lubrication.

1. Splash In the splash lubricating system, oil is splashed from the oil pan into the lower part of the crankcase. Usually, the connecting rod has a dipper that dips into the crankcase oil each time the piston reaches BDC. This splashes the oil. Some small engines also use oil slingers which are driven by the camshaft. These are gearlike parts that throw oil from the oil pan up into the moving engine parts. The splash system is used on most small four-cycle engines for power lawn mowers and similar applications.

2. Pressure feed In the pressure-feed lubricating system, many engine parts are lubricated by oil fed under pressure from the oil pump (Figs. 19-1 and 19-2). The oil from the pump enters an oil line (or a drilled header, or channel, or gallery, as it is variously called). From the oil line, it flows to the main bearings and camshaft bearings. The main bearings have oil-feed holes or grooves that feed oil into drilled passages in the crankshaft. The oil flows through these passages to the connecting-rod bearings. From there, on some engines, it flows through holes drilled in the connecting rods to the piston-pin bearings.

In I-head engines, oil is fed under pressure to the valve mechanisms in the head. For example, some engines have the rocker arms mounted on hollow shafts. The shafts feed oil to the rocker arms. Some engines with independently mounted rocker arms (Fig. 19-2) have hollow mounting studs. These studs feed oil from an oil gallery in the head to the rocker-arm ball pivots. The oil

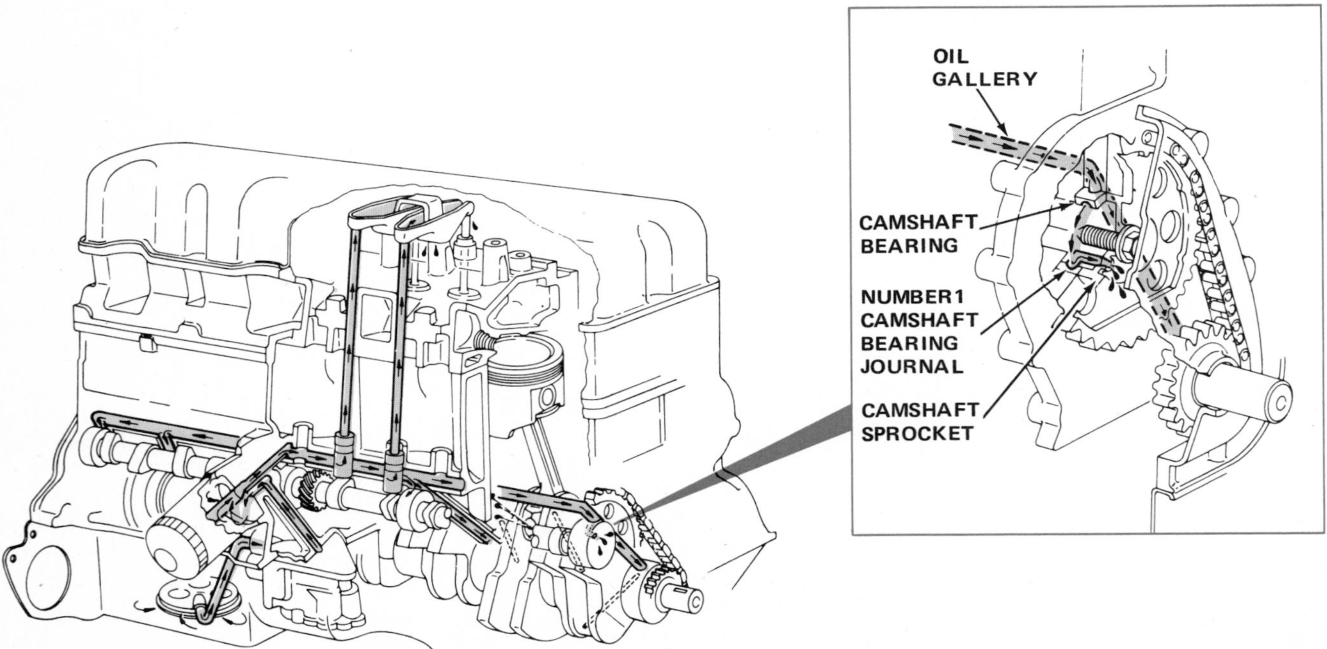

Fig. 19-1 Lubricating system for a six-cylinder, overhead-valve engine with seven main bearings. The boxed insert to the right shows how the camshaft sprocket and chain are lubricated. (*American Motors Corporation*)

spills off the rocker arms and provides lubrication for the valve stems and push-rod and valve-stem tips. Thus, all valve-mechanism parts are lubricated. On other engines, the oil flows up through hollow push rods to lubricate the valve stems and other valve-train parts.

Cylinder walls are lubricated by splashing oil thrown off by the connecting-rod bearings. Some engines have oil-spit holes or grooves in the connecting rods that align with drilled holes in the crankpin journals with each revolution. As this happens, a stream of oil is spit, or thrown, onto the cylinder walls (upper left, Fig. 19-2). On many V-8 engines, the oil-spit holes or grooves are arranged so that they lubricate opposing cylinders in the other cylinder bank. That is, the spit holes in the connecting rods in

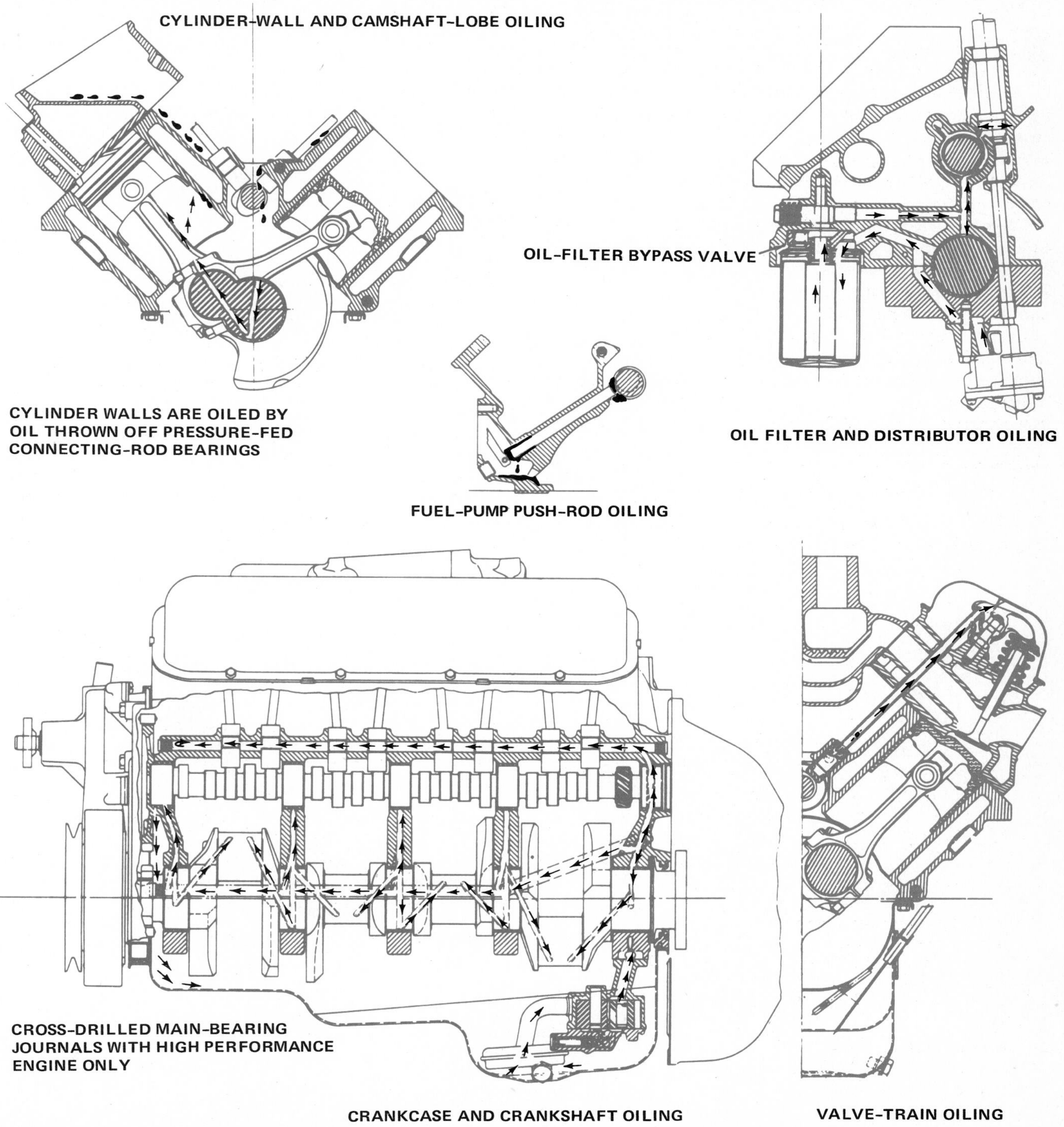

Fig. 19-2 Lubricating system of a V-8 overhead-valve engine. Arrows show the flow of oil to the moving parts in the engine. (*Chevrolet Motor Division of General Motors Corporation*)

the right-hand bank lubricate the cylinder walls in the left-hand bank, and vice versa. In many engines, the piston pins are lubricated with oil scraped off cylinder walls by the piston rings. The pistons have grooves, holes, or slots to feed oil from the oil-control-ring groove, or from oil scoops on the piston, to the piston-pin bushings.

● 19-8 Two-Cycle Engines

In two-cycle engines, the air-fuel mixture passes through the crankcase on its way from the carburetor to the engine cylinders. For this reason it is not possible to maintain a reservoir of oil in the crankcase. The oil would be picked up by the passing air-fuel mixture, carried to the engine cylinders, and burned. Therefore, to provide lubrication of two-cycle engine parts, the oil is mixed with the fuel. The air and oil-fuel mixture enter the crankcase. The fuel, being more volatile, evaporates and passes on to the engine as an air-fuel mixture. Some of the oil is carried along with the air-fuel mixture and is burned. But enough oil is left behind to keep the moving engine parts coated with oil.

Some two-cycle engines, for example, many used in motorcycles, have an oil feed to the carburetor (Fig. 19-3). The oil is not mixed with the gasoline for these engines. Instead, the oil enters the air-fuel mixture in the carburetor. The system meters the oil so that the proper amount is added to assure proper lubrication of the engine under all operating conditions. This is important for motorcycle engines which operate at greatly varying speed. A standard gasoline-oil mix such as is used for a power mower, for example, is not satisfactory. The system shown in Fig. 19-3, which is typical for motorcycle engines, varies the amount of oil entering the air-fuel mixture. It provides more oil at high engine speed, thus satisfying operating requirements.

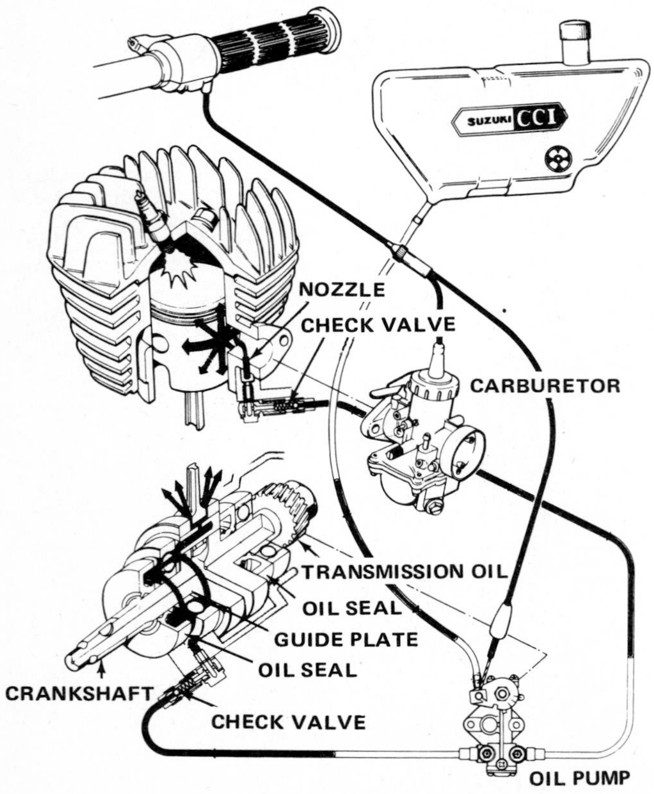

Fig. 19-3 Lubricating system for a one-cylinder, two-cycle motorcycle engine. (*Suzuki Motor Company, Ltd.*)

● 19-9 Overhead-Camshaft-Engine Lubrication

In the overhead-camshaft engine, additional lubrication must be furnished the cylinder head for the camshaft bearings. Figure 19-4 shows the lubricating system for the Opel overhead-camshaft engine. Note the oil gallery running the length of the cylinder head. It supplies oil to the camshaft bearings. The gallery also supplies oil to the hydraulic valve lifters and the valve-train parts.

● 19-10 Oil Pumps

The two general types of oil pumps used in pressure-feed lubricating systems are shown in Figs. 19-5 and 19-6. The gear-type pump uses a pair of meshing gears. As the gears rotate, the spaces between the gear teeth are filled with oil from the oil inlet. Then, as the teeth mesh, the oil is forced out through the oil outlet. The rotor-type pump uses an inner rotor and an outer rotor. The inner rotor is driven and causes the outer rotor to turn with it. As this happens, the spaces between the rotor lobes become filled with oil. When the lobes of the inner rotor move into the spaces in the outer rotor, oil is squeezed out through the outlet.

Oil pumps are usually driven from the engine camshaft, by the same spiral gear that drives the ignition distributor (Fig. 19-7). The oil intake for the oil pump is attached to a float in many engines. This floating intake takes oil only from the top of the oil in the oil pan. Since dirt particles sink, the top oil is cleanest.

● 19-11 Relief Valve

To keep the oil pump from building up too much pressure, a relief valve is included in the lubricating system. The valve is a spring-loaded ball (Fig. 19-5) or a spring-loaded plunger. When the pressure reaches the preset value, the ball or plunger is moved against its spring. It opens a port through which oil can flow back to the oil pan. Enough oil flows past the relief valve to prevent excessive pressure. The oil pump can normally deliver much more oil than the engine requires. This is a safety factor that assures delivery of enough oil under extreme operating conditions.

19-12 Oil Cooler

Some engine lubricating systems have oil coolers. Oil coolers are used on almost all automotive air-cooled engines. One type consists of a small radiator, mounted on the side of the engine block. Oil and water (coolant) circulate through the radiator. The coolant comes from the engine cooling system. As the coolant circulates through the oil cooler, it picks up heat. The heat is carried

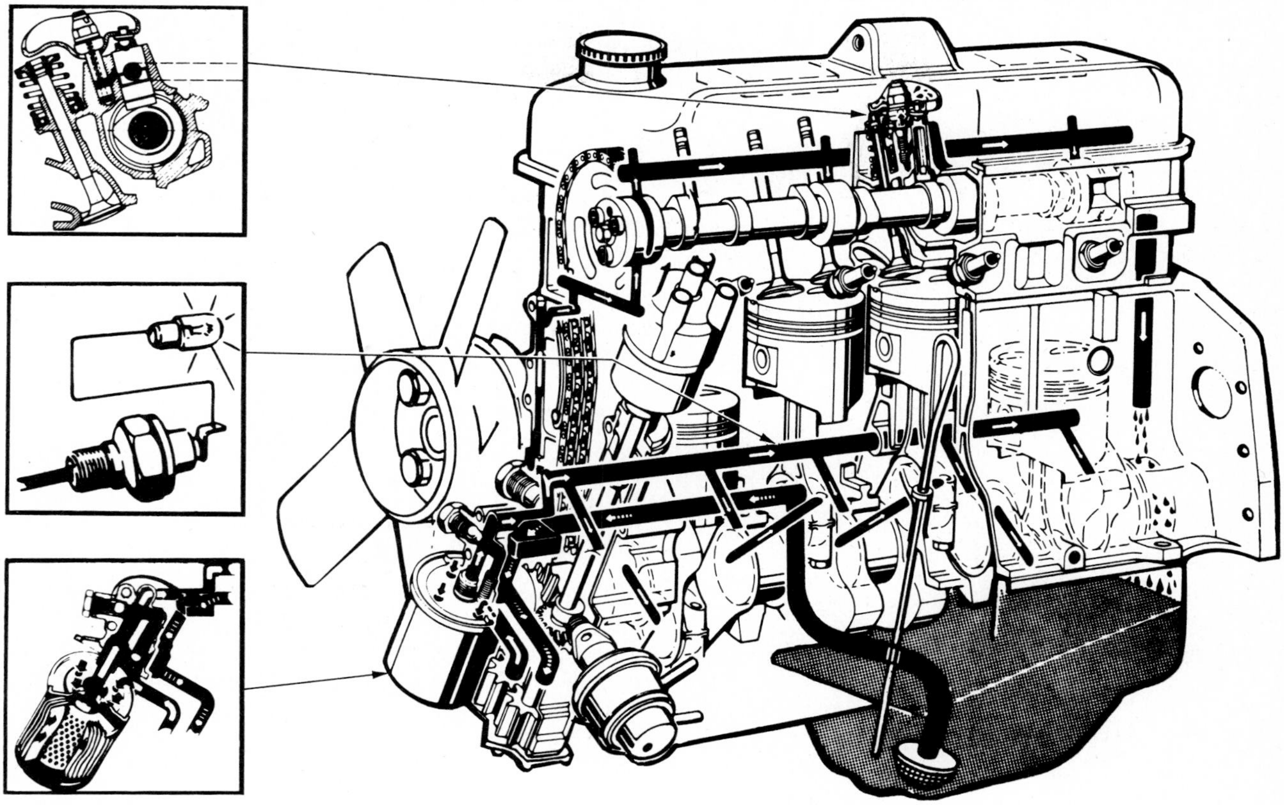

Fig. 19-4 Lubricating system for a four-cylinder, overhead-camshaft engine. Arrows show the flow of oil to the moving engine parts. (*Buick Motor Division of General Motors Corporation*)

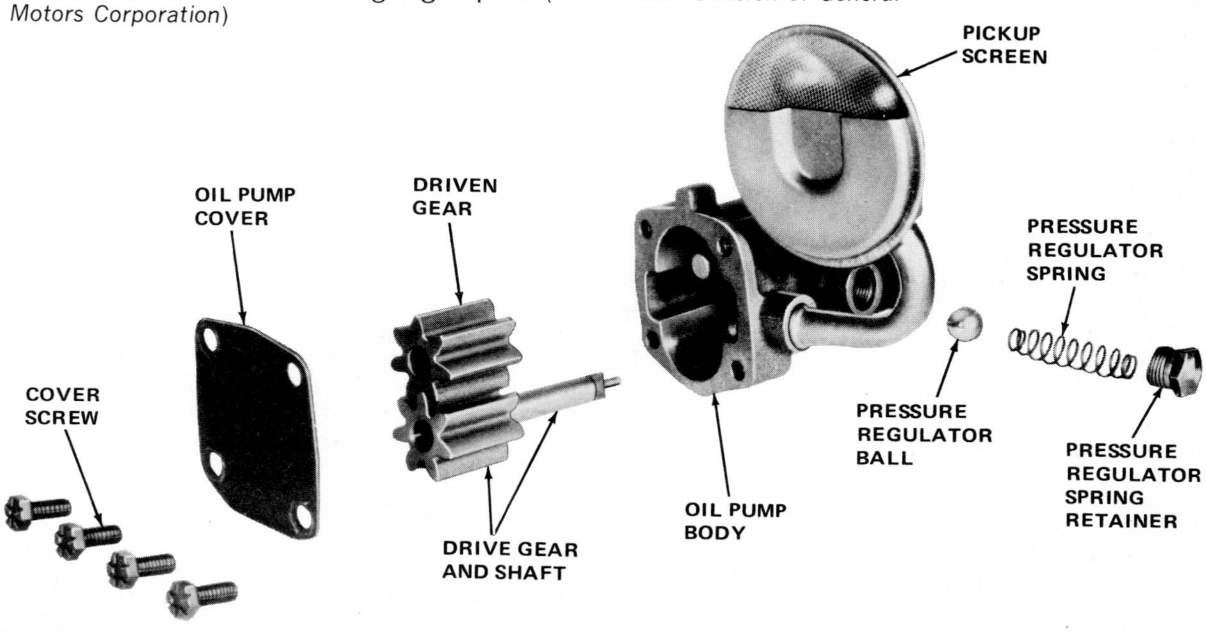

Fig. 19-5 Disassembled gear-type oil pump. (*Pontiac Motor Division of General Motors Corporation*)

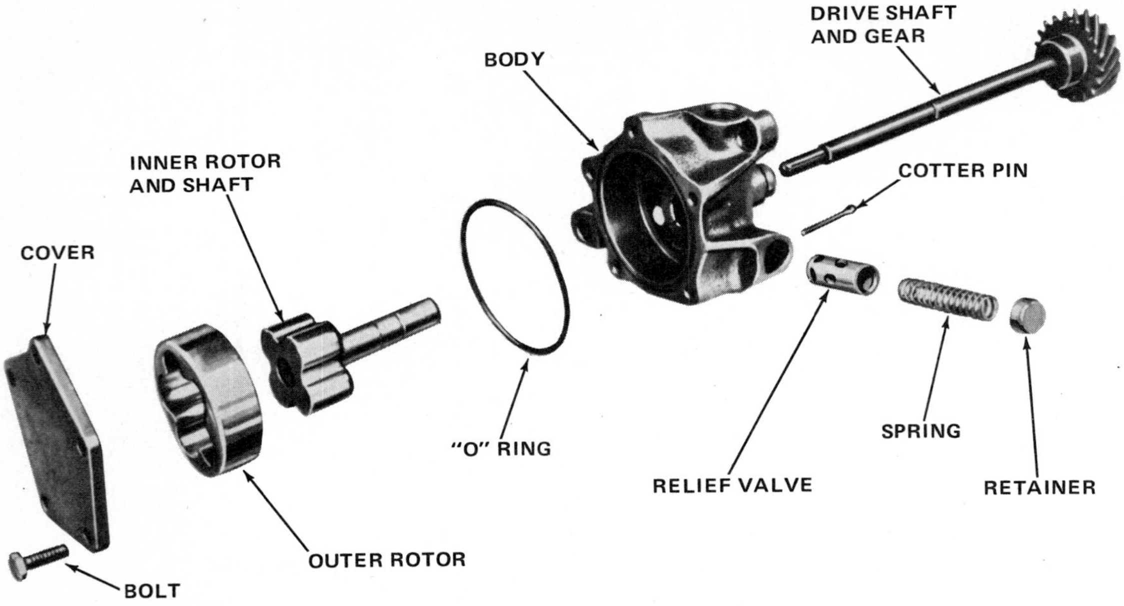

Fig. 19-6 Disassembled rotor-type oil pump. (*Chrysler Corporation*)

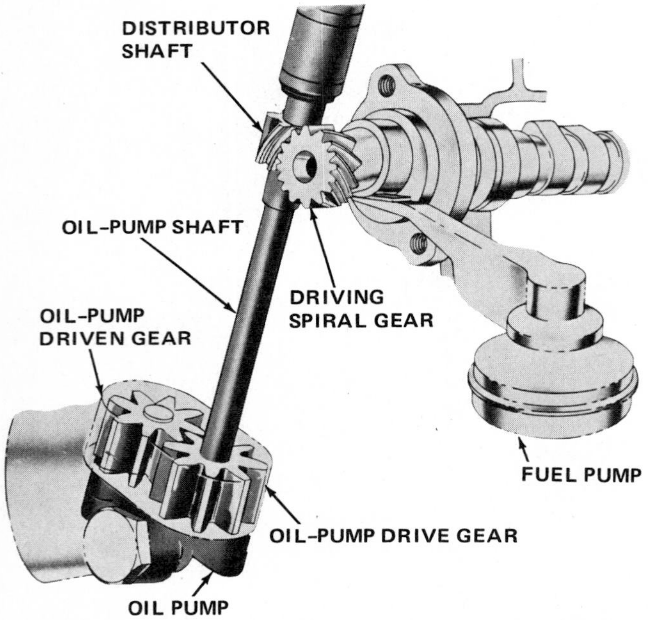

Fig. 19-7 Oil-pump, distributor, and fuel-pump drives. The oil pump is the gear type. A gear on the end of the camshaft drives the distributor. An extension of the distributor shaft drives the oil pump. The fuel pump is driven by an eccentric on the camshaft. (*Buick Motor Division of General Motors Corporation*)

to the radiator, where it is passed on to air circulating through the radiator. This process helps to cool the oil and keep it at a workable temperature. Another design uses a small section of the cooling system radiator, so that an extra radiator is not required. The type used in air-cooled engines consists of a small radiator much like the radiators used in liquid-cooling systems. (See Chapter 21.)

● 19-13 Oil Filters

All automotive-engine lubricating systems have an oil filter. Some or all of the oil from the oil pump circulates through this filter. In the filter is a cartridge of filtering material that traps particles of foreign matter. The filter thus helps to keep the oil clean and to prevent particles from entering the engine. Filters are of two types. Those which filter part of the oil from the oil pump are called *bypass* filters. Those which filter all the oil in circulation through the system are called *full-flow* filters. The full-flow filter includes a spring-loaded bypass valve. It protects the engine against oil starvation in case the filter becomes clogged. When this happens, the valve is opened by increased pressure from the pump trying to push oil through. With the valve opened, oil bypasses the filter. The engine is thus assured of sufficient oil. However, the filter element should be replaced every so often, so that the filter maintains filtering efficiency. Figure 19-8 is a cutaway view of a filter.

● 19-14 Oil-Pressure Indicators

The oil-pressure indicator tells the driver what the oil pressure is in the engine. This gives warning if something in the lubrication system prevents delivery of oil to vital parts. Oil-pressure indicators are of three general types: pressure-expansion, electric-resistance, and indicator light. The latter two are the more commonly used.

1. Pressure-expansion The pressure-expansion indicator uses a hollow Bourdon (curved) tube, fastened at one end and free at the other. The oil pressure is applied to the curved tube through an oil line from the engine. It causes the tube to straighten out somewhat as pressure in-

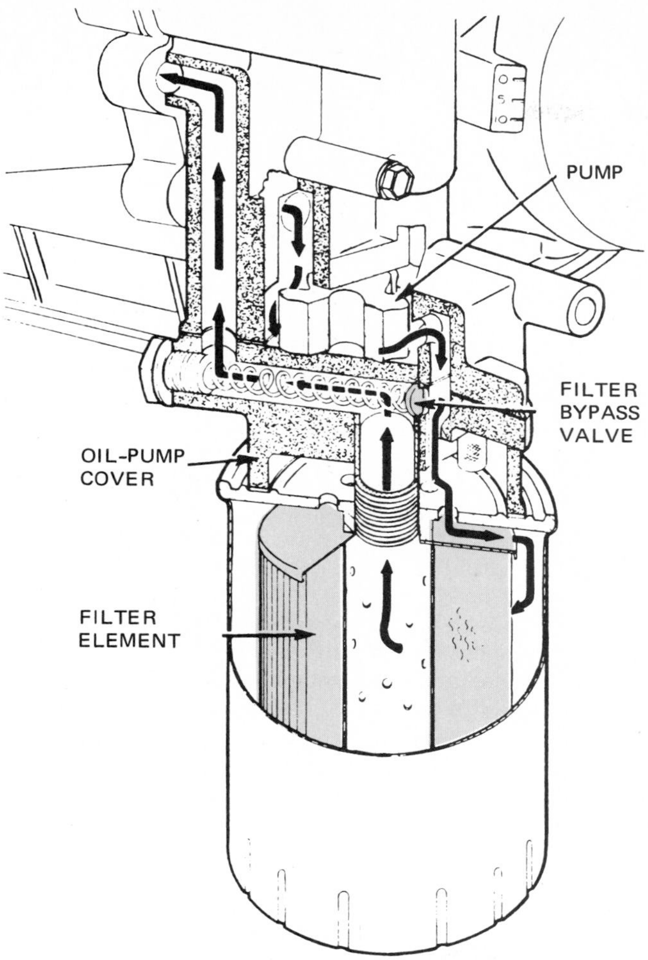

Fig. 19-8 Cutaway view of a full-flow oil filter with bypass valve. (*Buick Motor Division of General Motors Corporation*)

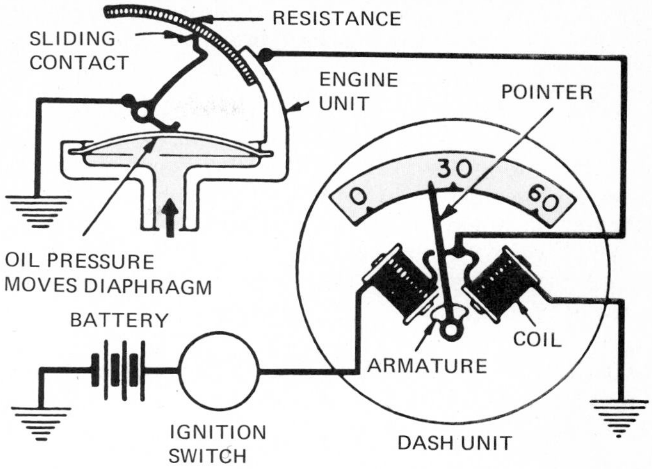

Fig. 19-9 Electric circuit of an electric-resistance oil-pressure indicator.

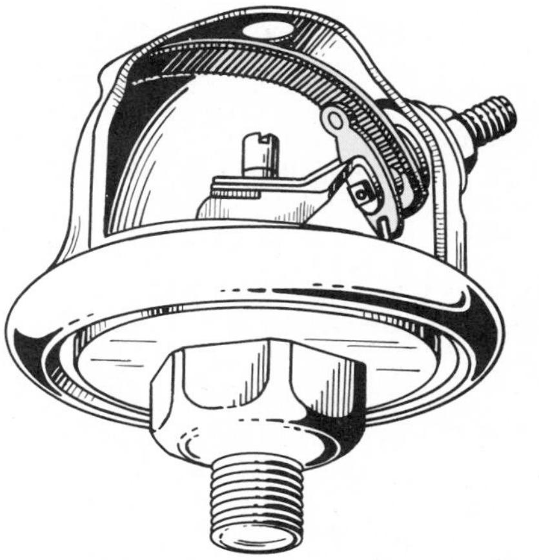

Fig. 19-10 Cutaway view of the engine unit of the oil-pressure indicator.

creases. This movement is transmitted to a needle by linkage and gears from the end of the tube. The needle moves across the face of a dial and registers the amount of oil pressure.

2. Electric Electrically operated oil-pressure indicators are of two types, the balancing-coil type and the bimetal-thermostat type. The balancing-coil type makes use of two separate units, the engine unit and the indicating unit (Fig. 19-9). It is similar to the balancing-coil fuel gauge, previously covered in ● 12-6. The engine unit consists of a variable resistance and a movable contact. The contact moves from one end of the resistance to the other as oil pressure against a diaphragm varies (Fig. 19-10). As pressure increases, the diaphragm moves inward. This causes the contact to move so that more resistance is placed in the circuit between the engine and indicating units. This reduces the amount of current that can flow in the circuit. The indicating unit consists of two coils that balance each other in a manner similar to electrically operated fuel gauges (● 12-6). In fact, this type of indicator operates in the same manner as the fuel indicator. The only difference is that the fuel indicator uses a float that moves up or down as the gasoline level changes in the gasoline tank. In the oil-pressure indicator, changing oil pressure operates a diaphragm that causes the resistance to change.

The bimetal-thermostat-type oil-pressure indictor is similiar to the bimetal-thermostat fuel gauge (● 12-6). The dash units for both types are practically identical. The engine unit is similar to the engine unit of the balancing-coil oil-pressure gauge.

3. Indicator light Instead of a gauge, many vehicles have an oil-pressure indicator light. The light comes on when the ignition is turned on and the oil pressure is low. Normally, after the engine has started and oil pressure has built up, the light goes off. If it does not, then the engine and lubricating system should be checked at once to find the cause of the low oil pressure. The light is connected to a pressure switch in the engine. The switch is closed except when oil pressure increases to normal values. The indicator light and the pressure switch are connected in series to the battery, through the ignition switch. When the ignition switch is turned on, the indictor light comes on. It stays on until the engine starts and the oil pressure builds up enough to open the pressure switch.

● 19-15 Oil-Level Indicators

To determine the level of the oil in the oil pan, an oil-level stick, or "dipstick," is used. The dipstick is placed so that it protrudes down into the oil. The oil level is determined by withdrawing the dipstick and noting how high the oil rises on the dipstick. In the closed or positive crankcase ventilating system (PCV), the dipstick tube is sealed at the top when the dipstick is in place. This keeps unfiltered air from entering the crankcase and crankcase gases from escaping.

● 19-16 Crankcase Ventilation

As mentioned in ● 12-14, air must circulate through the crankcase when the engine is running. This removes the water and liquid gasoline that appear in the crankcase when the engine is cold. Also, it removes blow-by gases from the crankcase. Unless the water, liquid gasoline, and blow-by gases are removed from the crankcase, there will be trouble. Sludge and acids will form. Sludge can clog oil lines and starve the lubricating system. This could mean a ruined engine. Acids corrode metal parts, and this, too, can ruin the engine.

The removal process requires that the engine must first heat up enough to vaporize the liquid water and gasoline. Then the circulating air can remove them, along with the blow-by gases.

In older engines, the crankcase was vented into the open air. The engine had an opening at the front into the crankcase and a vent tube at the back. The forward motion of the car and rotation of the crankshaft moved air through the crankcase so the crankcase was ventilated. This is no longer permitted. All modern engines have a closed, or positive, crankcase ventilation (PCV) system. This system carries air through the crankcase, as with the older engines. However, the air is then directed into the intake manifold instead of into the open air.

● 19-17 Other Automotive Lubricants

The modern automobile needs many other lubricants besides engine oil. There are chassis lubricants, automatic-transmission fluids, steering-gear lubricants, and more. Here, we take a brief look at various lubricants and fluids used in automotive vehicles. In your automotive shop work, you will be handling most, if not all, of these lubricants and fluids.

1. Grease This is a fluid lubricant such as mineral oil mixed with a thickening agent to make it a semisolid, or plastic. The thickening agent may be a metallic soap or a nonsoap substance such as clay. The soaps commonly used are lithium, calcium, sodium, aluminum, and barium. Each of these, alone or in combination, gives the grease special characteristics. Aluminum gives the grease good adhesion. Sodium gives the grease a thick, fibrous appearance. A number of additives are also mixed in to improve the performance of the grease.

Among the characteristics a good grease must have are consistency, stability, oxidation resistance, ability to protect against friction, wear, and corrosion, and feedability (ability to flow through dispensing equipment).

2. Fields of use Greases are commonly classified according to their use, as follows:

a. ***Wheel-bearing greases***—greases resistant to high temperatures and separating effect of centrifugal force.

b. ***Universal-joint greases***—greases compounded to stay in place as the universal joints spin and flex.

c. ***Chassis greases***—greases that can be applied with grease guns through fittings. They have the characteristics needed to keep them in place on the moving chassis surfaces without separating or losing lubricating effect.

d. ***Extended-lubrication-interval (ELI) chassis greases***—greases with the composition, structure, consistency, life, and antiwear and anticorrosion characteristics for use in "lifetime" applications, including suspension, driveline, and steering systems having sealed joints, prepacked during manufacture or assembly. They normally do not need relubrication for long intervals.

e. ***Multipurpose greases***—greases compounded to meet the performance requirements for chassis grease, wheel-bearing grease, universal-joint grease, and other automotive uses such as fifth-wheel service.

Some ELI greases are good for multipurpose uses.

f. ***Extreme-pressure (EP) greases***—greases suitable for high-load-carrying applications. Some have a surface-active additive that gives antiwear or antiseize properties beyond those of other greases. "Surface-active" means the agent bonds to metal surfaces to form a barrier that comes into play if normal lubricant film is pierced.

g. ***Other greases*** There are other special greases. Brake grease is specified for the moving parts in the drum-brake mechanisms. Distributor breaker cam grease is specified for the cam in ignition distributors. Speedometer-cable lubricant is another special lubricant.

3. Automatic-transmission fluid There are several types of automatic-transmission fluid. Each model of automatic transmission has special lubrication requirements. For example, there are at least two Dexron® fluids, specified for different transmission models.

4. Power-steering fluid This is another special fluid that meets the special needs of the power-steering unit.

5. Other fluids There are several other fluids used in automobiles, including antifreeze fluid (ethylene glycol) and brake fluid. In service work, a variety of fluids are used to clean or loosen parts. Carburetor cleaning fluid is one example. The manifold heat-control valve solvent is another, used to loosen up the valve if it gets stuck.

The list above may look long and complicated. But remember this—lubricant and fluid makers have tried to make your job easier by supplying the specific substances needed for each service. Also, the automotive manufacturers have done their part. Service manuals contain specific lubrication charts, such as shown in Fig. 19-11. These charts indicate the type of lubricant or fluid to use at every place needing lubrication and the intervals at which these services should be supplied.

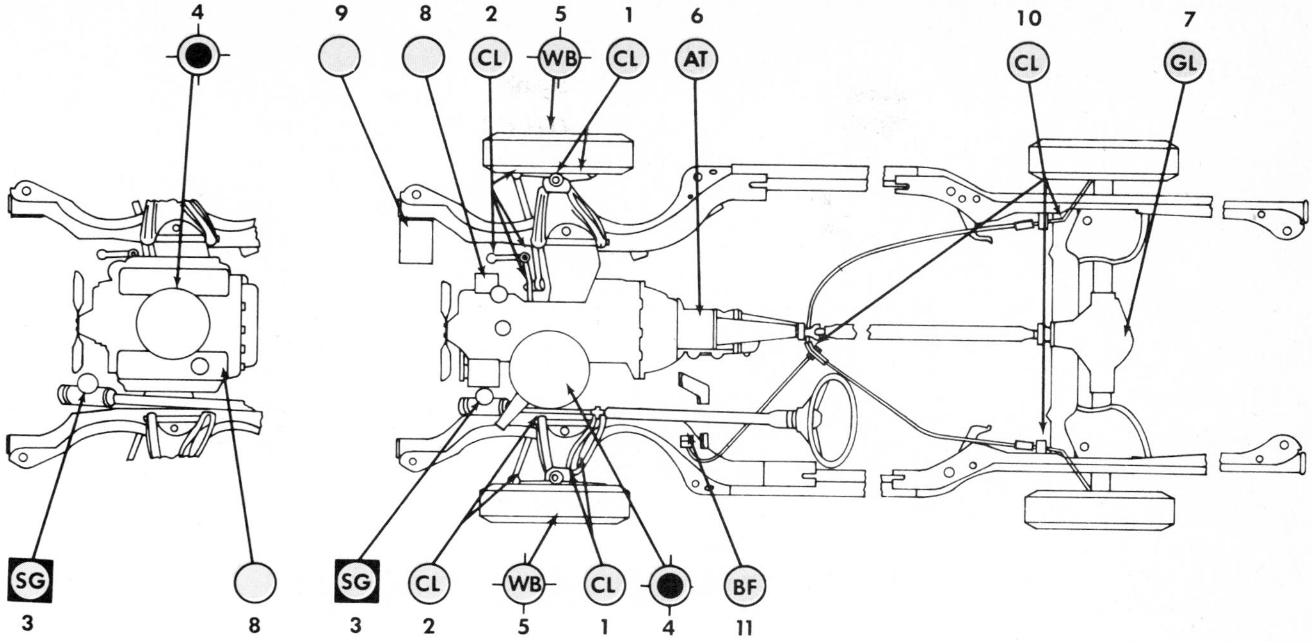

CHECK OR LUBRICATE EVERY 7500 MILES (12 000 km)

REPLACE EVERY 30 000 MILES (48 000 km)

REPACK EVERY 30 000 MILES (48 000 km)

CHECK FOR GREASE LEAKAGE
EVERY 30 000 MILES (48 000 km)

*REFILL POSITRACTION REAR AXLE
WITH SPECIAL LUBRICANT ONLY

GL – MULTI-PURPOSE OR UNIVERSAL GEAR LUBRICANT*
WB – WHEEL BEARING LUBRICANT
CL – CHASSIS LUBRICANT
AT – DEXRON-II® AUTOMATIC TRANSMISSION FLUID
OR EQUIVALENT
BF – BRAKE FLUID
SG – STEERING GEAR LUBRICANT

1. FRONT SUSPENSION
2. STEERING LINKAGE
3. STEERING GEAR
4. AIR CLEANER
5. FRONT WHEEL BEARINGS
6. TRANSMISSION
7. REAR AXLE
8. OIL FILTER
9. BATTERY
10. PARKING BRAKE
11. BRAKE MASTER CYLINDER

Fig. 19-11 Lubrication chart for some Chevrolet models. *(Chevrolet Motor Division of General Motors Corporation)*

REVIEW QUESTIONS

1. Name the six jobs that engine oil must do.
2. Explain how engine oil removes heat from the engine.
3. What is viscosity?
4. Does temperature influence oil viscosity? In what way?
5. What is viscosity index?
6. Explain how sludge forms in the crankcase.
7. Which is more apt to have crankcase sludge, the car driven 50 or more miles each time it is started, or the car that is used only for driving around town? Why?
8. What are the five service ratings for gasoline-engine lubricating oils?
9. Is there any difference between the viscosity rating and the service rating of lubricating oil?
10. What two main engine factors influence oil consumption?
11. Explain why wear of engine parts increases oil consumption.
12. What are the two general types of four-cycle-engine lubricating systems?
13. Explain how the two-cycle engine is lubricated.
14. What are the two general types of oil pumps?
15. What is the purpose of the relief valve in the lubricating system?
16. Describe the path that the oil takes to lubricate the valves in an overhead-valve engine.
17. Describe the path that the oil takes to lubricate the camshaft bearings in a typical overhead-camshaft engine.
18. What is the purpose of the oil filter?
19. Describe the operation of the two types of electric oil-pressure indicators.
20. What is the purpose of the crankcase ventilation system?

SELF PROJECTS

Cans of engine oil have information on the properties of and additives in the oil. Whenever you have a chance, copy this information on sheets of paper to file in your notebook. Check the cans containing oil of different viscosity numbers and service ratings. Build up as complete a file as you can. After you have compiled a file on one brand of oil (Gulf, for example), work on getting a file for other brands (Texaco, Shell, and so on).

CHAPTER 20
Lubricating-System Service

After reading this chapter, you should be able to:

- **Discuss troubles in the lubricating system and explain what might cause them.**
- **Describe lubricating-system servicing procedures, including changing oil and the oil filter, cleaning the oil pan, and the like. Perform these operations in the shop.**

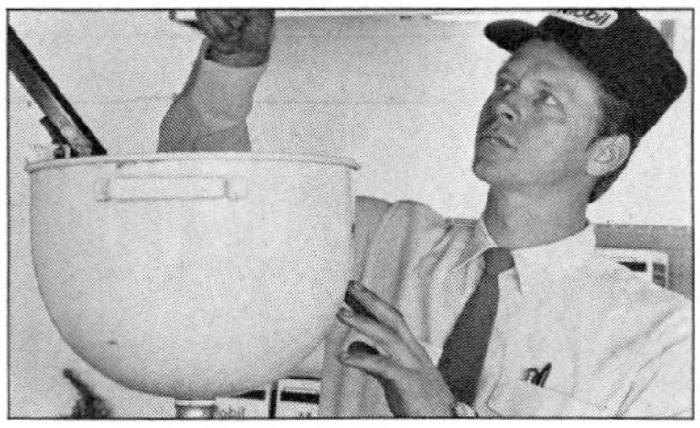

This chapter discusses troubles that may occur in the engine lubricating system. It explains the services required on lubricating-system components. Later chapters discuss engine troubleshooting and servicing procedures.

● 20-1 Trouble Tracing in the Lubricating System

The two complaints you will hear related to the engine lubricating system are:

1. The engine is using too much oil.
2. The indicator light is acting up (or the oil-pressure gauge shows low pressure).

We discussed in ● 19-6 the many possible causes of excessive oil consumption. Two main factors affect oil consumption, engine speed and the amount that engine parts have worn. If oil consumption is high, both of these factors must be considered. Figure 20-1 shows places in the engine where oil may be lost. If the oil is leaking from the engine, due to a faulty gasket, for example, then oil usage will be high.

As previously mentioned (● 19-6) worn bearings can cause high oil usage. A device sometimes used to check for worn bearings is the engine prelubricator (Fig. 20-2). The engine prelubricator is used by engine builders and rebuilders to lubricate all engine parts before starting the engine. The prelubricator sends oil under pressure through the lubricating system. If bearings are worn, then excessive amounts of oil will leak out of the engine.

If a bearing does not leak enough oil, then either the bearing clearance is too low, or an oil passage in the crankshaft or cylinder block is stopped up.

The second complaint was erratic indicator light or oil-pressure gauge action. If the light comes on part of the time, or the pressure gauge sometimes shows low pressure, either there is not enough oil in the crankcase, or else the oil pickup is not consistently picking up oil. This could be due to the oil pickup falling off or having been pushed up by a bent oil pan. The oil pan can become bent if it hits an object as the car is moving forward.

If the light stays on all the time, or the pressure gauge consistently reads low, then the first check is to see if the engine oil is low. If it is at normal height, check the oil-pressure sending unit. Remove it and install a pressure gauge to check for pressure with the engine running. If the pressure is okay, then the trouble is a defective sending unit. If the pressure is low there is other trouble—trouble in the engine itself, probably. Of course, trained mechanics can tell by listening if the oil pressure is low. The clatter of the hydraulic valve lifters and the bearings running without oil warns them that the oil pressure is low. Actually, an engine operating for any length of time without oil pressure is a ruined engine.

Causes of low oil pressure include:

1. A weak relief-valve spring
2. A worn oil pump
3. A broken or cracked oil line
4. Obstructions in the oil lines
5. Insufficient or excessively thin oil
6. Bearings that are so badly worn that they can pass more oil than the oil pump is capable of delivering
7. A defective oil-pressure indicator may be recording low

Excessive oil pressure may result from:

1. A stuck relief valve
2. An excessively strong valve spring
3. A clogged oil line
4. Excessively heavy oil

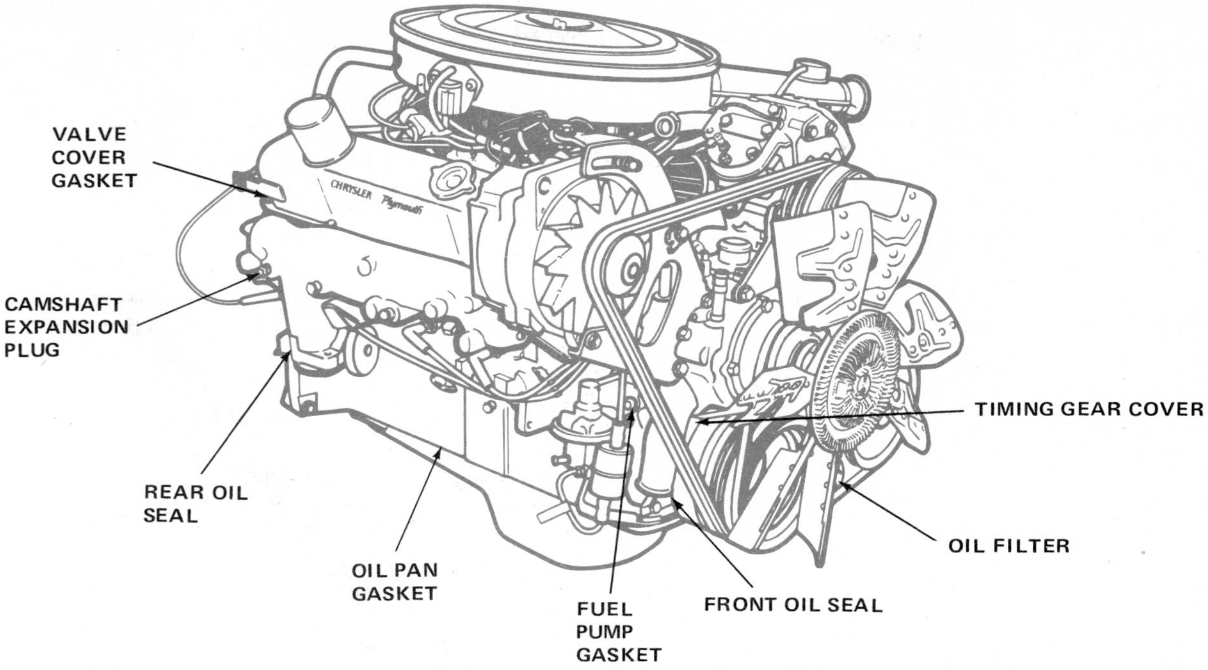

Fig. 20-1 Places where oil leaks might occur. (*Chrysler Corporation*)

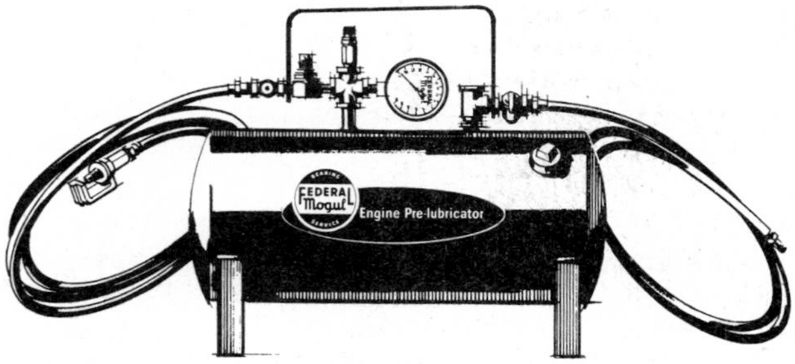

Fig. 20-2 Engine-bearing prelubricator to check main and connecting rods for wear. (*Federal-Mogul-Bower Bearings, Inc.*)

● 20-2 Lubricating-System Service

There are certain lubricating-system jobs that are done more or less automatically when an engine is repaired. For example, the oil pan is removed and cleaned during such engine-overhaul jobs as replacing bearings or rings. When the crankshaft is removed, it is the usual procedure to clean out the oil passages in the crankshaft. Also, the oil passages in the cylinder block should be cleaned out as part of the engine-block service job. All these services are described in later chapters on engine service. Sections that follow describe such lubricating-system service jobs as:

1. Checking oil level
2. Changing oil
3. Servicing the oil-pressure relief valve and the crankcase ventilator valve
4. Changing the oil filter
5. Servicing the oil pump and the oil-pressure indicator.

● 20-3 Checking Oil Level

Most engines use a bayonet type of oil-level gauge (the dipstick). It can be withdrawn from the crankcase to determine the oil level in the crankcase. The gauge should be withdrawn, wiped clean, reinserted, and again withdrawn so that the oil level on the gauge can be seen. The gauge is usually marked to indicate the proper oil level. The appearance of the oil should be noted to see whether it is dirty, thin, or thick. A few drops of oil can be placed between the thumb and fingers and rubbed to detect dirt. This will also tell you whether the oil has sufficient body, that is, whether it is sticky. If the oil level is low, oil should be added to the crankcase. If the oil is thin or dirty, it should be drained, and clean oil added.

The car should be on a level surface. If it is on a slope, you will get a false reading.

If the engine has just been shut off, wait for a few moments for the oil to drain back down into the oil pan before checking the oil level.

● 20-4 Changing Oil

Up to about 1960, standard practice called for changing the engine oil at 500-, 1,000-, or 2,000-mile [804-, 1,609-, or 3,218-km] intervals, according to the type of operation. With the development of improved lubricating oil and filters, automotive manufacturers have changed their recommendations, as explained in ● 19-5. Always follow the engine manufacturer's recommendations for the time or mileage intervals between oil changes. Note that the filter should be changed the first time the oil is changed for a new engine, and every other oil change thereafter.

Oil filters installed in the system tend to increase the time between oil changes. But oil changes are still needed. Oil should be changed more frequently during cold weather, particularly if most trips are short. With short-trip operation, the engine operates cold a greater part of the time. This increases the chances for water sludge to form. More frequent oil changes remove this sludge before dangerous amounts can collect.

When the car is operated on very dusty roads, the oil should be changed more frequently. Despite the air filters in the carburetor air cleaner and crankcase ventilator, dust does work its way into the engine. This is particularly true when the car operates in dusty areas. Changing the oil flushes this dust out so that it cannot harm the engine. Manufacturers recommend that a car that has been driven through a dust storm, for example, have the oil changed immediately. It does not matter, in such a case, how recently the last oil change was made. At the same time, the air filter should be cleaned, and the oil filter (if used) changed. When the engine oil is changed, various points in the engine accessories and chassis should be lubricated.

There is nothing special about changing oil. The car is raised on a lift, the oil drain pan is put into position, and the engine drain plug is removed (Fig. 20-3). After the oil is drained, the plug is replaced and the car is lowered. Then the recommended amount and grade of oil is put into the engine. Then start and run the engine and check for leaks.

Note: Always put on a new door-jamb sticker with the car mileage and the date the oil was changed.

● 20-5 Oil-Pan Service

Whenever the oil pan is removed for engine service, it should be cleaned. The oil screen and oil pump should also be cleaned. Before replacing the oil pan, make sure all the old gasket material has been scraped from the pan and block gasket surfaces. Apply new gasket cement (if specified). Lay the gasket, or gaskets, in place. Be sure the bolt holes in the gasket and pan line up. Install the pan, and tighten the bolts to the proper tension.

Fig. 20-3 Draining oil in preparation for adding fresh oil. *(Mobil Oil Company)*

● 20-6 Relief Valve

Relief valves are not usually adjustable, although springs of different tension may be installed to change the regulating pressure. This is not usually recommended, however, since a spring of the proper tension was originally installed on the engine. Any change of pressure is usually brought about by some defect that requires correction. For example, badly worn bearings, may pass so much oil that the oil pump cannot maintain normal pressure in the lines. Installing a stronger spring in the relief valve would not increase oil pressure. The relief valve does not operate under these circumstances.

● 20-7 Oil Filters

Oil filters are serviced by replacing the oil-filter element or the complete filter, according to the type. Oil screens are serviced by flushing out collected sludge and dirt. Where a floating type of oil intake is used, the float and screen should also be cleaned (● 20-5).

As the oil filter becomes clogged, it passes less and less oil. The condition of the oil filter can be determined by feeling it after the engine has been operated for a short time. If the filter is hot to the touch, oil is flowing through the filter. If it is cold, the filter is probably clogged and is not passing oil. However, the best procedure is to replace the filter at intervals. The usual recommendation is to replace the filter every other time the engine oil is changed. More frequent replacement should be made if the car is operated in unusually dusty conditions.

● 20-8 Filter Replacement

On most engines, the filter element and container are replaced as a unit. For example, on the type shown in Fig. 20-4, the old filter can be unscrewed and a new filter screwed into place by hand. A drip pan should be placed under the old filter as it is removed, to catch any oil that runs out. With the old filter off, the recess and sealing face of the filter bracket should be wiped with a clean shop towel. Then, the sealing gasket of the new filter should be coated with oil. Finally, the new filter should be hand-tightened until the gasket comes up against the bracket face. It should then be hand-tightened another half turn. After installation, the engine should be operated at fast idle to check for leaks. Check the oil level in the crankcase and add oil if necessary.

CAUTION: Engine oil should be changed before the new filter is installed. A new filter should always start out with new oil.

On other filters, oil lines must be disconnected and reconnected when they are changed. Some filters have replaceable elements. On these, the procedure is as follows: Remove the drain plug (if present) from the bottom of the housing. Take the cover off by loosening the center bolt or clamp. Lift out the element. If the filter housing has no drain plug, remove the old oil or sediment with a special pump. Wipe the inside of the housing with a clean cloth. Be sure that no traces of lint or dirt remain. Install the new filter element. Replace the plug and cover, using a new gasket. Start the engine, and check for leaks around the cover. Note whether the oil pressure has changed (with a new element, which passes oil more easily, it may be lower). Check the level of oil in the crankcase, and add oil if necessary. Installing a new filter element usually requires an additional quart of oil to bring the oil level up to the proper height in the crankcase. Note also the caution above about changing oil every time the filter is changed.

After a filter element or filter is replaced, the date and the mileage should be marked on the doorjamb sticker and the filter housing. Then, after the specified replacement time or mileage, the driver and technician will know that it is time to replace the filter element again.

● 20-9 Oil Pumps

Oil pumps are simple devices that require little service in normal operation. If a pump is badly worn, it will not maintain oil pressure and should be removed for repair or replacement. In such a case, refer to the manufacturer's shop manual for details of servicing.

● 20-10 Oil-Pressure Indicators

Oil-pressure indicators are discussed in ● 19-14. These units require very little service. Defects in either the dash unit or the engine unit usually require replacement of the defective unit. On the type of unit with a thermostatic blade, dirty contact points may cause incorrect readings. Points can usually be cleaned by pulling a strip of bond paper between them. Be sure that no particles of paper are left between the points. Never use emery cloth to clean the points. Particles of emery might embed and prevent normal indicator action. If the indicator is not functioning normally, a new engine unit may be temporarily substituted for the old one. This will determine whether the fault is in the engine unit or the dash unit.

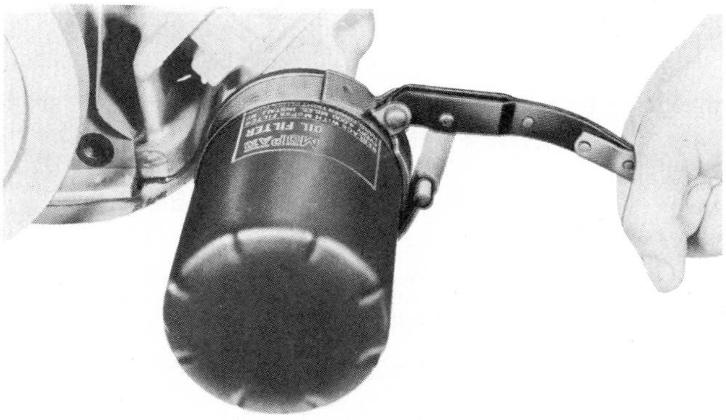

Fig. 20-4 Removing an oil filter. It is usually necessary to use a tool, as shown, to loosen the filter. (*Chrysler Corporation*)

● 20-11 Servicing the Crankcase Ventilator (PCV) Valve

A widely accepted recommendation is to replace this valve every one or two years. If the valve becomes clogged, it can cause engine loping (the engine speeds up and slows down) and rough idle. It cannot be cleaned satisfactorily—a clogged valve must be replaced. With engine loping and rough idle, install a new PCV valve, and see if the idling condition improves. If it does, leave the new valve in. If the loping or rough idle persists, check for restrictions in the lines. Procedures for testing the system with PCV testers are outlined in Part Five of the book, Automotive Emission Controls.

REVIEW QUESTIONS

1. Explain the purpose of the engine prelubricator.
2. Describe the procedure for checking the oil level in the crankcase.
3. Explain how to change engine oil.
4. Name five possible causes of low oil pressure.
5. Explain how to change the oil filter. How often should this be done?
6. Why should engine oil be changed periodically?
7. How will the engine probably react if the PCV valve becomes clogged?

SELF PROJECTS

Make a list of possible troubles in the lubricating system along with their possible causes. File this list in your notebook. Then make a list of the service operations required on the lubricating system, starting with "1. Checking oil level" and continuing through to "Servicing the PCV valve." Jot down any thoughts you have on how each of these services is performed. Refer to the book if you need to. The act of writing down these facts will help you remember them. File the list in your notebook.

CHAPTER 21
Engine Cooling System

After you read this chapter, you should be able to:

- **Discuss the purpose of the cooling system and explain how it works.**
- **Identify on cars and describe the construction and operation of the components in the cooling system, including the radiator, water pump, fan, and thermostat.**

This chapter discusses the construction and operation of automotive-engine cooling systems. The cylinder block and cylinder head have water jackets through which coolant (water plus antifreeze) can circulate. The circulation of coolant, between the water jackets and the cooling-system radiator, removes heat from the engine.

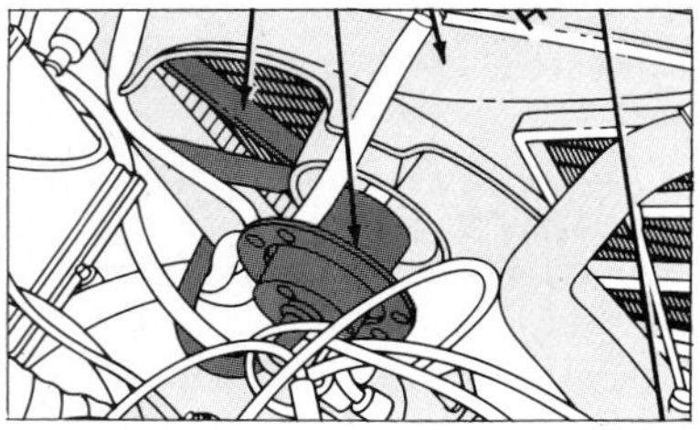

● 21-1 Purpose of Cooling System

The purpose of the cooling system is to keep the engine at its most efficient operating temperature at all engine speeds and all driving conditions. During combustion of the air-fuel mixture in the engine cylinders, temperatures reach as high as 6000°F [3316°C]. Some of this heat is absorbed by the cylinder walls, cylinder heads, and pistons. They, in turn, must be cooled so that their temperatures are not excessive. Cylinder-wall temperature must not increase beyond about 400 or 500°F [204 or 260°C]. Higher temperatures cause the lubricating-oil film to break down and lose its lubricating properties. But it is desirable to operate the engine at temperatures as close as possible to the limits imposed by oil properties. Removing too much heat through the cylinder walls and head would lower engine thermal efficiency (● 10-19). Cooling systems are designed to remove about 30 to 35 percent of the heat produced in the combustion chambers.

The engine is quite inefficient when cold. Therefore, the cooling system includes devices that prevent normal cooling action during engine warm-up. These devices allow the working parts to reach operating temperatures more quickly. They shorten the inefficient cold-operating time. Then, when the engine reaches operating temperature, the cooling system begins to function. Thus, the cooling system cools rapidly when the engine is hot. And it cools slowly or not at all when the engine is warming up or cold.

Two general types of cooling systems are used, air cooling and liquid cooling. Most automotive engines now employ liquid cooling although engines for motocycles, power lawn mowers, and the like, are all air-cooled. Air-cooled engines have metal fins on the heads and cylinders to help radiate the heat from the engine. Cylinders are usually partly or completely separated to improve air circulation around them. Special shrouds and blowers are used on many air-cooled engines to improve air circulation around the cylinders and heads. Figure 6-6 illustrates an air-cooled engine.

Liquid cooling systems usually have a water pump to maintain circulation in the system. Figure 21-1 shows the cooling system for a V-8 engine. The water pump is driven by a belt from the engine crankshaft. It circulates the cooling liquid (called the *coolant*) between the radiator and the engine water jackets. The coolant is water to which antifreeze has been added. Following sections describe the cooling-system components in detail.

● 21-2 Water Jackets

Just as we might put on a jacket to keep warm on a cool day, so are water jackets placed around the engine cylinders. There is this difference: Water jackets are designed to keep the cylinders cool. The water jackets are cast into cylinder blocks and heads. See Fig. 21-2.

● 21-3 Water Pumps

Water pumps are of the centrifugal type. They are mounted at the front end of the cylinder block between

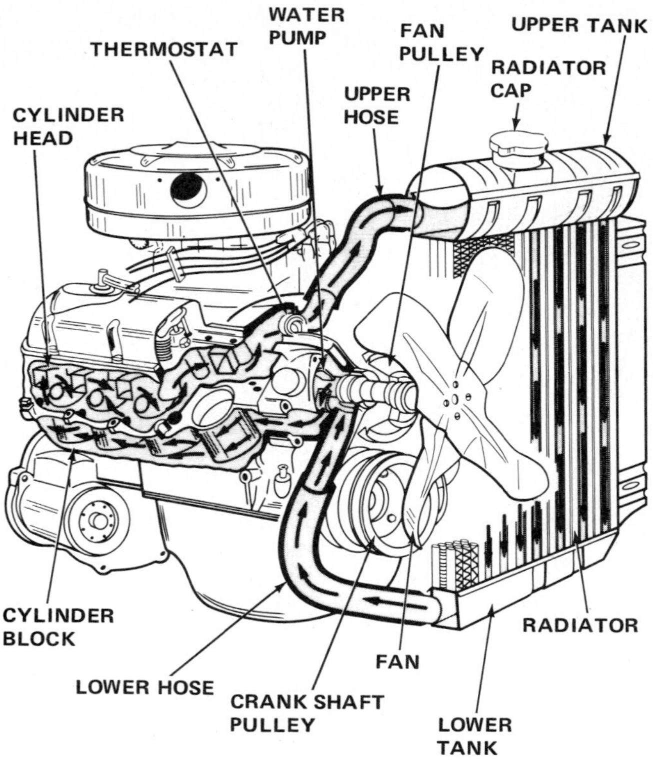

Fig. 21-1 Engine cooling system for a V-8 engine.

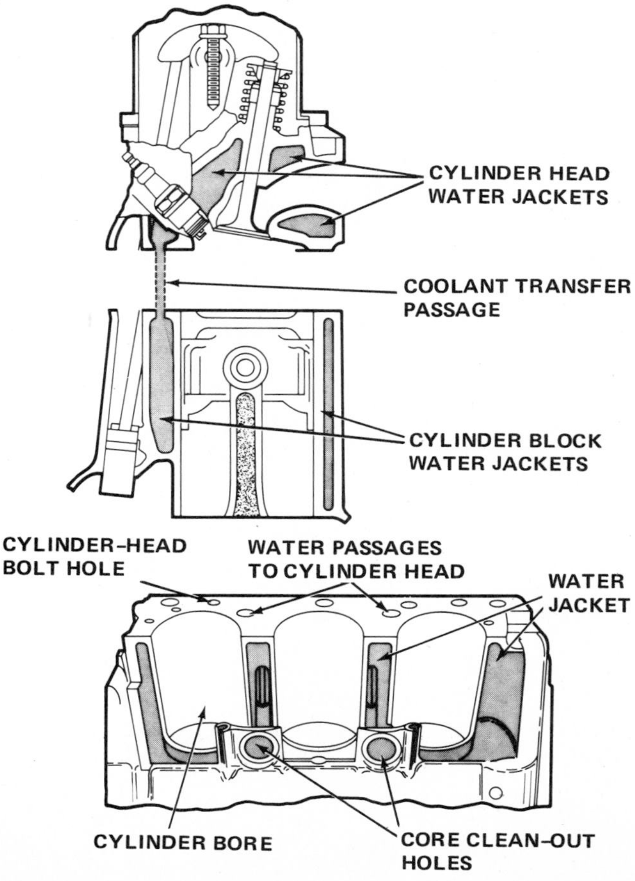

Fig. 21-2 Water jackets in cylinder head and cylinder blocks.

the block and the radiator (Fig. 21-1). The pump (Figs. 21-3 and 21-4) consists of a housing, with a water inlet and outlet, and an impeller. The impeller is a flat plate mounted on the pump shaft with a series of flat or curved blades, or vanes. When the impeller rotates, the coolant between the blades is thrown outward by centrifugal force. The coolant is forced through the pump outlet and into the cylinder block. The pump inlet is connected by a hose to the bottom of the radiator. Coolant from the radiator is drawn into the pump to replace the coolant forced through the outlet.

The impeller shaft is supported on one or more bearings. A seal prevents coolant from leaking out around the bearing. The pump is driven by a belt to the drive pulley, mounted on the front end of the engine crankshaft.

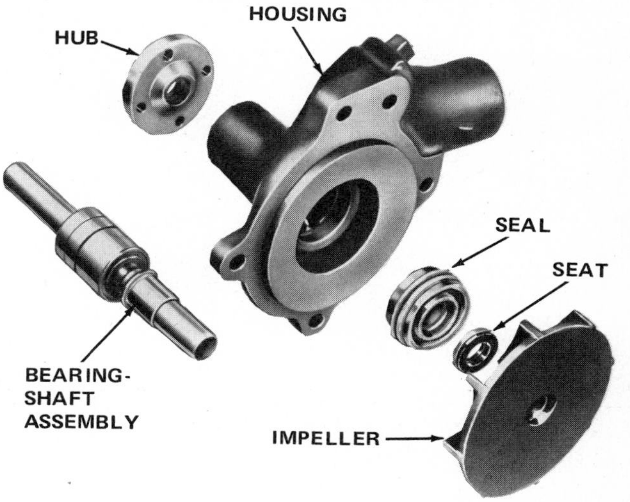

Fig. 21-3 Disassembled water pump. (*Pontiac Motor Division of General Motors Corporation*)

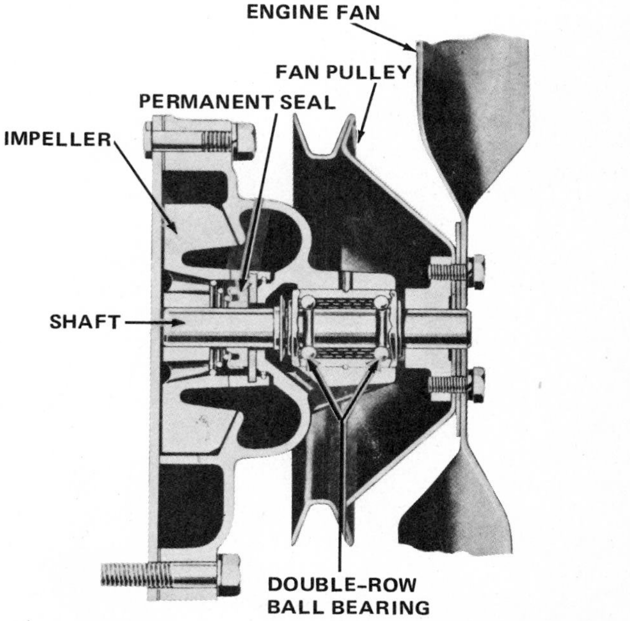

Fig. 21-4 Sectional view of a water pump. Note the double-row ball bearings which support the shaft. The fan and pulley are mounted on the shaft.

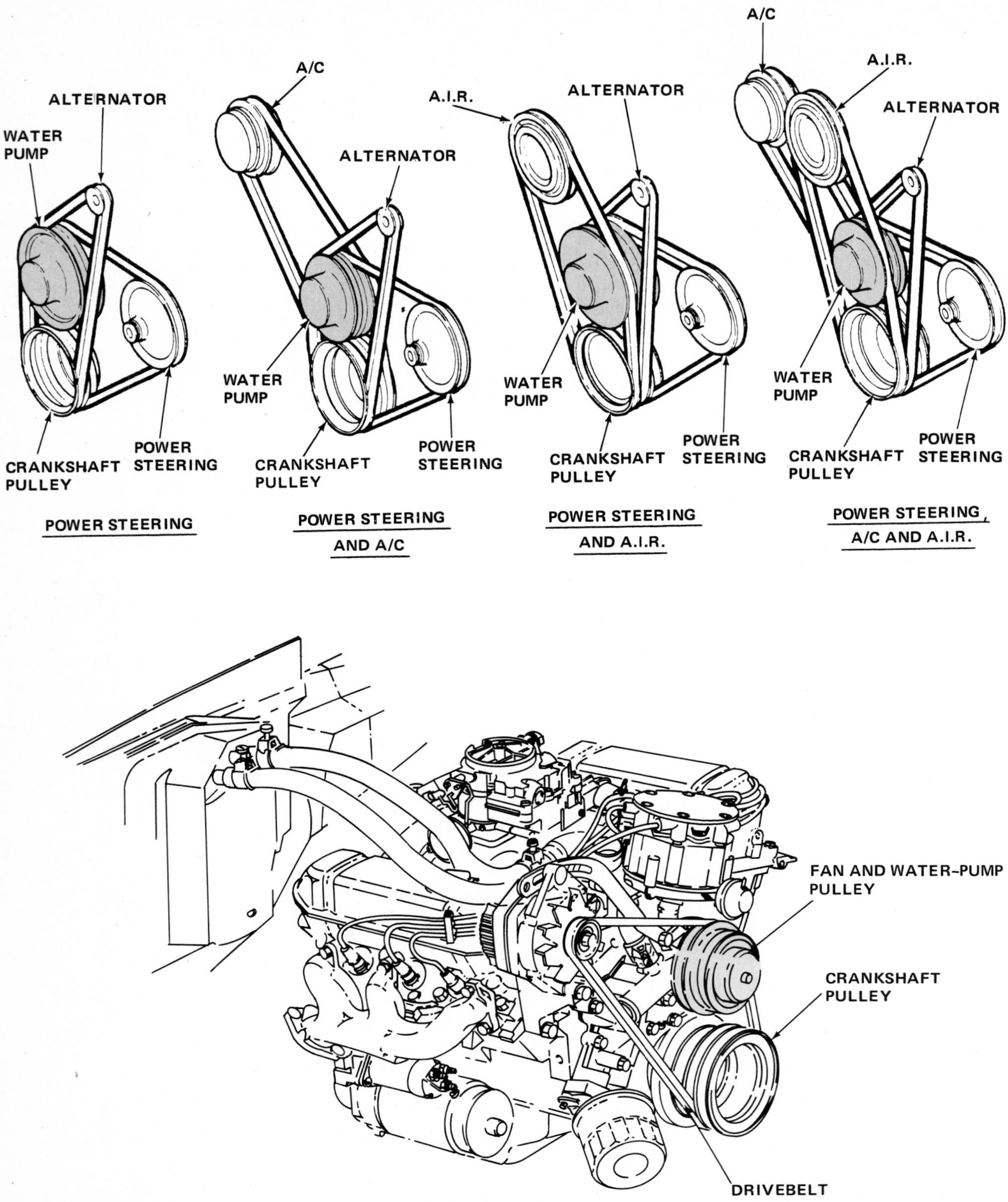

Fig. 21-5 The water pump and fan are driven by the same belt that drives the alternator, from the crankshaft pulley. The illustration shows various belt combinations of engines equipped with power steering, air conditioning (A/C), and air-injection reactor (AIR), an emission-control device. (*Pontiac Motor Division of General Motors Corporation*)

21-4 Engine Fan

The engine fan is usually mounted on the water-pump shaft. It is usually driven by the same belt that drives the pump and the alternator (Fig. 21-5). The purpose of the fan is to provide a powerful draft of air through the radiator. Some fans are equipped with a fan shroud that improves their performance. The shroud increases the efficiency of the fan. It assures that all air pulled back by the fan must first pass through the radiator.

1. Fan belt The belts are of the V type. There is friction between the sides of the belt and the sides of the grooves in the pulleys. The friction causes the driving power to be transmitted through the belt from one pulley to the other. The V-type belt provides a large area of contact, so that considerable power may be transmitted. The wedging action of the belt, as it curves into the pulley grooves, aids in preventing belt slippage.

Many engines use two V belts with double pulleys, that is, with pulleys that have two belt grooves. The added belt provides the power required to drive the alternator and water pump. These belts are matched. If one belt is to be replaced, then both should be replaced at the same time. Otherwise, the new belt will take most of the driving effort and will wear rapidly.

2. Variable-speed fan drive Many engines use a variable-speed fan drive which reduces fan speed. This conserves horsepower at high engine speed and when cooling requirements are low. At high speeds, a typical engine fan might use up several horsepower and might produce some noise. The variable-speed fan drive (Figs. 21-6 and 21-7) contains a small fluid coupling partly filled with a special silicone oil. When engine cooling requirements are severe, as during high-temperature, high-speed operation, more oil is injected into the fluid coupling as explained below. This causes more power to pass through the coupling. Fan speed therefore goes up. When cooling requirements are low, as during cool-weather, medium-speed operation, oil is withdrawn from the fluid coupling. Less power passes through, and fan speed drops off.

The amount of oil in the fluid coupling, and thus the fan speed, is controlled by a thermostatic strip or coil. (Figure 21-7 shows the coil.) If a strip is used, it is held at both ends by clips on the face of the fan drive. The strip bows outward with increasing under-the-hood temperatures. This motion allows a control piston to move outward. The outward-moving piston forces more oil into the fluid coupling. This increases fan speed and engine cooling. As under-the-hood temperatures drop, the thermostatic strip straightens out, forcing the control piston in. This action causes oil to leave the fluid coupling so that fan speed drops. The bimetal thermostatic coil works similarly to match fan speed with cooling requirements.

3. Flex fan Flexible blades reduce both the power needed to drive the fan and fan noise at high speed (Fig. 21-8). With this design, the pitch of the blades decreases as fan speed increases, owing to centrifugal force. The result is that each blade pushes less air. Thus, power needs and noise are lower at higher speeds.

4. Fan for crosswise engine Engines mounted crosswise at the front (and driving front wheels) need another method of driving the engine fan. Figure 21-8 shows a typical arrangement. The radiator is in front, as usual. The fan is driven by an electric motor. A thermostatic switch turns the motor on only when needed. For example, in one engine, the switch turns the motor on when the coolant reaches 193–207°F [89–97°C] and turns the motor off if the coolant temperature is below these figures. On cars with air conditioners, the thermostatic switch is bypassed and the motor runs all the time the air conditioner is on.

21-5 Radiator

The radiator (Fig. 21-1) is a device for holding a large volume of coolant in close contact with a large volume of air. This allows heat to transfer from the coolant to the

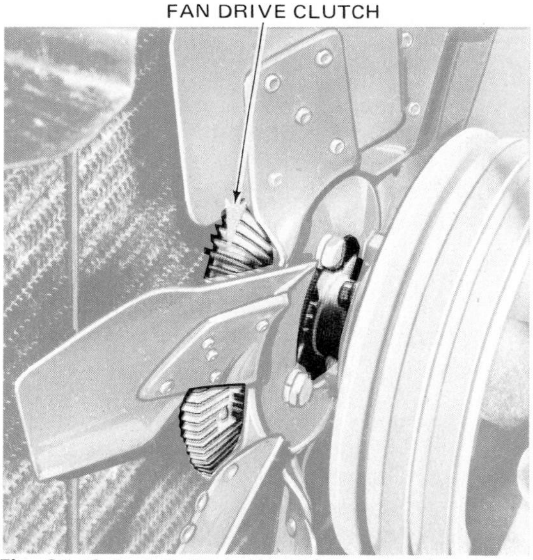

Fig. 21-6 Installation of variable-speed fan clutch drive. The drive is located between the fan hub and the pulley shaft. *(Ford Motor Company)*

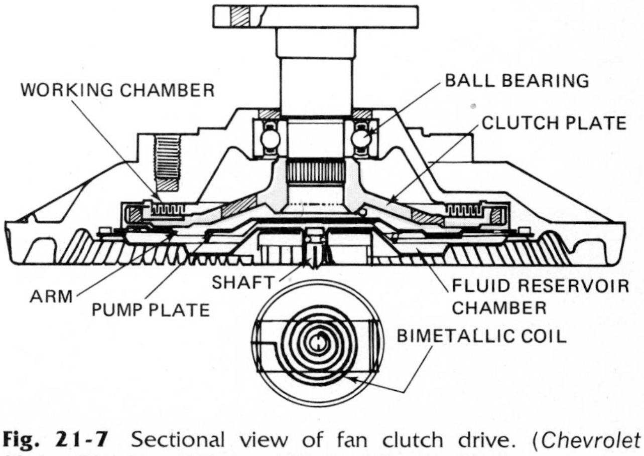

Fig. 21-7 Sectional view of fan clutch drive. *(Chevrolet Motor Division of General Motors Corporation)*

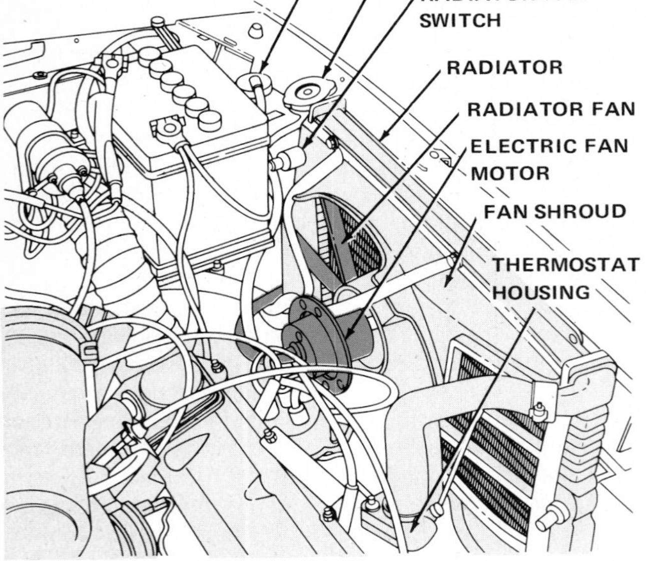

Fig. 21-8 When the engine is mounted crosswise at the front of the car, the engine fan is driven by an electric motor. (*Chrysler Corporation*)

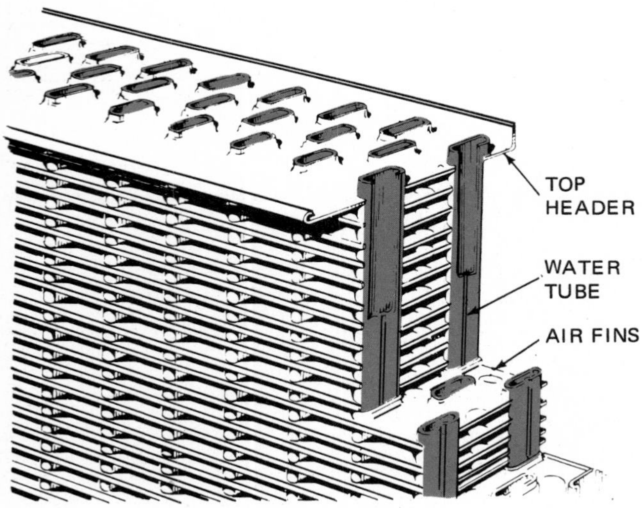

Fig. 21-9 Construction of a tube-and-fin radiator core.

air. The radiator core is divided into two separate and intricate compartments. Coolant passes through one, and air passes through the other. There are several types of radiator core. Two of the more commonly used types are the tube-and-fin (Fig. 21-9) and the ribbon-cellular (Fig. 21-10). The tube-and-fin type consists of a series of long tubes extending from the top to the bottom of the radiator (or from upper to lower tank). Fins are placed around the tubes to improve heat transfer. Air passes around the outside of the tubes, between the fins, absorbing heat from the coolant in passing.

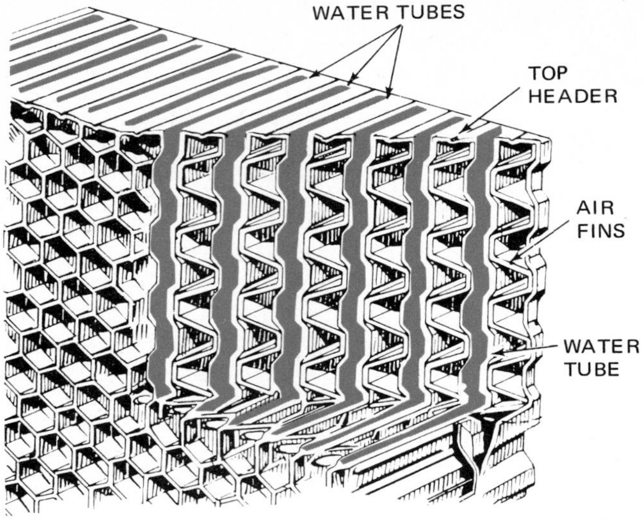

Fig. 21-10 Construction of a ribbon-cellular radiator core.

The ribbon-cellular radiator core (Fig. 21-10) is made up of a large number of narrow coolant passages. The passages are formed by pairs of thin metal ribbons soldered together along their edges, running from the upper to the lower tank. The edges of the coolant passages, which are soldered together, form the front and back surfaces of the radiator core. The coolant passages are separated by air fins of metal ribbon, which provide air passages between the coolant passages. Air moves through these passages from front to back, taking heat from the fins. The fins, in turn, absorb heat from the coolant moving downward through the coolant passages. As a consequence, the coolant is cooled.

Radiators can be classified in another way, according to the direction of coolant flow through them. In some, the coolant flows from top to bottom (down-flow type). In others, the coolant flows horizontally from an input tank on one side to another tank on the other side (cross-flow type). (See Figs. 21-11 and 21-12.)

The coolant tank above or to the side of the radiator serves two purposes. It provides a reserve supply of coolant. It also provides a place where the coolant can be separated from any air that might be circulating in the system. The tank has a filler cap which can be removed for addition of coolant as necessary.

1. Expansion tanks Many cooling systems have a separate expansion tank (Fig. 21-13). The tank is partly filled with coolant and is connected to the radiator filler neck. The coolant expands in the engine as it heats up. This sends part of the coolant into the expansion tank.

When the engine cools off, the coolant in the cooling system contracts. This produces a vacuum in the cooling system. Now coolant from the reservoir flows back into the radiator. The system works to keep the engine water jackets and radiator filled with coolant. This allows the system to operate at maximum efficiency at all times. The coolant flows from the cooling system to the expansion tank, and back again, through the radiator pressure cap. As we explain in ● 21-9, the pressure cap is designed to

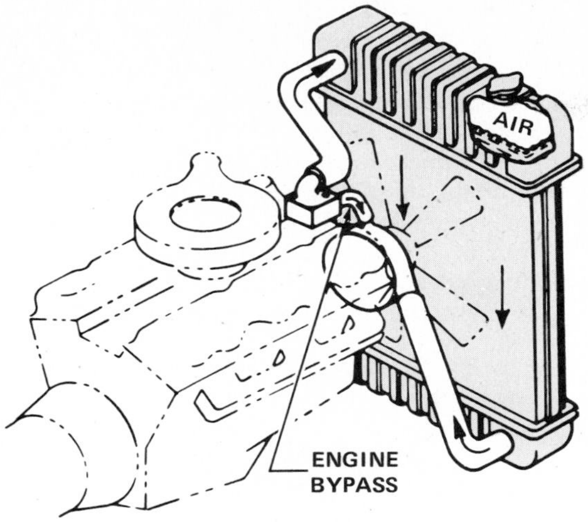

Fig. 21-11 Cooling system using a down-flow radiator. (*Harrison Radiator Division of General Motors Corporation*)

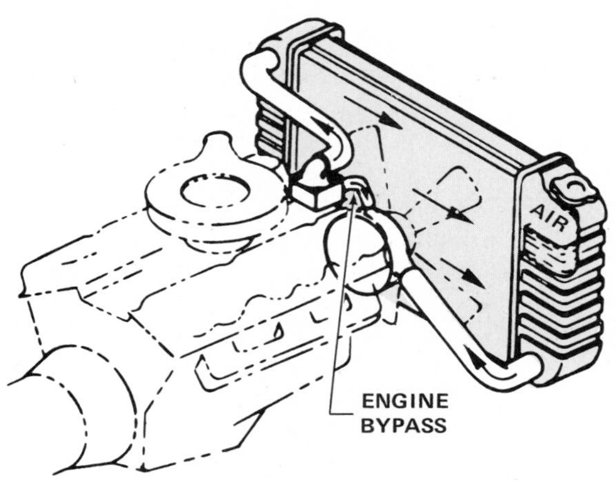

Fig. 21-12 Cooling system using a cross-flow radiator. (*Harrison Radiator Division of General Motors Corporation*)

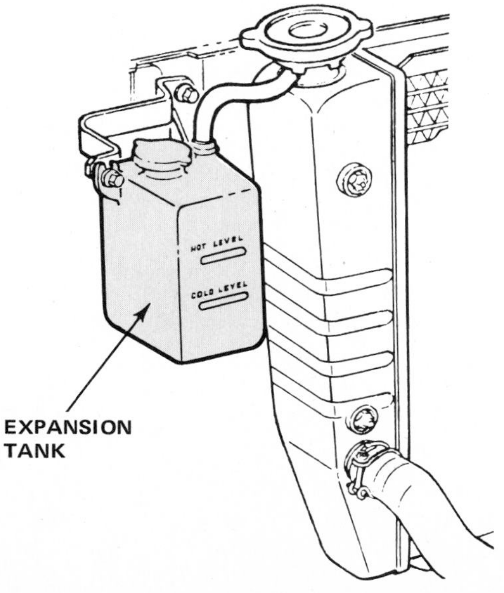

Fig. 21-13 Cooling system using an expansion, or constant-full, tank. (*Ford Motor Company*)

maintain some pressure in the cooling system which increases its cooling efficiency.

The radiator cap is not removed to check coolant level on cars with an expansion tank. Instead level is checked by looking at the tank. On many cars, the tank has markings which show where the level should be with the engine hot and with the engine cold.

2. Radiator grills Radiator grills add to the appearance of the car. But they also place an extra load on the cooling system. That is, they slow down the flow of cooling air through the radiator. However, the cooling system is designed to take this into account.

3. Radiator shutters Some radiators for heavy-duty applications (truck and buses) have automatically controlled radiator shutter systems (Fig. 21-14). The shutter is closed during cold starts and warm-up. As the engine reaches operating temperature, a thermostat (called a shutterstat in Fig. 21-14) operates a valve which admits compressed air to an air cylinder. The shutterstat is in the upper line to the radiator, so it senses the temperature of the coolant coming from the engine—that is, engine temperature. The compressed air forces a piston to move in the cylinder. This operates a lever which causes the shutter to open. You can compare this with the way venetian blinds work. When the shutter opens, more air can flow through the radiator to increase the cooling action. During cold-weather operation, the cooling system may be working too efficiently and overcooling the engine. In this case, the shutterstat, sensing the temperature of the coolant from the engine, causes the shutter to close.

The compressed air that operates the system comes from the air brake system, as shown in Fig. 21-14.

● 21-6 Transmission Oil Coolers

Many cars with automatic transmissions are equipped with cooler tubes or radiators to cool the transmission oil. The oil in automatic transmissions can get quite hot. Overheating reduces transmission performance and can damage the transmission. One system is shown in Fig. 21-15. The transmission is connected by two tubes to the oil-cooler tube in the side tank or bottom tank of the radiator (Fig. 21-16). The oil-cooler tube, being immersed in the cooler coolant in the radiator, is cooled. This cools the transmission oil passing through.

● 21-7 Hot-Water Car Heater

Most automobiles are equipped with car heaters of the hot-water type (Figs. 21-17 and 21-18). This device might be considered a secondary radiator. It transfers heat from the cooling system to the passenger compartment instead of to air passing through the main radiator. Hot coolant from the engine is circulated through the heater radiator. A small electric motor drives a fan that forces air through the radiator section of the heater. The air absorbs heat from the heater radiator.

A more complicated version is shown in Fig. 21-18. The principle is the same. Hot coolant from the engine circulates through a small heater radiator. An electric fan circulates air from the passenger compartment through the radiator. The system also includes a defrosting arrangement. The driver can operate controls that direct the heated air into the passenger compartment. Or, it can be sent up against the windshield to melt any frost, or evaporate any mist, that has formed.

In some cars, the passenger-compartment heating system has automatic controls. These controls turn the system on when the passenger compartment is cold and turn it off when the compartment is warm enough.

Some systems are fully automatic. They work with the air-conditioner system to maintain the desired temperature, summer or winter. The driver merely has to set a control to the temperature desired. The system either heats or cools as required to maintain that temperature.

● 21-8 Thermostat

The thermostat is placed in the coolant passage between the cylinder head and the top of the radiator (Fig. 21-1). Its purpose is to close off this passage when the engine is cold, so that coolant circulation is restricted. This allows the engine to reach operating temperature more quickly. The thermostat consists of a thermostatic device and a valve (Fig. 21-19). Various valve arrangements and thermostatic devices have been used. The bimetal type uses a thermostatic coil to operate the valve. Nowadays, most cooling-system thermostats are powered by a wax pellet which expands with increasing temperature to open the valve. The thermostat shown to the right in Fig. 21-19 is of the wax-pellet type. Figure 21-20 is a sectional view of a similar thermostat.

Thermostats are designed to open at specific temperatures. For example, a thermostat designated as a 195°F [90.6°C] unit will start to open between 192 and 197°F [88.9 and 91.7°C] and will be fully open at 207°F [97.2°C]. A 180°F [82.2°C] thermostat will operate at 15°F below these figures. Thermostats of the proper characteristics are selected to suit the operating requirements of the engine as well as the antifreeze used.

With the engine cold and the thermostatic valve closed, the coolant cannot pass between the engine and radiator. Instead, it recirculates through the cylinder block and head. This keeps the heat in the engine. The engine thus reaches operating temperature more rapidly. When the engine reaches operating temperature, the thermostatic valve begins to open. Then, coolant can

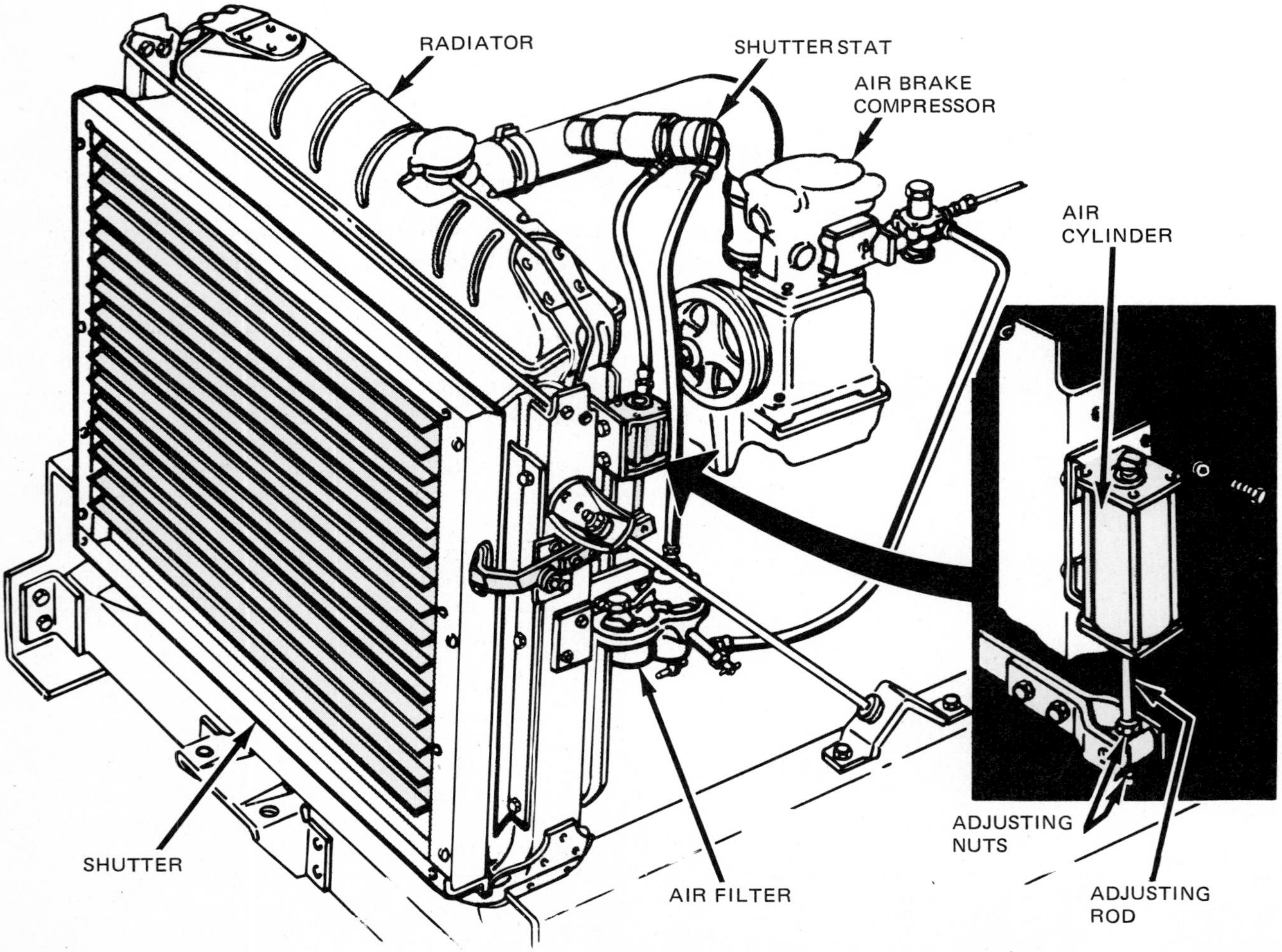

Fig. 21-14 Automatically controlled radiator shutter system. (*Ford Motor Company*)

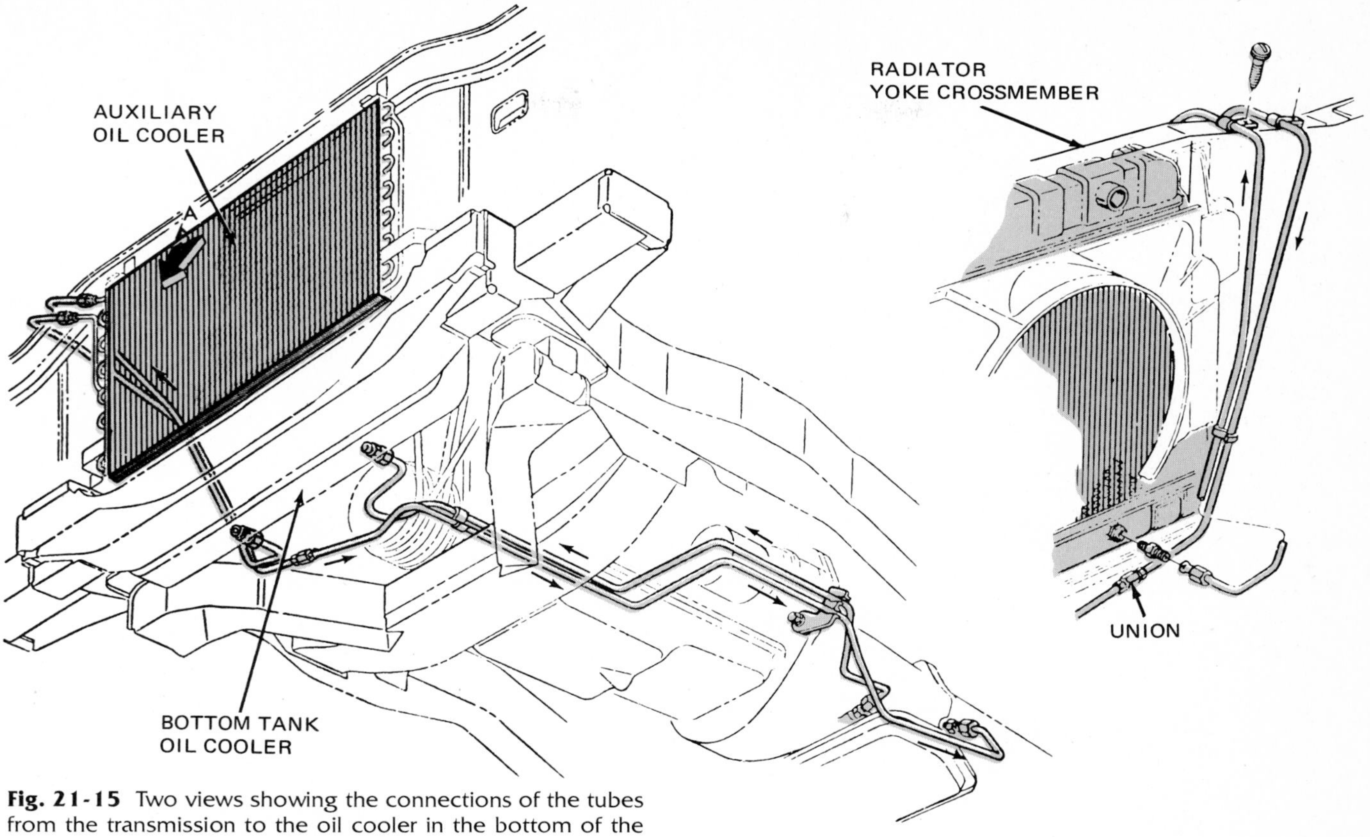

Fig. 21-15 Two views showing the connections of the tubes from the transmission to the oil cooler in the bottom of the cooling-system radiator. (*Chrysler Corporation*)

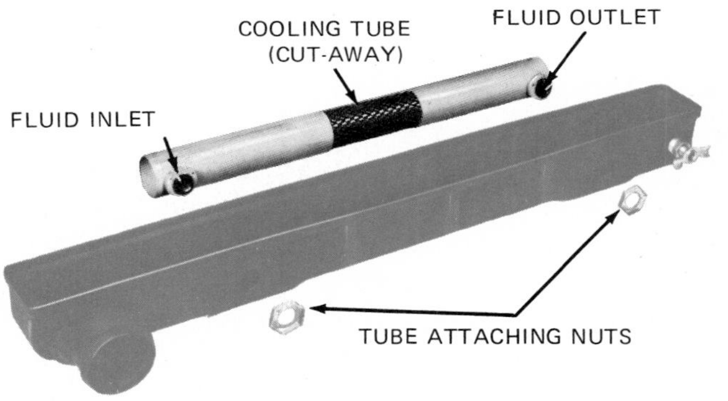

Fig. 21-16 Lower tank of the radiator and the oil-cooler tube, showing how the tube fits into the lower tank. (*Chrysler Corporation*)

circulate through the radiator, and operates the cooling system normally, as already described.

A bypass is required to permit coolant to circulate within the engine itself when the engine is cold. However, when the engine warms up, the bypass must close or become restricted. Otherwise, the coolant would continue to circulate within the engine and too little would go to the radiator for cooling.

One bypass system uses a small, spring-loaded valve. The valve is forced open by coolant pressure from the pump when the thermostat is closed. As the thermostat opens, the coolant pressure drops within the engine, and the bypass valve closes.

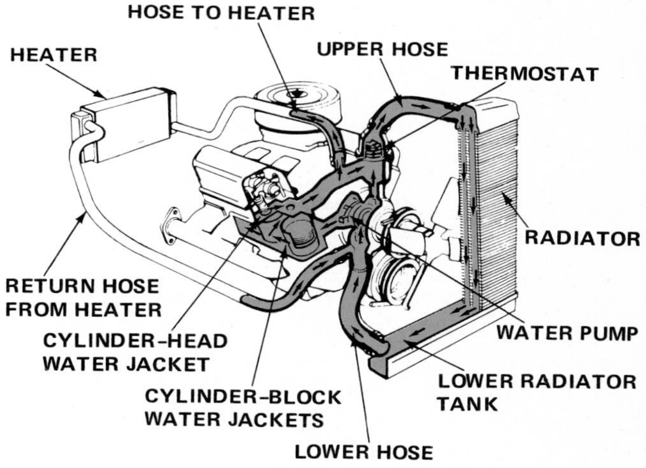

Fig. 21-17 Cutaway of a V-8 engine, showing the cooling system and the car hot-water heater. This is the simplest car heater system.

Another widely used system has a blocking bypass thermostat (Fig. 21-21). This thermostat operates like others already described, but it also has a secondary valve. When the primary valve is closed, the circulation to the radiator is shut off. However, at that time the secondary valve is open, permitting coolant to circulate through the bypass. But when the primary valve opens, permitting coolant to flow to the radiator, the secondary valve closes, blocking off the engine bypass.

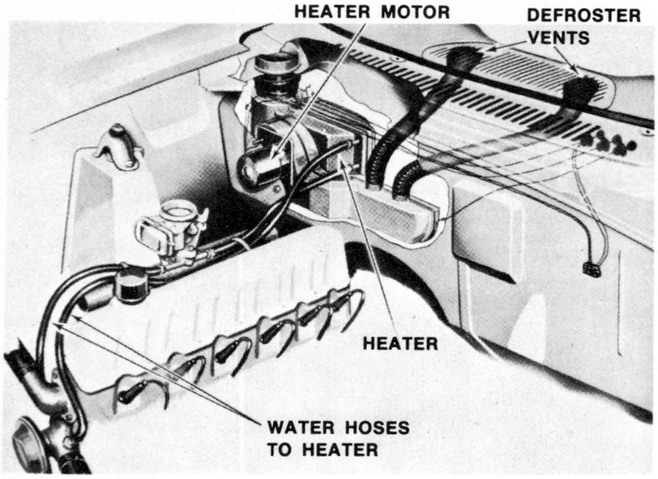

Fig. 21-18 Car heater system. Hot coolant from the engine cooling system circulates through a small radiator. The fan blows air through the radiator.

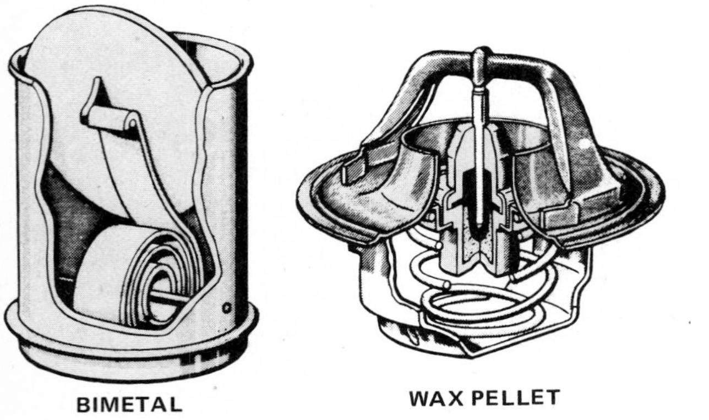

Fig. 21-19 Two types of cooling-system thermostats. (*Ford Motor Company*)

● 21-9 Radiator Pressure Cap

To improve cooling efficiency and prevent evaporation, many late-model automobiles use a pressure cap on the radiator (Fig. 21-22). At sea level, where atmospheric pressure is about 15 psi (pounds per square inch) [103.42 kPa], water boils at 212°F [100°C]. At higher altitudes, where atmospheric pressure is less, water boils at lower temperatures. Higher pressure increases the temperature required to boil water. Each added pound per square inch [0.0703 kg/cm^2] increases the boiling point of water about $3\frac{1}{4}$°F [1.8°C]. The use of a pressure cap on the radiator increases the air pressure within the cooling system several pounds per square inch. Thus, the water may be circulated at high temperatures without boiling. The water therefore enters the radiator at a higher temperature, so the difference in temperature between the air and the water is greater. Heat then is more quickly transferred from the water to the air, improving cooling efficiency. Evaporation of water is reduced by the higher pressure, because the boiling point of the water is higher. The pressure cap also prevents loss of water due to surging when the car is quickly braked to a stop.

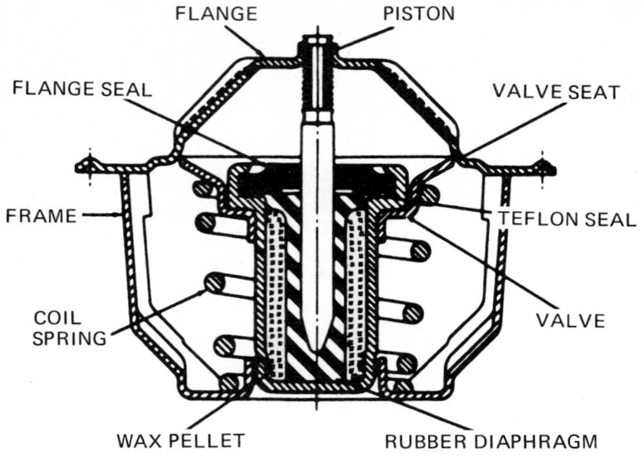

Fig. 21-20 Sectional view of a wax-pellet thermostat. (*Chevrolet Motor Division of General Motors Corporation*)

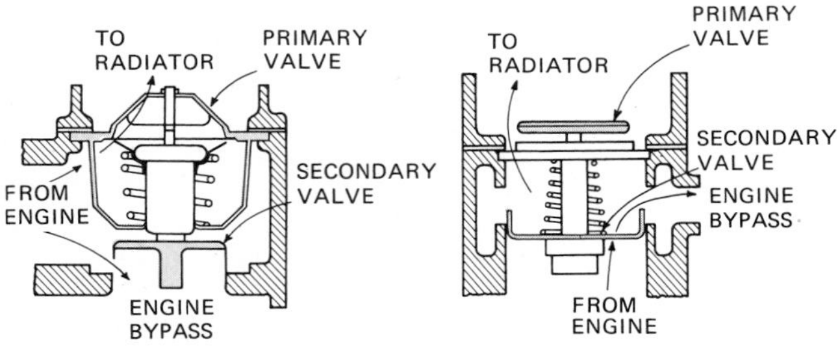

Fig. 21-21 Blocking bypass thermostats. (*Harrison Radiator Division of General Motors Corporation*)

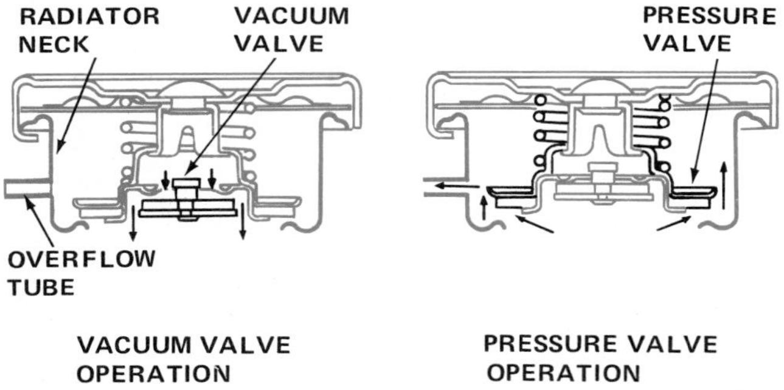

Fig. 21-22 Radiator pressure-vacuum cap, showing vacuum-valve and pressure-valve operation. When the pressure increases sufficiently in the cooling system, the pressure valve is raised to permit coolant to flow from the radiator into the expansion tank. (*American Motors Corporation*)

The pressure cap fits over the radiator filler tube and seals tightly around the edges. The cap contains two valves, the pressure valve and the vacuum valve. The pressure valve consists of a valve held against a valve seat by a calibrated spring. The spring holds the valve closed so that pressure is produced in the cooling system. If pressure rises above that for which the system is designed, the pressure valve is raised off its seat. This relieves the excessive pressure. Pressure caps are designed to add as much as 15 pounds of pressure per

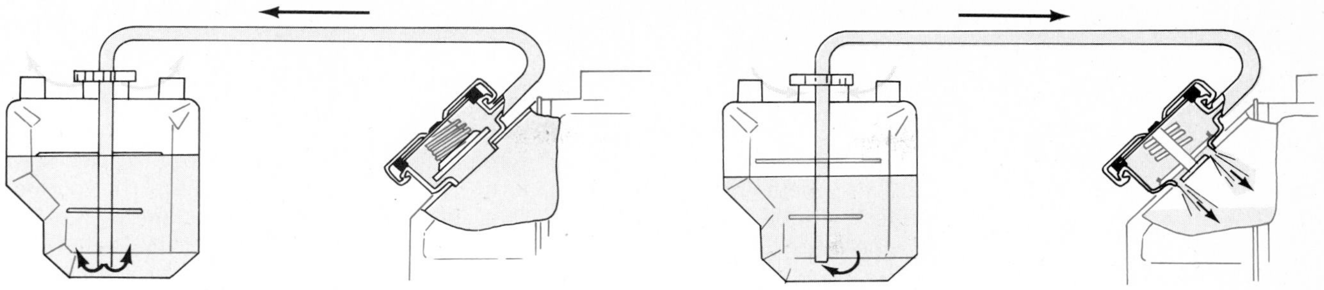

PRESSURE RELIEF VALVE LIFTED, ALLOWING COOLANT TO FLOW INTO THE EXPANSION TANK
VACUUM RELIEF VALVE OPEN, ALLOWING COOLANT TO SYPHON BACK INTO THE RADIATOR

Fig. 21-23 Operation of the radiator cap under pressure conditions (left) and vacuum conditions (right). This action, with the expansion tank, assures a full radiator and cooling system at all times. (*Ford Motor Company*)

square inch [1.0546 kg/cm^2] in the cooling system. This increases the boiling point of the water to as much as 260°F [126.6°C].

The vacuum valve prevents the formation of a vacuum in the cooling system when the engine has been shut off and begins to cool. If a vacuum forms, atmospheric pressure from the outside causes the small vacuum valve to open, admitting air into the radiator. Without a vacuum valve, the pressure within the radiator might drop so low that atmospheric pressure would collapse it.

In cooling systems with expansion tanks (● 21-5 and Fig. 21-13), the radiator pressure cap is more or less permanently installed. It should not be removed. If coolant must be added to the cooling system, it is put into the expansion tank. In these systems, the radiator overflow tube in the pressure cap is connected to the expansion tank. See Fig. 21-23.

● 21-10 Antifreeze Solutions

Water freezes at 32°F [0°C]. If water freezes in the engine cooling system, it stops coolant circulation. Some parts of the engine will overheat. This could seriously damage the engine. Worse, however, is the fact that water expands when it freezes. Water freezing in the cylinder block or cylinder head could expand enough to crack the block or head. Freezing water in the radiator could split the radiator seams. In either case, there is serious damage. A cracked block or head cannot be repaired satisfactorily. A split radiator is hard to repair.

To prevent freezing of the water in the cooling system, antifreeze is added to form the coolant. The most commonly used antifreeze is ethylene glycol. A mixture of half water and half ethylene glycol will not freeze above −34°F [−36.7°C]. This is 34° below zero, and it seldom gets that cold in any part of the United States. A higher concentration of antifreeze will prevent freezing of the coolant at temperatures as low as −84°F [−64.4°C].

Some antifreeze compounds also plug small leaks in the cooling-system radiator. These antifreeze compounds contain tiny plastic beads or inorganic fibers which circulate with the coolant. If a leak develops, the beads or fibers jam in the leak and plug it. Of course if the leak is too large, they cannot plug it. Also, they cannot stop leaks in hoses, cylinder-head gaskets, or pump seals.

Corrosion protection is also built into antifreeze solutions. That is, compounds are added that fight corrosion inside engine water jackets and the radiator.

Antifreeze solutions also serve a purpose during hot-weather operation. They improve the hot-weather efficiency of the cooling system and continue to fight corrosion.

Car manufacturers recommend that the cooling system be drained, flushed out, and refilled with a fresh mixture of water and antifreeze periodically. One recommendation is that this be done every 2 years. Another is that it be done every year, preferably in the late fall, just before freezing weather sets in.

Note: Cooling systems on modern cars should never be filled with only water. Indicator lights are designed not to come on to show excessive temperatures until well above the boiling point of water. Thus, just plain water could boil even though the indicator light does not come on. The water-antifreeze mixture will not boil until a higher temperature is reached.

● 21-11 Temperature Indicators

The operator should, at all times, know the coolant temperature. For this reason, a temperature indicator is installed in the car. An abnormal heat rise is a warning of abnormal conditions in the engine. The indicator thus warns the operator to stop the engine before serious damage is done. Temperature indicators are of three general types, vapor-pressure, electric, and indicator light.

1. Vapor pressure The vapor-pressure temperature indicator consists of a metal indicator bulb and a metal tube connecting the bulb to the indicator unit. The indicator unit contains a curved, or Bourdon, tube. One end of the tube is linked to the indicator needle. The other end is open and is connected through a tube to the bulb. The indicator bulb is usually placed in the water jacket of the engine. It is filled with a liquid that vaporizes at fairly low temperature. As the engine temperature increases, the liquid in the bulb begins to vaporize. This creates pressure that is conveyed through the connecting tube to the Bourdon tube in the indicating unit. The pressure tends to straighten out the tube. This movement causes the indicating needle to move across the dial face and indicate the temperature in the water jacket.

2. Electric indicators Electrically operated temperature indicators are of two types, the balancing-coil type and the bimetal-thermostat type. The balancing-coil oil-pressure indicator (Fig. 19-9), fuel gauge (Fig. 12-10), and temperature indicator all operate similarly. The instrument panel indicating units are, in fact, practically identical. Each consists of two coils and an armature to which a needle is attached (Fig. 21-24). The engine unit changes resistance with temperature. At higher temperatures it has less resistance and thus passes more current. When this happens, more current passes through the right-hand coil in the indicating unit. In this case, the armature to which the needle is attached is attracted by the increased magnetic field. The armature and the needle move around so that the needle indicates a higher temperature.

The bimetal-thermostat temperature indicator is similar to the bimetal-thermostat fuel gauge (Fig. 12-11). The instrument-panel units are practically identical.

3. Indicator light One indicator-light system is shown in Fig. 21-25. This system has a coolant-temperature sending unit mounted on the engine so that it is exposed to the cooling-system coolant. The sending unit is connected to two light bulbs and the battery through the ignition switch. When the igniton switch is first turned on, to start a cold engine, the sending-unit thermostatic blade is in the proper position to connect the COLD light to the battery. It comes on. The COLD light, which appears in blue on the instrument panel, remains on until the engine approaches operating temperature. As this happens, the thermostatic blade in the sending unit is bent by the increasing temperature. The blade therefore moves off the cold terminal, disconnecting the COLD light so that it turns off. If the engine overheats, the thermostat will warp further so that it moves under the hot terminal. This connects the HOT bulb to the battery so that it glows and appears in red on the instrument panel. This is a signal to the driver that the engine has overheated and should be stopped before damage results.

Some indicator-light systems do not have a COLD light. Instead, the HOT light comes on during cranking and then goes off when the engine starts. It stays off unless the engine overheats, when it comes on to signal that the engine is getting too hot.

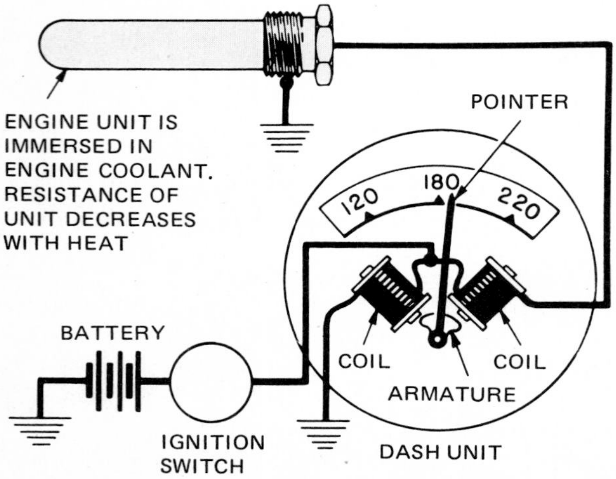

Fig. 21-24 Circuit diagram of a balancing-coil temperature-indicating system.

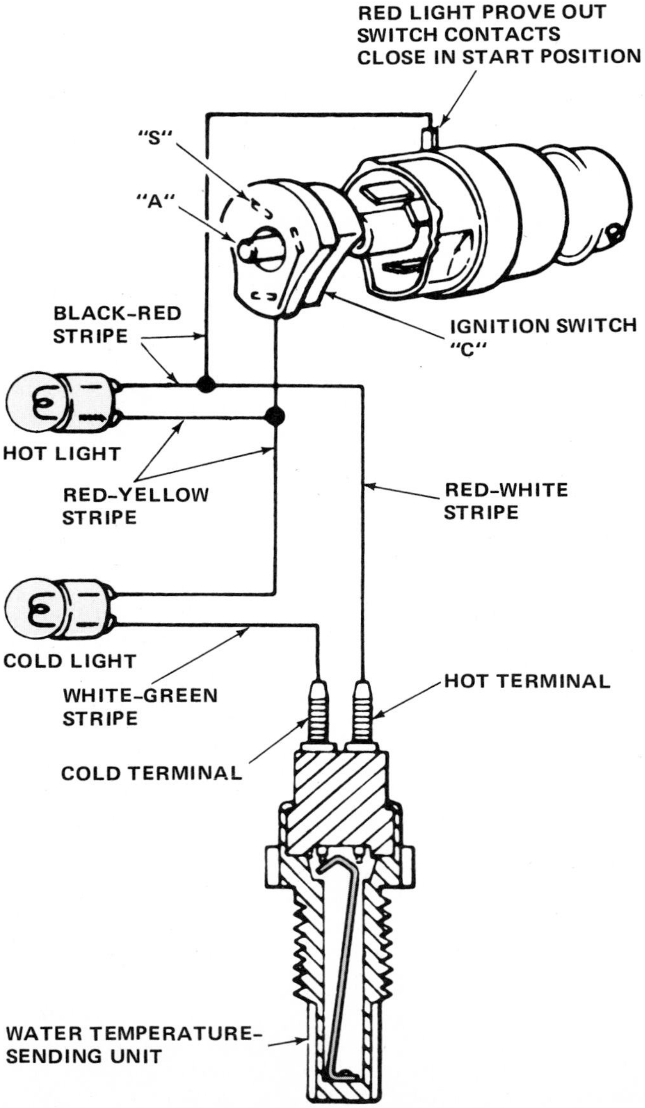

Fig. 21-25 Temperature-indicating system using COLD and HOT indicating lights. (*Ford Motor Company*)

REVIEW QUESTIONS

1. What is the purpose of the engine cooling system?
2. What are the two types of engine cooling systems?
3. Give two examples of air-cooled engines.
4. What are water jackets? What are they for?
5. What is the purpose of the water pump?
6. What is the coolant made of?
7. What is the purpose of the engine fan?
8. Explain how the variable-speed fan works.
9. Suppose an engine uses two V belts to drive the water pump and alternator. Is it all right to replace only one belt if only one is worn? Why?
10. What is the purpose of the expansion tank?
11. Explain how an automatic radiator shutter works.
12. What is the purpose of the transmission-oil cooler? Where is the oil-cooler tube located?
13. With regard to the direction of coolant flow, what are the two types of radiators?
14. What is the purpose of the thermostat? How does it work?
15. Explain how a blocking bypass thermostat works.
16. How many valves does the radiator pressure cap have? How do they work?
17. Explain how the radiator pressure cap works in the cooling system using an expansion tank.
18. Explain how a car heater works.
19. What is the basic purpose of antifreeze?
20. What are two recommendations regarding how often the cooling system should have fresh antifreeze?
21. Explain how the indicator-light system, described in the book, works.

SELF PROJECTS

1. Whenever you raise the hood of a car, note whether or not the cooling system has an expansion tank. If it does, make a quick sketch showing its location in the engine compartment. Add the make and year model to the sketch and file it in your notebook. As you build up your file, you will have complete information on expansion-tank locations on all late-model cars.
2. Examine cooling-system thermostats. Note the location of the wax pellet which is used in most modern thermostats to close or open the thermostatic valve.
3. Examine radiator pressure caps and find the locations of the pressure and vacuum valves.
4. Copy the information from a container of antifreeze and file it in your notebook.

CHAPTER 22
Cooling-System Service

After you read this chapter, you should be able to:

- **Discuss the cause and corrections of cooling-system troubles.**
- **Explain how to test the various components of the cooling system. Perform these tests.**
- **Explain how to clean the cooling system and add a fresh charge of antifreeze and water. Do this job on a car.**

This chapter describes trouble diagnosis, testing, care, and servicing of the automotive-engine cooling system and its component parts.

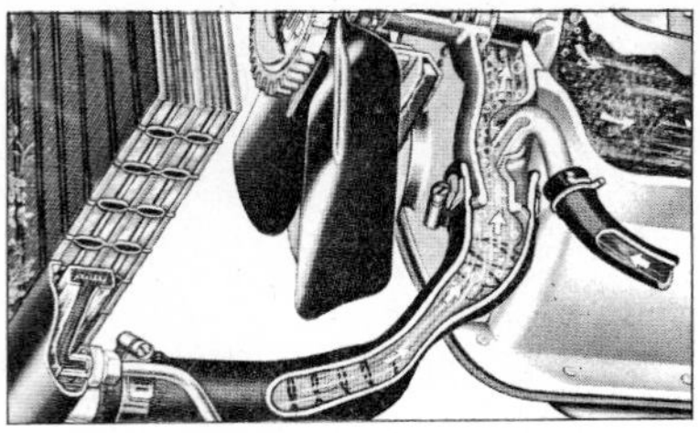

● 22-1 Cooling-System Trouble Diagnosis

Two common complaints related to the engine cooling system are engine overheating and cooling-system leaks. If the engine is slow to warm up, this could also be a fault of the cooling system. Possible causes of these complaints are discussed in following sections. Detailed testing of the system is covered later in the chapter.

● 22-2 Overheating

The driver may notice that the red light stays on or the temperature gauge registers in the overheating zone. Also, the driver may complain that the engine boiled over. Possible causes of engine overheating include:

1. Low coolant level due to leakage of coolant from the system, as described in ● 22-3.
2. Accumulation of rust and scale in the system which prevents normal circulation of coolant. This is less common today, because modern antifreeze compounds contain agents that fight rust.
3. Collapsed hoses which prevent normal coolant circulation.
4. Defective thermostat which does not open normally and thus blocks circulation of coolant. If the engine overheats without the radiator becoming normally warm, and if the fan belt is properly tightened, then the thermostat is probably at fault. Sometimes, on new cars, a grain of sand from the engine block or head core will lodge behind the thermostat, preventing it from opening.
5. Defective water pump which does not circulate enough coolant through the engine. Just as an experiment you could check the water-pump action by installing a clear plastic pipe in place of the upper radiator hose and then running the engine. You can then see how much coolant is circulating. Note that one of the more common causes of water-pump bearing failure is an overtight fan belt. Fan belts should always be tightened correctly with a belt tension gauge (Fig. 22-12). Bearing failure is usually fairly obvious because it is noisy. A quick check of the bearing can be made—with the engine *off*—by grasping the tips of the fan blades and attempting to move the fan toward and away from the radiator. Any movement indicates a worn bearing.

CAUTION: Do not stand directly in line with the engine fan with the hood up and the engine running. There have been cases of fan blades breaking off and hitting someone standing nearby, causing serious injury. Ford suggests you examine the fan, with the engine *not running*, for cracks. If you find cracks, or signs that the blades are beginning to separate from the hub, the fan should be replaced.

6. A loose or worn fan belt will not drive the water pump fast enough. It should be tightened or replaced. Where a pair of belts is used, both should be replaced at the same time, not just the one that appears most worn. If you replace only one belt, you put all the work on the new belt and it will wear rapidly. If both are replaced with a new matched pair, then each belt will handle half the job.
7. The trouble may be due to afterboil. That is, the coolant starts to boil after the engine has been turned off. This could happen, for example, after a long, hard drive. The engine has so much heat in it that, after the engine is turned off, the coolant starts to boil.
8. Boiling can also occur if the radiator is frozen. This hinders or stops the circulation of coolant. Then the

coolant in the engine becomes so hot that it boils. Freezing of the coolant in the radiator, engine block, or head may crack the block or head and can open up seams in the radiator. A frozen engine is apt to be damaged seriously.

It must be remembered that there are other causes of engine overheating which have nothing to do with the conditions in the cooling system. High-altitude operation, insufficient oil, overloading of the engine, hot-climate operation, improperly timed ignition, and long periods of slow-speed or idling operation—any of these can cause overheating of the engine. See Chap. 42 on engine trouble diagnosis.

● 22-3 Loss of Coolant

Loss of coolant is obvious if the system uses an expansion tank. If the coolant level is low or the tank is empty with the engine at operating temperature, the system has lost coolant and it must be examined for leakage points. Overheating is often a sign of coolant loss. In past years, it was a common service check to remove the radiator cap and check the level of the coolant in the radiator. Today, this is no longer recommended. In fact, Chrysler Corporation, for example, has this to say in their service manuals:

There is no need to remove the radiator cap at any time except to:

1. *Check strength of antifreeze in the coolant.*
2. *Add antifreeze to lower the freezing point.*
3. *Refill system with new antifreeze-water mix.*
4. *Conduct service procedures.*

If there are signs of leakage and you must remove the radiator cap to make some checks, note the following caution.

CAUTION: Never under any circumstances remove the radiator cap when the engine is running. Not only can this seriously damage the engine and cooling system, but also the gusher of steam and coolant that will spurt out can seriously scald you. To avoid being scalded, make sure the engine is turned off and has cooled down. If you must remove the radiator cap before the engine has cooled off, wrap a thick cloth around the cap. Then turn it slowly to the first stop. Step back while the pressure is being released. When you are sure all the pressure is gone, press down on the cap with the cloth and turn it further to remove it. This is hazardous and is *not* recommended.

Many radiator caps are marked with the warning words "Do not open hot." This reinforces what we have just said about the danger of removing the cap when the engine is hot.

If you do want to remove the radiator cap to check the coolant level, the best time to do it is the first thing in the morning before starting the engine.

Points of leakage are usually easy to spot. They show up as telltale scale or water marks below the leak. Also, most ethylene glycol antifreeze compounds have a dye color in them. This color shows up at any leakage point. Section 22-12 describes pressure testing the cooling system to help locate leaks. Figure 22-11 shows locations of possible leaks in the system.

If it is a gasket leak (cylinder head, water pump), the gasket may require replacement. Attaching bolts should be tightened to the correct tension.

If the leak is in the radiator, the radiator should be removed and either repaired (● 22-18) or replaced.

If the leak is at a hose connection, the hose connection should be tightened. If a hose is leaking, it should be replaced with a new hose of the correct type and specifications.

● 22-4 Slow Warm-Up

The probable cause of slow engine warm-up is a thermostat that is stuck open. This open position allows the coolant to circulate between the engine and radiator even though the engine is cold. The engine therefore has to run longer to reach operating temperature. As a result, engine wear will be greater because the engine operates cold for a longer time. The driver's complaint here would be that it takes a long time for the car heater to start putting out heat. If you get this complaint, suspect a defective thermostat that is stuck in the open position (see ● 22-7).

● 22-5 Cooling-System Tests

Cooling-system tests include:

1. Checking coolant level
2. Checking coolant antifreeze strength
3. Testing the thermostat
4. Checking the hose and hose connections
5. Testing the water pump
6. Checking for exhaust-gas leakage into the system
7. Pressure testing the system and cap
8. Checking the fan belt or belts for wear and tension
9. Checking system for accumulations of rust and scale

These are covered in detail in following sections.

CAUTION: Keep your hands away from the revolving engine fan. You can be seriously cut by the whirling blades. See also the caution in ● 22-2 about standing in line with the fan when the engine is running.

● 22-6 Testing Antifreeze Strength

The strength of the antifreeze solution must be great enough to protect against freezeup at the lowest expected temperatures. The strength of the antifreeze can be checked with any of the three testers. One is the hydrometer (Fig. 22-1). The higher the float rises in the coolant, the higher the percentage of antifreeze in the coolant. To

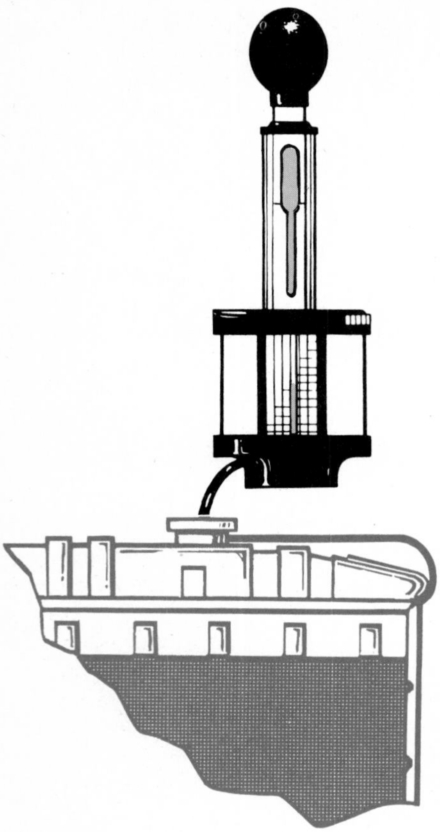

Fig. 22-1 Cooling-system hydrometer being used to check coolant to determine the amount of freezeup protection. *(Ford Motor Company)*

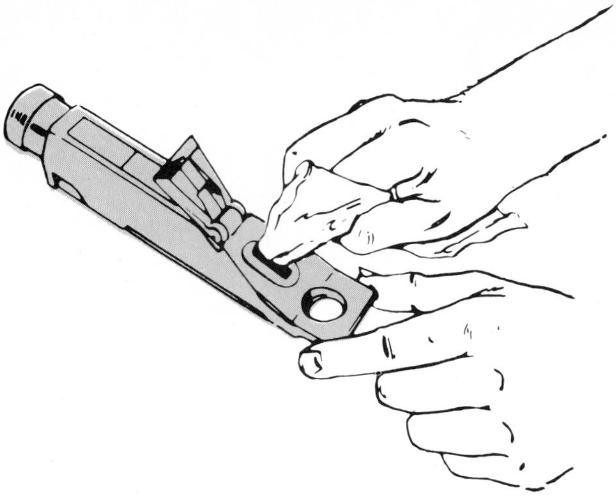

Fig. 22-2 Cleaning refractometer measuring window.

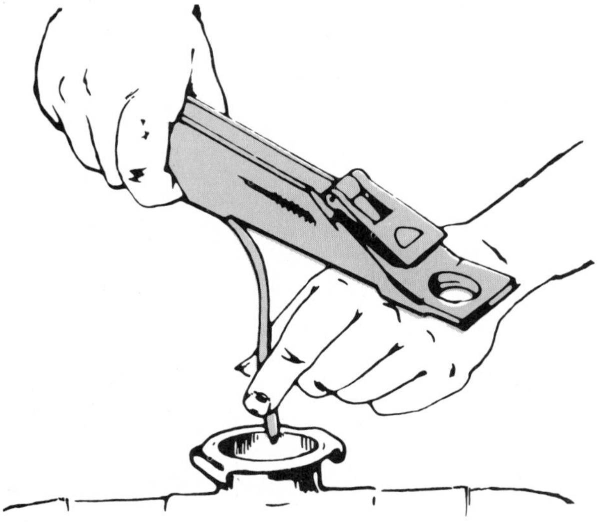

Fig. 22-3 Collecting a sample of coolant from the radiator.

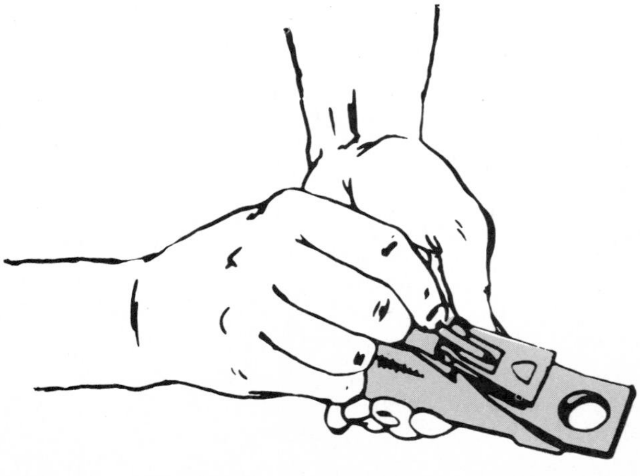

Fig. 22-4 Dropping a sample of coolant on the measuring window.

use the hydrometer, put the rubber tube into the coolant, and then squeeze and release the rubber bulb. Note how high the float rises in the coolant. Check the lower scale which shows the temperature of the coolant and the actual amount of protection the coolant gives against freezeup. That is, how low the temperature must go before the coolant will freeze.

A second tester has several balls in a glass tube. Coolant is sucked into the tube by squeezing and releasing a rubber bulb. The stronger the solution, the more balls will float. This is a relatively inexpensive tester that is small and easily handled with little danger of breakage. For these reasons, it is popular with service technicians.

A third tester, called a *refractometer,* uses the principle of light refraction (bending of light rays) as light passes through a drop of the coolant. To use the refractometer, open the plastic cover at the slanted end of the tester (Fig. 22-2). Wipe the measuring window and the bottom of the plastic cover with tissue or a clean cloth.

Close the plastic cover. Release the tip of the pump from the tester housing. Insert the tip of the tube into the radiator filler neck (Fig. 22-3). Be sure the end of the tube is well below the level of the coolant. Now press and release the bulb, so a sample of the coolant is sucked up. Bend the tube around, as shown in Fig. 22-4, so the tip can be inserted into the cover-plate opening. Squeeze the

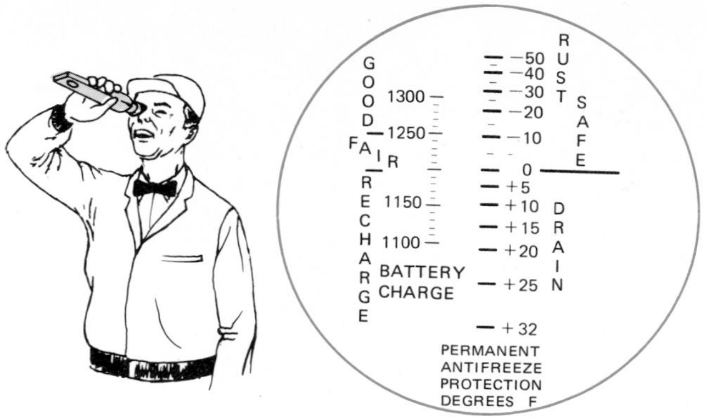

Fig. 22-5 Reading the refractometer.

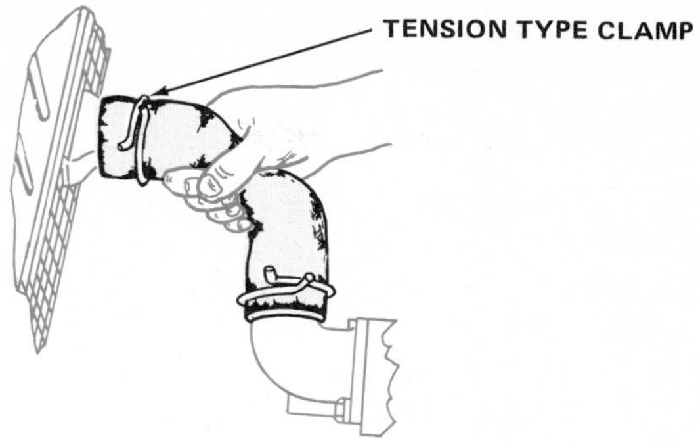

Fig. 22-6 Squeezing hose to determine its condition. (*Chrysler Corporation*)

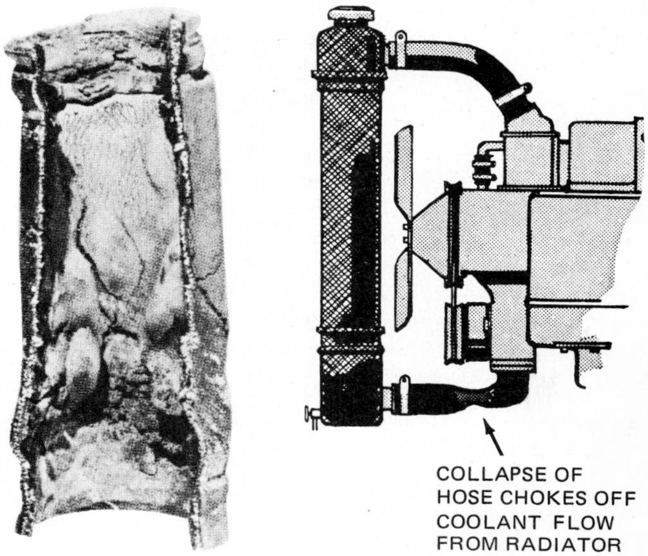

Fig. 22-7 Defective water hose. (*left*) Appearance when split open; (*right*) how collapsed hose restricts coolant flow.

bulb to put a few drops of coolant into the measuring surface.

Now point the tester toward the light, and look into the eyepiece (Fig. 22-5). The antifreeze-protection reading is where the dividing line between light and dark (the edge of the shadow) crosses the scale. Readings on the lower half of the scale indicate solutions without sufficient antifreeze protection.

The refractometer can also be used to check battery electrolyte to determine the battery state of charge (see ● 25-12).

CAUTION: The coolant is poisonous. It can cause serious illness and even death if it is swallowed.

● 22-7 Testing Thermostat

Different car manufacturers have different testing procedures. Chevrolet, for example, says to suspend the thermostat in a solution of one-third antifreeze and two-thirds water. The solution should be heated to 25°F above the temperature stamped on the thermostat. The thermostat should open. Then submerge the thermostat in a solution of one-third antifreeze and two-thirds water. This solution should be 10°F below the temperature stamped on the thermostat. The thermostat should close. If it does not open and close during the test, it is defective.

Note: Some thermostats are stamped in Celsius or Centigrade (C) instead of Fahrenheit (F).

Chrysler recommends the following test procedure. Remove radiator pressure cap. If engine is hot, wait 15 minutes. Then place a rag over the cap and turn it to the first stop. This allows the pressure to escape. After that, remove the cap. Drain one quart of coolant from the radiator. Warm the engine to operating temperature by idling it for 20 minutes. With a thermostat in the radiator filler neck, measure coolant temperature. Coolant should stabilize at no lower than 187°F [86.1°C]. This is 8°F below thermostat opening temperature.

Ford recommends immersing the thermostat in boiling water. If it does not open, it is defective. If the problem is slow warm-up, the thermostat may be leaking. Hold the thermostat up to the light to see if the valve is closing completely. If there is a gap between the valve and valve seat, replace the thermostat.

● 22-8 Checking the Hose and Hose Connections

The appearance of the hose and hose connections usually indicates their condition. If the hose is rotten and soft and collapses easily when squeezed (Fig. 22-6), it should be replaced. Figure 22-7 shows a badly decayed section of hose split open so you can see its internal appearance. The hose must be in good condition and connections should be properly tightened to avoid leaks.

● 22-9 Checking for Exhaust-Gas Leakage into the System

A defective cylinder-head gasket may allow exhaust gas to leak into the cooling system. This is very damaging.

Strong acids can form as the gas unites with the water in the coolant. These acids corrode the radiator and other cooling-system parts. A test for exhaust-gas leakage can be made with a Bloc-Chek tester which is installed in the radiator filler tube, as shown in Fig. 22-8. The test is made by putting the tester in the radiator opening with the engine running, and then squeezing and releasing the bulb. This draws an air sample from the cooling system up through the test fluid. The test fluid is ordinarily blue. But if combustion gas is leaking into the cooling system, the test fluid will change to yellow. If a leak is indicated, the exact location can be found by removing one spark plug wire at a time and retesting. When a leaking cylinder is firing, the liquid will change to yellow. When nonleaking cylinders only are firing, the liquid will remain blue in color.

Observe the cautions discussed previously in ● 22-2 and 22-3. These warn against standing in line with the fan and removing the radiator cap when the engine is running.

Undetected combustion leaks in the valve areas can cause cracked valve seats and cylinder heads. The coolant is forced away from the cracked area during heavy acceleration by the leakage of combustion gases through the leak. This causes excessive heat buildup. Then, when acceleration stops, the diverted coolant rushes back to the overheated area. The sudden cooling of the area can crack the head and valve seat.

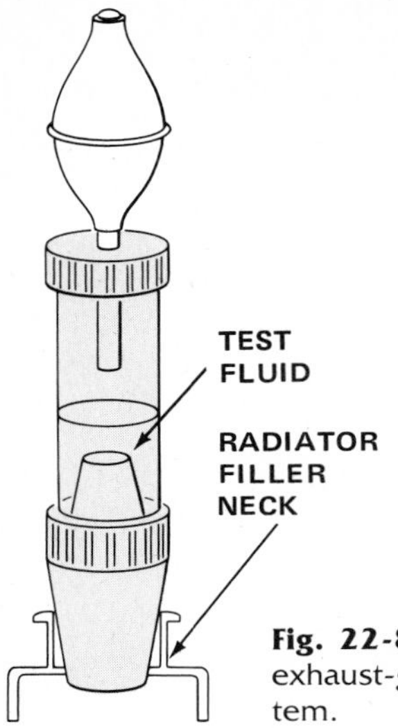

Fig. 22-8 Using Bloc-Chek to check for exhaust-gas leakage into the cooling system.

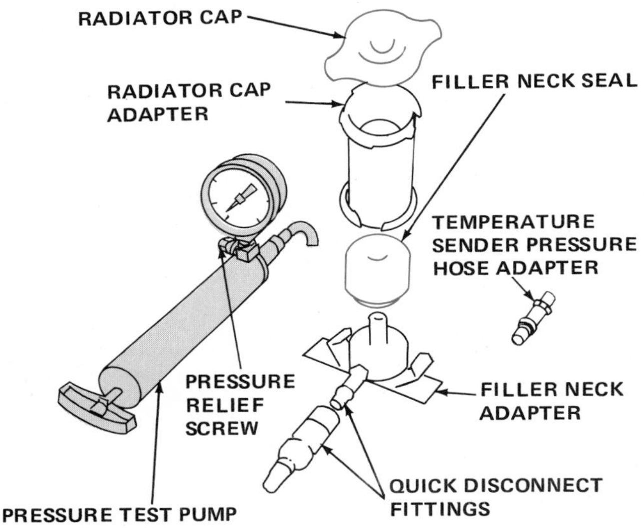

Fig. 22-9 Pressure tester with attachments. (*Ford Motor Company*)

● 22-10 Pressure Testing the System

To pressure-test the system, you apply pressure with a special tester (Figs. 22-9 and 22-10). The tester quickly shows a leaky cooling system. To use the tester, remove the radiator cap and spark plugs, fill the radiator within $\frac{1}{2}$ inch [12.7 mm] of the filler neck, wipe the neck sealing surface, and attach the tester. Then you operate the pump to apply a pressure of 15 psi (pounds per square inch) [1.0546 kg/cm^2]. If the pressure holds steady, the system is not leaking. But if the pressure drops, there are leaks. Look for external leaks at hose connections, hose, engine expansion plugs, water pump and cylinder-head gaskets, water-pump drive shaft, and radiator. See Fig. 22-11.

Fig. 22-10 Using a pressure tester to check a cooling system for leaks. (*Texaco Incorporated*)

● 22-11 Pressure Testing the Radiator Cap

The same pressure gauge as shown in Fig. 22-9 can be used to check the radiator pressure cap. If the cap will not hold its rated pressure, it should be discarded.

● 22-12 Testing the Fan Belt

Fan belts should be checked for wear and tension. If worn, fan belts should be replaced. On cars with two fan belts running in parallel, if one is worn and requires replacement, then both should be replaced. They come in

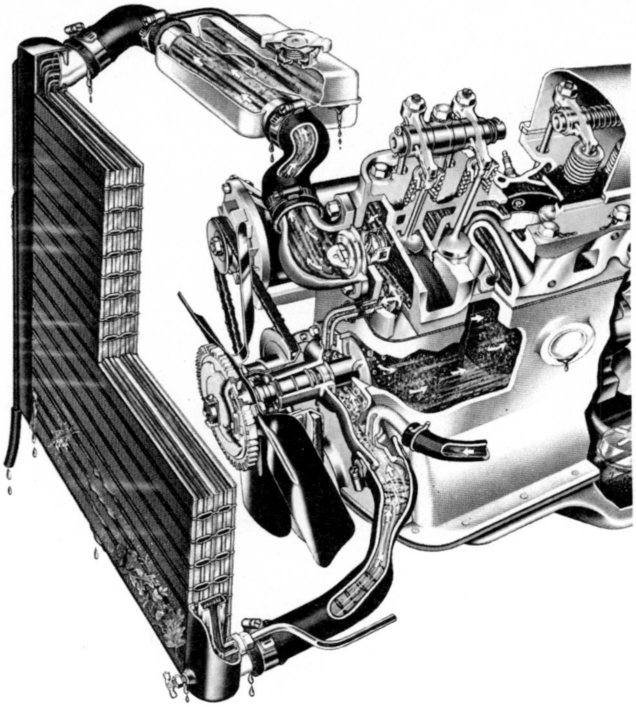

Fig. 22-11 Places in the cooling system where leaks might occur. (*Union Carbide Corporation*)

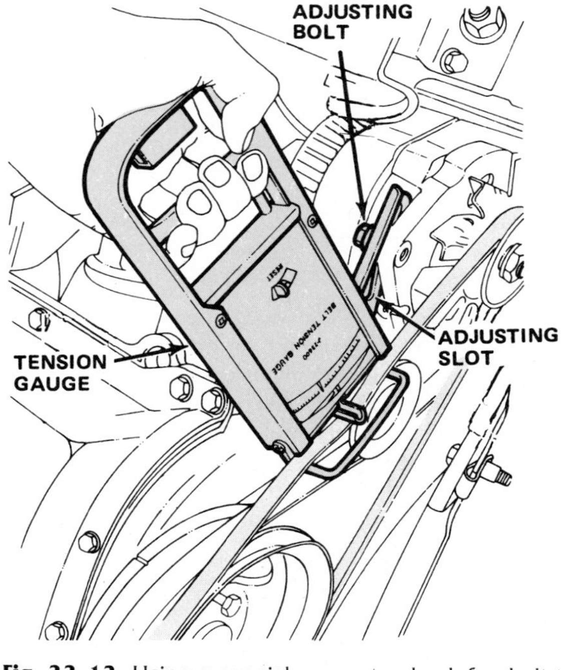

Fig. 22-12 Using a special gauge to check fan-belt tension. (*American Motors Corporation*)

matched sets. If only one is replaced, the new belt will take most of the wear because it is not stretched like the old belt. This means the new belt will wear rapidly while the old belt may slip, overheat, and also wear rapidly.

You should always use a tension gauge, such as shown in Fig. 22-12, to check and adjust fan-belt tension.

The fan belt should be checked every few thousand miles to make sure it is still in good condition. A fan belt that has become worn or frayed, or has separated plies, should be discarded. Remember that a faulty belt can not only cause engine overheating, but may also cause a run-down battery. The faulty belt cannot drive the water pump or alternator fast enough for normal operation.

● 22-13 Testing the System for Accumulations of Rust and Scale

At one time, it was common practice to use only water in the cooling system during the summer months. However, with pressurized systems and the use of antifreeze all year around, less rust collects in the cooling system. Some scale may accumulate from minerals in the water. This is the reason that many manufacturers recommend periodic cleaning of the cooling system. In modern high-temperature engines, the cooling system is more sensitive to accumulations of scale and rust. There is no easy way to determine the actual amount of scale and rust.

● 22-14 Replacing the Antifreeze

The antifreeze wears out and loses its effectiveness after a couple of years. In particular, the anticorrosion additives become used up. Before this happens, the old antifreeze should be drained and a new solution of water and antifreeze put into the system. At the same time, the usual recommendation is to flush or reverse-flush the system. If the system is filled with scale, it should also be cleaned out, as noted in ● 22-15.

● 22-15 Cleaning the Cooling System

The cooling system should be cleaned periodically to remove collected rust and scale and restore the system to top operating condition. Recommendations vary. Chevrolet, for example, recommends that the system be drained and flushed every 2 years. Coolant with new antifreeze should be put in at that time. Here is the procedure that Chevrolet recommends if the system is dirty enough to require cleaning:

1. Drain the cooling system. Remove the thermostat, and replace the thermostat housing. Close drain cocks, and add the liquid part of the cooling-system cleaner. Fill the system with water to about 3 inches [76.2 mm] below the top of the overflow pipe.
2. Cover the radiator, and run the engine at moderate speed until the water reaches 180°F [82.2°C]. Remove the cover from the radiator, and run the engine for another 20 minutes. Stop the engine if the water begins to boil.
3. With the engine running, add the powder part of the cooling-system cleaner. Run the engine for another 10 minutes.

CAUTION: Be careful not to scald your hands.

4. Stop the engine, and wait for it to cool. Then open the drain cocks again to drain the system. Remove the lower hose connection from the radiator.
5. Blow dirt and bugs from the radiator fins with compressed air, blowing from the engine side. Do not bend the radiator fins because this will decrease cooling-system efficiency.
6. Reverse-flush the radiator and engine block. In reverse flushing, water is forced through the system in the direction opposite to normal flow. This gets behind the scale and rust to loosen it so it will be flushed away.
7. To reverse-flush the radiator, remove the radiator upper and lower hoses, and replace the radiator cap.
8. Connect a leadaway hose to the top of the radiator. Connect a new hose to the radiator outlet at the bottom of the radiator (Fig. 22-13). Connect the water hose of the flushing gun to a water outlet, and the air hose to an air line, as shown. Connect the flushing gun to the new hose at the bottom of the radiator.
9. Turn on the water. When the radiator is full, turn on the air in short blasts, allowing the radiator to fill between blasts of air.

Careful: Apply the air gradually because the radiator will stand only about 20 psi [1.4062 kg/cm^2].

10. Continue until the water from the leadaway hose runs clear.
11. To reverse-flush the cylinder block and head, first disconnect the heater hose. Cap the connections at the engine. Then, with the radiator hose removed, attach a leadaway hose to the water-pump inlet. Connect a length of new hose to the coolant outlet at the top of the engine (Fig. 22-14). Connect the flushing gun to the new hose at the top of the engine.
12. Turn on the water. When the engine water jacket is full, turn on the air in short blasts.

NOTE: Excessive pressure can blow out the cylinder-block freeze plugs (also called expansion core plugs and water-jacket plugs). Servicing these plugs is covered in ● 39-22.

13. Continue flushing until the water from the leadaway hose runs clear.
14. The heater radiator core should also be reverse-flushed, but care must be taken to avoid too much air pressure. The core can be damaged by excessive air pressure.
15. Replace the thermostat and the radiator upper and lower hoses. Use new hoses if the old hoses are worn or damaged in any way. Make sure hoses are fully engaged on the tubes and that the clamps are properly tightened.
16. Add enough antifreeze to give full protection against freezeup in the lowest temperatures expected. Fill the system with water. Since the water is put in cold, the thermostat will close and prevent quick filling. With the thermostat closed, air is trapped back of the closed thermostat (Fig. 22-15). The thermostat has a small hole that will permit air to leak out, but this takes some time. So you may have to wait and refill the radiator a couple of times. The engine can be started and run for a few moments until the thermostat heats up and opens. Then completely fill the radiator with water.
17. Check the system for leaks after running the engine for a few minutes.

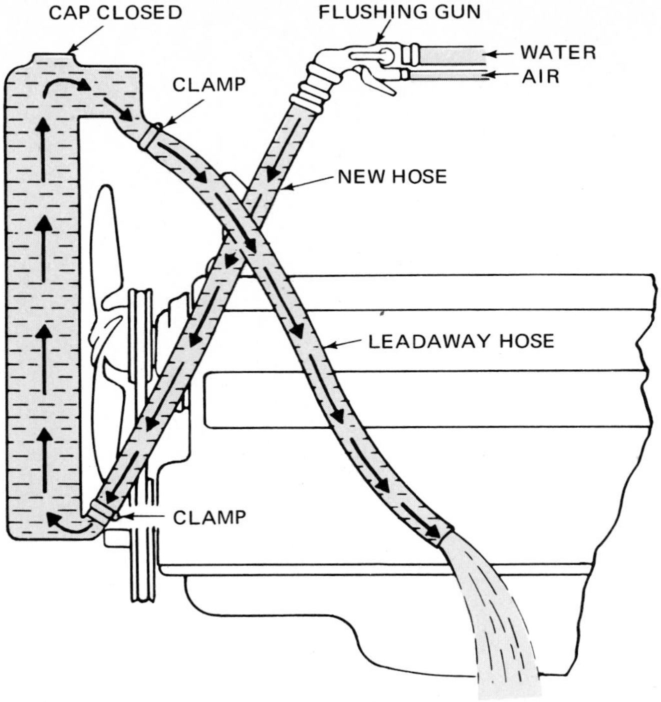

Fig. 22-13 Reverse flushing a radiator.

● 22-16 Fast-Flush Cleaning Method

The following procedure, recommended by Prestone, does not require removal of the engine thermostat and cylinder-block drain plugs. It is thus a quicker flushing procedure. With this method, the water supply is connected to the heater-supply nipple. It flows through the engine water jacket, the water pump, and the bottom of the radiator. It exhausts mainly through the filler neck of the top radiator tank. The specific directions, keyed to Fig. 22-16, follow.

1. Set the heater temperature control (1) to high. If the car has a vacuum-operated heater valve, start the engine and run it at idle during flushing. Be sure to turn it off before shutting off the water.
2. Open the radiator drain cock (2).
3. Remove the radiator cap, and install a leadaway hose or a deflection elbow (3).
4. Remove the hose from the heater-supply nipple at the engine block (4). Point the hose down so that it will drain. [If the nipple is hard to get at, remove the heater-supply hose at the heater. Connect the water supply to flow into this hose—not into the heater.

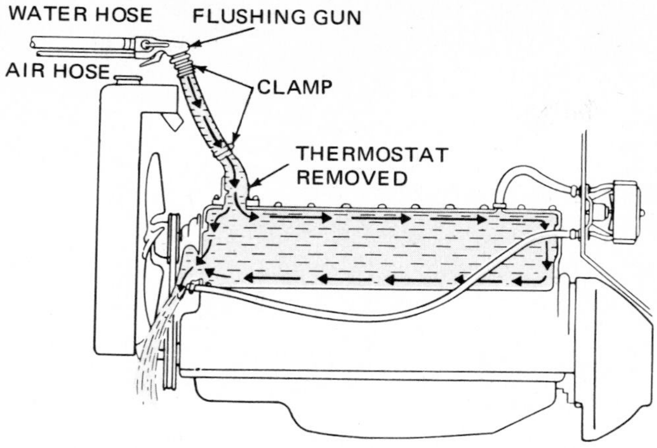

Fig. 22-14 Reverse flushing the engine water jackets.

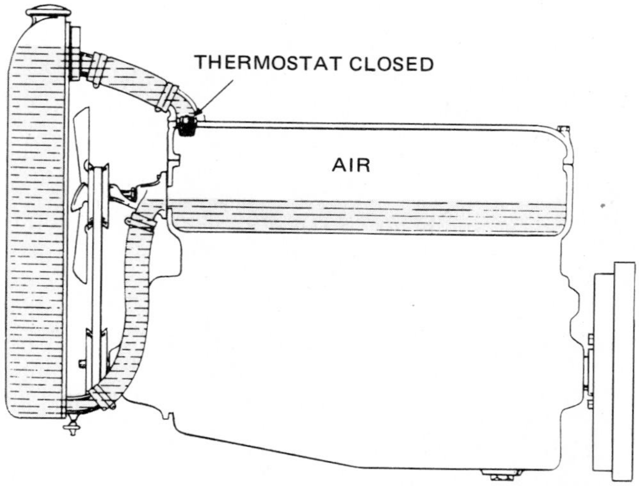

Fig. 22-15 Air trapped back of a closed thermostat as engine cooling system is filled.

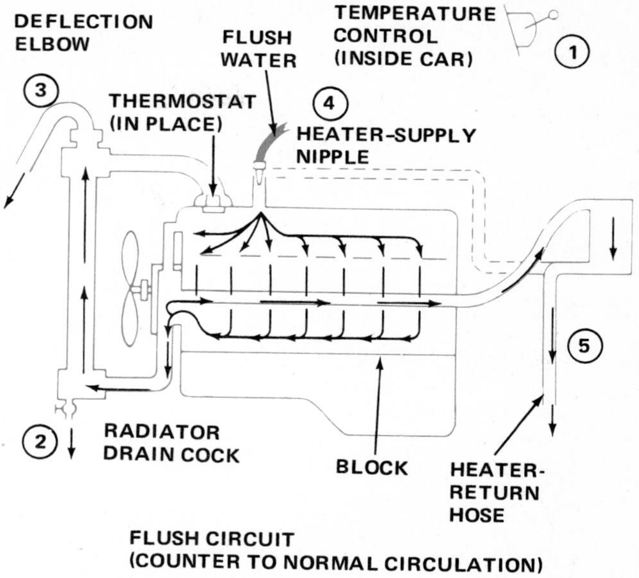

Fig. 22-16 Union Carbide or Prestone fast-flush method. *(Union Carbide Corporation)*

Attach a short piece of hose at the heater, and point it down (5). *Never connect the water supply directly to the heater!* This could damage the heater.]

5. After connecting the water supply either to the heater-supply nipple (4) or to the heater-supply hose, turn on the water. A flushing gun can also be used, as noted in the flushing instructions given earlier. Avoid excessive air pressure if air is used. During the last minute, squeeze the upper radiator hose to remove any trapped liquid.
6. Turn off the water. Reconnect the heater-supply hose, and disconnect the deflection elbow. Allow enough water to drain so there is room in the cooling system for antifreeze (or summer antirust compound). Then close the radiator drain cock. Add and adjust the antifreeze compound.

22-17 Fast-Flush-and-Fill Machine

This is a machine that flushes, tests, and refills the cooling system in about 10 minutes. It is not necessary to remove the thermostat, open the drains at the radiator and engine block, or remove and replace cooling-system hoses when using the machine. Here is the procedure:

1. Open the car hood, and remove the radiator cap. Attach a plastic tube to the radiator-filler neck. This is the leadaway hose through which the system is flushed.
2. Set the car temperature control to high.
3. Locate the heater inlet hose which is connected to the top of the engine. Apply temporary clamps on the two sides of the place you will cut. Cut the hose, and install the tee fitting with permanent clamps. The tee fitting now becomes a permanent part of the cooling system. Tighten the clamps enough to make leakproof joints. Remove the temporary clamps.
4. Attach the hoses from the flush-and-fill machine to the tee fitting. You are now ready to start the machine. The machine must be connected to an electric outlet and a water source.
5. Turn the selector switch on the machine to FLUSH. Operate the machine for 4 minutes, or until the flush water coming out the plastic tube is clean. This is a reverse-flushing operation.
6. While the machine is working, pour the required amount of antifreeze into the machine reservoir.
7. When flushing is finished, turn the selector switch off. Disconnect the plastic tube, and replace the radiator cap. Operate the pressure-test valve on the machine, and note the gauge reading. This tells you whether or not the system is tight. If the indication is that there are leaks, inspect the connections, hoses, radiator, water pump, and engine gaskets for signs of leakage.
8. If the system has no leaks, remove the radiator cap. Reinstall the plastic tube on the radiator-filler neck. Turn the selector switch to FILL so the machine will pump the antifreeze into the system. When the machine reservoir is empty, turn off the machine. Remove

the plastic tube from the radiator-filler neck, and reinstall the radiator cap.

9. Disconnect the coupling of the machine hose. Cap the tee fitting. The job is now complete.

● 22-18 Locating and Repairing Radiator Leaks

Leaks in a radiator are usually obvious. Telltale scale marks or water marks form on the outside of the core below the leaks. An accurate way to locate radiator leaks is to remove the radiator from the car and drain out all the water. Then close the openings at top and bottom, and immerse the radiator in water. It helps to locate leaks if the radiator is pressurized using the pressure tester (Figs. 22-9 and 22-10). Air bubbles will escape from the core through any leaks. Small leaks can sometimes be repaired without removing the radiator from the car. Certain liquid compounds, when poured into the radiator, seep through the leaks. They harden on contact with the air, sealing off the openings. A more effective way of repairing leaks is to solder them. If there are several leaks at various places in the core, it may not be worthwhile to attempt repair. The core is probably corroded to a point where other leaks will soon develop.

Radiator repair is usually done in a radiator shop that specializes in radiator service.

Removing a radiator is a simple job, although it takes a considerable amount of work. The procedure varies somewhat from car to car but, in general, is as follows:

1. Drain the engine and radiator by opening the drain cocks in the radiator and engine block.
2. Detach the upper and lower radiator hoses.
3. Remove any support bolts, horns, wiring harnesses, and so forth, that might interfere with radiator removal.
4. With these parts out of the way and the radiator loose, lift it straight up and off the car.

● 22-19 Water-Pump Service

The water pump is a simple mechanism that requires little service in normal operation. Some pumps require periodic lubrication. Most, however, have sealed ball bearings that require no lubrication. If the pump develops noise or leaks or becomes otherwise defective, it must be removed for repair. Refer to the manufacturer's shop manual for details of the servicing procedure.

REVIEW QUESTIONS

1. Describe two ways to test the thermostat.
2. Should you remove the radiator cap to check the coolant level on a system with an expansion tank? Why?
3. Explain how to use a pressure tester to test a cooling system for leaks.
4. Explain how to use the hydrometer to check antifreeze strength in the coolant.
5. What in the cooling system could cause slow warm-up?
6. What in the cooling system could cause engine overheating?
7. Describe the cooling-system cleaning procedure recommended by Chevrolet.
8. What does reverse flushing mean?
9. Explain the fast-flush method.
10. Explain how to use the fast-flush-and-fill machine.
11. What are the telltale signs of radiator leakage?
12. How can you detect leakage points in a radiator core that has been removed from a car?
13. Can leaks in a radiator core be fixed? How?
14. What is afterboil? What causes it?

SELF PROJECTS

1. Make a list of possible causes of engine overheating, including those not caused by cooling-system problems. File the list in your notebook. Making lists such as this helps you remember the essential information.
2. Make a list of places in the cooling system and engine where leaks can occur. You can refer to the illustration in the chapter while you make this list.
3. Make a list of cooling-system tests as outlined in the chapter. As you make the list, review in your mind the way that each test is made.
4. If you are able to go to a radiator shop, study the way that the specialists test and repair radiators. Write a description of this procedure and file it in your notebook.

Automotive Electrical and Electronic Equipment

4

Part Four of *Automotive Mechanics* discusses electricity and electronics. It explains how the components of the automotive electrical system work and how they are serviced. There are ten chapters in Part Four.

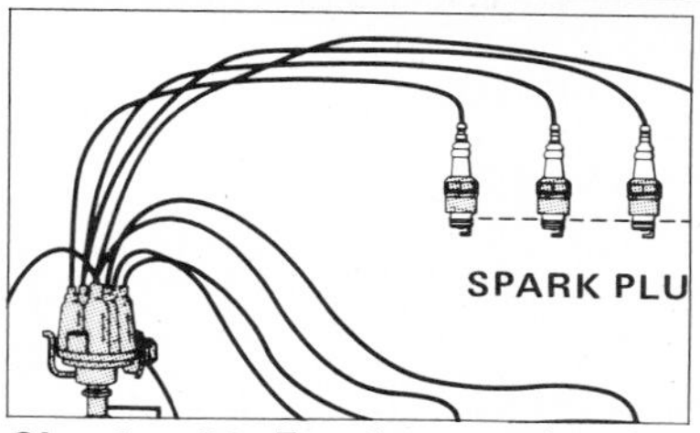

Chapter 23 Fundamentals of Electricity and Electronics

Chapter 24 Automotive Batteries

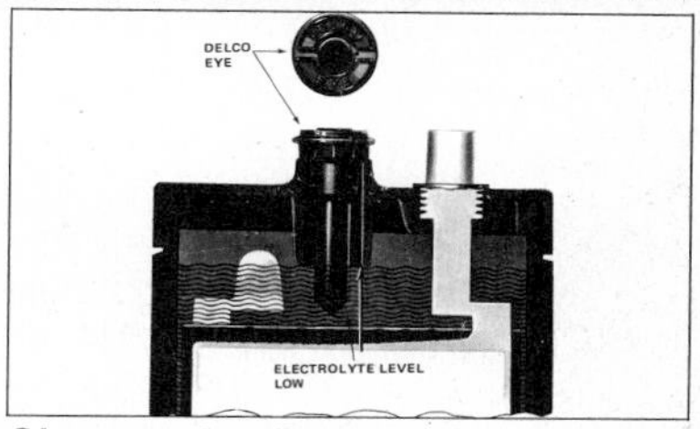

Chapter 25 Battery Service

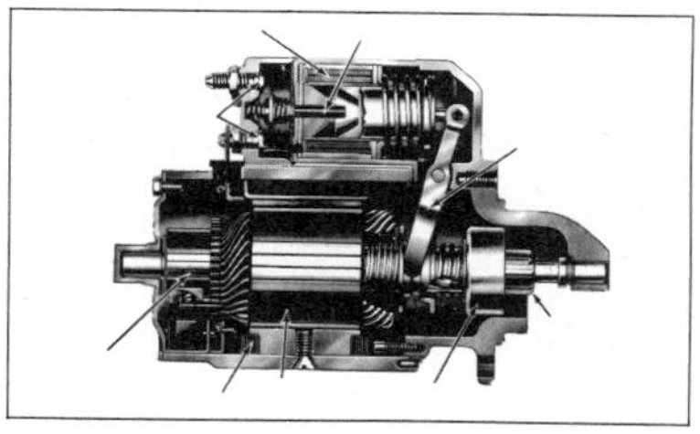
Chapter 26 Starting Motors

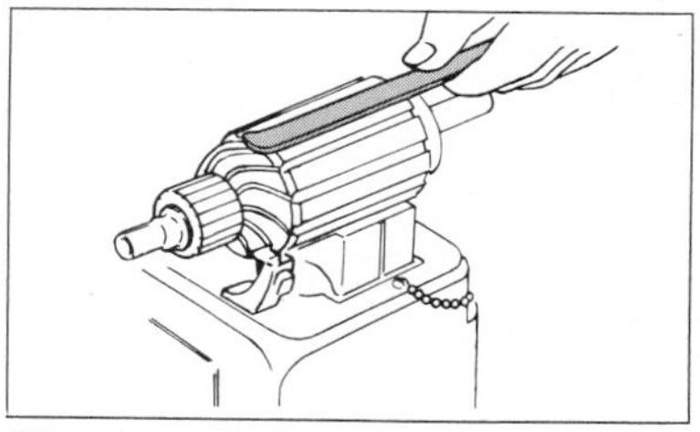
Chapter 27 Starting-System Service

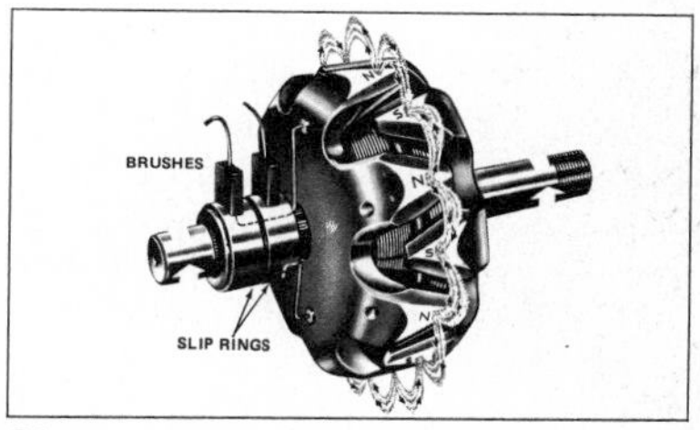

Chapter 28 Automotive Charging Systems

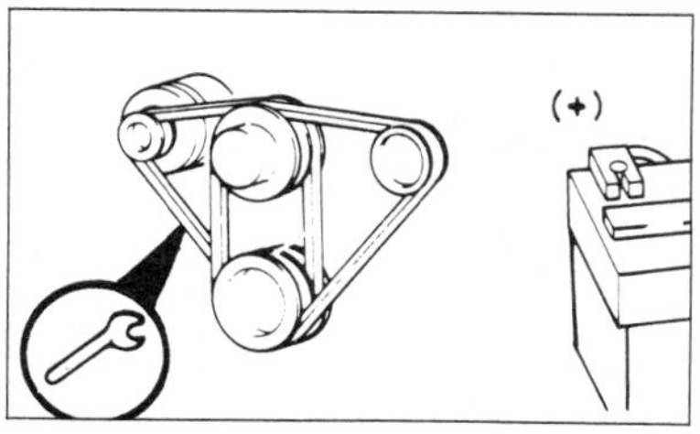

Chapter 29 Servicing Charging Systems

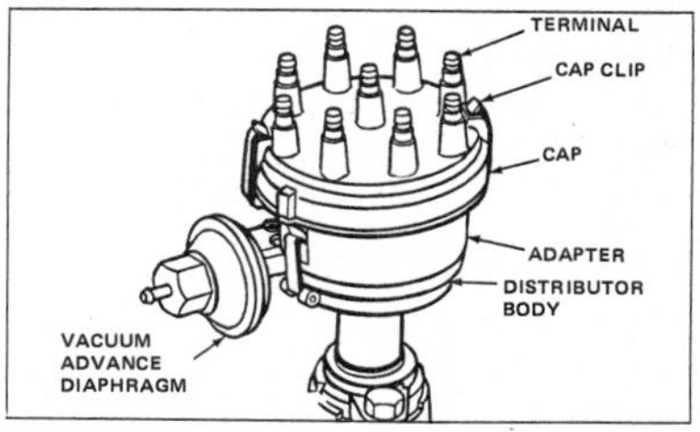

Chapter 30 Ignition Systems

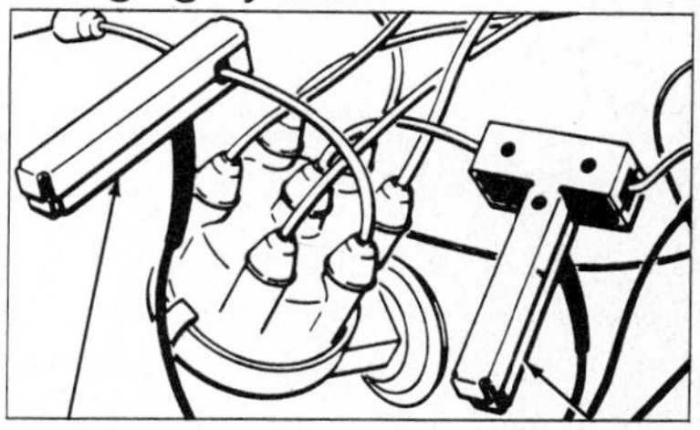
Chapter 31 Ignition-System Service

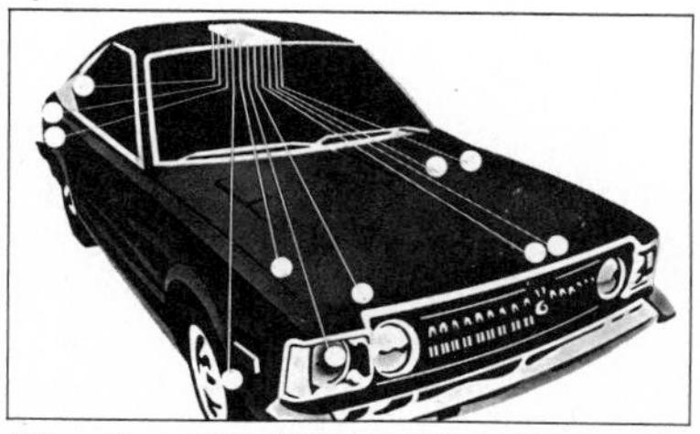
Chapter 32 Other Electric and Electronic Units

CHAPTER 23
Fundamentals of Electricity and Electronics

After you read this chapter, you should be able to:

- **Discuss electricity and explain what current is and how electrons are made to move in a conductor.**
- **Describe electromagnets and explain how they produce magnetism.**
- **Define amperes, voltage, and resistance, and explain how they are related in an electric circuit.**
- **Discuss electronics and explain how diodes and transistors work.**

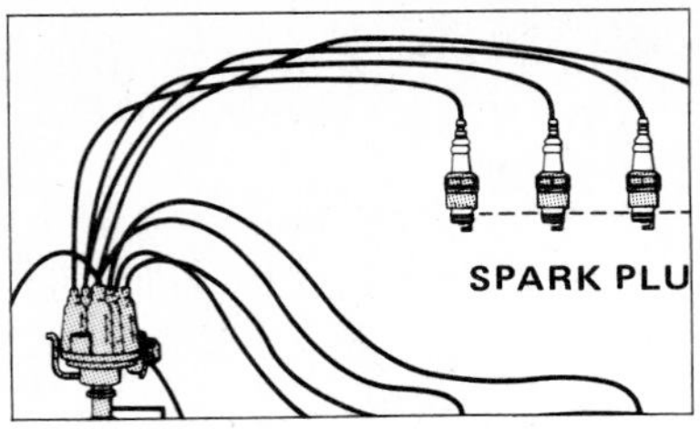

This chapter is your introduction to electricity and electronics. Electricity does many jobs in this world. It gives us light and runs much of the machinery around us. Shop and factory machine tools, refrigerators, television sets, and subway trains are only examples. In the car, it does several jobs. Figure 23-1 shows the electric system in an automobile, with the major components named. The automotive electric system starts the engine when you turn the ignition switch on. It makes the sparks that ignite the compressed air-fuel mixture. It operates the radio, electric gauges, and lights. We cover all of these in the pages that follow. First, however, let us find out what electricity is and how it does these jobs.

23-1 What Electricity Is

Recall from Chap. 4 that the atom has two basic particles that are electrically charged. These particles are:

1. The proton, which has a positive or plus (+) charge.
2. The electron, which has a negative or minus (−) charge.

Electrons can be detached from their atoms and moved along a conductor, such as a copper wire. We call this flow a *current* of electricity. It takes a fantastic number of electrons in motion to produce an electric current strong enough to do work. The reason is that electrons are so very small. In an ounce of iron, for example, there are something like 22 million billion billion electrons.

It takes an electrical device, such as a battery or an alternator, to get electrons moving in a conductor. Both the battery and the alternator can force the electrons to move all in the same direction. This electron flow, or electric current, can then light the car lights or turn the starting motor. When many electrons are moving, the current is high. When few are moving, the current is low. Chapter 24 covers batteries and Chap. 28 covers alternators.

23-2 Measuring Current

Current is measured in amperes, or *amps*. One ampere of electric current is a rather small amount of current. The battery can put out two or three hundred amps when it is operating the starting motor. Headlights draw 10 or more amps. A single ampere is 6 billion billion electrons flowing each second.

Obviously, nobody is going to count electrons to find out how many are flowing in a wire. We have to use an ammeter to measure amperes of current. The ammeter uses an interesting effect of electron flow.

This effect is that a flow of electrons, or an electric current, produces magnetism.

● 23-3 Magnetism

Magnetism comes in two forms—natural and electrical. Both act the same. Natural magnets are made of iron or certain other metals. Magnets attract iron objects. Two facts about magnets that are important are:

Magnets can produce electricity.

Electricity can produce magnets. Electrically produced magnets are called *electromagnets.*

We shall say more about electromagnets later.

● 23-4 The Ammeter

Now let us see how the ammeter measures electric current. The simplest kind of ammeter is shown in Fig. 23-2. This is the kind of ammeter found in many cars. Its purpose is to tell the driver whether the alternator is charging the battery or not. The battery will run down if the alternator is not working right—that is, if it does not charge the battery when it is supposed to. A run-down battery means no starting, no running engine, a useless car.

Here is how the ammeter works: The conductor is connected at one end to the battery. The pointer is mounted on a pivot. There is a small piece of iron, oval-shaped, mounted on the same pivot. This oval-shaped piece of iron is called the *armature*. A permanent magnet,

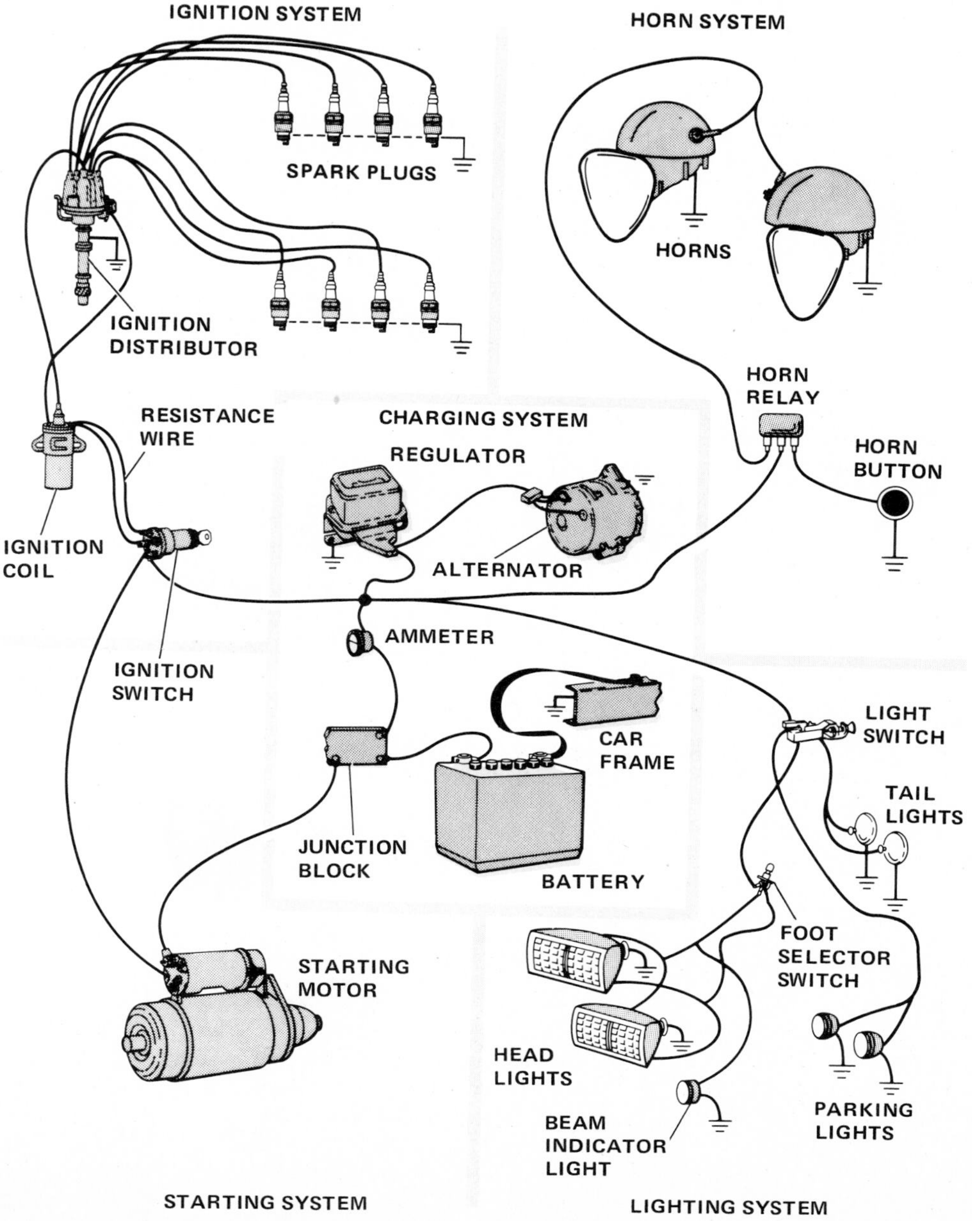

Fig. 23-1 Typical automobile electric system, showing the major electric units and the connections between them. The symbol ⏚ means ground, or the car frame or engine. Using the car frame or engine as the return circuit requires only half as much wiring. (*Delco-Remy Division of General Motors Corporation*)

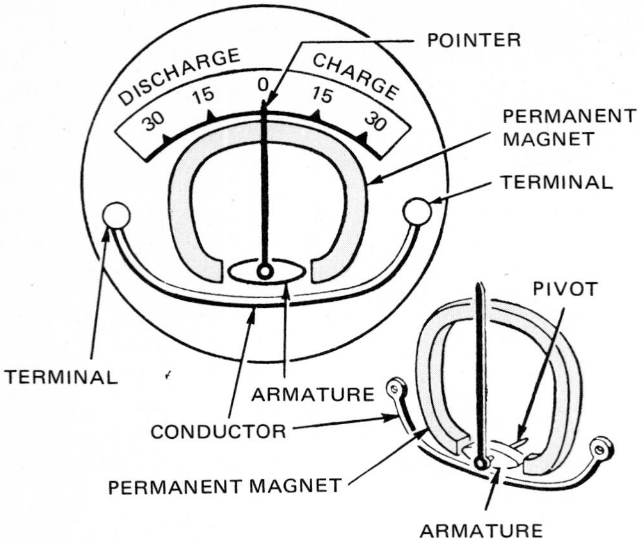

Fig. 23-2 Simplified drawing of a car ammeter and its interior construction.

almost circular in shape, is placed so its two ends are close to the armature. The permanent magnet attracts the armature and tends to hold it in a horizontal position. In this position, the pointer or needle points to 0. Nothing is happening. Now suppose the alternator starts sending current to the battery. This current passes through the conductor. The current produces magnetism. This magnetism attracts the armature and causes it to swing clockwise. This moves the pointer to the "charge" side. The more current that flows, the stronger the magnetism and the farther the pointer moves. The meter face is marked off to show the number of amperes flowing.

Now suppose the alternator is not working and you turn on the car lights. Current flows from the battery to the lights. It flows in the reverse direction through the conductor in the ammeter. Now the armature is attracted in the opposite direction, and it swings counterclockwise. This moves the pointer to the "discharge" side. The more current being taken out of the battery, the farther the pointer moves.

● 23-5 What Makes Electrons Move?

Electrons on the move make up electric current. But what makes the electrons move? Simply this: too many electrons in one spot. When electrons are gathered in one place, they try to move away. The battery and the alternator are devices that collect electrons. They collect electrons at one terminal by taking them away from the other. If we connect the two terminals with a conductor, electrons flow from the "too many" terminal to the "too few" terminal.

● 23-6 Voltage

Suppose there are a great many electrons at one terminal. And suppose the other terminal has a great shortage of electrons. With this great excess and great shortage, we say the electric pressure is high. That is, the pressure on electrons to move from the "too-many" terminal to the "too-few" terminal is high.

We measure this electric pressure in *volts*. High pressure is high voltage. Low pressure is low voltage. Car batteries are 12-volt units; 12 volts is low pressure. The spark at the spark-plug gap is a flow of electrons at high voltage. The voltage there can be 30,000 volts or more. That is high, but not nearly as high as the voltage on cross-country power lines. These are the wires that carry electricity from power plants to your home and to factories. The voltage in these lines is several hundred thousand volts. It could make a spark several feet long.

● 23-7 Insulation

Wires that carry electric current are covered with insulation. Power lines are hung from long insulators on the power poles or towers. We don't want those electrons to escape. If they escape, electricity is lost. Worse, electrons on the loose can cause serious trouble. For instance, damaged insulation on the wires of household appliances can start a fire. Or someone can be electrocuted.

Also, on the car, there are wires between the battery, the alternator, and other electrical devices. These wires are all covered with insulation. The insulation is a *nonconductor*. That is, it will *not* let electrons—electric current—flow through it. But if this insulation goes bad, the electric current will flow where it is not supposed to. It could take a short cut through the metal of the car frame and engine. This is called a *short circuit*. It can cause all sorts of trouble, as you will see later.

Just remember that insulation keeps the electric current moving in the proper paths, or circuits. These circuits include the wires and electrical devices on the car.

● 23-8 Magnets—Another Look

Let's take another look at magnets. Magnets act through *lines of force*. The lines of force stretch between the ends of the magnet. The two ends of the magnet are called the *poles*, or the *magnetic poles*. One pole is called the *north* pole, the other the *south* pole. The area surrounding the poles, where the lines of force are, is called a magnetic *field*.

● 23-9 Lines of Force

Lines of force have two properties. One is that they try to shorten up. If you hold the north pole of one magnet close to the south pole of another magnet, the two magnets pull together (Fig. 23-3). If we drew the lines of force between the two poles, the picture would look something like Fig. 23-4. The lines of force, stretching between the two poles, try to shorten up and so pull the two poles together.

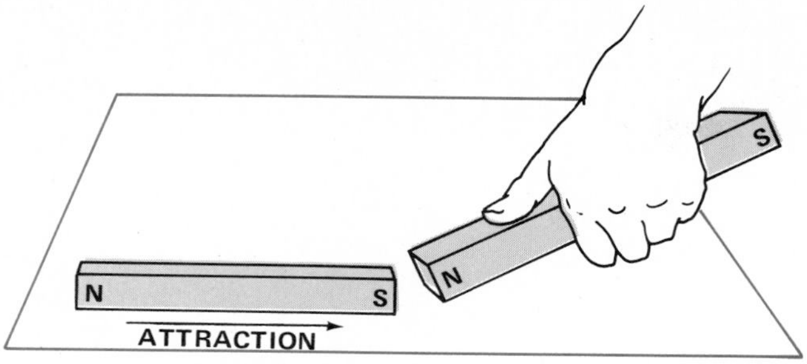

Fig. 23-3 Unlike magnetic poles attract each other.

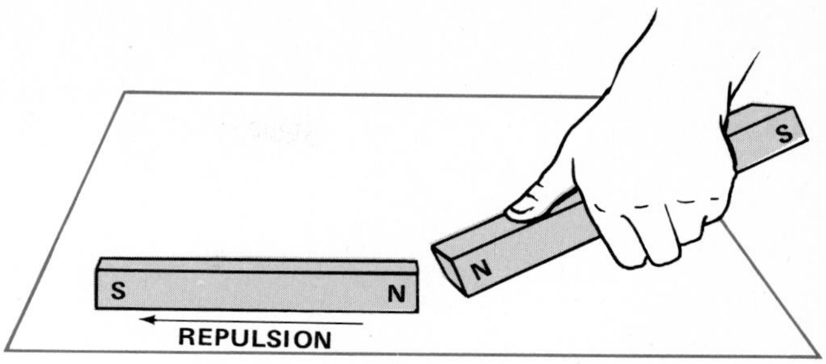

Fig. 23-5 Like magnetic poles repel each other.

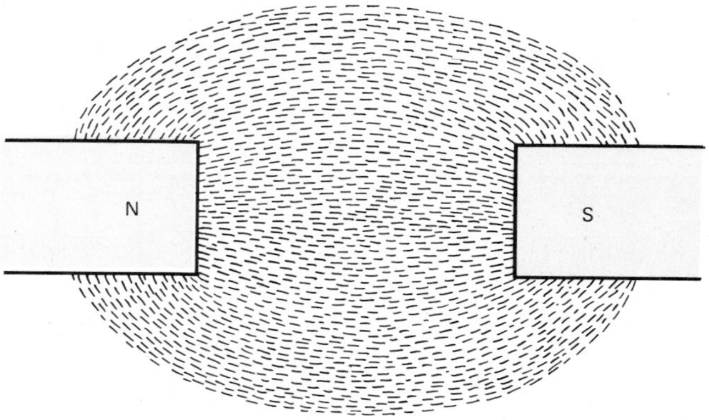

Fig. 23-4 Magnetic lines of force between two unlike magnetic poles. The magnetic lines of force tend to shorten, thus producing the attractive force that pulls the poles together.

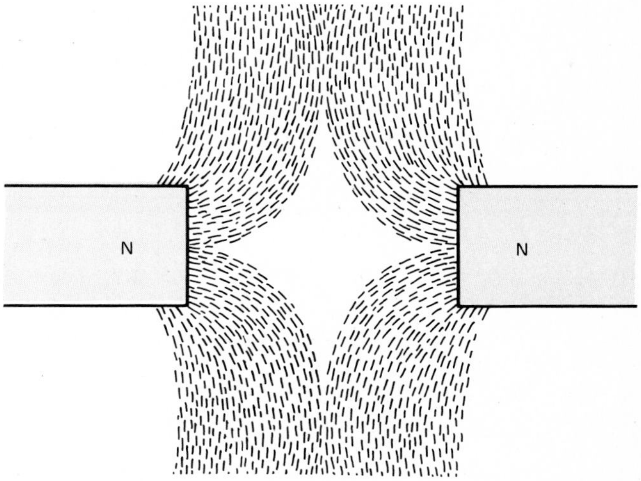

Fig. 23-6 Magnetic lines of force between two like poles. Magnetic lines of force tend to parallel each other, thus forcing the two like poles away from each other.

The other property is that lines of force run more or less parallel to each other. Also, they try to push away from each other. Suppose we bring two like poles together—two north poles, for example (Fig. 23-5). The lines of force run parallel to each other and try to push away. The magnet that is free actually moves away when the like pole of the other magnet is brought close (Fig. 23-6).

So now we can draw these conclusions:

Like magnetic poles repel each other. North repels north. South repels south.

Unlike magnetic poles attract each other. North and south attract.

23-10 Electromagnets

Electromagnets act just like magnets. An electromagnet can be made by wrapping wire around a tube. We saw what happened in the ammeter when current flowed one way or the other through the conductor. The current produced magnetism. That is, it produced magnetic lines of force.

Current flowing through a single conductor will not produce very much magnetism. But suppose we wind a conductor—a wire—around a tube. Then suppose we connect the ends of the wire to a source of electric current (or electrons). The winding produces strong magnetism. That is, there is a strong magnetic field around the coil of wire.

With current flowing through the winding, the winding acts just like a bar magnet. One end of it will either attract or repel a pole of a bar magnet. One end of the winding is a north pole; the other end is a south pole. You can change the poles by reversing the leads to the source of current. This shows that when electrons flow through in one direction, it makes one of the poles north. But when the electrons flow through in the reverse direction, the poles reverse. That is, the north pole becomes the south pole, and the south pole becomes the north pole.

An electromagnet, such as that made by winding wire around a tube, is also called a *solenoid*. It is used in several places in the electric system of the automobile. We shall find out more about this later, when we discuss the starting motor, alternator, and ignition system.

23-11 Resistance

An insulator has a high resistance to the movement of electrons through it. A conductor, such as a copper wire, has a very low resistance. Resistance is a fact of life in all electric circuits. We want resistance in some circuits so too much current (too many electrons) will not flow. In other circuits, we want as little resistance as possible so that a high current can flow.

Resistance is measured in *ohms.* For instance, a 1,000-foot [304.8-m] length of No. 10 wire (which is about 0.1 inch [2.54 mm] in diameter) has a resistance of 1 ohm. A 2,000-foot [605.6-m] length has a resistance of 2 ohms. If the wire is heavier, the resistance drops. For instance, a wire 0.2 inch [5.08 mm] in diameter (No. 4 wire) has only $\frac{1}{4}$ ohm resistance per 1,000 feet [304.8 m].

The explanation is simple. The longer the path, or circuit, the farther the electrons have to travel—that is, the higher the resistance to electric current. With the heavier wire, the path is wider. More electrons can flow, so the resistance is lower.

With most substances, copper for example, resistance goes up with temperature. A hot copper wire will carry less current than a cold wire. The reverse is true for a few materials. The substance in the engine unit of the cooling-system temperature indicator (● 21-11), for example, loses resistance as the temperature goes up.

● 23-12 Ohm's Law

There is a definite relation between current (electron flow), voltage (electric pressure), and resistance. As the electric pressure goes up, more electrons flow. That is, increasing the voltage increases the amperes of current. However, increasing the resistance decreases the amount of current that flows. These relationships can be summed up in a statement known as *Ohm's law,* which is shown below.

Voltage is equal to amperage times ohms:

$$E = IR$$

where E = voltage
I = current, in amperes
R = resistance, in ohms

The main thing to remember about Ohm's law is that increasing the resistance reduces the current. We shall discuss this again, several times, when we get to the automotive electrical system. A major cause of electrical troubles is excessive resistance in circuits. Such resistance can be due to poor connections, defective wires, or bad contacts.

● 23-13 One-Wire Systems

For electricity to flow, there must be a complete path, or circuit. That is, the electrons must flow from one terminal of the battery or alternator, through the circuit, and back to the other terminal. In the automobile, the engine and car frame are used as the return circuit. Therefore, no separate wires are required for returns from electrical devices to the battery or alternator. The return circuit is called *ground* and is indicated in wiring diagrams by the symbol ⏚. Just remember that ground—the engine and car frame—is the other half of the circuit. That is, it is the return circuit between the source of electricity (battery or alternator) and the electrical device.

● 23-14 Alternating and Direct Current

Most of the electricity generated and used in this world is alternating current (ac). The current flows first in one direction and then in the opposite direction. That is, it alternates. The current you use in your home is ac. It alternates 60 times per second and is therefore called 60-cycle [Hz] ac. (In the metric system of measurement, one cycle per second is called a hertz, abbreviated Hz.)

The automobile cannot use ac. The battery is a direct-current (dc) unit. When you discharge it—that is, connect electrical devices to it—you take current out in one direction only. The current does not alternate, or change directions. Likewise, the other electrical devices in the car operate on dc only.

● 23-15 What Electronics Is

The terms "electronics" and "solid-state" tend to scare some people. Actually, there is nothing so complicated about electronics. If you have learned about electricity, as covered in the first part of this chapter, you are ready for electronics and solid-state. You use electronic instruments every time you turn on your radio or television set, use your pocket calculator or electronic watch, or drive your recent-model car.

Before electronics, all electric circuits were turned on and off with mechanical switches of some sort. Now, thanks to electronics, circuits can be turned on and off electronically—with no moving parts. The primary circuit in the automotive ignition system is an example. The old-style system used a set of contact points. The new or electronic-style system uses electronics to do the job—faster, more accurately, and with no moving parts to wear. We describe ignition systems in Chap. 30.

Our special interest in this book is what electronics does in the automobile and how it does its job. First, we look at semiconductors, diodes, and transistors—the electronic devices that are the heart of electronic equipment. Basically, diodes are one-way valves that permit electricity to flow through in only one direction. Transistors are electrical switches that can start and stop a flow of current. Both diodes and transistors use materials called semiconductors.

Semiconductors are materials that are halfway between a conductor and a nonconductor. Sometimes they conduct, sometimes they act like insulators. You will see what we mean when we discuss diodes and transistors.

● 23-16 Diodes

The diode is a device that permits electricity to flow through in one direction but not in the other. Figure 23-7 shows this. The alternator is the device in the automotive electrical system that produces current to charge the battery and operate electrical devices that are turned on. However, the current from the alternator is alternating.

So diodes are used to change it to direct current. You cannot use alternating current to charge the battery or operate automotive electrical equipment. Direct current is required. A later chapter describes alternators and the diodes used in them.

23-17 Transistors

The transistor is a diode with some additional semiconductor material. This added material makes it possible for the transistor to amplify current. That is, a small signal current allows a large main current to flow. Figures 23-8 and 23-9 show this. In Fig. 23-8 a small current of 0.35 ampere is flowing to the base marked *n*. The *n* means this slice of semiconductor material has extra electrons which have negative charges. When the extra electrons flow in (after the switch is closed), they form a path through which current can flow from one *p* slice to the other, as shown by the arrows. The *p* means that the material lacks electrons and therefore, in effect, has a positive charge.

When the switch is opened, as shown in Fig. 23-9, no current (electrons) flows into the *n* slice. Thus there is no path through the slice for electrons to flow from one *p* to the other.

Just remember this—a small current to the controlling slice of semiconductor material allows a large current to flow through the transistor. When this small current—called the signal or trigger current—flows, the large current flows. When the signal current stops, the large current stops.

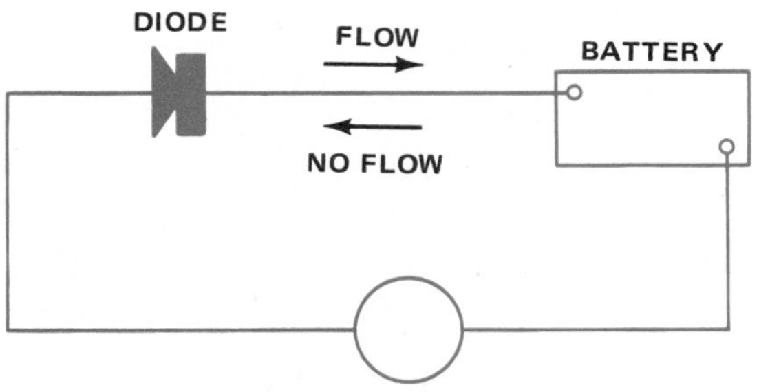

Fig. 23-7 Alternating current from an alternator can be rectified, or changed to direct current, by a diode so it can charge a battery.

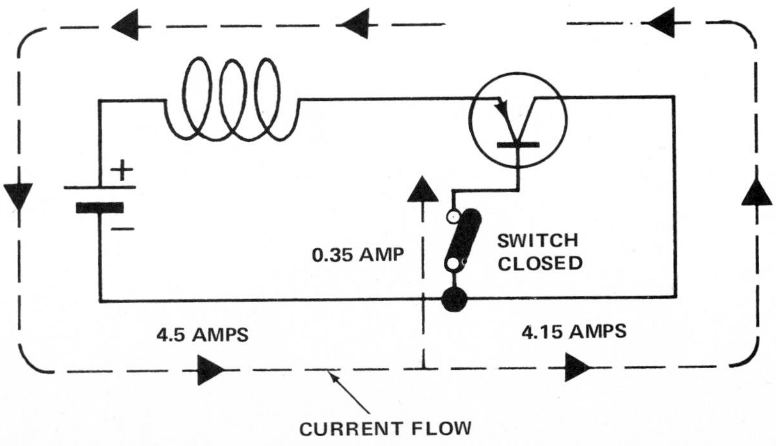

Fig. 23-8 When the switch is closed, current flows.

23-18 Integrated Circuits

Scientists and engineers have found ways to make diodes and transistors extremely small. This makes it possible to group large numbers of these semiconductor devices together in a small space. Such groups are called integrated circuits. That is, many components are put together, or integrated, into a single package. Such combinations are used in computers and complex controlling devices. Examples in the automobile are the electronic ignition system, electronic fuel injection, and electronic engine-system controls, all described later in the book.

Transistors, diodes, and similar devices are called *solid-state* devices because they are solids and have no moving parts (except electrons). That's a name you will hear many times—*solid state.*

23-19 Electronic Devices in the Automobile

A partial list of the automotive components using electronic controls follows:

1. Alternator voltage regulator
2. Electronic ignition system
3. Seat-belt interlock system (no longer used, although there may still be some on cars made in earlier years)
4. Air bags
5. Anticollision radar
6. Antilock braking system
7. Electronic fuel injection
8. Electronic engine-system control
9. Automatic level control
10. Accessories such as solid-state clocks, radios, and tape players, headlight dimmer, automatic on-off headlight control, speed controls, automatic temperature controls, and antitheft systems

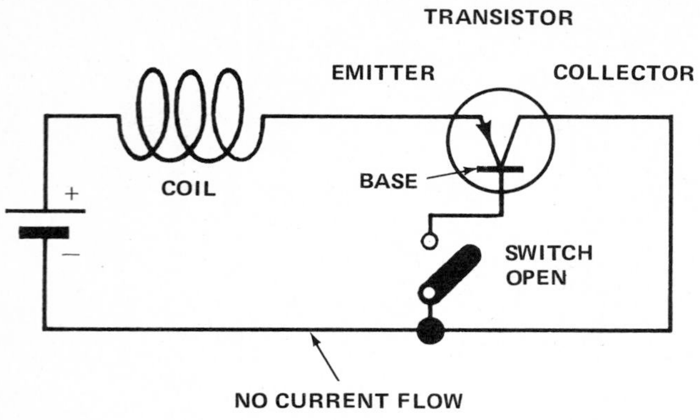

Fig. 23-9 When the switch is open, no current flows.

● 23-20 Electronics—The Mechanic's Friend

Many mechanics and technicians worried about the use of electronics in the automobile. It seemed that they would have to learn a whole new technology. But it did not turn out that way. An electronic circuit is a "go no-go" circuit. Either it works or it does not work. If it does not work, you replace the electronic component that is not doing its job. There are no adjustments. The testers are simple to use. They plug in and show, by lights or meter readings, what is working and what is not. So you quickly pinpoint causes of trouble and fix them by a simple replacement, as a rule.

Another advantage is that transistors and diodes never wear out, at least in theory. Connections may go bad over a period of time, but the solid-state device continues to do its job as long as it is not overheated or jarred excessively.

REVIEW QUESTIONS

1. What do you call a flow of electrons all moving in the same direction in a wire?
2. Name two sources of electric current in an automobile.
3. The ampere is a measurement of what?
4. Can magnets produce electricity?
5. Can electricity produce magnets?
6. What device is used to measure current?
7. What makes electrons move in a wire?
8. What is voltage?
9. What is the purpose of insulation?
10. What is a short circuit?
11. What are the two ends of a magnet called?
12. Do like magnetic poles attract or repel each other?
13. Do unlike magnetic poles attract or repel each other?
14. Explain how to make an electromagnet.
15. What is resistance? How is it measured?
16. What is Ohm's law?
17. What is one of the major causes of electrical troubles in automobile electric systems?
18. Explain what a one-wire system is. Why is it an advantage to have a one-wire system in the automobile?
19. What is dc?
20. What is ac?
21. Name ten electronic devices in the modern automobile.
22. What is the purpose of the diode?
23. Explain how a transistor works.
24. Why are transistors and diodes called solid-state devices?

SELF PROJECTS

1. You can perform many interesting experiments with magnets, a sheet of clear plastic, iron filings, some wire, and a small dry-cell battery. You can show magnetic lines of force and how electricity can produce magnetism. If you do any of these experiments, write down the procedures you followed and the conclusions you came to. File these reports in your notebook.
2. Refer to Fig. 23-1 and list all the electrical units you see in the illustration. File this list in your notebook.

CHAPTER 24
Automotive Batteries

After you have read this chapter, you should be able to:

- **Explain the construction and operation of storage batteries.**
- **Discuss battery efficiency and variations in terminal voltage.**
- **Explain the chemical actions in the battery during charge and discharge.**
- **Define the various battery ratings.**

In this chapter, we describe the construction and operation of automotive storge batteries. The battery is called an *electrochemical* device. This means that it produces electric current by chemical means, as we explain in the chapter.

24-1 Purpose of the Battery

The battery (Figs. 24-1 and 24-2) supplies current to operate the starting motor and the ignition system when the engine is being started. It also supplies current for lights, radio, and other electrical accessories when the alternator is not handling the electric load. The amount of current the battery can supply is limited by the "capacity" of the battery. This, in turn, depends on the amount of chemicals it contains.

24-2 Chemicals in the Battery

The chemicals in the battery are sponge lead (a solid), lead oxide (a paste), and sulfuric acid (a liquid). These three substances are made to react chemically to produce a flow of current. The lead oxide and sponge lead are held in *plate grids* to form positive and negative plates.

The plate grid (Fig. 24-3) is a framework of lead alloy with horizontal and vertical bars. The plate grids are made into plates by applying lead oxide paste. The horizontal and vertical bars hold the paste in the plate.

After the plates are assembled into the battery, the battery is given a "forming" charge. This changes the lead oxide paste in the negative, or minus, plate to sponge lead. It changes the lead oxide paste in the positive, or plus, plate to lead peroxide.

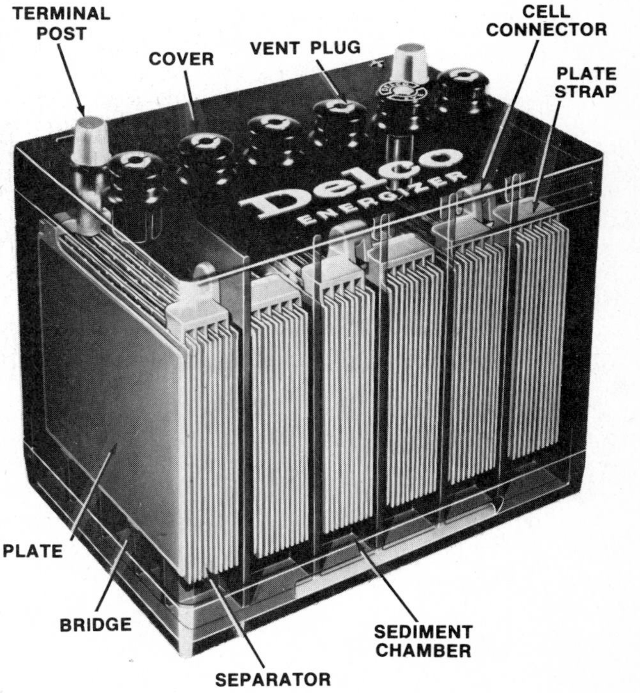

Fig. 24-1 Phantom view of a 12-volt battery. The case is shown as though it were transparent, so the insides can be seen. (*Delco-Remy Division of General Motors Corporation*)

24-3 Battery Construction

In the battery, several similar plates are properly spaced and welded, or *lead-burned,* to a strap. This forms a plate group (Fig. 24-4). Plates of two types are used, one for the positive plate group, the other for the negative plate group. A positive plate group is nested with a negative plate group. Separators are placed between the plates to form an element. The separators hold the plates apart so that they do not touch. At the same time the separators are porous enough to permit liquid to circulate between the plates. Wooden sheets, spun glass matted into sheets, and porous sponge-rubber sheets have been used as

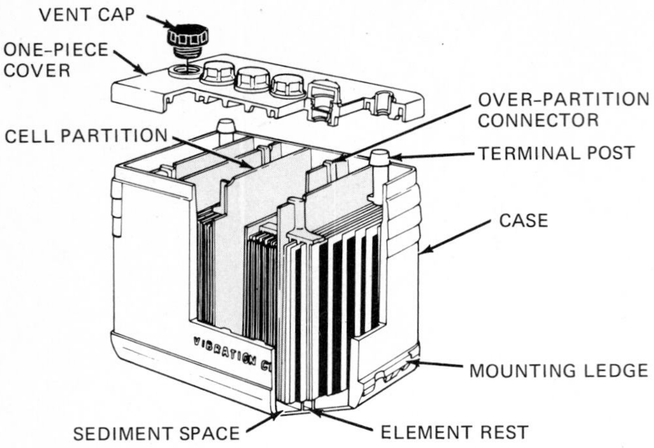

Fig. 24-2 Partly cut away and disassembled 12-volt battery. *(Ford Motor Company)*

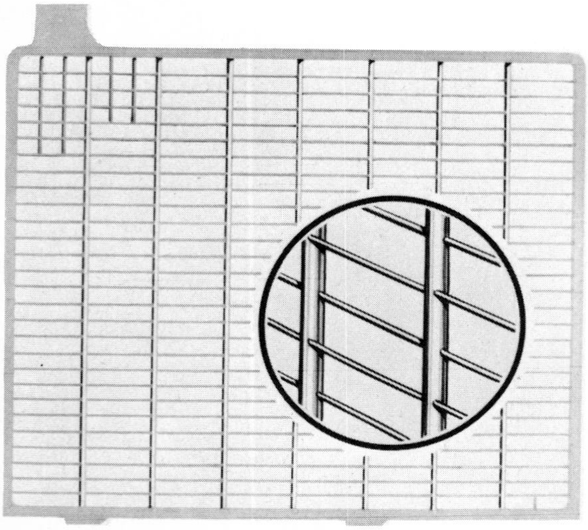

Fig. 24-3 Battery plate grid.

separators. Late-model batteries have separators made of acid-resistant polyvinyl chloride or polyethylene-saturated cellulose.

The elements are placed in cells in the battery case. Then heavy lead connectors are attached to the cell terminals to connect the cells in series. Many batteries have connectors that pass through the partitions, as in Fig. 24-5. Others have connectors that go over the partitions, as in Fig. 24-2. After the internal connectors are in place, the cover is put on (Figs. 24-1 and 24-2). The cover has openings through which liquid can be added when the filler plugs or vent caps are removed. After the liquid is added and the battery is given an initial charge, it is ready for operation.

Some batteries have the two main terminals on the battery cover, as in Figs. 24-1 and 24-2. Other batteries have the terminals in the side of the battery case, as in Fig. 24-6. This type of battery is called an ST (for side terminal or sealed terminal) battery by the manufacturer. Figure 24-7 shows how the cables are connected to the ST battery. It also shows the battery-mounting arrangement.

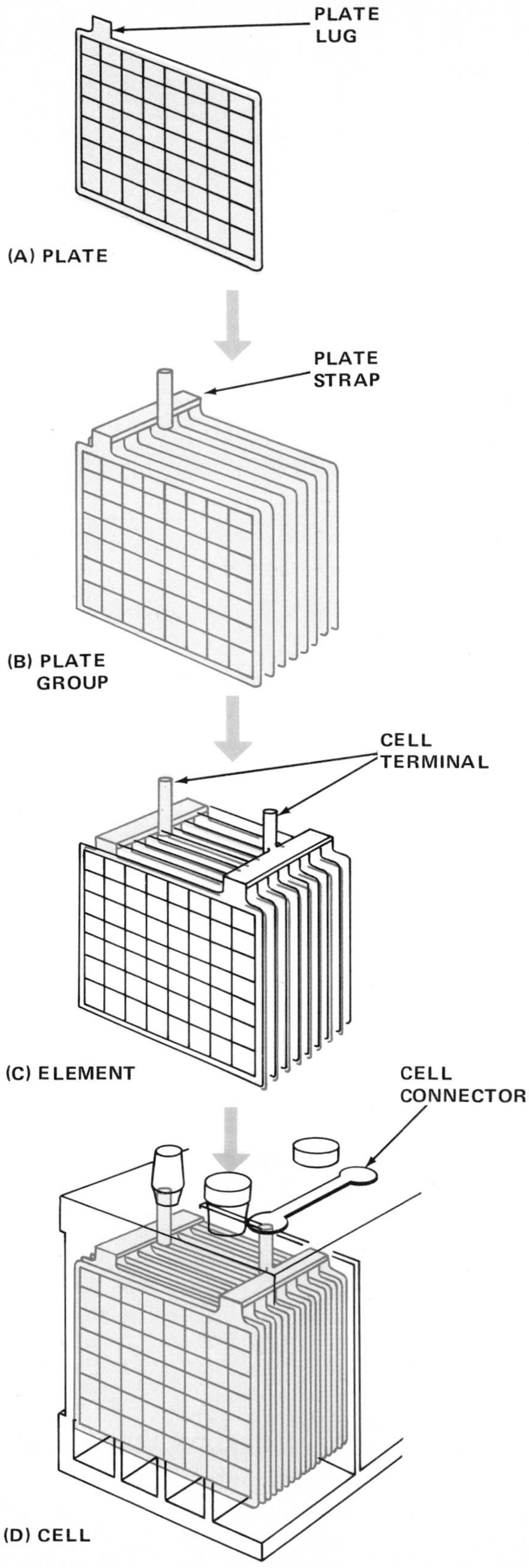

Fig. 24-4 Assembling of a battery cell.

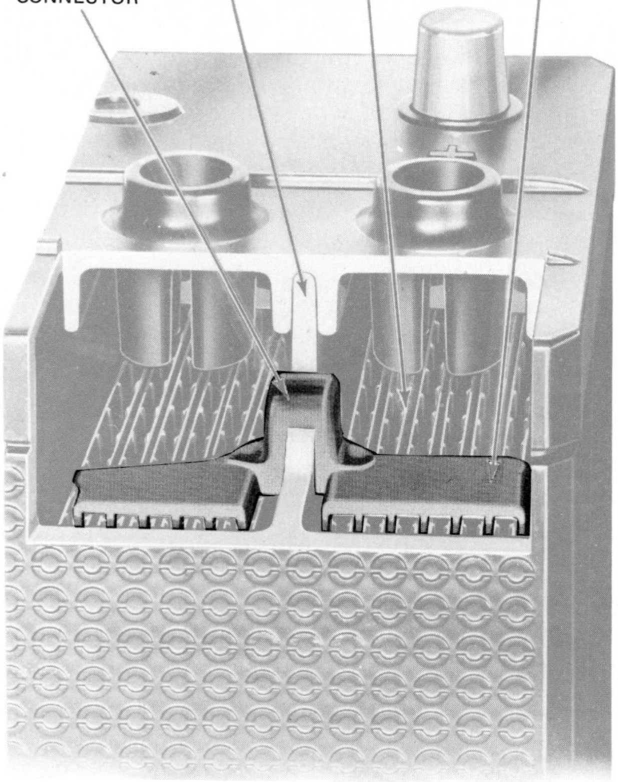

Fig. 24-5 Cutaway of two cells in a battery which has a one-piece cover and cell connectors that pass through the partitions between the cells. (*Delco-Remy Division of General Motors Corporation*)

There is also a "no-service" battery which is sealed. It never requires the addition of water, as other batteries occasionally do.

Many replacement batteries are the dry-charged type. They have fully charged positive and negative plates but no electrolyte. Because they are dry, they have a long shelf life and require no periodic servicing before they are sold. See ● 25-23. When a dry-charged battery is sold, electrolyte is added as explained in ● 25-23.

Delco-Remy calls their batteries "energizers." They are still, however, batteries.

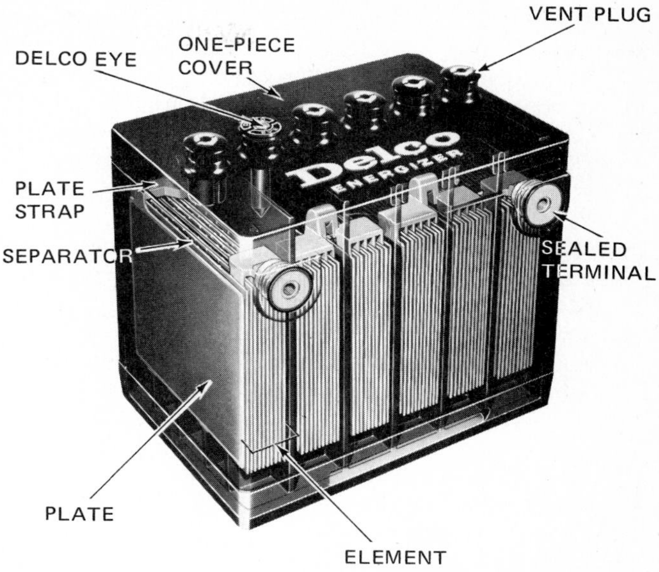

Fig. 24-6 Phantom view of a 12-volt battery with the terminals in the side of the battery case. (*Delco-Remy Division of General Motors Corporation*)

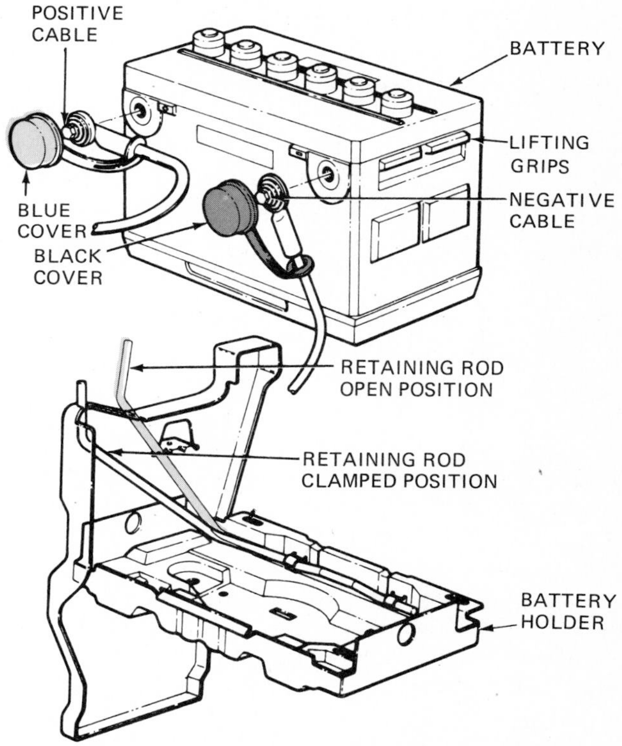

Fig. 24-7 Cable connections and battery-mounting arrangement for a side-terminal battery. (*Cadillac Motor Car Division of General Motors Corporation*)

● 24-4 Chemical Activities in the Battery

The liquid in a battery is called the *electrolyte.* It is made up of about 40 percent sulfuric acid and about 60 percent water (in a fully charged battery). When sulfuric acid is placed between the plates, chemical actions take place. These actions remove electrons from one group of plates and collect them at the other. This creates a 2.1-volt pressure between the two groups of plates. That is, there is a pressure of 2.1 volts between the two terminals of the battery cell. If the two terminals are not connected by any circuit, no further chemical activity takes place.

However, when the two terminals do become connected by an electric circuit, electrons (current) will flow. They flow from the terminal where chemical activity has collected them. They flow through the circuit to the other terminal, where the chemical activity has removed them. Chemical activities now begin again so the 2-volt pressure is maintained. The current flow continues. The chemical actions "use up" the sponge lead, lead peroxide, and sulfuric acid. Thus, after a certain amount of current has been withdrawn, the battery is *discharged* (or "run down," "dead," or "flat"). It is not capable of delivering any additional current. When the battery has reached this state, it may be *recharged.* This is done by supplying it with a flow of current from some external source. The external source forces current back through the battery. This reverses the chemical activities in the battery. The plates are restored to their original composition, and the battery becomes recharged. It is then ready to deliver additional current.

The chemical actions that take place are rather complicated. The sponge lead (negative plate) and lead peroxide (positive plate) change to *lead sulfate* during the discharge process. The sulfate comes from the sulfuric acid. The electrolyte loses acid and gains water as the sulfate goes into the plates. Thus, discharging the battery changes the two different chemicals in the battery plates to a third chemical, lead sulfate. Recharging the battery changes the lead sulfate back to sponge lead in the negative plates, and to lead peroxide in the positive plates. Meantime, the sulfuric acid reappears in the electrolyte of the battery.

24-5 Connecting Cells

Automotive batteries are usually 12-volt units. There are six cells in the 12-volt battery. The six cells are connected in series. In series connections, the voltages add. Some special applications use 24-volt batteries; these special-purpose batteries have 12 cells.

Although a battery cell at 80°F [26.7°C] will test on open circuit about 2.1 volts when fully charged, common practice is to call it 2.0 volts. Thus, a six-cell battery is said to be a 12-volt battery, rather than a 12.6-volt battery.

24-6 Battery Ratings

The amount of current that a battery can deliver depends on the total area and volume of active plate material. It also depends on the amount and strength of elecrolyte, that is, the percentage of sulfuric acid in the electrolyte. Factors that influence battery capacity—its ability to deliver current—include the number of plates per cell, the size and thickness of the plates, the cell size, and quantity of electrolyte. The ratings most commonly used in referring to battery capacity are discussed below.

1. Reserve capacity Reserve capacity is the length of time in minutes that a fully charged battery at 80°F [26.7°C] can deliver 25 amperes. A typical rating would be 125 minutes. This figure indicates the ability of a battery to carry the electrical operating load when the alternator is not operating.

2. Cold-cranking rate One of the two cold-cranking rates is the number of amperes that a battery can deliver for 30 seconds when it is at 0°F [−17.8°C] without the cell voltages falling below 1.2 volts. A typical rating for a battery with a reserve capacity of 125 minutes would be 430 amperes. This figure indicates the ability of the battery to crank the engine at low temperatures. The second cold-cranking rate is measured at −20°F [−28.9°C]. In this, the final voltage is allowed to drop to 1.0 volt per cell. A typical rating for a battery with a reserve capacity of 125 minutes would be 320 amperes.

3. Overcharge life units This is a measure of how well the battery will stand up when it is overcharged.

4. Charge acceptance This is a measure of how well the battery will accept a charge under normal operating conditions with a voltage-regulated automotive charging system.

5. Watts Delco is promoting an additional rating—watts. This is roughly equivalent to the battery cold-cranking rating.

24-7 Battery Efficiency

The ability of the battery to deliver current varies within wide limits. It depends on temperature and rate of discharge. At low temperature, chemical activities are greatly reduced; the sulfuric acid cannot work so actively on the plates. Thus the battery is less efficient and cannot supply as much current for as long a time. High rates of discharge will not produce as many ampere-hours as low rates of discharge. At high discharge rates, the chemical activities take place only on the surfaces of the plates. They do not have time to penetrate the plates and to use the materials below the plate surfaces.

Here are some figures that relate battery efficiency to battery temperatures. Note that these are only approximations.

Efficiency, percent	Battery temperature degrees F [C]
100	80 [26.7]
65	32 [0]
50	0 [−17.8]
10	−45 [−42.8]

24-8 Variations in Terminal Voltage

Because the battery produces voltage by chemical means, the voltage varies according to a number of conditions. These conditions and their effect on battery voltage may be summed up as follows:

1. Terminal voltage, battery being *charged,* increases with:
   a. Increasing charging rate. To increase charging rate (amperes input), the terminal voltage must go up.
   b. Increasing state of charge. As state of charge goes

up, voltage must go up to maintain charging rate. For example, a voltage of approximately 2.6 volts per cell is required to force a current through a fully charged battery. This is the reason that voltage regulators are set to operate at 15 volts—slightly below the voltage required to charge a fully charged battery. This setting protects the battery from overcharge.

c. Decreasing temperature. Lower battery temperatures require a higher voltage to maintain charging rate.

2. The terminal voltage of a battery that is being *discharged* decreases with:
   a. Increasing discharge rate. As the rate of discharge goes up, chemical activities increase and cannot penetrate plates so effectively. Therefore, voltage is reduced.
   b. Decreasing state of charge. With less of the active material and sulfuric acid available, less chemical activity takes place, and voltage drops.
   c. Decreasing temperature. With lower temperature, the chemical activities cannot go on so effectively, and the voltage drops.

REVIEW QUESTIONS

1. What is the purpose of the battery?
2. What are the three substances that react in the battery to produce current?
3. What are the two types of battery plates?
4. What is placed between the two sets of plates in a battery cell?
5. What are the two locations of the battery terminals?
6. Describe the chemical actions that take place during battery discharge.
7. Describe the chemical actions that take place during battery charge.
8. What is the reserve capacity of a battery?
9. Name three things that cause the battery terminal voltage to increase during battery charging.
10. Name three things that cause the battery terminal voltage to go down during battery discharging.

SELF PROJECTS

Make a collection of battery plates, separators, connectors, and covers. Find an old battery. Dump the electrolyte in a safe place such as on the ground a safe distance from any building. It will kill vegetation so don't dump it on grass or near trees or any place where there are growing plants. Then carefully remove the battery covers and take out the battery elements.

CAUTION: Remember that you are dealing with a highly corrosive substance—sulfuric acid. Wear rubber gloves. As you disassemble the battery, wash the various parts with lots of water. Let them soak in a bucket of clear water overnight. Don't get electrolyte on your skin or clothes. Read the cautions at the beginning of Chap. 25 about sulfuric acid. If you heed the cautions, you can safely tear down a battery.

Once you have removed the elements, cut the battery plates from the plate straps. Separate the plates and separators. Now, wash the plates and separators in clean water, rinsing them several times. Put them in a safe place to dry. Wash off the covers. When everything is dry, mount a negative plate, a positive plate, a separator, and other parts on a board for display in the school shop.

Some students make cutaway batteries such as shown in Fig. 24-1. This is more complicated. If you want to do this, discuss it with your instructor.

CHAPTER 25
Battery Service

After reading this chapter, you should be able to:

- **Discuss battery service and enumerate and explain the various battery checks. Safely make these checks on batteries.**
- **Explain the two cautions you must observe in working around batteries.**
- **Explain the various battery-charging procedures. Put a battery on charge and safely charge it.**
- **Explain how to put a dry-charged battery into service.**

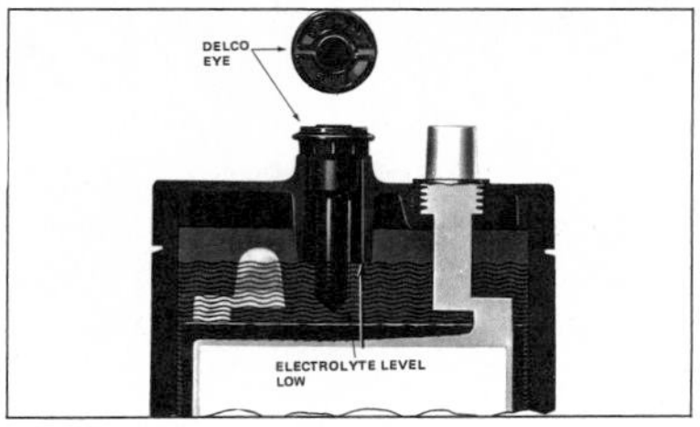

In this chapter, we look at battery maintenance and service. At one time, battery rebuilding was big business. In battery rebuilding, new plates and separators are installed in the old case and cover. Today, however, battery failure means discarding the old battery and installing a new one. Here is how to get full life out of your battery.

● 25-1 Battery Maintenance

Most people tend to forget about their car battery. That is, they forget it until one cold morning when the battery won't do its job and the engine won't start. Battery failure is one of the more common car troubles.

If people would check their batteries once in a while, much of this battery trouble could be avoided. Here are the things that should be done.

1. Visually inspect the battery.
2. Check electrolyte level in all cells periodically.
3. Add water if the level is low.
4. Clean off corrosion around battery terminals and top.
5. Check battery condition with a testing instrument. We describe battery-test instruments in later sections.
6. Recharge battery if it is low.

CAUTION: Sulfuric acid, the active ingredient in battery electrolyte, is very corrosive. It can destroy most things it touches. It will cause painful and serious burns if it gets on the skin. It can cause blindness if it gets into eyes. If you get battery acid (electrolyte) on your skin, flush it off at once with water. *Continue to flush for at least 5 minutes.* Put baking soda (if available) on the skin. This will neutralize the acid. If you get acid in your eyes, flush your eyes out with water, *over and over again* (Fig. 25-1). *Get to a doctor at once!* Do not wait!

CAUTION: The gases that form in the tops of battery cells during charging are very explosive. Never light a match or a cigarette near a recently charged battery. Never blow off

Fig. 25-1 If you get battery acid in your eye, flush it out immediately. Continue flushing for several minutes and see a doctor as quickly as possible.

a battery with an air hose. The compressed air could lift the cell cover and splash electrolyte all over you.

● 25-2 Visual Inspection of Battery

Look the battery over for signs of leakage, cracked case or top, corrosion, missing vent plugs, and loose or missing hold-down clamps. Leakage signs, which could indicate a cracked battery case, include white corrosion on the battery carrier, fender inner panel, or the car frame. If the top of the battery is covered with corrosion, and the owner complains that the battery needs water frequently, chances are the battery is being overcharged. This means a check of the charging system should be made.

The most common cause of a cracked top is improper installation. If the wrong wrench is used to remove or tighten the cable clamps, the battery top probably will be broken. See ● 25-18 on how to remove and replace cable clamps.

The most common cause of a cracked case is excessive tightening of the hold-down clamps. Also, a front-end crash, even if minor, so that little damage is done to the sheet metal, may cause the battery case to crack.

● 25-3 Checking Electrolyte Level and Adding Water

This is standard procedure at many service stations. The attendant starts the gas pump, and then opens the hood of the car to check engine oil level, the battery, and other things. The attendant removes the vent caps and looks down into the cells. If water is needed, it is added. Distilled water is recommended, but any water this is fit to drink may be used, provided it does not have excessive iron or other metallic chemicals in it. These chemicals can damage the battery, causing it to self-discharge rapidly and shortening battery life. Section 25-9 discusses self-discharge.

You can't check battery cells on sealed batteries, of course. But you can make sure connections are tight at the terminals. Also, many of these batteries have a state-of-charge indicator in the top and you can look at this to see if the battery needs a charge (Fig. 25-4).

CAUTION: If the charge indicator shows yellow, do not attempt to recharge the battery. The battery is on the verge of failure and should be replaced. Delco-Remy says a battery with a yellow charge indicator should *not be charged or tested.* Also, another battery should not be used to jump-start the car.

Many batteries have rings in the cell covers which show whether or not the battery needs water. The ring looks as shown in Fig. 25-2. The figure shows the ring when the level is too low and when it is correct.

Many batteries have "Delco Eye," a special vent cap or plug, in one of the six cells (Fig. 25-3). It has a transparent rod extending down into the cell. When the end of the rod is immersed, the exposed top of the rod shows black. When the level of the electrolyte falls below the tip, the top of the rod glows. This means water should be added. Thus vent caps do not need to be removed to check electrolyte level.

Careful: Don't add too much water. Too much water will cause the electrolyte to leak out. This will corrode, or eat away, the battery carrier and any other metal around.

● 25-4 Cleaning Corrosion Off the Battery

Battery terminals, especially those located on top of the battery, tend to corrode (Fig. 25-5). This corosion builds up around the battery and the cable clamp and also, unseen, between the terminal posts and clamps. To get rid of it, and to clean the battery top, mix some common baking soda in a can of water. Brush on the solution, wait until the foaming stops, and then flush off the battery top with water. If the buildup of corrosion around the terminals is heavy, detach the cables from the terminals (as explained later), and use the special wire brushes, shown in Fig. 25-6, to clean the terminal posts and cable clamps. Then coat the terminals with an anticorrosion compound to retard additional corrosion.

● 25-5 Checking Battery Condition

There are several ways to test battery condition. The most common way is with a battery hydrometer. Other methods use testing meters. In the shop, you will be shown how to use the instruments that are available and how to tell a good battery from a bad battery. Here, we cover the highlights of the tests.

● 25-6 Hydrometer Test

The hydrometer tests the *specific gravity,* or *gravity,* of the battery electrolyte. There are two types. One has a series of plastic balls, the other a glass float with a stem on top (Fig. 25-7). Both are used in the same way. The

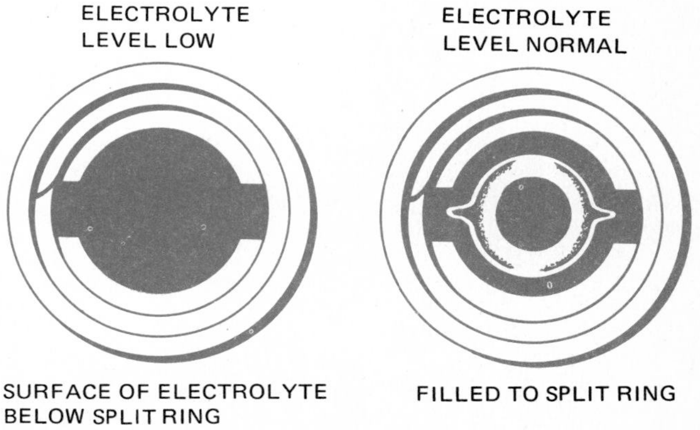

Fig. 25-2 Appearance of the electrolyte and split ring when the electrolyte is too low, and when it is correct. (*Delco-Remy Division of General Motors Corporation*)

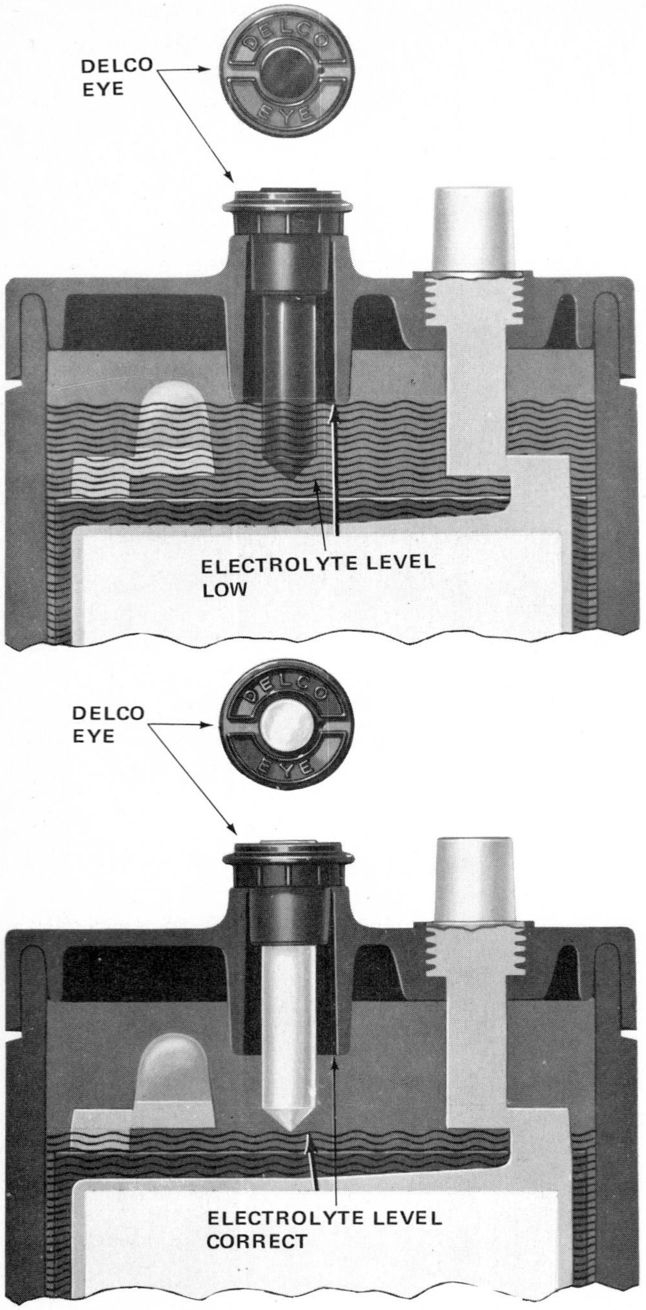

Fig. 25-3 Cutaway views of a battery cell with the Delco Eye. (*bottom*) The electrolyte is at the proper level. (*top*) It is low. (*Delco-Remy Division of General Motors Corporation*)

plastic-ball type is smaller and easier to use. You insert the end of the rubber tube into the battery cell so it is in the electrolyte. Then you squeeze and release the bulb. This sucks electrolyte up into the glass tube (Fig. 25-8). The number of balls that float tells you the state of charge of the battery cell. If all float, the cell is up to charge. If none float, the battery is run down, or flat. Test all cells.

To use the float type (right in Fig. 25-7), you suck electrolyte up into the glass tube in the same way, by squeezing and releasing the bulb. The float will float in this electrolyte. The amount the stem sticks out of the electrolyte tells you the battery state of charge. Take the reading at eye level, as shown in Fig. 25-9. Test all the cells.

Small batteries, such as those used in motorcycles, usually do not have enough electrolyte in the cells to float the float. Therefore, you must use the ball type to test these batteries.

CAUTION: Do not drip electrolyte on the car or on yourself! It will ruin the paint or the car and eat holes in your clothes! See Caution in ● 25-1.

If the float sticks out so the reading on the stem is between 1.260 and 1.290, the battery is fully charged. If the reading at the electrolyte level is between 1.200 and 1.230, the battery is only half charged. If the reading is around 1.140, the battery is about run down and needs a recharge. The following table of specific-gravity readings gives a general idea of battery conditions.

1.265–1.299	Fully charged battery
1.235–1.265	Three-fourths charged
1.205–1.235	One-half charged
1.170–1.205	One-fourth charged
1.140–1.170	Barely operative
1.110–1.140	Completely discharged

If some cells test significantly lower than others, it means there is something wrong with those cells. it could be a cracked case has allowed electrolyte leakage, or perhaps there is internal damage to the plates or separators. If the variation is only a few specific-gravity points, then there is probably no cause for alarm. But if the low cells measure 25 to 50 points lower, then those cells are defective and the battery should be replaced.

Some late-model 12-volt batteries for passenger-car service have a somewhat lower specific gravity when

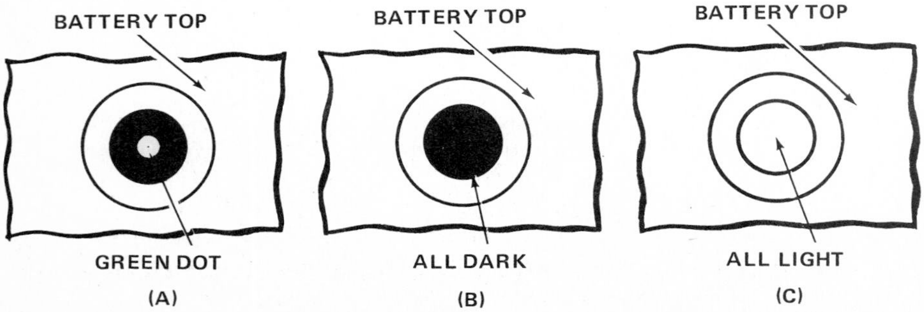

Fig. 25-4 Appearance of the charge indicator in the top of some sealed, or no-service, batteries. *A*, If the green dot shows, the battery is in a charged condition; *B*, if the indicator shows black, the battery is low and should be charged before any test; *C*, if the indicator shows light yellow, the battery is dead and should be discarded.

charged. For instance, one type is fully charged with a specific gravity of 1.270. Other batteries, for example, those used in hot climates, have a specific gravity of 1.225 when fully charged.

The decimal point is not normally referred to in a discussion of specific gravity. For example, "twelve twenty-five" means 1.225, and "eleven fifty" means 1.150. Also, the word "specific" is dropped, so that the term becomes just "gravity."

25-7 Variation of Gravity with Temperature

In addition to the effect of the state of charge on the gravity, temperature also changes the gravity. This effect is due to the fact that as a liquid cools it becomes thicker and gains gravity. As a liquid warms it becomes thinner and loses gravity. Thus, temperature must be considered when a gravity reading is taken with the float-type hydrometer (Fig. 25-9). On the ball type, this is not necessary because the balls take on the temperature of the electrolyte and thus they nullify the temperature effect. On the float type, however, a correction must be made if the temperature varies from standard. This correction involves the addition or subtraction of gravity points,

Fig. 25-5 Corroded battery cable and terminal post.

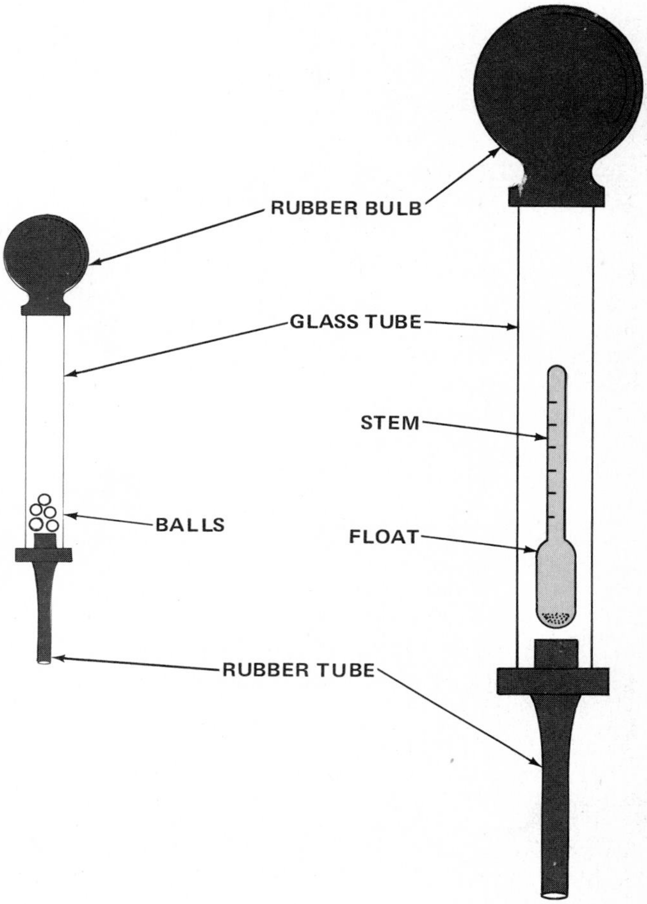

Fig. 25-7 Battery hydrometers.

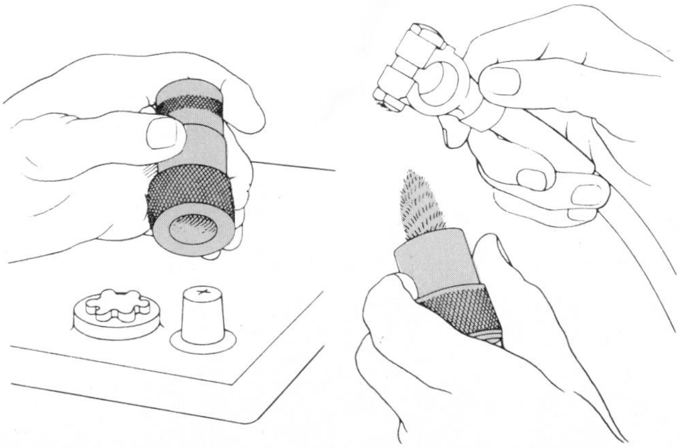

Fig. 25-6 Using special wire brushes to clean battery terminal posts and cable clamps. (*Buick Motor Division of General Motors Corporation*)

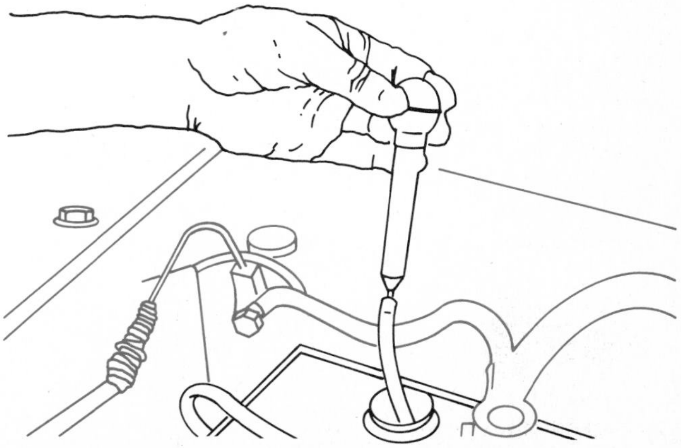

Fig. 25-8 Using ball-type hydrometer to check a motorcycle battery. (*K-D Manufacturing Company*)

according to whether the electrolyte temperature is above or below the 80°F [26.7°C] standard. The gravity of electrolyte changes about four points, or thousandths (0.004) for every 10°F temperature. To make temperature correction, four points must be subtracted for every 10°F below 80°F [26.7°C].

EXAMPLE 1.250 at 120°F. Add 0.016 (4 × 0.004). Corrected reading is 1.266. 1.230 at 20°F. Subtract 0.024 (6 × 0.004). Corrected reading is 1.206.

The battery hydrometer shown in Fig. 25-10 is compensated for temperature. No temperature corrections have to be made when this hydrometer is used.

25-8 Loss of Gravity from Age

As the battery ages, the electrolyte gradually loses gravity. This is because of the loss of active material from the plates (as it sheds and drops into the bottom of the cells). It is also lost due to gassing. Over a period of 2 years, for example, battery electrolyte may drop to a top gravity, when fully charged, of not more than 1.250. The original top gravity, when new, may have been 1.280. Little can then be done to restore gravity, since the loss is an indication of an aging battery.

25-9 Loss of Gravity from Self-Discharge

If a battery is allowed to stand idle for a long time, it will slowly self-discharge. This is brought about by internal chemical reactions between the battery materials. The higher the battery temperature, the more rapidly self-discharge will take place. The lead sulfate that forms on the battery plates as a result of self-discharge is difficult to reconvert into active material. Thus, a battery that is badly self-discharged may be ruined.

25-10 Battery Gravities for Hot Climates

In hot climates, chemical activities take place more readily in the battery. Thus, it is often desirable to adjust the gravity reading to as low as 1.210 (28.5 percent acid) for a fully charged battery. This reduces the amount of self-discharge and prolongs the life of the battery. On discharge, the electrolyte may be reduced to a gravity as low as 1.075 before the battery stops delivering current. Where there is no danger of freezing, low gravities can be used.

25-11 Freezing Point of Electrolyte

The higher the gravity of the electrolyte, the lower its temperature must be before it freezes. The battery must be kept in a sufficiently charged condition to prevent its freezing. Freezing usually ruins the battery (see Fig. 25-11).

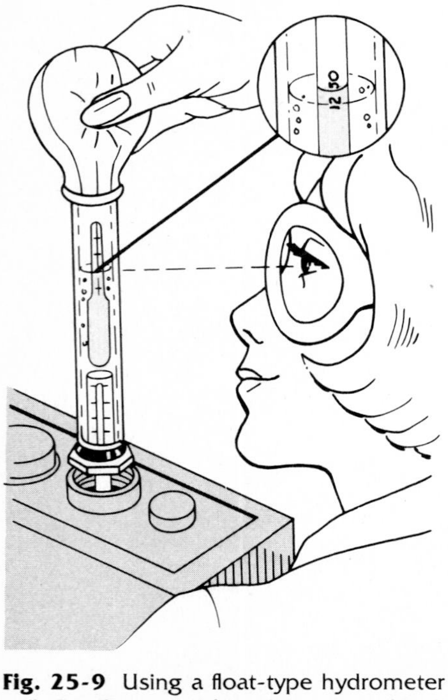

Fig. 25-9 Using a float-type hydrometer to check a battery cell. Reading must be taken at eye level.

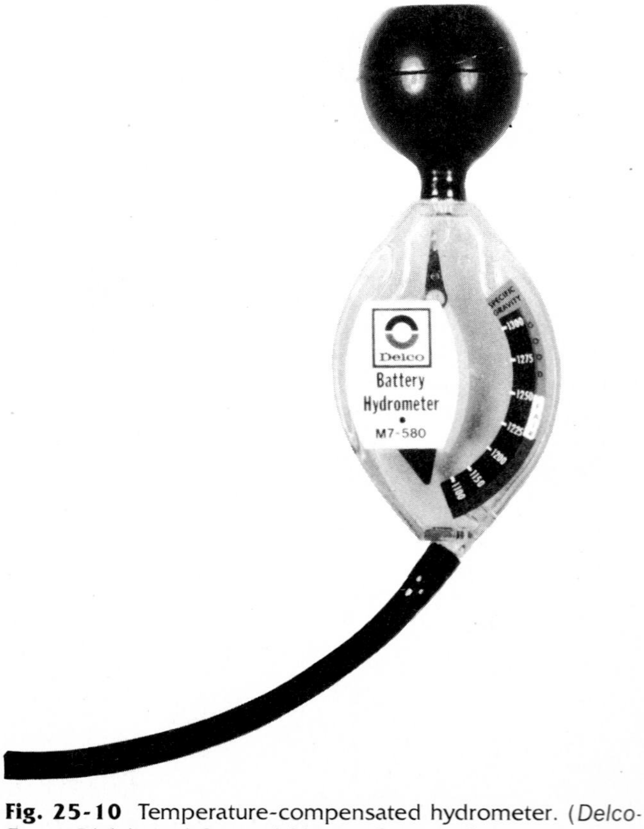

Fig. 25-10 Temperature-compensated hydrometer. (*Delco-Remy Division of General Motors Corporation*)

25-12 Refractometer Tester

This is the tester described in Chap. 22 and illustrated in Figs. 22-2 to 22-5. It is used in the same way for testing the antifreeze strength of engine coolant and the specific gravity of battery electrolyte. Refer to Figs. 22-2 to 22-5 and ● 22-6 for details of the use of the refractometer. The major difference is that, when testing the battery, you use the black dip stick. First, pick up a couple of drops of electrolyte from a battery cell with the black dip stick. Then put the end of the black dip stick into the cover-plate opening. This deposits the drops of electrolyte in the measuring window. Now you can read the electrolyte strength and determine the battery's state of charge. Test all battery cells in this manner. After each test, wipe the black dip stick, the measuring window, and the plastic cover of the refractometer (Fig. 22-2).

CAUTION: Be careful not to drip electrolyte on yourself or on the car. Remember that electrolyte contains sulfuric acid, and this acid is very corrosive. It will ruin the paint on the car. And, of course, it will burn your skin if you get electrolyte on it. See ● 25-1.

Specific gravity	Freezing temperature, degrees F [C]
1.100	18 [−8.2]
1.160	1 [−17.2]
1.200	−17 [−27.3]
1.220	−31 [−35]
1.260	−75 [−59.4]
1.300	−95 [−70.5]

Fig. 25-11 Table of specific gravities and freezing points.

25-13 High-Discharge or Capacity Test

For this test, the battery voltage is measured during a high discharge. The battery should be in good condition with no obvious defects, such as a broken cover or case. Specifications for the amount of high discharge vary. You should always check the manufacturer's manual. Figure 25-12 outlines the procedure recommended by Ford Motor Company.

25-14 The 421 Test

The 421 test is designed for use with batteries having a one-piece cover. The test requires a special tester which applies a series of timed discharge and charge cycles to the battery. The battery condition can thus be determined very accurately within a few minutes. When using the tester, follow the tester manufacturer's insturctions carefully.

25-15 Cadmium-Tip Test

This test requires a special tester. It has cadmium tips that are inserted into the electrolyte of neighboring cells

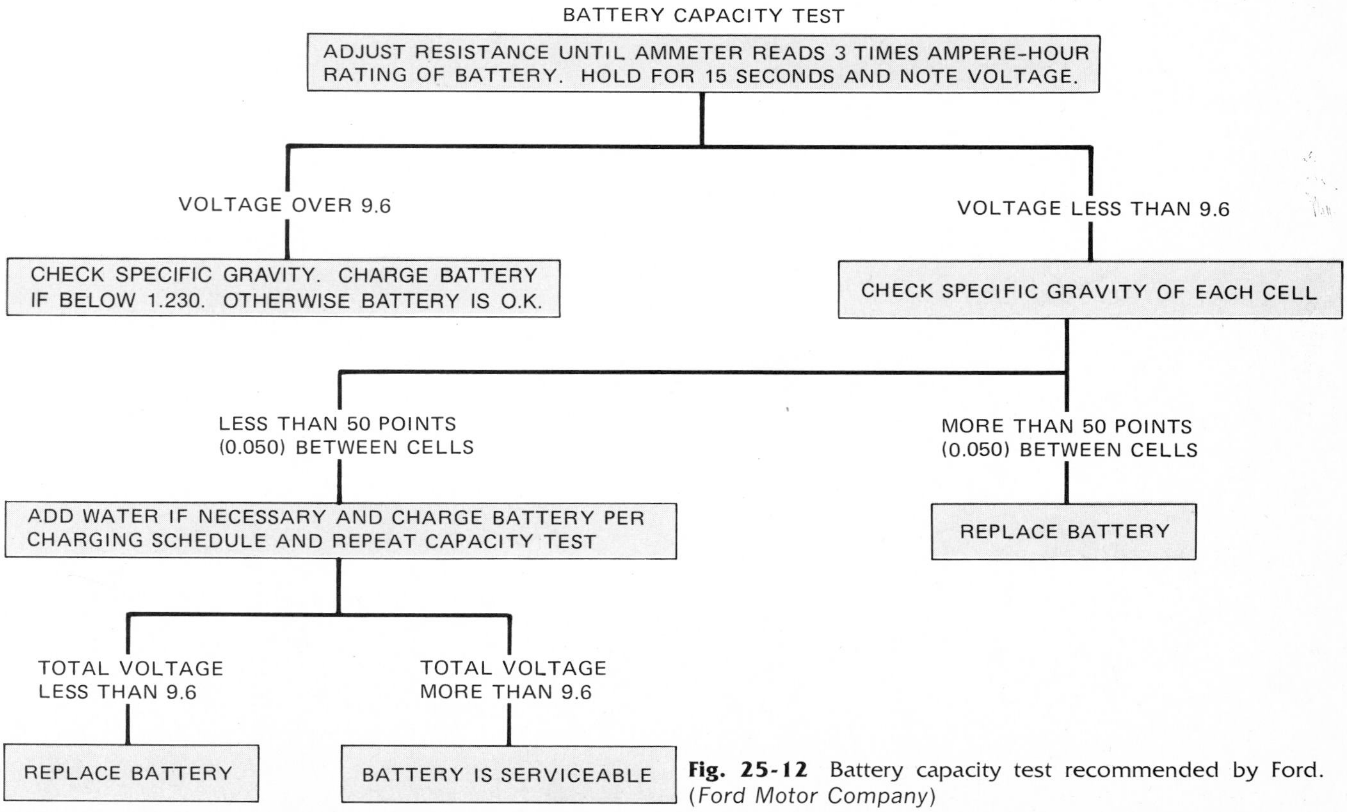
BATTERY CAPACITY TEST
ADJUST RESISTANCE UNTIL AMMETER READS 3 TIMES AMPERE-HOUR RATING OF BATTERY. HOLD FOR 15 SECONDS AND NOTE VOLTAGE.
VOLTAGE OVER 9.6
VOLTAGE LESS THAN 9.6
CHECK SPECIFIC GRAVITY. CHARGE BATTERY IF BELOW 1.230. OTHERWISE BATTERY IS O.K.
CHECK SPECIFIC GRAVITY OF EACH CELL
LESS THAN 50 POINTS (0.050) BETWEEN CELLS
MORE THAN 50 POINTS (0.050) BETWEEN CELLS
ADD WATER IF NECESSARY AND CHARGE BATTERY PER CHARGING SCHEDULE AND REPEAT CAPACITY TEST
REPLACE BATTERY
TOTAL VOLTAGE LESS THAN 9.6
TOTAL VOLTAGE MORE THAN 9.6
REPLACE BATTERY
BATTERY IS SERVICEABLE

Fig. 25-12 Battery capacity test recommended by Ford. *(Ford Motor Company)*

(Fig. 25-13), after filler plugs are removed. Electrolyte must be up to the proper level. If the car has been operated, or the battery charged, within 8 hours, turn on the headlights for 1 minute. Then turn the headlights off. Start the test by putting the red probe into the cell that has the positive terminal. Put the black probe into the next cell. Note the meter reading. Move the probes to cells 2 and 3, and so on, noting the meter readings. Compare the readings. Figure 25-14 shows various readings and the conditions they indicate. Note the following:

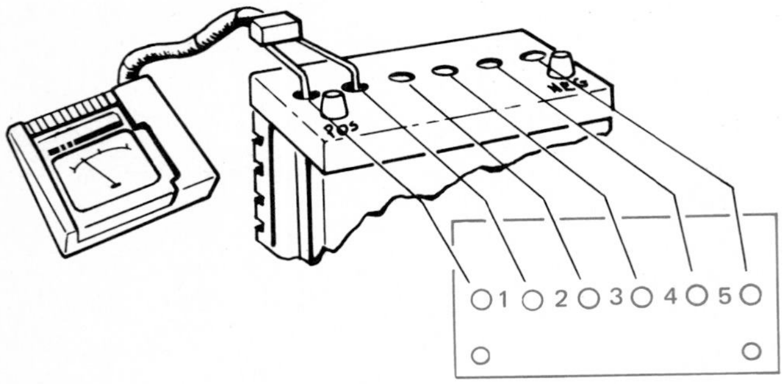

Fig. 25-13 Testing battery cells with a cadmium-tip battery-cell tester. (*Chrysler Corporation*)

1. If any two cells vary five scale divisions or more (top scale), the battery is at the point of failure and should be replaced.
2. If all cells vary less than five scale divisions and all read in the green section, the battery is charged and in good condition.
3. If all cells vary less than five scale divisions but some fall in the red section, the battery is in good condition but needs charging.
4. If any reading falls in the RECHARGE AND RETEST area, the battery is too low to make a good test. Recharge and retest it.

25-16 Battery Service

Battery service can be divided into four parts: visual inspection, testing, charging, and care of batteries in stock.

25-17 Battery Testing

Battery testing includes a check of the condition of the battery, as we have seen. It should also include analysis

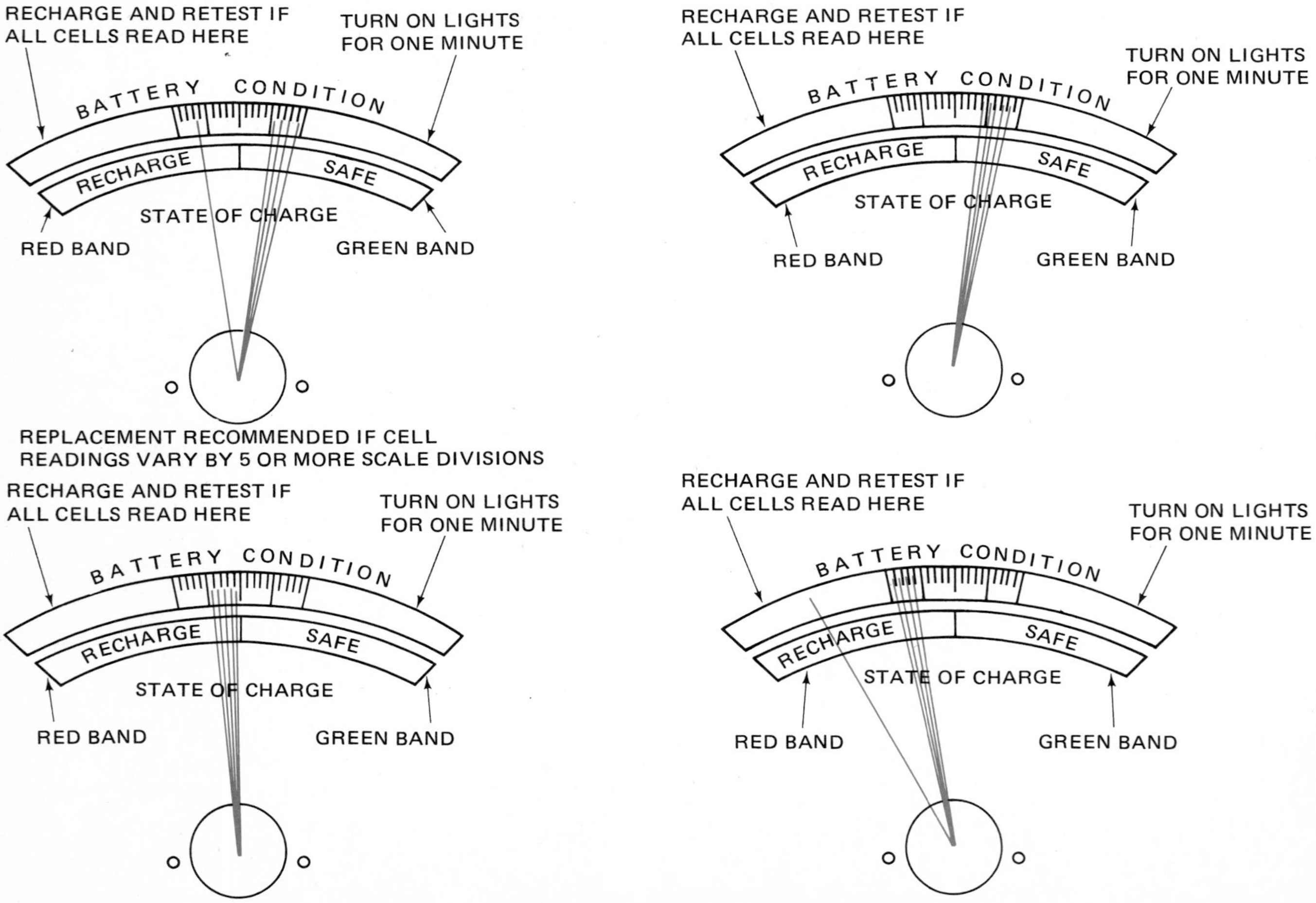

Fig. 25-14 Interpretation of meter readings from the cadmium-tip test. (*Chrysler Corporation*)

of any abnormality found, so that corrections can be made. This will prevent a repetition of trouble. Following are various battery troubles and their possible causes.

1. Overcharging If the battery requires a considerable amount of water, it is probably being overcharged. That is, too much current is being supplied to the battery. This is a damaging condition that overworks the active materials in the battery and shortens battery life. In addition, overcharging causes more rapid loss of water from the battery electrolyte. Unless this water is replaced frequently, the electrolyte level is likely to fall below the tops of the plates. This exposes the plates and the separators to the air and may ruin them. Also, battery overcharge causes the battery plates to buckle and crumble. Thus, a battery subjected to severe overcharging will soon be ruined. Where overcharging is experienced or suspected, the charging system should be checked. It should be serviced if necessary to prevent overcharging (Chap. 29).

2. Undercharging If the battery is discharged, it should be recharged, as outlined later in this chapter. In addition, an attempt should be made to determine the cause of the trouble. It could be caused by:

a. Charging-system malfunctioning
b. Defective connections in the charging circuit between the alternator and the battery
c. Excessive load demands on the battery
d. A defective battery
e. Permitting the battery to stand idle for long periods so that it self-discharges excessively

In addition, an old battery may have a low specific-gravity reading because it is approaching failure.

3. Sulfation The active materials in the plates are converted into lead sulfate during discharge, as has already been noted. This lead sulfate is reconverted into active material during recharge. However, if the battery stands for long periods in a discharged condition, the lead sulfate is converted into hard, crystalline substance. This substance is difficult to reconvert into active materials by normal charging processes. Such a battery should be charged at half the normal rate for 60 to 100 hours. Even though this long charging period may reconvert the sulfate to active material, the battery may still remain in a damaged condition. The crystalline sulfate, as it forms, tends to break the plate grids.

4. Cracked case A cracked case may result from excessively loose or tight hold-down clamps, from battery freezing, or from flying stones.

5. Bulged cases Bulged cases result from tight hold-down clamps or from high temperatures.

6. Corroded terminals and cable clamps This condition occurs naturally on batteries. You should be prepared to remove excessive corrosion periodically from terminals and clamps. Cable clamps should be disconnected from the terminal and the terminal posts and cables cleaned, as already explained (● 25-4).

7. Corroded battery holder Some spraying of battery electrolyte is natural as the battery is being charged. The battery holder may become corroded from the effects of the electrolyte. Such corrosion may be cleaned off, with the battery removed. Use a wire brush and common baking-soda solution.

8. Dirty battery top The top of the battery may become covered with dirt and grime mixed with electrolyte sprayed from the battery. This should be cleaned off periodically as already explained (● 25-4).

9. Discharge to metallic hold-down If the hold-down clamps are of the uncovered metallic type, a slow discharge may occur from the insulated terminal to the hold-down clamp. This is more apt to occur with a dirty battery top, across which current can leak. The remedy is to keep the battery top clean and dry.

● 25-18 Removing and Replacing a Battery

To remove a battery from a car, first take off the grounded-battery-terminal cable clamp. This prevents accidental grounding of the insulated terminal when it is disconnected. To remove a nut-and-bolt type of cable, loosen the clamp nut about $\frac{3}{8}$ inch [9.53 mm]. Use a box wrench or special cable pliers (Fig 25-15). Do not use ordinary pliers or an open-end wrench. Either of these might break a cell cover when swung around. If the clamp sticks, use a clamp puller (Fig. 25-16). Do not use a screwdriver or bar to pry on the clamp. This could damage the battery cell or cover. To detach the spring-ring type of clamp, Squeeze the ends of the rings apart with vise-grip or channel-lock pliers (Fig. 25-17).

After the grounded cable is disconnected, disconnect the insulated-terminal cable. Clean both battery terminals and cable clamps with special tools (Fig. 25-6). Loosen the battery hold-downs, and take out the battery. When installing a battery, do not reverse the terminal connections. (Some automobiles have the negative terminal

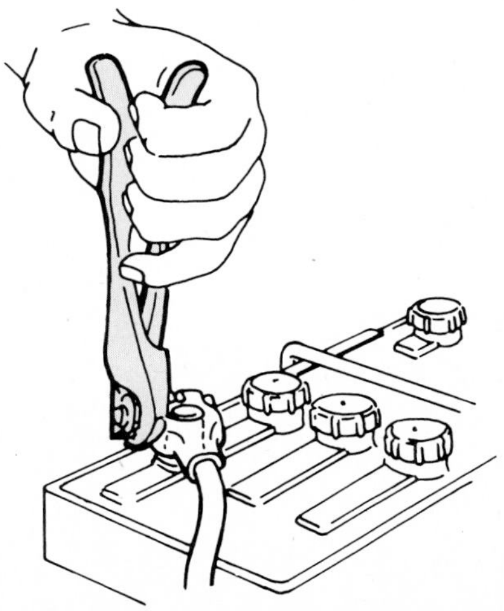

Fig. 25-15 Using battery pliers to loosen the nut-and-bolt type of battery-cable clamp.

grounded, others the positive terminal.) Reconnect the insulated-terminal cable first, and then reconnect the grounded-terminal cable. Apply corrosion inhibitor to clamps and terminals. Install and tighten the hold-downs. Avoid overtightening.

Careful: Make sure the cable clamps are tight and make good connections with the terminal posts. See Fig. 25-18. If the jaws of the clamp come together as shown at the left, chances are the clamp is not tight on the post. This could mean starting trouble. Correct the condition by disconnecting the clamp from the post. Shave the clamp jaws with a file so you get a gap as shown to the right (Fig. 25-18) when the clamp is installed.

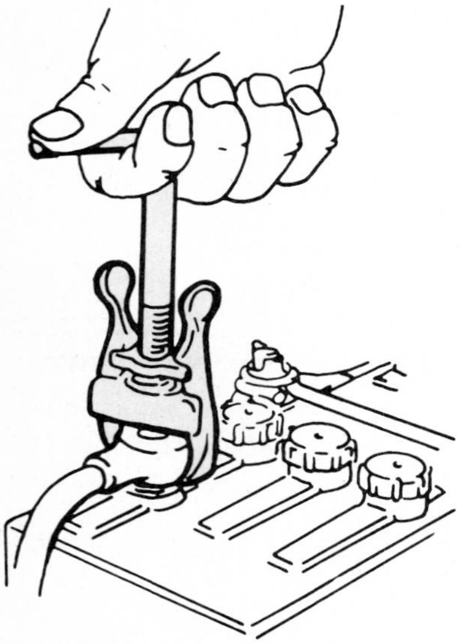

Fig. 25-16 Using special clamp puller to pull the cable from the battery terminal.

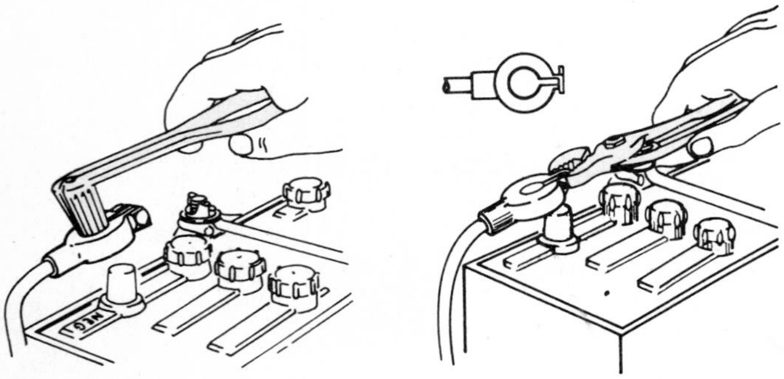

Fig. 25-17 Using pliers to loosen the spring-ring type of cable clamp from a battery terminal.

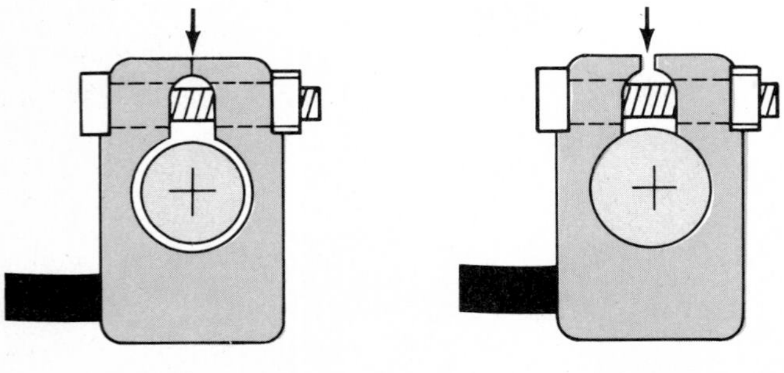

Fig. 25-18 If there is no gap between the jaws of the clamp (left), the clamp is probably loose on the terminal post.

25-19 Battery "Dopes"

"Dopes" is a name for certain chemical compounds that are supposed to restore a battery to a charged condition. Such chemicals should never be added to the battery. Their use may void the battery guarantee and cause battery failure.

25-20 Battery Slow Charging

Two methods of slow-charging batteries are in use, the constant-current and the constant-voltage (constant-potential) methods. In the constant-current method, the current input to the battery is adjusted to the manufacturer's specifications. The charging is continued until the battery is gassing freely and there is no rise in gravity for 2 hours.

In the constant-voltage method, the charging voltage is held at a constant value. The battery, as it approaches a charged condition, increases in resistance to the charging current. At the same time, the current input gradually tapers off. When the battery is fully charged, the current input has been reduced to a few amperes. We assume, in this, that the battery-electrolyte temperature remains within bounds. If the battery-electrolyte temperature increases greatly, the resistance of the battery will remain low. Then the battery will be damaged by overcharging, unless it is removed from the charging line in time.

25-21 Quick Chargers

Quick chargers charge the battery at a high rate (as much as 100 amperes) for a short time—30 to 45 minutes. The battery is thus brought to a fair state of charge before the battery temperature increases excessively. The quick-charger method of charging does not seem to harm batteries that are not subject to excessive temperatures. However, high charging rates combined with battery-electrolyte temperatures above 125°F [51.7°C] are very damaging to a battery.

Quick chargers cannot, as a rule, bring a battery up to *full* charge in a short time. Suppose a battery has been quick-charged for a short time. If the charging operation is finished by a slow-charging method, the battery will come up to full charge.

A battery with discolored electrolyte (from cycling) or with gravity readings more than 25 points apart should not be quick-charged. Likewise, a badly sulfated battery should not be quick-charged. Such batteries may be near failure, but they may give additional service if slow-charged. However, quick-charging might damage them further. During quick-charging, check the color of the electrolyte. Stop charging if it becomes discolored as a result of the stirring up of washed-out active material. Also, cell voltages should be checked every few minutes. Charging should be stopped if cell voltages vary more than 0.2 volt.

A very low battery may not accept a fast charge. The electrolyte in a very low battery does not have very much sulfuric acid in it. Therefore, the conductivity of the electrolyte is too low to allow a high current to flow through the battery. You might think a battery that refuses to take a high charge is worn out. However, it may be possible to restore the battery to a charged condition, as follows: First, slow-charge it for a few minutes to see if it starts coming up to charge. If it does, then it can be put on fast charge. Some fast chargers have a special circuit which will slow-charge a dead battery for a short time and then switch to fast charging.

● 25-22 Care of Batteries in Stock

Wet batteries (or batteries with electrolyte in them) are perishable. They are subject to self-discharge. If allowed to proceed for too long a time, this can completely ruin them. To prevent this, batteries in stock should be recharged at 30-day intervals. They should not be stacked on top of each other without some means of individual support. The weight of one battery is enough to collapse the plate assemblies and cause short circuits.

● 25-23 Dry-Charged Batteries

Dry-charged batteries contain fully charged positive and negative plates but no electrolyte. The batteries are sealed with rubber or plastic seals placed in the vent plugs. Since the batteries contain no moisture, practically no chemical action can take place in them. This means that they will remain in good condition for as long as 36 months, the manufacturers say, if they are properly stored.

Dry-charged-battery manufacturers supply ready-mixed electrolyte in special cartons. The carton contains an acid-proof plastic bag which holds the electrolyte. To activate a battery or get it ready for service, all that is necessary is the following:

1. Remove the vent plugs, and take out the plastic seals.
2. Remove the lid from the electrolyte container. Unfold the top of the plastic bag, and cut a small opening in one corner of the bag.
3. Use a glass or acid-proof plastic funnel and fill each battery cell. *Wear goggles, and observe all cautions already noted regarding sulfuric acid.* Wait a few minutes, and then add more electrolyte if necessary. Some electrolyte will probably be left; do not attempt to use it all. Do not overfill the battery.
4. Before discarding the container, empty it. Rinse the bag thoroughly with water. Otherwise, someone who handles the carton might be severely burned.

REVIEW QUESTIONS

1. What should you do if you get battery acid on your skin? In your eye?
2. Why is it dangerous to bring an open flame near a battery that is being charged?
3. How do you check the electrolyte level in batteries?
4. What is the Delco Eye? What is its purpose?
5. Explain how to clean corrosion off the top of a battery.
6. Explain how to use a hydrometer to test a battery.
7. Explain how to use a refractometer to test a battery.
8. Explain how to perform the cadmium-tip test on a battery.
9. Which will freeze more easily, a fully charged battery or a half-charged battery? Why?
10. Why will overcharging a battery damage it?
11. Describe the proper way to disconnect a cable clamp from a battery terminal.
12. What are the two basic charging methods?
13. Explain how to perform a high-discharge test on a battery.
14. Explain why a very low battery may not be able to accept a high charging rate.
15. Explain how to put a dry-charged battery into service.

SELF PROJECTS

1. Make a list of the six steps in battery maintenance and file it in your notebook.
2. Write down—so you will be sure to remember it—what you should do if you get electrolyte on your skin or in your eyes. Knowing what to do could save you from a severe burn or the loss of your eyesight.
3. Make a list of gravity readings and their meaning. This will help you remember them so that, when you use a hydrometer, you will know what the readings mean.
4. If your shop has a refractometer, copy the instructions that come in the package. File these instructions in your notebook.
5. If you are going to do any amount of battery work, you may want to get a pair of special cable-clamp pliers, such as shown in Fig. 25-15.

CHAPTER 26
Starting Motors

After you have read this chapter, you should be able to:

- **Explain the basic principles and construction of starting motors.**
- **Explain the construction and operation of inertia and overrunning clutch drives.**
- **Describe the construction and operation of magnetic switches and solenoids used with starting motors.**
- **Locate the starting motor and wiring in automobiles.**

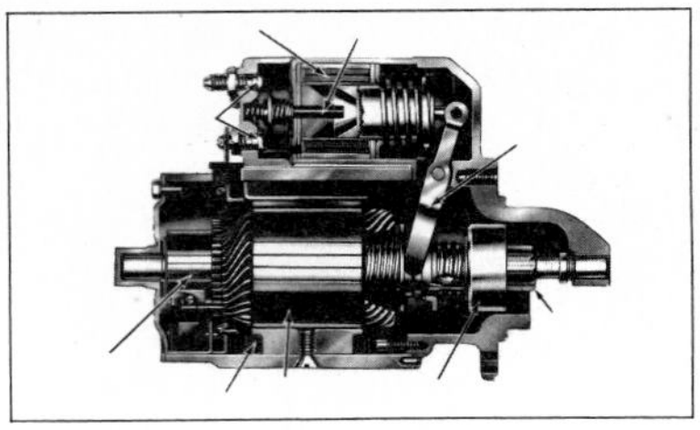

In this chapter, we describe the construction and operation of starting motors. The automotive starting motor is a high-capacity, sturdy electric motor. It is especially designed to provide the high torque required to spin the engine crankshaft and get the engine started. The starting motor is also called a *starter* and a *cranking* motor. We begin the chapter by looking into the basic principles of electric motors.

26-1 Basic Motor Principles

When current moves through a conductor, a magnetic field builds up around that conductor. If the conductor is in a magnetic field, as from a horseshoe magnet, force is exerted on the conductor. Figure 26-1 illustrates the conductor in a magnetic field. Figure 26-2 shows the conductor in end view, with the resulting magnetic field indicated. The cross in the center of the conductor indicates that the current is flowing away from the reader. This causes the magnetic field due to the current flow to circle the conductor in a counterclockwise direction. The circular magnetic field to the left of the conductor is in the same direction as the straight-line magnetic field from the magnet. To the right of the conductor, it is in the opposite direction. This weakens the magnetic field to the right of the conductor. It strengthens the magnetic field to the left of the conductor. Thus, the resulting magnetic field distorts around the conductor, as shown in Fig. 26-2.

Magnetic lines of force tend to shorten up to a minimum length. Thus, the bent lines of force in the magnetic-field pattern in Fig. 26-2 try to straighten out. As they do, they exert a push to the right on the conductor. The more current flowing, the more the lines of force will be distorted around the conductor—and the stronger will be the push. Increasing the straight-line magnetic field has a similar effect.

26-2 Motor Construction

Suppose we bend the conductor into a U and connect the two ends to the two halves of a split copper ring. We now have the elements of an electric motor (Fig. 26-3). Stationary brushes, connected to a battery and resting on the split ring, and two poles of a magnet complete the motor. The brushes are carbon blocks that form sliding contacts with the commutator. The U-shaped conductor loop and the split ring (which is called the *commutator*) are able to rotate together. Current flows from the battery, through the right-hand brush and segment of the commutator. Then it flows, through the conductor and left-hand segment of the commutator and brush, back to the battery, as shown. This causes the left-hand part of the conductor to be pushed upward, and the right-hand part to be pushed downward (see Fig. 26-2). Thus, the loop rotates in a clockwise direction. As the two sides of the loop reverse positions, the direction of the current flow through the two sides reverses. The force thus continues to rotate the loop clockwise.

The starting motor must use more than one loop to develop enough power. Actually, many loops or conductors are used, as shown in Fig. 26-4, which illustrates a starting-motor armature and field assembly. The ends of the conductors in the armature are connected to the commutator segments.

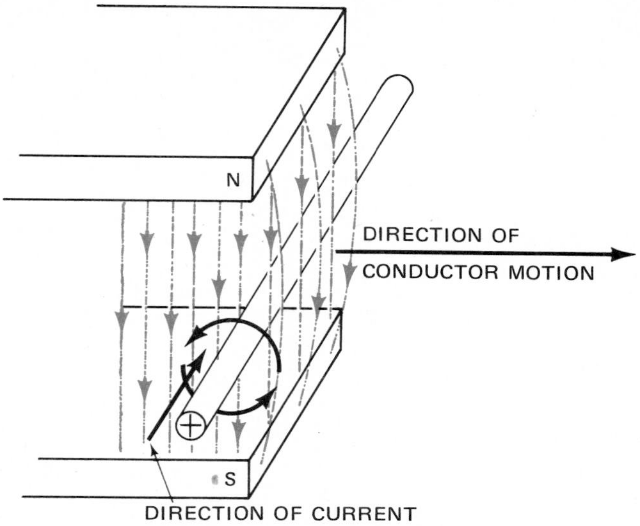

Fig. 26-1 Conductor held in the magnetic field of a magnet. The direction of current flow and the encircling magnetic field around the conductor are shown by arrows.

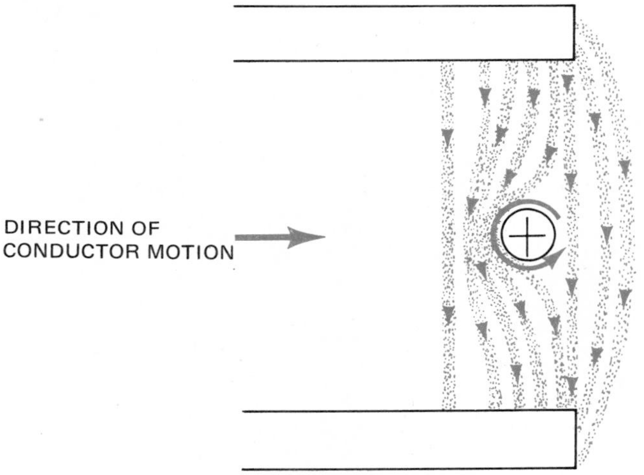

Fig. 26-2 End view of the conductor shown in Fig. 26-1.

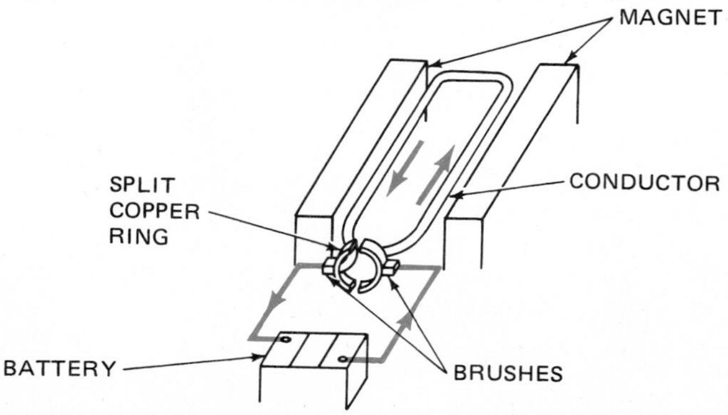

Fig. 26-3 Simple electric motor with a two-segment commutator.

High magnetic-field strength is needed for powerful starting motor actions. The natural magnetic strength of the magnetic poles is aided by field windings. Current flows through the field windings in such a direction as to increase the magnetic field between the two poles. Figure 26-5 is a simple wiring diagram of a starting motor. Current enters the motor and passes through the two field windings, then through the armature, and back to the battery. If the battery connections were reversed, the current would flow through the armature first, as shown in Fig. 26-6. This is a schematic drawing of a simple motor. This type of motor is called a *series-wound*, or *series*, motor. The armature and field windings are connected in series.

The wiring diagram in Fig. 26-5 is of a two-pole two-brush starting motor. Many starting motors have four brushes and four poles. Some also have one or two shunt windings (and are called *series-shunt*, or *compound*, units). The shunt windings prevent overspeeding (see Fig. 26-7). Note that the shunt winding (Fig. 26-7) is connected in parallel with (shunted across) the series windings.

A typical starting motor, with the main parts disassembled, is shown in Fig. 26-8. The motor consists of:

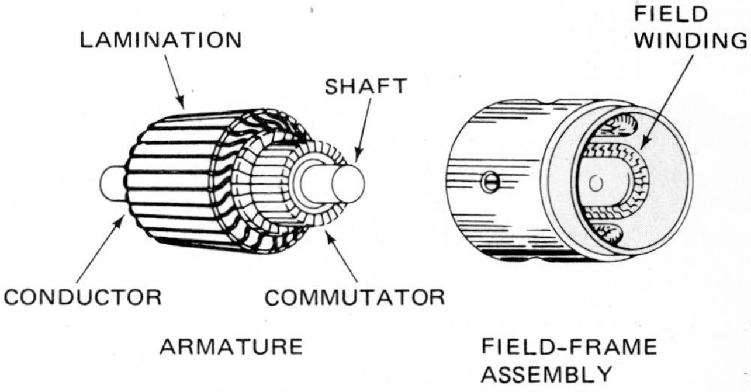

Fig. 26-4 Two major parts of a starting motor, the armature and the field assembly. (*Delco-Remy Division of General Motors Corporation*)

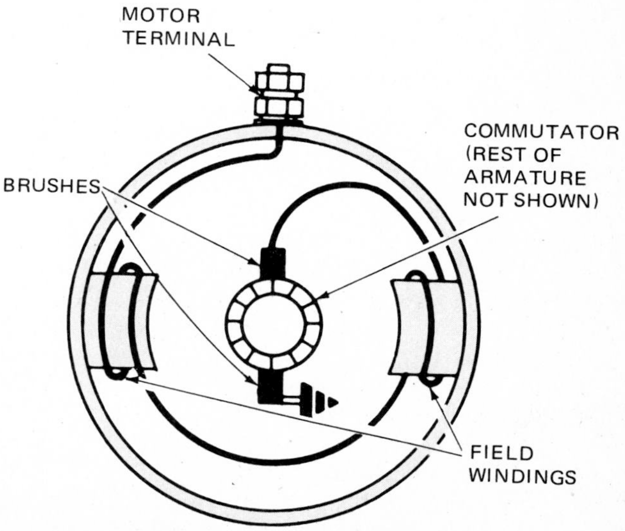

Fig. 26-5 Wiring diagram for a starting motor.

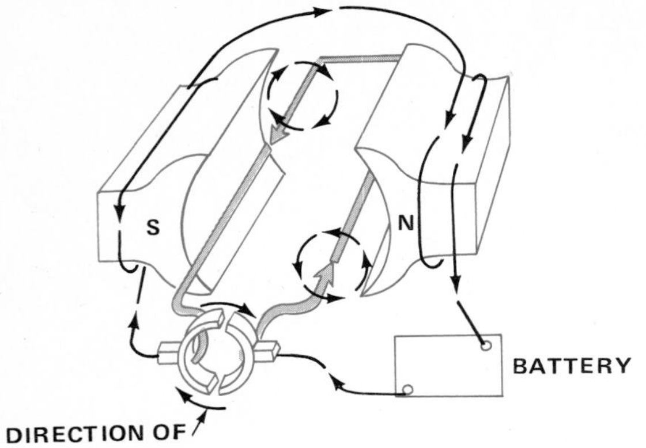

Fig. 26-6 Schematic drawing of a starting motor. Heavy arrows show direction of current flow. Light circular arrows show direction of magnetic field around conductors.

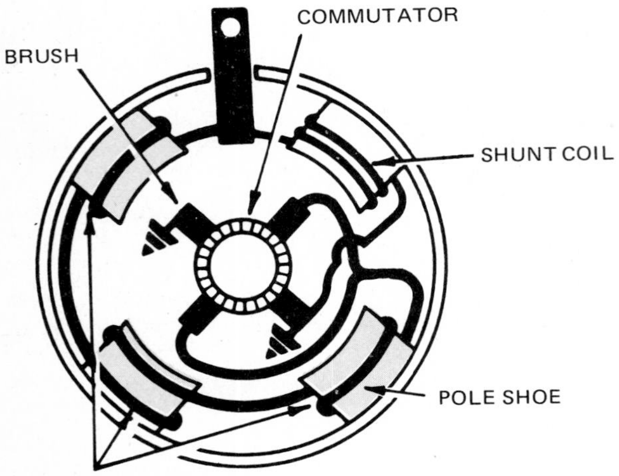

Fig. 26-7 Wiring diagram of a four-pole, series-shunt, or compound, starting motor. (*Delco-Remy Division of General Motors Corporation*)

1. The commutator end head, holding the brushes
2. The field frame, into which the field windings are assembled around iron pole shoes
3. The drive housing, which houses the drive assembly and supports the motor on the engine flywheel housing
4. The armature
5. The drive assembly

Most starting motors also have a solenoid that operates the shift lever (● 26-6).

● 26-3 Drive Arrangement

The drive assembly contains a small pinion that, in operation, meshes with teeth on a larger gear on the flywheel (Fig. 26-9) or torque converter. This provides gear reduction, so that the armature must rotate about fifteen times to cause the flywheel to rotate once. The armature may revolve about 2,000 to 3,000 rpm (revolutions per minute) when the starting motor is operated. This causes the flywheel to spin at speeds as high as 300 rpm. This is ample for starting the engine.

After the engine starts, it may increase in speed to 3,000 rpm or more. If the starting-motor drive pinion remained in mesh with the flywheel, it would be spun at 45,000 rpm because of the 15:1 gear ratio. This means that the armature would be spun at this terrific speed. Centrifugal force would cause the conductors and commutator segments to be thrown out of the armature, ruining it. To prevent such damage, automatic meshing and de-meshing devices are used. For passenger cars there are two general types, inertia and overrunning clutch, described below. The overrunning clutch is used on most automobiles made today.

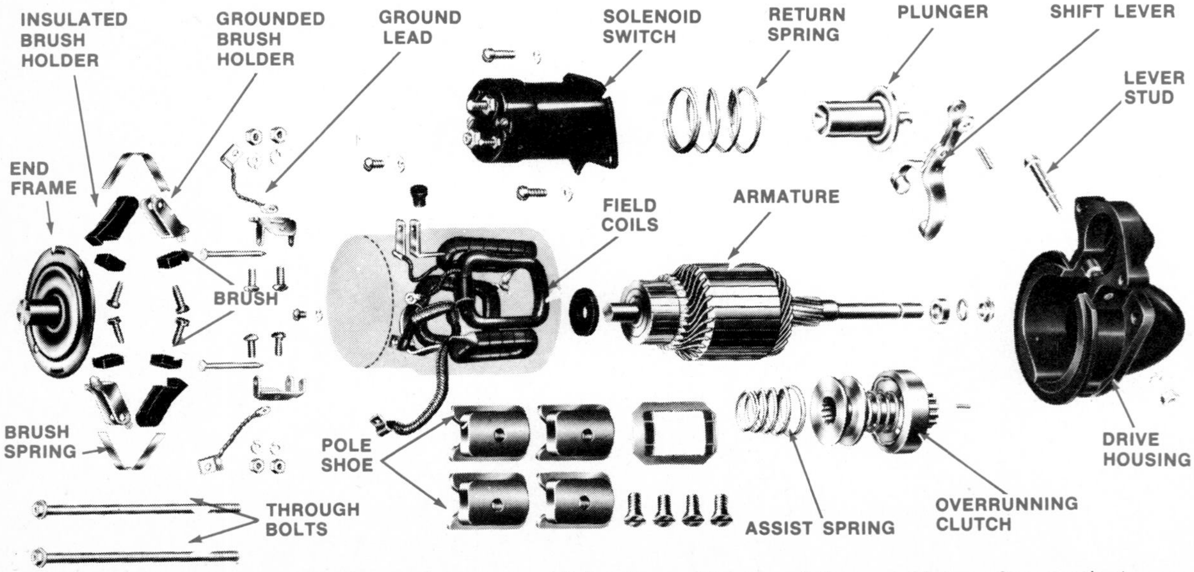

Fig. 26-8 Disassembled view of a starting motor. (*Delco-Remy Division of General Motors Corporation*)

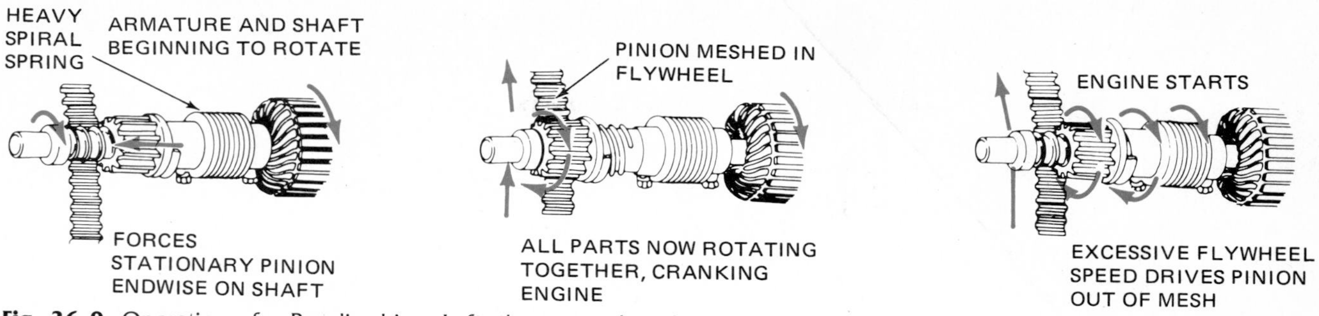

Fig. 26-9 Operation of a Bendix drive. Left, the screw threads turn in the pinion, forcing the pinion toward the flywheel. Center, the pinion meshes with the flywheel so cranking starts. Right, the flywheel spins the pinion faster than the screw threads, forcing the pinion to move back out of mesh. (*Delco-Remy Division of General Motors Corporation*)

26-4 Inertia Drive

The inertia drive depends on the inertia of the drive pinion to produce meshing. Inertia is the property that all things resist any change in motion. When the drive pinion is not rotating, it resists any force that attempts to set it into motion. Two types of inertia drive are the Bendix and the Folo-Thru (Figs. 26-9 and 26-10).

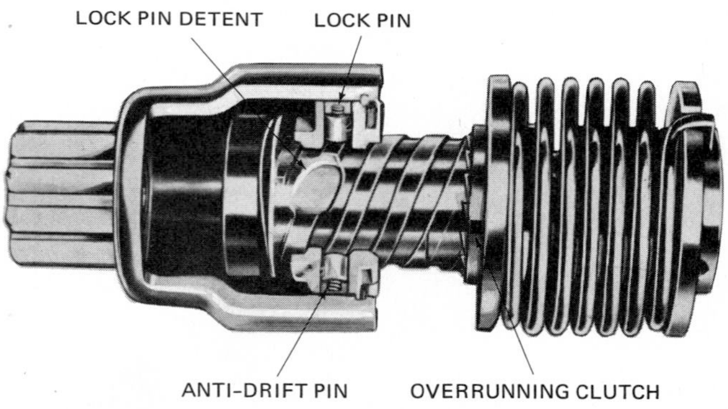

Fig. 26-10 Folo-Thru drive. The lock pin locks the pinion in the meshed position so it will not kick out during a false start. The antidrift pin prevents the pinion from drifting into mesh when the engine is running. (*Ford Motor Company*)

Neither the Bendix drive nor Folo-Thru drive is used on cars today, although both were widely used many years ago. Some variations of the Bendix drives may still be found on some small engines; those used in powered lawn equipment, for example.

26-5 Overrunning Clutch

The overrunning clutch (Fig. 26-11) is operated by a shift lever. The lever pushes the drive pinion into mesh with the flywheel teeth. As the shift lever completes its travel, it closes the starting-motor switch so that cranking takes place. Straight or spiral splines in the armature shaft and the clutch sleeve cause both to rotate together. A spiral spring is placed between the clutch housing and the shift-lever collar. This spring compresses if the pinion and the flywheel teeth should happen to butt instead of mesh. Then, after the starting-motor switch is closed and the armature starts to rotate, meshing is completed by the spring pressure.

The clutch (Fig. 26-12) consists of the outer shell and the pinion and collar assembly. The outer shell has four hardened-steel rollers fitted into four notches. The notches are not concentric, but are smaller in the end opposite to the plunger springs (Fig. 26-12). When the

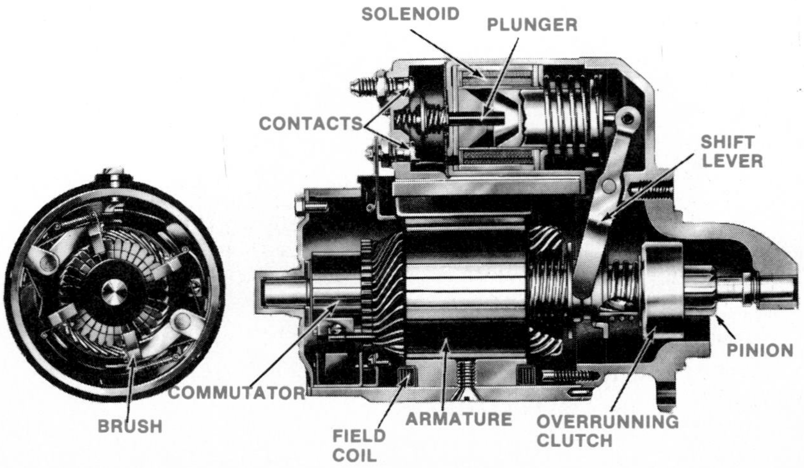

Fig. 26-11 Sectional view of an enclosed shift-lever starting motor with solenoid. (*Delco-Remy Division of General Motors Corporation*)

armature and the shell begin to rotate, the pinion does not. This causes the rollers to rotate into the smaller sections of the notches, where they jam tight. The pinion must now rotate with the armature, cranking the engine. After the engine starts, it spins the pinion faster than the armature is turning. The rollers are rotated into the larger sections of the notches, where they are free. This allows the pinion to spin independently of, or *overrun*, the remainder of the clutch. A spring on the shift lever pulls the pinion back out of mesh when the shift lever is released.

Gear reduction The starting-motor shown in Fig. 26-13 has a gear reduction which increases cranking torque. The shift lever (or fork) is enclosed. When it is actuated by the solenoid, it shifts the overrunning-clutch pinion into mesh with the flywheel. It also shifts the large driven gear on the clutch shaft into mesh with the smaller gear on the armature shaft. The gear ratio between the armature and the flywheel, due to the extra gears in the starting motor, is 45:1. That is, the armature turns 45 times to turn the flywheel, and the engine crankshaft, once. This provides a high cranking torque for starting.

● 26-6 Starting-Motor Controls

Starting-motor controls have varied from a simple foot-operated pedal to automatic devices that close the starting-motor circuit when the accelerator pedal is depressed. The system presently used in passenger cars and other

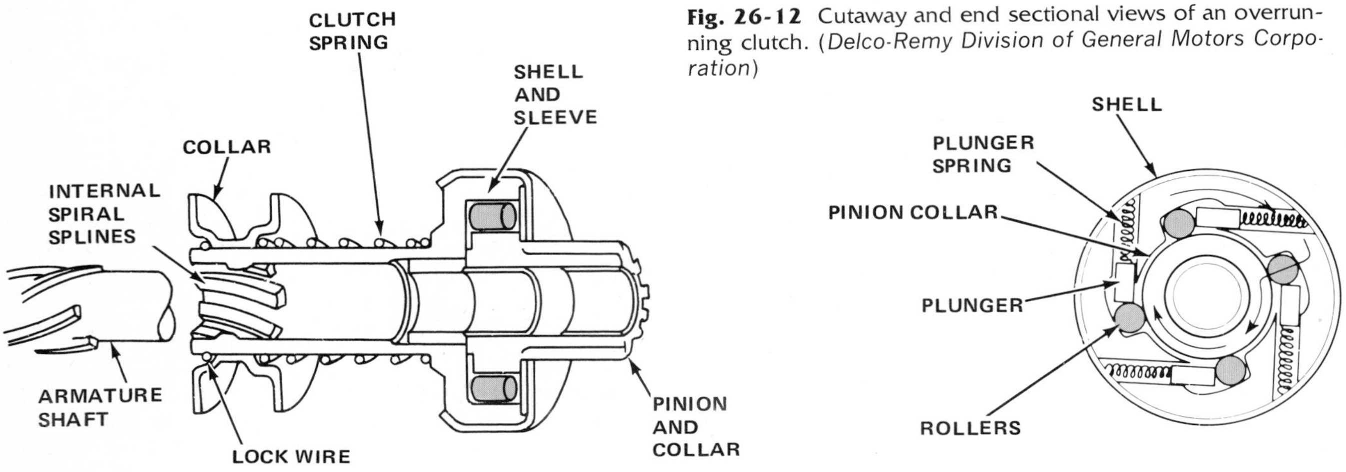

Fig. 26-12 Cutaway and end sectional views of an overrunning clutch. (*Delco-Remy Division of General Motors Corporation*)

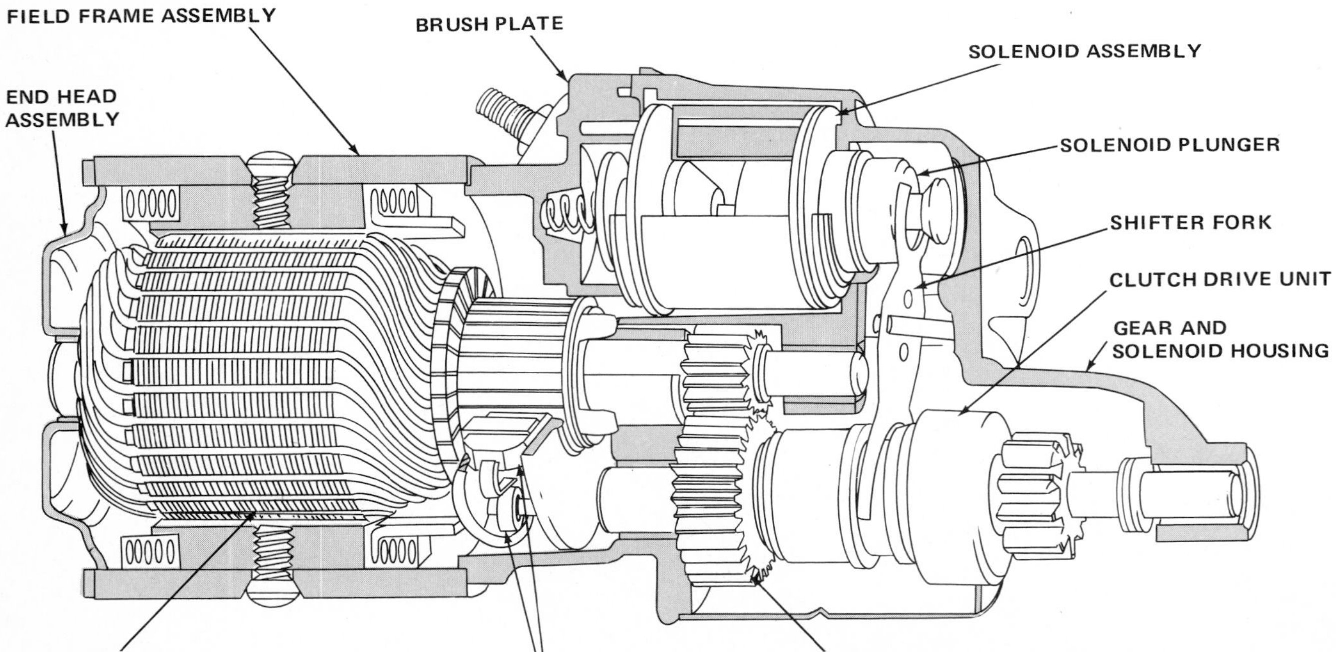

Fig. 26-13 Sectional view of a gear-reduction, overrunning-clutch starting motor. (*Chrysler Corporation*)

vehicles has starting contacts in the ignition switch. When the ignition key is turned against spring pressure past the ON position to START, the starting contacts close. This connects the starting-motor solenoid or magnetic switch to the battery. After the engine starts and the ignition key is released, spring pressure returns it to the ON position. This opens the starting contacts.

Bendix-type starting motors use a magnetic switch. When the starting contacts in the ignition switch are closed, the magnetic-switch winding is connected to the battery. The magnetism produced in the winding pulls the plunger in. This forces the contact disk against the magnetic-switch contacts. Now, current can flow through the contacts and disk to operate the starting motor. After the engine starts, the driver takes his hand off the ignition key. The key is returned by spring action to the ON position. This opens the magnetic-switch winding circuit. A spring moves the plunger and contact away from the magnetic-switch contacts. Now, the starting-motor circuit is opened, and the starting motor stops operating.

On starting motors with overrunning clutches, a solenoid is commonly used to produce the clutch-shifting action. The solenoid contains a pair of windings that are connected to the battery when the starting switch is closed. This produces a magnetic field that pulls a plunger in. The plunger movement causes a shift lever to move the overrunning clutch on the armature shaft. This shifts the overrunning-clutch pinion into mesh with the flywheel teeth. At the same time, the plunger movement forces a heavy switch to connect the starting motor directly to the battery. Now, cranking begins. Figure 26-14 shows the sequence of actions. Figure 26-15 is a wiring diagram of a starting-motor system.

Note that the solenoid has two windings, a pull-in winding and a hold-in winding. They work together to pull the plunger in. This combination of windings provides sufficient magnetic strength to mesh the pinion and close the starting-motor switch. After the pinion is meshed and the switch is closed, less magnetism is required to hold the core in. Consequently, as the switch closes, the pull-in winding is shorted out, since it is connected between the two solenoid terminals. This reduces the drain on the battery during the cranking operation.

On the Ford starting motor illustrated in Fig. 26-16, there is no separate solenoid. Instead, the starting-motor field windings produce the magnetic field which causes both the shifting action and the rotation of the armature. The magnetic field causes the pole shoe to slide in the frame. This action moves the shift lever so that the drive pinion is forced into mesh with the flywheel teeth. When the ignition switch is turned to START and the automatic-transmission neutral safety switch is closed, the magnetic-switch winding is connected to the battery. This causes the magnetic switch (starter relay) to close its contacts. This connects the starting motor directly to the battery.

One of the field windings acts as the winding to move the pole shoe. It has two parts, as shown, a pull-in winding and a hold-in winding. As the magnetic switch closes its contacts, these two windings are connected directly across the battery (through the contact points). This produces maximum magnetic strength. The other three field windings are connected in series with the armature. The magnetic field of the pole-shoe actuating windings moves the pole shoe and shifts the drive pinion into mesh. At the same time, the armature begins to turn so that the engine is cranked. As the pole shoe and shift lever move, they cause the contact points to open. Now, the pull-in winding is connected in series with the armature. Its magnetism drops. At the same time, however, the hold-in winding retains its full magnetism so that the pole shoe is retained in the cranking position. As soon as the ignition switch is released, the magnetic-switch winding is disconnected from the battery. It opens its contacts to disconnect the starting motor from the battery. Cranking stops. The drive pinion is demeshed from the flywheel by the return spring.

1. Automatic transmissions Years ago, cars with automatic transmissions had a special switch that prevented starting while the car was in gear. As shown in Fig. 26-15, the switch (neutral safety switch) is connected between the ignition switch and the solenoid. This switch is open at all transmission-lever positions except PARK or NEUTRAL. Late-model cars with the ignition switch and selector lever on the steering column do not need a safety

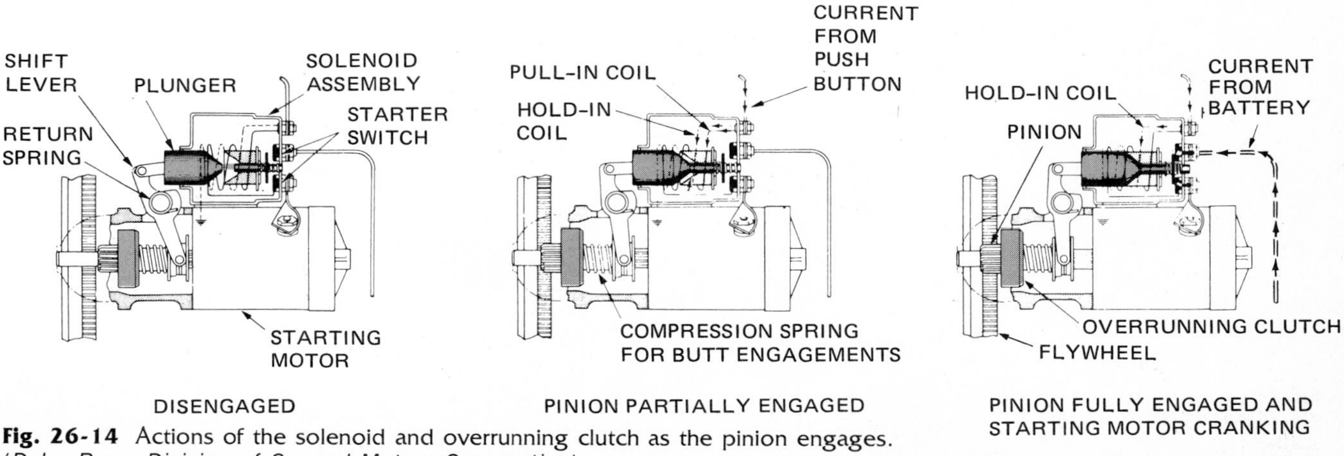

Fig. 26-14 Actions of the solenoid and overrunning clutch as the pinion engages. (*Delco-Remy Division of General Motors Corporation*)

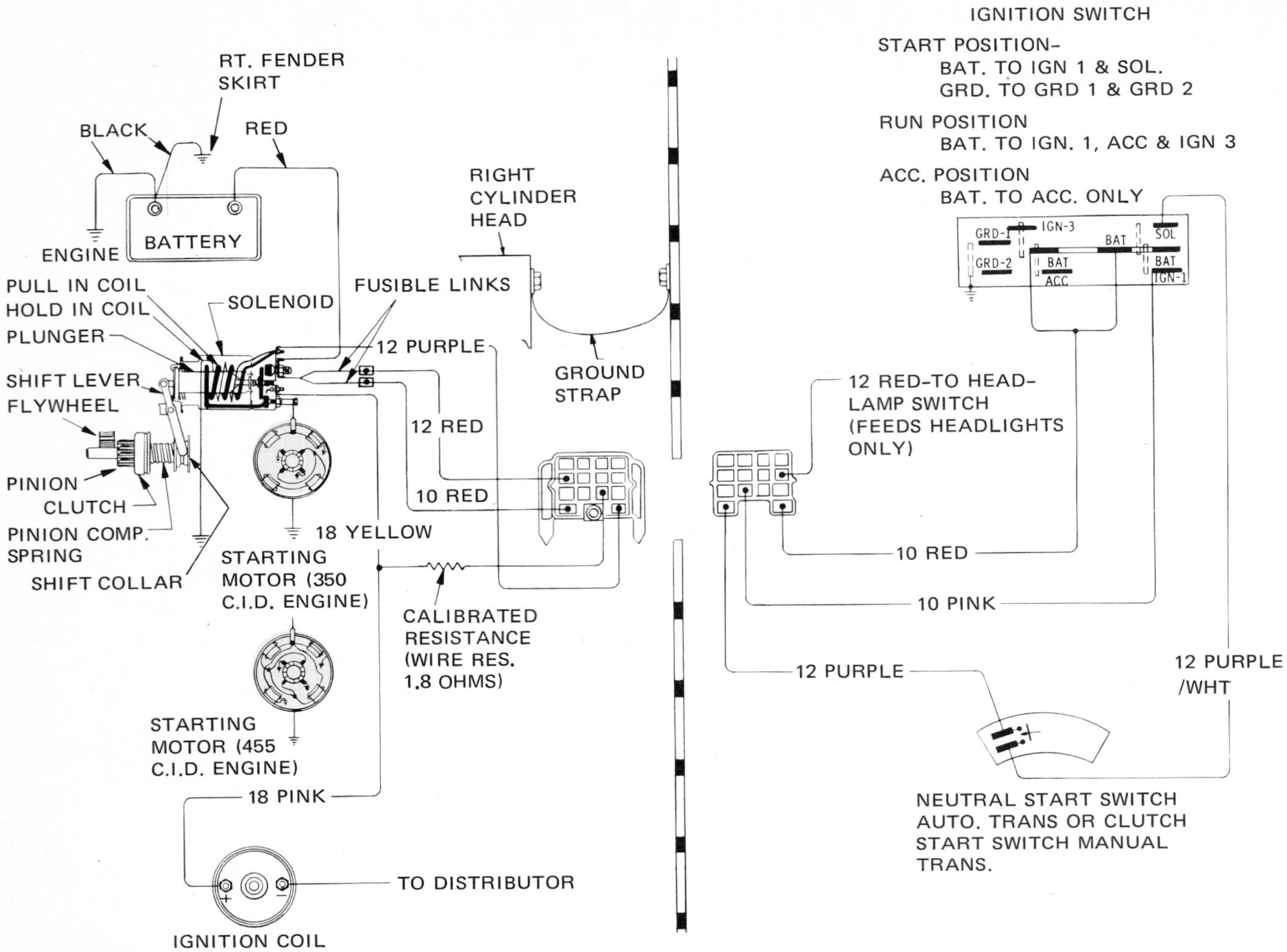

Fig. 26-15 Wiring diagram of a starting-motor system. *(Buick Motor Division of General Motors Corporation)*

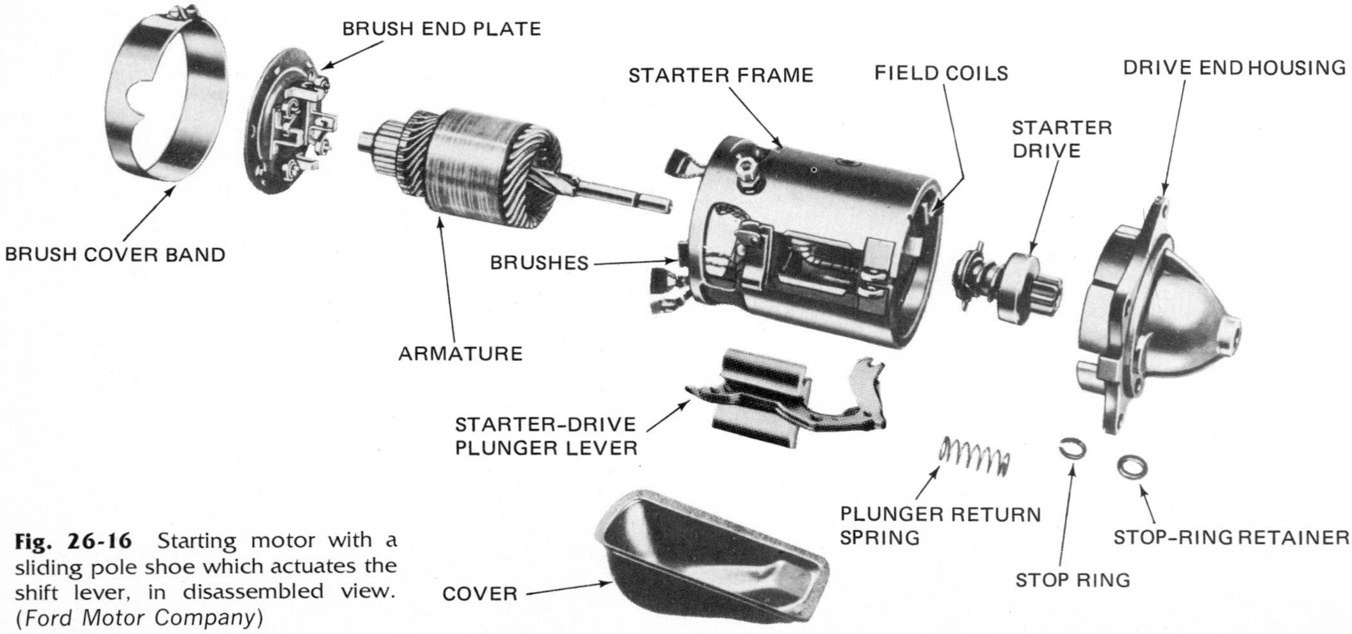

Fig. 26-16 Starting motor with a sliding pole shoe which actuates the shift lever, in disassembled view. *(Ford Motor Company)*

switch. On these, the ignition switch cannot be turned to ON or START unless the selector lever is in PARK or NEUTRAL.

2. Ignition resistance In many cars, the ignition system includes a resistance wire. The resistance is in series with the ignition-coil primary when the engine is running. This protects the ignition contact points from excessive current. However, during cranking, the ignition switch shorts out the resistance (when the key is turned to START). Now, full battery voltage is imposed on the ignition coil for good performance during cranking. The resistance is also called a *ballast* resistance. On some cars, the resistance is a separately mounted part. But most cars have the resistance in the wiring harness.

3. Other controls Other types of controls have been used to prevent starting-motor operation at all times except when starting. Vacuum switches, mounted on the intake manifold or in the carburetor, have been used extensively. This type of vacuum switch closes when the engine is not running. But the switch opens when the engine starts and a vacuum builds up in the intake manifold. Vacuum switches were usually linked to the throttle so that, at part to full throttle, the contacts were opened. This was an added safeguard against starting-motor operation while the engine was running. Sometimes a solenoid relay was used. This was a small relay with a single winding and a pair of contacts. When the control switch or circuit was closed, it would connect the relay winding to the battery. Then, the relay would close to connect the solenoid windings to the battery, so that the solenoid would operate.

For heavy-duty applications, a two-step control has been used. This control first imposes a relatively low voltage on the starting motor so as to get the armature started. Then, full voltage is imposed to crank the engine. A still different heavy-duty system uses a series-parallel switch with two 12-volt batteries. During normal operation with the engine running, the two batteries are connected in parallel, and the system is 12-volt throughout. But for starting, the two batteries are connected in series to supply the starting motor with 24 volts. This higher voltage causes the starting motor to develop a higher cranking torque.

Diesel engines, because of their high compression, are harder to start and require more cranking power. Thus, many automobiles using diesel engines have heavy-duty starting motors and a heavy-duty battery or two 12-volt batteries connected together. The V-8 diesel engine used in some General Motors cars, for example, has a heavy-duty starting motor and two 12-volt batteries connected in parallel. This provides the extra power needed for the glow plugs and the starting motor.

REVIEW QUESTIONS

1. What are the two major parts of the starting motor?
2. Why must the starting-motor pinion be demeshed from the flywheel after the engine starts?
3. Describe the construction of the overrunning clutch. Explain how it works.
4. Why does the starting-motor solenoid have two windings?
5. Explain how the Ford starting motor works to shift the pinion into mesh with the flywheel teeth.
6. What is the purpose of the switch, in the automatic transmission, that is tied in with the starting-motor control circuit?
7. What is the purpose of the resistance wire in the ignition circuit?

SELF PROJECTS

1. Refer to manufacturers' shop manuals and make a list of car models and the type of starting-motor drive they use.
2. You might like to make yourself a simple electric motor that will run off a dry-cell battery. Parts for making such motors can often be found in hobby stores. Your instructor may be able to tell you where to find the parts.
3. If you can find discarded and defective starting motors, disassemble them. Clean the parts and mount them on a wood panel for display in the school shop.

CHAPTER 27
Starting-System Service

After you read this chapter, you should be able to:

- **Discuss possible troubles in the starting motor and starting-motor system and describe possible causes of each.**
- **Explain how to remove, disassemble, assemble, and install a starting motor. Also perform this service.**
- **Explain how to jump start a car safely and then do this job.**

This chapter describes testing, troubleshooting, servicing, and repair procedures for starting motors and starting systems.

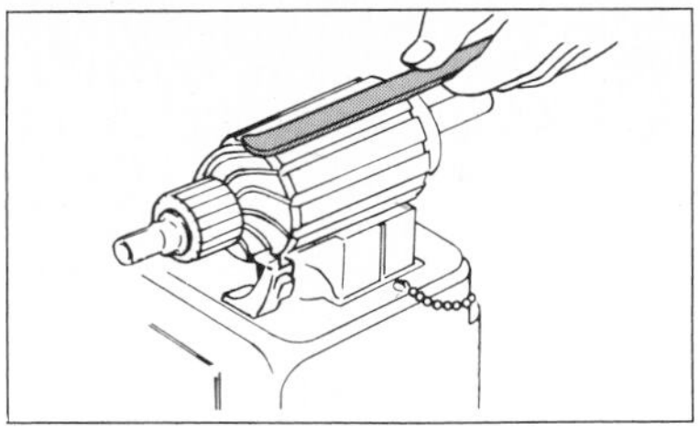

● 27-1 Troubleshooting the Starting System

The troubleshooting of starting motors can be divided into two parts.

1. Tests made on the car when trouble occurs. These are to determine whether the starting motor or some other component is at fault.
2. Tests made on the starting motor after it has been removed from the engine.

● 27-2 Testing the Starting Motor on the Car

There are three basic starting-motor troubles:

1. The starting motor does not turn over.
2. The starting motor turns over slowly, but the engine does not start.
3. The starting motor turns over and cranks the engine at normal speed, but the engine does not start.

This last trouble cannot be blamed on the starting motor. If it spins the engine at normal cranking speed, it has done its job. There is some problem in the fuel or ignition system, or the engine, that prevents starting. Let us look at the other two conditions.

● 27-3 The Starting Motor Does Not Turn Over

The most likely cause here is a run-down battery. But there could be other causes. Turn on the headlights, and try cranking. There are five possibilities.

1. No cranking, no lights This is probably due to a completely dead battery. It could be caused by a bad connection at the battery or starting motor or an open fusible link (which indicates a short or ground in the system).

2. No cranking, but lights go out as you turn the key to START This is an almost sure sign of a bad connection at the battery. It could also be due to a nearly dead (discharged) battery. Try wiggling the battery connections to see if this helps.

3. No cranking and lights dim only slightly as you try to start Chances are the trouble is in the starting motor. The pinion may not be engaging with the flywheel. If the starting-motor armature spins, then the overrunning clutch is slipping (or the inertia drive is stuck in the demeshed position).

4. No cranking, and lights dim heavily as you try to start This is most likely due to a run-down battery. It could be low temperature, too. Remember that a battery is much less efficient at low temperatures, and the engine oil is much thicker. The combination could prevent cranking, even though the battery is in fairly good condition. Also, there may be some sort of jam-up in the starting motor or engine.

5. No cranking, and lights stay bright Listen to hear if the solenoid is pulling the plunger in. You can hear this as a definite click. If nothing happens when you try to start, check the solenoid as follows: Connect one end of a jumper to the solenoid battery terminal. Connect the other end to the small terminal on the solenoid that is connected to the ignition switch. If nothing happens, the trouble is in the solenoid. If the solenoid and starting motor work with the jumper connected, the trouble is in the ignition switch, the transmission switch, or the wires connecting them.

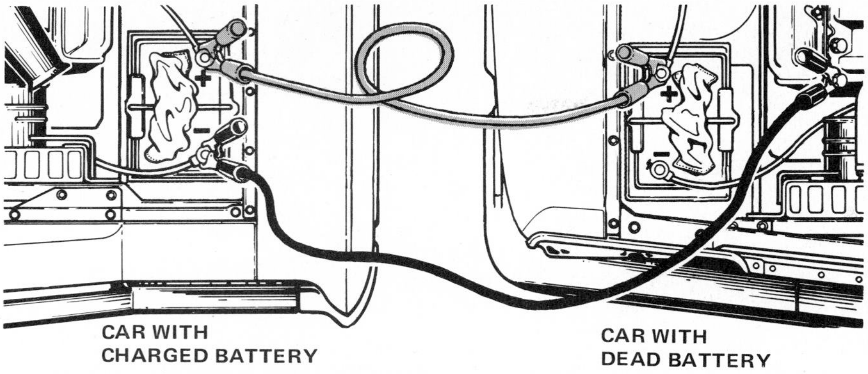

Fig. 27-1 Connections between the booster battery and the dead battery for starting a car with a dead battery.

● 27-4 Engine Cranks Slowly but Does Not Start

This is very likely due to a run-down or defective battery. The battery is unable to spin the starting motor at normal speed. Low temperature could also be a factor here, as we noted in the previous section.

It is also possible that the driver may have run the battery down trying to start. That is, some condition in the engine, or the fuel or ignition system, is preventing normal starting. The driver continued to try, however, until the battery ran down.

The procedure here is to test the battery and replace or recharge it if it is low. Or connect a booster battery, and then try to start (Fig. 27-1). See the caution below about using a booster battery. If the engine cranks normally and starts, the trouble is a low battery. If the engine cranks normally but does not start, the trouble is in the engine. If it still cranks slowly with a good battery, the trouble is either in the starting motor or the engine.

CAUTION: Use care in connecting a booster battery, to avoid hurting yourself or damaging the car's electrical equipment. Here is Ford's recommendation for using a jumper between two cars (for a negative-ground battery).

1. Remove the vent caps from both batteries. Cover the holes with cloths to prevent splashing of the electrolyte in case there is an explosion.
2. Shield your eyes.
3. Do not allow the two cars to touch each other. If the electrical systems of the two cars are not compatible, connecting them by allowing the two cars to touch could cause serious damage.
4. Make sure all electrical equipment except the ignition is turned off on the car you are trying to start.
5. Connect the end of one cable to the positive (+) terminal of the booster battery. Connect the other end of cable to the positive terminal of the dead battery.
6. Connect one end of the second cable to the negative (−) terminal of the booster battery.
7. Connect the other end of the second cable to the engine block of the car you are trying to start. *Do not connect it to the negative (−) terminal of the car battery!* If you did connect it to the negative (grounded) terminal, you could cause a spark. The spark could set off any hydrogen in the battery. This would produce a serious explosion. You could be badly injured by sulfuric acid or by parts of the battery.
8. Now start the car which has the booster battery. Then start the car which has the low battery. After the disabled car is started, disconnect the booster cables by first disconnecting the cable from the engine block. Then disconnect the other end of this (the negative) cable. Finally, disconnect the positive cable.

Careful: Never operate the starting motor for more than 30 seconds at a time. Pause for a few minutes to allow it to cool off. Then try again, if necessary. It takes a very high current to crank the engine. This can overheat the starting motor if it is used for too long a time. Overheating can ruin the starting motor.

● 27-5 Testing a Detached Starting Motor

No-load and stall tests are made on a starting motor that has been removed from the car. These tests, plus the use of a set of test points (Fig. 27-2), will tell you the condition of the starting motor.

1. No-load test To make the no-load test, connect the starting motor to a battery of the correct voltage, in series with a high-reading ammeter (Fig. 27-3). Measure the rpm (revolutions per minute) and current draw. These should be compared with the manufacturer's specifications for the motor.

2. Stall test To make this test, lock the drive pinion so that the armature cannot turn. Then apply the specified voltage to see what current the stalled motor will draw (Fig. 27-4). A high-reading ammeter is required for this test, as well as a high-capacity carbon-pile rheostat (variable resistance). However, service instructions from manufacturers no longer recommend this test.

3. Interpreting no-load and stall test results Following are the six most common combinations of conditions found in testing starting motors, along with further tests:

A. Rated current draw and no-load speed indicate normal condition of the cranking motor.

B. Low free speed and high current draw may result from: (1) Tight, dirty, or worn bearings; bent armature shaft; or loose field-pole screws, which allow the armature to drag on the pole shoes.

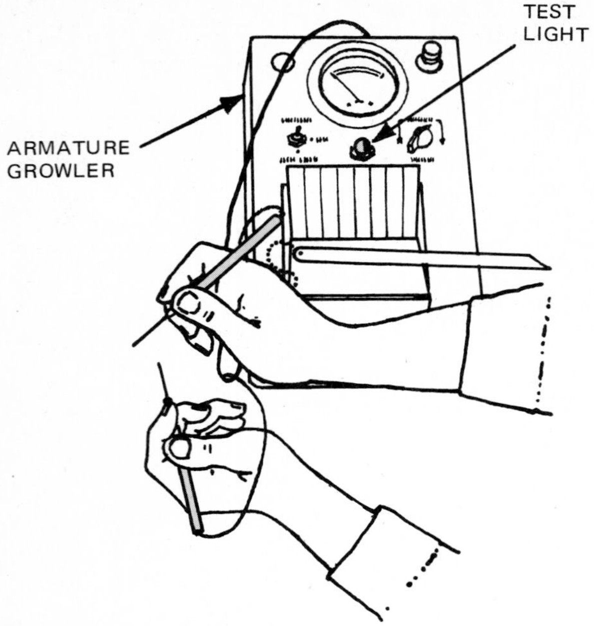

Fig. 27-2 Test points and lamp.

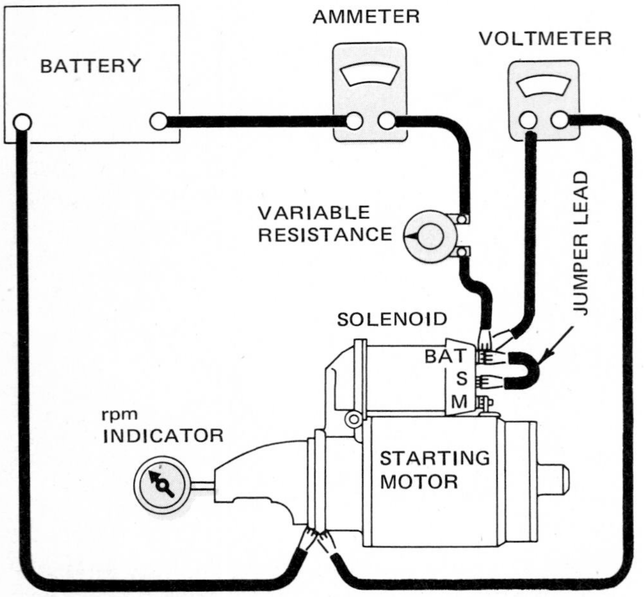

Fig. 27-3 Test setup for a no-load test. The variable resistance is used to adjust the applied voltage to the correct value. (*Delco-Remy Division of General Motors Corporation*)

(2) Grounded armature or fields. Raise the grounded brushes from the commutator, and insulate them with cardboard. Then check with the test lamp between the insulated terminal of the starting motor and the frame. If the test lamp lights, indicating a ground, raise the other brushes from the commutator. Check fields and commutator separately to determine which is grounded. On some units, one end of the field circuit is normally grounded. The ground screw or screws must be removed before the field can be tested for ground.
(3) Shorted armature. Check the armature further on a growler (● 27-8).

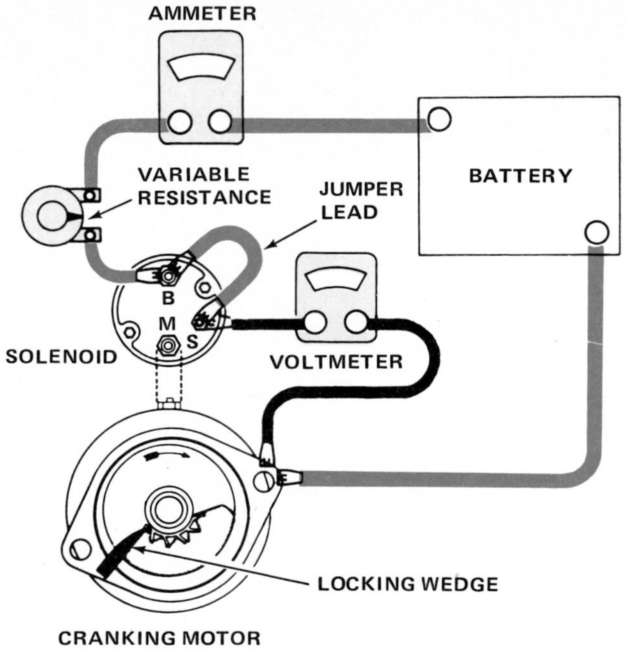

Fig. 27-4 Stall test of a starting motor. (*Delco-Remy Division of General Motors Corporation*)

C. Failure to operate at all with a high current draw indicates:
(1) Direct ground in the switch, terminal, or fields. This can be found with a test lamp by raising the grounded brushes as in (2) under item b above.
(2) Frozen shaft bearings, which prevent the armature from turning.
(3) Grounded armature windings, due, for instance, to thrown windings.
D. Failure to operate with no current draw indicates:
(1) Open field circuit. Inspect internal connections, and trace circuit with a test lamp, checking the brushes, armature, and fields.
(2) Open armature coils. This condition causes badly burned commutator bars.
(3) Broken or weak brush springs; worn brushes; high mica on the commutator; glazed or dirty commutator; or any other condition that prevents good contact between the commutator and the brushes. Most of these can be found by visual inspection.
E. Low no-load speed with low current draw indicates:
(1) An open field winding. Raise and insulate the ungrounded brushes from the commutator. Check the fields with a test lamp. Lamp should light as points are connected across each field.
(2) High internal resistance due to poor connections; defective leads; dirty commutator; or any other condition listed in (3) under item d above.
F. High free speed with a high current draw indicates shorted fields. Since the fields already have a low resistance, there is no practical way to test for this condition. If shorted fields are suspected, replace the fields and check for improvement in performance. But check the other components of the starting motor before going to this trouble.

27-6 Starting-Motor Service

Most starting motors require service only when the engine is overhauled. However, if a starting motor has heavy use, it may require more frequent service. Also, if it is damaged or defective, it will require service. Repair and servicing procedures are discussed below.

27-7 Damaged Starting-Motor Parts: Causes and Corrections

Several kinds of defects may develop in a starting motor, as follows:

1. Thrown armature windings This condition results from excessive armature speed that has thrown the windings from the armature. Improper adjustment of the throttle-opening linkage allows the starting motor to operate too long or spin the armature too fast before demeshing can take place. On the overrunning-clutch drive, the trouble could also be caused by a defective clutch. The clutch can become defective if spun at high speed before demeshing. This overheats the clutch. The inertia drive may be slow to demesh and this could spin the armature at high speed, causing thrown windings.

2. Burned commutator bars Burned commutator bars usually indicate an open-circuited armature. The open circuit normally will be found at one or more commutator connections. It is most often caused by excessively long cranking periods. Such long cranking periods overheat the starting motor and melt the solder at the connection. This not only throws solder (thrown solder may be found on the cover band), it also causes the connection to loosen. Arcing then takes place each time the bar with the bad connection passes under the brushes. The bar soon burns. If the bars are not too badly burned, the armature can be repaired. Resolder the connections at the riser bars (using rosin, not acid flux), and then turn the commutator. Some windings are attached by welding and these almost never need repair.

3. Broken or distorted Bendix spring The starting motor may have a broken or distorted Bendix spring or a broken drive housing. This is usually caused by an attempted meshing while the engine is on a rockback or by an engine backfire during cranking. On Bendix-drive starting motors, several seconds should elapse between attempts to start. Then the drive pinion will not go into mesh with the flywheel when the engine is rocking backward. If the ignition is out of time, it may cause the engine to backfire during cranking. This could cause a damaged drive or housing.

4. Dirty or gummy commutator The commutator sometimes becomes covered with a film of dirt or gum. Sometimes, this can be cleaned off with No. 00 sandpaper held against the commutator while the starting motor is being operated (not more than 30 seconds). This is somewhat difficult and hazardous. So it is best to correct this condition by turning the commutator with a lathe.

27-8 Starting-Motor Disassembly, Repair, and Assembly

The disassembly of the starting motor is usually simple. The solenoid or switch, where present, is removed first. Next, the cover band (where present) is removed, and the brush leads are disconnected. Where leads are soldered, the brushes are removed from the holders. Then, after the through bolts are taken out, the commutator end frame, field frame, and drive end can be separated. The Bendix drive can be removed from the armature shaft by taking out the drive-head attaching screw. The Folo-Thru drive is removed from the armature shaft by compressing the spring so that the end anchor plate clears the drive pin. The drive pin can then be pushed out of the shaft, and the drive slid off the shaft. On overrunning-clutch starting motors, the overrunning clutch can be slid off the shaft. Some models have a retainer and snap ring. These must be removed before the overrunning clutch can be slid off. Figure 26-8 is a disassembled view of a passenger-car starting motor.

1. Cleaning starting-motor parts The armature and fields should never be cleaned in any solution that dissolves or damages the insulation. They should be wiped off with a clean cloth. Never clean the overrunning clutch in a solvent tank. The solvent will dissolve the clutch lubrication and ruin the clutch.

2. Field-winding service Test for a grounded field with test-lamp points on the terminal stud and frame. If the lamp lights, the field is grounded. Test for open with points at two ends of the field circuit. The lamp should light. If field windings require replacement, use a pole-shoe screwdriver (Fig. 27-5). The pole-shoe screwdriver prevents damage to the pole-shoe screws and assures tight reassembly of the shoes. Rapping the frame with a plastic hammer while the screws are being tightened helps align the shoes properly. When resoldering connections, use rosin flux. For moist conditions, manufacturers recommend application of special insulating varnish to field windings after reassembly. This reduces the effects of moisture.

Many field windings are supplied permanently assembled into the field frame. The pole shoes are welded to the frame. On these, a defective field winding requires a new field-and-frame assembly.

3. Armature service Causes and correction of thrown armature windings and burned commutator bars have been discussed. Inspect the armature lamination for rub marks. These mean a worn bearing or a bent shaft has allowed the insulation to rub on the pole shoes. A check for a bent shaft can be made by putting the armature in V blocks. Rotate it while a dial indicator is placed in position to measure run-out. The run-out, or out-of-roundness, of the commutator can be checked at the same time.

The armature is tested for ground by placing one test point on the lamination and the other on the commutator. If the lamp lights, the armature is grounded. It is tested for short circuits on the growler (Fig. 27-6). The

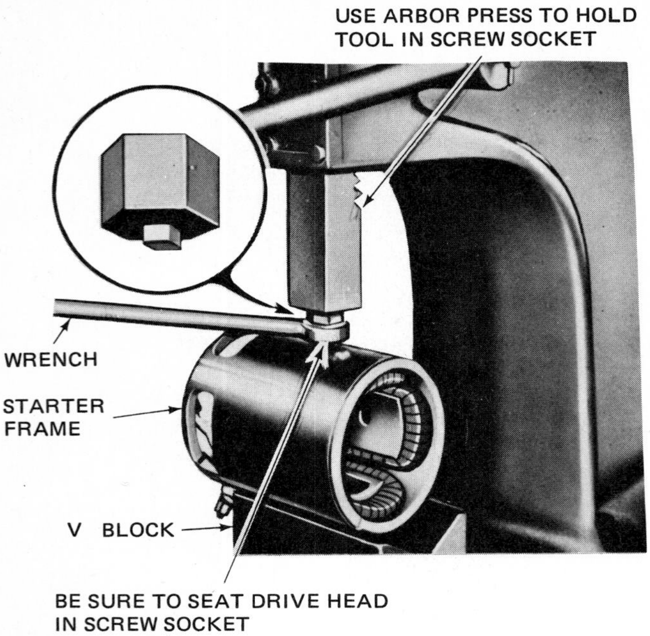

Fig. 27-5 Using a pole-shoe screwdriver to loosen or tighten a pole-shoe screw. (*Ford Motor Company*)

armature is placed on the growler and slowly revolved while a hacksaw blade is held above the armature core. The hacksaw blade vibrates against the core when it is above a slot containing a shorted winding. A shorted or grounded armature should be discarded.

If the commutator is out of round, or worn, or if it has high mica, it should be turned in a lathe (Fig. 27-7). The cut should be as smooth and as light as possible.

For moist conditions, manufacturers recommend application of special insulating varnish to the armature to reduce the effects of moisture. Varnish should be kept off the shaft and commutator.

4. Brush service Brushes that are worn to one-half their original length should be replaced. When the brush lead is soldered, unsolder it and unclinch the lead from the connector. Where the lead terminal clip is riveted to the frame, unsolder and unclamp the lead from the clip. Then the lead of a new brush can be clamped and soldered to the clip. With new brushes in place, put the armature into position so that the brushes rest on the commutator. If the brushes do not align with the commutator bars, the brush holders are bent. This requires replacement of the brush holders or the end frame. The brush-spring tension should be checked with a spring scale. Pull is required to raise the brushes, brush arms, or holders from the contact position. Replace the springs if the tension is not correct.

5. Starting-motor-drive lubrication The Bendix-type and Folo-Thru drives should be cleaned by washing in kerosene. They should not be lubricated. The overrunning-clutch type of drive must never be cleaned by a high-temperature or grease-dissolving method. This would remove the grease originally packed in the clutch. With the grease removed, the clutch would soon fail.

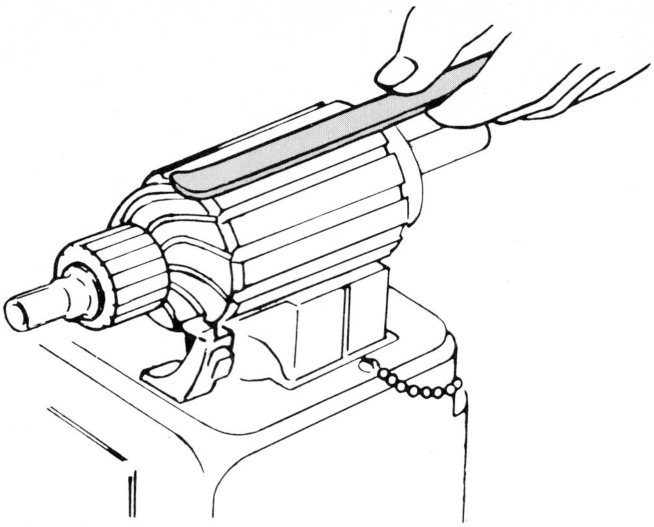

Fig. 27-6 Using a growler to test an armature for short circuits.

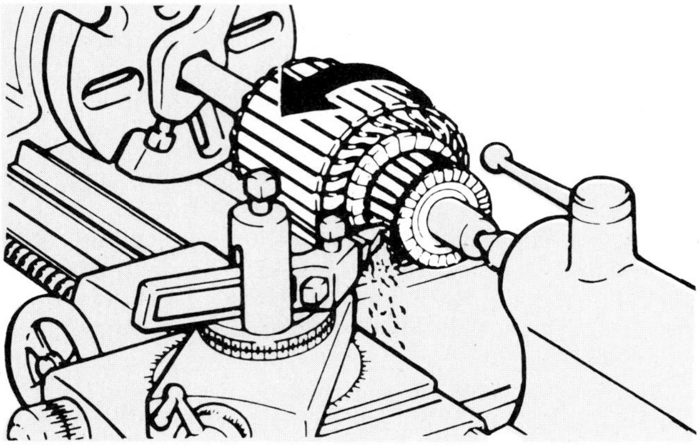

Fig. 27-7 Turning an armature commutator on a lathe.

6. Testing the overrunning clutch The overrunning-clutch pinion should turn freely and smoothly in the overrunning direction. It should not slip in the cranking position with normal cranking torque imposed on it. If the pinion turns roughly in the overrunning direction, the rollers are chipped or worn. The clutch should be replaced. If the pinion slips in the cranking direction, the clutch should be replaced.

7. Starting-motor lubrication During reassembly of the starting motor, all bearings should be lubricated with a few drops of light engine oil. Many starting motors have oilless bearings that have no provision for oiling. They should be lubricated, however, before reassembly, with a few drops of light engine oil.

8. Starting-motor assembly The assembly procedure is the reverse of disassembly. Soldered connections should be made with rosin, not acid, flux.

9. Overrunning-clutch-pinion clearance On the overrunning-clutch type of starting motor, the clearance between the pinion and the thrust washer, retainer, or housing should be measured after assembly. The pinion should be in the cranking position. Refer to the manufacturer's service manuals for details of checking.

10. Folo-Thru drive On the Folo-Thru drive, do not turn the pinion out to the extended, or cranking, position. In this position, a lockpin drops into a slot in the sleeve thread to lock the pinion (see Fig. 26-10). The only way the pinion can be unlocked is to mount the starting motor on the engine and start the engine. When engine speed increases to around 400 rpm, the lockpin will be retracted by centrifugal force. The pinion will then demesh and move back to the retracted position.

11. Testing assembled starting motor The starting motor should be submitted to no-load and torque tests, as outlined above, to make sure it operates according to specifications.

● 27-9 Installing Starting Motor

Whenever a starting motor is being installed or removed, the battery ground cable should be disconnected from the battery terminal. This avoids shorting the battery by an accidental grounding of the insulated cable. When installing the starting motor, connect the leads after the motor is bolted into place in the flywheel housing. Then check the throttle-cracker linkage on cars so equipped. Adjust it as necessary to obtain the proper throttle opening during cranking. This is particularly important on overrunning-clutch starting motors. An excessive throttle opening might spin the overrunning clutch at high speed during initial engine operation. This would cause the overrunning clutch to be overloaded.

● 27-10 Checking the Starting-Motor Circuit

After the starting motor has been reinstalled on the engine, the cables and connections can be tested. Use a low-reading voltmeter to measure voltage drop in the circuit, while a high current draw is taken through the circuit. This procedure locates any excessive resistance due to poor connections or bad cables. Excessive resistance prevents the delivery of normal amounts of current to the starting motor.

Two methods can be used to provide high current through the cables. In one method, the starting motor is operated. In the other, a heavy variable resistance is used. With either system, the high amperage flowing through the circuit from the battery shows up excessive resistance in the circuit: There is an excessive voltage drop. Too much voltage drop reduces the voltage at the starting motor, and you do not get normal operation.

● 27-11 Checking Circuit with the Starting Motor

A low-reading voltmeter is required for this check. Disconnect the ignition primary lead from the ignition distributor to the coil, so that the engine will not start. Operate the starting motor and, *very quickly,* check (1) from the insulated battery post to the solenoid battery terminal, (2) from the grounded battery post to the starting-motor housing, and (3) from the solenoid battery terminal to the solenoid motor terminal. Do not use the starting motor for more than 30 seconds. More than a 0.2-volt reading (0.1 volt on 6-volt circuits) on any of these indicates excessive resistance.

The remedy is to disconnect the cables and clean the cable clamps and terminal posts (see ● 25-18). Use new cables if the old ones seem to be in bad condition. Be sure that all connections are clean and tight. Always use cables of adequate size. Undersize cables have too much resistance and may prevent normal cranking, particularly during cold weather. If the excessive resistance is in the solenoid switch, disassemble the switch. Clean or replace the contact disk or contacts.

Extra-long battery cables are sometimes required because of the locations of battery and starting motor. This may result in somewhat higher voltage drops than the recommended 0.1 or 0.2 volt. In such cases, the normal voltage drop should be established by checking several vehicles. Then, when a voltage drop well above this normal value is found, excessive resistance must be located and eliminated.

REVIEW QUESTIONS

1. What are the three basic starting-motor problems?
2. A starting motor does not turn over. What are the five possible causes of the trouble?
3. What is the checking procedure if the engine cranks slowly but does not start?
4. Explain how to connect a booster battery to start a car.
5. Explain how to no-load-test a starting motor.
6. What is the most common cause of burned commutator bars?
7. Explain how to turn a commutator.
8. Explain how to test an armature on a growler.
9. Explain how to test the starting-motor circuit with a voltmeter.

SELF PROJECTS

1. Make a list of the tests to be made if the starting motor does not turn over when starting is attempted.
2. List the conditions that might cause slow cranking.
3. List the steps to take when using a booster battery in another car to get a car started.
4. If you can find defective starting-motor parts, examine them and decide what has caused the defects. Burned commutator bars, grounded field windings, loose conductor connections to the commutator bars, thrown armature windings, and bent armature shafts are samples. Write descriptions of the defects on tags and attach the tags to the defective parts. Note that this is the actual procedure used by automotive dealer service departments when they return defective parts to the factory under warranty.

CHAPTER 28
Automotive Charging Systems

After you read this chapter, you should be able to:

- **Describe the construction and operation of an alternator.**
- **Explain how the output of an alternator is controlled.**
- **Locate alternators and regulators (where separate) on cars and trace the charging circuit wiring.**

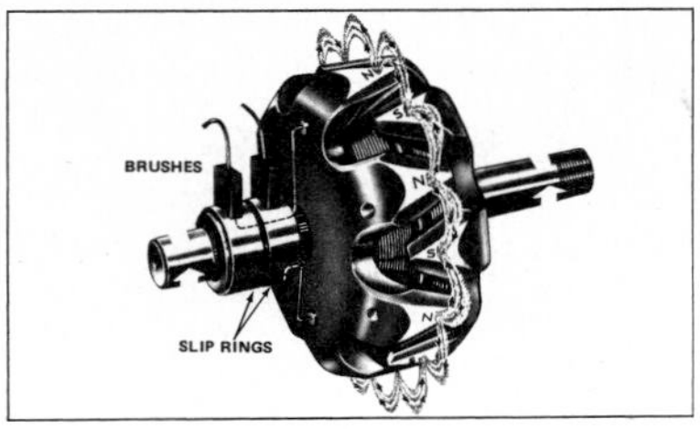

This chapter describes the various types of charging systems used on automobiles. The charging system has the job of keeping the battery in a charged condition. The starting motor takes current out of the battery when it cranks the engine. The charging system put the current back. In addition, the charging system handles electric loads, such as the ignition system and the lights, when the car is running. The charging system includes the generator or alternator, a regulator, and connecting wires (Fig. 28-1).

28-1 Function of Alternator

The alternator (or generator) converts mechanical energy from the engine into electric current. As already noted, it keeps the battery in a charged condition and handles electric loads when the engine is running. For many years, all cars used direct-current (dc) generators. In recent years, however, manufacturers have switched to alternators, or alternating-current (ac) generators.

Direct current (dc) flows in one direction. Alternating current (ac) flows in one direction for a moment, and then flows in the opposite direction. It alternates. The current in your home is ac. It alternates, or changes direction, 120 times per second. It is called 60-cycle [Hz] current. (In the metric system, one electrical cycle is called a hertz, abbreviated Hz.) The electric units on the car are all dc units and require dc.

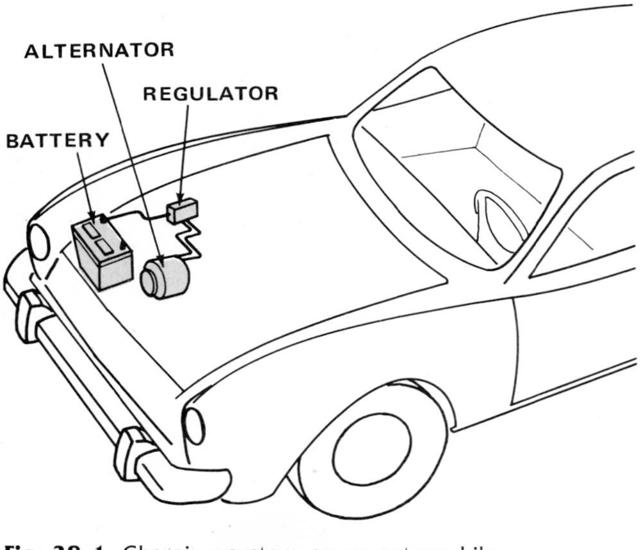

Fig. 28-1 Charging system on an automobile.

28-2 Generator

For many years, the direct-current generator was almost universally used in automotive vehicles. Today, however, you will seldom see this generator on anything except antique cars, some small engine applications, and some farm tractors. Figure 28-2 shows a generator. It is long and relatively small in diameter.

28-3 Alternator

The alternator (Fig. 28-3) is short and comparatively large in diameter. We show the two, generator and alternator (Figs. 28-2 and 28-3), so you can recognize them when you see them. The alternator has replaced the generator

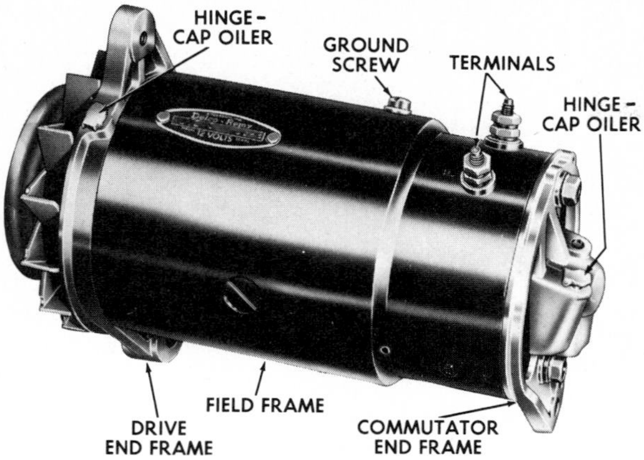

Fig. 28-2 A generator.

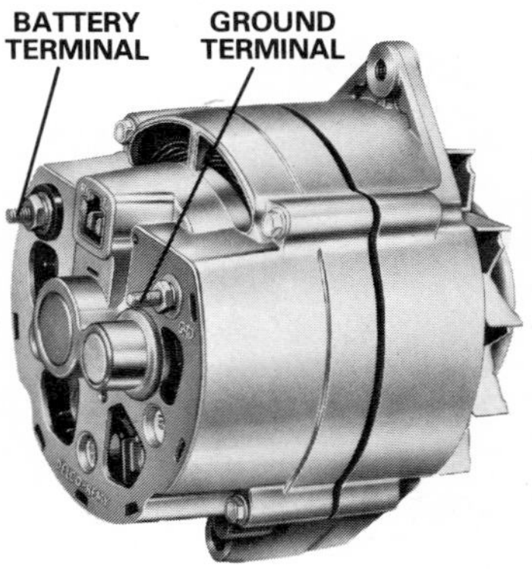

Fig. 28-3 An alternator.

in automotive vehicles for several reasons. It is lighter in weight and simpler in construction. It has fewer wearing parts and is easier to service. The regulators used with late-model alternators require no adjustments. Most are solid-state devices using diodes and transistors. The purpose of the regulator is to keep the alternator from producing too much current. Too much output from the alternator can damage the alternator and any electrical devices that are turned on. Also, this will overcharge the battery.

Some regulators are mounted inside the alternator. Others are separately enclosed in a metal box.

● 28-4 Producing Electricity

Before we discuss the alternator and regulator, let us review what we have learned about electricity. In Chap. 23, we said that electric current is a flow of electrons. We can make electrons move by moving a wire through a magnetic field. You can demonstrate this effect with a strong horseshoe magnet, a wire, and a sensitive ammeter, as shown in Fig. 28-4. As you move the wire through the magnetic field, the wire cuts lines of force. This causes the electrons (current) to move in the wire. So, as you move the wire through the magnetic field, the meter needle registers the current flow. As you move the wire back and forth, the needle moves first in one direction and then in the other. The same thing would happen if you held the wire stationary and moved the magnet. In

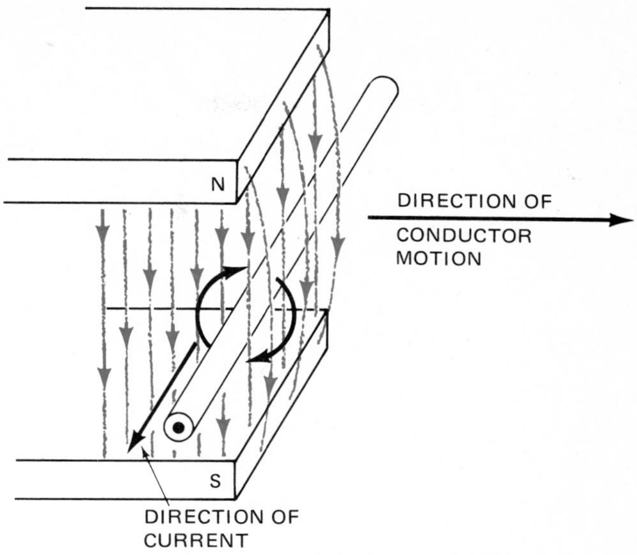

Fig. 28-4 A conductor, moving through a magnetic field as shown, has a flow of current induced in it.

either case, current flows through the wire. The key point is that the wire is cutting through the magnetic field. When it cuts in one direction, the current flows one way. When it cuts in the opposite direction, the current flows the other way through the wire.

● 28-5 Alternator Principles

In the alternator, the wires, or conductors, are held stationary and a magnetic field is moved through them. Actually, the alternator rotates the magnetic field so that the stationary conductors cut the moving magnetic lines of force.

Let us look at a simple alternator (Fig. 28-5). In the simple one-loop unit shown, the rotating bar magnet supplies the moving field. At the top, the north pole of the bar magnet passes the upper leg of the loop, and the south pole passes the lower leg of the loop. Current (electron flow) is induced in the loop in the direction shown by the arrows. At the bottom, the magnet has rotated half a turn. Its south pole is now passing the upper leg of the loop, and its north pole is passing the lower leg. Now, magnetic lines of force are being cut by the two legs in the opposite direction. So current (electron flow) is induced in the loop in the opposite direction. Thus, as the magnet spins and the two poles alternately pass the two legs of the loop, electrons in the loop are pushed first in one direction, and then in the other. In other words, the electrons alternate in direction; alternating current flows.

Three things will increase the current (number of electrons) moving in the loop. They are as follows:

1. Increasing the strength of the magnetic field.
2. Increasing the speed with which the magnetic field rotates.
3. Increasing the number of loops.

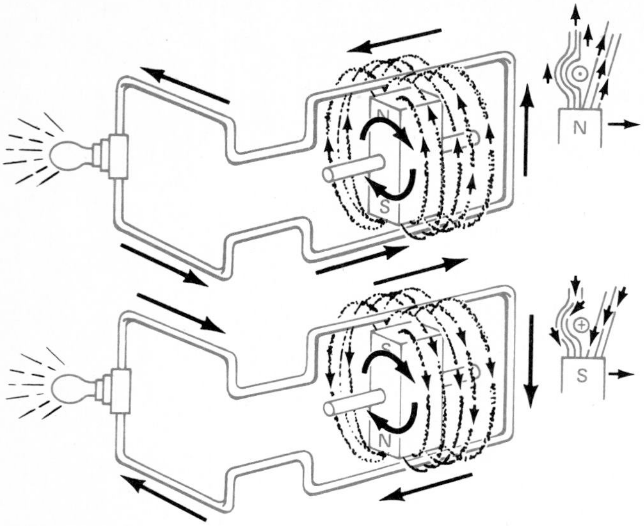

Fig. 28-5 Simplified alternator consisting of a single stationary loop of wire and a rotating bar magnet. The distortion of moving lines of force around a leg of the loop (conductor), and the direction of current (electron) flow, are shown at the right.

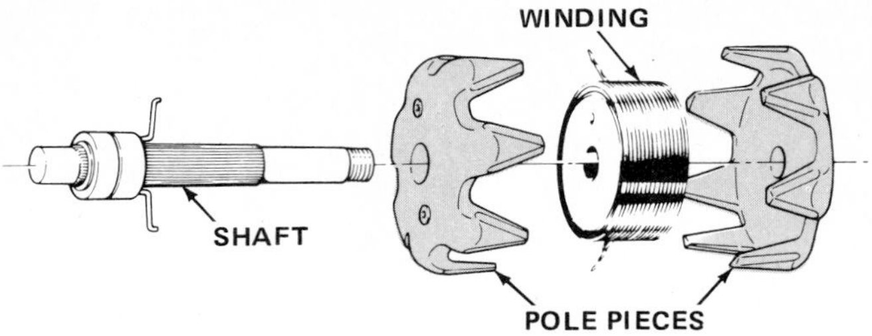

Fig. 28-6 Rotor of an alternator, partly disassembled. *(Delco-Remy Division of General Motors Corporation)*

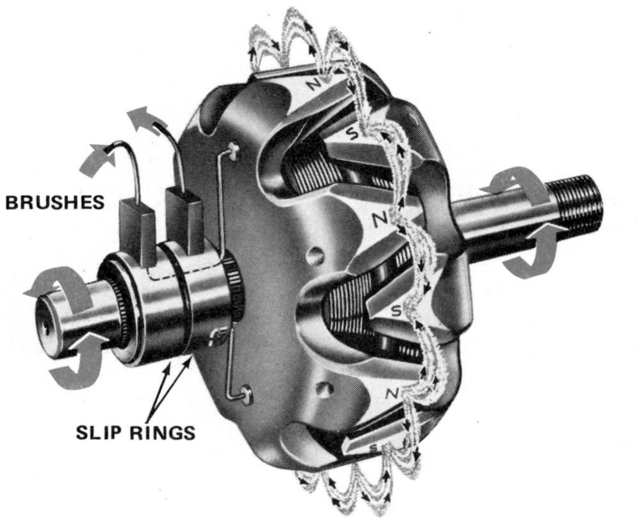

Fig. 28-7 Rotor of an alternator, showing brushes in place on slip rings. *(Delco-Remy Division of General Motors Corporation)*

In the actual alternator, both the strength of the magnetic field and the number of loops are increased. Instead of a simple bar magnet, the rotating part of the alternator is made up of two or more pole pieces. They are assembled on a shaft over an electromagnetic winding. The electromagnet is made up of many turns of wire. When current flows in the electromagnetic winding, a strong magnetic field is created. The pointed ends of the two pole pieces become, alternately, north and south poles (Fig. 28-6). The winding is connected to the battery through a pair of insulated rings that rotate with the shaft. A pair of stationary brushes ride on the rings. The two ends of the winding are attached to the rings, and the brushes make continuous sliding (or slipping) contact with the slip rings (Fig. 28-7).

Figure 28-8 shows the stationary loops of an ac generator assembled into a frame. The assembly is called a *stator*. The loops are interconnected as explained below so that the current produced in all loops adds together. Since this current is alternating, it must be *rectified*, or converted, into direct current.

28-6 The Alternator

The ac generator, or alternator, produces alternating current. The battery, ignition system, and other electrical components on the automobile cannot use ac, however. They are all dc units. The ac output must therefore be rectified, or changed to dc.

1. Rectifying ac Automotive alternators have built-in diode rectifiers. The *diode* is an electronic device that permits current to flow through it in one direction only. Figure 28-9 illustrates how four diodes can be used to change ac to dc. The four diodes are numbered 1 to 4 in

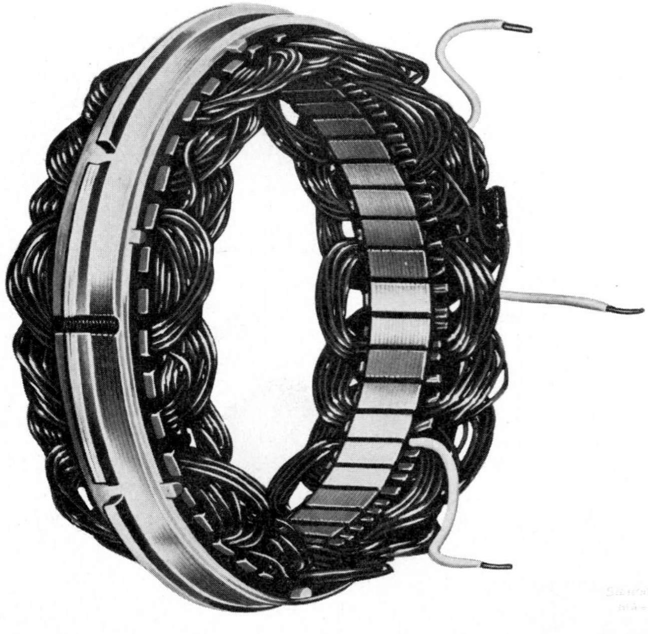

Fig. 28-8 Stator of an alternator. *(Delco-Remy Division of General Motors Corporation)*

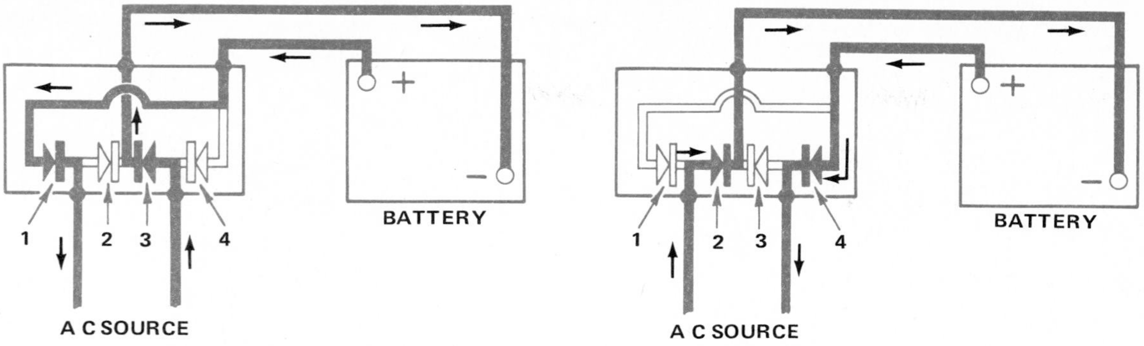

Fig. 28-9 Four diodes connected to an ac source. The diodes rectify the alternating current and change it to direct current to charge the battery.

the illustration. To the left, the current from the ac source follows the conductors shown solid. Diodes 1 and 3 permit the current to flow through. But diodes 2 and 4 will not, since the current is flowing in the wrong direction from them. However, when the direction of the current has reversed, as shown to the right in Fig. 28-9, diodes 2 and 4 will pass the current. But diodes 1 and 3 will not.

2. Three-phase The circuit in Fig. 28-9 is termed *single-phase,* since there is only a single ac source. Such a source would result in a pulsating current. This is like a single-cylinder engine. It does not provide a smooth flow of power, but rather a series of peaks between which no power is delivered. To provide a much smoother flow of current, ac generators are built with three stator circuits. These, in effect, give overlapping pulses of ac. When these are rectified, a smooth flow of dc is obtained.

The three stator circuits can be interconnected in either of two ways, with "Y" connections or with "delta" connections (Figs. 28-10 and 28-11). They operate similarly and are serviced similarly. The ac generated in the three legs of the stator passes through the six diodes and is converted into dc.

3. Diode heat sinks Diodes are usually mounted in the slip-ring end of the generator, in a metal bracket called a *heat sink.* The heat sink takes heat from the diodes, which become rather hot in operation. The heat sink has large radiating surfaces. They radiate the heat into the air surrounding the alternator. The diodes, therefore, do not overheat.

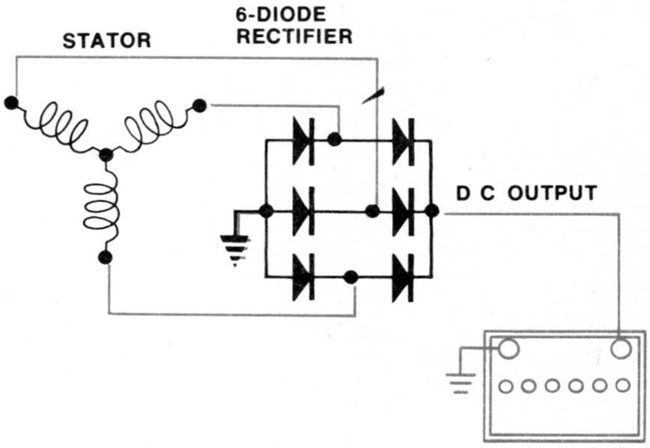

Fig. 28-10 Wiring diagram for an alternator with a six-diode rectifier and a Y-connected stator.

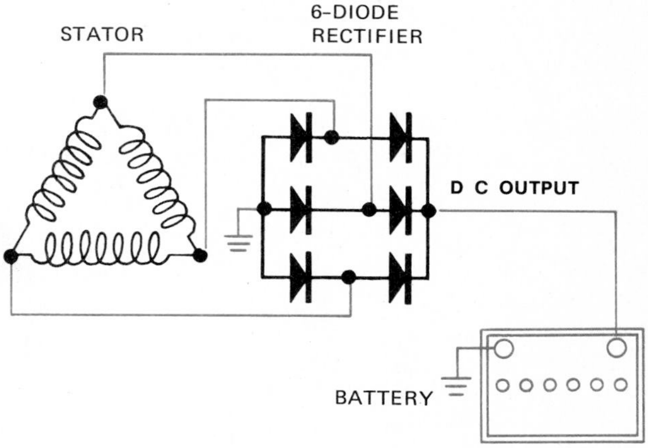

Fig. 28-11 Wiring diagram for an alternator with a six-diode rectifier and a delta-connected stator.

28-7 Alternator Regulation

A variety of devices has been used to regulate alternators. When alternators were first introduced, many of the regulators were very complex. They included a field relay, an indicator-light relay, and a voltage regulator. Recently, alternator-regulator systems have been much simplified. For instance, some of the latest types have the regulator built into the alternator (Fig. 28-12), so that the circuit looks like Fig. 28-13. Basically, the regulator limits the alternator field current as necessary to prevent excess alternator voltage. The stator remains permanently connected to the battery through the diodes. The diodes prevent the battery from discharging back through the stator when the alternator is not operating. The field (rotor) is connected to the battery only when the alternator is operating. The connection is made through either the field relay or the ignition switch.

28-8 Regulators for Alternators

Several types of alternator regulators that are mounted outside the alternator have been used. Ford Motor Com-

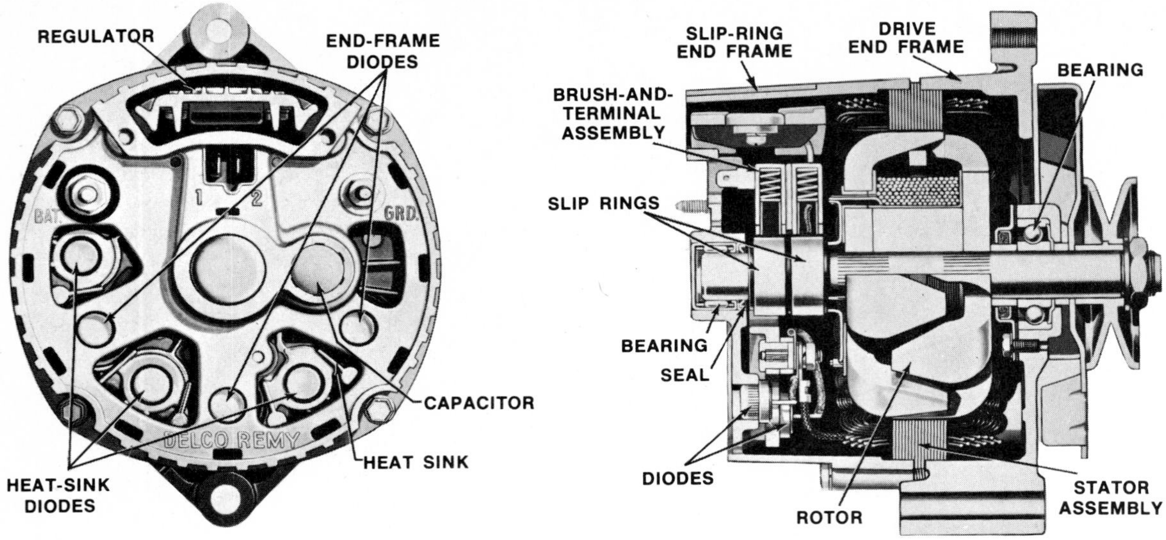

Fig. 28-12. End and sectional views of an alternator with built-in diodes and voltage regulator. This is called a Delcotron by Delco-Remy. (*Delco-Remy Division of General Motors Corporation*)

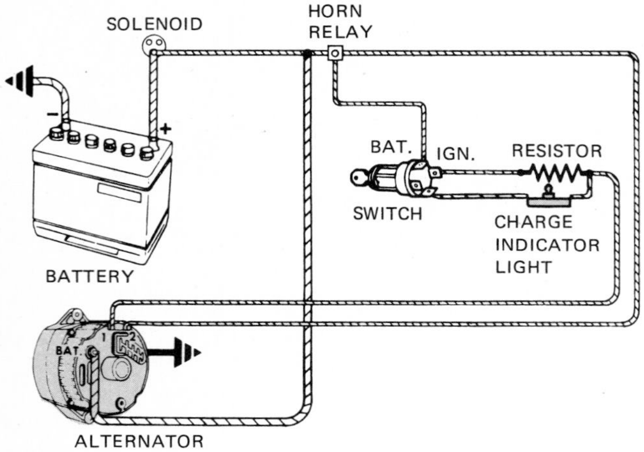

Fig. 28-13 Wiring diagram for a charging system using an alternator with an integral voltage regulator and a charge-indicator light. (*Delco-Remy Division of General Motors Corporation*)

pany has used three types of such regulators in recent years. Chrysler Corporation has used one type in late-model cars. We look at each in detail.

1. Chrysler external voltage regulator This regulator (Fig. 28-14) is a solid-state device which has diodes, transistors, and other electronic components. Figure 28-15 is the wiring diagram for the system. The regulator operates as previously mentioned, by limiting the current to the field (rotor) to prevent excessive alternator voltage. Note that one end of the field is connected through the regulator. The other end is connected through the ignition switch. If the voltage starts to rise above the specified maximum, the electronic circuitry cuts down the amount of current flowing to the field (rotor). This weakens the magnetic field cutting through the stator windings. The voltage is therefore kept from rising any further.

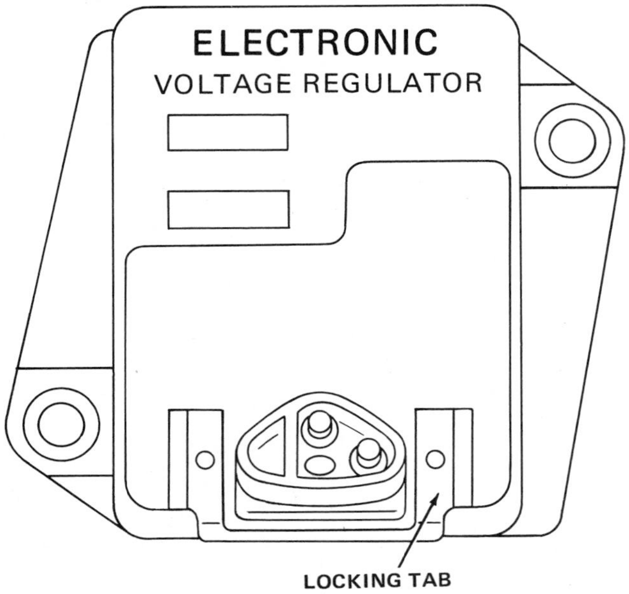

Fig. 28-14 Chrysler electronic voltage regulator. (*Chrysler Corporation*)

2. Ford external voltage regulators In recent years, Ford Motor Company cars have had three different types of voltage regulators: electromechanical, transistorized, and electronic.

The electromechanical regulator has two mechanical units operated by electromagnetism. These are a vibrating voltage regulator and a field relay. The voltage regulator has a winding that is connected across the alternator so it senses alternator voltage. When that voltage reaches the specified maximum, the magnetism is strong enough to pull down a flat plate, called the armature. This opens contact points. When the points open, the field current must flow through a resistance. This reduces the field current and the alternator voltage. With a reduced voltage, the voltage regulator releases the armature and the points close. Voltage goes up and the cycle is repeated, many times a second. The points stay open enough of the time to prevent excessive voltage.

The field relay is open until the engine starts and the alternator rotor begins to spin. This produces a voltage in the stator. The voltage causes the field relay to close, directly connecting the field to the battery.

The transitorized voltage regulator (Fig. 28-16) has a field relay just like the one described above. This regulator is electronic, including diodes and transistors. It works like the one used with Chrysler alternators.

The electronic voltage regulator comes in two variations, one for ammeter-equipped cars, the other for cars equipped with indicator lamps. The circuits for these two are shown in Figs. 28-17 and 28-18. Note that none of the Ford regulators are interchangable. That is, you cannot use the electronic voltage regulator in place of the other regulators. The same goes for the other two. The alternators used with the three regulators are different, as are the connectors.

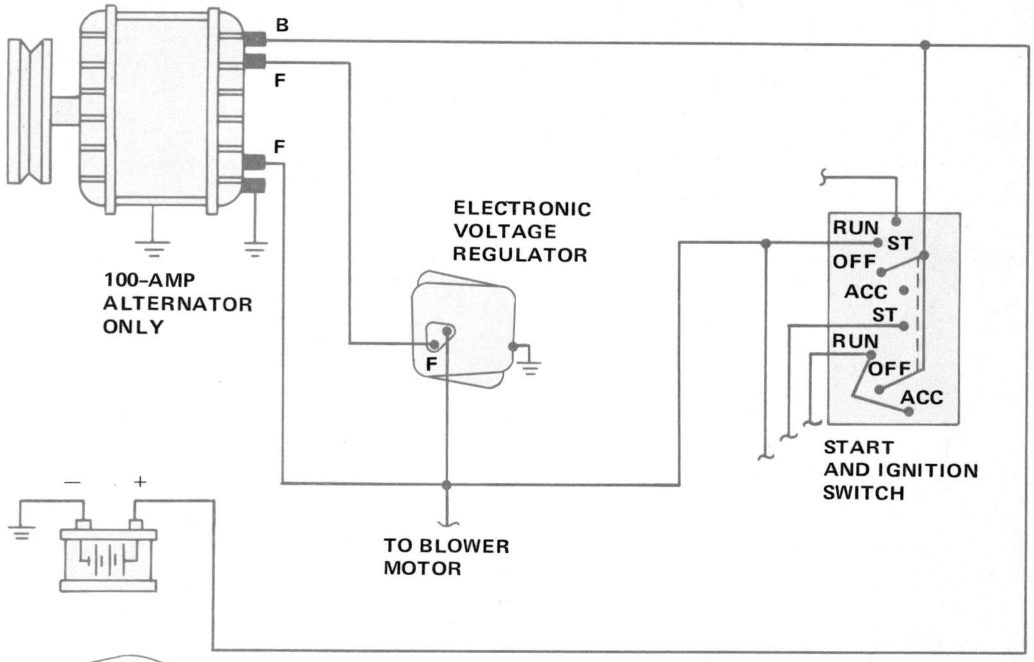

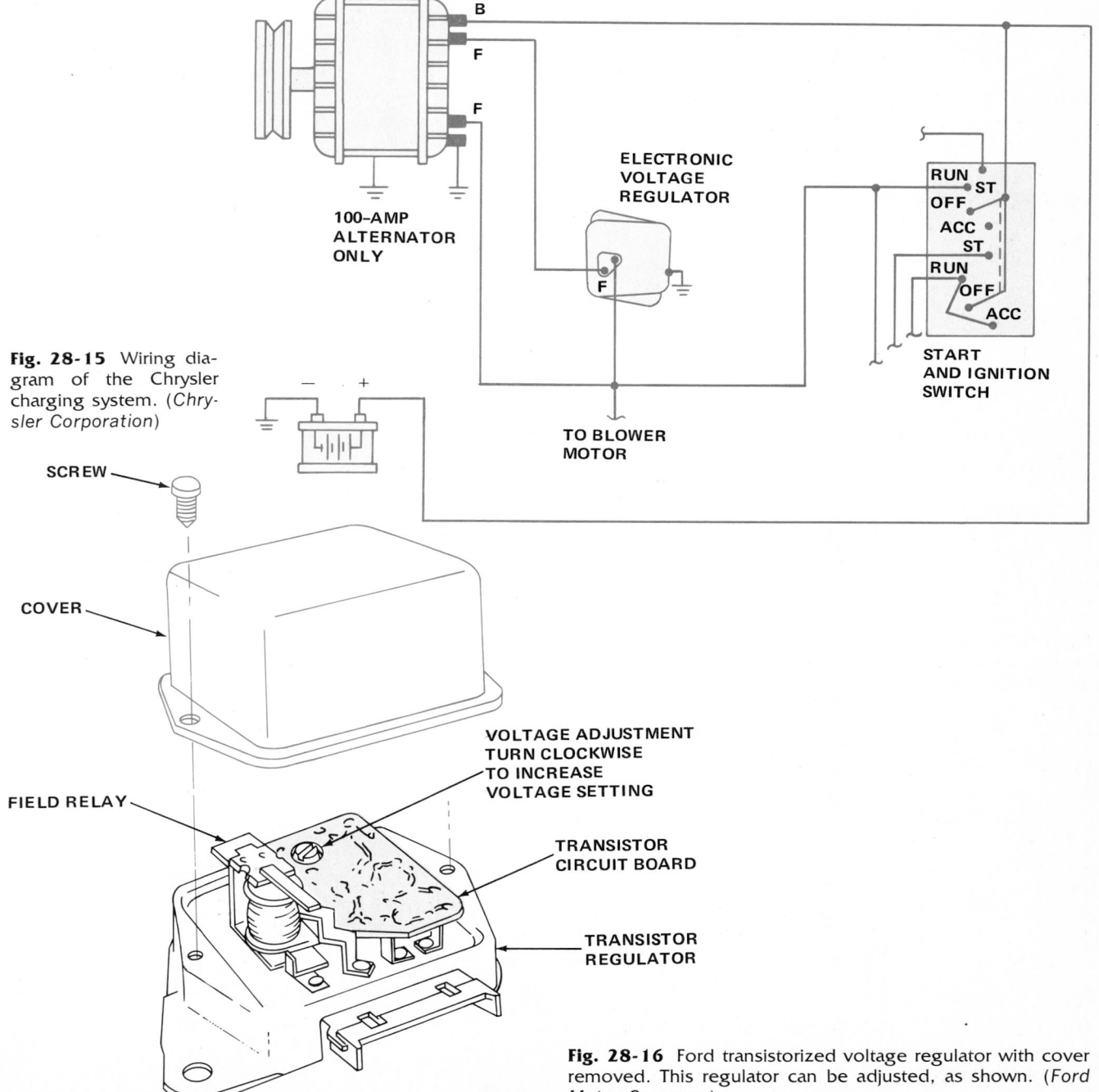

Fig. 28-15 Wiring diagram of the Chrysler charging system. (*Chrysler Corporation*)

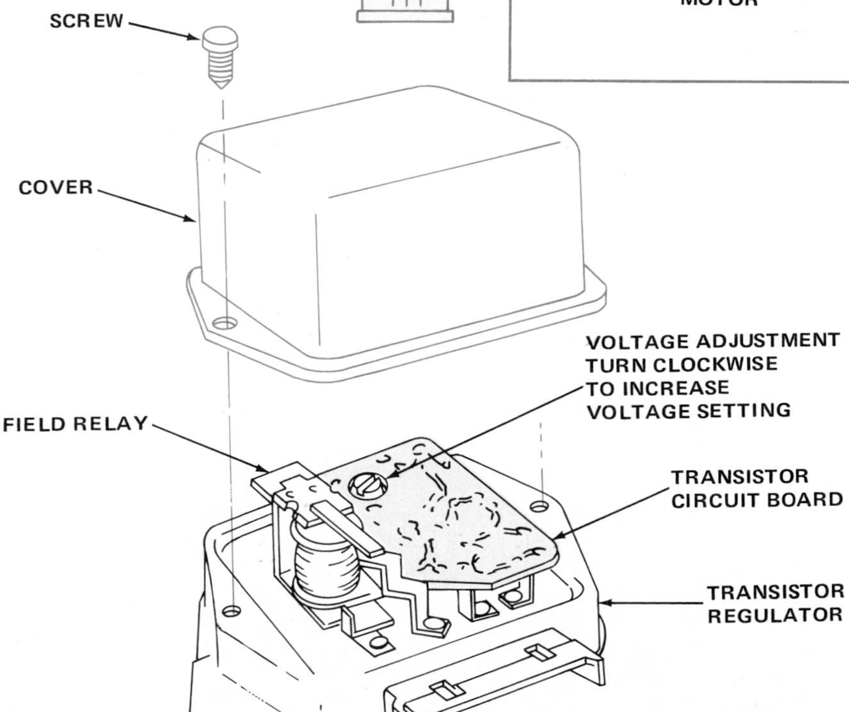

Fig. 28-16 Ford transistorized voltage regulator with cover removed. This regulator can be adjusted, as shown. (*Ford Motor Company*)

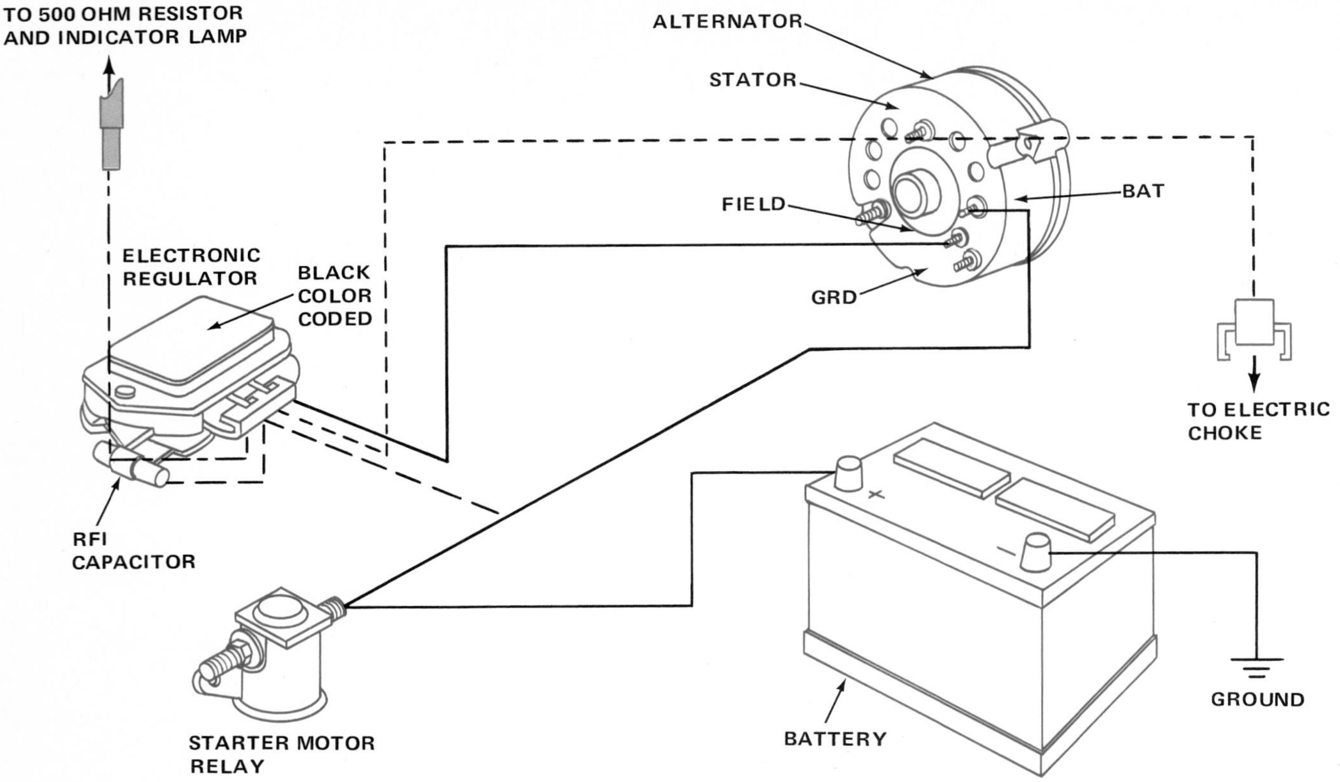

Fig. 28-17 Wiring circuit of charging system with warning-lamp indicator. (*Ford Motor Company*)

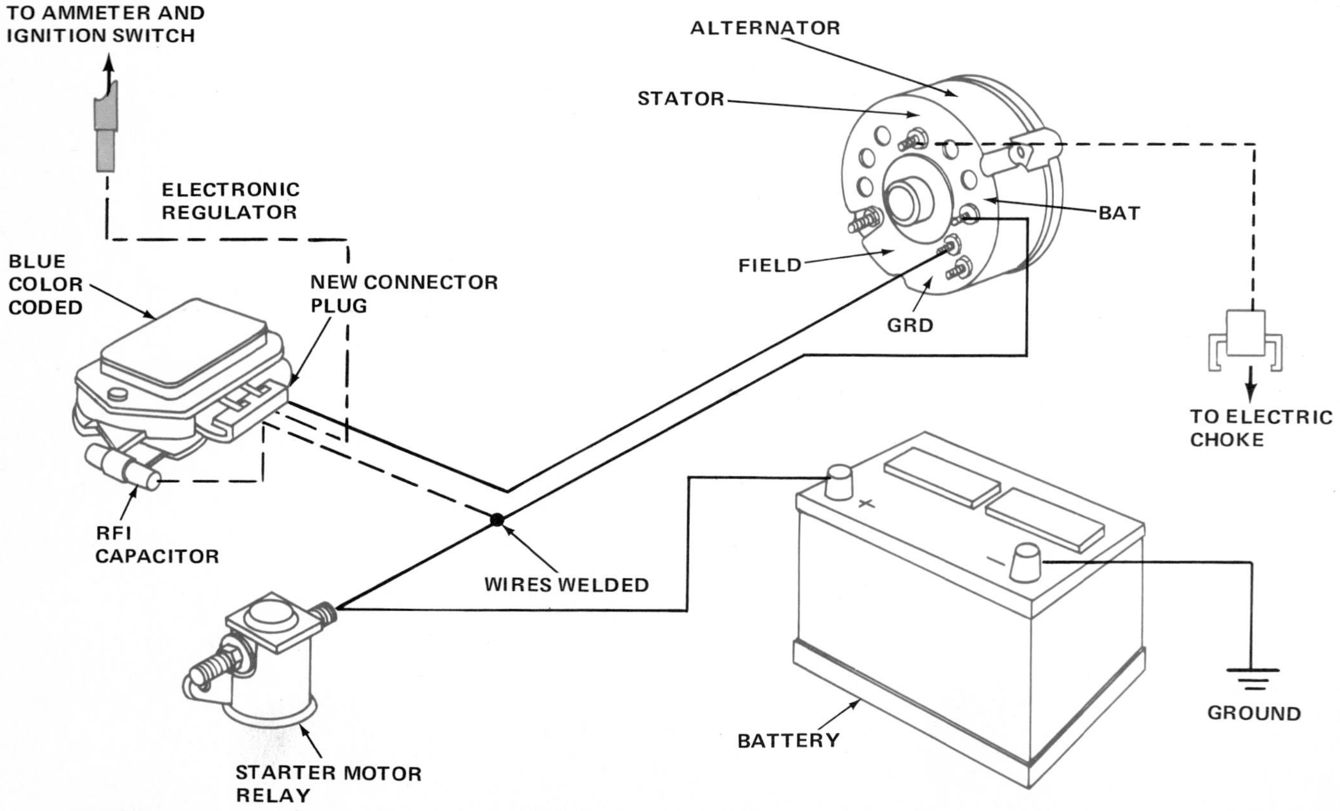

Fig. 28-18 Wiring circuit of a charging system with ammeter. (*Ford Motor Company*)

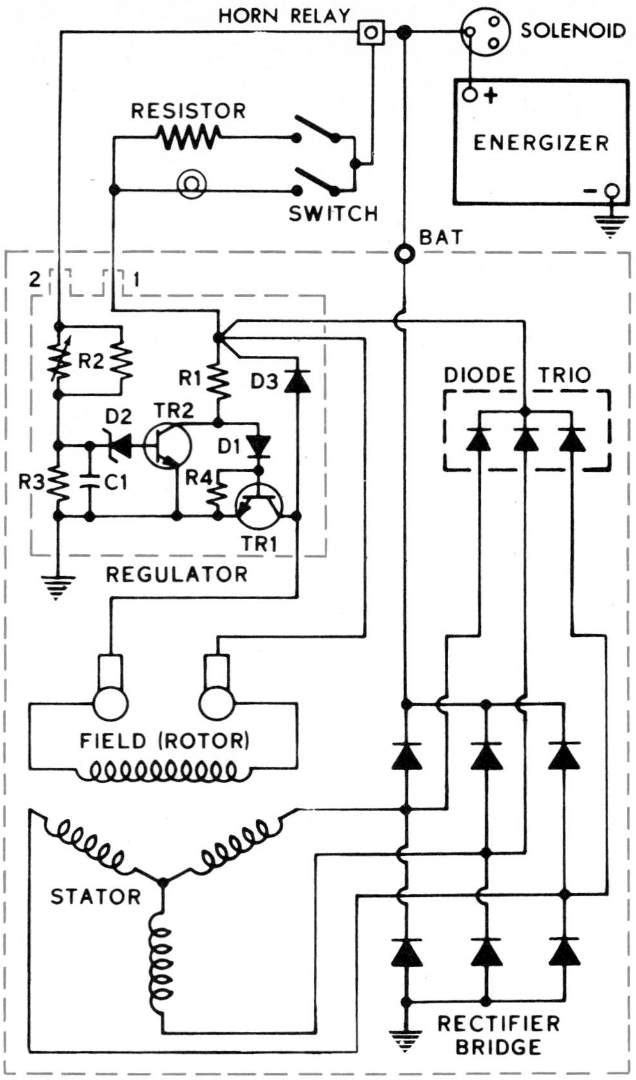

Fig. 28-19 Internal wiring diagram for an alternator with an integral voltage regulator. The rectifier and external circuits are also shown. (*Delco-Remy Division of General Motors Corporation*)

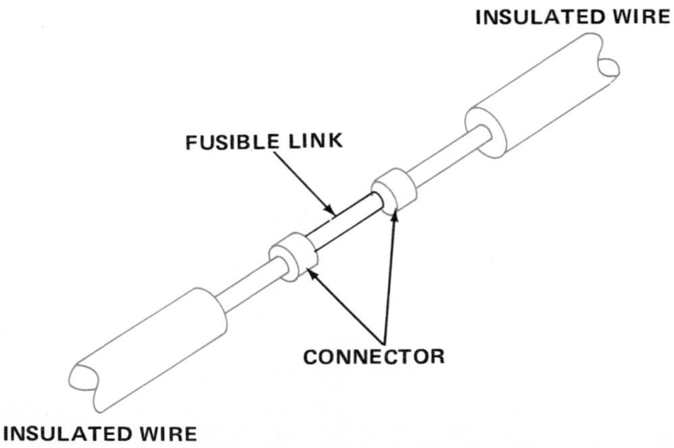

Fig. 28-20 Fusible link connected into a hot-wire circuit. Insulation is removed so link can be seen.

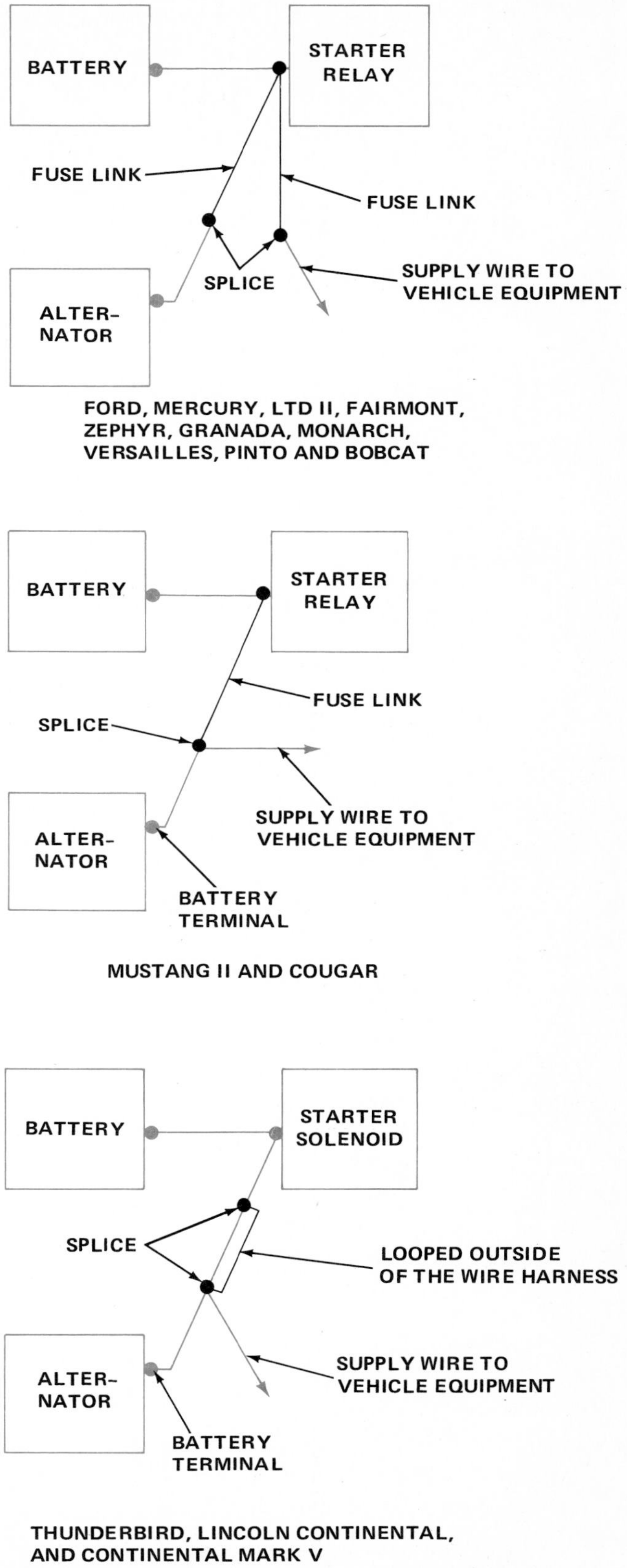

Fig. 28-21 Locations of fusible links in various charging systems of Ford cars. (*Ford Motor Company*)

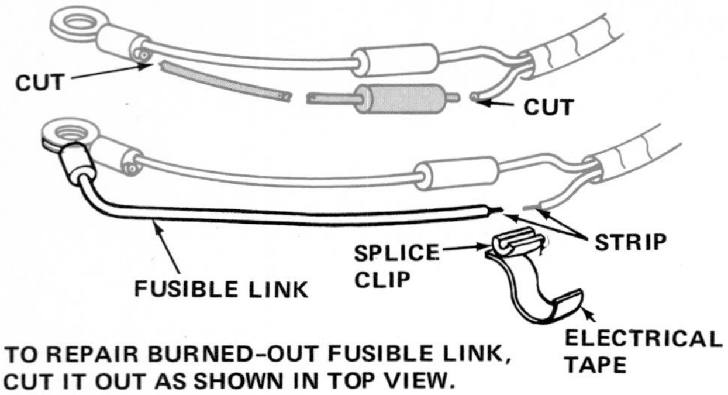
CUT
CUT
STRIP
SPLICE CLIP
FUSIBLE LINK
ELECTRICAL TAPE
TO REPAIR BURNED-OUT FUSIBLE LINK, CUT IT OUT AS SHOWN IN TOP VIEW.
STRIP BACK INSULATION. SPLICE WIRES WITH SPLICE CLIP, AND SOLDER IN THE NEW FUSIBLE LINK. TAPE SPLICE WITH DOUBLE LAYER OF ELECTRICAL TAPE.

Fig. 28-22 Repairing a burned-out fusible link. (*Buick Motor Division of General Motors Corporation*)

3. General Motors and American Motors voltage regulators In recent years, these voltage regulators have been built into the alternators, as shown in Figs. 28-12 and 28-13. Figure 28-19 shows the complete wiring circuit, including the internal diagrams for the regulator and the alternator.

● 28-9 Charge Indicators

There are two types of charge indicator, the ammeter, described in ● 23-4 (see Fig. 23-2), and an indicator light. The ammeter needle moves off center (or zero) to indicate a flow or current to the battery (charging) or a flow of current from the battery (discharging). The charge indicator light comes on when first starting to indicate the alternator is not charging the battery. Then, when the alternator speed increases enough to send a charge to the battery, the indicator light goes off.

The ammeter is connected into the circuit from the alternator to the battery so it can show what's happening. The charge indicator light has a different circuit (See Fig. 28-13). When the ignition switch is turned on, current can flow through the resistor and charge the indicator light. It comes on. As soon as the alternator builds up voltage, however, the voltage on the two sides of the resistor and light becomes the same. The light goes out and this indicates that the alternator voltage has built up so the alternator is charging the battery. If the alternator should fail for any reason, its voltage will fall and the light will come on to indicate the failure.

● 28-10 Fuse Links

The fuse link, or fusible link as it is also called, is a short length of insulated wire within the wiring harness. It is smaller than the circuit that it protects (Fig. 28-20). The illustration shows it uncovered. In the actual harness, it is covered with insulation. The fuse links are identified in the harness by a special tag or color code. On some cars, the fuse link is looped outside the wiring harness. Figure 28-21 shows the locations of fusible links in various charging systems. Their purpose is to burn or blow out if the current goes too high. This protects the rest of the system from damage. A blown fuse link is usually easy to spot. Bare wire may be sticking out of the insulation. Or the insulation may be bubbled or burned from the heat. Figure 28-22 shows how to replace a fusible link.

REVIEW QUESTIONS

1. What is the purpose of the alternator?
2. How can you tell the difference between a generator and an alternator?
3. Discuss alternator principles and explain how current is induced in the conductors, or stator.
4. What three things will increase the current from an alternator?
5. What is the rectification and why does an alternator require it?
6. Explain how the set of diodes in the alternator change the alternating current to direct current.
7. How does the alternator regulator work?
8. What are the three types of external voltage regulators that Ford has used in recent years?
9. What are the two types of charge indicator?
10. What is a fusible link and how can you tell if it is blown?

SELF PROJECTS

1. If you are able to get junked generators and alternators, disassemble them. Examine the internal parts and note how the connections are made to the generator brushes and to the alternator brushes and diodes. See if you can trace the connections between the diodes and the stator of the alternator.
2. Write a short explanation of the differences in design and construction of the generator and alternator. Explain why, in your opinion, all automotive manufacturers have switched to alternators in recent years.
3. Refer to automotive manufacturers' shop manuals and find the wiring diagrams they show for the charging systems in their cars. Copy these diagrams for your notebook. You can use different colored pencils to show the circuits. For example, you could use a red pencil to show the rotor circuit. Use a blue pencil to show the stator and diode circuit.

CHAPTER 29
Servicing Charging Systems

After you read this chapter, you should be able to:

- **Troubleshoot ailing charging systems to find trouble causes.**
- **Test alternators and regulators.**
- **Replace defective alternators and regulators.**

This chapter discusses troubles in the charging system, how to locate the causes, and how to correct them, including the servicing of alternators.

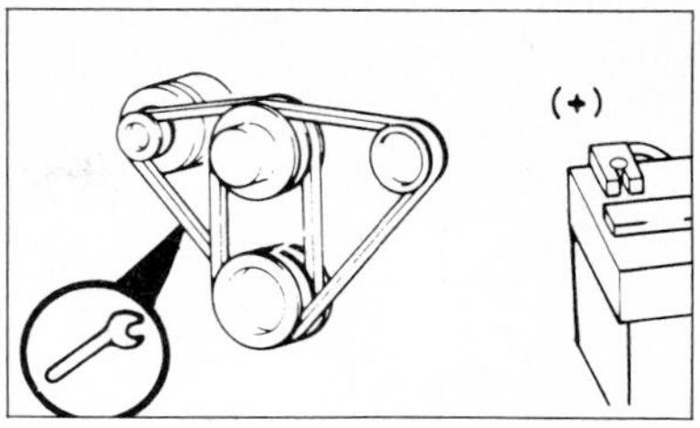

● 29-1 Testing Alternator-Regulator Systems

Many late-model cars use solid-state alternator regulators. As we mentioned in the previous chapter, General Motors cars are equipped with alternators inside of which a solid-state regulator is installed. Chrysler Corporation cars have an external solid-state voltage regulator. Ford Motor Company cars have, in recent years, used three different types of voltage regulators; none are interchangeable. These are the electromechanical type (vibrating points), the transistorized (transistorized voltage unit with a mechanical field relay), and the full electronic regulator. All of these require their own special tests. All have the same remedy for regulator or alternator troubles—replace the unit. Defective regulators cannot be serviced. Alternators, however, can be torn down for repair and replacement of defective parts.

The tests include:

1. Charging-circuit resistance test
2. Check of regulator operating voltage
3. Testing alternator for current output

● 29-2 General Instructions

There are four kinds of trouble that might bring a car into the shop for service:

1. Low battery
2. Overcharged battery
3. Faulty indicator-lamp action
4. Alternator noisy

Regardless of the type of trouble, there are certain general precautions you should observe:

1. Never attempt to polarize the alternator. This can seriously damage the system. The old-style dc generator (Fig. 28-2) did require this service after any test, but it must not be done on the alternator.
2. Do not short across or ground any terminal in the charging circuit. This can cause great damage. There are one or two exceptions to this general rule, however, but these are *specifically spelled out in the manufacturer's instructions*.
3. Never operate the alternator with the output terminal disconnected. The alternator voltage can go high enough to burn out the alternator.
4. Make sure the alternator and battery are connected correctly, according to their polarity. Installing a battery backwards, for example, will allow a high current to flow through the system and everything will burn up (if fuses or fuse links do not burn out first).
5. When connecting a booster battery, connect negative to negative and positive to positive, as explained in ● 27-4. Observe the cautions outlined there.

● 29-3 Charging-Circuit Resistance Test

A typical charging-circuit test is the one specified by Chrysler Corporation. See Fig. 29-1. The procedure is as follows.

1. Disconnect the battery ground cable and connect the test ammeter, voltmeter, and carbon-pile rheostat, as shown in Fig. 29-1. Note that the lead to the regulator has been disconnected from the alternator and a jumper wire connected from the alternator terminal to ground instead. This eliminates any regulator action that could influence the test. It also can permit the alternator voltage to go too high if you over-rev the engine. So do not increase engine speed any higher than necessary to make the test. Reconnect the battery ground cable.
2. Start engine and immediately reduce engine speed to idle.

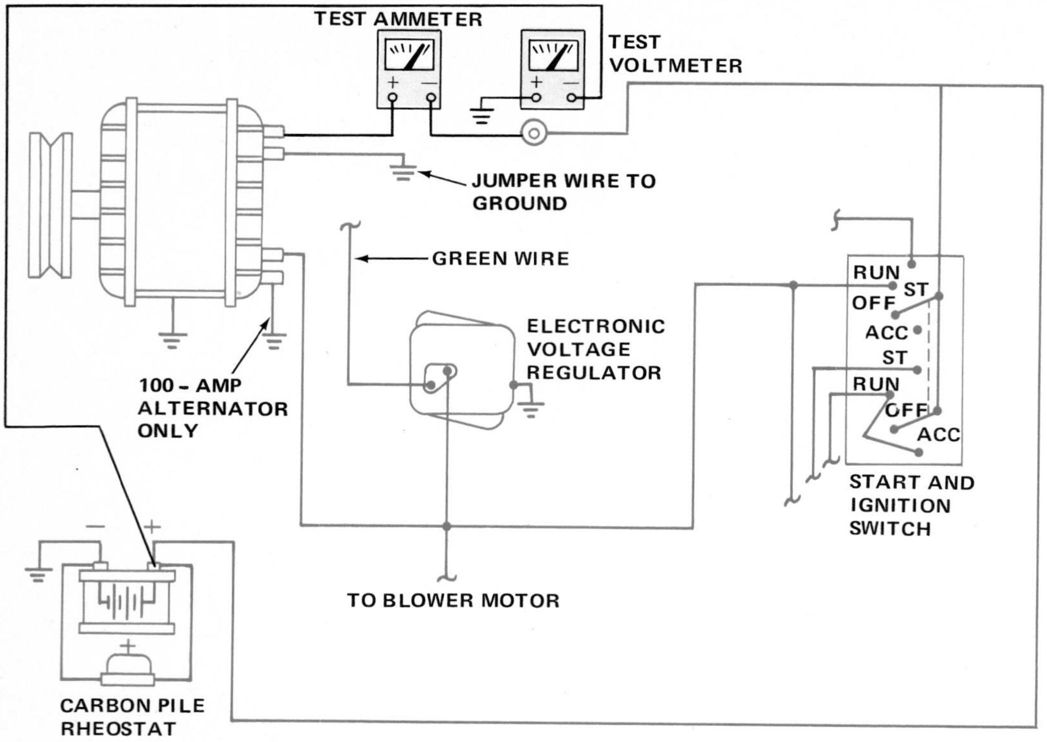

Fig. 29-1 Charging-circuit resistance test. (*Chrysler Corporation*)

3. Adjust engine speed and carbon pile to obtain 20-ampere flow in the circuit. The voltage reading should not exceed 0.7 volt. If it does, there is excessive resistance and all connections and leads should be checked to eliminate it.

CAUTION: Whenever making or breaking connections in the electric circuit, always temporarily disconnect the battery ground cable first. Then, if you should accidentally ground another lead, nothing will happen. However, if the ground cable is still connected, you could get a direct short across the battery and the sparks would fly! This could be dangerous. You could burn your hand. Or the sparks might cause a fire, especially if there is any gasoline or gasoline vapor in the engine compartment.

29-4 Current-Output Test

The current-output test determines if the alternator can produce its rated maximum. The way Chrysler sets up the test is shown in Fig. 29-2. Connections are made as for the circuit-resistance test except that the test voltmeter is connected from the alternator hot terminal to ground. Disconnect the battery ground cable before making any connections and then reconnect it after all other connections have been made. Connect an engine tachometer to measure engine speed.

Run the engine at idle. Then slowly increase engine speed and readjust the carbon pile until you get an rpm of 1,250 and a voltage of 15 volts. (Chrysler says 900 rpm and 13 volts for their 100-ampere alternator.) Note the amperage output. It should be within the limits as specified by the alternator specification chart in the manufacturer's shop manual.

If the alternator does not come up to rated output, there is something wrong and it should be serviced.

29-5 Voltage-Regulator Test

Manufacturers supply special testing instruments to check the operation of their voltage regulators. However, most can be checked with a voltmeter (Fig. 29-3). First, check the battery gravity. If it is below 1.200, disconnect it and connect a fully charged battery in its place. Then, with the voltmeter connected, as shown in Fig. 29-3, operate the engine at 1,250 rpm with all lights and accessories off. Voltmeter should read within the limits specified by the manufacturer. The voltage will go down as the temperature goes up. This is a built-in temperature compensation that helps to overcome the increased battery resistance to charge as its temperature goes down. That is, at low temperature, the charging voltage is higher. Chrysler supplies the following specifications for late-model vehicles:

Temperature near regulator	Voltage range
−20°F [−28.9°C]	14.9 to 15.9
80°F [26.7°C]	13.9 to 14.6
140°F [60.0°C]	13.3 to 13.9

If the shop has one of the special testers designed to test voltage regulators, follow the instructions that come with it to check regulator operating voltages.

29-6 Troubles and Their Symptoms

Each type of trouble in the charging system has its own set of symptoms, as follows:

1. Low battery A battery that is low, or flat, does not have enough reserve to crank the engine at normal speed, if at all. If the trouble is slow or no cranking, the trouble could be due to:

a. Start-and-stop driving during which the alternator never operates long enough to put back into the battery the current taken out by the starting motor.
b. Accessories or lights left on when the engine is turned off.

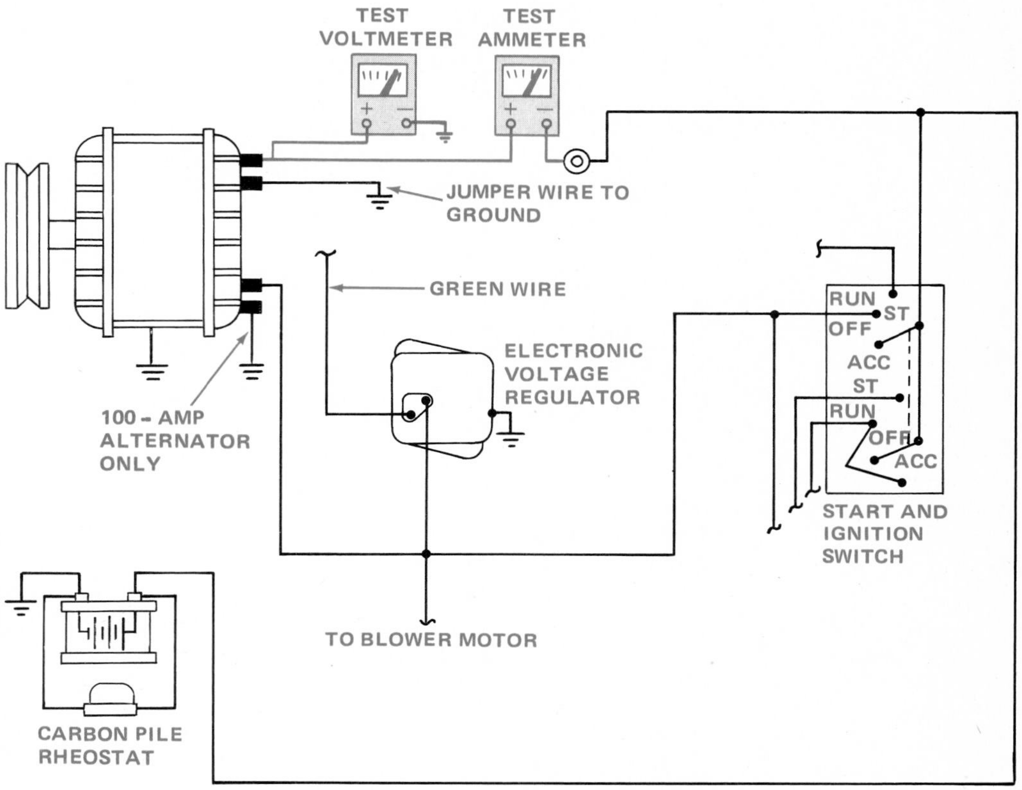

Fig. 29-2 Alternator output test. *(Chrysler Corporation)*

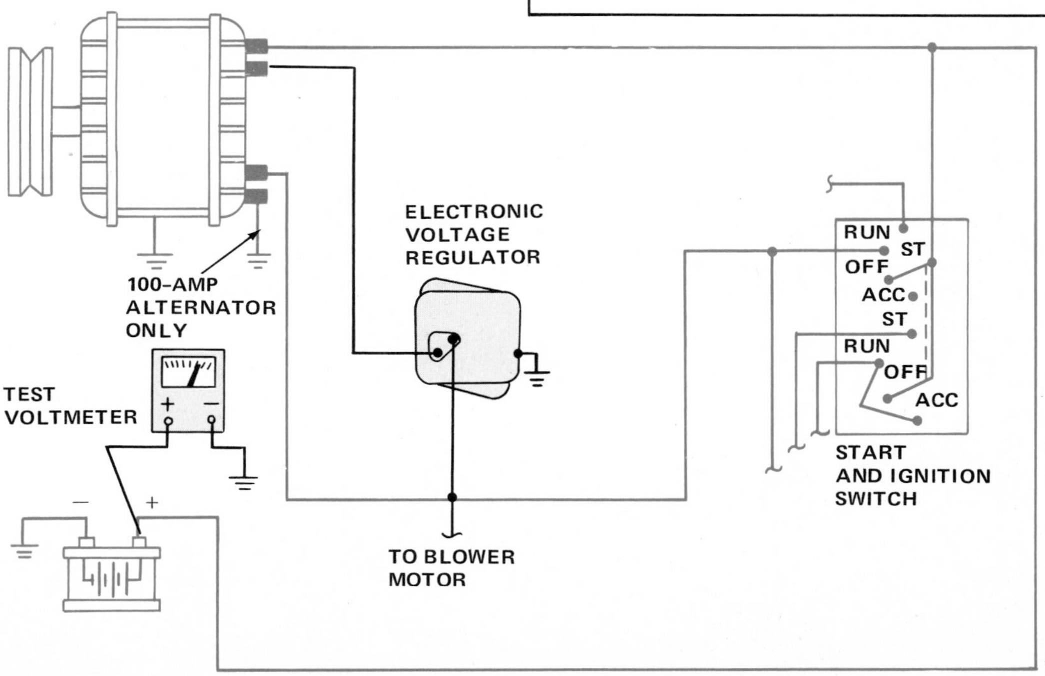

Fig. 29-3 Voltage regulator operating-voltage test. *(Chrysler Corporation)*

c. Loose or defective drive belt.
d. An old battery that will not accept a charge normally.
e. Defective wiring or connections.
f. Defective alternator or regulator.

Sections 27-3 and 27-4 describe in detail procedures to use to check the starting system when there is no or slow cranking.

2. Overcharged battery An overcharged battery needs water frequently because the continued high charging rate to a charged battery causes water loss. It also damages the battery plates and shortens battery life. In addition, the condition usually means the charging voltage is high, and this imposes a high voltage on all electrical accessories such as lights which will shorten their lives. To check for high voltage, connect a voltmeter across the battery with the engine operating on fast idle and warmed up. If the voltage is above the specified maximum, the field may be grounded or shorted inside the alternator, or else the regulator is defective and is not holding down the voltage.

3. Faulty indicator lamp action Three different kinds of trouble can occur with the alternator charge-indicator lamp:

a. Alternator indicator lamp on with engine running. It should go off when the engine starts.
b. Alternator lamp off with ignition on and engine not running. The lamp should come on when the ignition switch is turned to START or ON with the engine not running.
c. Alternator lamp on with ignition off. The lamp should be off when the ignition is turned off.

4. Noise It often happens that the noise which is being blamed on the alternator is actually originating elsewhere—in the water pump, power-steering pump, or other belt-driven accessory. Noise from the alternator can be caused by a worn drive belt, loose alternator mounting, rotor or fan damaged, bearings worn, one or more diodes in the rectifier open or shorted, or opens, grounds, or shorts in the stator.

● 29-7 Troubleshooting Procedures

As previously noted, the three possible troubles in a charging system include low battery, overcharged battery, and faulty indicator-lamp action. In addition, there may be complaints of a noisy alternator. Different manufacturers supply different troubleshooting procedures for each of these conditions. For example, Chrysler has a chart, shown in Fig. 29-4. Following this chart in the Chrysler shop manual are explanations relating to the tests indicated in the chart.

The shop manuals for General Motors and American Motors Corporation include DARS charts (for Diagnosis and Repair Simplification). These are picture charts that show, step by step, how to run down troubles and fix them. See Fig. 29-5, which shows a part of the chart for the condition "Battery Undercharged." The column to the right—"RESULT"—shows what has been accomplished, or what the next step should be. For example, at the end of step 1, the Result column has two numbers. If the test light is on, you go to step 2. If the test light is off, you go to step 3. If there is a Stop in the right column, the problem has been solved. The complete set of charts covering trouble diagnosis of the three major problems noted above (low battery, overcharged battery, faulty indicator-light action), takes up six pages in the manufacturer's shop manual. Refer to one of these shop manuals and follow the instructions through for each type of trouble.

● 29-8 Fuse-Link Repair

We discussed fuse links and their purpose in ● 28-10. They burn out or blow when high current starts to flow in a circuit, thus protecting the circuit from damage. When a fuse link has blown, the insulation covering it is usually bubbled or burned from the heat. Also, bare wire may be sticking out of the insulation. To make a repair, the blown fuse link is cut out and replacement fuse link and butt connectors installed. Figure 29-6 shows various procedures recommended by Ford Motor Company.

Before completing the repair, you must find out what caused the fuse link to blow and then fix the trouble. If you don't, the new fuse link will also blow.

● 29-9 Alternator Service

When removing or replacing ball or roller bearings, be sure to exert pressure on the race that is tight (on the shaft or in the frame), not on the race that is free. Service procedures for some typical units follow.

1. Chrysler To disassemble the alternator, remove the brushes (insulated brush first) and the through bolts. Then separate the stator and the slip-ring end frame from the drive end frame by prying with a screwdriver. Take off the pulley with a puller. Pry the bearing retainer (three places) off the end frame, to remove the rotor and bearing from the frame.

If diodes are defective, replace the rectifier assembly. Years ago, the Chrysler manual had instructions for servicing individual diodes, as follows.

A special tool is required to press out defective rectifier diodes and press in the new units. Hold the diode wire close to the diode with pliers when soldering the wire. This prevents heat from working back into the diode.

Careful: The diode must not be overheated, jarred, dropped, hammered on, or otherwise treated roughly. Such treatment may ruin it.

New bearings or slip rings can be installed with special puller and replacer tools.

2. Delco-Remy There are several models of Delco-Remy alternators. In general, to disassemble, remove the through bolts and separate the end frames from the stator. To avoid damaging the brushes, use through bolts to hold the brushes up when separating the end frame from the stator. Take off the pulley to separate the rotor from the end frame. Bearings are replaced with a special puller.

If diodes are defective, replace the rectifier assembly. Years ago, General Motors manuals carried instructions on servicing individual diodes, as follows.

To service diodes, heat the end frame to 150°F [65.6°C] to make diode removal easier. New diodes have long leads. Cut them to length, holding the lead (not the

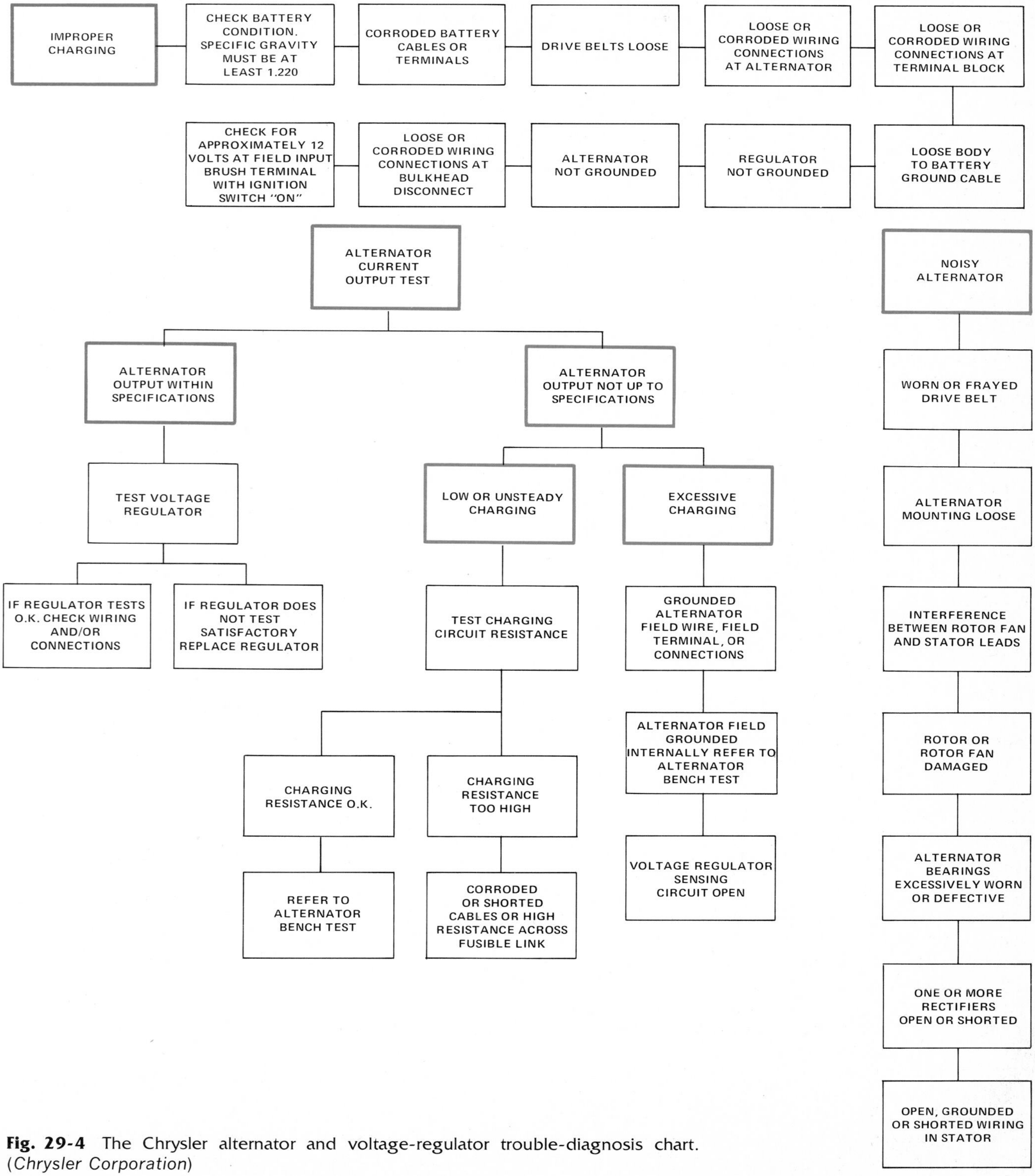

Fig. 29-4 The Chrysler alternator and voltage-regulator trouble-diagnosis chart. (*Chrysler Corporation*)

diode case) in pliers. Cover the diode threads with silicone grease, and tighten in the end frame to the specified torque. Crimp the lead to flexible leads, and solder with 360°F [182.2°C] solder. Use rosin flux. Avoid excessive heat on the diode.

3. Ford This is a typical procedure for Ford alternators. Remove the pulley (with a puller). Then remove the nuts, washers, and insulators from the terminals. Then remove the screws and terminal cover. Next remove the brush-holder assembly. Pull the slip-ring end housing and stator assembly from the rotor with a puller. If bearings need replacement, pull them with a puller. If diodes are defective, replace the complete rectifier assembly in the slip-ring end housing. (Diodes are not separately replaceable.)

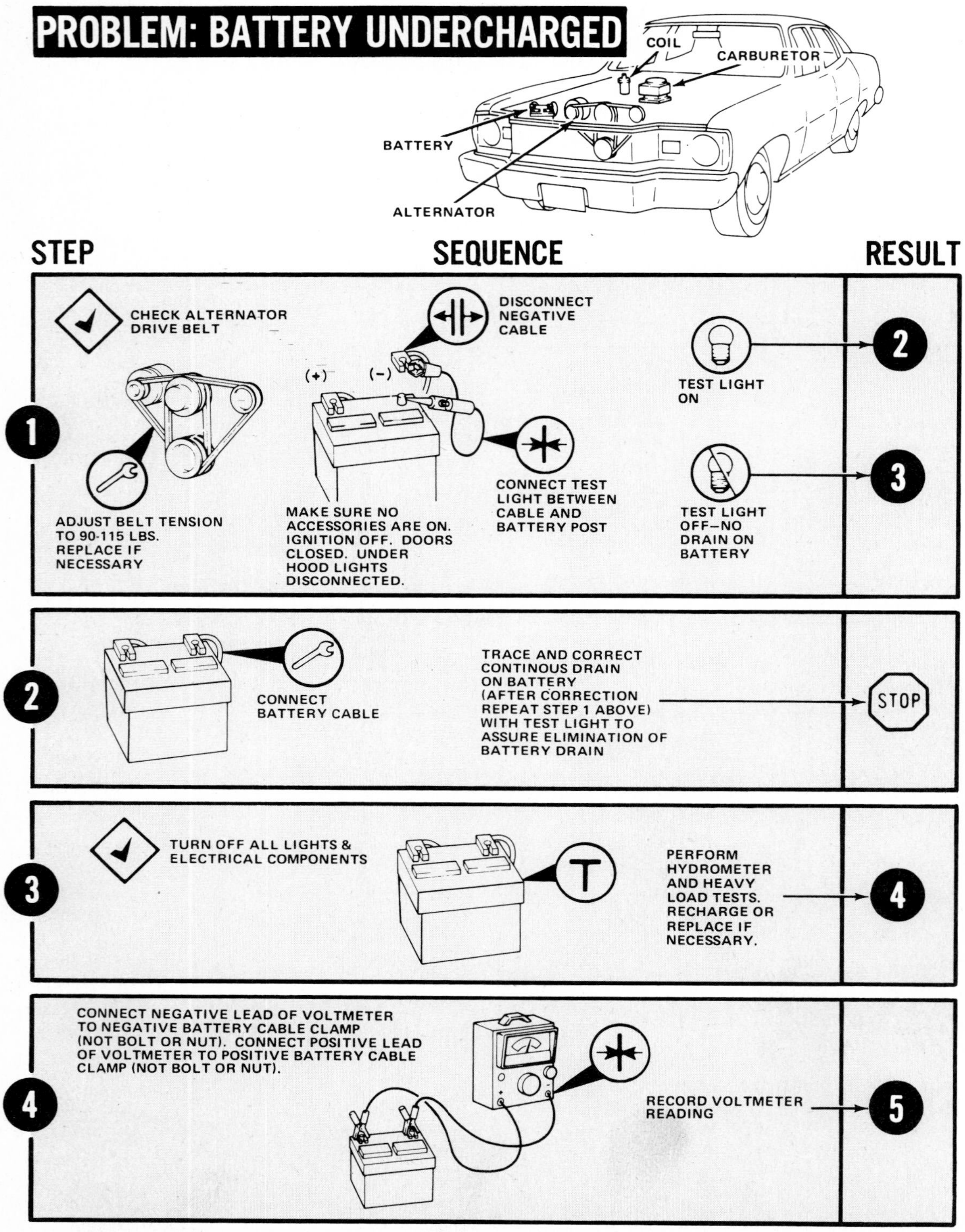

Fig. 29-5 First part of the DARS chart for battery undercharged problem. (*American Motors Corporation*)

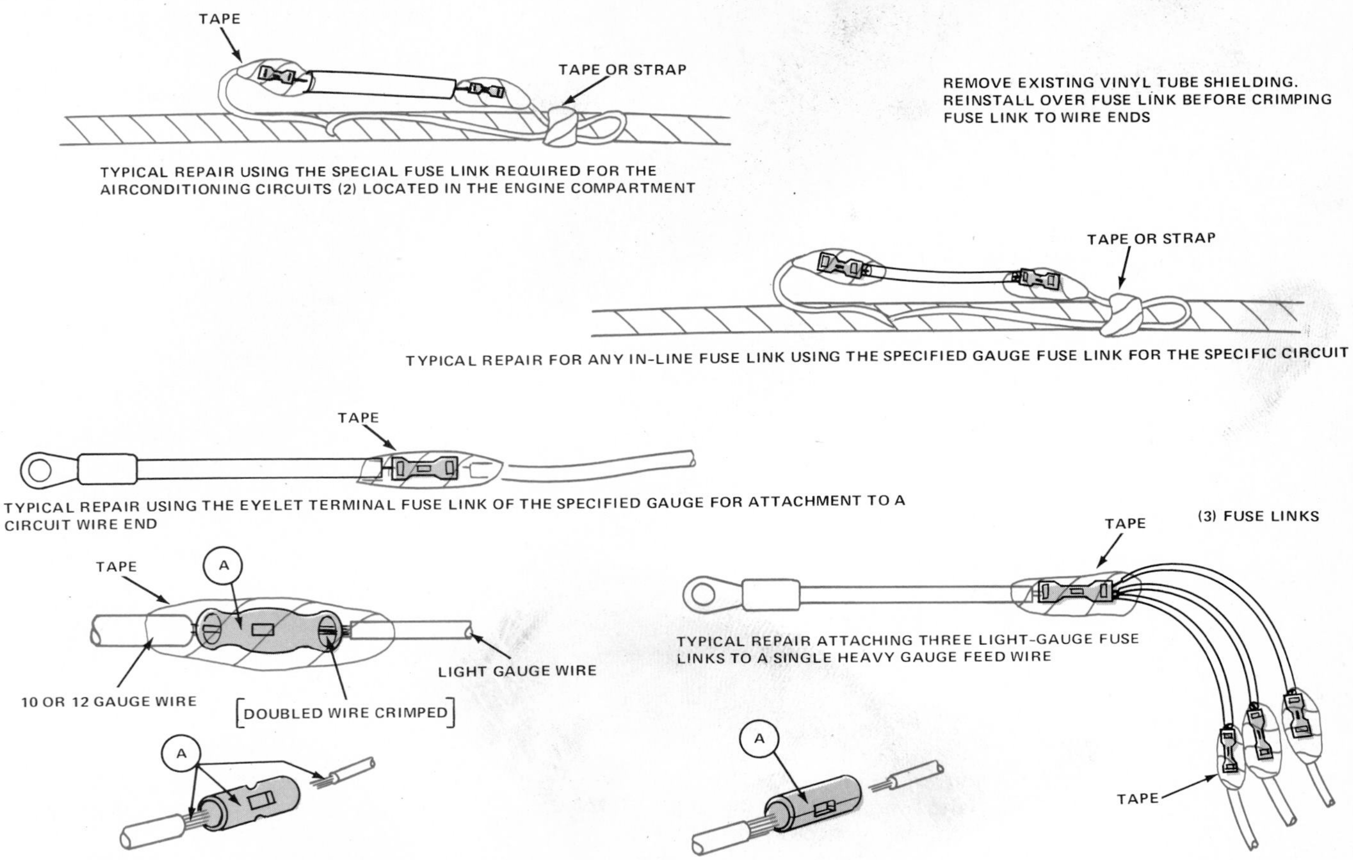
TAPE
TAPE OR STRAP
TYPICAL REPAIR USING THE SPECIAL FUSE LINK REQUIRED FOR THE AIRCONDITIONING CIRCUITS (2) LOCATED IN THE ENGINE COMPARTMENT
REMOVE EXISTING VINYL TUBE SHIELDING. REINSTALL OVER FUSE LINK BEFORE CRIMPING FUSE LINK TO WIRE ENDS
TAPE OR STRAP
TYPICAL REPAIR FOR ANY IN-LINE FUSE LINK USING THE SPECIFIED GAUGE FUSE LINK FOR THE SPECIFIC CIRCUIT
TAPE
TYPICAL REPAIR USING THE EYELET TERMINAL FUSE LINK OF THE SPECIFIED GAUGE FOR ATTACHMENT TO A CIRCUIT WIRE END
TAPE
A
10 OR 12 GAUGE WIRE
DOUBLED WIRE CRIMPED
LIGHT GAUGE WIRE
A
TAPE
(3) FUSE LINKS
TYPICAL REPAIR ATTACHING THREE LIGHT-GAUGE FUSE LINKS TO A SINGLE HEAVY GAUGE FEED WIRE
A
TAPE

Fig. 29-6 Fusible-link repair procedures. (*Ford Motor Company*)

● 29-10 Belt-Tension Adjustment

The drive belt must be properly tightened. A loose belt slips and soon wears out. It does not drive the alternator fast enough to keep the battery charged. An excessively tight belt causes rapid bearing wear. Figure 22-12 shows a belt-tension tool in place. This tool applies a measured amount of tension to the belt to check the amount of deflection. If the deflection is too great, the belt tension is low and should be increased. To do this, loosen the alternator-mounting and adjusting-bracket bolts. Move the alternator outward to increase the tension. Tighten the bolts after adjustment is complete.

REVIEW QUESTIONS

1. Should you polarize an alternator?
2. Why should you never operate the alternator with the output terminal disconnected?
3. When connecting a booster battery to a car battery, should you connect negative to negative, or negative to positive?
4. In what three ways do most charging-system troubles show up?
5. Name five possible causes of a run-down battery.
6. Can the built-in regulators, used in late-model General Motors cars, be adjusted?
7. Can the late-model Ford regulator be adjusted?
8. What are the special cautions to observe when working with diodes?

SELF PROJECT

Here is a self project that is both interesting and instructive. Get several manufacturers' shop manuals on late-model cars. Study the charging-system checks for the different makes of cars. Write out the specific checking procedures for the various models. Put your sheets of paper with the checking procedure into your notebook. This not only helps you understand the various procedures, but also gives you a permanent record of the procedures. Another project you will find interesting is to disassemble an old alternator and mount the parts on a board for a shop exhibit. Neatly letter cards naming each part and glue them under the alternator parts.

NOTE: As mentioned previously (● 30-5), secondary wiring used today has resistance that takes the place of the spark-plug resistor.

CHAPTER 30
Ignition Systems

After reading this chapter, you should be able to:

- **Explain how the contact-point ignition system works.**
- **Explain how the electronic ignition system works.**
- **Locate and identify the components of the ignition system on various cars.**
- **Explain how the advance mechanisms work and why they are necessary.**
- **Explain how the Chrysler electronic lean-burn ignition system, the Ford EEC system, and the General Motors MISAR electronic spark-timing system work.**

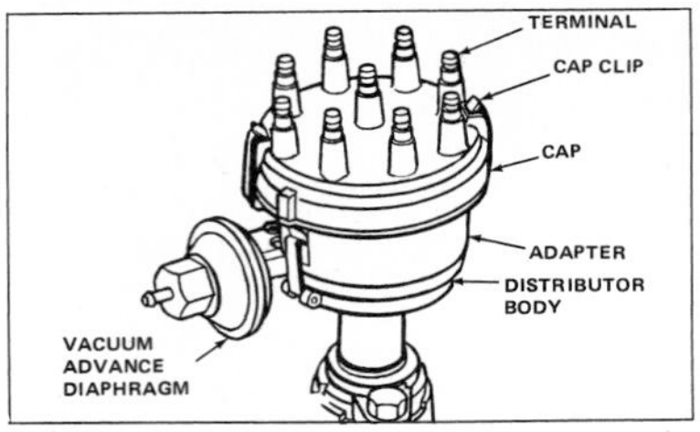

This chapter describes the construction and operation of automotive ignition systems and their component parts. There are two general types of ignition system, the type using contact points and the electronic type. Although they are somewhat different in construction and operation, both do the same job. That job is to produce and distribute high-voltage surges to the spark plugs, timing them to arrive at the correct instant. We cover the contact-point ignition system in the first part of the chapter and the electronic ignition system, later in the chapter. First, we explain the function of the ignition system.

● 30-1 Function of Ignition System

The ignition system (Figs. 30-1 and 30-2) supplies high-voltage surges (as high as 35,000 volts) to the spark plugs in the engine cylinders. These surges produce electric sparks at the spark-plug gaps. The sparks ignite, or set fire to, the compressed air-fuel mixture in the combustion chambers. Each spark appears at the plug gap just as the piston approaches top dead center on the compression stroke, when the engine is idling. At higher speed or during part-throttle operation, the spark is advanced. It occurs somewhat earlier in the cycle. The mixture thus has ample time to burn and deliver its power. The ignition system consists of the battery, ignition switch, ignition distributor, ignition coil, spark plugs, and wiring (Fig. 30-2). Also, the electronic ignition system has an electronic control module, as we explain later.

Contact-Point Ignition System

This ignition system uses contact points to open and close the circuit between the battery and ignition coil.

● 30-2 Ignition Distributor

The ignition distributor using contact points has two jobs. First, it closes and then opens the contact points to complete and then interrupt the circuit between the battery and the ignition coil. When the circuit is completed through the closed contact points, current flows in the ignition coil and builds up a magnetic field. When the points open, the circuit is open and current stops flowing. The magnetic field collapses and this produces a high-voltage surge in the ignition coil. The distributor's second job is to distribute each high-voltage surge to the correct spark plug at the correct instant. It does this with the distributor rotor and cap and the secondary cables.

Figures 30-3 and 30-4 shows various views of distributors using contact points. The contact-point distributor

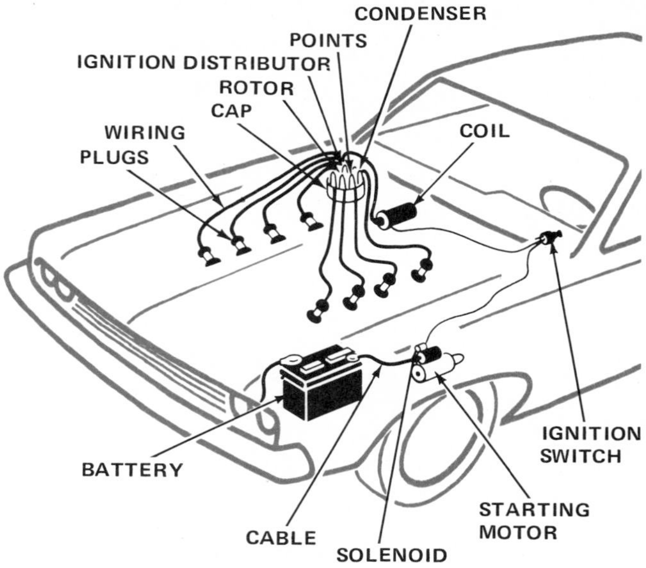

Fig. 30-1 Simplified drawing of an ignition system using contact points.

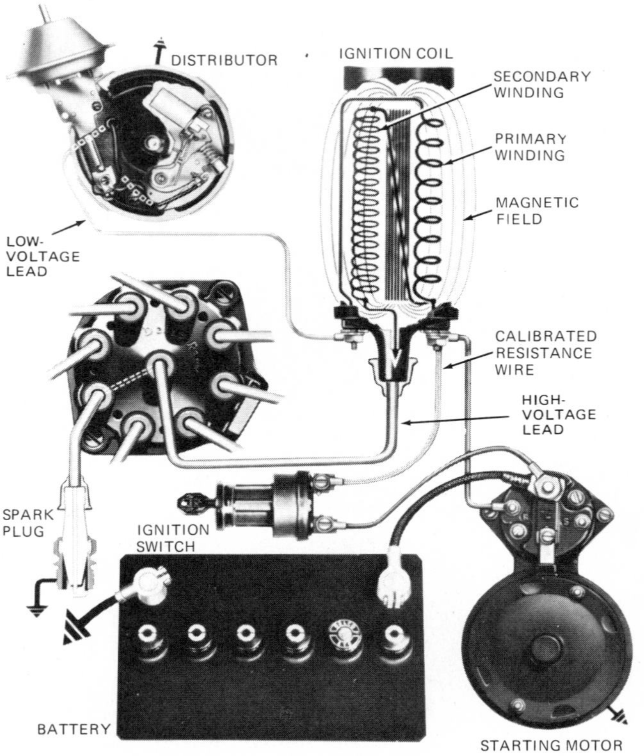

Fig. 30-2 Schematic layout of an ignition system using contact points. It includes the battery (source of power), ignition switch, ignition coil (shown schematically), distributor (shown in top view with its cap removed and placed below it), spark plugs (one shown in sectional view), and wiring. (*Delco-Remy Division of General Motors Corporation*)

consists of a housing, a drive shaft with breaker ca advance mechanism, a breaker plate with contact poi and a condenser, a rotor, and a cap. The shaft is usually driven by the engine camshaft through spiral gears (Fig. 19-7). It rotates at one-half crankshaft speed. Usually, the distributor drive shaft is coupled with a shaft that drives the oil pump.

Rotation of the shaft and breaker cam causes the distributor contact points to open and close. The breaker cam usually has the same number of lobes as there are cylinders in the engine. It rotates at one-half crankshaft speed. The contact points close and open once for each cylinder with every breaker-cam rotation. Thus one high-voltage surge is produced by the coil for each cylinder every two crankshaft revolutions. This ignites the air-fuel mixture compressed in each cylinder every other crankshaft revolution.

The rotor rotates with the breaker cam on which it is mounted. As it does, a metal spring and segment (or blade) on the rotor connect the center terminal of the cap with each outside terminal in turn. Thus the high-voltage surges from the coil are directed first to one spark plug, then to another, and so on, according to the firing order.

● 30-3 Ignition Coil

The ignition coil transforms, or steps up, the 12 volts of the battery to the high voltage required to make the current jump the spark-plug gap. The air-fuel mixture between the two electrodes presents a high resistance to the passage of current. The voltage (pressure) must be very high in order to push current (electrons) from the center to the outside electrode.

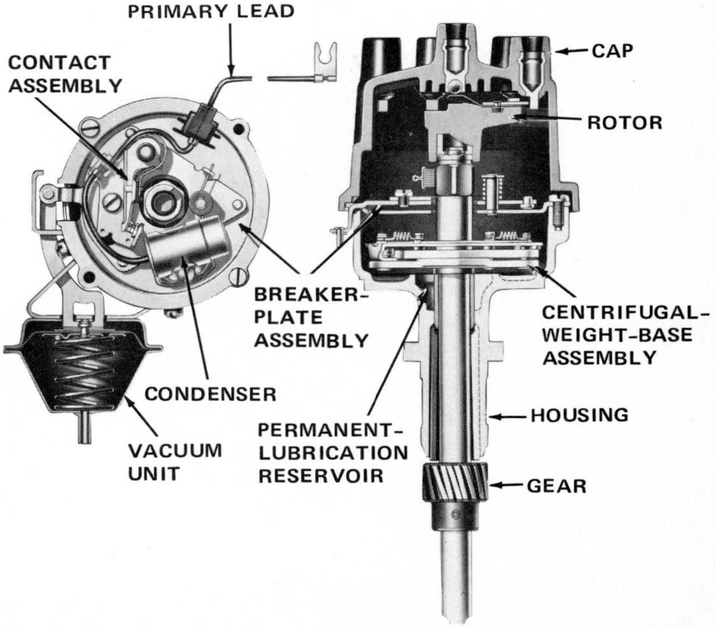

Fig. 30-3 Top and sectional views of an ignition distributor using contact points. In the top view (to left) the cap has been removed so the breaker plate can be seen. (*Delco-Remy Division of General Motors Corporation*)

wo windings, or circuits, a pri-
ondary circuit (Fig. 30-2). The
de up of many thousands of turns
mary circuit is made up of a few
avier wire. The wire is wrapped or
utside of the secondary winding, as
When the distributor contact points
ci… lows in the primary circuit, a magnetic
field bu… hen the distributor contact points open
and current stops flowing, the magnetic field collapses. The collapsing magnetic field induces high voltage in the secondary winding. This creates the high-voltage surge that is conducted through the distributor rotor and cap to a spark plug.

30-4 Primary and Secondary Circuits

In order to get a clearer picture of the two circuits in the ignition system, let us look at each one separately. Figure 30-6 shows the primary circuit. It consists of the battery, the contact points in the distributor, the primary winding in the ignition coil, the ignition switch, and the wiring.

Figure 30-7 is the same illustration with the secondary circuit added. The secondary circuit includes the secondary winding in the ignition coil, the distributor cap and rotor, the spark plugs, and the connecting wires.

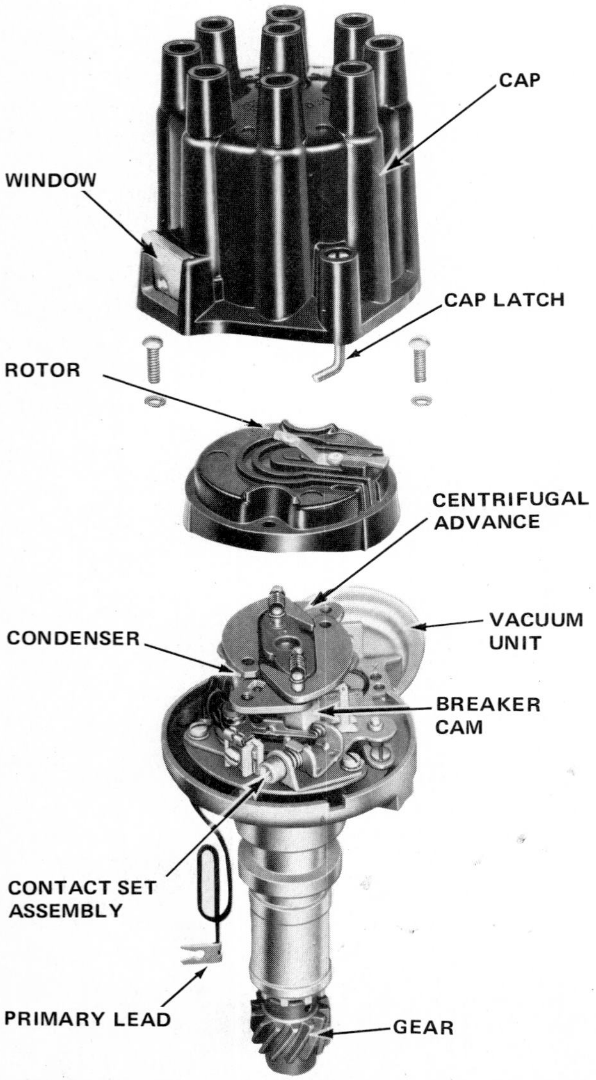

Fig. 30-4 Partly disassembled distributor. (*Delco-Remy Division of General Motors Corporation*)

30-5 Secondary Wiring

The secondary wiring consists of the high-voltage cables connected between the distributor cap, the spark plugs, and the high-voltage terminal of the ignition coil. These cables carry the high-voltage surges that produce the sparks at the plug gaps. Thus, they must be heavily insulated to contain the high voltage. The insulation must be able to withstand the effects of high temperature, oil, and high voltage.

Before 1961, the cores of the cables were of copper or aluminum wire. Later, in 1961, automotive manufacturers in the United States began to use carbon-impregnated

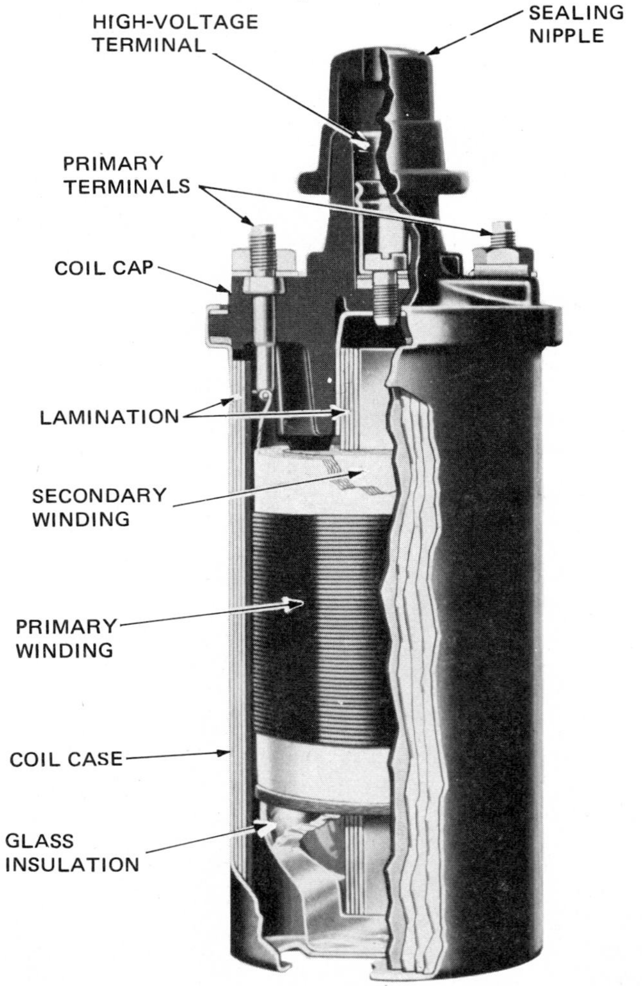

Fig. 30-5 Ignition coil with case partly cut away so the windings can be seen. The primary winding is on the outside of the secondary winding. (*Delco-Remy Division of General Motors Corporation*)

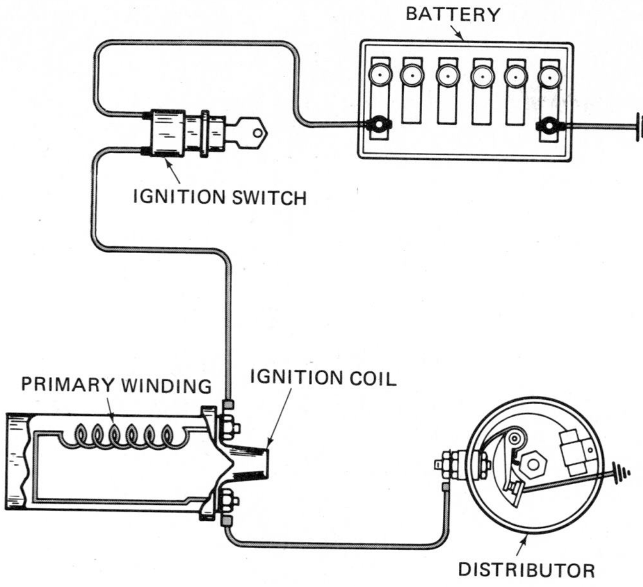

Fig. 30-6 Simplified primary circuit of the ignition system.

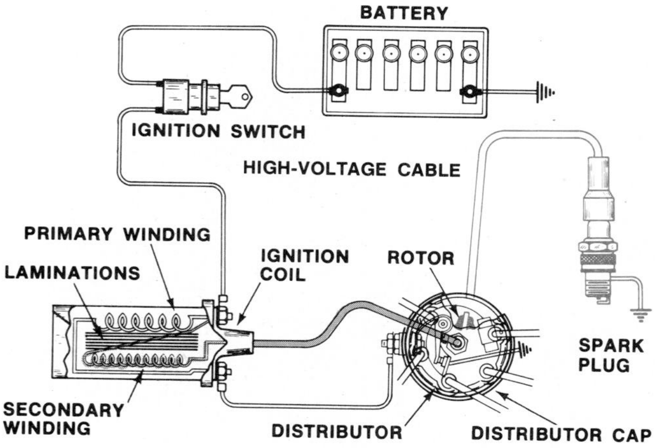

Fig. 30-7 Secondary circuit added to the primary circuit of the ignition system. Only one spark plug is shown.

linen cores. The carbon-impregnated linen forms a resistance path for the high-voltage surges. It produces the same effect as the resistors in the spark plugs as we explain later. These cables thus prevent the ignition system from interfering with radio and television.

In 1963, manufacturers began using cables with graphite-saturated fiber-glass cores. These worked like the carbon-impregnated linen-core cables. However, it is claimed that they resist breakage when pulled off spark plugs. Also, they have less tendency to char from high temperatures.

Today, some spark-plug wires are made of special string that is coated with carbon. A jacket of silicone is placed around the string. This is the insulation. Silicone wires are easily identified by their size. Older spark-plug wires are 7 mm in diameter. Silicone wires are 8 mm in diameter. Handle silicone wires carefully; they are more easily damaged than the older type of spark-plug wire.

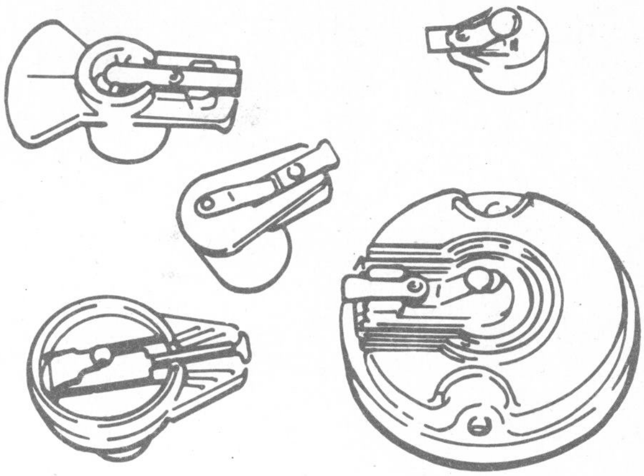

Fig. 30-8 Distributor rotors. The one at the lower left has a carbon resistor. The one at the lower right is attached to the advance mechanism by screws.

30-6 Distributor-Cap and Rotor Action

As you can see from Figs. 30-3 and 30-4, the rotor sits on top of the cam in the distributor. Figure 30-8 shows several rotors. The purpose of the rotor is to connect the center terminal of the distributor cap to the outside terminals of the cap.

The terminals are insulated from one another and are molded in place in the cap. You can see three of them cut away in Fig. 30-9. The center terminal of the cap has a carbon button on its lower end. This button rests on one end of the rotor blade. A small spring holds the carbon button and rotor blade in continuous contact. Therefore, the rotor blade is always connected to the secondary winding of the ignition coil. Whenever the coil secondary winding produces a high-voltage surge, we have this condition. The rotor blade is pointing at the side terminal which is connected to the spark plug that is ready to fire (Fig. 30-7).

Let's review the actions: The contact points open. The magnetic field created by current flowing in the coil primary winding collapses. This collapse produces a high-voltage surge in the coil secondary winding. The high-voltage surge is led from the center terminal of the ignition coil to the center terminal of the distributor cap. From there it goes through the rotor blade to one side terminal. The side terminal is connected to the spark plug in the cylinder in which the compression stroke is ending. The spark produced at the spark plug by the high-voltage surge ignites the compressed air-fuel mixture. It burns, and the power stroke follows.

30-7 Condenser Effect

As the distributor contact points open, the current from the battery through the primary winding of the coil is stopped. Instantly, the magnetic field begins to collapse. This collapse tends to reestablish the flow of current. If it were not for the condenser (also called a *capacitor*), the flow of current would be reestablished. This means a

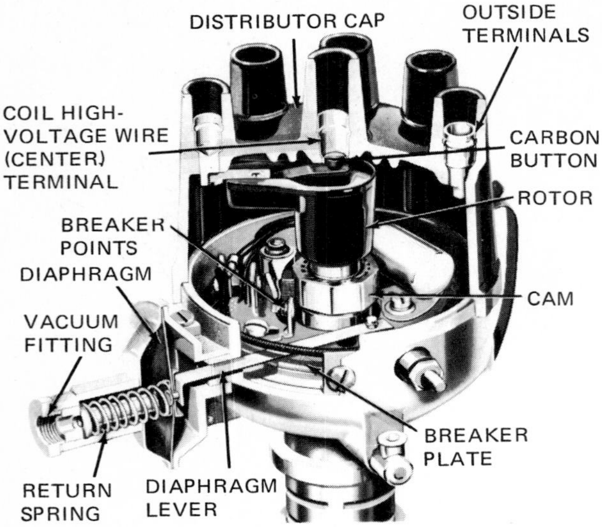

Fig. 30-9 Cutaway view of a distributor showing how the rotor is mounted on top of the cam. The picture also shows the construction of the vacuum advance mechanism. (*Ford Motor Company*)

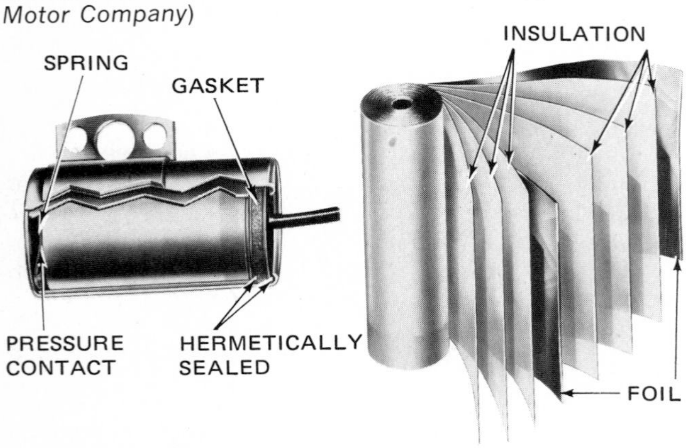

Fig. 30-10 Condenser assembled and with the winding partly unwound.

heavy electric arc would take place across the separating contact points. The points would burn, and the energy stored in the ignition coil as magnetism would be consumed by the arc. The condenser prevents this, however. It provides a place for the current to flow as the points begin to move apart.

The condenser (or *capacitor*) is made up of two thin metallic plates separated by an insulator. The plates are two long, narrow strips of lead or aluminum foil. They are insulated from each other by special condenser paper and wrapped to form a winding. The winding is then installed in a container. A condenser is shown in Fig. 30-10. The two plates provide a large surface area onto which the electrons (flow of current) can move at the instant the contact points separate. Remember, it is the massing of electrons in one place in a circuit that causes them to move and produce a current. The condenser provides a large surface area. Thus, many electrons can flow into it without producing an excessive massing of electrons in one spot.

The number of electrons the condenser can accept is, however, limited. It quickly becomes charged. But, by this time, the contact points are sufficiently far apart to prevent an arc from forming between them. In effect, the condenser acts as a reservoir into which electrons flow at the instant the points begin to separate. By the time the reservoir is filled, the points are too far apart for the electrons to jump across them. The electrons, or current, must stop flowing in the primary circuit. It is a current flow that induces the magnetic field. Thus, the quick stoppage of the current causes the magnetic field to collapse rapidly. It is this rapid collapse that induces the high voltage in the secondary winding of the ignition coil.

● 30-8 Effect on Secondary Winding of Magnetic-Field Collapse

The rapid collapse of the magnetic field causes the magnetic lines of force to move rapidly across the thousands of turns of wire in the secondary winding. This means that each turn has a voltage induced in it. All turns are connected in series, so that the total voltage induced is the sum of the voltages in all the turns. Thus, the winding will supply a high voltage during the magnetic-field collapse. One end of the secondary winding is connected through ground (by way of the cylinder block and head) to the side electrode in the spark plug. The other end of the secondary winding is connected through the cap and the rotor of the distributor, and through the wiring to the center electrode in the spark plug. This high voltage, suddenly imposed on the spark plug, causes electrons (current) to jump across the gap, producing an electric spark. The spark is timed by the spark-advance mechanisms located in the distributor (● 30-12).

The ignition coil voltage output varies with different operating conditions. The coil produces only enough voltage to jump the spark-plug gap. The different plugs in an engine will have different voltage requirements because of differences in their gaps. Also, air-fuel mixture richness, and the amount of mixture compressed each compression stroke, will vary. These variations result from changes in throttle opening and engine speed. The coil voltage must change to meet these different conditions.

● 30-9 Ignition-Coil Resistor

In many passenger cars with 12-volt systems, there is a resistance wire in the ignition-coil primary circuit (see Figs. 26-15 and 30-2). This wire is shorted out by the ignition switch when it is turned to START. Now, full battery voltage is imposed on the ignition coil for good performance during cranking. After the engine is started and the ignition switch is turned to ON, the resistance is in the ignition primary circuit. It thus protects the contact points from excessive current.

● 30-10 Summary of Actions

Let us review briefly the action taking place in the ignition system. The piston in one of the engine cylinders starts up on the compression stroke. At the same time, one of the distributor breaker-cam lobes moves away from the contact-point breaker arm. The contact points close. Current flows through the primary winding of the ignition coil, and a magnetic field builds up. Then, the piston reaches the position in the cylinder at which ignition of the compressed air-fuel mixture should take place. At this instant the next cam lobe has moved around to where it strikes against the contact-point breaker arm, so that the contact points separate. The current stops flowing in the primary circuit, and the magnetic field collapses. This induces high voltage in the secondary winding. The rotor on top of the breaker cam, in the meantime, has moved into position. It is now opposite the outside distributor-cap terminal connected to the cylinder spark plug. The spark plug is thus connected to the secondary winding of the ignition coil through the cap and rotor at the instant that the high voltage is induced. A spark therefore occurs at the spark-plug gap.

● 30-11 Spark Plugs

The spark plug (Fig. 30-11) is a metal shell in which a porcelain insulator is fastened. An electrode extends through the center of the insulator. A second electrode is attached to one side of the shell. This electrode is bent in toward the center electrode. Threads on the metal shell allow it to be screwed into a tapped hole in the cylinder head. This grounds the electrode that is attached to the shell. The two electrodes are of special heavy wire. There is a gap of up to 0.040 inch [1.02 mm] between them (0.080 inch [2.03 mm] on some electronic ignition systems). The electric spark jumps this gap to ignite the air-fuel mixture in the combustion chamber. The spark jumps from the center, or insulated, electrode to the grounded, or outer, electrode. Some spark plugs have a built-in resistor (Fig. 30-11) which is part of the center electrode. This resistor reduces radio and television interference from the ignition system. It also reduces electrode erosion caused by over-long sparking. We have been talking of the high-voltage surge from the ignition-coil secondary as if it were a single powerful surge. Actually, the action is more complex than that. There may be a number of early surges before a full spark forms. At the end of the sparking cycle, the spark may die and re-form several times. All this takes place in only a few ten-thousandths of a second. The effect is that the ignition wiring acts like a radio transmitting antenna. The surges of high voltage send out static that causes radio and television interference. However, the resistors in the spark plugs tend to reduce the number of surges. They thus reduce interference and the wear on the electrodes. NOTE: As mentioned previously (●30-5), secondary wiring used today has resistance that takes the place of the spark plug.

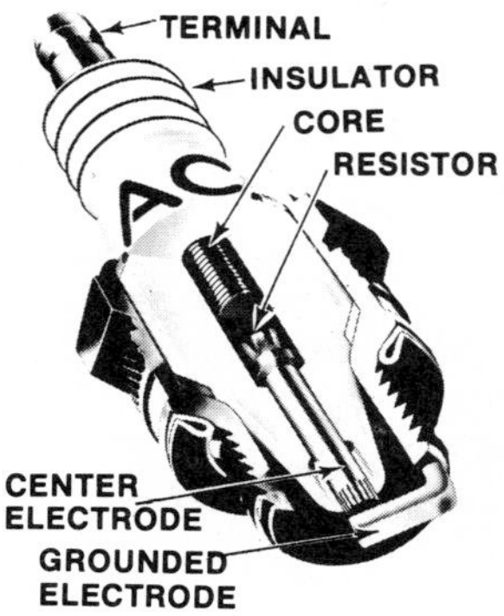

Fig. 30-11 Cutaway resistor-type spark plug. (*AC Spark Plug Division of General Motors Corporation*)

● 30-12 Spark-Plug Heat Range

The heat range of a spark plug tells how hot the plug gets in operation (Fig. 30-12). The temperature that a plug reaches depends on how far the heat must travel. The heat path is from the center electrode to the cooler outer shell of the plug and then to the cylinder head. If the path is long, the plug will run hotter than if the path is short. When a plug runs too cold, sooty carbon will deposit on the insulator around the center electrode. A hotter-running plug burns this carbon away or prevents its formation. Carbon deposits can also be caused by too-rich air-fuel mixtures or by too much oil in the cylinder.

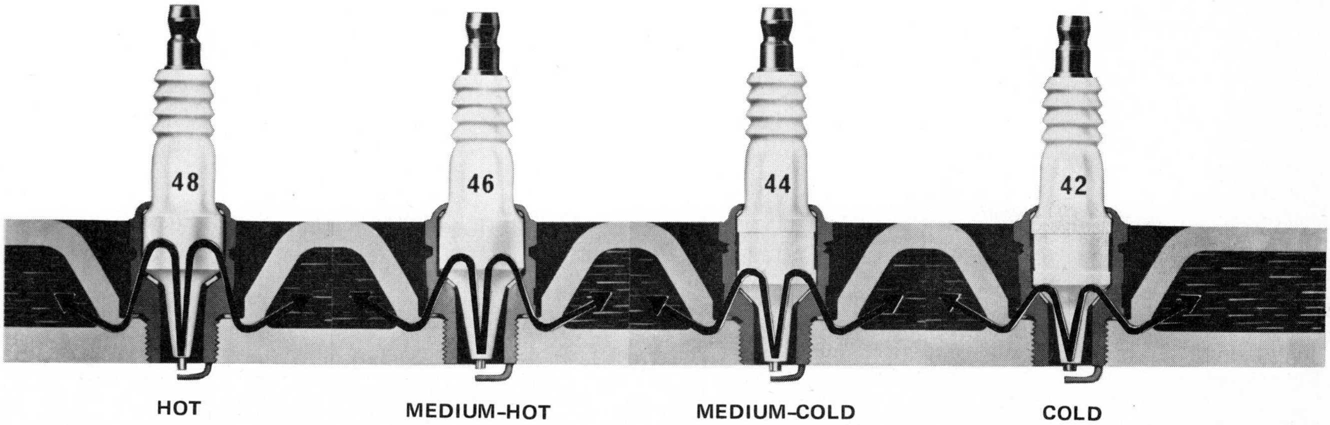

Fig. 30-12 Heat range of spark plugs. The longer the heat path (indicated by arrows) the hotter the plug runs. (*AC Spark Plug Division of General Motors Corporation*)

If the plug runs too hot, the insulator may take on a white or grayish cast and may appear blistered. A plug that runs hot wears more rapidly. The higher temperatures cause the electrodes to burn away more quickly. In addition, with a hot-running plug, there is always danger of preignition (see ● 11-9).

● 30-13 Spark-Advance Mechanisms

There are two general types of spark-advance mechanisms, centrifugal and vacuum. These mechanisms vary spark timing for different engine-operating conditions.

● 30-14 Centrifugal Advance

When the engine is idling, the spark is timed to occur just before the piston reaches top dead center on the compression stroke. At higher speeds, it is necessary to deliver the spark to the combustion chamber somewhat earlier. This gives the mixture time to burn and deliver its power to the piston. To provide this advance, a centrifugal advance mechanism is used (Fig. 30-13). It consists of two weights that are thrown out against spring tension as engine speed increases. This movement is transmitted through a toggle arrangement to the breaker cam. This causes the cam to advance or move ahead, with respect to the distributor drive shaft. This advance causes the cam to open and close the contact points earlier in the compression stroke at high speeds. Since the rotor, too, is advanced, it comes into position earlier in the cycle. The timing of the spark to the cylinder thus varies from no advance at low speed to full advance at high speed (when the weights have reached the outer limits of their travel). Maximum advance may be as much as 45 degrees of crankshaft rotation before the piston reaches top dead center. It varies with different makes of engines. The toggle arrangement and springs are designed to give the correct advance for maximum engine performance.

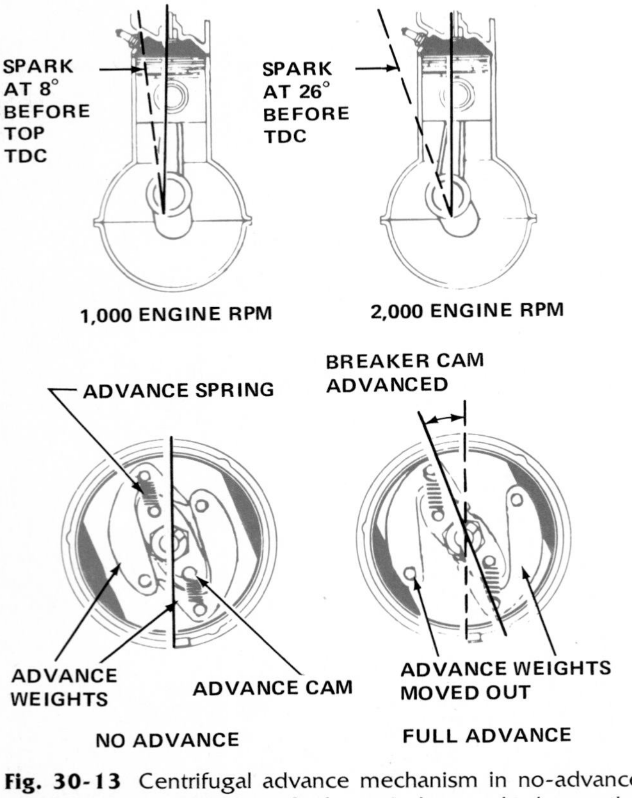

Fig. 30-13 Centrifugal advance mechanism in no-advance and full-advance positions. In the typical example shown, the ignition is timed at 8 degrees before top dead center on idle. There is no centrifugal advance at 1,000 engine rpm. There is 26 degrees total advance (18 degrees centrifugal plus 8 degrees due to original timing) at 2,000 engine rpm. (*Delco-Remy Division of General Motors Corporation*)

● 30-15 Vacuum Advance

Under part throttle, a partial vacuum develops in the intake manifold. This means that less air and fuel will be admitted to the cylinder (volumetric efficiency is lowered). Thus, there will be less mixture to be compressed. The mixture will burn more slowly when ignited. In order to realize full power from it, the spark should be somewhat advanced. To obtain this spark advance, a vacuum advance mechanism is used.

Figure 30-9 shows a type of vacuum advance mechanism used on contact-point distributors. It contains a spring-loaded, airtight diaphragm. The diaphragm is connected by a linkage, or lever, to the breaker plate. The breaker plate is supported on a bearing so it can turn with respect to the distributor housing. It actually turns only a few degrees. The linkage to the spring-loaded diaphragm prevents any greater rotation than this.

The spring-loaded side of the diaphragm is connected through a vacuum line to an opening in the carburetor (Fig. 30-14). This opening is on the atmospheric side of the throttle valve when the throttle is in the idling position. There is no vacuum advance in this position.

As soon as the throttle is opened, however, it moves past the opening of the vacuum passage. The intake-

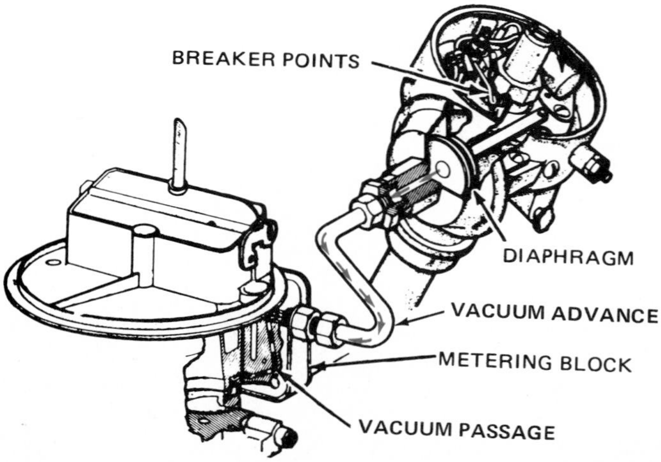

Fig. 30-14 Connections of the vacuum line between the carburetor and the vacuum advance mechanism on the distributor. (*Ford Motor Company*)

manifold vacuum can then draw air from the vacuum line and the airtight chamber in the vacuum advance mechanism. This causes the diaphragm to move against the spring. The linkage to the breaker plate then rotates the breaker plate. This movement carries the contact points around. Thus, the cam, as it rotates, closes and opens the points earlier in the cycle. The spark then appears at the spark-plug gap earlier in the compression stroke. As the throttle is opened wider, there is less vacuum in the intake manifold, and less vacuum advance. At wide-open throttle, there is no vacuum advance at all. The spark advance under this condition is provided entirely by the centrifugal advance mechanism.

30-16 Combination of Centrifugal and Vacuum Advances

At any particular engine speed, there will be some centrifugal advance due to engine speed. There may be an additional spark advance due to the operation of the vacuum advance mechanism. Figure 30-15 illustrates this. At 40 mph (miles per hour) [64.37 km/h], the centrifugal advance mechanism provides 15 degrees of spark advance in this example. The vacuum mechanism will supply up to 15 degrees of additional advance under part-throttle conditions. However, if the engine is operated at wide-open throttle, no vacuum advance will be obtained. The advance usually varies between the straight line (centrifugal advance) and the curved line (centrifugal advance plus total possible vacuum advance) as the throttle is closed and opened.

Some late-model electronic ignition systems do not use mechanical devices to produce spark advance. Instead, spark advance is produced electronically, as explained later in the chapter.

30-17 Full Vacuum Control

The distributor in Fig. 30-16 does not contain a centrifugal advance mechanism. Instead, it utilizes vacuum from the carburetor venturi and intake manifold to produce the proper advance. Full control by vacuum alone is possible because air speed through the carburetor air horn, and thus the vacuum in the venturi, is directly related to engine speed. Let us see how the system works.

In the carburetor shown in Fig. 30-16, there are two vacuum openings in the air horn. One is at the venturi, and the other is just above the throttle when it is closed. The lower, or throttle, vacuum-takeoff opening may have two ports on some models, as in Fig. 30-16. These openings are connected by vacuum passages to each other. They are connected to the distributor vacuum advance mechanism by a vacuum line. Vacuum imposed on the diaphragm in the vacuum advance mechanism causes the breaker-plate assembly to rotate. This is very similar to the action of the vacuum advance devices discussed earlier. Rotation of the breaker-plate assembly causes an advance of the spark.

As engine speed increases, the vacuum at the venturi in the carburetor increases. This is due to the increase of air speed through the venturi. This causes an increasing spark advance which is related to engine speed. At the same time, under part-throttle operating conditions, there will be a vacuum in the intake manifold. This acts at the throttle vacuum ports in the carburetor to produce a further vacuum advance. Thus, the vacuum conditions at the two points in the carburetor produce, in effect, a

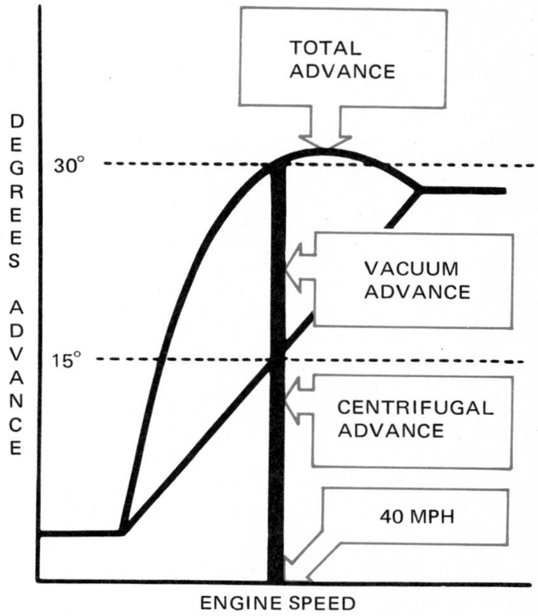

Fig. 30-15 Centrifugal and vacuum advance curves for one application.

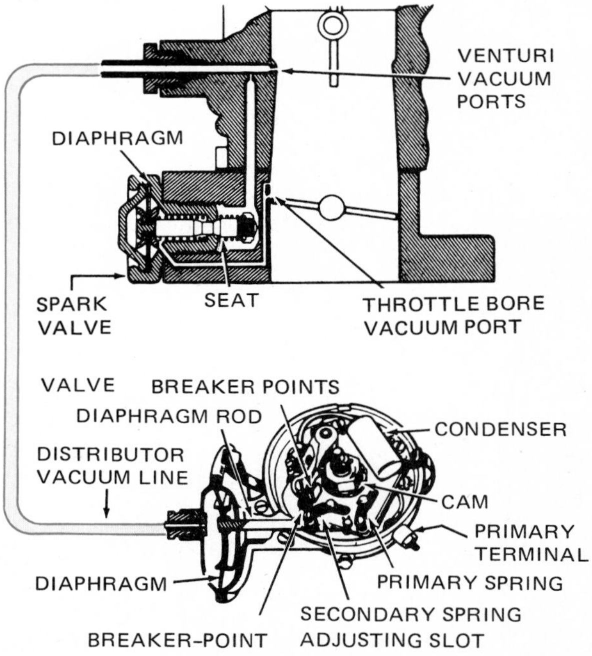

Fig. 30-16 Vacuum-line connections between a carburetor and a distributor having full vacuum control. (*Ford Motor Company*)

combined speed advance (as with a centrifugal device) and vacuum advance.

Vacuum advance disconnect devices, such as are installed on used cars for exhaust emission control, should not be installed on engines having this type of distributor. The device would eliminate, at times, any spark advance at all and this could cause very poor engine performance.

30-18 Vacuum Advance Controls for Emission Reduction

During some operating conditions, vacuum advance can increase the formation of nitrogen oxides during combustion. As we explain in Chap. 33, nitrogen oxides (NO_x) form during combustion at high temperatures. Thus, part-throttle operation in the lower gears can cause an increase in NO_x in the exhaust gases. To prevent this, automobiles are equipped with control systems that prevent vacuum advance under some conditions.

Car manufacturers achieve vacuum advance control by different methods. Figure 30-17 shows one system. It allows vacuum advance only in high gear—with certain exceptions noted below. The system includes a transmission switch, a solenoid vacuum switch, and a temperature override system. For normal operation in any gear but high, the transmission switch is closed. This connects the solenoid vacuum switch to the battery. The solenoid pulls in its plunger, closing off the vacuum connection to the lower part of the carburetor (that is, to manifold vacuum). At the same time, the solenoid opens a connection to the upper part of the carburetor through a clean-air vent. This releases any vacuum on the vacuum advance mechanism at the distributor. There is no vacuum advance.

When the transmission goes into high gear, the transmission switch is opened. This allows the solenoid plunger to be pushed up by a spring. This shuts off the clean-air vent. It opens the line from the manifold vacuum vent to the vacuum advance unit. Now, normal vacuum advance can take place.

The temperature override system provides full advance

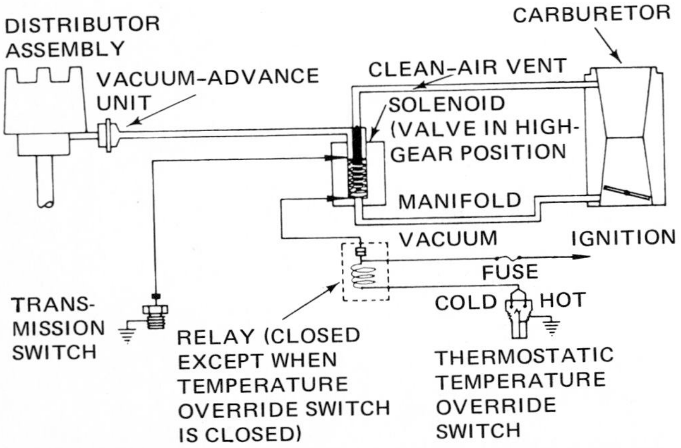

Fig. 30-17 Schematic view of the transmission-controlled spark (TCS) system. (*Chevrolet Motor Division of General Motors Corporation*)

for better performance in all gears when the engine is cold. A thermostatic water-temperature switch is closed when the engine is cold. This connects the relay winding to the battery (through the ignition switch). The relay contact points open, thus opening the circuit to the solenoid. Now, regardless of the position of the transmission switch, the solenoid cannot operate. The vacuum advance unit remains connected to the manifold vacuum vent in the carburetor. Thus, the system provides normal vacuum advance.

When the engine warms up, the thermostatic override switch opens. This opens the relay winding. The relay points close. The system now operates so that vacuum advance is obtained only in high gear.

The system also has a hot override position. This provides vacuum advance at all gear positions if the engine is overheating. The vacuum advance improves engine cooling. See Chaps. 33 to 35 for the details of this and other vacuum advance control systems.

30-19 Ignition Switch

In late-model cars, the ignition switch is mounted on the steering column, as shown in Fig. 30-18. This arrangement locks the steering shaft when the ignition switch is turned off and the ignition key is removed. A small gear on the end of the ignition switch rotates and releases a plunger. The plunger enters a notch in a disk on the steering shaft to lock the shaft. If a notch is not lined up with the plunger, the plunger rests on the disk. When the steering wheel, shaft, and disk are turned slightly, the plunger drops into a notch.

When the ignition key is inserted and the ignition switch is turned on, the plunger is withdrawn from the disk to unlock the steering shaft.

The ignition switch has an extra set of contacts that are used when the switch is turned past ON to START. The contacts connect the starting-motor solenoid to the battery so that the starting motor can operate. When the engine is started and the switch released, it returns to ON. The starting motor is then disconnected from the battery.

The alternator field circuit is connected to the battery through the ignition switch when it is turned to ON. When the ignition switch is turned to OFF, the alternator field circuit is disconnected. This keeps the battery from running down through the field circuit.

The ignition switch also operates a buzzer if the key is in the lock when the driver's car door is open. This reminds the driver to remove the key from the lock when the car is parked. It helps guard against theft of the car.

Such accessories as the radio and the car heater are connected to the battery through the ignition switch. This arrangement prevents drivers from leaving these units running when they turn off the engine and leave the car.

Electronic Ignition System

This ignition system uses electronic components to open and close the circuit between the battery and the ignition coil. All other parts of the electronic ignition system are practically the same as in the contact-point system. The ignition switch, ignition coil, wiring and cables, centrifugal and vacuum advance, and cap and rotor are the same for both systems. Now, we look at the special features of electronic ignition systems.

There are some new ignition systems which do not use mechanical centrifugal and vacuum advance controls.

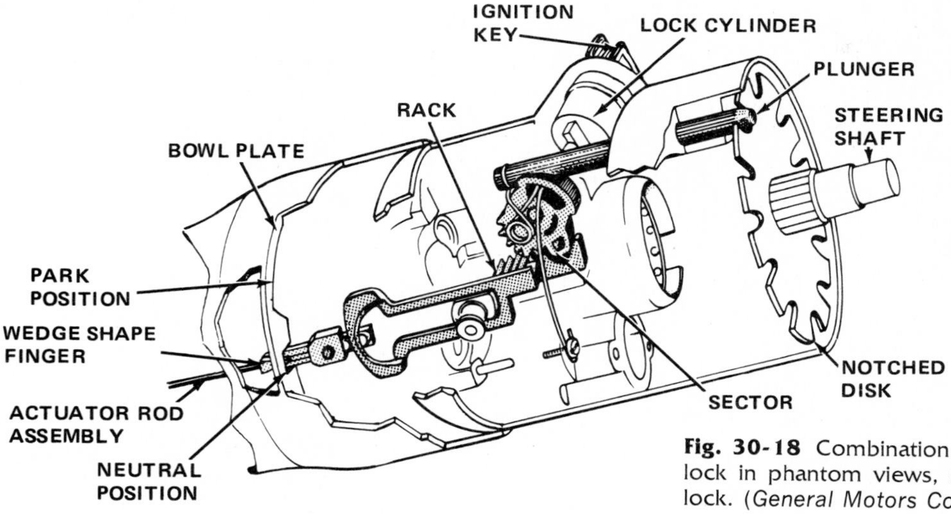

Fig. 30-18 Combination ignition switch and steering-wheel lock in phantom views, showing the two positions of the lock. (*General Motors Corporation*)

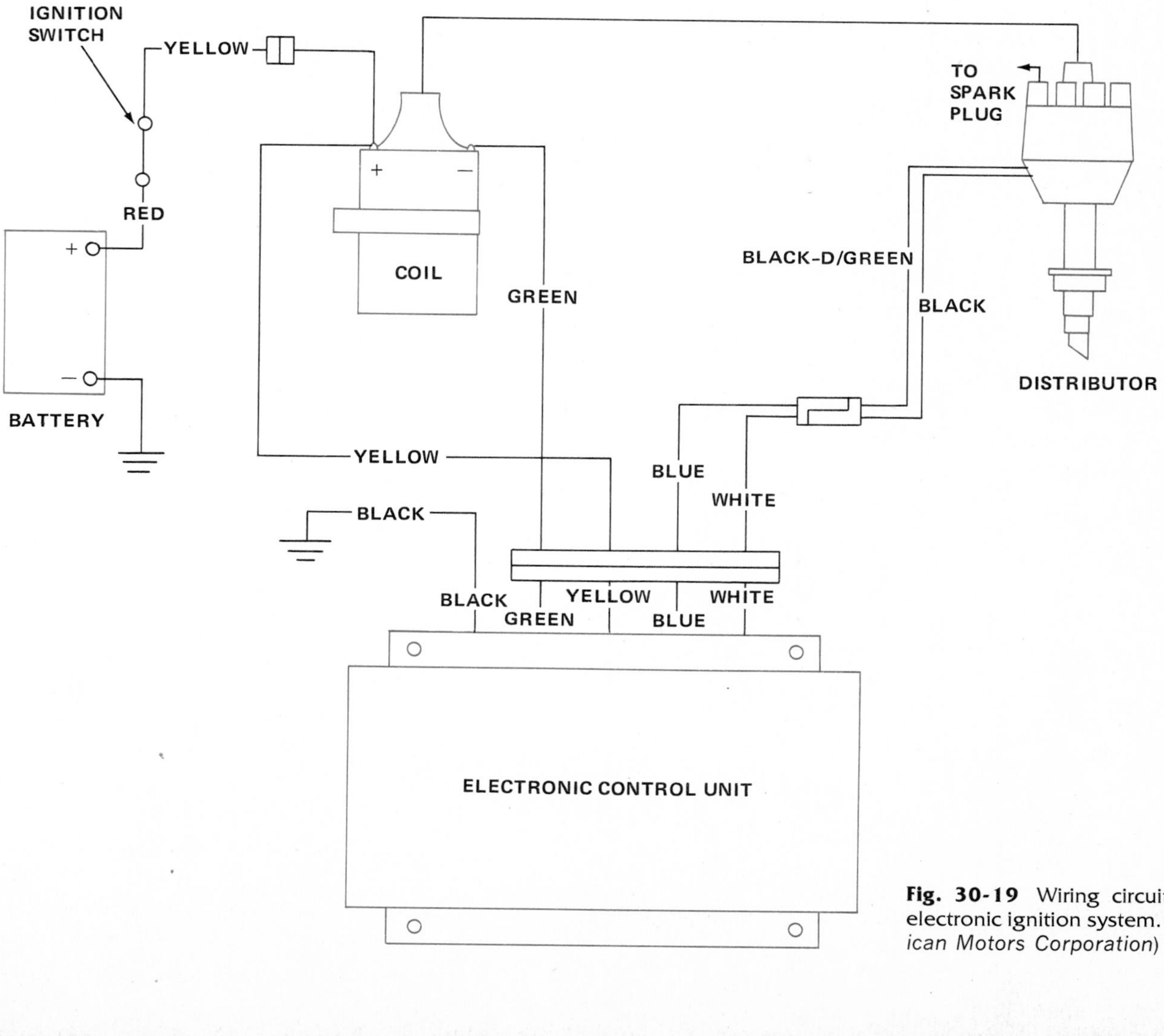

Fig. 30-19 Wiring circuit of an electronic ignition system. (*American Motors Corporation*)

Instead, these systems use electronic controls and sensing devices to produce accurate control of advance. They advance or retard the spark as engine temperature, manifold vacuum, engine speed, atmospheric pressure, and other conditions change. Three such systems, the Chrysler Electronic Lean Burn system, the General Motors Oldsmobile MISAR system, and the Ford Lincoln Versailles Electronic Engine Control (EEC) system, are discussed at the end of this chapter. First, however, we look at the electronic ignition systems with mechanical spark-advance controls.

30-20 American Motors Corporation Electronic Ignition System

The wiring circuit for the American Motors Corporation electronic ignition system (Fig. 30-19) looks very much like that for the contact-point system. There is one addition and this is the electronic control unit. Actually, the wiring circuits for all electronic ignition systems look pretty much alike.

Figure 30-20 shows a disassembled distributor for the electronic ignition system. It is the same as the distributors for contact-point systems except that the contact points have been replaced by a sensor and trigger wheel. It uses mechanical centrifugal and vacuum advance controls. The distributor is mounted and is driven in the same way as the contact-point distributors. As the shaft and trigger wheel rotate, the spokes on the wheel swing past the sensor assembly. A coil in the sensor magnetically detects the movement of each spoke because it causes a change in the magnetic field around the coil. The coil therefore almost instantly sends a voltage pulse to the control unit as a spoke passes by. The control unit then opens the circuit from the battery to the ignition coil. The magnetic field in the coil collapses and a high-voltage surge is produced in the coil secondary. This surge is carried by the high-voltage cables to the distributor cap, through the rotor and spark-plug cable to the plug that is ready to fire. Thus, you can see that the ignition coil, distributor cap, rotor, and cables work exactly the same as in the contact-point system.

Notice, in Fig. 30-20, that the trigger wheel has six spokes, one for each cylinder in the six-cylinder engine it is designed for. This corresponds to the six lobes on the cam in the contact-point cylinder. For an eight-cylinder engine, there would be eight spokes on the trigger wheel.

The electronic control unit has diodes, transistors, and other electronic components which work together as an electronic switch. When the voltage pulse comes from the sensor coil, they switch off the circuit between the battery and ignition coil. Then, as soon as the spoke on the trigger wheel passes the coil, the voltage pulse from the sensor coil dies out. The electronic control unit immediately switches the circuit on between the battery and ignition coil. Current flows and a magnetic field builds up in readiness to produce the next high-voltage surge. This happens once for each spoke on the trigger wheel every time the wheel completes one revolution.

30-21 Chrysler Electronic Ignition System

Figure 30-21 is the wiring diagram of the electronic ignition system used on Chrysler Corporation cars. Figure

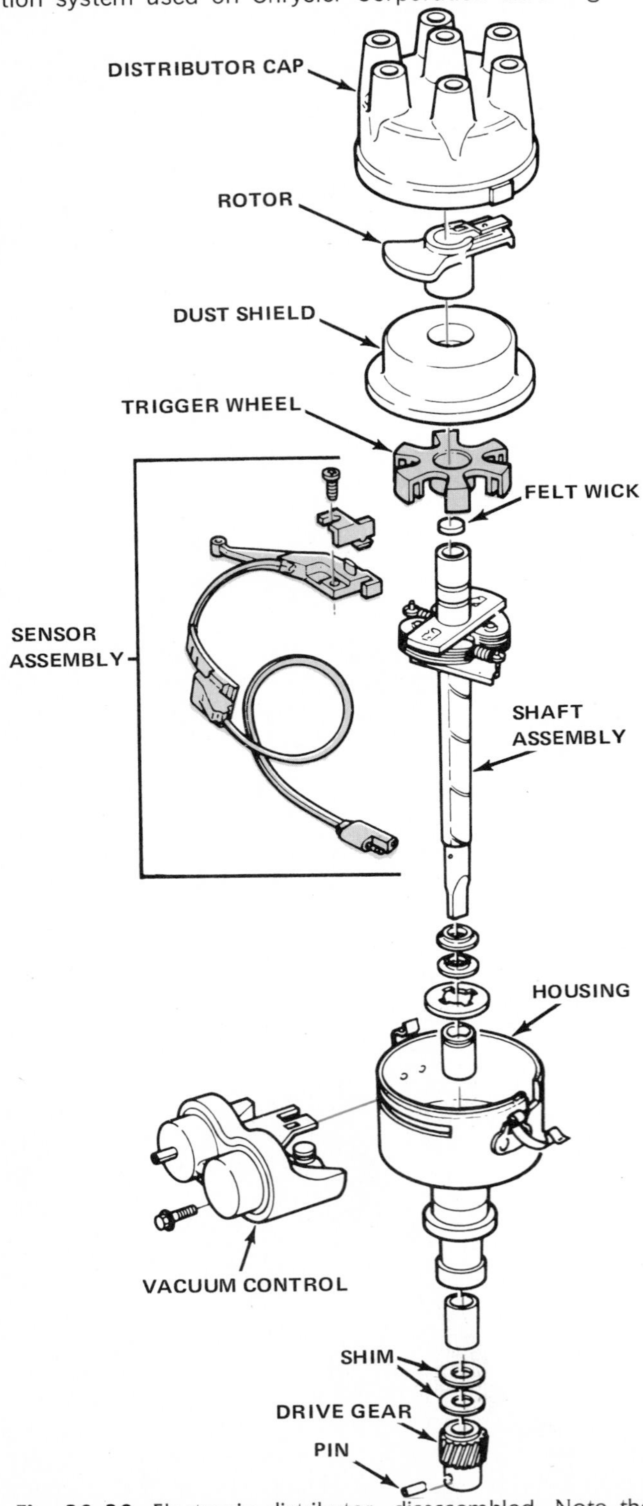

Fig. 30-20 Electronic distributor, disassembled. Note the trigger wheel and the sensor assembly (a coil). (*American Motors Corporation*)

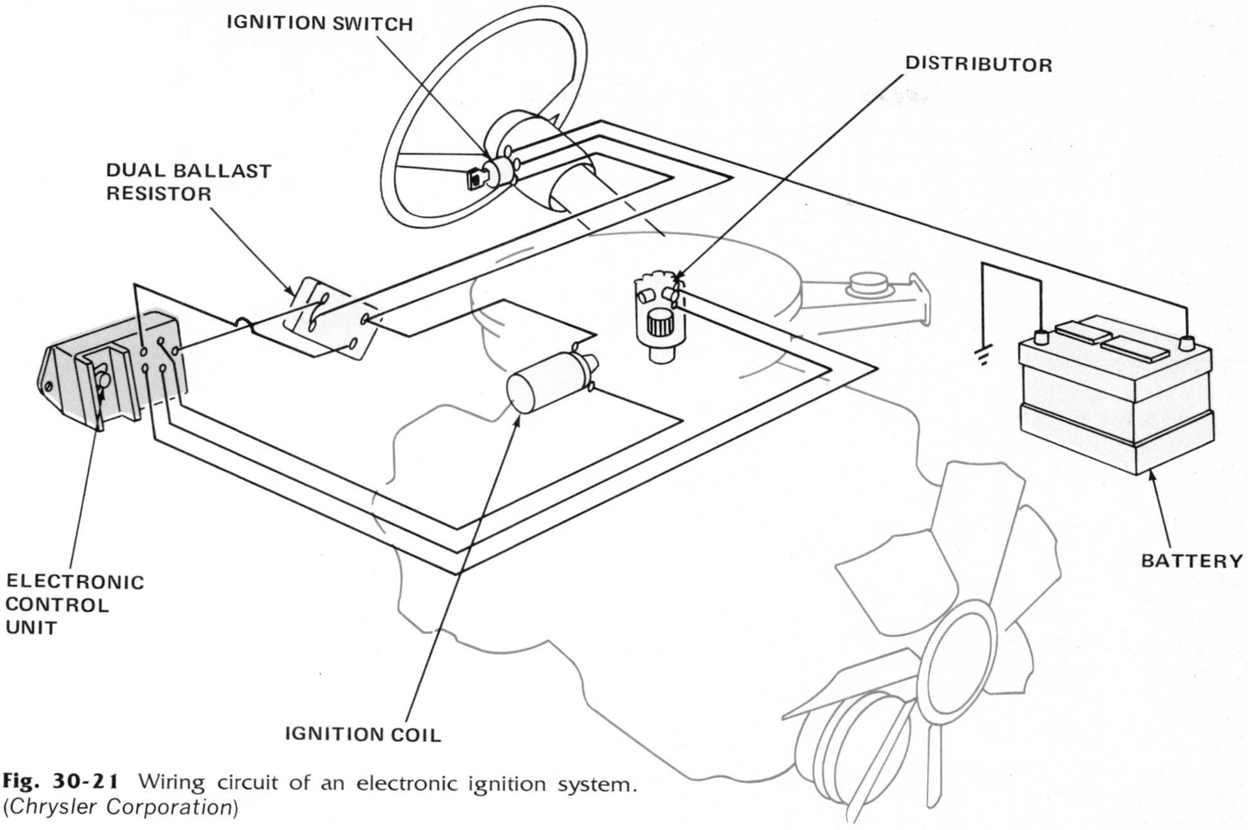

Fig. 30-21 Wiring circuit of an electronic ignition system. *(Chrysler Corporation)*

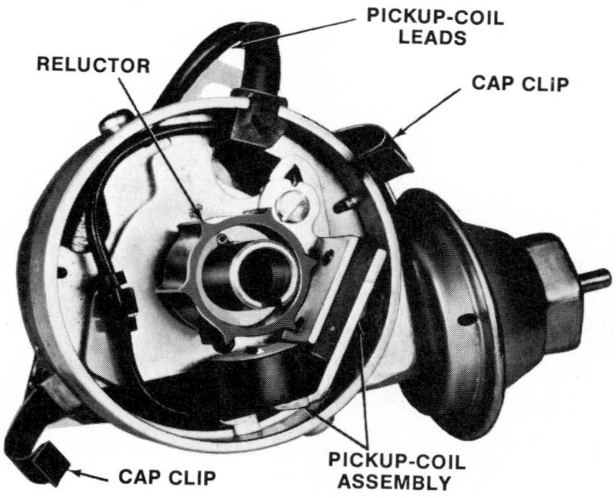

Fig. 30-22 Top view of the Chrysler electronic ignition distributor. The cap and rotor have been removed to show the reluctor and the pickup coil. *(Chrysler Corporation)*

30-22 is a top view, with the cap and rotor removed, of the distributor used with this system. Note that this distributor is somewhat different in design from the distributor used with the American Motors Corporation electronic ignition system. Instead of the trigger wheel, it has what Chrysler calls a reluctor. The reluctor has the same number of tips as there are cylinders in the engine. The sensor (American Motors name) is called the pickup-coil assembly in the Chrysler unit. The effect is exactly the same. As each tip passes the pickup coil, the pickup coil sends a voltage pulse to the electronic control unit. The control unit then switches off the circuit between the battery and ignition coil. The coil produces a high-voltage surge and the distributor cap, rotor, and high-voltage cables send it to the spark plug that is ready to fire.

After the tip passes the pickup coil, the voltage pulse dies and the control unit switches the circuit between the battery and ignition coil on again. The whole procedure occurs again as the next reluctor tip passes the pickup coil.

Note that this distributor also has mechanical centrifugal and vacuum advance controls. The dual ballast resistor (Fig. 30-21) maintains constant primary current even though engine speed varies. This provides a uniformly strong spark. The resistance is bypassed during starting so full battery voltage is applied to the ignition coil. This assures a strong spark for starting.

30-22 Ford Electronic Ignition System

Ford calls their electronic ignition system the "Dura Spark" system. There are two variations, "Dura Spark I" and "Dura Spark II." Basically, the systems work the same as the American Motors and Chrysler systems previously described. Figure 30-23 is a wiring diagram of the Ford system and Fig. 30-24 is a top view of the distributor with the cap and rotor removed. Notice the distributor has the same general structure as the other electronic distributors previously discussed. This distributor and the system work the same as the other systems. Note, however, that Ford calls the rotating element with spokes on it the "armature." Also, the sensor is called the magnetic pickup assembly. There is another difference worth noting. Shown in Fig. 30-25 is an external view of the distributor. Note that the cap has solid-spark-plug-type terminals instead of the push-in type used in many other distributor caps (such as shown in Fig. 30-20).

On some applications, a special Dura Spark II electronic module is used which has altitude and load compensation. That is, the module provides a change in spark advance as altitude and atmospheric pressure change. The module uses an atmospheric-pressure switch which signals the module as pressure changes. The module then modifies ignition timing to suit. For example, it provides a more advanced spark timing at high altitudes to compensate for the thinner mixture. It then retards the spark at low altitude to prevent spark knock. The system also includes a vacuum switch which senses intake-manifold vacuum. The signals from this switch cause the module to change the spark advance as the manifold vacuum changes. Thus, when the throttle is open and the engine is under heavy load, the module retards the spark to prevent spark knock. At light loads and part throttle, the module advances the spark.

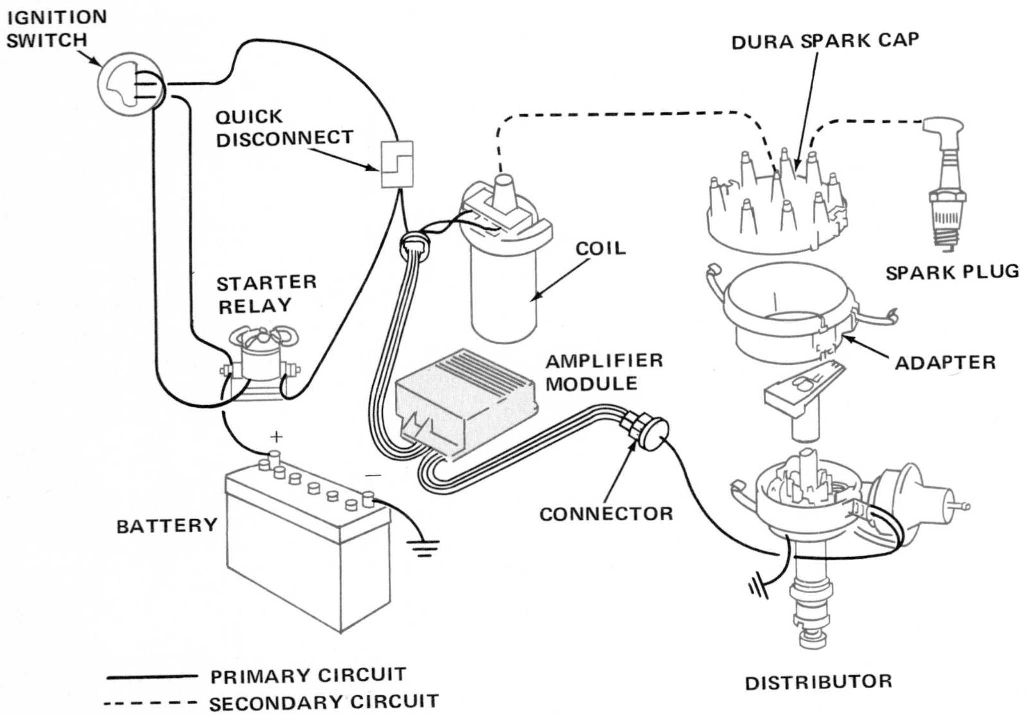

Fig. 30-23 Wiring circuit of the Ford electronic ignition system. (*Ford Motor Company*)

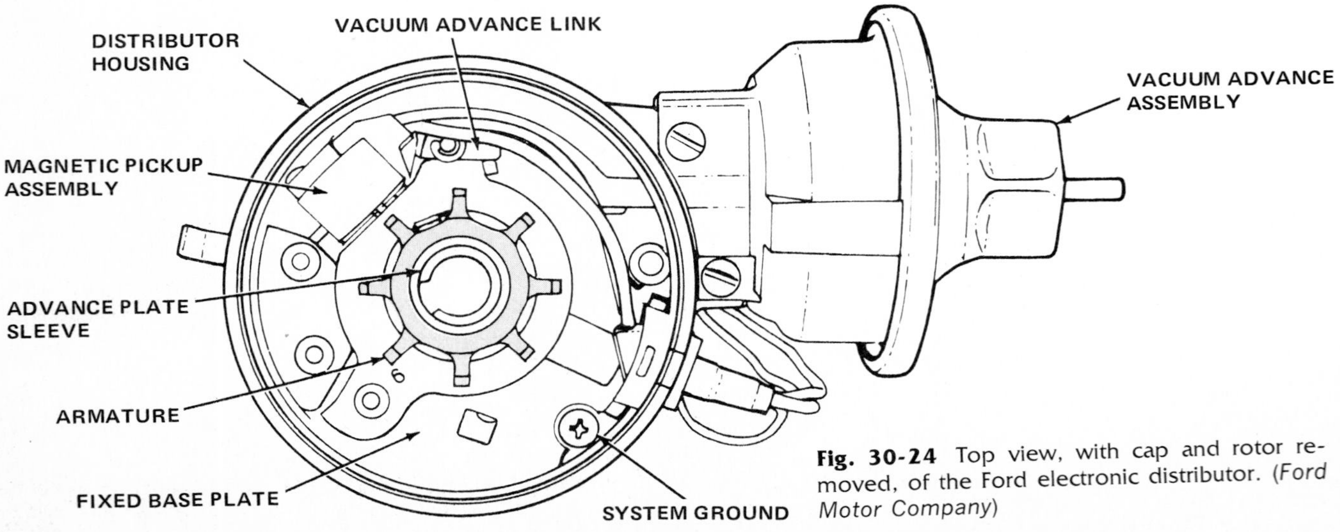

Fig. 30-24 Top view, with cap and rotor removed, of the Ford electronic distributor. (*Ford Motor Company*)

30-23 General Motors Electronic Ignition System

Figure 30-26 is the wiring diagram for the General Motors electronic ignition system. Figure 30-27 is a top view of the distributor with the cap and rotor off. This system uses a somewhat different design. General Motors calls the distributor a magnetic-pulse distributor. It has a permanent magnet and pickup coil on top of which is mounted a pole piece. The pole piece has a series of teeth pointed inward. There are the same number of teeth as there are cylinders in the engine. The rotating trigger is called the timer core and it also has the same number of teeth, pointing outward, as there are engine cylinders. Each time the teeth on the timer core and the teeth on the pole piece align and then move out of alignment, a magnetic field sweeps through the pickup coil. The pickup coil therefore sends a voltage pulse to the ignition-pulse amplifier. The amplifier shuts off the flow of current to the ignition coil. The coil then produces a high-voltage surge which passes through the cables, cap, and rotor to the spark plug that is ready to fire.

The distributor includes mechanical centrifugal and vacuum advance controls, as do the other distributors previously described.

30-24 General Motors High-Energy Ignition System

This ignition system (Fig. 30-28) has a distributor which, on eight-cylinder models, has the ignition coil as part of the assembly (Fig. 30-29). The six-cylinder model has a separately mounted ignition coil (Fig. 30-30). Both distributors have the electronic amplifier or "electronic module" as General Motors calls it, mounted inside the distributor. In the eight-cylinder model, the coil is mounted on top of the distributor cap, as shown in Fig. 30-29. Note how this arrangement simplifies the wiring system (Fig. 30-28).

The name of the system, "high energy," comes from its producing high voltages of up to 35,000 volts. This is considerably more than the standard systems produce.

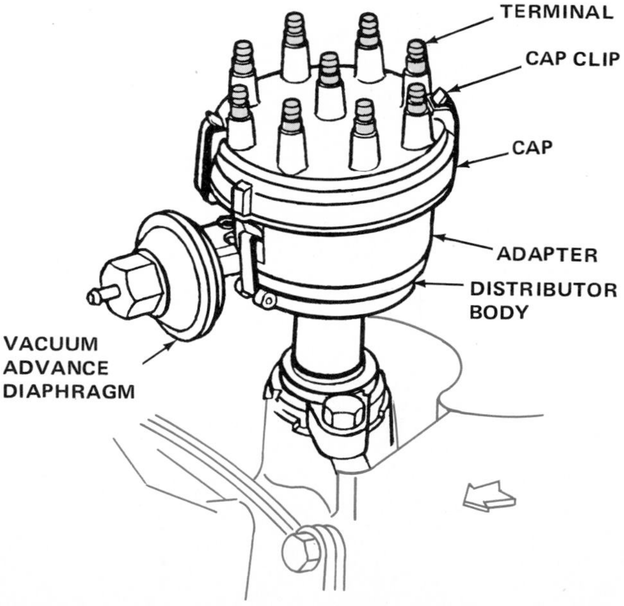

Fig. 30-25 Ford electronic distributor. Note the spark-plug-type terminals in the cap. (*Ford Motor Company*)

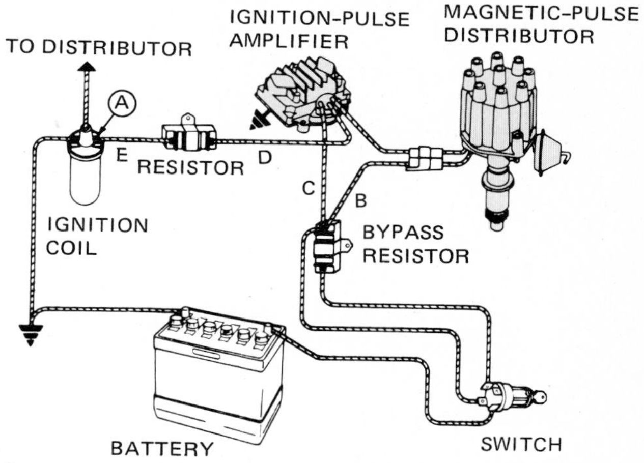

Fig. 30-26 Wiring circuit for the General Motors electronic ignition system using a separately mounted ignition-pulse amplifier (electronic control unit). (*Delco-Remy Division of General Motors Corporation*)

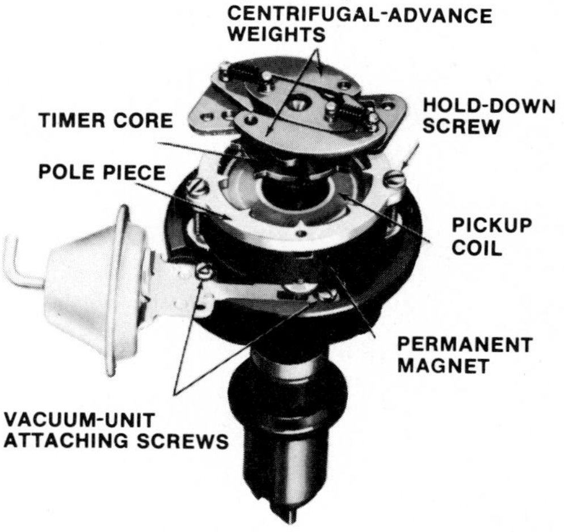

Fig. 30-27 Electronic distributor with cap and rotor removed. (*Delco-Remy Division of General Motors Corporation*)

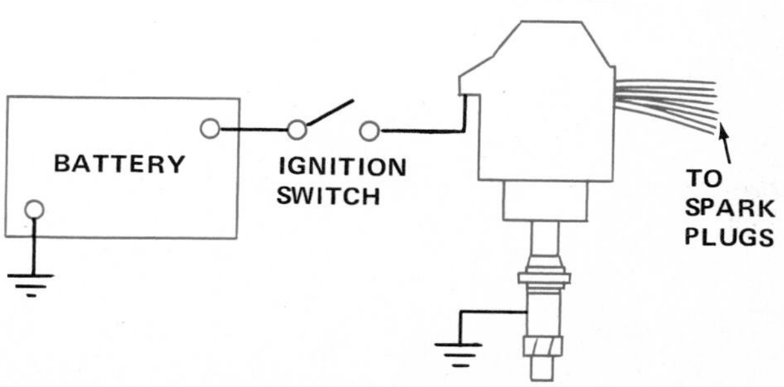

Fig. 30-28 Wiring diagram of high-energy distributor. (*Delco-Remy Division of General Motors Corporation*)

The higher voltage permits the use of spark plugs with wide gaps—up to 0.080 inch [2.03 mm]. The longer spark can more easily fire lean mixtures. Lean mixtures produce a more economical engine with fewer emission problems (see Part Five on emission controls).

Because of the higher voltages, special high-voltage cables must be used in the secondary system to handle these voltages. The cables are larger in diameter because of the heavier insulation, have greater heat resistance, and are gray in color.

The distributor uses the magnetic-pulse principle, the same as the other General Motors Electronic distributors.

All of the electronic ignition systems described so far are very similar in operation. A sensing device senses the rotation of a trigger wheel, armature, or timer core. The sensing device sends voltage pulses to the electronic module. These pulses close and open the battery-to-coil

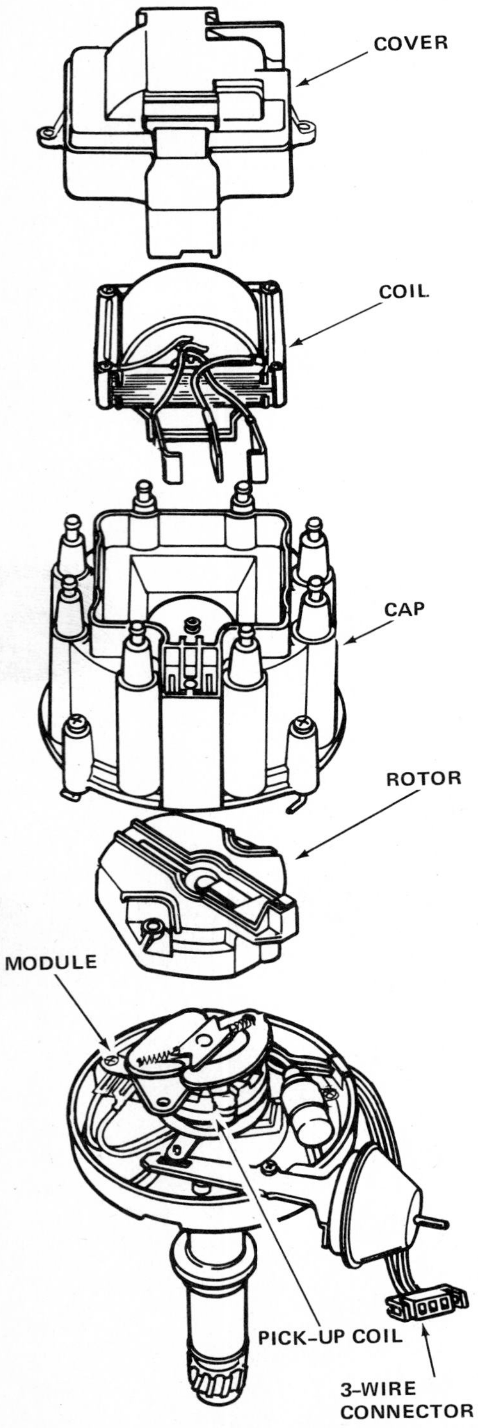

Fig. 30-29 Partly disassembled high-energy distributor. (*Delco-Remy Division of General Motors Corporation*)

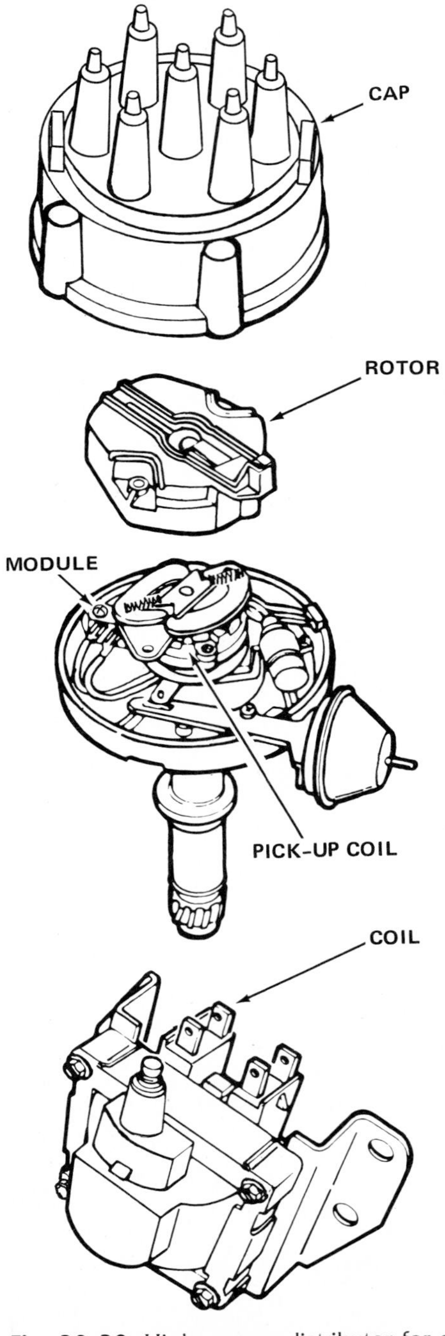

Fig. 30-30 High-energy distributor for six-cylinder engine. The ignition coil is separately mounted. (*Delco-Remy Division of General Motors Corporation*)

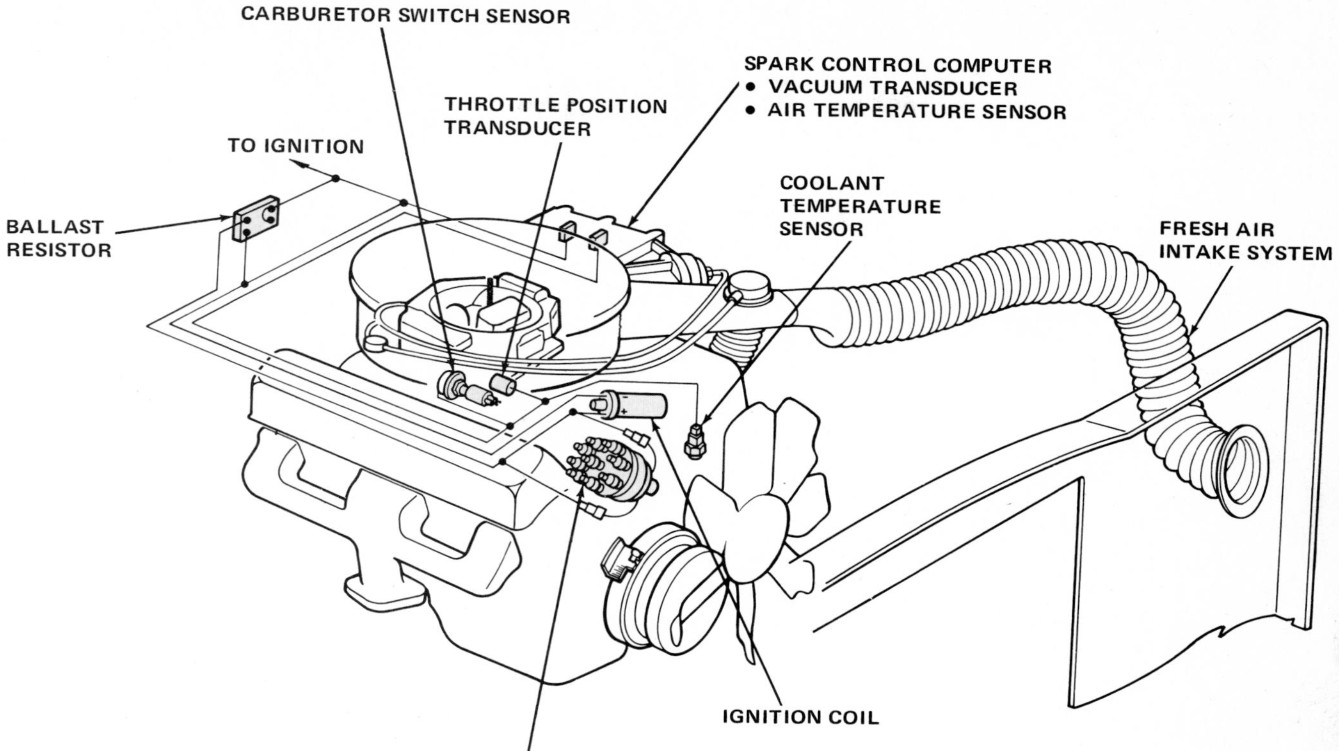

Fig. 30-31 Chrysler Lean Burn system, showing locations of components. (*Chrysler Corporation*)

circuit (just like the points in the contact-point system). This causes the coil to produce the high-voltage surges which make the sparks at the spark plugs.

Electronic Ignition Systems with Electronic Spark-Advance Control

Electronic ignition systems with electronic spark-advance controls are the same as the electronic systems previously described, with one major exception. They have no mechanical centrifugal or vacuum advance controls. Instead, various sensors feed information to a central computer which then produces the proper advance for the operating conditions. We describe three variations of this system, the Chrysler Lean Burn system, the General Motors Oldsmobile MISAR system, and the Ford Electronic Engine Control (EEC) system.

● 30-25 Chrysler Lean Burn System

This system (Fig. 30-31) consists of a spark-control computer, several engine sensors, and a special lean-calibration carburetor. The system is designed to provide voltage surges of sufficient strength and duration to burn a lean air-fuel mixture. Before we describe the operation of the system, we look at the computer and the sensors.

1. Spark-control computer The spark-control computer includes two interrelated modules, the program-schedule module and the ignition-control module. The program-schedule module receives signals from the sensors and almost instantly computes the correct spark advance for the operating condition. It then "tells" the ignition-control module how much to advance the spark.

2. Sensors There are seven sensors that provide information to the spark-control computer, as follows.

a. Coolant-temperature sensor This sensor is located in the water-pump housing. Its function is to signal the computer when the engine-coolant temperature is below 150°F [65.6°C].

b. Air-temperature sensor This sensor is inside the computer, and it supplies a signal based on the temperature of the air coming into the air cleaner. This signal also affects the amount of additional spark advance produced by the computer as related to the throttle-position transducer.

c. Throttle-position transducer "Transducer" is a fancy term for a device that receives a signal from some device and changes it to another related signal which it sends to a second device. The throttle-position transducer, for example, is located in the carburetor. It signals to the computer the position and rate of change of the throttle valves. As the throttle valves start to open, they "tell" the computer and the computer provides additional advance. The amount the throttle opens determines the amount of additional advance. If the throttle is opened quickly, maximum advance is provided for about one second. However, the air-temperature-sensor signal controls how much of this maximum advance will be allowed. If the air temperature is hot, it will allow less advance. If cold, it allows more advance.

d. Carburetor-switch sensor This sensor, located on the right side of the carburetor, tells the computer if the engine is operating at idle or off idle.

e. Vacuum transducer This device is located on the computer and it is connected to a port in the carburetor so it senses intake-manifold vacuum. As vacuum increases, its signal to the computer produces more advance. The lower the vacuum, the less the spark advance. Thus, as you can see, the vacuum transducer works about the same as the mechanical vacuum advance unit on the earlier distributors.

f. Start pickup This is one of two sensors located inside the distributor (Fig. 30-32). During cranking, this sensor supplies a signal to the computer that causes it to provide a fixed spark advance. All other advance possibilities are canceled out.

g. Run pickup This sensor (Fig. 30-32) supplies the basic timing signal to the computer. This signal "tells" the computer how fast the engine is running and also when the pistons are moving up on their compression strokes. Thus, as you can see, this sensor—the run pickup—does the same job as the sensor in the other electronic ignition systems previously described.

30-26 Operation of Chrysler Lean Burn System

When the ignition key is turned on, the start pickup signals the computer to supply additional spark advance during cranking. Then, when the engine starts, the run pickup takes over and begins to inform the computer of the speed and piston locations of the engine. During the first minute of running, the computer also supplies additional advance to give the cold mixture time to burn. This additional advance is phased out over the first minute.

With the coolant temperature below 150°F [65.6°] the coolant-temperature sensor supplies a signal that will prevent additional spark advance due to the vacuum-transducer signal. This prevents excessive spark advance with a high intake-manifold vacuum which could cause engine stumble.

After the engine reaches operating temperature, the warm-engine operating mode is reached. The run pickup signals the speed and piston positions to the computer so it can provide the correct advance for any engine speed. At the same time, the input signals from the air-temperature sensor, throttle-position transducer, carburetor-switch sensor, and vacuum transducer are entering the computer. The computer combines all these signals and "decides" the exact spark advance required for best engine operation, based on all these signals.

30-27 Ford Electronic Engine Control (EEC) System

This system (Fig. 30-33) also uses a computer and is similar in many ways to the Chrysler Lean Burn system.

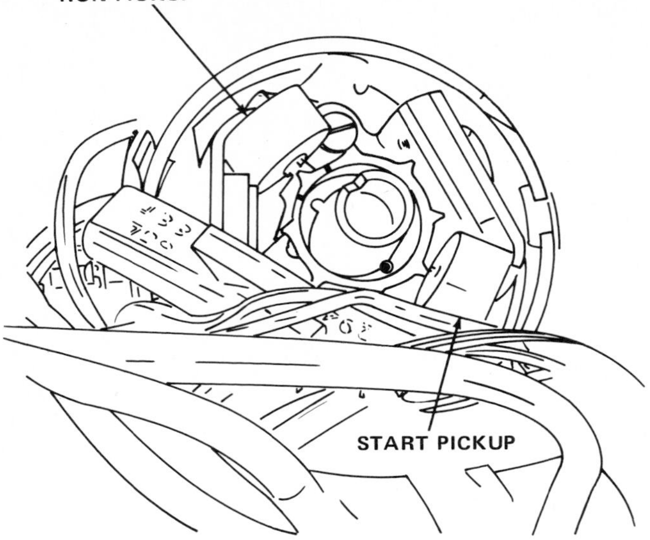

Fig. 30-32 Locations of start-pickup and run-pickup sensors in distributor. (*Chrysler Corporation*)

In addition to controlling the spark advance, the Ford EEC system also controls the exhaust-gas-recirculation valve and the air-injection system, as we explain later.

The computer is fed information from sensors that

- Sense the inlet air temperature entering the air cleaner.
- Sense throttle position.
- Sense coolant temperature.
- Sense air pressure which changes as altitude changes.
- Sense intake-manifold vacuum.
- Sense crankshaft and thus piston positions. This signal originates at a disk with four teeth 90° apart on the rear of the crankshaft. When each tooth lines up with the sensor, it produces a signal that tells the computer the crankshaft position.
- Sense engine speed. The signal from the crankshaft disk and sensor also tells the computer how fast the crankshaft is turning.

The exhaust-gas recirculation (EGR) system sends some exhaust gas back through the engine to reduce the amount of nitrogen oxide formed during combustion. This gas is considered a pollutant, and the amount must be kept low. We mention the system in ● 12-13 and describe it in detail in Part Five of the book. The Ford EEC system also controls the operation of the valve that allows some exhaust gas to flow back into the intake manifold. It uses two solenoid valves to produce this control. The purpose of this control is to allow the EGR system to work only when the engine temperature and operating condition can tolerate the exhaust gas. That is, allow the exhaust gas to enter the intake manifold without reducing engine performance.

A second emission control—exhaust manifolds with air-injection systems—is mentioned in ● 12-15 and covered in detail in Part Five of the book. The air-injection

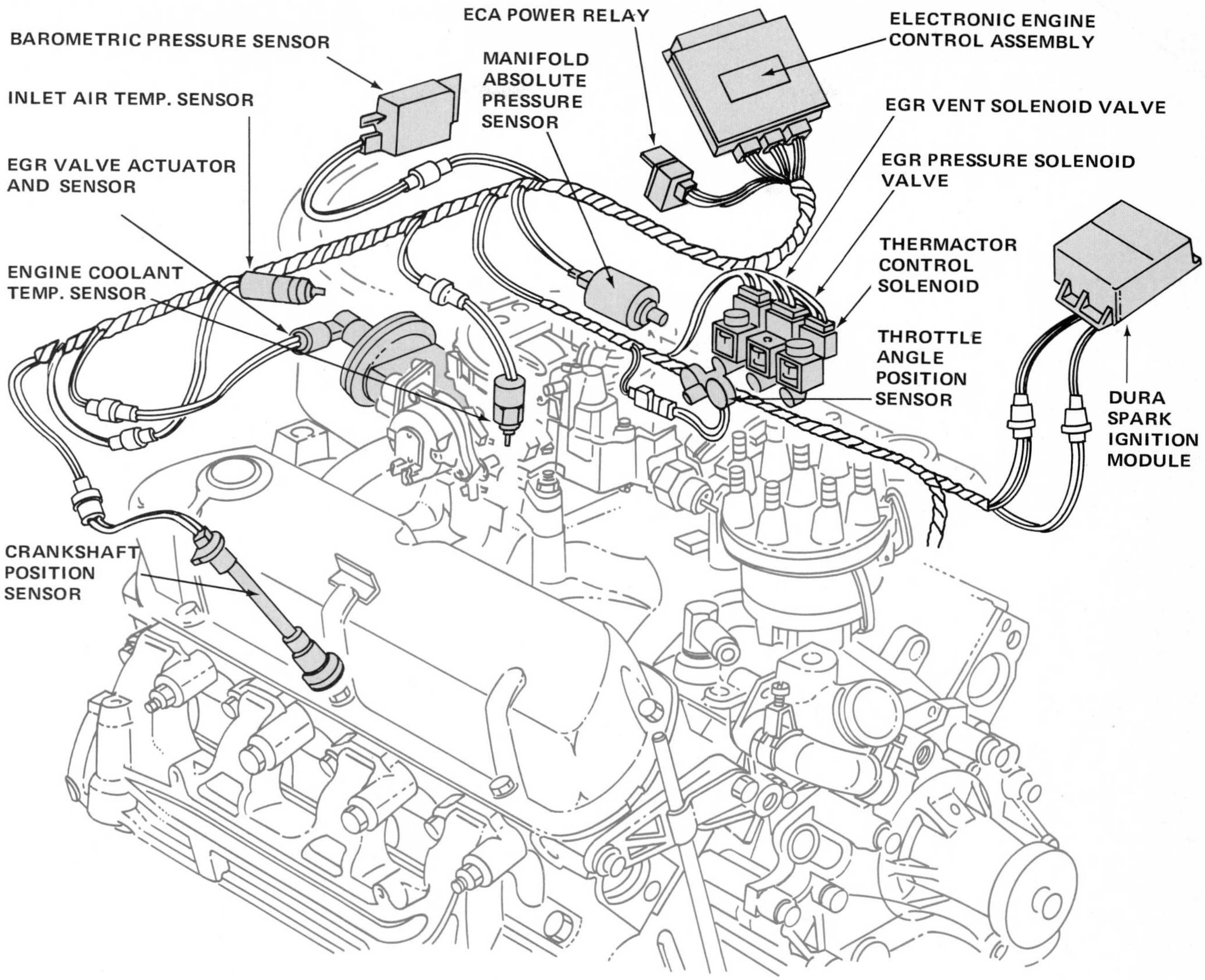

Fig. 30-33 Ford electronic engine control (EEC) system. (*Ford Motor Company*)

system sends air into the exhaust manifold to help burn up any unburned gasoline (HC) or partly burned gasoline (CO) in the exhaust gases. The Ford EEC system controls the air pump that is part of the air-injection system, cutting off the air flow to the exhaust manifold when it could cause reduced engine performance.

30-28 General Motors Microprocessing Sensing Automatic Regulation (MISAR)

This system is similar in many ways to the two systems previously described. It controls spark advance electronically, based on the information fed to the electronic control unit by four engine sensors. These sensors are:

1. Crankshaft sensor (in 1977 cars; in 1978, the sensor is located inside the distributor)
2. Engine-coolant sensor
3. Manifold-vacuum sensor
4. Atmospheric-pressure sensor

The 1977 version of MISAR has a crankshaft sensor mounted to the front engine mounting brackets (Fig. 30-34). A disk with teeth is mounted between the pulley and harmonic balancer on the front of the crankshaft (Fig. 30-35). As the engine runs and the disk rotates, it causes the crankshaft sensor to send signals to the electronic control unit. These signals "tell" the control unit how fast the engine is running and the position of the crankshaft (and thus the pistons on the compression strokes). Using this information, the control unit adjusts the spark advance to suit engine speed.

The 1978 version of MISAR places the speed and piston position sensor in the distributor. In this position, it provides the same information as the disk on the crankshaft and its sensor.

On the 1977 version, timing is adjusted by moving the crankshaft sensor (Fig. 30-35). The 1978 version is adjusted by shifting the distributor in its mounting.

The engine-coolant sensor is located in the engine cooling system and senses engine-coolant temperature. It sends a continuous signal to the electronic control unit to

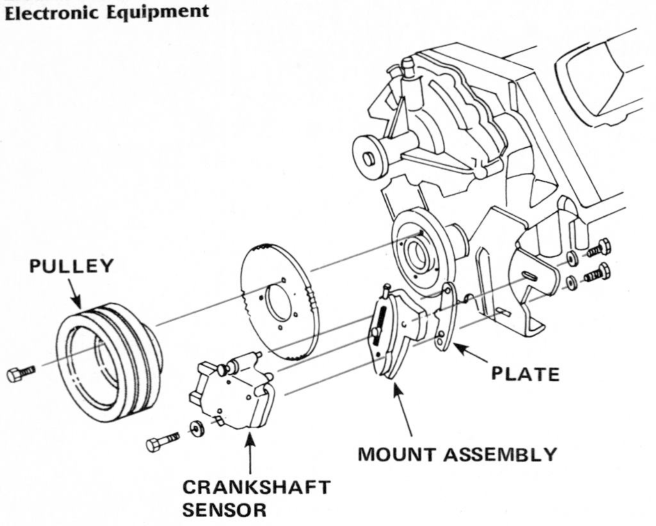

Fig. 30-34 Crankshaft sensor mounting for the 1977 MISAR ignition system. (*Oldsmobile Division of General Motors Corporation*)

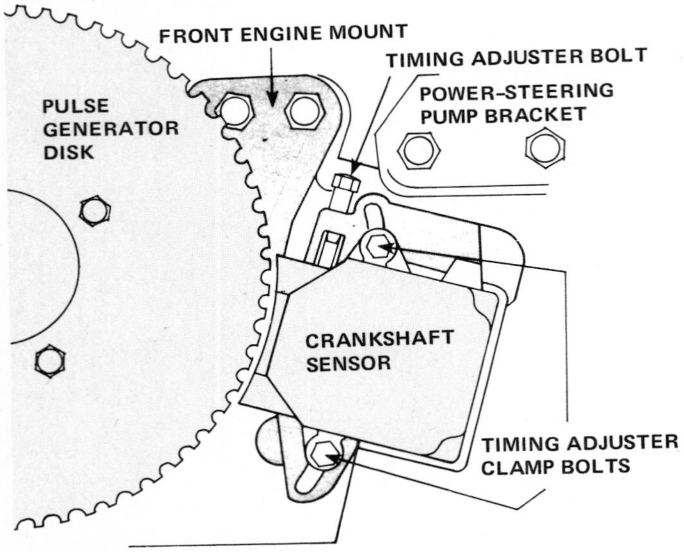

Fig. 30-35 Relationship of crankshaft sensor to pulse generator disk on the engine crankshaft. (*Oldsmobile Division of General Motors Corporation*)

tell it the engine temperature. The control unit then adjusts the spark advance as needed.

The manifold-vacuum sensor reacts to intake-manifold vacuum, providing a continuous signal to the control unit. This causes the control unit to adjust the spark advance to suit vacuum conditions. For example, when the throttle is partly closed and there is intake-manifold vacuum, the control advances the spark. This gives the thinner air-fuel mixture more time to burn. This is the same function that the vacuum advance unit on the mechanical distributor performs.

The atmospheric pressure sensor adjusts the spark advance to suit the atmospheric pressure. Thus, at higher altitudes, where the pressure is lower, the sensor tells the electronic control to advance the spark. Less air-fuel mixture gets into the cylinders when the atmospheric pressure is lower. The spark advance gives this thinner mixture enough time to burn.

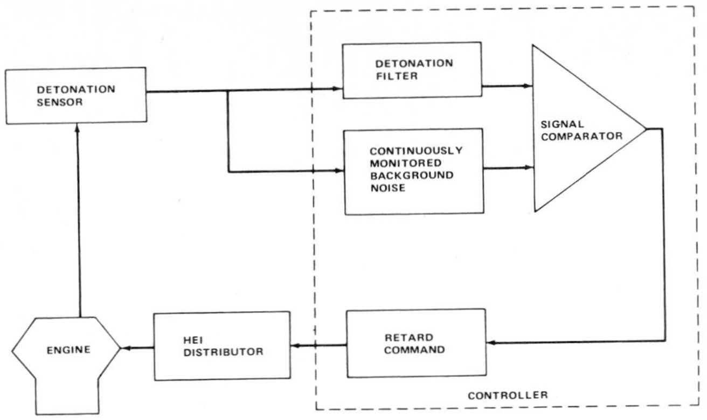

Fig. 30-36 Block diagram of the electronic logic system for the ESC system which retards the spark advance if detonation begins. (*Buick Motor Division of General Motors Corporation*)

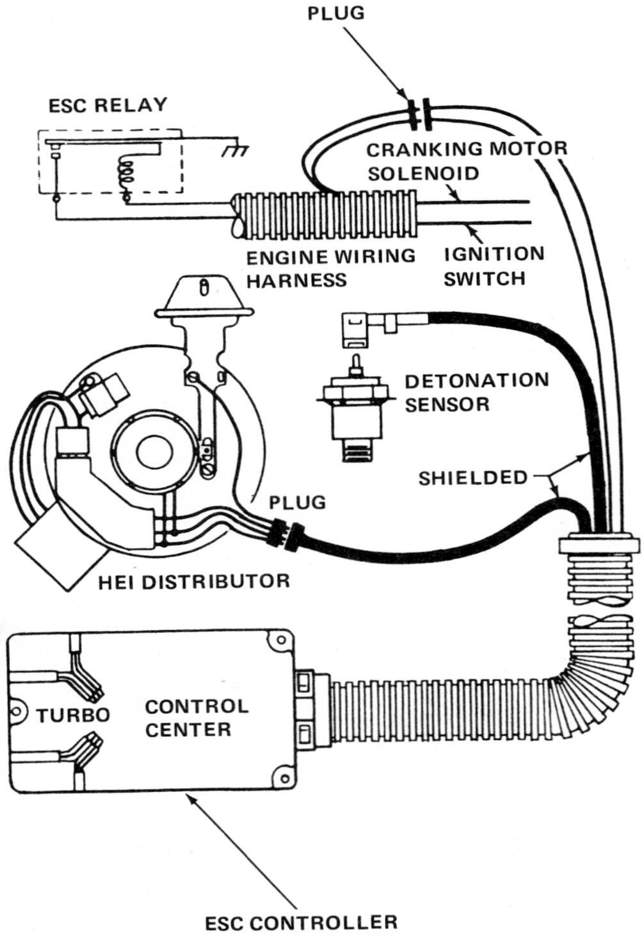

Fig. 30-37 Layout of the ESC system. (*Buick Motor Division of General Motors Corporation*)

30-29 Electronic Spark Control

Electronic spark control (ESC) is used on some turbocharged engines to guard against detonation. Turbochargers are described in ●12-18. The turbocharger packs additional air-fuel mixture in the engine cylinders to increase engine power. However, if too much is packed in, with normal spark advance, detonation can occur. The ESC prevents this by backing off the spark advance just before detonation starts to occur. Figure 30-36 shows the logic system of the ESC. The detonator sensor is mounted on the intake manifold at the thermostat housing. When detonation starts, the signal from the sensor activates the electronic control. The electronic control then signals the distributor so it overrides the distributor advance mechanisms (mechanical type) and causes the spark to retard. Figure 30-37 is a layout of the system. Note that a High Energy Ignition (HEI) distributor is used.

REVIEW QUESTIONS

1. What is the basic purpose of the ignition system?
2. Does the electronic ignition system use contact points?
3. What are the two circuits through the contact-point-type ignition distributor?
4. How many tips are there on a reluctor for the distributor of an eight-cylinder engine?
5. When a tip of the reluctor passes the pickup coil, does the control unit open or close the circuit to the ignition-coil primary?
6. What is special about the Delco-Remy high-energy ignition system?
7. What is the purpose of the resistor used in some spark plugs?
8. What is meant by the spark-plug heat range? Describe a cool-running plug.
9. Explain how the starting and stopping of a flow of current in the ignition-coil primary winding produces a high-voltage surge in the secondary winding.
10. Name the essential parts in the primary circuit. In the secondary circuit.
11. Describe the cap and rotor action.
12. Explain the condenser effect.
13. What are the two general types of spark-advance mechanisms?
14. Describe the operation of the centrifugal advance mechanism.
15. Describe the operation of the vacuum advance mechanism.
16. What is the purpose of the transmission switch in the emission-reduction vacuum advance control system?
17. Explain the operation of the solenoid vacuum switch in the emission-reduction vacuum advance control system.
18. What is the purpose of the temperature override system?
19. Where is the ignition switch located in late-model cars?
20. Name four jobs that the ignition switch does in the modern car.
21. Explain how an electronic ignition system works.
22. What are three names for the rotating part that magnetically influences the sensor or pickup coil?
23. Where is the electronic module mounted in the General Motors high-energy ignition system?
24. Explain how the electronic ignition system with electronic spark-advance control, such as the EEC or MISAR system, works.

SELF PROJECTS

1. Refer to the line drawings, in the chapter, showing the primary and secondary circuits. Draw your own versions of these circuits. Use different colors for the different circuits and for the system components. For instance, you could use blue for the secondary circuit, red for the primary, and black for the components. File your drawings in your notebook.
2. If you can find junk ignition coils and condensers, you might be interested in tearing them down. Use a hacksaw to cut the top off the condenser can. If the condenser won't come out, split the can open by sawing along one side. You can probably remove the top from the ignition coil by opening the seam at the top. Then you may have to split the can to get to the windings.
3. As you run across articles in magazines on the new electronic systems and controls, clip them and file them in your notebook.

CHAPTER 31
Ignition-System Service

After you read this chapter, you should be able to:

- **Use the oscilloscope under the instructor's guidance, to check ignition coils and ignition systems.**
- **Install and adjust contact points.**
- **Use a timing light to adjust timing on an engine.**
- **Remove and replace an ignition distributor, spark plugs, and ignition cables.**
- **Perform the other ignition service jobs discussed in the chapter.**

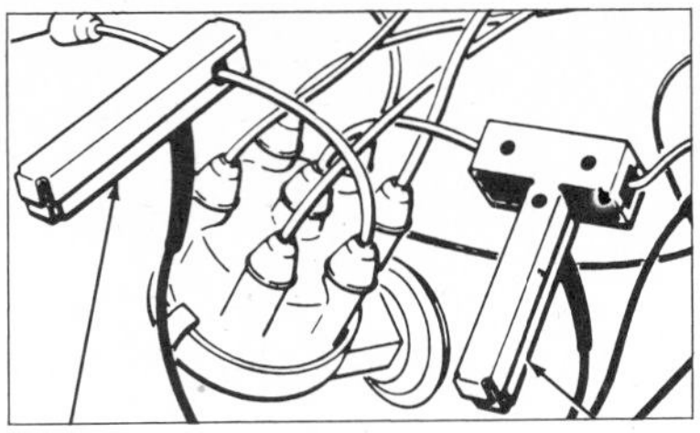

In this chapter we discuss the testing and servicing of the ignition system and its component parts. The ignition system includes the distributor, spark plugs, ignition coil, switch, and wiring. Of these, only the distributor and spark plugs require periodic service. If there is difficulty with the coil, switch, wiring, or electronic components, the correction is replacement.

● 31-1 Ignition Testing Equipment

Complete testing of the ignition system and its component parts requires a variety of testing equipment. These include the oscilloscope, coil tester, condenser tester, distributor tester, contact-point-opening testers, and contact-pressure gauge. In addition, electronic ignition systems can be tested with a special tester developed for this purpose. We discuss all these testers in sections that follow.

● 31-2 Oscilloscope

The oscilloscope is a high-speed voltmeter that uses a television-like picture tube to show ignition voltages. Figure 31-1 shows an electronic engine tester which includes an oscilloscope. The oscilloscope, or "scope," is to the upper left in the picture.

The oscilloscope draws a picture of the ignition voltages on the face of the tube. The picture shows what is happening in the ignition system. If something is wrong, the picture will show what it is.

To understand the pictures, we first review the ignition-system action. When current flows in the coil primary winding, a magnetic field builds up. Then, when the current flow stops, the magnetic field collapses and a high-voltage surge is produced in the secondary winding. The current flow is started and stopped by either contact points or the electronic control unit, according to the type of ignition system.

The secondary voltage jumps up to thousands of volts. This high voltage surges to a spark plug and produces a

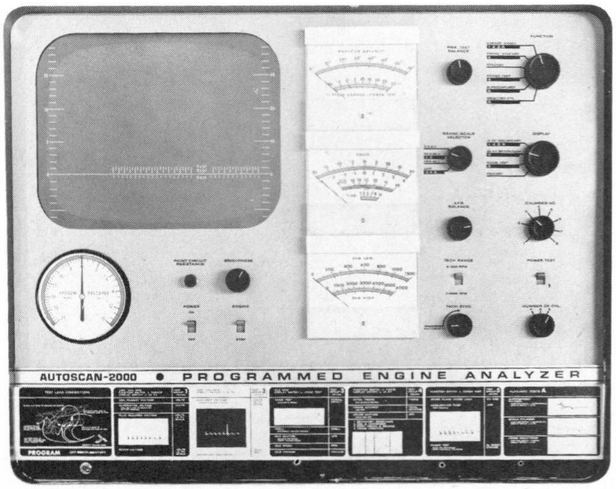

Fig. 31-1 An electronic engine tester with an oscilloscope. (*Autoscan, Inc.*)

spark. That is, the high voltage jumps the gap between the insulated and grounded electrodes of the spark plug. It takes a high voltage to start the spark. But after the spark is established, much less voltage is needed to keep the spark going. The scope can, among other things, draw a picture of how and when this voltage goes up and down.

The picture is drawn on the face of the tube by a stream of electrons. This is exactly the way the picture tube in a television set works. In the scope, however, the stream of electrons draws a picture of just one thing—ignition-system voltages feeding into the scope. Figure 31-2 shows the face of the picture tube and helps to explain what we mean.

When a voltage is detected by the scope, a "spike," or vertical line, appears on the face of the tube. This is shown in Fig. 31-2. The higher the spike, the higher the voltage. If the voltage spike points down, it indicates that the ignition coil or the battery is connected backward.

To see how the scope picks up the voltages, and what the pictures mean, let us first study what is called the *basic pattern* (see Fig. 31-3). The basic pattern is what the scope would show if it were drawing the voltage pattern for one spark plug. First the current stops flowing in the coil primary. The high-voltage surge from the coil arrives at the spark plug. The voltage goes up, from A to B, as shown. This is called the *firing line.* After the spark is established, the voltage drops off and holds fairly steady, from C to D. This is a very short time, measured in hundred-thousandths of a second. But the spark lasts for as long as 20 degrees of crankshaft rotation. This is long enough to ignite the compressed air-fuel mixture in the cylinder.

After most of the magnetic energy in the coil has been converted into electricity to make the spark, the spark across the spark-plug gap dies. However, there is still some energy left in the coil, and this produces a wavy line, from D to E. This line is called the *coil-condenser oscillation line.* What this wavy line means is that the remaining energy is pushing electricity back and forth in the ignition secondary circuit. The voltage alternates, but it is no longer high enough to produce a spark. After a very short time, the voltage dies out. Then, at E, the current again starts to flow in the primary winding of the ignition coil. Now an alternating voltage is produced in the secondary. This is the result of the buildup of current in the coil primary winding. This is shown by the oscillations that usually follow E. The section from E to F is called the *dwell* section. This is the time during which the contact points are closed or the electric control unit maintains a complete battery-to-coil circuit. During this time, the magnetic field is building up in the ignition-coil primary. Then, at F, we are back to A again and the whole process is repeated.

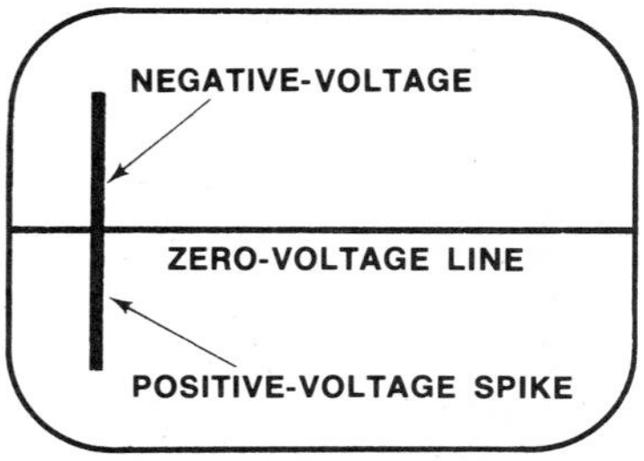

Fig. 31-2 The oscilloscope draws a horizontal zero-voltage line until a negative or positive voltage pulse enters. This causes the trace to kick up or down, as shown. The higher the voltage, the farther the trace moves up or down. The sharp up and down movements of the trace are called *spikes.*

31-3 Oscilloscope Patterns

The curves that the scope draws on the tube face are called *patterns.* The patterns can be drawn on the tube

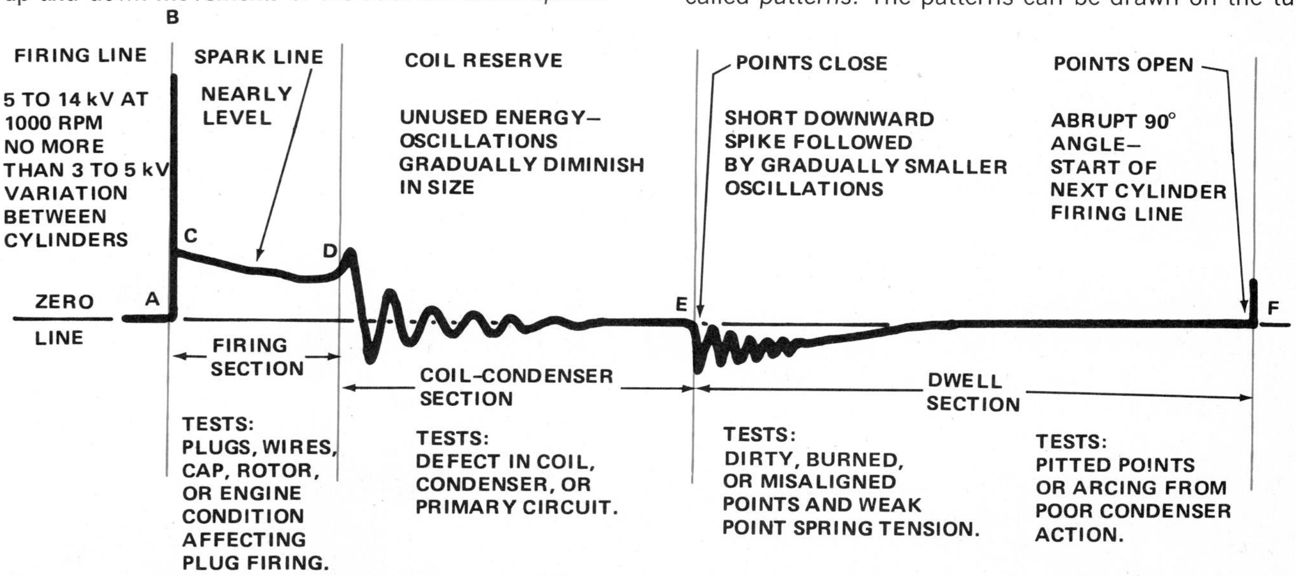

Fig. 31-3 A waveform, or trace, showing one complete spark-plug firing cycle. Note that the "dwell section" is the period during which the points are closed or the electronic control unit is maintaining a complete circuit between the battery and ignition coil. (*Sun Electric Corporation*)

face in different ways. For example, the scope can be adjusted to draw a *parade* pattern, as shown in Fig. 31-4. It is called a parade pattern because the traces for the separate cylinders follow one another across the tube face, like marchers in a parade. Note that they follow from left to right across the screen, in normal firing order, with No. 1 cylinder on the left.

By adjusting the scope in a different way, the traces can be stacked one above the other, as shown in Fig. 31-5. This is called a *raster* pattern. It lets you compare the traces, so you can see if something is wrong in a cylinder. The pattern is read from the bottom up in the firing order, with No. 1 cylinder at the bottom.

A third way to display the traces is to superimpose them (Fig. 31-6). That is, put them one on top of another. This gives a quick comparison and shows whether the voltage pattern from any one cylinder differs from that of the others. If everything is okay in the cylinders, only one curve appears on the tube face. This is because all the curves fall on top of one another.

31-4 Using the Scope

There are several makes of oscilloscopes. Many are combined in consoles with other instruments for testing the separate ignition components, engine rpm, intake-manifold vacuum, and so on. Figure 31-7 shows a complete tester of this type. Figure 31-1 shows the face of a similar tester. Scopes have pickup sensors that can be clamped onto the ignition wires, as shown in Fig. 31-8. It is not necessary to disconnect and reconnect the ignition circuits. The pattern-pickup sensor is clamped onto the wire that goes from the ignition coil to the distributor-cap center terminal. The sensor senses the high-voltage surges going to all the spark plugs. The trigger-pickup

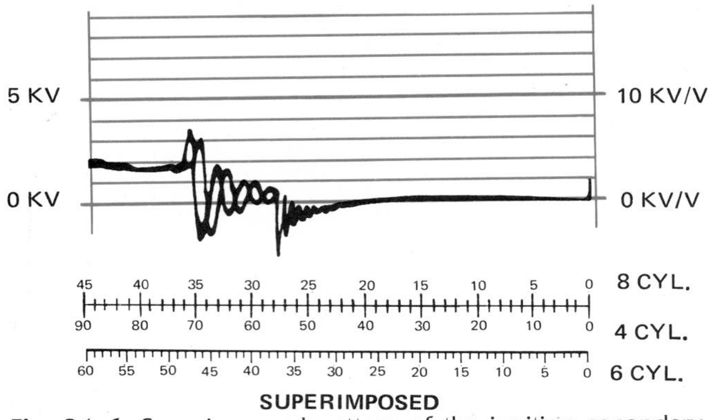

Fig. 31-6 Superimposed pattern of the ignition secondary voltages in a six-cylinder engine. (*Sun Electric Corporation*)

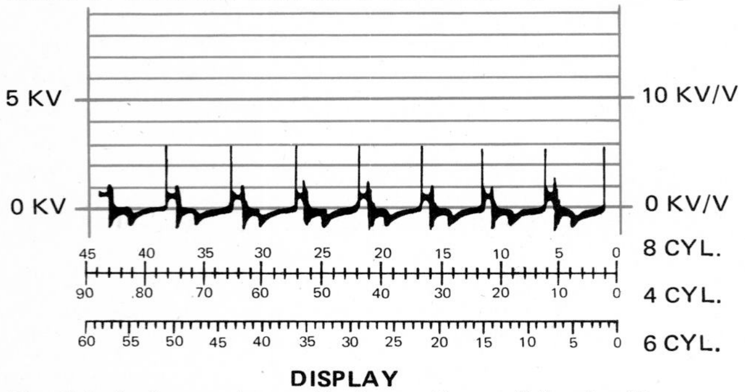

Fig. 31-4 A parade or display pattern of the ignition secondary voltages in an eight-cylinder engine. (*Sun Electric Corporation*)

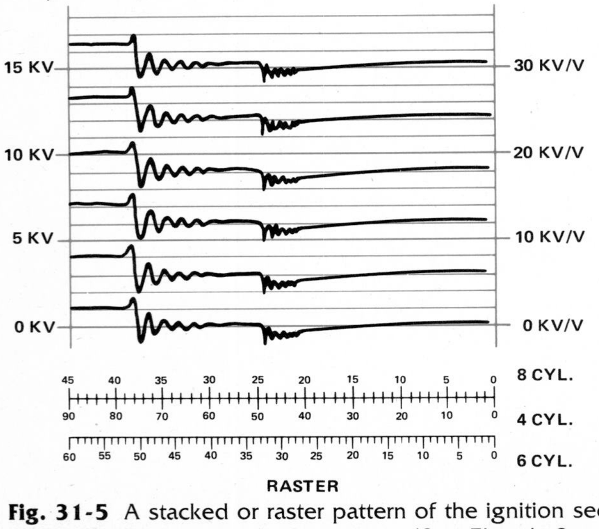

Fig. 31-5 A stacked or raster pattern of the ignition secondary voltages in a six-cylinder engine. (*Sun Electric Corporation*)

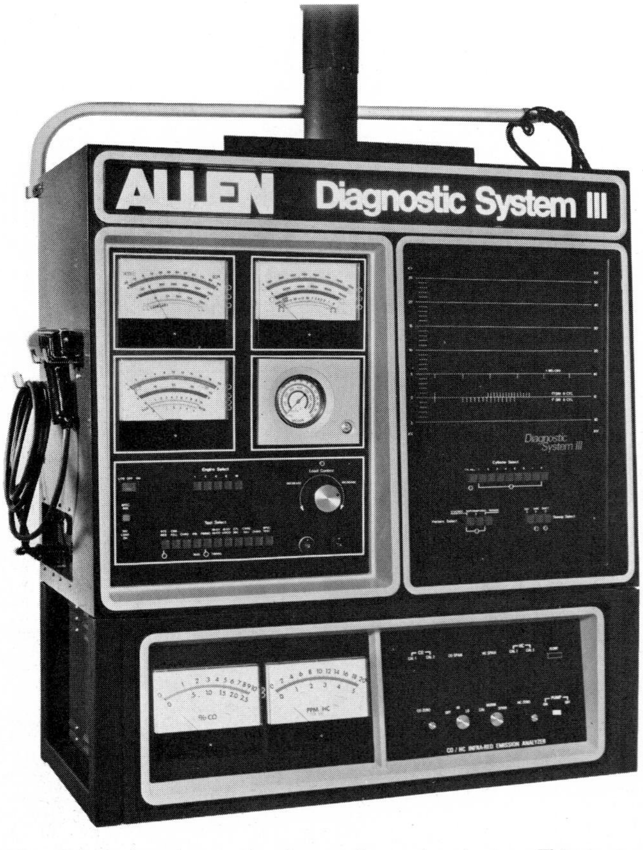

Fig. 31-7 An electronic-diagnosis engine tester. This tester includes an oscilloscope (upper right) and other testing devices to check engine vacuum, contact-point dwell, engine speed, HC and CO content in exhaust gases (at bottom), and other data. (*Allen Testproducts Division, The Allen Group, Inc.*)

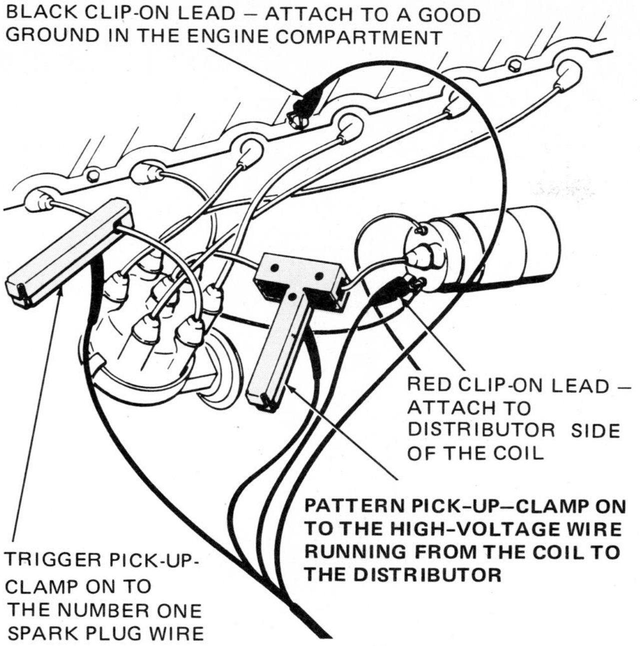
BLACK CLIP-ON LEAD – ATTACH TO A GOOD GROUND IN THE ENGINE COMPARTMENT
RED CLIP-ON LEAD – ATTACH TO DISTRIBUTOR SIDE OF THE COIL
PATTERN PICK-UP—CLAMP ON TO THE HIGH-VOLTAGE WIRE RUNNING FROM THE COIL TO THE DISTRIBUTOR
TRIGGER PICK-UP-CLAMP ON TO THE NUMBER ONE SPARK PLUG WIRE

Fig. 31-8 Test leads are clipped to terminals, and pickup sensors are clamped on high-voltage leads, to test an ignition system. The tester is of the oscilloscope type. (*Autoscan, Inc.*)

sensor is clamped onto the wire that goes to the plug in No. 1 cylinder. The trigger pickup senses when the plug fires. This is the signal to the scope to start another round of traces.

● 31-5 Reading the Patterns

The patterns in Fig. 31-9 show different troubles that occur in the ignition system. The pattern of any cylinder's ignition-circuit voltage shows what voltages are occurring in that circuit. The way that the voltage varies from normal shows you where the electrical problem exists. For example, the scope can detect wide or narrow spark-plug gaps, open spark-plug wires, shorted coils or condensers, arcing contact points, improper contact-point dwell, and so on. Many abnormal engine conditions change the voltage needed to fire the plug. This, too, shows up on the scope. When you work in a shop that has an oscilloscope, you will have instructions on how to use it.

● 31-6 Ignition-Coil Testers

There are three types of ignition-coil testers, the spark gap, the high-frequency, and the oscilloscope. The oscilloscope gives an accurate picture of coil condition (Fig. 31-10).

● 31-7 Condenser Tester

A good condenser tester should be able to test for grounds or shorts. It should also test for insulation resistance, series resistance, and capacity. These conditions all affect ignition performance.

● 31-8 Distributor Tester

Distributor testers are variable-speed devices. They drive the distributor at various speeds so that the operation of the centrifugal advance can be checked (Fig. 31-11). They also may include a dwell meter to measure dwell, or cam angle (Fig. 31-12). In addition, they usually have a vacuum device to test the operation of vacuum advance mechanisms.

● 31-9 Contact-Point-Opening Testers

The opening of the distributor contact points must be correct to assure good ignition performance. Thus, their adjustment is of great importance. There are three methods of testing the amount of contact-point opening. One makes use of a feeler gauge placed between the points. The breaker cam must be positioned so that the lever-arm rubbing block rests on the high point of one cam lobe. This method, although satisfactory for new points, should not be used for worn points. Points that have been used are likely to be rough, even though they are still good for many more miles of service. To test such points, a dial indicator or a dwell meter should be used. The dwell is the number of degrees of cam rotation from the instant the contact points close until they open again (Fig. 31-12). Decreasing the contact-point opening increases the dwell. Increasing the contact-point opening decreases the dwell. Adjustment is made by loosening a locking screw and turning an eccentric. Some contacts are adjusted by loosening a locking nut and turning the contact screw. On the unit shown in Fig. 30-4, the contacts are adjusted without removing the cap. The window can be raised to expose the adjustment screw.

● 31-10 Contact-Pressure Gauge

The contact-point pressure must be within specifications. Low point pressure will allow the points to bounce and burn. High pressure will cause rapid wear of the points, cam, and rubbing block. A spring gauge can be used to measure the spring pressure. Adjustment is made by bending the breaker-lever spring, or by loosening the spring attachment and sliding the spring in or out as required. Tighten the screw and retest the adjustment.

The contact-point sets in many late-model distributors are preadjusted so that it is not necessary to check or adjust spring tension. On these, only the point opening, or dwell, requires checking.

● 31-11 Electronic Ignition Testing

Electronic ignition systems (● 30-20 to 30-28) require special testing procedures and equipment. Figure 31-13 shows the special electronic ignition tester recommended by Chrysler. Its use is simple. You plug the tester into the wiring harness between the distributor connector and the control-unit connector. Then, with the ignition switch turned on, you check the ignition system with the tester. The green lights come on if everything is okay. The red lights come on to signal trouble. When you use the tester, follow the special instructions that explain the testing procedure.

● 31-12 Ignition Timing

The sparks must reach the spark plugs in the cylinders at exactly the right time. They must arrive a specific number of degrees before TDC (top dead center) on the compression stroke. Adjusting the distributor to make the sparks arrive at the right time is called *ignition timing*. On most ignition systems, you adjust the timing by turning the

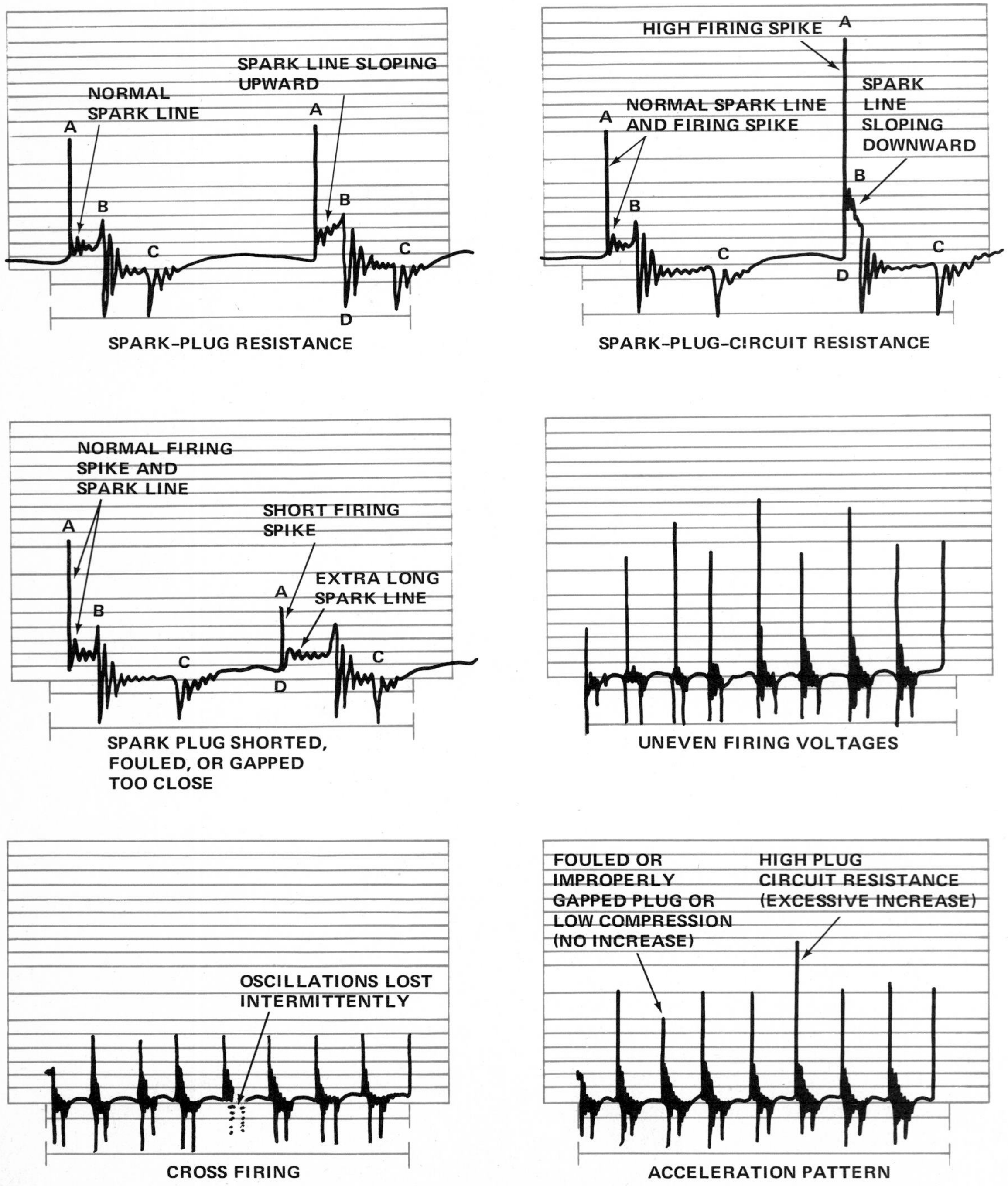

Fig. 31-9 Abnormal traces and their causes. (*Ford Motor Company*)

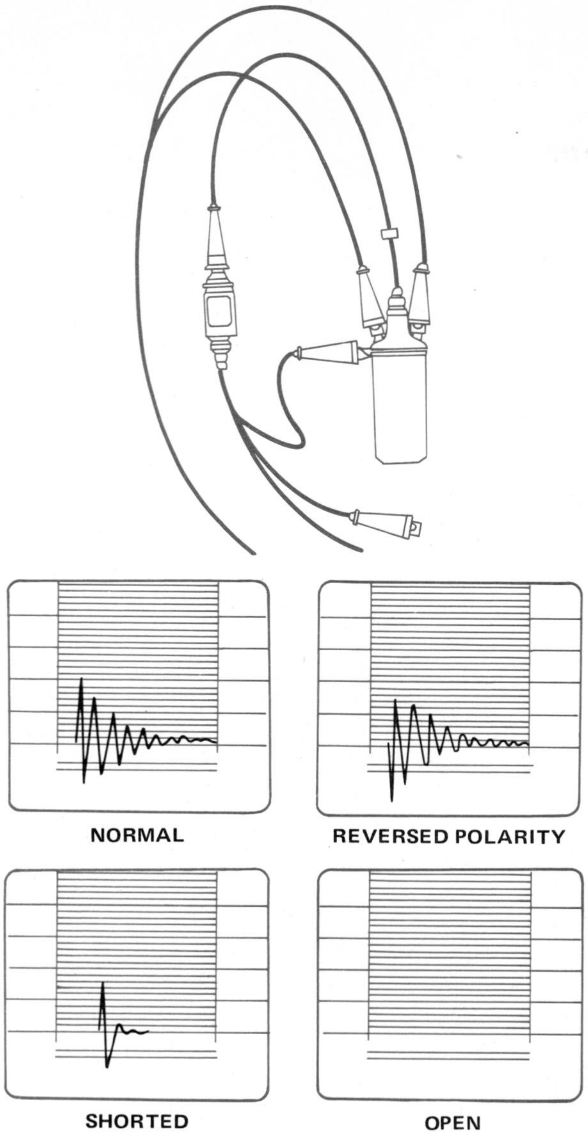

Fig. 31-10 Testing ignition coil with oscilloscope. Top, leads connected to the coil. Bottom, scope patterns for various coil conditions. (*Sun Electric Corporation*)

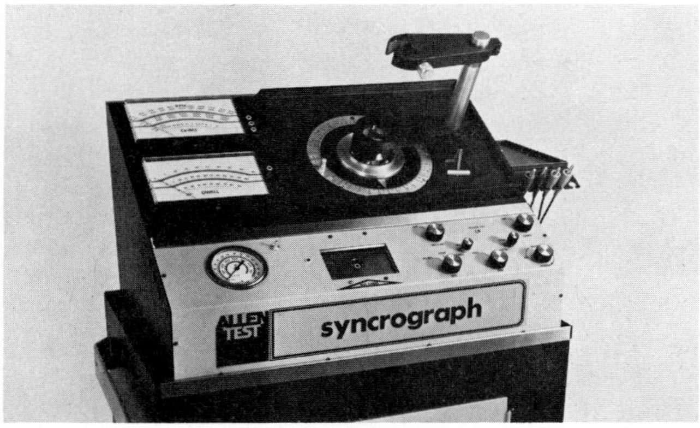

Fig. 31-11 Ignition-distributor tester. (*Allen Testproducts Division, The Allen Group, Inc.*)

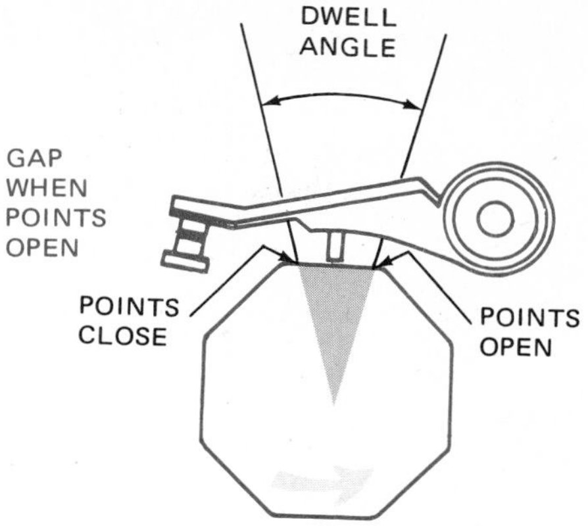

Fig. 31-12 Dwell angle.

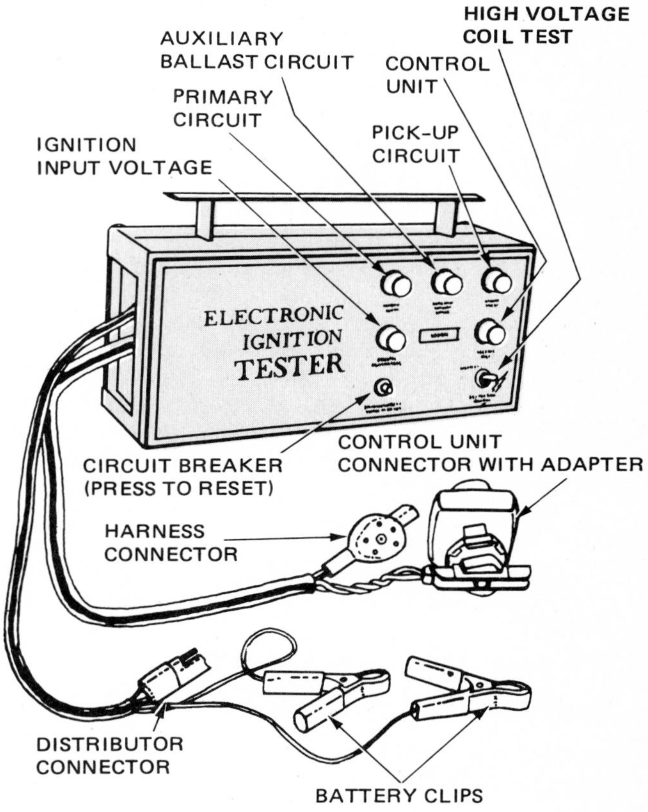

Fig. 31-13 Electronic ignition tester. (*Chrysler Corporation*)

distributor in its mounting. If you rotate the distributor in the direction opposite to normal cam rotation, you move the contact points ahead. That is, the points will close and open earlier. This advances the spark, so the sparks appear at the spark plugs earlier. Turning the distributor in the direction of normal cam rotation retards the sparks. The sparks appear at the plugs later.

To time the Ford EEC and the 1977 General Motors MISAR systems (● 30-27 and 30-28), you change the position of the crankshaft sensor.

To time the ignition, check the markings on the crankshaft pulley with the engine running. Since the pulley turns rapidly, you cannot see the markings in normal light. But by using a special timing light, you can make the pulley appear to stand still. The timing light is a *stroboscopic* light. You use it by clamping an inductive pickup around the cable to the No. 1 spark plug. The inductive pickup senses the voltage surge and triggers the timing light. Thus, every time the plug fires, the timing light gives off a flash of light (Fig. 31-14). The light lasts only a small fraction of a second. The repeated flashes of light make the pulley seem to stand still.

Careful: Do not puncture cables or nipples with test probes. This will ruin the cables. See the **Careful!** in ● 31-14.

To set the ignition timing, loosen the clamp screw that holds the distributor in its mounting. Then turn the distributor one way or the other. As you turn the distributor, the markings on the pulley will move ahead or back. When the timing is correct, the markings will align with a timing pointer, or timing mark, as shown in Fig. 31-15. Tighten the distributor clamp.

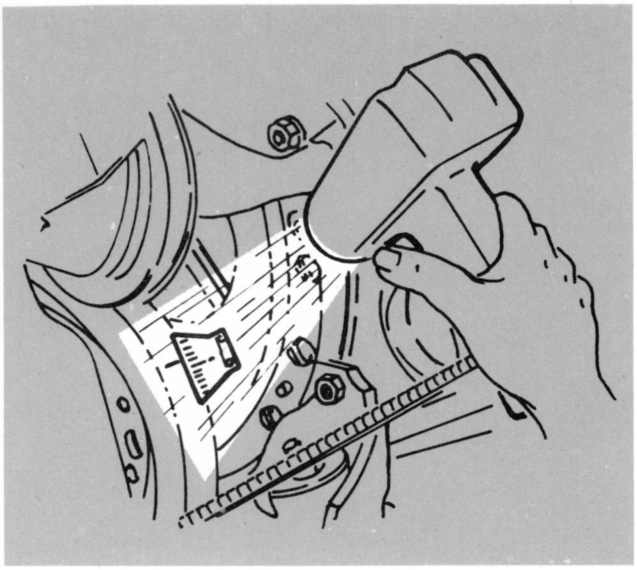

Fig. 31-14 The timing light flashes every time No. 1 spark plug fires.

● 31-13 Spark-Plug Service

Spark plugs will foul or the electrodes will wear rapidly if their heat range is wrong for the engine. See Fig. 30-12, which illustrates spark-plug heat range. Figure 31-16 relates spark-plug appearance to various conditions in the engine. Figure 31-17 shows a spark-plug cleaner. The spark plug is put into the cleaner. The cleaner sends a blast of grit against the electrodes and insulator to clean them. After the cleaning, the spark-plug electrodes should be filed flat with an ignition file. Then a special tool is used to adjust the electrode gap (Fig. 31-18).

The cost of labor is high, and the cost of spark plugs is relatively low. This has caused many service experts to recommend installation of new plugs rather than cleaning and regapping the old plugs. Also, sandblasting roughens the porcelain insulation so that contaminants adhere. This can set up a leakage path which causes the plug to misfire. That is, the high voltage leaks across the porcelain rather than producing a spark at the gap.

Fig. 31-15 Ignition timing marks on the crankshaft pulley.

● 31-14 Ignition Wiring

An important part of ignition service is to inspect the wiring to make sure it is in good condition. Cracks or punctures in the secondary-cable insulation can allow high-tension leakage and engine miss, particularly under heavy load when voltages peak.

Visually inspect the secondary cables for cracks, burned spots caused by being too close to the exhaust manifold, and brittleness. Feel the cables to see if they are hard or crumbly. You can make a secondary-insulation check with the oscilloscope. If you do not have an oscilloscope, you can check secondary-cable insulation as follows: With the engine not running, connect one end of a test probe to a good ground such as the engine block. This leaves the other end with the test point free to probe. Disconnect the cable from a spark plug, and insulate the clip end from ground. Now start the engine, and move the test probe along the entire length of the cable. If there are punctures or cracks, a spark will jump through the insulation to the end of the test point.

Here is the recommended way to install new cable assemblies. Grasp the nipple and clip end of the cable, as shown in Fig. 31-19. Gently push the cable clip into the cap tower. Pinch the larger diameter of the nipple to release trapped air. Then push the cable and nipple until the cable clip is fully entered into the cap terminal, and the nipple is all the way down around the terminal.

If you are replacing a set of ignition cables, replace one cable at a time. This avoids getting mixed up and connecting a cable from the distributor cap to the wrong spark plug. If all cables have been removed, first determine which direction the rotor turns and the firing order. From these, you will be able to figure out how the cables are to be connected.

Careful: Never remove cable and nipple assemblies from the distributor or coil towers unless (1) the nipples are damaged, or (2) cable testing shows the cables are bad and must be replaced. You can ruin a cable by careless removal and installation.

Do not puncture cables or nipples with test probes. Puncturing cable insulation or a nipple can ruin the cable. The probe can separate the conductor and cause high resistance. Also, breaking the insulation can result in high-voltage leakage to ground. Either of these can cause engine miss.

● 31-15 Causes of Ignition Failure

Most often, ignition failure results from normal wear of the contact points, spark plugs, and other components. Ignition failure can be classified under three headings:

NORMAL

Brown to grayish tan color and slight electrode wear. Correct heat range for engine and operating conditions.

RECOMMENDATION: Properly service and reinstall. Replace if over 10,000 miles of service.

SPLASHED DEPOSITS

Spotted deposits. Occurs shortly after long-delayed tune-up. After a long period of misfiring, deposits may be loosened when normal combustion temperatures are restored by tune-up. During a high-speed run, these materials shed off the piston and head and are thrown against the hot insulator.

RECOMMENDATION: Clean and service the plugs properly and reinstall.

CARBON DEPOSITS

Dry soot.

RECOMMENDATION: Dry deposits indicate rich mixture or weak ignition. Check for clogged air cleaner, high float level, sticky choke,or worn breaker contacts. Hotter plugs will temporarily provide additional fouling protection.

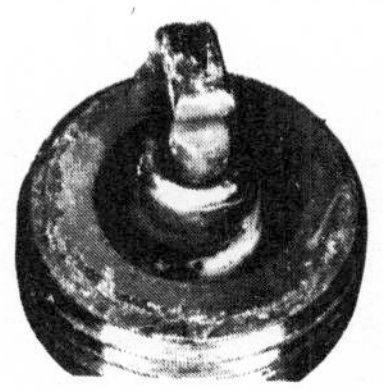

HIGH-SPEED GLAZING

Insulator has yellowish, varnish-like color. Indicates combustion chamber temperatures have risen suddenly during hard, fast acceleration. Normal deposits do not get a chance to blow off, instead they melt to form a conductive coating.

RECOMMENDATION: If condition recurs, use plug type one step colder.

OIL DEPOSITS

Oily coating.

RECOMMENDATION: Caused by poor oil control. Oil is leaking past worn valve guides or piston rings into the combustion chamber. Hotter spark plug may temporarily relieve problem, but positive cure is to correct the condition with necessary repairs.

MODIFIER DEPOSITS

Powdery white or yellow deposits that build up on shell, insulator, and electrodes. This is a normal appearance with certain branded fuels. These materials are used to modify the chemical nature of the deposits to lessen misfire tendencies.

RECOMMENDATION: Plugs can be cleaned or, if replaced, use same heat range.

TOO HOT

Blistered, white insulator, eroded electrodes and absence of deposits.

RECOMMENDATION: Check for correct plug heat range, overadvanced ignition timing, cooling system level and/or stoppages, lean air-fuel mixtures, leaking intake manifold, sticking valves, and if car is driven at high speeds most of the time.

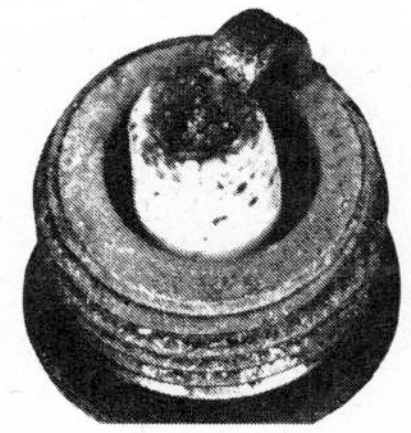

PREIGNITION

Melted electrodes. Center electrode generally melts first and ground electrode follows. Normally, insulators are white, but may be dirty due to misfiring or flying debris in combustion chamber

RECOMMENDATION: Check for correct plug heat range, overadvanced ignition timing, lean fuel mixtures, clogged cooling system, leaking intake manifold, and lack of lubrication.

Fig. 31-16 Appearance of spark plugs related to causes. (*Ford Motor Company*)

1. Loss of energy in the primary circuit. This, in turn, may be caused by several conditions.
   a. Resistance in the primary circuit due to defective leads, bad connections, burned distributor contact points or switch, or open coil primary
   b. Points not properly set
   c. Discharged battery or defective alternator
   d. Defective condenser (shorted, low insulation resistance, high series resistance)
   e. Grounded primary circuit in coil, wiring, or distributor
2. Loss of energy in the secondary circuit.
   a. Plugs fouled, broken, or out of adjustment
   b. Defective high-voltage wiring, which allows high-voltage leaks
   c. High-voltage leakage across coil head, distributor cap, or rotor
   d. Defective connections in high-voltage circuits
   e. Defective ignition coil
3. Out of time.
   a. Timing not set properly
   b. Distributor bearing or shaft worn, or shaft bent
   c. Vacuum advance defective
   d. Centrifugal advance defective
   e. Preignition, due to plugs of wrong heat range, fouled plugs, etc.

Fig. 31-17 Spark-plug cleaner. *(AC Spark Plug Division of General Motors Corporation)*

● 31-16 Quick Checks of Ignition System

A number of quick checks can be made to determine whether the ignition is at fault if the engine does not operate normally. The following applies mainly to the contact-point-type ignition system. Some of the items listed may apply to electronic ignition systems. But, generally speaking, the electronic systems should be checked in the special way outlined in the manufacturer's shop manuals.

1. Engine does not run If the engine cranks at normal speed but does not start, the trouble could be in either the ignition or the fuel system. To check the ignition system, disconnect the lead from one spark plug (or from the center distributor-cap terminal). Use insulated pliers to hold it about $\frac{3}{16}$ inch [4.76 mm] from the engine block. Crank the engine. If a good spark occurs, chances are that the ignition system is in reasonable condition (although the timing could be off). If no spark occurs, check the ignition system further.

CAUTION: Do not try this with the General Motors high-energy ignition system. The 35,000 volts this system can produce can jump considerable distances and can give you a very painful shock.

Connect a test ammeter into the ignition-coil primary circuit, and watch it while cranking the engine. If there is a small, steady reading that fluctuates somewhat, the primary circuit is probably all right. The trouble is probably a defective coil secondary or secondary leads, a defective condenser, or high-voltage leakage across the cap, rotor, or coil head.

If the ammeter shows a fairly high and steady reading:
a. The contact points are out of adjustment.
b. The condenser is shorted.
c. The coil primary circuit is grounded.

If there is no ammeter reading, the primary circuit is open. This could be due to out-of-adjustment contact points, a loose connection, defective wiring or switch, or an open coil primary winding.

2. Engine misses Missing is caused by such defects in the ignition system as:
a. Worn or out-of-adjustment contact points
b. Defective condenser
c. Centrifugal or vacuum advance malfunctioning
d. Defective secondary wiring
e. Defective ignition coil
f. Poor connections
g. High-voltage leakage across ignition-coil head, rotor, or cap
h. Defective spark plugs

The wrong ignition coil for the engine, or reversed connections to the ignition coil, may also cause missing. Putting a battery in backwards can also cause missing, because this also reverses the polarity of the coil.

Careful: If a battery is put in backwards in a car with an electronic ignition system, the system will probably burn

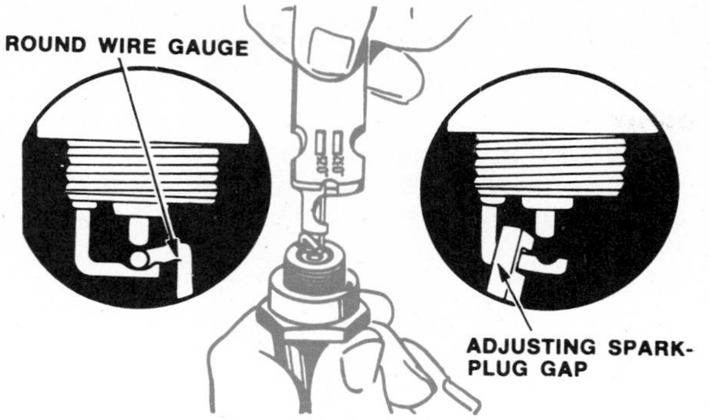

Fig. 31-18 Using a special gauge and adjusting tool to adjust plug gap.

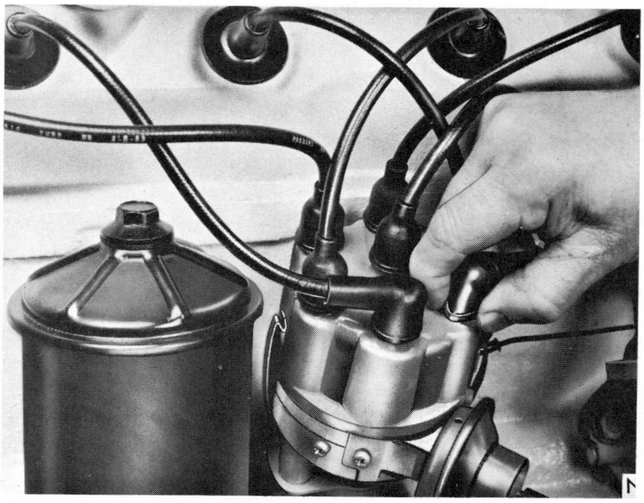

Fig. 31-19 Installing cable and nipple on distributor tower. *(Chrysler Corporation)*

up. A high current will flow and, unless there is a fuse link that will blow, the current can ruin the electronic control module. Also, a battery that is connected backwards in a system using an alternator can burn up the alternator.

With reversed coil polarity, the electrons have to jump from the relatively cool outer electrode to the center electrode. This requires a considerably higher secondary voltage. It increases the possibility of the engine missing, especially at high speed. Normally, the coil and battery are connected so that electrons jump from the hot center electrode to the outer electrode. With the emitting electrode hot, the electrons can jump the spark-plug gap more easily. Voltage requirements are considerably lower. Coil terminals are usually marked to prevent incorrect connections.

Reversed polarity can be easily detected with the oscilloscope. If the polarity is reversed, the pattern will be upside down. If an oscilloscope is not available, the reversed-polarity test can be made as follows. Hold an ordinary pencil tip between the high-voltage wire clip and the spark-plug terminal (Fig. 31-20). The spark should flare out between the pencil tip and the spark plug, as shown. If it flares out between the pencil tip and the wire clip, the polarity is reversed. Another test uses a neon bulb (NE-2 or similar) between the spark-plug terminal (high-voltage lead connected) and ground. With the engine running, the electrode in the neon bulb connected to the spark-plug terminal should glow. If the terminal connected to ground glows, the polarity is reversed. Reverse the coil primary leads to correct the polarity.

3. Overheating and spark These conditions may be caused by improper ignition timing.

● 31-17 Ignition Service

With the oscilloscope, you can fully check the operating condition of the ignition system. The scope face shows the pattern of voltages in the ignition system. Any variation from the normal pattern indicates trouble. The way the pattern varies from normal indicates the type and source of trouble.

In addition, a mechanical check should be made, to determine the condition of the battery and cables, ignition coil, distributor, ignition wiring, and spark plugs. Check ignition wiring as explained in ● 31-14. The distributor cap and rotor, and the ignition-coil head, should be examined for cracks or carbonized paths that could allow high-voltage leakage. The distributor contact points should be examined, checked, and adjusted as necessary. If the points are burned, pitted, or oxidized, they should be replaced. They can be cleaned in an emergency with a fine-cut contact file, but most technicians prefer to install new points. Emery cloth must never be used to clean points. Particles of emery will embed and cause rapid burning away of the points. Distributors requiring lubrication should be lubricated, as explained in the factory shop manual, at periodic intervals. See also ● 31-19.

Summary of ignition troubles Various troubles that may occur in the components of the ignition system are discussed below. Once again, let us emphasize that the items listed below apply mostly to the contact-point ignition systems.

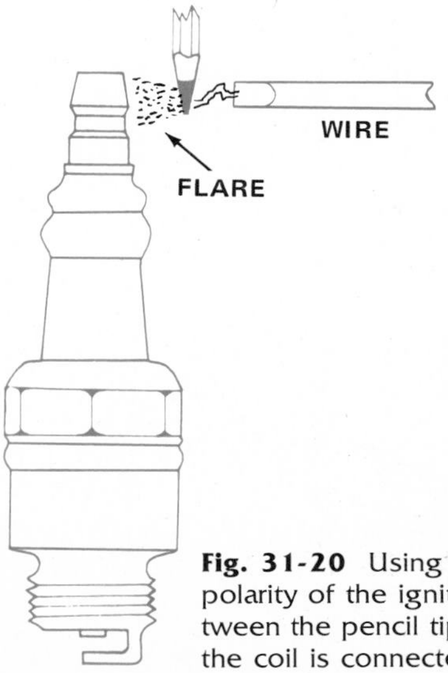

Fig. 31-20 Using a pencil tip to check the polarity of the ignition coil. If the flare is between the pencil tip and the plug, as shown, the coil is connected correctly.

1. Burned or oxidized contact points. It is normal for ignition-distributor contact points gradually to burn away over a long period of time. Rapid burning or oxidizing of the points may be due to several conditions.
   a. Excessive resistance in the condenser circuit caused by high series resistance in the condenser or by a loose mounting or connection.
   b. High voltage produced by a high-voltage-regulator setting.
   c. Dwell too large (point opening too small). The points, closed too much of the total operating time, burn away rapidly.
   d. Weak spring tension that causes the contact points to bounce and arc.
   e. Oil or crankcase vapors entering the distributor housing are deposited on the point surfaces, causing them to burn rapidly. A glance at the breaker plate usually discloses this condition. It causes a black smudge on the breaker plate under the points. Clogged engine crankcase PCV valves or hoses and worn distributor bearings can produce this trouble.
2. Sooty, burned, or cracked spark-plug insulator. Spark plugs may fail for a variety of reasons. Spark-plug manufacturers usually recommend replacement of spark plugs at 10,000-mile [16,090-km] intervals. This will help prevent failure and maintain the engine at good operating efficiency. One cause of spark-plug trouble is the installation of plugs of the wrong heat range. See ● 31-13. Cracked insulators are usually caused by careless installation of the plug or by careless adjustment of the plug gap.

● 31-18 Distributor Disassembly and Reassembly

Instead of disassembling, servicing, and reassembling distributors, many shops prefer to turn the old units in for new units. Installing a new or properly rebuilt distributor relieves the shop of the rebuilding chore. The *Workbook for Automotive Mechanics* has a step-by-step servicing procedure on a distributor. Also, the manufacturer's shop manuals cover this job in detail.

Disassembly procedures vary with different distributor designs. A typical procedure starts with removal of the cap, rotor, terminal parts, and breaker plate. The coupling or gear is taken off after the head of the pin has been filed off, and the pin driven out. This permits removal of the shaft from the housing.

Reassembly is the reverse of disassembly. Replace the bearing in its housing if it is worn, that is, if it allows excessive side play or wobble of the shaft. The old bearing can be pressed out, and the new one pressed in, with an arbor press. Some bearings will require reaming to size after installation. When installing the coupling or the gear, add or remove shims (between the coupling or gear and the housing), in order to get the correct shaft end play. Then peen over the end of the pin.

● 31-19 Distributor Lubrication

Many distributors have built-in lubrication and need no further lubrication. This is not so for the contact-point types. Many of these have a cam lubricator, as shown in Fig. 31-21. On these, the specifications call for turning the lubricator 180 degrees every 12,000 miles [19,308 km], and replacing it every 24,000 miles. Some Ford distributors are equipped with an oil cup which should have a few drops of SAE10W oil periodically.

● 31-20 Distributor Removal and Replacement

Distributor removal and replacement is a simple job if the engine is left undisturbed while the distributor is out. However, if the engine is cranked so that crankshaft and camshaft are turned with the distributor out, then the replacement job is a little more complicated.

1. Distributor removal Remove the air cleaner, and disconnect the vacuum hose or hoses from the distributor. Disconnect the primary lead running from the ignition coil to the distributor. Remove the distributor cap, and push the cap and wire assembly aside.

Scratch a mark on the distributor housing. Scratch another mark, which lines up with the first, on the engine block. These marks locate the position of the distributor housing in the block. Scratch a third mark on the distributor housing exactly under the rotor tip. This mark locates the position of the rotor in the housing.

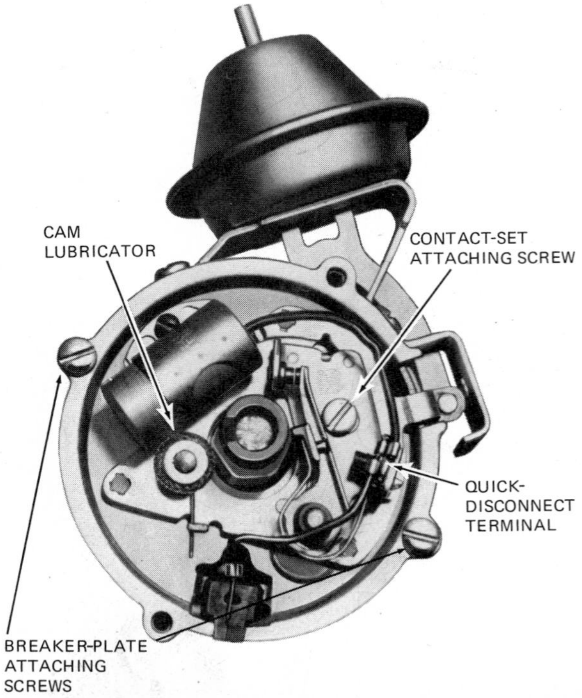

Fig. 31-21 Ignition distributor with cap and rotor removed to show the location of the cam lubricator. (*Delco-Remy Division of General Motors Corporation*)

Remove the distributor hold-down bolt and clamp. Lift the distributor out of the block.

If the engine is not cranked while the distributor is out, the distributor can be easily installed in the correct position. Simply align the marks on the distributor housing and cylinder block. Note, however, that as you push the distributor down into place, the shaft and rotor will turn as the spiral gears mesh. Therefore, you will have to start with the rotor turned back from the installed position so that when you install the distributor, the rotor will turn into the proper position. That is, the tip will line up with the mark on the distributor housing. When correctly installed, the tip will point to the cap terminal connected to the spark plug in No. 1 cylinder.

2. Distributor installation If the engine has been cranked with the distributor out, timing has been lost. The engine must be retimed. This is necessary to establish the proper relationship between the distributor rotor and the No. 1 piston.

Remove No. 1 spark plug from the cylinder head. Place a shop towel over your finger, and cover the spark-plug hole. Crank the engine until you feel compression pressure on your finger.

Bump the engine with the starting motor until the timing marks on the crankshaft pulley and timing cover are aligned. This means that No. 1 piston is in firing position.

Now, the distributor can be installed in the cylinder block. Make sure to align the marks you made on the distributor housing and cylinder block. Check to make sure that the distributor gasket or rubber O ring is in place when you install the distributor.

Three different distributor drives are shown in Fig. 31-22. You may have to turn the rotor slightly to engage the drive.

Make sure the distributor housing is fully seated against the cylinder block. If it is not, the oil-pump shaft is not engaging. Hold the distributor down firmly, and bump the engine a few times until the distributor housing drops into place. Then bump the engine again to realign the timing marks.

Install, but do not tighten, the distributor clamp and bolt. Rotate the distributor until the contact points just start to open to fire No. 1 cylinder. Hold the distributor cap in place above the distributor. Make sure that the rotor tip lines up with No. 1 terminal on the cap. Install the cap with wires. Connect the primary wire from the ignition coil to the distributor.

Start the engine. Set the ignition timing (● 31-12). Connect the vacuum hose or hoses to the distributor. Replace the air cleaner.

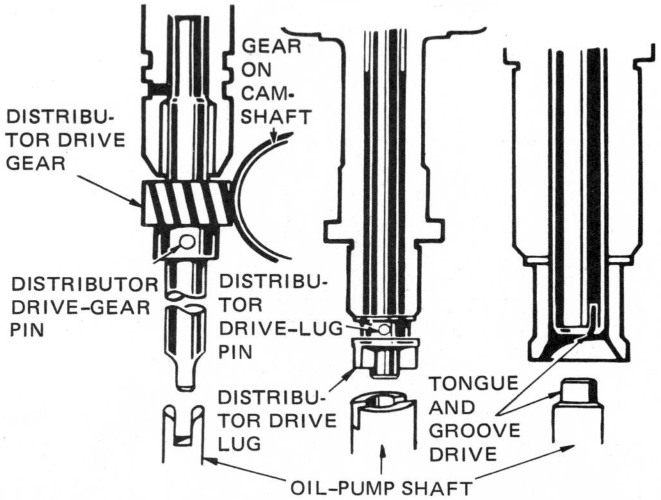

Fig. 31-22 Three distributor-drive methods.

REVIEW QUESTIONS

1. What is a parade pattern? A stacked pattern?
2. What is contact-point dwell?
3. How is contact-point dwell adjusted?
4. Considering possible causes of ignition trouble, name four conditions that could cause loss of energy in the primary circuit.
5. Explain how to check the ignition system with the spark test when the engine will crank but not start.
6. Explain how to check the ignition system to determine if the ignition coil has been connected backwards.
7. Explain how to adjust ignition timing on an engine.
8. How does the timing light work?
9. Explain how to clean and adjust a spark plug.
10. Explain how to check secondary-wiring insulation with a test probe.
11. Describe the proper way to install a cable and nipple.
12. Why must you not puncture cables or nipples with test probes?
13. What is the recommended service procedure for contact-point distributors having cam lubricators?
14. What is the usual cause of cracked spark-plug insulators?

SELF PROJECT

It is important for you to know how to use different kinds of ignition-system testing instruments. Here is one way to learn more about them. Get hold of the instruction sheet or booklet that accompanies each testing instrument in the shop. For example, the oscilloscope has a comprehensive instruction booklet. Study these booklets. Write brief summaries of how each test instrument is used. Note especially the cautions and the various test results, along with their meanings. File all your summaries in your notebook. You now have a permanent record that tells you how to use the different test instruments. Furthermore, because you wrote these summaries yourself, you will remember the instructions better.

CHAPTER 32
Other Electric and Electronic Units

After you have studied this chapter, you should be able to:

- **Explain the purpose of fuses, circuit breakers, and fuse links.**
- **Explain how to aim headlights and, under the supervision of an instructor, do this job.**
- **Explain the operation of horns and horn relays.**

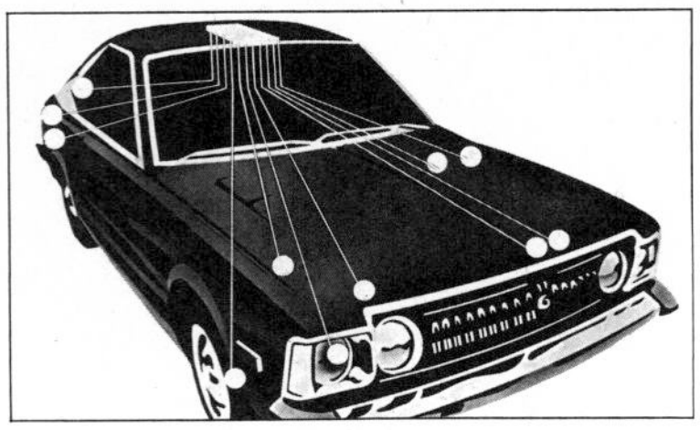

We covered the major units in the electrical and electronic systems in earlier chapters. These include the battery, alternator, starting motor, contact-point and electronic ignition systems, and electronic fuel-injection systems. Now, in this chapter, we look at other units in the electrical and electronic systems. These include the lights, horn, windshield wipers, window and seat adjusters, automatic level controls, and brake antilock system.

32-1 Wiring Circuits

The electric units in the automobile are connected by wires of different sizes. The size of each wire depends on the amount of current the wire must carry. The heavier the current, the larger the wire must be. The wires are gathered together to form wiring harnesses. Each wire is identified by the color of its insulation. For example, wires are light green, dark green, blue, red, black with a white tracer, and so on. The car manufacturers' shop manuals have illustrations that show the various wires and their colors. If you ever have to trace a particular wire, refer to the shop manual to determine its color. Figure 32-1 shows the wiring diagram for the front lighting and engine compartment for one car. Figure 32-2 shows the wiring harness in the instrument panel for another. Note that each wire has its own distinctive color.

32-2 Printed Circuits

The instrument panel has a number of indicating devices, switches, and controls (Fig. 32-3). Because the panel is crowded, there can be problems in making connections between the instruments. One solution is the use of printed circuits.

A printed circuit is a flat piece of insulating material on which a series of conduction strips are printed. Figure 32-4 shows part of a printed circuit. When a printed circuit is installed on the instrument panel, the conduction strips carry current between the units. For example, when indicator lamps are installed, the contacts on the lamps rest on the metallic strips to complete the circuit. Likewise, when a switch is installed, its contact terminals connect to the eight strips feeding into the switch. Figure 32-5 shows how a printed circuit is installed behind the instrument panel.

32-3 Fuses

Fuses, fusible links, and circuit breakers are installed in circuits to protect the electrical devices in the circuits. Their purpose is to open the circuit in case a short or ground develops and dangerously high currents start to flow. If this should happen, the fuse "blows" or the circuit breaker opens.

A typical "old-style" cartridge fuse is shown partly cut away in Fig. 32-6. It contains a soft metal strip, connected at the ends to the fuse caps. It is connected in series in the circuit. All current in the circuit flows through the fuse. If excessive current flows, the metal strip overheats and melts or "blows," thus opening the circuit. This protects the rest of the circuit and connected electrical units from damage. When a fuse blows, the circuit should be checked to see what caused it. Then, after the trouble is fixed, a new fuse should be installed. Figure 32-7 shows the type of fuse block that takes the cartridge fuse.

Instead of the cartridge fuse, shown in Fig. 32-6, many cars today use a U-shaped fuse (Fig. 32-8). These were engineered for compactness and ease of service. The type of fuse block in which these fuses are installed is shown in Fig. 32-9.

32-4 Circuit Breakers

Circuit breakers are used in some circuits, such as the headlight circuit. The circuit breaker has a small winding

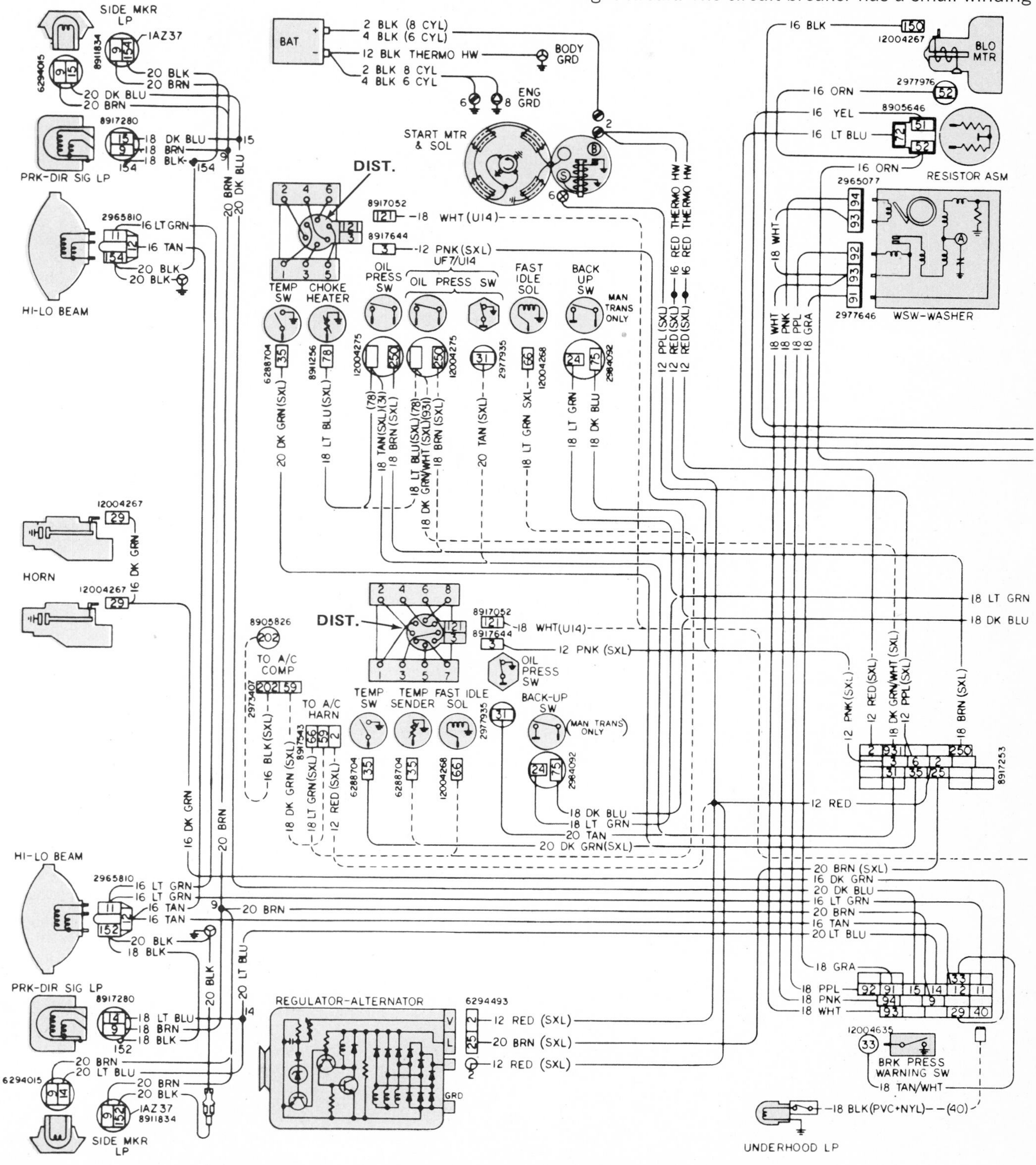

Fig. 32-1 Front-lighting and engine-compartment wiring for V-6 and V-8 engines. *(Chevrolet Motor Division of General Motors Corporation)*

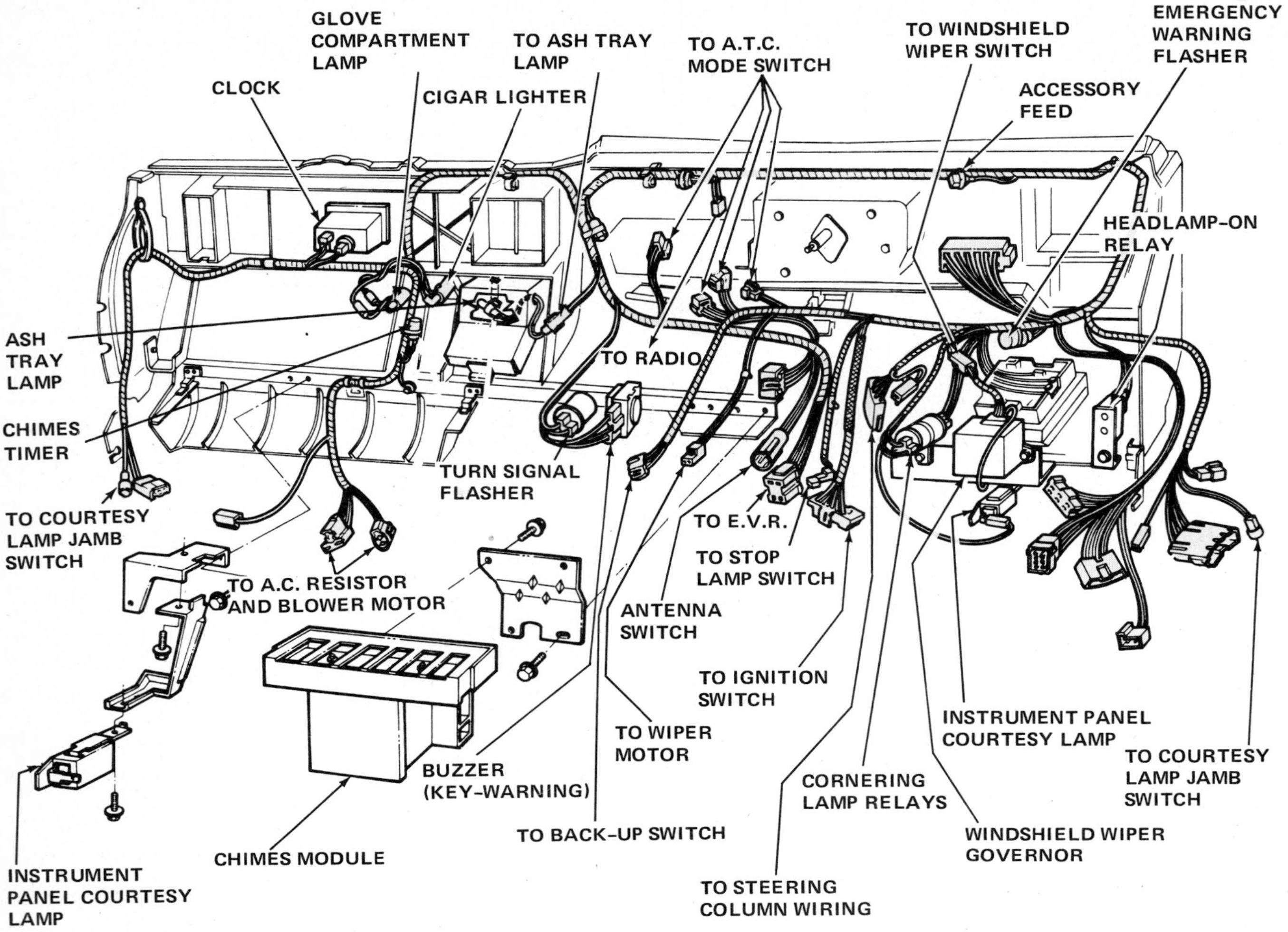

Fig. 32-2 Instrument-panel wiring harness for some models. (*Ford Motor Company*)

that carries the current in the circuit. When the current is too high, the winding magnetism opens points to open the circuit. The advantage of the circuit breaker is that it keeps resetting itself. Thus it gives a warning of trouble but does not completely kill the circuit. For instance, if excessive current starts to flow in the headlight circuit, the circuit breaker will operate. The lights will come on and off, warning the driver of trouble. The flashing light gives the driver enough time to pull over to the side of the road and stop.

● 32-5 Fusible Link

For added protection, many cars have fusible links in the insulated battery cable and in the larger high-current-carrying wires. The fusible link is simply a wire several gauges smaller than the wire it is protecting. If a short or ground occurs, the fusible link will burn in two before the larger wire. As a result the other parts of the circuit will not be damaged. See ● 28-10 for a more detailed discussion, with pictures, of fusible links.

● 32-6 Headlights

A typical lighting system for a car is shown in Fig. 32-10. The complete system includes headlights, parking lights, turn signals, side marker lights, stoplights, backup lights, taillights, and interior lights. The interior lights include instrument-panel lights, various warning and indicator lights, and courtesy lights that turn on when a car door is opened.

Figure 32-11 shows a headlight. It has a reflector and a filament at the back, and a lens at the front. When the filament is connected to the battery through the light switch, current flows through the filament. It glows white hot. The light is concentrated by the reflector into a forward beam and is focused by the lens.

Headlights are made in two types and four sizes (two round and two rectangular, see Fig. 32-12). The round sizes are $5\frac{3}{4}$ inches [146.05 mm] in diameter and 7 inches [177.80 mm] in diameter. The rectangular sizes are 4 by $6\frac{1}{2}$ inches [100 by 165 mm] and 5.6 by 9.7 inches [142 by 200 mm]. Both sizes are identified by the number 1 or 2 molded into the glass at the top of the

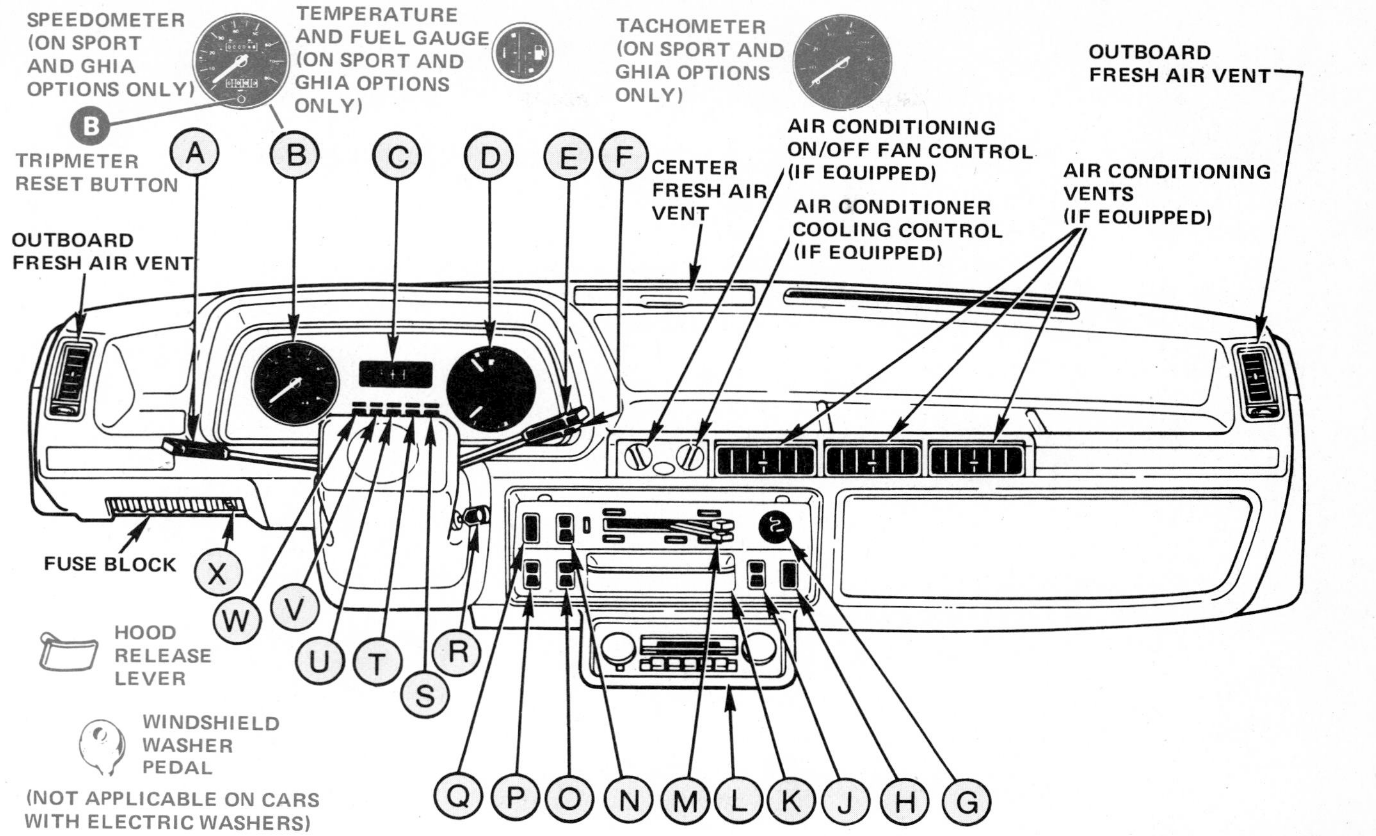

A—Multifunction Switch
B—Speedometer
C—Shift Pattern
D—Fuel/Temperature Guages
E—Windshield Wash-Wipe Switch
F—Headlight Switch
G—Cigarette Lighter
H—Seat Belt Warning Light
J—Hazard Flasher Switch
K—Ashtray
L—Radio
M—Heater Controls
N—Heater Fan Switch
O—Tailgate Wash/Wipe Switch
P—Heated Rear Window Switch
Q—Plug
R—Ignition Switch
S—Charging System Warning Light
T—High Beam Warning Light
U—Turn Signal Warning Light
V—Parking Brake Warning Light
W—Oil Pressure Warning Light
X—Rheostat Switch

Fig. 32-3 Instrument panel for some models. (*Ford Motor Company*)

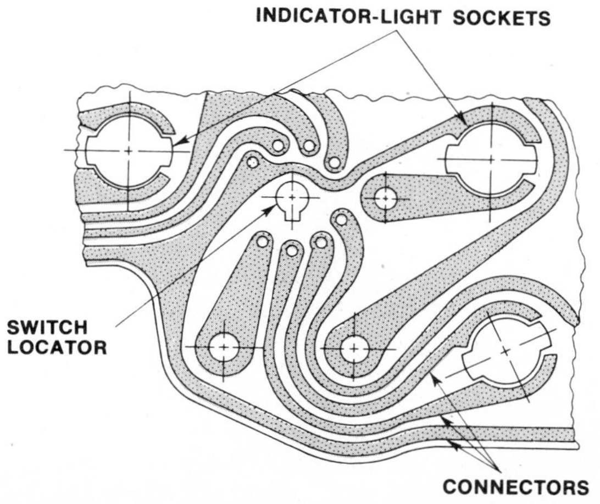

Fig. 32-4 Part of a printed circuit.

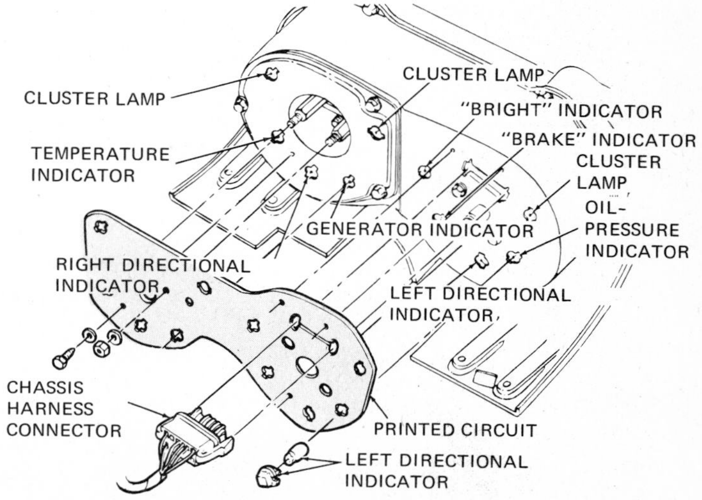

Fig. 32-5 Instrument cluster assembly with printed circuit. (*Chevrolet Motor Division of General Motors Corporation*)

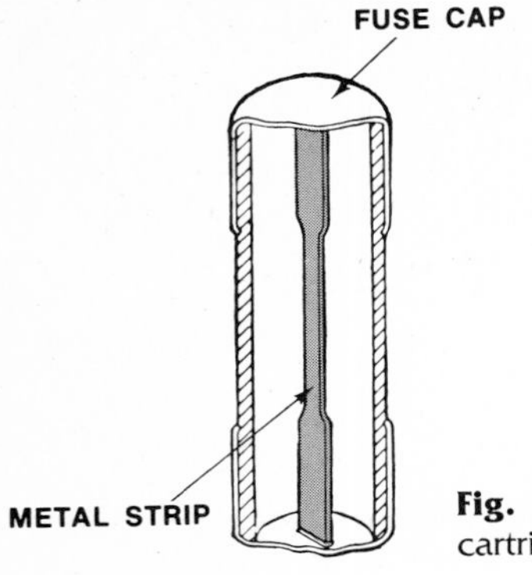

Fig. 32-6 Sectional view of a cartridge fuse.

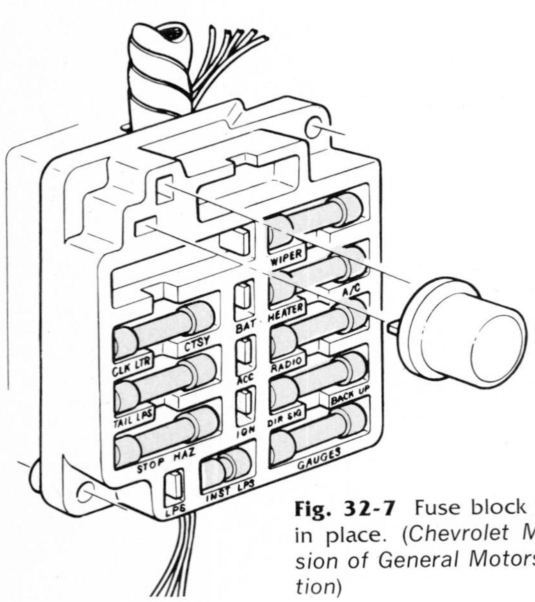

Fig. 32-7 Fuse block with fuses in place. (*Chevrolet Motor Division of General Motors Corporation*)

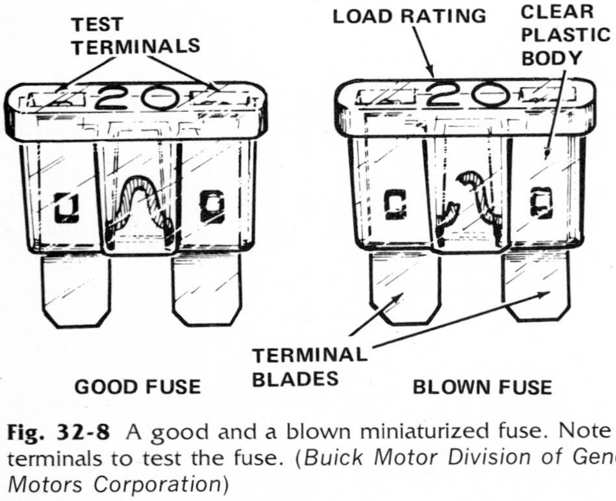

Fig. 32-8 A good and a blown miniaturized fuse. Note the terminals to test the fuse. (*Buick Motor Division of General Motors Corporation*)

lens. Type 1 has only one filament. Type 2 has two filaments, one for the high beam and the other for the low beam. The high beam is for driving on the highway when there is no car approaching from the other direction. The low beam is for city driving and for passing a car coming in the opposite direction. The use of the low beam in passing prevents the oncoming driver from being temporarily blinded by the high beam.

The rectangular headlights have become increasingly popular in recent years. They fit the tapered front-end style of the modern car.

Some cars have only one pair of headlights. These are type 2. Other cars have two pairs of headlights—one pair of type 1 and one pair of type 2.

The driver uses the dimmer switch to select the filaments that will glow. For example, on a car having only one pair of headlights (type 2), the driver operates the dimmer switch to select either the high or the low beam. On a car having two sets of headlights (one set of type 1 and one set of type 2), the arrangement is different. When the driver operates the dimmer switch for low-beam driving, one of the filaments in the type-2 lights comes on. When drivers change the dimmer switch to high beam, the other filament of the type-2 lights comes on. At the same time, the single filament of type-1 lights comes on.

The backup lights come on when the driver shifts into reverse. This closes a switch, linked to the selector lever, which connects the backup lights to the battery.

Blinker lights provide a means of signaling when a car is stalled on the highway or has pulled off to the side. The blinking is much more noticeable than a steady light and provides a warning to approaching cars.

32-7 Headlight Cover

Some cars have vacuum-operated headlight covers that move upward to expose the headlights when they are turned on. The system lowers the covers when the headlights are turned off. Figure 32-13 shows one such system. There is a separate vacuum motor, or actuator, at each headlight, linked to a cover. In operation, pulling the light switch all the way out to turn on the headlights operates a distribution valve. The valve is mounted on the back end of the light switch. It directs vacuum to the two vacuum motors. Vacuum applied to the diaphragms in the motors causes them to move and thus lift the headlight covers. The distribution valve also has an opening through which the atmosphere side of the motor diaphragms is vented. Thus, atmospheric pressure is always applied to this side of the diaphragms.

When the headlights are turned off, the distribution valve allows atmospheric pressure to enter the vacuum side of the vacuum motors. Now, springs on the headlight covers cause them to drop to cover the headlights.

A vacuum reservoir holds sufficient vacuum for several cover operations. It is used if the headlights are turned on and off when the engine is not running. Also, the headlight covers can be operated manually in case the vacuum system fails. This is done by turning on the headlights and then lifting the covers by hand.

Fig. 32-9 Fuse panel using the miniaturized fuses. (*Chevrolet Division of General Motors Corporation*)

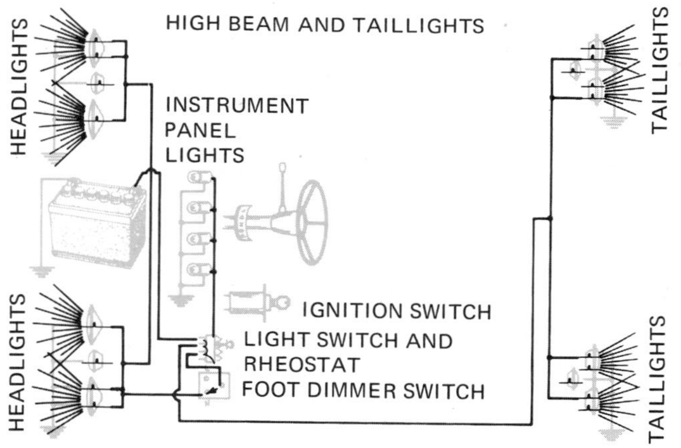

Fig. 32-10 Typical lighting system.

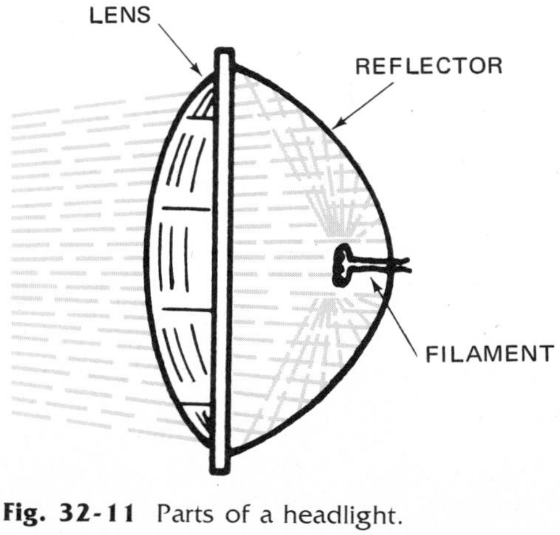

Fig. 32-11 Parts of a headlight.

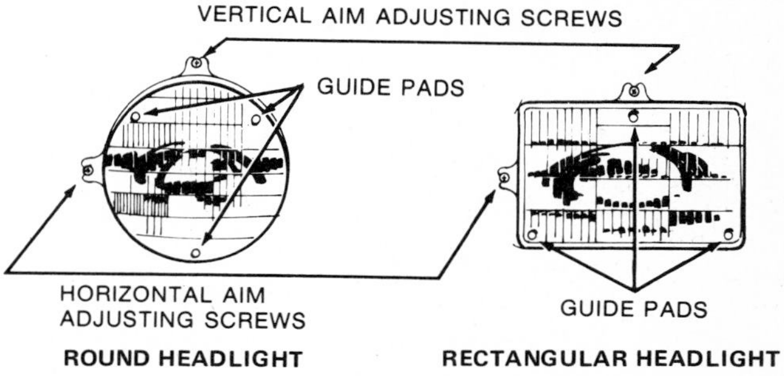

Fig. 32-12 Shapes of round and rectangular headlights.

● 32-8 Headlight Aiming

It is important for the headlights to be aimed correctly. If they are aimed too high or to the left, they might blind an oncoming driver and cause a serious accident. Incorrect aiming can also reduce the driver's ability to see the road properly. This could also lead to an accident.

Headlights have three aiming buttons on the front lens. Adjustments are made by turning spring-loaded screws. There is one spring-loaded screw at the top for up-and-down adjustment, and one at the side for left-to-right adjustment (Fig. 32-14).

There are several methods of checking the aiming of headlights. The simplest method uses a screen set 25 feet [7.60 m] in front of the vehicle and a perfectly level floor. With the car aligned perpendicular to the screen, the low beam and high beam are checked separately (Fig. 32-15). Note that the centerlines of the lights and the

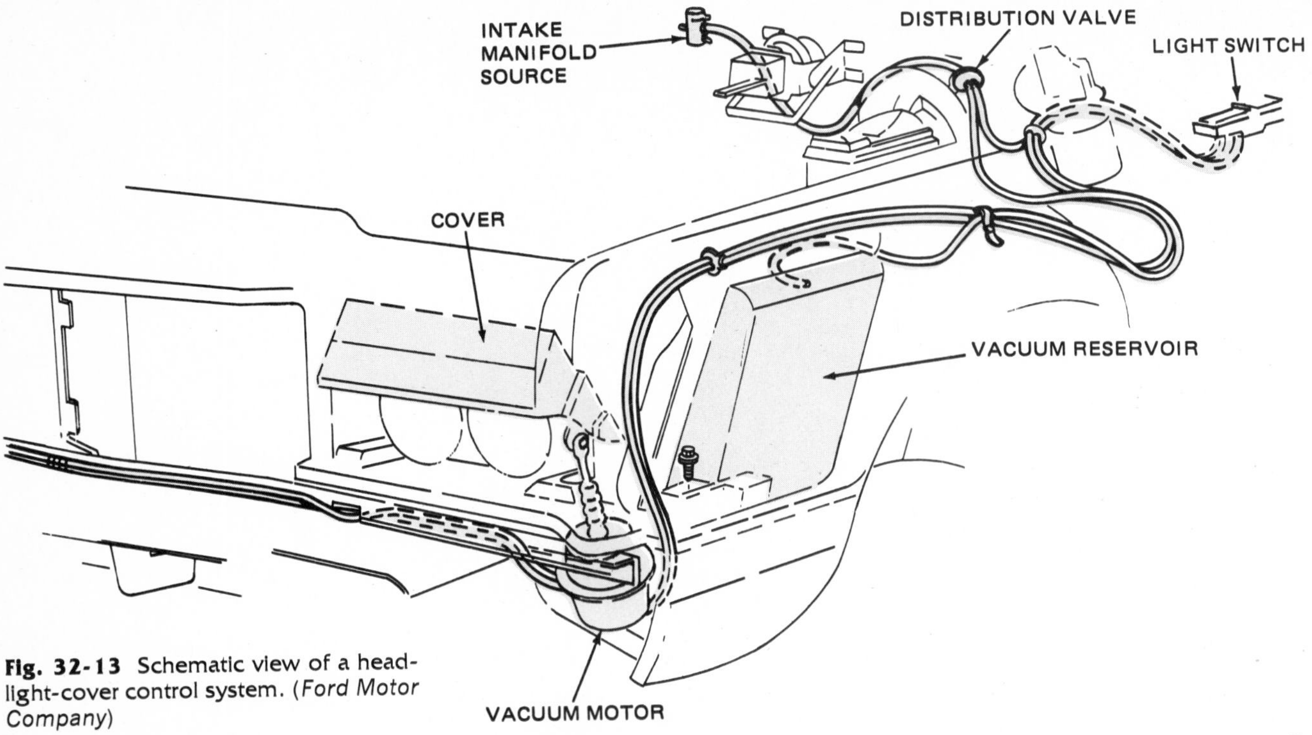

Fig. 32-13 Schematic view of a headlight-cover control system. (*Ford Motor Company*)

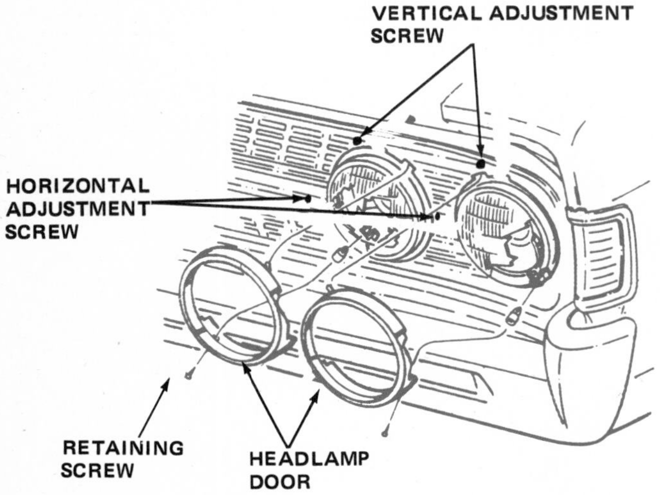

Fig. 32-14 Locations of headlight-aiming screws. (*Ford Motor Company*)

centerline of the car are indicated. Also shown are the hot-spot (or high-intensity area) reference lines. The low beams are adjusted so that the hot spot is 2 inches [50.80 mm] to the right of the reference lines.

Some manufacturers, in their aiming instructions, call for a full fuel tank and an empty car. Others call for a partly full tank and two people in the front seat. Tires must be inflated to the specified pressure. Just before checking the aim, after the car has been positioned, bounce each corner of the car a couple of times to equalize the suspension system.

A variety of headlight-aiming instruments are available. You should learn to use as many different kinds as you can. Figure 32-16 shows the type of instrument recommended by Chrysler. Note that there are two of them, installed by vacuum cups onto the two front headlights. Adapters must be used to install them on the two different size round headlights and the rectangular headlights

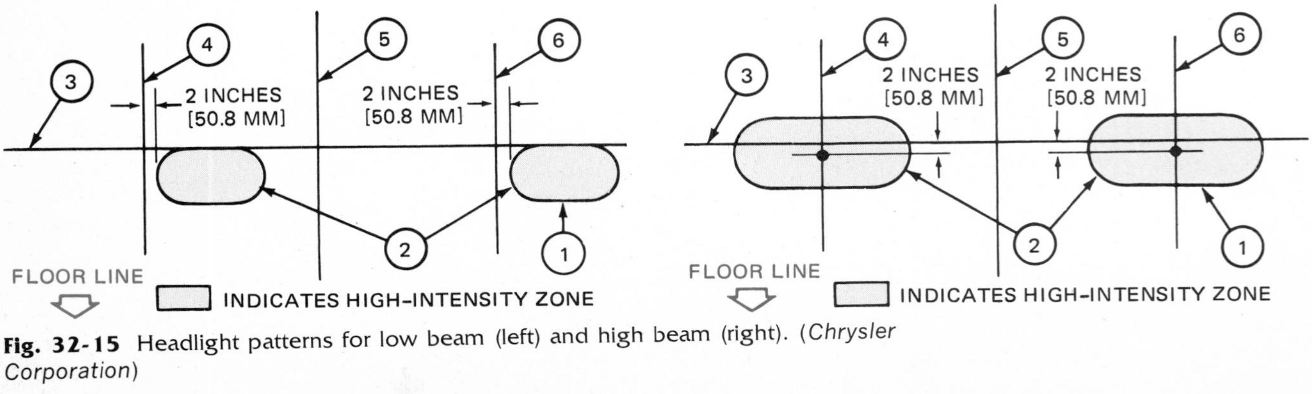

Fig. 32-15 Headlight patterns for low beam (left) and high beam (right). (*Chrysler Corporation*)

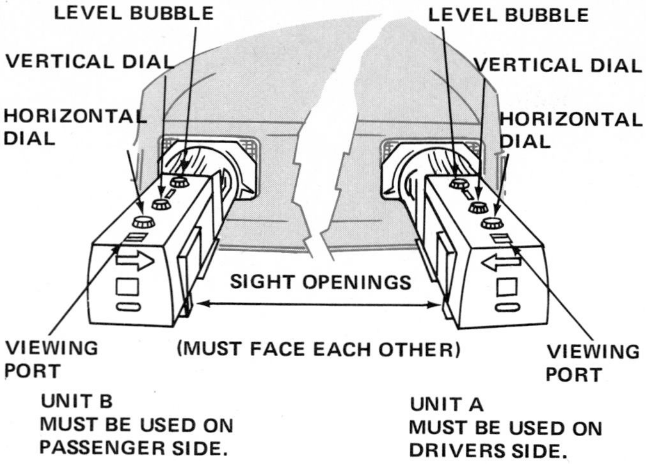

Fig. 32-16 Headlight aimers in position on the two headlights. *(Chrysler Corporation)*

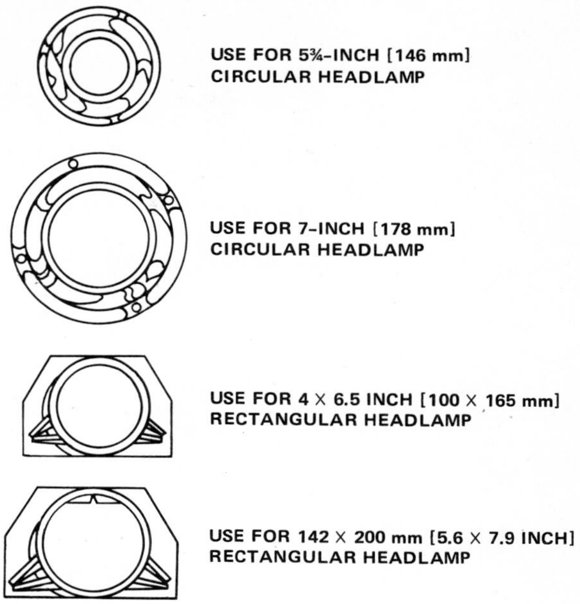

Fig. 32-17 Adapters to fit headlight aimers on the headlights. *(Chrysler Corporation)*

(Fig. 32-17). The instruments shown (two are needed) are attached, leveled, and then sighted through to determine the aim of the headlights. Adjustments are made as shown in Fig. 32-15.

32-9 Phototube Headlight Control

This is an electronic device that automatically selects the proper headlight beam. It holds the lights on high beam until a car approaches from the other direction. The headlights of an approaching car trigger the phototube to switch automatically from high to low beam. When the other car has passed, it switches back to high beam. The device has several names: Autronic Eye, Guide-Matic, Automatic Headlight Dimmer, and so on.

The phototube unit is mounted either on the top left side of the instrument panel or behind the radiator grill (see Fig. 32-18). In either place the unit is in line with the lights of oncoming cars. The phototube is an electronic tube that releases electric current when light strikes it. The current is very small, but it is amplified by a special amplifier unit. This provides enough current to operate a power relay and shift the headlights from high to low beam. The driver can still use the manually operated dimmer switch to override the phototube.

32-10 Other Lights

In addition to headlights and stop lights, cars have courtesy lights, turn signals, and warning-blinker systems. The courtesy lights are operated by switches in the doors. When a door is opened, the switch closes to connect the internal lights in the dome or side. The lights come on so a passenger or driver can see.

Turn signals are operated by a separate lever on the steering column. When the driver flips this lever for a turn, lights at the front and rear of the car flash on and off to indicate that the car is about to make a turn. If it is a left turn, the lights on the left front and rear flash on and off. If a right turn, the lights on the right front and rear flash on and off. Some cars are equipped with sequential turn-signal lights. That is, there are several separate bulbs that come on in sequence, starting near the rear center of the car and moving to the side. In addition, many cars are equipped with side lights at the front of the car. These come on and remain on when the turn-signal lever is moved. This provides additional light to the driver. The area the car is turning into is lighted. The turn-signal lever is returned to neutral when the driver turns the steering wheel back to straight ahead.

The warning-blinker system is operated by a separate switch usually located on the steering column. When there is an emergency of some sort, the driver can actuate the system by closing this switch. Then all the outside car lights flash on and off every few seconds. This system is used when the car has stopped. It is not normally used when the car is operating on the highway.

32-11 Stoplight Switch

The stoplight switch operates lights at the rear of the car. The lights warn trailing drivers that the brakes are being applied. In older cars, the switch was connected into the brake hydraulic system. When the brakes were applied, the pressure on the brake fluid operated the switch. Today, with dual-braking systems, this type of switch cannot be used. Dual-braking systems are really two braking systems in one. One operates the front brakes; the other operates the rear brakes. If one system fails, the other system can still stop the car.

The stoplight switch used on cars today is shown in Fig. 32-19. It is a mechanical switch that is operated by the brake pedal. When the brakes are applied, the switch contacts close, and the stoplights come on.

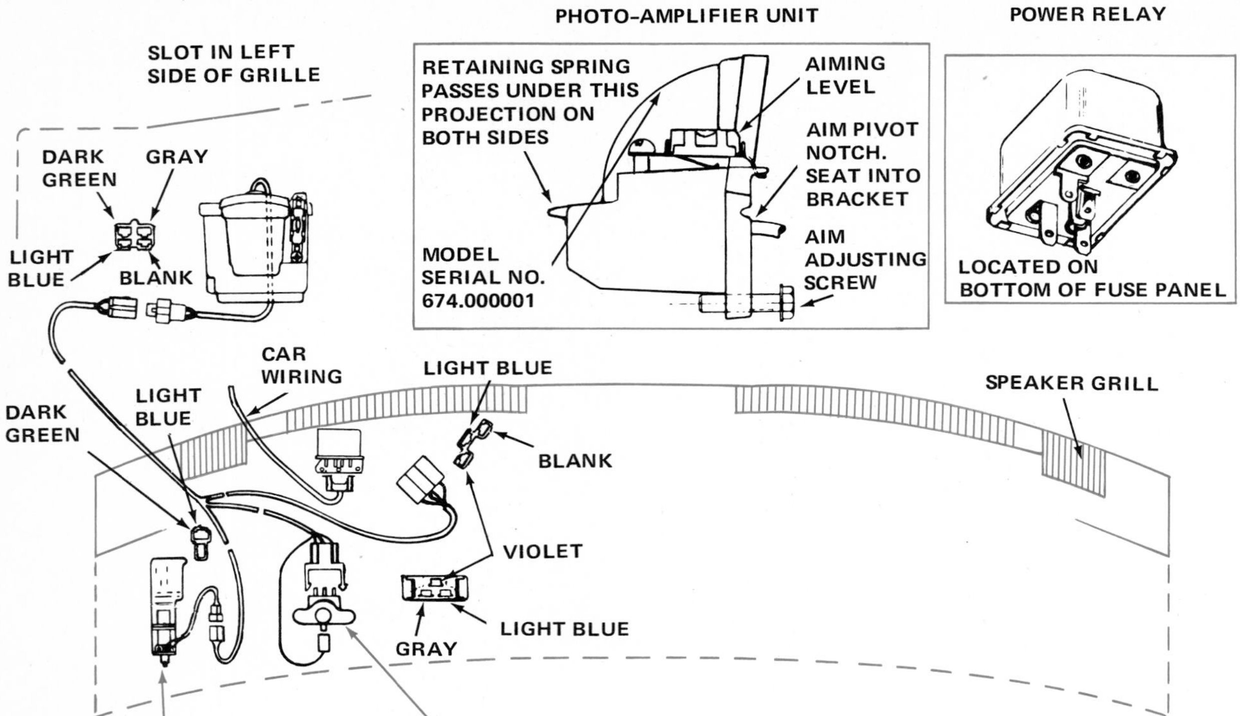

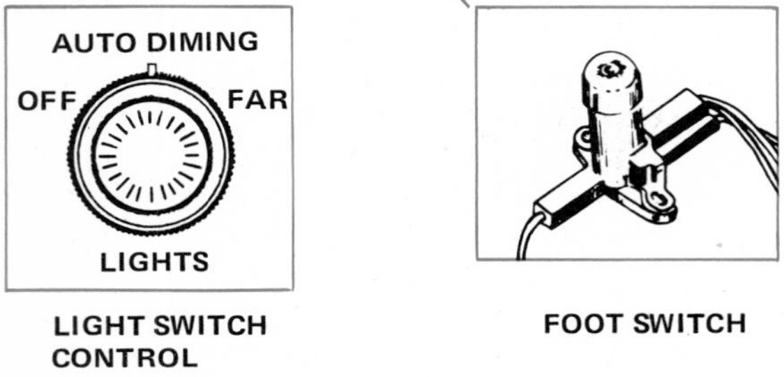

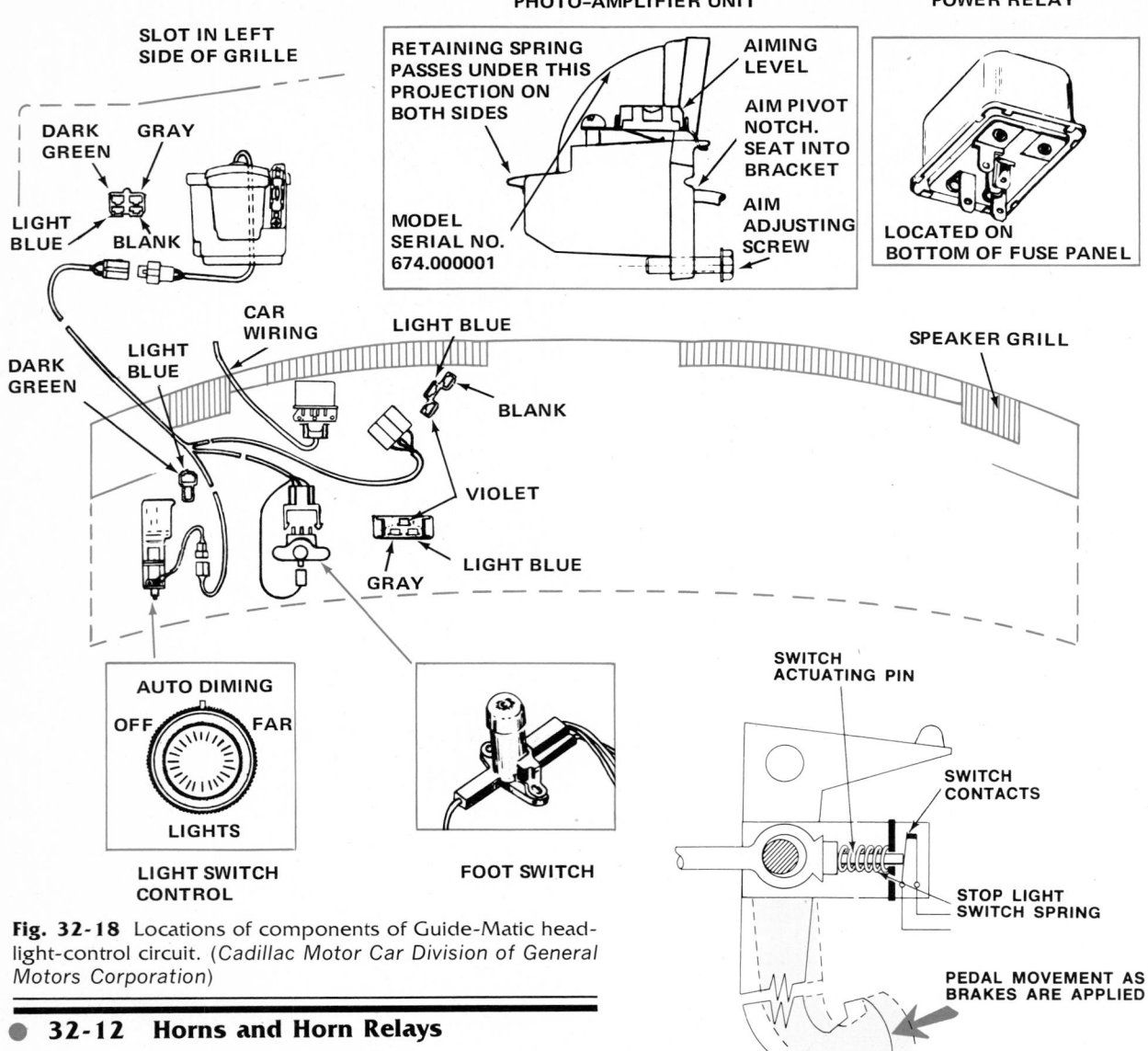

Fig. 32-18 Locations of components of Guide-Matic headlight-control circuit. (*Cadillac Motor Car Division of General Motors Corporation*)

32-12 Horns and Horn Relays

The automotive horn is of the vibrating type. Figure 32-20 is a sectional view of a typical horn. It has a field coil, a set of contact points, and a metal diaphragm. When the horn button is pressed, this closes the circuit and current starts flowing through the field coil. This produces a magnetic field that pulls the diaphragm down. The diaphragm movement produces a click. As the diaphragm moves down, the contacts separate so no current can flow. The magnetic field in the field coil dies, and the diaphragm is released. It moves up with another click. This action is repeated, so rapidly that the separate clicks blend to form the sound you hear.

The horn relay (Fig. 32-21) has a single winding that is connected through the horn button to the battery. Current flowing through the winding produces a magnetic field that pulls the armature down and closes the contact points. This connects the horns directly to the battery so the horns sound (see Fig. 32-21).

In today's cars, the horn relay has a second job. If the driver leaves the ignition key in the ignition switch and then opens the car door, the horn relay buzzes. The circuit for this arrangement is shown in Fig. 32-22. When the ignition key is left in the ignition switch, the warning switch remains closed. The warning switch is located in the ignition switch and is connected to the door switch, as shown. The door switch is closed as the door is opened. This completes the circuit to the horn-relay winding. The circuit runs through a special set of contact points above the armature. When the circuit to the winding is completed, the winding magnetism pulls the armature down. This opens the upper points to open the winding circuit. The magnetism dies and the armature moves back up. The points close, and the action is repeated. This action produces a buzzing sound that warns the driver to remove the ignition key.

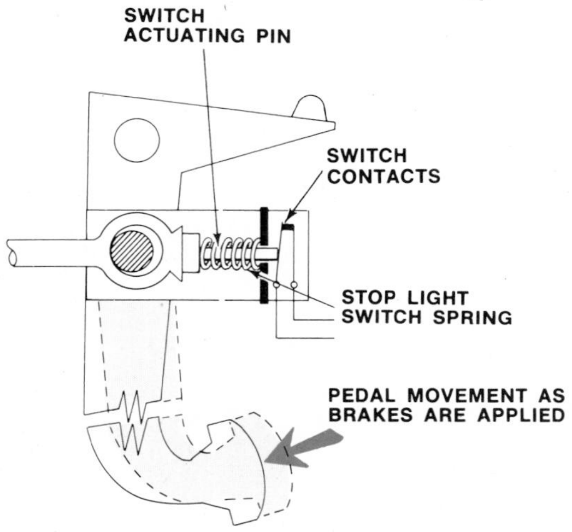

Fig. 32-19 Mechanical stoplight switch. When brakes are applied, as shown, the brake-pedal movement allows the switch contacts to close. (*Ford Motor Company*)

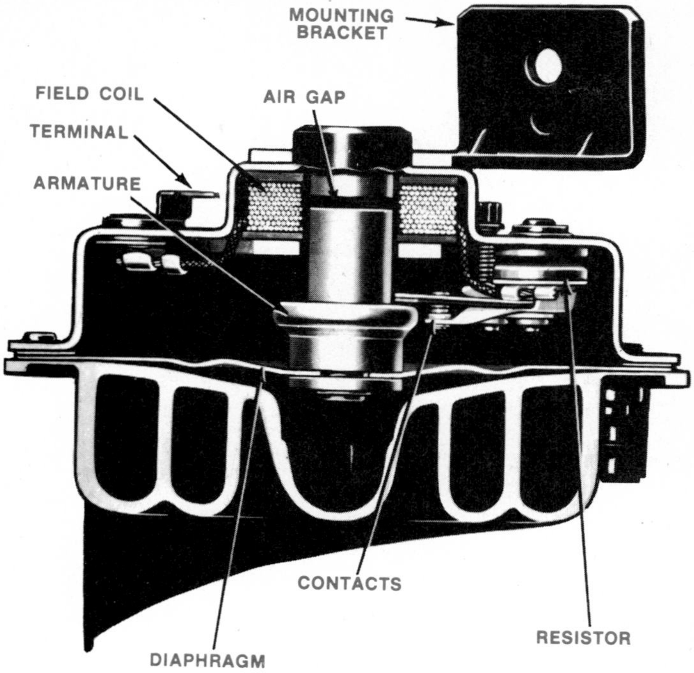

Fig. 32-20 Sectional view of a horn.

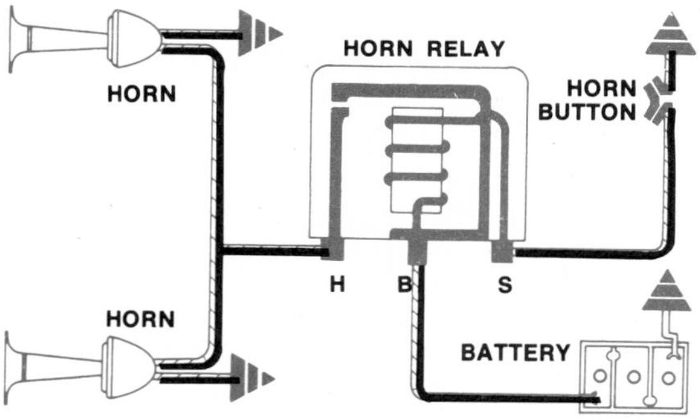

Fig. 32-21 Horn relay and horn circuit.

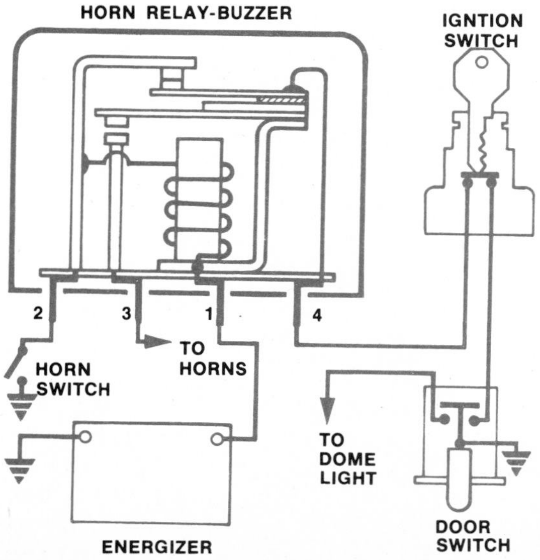

Fig. 32-22 Horn relay which includes a warning system.

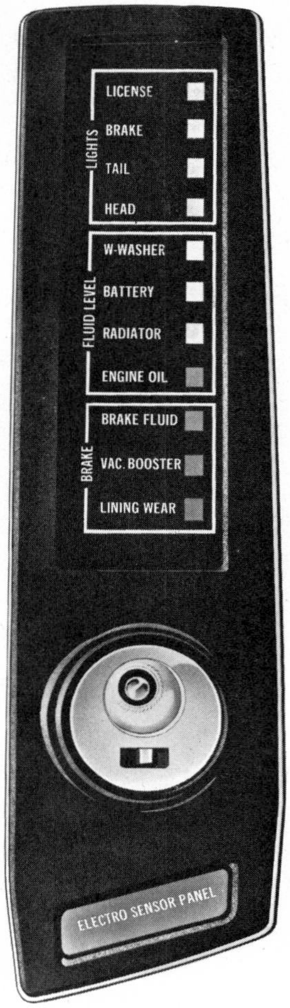

Fig. 32-23 Sensor panel, called the Electro Sensor Panel or ESP by the manufacturer. (*Toyota Motor Sales, Limited*)

In some cars, the buzzer will also sound if the driver leaves the car with the headlights on. This is a reminder to turn off the headlights so the battery will not run down while the driver is absent from the car.

● 32-13 Indicating Devices

Many cars have an ammeter or charge indicator, a fuel gauge, an oil-pressure gauge, and an engine-temperature indicator. These instruments keep the driver informed of the operating condition of the engine. For example, if the oil pressure drops too low, the indicator tells the driver that something is wrong. This gives the driver time to stop the car before the engine is damaged. These gauges are described in other chapters: fuel gauges in Chap. 18; oil-pressure gauges in Chap. 19; and engine-temperature indicators in Chap. 21.

An innovation in indicating devices is the sensor panel introduced by Toyota in 1974 (Fig. 32-23). This panel is installed on the roof of the car above the driver (Fig. 32-24). It is connected to sensors in the light circuits, the brakes, the windshield washer, battery, cooling-system radiator, and engine crankcase (see Fig. 32-25). The sensor panel has eleven warning lights which come on if something needs attention. For example, if any of the four lights at the top of the panel light (license, brake, tail, head), it indicates trouble in that circuit. If one headlight burns out, "head" would come on to warn the driver of the trouble. The four "fluid level" lights (W-washer, battery, radiator, engine oil) indicate low fluid level in any of these four areas. That is, if the engine oil drops to a low level, the "engine oil" light would come on. The "brake" section of the panel warns of low brake fluid, loss of vacuum in the power-brake unit, or excessive brake lining wear. Figure 32-25 shows how the eleven warning lights are connected by sensors to the service areas. Figure 32-26 shows how the brake fluid-level sensor is installed on the brake-master-cylinder cap.

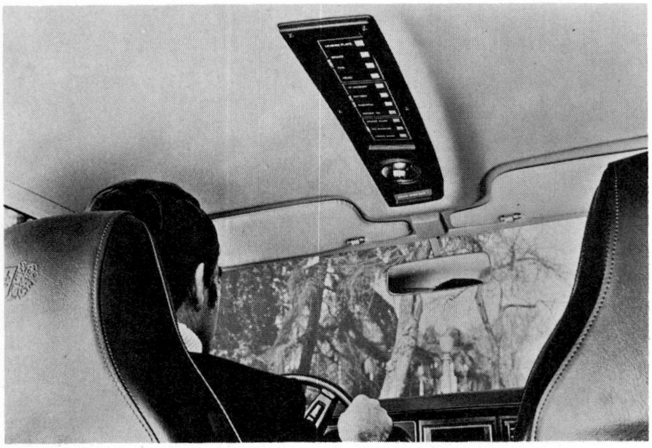

Fig. 32-24 Location of sensor panel in car. (*Toyota Motor Sales, Limited*)

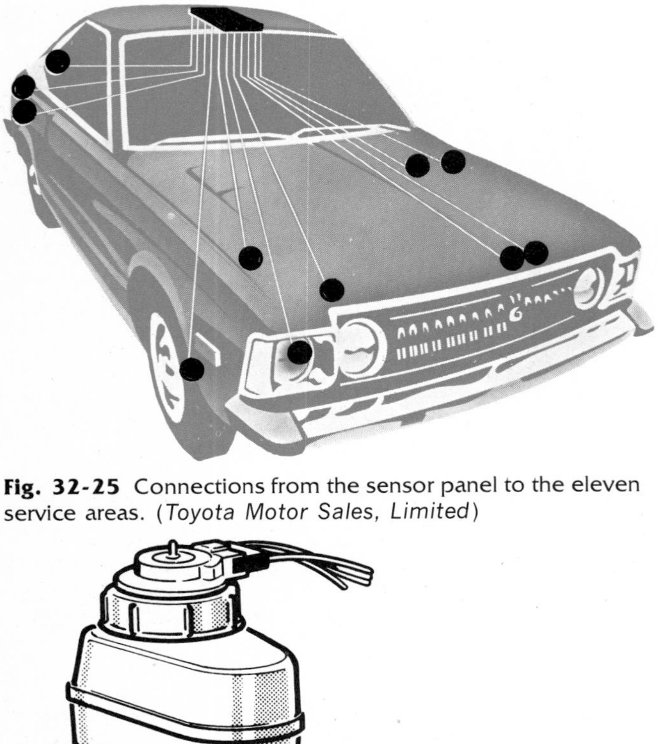

Fig. 32-25 Connections from the sensor panel to the eleven service areas. (*Toyota Motor Sales, Limited*)

Fig. 32-26 Brake fluid-level sensor installed on master cylinder. (*Ford Motor Company*)

● 32-14 Ammeter

In Chap. 29 we described the ammeter in detail. The ammeter tells the driver which way the current is flowing—to the battery or from the battery—and how much current is flowing (Fig. 29-2).

● 32-15 Charge-Indicator Light

Many cars have an indicator light rather than an ammeter. When the alternator is not charging the battery, the light comes on. When the alternator is charging the battery, the light stays out.

● 32-16 Electric Fuel-Pump Control System

The tank-mounted electric fuel pump, described in ● 12-10, is connected through a hydraulic switch to the ignition switch. The hydraulic switch is operated by pressure in the engine lubricating system. This switch has two pairs of contacts. During cranking, the first set of contacts is closed. Current is fed through these contacts from the solenoid on the starting motor. This current flows to the fuel pump so that it operates. These connections continue as long as the engine is being cranked. As soon as the engine starts, the solenoid is disconnected from the battery so that this source of current is cut off. However, as soon as the engine starts, the oil pressure builds up. This causes the control switch to open the first set of contacts and close the second set of contacts. Now,

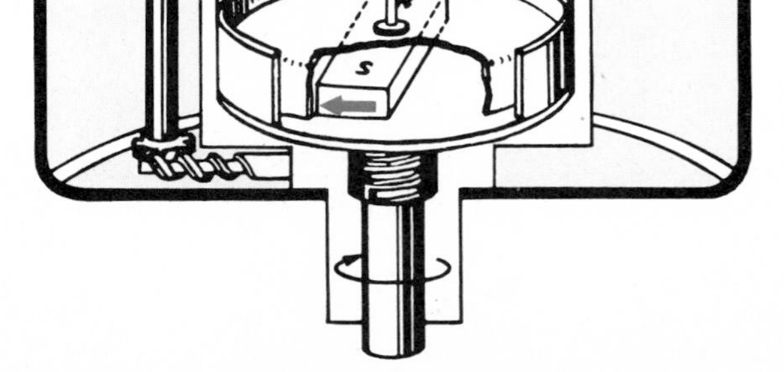

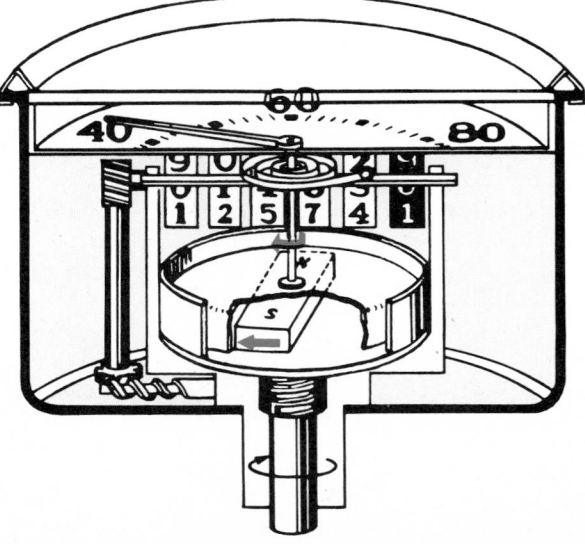

Fig. 32-27 Cutaway view of a speedometer-odometer assembly.

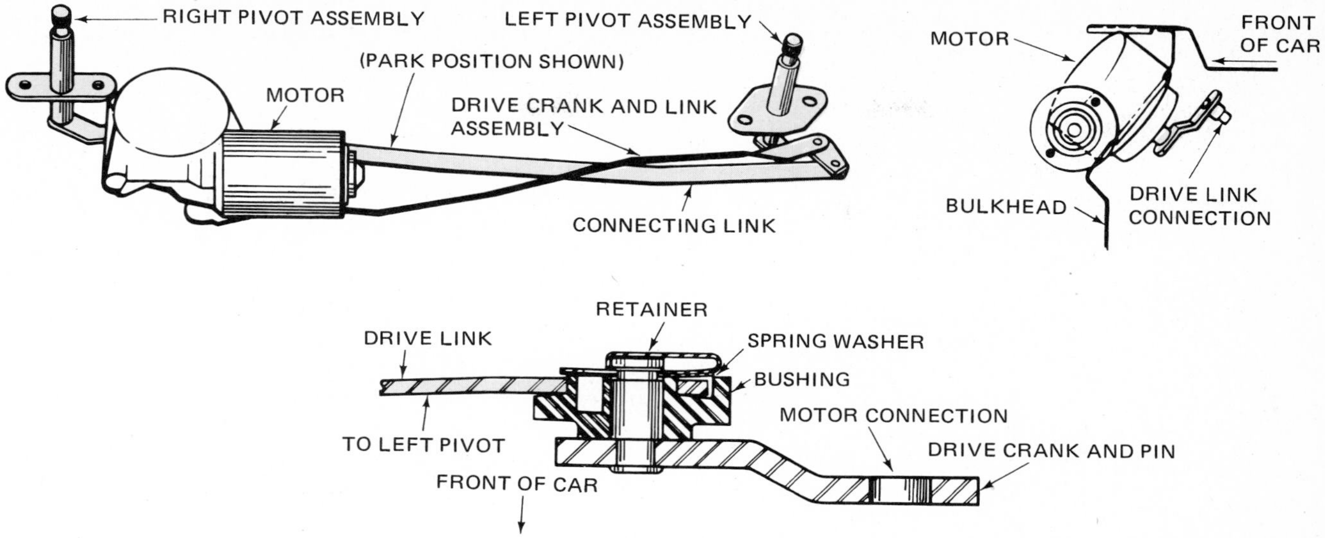

Fig. 32-28 Three-speed windshield-wiper assembly and mounting arrangement. *(Chrysler Corporation)*

current is fed to fuel pump through ignition switch and the second set of contacts in the control switch.

If the engine stalls during operation, the pressure becomes too low to keep the second set of contacts in the control switch closed. They open, and the fuel pump stops. If the engine suddenly loses oil pressure, the second set of contacts in the control switch opens so that the fuel pump stops. Now, the engine will run only until the fuel in the carburetor float bowl is used up. This protects the engine against serious damage from operating without oil pressure. The fuel pump also stops when the ignition switch is turned off.

● 32-17 Speedometer and Odometer

The speedometer and the odometer are not electrical. We discuss them here, however, because they are mounted on the car dash, along with the other instruments. The speedometer tells the driver how fast the car is going. The odometer tells the driver how many miles the car has gone. Figure 32-27 is a cutaway view of the assembly.

There is a small magnet mounted on a shaft inside the speedometer. This magnet is driven by a flexible cable from the transmission. The faster the car goes, the faster the magnet spins. This action produces a rotating magnetic field that drags on the aluminum ring surrounding the magnet. The faster the spinning, the more drag on the ring. The spinning causes the ring to swing around against the tension of a spring. This, in turn, moves a pointer attached to the ring, which indicates car speed.

The odometer is operated by a pair of gears from the same rotating flexible cable that drives the speedometer. The motion is carried through the gears to the mileage rings on the odometer indicator. These rings turn to show how many miles the car has been driven.

The cable is driven from a pair of gears in the rear extension housing of the transmission. One of these gears is on the main shaft of the transmission. The other is on the end of the flexible cable.

● 32-18 Windshield Wiper

Windshield wipers are driven by an electric motor. A typical system is shown in Fig. 32-28. The motor, through gearing, causes the wiper blades to move back and forth on the windshield. Many cars have a windshield washer as part of the windshield-wiper system. When the driver presses a button, a squirt of liquid covers the windshield and the blades can clean more effectively.

● 32-19 Antilock Braking System

This system reduces the tendency of the tires to skid by preventing wheel lock. That is, when the car is braked hard, the wheels can lock and the tires will skid on the pavement. As we explain in the chapter on brakes, when tires skid they are less effective in stopping the car. The antilock system uses an electronic module which monitors the wheel rpm. If a wheel starts to slow down too fast, or faster than other wheels, during braking, the electronic module does this: It eases up on the brake at the wheel that is slowing down too fast. This keeps the wheel below the lock point. This system, which has electronic components, is covered in detail later in the book.

● 32-20 Seat Adjuster

Many cars have a motor-powered seat adjuster. Figure 32-29 shows one type. It is a six-way adjuster, which moves the seat forward or backward, or up or down, and tilts the seat forward or backward. The mechanism includes a drive motor, drive cables, jack screws, and a transmission. Depending on which lever the driver operates, the movement of the motor puts into action one or another of the drive cables. The mechanism causes the seat to move in the direction selected by the driver.

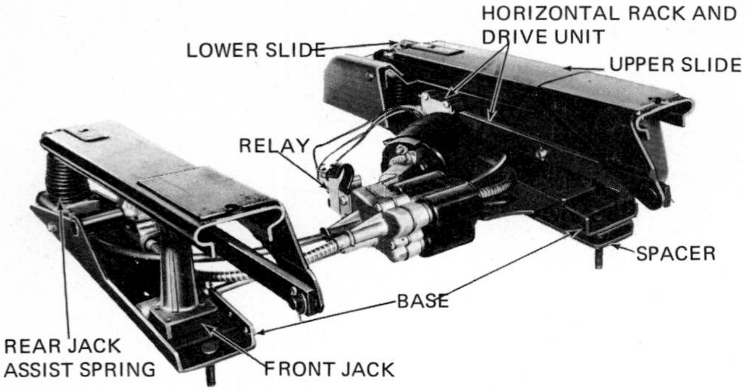

Fig. 32-29 Six-way power seat adjuster. (*Chrysler Corporation*)

32-21 Window Regulators

Many cars have power window regulators. They are operated by an electric motor that causes levers to move so that the window can be raised or lowered. Figure 32-30 shows a power window regulator.

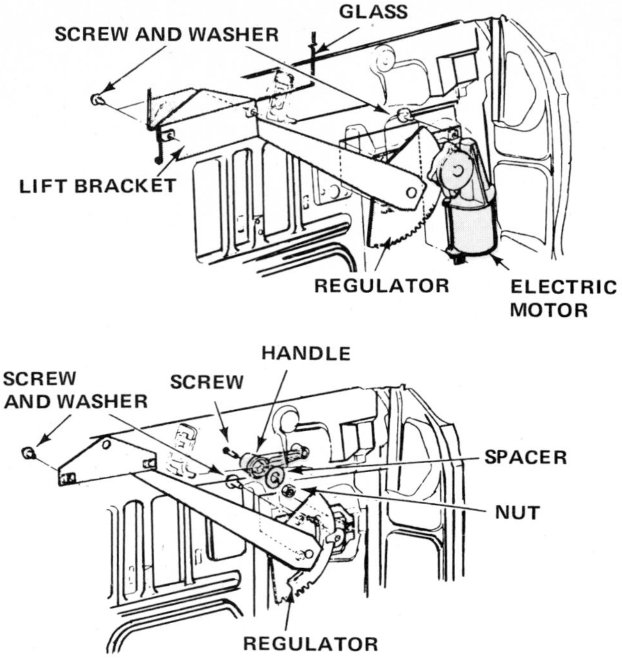

Fig. 32-30 Window regulators; mechanical at bottom, electrical at top. (*Chrysler Corporation*)

REVIEW QUESTIONS

1. How are the wires in a wiring harness identified?
2. What is a printed circuit?
3. Describe a fuse.
4. Explain how a circuit breaker works.
5. Describe the construction of a headlight.
6. Explain how a headlight cover works.
7. Explain how headlight aiming is checked. How it is adjusted.
8. Explain how a phototube headlight control works.
9. Describe the operation of a horn.
10. Explain how the horn relay works.
11. Explain how the speedometer works. The odometer.
12. Explain how the electric fuel-pump control system works.
13. What provision does the electric fuel-pump control system have to stop the fuel pump if engine oil pressure is lost?
14. Explain how the windshield wiper works.
15. Explain how the power seat adjuster works.

SELF PROJECT

Examine and tear down junk parts so you will get a better idea of what is inside them and how they work. For instance, take the cover off a horn relay, and examine what is inside.

Automotive Emission Controls

5

Part Five of this book discusses the special devices installed on engines and automobiles, as well as changes made in engines, to reduce automotive pollution.

Tampering with automotive emission controls is against the law. If an employer or any employee is caught tampering, the employer can be fined for each vehicle or vehicle engine that has been tampered with. Tampering is defined as removing, disconnecting, damaging, or in any way rendering ineffective any emission-control device installed on a motor vehicle or motor-vehicle engine.

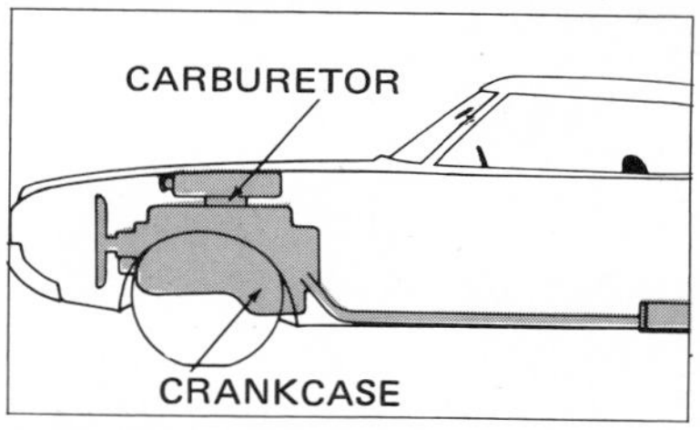

Chapter 33 Air Pollution, Smog, and the Automobile

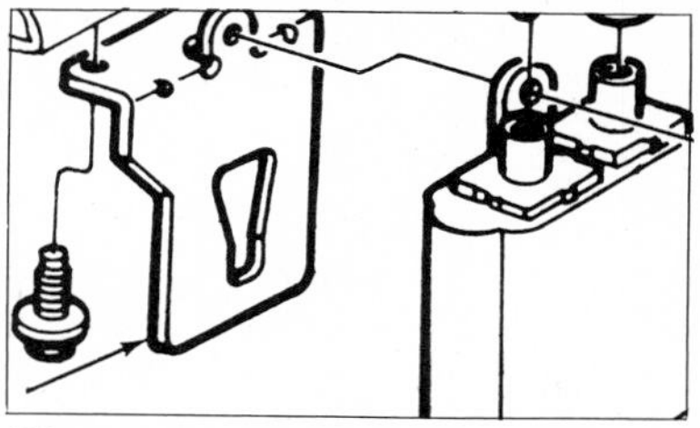

Chapter 34 PCV and Fuel-Vapor Emission-Control Systems

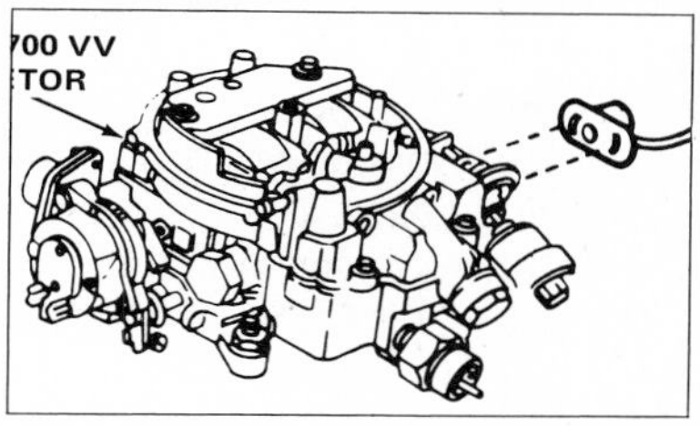

Chapter 35 Cleaning Up the Exhaust Gas

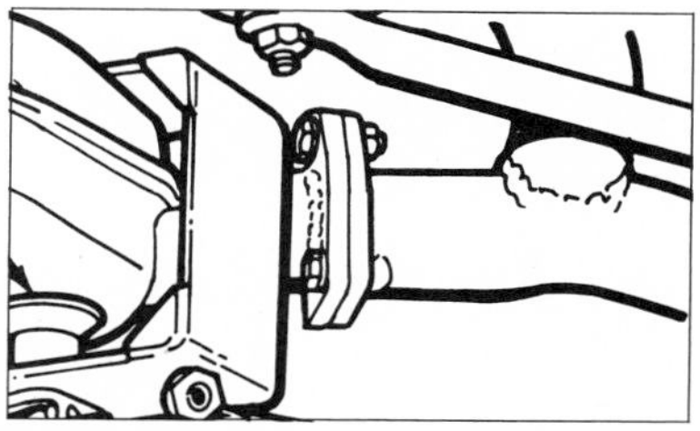

Chapter 36 Servicing Emission-Control Systems

CHAPTER 33
Air Pollution, Smog, and the Automobile

After you study this chapter, you should be able to:

- **Define smog and explain what produces it.**
- **Describe the three major pollutants from the automobile and explain how they are formed.**

In this chapter, we look at sources of atmospheric pollution, the formation of smog, and the automobile's role in air pollution.

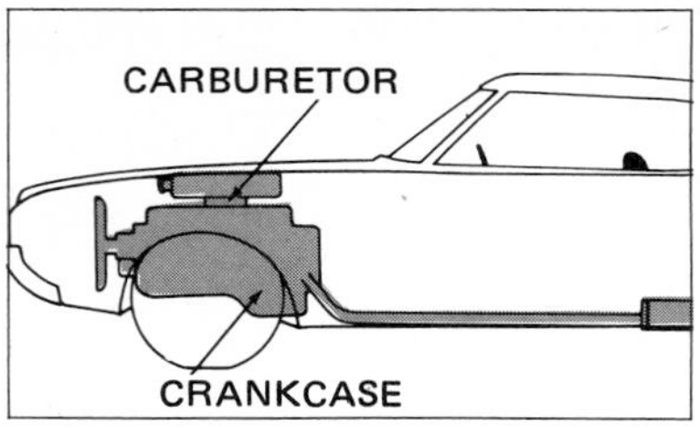

33-1 Smog

The word "smog" comes from "smoke" and "fog." Smog is a sort of fog with other substances mixed in. Smog has been here a long time. Billions of years ago, volcanos sent millions of tons of ash and smoke into the air. Winds whipped up dust clouds. Animal and vegetable matter decayed, adding polluting gases.

When people came along, they began to produce their own kind of air pollution. They discovered fire. In the Middle Ages, people in cities such as London used soft coal to heat their homes. The smoke from these fires, combined with moisture in the air, produced dense layers of smog. The smog would blanket the city for days, particularly in winter. The heat generated in large cities tends to circulate air within a dome-like shape, as shown in Fig. 33-1. This traps the smog and holds it over the city.

Smog, and the chemicals and other substances in it, can be harmful, even deadly. Smog blurs vision (Figs. 33-2 and 33-3). It irritates the eyes, the throat, and the lungs. Eyes water, throats get sore, people cough. Smog can make people ill. And it can make sick people sicker. Air pollution has been linked to eczema, asthma, emphysema, cardiovascular difficulties, and lung and stomach cancer. It also has a harmful effect on the environment.

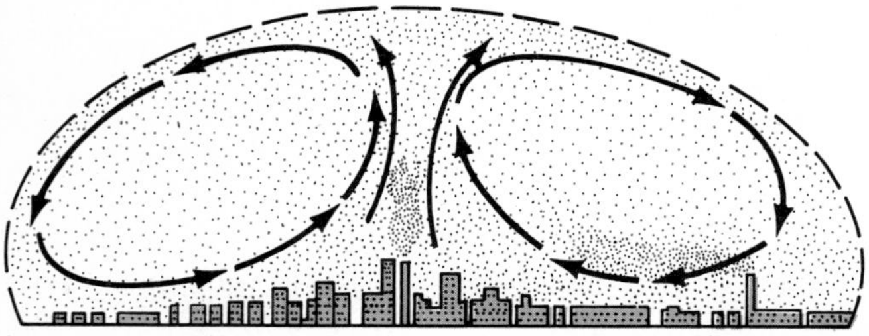

Fig. 33-1 The heat generated within a large city tends to produce a circulatory air pattern which traps smog within a dome.

Food crops and animals suffer. Paint may peel from houses. It is obvious that we must do everything possible to reduce man-made atmospheric pollutants and smog.

33-2 Not All Air Pollution Is Smog

Smog, along with smoke, is the most visible evidence of atmospheric pollution. But some atmospheric pollution is not visible and may not become visible until it is mixed with moisture. Lead compounds from leaded gasoline, hydrocarbons (unburned gasoline), carbon monoxide, and other gases may pollute the air without being seen.

33-3 Air Pollutants

All air is polluted to some extent. That is, all air carries some polluting substances. Much of it is natural: smoke and ash from volcanos, dust stirred up by the wind, compounds given off by growing vegetation, gases given off by rotting animal and vegetable matter, salt particles from the oceans, and so on.

Man adds to these pollutants by burning coal, oil, gas, gasoline, and many other things. It is these added substances that we are concerned with in this book—especially those that come from the automobile.

Before we get to the automobile, however, let us review what we have learned about combustion. Most fuels, such as coal, gasoline, and wood, contain hydrogen and carbon in various chemical combinations. During combustion, oxygen unites with the hydrogen and carbon to form water (H_2O), carbon monoxide (CO), and carbon dioxide (CO_2).

In addition, many fuels contain sulfur; this burns to produce sulfur oxides. Also, in the heat of combustion, some of the nitrogen in the air combines with oxygen to form nitrogen oxides (NO_x). Some of the fuel may not burn completely, so that smoke and ash are formed.

Fig. 33-2 View of Los Angeles during a clear day. *(Los Angeles County Air Pollution Control District)*

Fig. 33-3 Same view of Los Angeles, during a typical smoggy day. Note that many buildings are hidden. *(Los Angeles County Air Pollution Control District)*

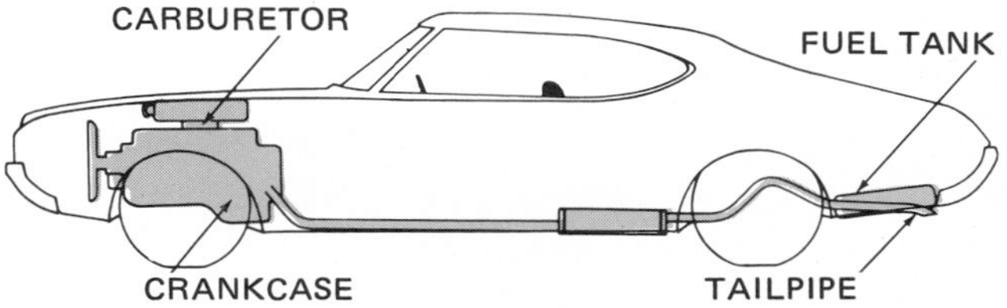

Fig. 33-4 Four possible sources of atmospheric pollution from the automobile.

Smoke is simply particles of unburned fuel and soot, called *particulates,* mixed with air.

Altogether, it is estimated that 200 million tons of man-made pollutants enter the air every year in the United States alone. This is about a ton for every man, woman, and child in the country!

This man-made pollution is what clean-air laws are aimed at.

33-4 Los Angeles

Consider Los Angeles, a large city set in a basin, with about 7,000,000 inhabitants. It is surrounded on three sides by mountains, and on the fourth by the Pacific Ocean. When the wind blows out over the ocean, it sweeps away pollutants. But at other times, the air is stagnant. Smoke and other pollutants from industry and automobiles do not blow away (Figs. 33-2 and 33-3). They just build up into a thick, smelly, foggy layer of smog. The location of Los Angeles, plus all the people and industry there, make it one of the biggest "smog centers" in the country. And it is Los Angeles which has led in measures to reduce smog.

Los Angeles has banned unrestricted burning, for example, burning trash. Incinerators without pollution controls were outlawed. Industry was forced to change combustion processes and add controls to reduce pollutants coming from their chimneys. Laws were passed that required the addition of emission controls on automobiles. We shall discuss these controls in Chaps. 34 and 35. All these measures have significantly reduced atmospheric pollution in the Los Angeles area.

33-5 Pollution from Automobiles

If not controlled, the automobile can give off pollutants from four places, as shown in Fig. 33-4. Pollutants can come from the fuel tank, the carburetor, the crankcase, and the tail pipe. Pollutants from the fuel tank and carburetor consist of gasoline vapors. Pollutants from the crankcase consist of unburned and partly burned air-fuel mixture that has blown by the piston rings. Pollutants from the tail pipe consist of partly burned gasoline (HC), carbon monoxide (CO), nitrogen oxides (NO_x), and—if there is sulfur in the gasoline—sulfur oxides (SO_x). In the chapters that follow, we discuss the causes and cures for these pollutants.

REVIEW QUESTIONS

1. What is smog?
2. What are some of the effects of smog on people?
3. What are some of the causes of natural air pollution?
4. Name four products of combustion.
5. Name the four places on the automobile from which pollutants can come.

SELF PROJECTS

Go to your local library and ask the librarian for help in looking up the latest laws (state and federal) on automotive pollution controls. See if you can find some short explanations, since the full laws are complicated and sometimes hard to understand. Find out what laws call for reduction of the pollutants coming from cars. Write a short essay on what you learn.

CHAPTER 34
PCV and Fuel-Vapor Emission-Control Systems

After you have studied this chapter, you should be able to:

- **Identify the components of the PCV and vapor emission-control systems, and point them out on a vehicle.**
- **Explain the purpose of and operation of these systems.**

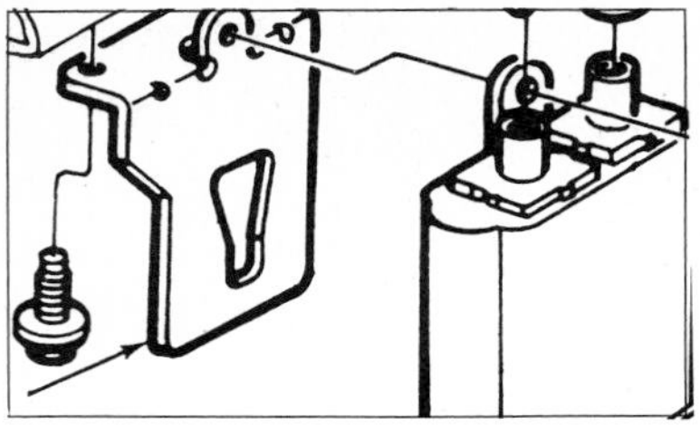

This chapter describes the positive crankcase-ventilation (PCV) system and the fuel-vapor emission-control systems. These latter systems capture gasoline vapors from the fuel tank and carburetor. The next chapter discusses methods of cleaning up exhaust gases. Thus, these two chapters examine the four sources of automotive pollutants (see Fig. 33-4). They explain what has been done to reduce or prevent pollution from these sources.

● 34-1 Positive Crankcase Ventilation

We explained, in ● 12-14, that the engine crankcase must be ventilated. That is, outside air must flow through the crankcase to remove blow-by. This blow-by gets past the piston rings during the compression and power strokes. Unless it is cleared from the crankcase, it will cause trouble. It can form sludge and acids. The sludge can clog oil lines and starve the lubricating system. This could mean a ruined engine. Acids corrode metal parts, and this, too, can ruin the engine.

The removal process requires that the engine must first heat up enough to vaporize the liquid gasoline and water that has collected in the crankcase. Then the circulating air can remove them, along with the blow-by gases.

In older engines, the crankcase was ventilated by an opening at the front of the engine and a vent tube in the back. The forward motion of the car and the rotation of the crankshaft moved air through the crankcase, as shown in Fig. 34-1. The air passing through removed the water, fuel vapors, and blow-by. However, discharging these gases into the atmosphere caused air pollution.

To prevent this pollution, modern engines have a closed, or positive, crankcase-ventilating (PCV) system. A typical system for a six-cylinder engine is shown in Fig. 34-2. Figure 34-3 shows the system for a V-8 engine. The idea is simple. Filtered air from the carburetor air cleaner is drawn through the crankcase. In the crankcase it picks up the water, fuel vapors, and blow-by. The air then flows back up to the intake manifold and enters the engine. There, unburned fuel is burned.

Too much air flowing through the intake manifold during idling could upset the air-fuel ratio. This could cause poor engine idling and even stalling. To prevent this, a regulator valve is used. The valve is called a posi-

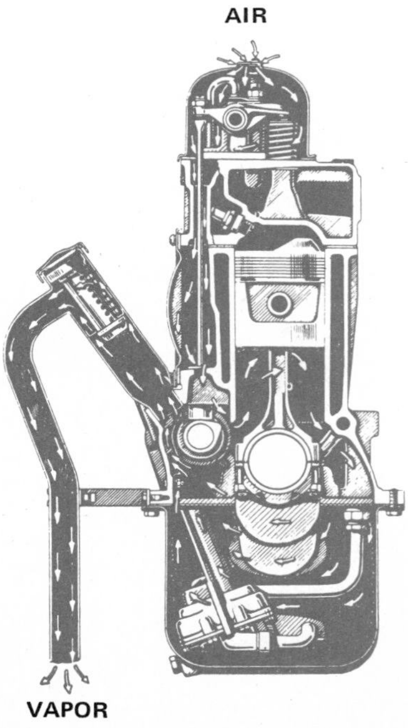

Fig. 34-1 Open crankcase-ventilating system.

tive-crankcase-ventilation (PCV) valve. The PCV valve allows only a small amount of air to flow through during idle. But as engine speed increases, reduced intake-manifold vacuum allows the valve to open more. This, in turn, allows more air to flow through. Figure 34-4 shows the valve in the two positions.

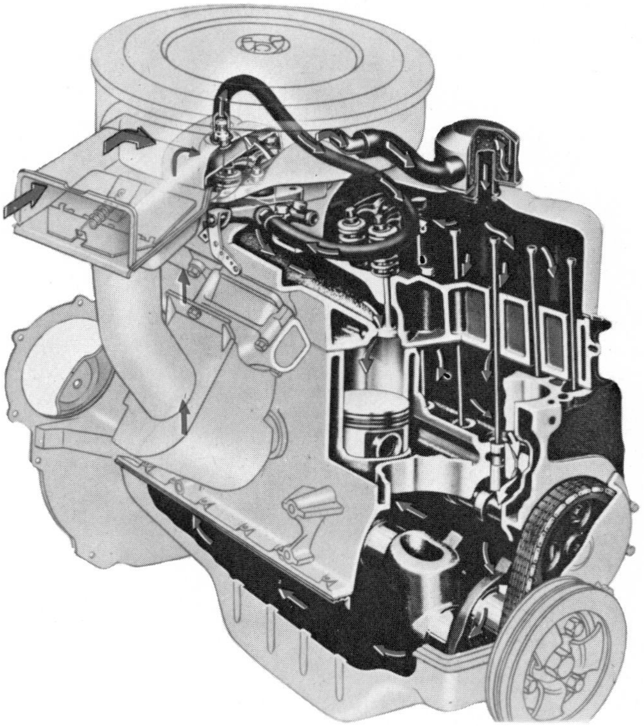

Fig. 34-2 Positive-crankcase-ventilating system for a six-cylinder engine. (*Ford Motor Company*)

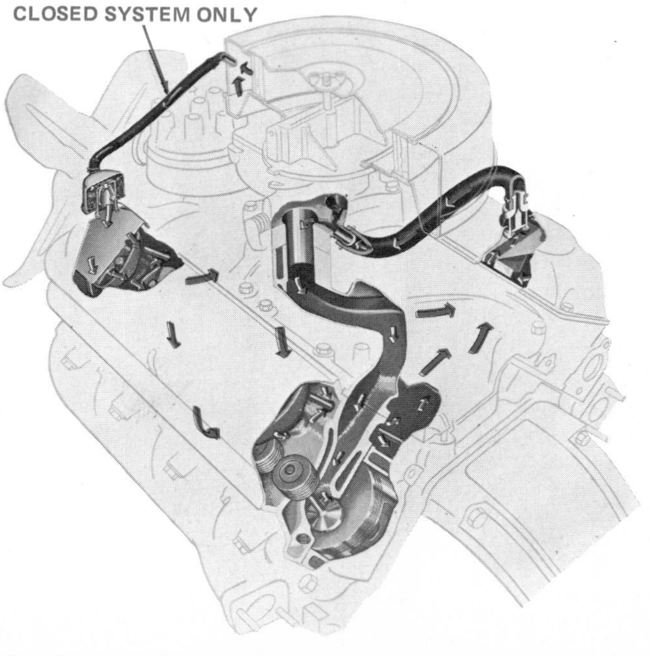

Fig. 34-3 Positive-crankcase-ventilating system for a V-8 engine. (*Ford Motor Company*)

● 34-2 Need for Fuel-Vapor Emission-Control System

Both the fuel tank and carburetor can lose gasoline vapor to the atmosphere, causing pollution, if the car does not have a vapor emission-control system. The fuel tank "breathes" as temperature changes. That is, as the tank heats up, the air inside it expands. Part of the air is forced out through the tank vent tube, or through the vent in the tank cap. This air is loaded with gasoline vapor. Then, when the tank cools, the air inside contracts. More air enters the tank from outside. This breathing of the tank causes a loss of gasoline. The higher the tank temperature goes (for instance, when the car is parked in the sun), the more gasoline vapor is lost.

The carburetor also can lose gasoline by evaporation. The carburetor float bowl is full whenever the engine is running. When the engine stops, engine heat evaporates some or all of the gasoline stored in the float bowl. Without a vapor-recovery system, this gasoline vapor would pass into the atmosphere.

A fuel-vapor emission-control system captures these gasoline vapors and prevents them from escaping into the air. It thus tends to reduce atmospheric pollution. All modern cars are equipped with such systems. They are called by various names: ECS (Evaporation Control System), EEC (Evaporation Emission Control), VVR (Vehicle Vapor Recovery), and VSS (Vapor Saver System). All work in the same general way.

● 34-3 Fuel-Vapor Emission-Control Systems

Figures 34-5 and 34-6 show typical vapor emission-control systems. The canister is filled with activated charcoal. Just after the engine is shut off, heat continues to enter the carburetor. This vaporizes gasoline in the carburetor float bowl. The vapor passes through the control line and

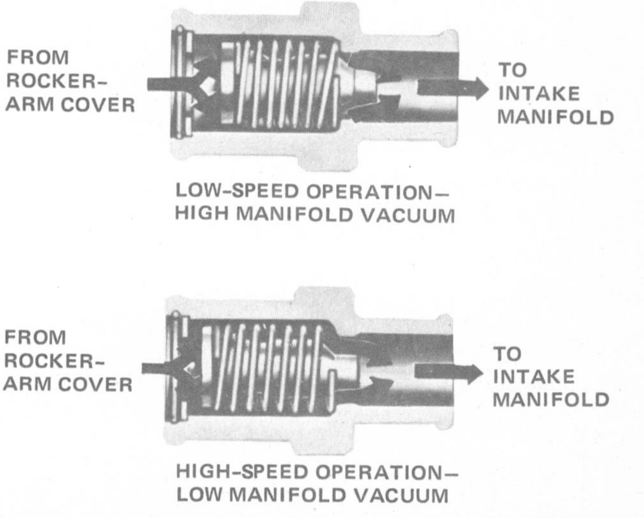

Fig. 34-4 The two extreme operating conditions of the PCV valve. (*Ford Motor Company*)

into the canister where it is adsorbed by the charcoal. "Adsorbed" means that the gasoline vapor is trapped by the charcoal particles. (This is somewhat like the charcoal filters on cigarettes. Their purpose is to trap particles of tar and other substances to prevent their entering the mouth and lungs of the smoker.) Some carburetor float bowls have a special vent (see Fig. 13-11), connected by a tube to the charcoal canister. The vent and tube carry the float-bowl vapor directly to the canister.

At the same time, vapor-laden air from the fuel tank is carried by a special emission-control pipe to the canister. As the air passes down through the canister, the gasoline

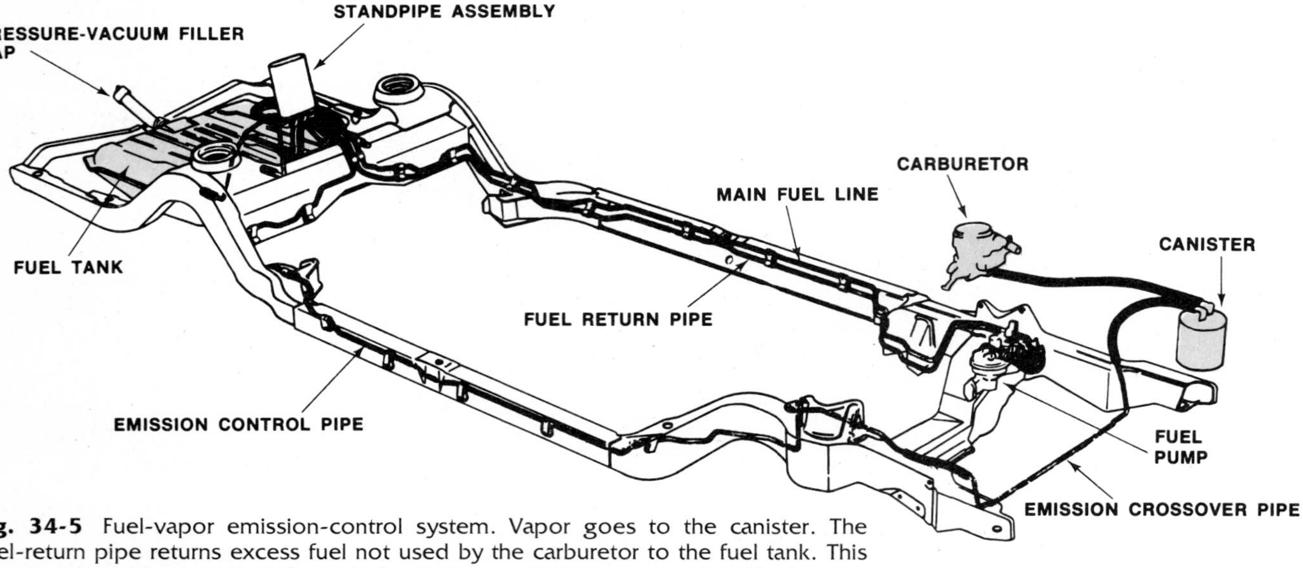

Fig. 34-5 Fuel-vapor emission-control system. Vapor goes to the canister. The fuel-return pipe returns excess fuel not used by the carburetor to the fuel tank. This constant flow of excess fuel through the fuel pump helps prevent vapor lock.

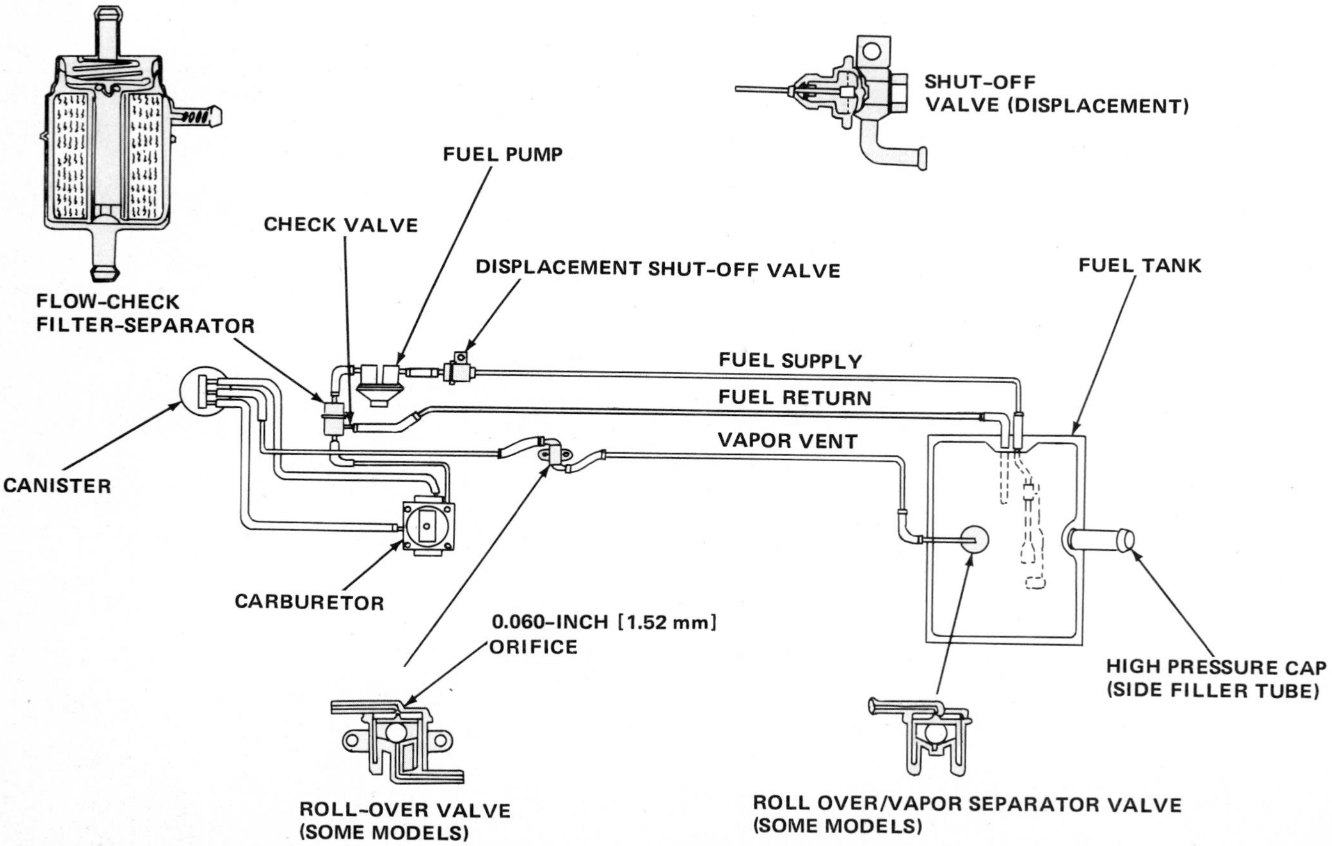

Fig. 34-6 Late-model fuel-vapor emission-control system. Note that the system has roll-over valves that shut off fuel flow in case the car rolls over. (*Chrysler Corporation*)

vapor is trapped by the charcoal particles. The air exits from the bottom of the canister, leaving the HC (hydrocarbon) vapor behind. There is a filter at the bottom of the canister. It comes into action during the *purge* phase of operation. This occurs when the engine is started. Now, intake-manifold vacuum draws fresh air up through the canister. This fresh air removes, or purges, the gasoline vapor from the canister. It takes the HC through a purge line to a connection at the carburetor.

● 34-4 Fuel-Vapor-Return Line

Note, in Figs. 34-5 and 34-6, that a fuel-return line parallels the main fuel line. This return line connects the pressure side of the fuel pump to the fuel tank. Thus, any excess gasoline being pumped by the fuel pump is returned to the fuel tank. This action removes any vapor that might develop in the fuel pump. It also maintains a flow of fuel through the fuel pump. This keeps the fuel pump relatively cool and helps prevent vapor lock. In some systems, there is a check valve in the fuel-vapor return line (Fig. 12-15). Its purpose is to prevent fuel from feeding back to the carburetor from the fuel tank through the return line. There is more on vapor-return lines in ● 12-8.

● 34-5 Charcoal Canister

Figure 34-7 shows a charcoal canister in sectional view. This canister is used on some V-6 General Motors engines. The downward-pointing arrows to the left show the flow of fuel vapor from the carburetor float bowl and from the fuel tank during the time that the engine is not running. When the engine is running, the action is as shown on the right-hand side. That is, air is pulled up through the charcoal, as shown by the upward-pointing arrows. This purges the gasoline vapor from the charcoal. Ford calls the canister a *carbon* canister. Charcoal is a special form of carbon.

Some of the early charcoal canisters did not have the connection for a vapor line from the carburetor float bowl. They had just two connections at the top, one for the tube from the fuel tank, the other for the purge line to the carburetor.

Figure 34-8 shows a typical canister-hose routing. Hose routings vary somewhat from car model to car model. The arrangement in Fig. 34-8 is for a V-8 engine.

Figure 34-9 is a canister for a four-cylinder engine. Note that it has something added—a purge valve (to upper left). This valve limits the flow of vapor and air to the carburetor during idle. But it allows full air-vapor flow during part- to full-throttle operation. In the four-cylinder engine, full fuel-vapor flow could upset engine idle. At higher engine speeds, however, full air-fuel vapor can be tolerated. The valve is operated by a vacuum signal from the carburetor. During idle, a small amount of purging takes place through a small "constant purge hole." When the throttle is opened, it passes a drilled hole in the carburetor that allows intake-manifold vacuum to lift the purge valve off its seat. Now, additional air vapor can flow through the purge hole. Note that this hole is connected by a tube to the PCV system. The air vapor flows through a tube that is connected to the positive-crankcase-ventilation hose on the engine.

Figure 34-10 shows in sectional view a charcoal canister similar to the one in Fig. 34-7, except that a vapor-vent valve has been added. This valve is operated by a vacuum signal from the carburetor. When the engine is off, there is no manifold vacuum and no vacuum signal. Therefore, the spring pushes the valve down so it is open. Now, any

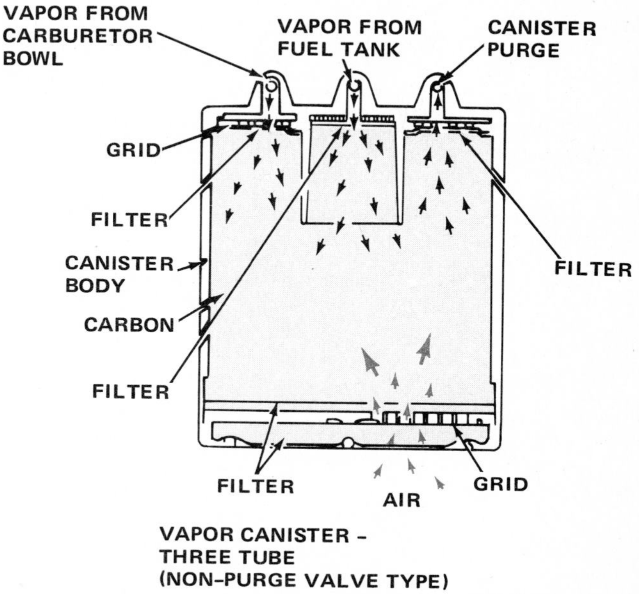

Fig. 34-7 Charcoal canister for a V-6 engine. (*Buick Motor Division of General Motors Corporation*)

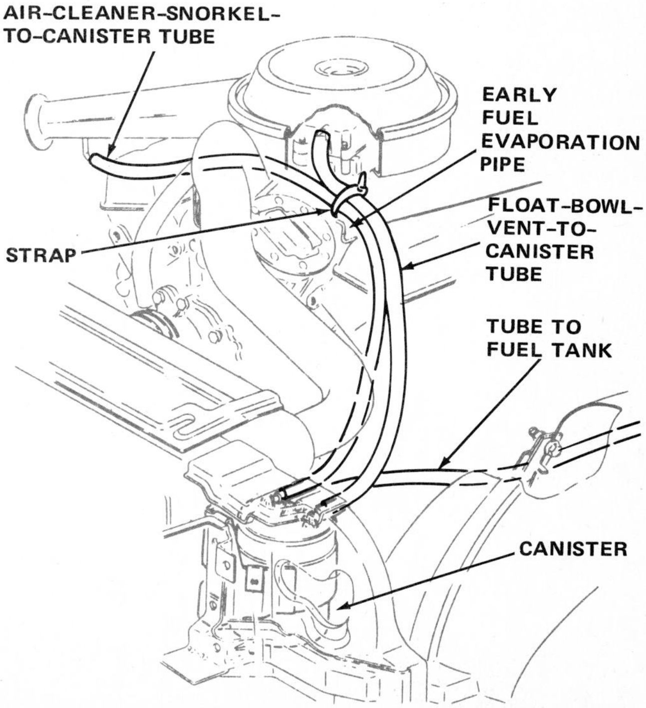

Fig. 34-8 Hose routing for a fuel-vapor emission-control system. (*Buick Motor Division of General Motors Corporation*)

vapor forming in the carburetor float bowl can flow through a tube and into the canister. When the engine starts, vacuum from the intake manifold pulls the vapor-vent valve closed. This prevents any air leak from the canister back into the float bowl. Such a leak could upset the carburetor calibration. Recall, from our discussions of carburetors, that most are balanced (● 13-12). Allowing air leaks to the float bowl could unbalance the carburetor and result in an excessively rich air-fuel mixture. This vapor-vent valve has a more positive action than the pressure-relief valve used in some carburetors (Fig. 13-11).

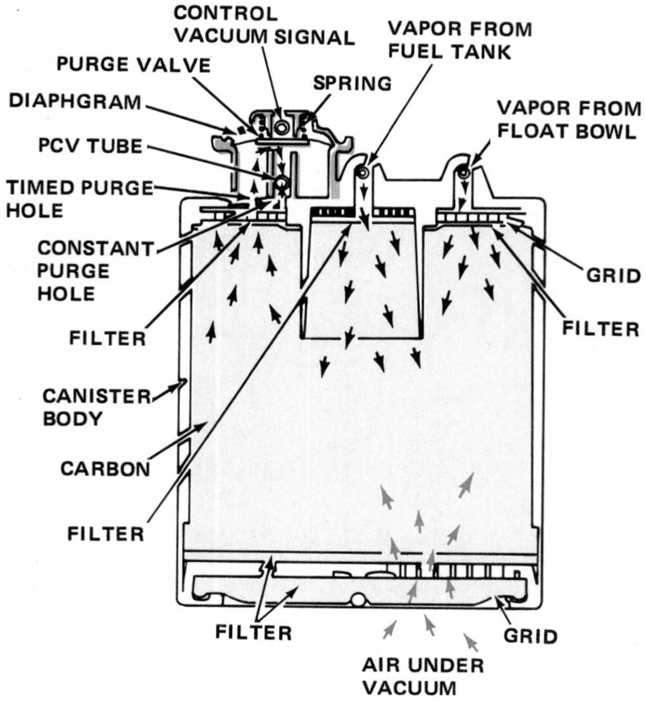

Fig. 34-9 Charcoal canister for a four-cylinder engine. *(Pontiac Motor Division of General Motors Corporation)*

Figure 34-11 is a sectional view of a charcoal canister having both the purge valve, shown in Fig. 34-9, and the vapor-vent valve, shown in Fig. 34-10. This canister is used on cars sold in areas demanding the most rigid fuel-vapor control.

Many of the canisters we have described so far are cylindrical, as shown in Fig. 34-8. Recent Ford Motor Company cars have used a rectangular canister (Fig. 34-12). Some Ford models with big engines and large fuel tanks use two canisters (Fig. 34-13). One canister handles the fuel-tank vapors, the other handles the carburetor float-bowl vapors. Float-bowl venting on many Ford models is handled by a purge control valve, a solenoid vent valve, or a thermal vent valve.

The purge control valve (Fig. 34-14) is similar to the unit previously described (Fig. 34-9). But it also handles the vapor from the fuel tank. When the engine is running above idle, the vacuum opens the purge valve to allow a free flow of air through the canister. When the engine is idling, the valve closes to reduce the air flow.

The float-bowl solenoid vent valve (Fig. 34-15) is located in the line from the bowl to the canister. When the engine is turned off, the valve is open to allow vapor to flow from the bowl to the canister. When the engine is turned on, the solenoid is connected to the battery through the ignition switch. This causes the solenoid to

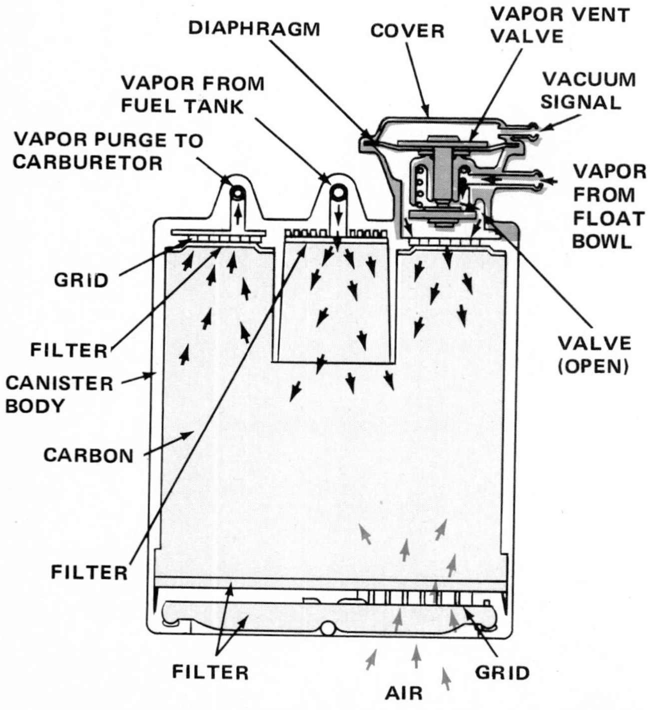

Fig. 34-10 Charcoal canister with added vapor-vent valve. *(Pontiac Motor Division of General Motors Corporation)*

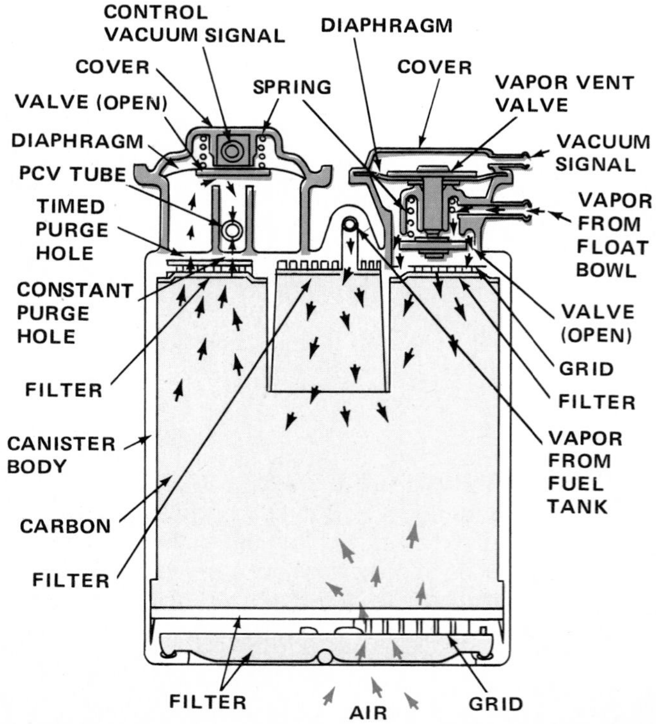

Fig. 34-11 Charcoal canister with purge valve and vapor-vent valve. *(Pontiac Motor Division of General Motors Corporation)*

close the valve. It does the same job as the vapor-vent valve, shown in Fig. 34-10. However, it does the job electrically. The valve in Fig. 34-10 does the job by vacuum. Figure 13-11 shows another solenoid vent valve installed in the carburetor. See also ● 13-12.

The float-bowl thermal vent valve is connected in the line from the float bowl and the canister (Fig. 34-16). Its purpose is to prevent fuel-tank vapors from being vented through the carburetor float bowl. This tends to happen when the engine is cold and the fuel tank is hot. For example, a car sitting out all night will be cold. But when the sun strikes it, the fuel tank will warm up much faster than the engine. This could send some fuel vapor from the tank through the float bowl. The bimetal plate in the thermal vent valve, however, distorts when cold and seals off the line. When the engine warms up, the bimetal plate opens the valve so normal venting can take place.

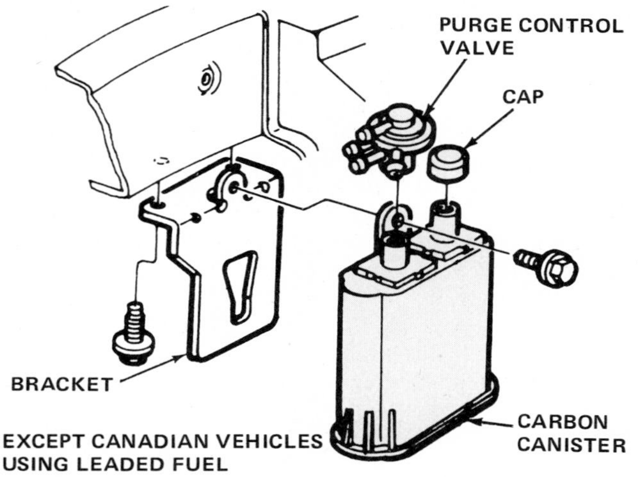

Fig. 34-12 Charcoal canister used by Ford. (*Ford Motor Company*)

● 34-6 Separating Vapor from Fuel

A variety of devices have been used to prevent liquid gasoline from leaking back through the vapor-vent line from the tank to the canister. An early system used a standpipe assembly (Figs. 34-5 and 34-17). It contains a series of pipes that are connected to three vents in the fuel tank (Fig. 34-18). One of the vents is always above the fuel level so it can feed vapor through the standpipe assembly to the vent pipe connected to the canister.

Another type of vapor-fuel separator is shown in Fig. 34-19. In this system, a liquid check valve is used. The liquid check valve will pass air or vapor but not liquid. This prevents liquid gasoline from getting to the canister.

Another type of vapor separator is shown in Fig. 34-20. It is mounted on top of the tank. It is filled with filter material that will pass vapor but not liquid.

Figure 34-21 shows another type of vapor-fuel separator. As long as vapor flows through, the float stays down. But if liquid enters, the float goes up and closes the orifice to the tube leading to the canister.

Many fuel tanks that mount horizontally have a dome, or else the filler pipe on the tank is slightly below the top of the tank. In either case, the tank cannot be completely filled. If it were possible to completely fill the tank, then expansion of the gasoline as it warmed up would send it spilling out through the fuel-tank cap or through the canister. Some fuel tanks also used an internal expansion tank which served the same purpose—gave the gasoline space in which to expand as it warms up. (Fig. 34-17).

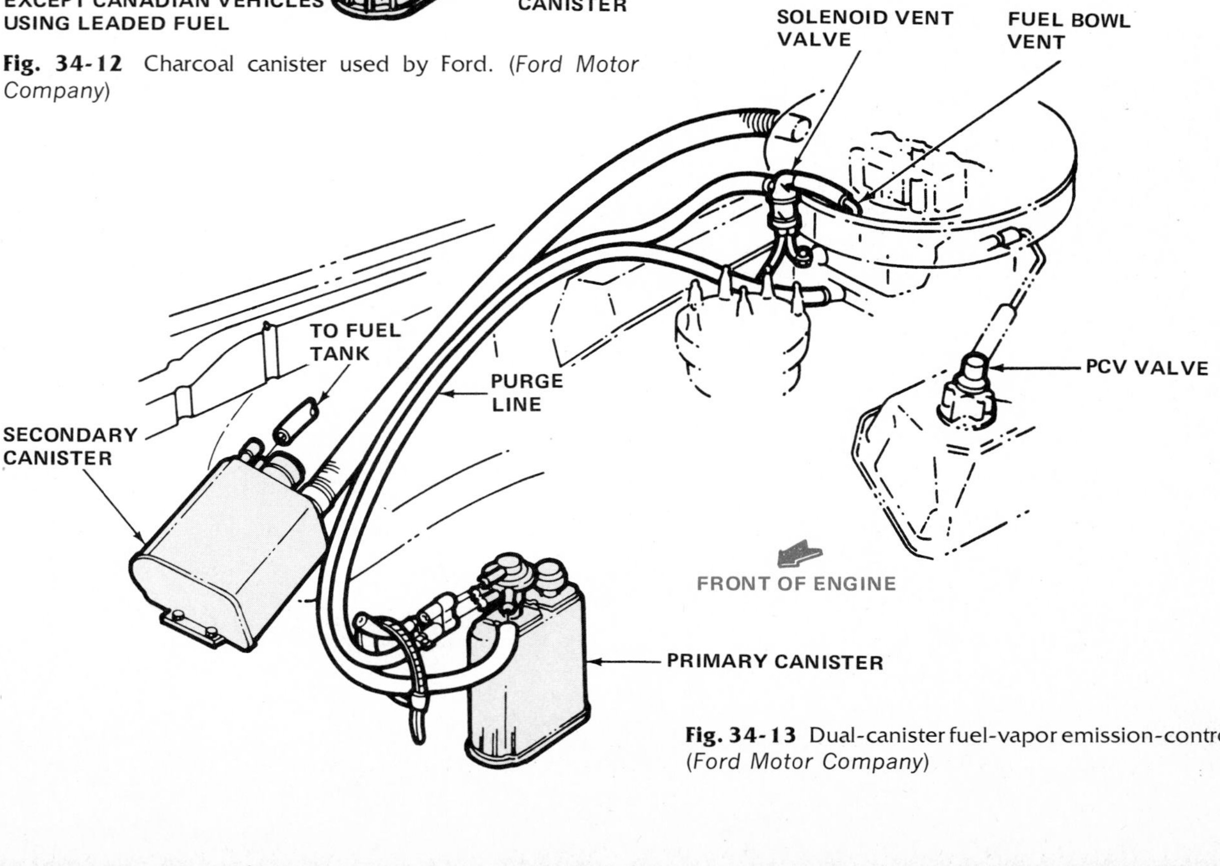

Fig. 34-13 Dual-canister fuel-vapor emission-control system. (*Ford Motor Company*)

Many vapor-vent lines have a rollover check valve (Figs. 12-16 and 34-6). Its purpose is to block the line if the car rolls over and is upside down. This prevents gasoline leakage back through the line to the carburetor or canister. This is described in ● 34-9. See also the two locations of the rollover check valves in many late-model Chrysler Corporation cars (Fig. 34-6).

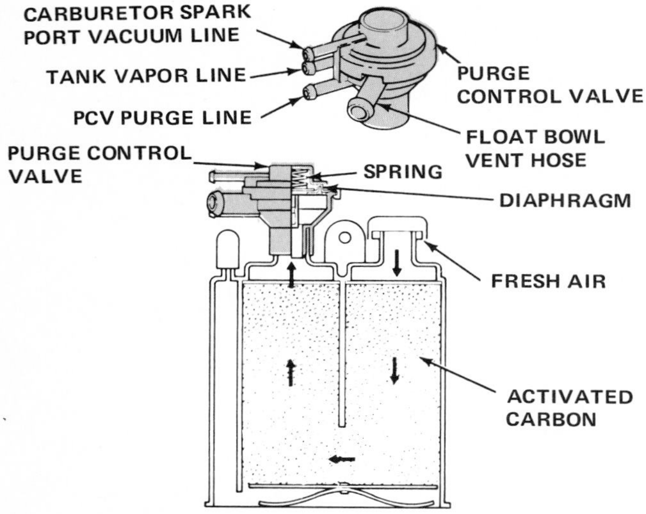

Fig. 34-14 Canister with purge control valve. (*Ford Motor Company*)

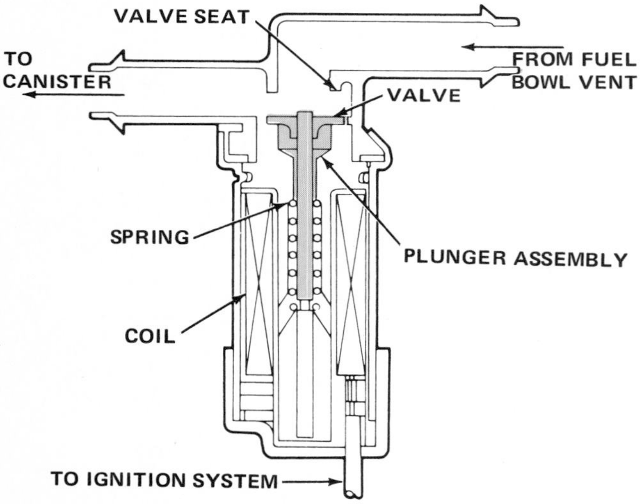

Fig. 34-15 Float-bowl solenoid vent valve. (*Ford Motor Company*)

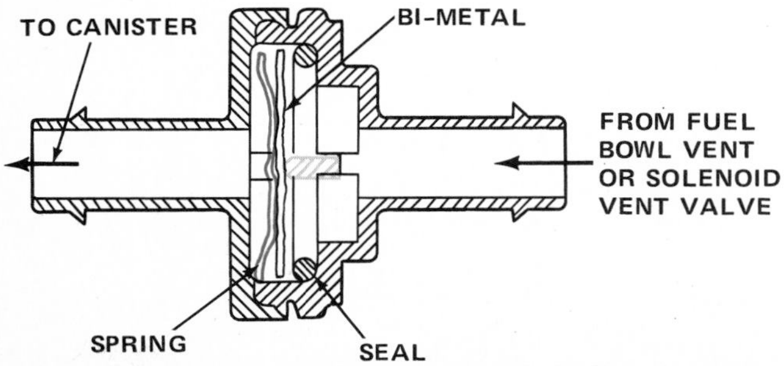

Fig. 34-16 Float-bowl thermal vent valve. (*Ford Motor Company*)

● 34-7 Sealed Fuel Tank

The fuel tank must be sealed to prevent the escape of gasoline vapor into the atmosphere as the tank "breathes" (● 34-2). Figure 34-22 shows the special fuel-tank filler cap used. It has a two-way relief valve, operating on both pressure and vacuum. When gasoline is withdrawn from the tank, a slight vacuum develops. The vacuum valve opens to admit air. If the pressure builds up excessively, the pressure valve opens to relieve the pressure. Normally, excessive pressure would not develop because any pressure is relieved through the fuel-vapor emission-control system. Some filler caps have a rollover check valve which closes if the car is in an

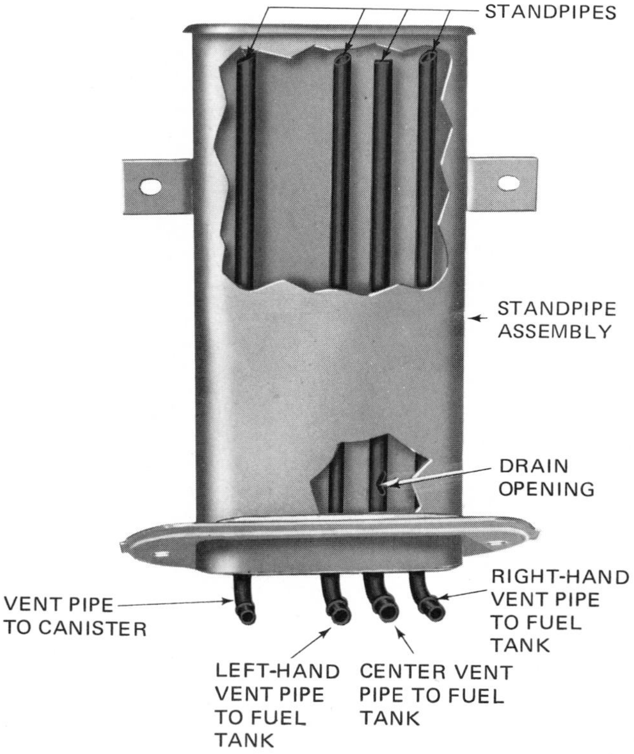

Fig. 34-17 Cutaway view of a standpipe assembly. (*Oldsmobile Division of General Motors Corporation*)

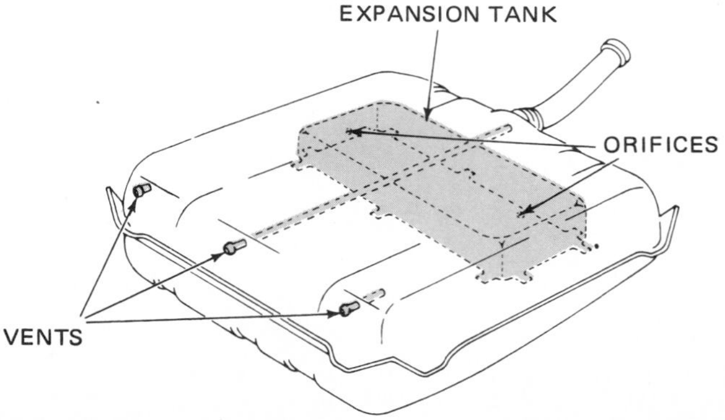

Fig. 34-18 Fuel tank for a fuel-vapor emission-control system. (*Pontiac Motor Division of General Motors Corporation*)

accident that rolls it over. When the car, and cap, are upside down, the check valve closes to prevent leakage of gasoline from the tank. See Fig. 12-6.

34-8 Carburetor Insulation

Some carburetors use an insulator (Fig. 34-23) to reduce heat flow from the engine to the float bowl. The insulator is placed between the carburetor and the intake manifold. Also, some carburetors have an insulator between the throttle body and the float bowl. Either position forms a heat barrier to the float bowl. This reduces fuel evaporation from the float bowl after the engine is turned off. Another arrangement uses an aluminum heat-dissipating plate which sticks out, as shown in Fig. 34-24.

34-9 Vapor Storage in Crankcase

Some Chrysler Corporation cars have used the crankcase to store gasoline vapors from the fuel tank and carburetor. When the engine is stopped, gasoline vapors from the vapor separator at the fuel tank flow to the crankcase air cleaner. From there, they flow down into the crankcase. At the same time, fuel vapors from the carburetor float bowl flow down into the crankcase. The vapors are two to four times as heavy as air. Thus, they sink to the bottom of the crankcase. Then, when the engine is started, the positive-crankcase-ventilating system clears the crankcase of the vapors. The vapors are carried up into the intake manifold and then into the engine, where they are burned.

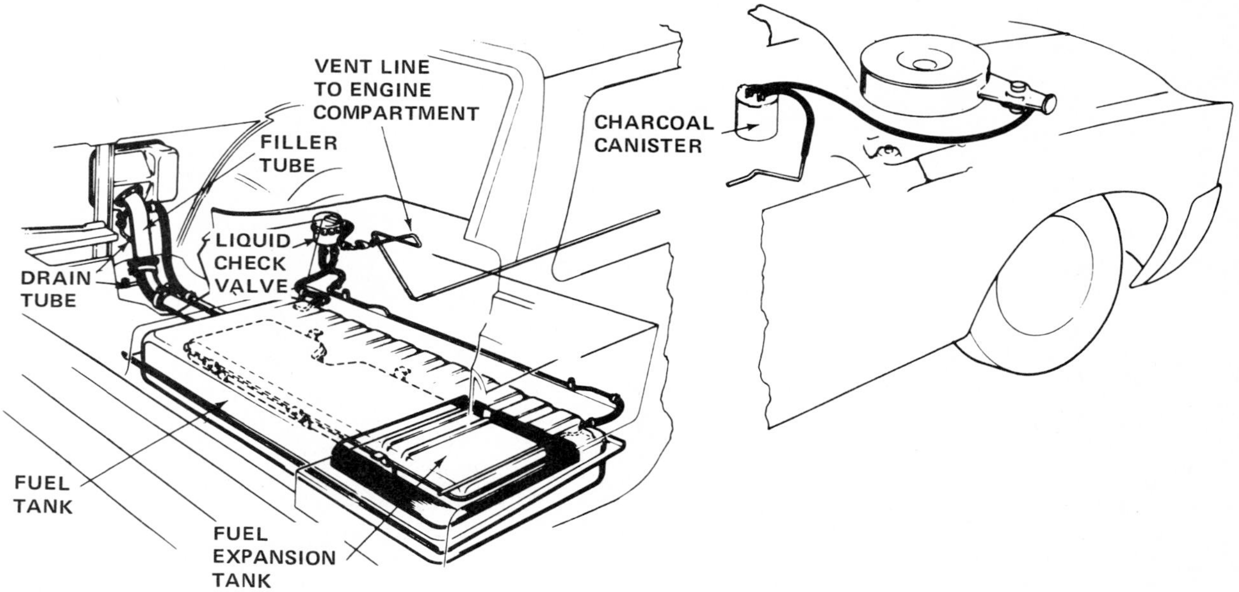

Fig. 34-19 Fuel-vapor emission-control system using a liquid check valve. (*American Motors Corporation*)

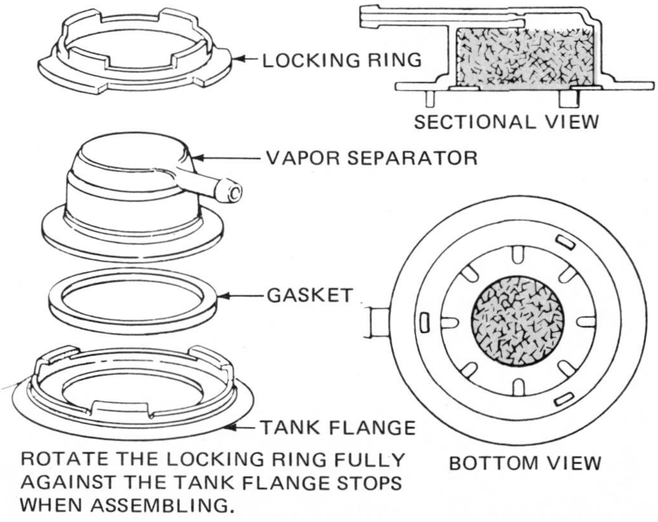

Fig. 34-20 Vapor separator using filter material. (*Ford Motor Company*)

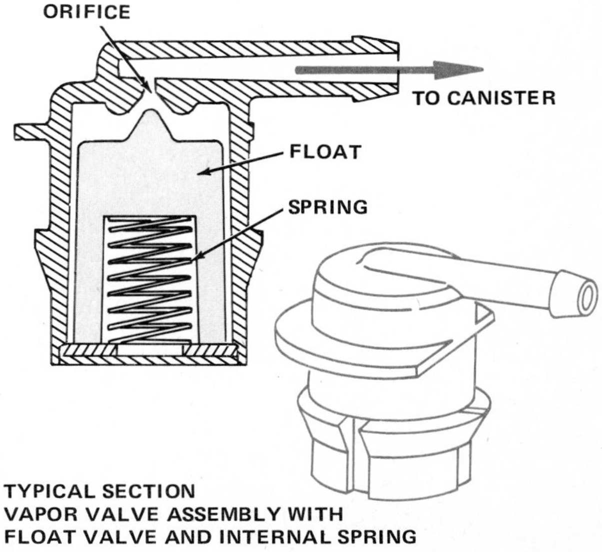

Fig. 34-21 Vapor separator using a float. (*Ford Motor Company*)

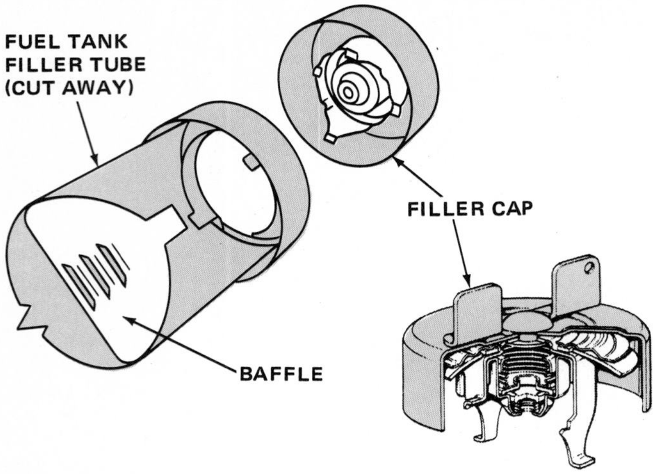

Fig. 34-22 Fuel-tank filler cap and filler tube cut away to show baffle. *(Chrysler Corporation)*

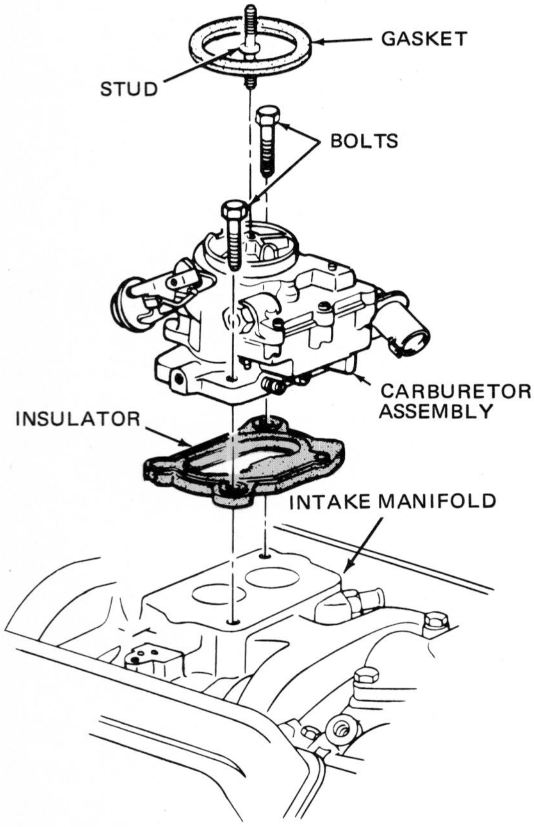

Fig. 34-23 Carburetor insulator placed between the carburetor and the intake manifold. The insulator blocks passage of heat to the carburetor. It thus reduces evaporation of fuel from the float bowl. *(Chevrolet Motor Division of General Motors Corporation)*

REVIEW QUESTIONS

1. What is the purpose of the PCV system? How does it work?
2. Explain how the PCV valve works.

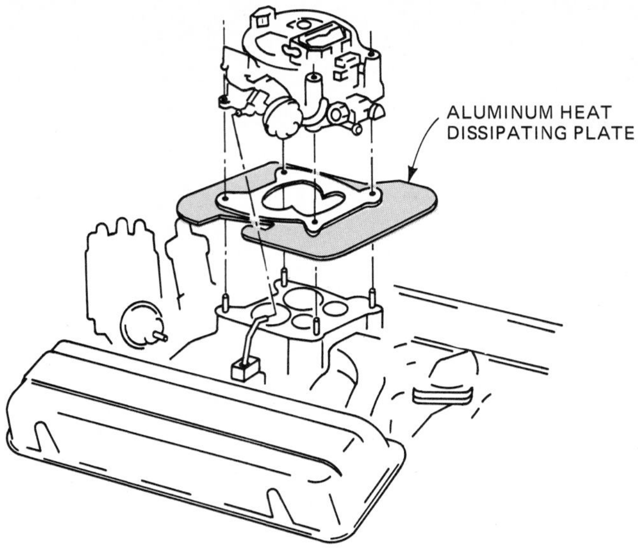

Fig. 34-24 Insulator and aluminum heat-dissipating plate between the carburetor and intake manifold, to reduce heat flow to the carburetor.

3. Explain the purpose and operation of a fuel-vapor emission-control system.
4. From what two places can gasoline vapors escape?
5. In which direction does the air-vapor mixture from the fuel tank pass through the charcoal canister, up or down?
6. When the engine starts, in which direction does fresh air pass through the charcoal canister, up or down?
7. What is the purpose of the fuel-return line?
8. Explain the purpose and operation of the float-bowl solenoid vent valve.
9. Describe three methods of separating vapor from liquid fuel.
10. What is special about the fuel-tank cap used on a fuel tank for a vapor-recovery system?
11. What is the purpose of the carburetor insulator?

SELF PROJECT

It is always instructive to look inside components to see how they are made. It is difficult to tear open a PCV valve. And, if you did, there wouldn't be much to see except the valve and spring. Likewise, the charcoal canister is little more than a plastic container filled with charcoal grains. *Don't try to open a fuel tank!* It is very difficult to remove all gasoline and gasoline vapor from a fuel tank. An empty fuel tank that still has gasoline vapor in it is a potential bomb. A spark from a chisel or hacksaw, or from an acetylene torch, could cause a terrible explosion. But you can examine these units and note their locations in different cars. Notice especially the PCV hoses and connections and the various ways the canisters and other components of the vapor emission-control systems are arranged. Make sketches of these on different cars for filing in your notebook.

CHAPTER 35
Cleaning Up the Exhaust Gas

After studying this chapter, you should be able to:

- **Describe the three methods of cleaning up the exhaust gas.**
- **Identify and point out the components of, and connections for, the exhaust-gas recirculation system, the air-injection or thermactor system, vacuum advance-control system, and catalytic converters.**
- **Explain how each system works.**

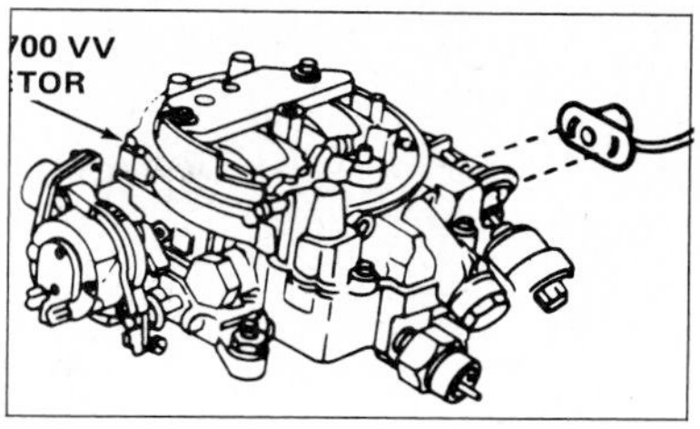

In the previous chapter, we described the positive-crankcase-ventilating system and the fuel-vapor emission-control system. These cover three of the four sources of automotive pollution (see Fig. 33-4). In this chapter, we consider the steps that have been taken to clean up the fourth source of pollution—exhaust gas coming from the tail pipe.

There are three ways of cleaning the exhaust gas: (1) controlling the air-fuel mixture, (2) controlling combustion, and (3) treating the exhaust gas.

35-2 Controlling the Air-Fuel Mixture

Gasoline has been changed to make it burn cleaner and more completely. One of the changes is the elimination of lead. Lead is now being removed to permit the use of catalytic converters, which we shall discuss later in the chapter. Removing the lead has required a reduction of compression ratios.

Basically, controlling the air-fuel mixture has meant (1) modifying the carburetor to deliver a leaner air-fuel mixture, and (2) faster warm-up and quicker choke action.

We have covered some of the items we now discuss before. But we review them here so the complete story on cleaning up the exhaust gas is in this chapter.

35-3 Leaner Idling Air-Fuel Mixture

Modern carburetors have an idle limiter (Fig. 35-1). The idle-mixture adjustment screw is adjusted at the factory. Then the idle-limiter cap is installed. The cap permits a small amount of adjustment of the idle-mixture adjustment screw. The cap can be removed, of course, if the carburetor requires a major overhaul. But it must then be reinstalled, as explained in Chap. 14. Some carburetors have the screw sealed with a steel plug. The carburetor must be removed from the engine and the plug knocked out to adjust the screw.

35-4 Faster Warm-Up

If the air-fuel mixture coming from the carburetor is cold, only part of the fuel will vaporize. This means that an extra-rich mixture is needed. Otherwise, the engine will not get enough gasoline vapor for it to run. Of course, the situation changes as soon as the engine begins to run. Then, the hot exhaust gas circulating around the manifold heat-control valve begins to heat the intake manifold. (See ● 13-26.) However, this is too slow for the new systems. So a thermostatically controlled air cleaner is used to provide heated air quickly to the carburetor when the engine is cold. This system, called the *heated-air*

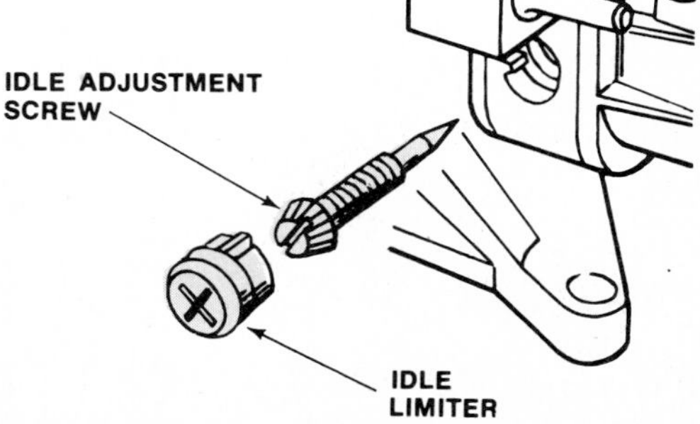

Fig. 35-1 Idle limiter on a carburetor.

system, is discussed in detail in ● 12-12. Figure 35-2 shows the system installed on a V-8 engine.

As you will recall, this system includes a control-damper assembly in the snorkel of the air cleaner. Also, it has a heat stove that surrounds the exhaust manifold. When the engine is cold, the temperature sensor in the air cleaner closes the damper. In this position, all air has to come from the heat stove. When the engine starts and the exhaust manifold begins to warm up, hot air is delivered to the carburetor. This improves cold and warm-up operation.

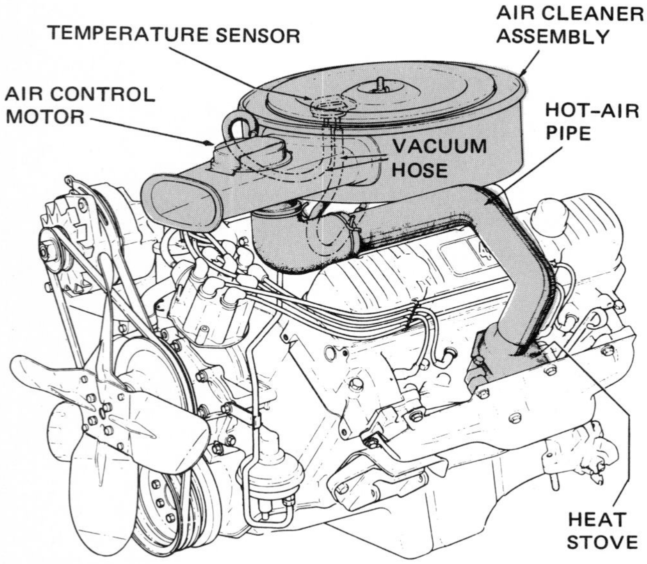

Fig. 35-2 Heated-air system on a V-8 engine. (*Buick Motor Division of General Motors Corporation*)

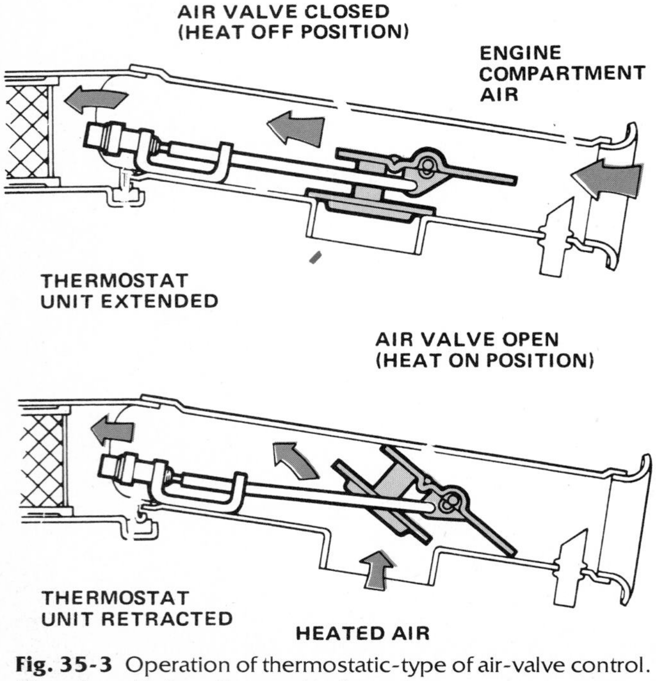

Fig. 35-3 Operation of thermostatic-type of air-valve control. (*American Motors Corporation*)

As the engine begins to warm up, the ingoing-air temperature rises above 100°F [37.8°C]. This causes the temperature sensor to open the air valve (control damper) so air can enter from the engine compartment. There are two kinds of temperature-sensor controls (Figs. 35-3 and 35-4). One type uses a thermostat unit (Fig. 35-3). It elongates as it warms up and this shuts off the duct to the heated air and opens the passage to the engine compartment. The other temperature-sensor control uses an air bleed that is opened or closed by a thermal sensor (Fig. 35-4). When the air bleed is closed, manifold vacuum can operate the vacuum motor, as shown, to admit heated air.

The system allows the engine to start and operate satisfactorily when cold, even though the idle mixture is lean.

● 35-5 Faster-Acting Choke

In earlier systems, the choke operated only by engine heat (● 19-25). Late-model cars have electric-assist chokes (Fig. 35-5). The choke thermostat is subjected to

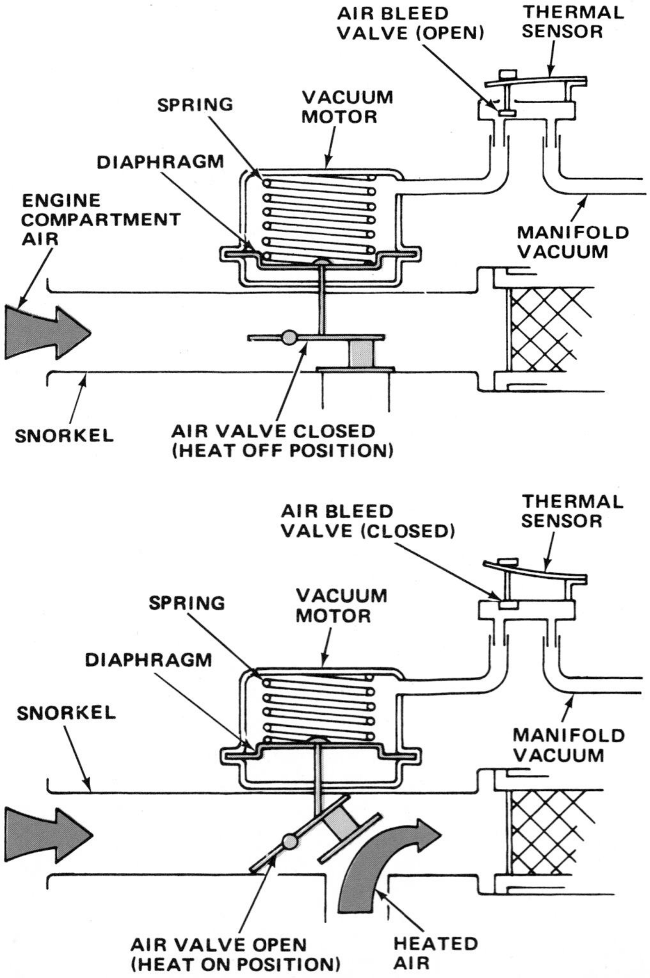

Fig. 35-4 Operation of thermal-sensor type of air-valve control. (*American Motors Corporation*)

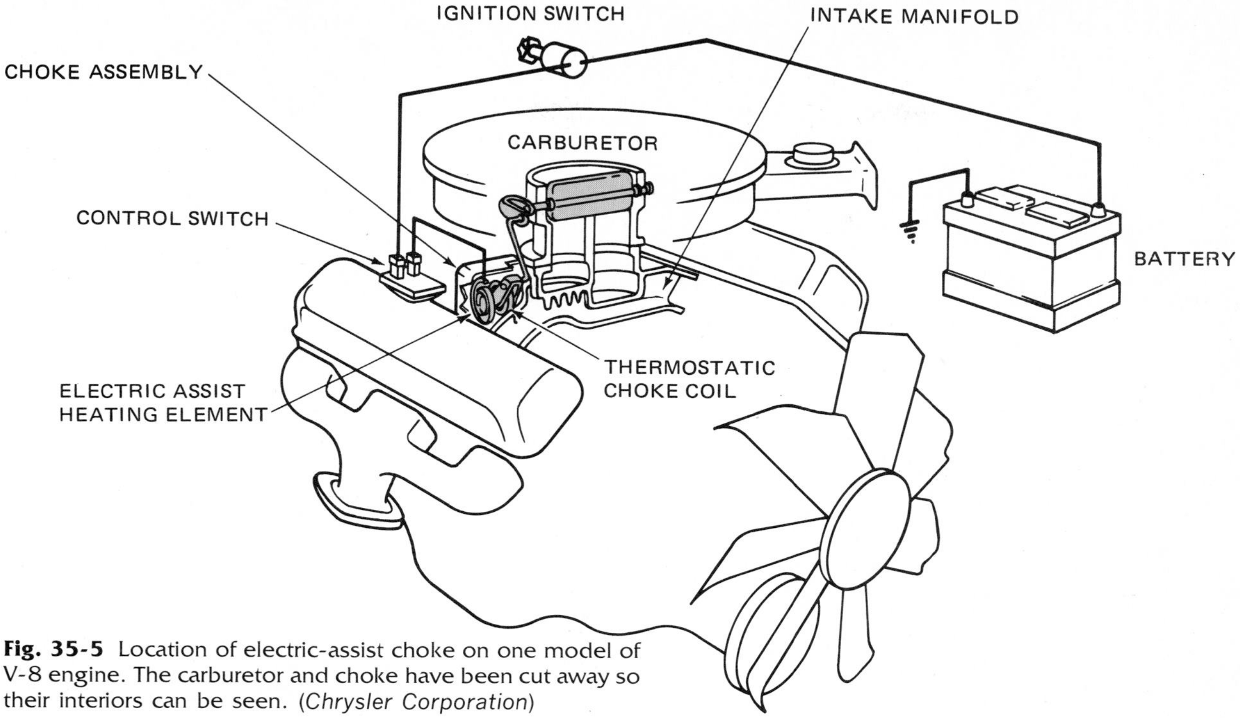

Fig. 35-5 Location of electric-assist choke on one model of V-8 engine. The carburetor and choke have been cut away so their interiors can be seen. *(Chrysler Corporation)*

heat from a heating element in the choke, and heat from the exhaust manifold. Therefore, the choke acts faster and opens more quickly. This reduces the amount of time during which the engine operates in a choked condition. With the choke valve closed, the engine is being fed a very rich air-fuel mixture. The exhaust gas is thus loaded with unburned HC and with CO. The electric-assist choke reduces the length of time during which these pollutants are fed into the atmosphere. There is more information on the electric-assist choke in ● 13-25.

● 35-6 Controlling the Combustion Process

The combustion process seems simple at first glance. A mixture of air and gasoline vapor is compressed in the combustion chamber. A spark ignites it. It burns and produces the high pressure that pushes the piston down. However, the process is complicated. Here are some of the factors involved:

1. The layers of air-fuel mixture next to the relatively cool cylinder head and pistonhead do not burn. The metal surfaces chill these layers below the combustion point. So the unburned fuel is swept out of the cylinder on the exhaust stroke. This adds polluting HC to the atmosphere. There are two methods of combating this problem. One is to use stratified charge or fuel injection. The other is to reduce the surface area surrounding the combustion chamber. We shall come back to these two methods later.
2. Increasing the combustion temperature improves combustion of the fuel. But the higher temperature produces more nitrogen oxides (NO_x), and this produces another problem. More on this later.

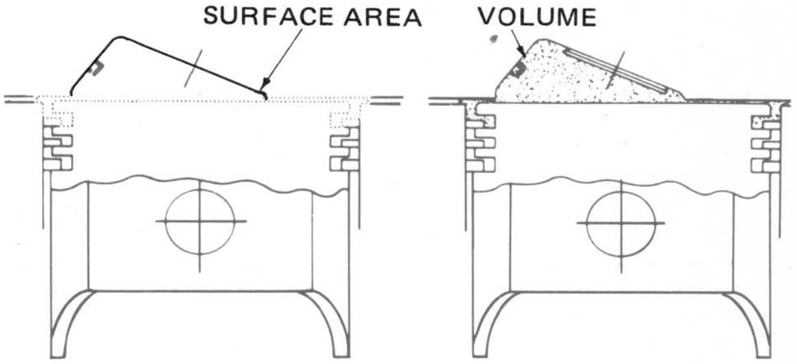

Fig. 35-6 The ratio of the surface area to the volume of the combustion chamber, or *S/V*. This ratio has an effect on the amount of unburned hydrocarbons in the exhaust gas.

3. Vacuum advance gives the air-fuel mixture a longer time to burn when the engine operates at part throttle. But it also gives more time for NO_x to form under certain operating conditions. So a means must be provided to kill the vacuum advance during these special operating conditions. We cover this in detail later.

● 35-7 Reducing Combustion-Chamber Surface Area

Actually, what we are talking about here is reducing the *S/V* ratio (Fig. 35-6), that is, the ratio between the surface area and the volume of the combustion chamber. A sphere has the lowest possible *S/V* ratio. The wedge combustion chamber (Fig. 35-7) has a higher *S/V* ratio. Thus, the hemispheric combustion chamber has a lower surface area. It has less surface to chill the air-fuel mixture and, therefore, produces a lower percentage of unburned HC in the exhaust.

35-8 Stratified Charge

The stratified-charge engine has a means of concentrating rich mixture in the center of the compressed air-fuel mixture (Fig. 35-8). During combustion, the burning rich mixture spreads outward and moves into areas where the mixture is lean and harder to ignite. With stratified charging, a much leaner air-fuel mixture, on the average, can be used. The combustion takes place largely in and around the concentration of rich mixture. This means that the fuel is more completely burned. The amount of pollutants, such as carbon monoxide, unburned gasoline, and nitrogen oxide, is reduced.

One way to achieve stratified charging is to give the air-fuel mixture a swirling motion as it enters the cylinder. This can be done by careful placement of the intake port.

Another method of centering the combustion is to use a turbulence generating pot (TGP) as developed by Toyota. It is described in ● 7-9 and illustrated in Fig. 7-17.

There is also the so-called Honda system. Here, a separate small precombustion chamber is used. This precombustion chamber has the spark plug and its own intake valve. Figure 35-9 is an outline view of the engine showing the valves, spark plug, and pistons for one cylinder. Figure 35-10 shows how the arrangement works. In operation, the carburetor delivers a very lean mixture to the main combustion chamber, and a very rich mixture to the precombustion chamber. Ignition takes place in the precombustion chamber. The rich mixture, under the high pressure of combustion, streams out into the main combustion chamber, as shown in 3 and 4 of Fig. 35-10. There, it mixes with the lean mixture, and combustion continues. This assures good burning of the fuel so that polluting gases—carbon monoxide, unburned fuel, and nitrogen oxides—are kept to a low level. Figure 35-10 shows the sequence of actions.

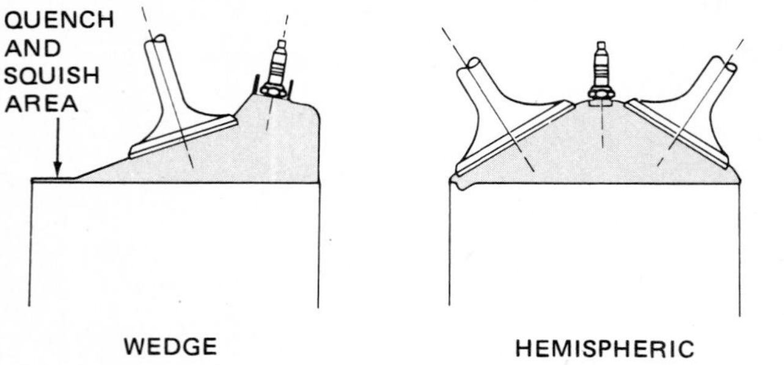

Fig. 35-7 Wedge and hemispheric combustion chambers. (*General Motors Corporation*)

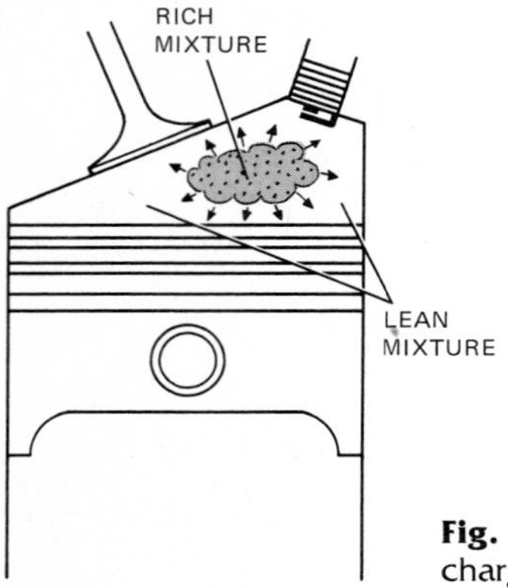

Fig. 35-8 Principle of stratified charging.

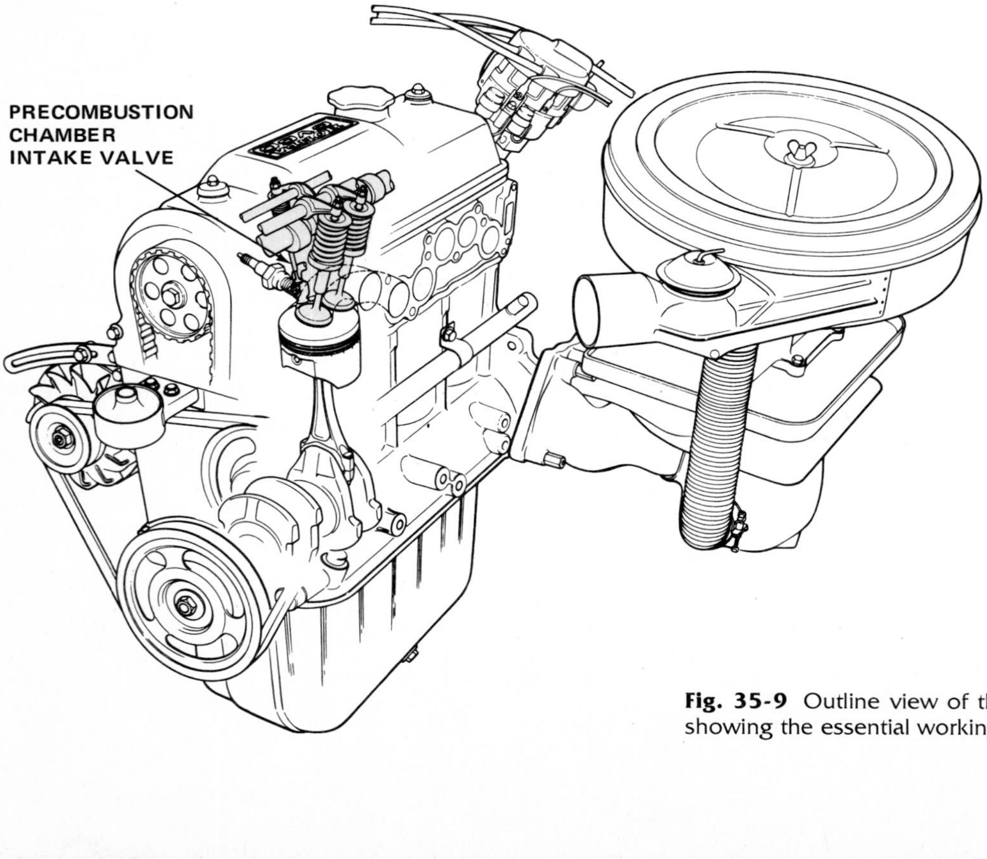

Fig. 35-9 Outline view of the Honda four-cylinder engine, showing the essential working parts of one cylinder. (*Honda*)

● 35-9 Fuel Injection

We have already described fuel-injection systems for gasoline engines in Chap. 15. Fuel injection can improve combustion, and reduce HC and CO in the exhaust. For one thing, the fuel-injection system more accurately meters the fuel. It supplies the same amount of fuel to each cylinder. In the carburetor-type system, some cylinders can get a richer mixture than others.

● 35-10 Increasing Combustion Temperature

Increasing the combustion temperature reduces CO and unburned HC in the exhaust. But it increases the formation of NO_x. One method of reducing NO_x is to reduce the compression ratio. This reduces top combustion temperatures and thus the amount of NO_x that is formed.

Other NO_x-reduction methods are used. One of these reduces NO_x during normal running of the engine (the exhaust-gas recirculation, or EGR system). A second method reduces NO_x during acceleration in lower gears and during part-throttle, high-vacuum conditions. A third method uses a three-way catalytic converter that converts the NO_x into nitrogen (N) and oxygen (O). We discuss all of these in following sections.

● 35-11 Exhaust-Gas Recirculation

If a small part of the exhaust gas is sent back through the engine, it reduces the combustion temperature and lowers the formation of NO_x. The amount sent through the engine should vary according to operating conditions. The simplest system is shown in Fig. 35-11. There, a special passage connects the exhaust manifold with the intake manifold. This passage is opened or closed by a special exhaust-gas recirculation (EGR) valve. The upper part of this valve is sealed. It is connected by a vacuum line to a signal port in the carburetor, as shown. When there is no vacuum at work on the signal port, there is no vacuum in the EGR valve. The spring holds the valve closed. No exhaust gas recirculates. This is the situation during idling, when NO_x formation is near minimum.

However, when the throttle is opened, it passes the signal port. This allows the intake-manifold vacuum to operate the EGR valve. The vacuum raises the diaphragm in the valve. This lifts the valve off the seat. Now exhaust gas can pass into the intake manifold. There, it mixes with the air-fuel mixture and enters the engine cylinders. The exhaust gas lowers the combustion temperature and thus reduces the formation of NO_x. Note that, at wide-open throttle, there is little vacuum in the intake manifold. Thus, the EGR valve is nearly closed. At wide-open throttle, there is less need for exhaust-gas recirculation.

Figure 35-12 shows the EGR valve in the fully open position. A thermal vacuum switch on many late-model cars prevents exhaust-gas recirculation until engine temperature reaches about 100°F [37.8°C]. The thermal vacuum switch is also called a coolant-temperature override (CTO) switch. It is connected into the vacuum line between the carburetor and the EGR valve. The CTO switch is mounted in a cooling-system water jacket, so it senses coolant temperature. If this temperature is below 100°F [37.8°C], the switch remains closed. This prevents the vacuum from reaching the EGR valve, so exhaust gas does not recirculate. This improves cold-engine performance for the first few moments of operation. After the engine warms up to where it can tolerate exhaust-gas recirculation, the CTO valve opens. Now vacuum can get to the EGR valve, so that exhaust gas can recirculate.

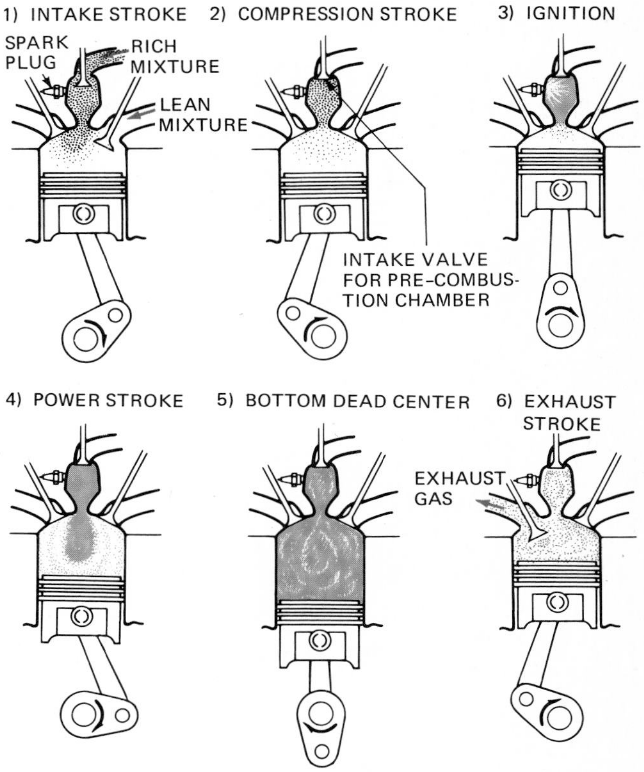

Fig. 35-10 Sequence of actions in the Honda system. (*Honda*)

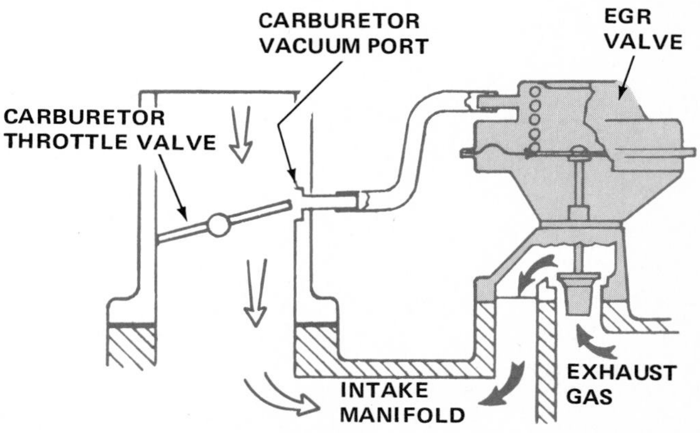

Fig. 35-11 Schematic view of an exhaust-gas recirculation system. (*Chevrolet Motor Division of General Motors Corporation*)

Figures 35-13 and 35-14 show the EGR system for a six-cylinder and an eight-cylinder engine. There are other mounting arrangements for the EGR valve and coolant-temperature override (CTO) switch. Note that the EGR valve is a combination EGR valve and back-pressure sensor. The valve functions as previously described (see Fig. 35-12). The back-pressure sensor has been added to control the flow of exhaust gas into the intake manifold in accordance with the exhaust-gas (or back) pressure in the exhaust manifold.

When the engine load is light and back pressure is low, the back-pressure sensor keeps the EGR valve closed so there is no exhaust-gas recirculation. But when power demands are made on the engine, the exhaust back pressure goes up. Now the back-pressure sensor allows the EGR valve to open so recirculation can proceed.

Figure 35-15 shows how it works. To the left, we see the condition when the engine load is light. The back pressure from the exhaust is low. It enters through the small hole in the stem of the EGR valve. It works on the space between the diaphragm and the plate under it. But it is not great enough to have any influence. Air can flow, as shown by the arrows, and satisfy the vacuum. To the right, we see what happens when the engine load goes up. The back pressure is higher—high enough to lift the diaphragm above the plate. As it moves up, it closes the control valve. Now, no air can flow up to the vacuum chamber. This vacuum pulls up the actuating diaphragm and plate. This lifts the EGR valve off its seat, as shown, so exhaust gas can flow to the intake manifold.

There are several variations of this basic system. For instance, some EGR valves have a second diaphragm. Its purpose is to produce increased exhaust-gas recirculation when the engine is heavily loaded, as during hard acceleration. Also, some high-performance engines use an additional modulator system to provide additional control based on car speed. One system of this type is shown in Fig. 35-16. The modulator system is enclosed in dashed lines. It includes a solenoid valve that is normally open, allowing intake-manifold vacuum to pass through it. When engine temperature is high enough to open the thermal switch, and the throttle is partly opened, intake manifold can operate the EGR valve. Exhaust-gas recirculation results. However, when car speed reaches a certain level, the speed sensor sends a signal to the electronic amplifier. This causes the amplifier to close the solenoid valve. Now the vacuum line is closed, and exhaust-gas recirculation stops.

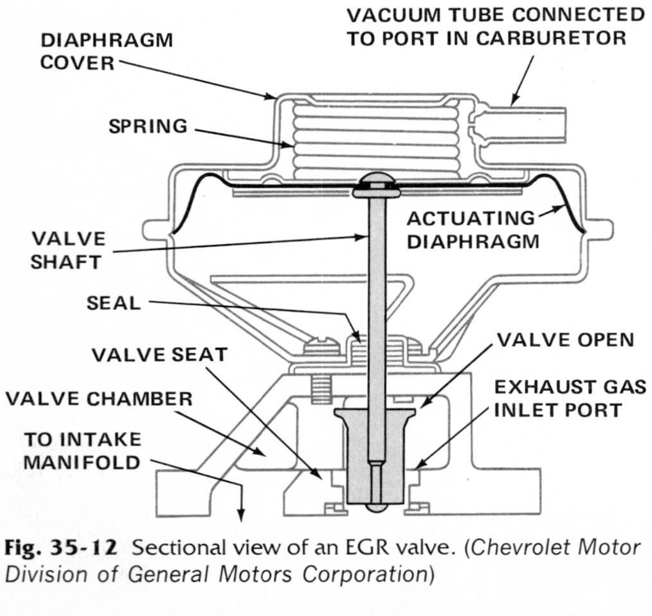

Fig. 35-12 Sectional view of an EGR valve. (*Chevrolet Motor Division of General Motors Corporation*)

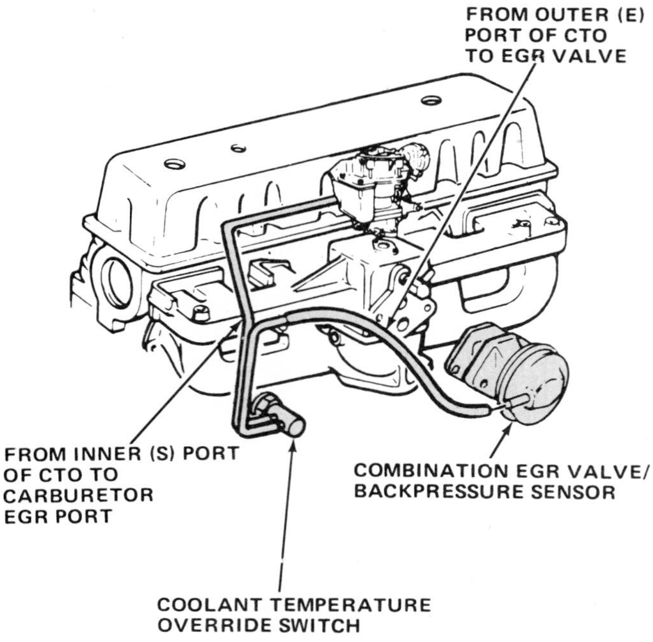

Fig. 35-13 EGR system on a six-cylinder engine. (*American Motors Corporation*)

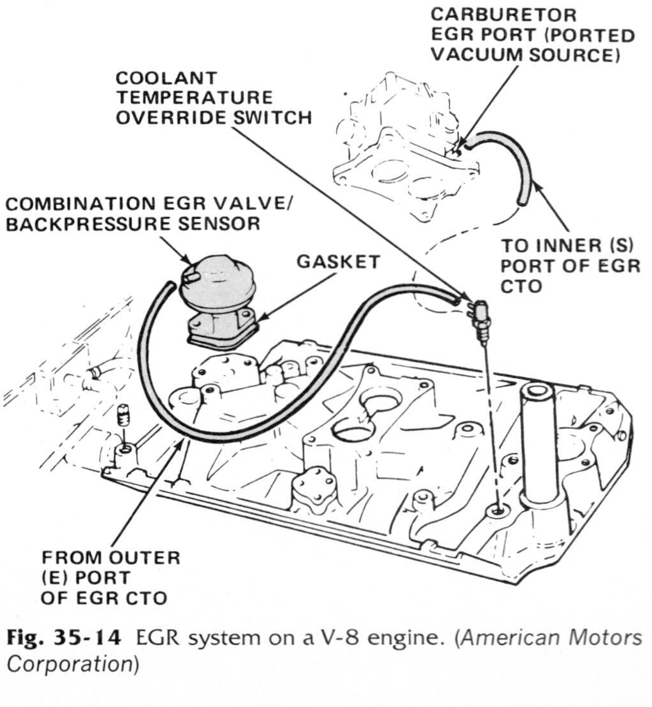

Fig. 35-14 EGR system on a V-8 engine. (*American Motors Corporation*)

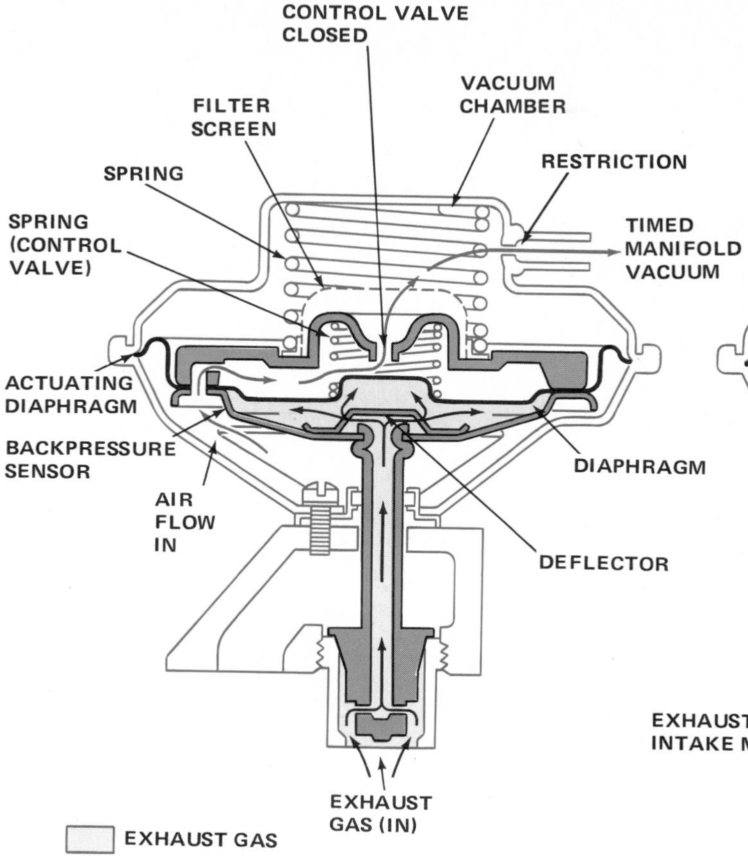

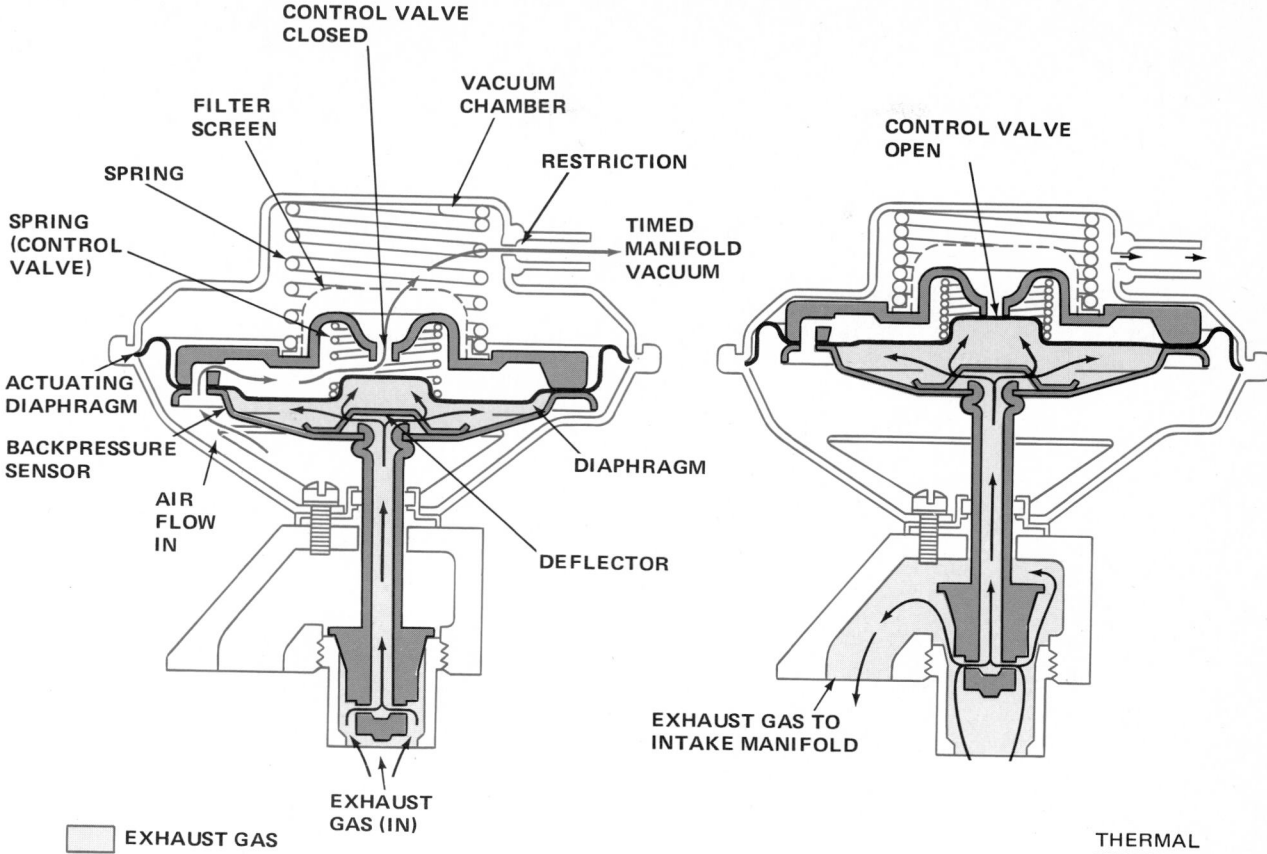

Fig. 35-15 Combination EGR and back-pressure sensor valve. *(Chevrolet Motor Division of General Motors Corporation)*

Figure 35-17 shows a still different arrangement. Here, a vacuum amplifier is used to increase the vacuum enough to operate the EGR valve. At wide-open throttle, recirculation is eliminated by a dump diaphragm inside the amplifier. The amplifier continually compares the vacuum signal going to the EGR valve and the vacuum in the intake manifold. When the comparison shows that the throttle is wide open (vacuum practically the same), the amplifier stops amplification of the vacuum signal. Now, it is not great enough to hold the EGR open so it closes, halting recirculation.

The EGR delay timer and EGR delay solenoid, to the upper left in Fig. 35-17, delay EGR action for 35 seconds after starting a cold engine. This assures reliable starting and initial operation of the engine. After the engine has begun to warm up, EGR can commence without upsetting the operation of the engine. If EGR started immediately after the engine began to run, the engine could stumble and even stall. The EGR delay prevents this.

The reasons for the variety of EGR systems lie in the differing characteristics of the engines and exhaust systems with which they are used. Each EGR system must be tailored to fit the engine it will work with. We have covered several variations of the basic system, but there are still others. However, if you know how the basic system works, you will have no trouble understanding any variation you run across.

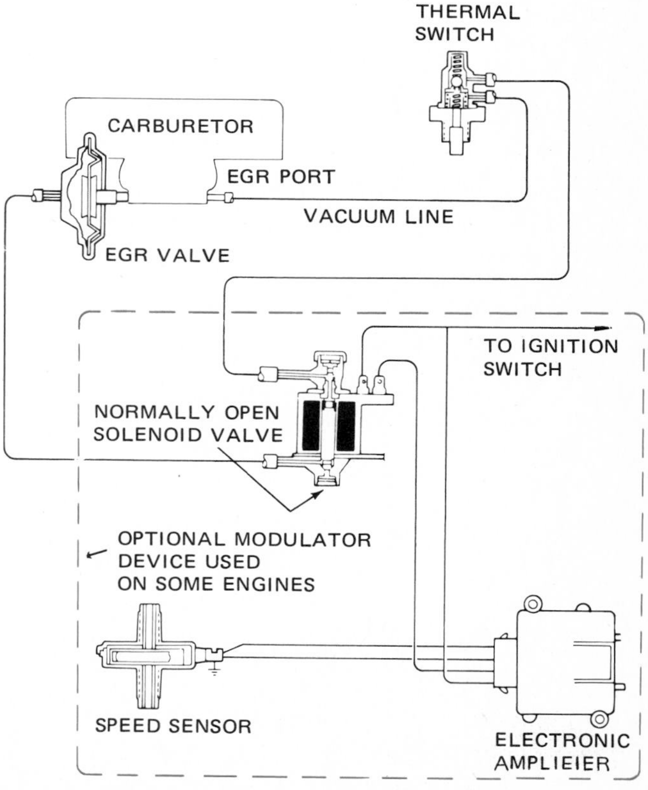

Fig. 35-16 Exhaust-gas recirculation system showing optional modulator device for some engines (shown inside dashed line). *(Ford Motor Company)*

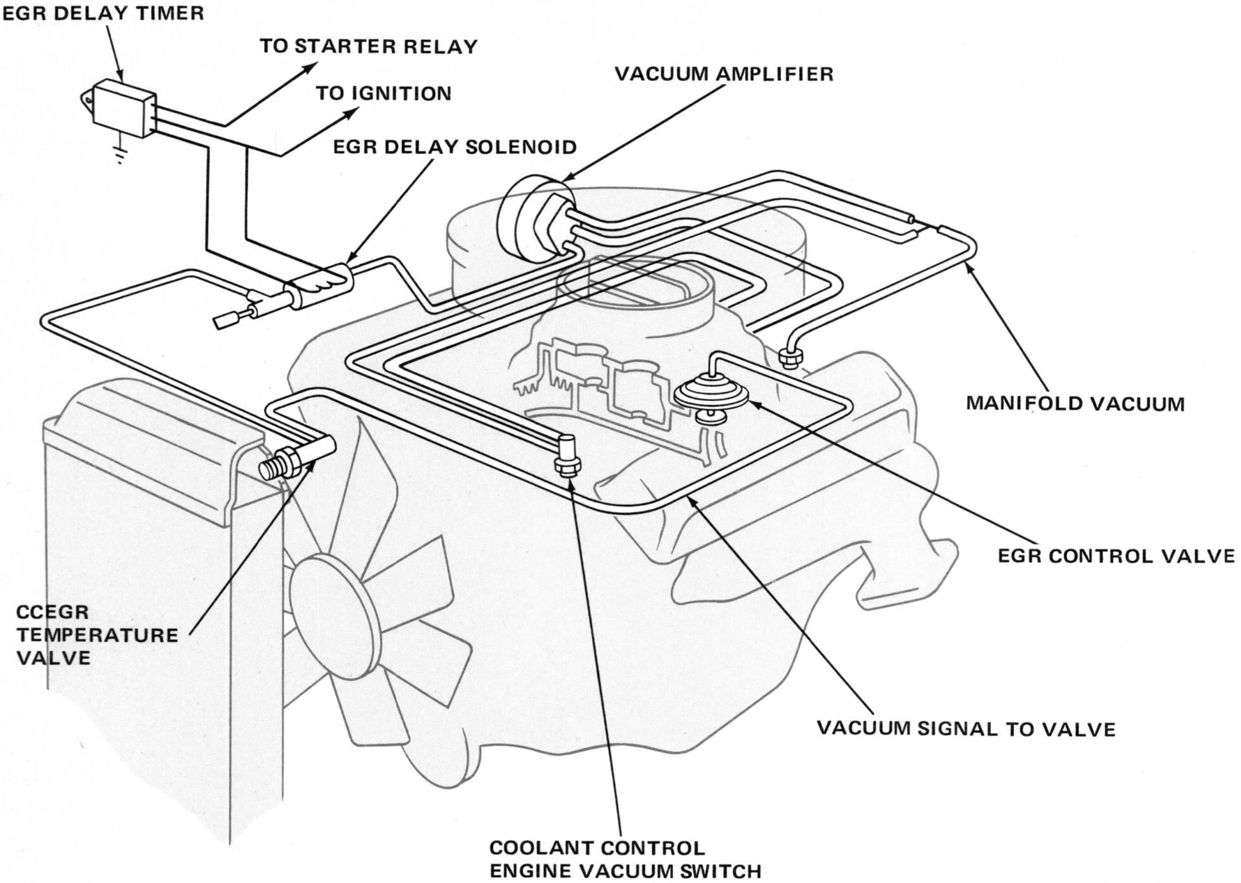

Fig. 35-17 Exhaust-gas recirculation system with vacuum amplifier and EGR delay timer and solenoid. (*Chrysler Corporation*)

35-12 Valve Overlap

One of the features of the Chrysler complete emission-control systems is additional valve overlap. The complete set of systems is shown in Fig. 35-18. Additional valve overlap does the same thing as the EGR system, but in a different way. Increased valve overlap leaves more of the exhaust gas in the cylinders. That is, the intake valve opens while there is still quite a bit of exhaust gas in the cylinder. So exhaust gas mixes with the air-fuel mixture entering the cylinder. The result is that the top combustion temperatures are reduced, and there is less NO_x formation. However, increased valve overlap can cause rough idling.

35-13 Control of Vacuum Advance: TCS System

During part-throttle operation, the distributor vacuum advance operates. This provides more time for the leaner air-fuel mixture to burn. However, this added time also allows more NO_x to develop. Thus, a variety of controls have been used to prevent vacuum advance under certain conditions. For example, Chevrolet uses a transmission-controlled spark (TCS) system on cars with manual transmissions. The TCS system prevents vacuum advance when the car is operated in reverse, neutral, or low forward gears. Under these special conditions vacuum advance could greatly increase the formation of NO_x.

Figure 35-19 shows the Chevrolet TCS system for a six-cylinder engine in a manual-transmission car. The diagram also shows the engine temperature switch (lower left) and the idle-stop solenoid. Figure 35-19 shows the situation during starting. Turning on the ignition switch energizes the idle-stop solenoid. The plunger extends to contact the throttle lever. This prevents the throttle from closing completely, so that idle speed stays high enough. When the engine is turned off, the idle-stop solenoid allows the throttle to close completely. This prevents "dieseling," or the engine running with the ignition off.

Note that turning on the ignition switch completes the circuit through the vacuum advance solenoid and temperature-switch cold terminal. At the same time, the circuit to the 20-second time relay is completed. With either of these circuits complete, the vacuum advance solenoid is energized. Vacuum is admitted to the distributor vacuum advance mechanism so vacuum advance is obtained.

If the engine temperature has gone up enough in low-gear operation, the temperature-switch cold points

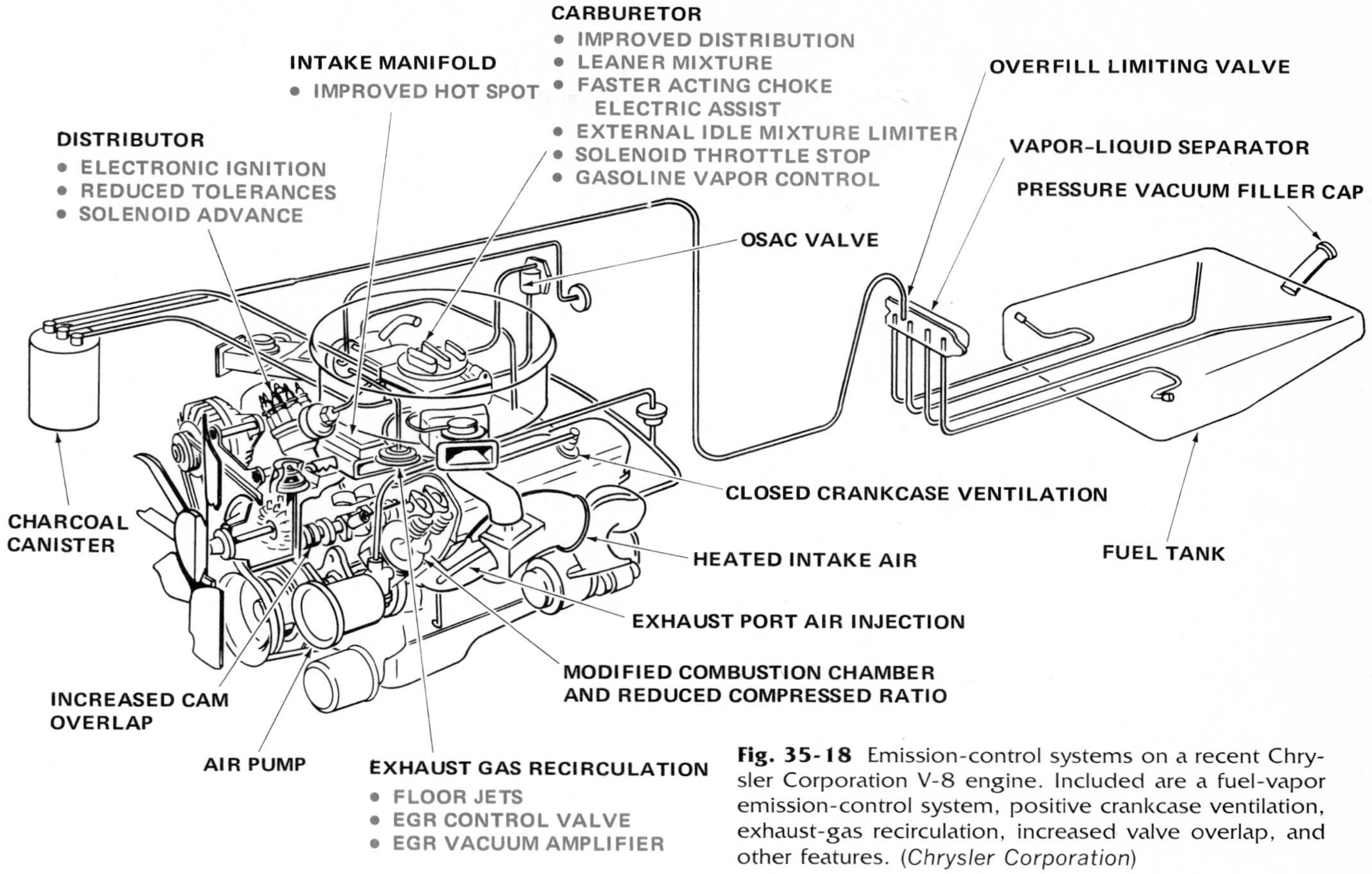

Fig. 35-18 Emission-control systems on a recent Chrysler Corporation V-8 engine. Included are a fuel-vapor emission-control system, positive crankcase ventilation, exhaust-gas recirculation, increased valve overlap, and other features. (*Chrysler Corporation*)

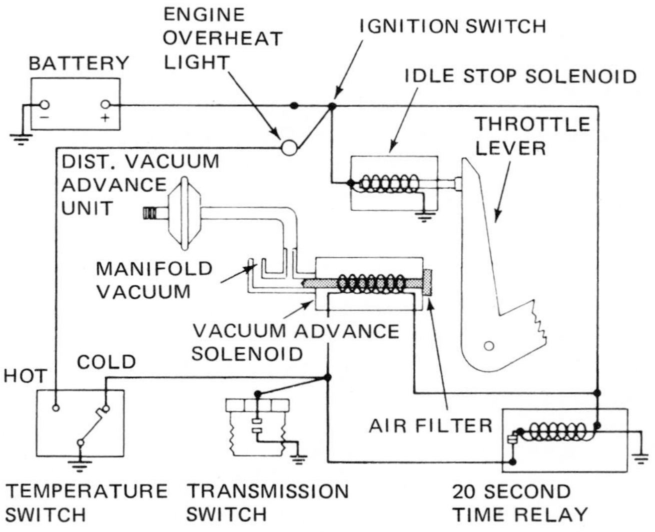

Fig. 35-19 TCS system with cold engine running. (*Chevrolet Motor Division of General Motors Corporation*)

have opened. Also, after 20 seconds, the time-relay switch points open. Thus, the circuit to the vacuum advance solenoid is opened by either of these conditions. The solenoid plunger moves to block vacuum to the distributor vacuum advance. No vacuum advance results.

The transmission switch closes its points when the transmission is shifted into high. This energizes the vacuum advance solenoid so that vacuum is admitted to the distributor vacuum advance mechanism. Vacuum advance can then result.

Some systems have a temperature override switch. This switch causes the system to provide vacuum advance under any condition if the engine begins to overheat. If the engine becomes too hot, the hot points in the temperature override switch close. This energizes the solenoid so that vacuum is admitted to the distributor vacuum advance. With vacuum advance, engine speed increases, and improved cooling results.

● 35-14 Control of Vacuum Advance: TRS System

A Ford transmission-regulated spark (TRS) system is shown in Fig. 35-20. It is for both manual and automatic transmissions. The system works in about the same way as the Chevrolet TCS system described in ● 35-13. The solenoid valve is normally open, allowing vacuum advance when the transmission is in high gear. In the lower gears, the transmission switch is closed. This closes the solenoid valve. With the solenoid valve closed, vacuum is shut off from the distributor vacuum advance. Thus, there is no vacuum advance.

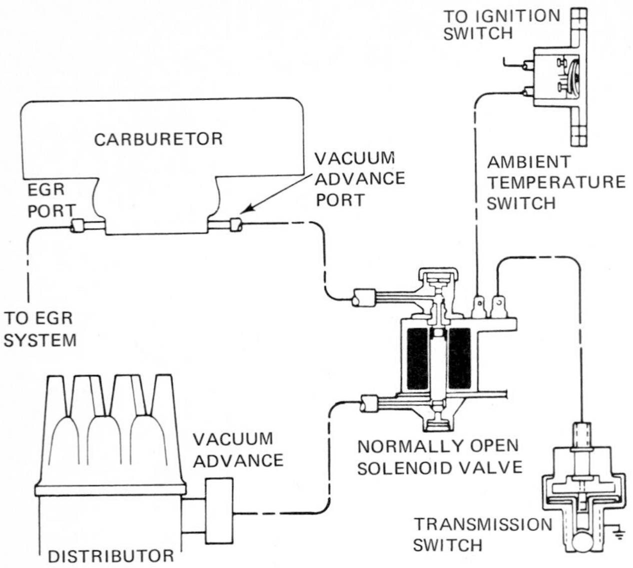

Fig. 35-20 Transmission-regulated spark (TRS) system. *(Ford Motor Company)*

● 35-15 Other Vacuum Advance Control Systems

There are other vacuum advance controls. Most are specially designed for the engines and vehicles with which they are used. Some cars produced by Chrysler Corporation use an orifice spark-advance control (OSAC). It includes a very small hole, or orifice. This delays any change in the application of vacuum to the distributor by about 17 seconds, between idle and part throttle. Therefore, there is a delay in vacuum advance until acceleration is well under way. This is a critical time, during which vacuum advance could produce high NO_x.

Ford has a somewhat similar system called the spark-delay valve system. This system delays vacuum advance during some vehicle-acceleration conditions. The spark-delay valve is connected in series with the vacuum supply from the vacuum advance port in the carburetor and the distributor vacuum advance. During mild acceleration, the vacuum signal to the distributor can increase only gradually. This is because the spark-delay valve only allows the vacuum to pass through slowly. During deceleration or heavy acceleration, the change in vacuum is great enough to open a check valve. This valve allows the vacuum to bypass the spark-delay valve. This produces vacuum advance during these critical times, for better engine performance. If engine temperatures are low, the temperature switch actuates the solenoid valve. The actuated valve then passes vacuum directly to the distributor vacuum advance (through the check valve). This provides vacuum advance when the engine is cold.

● 35-16 Electronic Spark Advance

As we learned in Chap. 30 (● 30-25 to 30-28) some late-model cars have electronic ignition systems which utilize electronic spark advance. Electronic control of spark advance provides more accurate spark advance for every operating condition. This, combined with the new electronic ignition systems, permits the engines to burn leaner mixtures. This increases engine efficiency (more miles per gallon) and also reduces the amount of pollutants in the exhaust gas.

● 35-17 Electronic Control of Air-Fuel Ratio

There are electronic controls of the spark advance, as noted in the previous section. Also, new systems have been developed which control the air-fuel ratio of the mixture being delivered from the carburetor. What these new systems do is monitor the exhaust gas in the exhaust manifold. If the exhaust gas is low in oxygen, the air-fuel mixture is rich. If it is high in oxygen, the air-fuel mixture is lean. By continuous measurement of the oxygen content, the system determines if the air-fuel ratio needs adjustment. If it does, and the mixture is rich, the system automatically reduces the amount of fuel entering the air going through the carburetor. This leans out the mixture. If the mixture becomes too lean (more oxygen in the exhaust), the engine will not run well. In this case, the electronic system allows more fuel to flow so the mixture becomes richer. We discuss two systems that control air-fuel ratio electronically, the General Motors Electronic Fuel Control (EFC) system, and the Ford Feedback-Carburetor Electronic-Engine-Control system. All such systems reduce the pollutants in the exhaust gas.

● 35-18 General Motors Electronic Fuel Control (EFC) System

This system is shown in Fig. 35-21. It includes an exhaust-gas oxygen sensor, an electronic control unit (ECU), an engine temperature switch, three vacuum devices, and a three-way catalytic converter of advanced design that handles NO_x as well as CO and HC. The sensor generates a voltage that varies as the oxygen in the exhaust gas varies. It sends this voltage signal to the electronic control unit. At the same time, the temperature switch is reporting the engine temperature.

If the oxygen content of the exhaust gas is low, indicating a rich mixture, the ECU signals the vacuum switch and vacuum modulator. They allow vacuum to work the carburetor feedback diaphragm. These are vacuum motors that provide a mechanical movement as the vacuum on them changes. One of the feedback diaphragms operates an idle needle that allows more or less air to bleed into the idle system. As more air bleeds in, the idle mixture is leaned out. The other feedback diaphragm meters the fuel flow from the main fuel valve. If the main metering system is feeding too much fuel so the mixture is too rich, the feedback diaphragm reduces the amount of fuel so the mixture is leaned out.

The system constantly keeps adjusting the air-fuel ratio to the lean side so that pollutants are kept to a

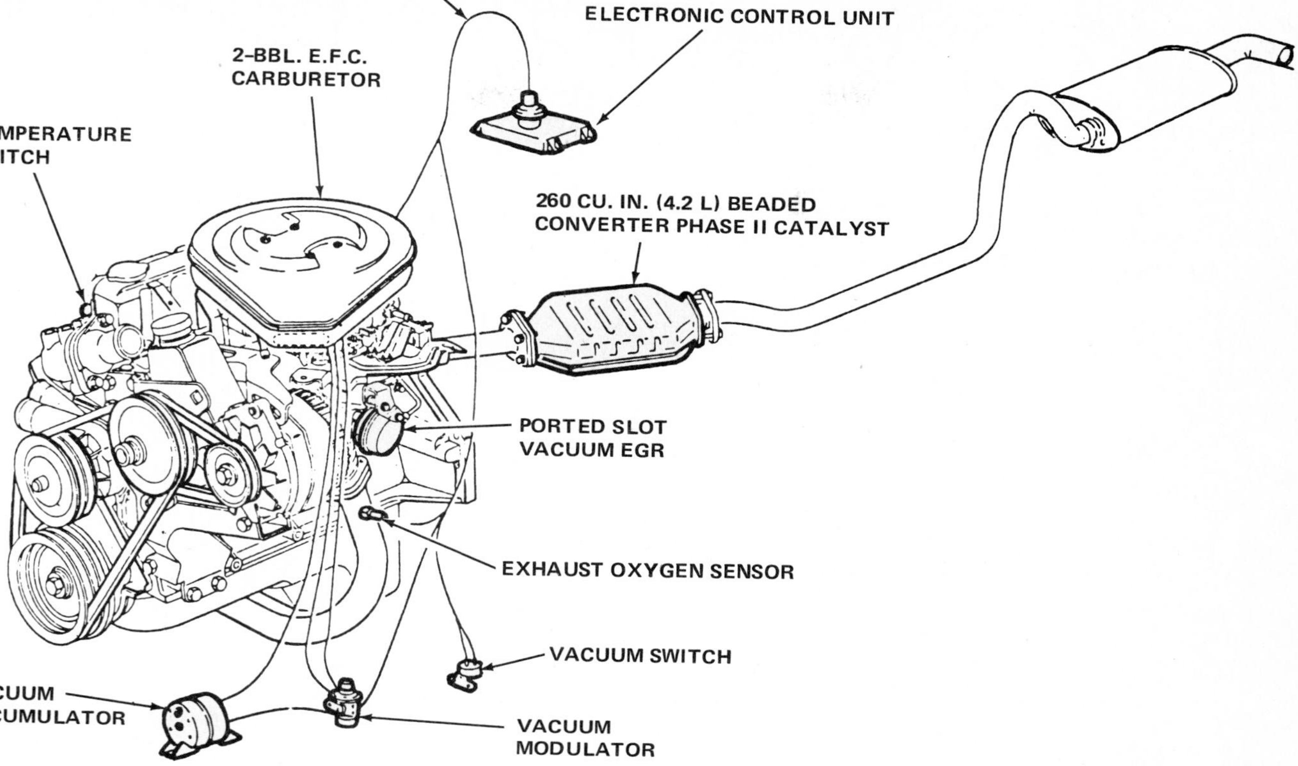

Fig. 35-21 Electronic fuel control (EFC) system on a V-8 engine. (*Chevrolet Motor Division of General Motors Corporation*)

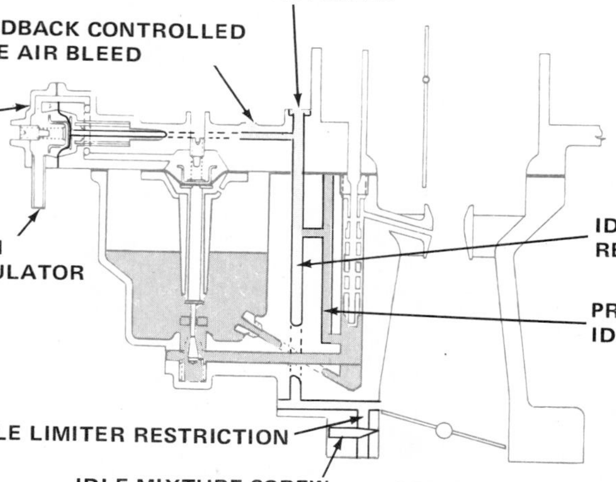

Fig. 35-22 Sectional view of the carburetor for the EFC system, showing the idle system and idle feedback diaphragm. (*Pontiac Motor Division of General Motors Corporation*)

minimum. Note that the catalytic converter, discussed later, is a three-way unit that handles HC, CO, and NO_x.

The engine-temperature switch sends a signal that indicates whether the engine is cold or warmed up. If cold, the signal tells the ECU not to lean out the air-fuel mixture too much. A cold engine needs a somewhat richer mixture to run well. As soon as the engine approaches operating temperature, the temperature switch "tells" the ECU. The ECU then switches to the hot-engine mode and the mixture goes more to the lean side.

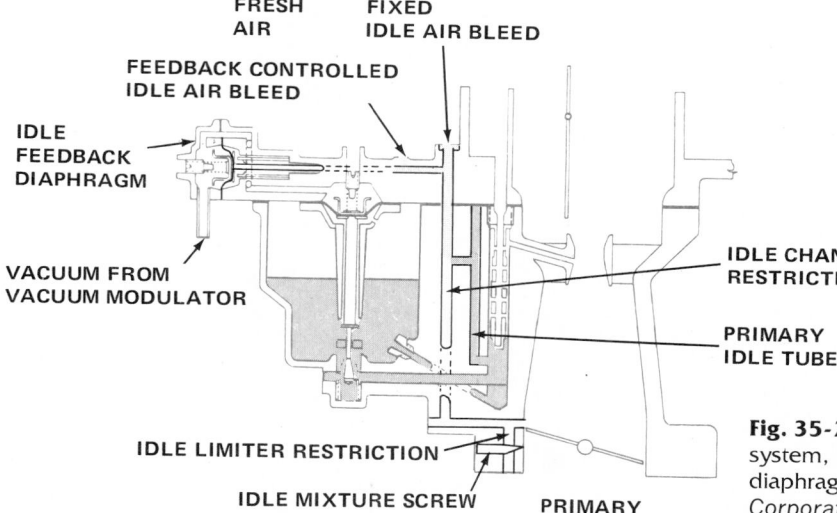

Figure 35-22 is a sectional view of the carburetor showing the idle system. Note that the vacuum from the vacuum modulator works on the idle feedback diaphragm. If the vacuum is sufficient, it causes the diaphragm to pull back the needle to the feedback controlled idle air bleed. This allows more air to feed in so the idle mixture is leaned out.

Figure 35-23 shows the main metering system feedback diaphragm. It is connected to the tapered needle centered in the feedback-controlled main metering ori-

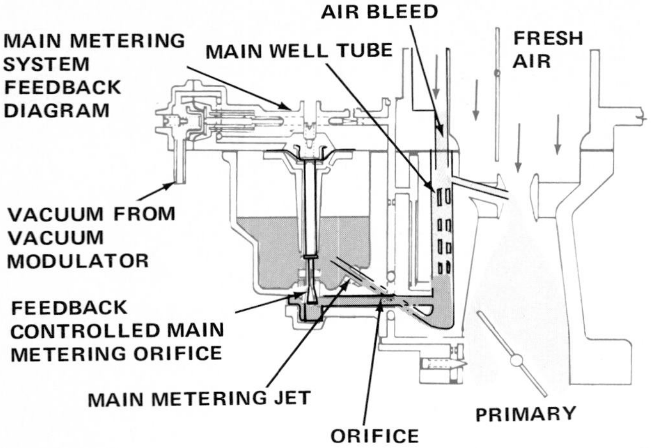

Fig. 35-23 Sectional view of the carburetor for the EFC system, showing the main metering system with feedback diaphragm. (*Pontiac Motor Division of General Motors Corporation*)

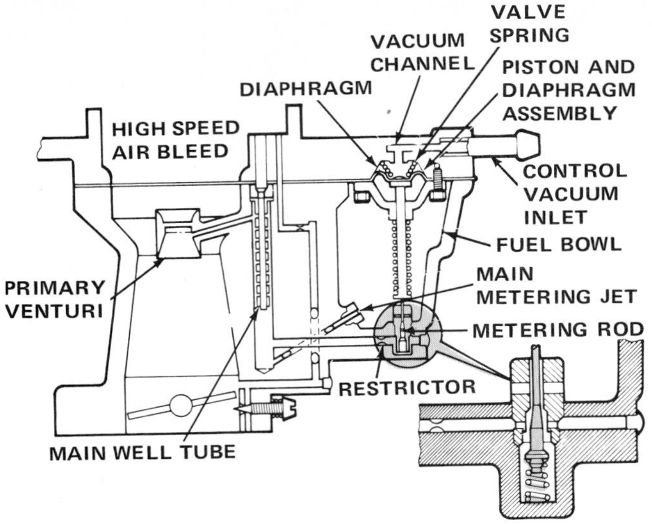

Fig. 35-24 Sectional view of the feedback carburetor for an electronic engine control system with diaphragm control of the main metering system. (*Ford Motor Company*)

fice. If the mixture is too rich, there will be a sufficient vacuum on the diaphragm so it lifts the needle. This reduces the amount of fuel that can flow to the main system, so the mixture is leaned out.

● 35-19 Ford Feedback-Carburetor Electronic-Engine-Control System

The Ford system for air-fuel-mixture control is somewhat more complex and is tied in with the thermactor or air-injection system. A later section describes the thermactor system. What it does is blow fresh air into the exhaust manifold to help burn any unburned gasoline (HC) or partly burned gasoline (CO).

The system consists of three subsystems:

1. Dual catalytic converter
2. Thermactor air control
3. Electronic feedback carburetor

The electronic feedback carburetor is very similar to the General Motors unit (Figs. 35-22 and 35-23). However, it has only one feedback diaphragm and this controls the main metering system (Fig. 35-24). The system uses an oxygen sensor in the exhaust manifold. When the oxygen content of the exhaust gas goes down, indicating a rich mixture, the sensor reports this to the electronic control unit (ECU). The ECU then signals the vacuum switch system to send more vacuum to the feedback diaphragm. The diaphragm therefore lifts the metering rod to restrict the flow of fuel. Less fuel gets to the main fuel nozzle so the mixture is leaned out.

The thermactor system has two operating modes (Figs. 35-25 and 35-26). When the engine is cold (Fig. 35-25), the fuel system is feeding a rich mixture to the engine. This is due to the choke action, as we explained in an earlier chapter. The exhaust gas is therefore rich in unburned and partly burned fuel (HC and CO). With this condition, the thermactor sends air into the exhaust manifold to help complete the combustion of these pollutants. The additional oxygen in the air helps turn the HC into H_2O and CO_2. It helps turn the CO into CO_2. In doing this, it guards the catalytic converter from overload, as we explain in a later section.

When the engine warms up (Fig. 35-26), the vacuum switch, which senses engine-coolant temperature, shuts off the vacuum to the air-control valve. As a result, the thermactor system sends the air it is pumping to the catalytic converter (Fig. 35-26). Here, it aids the converter in changing the pollutants to harmless gases, as we explain in a later section.

● 35-20 Treating the Exhaust Gas by Air Injection

After the exhaust gases leave the engine cylinders, they can be treated to reduce the HC, CO, and NO_x in the gas. One method is to blow fresh air into the exhaust manifold. This system is called a thermactor or air-injection system. It provides the additional oxygen needed to burn the HC and CO. Figure 35-27 shows one system.

The air pump pushes air through the airlines and air manifold into a series of air-injection tubes. These tubes are located opposite the exhaust valves. The oxygen in the air helps to burn any HC or CO in the exhaust gas. The check valve prevents any backflow of exhaust gas to the air pump, in case of a backfire. The air bypass valve operates during engine deceleration. During deceleration, intake-manifold vacuum is high. The bypass valve momentarily diverts air from the air pump to the air cleaner, instead of to the exhaust manifold. This tends to prevent backfiring in the exhaust manifold.

We have already discussed the special application of the air-injection or thermactor system in the Ford electronic engine control system using a feedback carburetor (● 35-19). Note that this thermactor system sends air into the exhaust manifold only part of the time. The rest of the time, when the engine reaches operating temperature, the air is sent to the catalytic converter.

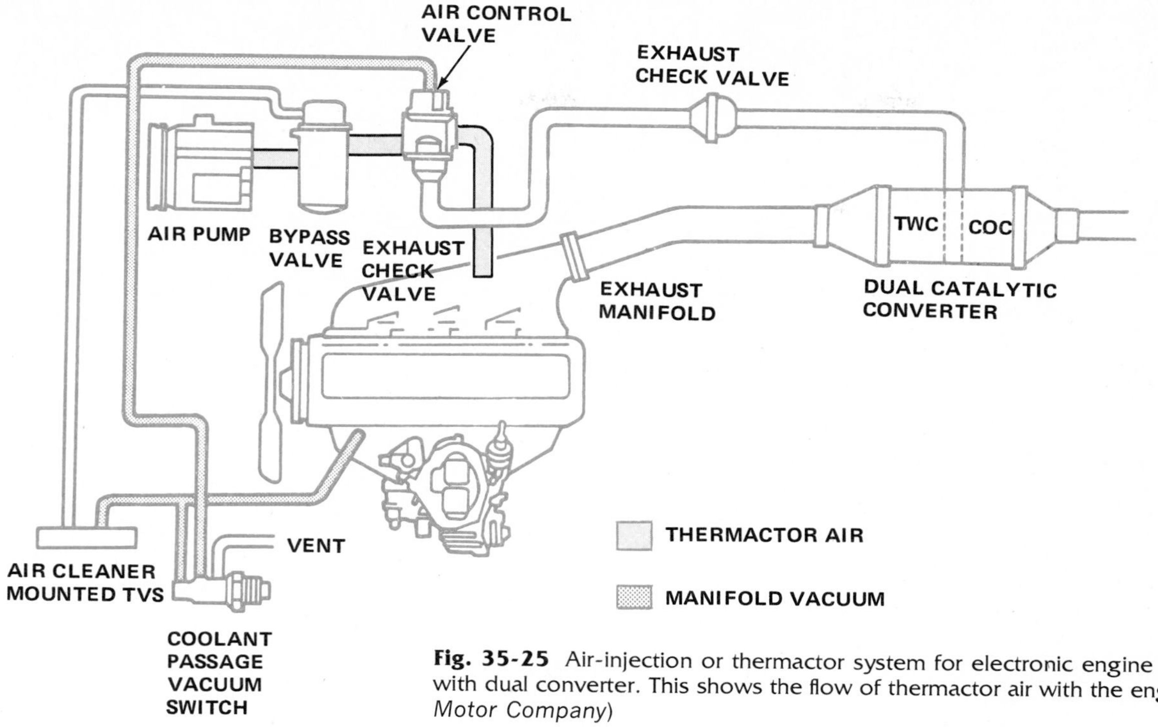

Fig. 35-25 Air-injection or thermactor system for electronic engine control system, with dual converter. This shows the flow of thermactor air with the engine cold. *(Ford Motor Company)*

35-21 Aspirator Air System

This system is used on some Chrysler Corporation cars instead of the air-injection system. It works in about the same way. It uses a valve connected in a line between the air cleaner and the exhaust manifold. The valve (Fig. 35-28) is a simple pressure-vacuum valve. It uses the exhaust pressure pulsations to draw air into the exhaust system. This air enters the exhaust manifold and mixes with exhaust gases. Oxygen in the air helps convert HC and CO into harmless gases, just as in air-injection systems.

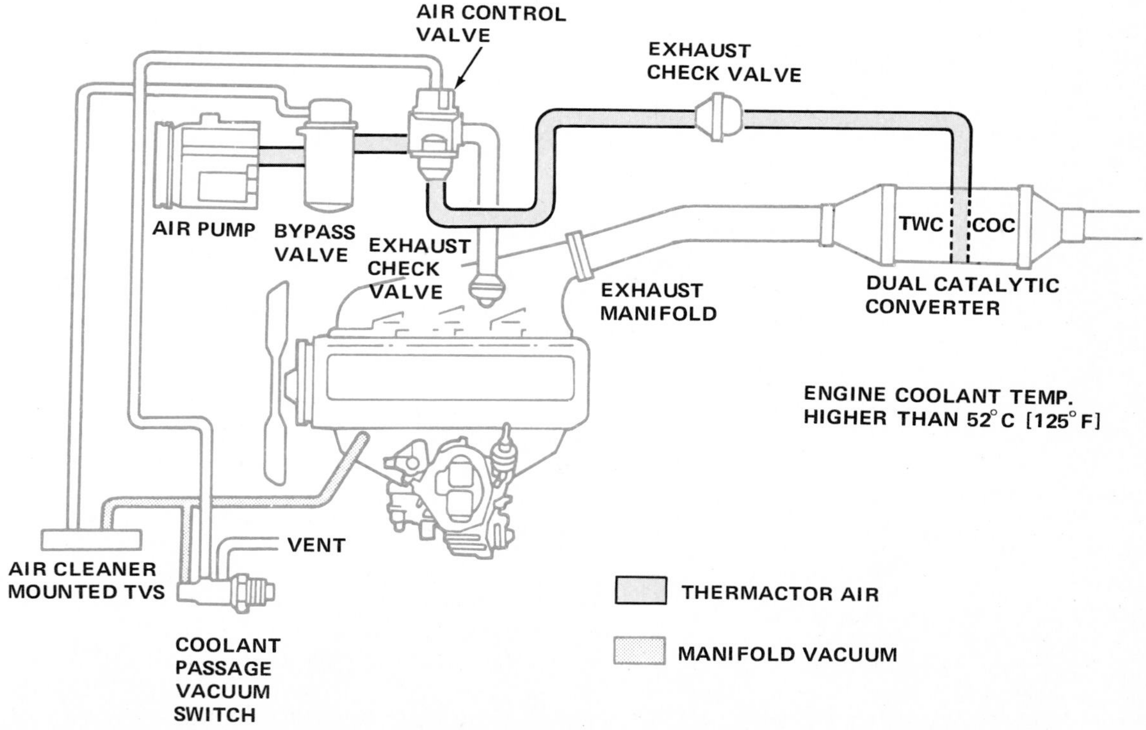

Fig. 35-26 Air-injection or thermactor system with the engine hot. Thermactor air flows to the dual catalytic converter. *(Ford Motor Company)*

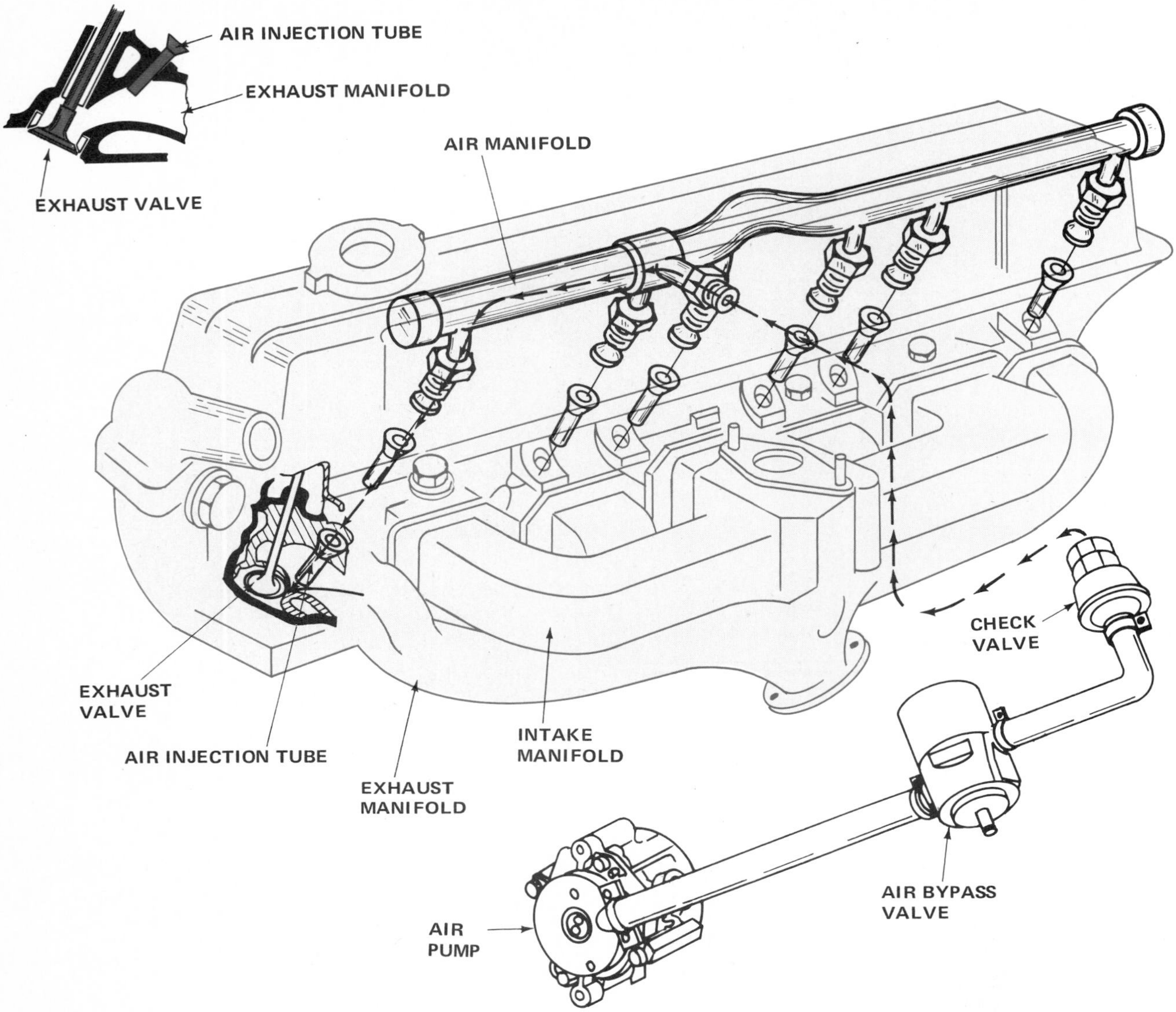

Fig. 35-27 Air-injection system. Air manifold and other parts of the system are shown detached so they can be seen better. The cylinder head has been cut away at the front to show how the air-injection tube fits into the head.

When the exhaust pressure is lower than the pressure in the air cleaner, the valve opens, as shown in view A (Fig. 35-28). Fresh air can then flow into the exhaust manifold. When the exhaust pressure goes up, the valve closes, as shown in view B. This prevents any backflow of exhaust gas into the air cleaner.

● 35-22 Treating the Exhaust Gas by Catalytic Converter

A second method of treating exhaust gas uses catalytic converters. These convert the gaseous pollutants into harmless gases. A catalyst is a material that causes a chemical change without entering into the chemical reaction. In effect, the catalyst stands by and encourages two chemicals to react with each other. For example, in the HC/CO catalytic converter, the catalyst encourages the HC to unite with oxygen to produce H_2O (water). It encourages the CO to change to CO_2 (carbon dioxide). The catalyst in the NO_x converter splits the nitrogen from the oxygen. NO_x becomes harmless nitrogen and oxygen.

Figure 35-29 shows a recent General Motors emission-control system. The catalytic converter is the two-way type that converts only HC and CO. Figure 35-30 is a cutaway view of the converter and Fig. 35-31 shows the flow of the exhaust gas through the converter. The converter is filled with BB-shot-size metal pellets. They are coated with a thin layer of platinum or similar catalytic metal. The pellets form a matrix through which the exhaust gas must pass. As the exhaust gas flows through, the catalyst produces the chemical reaction. Another type of catalytic converter uses a honeycomb arrangement through which the gas must pass.

Cars with catalytic converters must use nonleaded gasoline, as we have mentioned in previous chapters. If the gasoline contains lead, the lead will coat the catalyst and the converter will stop working. If this happens to the pellet-type converter, there is a way to remove the old pellets and put in a charge of fresh pellets (● 36-16). But on the honeycomb type, the complete converter unit must be replaced.

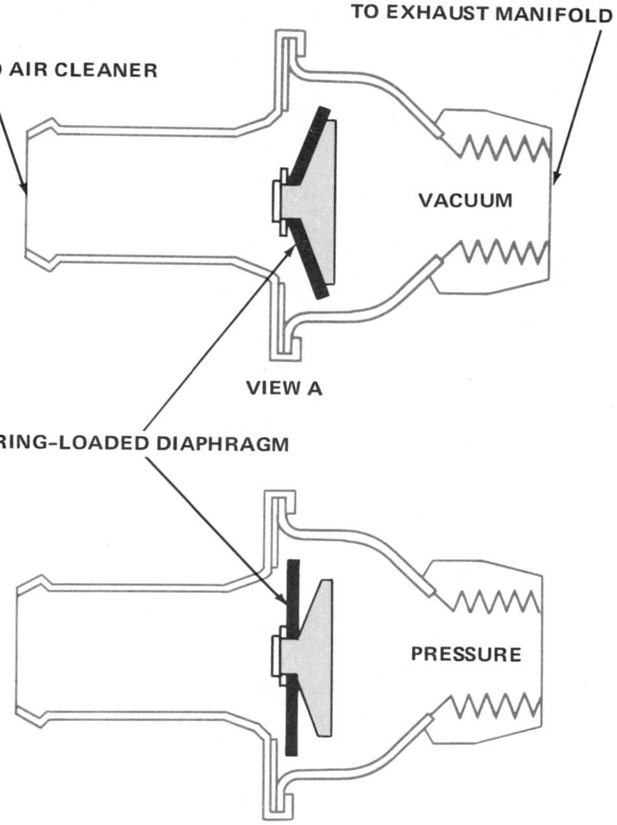

Fig. 35-28 Two views of aspirator valve, showing it open (view A), and closed (view B). *(Chrysler Corporation)*

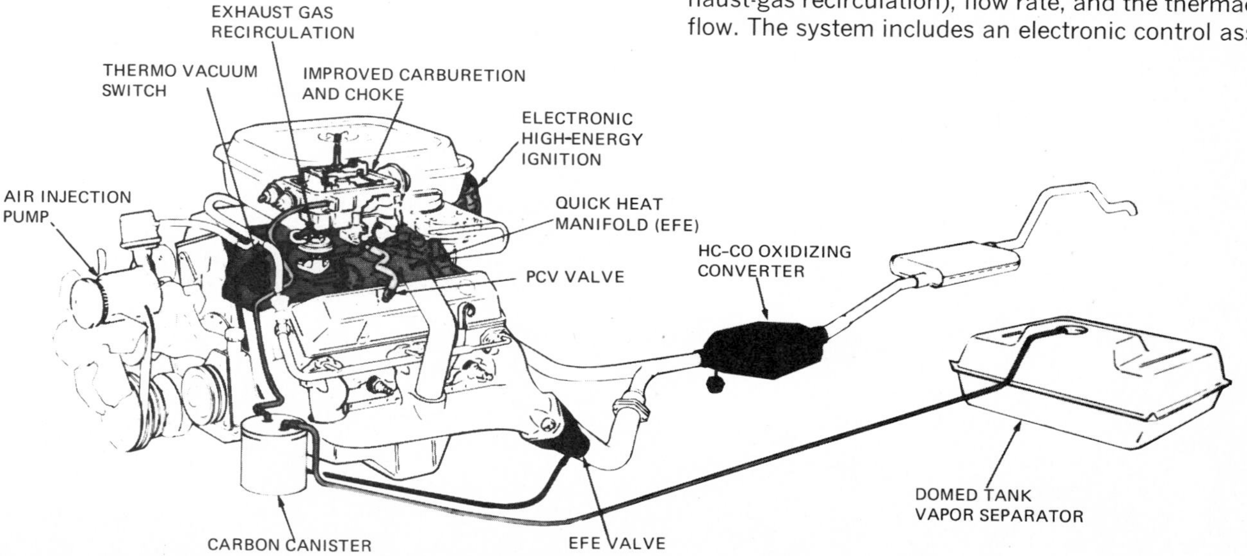

Fig. 35-29 Emission-control system using an under-the-floor catalytic converter. Note that the system uses air injection and other emission-control features described previously. *(General Motors Corporation)*

The converter gets hot. Therefore, the floor pan above it must be insulated to prevent this heat from working up into the passenger compartment (Fig. 35-32).

● 35-23 Three-Way Catalytic Converter

The three-way catalytic converter handles not only HC and CO, but NO_x as well. We mentioned in our discussions of the General Motors EFC system and the Ford feedback carburetor electronic-engine-control system that they both used three-way catalytic converters. The General Motors version includes additional catalytic metals which handle the NO_x. The Ford system has an additional feature. It is tied in with the thermactor or air-injection system (Figs. 35-25 and 35-26). The catalytic converter used with this system is shown in cutaway view in Fig. 35-33. The air from the thermactor pump feeds into the space between the three-way catalyst (TWC) and the two-way or conventional oxidation catalyst (COC). The front converter is coated with the metals rhodium and platinum. The combination handles NO_x and partly handles the HC and CO. The partly treated exhaust gas then flows into the rear converter. As it does so, it mixes with the air being pumped in from the air pump. This puts more oxygen into the exhaust gas so the two-way catalyst can take care of the CO and HC. Remember that the thermactor system feeds air into the converter only when the engine is at operating temperature. When the engine is cold, the air from the thermactor goes into the exhaust manifold (see ● 35-19).

● 35-24 Ford Electronic Engine Control (EEC) System

This system, introduced on one Lincoln model in 1978, is an electronic system to control ignition timing, EGR (exhaust-gas recirculation), flow rate, and the thermactor air flow. The system includes an electronic control assembly

Fig. 35-30 Cutaway view of the catalytic converter. (*General Motors Corporation*)

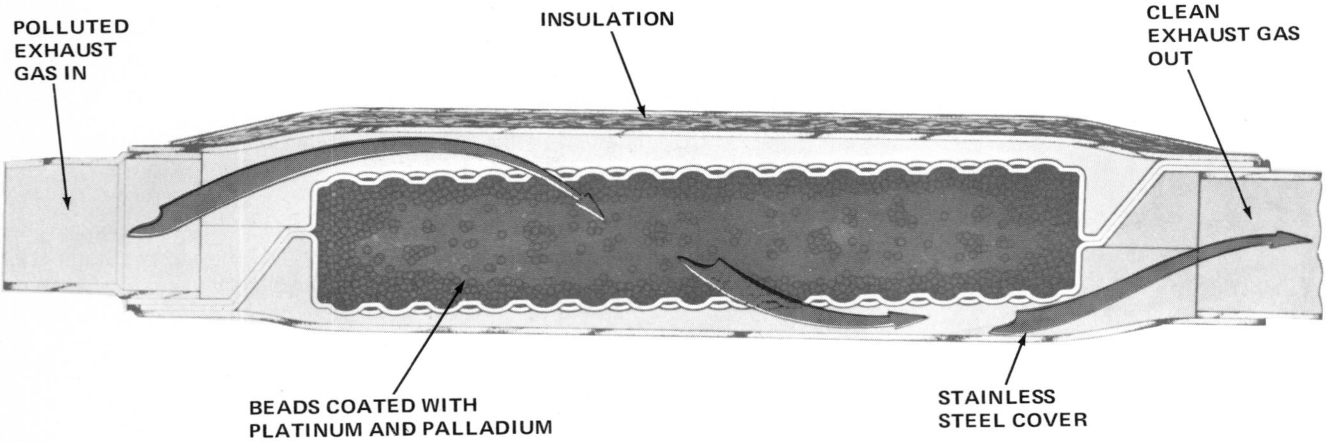

Fig. 35-31 Flow of exhaust gas through the converter, shown by arrows. (*General Motors Corporation*)

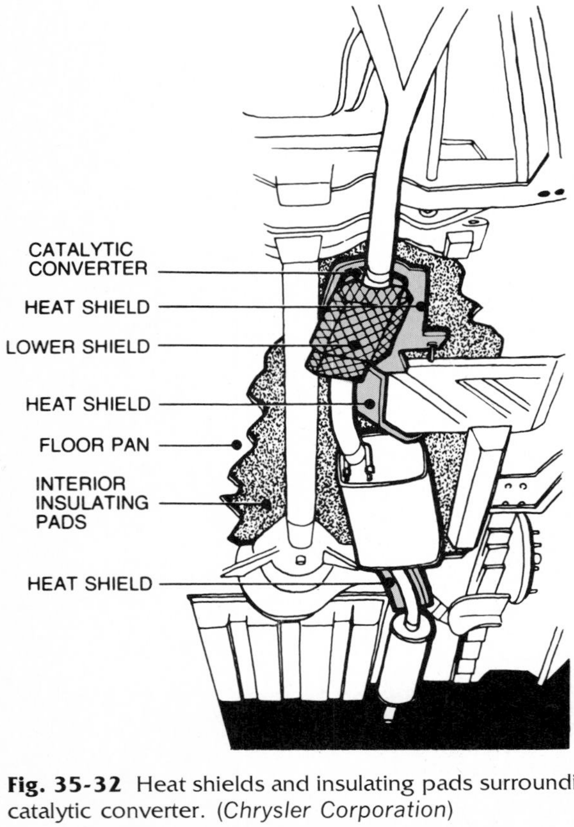

Fig. 35-32 Heat shields and insulating pads surrounding the catalytic converter. (*Chrysler Corporation*)

(ECA), seven sensors, a special ignition module and coil (Dura-Spark II), a new design of distributor, and an air-pressure-operated EGR system (Fig. 35-34). The seven sensors sense the following:

- Manifold absolute pressure (or vacuum)
- Barometric (atmospheric) pressure
- Engine-coolant temperature
- Inlet-air temperature
- Crankshaft position
- Throttle position
- EGR valve position

The sensors report these seven engine conditions continuously to the ECA. The ECA puts the information together and then computes the correct ignition timing, EGR flow rate, and thermactor air flow for the best engine performance and emission control. The ECA does this continuously, instantly adjusting its computations as the sensors report changes.

Exhaust-gas recirculation (EGR) is controlled by the EGR-valve-position sensor, two control solenoids, and a cooler assembly. The cooler assembly is mounted over the right valve cover. It consists of a long tube through which the exhaust gases flow on their way to the intake manifold, and a water jacket surrounding the tube. Coolant from the engine cooling system flows through the water jacket, thereby cooling the exhaust gas. The two control solenoids operate on signals from the ECA assembly. They change the amount of exhaust gas flowing

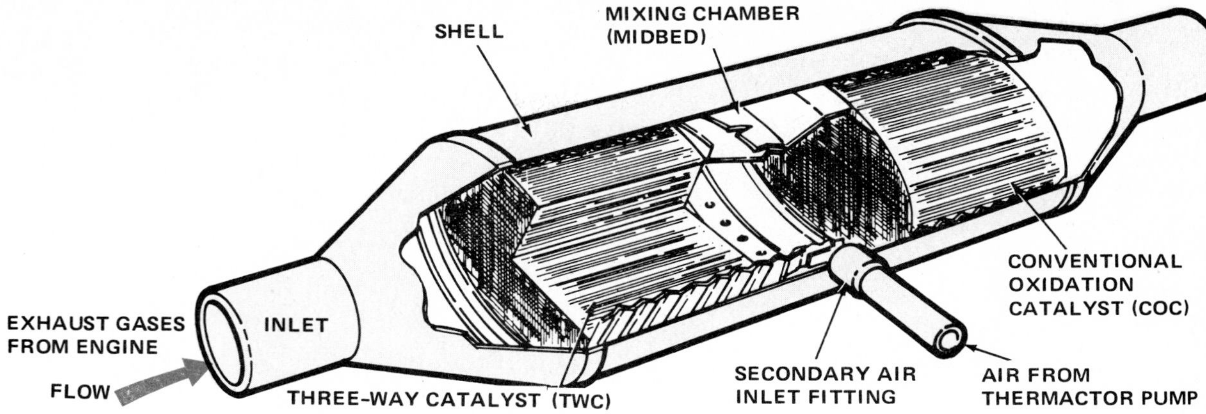

Fig. 35-33 Partial cutaway view of dual catalytic converter. (*Ford Motor Company*)

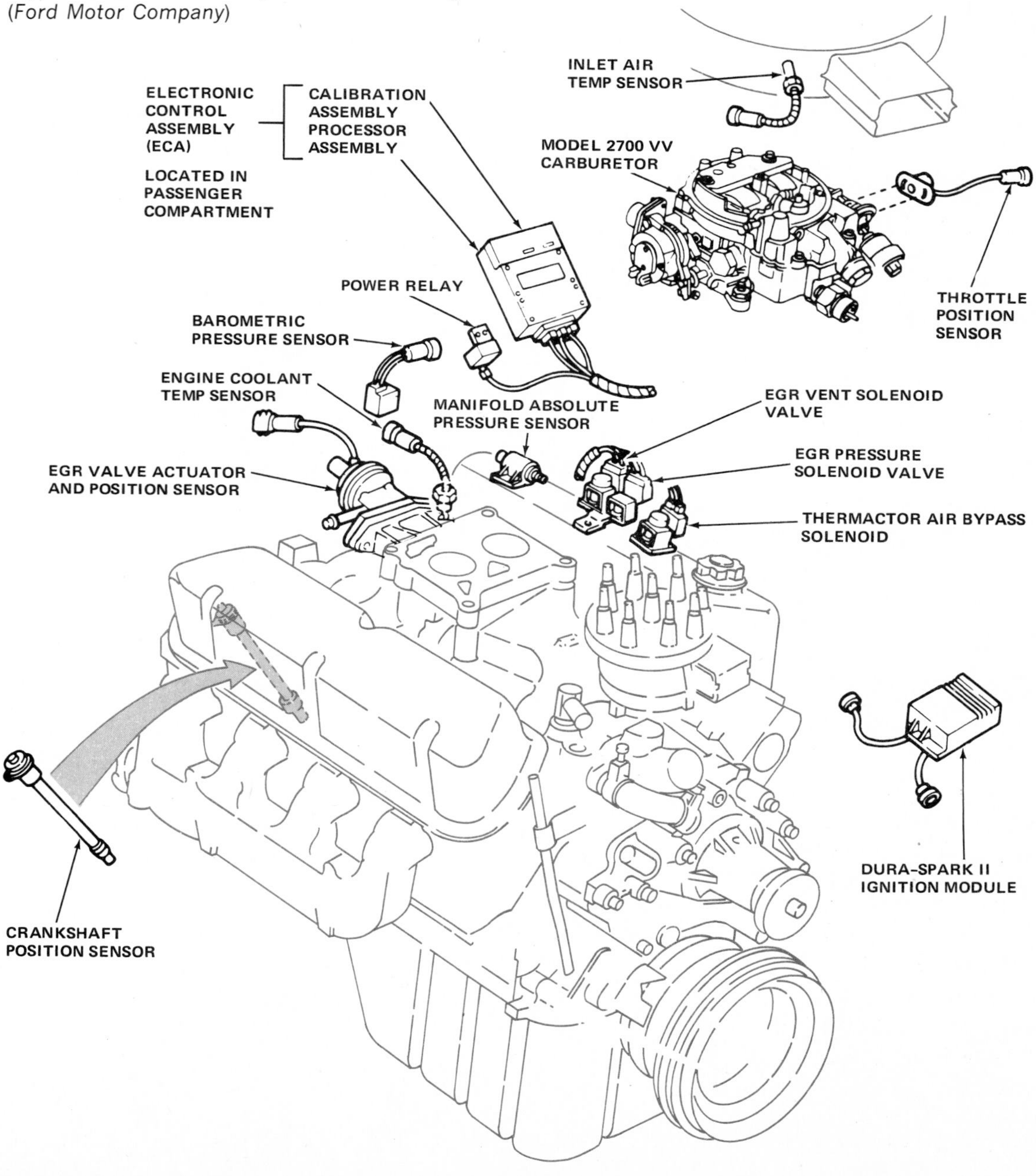

Fig. 35-34 Details of the Ford Electronic Engine Control (EEC) system, showing major components. (*Ford Motor Company*)

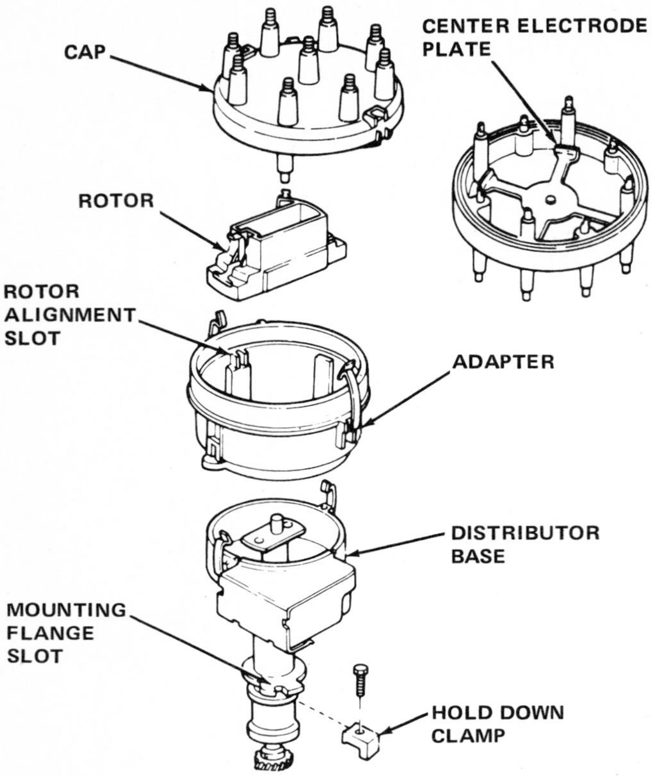

Fig. 35-35 Distributor for the Ford EEC system. The rotor is a two-level unit, alternately sending high-voltage surges to the long and the short terminals inside the cap. (*Ford Motor Company*)

to the intake manifold according to engine operating conditions.

Thermactor (air-injection) action is controlled by a solenoid valve which is actuated by the ECA assembly. When this valve is energized, air is injected into the exhaust ports. When it is deenergized, the valve dumps the thermactor air into the atmosphere.

The ignition system uses a special two-level distributor rotor and cap (Fig. 35-35). The purpose is to increase the distance between the terminals inside the cap. Because the system produces a very high voltage, it is desirable to position the terminals as far apart as possible. The high voltage entering the center electrode is carried to the center electrode plate. As the rotor segment passes the bent-up end of one spoke of the plate, the voltage jumps to the rotor segment. From there it jumps to the cap terminal aligned with either the upper or the lower rotor blade, according to which is aligned. As added protection against crossfiring due to sparks jumping the gaps between the terminals, the rotor tips and tips of the center electrode plate are coated with silicone grease. This grease must not be removed.

The spark advance is controlled by the ECA in a manner similar to other electronically controlled systems previously discussed. This system is unique, however, in that it is the first system to control three separate but interrelated variables—ignition timing, EGR flow rate, and thermactor air flow.

REVIEW QUESTIONS

1. What is meant by the expression "Cleaning up the exhaust gas"?
2. What are the three ways of cleaning up the exhaust gas?
3. What are the two ways of controlling the air-fuel mixture to achieve lower HC and CO in the exhaust?
4. Explain how the thermostatically controlled air cleaner works.
5. Explain how the electric-assist choke works.
6. The layer of air-fuel mixture next to the metal surfaces in the combustion chamber does not burn. Why not?
7. What is meant by the *S/V* ratio? Why is it important as regards the amount of HC and CO in the exhaust?
8. What is meant by stratified charge? How does it work?
9. How does the Honda system work?
10. What is EGR? How does it work? Why does it reduce NO_x in the exhaust?
11. Why does additional valve overlap (Chrysler system) reduce NO_x in the exhaust?
12. What is the TCS system? How does it work?
13. What is the purpose of the air-injection system? How does it work?
14. What is a catalytic converter? How does it work in the car exhaust system?
15. Explain how the General Motors EFC system works.
16. Explain how the Ford feedback-carburetor electronic-engine-control system works.

SELF PROJECTS

1. Disassemble a thermostatically controlled air cleaner to see how the parts go together. This will help you understand how the system works.
2. Examine cars with EGR systems. Make drawings showing how the system is connected. File the drawings in your notebook.
3. Examine a car with a TCS system. List the parts involved, and make a drawing of the system. File the drawing in your notebook.
4. Examine cars with air-injection systems and trace out the hose connections.

CHAPTER 36
Servicing Emission-Control Systems

After you have read this chapter, you should be able to:

- **Explain how to do the various basic services required on the emission controls described in the chapter. Also, perform these services.**

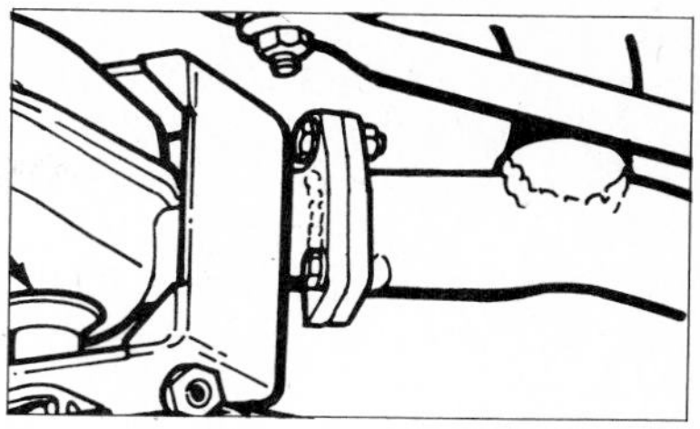

This chapter outlines the fundamental services required on the various systems for controlling automotive emissions. We have described these systems in previous chapters. They include:

1. Positive-crankcase-ventilation system
2. Fuel-evaporation emission-control system
3. Air-injection system
4. Heated-air system
5. Exhaust-gas recirculation system
6. Transmission-controlled spark vacuum advance (TCS)
7. Electronic spark-advance controls
8. Catalytic-converter systems

We now look at the services each of these systems require.

It is against the law to tamper with emission controls on vehicles or vehicle engines. The penalty is a fine for every vehicle tampered with. When servicing engines and emission controls, do *only* those jobs specified in the manufacturer's shop manuals. If you remove, disconnect, damage, or render ineffective any emission control, you are breaking the law.

● 36-1 Positive-Crankcase-Ventilation Troubles

Several engine troubles can result from defective conditions in the PCV system. These troubles can also result from faults in other systems as we explain in Part Seven, Engine Trouble Diagnosis and Tune-Up. Here, we look only at troubles arising from the PCV system.

Rough idle and frequent stalling could result from a plugged or stuck PCV valve or to a clogged PCV air filter. In either case, the remedy is to replace the valve or filter.

Vapor flow into the air cleaner and oil in the air cleaner; this can result from backflow. That is, instead of filtered air flowing into the crankcase, vapors from the crankcase are flowing into the air cleaner. The cause is a plugged PCV valve or plugged or leaking condition somewhere in the PCV system. It would also be caused by worn piston rings or cylinder walls which allow more blow-by than the PCV system can handle.

Sludge or oil dilution in the crankcase can result from a plugged condition—valve or line—that prevents normal circulation.

● 36-2 Positive-Crankcase-Ventilation Service

The PCV valve should be checked periodically and replaced at specified intervals. Typical recommendations are:

American Motors Corporation Replace the PCV valve every 15,000 miles [24,140 km].

Chevrolet and Ford Replace the valve every 24,000 miles [38,624 km] or 24 months, whichever comes first.

Chrysler Check the valve every 12 months or 12,000 miles [19,312 km]. Replace it every 24,000 miles [38,624 km] or 24 months.

There are special testers that can be used to check the operation of the PCV valve. A simple way to check the system and the valve is to remove the valve or valve connection, with the engine running. Place your hand over the opening. You should feel a slight vacuum pull against your hand. If there is no vacuum, or if you can feel a positive pressure, then something is wrong. Check the PCV valve, hoses, and connections.

Chevrolet offers a different method. With the engine running at idle, remove the PCV valve from the rocker-arm-cover grommet, with hose attached. Block the opening of the valve, and note the change in engine speed. A decrease of less than 50 rpm indicates a plugged PCV valve. You should use a tachometer when making this test, to get an accurate reading of the engine rpm.

Chrysler recommends the following test procedure. Remove the PCV valve from the rocker-arm cover, with the engine idling. The valve should hiss. You should be able to feel a strong vacuum when your finger is placed over the valve inlet. Reinstall the PCV valve, and remove the crankcase inlet air cleaner. Hold a piece of stiff paper over the opening of the rocker-arm cover. After a few seconds, the paper should be sucked against the opening. Then stop the engine. Remove the PCV valve from the rocker-arm cover, and shake it. It should click, showing that the valve is free. If the system does not meet these tests, replace the PCV valve and try again. If the system still does not pass the tests, the hose may be clogged. It should be cleaned out or replaced. It may be necessary to remove the carburetor and clean the vacuum passage with a $\frac{1}{4}$-in [6.35-mm] drill. Also, clean the inlet vent on the crankcase inlet air cleaner that is connected by the hose to the carburetor air cleaner.

36-3 Fuel-Vapor Emission-Control Troubles

Here are troubles that might be caused by conditions in the fuel-vapor emission control or fuel system.

Fuel odor or loss of fuel could be caused by a number of conditions which we list here. Correction is obvious in each case:

1. Overfilled fuel tank
2. Leaks in fuel, vapor, or vent line
3. Wrong or faulty fuel-tank cap
4. Faulty liquid-vapor separator
5. Excessively high fuel volatility
6. Vapor-line restrictor missing
7. Canister drain cap or hose missing

A collapsed fuel tank can result if the wrong fuel-tank cap is installed or if the vacuum valve in the cap sticks. In either case, no air can enter to replace gasoline being withdrawn by the fuel pump. The result could be a vacuum in the fuel tank great enough to allow atmospheric pressure to crush the tank.

Excessive pressure in the fuel tank could result from a combination of high temperatures and a plugged vent line, liquid-vapor separator, or canister. Pressure can be released by turning the tank filler cap just enough to allow the pressure to slowly escape.

Many different engine-idling problems can result from faulty or improper connections of a hose in the control system. A plugged canister, vapor-line restrictor missing, or high-volatility fuel can also cause poor idle.

36-4 Fuel-Vapor Emission-Control Service

These systems require little in the way of service. About the only troubles that occur are restriction of the fuel flow (so the engine is starved and stalls) and a collapsed fuel tank, as we have mentioned in the previous section. About the only periodic service required is to replace the filter in the bottom of the canister every year or every 12,000 miles [19,312 km] of operation (Fig. 36-1).

36-5 Air-Injection-System Troubles

Troubles related to the air-injection system (Fig. 36-2) include noise, no air supply, backfire, and high HC and CO levels in the exhaust.

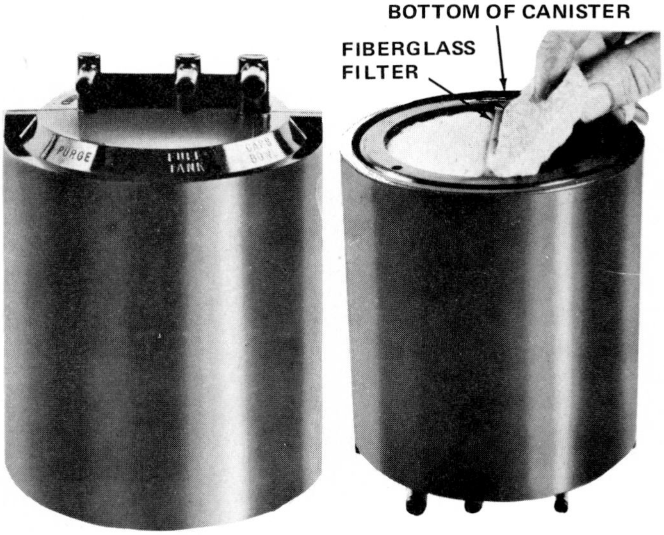

Fig. 36-1 Replacing the air filter in the charcoal canister.

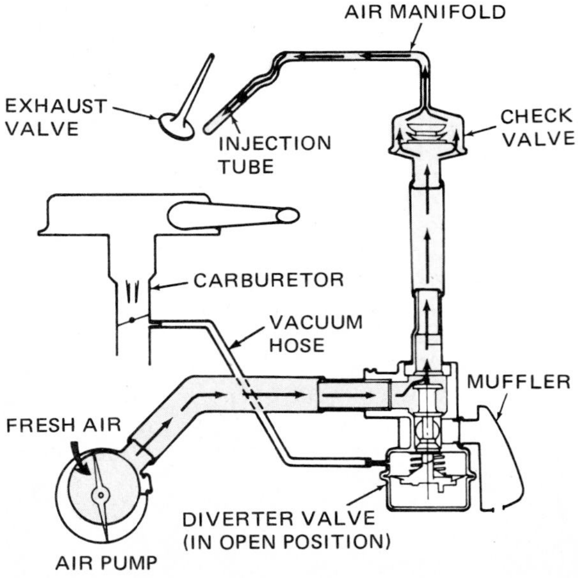

Fig. 36-2 Schematic view of the air-injection system. *(Chevrolet Motor Division of General Motors Corporation)*

Noise from the belt or air pump could result from a loose belt, loose air-pump mounting bolts, worn pump bearings or other internal trouble, or air leaks from the system. The pump is not repairable; it must be replaced if damaged. The air pump should be adjusted so the belt has the correct tension. Do not pry on the pump housing because this can ruin the pump. If you have to use a pry bar, pry as close to the pulley end as possible. Some manufacturers recommend using a belt tensioner and a belt-tension gauge (Fig. 36-3).

Air leaks should be stopped by tightening hose connections and replacing any hose that is defective.

If no air is getting to the air manifold, the exhaust gas will probably be high in HC and CO. Causes of no or inadequate air include a loose belt, frozen pump, leaks in the hoses or connections, and failure of the diverter or check valve. Remedies here are obvious. A defective pump or valve must be replaced. They are not repairable.

Backfire is usually caused by a defective diverter valve which fails to block off the air supply under conditions of high intake-manifold vacuum. The same thing happens if the vacuum hose becomes disconnected or blocked.

● 36-6 Air-Injection-System Service

No routine service is required on the air-injection system except to replace the air filter every 12,000 miles [19,312 km] on those systems using a separate air filter. All late-model air pumps use a centrifugal filter which requires no separate service.

The air-pump drive belt should be checked periodically, as all the belts at the front of the engine should be. Check for wear and tension. Replace any belts that are worn, cracked, or brittle. Do not overtighten the air-pump drive belt. Overtightening can cause fast pump-bearing wear.

● 36-7 Heated-Air-System Troubles

The heated-air system is designed to let the engine run better when cold, with a relatively lean mixture. It does this by adding heat to the air going into the carburetor almost as soon as the engine is started. This means the mixture is warm and the gasoline vaporizes better. If the heated-air system is not working to add heat, it means the control damper has closed off the hot-air pipe. As a result, the mixture is not getting warmed. The cold engine will hesitate or stumble, or even stall, because the mixture is too cold and lean to fire consistently. On the other hand, if the damper does not shut off the hot-air pipe when the engine is hot, the mixture will be overheated so not enough mixture gets into the cylinders for full power. The engine will not develop full power.

You can check the operation of the damper by starting the engine cold. Note first whether the damper in the snorkel is closed or open. It should be wide open. Then, when the engine starts, it should close. This opens the hot-air pipe. As the engine warms up, the damper should move to the open position. An accurate check can be made with a temperature gauge or thermometer. If the damper does not perform properly, the trouble could be in the damper-operating mechanism. You may need a small mirror to observe the damper action. The damper-operating mechanism may be either a thermostat or a vacuum motor operated by intake-manifold vacuum. See ● 18-12.

The heated-air system requires no routine service.

● 36-8 Exhaust-Gas-Recirculation Troubles

Trouble in the EGR system is reflected in poor engine performance. For example, rough engine idle and stalling could be caused by a leaky EGR valve or valve gasket that allows exhaust gas or air to enter the intake manifold during idling. A defective thermal vacuum switch (on engines having this valve) could cause vacuum to operate the EGR valve when it should not.

Poor part-throttle performance, poor fuel economy, and rough running on light acceleration could also be caused by the vacuum valve being defective. In addition, a sticking or binding EGR valve, or deposits in the EGR passages, could cause these conditions. If deposits have clogged the EGR passages, remove the manifold to clean them away.

If the engine stalls on deceleration, it could be due to a restricted vacuum line that is preventing the EGR valve from closing promptly. Detonation at part throttle could be caused by insufficient exhaust-gas recirculation. This could be due to clogged or damaged hoses, an EGR valve that is stuck, or a defective thermal vacuum switch.

● 36-9 Testing the EGR System

Some EGR valves have the stem visible under the diaphragm (Fig. 36-4). You can check this type valve with the engine warmed up and idling. With the transmission in neutral, snap the throttle open to bring the engine rpm up to about 2,000. The EGR stem should move up indicating the valve has opened. If it does not, connect a vacuum tester to the vacuum tube of the valve. With the engine warmed up and idling, apply about 8 inches Hg [203.2 mm Hg] to the valve. The valve should operate. If it does not, it is either defective or dirty. You can tell when the valve operates because the engine will idle roughly and may even stall.

To test the thermal vacuum switch, you need a vacuum gauge and a vacuum tester, connected as shown in Fig. 36-5. With the engine cold, no vacuum should pass through the switch. When the engine warms up, vacuum should pass through.

● 36-10 EGR-System Service Intervals

Typical service intervals for systems are to check the system every 12 months or 12,000 miles [19,312 km] if

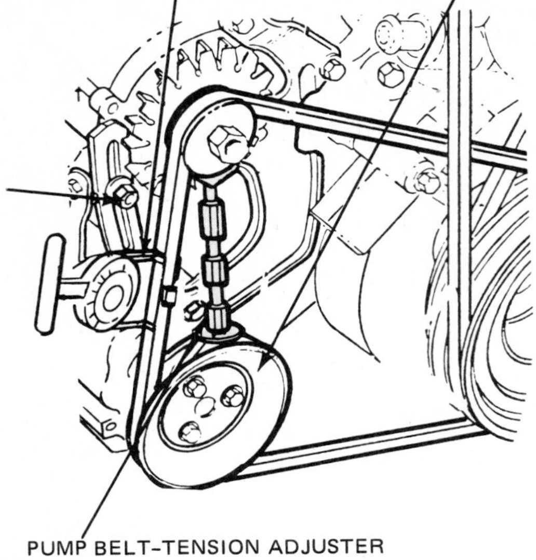

Fig. 36-3 A pump belt-tension adjuster being used to hold the air pump in place while the adjustment bolts are tightened. *(Ford Motor Company)*

the car is running on leaded gasoline. If the car is running on unleaded gasoline, check the system half as often (24 months or 24,000 miles [38,624 km]). Some cars made by Chrysler Corporation have an EGR-maintenance reminder light that comes on at 15,000 miles [24,140 km] to remind the driver to have the system checked. However, many late-model cars do not require any regular check. Instead, if trouble develops, the system should be checked, as we have explained.

● 36-11 Transmission-Controlled Spark-Vacuum-Advance (TCS) Troubles

This system, used on cars with manual transmissions, allows vacuum advance only when the transmission is in high gear.

Engine stall at idle, car creeping excessively in idle, and engine dieseling can all be due to a defective or improperly adjusted idle-stop solenoid.

Poor high-gear performance, stumble or stall on cold starts, poor fuel economy, and backfiring during deceleration could be due to an inoperative vacuum advance solenoid, to a defective temperature or transmission switch, or to failure of the time relay to energize.

If vacuum advance is obtained in all gears and HC and NO_x are high in the exhaust, you have the same four items to consider as mentioned in the previous paragraph. That is, a defective transmission or temperature switch, or time relay. Or the vacuum advance solenoid is defective.

● 36-12 Other Vacuum Advance Controls

The Ford TRS and the Chrysler OSAC systems work in a similar manner to the TCS system (see ● 35-13 to 35-15).

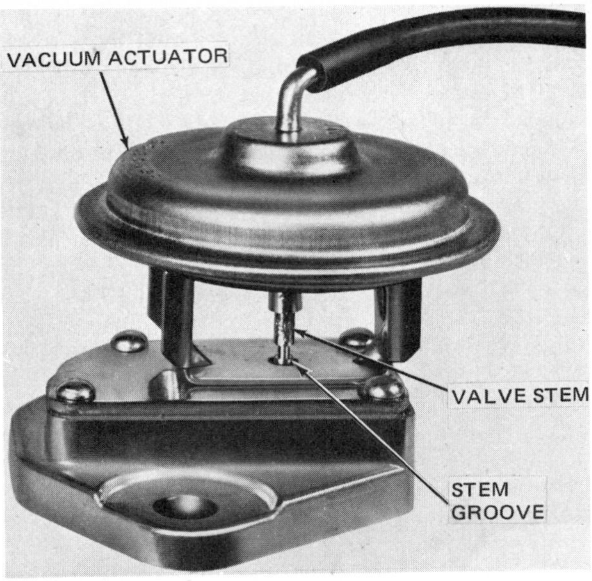

Fig. 36-4 EGR valve with exposed valve stem. *(Chrysler Corporation)*

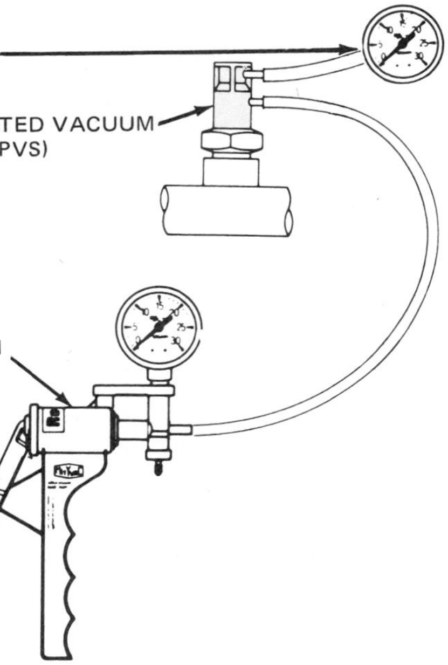

Fig. 36-5 Testing the EGR thermal vacuum switch. *(Ford Motor Company)*

Similar troubles can occur in all these systems. Failure of the system to work will give you the conditions described in ● 36-11.

● 36-13 Service Intervals for Vacuum Advance Controls

Generally, no special servicing checks are called for by the manufacturers. However, whenever a tune-up is performed, the system operation should be checked and the idle-stop solenoid adjusted. Inspection includes checking hoses for cracks, brittleness, or poor connections.

● 36-14 Electronic Control of Spark Advance

As we explained in Chap. 30 (● 30-25 to 30-28), electronic control of the spark advance is achieved by an electronic computer which receives data from various sensors. These sensors report coolant temperature, air temperature, throttle position, intake-manifold vacuum, engine speed, and so on. The computer puts all of these data together and then advances the spark the right amount for best engine operation.

These systems should develop very little trouble because everything is done electronically and, theoretically, transistors and diodes do not wear out. The manufacturers have developed special testers to check out these systems. Shops handling service to cars with electronic spark control have the special testers.

● 36-15 Catalytic-Converter Troubles

Catalytic-converter troubles would show up as noise, BB-size particles coming out the tail pipe, a rotten-egg smell, or high CO and HC levels in the exhaust.

Noise could be due to loose exhaust-pipe joints, a ruptured converter, or a loose or missing catalyst replacement plug (GM and AMC).

BB-size particles coming out the tail pipe means the converter has been overheated so the catalyst support has warped. This allows the beads to be blown out by the exhaust gas. The condition can happen only on the type of converter using the beads (GM and AMC). The remedy is converter replacement.

A rotten-egg smell comes from hydrogen sulfide (H_2S) that the catalytic converter is producing. The S, or sulfur, is in the gasoline. Some gasolines have more than others. Advise the driver to try a different brand of gasoline. Also, check the carburetor adjustments. The smell is more noticeable when a momentarily rich mixture enters a hot converter.

● 36-16 Catalytic-Converter Service

Damaged or overheated converters must be replaced. They are not repairable. However, on the bead-type converter, the old beads can be removed and a fresh charge of beads installed. Figures 36-6 and 36-7 show the special devices required. The vacuum pump is turned on while the vibrator and can are attached. It keeps the beads from falling out when the converter filler plug is removed. After the vibrator and can are attached, the vacuum is turned off. The air supply to the vibrator is turned on. Beads will now start falling in the can. It takes about ten minutes to clear the converter.

To install new beads, dump the old beads and fill the can with new beads. Attach the can to the vibrator. Turn on the air and vacuum lines. After the beads stop flowing, disconnect the air hose and remove the vibrator. The

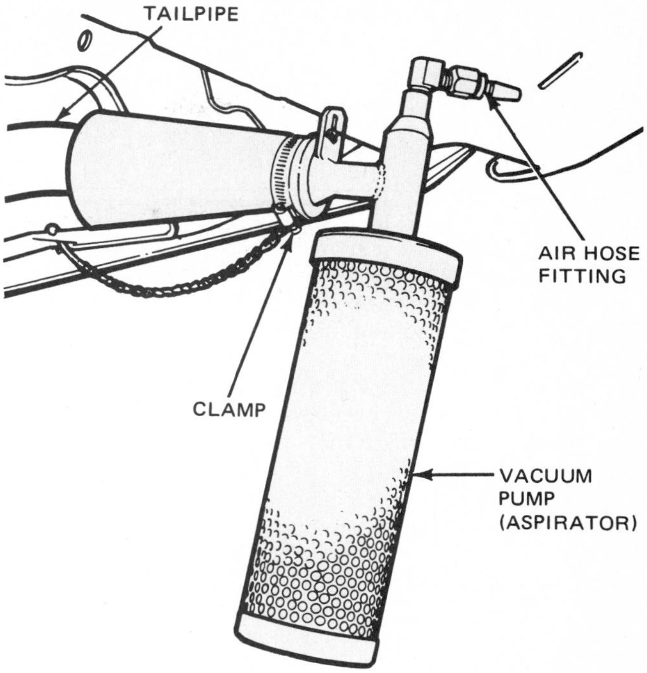

Fig. 36-6 Vacuum pump, or aspirator, mounted on the tail pipe of a car to change the beads in a pellet-type catalytic converter. (*American Motors Corporation*)

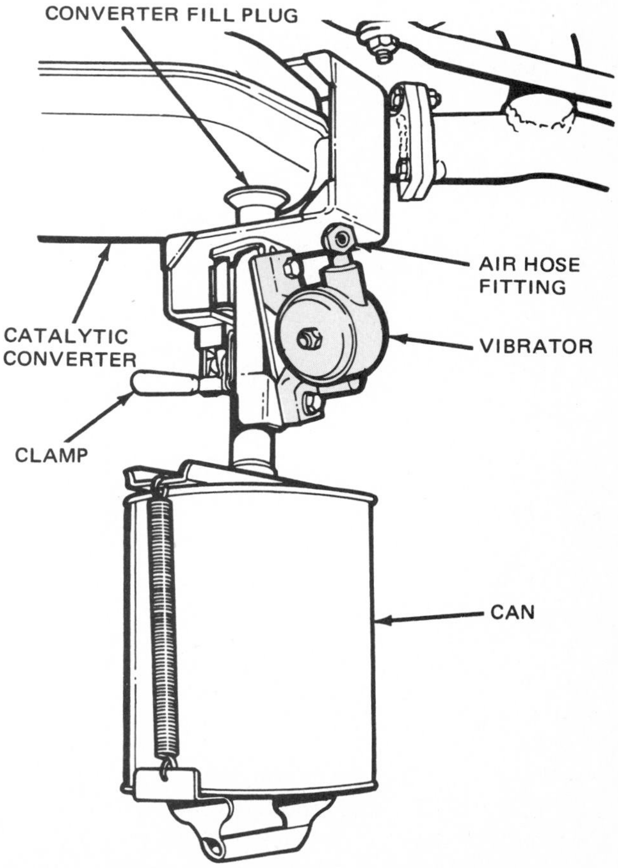

Fig. 36-7 Vibrator mounted on a catalytic converter. (*American Motors Corporation*)

converter shoud be filled flush with the full-plug hole. The vacuum pump is keeping them from falling out. Install the plug and remove the vacuum pump. Use antiseize compound on the plug threads.

REVIEW QUESTIONS

1. Can you legally disconnect a PCV system if a customer asks you to do so?
2. Describe the conditions that could result from a faulty PCV system.
3. Explain how to check the operation of the PCV valve.
4. Describe various conditions in the fuel-vapor emission-control system that could cause a noticeable fuel odor.
5. What could cause a collapsed fuel tank?
6. Name four troubles that might originate in the air-injection system.
7. How might a cold engine perform on idle if the control damper in the heated-air system is stuck in the hot-engine position?
8. How might a hot engine perform if the control damper in the heated-air system is stuck in the cold-engine position?
9. Discuss exhaust-gas-recirculation troubles and their effects on engine performance.
10. Explain how to test the EGR valve and the thermal vacuum switch.
11. Discuss various defects in the TCS system and their effects on engine performance.
12. What could cause a rotten-egg smell in the exhaust? What can be done to eliminate it?
13. What would cause BB-size particles to come out the tail pipe?
14. Explain how to install new beads in the General Motors converter.

SELF PROJECTS

Whenever you run across an article in a magazine or paper about automotive emission controls, clip it and put it in your notebook. Study the emission-control sections in every automotive shop manual you can find. You will not be able to clip this material from these manuals. However, if you study the material carefully, and make notes of what you learn, you will soon accumulate a valuable store of knowledge on the subject. Be sure to copy any wiring or vacuum diagrams for filing in your notebook.

Every time you raise the hood of a car, identify all emission controls. Trace the various hoses and note where they are connected. If you continue to do this, you will soon be familiar with all the different controls and variations of these controls. You will soon know what every component of every control is for, how it works, and what can happen to engine performance if it stops working.

Automotive Engine Service

6

Part Six of this book describes service operations on automotive engines. Included in Part Six are servicing details on engine valves and valve trains, pistons and related parts, crankshafts, and cylinder blocks.

There is another method of engine service called engine rebuilding. This results in an engine that is just about "as good as new."

Another way to go in engine service is to purchase, from the manufacturer, a "short block." This includes the cylinder block with all related parts such as pistons, piston pins, rings, connecting rods and bearings, crankshaft and main bearings, and so on.

There are four chapters in Part Six.

Chapter 37 Engine Service: Valves and Valve Mechanisms

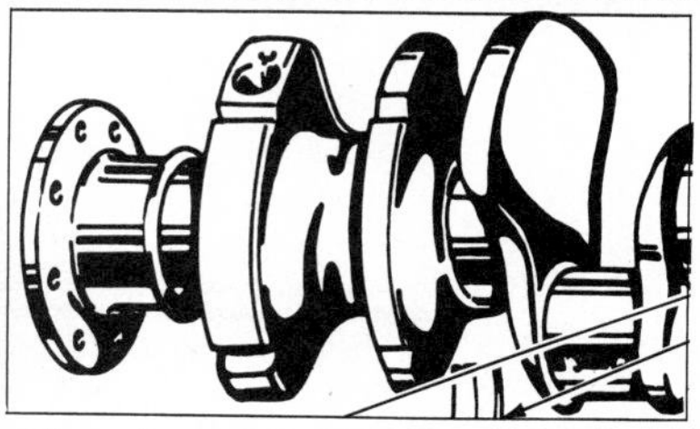

Chapter 38 Engine Service: Connecting Rods, Rod Bearings, Pistons, and Rings

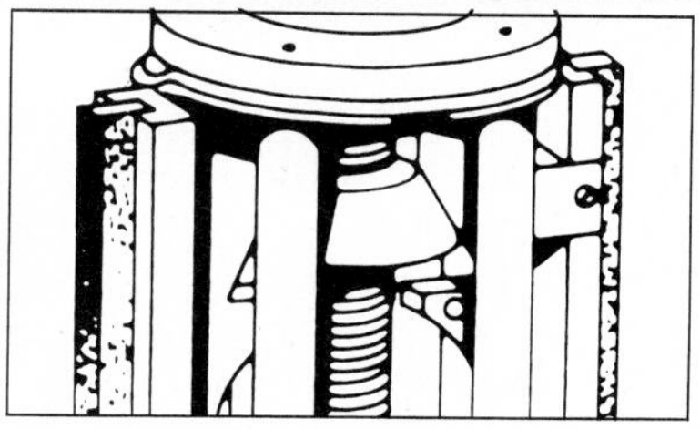

Chapter 39 Engine Service: Crankshafts and Cylinder Blocks

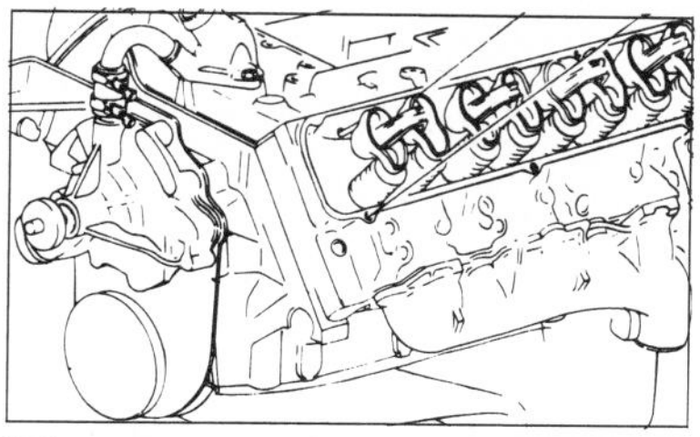

Chapter 40 Diesel-Engine Service

CHAPTER 37
Engine Service: Valves and Valve Mechanisms

After studying this chapter, you should be able to:

- **Discuss valve troubles and their possible causes.**
- **Explain how to perform the various valve jobs discussed in the chapter.**
- **Under the instructor's supervision, perform the valve jobs discussed in the chapter.**

This chapter describes the servicing of valves and valve mechanisms, including valves, valve lifters, push rods, rocker arms, crankshafts, and other valve-train parts. Figure 37-1 shows a cylinder head with valve-train parts for a V-8 engine.

● 37-1 Cleanliness

The major enemy of good engine-service work is dirt. A speck of dirt left on a bearing or cylinder wall can ruin a perfect service job. Thus, you must be absolutely sure you do not leave abrasive or dirt in the engine or on engine parts after a service job.

Before any major service job, the engine should be cleaned. Electrical and air-conditioning units should be removed or protected if the engine is to be steam-cleaned. Otherwise, they could be damaged.

● 37-2 Valve Troubles

Valves must be properly timed. They must seat tightly and operate without lag. On engines with mechanical valve lifters, the clearance between the rocker arm and valve stem must be correct. Clearance between the valve stems and guides must be within specifications. Hydraulic valve lifters must perform properly to take up clearance in the valve train when the valve is closed. Failure to meet any of these requirements means valve and engine trouble.

As an example, suppose there is too much clearance between valve stems and guides. This means that, on every intake stroke, oil can be pulled past the valve stems and guides (Fig. 37-2). It enters the combustion chamber, where it burns. This, in turn, leads to excessive oil consumption, engine deposits, preignition, clogged piston rings, fouled spark plugs, and probably burned or otherwise damaged valves and valve seats. Thus, what seems like a minor fault can lead to serious engine trouble. This shows the importance of doing every engine job correctly. Being slightly "off" on one measurement can cause serious engine trouble.

Oil leakage past the exhaust-valve stem can also occur.

Types of trouble Valve troubles include sticking, burning, breakage, wear, deposits, and valve-seat recession. We discuss each of these in following sections.

● 37-3 Valve Sticking

Gum or carbon deposits on the valve stem (● 37-9) cause valve sticking (Fig. 37-3). Excessive valve-stem clearance speeds up valve deposits. Another cause of valve sticking is warped stems. These could result from overheating, an eccentric seat (which throws side pressure on the valve), or a cocked spring or retainer (which tends to bend the stem). Insufficient oil also causes valve sticking. Sometimes valves stick when cold, but work free as the engine warms up.

● 37-4 Cleaning Valve Stems with Special Cleaner

When valves and piston rings are badly clogged with deposits, an engine overhaul is usually required. However, there are certain compounds that can be put in the fuel or oil to help free rings and valves. One of these comes in a pressure can. It is sprayed into the running engine through the carburetor. This is done after the air cleaner has been removed. When parts are not too badly worn, and the major trouble seems to be from deposits, use of these compounds often postpones engine overhaul.

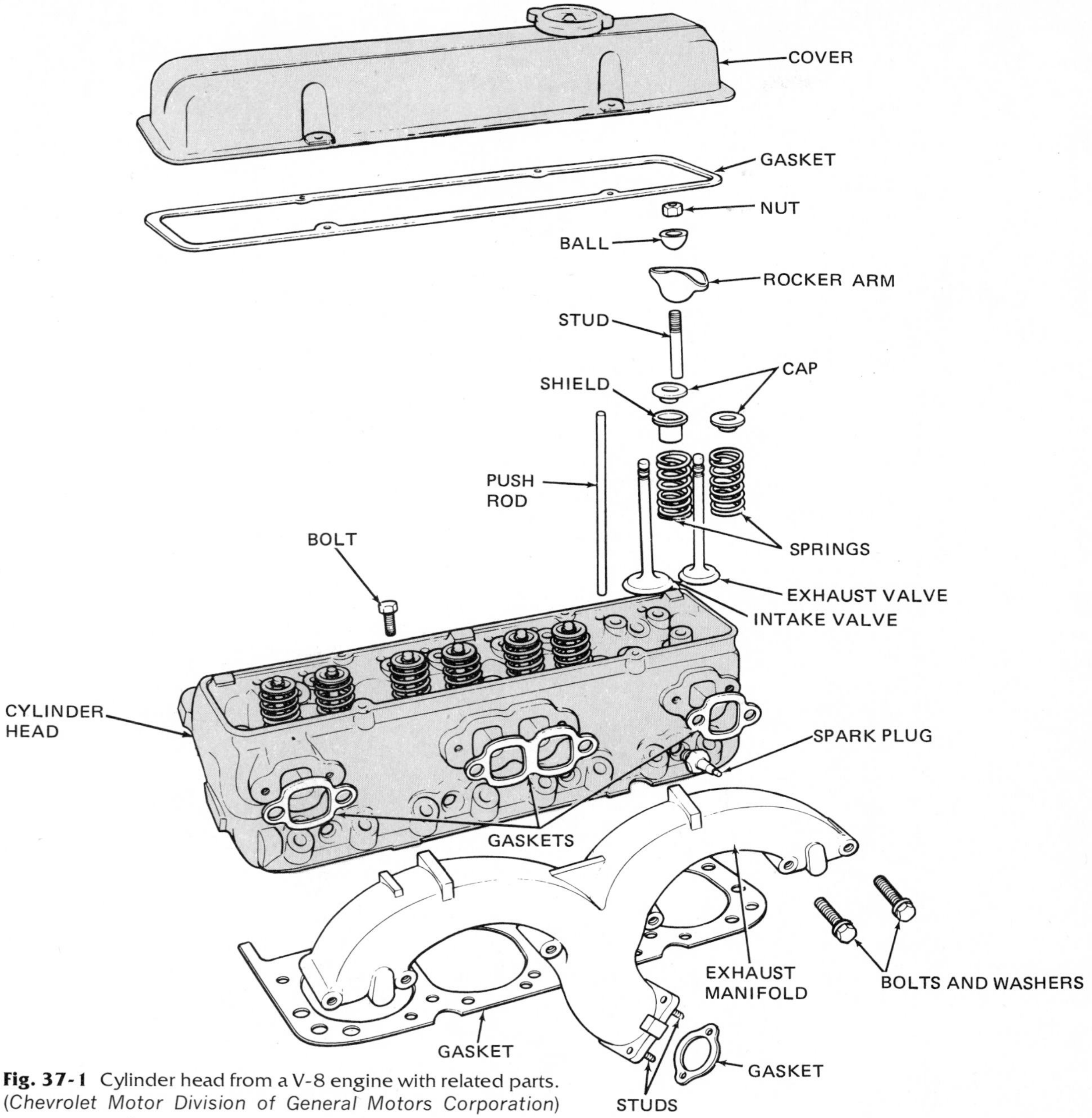

Fig. 37-1 Cylinder head from a V-8 engine with related parts. *(Chevrolet Motor Division of General Motors Corporation)*

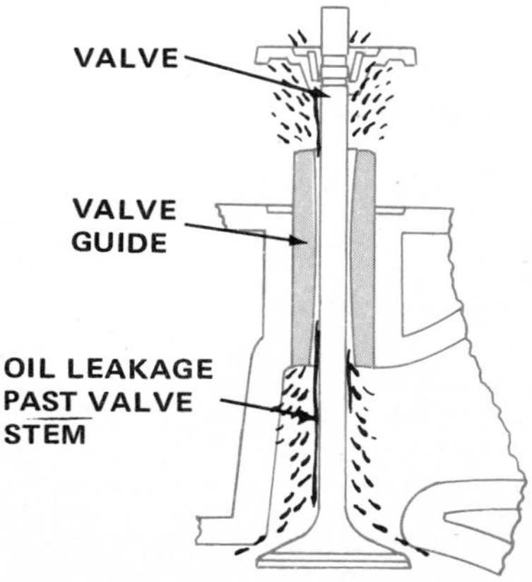

Fig. 37-2 Leakage of oil will result from a worn valve guide.

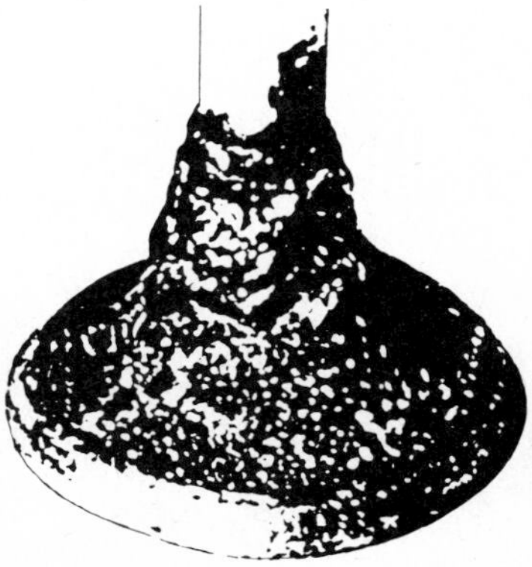

Fig. 37-3 Gummed intake valve. Note deposits under valve head. *(Clayton Manufacturing Company)*

Be sure to read the instructions on the can the cleaner comes in, to make sure it will not harm the catalytic converter on the car (if the car is so equipped). Some of these compounds can have a very harmful effect on the converter. Do not use a compound unless you are sure it will not damage the converter.

● 37-5 Valve Burning

This is usually an exhaust-valve problem. However, intake valves may also burn. For example, one cause could be the manifold heat-control valve sticking in the closed position. This increases the air-fuel mixture temperature which can lead to intake-valve burning. Any condition that prevents normal valve seating may lead to valve burning (Figs. 37-4 and 37-5). The poor seating prevents normal valve cooling through the valve seat. It also allows hot gases to blow by, further heating the valve. A worn guide also prevents normal valve cooling. Also water circulation around the valve seat may be slowed, by clogged distribution tubes or jackets, for example. Then local hot spots may develop. This could cause seat distortion, poor seating, and overheated valves. Seat distortion can also result from improper cylinder-head-bolt tightening. Other conditions that could prevent normal seating include a weak or cocked valve spring and insufficient valve-tappet clearance.

Engine overloading or overheating will cause hot valves. A lean air-fuel mixture may cause valve burning. In this case, the fuel system should be serviced. Preignition and detonation produce high combustion pressures and temperatures. These are hard on valves as well as other engine parts. Correction is to clean out carbon, retime the ignition, and use higher-octane fuel.

In some cases of seat leakage (especially where deposits on the valve seat and face prevent normal seating), an "interference angle" has proved helpful. The valve is faced at an angle $\frac{1}{4}$ to 1 degree flatter than the seat angle (Fig. 37-6). This produces greater pressure at the outer edge of the valve seat. The valve-seat edge tends to cut through any deposits that have formed, and thereby establish a good seal. The difference in angles between the valve face and seat gradually disappears as the valve face and seat wear. That is, the contact between the two changes from line contact (interference angle) to area contact. This improves the seating.

Figure 37-6 illustrates one manufacturer's recommendations for the valve and seat interference angles. This manufacturer does not recommend interference on stellite-faced exhaust valves and induction-hardened exhaust-valve seats. These surfaces are so hard that seating would not be improved by interference.

Sometimes valve stems stretch, due to overheating and to heavy valve springs.

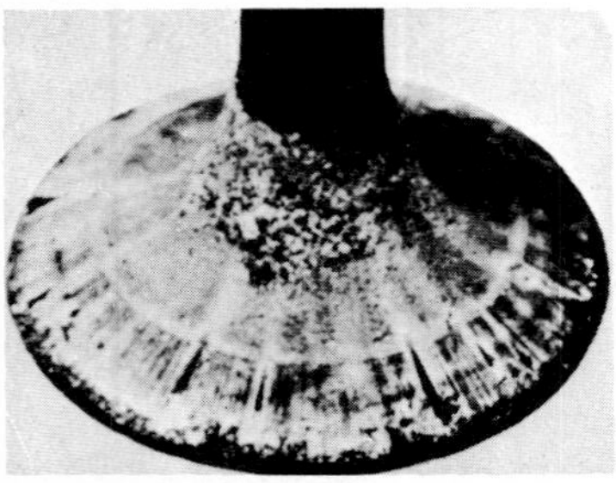

Fig. 37-4 Valve burning due to failure of the valve to seat fully. Note that the valve is uniformly burned all the way around its face. (*TRW Valve Division of Thompson, Ramo, Wooldridge, Inc.*)

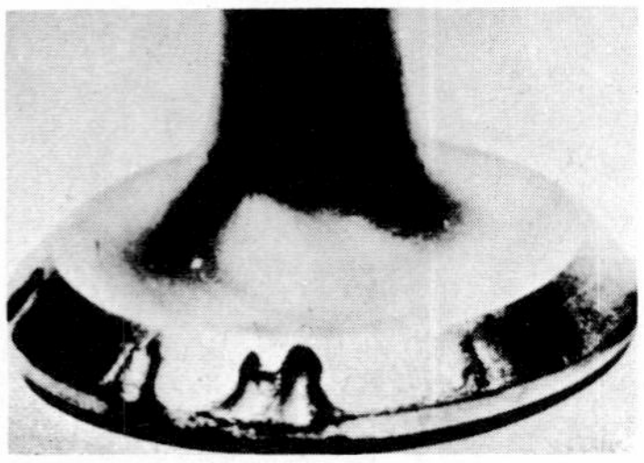

Fig. 37-5 Valve burning due to guttering caused by deposits on valve face and seat. Parts of the deposits break off. They leave a path through which exhaust gas can pass when valve is closed. These burn channels in the valve face. (*TRW Valve Division of Thompson, Ramo, Wooldridge, Inc.*)

● 37-6 Valve Breakage

Any condition that causes the valve to overheat (● 37-5) may cause it to break. Heavy pounding (as from excessive tappet clearance or from detonation) may cause valves to break. Excessive tappet clearance permits heavy impact seating. An off-center seat or cocked valve spring or retainer will put side pressure on the valve every time it seats. This may cause it to fatigue and break. A scratch on the stem may serve as a starting point for a crack and a break in the stem.

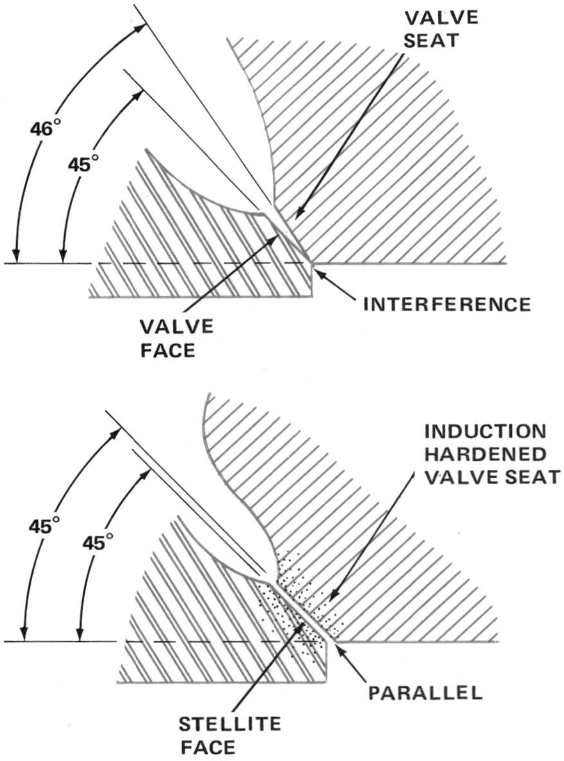

Fig. 37-6 Valve and valve-seat angles. Top, interference angle recommended for many intake and exhaust valves and seats. Bottom, parallel faces recommended for stellite-faced exhaust valves and induction-hardened exhaust-valve seats. (*Chevrolet Motor Division of General Motors Corporation*)

37-7 Valve-Face Wear

Excessive tappet clearance or dirt on the valve face or seat can cause valve-face wear. Excessive tappet clearance causes heavy impact seating. This wears the valve and may cause valve breakage (● 37-6). Dirt may cause valve-face wear if the engine operates in dusty conditions, or if the carburetor air cleaner is not functioning properly. The dust enters the engine with the air-fuel mixture, and some of it settles on the valve seat. Dust also causes bearing, cylinder-wall, and piston and ring wear.

37-8 Valve-Seat Recession

Valve-seat recession is the gradual wearing down of the valve seat so that it recedes away from the combustion chamber. This decreases valve-tappet clearance. That is, as the valve seat recedes, the clearance in the valve train is reduced. In an engine with mechanical valve lifters, the result can be a complete loss of clearance. The valve can no longer close completely, and valve and seat burning result.

More attention has been given to valve-seat recession recently. This is because lead has been removed from the gasoline used in converter-equipped cars. Lead additives in the gasoline form a lubricant between the valve face and valve seat. This prevents iron particles that flake off the valve seat from sticking to the valve face. However, without this lead coating, particles of iron do tend to stick on the valve face. Gradually, as these particles embed in the valve face, they build up into tiny bumps, or warts. This turns the valve face into a cutting surface. Therefore, the valve seat is gradually cut away and seat recession results. The warts also cause leakage, and loss of compression and power.

To prevent such seat recession in engines run on lead-free gasoline, the valves are given a very thin coating of aluminum, nickel, or other metal. This coating is less than 0.002 inch [0.05 mm] in thickness. It keeps particles from sticking to the valve face and thus causing valve-seat recession. The coating gives the valve face a dull, almost rough, appearance. The automotive mechanic might think the valve needs refacing. However, this must not be done. *Coated valves should not be refaced or lapped.* These would remove the coating and deny the valve seat the protection of the coating. The result could be very short valve and seat life. There is more on this later in the chapter, when we cover valve and seat service.

37-9 Valve Deposits

A fuel with excessive amounts of gum may deposit some of the gum on the intake valve. This happens as the air-fuel mixture passes the valve on the way to the engine cylinder. Carbon deposits may form because of an excessively rich mixture or because of oil passing a worn valve guide (in the intake valve). Improper combustion may be due to a rich mixture, defective ignition system, loss of compression in the engine, a cold engine, and so forth. Whatever its cause, it will result in carbon deposits on the exhaust valves. Dirty or improper oil will cause deposits to form on the valves.

37-10 Valve Service

Valve service includes adjusting valve-tappet clearances (also called *adjusting valve lash*), grinding valves and valve seats, installing new seat inserts, cleaning and knurling or replacing valve guides, servicing the camshaft and camshaft bearings, and timing the valves. A complete valve-service job, including grinding valves and seats, checking springs, cleaning guides, and tuning the engine, requires about 5 hours (for an overhead-valve six-cylinder engine). Replacing the camshaft requires about 8 hours; 4 additional hours are required for replacing camshaft bearings. These times are only approximate and vary from engine to engine.

37-11 Valve-Lifter Clearance

The procedure for checking and adjusting valve-tappet (or valve-lifter) clearance depends on the type and model of engine. Some engines with hydraulic valve lifters normally require no clearance adjustment. Others require checking and adjustment whenever valve-service work has been performed. The following procedures are typical.

37-12 L-Head Engine with Mechanical Valve Lifters

Remove the valve-cover plates. Use a feeler gauge to check the clearance between the valve stem and the adjusting screw in the valve lifter. Adjustment is correct when the feeler gauge can be moved between the screw and valve stem with some drag when the valve is closed. When a locking nut is used, it should be tightened after the adjustment is made. The clearance should be checked again. After the adjustment is completed, replace the cover plates, using new gaskets.

Valves on some L-head engines are set with the engine running.

37-13 I-Head Engine with Mechanical Valve Lifters

First remove the valve cover. Measure the clearance between the valve stem and rocker arm, as shown in Fig. 37-7. Most specifications call for making the check with the engine hot and not running. The clearance is measured with the valve lifter on the base circle of the cam. Turn the crankshaft by bumping the engine with the starting motor until the base circle of the cam is under the valve lifter.

There are two kinds of rocker arms. One is shaft mounted, and the other ball-stud mounted. The shaft-mounted type (Fig. 37-7) usually has an adjustment screw. This screw is usually self-locking and does not require a locking nut. Use a box wrench to turn the adjustment screw and adjust the clearance to specifications. Do not use an open-end wrench. This could damage the screw head.

On ball-stud-mounted rocker arms (Fig. 37-8), turn the self-locking rocker-arm stud nut to make the adjustment. Turning the nut down reduces clearance.

● 37-14 Free-Type Valve Rotator

Free-rotator valves are checked in the same way as the mechanical-lifter type. The clearance is checked between the tip cup on the valve stem and the adjusting screw in the valve lifter.

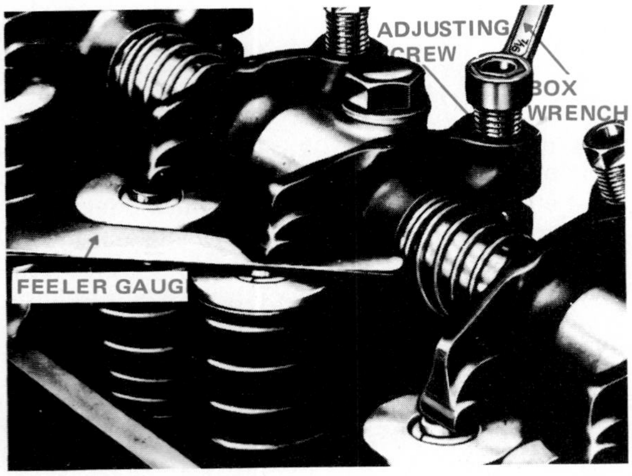

Fig. 37-7 Adjusting valve-tappet clearance on an I-head engine. (*Ford Motor Company*)

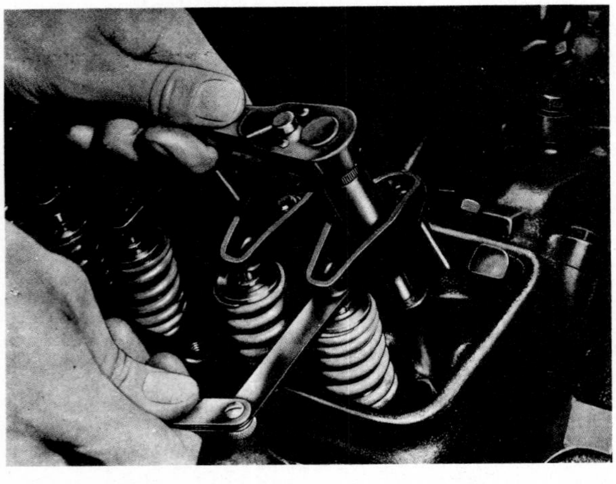

Fig. 37-8 Adjusting valve-tappet clearance on an engine with rocker arms independently mounted on ball studs. Backing the stud nut out increases clearance. (*Chevrolet Motor Division of General Motors Corporation*)

● 37-15 I-Head Engine with Hydraulic Valve Lifters

On some engines with hydraulic valve lifters, no adjustment is provided in the valve train. In normal service, no adjustment is necessary. The hydraulic valve lifter takes care of any small changes in valve-train length. However, adjustment may be needed if valves and valve seats are ground. Unusual and severe wear of the push-rod ends, rocker arm, or valve stem may also require adjustment. Then some correction may be required to reestablish the correct valve-train length. Typical checking and correcting procedures follow.

● 37-16 Ford Engines with Hydraulic Valve Lifters

Ford engines use two types of rocker arms, the shaft mounted (Fig. 37-9) and the ball-stud mounted (Fig. 37-10). On both types the clearance in the valve train is checked with the valve lifter bled down so that the valve-lifter plunger is bottomed. First, the crankshaft must be turned so that the lifter is on the base circle or low part of the cam (rather than on the lobe). This is done by setting the piston in No. 1 cylinder at TDC (top dead center) at the end of the compression stroke. Then check both valves in No. 1 cylinder. The crankshaft can then be rotated to put other lifters on the base circles of their cams so they can be checked.

To make the check, a special tool is used to apply slow pressure on the rocker arm (Figs. 37-9 and 37-10). This gradually forces oil out of the valve lifter, so that the

Fig. 37-9 Checking valve-train clearance. (*Ford Motor Company*)

plunger bottoms. Then, the clearance gauge is used to check the clearance between the valve stem and rocker arm. If the clearance is too small, a shorter push rod should be installed. If the clearance is excessive, install a longer push rod. The clearance might be too small if valves and seats have been ground. The clearance might be excessive due to wear of the valve-train parts. This includes wear of the push-rod ends, valve stem, and rocker arm.

37-17 Plymouth Engines with Hydraulic Valve Lifters

The procedure for setting Plymouth valves is typical of Chrysler-manufactured engines. It is necessary only when valves and valve seats have been ground. When this happens, the increased height of the valve stem above the cylinder head should be checked. With the valve seated, place the special gauge over the valve stem. If the height is excessive, the end of the valve stem must be ground off to reduce the height to within limits. The hydraulic valve-lifter plunger will now be near its center position rather than near the bottom, as it would with an excessively high valve stem.

37-18 Chevrolet Engines with Hydraulic Valve Liftes

The procedure for Chevrolet engines is typical of General Motors engines using the ball-pivot type of rocker arm (Fig. 37-11). With the valve lifter on the base circle of the cam, back off the adjustment nut until the push rod is loose. Then slowly turn the adjustment nut down. At the same time, rotate the push rod with your fingers until the push rod is tight, that is, until you cannot easily rotate the push rod. Then turn the adjustment nut down one additional full turn. This places the plunger of the valve lifter in its center position.

Fig. 37-10 Checking valve-train clearance on ball-stud-mounted rocker arm. The hydraulic valve lifter has been bled down with pressure from the special tool. (*Ford Motor Company*)

37-19 Overhead-Camshaft Engine

Overhead-camshaft engines have several arrangements for carrying the cam action to the valve stems. In some engines, cam action is carried directly to the valve stem through a cap, called the *valve tappet*. This cap fits over the valve stem and spring. In others, the cam action is carried through a rocker arm. We shall look at both types.

1. Chevrolet Vega This engine is pictured in Figs. 6-19 and 9-15. Figure 37-12 shows the valve-train parts. Adjustment of valve clearance is made by turning the adjustment screw located in the valve tappet. The adjustment screw has a flat on one side. Therefore, adjustment must be made by turning the screw *full turns only*. The checking and adjusting procedure follows.

Turn the camshaft so the valve tappet is on the base circle. Then measure the clearance between the cam and the valve tappet with a feeler gauge (Fig. 37-13). Use the special tool, as shown, to turn the adjustment screw. The screw must be turned complete revolutions, so the flat on the screw ends up directly above the valve stem. Turning the screw in, or clockwise, decreases clearance. Each full turn of the screw changes the clearance 0.003 inch [0.076 mm]. Check and adjustment are made with the engine cold.

2. Chevrolet LUV engine This engine has rocker arms which are held in place by springs. The rocker arms have

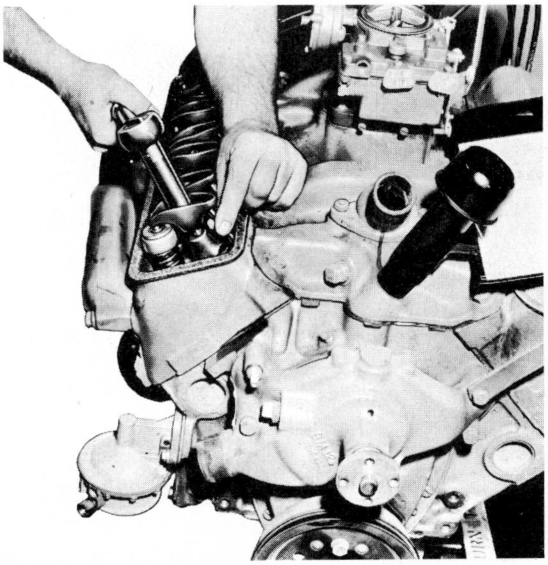

Fig. 37-11 Adjusting valve rocker-arm stud nut to properly position the plunger of the hydraulic valve lifter. (*Chevrolet Motor Division of General Motors Corporation*)

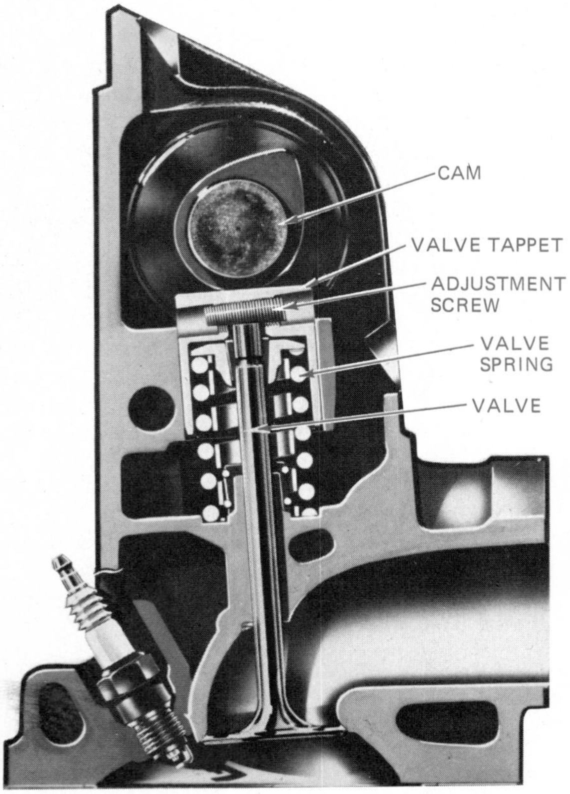

Fig. 37-12 Valve train for the Vega engine. (*Chevrolet Motor Division of General Motors Corporation*)

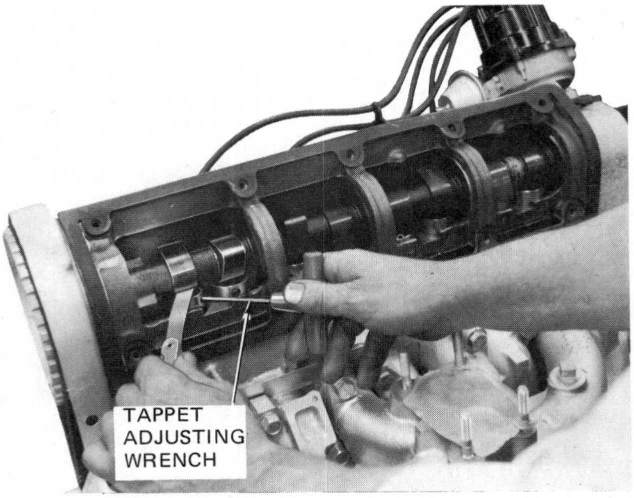

Fig. 37-13 Adjusting valve-tappet clearance on the Vega engine. (*Chevrolet Motor Division of General Motors Corporation*)

dome-shaped ends which fit over ball studs in the cylinder head. The valve ends of the rocker arms fit into a depression in the valve-spring retainer and rest on the valve stems. Measure the clearance between the cam surface of the rocker arm and the base circle of the cam. Use a flat feeler gauge. Adjust with a Phillips-head screwdriver inserted through the hole in the rocker-arm dome.

Fig. 37-14 Checking valve-tappet clearance on the Ford 2,000-cc engine. (*Ford Motor Company*)

3. Ford 2,000-cc four-cylinder engine This engine has rocker arms which float between a stationary stud on one side and the valve stem on the other. The center of the rocker arm rests on the cam. This is similar to the arrangement in the Chevrolet LUV engine. Figure 37-14 shows the use of a feeler gauge to check the clearance between the base circle of the cam and the rocker arm. Adjustment is made by loosening the lock nut. Use a 15-mm (millimeter) [0.59-in.] open-end wrench to turn the adjustment screw in or out. Turning the screw in increases the clearance. Tighten the lock nut securely after the adjustment, and recheck the clearance.

4. Pontiac overhead-camshaft engine In the valve-train arrangement, the rocker arm floats between the end of the valve stem and an automatic valve-lash adjuster. The automatic valve-lash adjuster is a special type of hydraulic valve lifter. It automatically takes up any clearance between the cam and the rocker arm. Normally, no valve adjustment is necessary on this engine.

● 37-20 The Complete Valve Job

A complete valve job requires the steps listed below. Valve and valve-seat servicing are described in detail in sections that follow:

1. Drain cooling system. Disconnect upper radiator hose from engine.
2. Remove air cleaner. Disconnect accelerator rod, fuel line, air and vacuum hoses from carburetor.
3. Remove or move aside lines and hoses as necessary to get at the cylinder head.
4. Disconnect spark-plug cables and temperature-sending-unit wire.
5. Remove crankcase ventilating system. On air-injec-

tion systems, disconnect the air hose at the check valve. Then remove the air-supply-tube assembly.

6. On many in-line engines, it is not necessary to remove manifolds. But in V-8 engines, the carburetor and intake manifold must be removed.
7. Remove rocker-arm cover or covers.
8. On engines with ball-stud-supported rocker arms, the rocker arms and push rods can be removed at this time. If they are left on, the nuts should be loosened so the rocker arms can be moved aside and the push rods removed. Push rods should be placed in a rack, in order. Then they can be reassembled in their proper positions.
9. On engines with rocker arms supported on shafts, remove the shaft assembly or assemblies (● 37-23). Then remove the push rods, in order.
10. Remove head bolts. Take the head off the engine.
11. Remove the valves and springs from the head. (Keep them in order so they can be put back in their proper positions.)

Careful: If a valve-stem end has mushroomed, the valve cannot be pulled out by hand. The mushroom must be removed, as explained in ● 37-25. Otherwise, if you try to pull the valve, you can break the valve guide.

12. Check valve guides for wear. Clean, replace, or knurl and ream for same-size valve stem if necessary. Or ream for a larger-diameter valve stem.
13. Check valves and valve seats. Clean valve heads and stems on a wire wheel. Grind valve seats, and reface valves as necessary. Check valve seating. Touch up valve-stem ends if necessary.

    If you are installing new valves of the coated type, do not reface them. Refacing or lapping coated valves removes the protective coating and greatly shortens valve and seat life.
14. Check rocker arms for wear. Service or replace as necessary.
15. Replace valves and springs in head.
16. Install head, push rods, rocker arms, rocker-arm cover, and other parts removed during head removal.
17. Check and adjust valve-stem clearance as necessary.

● 37-21 Removing, Cleaning, and Replacing Cylinder Heads

On some cars, the manifolds must be removed before the cylinder heads can be taken off (● 37-34). On other cars, the manifolds may be left in place.

1. Removing the cylinder head Follow the general instructions in ● 37-20 to remove the cylinder head. Slightly loosen all cylinder-head bolts first, to ease the tension on the head. Then remove the bolts. If the head sticks, carefully pry it loose. Do not pry hard. Do not insert the pry bar too far between the head and block. This could mar the mating surfaces and lead to leaks. Lift the head off, and place it in a head-holding fixture.

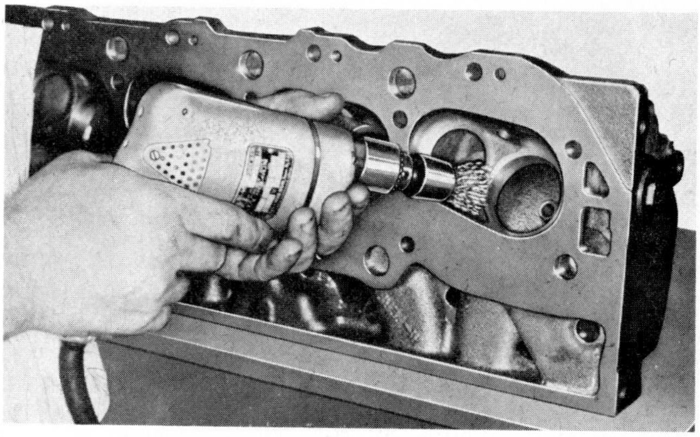

Fig. 37-15 Cleaning combustion chamber and valve ports with wire brush. (*Chevrolet Motor Division of General Motors Corporation*)

Careful: Never remove a cylinder head from a hot engine. Wait until the engine cools. If the head is removed hot, it can be distorted so that it cannot be used again.

2. Cleaning the cylinder head After the valves and other parts are removed from the head (as explained later), it should be cleaned and inspected. Clean carbon from the combustion chambers and valve ports. Use a wire brush driven by a drill motor (Fig. 37-15). An air-powered motor is best. Keep the wire brush away from valve seats, because it could scratch the seating surfaces. Scratched seats can cause poor valve seating and serious engine trouble. Blow out all dust with an air hose.

Some engine experts recommend temporarily reinstalling the valves before cleaning the combustion chambers. This protects the valve seats and at the same time cleans the valve heads.

CAUTION: Always use goggles when using a wire brush, compressed air, or similar equipment. They will protect your eyes from flying particles.

Clean gasket surfaces with a flat scraper. Be very careful not to scratch the gasket surface. All traces of gasket material and sealer should be removed.

Remove dirt and grease from the cylinder head. Then clean the water jackets and passages by soaking the head in a boil tank. Flush the water jackets as recommended by the manufacturer of the cleaning agent.

3. Inspecting the cylinder head As you remove the head, examine the gasket and mating surface for traces of leakage or cracks. If cracks are suspected, have the head checked with Magnaflux or similar equipment. A blown gasket or coolant leakage could result from a warped head or improper gasket installation. In the head, cracks usually occur between valve seats. If they are not too bad, they can be repaired by "cold welding." This is a job for an expert. First, drill a small hole in the crack and thread it. Then screw in a threaded rod and cut it off. Next, drill a second hole overlapping the first one and thread it. Screw in the threaded rod and cut it off. Repeat until the crack is completely treated. Make sure to get to both ends of the crack, to relieve the stress that caused the crack. Sometimes it is best to install a seat insert when the crack runs into the valve seat. This is done by making an undercut in the head and pressing in the insert (● 37-30).

Clean and inspect valve guides. Note the condition of valve seats and ball studs (on heads using them). We cover servicing of valve guides, ball studs, and seats later.

Warpage is detected by laying a straightedge against the gasket surface of the head (Fig. 37-16). Check crossways and longways. One specification calls for 0.005 inch [0.127 mm] maximum out of straight. More than this requires either a new head or machining the head so that the gasket surface is straight again.

Check the gasket surface of the head for nicks or rough spots. These can be removed with a fine-cut mill file.

If one head from a V-8 engine requires machining to remove gasket-surface roughness, then the other head should be machined a like amount. Otherwise, uniform compression and manifold alignment will be lost. Also, remember that removing metal from the gasket surface lowers the head with respect to the intake manifold. Therefore, a compensating amount may have to be machined from the manifold to restore alignment.

4. Installing the cylinder head Reassemble the cylinder head as explained later, so that valve springs, rocker arms, and other parts are in place. Then install the head, as follows. Always use a new gasket.

Before installing the head, check the cylinder block. Make sure the gasket surface is flat and in good condition. Make sure all traces of gasket material are removed from the block. Bolt holes in the block (where present) should be cleaned out. Cylinder-block studs (where present), should be in good condition. Cylinder-head bolts should be cleaned with a wire brush or wheel. Cylinder-block studs (where present) can be cleaned up with a thread chaser if the threads are damaged.

Use care when handling the gasket. If it is of the lacquered type, do not chip the lacquer. If the block has studs, put the gasket into place, right side up. Some are marked TOP so you know which side goes up. Use gasket cement only if specified by the manufacturer. For example, one manufacturer says to use cement on steel gaskets, but not on composition steel-asbestos gaskets.

If the block does not have studs, use two pilot pins set into two bolt holes to assure gasket alignment. Then lower the head into position. Substitute bolts for pilot pins (if used). Run on the nuts or bolts, finger tight.

Make sure that all bolt holes in the block have been cleaned out. If they are not clean, and the bolts bottom on foreign material, then the head will not be tight.

Use a torque wrench to tighten the nuts or bolts. They must be tightened in the proper sequence and to the proper tension. If they are not, head or block distortion, gasket leakage, or bolt failure may occur. Refer to the sequence chart for the engine being serviced, and note the torque called for. Figure 37-17 shows the sequence for one head of a V-8 engine. Each bolt should be tightened in two or more steps. That is, the complete circuit should be made at least twice, with each bolt or nut being drawn down little by little. After engine assembly is completed, the engine should be run until it is warm. Then the torques should be checked. Also, some engines using aluminum heads must be turned off and allowed to cool. Then the bolt or nut torques must be checked again.

Fig. 37-16 Checking head for warpage with a straightedge and feeler gauge. (*TRW, Inc.*)

Some torque specifications call for clean, dry threads. Others call for lightly lubricated threads. Antiseize or sealing compounds are often used on bolts in aluminum blocks.

Careful: If the rocker arms are in place, tighten the bolts slowly. This gives the hydraulic valve lifters time to bleed down to their operating length. If the bolts are tightened too rapidly, excessive pressure will be put on the lifters. They could be damaged, and the push rods could be bent. On most engines, the head bolts cannot be tightened if the rocker arms are in place.

On overhead-valve cylinder heads with the rocker arms and shaft in place, make sure the push rods are in position. That is, make sure the lower ends of the push rods are in the valve-lifter sockets.

● 37-22 Ball-Stud Service

If a ball stud is loose in the cylinder head, has damaged threads, or has begun to pull out, it should be replaced. The old stud is removed with a special puller. The stud hole is then reamed to a larger size to take the oversize stud. Some studs screw into tapped holes.

● 37-23 Servicing Rocker-Arm Assemblies

There are two methods of attaching the rocker arms to the cylinder head. In one, the rocker arms are mounted on a common shaft. In the other, the rocker arms are mounted individually on separate studs. They move on ball pivots instead of a shaft. On these (Fig. 37-11) removing the adjusting nut permits removal of the rocker arm. The rocker-arm studs can be replaced if they are loose or if the stud threads are damaged (● 37-22).

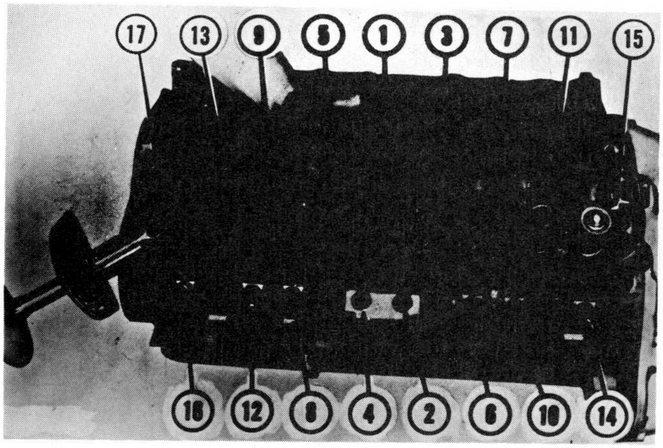

Fig. 37-17 Sequence for tightening cylinder-head bolts on a V-8 engine. (*Chrysler Corporation*)

On some shaft-mounted rocker arms, the shaft and rocker arms and brackets all come off as an assembly. On others, you remove the shaft lock plug and slip the shaft off the head. The rocker arms can then be removed from the head.

After rocker arms are removed, they should be inspected for wear or damage. Rocker arms with bushings can be rebushed if the old bushings are worn. On some rocker arms, the valve ends, if worn, can be ground down on the valve-refacing machine. Excessively worn rocker arms should be discarded.

When reinstalling rocker arms and shafts on the cylinder head, make sure that the oil holes (in shafts so equipped) are on the underside. Otherwise, they will not feed oil to the rocker arms. Be sure all springs and rocker arms are in their original positions when the shafts are reattached to the head.

● 37-24 Push-Rod Service

Push rods should be inspected for wear at the ends. Roll the rods on a flat surface to check for straightness. Replace defective rods. Rods on some engines have one tip hardened and marked with a stripe of color. The push rod should be installed so that the hardened end is toward the rocker arm. Always make sure that the lower end of the push rod is seated in the valve-lifter socket.

Special short-length push rods are available for some engines. These may be used in some engines after valves and valve seats have been reground.

● 37-25 Valve Removal

After the head is off, and rocker-arm mechanisms are removed from the head, the valves are taken out. Valves and valve parts must not be interchanged. Each valve, with its own spring, retainer, and lock, should be put back in the valve port from which it was removed. For this reason, a special valve rack is recommended (Fig. 37-18).

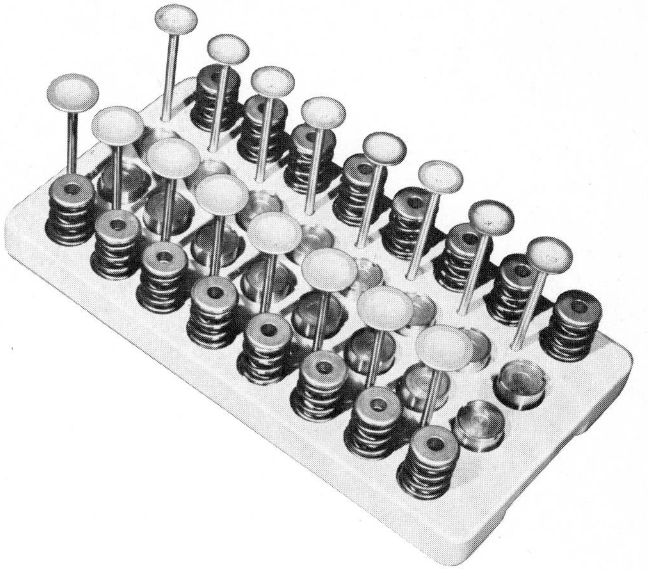

Fig. 37-18 Valve rack for holding valves and associated parts. (*Chevrolet Motor Division of General Motors Corporation*)

Likewise, each rocker arm and push rod should go back to its original position.

Before removing a valve from the cylinder head, examine the valve stem. Look for burrs at the retainer-lock grooves and for mushrooming on the end. Burrs and mushrooming must be removed with a file or a small grinding stone in a drill motor. If they are not removed, the valve guide could be badly damaged or broken when the valve is forced out. This would mean extra work in cleaning up the valve guide.

After everything else is off the head, the valves can be removed. This requires a spring compressor. Figure 37-19 shows how a spring compressor is used. As the handle is pressed, the spring is compressed. This allows removal of the retainer lock. Then the spring can be released so the retainer and spring can be removed. Valve-stem seals or shields (Figs. 9-13 and 9-14) should always be replaced. Many manufacturers recommend installation of new seals or shields whenever valves or valve springs are removed.

A single valve spring, stem seal, or shield can be replaced without removing the head on many models. An example is the engine with ball-pivot-type rocker arms. A special spring compressor is installed in place of the rocker arm to compress the spring. To hold up the valve while the spring is being compressed, compressed air from the shop supply is introduced into the cylinder through the spark-plug hole (Fig. 37-20). A special air-hose adapter, which can be screwed into the spark-plug hole, is required. The pressure holds the valve closed while the spring is compressed. If the air pressure does not hold the valve closed, then the valve is stuck or damaged. The head must be taken off for a closer look.

Careful: The air pressure may push the piston to BDC (bottom dead center). If it does, the valve could drop into

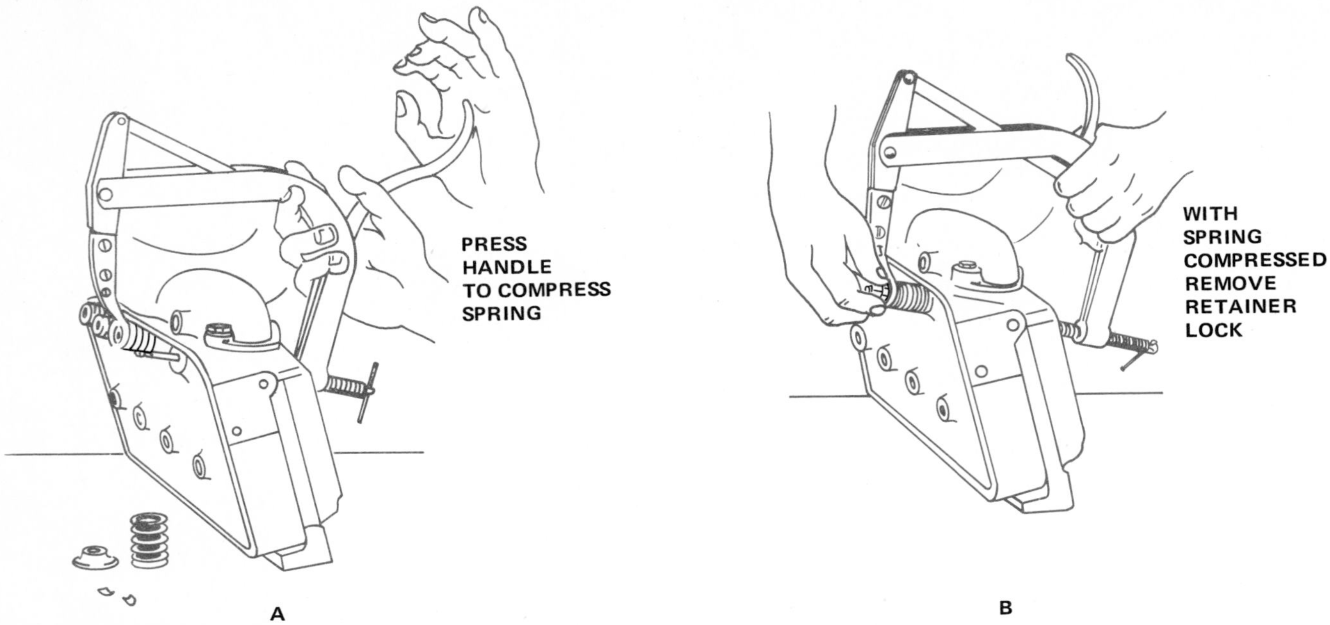

Fig. 37-19 Using a spring compressor. (*A*) Compressing the spring; (*B*) removing the retainer lock.

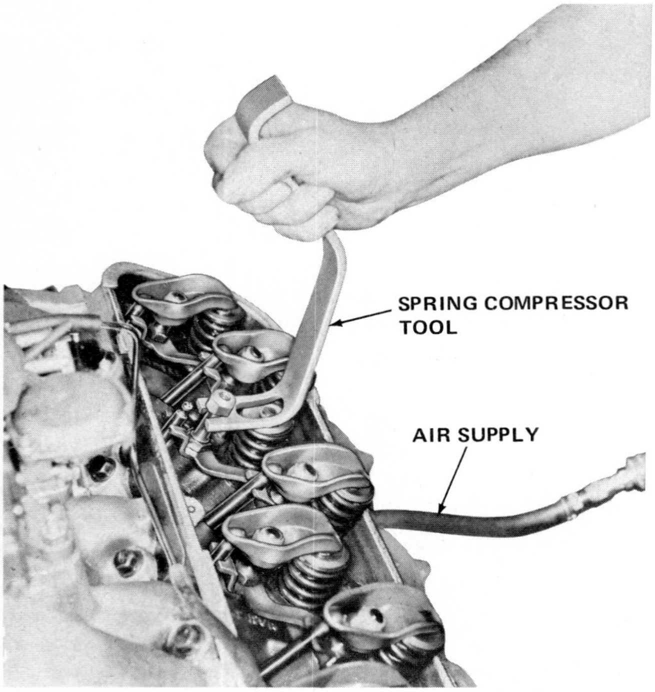

Fig. 37-20 Compressing a valve spring while the valve is held closed by air pressure. (*Chevrolet Motor Division of General Motors Corporation*)

the cylinder if the air pressure is released. To prevent this, wrap a rubber band or tape around the valve stem.

On some engines with the rocker arms mounted on a shaft, it is possible to bleed down the hydraulic valve lifter. To do this, pressure is applied with a special tool (Fig. 37-9). Then you can remove the push rod, and move the rocker arm to one side. With the rocker arm wired out of the way, a valve-spring compressor can be used to compress the spring. Now you can remove the retainer, spring, and seal. Air pressure must be applied to the cylinder, as explained in the previous paragraph, to hold the valve on its seat when the spring is compressed.

● 37-26 Valve Inspection

As you take the valves out of the head, inspect each one. Decide whether or not it can be serviced and used again. (See ● 37-1 to 37-9.) If it looks good enough to use again, put it into its proper place in the valve rack (Fig. 37-18). If it looks too bad to be cleaned up for further service, discard it. Put a new valve in the appropriate place in the valve rack.

● 37-27 Servicing Valves

Once all the valves are out of the head, remove them one by one from the valve rack. Clean each of them. Clean the carbon off the valves with a wire wheel. (Wear goggles to protect your eyes from flying particles of metal and dirt!) Polish the stems, if necessary, with a fine grade of emery cloth. Do not take off more than the dirty coating on the surface. Do not take metal off the stems.

Careful: Do not scratch the valve-seating surface or valve stem with the wire brush or emery cloth.

As you clean the valves, reexamine them to make sure all are usable. Small pits or burns in the valve seating face can be removed by grinding the valve. Larger pits or burns are hopeless; new valves will be required. Figure

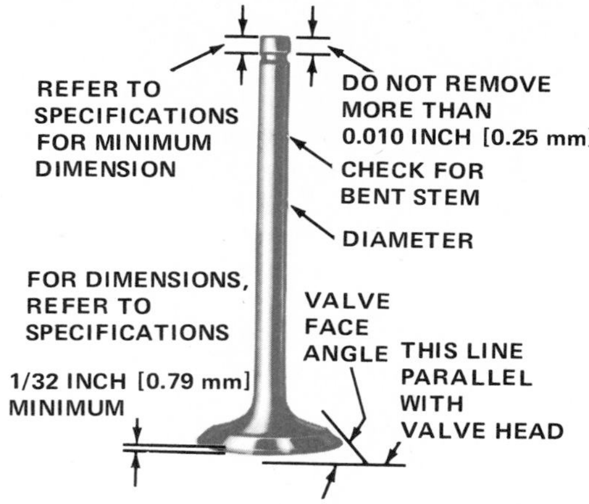

Fig. 37-21 Valve parts to be checked. On the valve shown, the stem is hardened at the end. Therefore, not more than 0.010 inch (0.254 mm) should be removed. *(Ford Motor Company)*

37-21 shows specific parts of the valve to be examined. Some engine manufacturers recommend the use of a run-out gauge to check for a bent valve stem. Eccentricity can also be checked in the valve grinder. If the run-out, or eccentricity, is excessive, discard the valve.

After cleaning the valves, replace them temporarily in their valve guides to check for guide wear. This procedure, and valve-guide service, are covered in ● 37-29.

1. Refacing or grinding valves The next step, if the valves are good enough to reuse, is to reface, or grind, them. This requires a valve-refacing machine (Fig. 37-22). The machine has a grinding wheel, a coolant-delivery system, and a chuck which holds the valve for grinding. Set the chuck to grind the valve face at the specified angle. This angle must just match the valve-seat angle, or else make an interference angle of $\frac{1}{4}$ to 1 degree (Fig. 37-6). Then put the valve into the chuck and tighten the chuck. The valve should be placed in the chuck so that the part of the stem that runs in the valve guide is gripped by the chuck.

To start the operation, align the coolant feed so it feeds coolant on the rotating valve face. Then start the machine. Move the lever to carry the valve face across the grinding wheel. The first cut should be a light one. If this cut removes metal from only one-half or one-third of the face, the valve may not be centered in the chuck. Or the valve stem is bent, and the valve should be discarded. Cuts after the first should remove only enough metal to true up the surface and remove pits. Do not take heavy cuts. If so much metal must be removed that the margin is lost, discard the valve. Loss of the margin causes the valve to run hot, and it will soon fail.

If new valves are required, reface them lightly, provided they are not of the coated type. *Never reface or lap coated valves!*

Follow the operating instructions of the refacer manufacturer. In particular, dress the grinding wheel as necessary with the diamond-tipped dressing tool. As the diamond is moved across the rotating face of the grinding wheel, it cleans and aligns the grinding face.

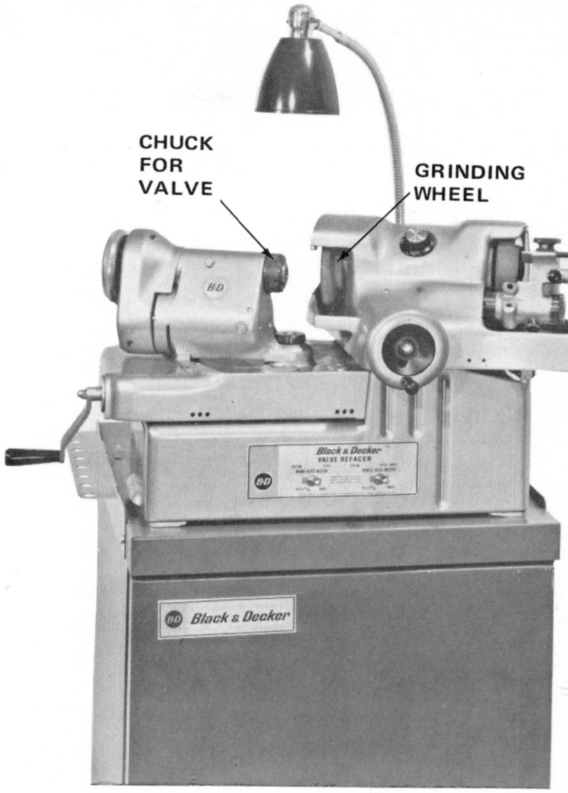

Fig. 37-22 Valve-refacing machine. *(Black and Decker Manufacturing Company)*

2. Refacing valve-stem tips The tip of a valve stem should be ground lightly. Use the special attachment furnished with the valve-refacing machine. One recommendation is to grind off as much from the stem end as you ground off the valve face. That way, you make up for the amount the valve sinks into the seat.

The ends of some valve stems are hardened. These should have no more than a few thousandths of an inch ground off (Fig. 37-21). Excessive grinding exposes soft metal so that the stem wears rapidly in service.

● 37-28 Valve Installation

As the valves are refaced and cleaned, they should be returned to the valve rack (Fig. 37-20). They are now ready for installation in the cylinder head. First, however, the valve guides and valve seats must be serviced (● 37-29 and 37-30). Also, the other components of the valve train—push rods, rocker arms, and valve lifters—must be checked and serviced as necessary.

New shields or seals should be installed if the old ones are worn or if the manufacturer recommends them. To avoid damage to the seals, special plastic caps can be placed over the ends of the valve stems. The seals will then slip on without being damaged by the sharp edges of the stem and/or lock grooves.

If valves and seats have been ground, the effective length of the valve spring will not be great enough. In order to restore normal spring tension, spring shims will be required.

Using a spring compressor (Fig. 37-19), install the springs, spring retainers, and locks. Measure the installed spring height (Fig. 37-23). Note that the end of the steel scale has been cut away. If the spring height is excessive, a spring shim is required. The shim is installed between the spring and cylinder head.

Do not install a shim that reduces the spring height below the specified minimum. This results in excessive spring pressure and rapid wear of valve-train parts.

Install the valve springs with the proper side against the cylinder head. As a rule, the close-spaced coils go next to the head (on springs with differential spacing of coils). Also, a damper spring (where used) must be placed inside the valve spring in an exact relationship with the spring coils. One typical example is that the coil end of the damper spring should be 135 degrees counterclockwise from the coil end of the valve spring.

● 37-29 Valve-Guide Service

The valve guide must be clean and in good condition for normal valve seating. It must be serviced before the valve seats are ground, if that is required. As a first step, the valve guide should be cleaned with a wire brush or adjustable-blade cleaner. Then, it should be checked for wear. If it is worn, it requires service. The type of service depends on whether the guide is integral or replaceable. If replaceable, the old guide should be pressed out. Then a new guide is installed and reamed to size. If integral, it can be serviced in either of two ways. One, ream it to a larger size, and install a valve with an oversize stem. Two, knurl and ream the guide. All these services are covered in the following paragraphs.

1. Testing the guide for wear One method of testing the guide uses a dial indicator (Fig. 37-24). With the valve in place, the dial indicator is attached as shown. The button just touches the edge of the valve head. Use a special tool, as shown in Fig. 37-25, to hold the valve off its seat. Then rotate the valve and move it sideways to determine the amount of guide wear. On some engines, the recommendation is to check valve movement from the stem end with the valve seated.

Another checking method is to insert a tapered pilot into the guide until it is tight. Then, pencil-mark the pilot at the top of the guide, and remove it. Measure the pilot diameter $\frac{1}{2}$ inch [12.7 mm] below the pencil mark. This gives the guide diameter, which can then be compared with the valve-stem diameter.

Another method that is more accurate than the two described above uses a special gauge (Fig. 37-26). The gauge is preset to register zero. There are two buttons near the end of the probe. When the probe is entered into a valve guide, the buttons will move in and out of the probe to show variations in guide diameter, that is, wear. Any movement of the buttons moves the needle to show how much wear there is.

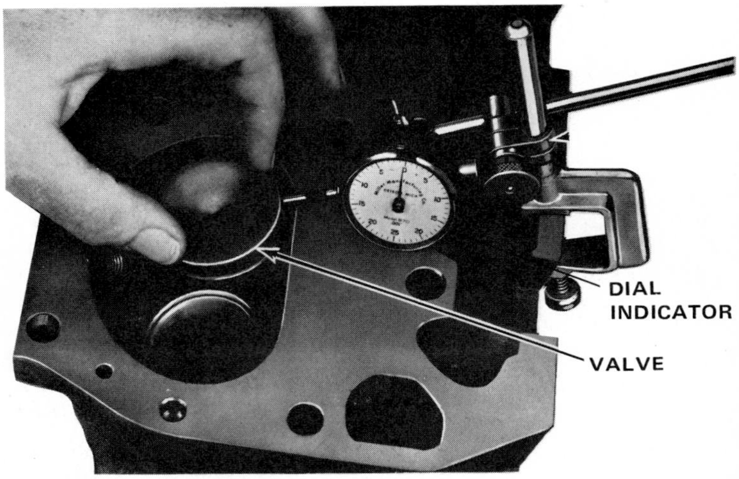

Fig. 37-24 Dial indicator set up to measure valve-guide wear. *(Chrysler Corporation)*

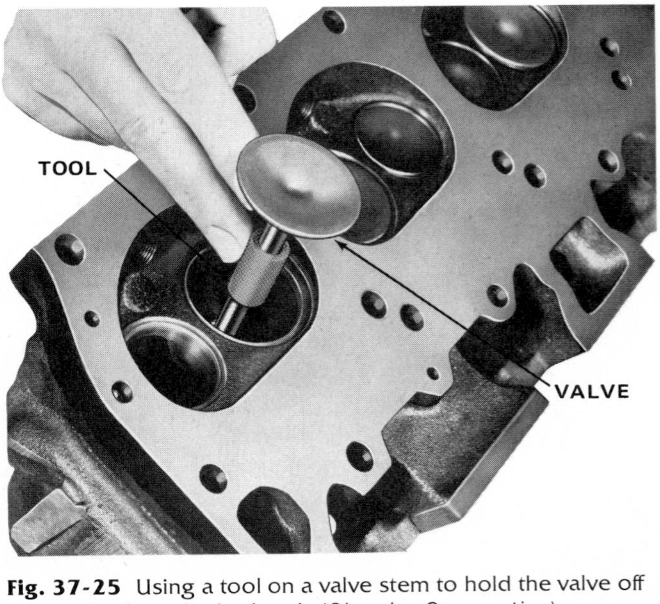

Fig. 37-25 Using a tool on a valve stem to hold the valve off its seat in the cylinder head. *(Chrysler Corporation)*

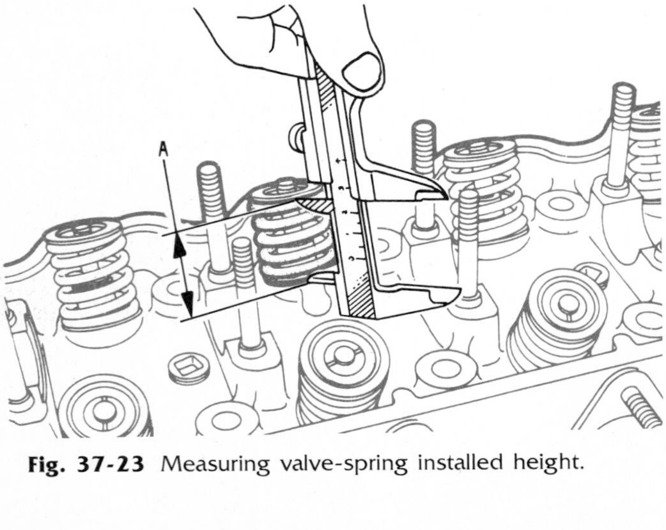

Fig. 37-23 Measuring valve-spring installed height.

2. Removing valve guide (replaceable type) On I-head engines, the valve guide can be pressed out of the head with an arbor press.

3. Installing valve guide (replaceable type) Guides can be installed in I heads with an arbor press. Guides must be installed to the proper depth in the block or head. Then, they must be reamed to size.

4. Reaming the guide The guide is reamed to take a valve with an oversize stem. After reaming, the sharp edge at the top of the guide should be broken (removed) with a scraper.

5. Knurling the guide With this procedure, you end up with a guide of the same size as originally, so you use a standard valve. In the guide-knurling operation, a knurling tool is run down into the guide. Then the guide is reamed. The procedure usually takes several steps: first knurl, first ream, second knurl, final ream, for example.

6. Checking concentricity with seat After the valve guide is serviced and checked for size, check its concentricity with the valve seat. The seat is always ground whenever the guide is serviced (● 37-30).

● 37-30 Valve-Seat Service

For effective valve seating and sealing, the valve face must be concentric with the valve stem. Also, the valve guide must be concentric with the valve face. In addition, the valve-face angle must match the valve-seat angle (or have an interference angle). Thus, as a first step in valve-seat service, the valve guides must be cleaned and serviced (● 37-29).

Valve seats are of two types. The *integral* type is actually the cylinder block or head. The *insert* type is a ring of special metal set into the block or head (Fig. 5-13). Replacing seat inserts and grinding seats are described below.

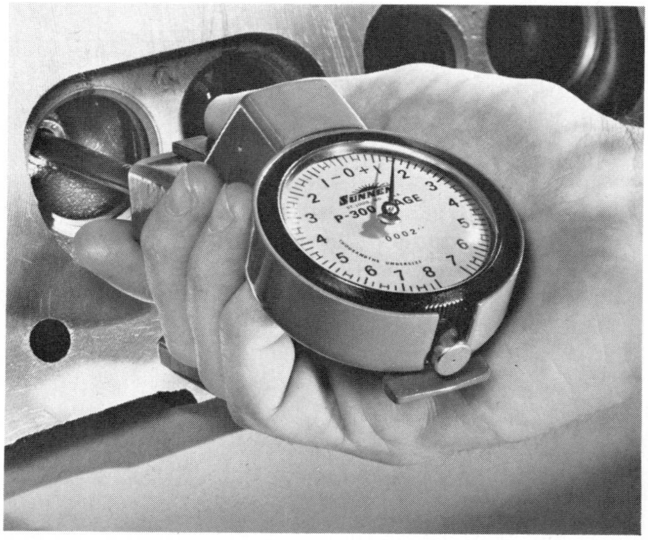

Fig. 37-26 Using special valve-guide gauge to check guide for wear. Movement of the probe in and out of the guide will cause the needle to move if the guide is irregularly worn. *(Sunnen Products Company)*

1. Replacing valve-seat inserts A valve-seat insert may be badly worn. Or, it may have been ground down on previous occasions so that there is insufficient metal for another grind. In either case, it must be replaced. The old seat must be removed with a special puller. If a puller is not available, the insert is punch-marked on two opposite sides. An electric drill is then used to drill holes almost through the insert. Then, a chisel and hammer can be used to break the insert into halves so that it can be removed. Care must be used so that the counterbore is not damaged. If the new insert fits too loosely, the counterbore must be rebored oversize. Then an oversize insert is installed. The new insert should be chilled in dry ice for 15 minutes to shrink it so it can be driven into place. Then, the valve seat should be ground.

2. Grinding valve seats Two types of valve-seat grinders are used, the concentric grinder and the eccentric grinder. The concentric grinder rotates a grinding stone of the proper shape on the valve seat (Fig. 37-27). The stone is kept concentric with the valve seat by a pilot installed in the valve guide (Fig. 37-28). This means that the valve guide must be cleaned and serviced (● 37-29) before the seat is ground. In the unit shown in Fig. 37-27, the stone is automatically lifted about once a revolution. This permits the stone to clear itself of grit and dust by centrifugal force. After the seat is ground, it may be too

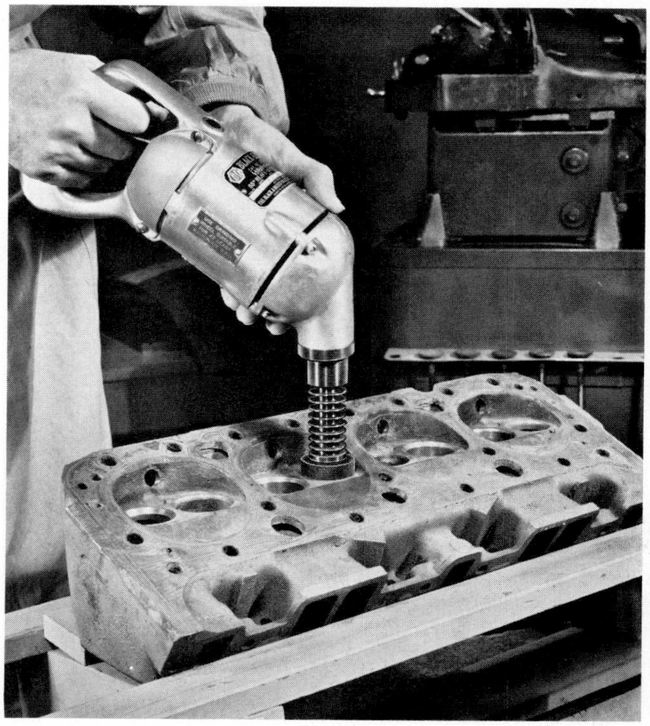

Fig. 37-27 Concentric valve-seat grinder using the patented Vibrocentric principle. The stone is rotated at high speed. About once every revolution, it is automatically lifted off the valve seat so it can throw off loosened grit and grindings. *(Black and Decker Manufacturing Company)*

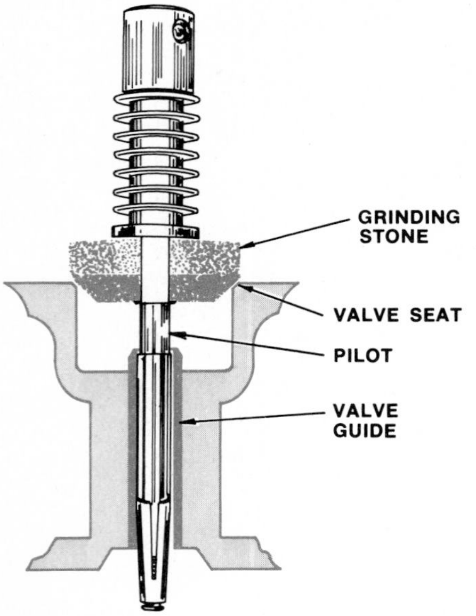

Fig. 37-28 Pilot on which grinding stone rotates. The pilot keeps the stone concentric with the valve seat. (*Black and Decker Manufacturing Company*)

Fig. 37-29 Testing a valve spring for the proper tension in a special fixture. (*Ford Motor Company*)

wide. It must be narrowed by using upper and lower grinding stones to grind away the upper and lower edges of the seat. A steel scale can also be used to measure seat width.

In the eccentric valve-seat grinder, the grinding stone is offset from the center of the valve seat. It makes only line contact with the valve seat. As the stone revolves, its center rotates slowly on an eccentric shaft. This permits the grinding line to progress evenly around the entire valve seat. This valve-seat grinder also pilots in the valve guide.

Careful: Be sure to follow instructions furnished by the grinder manufacturer. Note that the grinding stone must be dressed frequently with the diamond-tipped dressing tool.

3. Checking valve seats for concentricity After the valve guides are serviced and valve seats ground, the concentricity of the two can be checked with a valve-seat dial gauge. The gauge is mounted in the valve guide and is rotated so the indicator finger sweeps around the valve

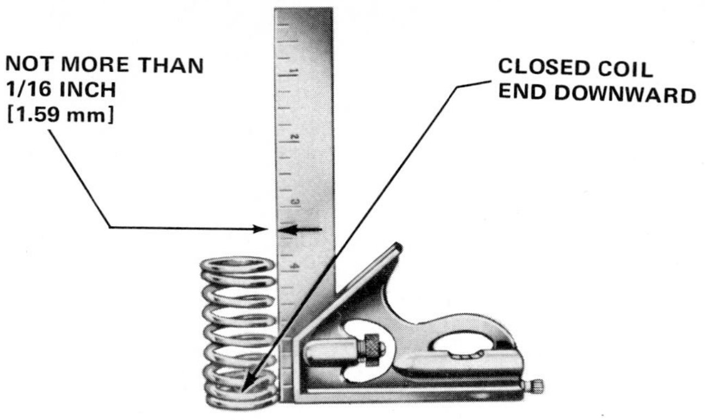

Fig. 37-30 Checking spring squareness. (*Ford Motor Company*)

seat. Any eccentricity (or run-out) of the seat is thus registered on the gauge dial.

4. Testing valve seating Contact between the valve face and seat may be tested as follows: Mark lines with a soft pencil about $\frac{1}{4}$ inch [6.35 mm] apart around the entire valve face. Then put the valve in place. With light pressure, rotate it half a turn to the left and then half a turn to the right. If this removes the pencil marks, the seating is good.

The seating can also be checked with prussian blue. Coat the valve face lightly with prussian blue. Put the valve on its seat, and turn it with light pressure. If blue appears all the way around the valve seat, the valve seat and guide are concentric with each other. Now, check the concentricity of valve face with valve stem. Remove the prussian blue from valve and seat. Lightly coat the seat with prussian blue, and then lightly rotate the valve on the seat. If blue transfers all the way around the valve face, the valve face and stem are concentric. This check is similar to the check that uses a run-out gauge.

37-31 Valve-Spring Inspection

Valve springs should be checked for proper tension and for squareness. A special fixture, such as shown in Fig. 37-29, checks tension. To check for squareness, stand the spring, closed-coil end down, on a flat surface. Hold a steel square next to it as shown in Fig. 37-30. Rotate the spring slowly to see if the top coil moves away from the square more than $\frac{5}{64}$ inch [1.984 mm] (Ford). If the spring is excessively out of square, or has lost tension, discard it. One manufacturer's recommendation is to replace all valve springs during the complete valve job. Then you are sure of good spring action.

37-32 Camshaft Service

Camshaft removal varies somewhat from engine to engine. It is less complex in an overhead-camshaft engine. The general procedure in an overhead-valve engine begins with removal of the radiator. Then take the pulley from the crankshaft. Remove the gear or timing-chain

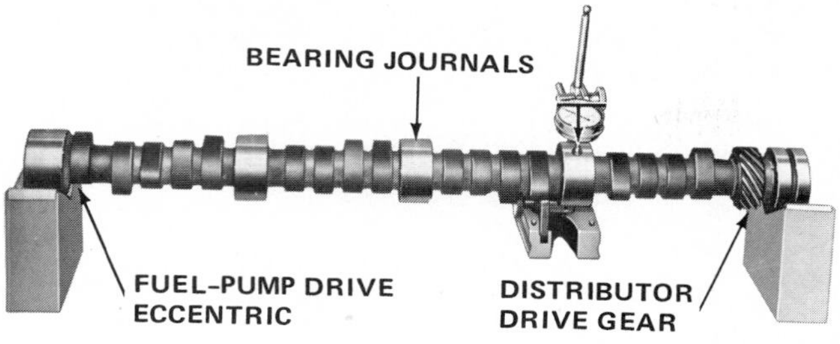

Fig. 37-31 Checking alignment of a camshaft. *(Chevrolet Motor Division of General Motors Corporation)*

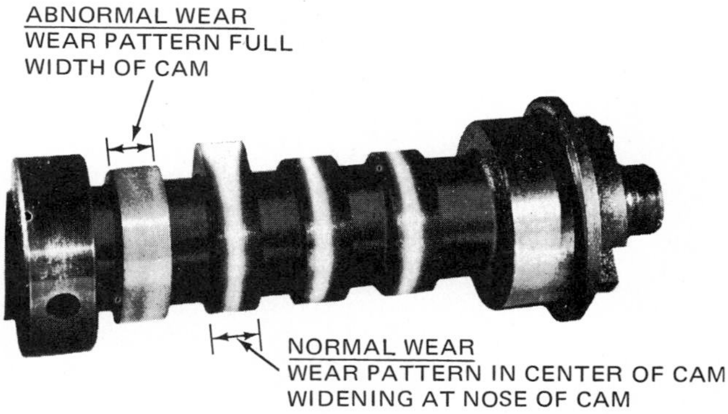

Fig. 37-32 Normal and abnormal cam wear. *(Oldsmobile Division of General Motors Corporation)*

cover. Detach the camshaft thrust plate (where present). Take off the camshaft sprocket and chain (where used). The distributor or oil pump (whichever has the driven gear) must be removed so the gear will not interfere with camshaft removal.

The push rods must be removed so the valve lifters can be raised up out of the way. Now, the camshaft is free and can be pulled forward and out. Be very careful to keep the journals and cams from scratching the camshaft bearings. Support the rear of the camshaft as it is pulled out, so the bearings are not damaged.

Supporting the camshaft is easier said than done. On most engines, it is necessary to remove the crankshaft in order to get to the camshaft and support it.

1. Checking the camshaft Check for alignment by rotating the camshaft in V blocks and using a dial indicator (Fig. 37-31). Journal diameters should be checked with a micrometer, and the bearings with a telescope gauge. The two dimensions can be compared to determine whether bearings are worn and require replacement.

2. Checking for cam wear Figure 37-32 shows normal and abnormal cam wear. Normal cam wear is close to the center of the cam, as shown. The reason for this is that the cam, in most engines, is slightly tapered. Also, the lifter foot is slightly spherical, or crowned, in shape. Therefore, the contact pattern, when all is well, is as shown in Fig. 37-33. If wear shows across the full width of the cam, a new camshaft is required. The lifter should also be checked (● 37-33). The general rule is: if new lifters are required, install a new camshaft. If a new camshaft is required, install new lifters.

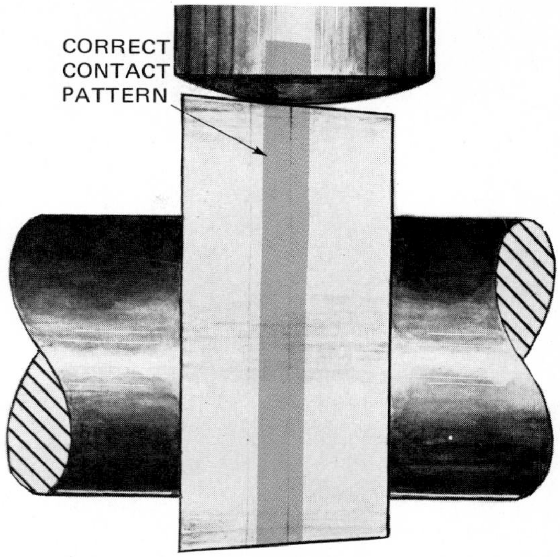

Fig. 37-33 Offset between the cam lobe and lifter face (which is crowned) gives a wide, centered, contact area. Taper and crown are shown exaggerated. *(Dana Corporation)*

The cam-lobe lift can be checked with the camshaft in or out of the engine. Figure 37-34 shows the check with the camshaft in the engine. The setup in Fig. 37-31 can be used to measure the lobe lift.

3. Replacing camshaft bearings A special bearing remover-and-replacer bar is required to do this job. For some engines, the bearing-puller bar is threaded, and a nut is turned to remove the bearings. For others, a hammer is used to drive against the bar and force the bearings out. Oil holes in the new bearings should align with the oil holes in the block. Also, new bearings should be staked in place if the old bearings were staked. If the new bearings are not of the precision type, they will require reaming to establish the proper fit.

4. Timing the valves The timing gears, or sprockets, and chain are marked for proper positions and correct valve timing (Figs. 9-2 to 9-4). To get to these markings, however, the front of the car has to be partly torn down. Thus, some engines have another marking system for checking valve timing. This marking is on the flywheel or vibration damper, near the ignition-timing markings. When this marking is visible or registers with a pointer, a designated valve should be just opening. Or it should have opened a specified amount. Valve action is observed by removing the valve cover.

When the flywheel or vibration damper is not marked, piston position can be measured with a special gauge. The gauge is inserted through a special hole in the head. The relationship of the piston with the valves can therefore be established.

5. Timing gear and chain Gear run-out can be checked by mounting a dial indicator on the block. The indicating finger should rest on the side of the gear (Fig. 37-35). Run-out will then be indicated as the gear is rotated. Gear backlash is measured by inserting a narrow feeler gauge between the meshing teeth. Excessive run-out or backlash requires gear replacement. Excessive slack in the timing chain indicates a worn chain, and possibly worn sprockets.

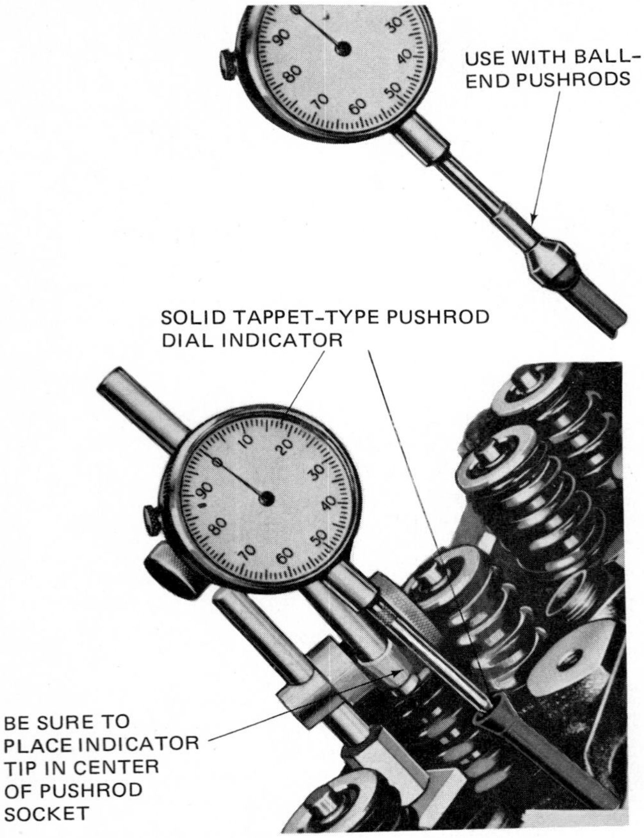

Fig. 37-34 Cam-lobe lift check on one engine. (*Ford Motor Company*)

37-33 Valve Lifters

The solid and the hydraulic valve lifters require different servicing procedures.

1. Solid lifter Solid lifters are removed from the camshaft side on some engines. This requires camshaft removal as a first step. In most engines, lifters are removed from the valve or push-rod side. Lifters should be kept in order so they can be restored to the bores from which they were removed. Oversize valve lifters may be installed on many engines if the lifter bores have worn. Before this is done, the lifter bores must be reamed oversize.

2. Hydraulic valve lifter On some engines, a "leakdown" test is used to determine the condition of the hydraulic valve lifters. One way to make this test is to insert a feeler gauge between the rocker arm and valve stem. Then, note the time it takes the valve lifter to leak enough oil to seat the valve. As the valve seats, the feeler gauge becomes loose. This indicates the end of the test. If the leak-down time is too short, the valve lifter is defective.

A more accurate leak-down test is made with the lifter out of the engine and installed in a special tester. With this tester, the time required for a uniform pressure (from the weight on the end of a lever) to force the lifter plunger to bottom is measured. If short, the lifter is defective.

To remove the hydraulic valve lifters from some engines, the push-rod cover and rocker-arm assembly must be removed. Then the push rod is taken out. On some engines with shaft-mounted rocker arms, the rocker arm can be moved by compressing the spring. This allows the push rod to be removed. Thus, the rocker-arm assembly does not have to be taken off these engines.

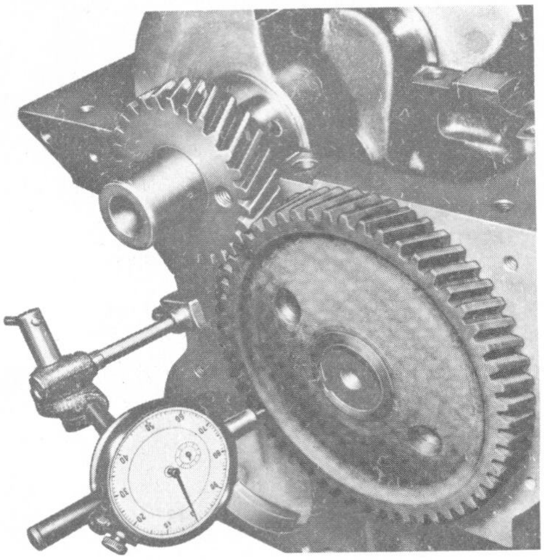

Fig. 37-35 Checking timing-gear run-out, or eccentricity, with a dial indicator. (*Buick Motor Division of General Motors Corporation*)

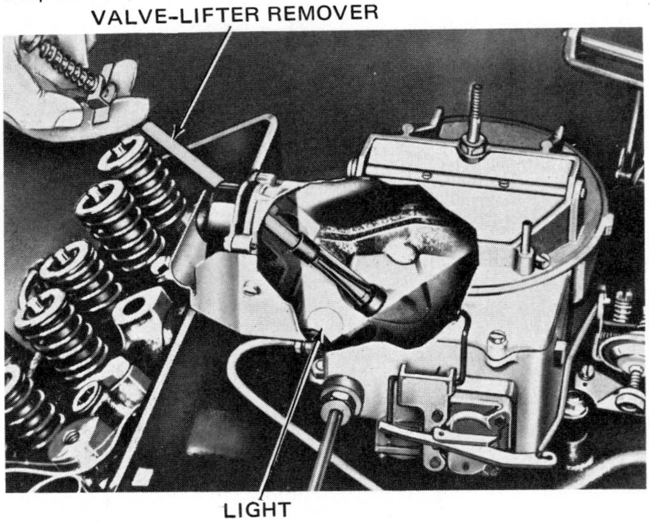

Fig. 37-36 Removing a valve lifter with a special tool. (*Ford Motor Company*)

The use of one lifter-removing tool is shown in Fig. 37-36. The tool is inserted through the push-rod opening in the block and is seated firmly on the end of the lifter. The lifter is then removed through the push-rod opening.

3. Servicing hydraulic valve lifters Usually, if a hydraulic valve lifter is defective, it is cheaper to replace it than to disassemble and service it. The labor would probably cost more than the new lifter. However, if you prefer to spend the time, you can service the lifter as follows. Disassemble the lifter, and clean all parts in solvent. If any part is defective, the lifter should be replaced. On reassembly, fill the lifter with clean, light engine oil.

Work on only one lifter at a time so you do not mix lifter parts. Also, make sure each lifter goes back into the bore from which it was removed.

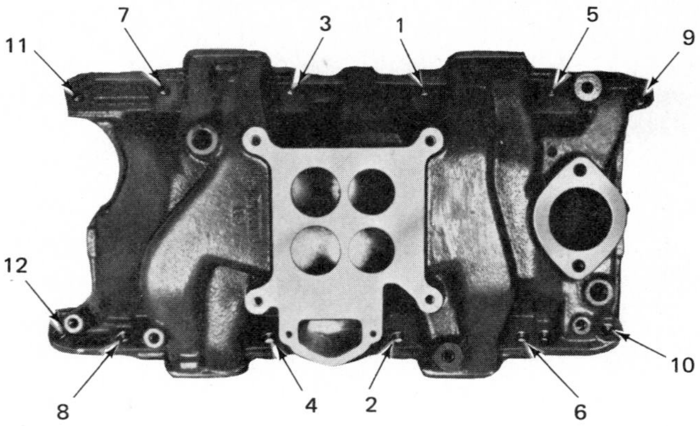

Fig. 37-37 Sequence chart for tightening the intake-manifold attaching bolts on a V-8 engine. (*Chevrolet Motor Division of General Motors Corporation*)

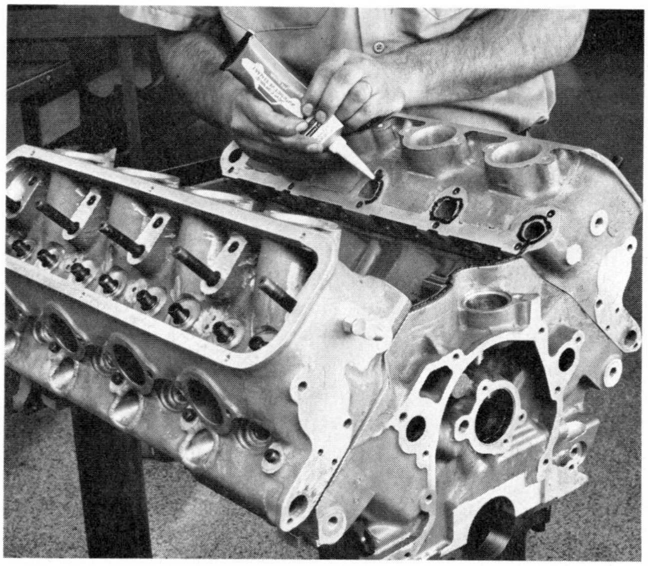

Fig. 37-38 Using plastic rubber to form a gasket. (*Dow Corning Corporation*)

Careful: Be extremely careful to keep everything clean when servicing and handling hydraulic valve lifters. It takes only one tiny particle of dirt to cause a lifter to malfunction.

4. Checking lifter foot As shown in Fig. 37-33, the foot of the lifter should be slightly spherical, or crowned. If it is worn or pitted, it can sometimes be reground and reused.

● 37-34 Removing and Replacing Manifolds

Take the carburetor off. Handle it with care to avoid damaging it or spilling gasoline from the float bowl. Disconnect vacuum lines, exhaust pipes, pollution-control hoses, and any other pipe or wire connected to the manifold. Remove the nuts or bolts, and take the manifolds off.

When reinstalling manifolds, be sure all old gasket material has been removed from the manifolds and cylinder head. Use new gaskets. Tighten nuts or bolts to the proper tension and in the proper sequence (Fig. 37-37).

Some mechanics say they have success with a plastic gasket compound (silicone rubber) that comes in a tube. It is squeezed out of the tube onto the gasket surfaces (Fig. 37-38), and it spreads to form the gasket.

REVIEW QUESTIONS

1. Explain some of the troubles that could result from excessive intake-valve-stem clearance.
2. What are five valve troubles?
3. What is an interference angle, as applied to valves and valve seats?
4. Describe the causes of valve burning. Of valve breakage.
5. What is valve-seat recession? What causes it?
6. Describe the procedure for adjusting valves in an I-head engine.
7. Describe the procedure for adjusting valves in an overhead-camshaft engine.
8. What are the two basic kinds of rocker arms?
9. Describe the procedure for adjusting valves on Chevrolet engines with hydraulic valve lifters.
10. Explain what coated valves are. Why shouldn't they be refaced before installation?
11. Explain the procedure for removing a cylinder head. For installing a cylinder head.
12. Describe a typical procedure for removing valves from a cylinder head.
13. Explain how to clean and inspect a cylinder head.
14. Explain how to replace a ball stud in a cylinder head.
15. Why should you examine the valve stem before removing the valve from the cylinder head?
16. Describe the procedure for cleaning, inspecting, and servicing valves, including valve grinding.
17. Describe a typical procedure for replacing valves in a cylinder head.
18. Describe the service procedure for a valve guide, including knurling.
19. Describe the procedure for removing and checking a camshaft.
20. Explain how to check and service a valve seat, including valve-seat grinding.

SELF PROJECTS

1. As you look through the junk box in the shop, and do valve-service jobs on engines, save valves showing different kinds of troubles. Mount the valves on a board. Add a label under each, identifying the trouble.
2. Study valve-adjusting procedures in various manufacturers' shop manuals. Write short summaries on how the various manufacturers want the valves in their engines adjusted. File these in your notebook.
3. Study the instruction manuals for the valve grinder and valve-seat grinder. Prepare, for your notebook, brief notes on how to use these two machine tools.

CHAPTER 38
Engine Service: Connecting Rods, Rod Bearings, Pistons, and Rings

After studying this chapter, you should be able to:

- **Discuss servicing procedures on connecting rods, pistons, piston rings, and rod bearings.**
- **Under the instructor's supervision, perform the service jobs discussed in the chapter.**

This chapter continues the discussion of engine service. It covers the servicing of pistons, piston rings, connecting rods, and connecting-rod bearings. Figure 38-1 is a disassembled view of a cylinder block for a V-8 engine. Only one rod-and-piston assembly is shown. Note that the piston pin is of the free-floating type.

Connecting Rods and Rod Bearings

● 38-1 Engine-Bearing Prelubricator

The engine-bearing prelubricator (Fig. 38-2) provides initial lubrication to all bearings after a service job. That is, it is used to charge the reconditioned engine with oil before the engine is operated. This prevents damage to the bearings during initial startup.

To use the bearing prelubricator, fill it with SAE20 or SAE30 oil and connect it to an air hose. Then connect the prelubricator hose to the engine lubricating system—in the oil-pressure indicator hole, for example. Then, when air pressure is applied, oil is forced through the lubricating system. The bearing prelubricator is used in many engine-building and rebuilding shops.

Many automotive technicians believe that the best prelubrication is to put plenty of oil on engine parts during assembly. Additional prelubrication may be required, for example, to fill the hydraulic lifters and oil filter, and to prime the oil pump. Then this should also be done: After the engine is assembled, add oil to the oil pan. Then use a slow-speed drill motor to turn the oil pump until the system pressurizes. Then install the ignition distributor.

● 38-2 Using the Prelubricator to Check Bearings

The bearing prelubricator can be used to check for bearing wear before starting a service job. It is used with the oil pan off (Fig. 38-2). Worn bearings pass much more oil. This means that more oil gets on the cylinder walls, and more oil works up into the combustion chambers where it is burned. A normal bearing will leak between 20 and 150 drops of oil per minute when the prelubricator is used. If the bearing leaks more, it is worn. If it leaks less than 20 drops per minute, the bearing clearance is too small, or the oil line to the bearing is stopped up.

When the oil holes in the crankshaft and in the bearing align, considerable oil is forced through the bearing. This will give the appearance of excessive wear. In such a case, the crankshaft should be turned a few degrees to move the oil holes out of register.

● 38-3 Preparing to Remove Rods

Connecting rods and pistons are removed from the engine as assemblies. Removing, servicing, and replacing connecting rods requires about 5 to 8 hours, depending on the type of engine. About 3 additional hours are required to install new piston rings. Additional time is needed for such services as piston-pin bushing replacement. On most engines, the piston-and-rod assemblies are removed from the top of the engine. Thus, the first step is to remove the cylinder head (● 37-19). Cylinders should be examined for wear. If wear has taken place, there will be a ridge at the top of the cylinder. This ridge, called the *ring ridge,* marks the upper limit of piston-ring travel. If this ridge is not removed, the top ring could jam under it as the piston is moved upward. This could break the rings or the piston-ring-groove lands (Fig. 38-3). Thus, the ridge, if present, must be removed.

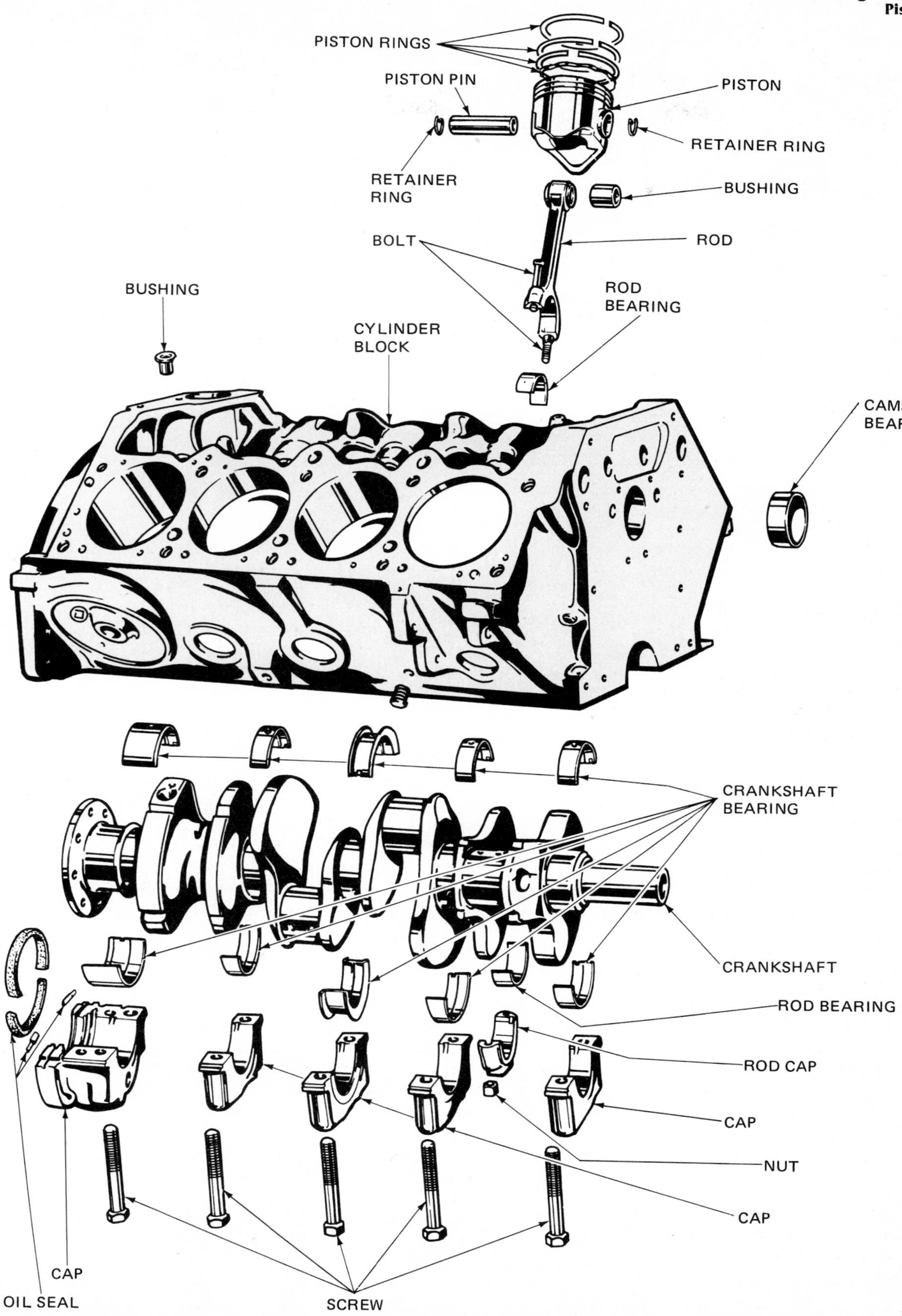

Fig. 38-1 Cylinder block with associated parts. (*Chrysler Corporation*)

Fig. 38-2 Oil-leakage test using the engine prelubricator. *(Federal Mogul Corporation)*

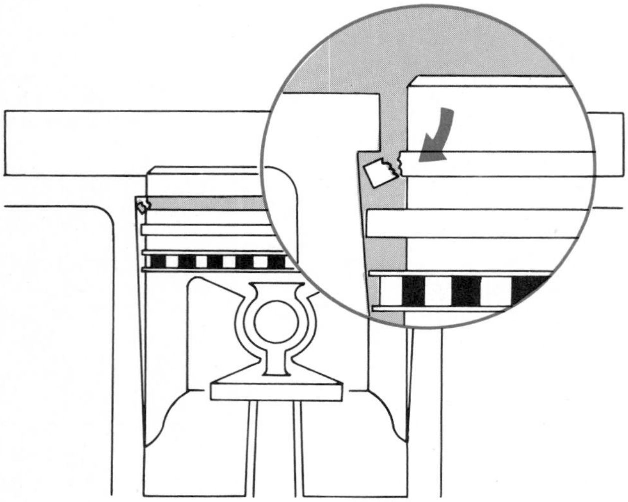

Fig. 38-3 How a ring ridge caused by cylinder wear might break the ring if the piston is pulled without removing the ridge. *(Sealed Power Corporation)*

A quick way to check for a ring ridge is to see if your fingernail catches under it. If your fingernail catches on the ring ridge, so will the piston rings. A more accurate way to check is to use an inside micrometer. Measure the diameter on the ring ridge and then immediately below the ring ridge. If the difference is more than 0.004 inch [0.102 mm], the ridge must be removed.

● 38-4 Removing Ring Ridge

To remove the ring ridge, use a special ring-ridge remover, as shown in Fig. 38-4. With the piston near BDC, stuff a cloth into the cylinder, and install the ridge remover.

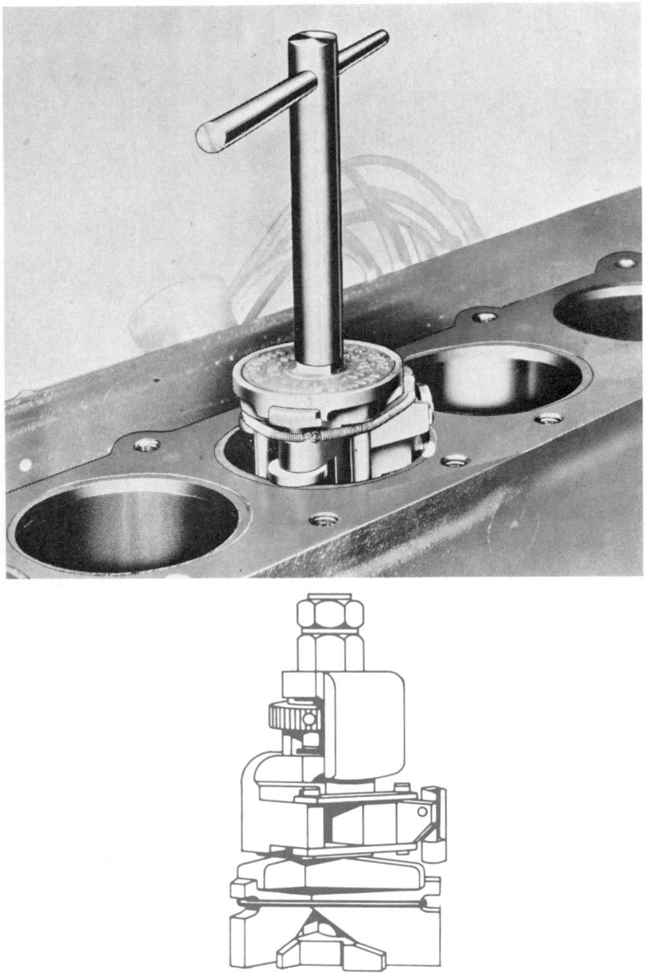

Fig. 38-4 Ridge-removing tool in place in the top of a cylinder. Cutters remove the ridge as the tool is turned in the cylinder.

There are several different kinds of ridge removers. Be sure to read carefully and follow the instructions for the ridge remover you use.

Adjust the cutter blades to take off just enough metal to remove the ridge. Cover the other cylinders to keep cuttings from getting into them. Rotate the tool to cut the ridge away.

Careful: Turn the ridge remover by hand, not with an impact wrench! Do not remove too much metal. That is, do not undercut the top of the cylinder deeper than the material next to the ring ridge. Do not run the cutting tool above the cylinder. This would taper the edge.

Remove the tool, take the cloth out, and wipe the cylinder clean. Repeat the process for the other cylinders.

● 38-5 Removing Oil Pan

The oil pan must be removed so that the connecting rods can be detached from the crankshaft. First, remove the drain plug to drain the engine oil (see Fig. 20-3). On many cars, the steering idler or other steering linkage must be removed. In such cases, note how the linkage is attached. Note the number and location of shims (if used). On some cars, the oil pan is easier to remove if the engine mounting bolts are removed and the engine is raised slightly. Other parts may require removal before the oil pan can be taken off. These include the exhaust pipe, oil-level tube, brake-return spring, and starting motor. Then, the nuts or bolts holding the oil pan to the engine cylinder block can be removed. Steady the pan as the last two nuts or bolts are removed, so that it does not drop. If the pan strikes the crankshaft so that it does not come free, turn the crankshaft over a few degrees.

If the oil pan does not break loose when the last bolt is out, tap the sides of the oil pan with a rubber mallet. If it still doesn't come loose, carefully force the claw or flat edge of a pry bar or scraper between the edge of the pan and cylinder block. Try to get the flat edge on the pan side of the gasket to avoid scratching the block. You can tap the pry bar with a hammer to help free the oil pan.

Many engines have metal reinforcements under the corner bolts of the oil pan. They help seal around the rear main bearing and the bottom of the timing cover. Don't let these get away from you when you remove the bolts.

Clean the oil pan, oil screen, and oil pump thoroughly before replacing the oil pan. Make sure that the gasket material is scraped off the pan and block gasket surfaces. Check the flatness of the oil-pan gasket surfaces. Make sure that the bolt holes have not been dished out by overtightening of the bolts. The gasket surfaces can be straightened by laying the oil pan on a flat surface and tapping the gasket-surface flanges with a hammer.

Apply new gasket cement, if specified. Lay the gasket, or gaskets, in place on the oil pan. Be sure that the bolt holes in the gasket and pan line up. Install the pan, and tighten the bolts or nuts to the proper tension.

Some oil pans are installed with plastic gasket material (see Fig. 37-38).

38-6 Removing Piston-and-Rod Assemblies

Now you are ready to remove the piston-and-rod assemblies.

Careful: Handle pistons and rods with care because they can be easily damaged. Never clamp a rod tightly in a vise. This can bend the rod and ruin it. Never clamp a piston in a vise. This can nick or break the piston. Do not allow the pistons to hit against each other or against other hard objects or bench surfaces. Distortion of the piston or nicks in the soft aluminum piston material may result from careless handling. These will ruin the piston.

With the head and pan off, crank the engine so the piston of No. 1 cylinder is near the bottom. Examine the rod and rod cap for identifying marks (Fig. 38-5). If none can be seen, use a small water-color brush and a little white metal paint, or a marking pen, to mark a "1" on the rod cap and the rod. Marks are needed to make sure that the parts go back into the cylinders from which they were removed. Each piston should also be numbered.

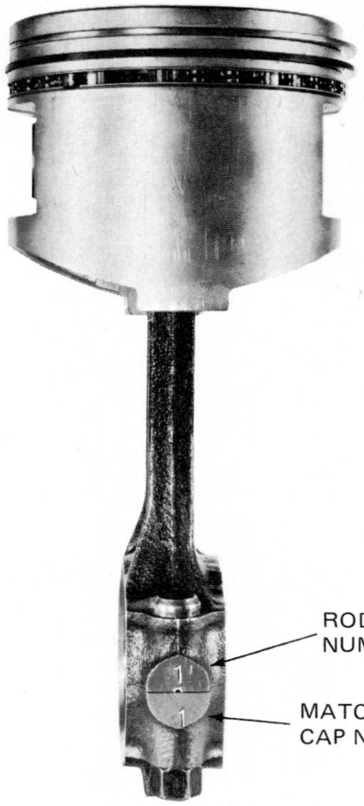

Fig. 38-5 Piston-and-connecting-rod assembly. The numbers on the rod and rod cap indicate the assembly goes into No. 1 cylinder. (*American Motors Corporation*)

Careful: Do not mark the rod caps and rods with metal numbering dies or a center punch and hammer. This can distort and ruin the rods and caps.

Remove the rod nuts and caps. Slide the rod-and-piston assembly up into the cylinder, away from the crankshaft. Use guide sleeves on the rod bolts, as shown in Fig. 38-6. They prevent the bolt threads from scratching the crankshaft journals. Also, the long handle permits easy removal and replacement of the piston-and-rod assembly. Short pieces of rubber hose, split and slipped over the rod bolts, will protect the crankshaft journals.

Turn the crankshaft as you go from rod to rod, so you can reach the rod nuts. The best procedure, recommended by engine specialists, is to remove a rod cap, and then take the rod-and-piston assembly out from the top of the cylinder block. That is, you remove each cap and rod in sequence. As you do so, lay the cap and rod out on a cloth spread on the bench. Or put them in a wooden piston box. Make sure each rod, cap, and piston is marked with the number of the cylinder from which they were removed.

38-7 Separating Rods and Pistons

There are five basic piston-rod-bushing arrangements, as shown in Fig. 8-2. The rods and pistons are separated by removing the piston pins. If the pin is free-floating, the pin is removed by removing the retainer ring and sliding the pin out. If the pin is locked to the connecting rod or

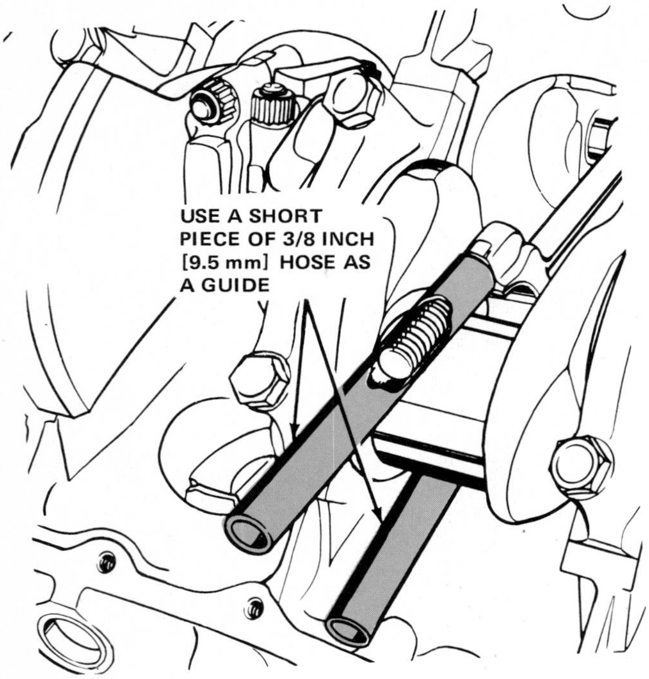

Fig. 38-6 Using guides to remove a connecting rod. *(Ford Motor Company)*

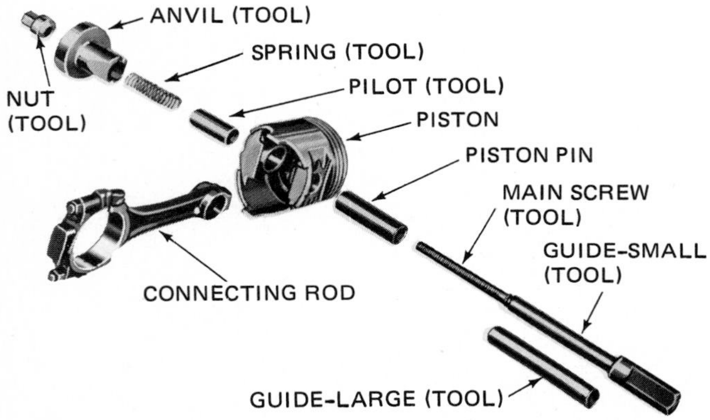

Fig. 38-7 Piston-pin-removal and installation tool, laid out to remove a pin. When installing a pin, the pilot is put on the other side of the piston with a spring between it and the anvil. *(Chrysler Corporation)*

piston with a lock bolt, loosen the lock bolt and slide the pin out.

If the pin is a press fit in the connecting rod, it must be pressed out. This requires a special tool assembly, as shown in Fig. 38-7. The tool is put together in the arrangement shown. Then an arbor press is used to press the pin out.

It is usually not necessary to separate the rods and pistons unless new pistons are to be installed and the old rods are to be reused.

Careful: Be very careful to avoid nicking or scratching the pistons, piston rings, or rods (see the Careful! at the start of ● 38-6). Check the rods, rings, and pistons, as explained in following sections.

● 38-8 Attaching Rods and Pistons

After rods and pistons have been cleaned, serviced, and checked, place them on a clean bench in their engine order. Make sure parts match as in the original assembly. The pistons should then be attached to the rods with the piston pins, as follows. Make sure the piston is in the correct position on the rod as the two are attached. On many engines, the piston notches face to the front of the engine. Also, the rod oil hole faces toward the outside of the block.

To attach the piston and connecting rod where a lock bolt is used, simply put the pin through the piston and rod. Tighten the lock bolt to hold the parts together. On the free-floating type, install the pin and retainer rings.

On the type with a press fit of the pin in the rod, a special tool assembly (Fig. 38-7) is required to press the pin in. Pressure is applied to the tool in an arbor press to push the pin into place. Plymouth recommends a fit test after the pin has been installed. This is done by placing the assembly in a vise. Then a torque wrench is used to apply 15 pound-feet [2.073 kg-m] of torque to the nut on the end of the tool. If this amount of torque causes the connecting rod to move down on the piston pin, the press fit is too loose. The connecting rod must be discarded. If the rod does not move, the fit is satisfactory.

● 38-9 Reinstalling Piston-and-Rod Assemblies

After rods are reattached to their pistons, the piston rings are installed (● 38-26). Then the assemblies go back into the engine. Rings should be positioned so that the ring gaps are uniformly spaced around the piston, or as specified by the manufacturer. For example, Fig. 38-8 shows a recent Ford recommendation. Note that the gaps in the compression rings are toward the front, while the gaps in the oil-ring parts are toward the back.

Dip the piston assembly above the piston pin in SAE30 oil. Drain excess oil. Use a piston-ring compressor to compress the rings into the piston-ring grooves (Fig. 38-9). Install guide sleeves on the rod bolts (Fig. 38-6), or cover the rod bolts with rubber hose. Then push the piston down into the cylinder. Tapping the head of the piston with the wooden handle of a hammer helps get the piston started. Make sure the assembly is installed with the piston facing in the right direction. Many pistons have a notch or other mark that should face toward the front of the engine (Figs. 38-8 and 38-9).

Attach the rod cap with the nuts turned down lightly. Then tap the cap on its crown lightly, to help center it. Tighten the nuts to specifications with a torque wrench.

Bearing clearances must be checked (● 38-15).

● 38-10 Checking Rod Side Clearance

Make sure that the rods are centered on the crankshaft crankpins. If a rod is offset to one side, the rod-and-

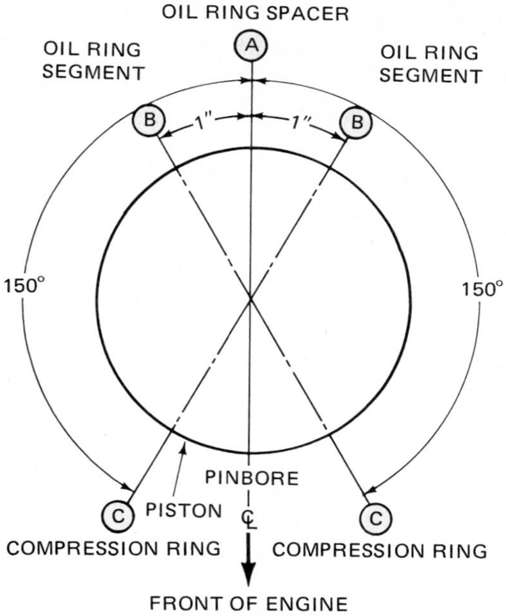

Fig. 38-8 Proper arrangement of ring gaps for one model engine. (*Ford Motor Company*)

piston assembly has probably been put in backward. That is, it has been turned 180 degrees from its correct position. Also, offset could mean a bent rod (see ● 38-11). Clearance between connecting rods on V-8 engines should also be checked (Fig. 38-10). Incorrect side clearance means a bent rod.

● 38-11 Checking Connecting Rods

After rods are detached from the pistons, the rod and rod caps should be cleaned and inspected. Make sure to clean out the oil holes in the rods. Blow them out with compressed air.

Inspect the rod big-end bearings (see ● 38-14). If the rod has a bushing in the small end, check its fit with the piston pin. If it is not correct, service is required (see ● 38-12).

Check rod alignment. Figure 38-11 is an exaggerated view showing the effects of a misaligned connecting rod. Heavy loading at points *A* and *B* on the bearing would cause bearing failure at these points. The heavy-pressure spots *C* and *D* on the piston cause heavy wear and possibly scoring of the piston and cylinder wall. This is a basic inspection check recommended by engine manufacturers. That is, look for uneven wear or shiny spots on the pistons. If any are found, the piston, pin, and rod should all be discarded.

A rough check for rod alignment can be made by detaching the oil pan and watching the rod while the engine is cranked. If the rod moves back and forth on the piston pin, or is not centered, the rod is out of line. It should stay centered on the pin.

Examine the rod bearings for side wear. If a rod is out of line, the upper bearing half will wear on one side and the lower half will wear on the other side (see Fig. 38-11).

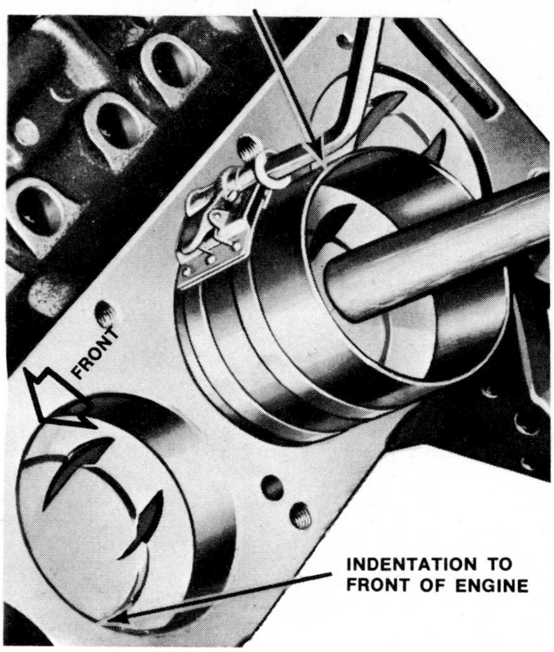

Fig. 38-9 Using a piston-ring compressor tool to install a piston-and-ring assembly. (*Chevrolet Motor Division of General Motors Corporation*)

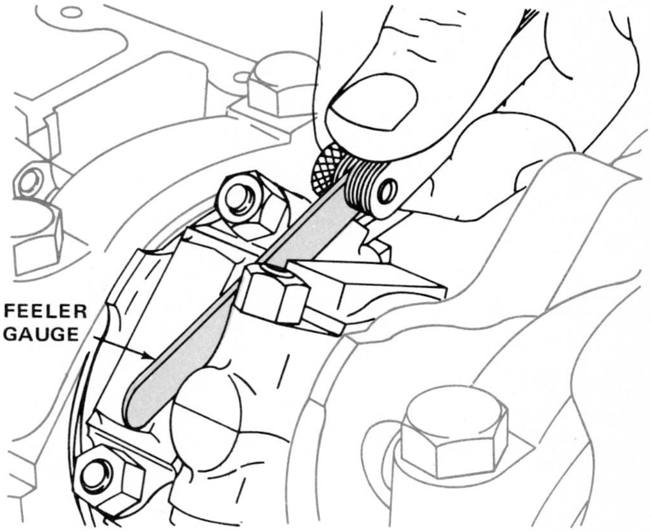

Fig. 38-10 Checking connecting-rod side clearance on a V-type engine. (*Ford Motor Company*)

To accurately check rod alignment out of the engine, a special alignment fixture is required. This fixture has an arbor and a faceplate. With the rod on the arbor and the piston pin in the rod, a V block is placed on the pin and the rod is moved back and forth across the face plate. This shows up any lack of alignment.

If the rod is out of line, check the crankpin for taper (● 38-16). A tapered crankpin can cause the rod to bend. Bent rods should be replaced. They cannot be satisfactorily straightened.

● 38-12 Piston-Pin Bushings in Rods

When the rod has a piston-pin bushing (Fig. 8-2), check the fit of the pin. If the fit is correct, the pin will not drop through the bushing of its own weight when held vertical. It will require a light push to force it through. If the fit is too loose, the bushing should be reamed or honed for an oversize pin, or else replaced.

Aluminum pistons usually have no bushings. They are supplied with prefitted piston pins as a matched set. If the pin is worn, or is too loose a fit in the piston, a new pin-piston set is required.

On some rods, the bushing cannot be replaced. If the bushing is so worn that it cannot be reamed or honed for an oversize pin, the complete rod must be replaced. On other rods, worn bushings can be replaced. The new bushings can be reamed or honed to fit the present pins (if they are in good condition) or new standard-sized pins.

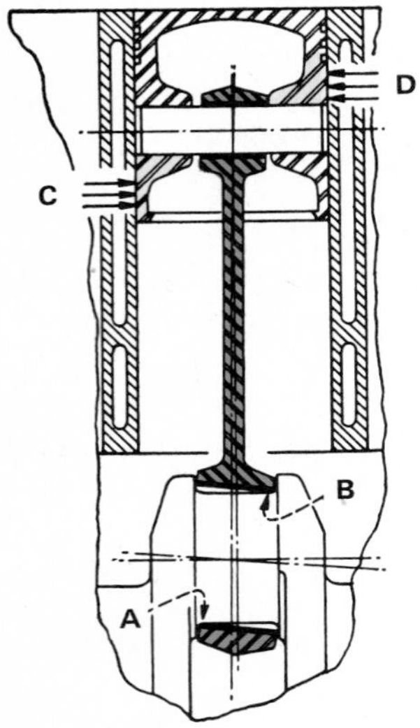

Fig. 38-11 Heavy-pressure areas due to a bent rod. The bent condition is exaggerated. Areas of heavy pressure (*A*, *B*, *C*, and *D*) wear rapidly so that early bearing failure results.

● 38-13 Connecting-Rod Bearings

Connecting-rod big-end bearings are of two types, direct-bonded and precision-insert. Some adjustment is possible on the direct-bonded type (● 38-15). But if this type of bearing is worn, the complete rod and cap must be replaced. The precision-insert type is not adjustable. However, this type of bearing can be replaced without difficulty, provided the rod, crankpin, and other engine components are in good condition. Whenever a rod bearing fails, an analysis should be made to determine the cause. Then the cause can be eliminated so that the failure will not be quickly repeated (see ● 38-14).

● 38-14 Analysis of Bearing Failures

Types of bearing failure are discussed below.

1. Bearing failure due to lack of oil (*A* in Fig. 38-12) When insufficient oil flows to a bearing, actual metal-to-metal contact results. The bearing overheats, and the bearing metal melts or is wiped out of the bearing shell. Welds may form between the rotating journal and bearing shell. There is a chance that the engine will "throw a rod." That is, the rod will "freeze" to the crankpin and break, and parts of the rod will go through the engine block. Oil starvation of a bearing could result from clogged oil lines, a defective oil pump or pressure regulator, or insufficient oil in the crankcase. Also, bearings with excessive clearance may pass all the oil from the pump, so that other bearings are starved and thus fail.

2. Fatigue failure of bearings (*B* in Fig. 38-12) Repeated application of loads on a bearing fatigue the bearing metal. It starts to crack and flake out. Craters, or

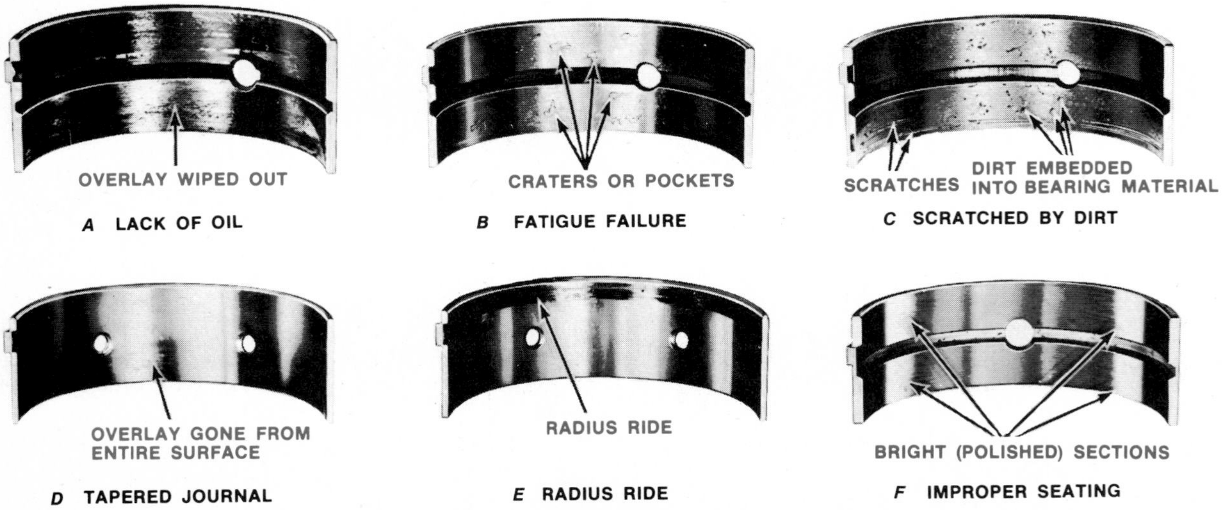

Fig. 38-12 Types of engine-bearing failure. The appearance of a failed bearing usually indicates the cause of its failure. (*Ford Motor Company*)

pockets, form in the bearing. As more and more of the metal is lost, the remainder is worked harder and fatigues at a faster rate. Then complete bearing failure occurs.

Fatigue failure seldom occurs under average operating conditions. However, certain special conditions will cause this type of failure. For instance, if a journal is worn out of round, the bearing will be overstressed with every crankshaft revolution. Also, if the engine is idled or operated at low speed much of the time, the center part of the upper rod-bearing half will carry most of the load and will "fatigue out." On the other hand, if the engine is operated at maximum torque with wide-open throttle (that is, if the engine is "lugged"), then most or all of the upper bearing half will fatigue out. High-speed operation tends to cause fatigue failure of the lower bearing half.

3. Bearing scratched by dirt in the oil (*C* in Fig. 38-12) Embeddability (● 7-21) enables a bearing to protect itself by allowing particles to embed in the bearing. Then they will not gouge out bearing material or scratch the rotating journal. Figure 38-13 shows, in exaggerated view, what happens when a particle embeds. The metal is pushed up around the particle, reducing oil clearance in the area. Usually the metal can flow outward enough to restore adequate oil clearance. However, if the dirt particles are too large, they do not embed completely. They are carried with the rotating journal, gouging out scratches in the bearing. Also, if the oil is very dirty, the bearing becomes overloaded with particles. In either case, bearing failure soon occurs.

4. Bearing failure due to tapered journal (*D* in Fig. 38-12) If the journal is tapered, one side of the bearing carries most or all of the load. This side will overheat and lose its bearing metal. Do not confuse this type of failure with the failure that would result from a bent connecting rod. With a tapered journal, both bearing halves will fail on the same side. With a bent rod, failure will be on opposite sides (*A* and *B* in Fig. 38-11).

5. Bearing failure from radii ride (*E* in Fig. 38-12) If the journal-to-crank-cheek radius is not cut away sufficiently, the edge of the bearing rides on this radius. This causes cramming of the bearing, possibly poor seating, rapid fatigue, and early failure. This trouble would be most likely to occur after a crankshaft-grind job during which the radii were not sufficiently relieved.

6. Bearing failure from poor seating in bore (*F* in Fig. 38-12) Poor seating of the bearing shell in the bore causes local high spots where oil clearances are too low. Figure 38-14 shows, in exaggerated view, what happens when particles of dirt are left between the bearing shell and the counterbore. This reduces oil clearance (as at *X*). Also, an air space exists which prevents proper cooling of the bearing (*A*). The combination can lead to quick bearing failure.

7. Bearing failure from ridging Crankpin ridging, or "camming," may cause failure of a partial-oil-groove type of replacement bearing installed without removal of the ridge. The ridge forms on the crankpin because of uneven wear between the part of the crankpin in contact with the partial oil groove and the part that runs on the solid bearing. The original bearing wears around this ridge. However, when a new bearing is installed, the center zone may be overloaded (at the ridge) and may soon fail. A ridge so slight that it can hardly be detected (except with a carefully used micrometer) may be enough to cause this sort of failure. Failures of this sort have been reported in engines having ridges of less than 0.001 inch [0.0254 mm].

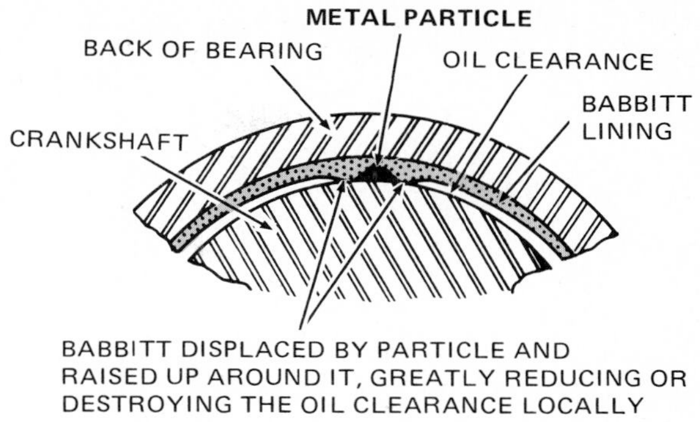

Fig. 38-13 Effect of a metallic particle embedded in bearing metal (babbitt lining). (*Federal-Mogul Service Division of Federal-Mogul-Bower Bearings, Incorporated*)

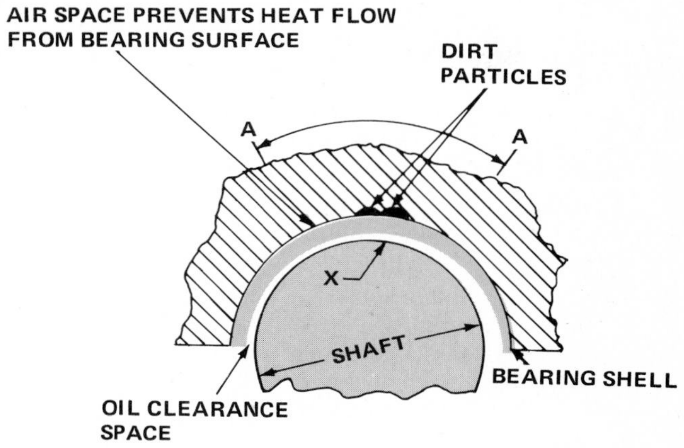

Fig. 38-14 Effect of dirt particles under the bearing shell, due to poor installation of the bearing. (*Federal-Mogul Service Division of Federal-Mogul-Bower Bearings, Incorporated*)

● 38-15 Checking Connecting-Rod Bearing Clearance

Precision-insert bearings are checked in one way, the direct-bonded type in another, as noted below.

Careful: Before installing new bearings, the crankpins should always be checked for taper or out-of-roundness (● 38-16).

1. Precision-insert bearings The clearance of these bearings can be checked with Plastigage, shim stock, or with micrometer and telescope gauge.

a. Plastigage Plastigage is a plastic material that comes in strips and flattens when pressure is applied to it. A strip of the material is put into the bearing cap. The cap is installed, and the rod nuts are tightened to the specified tension. Then, the cap is removed, and the amount of flattening is measured. If the Plastigage is flattened only a little, then oil clearances are large. If it is flattened considerably, oil clearances are small. Actual clearance is measured with a special scale supplied with the Plastigage (Fig. 38-15).

The bearing cap and crankpin should be wiped clean of oil before the Plastigage is used. The crankshaft should be turned so that the crankpin is about 30 degrees back of BDC (bottom dead center). Do not move the crankshaft while the cap nuts are tight. This would further flatten the Plastigage and throw off the clearance measurement.

b. Shim stock The shim-stock method is seldom used today because the Plastigage method is faster and easier. When using shim stock, you lay a strip of 0.001-inch [0.0254-mm] stock in the cap, replace the cap, and tighten the nuts to see if the rod tightens up on the crankpin. You repeat the procedure, adding strips until the rod tightens up. This tells you the amount of clearance.

c. Micrometer and telescope gauge Check the crankpin diameter with a micrometer. Check the bearing diameter (cap in place) with a telescope gauge and micrometer (or an inside micrometer). Compare the two diameters to determine the difference, or bearing clearance. At the same time, the crankpin can be checked for taper or eccentric wear. Measure the diameter at several places along the crankpin (to check for taper). Also measure around the crankpin (to check for eccentricity, or out-of-roundness).

2. Direct-bonded connecting-rod bearings On these, adjustment is made by installation or removal of shims under the cap.

● 38-16 Installing Precision Rod Bearings

New precision connecting-rod bearings are required if the old ones are defective (● 38-14), or have worn so much that clearances are excessive. They are also required if the crankpins have worn out of round, or tapered so much that they have to be reground. In this case, new undersize bearings are required. Engine rebuilders usually replace the bearings in an engine when it is torn down, whether or not the old bearings are in bad condition. Their reasoning is that it costs little more to put in new bearings when the engine is torn down for rebuilding. However, if the engine had to be torn down especially for bearing installation, the cost would be high. They believe it is cheap insurance against failure to install new bearings during the engine-rebuilding job.

1. Checking crankpins Crankpins should always be checked with a micrometer for taper and concentricity. If crankpins are out of round or tapered more than 0.0015 inch [0.037 mm], the crankshaft must be replaced or the crankpins reground (● 39-1). Bearings working against

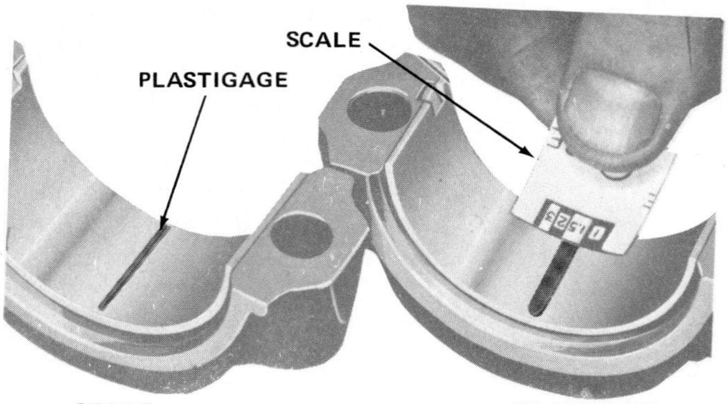

Fig. 38-15 Bearing clearance being checked with Plastigage. Left, Plastigage in place before tightening the cap. Right, measuring the amount of flattening (or bearing clearance) with a scale. (*Buick Motor Division of General Motors Corporation*)

taper or out-of-roundness of more than 0.0015 inch [0.037 mm] will not last long. And when bearings go, there is the chance that the engine will be severely damaged. Measurements should be taken at several places along the crankpin to check for taper. Diameter should be checked all the way around for out-of-roundness.

2. Installing new bearings When new bearings are to be installed, make sure your hands, workbench, tools, and all engine parts are clean. Keep the new bearings wrapped up until you are ready to install them. Then handle them carefully. Wipe each with a fresh piece of cleaning tissue just before installing it. Be very sure that the bores in the cap and rod are clean and not excessively out of round. (Some manufacturers recommend a check of bore symmetry with the bearing shells out. The cap should be attached with nuts drawn up to specified tension. Then a telescope gauge and micrometer or a special out-of-round gauge can be used to check the bore.) Put the bearing shells in place. If they have locating tangs, make sure that the tangs enter the notches in the rod and cap. Note comments about bearing spread and crush, below. Check clearance after installation (● 38-15).

Do not attempt to correct clearance by filing the rod cap. This destroys the original relationship between cap and rod and leads to early bearing failure.

3. Bearing spread Bearing shells are usually manufactured with "spread." That is, the shell diameter is somewhat greater than the diameter of the rod cap or rod bore into which the shell will fit (Fig. 38-16). When the shell is installed into the cap or rod, it snaps into place and holds its seat during later assembly.

4. Bearing crush In order that the bearing shell will "snug down" into its bore in the rod cap or rod when the cap is installed, the bearings have "crush" (Fig. 38-17). That is, they are manufactured to have some additional height over a full half. This additional height must be crushed down when the cap is installed. Crushing down the additional amount forces the shells into the bores in the cap and rod. It assures firm seating and snug contact with the bores.

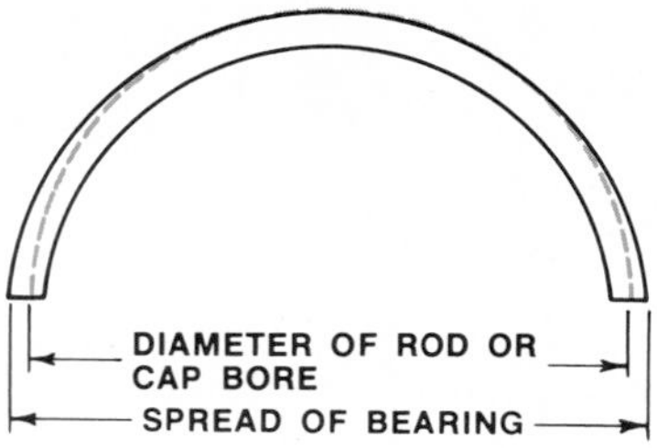

Fig. 38-16 Bearing spread.

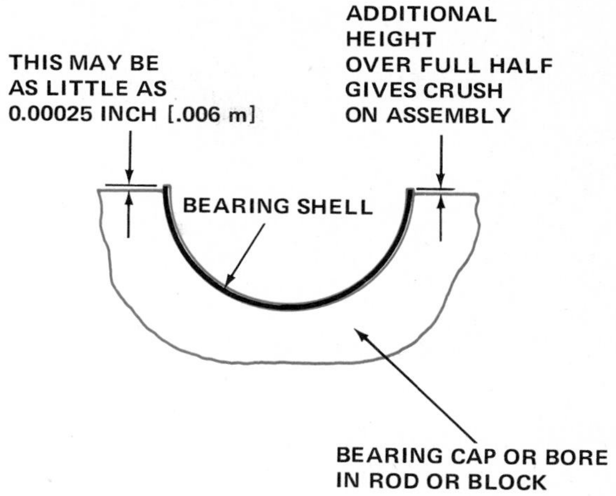

Fig. 38-17 Bearing crush.

Never file off the edges of the bearing shells in an attempt to remove crush. When you select the proper bearings for an engine (as recommended by the engine manufacturer), you will find that they have the correct crush. Precision-insert bearings must not be tampered with in any way to make them "fit better." This usually leads only to rapid bearing failure.

Pistons and Rings

● 38-17 Piston Service

After the piston-and-rod assemblies are removed from the engine, the pistons and rods should be separated (● 38-3 to 38-7). Then the rings can be removed from the pistons. The rings can also be removed from the pistons before the pistons and rods are separated. A special ring-expander tool can be used for ring removal. The tool has two small claws that catch under the ends of the ring (Fig. 38-18). When pressure is applied to the tool handles, the ring is sprung enough so it can be lifted out of the ring groove and off the piston. Discard the old rings. As a rule, expert mechanics replace the old rings with new ones during an engine overhaul. Once the ring break-in coating and tool marks are worn off, the ring will not reseat itself if it is reinstalled.

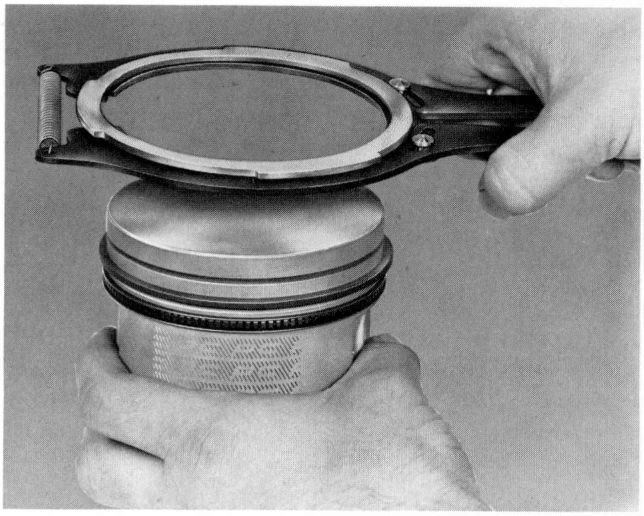

Fig. 38-18 Using a piston-ring-expander tool to remove or install a compression ring on a piston. (*Service Parts Division, Dana Corporation*)

● 38-18 Piston Cleaning

Remove carbon and varnish carefully from piston surfaces. Do not use a caustic cleaning solution or wire brush! These could damage the piston-skirt finish. You may decide to reinstall the pistons in the engine; therefore, you should not damage them. Use the cleaning method provided in your shop to clean the pistons. Clean ring grooves with a clean-out tool. You can also use the end of a broken piston ring filed to a sharp edge. Oil-ring slots, or holes, must be clean, so oil can drain back through them. Use a drill of the proper size. Do not remove metal when cleaning the slots or holes.

● 38-19 Piston Inspection

Examine the pistons carefully for wear, scuffs, scored skirts, worn ring grooves, and cracks. Look for cracks at the ring lands, skirts, pin bosses, and heads. Any defects require replacement of the piston, with these exceptions: Worn ring grooves can sometimes be repaired by cutting the grooves larger and using ring-groove spacers (● 38-20). Piston skirt wear or collapse (reduction in skirt diameter) can sometimes be corrected by knurling the piston skirt (● 38-21).

Check the fit of the piston pins to the pistons or piston bushings. One way of doing this is to use a small-hole gauge to check the piston bushing bores, and a micrometer to measure the pin diameter. On the type of piston without a bushing in which the pin oscillates (Fig. 8-2), the piston and pin are supplied in matched sets. If the fit is too loose, or there are other pin or piston defects, the pin and piston are replaced as a matched set. Chevrolet specifies a fit no looser than 0.001 inch [0.0254 mm].

Check the piston with a micrometer (Fig. 38-19). Compare the piston measurement with the cylinder diameter, measured 90 degrees from the piston pin. This measurement may be made with a telescope gauge and microme-

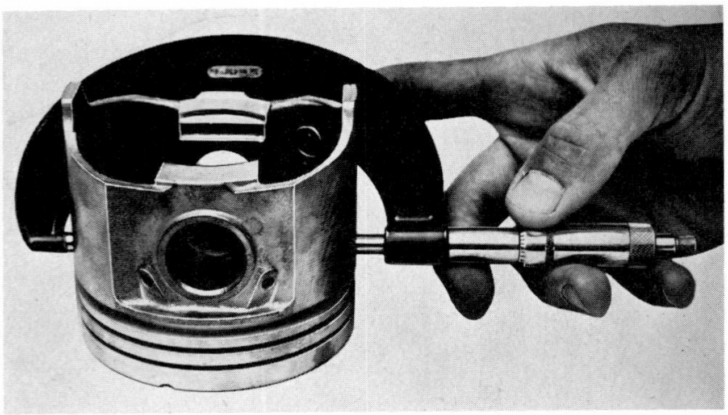

Fig. 38-19 Using a micrometer to measure piston diameter. *(Pontiac Motor Division of General Motors Corporation)*

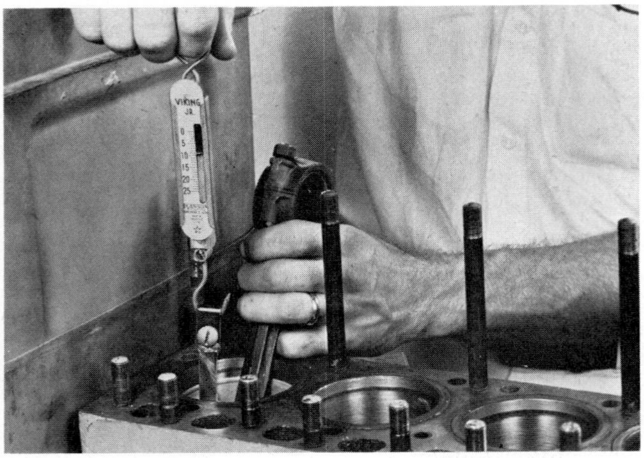

Fig. 38-20 Using a feeler ribbon and a spring scale to check piston clearance. *(Service Parts Division of Dana Corporation)*

ter. Or, it may be measured with a dial indicator and a micrometer.

If the cylinder wall is excessively worn or tapered, it will require refinishing (Chap. 39). If the cylinder wall is refinished, then a new, oversized piston will be required.

Prior to the introduction of the modern piston, the fitting process required the use of a feeler ribbon and spring scale (Fig. 38-20). If only a light pull is required to pull the ribbon out, the fit is the thickness of the ribbon. If the feeler ribbon comes out too easily, the fit is too loose. Some manufacturers still recommend using the feeler ribbon.

38-20 Ring-Groove Repair

If a piston is in good condition except for excessive ring-groove wear, it can often be repaired. The top ring groove is the one that wears most, because it gets the highest temperatures and pressures. One piston-ring manufacturer states that almost all aluminum pistons checked at overhaul have excessively worn top ring grooves. The ring groove may be checked with a special gauge. If the ring groove is excessively worn (as much as 0.006 inch [0.152 mm] or more), it can be machined to a larger width with a special hand-operated lathe. This squares up

Fig. 38-21 Top ring-groove spacer in place above the ring. *(Perfect Circle)*

the top and bottom sides of the ring groove. Then the ring is installed with a spacer, as shown in Fig. 38-21.

38-21 Piston Resizing

Resizing of modern pistons is not recommended by automotive manufacturers. The procedure can damage the piston finish. One piston-ring manufacturer has developed a knurling procedure, called "nurlizing." The piston skirt is run between a supporting wheel and a nurlizing tool. The procedure displaces metal and expands the diameter of the piston skirt. Also, the indentations form little pockets that can hold lubricating oil.

38-22 New Pistons

New pistons are of the finished type, ready for installation. They are available in a number of sizes. When these are used, the cylinders are finished to fit the pistons. Engine manufacturers supply oversize pistons of the same weight as the original pistons. Thus, it is not necessary to replace all pistons when only some cylinders require service. There is no problem of balance if all pistons are of the same weight, even if some are oversize.

Aluminum pistons are usually supplied with piston pins already fitted. This assures factory specifications on the pin fit to the piston.

Finished pistons have a special finish. They must not be buffed with a wire wheel or finished to a smaller size. This would remove the finish and cause rapid piston wear after installation.

38-23 Fitting Piston Pins in Pistons

On pistons with piston-pin bushings, worn bushings may be replaced. The new bushings are honed to size to fit the piston pins. Figure 8-2 shows this type of piston.

38-24 Rod and Piston Alignment

After the rod and piston have been reassembled, but before the rings are installed on the piston, alignment should be checked with the special alignment gauge. If the V block does not line up with the faceplate as the piston is moved to various positions, the connecting rod is twisted.

● 38-25 Piston-Ring Service

If an engine is torn down for service, the old rings should be discarded. Rings that have been used, even for only short mileage, will not seat properly to provide sealing.

If the engine trouble is due to rings sticking in the piston-ring grooves, a special compound can be introduced into the intake manifold and engine oil. This will sometimes free the rings (see ● 37-4). However, no compound should be used on cars equipped with catalytic converters unless you are sure the compound will not damage the converter.

Proper selection of new rings depends on the condition of the cylinder walls and whether or not they are to be reconditioned. In Chap. 39, we describe the checking of cylinder walls for wear and taper. If they are only slightly tapered or out of round, then standard-type rings can be installed. (Consult manufacturer's specifications for maximum allowable deviations.) The walls may have some taper, but not enough to warrant the extra expense of a rebore or hone job. Then special "severe," or "drastic," rings should be used. These rings have greater tension and are more flexible. This enables them to expand and contract as they move up and down in the cylinder. Thus, they follow the changing contours of the cylinder wall and provide adequate sealing (preventing blow-by) and oil control. Figure 38-22 shows a set of replacement rings for tapered cylinder walls.

Severe rings increase ring friction on the cylinder walls and thus reduce engine power output.

Automotive manufacturers generally recommend honing the cylinder walls lightly before ring installation, to "break the glaze." Cylinder walls take on a hard, smooth glaze after the engine has been in use for a while. Some engine rebuilders knock off this glaze by running a hone up and down the cylinder a few times before putting in new rings. However, at least one ring manufacturer says this does not need to be done on cast-iron cylinder walls, *provided the walls are not wavy or scuffed.* The glaze is a good antiscuff material and will not retard the wear-in of new rings. The walls should be reasonably concentric and in good condition.

The best honing job leaves a cross-hatch pattern (Fig. 38-23) with hone marks intersecting at a 60-degree angle. This leaves the best surface for seating of new rings.

● 38-26 Fitting Piston Rings

Piston rings must be fitted to the cylinder and to the ring grooves in the piston. Rings come in packaged sets in graduated sizes to fit various sizes of cylinders. All packages have instruction sheets that describe how to install the rings. Follow these instructions carefully.

Careful: Never throw the instructions away until you have finished the ring-installation job.

Fig. 38-22 Disassembled view of a set of replacement rings. 1, top compression ring; 2, second compression ring which includes an expander spring; 3, oil-control ring, which includes an expander spring. (*Grant Piston Rings*)

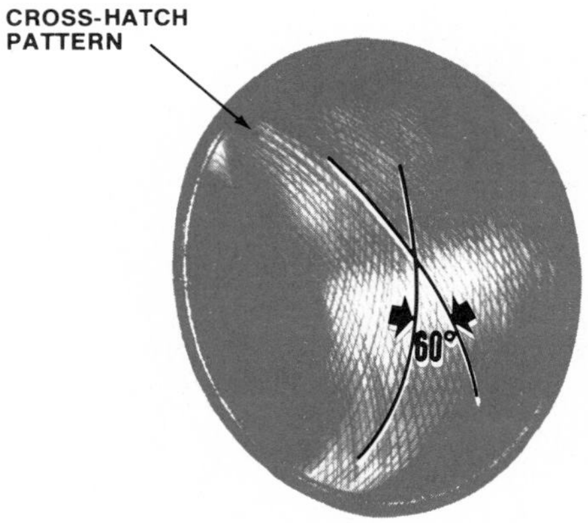

Fig. 38-23 Cross-hatch pattern left on a cylinder wall after a good honing job. (*Chrysler Corporation*)

As a first step, the ring should be pushed down into the cylinder with a piston, and the ring gap measured. The ring gap is the space between the ends of the ring. It is measured with a feeler gauge (Fig. 38-24). Figure 38-24 shows the gap being measured with the ring pushed down to the lower limit of ring travel. If the cylinder is worn, that is where the ring gap will be smallest. If the ring gap is too small, check the package the ring came in. The ring set may be wrong for the job. Rings come in sets in graduated sizes. If the ring gap is wrong, then either you have the wrong rings, you have incorrectly measured the cylinder diameter, or the wrong size rings were packaged in the box.

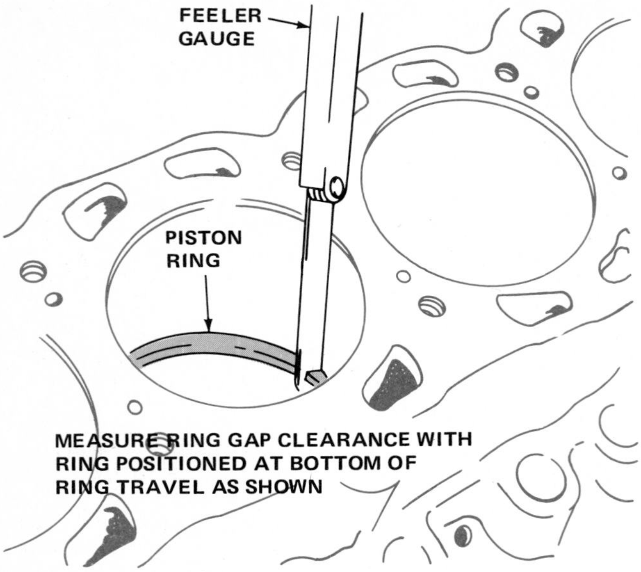

Fig. 38-24 Ring gap being measured with the ring in the cylinder. (*Ford Motor Company*)

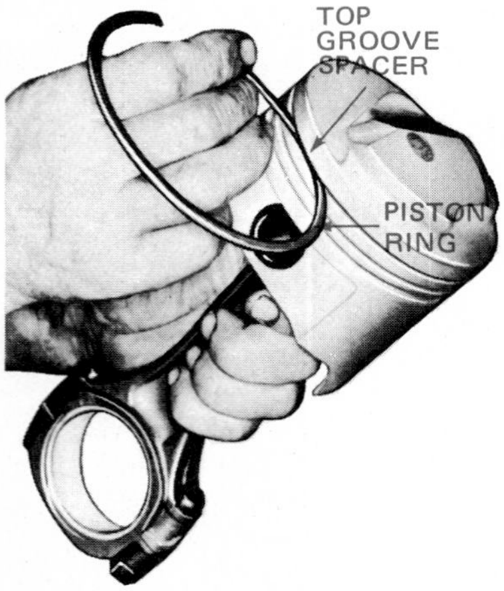

Fig. 38-25 Checking the fit of the ring in the ring groove. (*Chevrolet Motor Division of General Motors Corporation*)

On earlier engines, the recommendation was to file the ends of the ring with a fine-cut file. The file was first clamped in a vise. Then the ring was worked back and forth on the file (with the ring ends on the two sides of the file). This is no longer recommended. Filing the ring ends can remove some of the ring coating and cause early ring failure.

Careful: Remember, if the cylinder is at all tapered, the diameter at the lower limit of ring travel (in the assembled engine) will be smaller than the diameter at the top. This means the ring must be fitted to the diameter at the lower limit of ring travel. If it is fitted to the upper part of the cylinder, the ring gap will not be large enough at the lower limit of travel. This means that the ring ends will come together. The ring will be broken, and the cylinder wall scuffed. Always measure the ring gap with the ring pushed down to the point of minimum diameter—at the lower limit of ring travel.

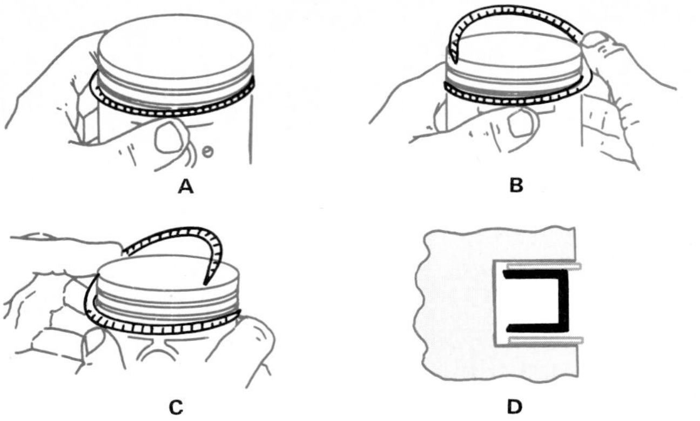

Fig. 38-26 Installation of three-piece oil-control ring. **A** Place expander spacer in the oil-ring groove with ends of spacer above a solid part of the groove bottom. **B** Hold ends of spacer together and install steel rail above the spacer. **C** Install other spacer on lower side of the spacer. Make sure the ends of the spacer are not overlapping. **D** Sectional view of the three parts fitted into the groove. (*Perfect Circle*)

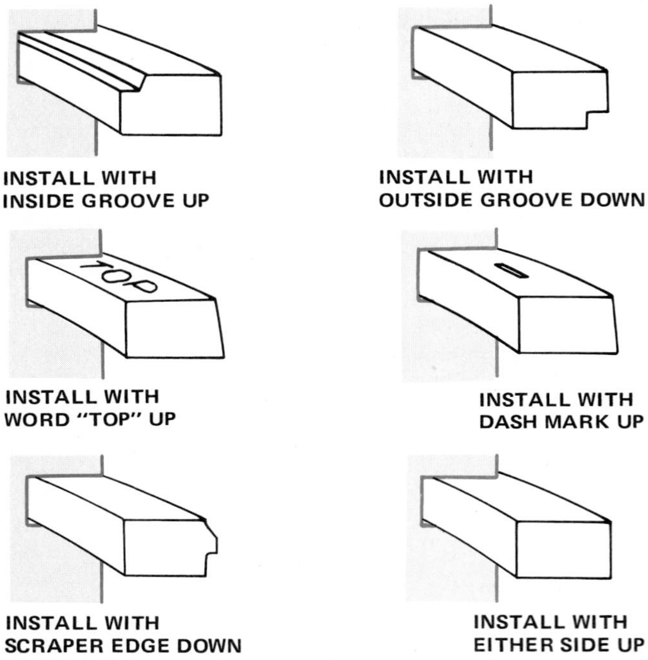

Fig. 38-27 Types of compression rings, and the proper way to install them. (*Perfect Circle*)

If the ring gap is correct, insert the outside surface of the ring into the proper groove in the piston (Fig. 38-25). Then roll the ring around in the groove, to make sure the ring has a free fit around the entire piston. An excessively tight fit probably means that the ring groove is dirty. Or else, the ring groove has been nicked or burred with the blade of the ring tool. Some authorities recommend using the end of a broken ring, filed to a sharp edge, to clean the ring grooves. This is preferred by many technicians because the piece of ring will not cause nicks or burrs.

Install the rings in the ring grooves, using the ring tool as shown in Fig. 38-18. Then recheck the fit. Insert a feeler gauge of the proper size between the ring and the side of the groove.

● 38-27 Cautions on Installing Rings

The three-part oil-control ring is installed one part at a time, as shown in Fig. 38-26. Various types of compression rings and their proper installation are shown in Fig. 38-27. One special caution to observe is never to spiral the compression rings into the grooves. (We show spiraling of the rails of the *oil-control* rings in Fig. 38-26.) This could distort or break the compression ring and cause loss of compression and blow-by. Instead, *always* use a ring-expander tool, as shown in Fig. 38-18. Also, never overexpand the compression rings.

● 38-28 Installing the Piston-and-Rod Assembly

We have already covered this procedure in ● 38-9. Be sure to use a ring compressor. Install the piston-and-ring assembly with the correct side facing forward (Fig. 38-9).

REVIEW QUESTIONS

1. Explain how to remove a ring ridge. Why is this necessary?
2. Describe the removal procedure for piston-and-rod assemblies.
3. What are the five basic arrangements for attaching the rod and piston? How is the pin removed from each type?
4. Explain how to remove and replace the piston pin that is a press fit in the rod.
5. Describe the procedure for installing piston-and-rod assemblies.
6. What is a piston-ring compressor? How is it used?
7. Explain how to use a piston-ring expander tool.
8. Describe, step-by-step, how to clean and inspect a piston.
9. What is the purpose of ring-groove repair? Explain how it is done.
10. What is nurlizing? What is its purpose?
11. What are the six basic types of bearing failure? What causes them?
12. Explain how to check connecting-rod fit with Plastigage. With shim stock.
13. Describe the procedure for removing an old rod bearing and installing a new one.
14. What are "severe" rings? What is their purpose?
15. Describe the procedure for fitting a piston ring to the cylinder. To the piston.
16. How is piston-ring gap measured in the cylinder?
17. Why should you never spiral a compression ring into the ring groove?
18. Describe the procedure for installing a piston-and-rod assembly, including the cautions to observe.

SELF PROJECTS

1. As you work on engines, and watch engines go through the service shop, try to collect rod bearings showing the six different kinds of bearing failure discussed in the chapter. Mount them on a board, with a label under each identifying the type of failure.
2. Save the instructions from the ring set you installed. Sometimes the instructions are printed on the envelopes the rings come in. Sometimes they are on a separate sheet. Sometimes they are printed on the package the ring envelopes come in. Cut them out, and paste them on a sheet of paper which you can put in your notebook. Add the year, make, mileage, conditions found, and so on, on the engine you serviced. Then, if you ever see the engine again, you can check on the performance and life of the job you performed.
3. Make ring-groove cleaners from broken rings. Put them in your toolbox.
4. Get four pieces of vacuum hose slightly smaller than the rod bolts. Trim the pieces to about three inches long. Then slit them from one end to the other. You now have rod-bolt covers that will protect the crankshaft when you remove piston-and-rod assemblies (Fig. 38-6). Put the covers in your toolbox.
5. Remove a connecting-rod bolt and nut from a rod. Examine them carefully. Compare them with the same size ordinary nut and bolt. Look for reasons why the rod bolt and nut are special.

CHAPTER 39
Engine Service: Crankshafts and Cylinder Blocks

After studying this chapter, you should be able to:

- **Describe the procedure of installing main bearings.**
- **Explain how to replace the main-bearing oil seal.**
- **Explain how to remove, check, and service crankshafts.**
- **Describe how to check and service cylinder walls.**
- **Under the instructor's supervision, perform the above services.**

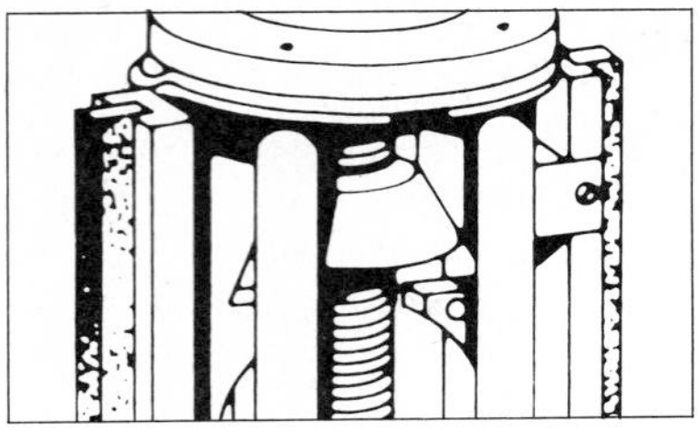

This chapter concludes the discussion of piston-engine service. It discusses the servicing of crankshafts, main bearings, and cylinder blocks. If the cylinder block, crankshaft, and main bearings all require service, it is often cheaper to buy a short block than to invest the time required for all the services.

Crankshafts and Main Bearings

● 39-1 Precision-Insert Main Bearings

Modern automotive engines have precision-insert main bearings that can be replaced without removing the crankshaft. Many main-bearing difficulties can be taken care of by this method of bearing replacement. However, bearing replacement will not fix stopped-up oil passages, worn crankshaft journals, a damaged crankshaft, or a block in which a bearing has spun. Owing to lack of oil, the bearing and crankshaft journal can become so hot that they weld momentarily. Then the bearing spins with the crankshaft and gouges the bearing bore in the cylinder block. Bearing spin will damage the cylinder block and require block replacement.

If all bearings have worn fairly evenly, then probably only crankshaft-journal checks and bearing replacement are required. Usually all bearings do not wear the same amount. Some bearings wear more than others. This is acceptable, provided none of the bearings wear beyond manufacturer's specifications. The lower half of the bearing wears the most. It takes the weight of the crankshaft and combustion pressures through the rods and cranks. Uneven wear can result from oil-pump wear. That is, as the oil pump wears, oil pressure and circulation drop. The main and rod bearings farthest from the oil pump get less oil. Thus, they wear the most. Also, a clogged oil passage will starve bearings. When this happens, the bearings fail. They may also spin, as noted above.

If main bearings have worn very unevenly, the best procedure is to remove the crankshaft from the engine block. Then the block and crankshaft can be checked separately for damage and clogged oil lines.

However, as we said, many main-bearing problems can be taken care of by replacing the bearings with the crankshaft still in the engine. We shall explain this in detail later. Replacing the main bearings (or mains) without removing the crankshaft requires about 5 hours. Removing and replacing the crankshaft requires added hours.

● 39-2 Checking Crankshaft Journals

Both the crankpins and crankshaft main journals should be checked whenever the bearing caps are removed. Checking crankpins was discussed in ● 38-16. Main journals can be checked with a special crankshaft gauge or with a special micrometer. Measurements should be taken in several places along the journal, to check for taper. Also, the crankshaft should be rotated by one-quarter or one-eighth turns to check for out-of-round wear. See ● 38-14 for a discussion of what a tapered, ridged, or out-of-round journal will do to the bearing. If journals are tapered or out of round by more than 0.003 inch [0.076 mm], they should be reground. Some authorities consider 0.0015 inch [0.037 mm] the outside tolerable limit for wear. They point out that *any* noticeable out-of-round or taper shortens bearing life.

To check journals, remove the oil pan (● 38-5) and bearing caps. It is not necessary to detach the connecting rods from the crankshaft. However, the spark plugs should be removed so that the crankshaft can be turned over easily.

● 39-3 Removing Bearing Caps

Sometimes, it is difficult to remove all bearing caps with the engine in the car. The front cross member is so close to the engine, in some cars, that the engine must be lifted to get enough clearance to remove the cap. Also, the rear main-bearing cap may be hard to remove because of interference with other parts.

If you are planning to check only journals and bearings, then remove bearing caps one at a time to make the checks. If the crankshaft is to come out of the engine, the connecting rods should be detached. All main-bearing caps should be removed.

Caps should be marked so they can be replaced on the journals from which they were removed. To remove a cap, remove the nuts or bolts. Bend back the lock-washer tangs (if used). Disconnect oil lines where necessary. Use new lock washers when reassembling.

If a cap sticks, work it loose carefully to avoid nicking or cracking it. In some engines, a bearing-cap puller can be used. The puller is screwed into the oil-coupling hole. In other engines, a screwdriver or pry bar can be used to work the cap loose. Sometimes, tapping the cap lightly on one side and then the other with a brass hammer will loosen it.

Careful: Heavy hammering or prying can nick or crack the cap, bend the dowel pins, or damage the dowel holes. In such a case, the bearing may not fit when the cap is replaced, and early bearing failure will occur. Also, remember that the bearing caps are cast iron, and so are brittle. A hard blow can crack or break a cap. A damaged cap will have to be discarded, and a new cap used. A cracked cap can break if it is reinstalled on the engine. Such a break usually means a ruined engine.

When a bearing cap is damaged or lost, a new cap is required. The new caps are supplied with a shim pack so the cap can be shimmed into alignment. However, it is difficult to get a good fit this way. So it may become necessary to take the engine out of the car, disassemble it, and install the caps, without bearings, back in the block. Then the block will have to be line bored to reestablish bearing-bore alignment.

● 39-4 Measuring Main Journals with Crankshaft Gauge

The special gauge is used as shown in Fig. 39-1. The journal and the gauge pads and plunger must be clean. Then, the plunger is retracted, and the gauge is held tightly against the journal (Fig. 39-1). Next, the plunger is released so that it moves out into contact with the journal. The plunger is then locked in this position by tightening the thumbscrew. Finally, an outside micrometer is used to measure the distance between *D*, or the end of the plunger, and *C*, or the button on the bottom of the gauge. This measurement, multiplied by 2, is the diameter of the journal.

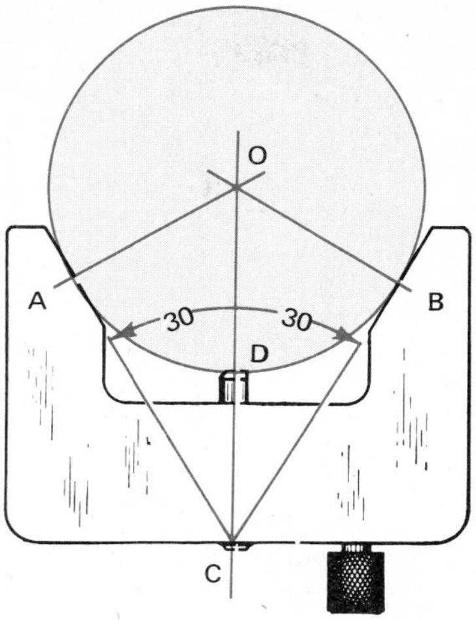

Fig. 39-1 Checking crankshaft main journal with a crankshaft gauge. (*Federal-Mogul*)

Take the measurements from one end of the journal to the other. Rotate the crankshaft by one-eighth turns to repeat the checks. This repeated measurement will detect journal taper and out-of-round. Write down the readings.

● 39-5 Measuring Main Journals with a Micrometer

To use the micrometer, the upper bearing half must be removed. This is done with a special roll-out tool, as explained in ● 39-8. Take measurements from one end of the journal to the other. Rotate the crankshaft by one-eighth turns to repeat the check. This procedure will detect journal taper and out-of-round.

● 39-6 Checking Main Bearings

Main, or crankshaft, bearings should be replaced if they are worn, burned, scored, pitted, rough, flaked, cracked, or otherwise damaged. (See ● 38-14 on bearing failures.) It is important to check the crankshaft journals (● 39-2 to 39-5) before installing new bearings. If the journals are not in good condition, the new bearings may soon fail. Also, bearings may have worn unevenly; that is, some bearings may be worn a lot more than others. If so, or if a bearing is damaged, the possibility of a bent crankshaft or clogged oil passages should be considered. Following sections describe the checking of bearing clearance, replacement of bearings, and servicing of crankshafts.

● 39-7 Checking Main-Bearing Clearances

Bearing fit (or oil clearance) should always be checked after new bearings are installed. The fit should also be

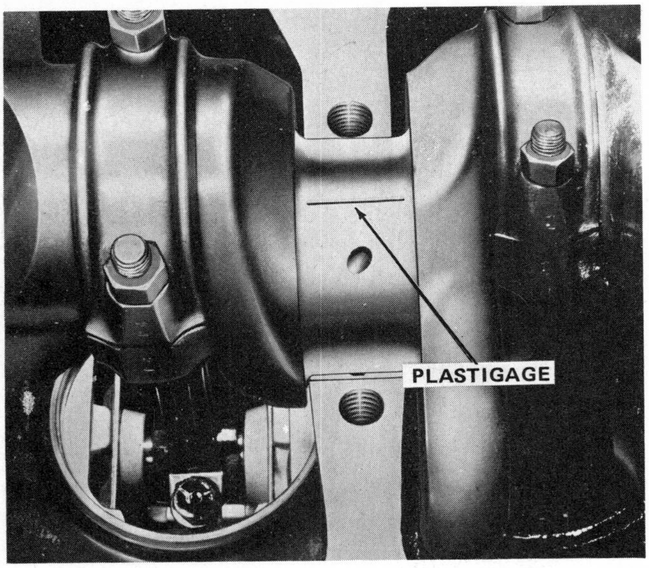

Fig. 39-2 Plastigage strip in place for bearing-clearance check. (*Chevrolet Motor Division of General Motors Corporation*)

Fig. 39-3 Checking flattening of Plastigage to determine bearing clearance. (*Chevrolet Motor Division of General Motors Corporation*)

checked whenever the condition of the bearings is being determined. Crankshaft-journal condition should be checked at the same time.

1. Precision-insert type Bearing clearance can be checked with shim stock or Plastigage.

a. With Plastigage Wipe the journal and the bearing clean of oil. Put a strip of Plastigage lengthwise in the center of the journal (Fig. 39-2). Replace and tighten the cap. Then remove the cap, and measure the amount the Plastigage has been flattened (Fig. 39-3). Do not turn the crankshaft with the Plastigage in place. (See ● 38-15 for more detailed information on Plastigage.)

b. Shim stock The shim-stock method is seldom used today because the Plastigage method is faster and easier. To use shim stock, you lay a strip of shim stock on the journal and replace the cap. Tighten the cap nuts and note the ease with which the crankshaft can be turned. Add strips of shim stock until the crankshaft tightens up. Count the strips and subtract one. That is the clearance.

The crankshaft must be supported so that its weight will not cause it to sag. (This could result in an incorrect measurement.) One way to do this is to position a small jack under the crankshaft. Let it bear against the counterweight next to the bearing being checked. Another method is to put shims in the bearing caps of the two neighboring main bearings. Then tighten the cap bolts. This lifts and supports the crankshaft. Of course, if the engine is off the car and inverted, this is not necessary.

2. Shim-adjusted main bearings Loosen all bearing caps just enough to permit the crankshaft to turn freely. Take off the rear main-bearing cap. Remove one shim from each side of the cap. Replace and tighten the cap bolts or nuts to the specified tension. Rotate the crankshaft to see whether it now drags. If it does not drag, remove additional shims (in pairs). Check for drag after each pair is removed. When a drag is felt, replace one shim on each side of the cap. If the crankshaft now turns freely when the cap is tightened, the clearance is correct. Loosen the cap bolts or nuts. Go to the next bearing, and adjust its clearance in the same way. Finally, when all bearings are adjusted, tighten all cap bolts or nuts to the proper tension. Then recheck for crankshaft drag as it is turned. If it drags, then recheck and readjust the bearings.

3. Checking crankshaft end play Crankshaft end play will become excessive if the end-thrust bearings are worn. This produces a sharp, irregular knock. If the wear is considerable, the knock will occur every time the clutch is released and applied. This action causes sudden endwise movements of the crankshaft. Check end play by forcing the crankshaft endwise as far as it will go. Then measure the clearance at the end-thrust bearing with a feeler gauge (Fig. 39-4). Consult the engine manufacturer's shop manual for allowable end play.

● 39-8 Replacing Precision Main Bearings

Before replacing bearings, crankshaft journals should be checked (● 39-2). Also, after bearings are installed, bearing fit should be checked (● 39-7). Precision-insert main bearings can be replaced without removing the crankshaft. However, some authorities do not suggest this. They say that you are working blind. You cannot be sure that the counterbore in the cylinder block is perfectly clean and that the shell is seating tightly. Furthermore, neither the crankshaft nor the block can be checked for alignment. As previously noted (● 39-1), with uneven bearing wear the crankshaft should be removed for further checking.

To install a precision-insert main bearing without removing the crankshaft, use a special roll-out tool as shown in Fig. 39-5. The tool is inserted into the oil hole in the crankshaft journal, as shown. Then, the crankshaft is rotated. The tool forces the bearing shell to rotate with the crankshaft so that it is turned out of the bore. The crankshaft must be rotated in the proper direction so that

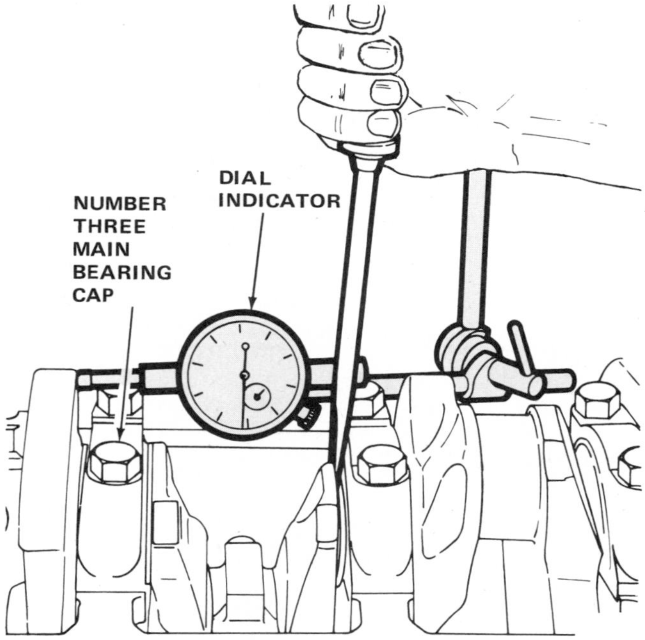

Fig. 39-4 Checking crankshaft end play at end-thrust bearing with a feeler gauge (*top*), and with a dial indicator (*bottom*). (*Chevrolet Motor Division of General Motors Corporation*)

the lock, or tang, in the bearing is raised up out of the notch in the cylinder block.

To install a new bearing half, coat the bearing surface with engine oil. Leave the outside of the bearing dry. Make sure that the bore, or bearing seat, in the block is clean. Do not file the edges of the shell (this would remove its crush). Use the tool, as shown in Fig. 39-5, to slide the bearing shell into place. Make sure that the tang on the bearing shell seats in the notch in the block. Then place a new bearing shell in the cap. Install the cap, and tighten the cap bolts or nuts to the specified tension. Tap the crown of the cap lightly with a brass hammer while tightening it. This helps to align the bearings properly. After all bearings are in place, check bearing fit.

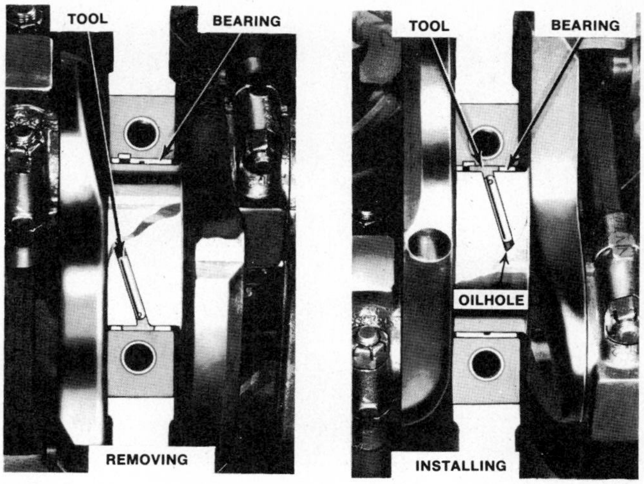

Fig. 39-5 Removing and installing upper main bearing with roll-out tool. The crankshaft journal is partly cut away to show the tool inserted in the oil hole in the journal. (*Chrysler Corporation*)

If the crankshaft is removed, it is easier to install main bearings. Also, you can wipe the bearing bores in the cylinder block and make sure they are in good condition. Then, the bearing inserts can be slid into position.

Some bearing sets have annular (or ring) grooves in only one bearing half. Others have grooves in both halves. Some do not use grooves. Be sure to check the service manual for the engine you are servicing, to determine what kind of bearing half goes where.

Some crankshaft journals have no oil hole. For example, the rear main journals of many in-line engines do not. To remove and replace the upper bearing half on these, first start the bearing half with a small pin punch and hammer. Then use a pair of pliers with taped jaws to hold the bearing half against the oil slinger. Rotate the crankshaft (Fig. 39-6). This will pull the old bearing out. The new bearing is put into position in the same manner. The last fraction of an inch can be pushed into place by holding only the oil slinger with the pliers while rotating the crankshaft. Or the bearing may be tapped down with a pin punch and hammer. Be careful that you do not damage the bearing.

While removing and replacing the upper bearing shell of a rear main bearing, hold the oil seal in position in the cylinder block. Otherwise, it may come out of position (see ● 39-9 on oil-seal replacement).

On all but a very few engines, precision-insert bearings are installed without shims. Never use shims on these bearings unless the engine manufacturer specifies them. Similarly, bearing caps must not be filed in an attempt to improve bearing fit.

● 39-9 Replacing Main-Bearing Oil Seal

An oil seal is required at the rear main bearing to prevent oil leakage at that point. When main-bearing service is being performed, or whenever leakage is noted at the rear main bearing, the oil seal must be replaced.

Fig. 39-6 Replacing the rear main-bearing half with pliers. On this engine, the crankshaft has no oil hole in the rear journal. *(Chevrolet Motor Division of General Motors Corporation)*

The replacement procedure varies with different constructions. On some engines using a split-type oil seal, the crankshaft must be removed. A special oil-seal compressor or installer is then used to insert the new seal in the cylinder-block bearing. The seal should then be trimmed flush with the block, as shown in Fig. 39-7. The oil seal in the cap can be replaced by removing the cap, installing the oil seal, and trimming it flush. On other engines, it is not necessary to remove the crankshaft. Removal of the flywheel permits access to the upper oil-seal retainer. The retainer cap screws can then be removed along with the retainer for oil-seal replacement. Some engines use a one-piece rubber-type oil seal. It can be pulled from around the crankshaft with a pair of pliers. Then a new oil seal can be worked into place. It should be coated with cup grease (except on the ends, since this would prevent the ends from meeting tightly). Then, one end of the seal should be forced up into the slot on one side until it is at the top of the bearing. Next, the other end can be forced up into the slot on the opposite side, so that the ends meet at the top of the bearing.

● 39-10 Removing the Crankshaft

Such parts as the oil pan, timing-gear or timing-chain cover, crankshaft timing gear or sprocket, interfering oil lines, and oil pump must be removed before the crankshaft can be taken off. Also, on some engines, the flywheel must be detached from the crankshaft. With other parts off, the bearing caps are removed to release the crankshaft.

CAUTION: The crankshaft is heavy. Support it adequately as you remove the bearing caps.

For a complete engine overhaul, the cylinder head and piston-and-rod assemblies have to be removed. However, if only the crankshaft is coming out, the piston-and-rod assemblies need not be removed. Instead, they can be detached from the crankshaft and pushed up out of the way. Be very careful not to push them up too far. If you do, the top piston ring may move up beyond the cylinder block. In this case, it can catch on the top of the block. You will then be unable to pull the piston-and-rod assembly back down. You'll have to take the head off so a ring compressor can be used to get the ring back down into the block again.

● 39-11 Servicing the Crankshaft

Check the crankshaft for alignment and for main-journal and crankpin wear. If the crankshaft is out of line, a new or reground crankshaft should be used. If journal or crankpin taper or out of round exceed safe limits, or if they are rough, scratched, pitted, or otherwise damaged, they must be ground. Then new undersize bearings must be installed.

● 39-12 Finishing Main and Crank Journals

A special grinder, or lathe, is required to service main and crank journals. This is a job that is done in automotive machine shops specializing in crankshaft service, and in engine-rebuilding shops.

Cylinder Blocks

● 39-13 Cylinder Wear

The piston and ring movement, the high temperatures and pressures of combustion, the washing action of gasoline entering the cylinder—all these tend to cause cylinder-wall wear. At the start of the power stroke, pressures are the greatest. The compression rings are forced with the greatest pressure against the cylinder wall. Also, at the same time, the temperatures are highest. The oil film is therefore least effective in protecting the cylinder walls. Thus, the most wear takes place at the top of the cylinder. As the piston moves down on the power stroke, the combustion pressure and temperature decrease. Thus, less wear takes place. The cylinder thus wears irregularly, as shown in Fig. 39-8.

The cylinder also tends to wear somewhat oval-shaped. This is due to the side thrust of the piston as it moves down in the cylinder on the power stroke. The side thrust results from the swing, from vertical, of the connecting rod. Another factor is the washing action of the gasoline. At times the air-fuel mixture is not perfectly blended. Small droplets of gasoline, still unvaporized, enter the cylinder. They strike the cylinder wall (at a point opposite the intake valve) and wash away the oil film. Therefore, this area wears somewhat more rapidly.

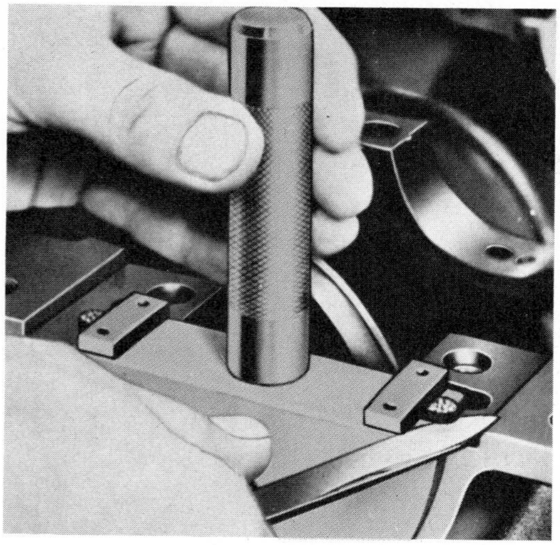

Fig. 39-7 Using a special tool to install a rear main-bearing oil seal in a cylinder block. (*Ford Motor Company*)

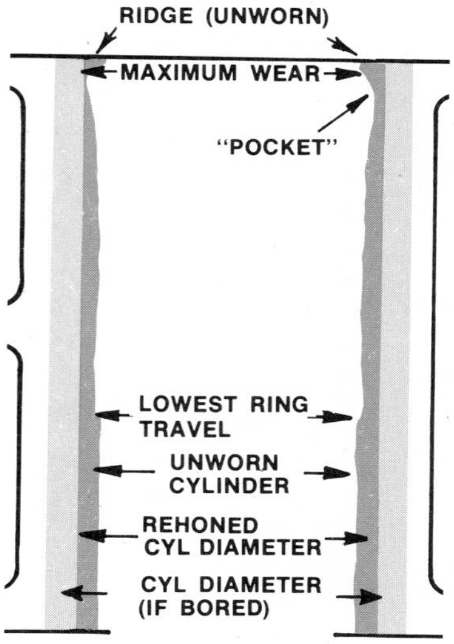

Fig. 39-8 Taper wear of engine cylinder (shown exaggerated). Maximum wear is at the top, just under the ring ridge. Honing the cylinder removes less material than boring, as indicated. Material to be removed by honing is shown solid. Material to be removed by boring is shown both solid and shaded. (*Sunnen Products Company*)

● 39-14 Cleaning and Inspecting the Cylinder Block

As a first step, make a visual inspection of the block. Major damage, resulting from a main bearing spinning or a broken connecting rod going through the block, means discarding the block. If the engine has overheated and there are cracks in the cylinder walls, discard the block.

Check the cylinder bores for cracks, grooves, scratches, or discoloration. Check for cracks across the top of the block between cylinders, between bolt holes, on the outside of the block, and in the main-bearing-bore webs. (See ● 39-21 on block repair.) If everything looks okay, measure the cylinder bores with a micrometer. If the bores are not too badly worn, and can be honed or rebored within specified limits, then clean and service the block. As a rule, cylinder block service is a machine shop job.

● 39-15 Bearing-Bore Check

Very uneven bearing wear, with some bearings wearing much more than others, may mean out-of-round bearing bores or a warped block. If the bores are out of round or out of line they must be line-bored to restore roundness.

● 39-16 Cylinder Service

Up to certain limits, cylinders may wear tapered or out of round and not require refinishing. As mentioned in ● 38-25, drastic replacement rings control compression and oil in cylinders with some taper and out-of-round wear. But when wear goes beyond a certain point, even the severest rings cannot hold compression and control oil. Loss of compression, high oil consumption, poor performance, and heavy carbon accumulations in the cylinders will result. Then the only way to get the engine back into good operating condition is to refinish the cylinders. New pistons (or resized pistons) and new rings must be installed at the same time.

● 39-17 Checking Cylinder Walls

Wipe walls, and examine them for scores and spotty wear (which shows up as dark, unpolished spots). Hold a light at the opposite end of the cylinder so you can see the walls better. Scores or spots mean the walls must be refinished. Even drastic rings cannot give satisfactory performance on such walls.

Next, measure the cylinders for wear, taper, and out-of-roundness (see Fig. 39-9). This can be done with an inside micrometer, with a telescope gauge and an outside micrometer, or with a dial indicator. The dial indicator is moved up and down in the cylinder and rotated at various positions to detect wear.

● 39-18 Refinishing Cylinders

As a first step, the block should be cleaned. A decision must be made on whether the cylinders are to be honed or bored. This decision depends on the amount of cylinder wear. Figure 39-8 shows the amount of metal removed by the hone and by boring. The hone (Figs. 39-10 and 39-11) uses a set of abrasive stones which are turned in the cylinder. The boring machine (Fig. 39-12) uses a revolving cutting tool. Where cylinder wear is not too great, only honing is necessary. But if wear has gone so far that much material must be removed, then honing will not do the job. The cylinder must be rebored and oversize pistons installed.

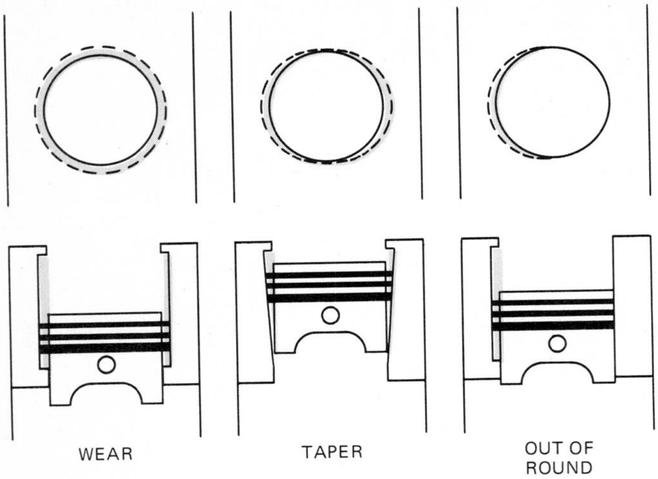

Fig. 39-9 Cylinder wear, taper, and out-of-round.

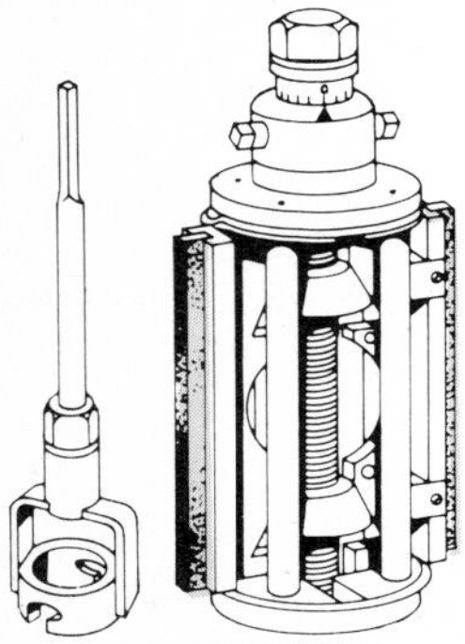

Fig. 39-10 Cylinder hone.

Fig. 39-11 Cylinder hone in place in engine cylinder. In operation, the hone revolves in the cylinder. The abrasive stones in the hone remove material from the cylinder wall. *(Hall Manufacturing Company)*

39-19 Cleaning Cylinders

Cylinders must be cleaned thoroughly after the honing or boring operation. Even slight traces of grit or dust on the cylinder walls may cause rapid ring and wall wear and early engine failure. As a first step, some engine manufacturers recommend wiping down the cylinder walls with very fine crocus cloth. This loosens embedded grit and knocks off "fuzz" left by the honing stones or cutting tool. Then use a stiff brush and hot soapy water to wash down the walls. It is absolutely essential to clean the walls of all abrasive material. If not removed, such material causes rapid wear of pistons, rings, and bearings.

After washing the walls, swab them several times with a cloth dampened with light engine oil. Wipe off the oil each time with a clean, dry cloth. At the end of the cleaning job, the cleaning cloth should come away from the walls showing no trace of dirt.

Gasoline and kerosene will not remove all the grit from cylinder walls. They are not recommended for cleaning grit or dust off cylinder walls.

39-20 Replacing Cylinder Sleeves

There are two types of cylinder sleeves, wet and dry. The wet sleeve is sealed to the block at the top and bottom. It is in direct contact with the coolant. The dry type is pressed into the cylinder. It is in contact with the cylinder wall from top to bottom.

Fig. 39-12 Cylinder boring bar. The cutting tool is carried in a rotating bar that feeds down into the cylinder as it rotates. This causes the rotating tool to remove material from the cylinder wall. *(Rottler Boring Bar Company)*

Cracked blocks, scored cylinders, cylinders worn so badly that they must be rebored to an excessively large oversize—all these can often be repaired by the installation of cylinder sleeves. As a first step, the cylinders are bored oversize to take the sleeves. Then the sleeves are pressed into place.

Figure 39-13 shows an operator using a pneumatic hammer to drive sleeves into place. The cylinder block

Fig. 39-13 Installing sleeves in a previously prepared V-8 engine block. (*Automotive Rebuilders, Inc.*)

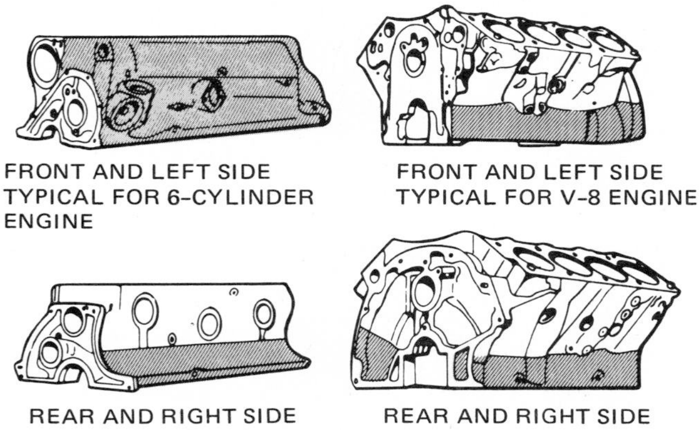

Fig. 39-14 Areas of cylinder block that can be repaired with epoxy. (*Ford Motor Company*)

has been prepared, and the sleeves have been positioned on the cylinder block. Then the pneumatic hammer is used. It hammers the sleeves down into place in the cylinder block. The hammer uses compressed air to operate the driving head. The sleeves are then finished to the proper size to take a standard piston and rings.

39-21 Repairing Cylinder-Block Cracks or Porosity

Sometimes, a block is in good condition except for some cracks or sand holes (left in the block during casting). It then may be worthwhile to repair it. Areas not subject to temperatures of more than 500°F [260°C] or pressure (from coolant, oil, or cylinder) can be repaired with a metallic plastic or epoxy. Permissible repair areas for one manufacturer's engines are shown in Fig. 39-14.

39-22 Expansion-Core Plugs

You might have to remove an expansion plug from the block (because of coolant leakage, for example). To do this, put the pointed end of a pry bar against the center

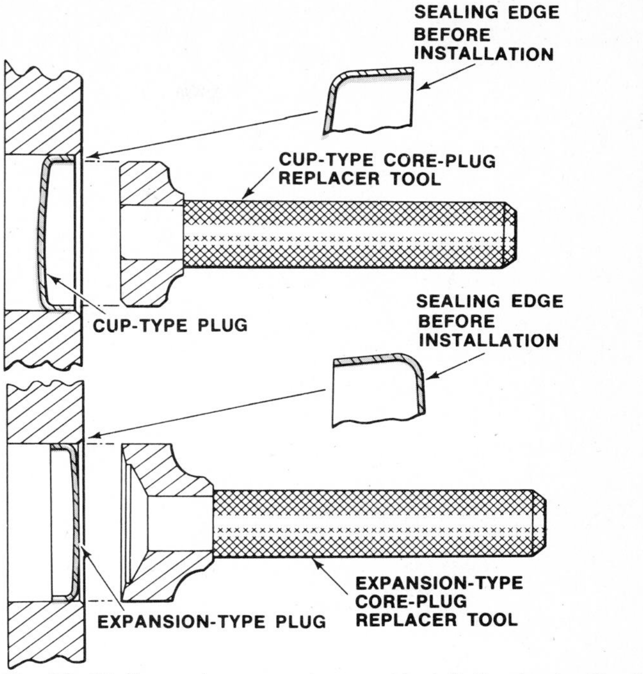

Fig. 39-15 Expansion-core plugs and installation tools. (*Ford Motor Company*)

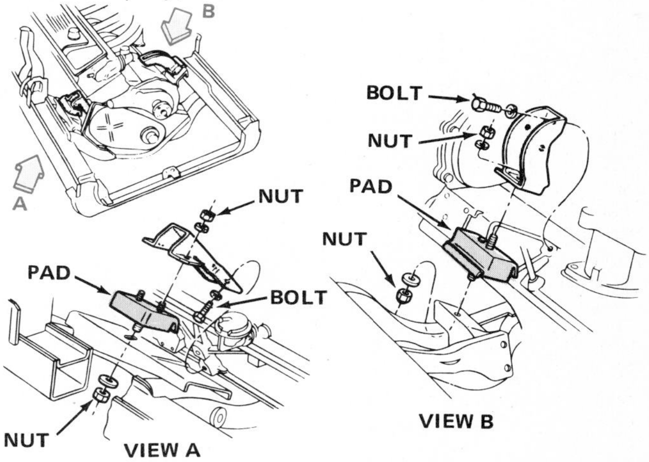

Fig. 39-16 Flexible mounts at front of engine. (*Chrysler Corporation*)

of the plug. Tap the end of the bar with a hammer until the point goes through the plug. Then press the pry bar to one side to pop the plug out. Another method is to drill a small hole in the center of the plug and then pry the plug out. Figure 39-15 shows core plugs and installation tools.

39-23 Replacing Engine Mounts

Damaged engine mounts should be replaced without delay. Figures 39-16 and 39-17 show one type of flexible engine mount. To replace mounts, support the engine with a wood block under the oil pan. Raise the engine slightly to relieve the weight on the mount. Then remove the mount and replace it with the new parts. If self-locking bolts or nuts are used, do not reuse the old nuts or bolts. Instead, use new ones.

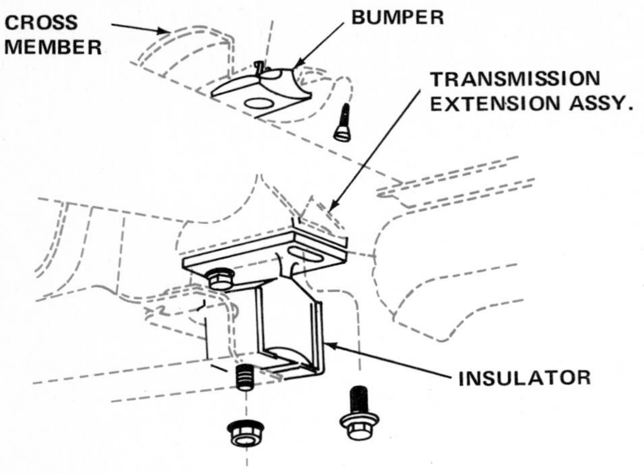

Fig. 39-17 Rear support for engine. (*Chrysler Corporation*)

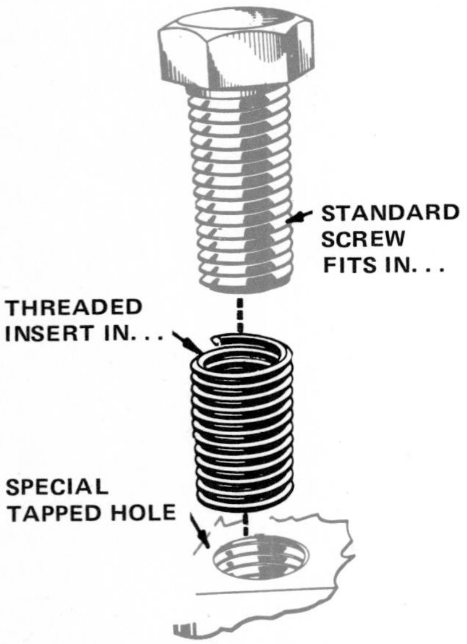

Fig. 39-18 Heli-Coil installation. (*Chrysler Corporation*)

● 39-24 Threaded Inserts

Damaged or worn threads in the block or head can often be repaired with a threaded insert. One such is the Heli-Coil (Fig. 39-18). First, drill out the worn threads. Tap the hole with the special Heli-Coil tap to make new threads. Then screw a Heli-Coil insert into the new threads, to bring the hole back to its original thread size. The original bolt can then be used in the hole.

REVIEW QUESTIONS

1. Is it safe to replace precision-insert bearings without removing the crankshaft if bearing wear is uneven? Explain your answer.
2. Explain two ways to check main journals on a crankshaft without removing the crankshaft.
3. Explain how to check main-bearing fit with Plastigage. With shim stock.
4. Explain how to install precision-insert main bearings without removing the crankshaft.
5. Explain how to remove the rear main-bearing half from block when journal does not have an oil hole.
6. Explain how to remove and replace a main-bearing oil seal.
7. Describe the procedure for removing the crankshaft.
8. Describe the procedure for refinishing main and crankpin journals.
9. Describe the procedure for checking and inspecting cylinder blocks.
10. When would you check bearing bores in the cylinder block for alignment? How is this done?
11. Describe the procedure for checking cylinder walls.
12. What is the quick way to measure the cylinder walls for taper, using a compression ring?
13. Describe the procedure for honing cylinders.
14. Describe the procedure for boring cylinders.
15. Why must the top of the cylinder block be smooth and free of nicks if the cylinders are to be bored?
16. Why must all main bearing caps be in place, with the cap bolts tightened to specifications, before cylinders are bored?
17. Describe the procedure for cleaning cylinders after honing or boring.
18. Describe the procedure for installing cylinder sleeves.
19. Explain how to repair cylinder-block cracks with epoxy.
20. Explain how to remove and replace both types of expansion-core plugs.

SELF PROJECTS

1. As you work on and watch engines in the service shop, try to collect main bearings showing the six basic types of bearing failure. Mount them on a board, with labels identifying each type of failure.
2. Study the instruction booklet for the cylinder boring bar. Write the procedure for using the boring bar.
3. Study the instruction booklet for the cylinder hone. Write the procedure.
4. Make a bearing roll-out tool from a cotter pin. (See Fig. 39-19.) Keep it in your toolbox.
5. Make a set of shim-stock bearing-clearance checkers by cutting $10\frac{1}{4}$- by $\frac{3}{4}$-inch [260.35- by 19.05-mm] pieces of brass shim stock. Be careful not to crinkle the edges when cutting the shims.

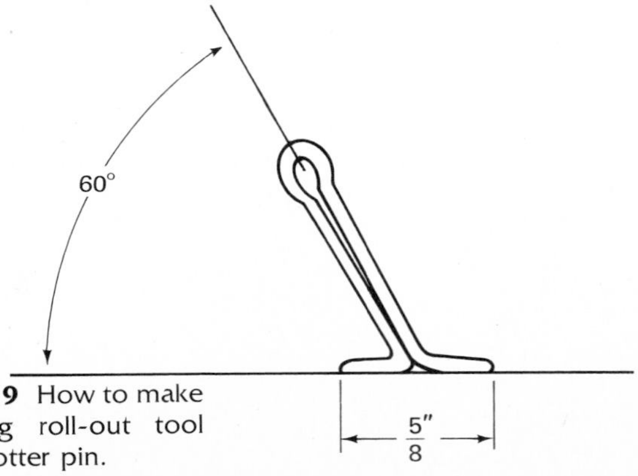

Fig. 39-19 How to make a bearing roll-out tool from a cotter pin.

CHAPTER 40
Diesel-Engine Service

After studying this chapter, you should be able to:

- **Discuss the special troubles diesel engines have and their causes.**
- **Discuss the special services diesel engines require beyond those gasoline engines need.**
- **Under instructor's supervision, perform diesel-engine service jobs.**

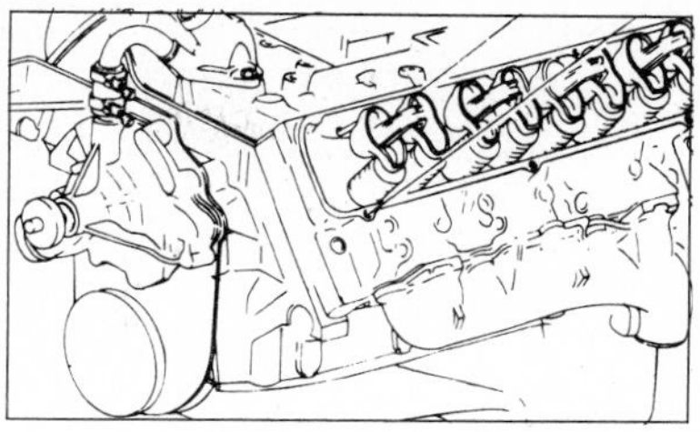

In this chapter, we look at the special problems and service needs of automotive-type diesel engines. Many service jobs are the same, or similar, for gasoline and diesel engines. For example, the valve trains and piston-connecting-rod assemblies are similar and serviced in a similar manner. The crankshaft and cylinder bores are also serviced in a similar way. We have already covered, in Chap. 18, the servicing of diesel-engine fuel-injection systems, including hydraulic lines and fittings, fuel filter, injector nozzles, and injector pump. We cover some of these again briefly in this chapter so we have more complete coverage of the diesel.

● 40-1 Diesel-Engine Troubles

Chapter 42 has a comprehensive trouble-diagnosis chart for automotive-type internal-combustion engines. Many of the troubles listed can happen to both gasoline and diesel engines. As examples, noisy lifters, worn rod or main bearings, troubles with valves, pistons, and rings are the same for both types of engines. Some troubles can occur only to diesels, as for example cold-weather starting trouble caused by the glow plugs not turning on.

In the chart that follows, we list the conditions that, in general, could occur only in automotive diesels.

● 40-2 Diesel-Engine Trouble Chart

The causes of such troubles as *engine will not turn over,* and *engine turns over slowly but does not start* are the same for both engine types. The diesel engine should turn over at least 100 rpm cold and 240 rpm hot when cranking.

DIESEL-ENGINE TROUBLE CHART

Complaint	Possible Cause	Check or Correction
1. Engine cranks normally but does not start	a. Incorrect starting procedure	Use correct procedure (see Chap. 17)
	b. Fuel solenoid or glow plugs inoperative	Use test lamp to check out circuits. Replace defective items
	c. Injector-pump timing off	Check, adjust timing
	d. Fuel system not delivering fuel	Check filter, fuel pump, injection pump, lines
	e. Loss of compression	Check compression

DIESEL-ENGINE TROUBLE CHART (Continued)

Complaint	Possible Cause	Check or Correction
2. Engine starts but stalls on idle	a. Slow idle incorrect	Adjust
	b. Fast-idle solenoid not working	Check solenoid for action, replace if defective
	c. Fuel system not working right	Check for restrictions, pump timing and operating condition, contaminated fuel, wrong fuel
	d. Loss of compression	Check compression
3. Engine starts, idles rough without abnormal noise or smoke	a. Slow idle incorrect	Adjust
	b. Fuel system not working right	Check components for restrictions, action, pump timing, dirty or wrong fuel
4. Engine starts, idles rough with noise and smoke	a. Pump timing off	Adjust
	b. Nozzles not working right	Loosen lines to nozzles, one at a time with engine running. If idle does not change, you have found a bad nozzle
5. Engine misses, one cylinder	a. Nozzle clogged	Clean
	b. Air in line	Bleed air out
	c. Pump trouble	Replace pump
6. Engine misfires above idle but idles okay	a. Plugged fuel filter	Replace filter
	b. Incorrect pump timing	Time correctly
	c. Wrong or dirty fuel	Flush system and use right fuel
7. Loss of power	a. Restricted air intake	Check air cleaner
	b. Restricted exhaust	Check and correct
	c. Plugged fuel filter	Replace
	d. Plugged vacuum vent in fuel-tank cap	Replace cap
	e. Restricted fuel-return line	Make correction
	f. Restricted fuel-injection system	Check lines, pump, nozzles, to find and eliminate cause
	g. Wrong or dirty fuel	Flush system, use right fuel
	h. External compression leaks	Check for leaks around nozzle seals, glow plugs. Tighten, replace seals
	i. Low compression	Check compression to find cause
8. Rap (like a rod-bearing knock) from one or more cylinders	a. Air in fuel system	Check for air leaks and make corrections
	b. Nozzle or nozzles sticking open or opening at low pressure	Remove nozzles for repair or replacement
9. Combustion noise with black smoke	a. Timing off	Set pump timing
	b. Internal engine problem	Check compression, engine condition
	c. Injection-pump internal problems	Remove for repair
10. Engine runs with key off	a. Injection-pump solenoid not returning fuel valve to the off position. Check electrical system solenoid action	

● 40-3 Diesel-Engine Service

We now cover briefly the special procedures required on automotive diesel engines. As noted previously, many service procedures are the same for both gasoline and diesel engines. Diesel-engine fuel-injection systems, and servicing these systems are covered in Chaps. 17 and 18.

The engine we use in the discussion of service procedures is the Oldsmobile General Motors V-8 diesel, one of the newer and more advanced diesel engines. Before working on any engine, always have the factory shop manual handy so you can follow the directions and meet the specifications.

● 40-4 Timing the Injection Pump

The marks on the top of the injection-pump adapter and the flange of the injection pump must align (Fig. 40-1).

This corrects the timing of the fuel delivery to the nozzles. The adjustment is made with the engine not running. To make the adjustment, loosen the three pump retaining nuts and align the marks. Torque nuts to 35 pound-feet [4.8 kg-m]. Use a $\frac{3}{4}$-inch-end wrench on the boss at the front of the pump to help when you rotate the pump into alignment.

40-5 Exhaust-Manifold Removal and Replacement

Figure 40-2 shows how the exhaust manifolds are attached. There is nothing special about removing and replacing them except that bolt locks are used to prevent the bolts from loosening. The diesel engine is subjected to greater vibrational stress. The bolts therefore must be secure.

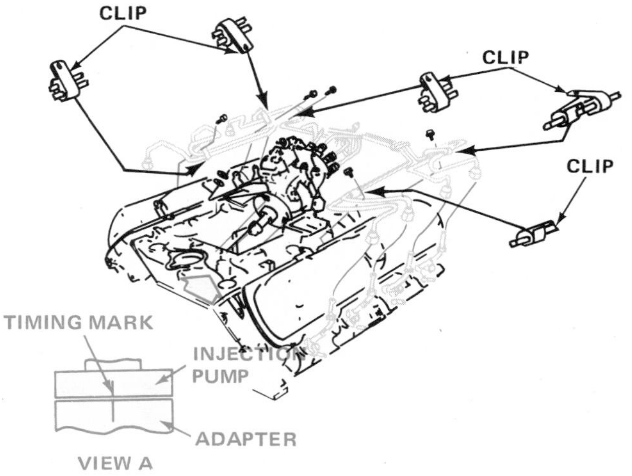

Fig. 40-1 Timing marks and injection-pump lines. *(Oldsmobile Division of General Motors Corporation)*

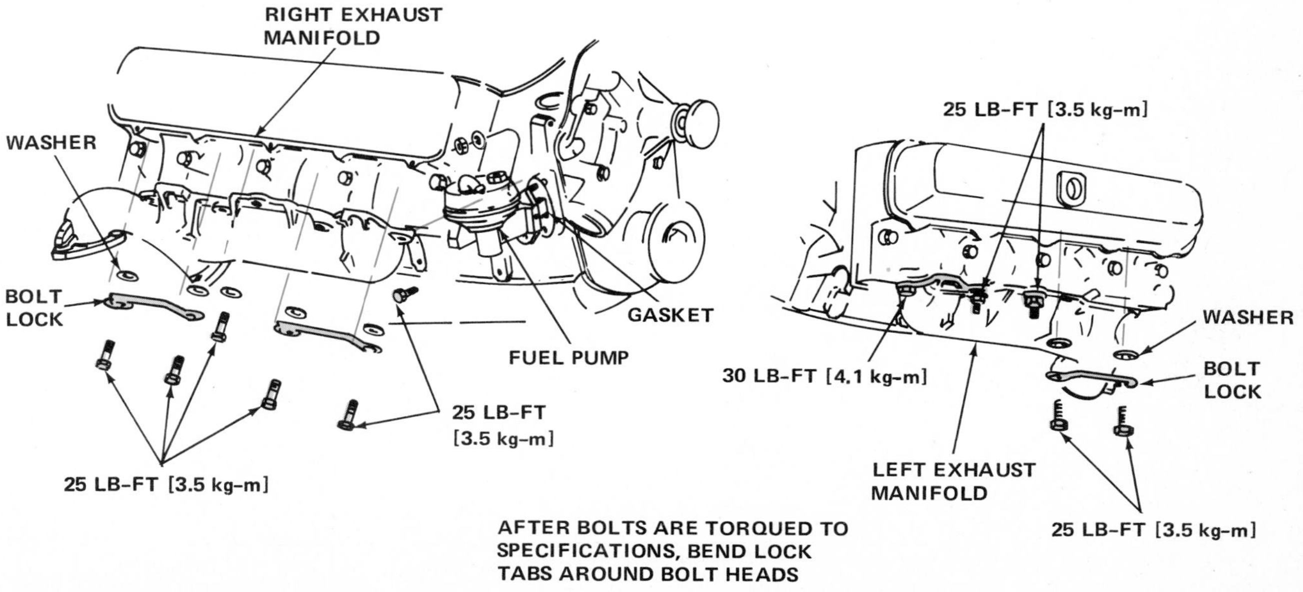

Fig. 40-2 Exhaust manifold attaching bolts and bolt locks. *(Oldsmobile Division of General Motors Corporation)*

40-6 Linkage Adjustments

Timing must be correct before linkage is adjusted. The following may require adjustment:

1. Throttle rod (Fig. 40-3).
2. Transmission TV or detent cable (Fig. 40-4).
3. Transmission vacuum valve.
4. Slow-idle speed. This requires a magnetic tachometer inserted in the tach hole (Fig. 40-5). The slow-idle adjustment screw is shown in Fig. 40-6.
5. Fast-idle solenoid. The fast-idle speed is controlled by the plunger in the fast-idle solenoid. It should be checked with the air conditioner on and the compressor wires disconnected.
6. Cruise-control servo relay rod.

Details are in the manufacturer's shop manual.

40-7 Checking Injection-Pump-Housing Fuel Pressure

Remove the air cleaner and crossover (Fig. 40-7). Cover the manifold openings with screened covers. From the pump (Fig. 40-6) remove the pressure tap plug. Attach low-pressure gauge by screwing adapter into the tap-plug hole. Install the magnetic-pickup tachometer (Fig. 40-5). Check pressure at 1,000 rpm (transmission in PARK). It should be 8 to 12 psi [0.5 to 0.7 kg/cm^2] with not more than 2 psi [0.1 kg/cm^2] fluctuation. If incorrect, remove the pump for repair.

Reinstall air crossover after removing the screened covers. Install the air cleaner.

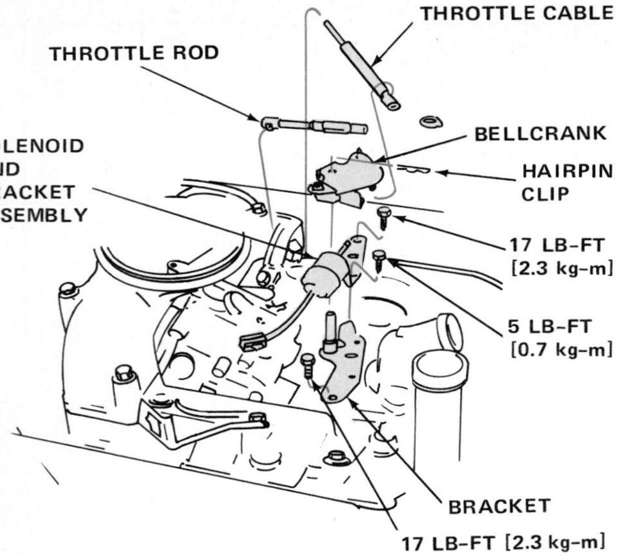

Fig. 40-3 Throttle linkage. (*Oldsmobile Division of General Motors Corporation*)

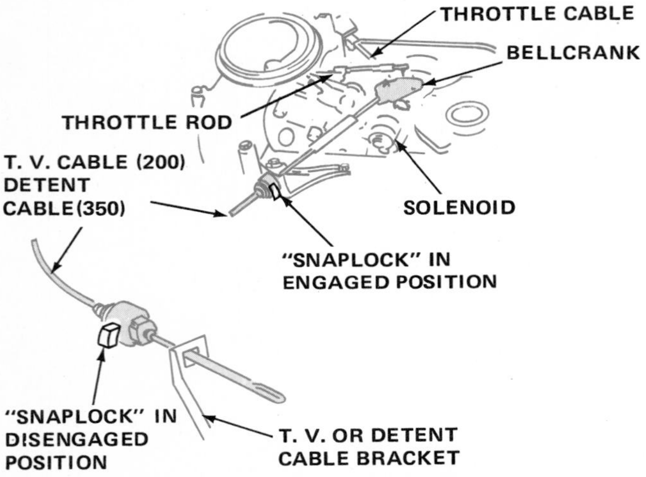

Fig. 40-4 Transmission TV cable adjustment. (*Oldsmobile Division of General Motors Corporation*)

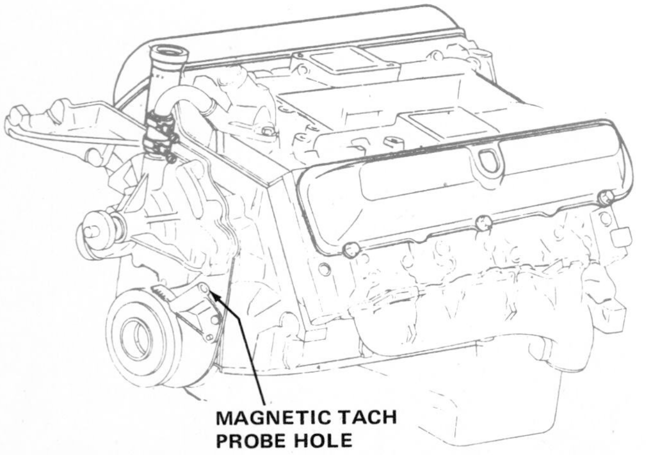

Fig. 40-5 Probe hole into which the magnetic tachometer probe is inserted to check engine rpm. (*Oldsmobile Division of General Motors Corporation*)

● 40-8 Injection-Pump Fuel Lines

When these lines (Fig. 40-1) are removed, the lines, nozzles, and pump fittings must be capped to keep dirt out.

To remove the lines, remove air cleaner, filters, and pipes from valve covers, and the air crossover. Take off the line clamps and disconnect the fuel lines, capping open lines, nozzles, and pipe fittings. (See ● 18-1.)

On reassembly, do not bend or twist the lines. Torque pump ends of lines to 35 pound-feet [4.8 kg-m] and the nozzle ends to 25 pound-feet [3.4 kg-m].

● 40-9 Removing Injection Pump

After removing the fuel lines (● 40-8), disconnect the throttle rod and return springs. Take off the bellcrank,

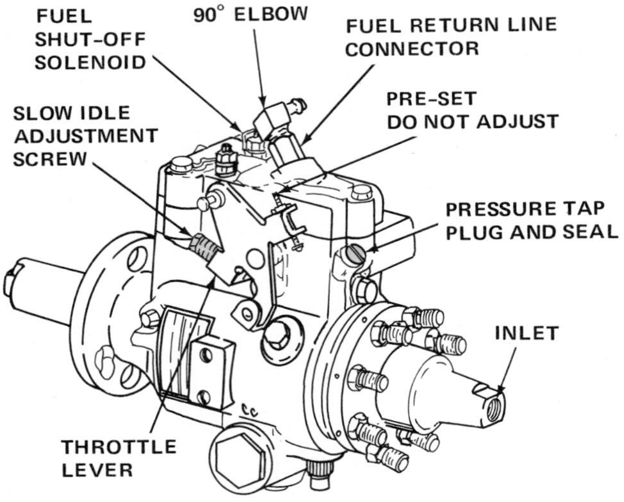

Fig. 40-6 Injection pump, showing locations of the slow-idle adjustment screw and the pressure tap plug. (*Oldsmobile Division of General Motors Corporation*)

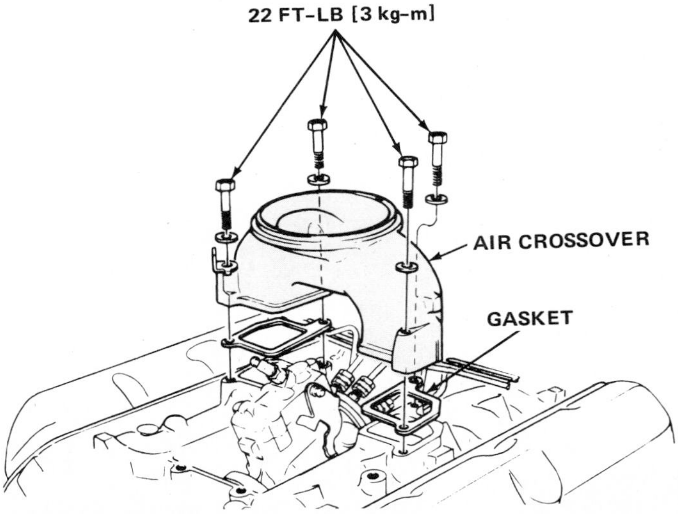

Fig. 40-7 Removing air crossover from intake manifolds. (*Oldsmobile Division of General Motors Corporation*)

throttle, and TV cables. Remove fuel filter. Disconnect the fuel line at the fuel pump and the fuel-return line from the injection pump.

With everything disconnected from the pump, remove the three nuts holding the pump and lift pump off the engine. Cap all open lines and nozzles. Section 18-4 covers the procedure in detail.

On reinstallation, remove the protective caps. Line up the offset tang on the pump drive shaft with the pump-driven gear (Fig. 18-5) and install pump. Adjust timing (● 40-4). Reconnect lines and linkages and adjust linkages as necessary.

● 40-10 Valve Cover

With everything out of the way (injection pump, lines, air crossover) remove the attaching screws and lift the valve cover off (Fig. 40-8). After cleaning the valve compartment and cover, run a bead of sealer around the edge of the valve cover. Replace the cover and tighten the screws. Replace all parts that were removed.

● 40-11 Intake Manifold

Figure 40-9 shows the intake manifold with related parts, raised above the cylinder block. Note that to take off the manifold the radiator must be drained and everything else above the manifold (injection pump, air crossover, vacuum pump, all interfering lines, etc.) must be removed. The air-conditioner lines at the compressor are flexible and can be moved to one side without disconnecting them from the compressor.

Do not bend the injection-pump lines when removing or replacing them.

Figure 40-10 shows the intake-manifold torque sequence or the order in which the bolts should be tightened.

● 40-12 Valves, Valve Trains, and Cylinder Heads

Figure 40-11 shows a cylinder head removed and the rocker-arm attachment method. Figure 40-12 shows the valves, springs, and rocker arms for one cylinder. Note that valve rotators are used.

To take off a cylinder head, you must remove the intake manifold (● 40-11). This means that you must also remove everything above the manifold, as already explained.

When reinstalling the cylinder head, make sure the gasket surfaces are in good condition and clean. Use care in handling the gasket so you do not damage the finish. Dip all head bolts in oil and torque to 130 pound-feet [18 kg-m].

Valves, valve seats, springs, push rods, and valve lifters are all checked and serviced as for the same parts in gasoline engines (Chap. 37).

● 40-13 Precombustion Chamber

The precombustion chambers (or pre-chambers as Oldsmobile calls them) can be removed if necessary (Fig. 40-13). First remove the glow plug and nozzle. Then tap out the pre-chamber with a small blunt drift. The pre-chamber can be installed in only one position (Fig. 40-14). Tap it into place with a small soft-head mallet. Install glow plug and nozzle.

● 40-14 Measuring Valve-Stem Height

Removing, servicing, and installing valves is the same as for gasoline engines (Chap. 37). Also, valve seats are serviced in the same way. The valve face and seat are finished to provide an interference angle (Fig. 37-6). Figures 40-15 and 40-16 show a gauge being used to check the installed height of the valve stem and the valve rotator. If the height is excessive, grind the valve tip. If grinding the tip reduces the distance between the tip and rotator excessively, replace the valve.

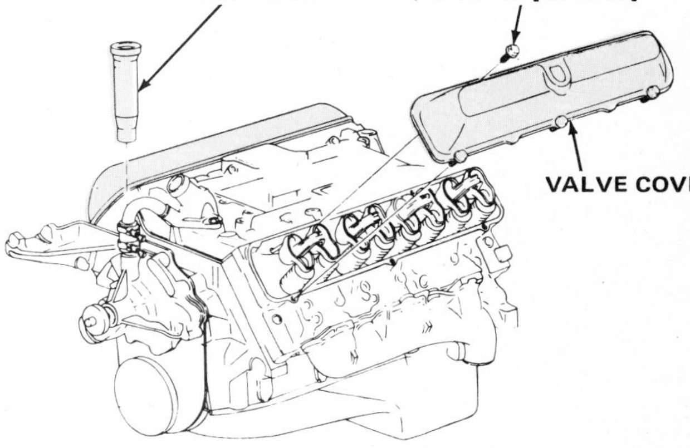

Fig. 40-8 Valve cover and oil-filler tube removed. (*Oldsmobile Division of General Motors Corporation*)

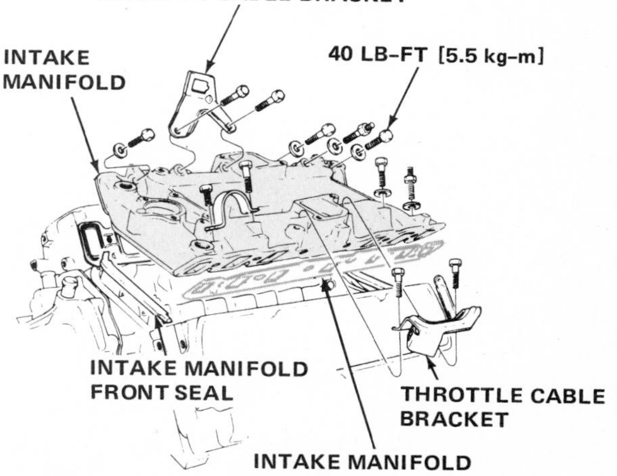

Fig. 40-9 Intake manifold and gasket. (*Oldsmobile Division of General Motors Corporation*)

40-15 Rod Bearings and Side Clearance

Connecting rods are checked and replaced as for gasoline engines (Chap. 38). It is especially important for the crankpins to be in good condition because of the high bearing pressure. This is due to the high compression ratio of the engine and the high combustion pressures developed in the cylinders.

Figure 40-17 shows the procedures for checking rod side clearance. If it is not correct, the probable cause is a bent rod. Bent rods must be discarded. They cannot be straightened satisfactorily.

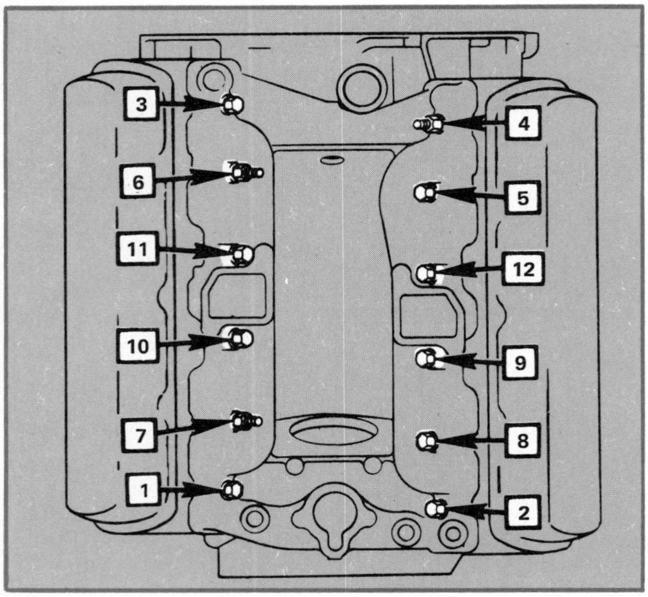

Fig. 40-10 Intake manifold attaching-bolt torque sequence. *(Oldsmobile Division of General Motors Corporation)*

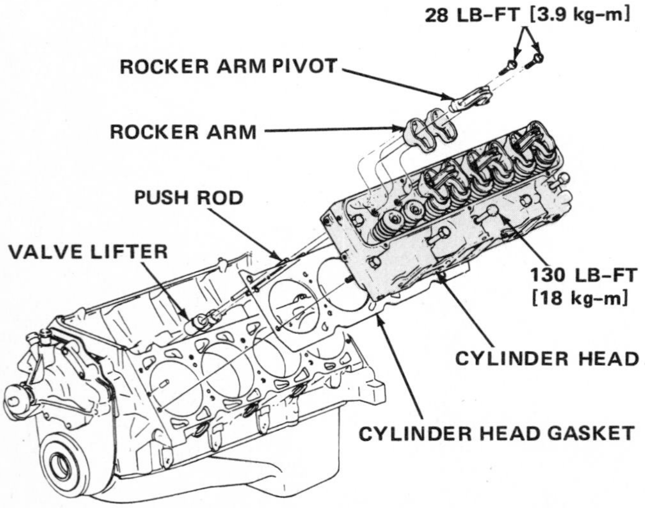

Fig. 40-11 Cylinder head removed from block with two rocker arms detached from head. *(Oldsmobile Division of General Motors Corporation)*

40-16 Pistons and Rings

The basic piston and ring servicing and replacement procedures used in gasoline engines also apply to diesel engines. When installing the pistons, correctly locate the larger valve depressions (Fig. 40-18). The large valve depression on cylinders 1, 2, 3, and 4 goes to the front. On cylinders 5, 6, 7, and 8, it goes to the back.

40-17 Camshaft, Injection-Pump Drive, and Driven Gears

The injection pump is driven by a pair of gears from the camshaft. See Fig. 17-8. This compares with the way the ignition distributor is driven. To remove the camshaft from the diesel engine, you take off the same parts as

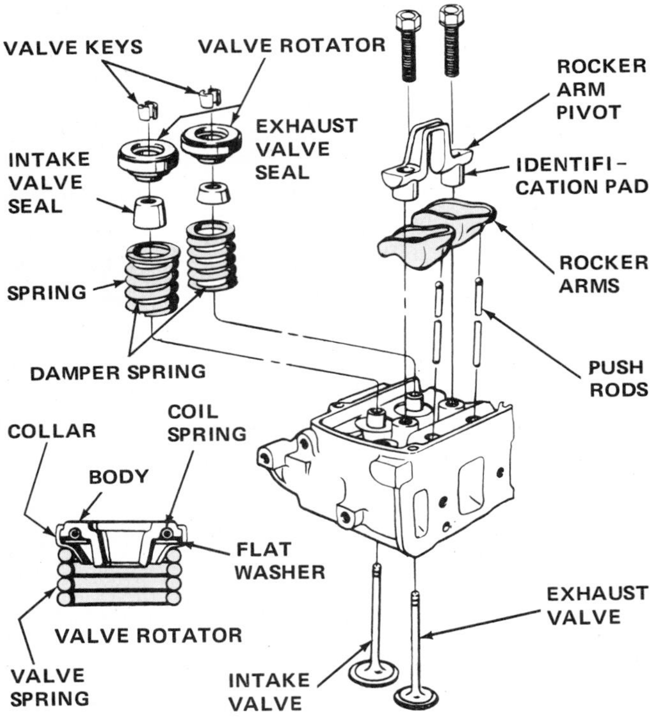

Fig. 40-12 Valve springs and rocker arms detached from head. *(Oldsmobile Division of General Motors Corporation)*

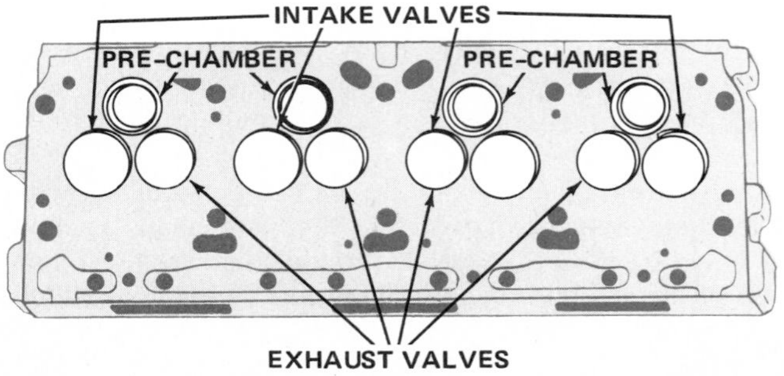

Fig. 40-13 Valve and pre-chamber locations. *(Oldsmobile Division of General Motors Corporation)*

with the gasoline engine. Figure 40-19 shows the engine front cover and other parts removed so you see the timing chain and camshaft and crankshaft sprockets.

During reassembly, line up the timing marks on the two sprockets. Also, line up "0" marks on the pump drive and driven gears. Then time the injection pump.

● 40-18 Crankshaft and Main Bearings

The service and replacement procedure for the crankshaft and main bearings is the same as for gasoline engines (Chap. 39). Before the bearing-cap bolts are tightened, the flanges of the thrust bearings are aligned with a block of wood (Fig. 40-20). The block of wood is used to bump the shaft in each direction. This centers the thrust bearing. Then tighten all bearing-cap bolts.

● 40-19 Complete Instructions

Even though you have learned the complete gasoline-engine servicing procedure, you have still more to learn

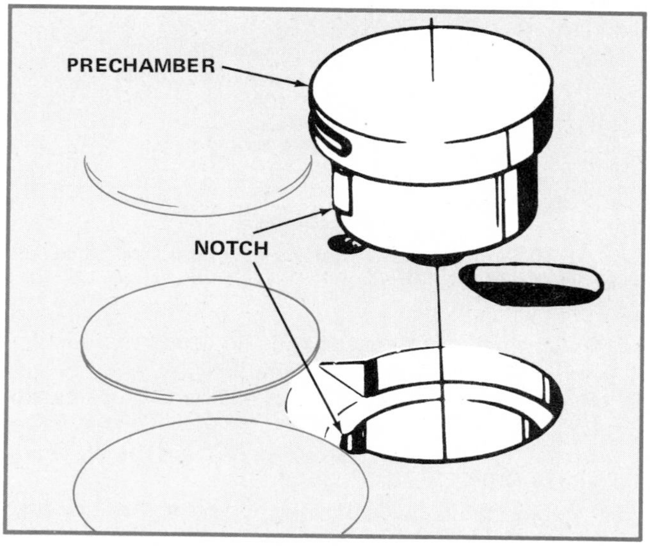

Fig. 40-14 Installation of pre-chamber in cylinder head. (*Oldsmobile Division of General Motors Corporation*)

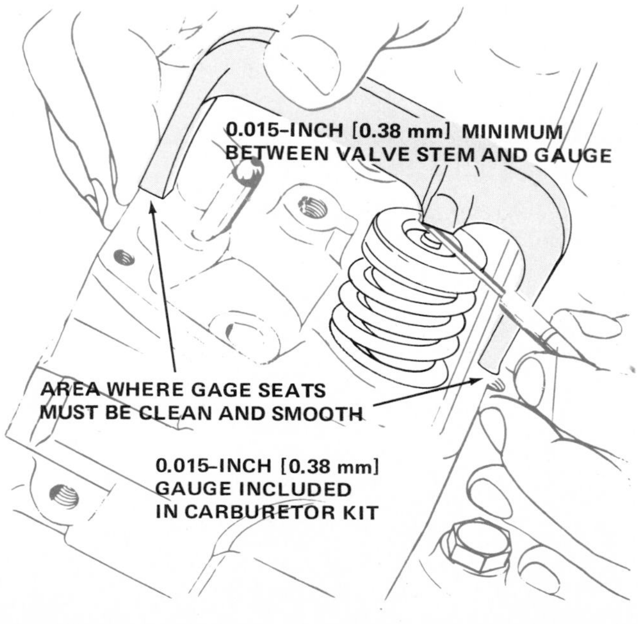

Fig. 40-15 Measuring valve-stem height with special gauge. (*Oldsmobile Division of General Motors Corporation*)

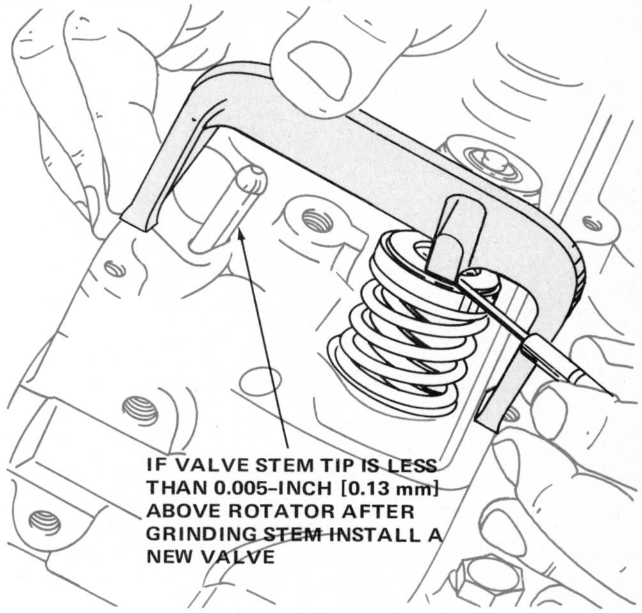

Fig. 40-16 Measuring distance valve stem rises above rotor. (*Oldsmobile Division of General Motors Corporation*)

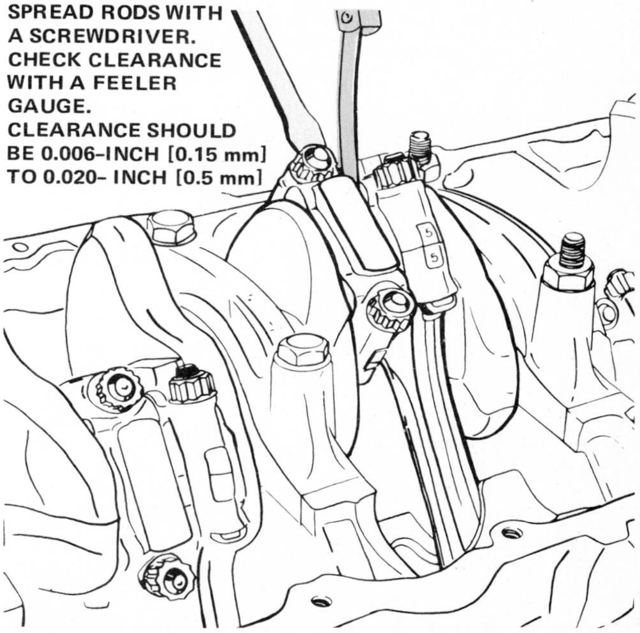

Fig. 40-17 Checking side clearance between connecting rods. (*Oldsmobile Division of General Motors Corporation*)

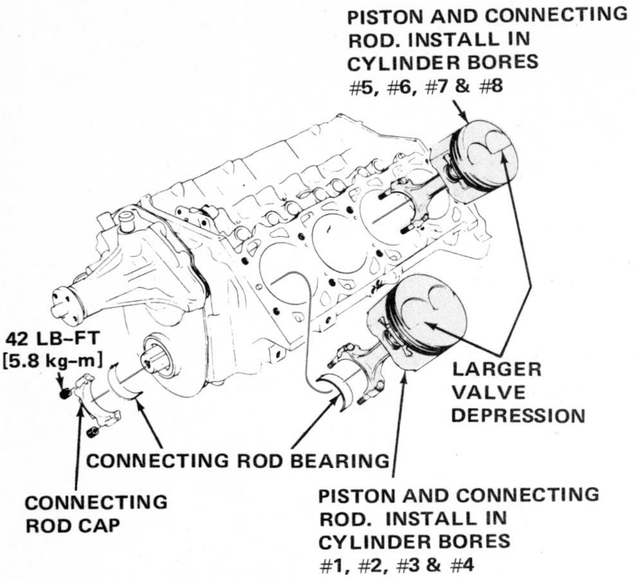

Fig. 40-18 Proper positioning of pistons in cylinder block. *(Oldsmobile Division of General Motors Corporation)*

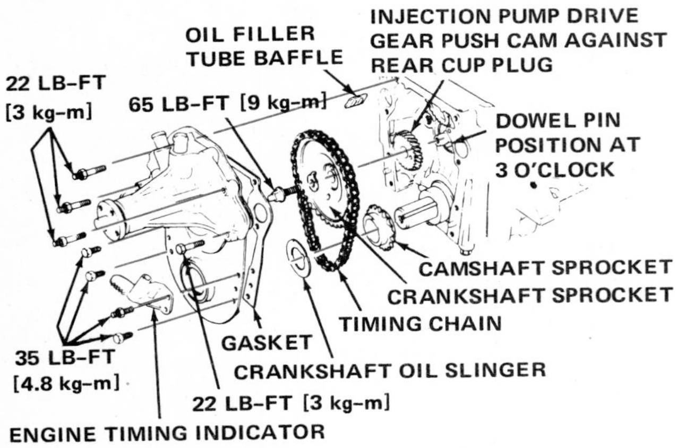

Fig. 40-19 Relationship of front cover, sprockets and timing chain, and injection-pump drive gear on camshaft. *(Oldsmobile Division of General Motors Corporation)*

on diesels. We have covered the highlights of the special procedures required on diesels in this chapter. However, before you tackle a diesel service job, you should have the shop manual issued by the diesel-engine manufacturer before you. And you should follow the specific instructions in that manual. Working on diesel engines is no harder than working on gasoline engines. It is just a little different for some jobs.

REVIEW QUESTIONS

1. What is the correct procedure for starting a diesel with glow plugs when the temperature is around zero?
2. Name four conditions in the diesel engine that could prevent starting.
3. Name eight conditions that could cause loss of power.

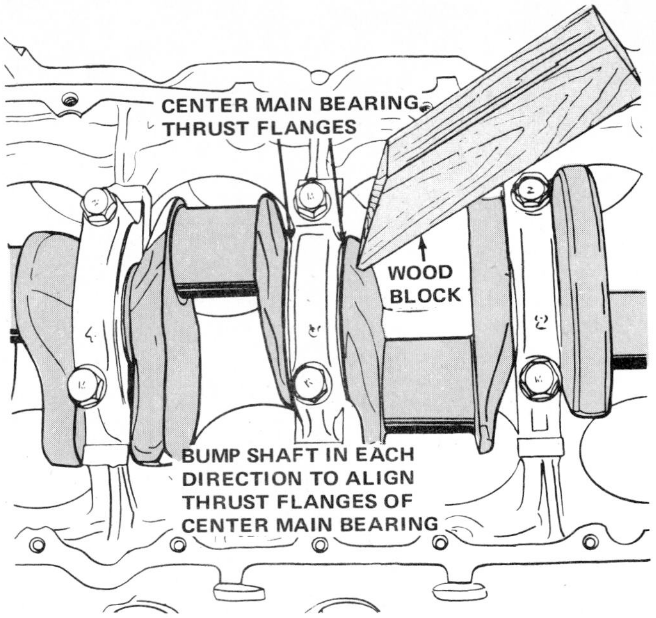

Fig. 40-20 Aligning thrust bearings by bumping crankshaft with wood block. *(Oldsmobile Division of General Motors Corporation)*

4. Name three conditions that could cause combustion noise with black smoke in the exhaust.
5. Marks on the camshaft and crankshaft sprockets are lined up during assembly to assure correct valve timing. How are the injector pump and driven gears aligned?
6. Explain how to check the injection-pump-housing fuel pressure.
7. Describe the procedure of removing the fuel lines and injection pump.
8. Explain how to remove the intake manifold.
9. After a valve refacing job, the installed height of the valve is excessive. What should be done?
10. If the rod side clearance is excessive, what should be done?

SELF PROJECTS

Get some 3 x 5 inch cards. On one side of a card, write a diesel-engine trouble, such as normal cranking but not starting. Then on the other side write the possible causes as listed in the chapter. Go through the trouble chart and prepare a card on each complaint. Carry the cards around with you. Every chance you get, take them out and study them. Soon, you will be able to relate complaints and causes without having to puzzle over them. In Chap. 42, you will study the complete trouble chart on gasoline engines. When you come to this chart, you will already know the story on diesels. So you will find it easier going through the big chart on gasoline engines.

Whenever you run across articles in magazines on diesel engines, clip them and file them in your notebook. Study the detailed servicing procedures in factory shop manuals on diesel engines. Make notes of important points for filing in your notebook.

Engine Trouble Diagnosis and Tune-up

7

Part Seven of this book explains how to diagnose troubles in the engine. It describes engine testing instruments and how to use them, and how to perform a complete tune-up on an engine. Some of the testing instruments were discussed in previous chapters. We review them here, however, so that this part of the book tells the complete story of engine trouble diagnosis and tune-up. There are three chapters in Part Seven.

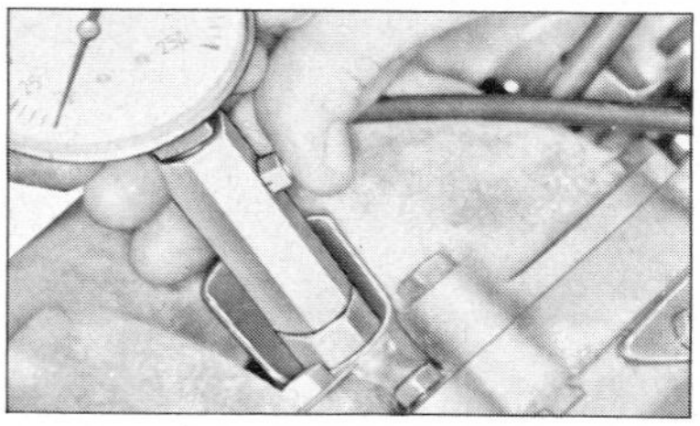

Chapter 41 Engine Testing Instruments

Chapter 42 Engine Trouble Diagnosis

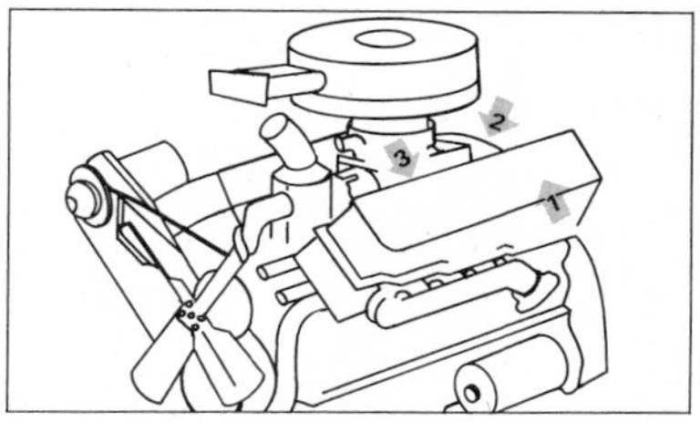

Chapter 43 Engine Tune-up

CHAPTER 41
Engine Testing Instruments

After studying this chapter, you should be able to:

- **List and describe the test instruments covered in the chapter.**
- **Explain how to use the test instruments.**
- **Under the supervision of an instructor, use the test instruments to make actual performance tests in the shop.**

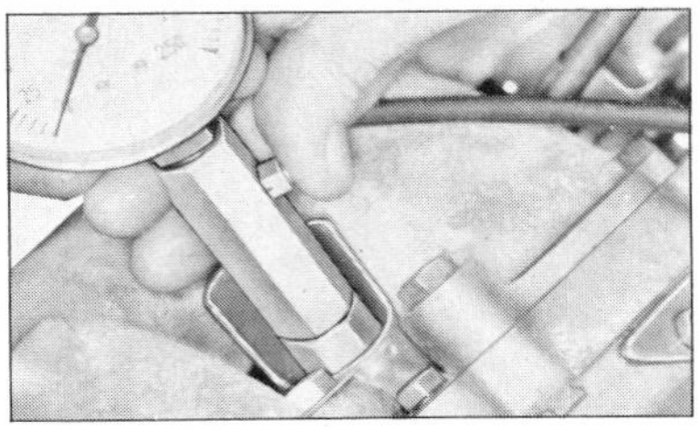

In this chapter we describe the testing instruments used by automotive technicians to check the engine and its systems. You will find that we review some of the test instruments covered in earlier chapters. We do this to put together, in this chapter, the complete story of engine testing instruments.

● 41-1 Engine Testing Instruments

The testing instruments we cover in this chapter are:

1. Tachometer, which measures engine speed in revolutions per minute (rpm)
2. Cylinder-compression tester, which measures the ability of the cylinders to hold compression
3. Cylinder-leakage tester, which finds places where there is compression leakage
4. Vacuum gauge to measure intake-manifold vacuum
5. Exhaust-gas analyzer, which measures the amount of pollutants in the exhaust gas
6. Ignition timing light, which is used to set the ignition timing and check the spark advance
7. Oscilloscope, which shows the overall operating condition of the ignition-system circuits
8. Chassis dynamometer, which checks the engine and its operating parts under operating conditions
9. Engine analyzer, which combines several testing instruments and makes several tests at once, or in sequence. Some are computerized so they are automatic in operation. Also, some have print-out capability and provide a printed record of the test results. See ● 43-4.

There are also instruments to test the battery, starting motor, charging system, and cooling system. There are other instruments to test ignition coils, condensers, spark plugs, distributor contact-point dwell, and distributor advance mechanisms.

● 41-2 Tachometer

The tachometer measures engine speed in revolutions per minute (rpm). It is a necessary test instrument because the idle speed must be adjusted to a specific rpm. Also, many tests must be made at specific engine speeds. Most tachometers are connected to the ignition system and operate electrically. Some are triggered by a magnet on the crankshaft which is sensed by a magnetic probe (Fig. 40-5).

The electric tachometer measures the number of times the primary circuit is interrupted each minute. It translates this into engine rpm. The tachometer selector knob can be turned to 4, 6, or 8, the number of cylinders in the engine being tested. Figure 41-1 shows a tachometer connected to an engine.

Many high-performance cars have tachometers mounted on the instrument panel. They tell the driver how fast the engine is turning. Then he or she can keep the rpms where the engine develops maximum torque. This lets the driver get the best performance from the engine. Many of these tachometers have a red line at the top rpm on the dial. The red line marks the danger point for engine speed. The driver should keep below this engine speed.

Some car tachometers are mechanical instead of electrical. They are driven off a gear on the ignition distributor shaft. They operate like the speedometer (● 32-17).

● 41-3 Cylinder-Compression Tester

The cylinder-compression tester measures the ability of the cylinders to hold compression. Pressure operates on a diaphragm in the tester. It causes the needle on the face of the tester to move around to indicate the pressure being applied. Figure 41-2 shows a compression tester being used to measure pressure in an engine cylinder.

To use the tester, first remove all the spark plugs. A recommended way to do this is to disconnect the wires and loosen the plugs one turn. Then reconnect the wires, and start the engine. Then, run the engine for a few moments at 1,000 rpm. Turn off the engine and remove the plugs. The combustion gases will blow out of the plug well any dirt that could fall into the cylinder when the spark plugs are removed. The gases also blow out of the combustion chamber any loosened carbon that was caked around the exposed threaded end of the plug. This prevents carbon and dirt from getting under a valve and holding the valve open during the compression test.

Next, screw the compression-tester fitting into the spark-plug hole of cylinder No. 1, as shown in Fig. 41-2. To protect the coil from high voltage, disconnect the primary lead from the negative terminal of the coil. This is the primary lead that goes to the distributor. Then, hold the throttle wide open, and operate the starting motor to crank the engine for several revolutions. The needle will move around to show the maximum compression pressure the cylinder is developing. Write down this figure. Test the other cylinders the same way.

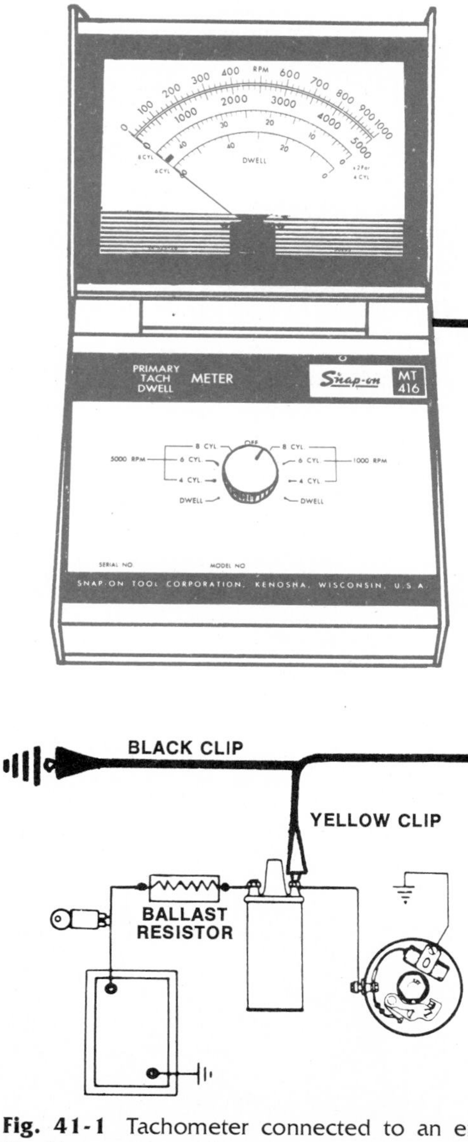

Fig. 41-1 Tachometer connected to an engine. (*Snap-on Tools*)

● 41-4 Diesel-Engine Compression Test

The compression test is different for diesel engines. First, remove the air cleaner and install a manifold cover. Disconnect the wire from the fuel-shutoff solenoid terminal of the injector pump (see Fig. 40-6). This prevents delivery of fuel during the test. Disconnect glow-plug wires and remove glow plugs. Screw the compression-tester fitting into the glow-plug hole of the cylinder to be checked. Then crank the engine for at least 12 crankshaft revolutions (six "puffs").

Check all cylinders the same way. The lowest compression reading should be not less than 70 percent of the highest. No cylinder should read less than 275 pounds. (Pounds per square inch or psi. In the metric measurement system, this would be 1,892 kPa.)

In addition to the pressure reached, note the following. If everything is normal, the compression builds up quickly and evenly. If there is leakage past the piston

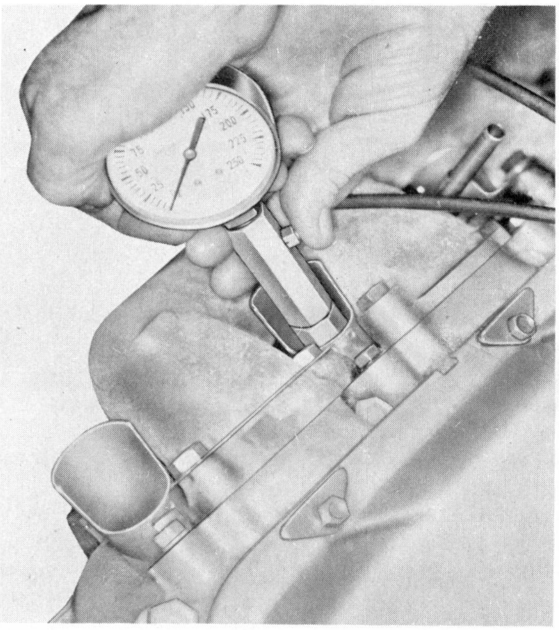

Fig. 41-2 Using cylinder-compression tester. (*Chevrolet Motor Division of General Motors Corporation*)

rings, the compression is low on the first strokes but will tend to build up toward normal with later strokes. However, it does not reach normal and the pressure is rapidly lost after cranking stops.

41-5 Results of the Compression Test

The manufacturer's specifications tell you what the compression pressure of the cylinders should be. If the test shows that the compression is low, there is leakage past the piston rings, valves, or cylinder-head gasket. To correct the trouble, you must remove the cylinder head and inspect the engine parts.

Before you do this, you can make one more test to pinpoint the trouble. Pour a small quantity of heavy oil into the cylinder through the spark-plug hole. Then retest the compression. If the pressure increases to a more normal figure, the low compression is due to leakage past the piston rings. Adding the oil helps seal the rings temporarily so they can hold the compression pressure better. The trouble in this case is caused by worn piston rings, a worn cylinder wall, or a worn piston. The trouble could also be caused by rings that are broken or stuck in the piston-ring grooves.

If adding the oil does not increase the compression pressure, the leakage is probably past the valves. This could be caused by:

1. Broken valve springs
2. Incorrect valve adjustment
3. Sticking valves
4. Worn or burned valves
5. Worn or burned valve seats
6. Worn camshaft lobes
7. Dished or worn valve lifters

It may also be that the cylinder-head gasket is "blown." This means the gasket has burned away so that compression pressure is leaking between the cylinder head and the cylinder block. Low compression between two adjacent cylinders is probably caused by the head gasket blowing between the cylinders.

Whatever the cause—rings, pistons, cylinder walls, valves, or gasket—the cylinder head has to be removed (except if valves need adjustment) so the trouble can be fixed. We discussed engine service earlier in the book.

41-6 Cylinder-Leakage Tester

The cylinder-leakage tester does about the same job as the compression tester, but in a different way. It applies air pressure to the cylinder with the piston at top dead center (TDC) on the compression stroke. In this position, both valves are closed. Very little air should escape from the combustion chamber. Figure 41-3 shows a cylinder-leakage tester. Figure 41-4 shows the tester connected to an engine cylinder and how it pinpoints places where leakage can occur.

To use the tester, first remove all plugs, as we have explained. Then remove the air cleaner, the crankcase filler cap or dipstick, and the radiator cap. Set the throttle wide open, and fill the radiator to the proper level. You are now ready to begin.

Connect the adapter, with the whistle, to the spark-plug hole of cylinder No. 1. Turn the engine over until the whistle sounds. When the whistle sounds, the piston is moving up on the compression stroke. Continue to rotate the engine until the TDC timing marks on the engine align. When the marks align, the piston is at TDC. Disconnect the whistle from the adapter hose and connect the tester, as shown in Figs. 41-3 and 41-4. Apply air pressure from the shop supply. Note the gauge reading, which shows the percentage of air leakage from the cylinder. Specifications vary, but if the reading is above 20

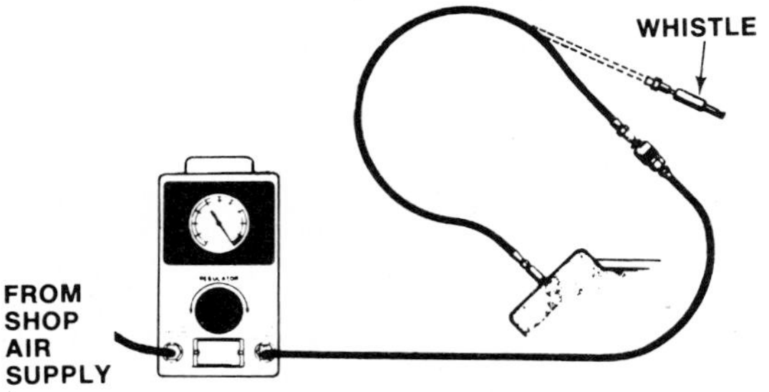
WHISTLE
FROM
SHOP
AIR
SUPPLY

Fig. 41-3 Cylinder-leakage tester. The whistle is used to locate TDC in cylinder No. 1. *(Sun Electric Corporation)*

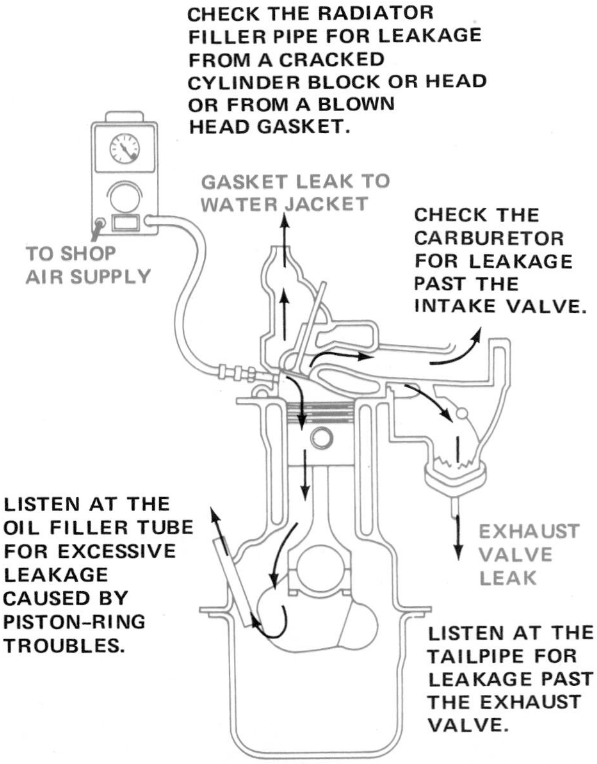
CHECK THE RADIATOR FILLER PIPE FOR LEAKAGE FROM A CRACKED CYLINDER BLOCK OR HEAD OR FROM A BLOWN HEAD GASKET.
GASKET LEAK TO WATER JACKET
CHECK THE CARBURETOR FOR LEAKAGE PAST THE INTAKE VALVE.
TO SHOP AIR SUPPLY
LISTEN AT THE OIL FILLER TUBE FOR EXCESSIVE LEAKAGE CAUSED BY PISTON-RING TROUBLES.
EXHAUST VALVE LEAK
LISTEN AT THE TAILPIPE FOR LEAKAGE PAST THE EXHAUST VALVE.

Fig. 41-4 How the cylinder-leakage tester works. It applies air pressure to the cylinder through the spark-plug hole with the piston at TDC and both valves closed. Points where air is leaking can then be pinpointed, as shown. *(Sun Electric Corporation)*

percent, there is excessive leakage. If the air leakage is excessive, check further by listening at the carburetor, tail pipe, and crankcase filler pipe. If the air is blowing out of an adjoining cylinder spark plug, it means that the head gasket is blown between the cylinders.

Figure 41-4 shows what it means if you can hear air escaping at any of the three listening points. If air bubbles up through the radiator, then the trouble is a blown cylinder-head gasket or a cracked cylinder head. This allows leakage from the cylinder to the cooling system.

Check the other cylinders in the same manner. A special adapter supplied with the tester lets you quickly find TDC on the other cylinders. When you use the tester, follow the instructions that explain how to use the adapter.

41-7 Engine-Vacuum Gauge

The engine-vacuum gauge is important for tracking down troubles in an engine that does not run as well as it should. This gauge measures intake-manifold vacuum. The intake-manifold vacuum changes with different operating conditions and with different engine defects. The way the vacuum varies from normal shows you what is wrong inside the engine.

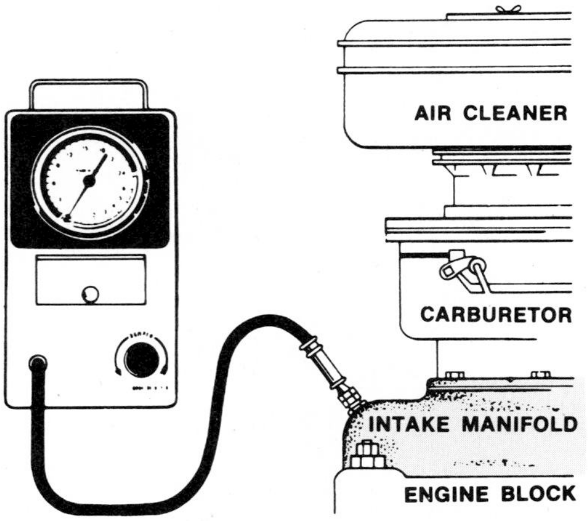
AIR CLEANER
CARBURETOR
INTAKE MANIFOLD
ENGINE BLOCK

Fig. 41-5 Vacuum gauge connected to the intake manifold to check manifold vacuum. (*Sun Electric Corporation*)

Figure 41-5 shows the vacuum gauge connected to the intake manifold. With the gauge connected, start the engine. Operate it at idle and at other speeds, as explained in the following paragraphs. The test must be made with the engine at operating temperature. The meanings of various readings are explained in the following paragraphs (see Fig. 41-6).

A steady and fairly high reading on idle indicates normal performance. Specifications vary with different engines, but a reading somewhere between 17 and 22 inches [432 and 559 mm] of mercury indicates the engine is okay. The reading will be lower at higher altitudes because of the lower atmospheric pressure. For every 1,000 feet [305 m] above sea level, the reading will be reduced about 1 inch [25.4 mm].

"Inches of mercury" refers to the way the vacuum gauges are set up. There is no mercury in the gauge; it is just a way of measuring vacuum.

A steady and low reading indicates late ignition or valve timing or possibly leakage around the pistons. Leakage around pistons—excessive blow-by—could be due to worn or stuck piston rings, worn cylinder walls, or worn pistons. Each of these conditions reduces engine power. With reduced power, the engine does not "pull" as much vacuum.

A very low reading indicates a leaky intake manifold or carburetor gasket, or possibly leaks around the carburetor throttle shaft. Air leaking into the manifold reduces the vacuum and engine power.

Late-model engines, with high-lift cams and more valve overlap, may have a lower and more uneven intake-manifold vacuum. Also, certain automotive-engine emission controls lower intake-manifold vacuum.

Back-and-fourth movement of the needle that increases with engine speed indicates weak valve springs.

Gradual falling back of the needle toward zero with the engine idling indicates a clogged exhaust line.

Regular dropping back of the needle indicates a valve sticking open or a plug not firing.

Irregular dropping back of the needle indicates that valves are sticking only part of the time.

Floating motion or slow back-and-forth movement of the needle indicates an air-fuel mixture that is too rich.

A test can be made for loss of compression due to leakage around the pistons. This would be the result of

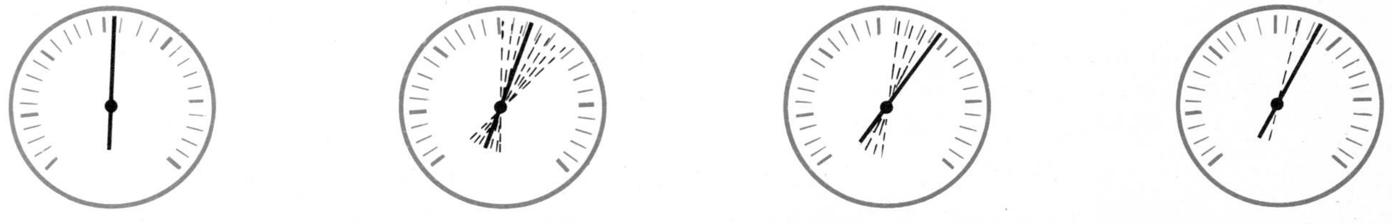
LOW AND STEADY READING INDICATES LOW COMPRESSION, AIR LEAKS, OR LATE IGNITION TIMING
RAPID VIBRATION WHEN ENGINE IS ACCELERATED INDICATES WEAK VALVE SPRINGS
INTERMITTENT DROP OF NEEDLE INDICATES STICKY VALVES
FLOATING MOTION OF NEEDLE INDICATES RICH MIXTURE

Fig. 41-6 Vacuum-gauge readings and their meaning.

stuck or worn piston rings, worn cylinder walls, or worn pistons. Race the engine for a moment, and then quickly release the throttle. The needle should swing around to 23 to 25 inches [584 to 635 mm] as the throttle closes, indicating good compression. If the needle fails to swing around this far, there is loss of compression. Further checks should be made.

● 41-8 Exhaust-Gas Analyzer

At one time, the major use of the early type of exhaust-gas analyzer was to adjust the carburetor. It is still used for that purpose. However, the new type of exhaust-gas analyzer (infrared type) is also used to check out the emission controls on the car. We covered emission controls in Chaps. 33 to 36. The main purpose of the emission controls is to cut down on the carbon monoxide (CO), hydrocarbons or unburned gasoline (HC), and oxides of nitrogen (NO_x) in the exhaust gas.

Figure 41-7 shows the infrared type of exhaust-gas analyzer. To use it, you stick a probe into the tail pipe of the car (Fig. 41-8). The probe draws out some of the exhaust gas and carries it through the analyzer. Two dials on the face of the analyzer, shown in Fig. 41-9, tell you how much HC and CO are in the exhaust gas. The HC meter reports in parts per million. The CO meter reports a percentage. Federal and state laws set the maximum legal limits on the amount of HC and CO permitted in the exhaust.

A different kind of tester is required for NO_x, but it works in the same general way. It draws exhaust gas from the tail pipe and runs the gas through the analyzer. The meter reports the amount of NO_x in the exhaust gas. Generally, NO_x testers are available only in testing laboratories. They are not widely used in the automotive service shop. Authorities say, however, that some day many well-equipped shops will have them.

Fig. 41-7 Exhaust-gas analyzer. (*Allen Test Products Division, The Allen Group, Inc.*)

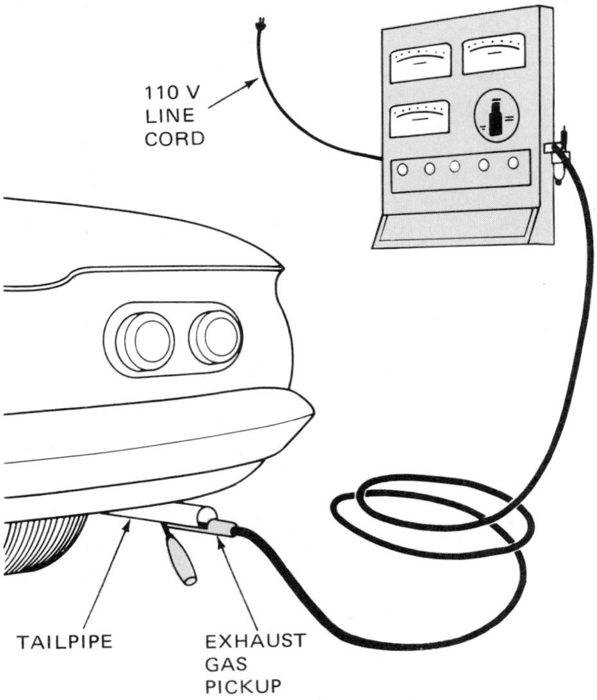

Fig. 41-8 Exhaust-gas analyzer connected for exhaust-gas test.

● 41-9 Ignition Timing Light

As you know, the sparks must reach the plugs in the cylinders at exactly the right time. Adjusting the distributor to make the sparks do this is called "ignition timing." You can adjust the distributor by turning the distributor in its mounting. The procedure is described in ● 31-12. You use a timing light (Fig. 31-14) to check the timing.

● 41-10 Oscilloscope

The oscilloscope, described in ● 31-2 to 31-5, is a high-speed voltmeter. It has a picture tube like that found in a television set. On the tube, it draws the voltages in the ignition-system primary or secondary circuit. If there is trouble, the picture, or traces, show what is wrong. Besides showing what is wrong in the ignition system, the oscilloscope can also detect many engine troubles. For example, with loss of compression in a cylinder, the secondary voltage does not have to go as high. This shows up on the trace for that cylinder, because the spike does not go as high as it would with normal compression.

● 41-11 Dynamometer

The chassis dynamometer can test the engine power output under various operating conditions. It can duplicate any kind of road test at any load or speed desired by the operator. The part of the dynamometer that you can see consists of two heavy rollers mounted at or a little above floor level (Fig. 41-10). The car is driven onto these rollers, as shown in Fig. 41-11, so that the car wheels can drive the rollers. Next, the engine is started, and the transmission is put into gear. The car is then operated as if it were out on an actual road test.

Under the floor is a device that can place loads on the rollers. This allows the technician to operate the engine

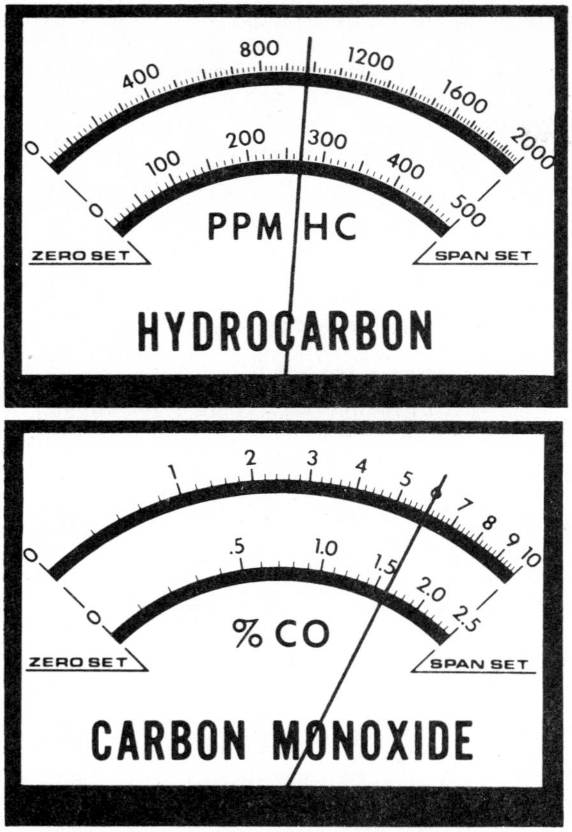

Fig. 41-9 HC and CO meter faces. (*Sun Electric Corporation*)

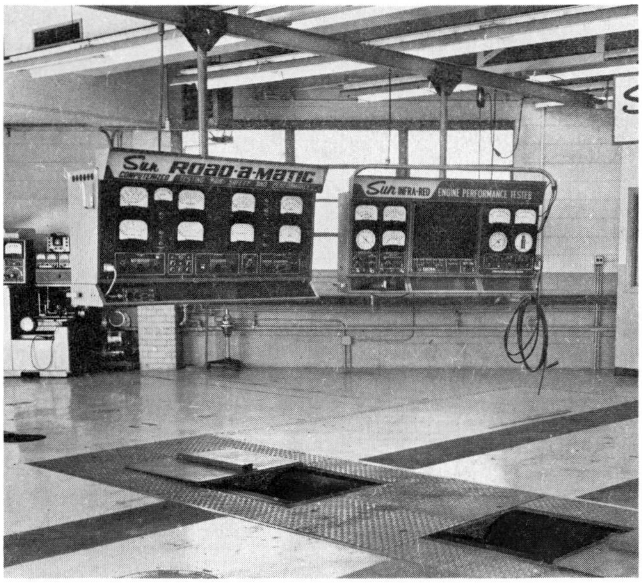

Fig. 41-10 Chassis dynamometer of the flush-floor type. The rollers are set at floor level. (*Sun Electric Corporation*)

under various conditons. You can find out how the engine would do during acceleration, cruising, idling, and deceleration. The test instruments, such as the scope, dwell tachometer, and vacuum gauge, are hooked into the

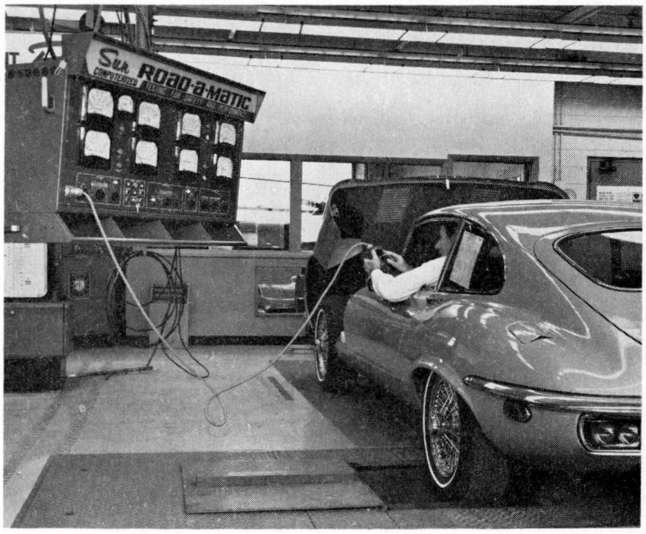

Fig. 41-11 Automobile in place on a chassis dynamometer. The rear wheels drive the dynamometer rollers. At the same time, instruments on the test panel measure car speed, engine power output, engine vacuum, and so on. (*Sun Electric Corporation*)

engine. These instruments then show the condition of the engine as it operates.

The dynamometer can also be used to check the transmission and the differential. For example, the shift points and other operating conditions of an automatic transmission can be checked on the dynamometer. Special diagnostic dynamometers are becoming more popular. These units have many instruments attached and have motored rollers that permit testing of wheel alignment, suspension, brakes, and steering.

● 41-12 Cooling-System Testers

There are three basic types of tester for the cooling system: antifreeze tester, pressure tester, and belt-tension tester. Let us look at each of these.

1. Coolant antifreeze tester There are three kinds of antifreeze testers, as illustrated in Figs. 22-1 and 22-2. As you know, the antifreeze protects the engine from damage resulting from coolant freeze-up. Thus, it is important to check the coolant during cold weather, to find out how much protection it has against freezing. That is, you measure the concentration of antifreeze.

2. Pressure tester The pressure tester is a small pump with a pressure gauge (Figs. 22-9 and 22-10). It is attached to the radiator filler neck, as shown, and the pump is operated to apply pressure. If the pressure holds steady and there are no signs of leaks, the cooling system is tight.

3. Belt-tension tester One type of belt-tension tester is shown in Fig. 22-12. If there is not enough tension, the belt will slip. The fan and water pump will not be driven fast enough, and the engine will overheat. Also, the slip-

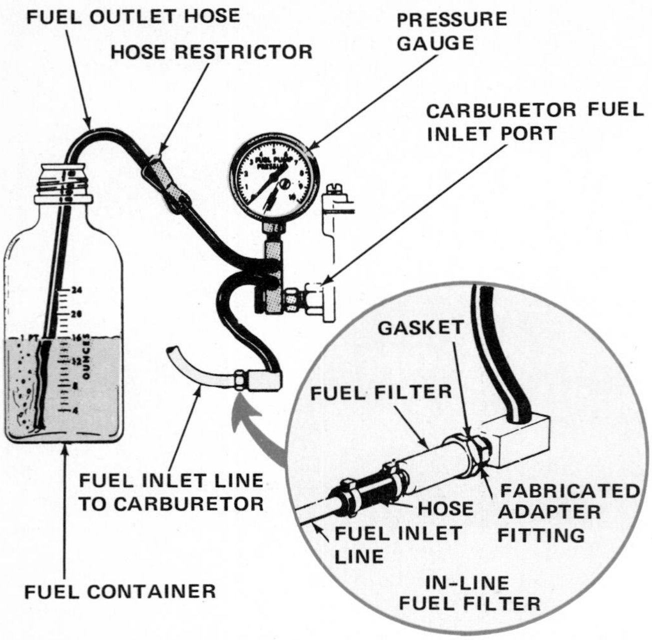

Fig. 41-12 Fuel-pump pressure and capacity tests. *(Ford Motor Company)*

ping belt will wear out rapidly. On the other hand, too much tension will wear the bearings of the devices being driven (alternator, air-injection or steering pump, etc.) Adjustment is made by moving the alternator or other driven unit in or out slightly.

● 41-13 PCV-Valve Tester

In Chap. 34 we mentioned that modern engines are equipped with a positive-crankcase-ventilating (PCV) system. This system passes air through the crankcase and then up to the intake manifold. The air passing through picks up any blow-by or hydrocarbon that has leaked down past the piston rings. The unburned and partly burned gasoline then passes through the engine again, where it has another chance to burn.

The PCV valve in the system is designed to prevent too much air from flowing during idle. If the valve sticks open, too much air will flow. The idle mixture will be upset, and the engine will idle roughly and may even stall. If the valve sticks closed, not enough air will get through, and blow-by products can gather in the cankcase. This could seriously damage the engine, because these products include corrosive acids and goo. They could plug oil lines and cause the engine to fail from oil starvation. Section 36-2 describes the testing procedure.

● 41-14 Fuel-System Testers

We have already covered various testers that check fuel-system performance. The exhaust-gas analyzer, engine-vacuum gauge, oscilloscope, and other instruments are used to check engine performance. An important part of the engine is the fuel system, and these test instruments also report on the fuel system. There are also fuel-pump pressure and capacity testers that check on how well the fuel pump is doing its job. Figure 41-12 shows a fuel-pump pressure and capacity tester.

● 41-15 Electric-System Testers

A variety of testers are required to test the electrical equipment on the car. These include the distributor, coil, and condenser testers for ignition-system components. To check the charging system, ammeters and voltmeters are required. Any of several instruments can be used to check the battery, including a hydrometer, a voltmeter, a 421 tester, and a cadmium-tip tester. We have described all these testers in earlier chapters.

REVIEW QUESTIONS

1. What is the purpose of the tachometer?
2. What is the purpose of the cylinder-compression tester? How is it used?
3. What is the purpose of the cylinder-leakage tester? How is it used?
4. If the compression is low in a cylinder, what would be the purpose of pouring some heavy oil into the cylinder and rechecking it?
5. If adding oil does not change the compression pressure, what are five possible causes of low compression?
6. What are the three listening points when you use the cylinder-leakage tester?
7. Name three abnormal vacuum-gauge readings. Explain the trouble that could cause each.
8. What is the purpose of the ignition timing light? How is it used?
9. What is the purpose of the oscilloscope? How is it used?
10. What is the purpose of the dynamometer? How is it used?
11. What are the three basic types of cooling-system testers?
12. Explain how to use the cooling-system pressure tester.
13. Explain how to use the PCV tester.
14. Name three fuel-system testers, and explain how to use them.
15. Name four electric-system testers.

SELF PROJECTS

1. Collect your own file of instructions on how to use various engine testing instruments.
2. The packages, boxes, cans, and instruction sheets that come with automotive parts are good sources of servicing information. Collect this information for your own notebook.

CHAPTER 42
Engine Trouble Diagnosis

After studying this chapter, you should be able to:

- **List and describe the basic engine troubles discussed in the chapter.**
- **For each trouble, explain possible causes.**
- **Perform the preliminary checks outlined in the chapter to help pinpoint trouble causes.**

This chapter discusses various engine troubles and explains what may cause them. It also indicates cures for these troubles. In other words, the chapter covers engine trouble diagnosis or, as many call it, engine "troubleshooting." Earlier chapters describe servicing and repair procedures to use after a trouble has been traced to its cause.

● 42-1 Trouble-Diagnosis Chart

The chart that follows lists various engine complaints, their possible causes, and checks or corrections to be made. The information in this chart will shorten the time you need to correct a trouble. If you follow a logical procedure, you can usually find the cause of trouble quickly. On the other hand, haphazard guesswork wastes time and effort. The trouble-diagnosis chart makes several references, where appropriate, to the Diesel-Engine Trouble Chart (● 40-2), that is, where certain additional or special trouble occurs only to diesel engines.

Engine Trouble-Diagnosis Chart

See ● 42-2 to 42-19 for detailed explanations of the trouble causes and corrections listed below.
Note: The troubles and possible causes are not listed according to how often they occur. That is, item 1 does not necessarily occur more often than item 2.

ENGINE TROUBLE-DIAGNOSIS CHART (Continued)

Complaint	Possible Cause	Check or Correction
1. Engine will not turn over (● 42-2)	a. Run-down battery	Recharge or replace; start engine with jumper battery and cables
	b. Starting circuit open	Find and eliminate the open; check for dirty or loose cables
	c. Starting-motor drive jammed	Remove starting motor, and free drive
	d. Starting motor jammed	Remove for teardown and correction
	e. Engine jammed	Check engine to find trouble
	f. Transmission not in neutral, or neutral switch out of adjustment	Check and adjust if necessary
	g. Seat belt not fastened, or interlock faulty	Check interlock
	h. Also causes listed under item 3 below, "Engine turns over at normal speed but does not start"; driver may have run battery down trying to start	

ENGINE TROUBLE-DIAGNOSIS CHART (Continued)

Complaint	Possible Cause	Check or Correction
2. Engine turns over slowly but does not start (● 42-3)	a. Run-down battery	Recharge or replace; start engine with jumper battery and cables
	b. Defective starting motor	Repair or replace
	c. Bad connections in starting circuit	Check for loose or dirty cables; clean and tighten
	d. Also causes listed under item 3 below; driver may have run battery down trying to start	
3. Engine turns over at normal speed but does not start (● 42-4)	a. Ignition system defective	Try spark test; check timing, ignition system.
	b. Fuel pump defective or overchoking	Prime engine; check accelerator-pump discharge, fuel pump, fuel line, choke, carburetor
	c. Air leaks in intake manifold or carburetor	Tighten mounting; replace gaskets as needed
	d. Defect in engine	Check compression or leakage (● 42-5), valve action, timing
	e. Ignition bypass resistor burned out	Replace
	f. Plugged fuel filter	Clean or replace
	g. Plugged or collapsed exhaust system	Replace collapsed parts
	h. See also item 1 in ● 40-2	
4. Engine runs but misses on one cylinder (● 42-6)	a. Defective spark plug	Clean or replace
	b. Distributor cap or spark-plug wire defective	Replace
	c. Stuck valve	Free valve; service guide
	d. Broken valve spring	Replace
	e. Burned valve	Replace
	f. Bent push rod	Replace
	g. Flat cam lobe	Replace camshaft
	h. Defective piston or rings	Replace; service cylinder wall as necessary
	i. Defective head gasket	Replace
	j. Intake-manifold leak	Replace gasket; tighten manifold bolts
	k. See also item 5 in ● 40-2	
5. Engine runs but misses on different cylinders (● 42-6)	a. Defective distributor advance, coil, condenser	Check distributor, etc.
	b. Defective fuel system	Check fuel pump, flex line, carburetor
	c. Crossfiring plug wires	Replace; relocate
	d. Loss of compression	Check compression, leakage
	e. Burned valves	Service
	f. Worn pistons and rings	Service
	g. Overheated engine	Check cooling system
	h. Manifold heat-control valve sticking	Free valve
	i. Restricted exhaust	Check tail pipe, muffler; eliminate restriction
6. Engine lacks power, acceleration, or high-speed performance, hot or cold (● 42-8)	a. Ignition defective	Check timing, distributor, wiring, condenser, coil, plugs
	b. Fuel system defective; secondary throttle valves not opening	Check carburetor, choke, filter, air cleaner, fuel pump
	c. Throttle valve not opening fully	Adjust linkage
	d. Restricted exhaust	Check tail pipe, muffler; eliminate restriction
	e. Loss of compression	Check compression or leakage (● 42-5)

ENGINE TROUBLE-DIAGNOSIS CHART (Continued)

Complaint	Possible Cause	Check or Correction
	f. Excessive carbon in engine	Remove carbon
	g. Defective valve action	Check with compression, leakage, or vacuum tester (● 42-5 and 42-7)
	h. Excessive rolling resistance from low tires, dragging brakes, wheel misalignment, etc.	Correct the defect causing rolling resistance
	i. Heavy oil	Use lighter oil
	j. Wrong or bad fuel	Use good fuel of correct octane
	k. Transmission not downshifting, or torque converter defective	Check transmission
	l. See also item 7 "Loss of Power" in ● 40-2	
7. Engine lacks power, acceleration, or high-speed performance, hot only (● 42-8)	a. Engine overheats	Check cooling system (see item 9 below)
	b. Choke stuck partly open	Repair or replace
	c. Sticking manifold heat-control valve	Free valve
	d. Vapor lock	Use different fuel, or shield fuel line
8. Engine lacks power, acceleration, or high-speed performance, cold only (● 42-8)	a. Automatic choke stuck open	Repair or replace
	b. Manifold heat-control valve stuck open	Free valve
	c. Cooling-system thermostat stuck open	Repair or replace
	d. Engine valves stuck open	Free valves, service valve stems and guides as needed
9. Engine overheats (● 42-10)	a. Lack of coolant	Add coolant; look for leak
	b. Ignition timing late	Adjust timing
	c. Loose or broken fan belt	Tighten or replace
	d. Thermostat stuck closed	Replace
	e. Clogged water jackets	Clean out
	f. Defective radiator hose	Replace
	g. Defective water pump	Repair or replace
	h. Insufficient engine oil	Add oil
	i. High-altitude, hot-climate operation	Drive more slowly; keep radiator filled
	j. Defective fan clutch	Replace
	k. Valve timing late, slack timing chain has allowed chain to jump a tooth	Retime; adjust or replace
	l. No vacuum advance in any gear	TCS system or distributor trouble
10. Rough idle (● 42-11)	a. Carburetor idle adjustment incorrect	Readjust idle mixture and speed
	b. PCV valve stuck open	Replace
	c. Other causes, listed under items 6 to 8 above, "Engine lacks power"	
11. Engine stalls cold or as it warms up (● 42-12)	a. Choke valve stuck closed, or will not close	Open choke valve; free or repair automatic choke
	b. Fuel not getting to or through carburetor	Check fuel pump, lines, filter, float, idle circuits
	c. Manifold heat-control valve stuck	Free valve
	d. Throttle solenoid improperly set	Adjust
	e. Engine idling speed set too low	Increase idling speed to specified value

ENGINE TROUBLE-DIAGNOSIS CHART (Continued)

Complaint	Possible Cause	Check or Correction
	f. Malfunctioning PCV valve	Replace
	g. Damper in thermostatic air cleaner stuck closed	Free up; replace motor
	h. See also item 2 in ● 40-2	
12. Engine stalls after idling or slow-speed driving (● 42-12)	a. Defective fuel pump	Repair or replace fuel pump
	b. Overheating	See item 9 above, "Engine overheats"
	c. High carburetor-float level	Adjust
	d. Idling adjustment incorrect	Adjust
	e. Malfunctioning PCV valve	Replace
	f. Throttle solenoid improperly set	Adjust
13. Engine stalls after high-speed driving (● 42-12)	a. Vapor lock	Use different fuel or shield fuel line
	b. Carburetor venting or idle compensator valve defective	Check and repair
	c. Engine overheats	See item 9 above, "Engine overheats"
	d. Malfunctioning PCV valve	Replace
	e. Improperly set throttle solenoid	Adjust
14. Engine backfires (● 42-13)	a. Ignition timing off	Adjust timing
	b. Spark plugs or wrong heat range	Install correct plugs
	c. Excessively rich or lean mixture	Repair or readjust fuel pump or carburetor
	d. Overheating of engine	See item 9 above, "Engine overheats"
	e. Carbon in engine	Clean out
	f. Valves hot or sticking	Adjust; free; clean; replace if bad
	g. Cracked distributor cap	Replace cap
	h. Inoperative antibackfire valve	Replace
	i. Crossfiring plug wires	Replace
15. Engine run-on or dieseling (● 42-14)	a. Idle-stop or solenoid adjustment not correct	Adjust; fix solenoid
	b. Engine overheating	See item 9 above, "Engine overheats"
	c. Hotspots in cylinders	Check plugs, pistons, cylinders for carbon; check valves for defects and faulty seating
	d Timing advanced	Adjust
	e. See also item 10 in ● 40-2	
16. Too much HC and CO in exhaust (● 42-15)	a. Ignition miss	Check plugs, wiring, cap, coil, etc.
	b. Incorrect ignition timing	Time ignition
	c. Carburetor troubles	Check choke, float level, idle-mixture adjustment screw, etc., as listed in item 20 below
	d. Faulty air injection	Check pump, hoses, manifold
	e. Defective TCS system	Check system
	f. Defective catalytic converters	Replace converters or catalyst
17. Smoky exhaust:		
a. Blue smoke	Excessive oil consumption	See item 18 and ● 42-16
b. Black smoke	Excessively rich mixture	See item 20 and ● 42-18
c. White smoke	Steam in exhaust	Replace gasket; tighten cylinder-head bolts to eliminate coolant leakage into combustion chambers
d. See also item 9 in ● 40-2		

ENGINE TROUBLE-DIAGNOSIS CHART (Continued)

Complaint	Possible Cause	Check or Correction
18. Excessive oil consumption (● 42-16)	a. External leaks	Correct seals; replace gaskets
	b. Burning oil in combustion chamber	Check valve-stem clearance, piston rings, cylinder walls, rod bearings
	c. High-speed driving	Drive more slowly
19. Low oil pressure (● 42-17)	a. Worn engine bearings	Replace
	b. Engine overheating	See item 9 above
	c. Oil dilution or foaming	Replace oil
	d. Lubricating-system defects	Check oil lines, oil pump, relief valve
20. Excessive fuel consumption (● 42-18)	a. Jackrabbit starts	Drive more reasonably
	b. High speed	Drive more slowly
	c. Short-run operation	Make longer runs
	d. Excessive fuel-pump pressure or pump leakage	Reduce pressure; repair pump
	e. Choke partly closed after warm-up	Open; repair or replace automatic choke
	f. Clogged air cleaner	Clean
	g. High carburetor float level	Adjust
	h. Stuck or dirty float-needle level	Free and clean
	i. Worn carburetor jets	Replace
	j. Stuck metering rod or full-power piston	Free
	k. Idle too rich or too fast	Adjust
	l. Stuck accelerator-pump check valve	Free
	m. Carburetor leaks	Replace gaskets; tighten screws, etc.
	n. Cylinder not firing	Check coil, condenser, timing, plugs, contact points, wiring
	o. Automatic transmission slipping or not up-shifting	Check transmission
	p. Loss of engine compression (worn engine)	Check compression or leakage (● 42-5)
	q. Defective valve action (worn camshaft, chain slack, jumped tooth)	Check with compression, leakage, or vacuum tester (● 42-5 and 42-7)
	r. Excessive rolling resistance from low tires, dragging brakes, wheel misalignment, etc.	Correct the defects causing the rolling resistance
	s. Clutch slippage	Adjust or repair
21. Engine is noisy		
a. Regular clicking	Valve and tappet	Readjust valve clearance, or replace noisy hydraulic lifters
b. Ping or chatter on load or acceleration	Detonation due to low-octane fuel, carbon, advanced ignition timing, or causes listed under item 14 above, "Engine backfires"	Use higher-octane fuel; remove carbon; adjust ignition timing
c. Light knock or pound with engine floating	Worn connecting-rod bearings or crankpin, misaligned rod, lack of oil	Replace or adjust bearings; service crankpins; replace rod; correct lack of oil
d. Light metallic double knock, usually most audible during idle	Worn or loose piston pin or lack of oil	Service pin and bushing; correct lack of oil
e. Chattering or rattling during acceleration	Worn rings, cylinder walls, low ring tension, broken rings	Service walls; replace rings

ENGINE TROUBLE-DIAGNOSIS CHART (Continued)

Complaint	Possible Cause	Check or Correction
f. Hollow, muffled, bell-like sound, engine cold	Piston slap due to worn pistons, walls, collapsed piston skirts, excessive clearance, lack of oil, misaligned connecting rods	Replace or resize pistons; service walls; replace rods; correct lack of oil
g. Dull, heavy, metallic knock under load or acceleration, especially when cold	Regular noise: worn main bearings; irregular: worn end-thrust bearing knock on clutch engagement or on hard acceleration	Replace or service bearings and crankshaft
h. Miscellaneous noises	Rattles etc., from loosely mounted accessories; alternator, horn, oil pan, front bumper, water pump, etc.	Tighten mounting
i. See also items 8 and 9 in ● 40-2		

● 42-2 Engine Will Not Turn Over

If the engine will not turn over (no cranking) when starting is attempted, make sure the control lever is in neutral (N) or park (P). Or, if the car has a manual transmission, make sure the clutch is depressed when starting is attempted. Check the battery and cables. A low battery may cause the solenoid plunger to pull in and release repeatedly (making a clattering noise) but will not crank the engine.

If the starter spins and the drive pinion engages, the starting-motor overrunning clutch is slipping. If the solenoid plunger pulls in (a loud click) but nothing else happens, there is trouble in the solenoid (poor contacts), starting motor, or the circuit. If the solenoid plunger does not pull in, the solenoid circuit may be at fault. Connect a jumper lead between the solenoid battery terminal and the solenoid switch terminal. If the starting motor operates after you connect the jumper lead, the solenoid is okay. The trouble is in the ignition switch, the neutral starting switch, or the circuit between these two switches. If the starting motor does not operate, remove it for service.

Another method of checking for the cause of trouble when the engine will not turn over uses the headlights or dome light. This is a preliminary, or "instant," check. It is not as accurate as the preceding method, but some technicians use it as a first step in the diagnosis. Here is how to do this quick check: Turn on the headlights or dome light, and try to start the engine. The lights will (1) stay bright, (2) dim considerably, (3) dim slightly, (4) go out, or (5) not burn at all.

1. If the lights stay bright, there is an open circuit in the starting motor or starting-motor circuit. Check as outlined in ● 42-3. Also, the transmission may not be in neutral, or the neutral switch is out of adjustment. In addition, on late-model cars, the ignition-interlock safety belts may not be fastened properly, or the system may be defective.
2. If the lights dim considerably, the battery may be run down. Or there may be mechanical trouble in the starting motor or engine. If the battery tests okay, remove the starting motor for further checks. Try to turn the engine flywheel in the normal direction of rotation to see if the engine is jammed.
3. If the lights dim only slightly, listen for cranking action (sound of an electric motor running). If the starting motor runs, the pinion is not engaging the flywheel (Bendix type), or the overrunning clutch is slipping. If the solenoid clicks but the starting motor does not rotate, it could be a low battery. However, it is probably trouble in the starting motor. Remove it for service.
4. If the lights go out as cranking is attempted, there may be a bad connection in the main circuit, probably at a battery terminal.
5. If the lights burn dimly or not at all when they are turned on, even before cranking is attempted, the battery is probably run down.

● 42-3 Engine Turns Over Slowly but Does Not Start

Causes of this condition could be a run-down battery, a defective starting motor, or mechanical trouble in the engine. Check the battery, starting motor, and circuit as outlined in Chaps. 25 and 27. If they are normal, the trouble probably is in the engine (defective bearings, rings, and so on, that could produce high friction). Remember that, in cold weather, cranking speed is reduced by thickening of the engine oil and reduction of battery efficiency.

If the battery is run down, it could be that the driver has discharged the battery in attempting to start. The cause of starting failure could be as noted in following sections.

● 42-4 Engine Turns Over at Normal Cranking Speed but Does Not Start

This means the battery and starting motor are in normal condition. The cause of trouble is probably in the ignition or fuel system. The difficulty could be due to overchoking. This applies to a cold engine. Failure to start with a hot engine may be due to a defective choke that fails to open properly as the engine warms up. This would cause flooding of the engine (delivery of too much gasoline). Open the throttle wide while cranking (this dechokes the engine), or open the choke valve by hand and then crank the engine. Try cranking with the throttle wide open. If the engine does not start, disconnect the lead from one spark plug (or from the center distributor-cap terminal). Use insulated pliers to hold the lead clip about $\frac{3}{16}$ inch [4.76 mm] from the engine block (see Fig. 42-1). Be careful so you do not damage the insulation. Crank the engine to see if a good spark occurs. If no spark occurs, check the ignition system. If a spark does occur, the ignition system is probably okay (the timing could be off, however).

If the ignition system operates normally, the fuel system should be analyzed. First, prime the engine by operating the carburetor accelerator pump several times.

CAUTION: Gasoline is highly explosive. Keep back out of the way while priming the engine; the engine might backfire through the carburetor. Replace the air cleaner before cranking.

If the engine now starts and runs for a few seconds, the fuel system is probably faulty. It is not delivering fuel to the engine. Temporarily disconnect the fuel inlet to the carburetor. Also, disconnect the primary lead from the negative terminal of the ignition coil so the engine will not start. Hold a container under the fuel line to catch fuel, and crank the engine to see whether fuel is delivered. If it is not, the fuel pump is defective or the fuel line is clogged. If fuel is delivered, the fuel filter is probably at fault, the automatic choke is not working correctly, or possibly there are air leaks into the intake manifold or carburetor.

If the fuel and ignition systems seem okay on preliminary checks, check the mechanical condition of the engine with the compression and leakage tests (● 42-5).

● 42-5 Cylinder Compression and Leakage Testers

These testers are used to determine whether or not the cylinder can hold compression, or whether there is excessive leakage past the rings, valves, or head gasket. The compression tester (Fig. 41-2) has been a basic engine-testing instrument for many years. Recently, the cylinder-leakage tester has come into use. Some mechanics believe that it is more accurate in pinpointing defects in the cylinder.

The use of compression and leakage testers has been described in detail (compression tester in ● 41-3 and 41-4, and leakage tester in ● 41-5).

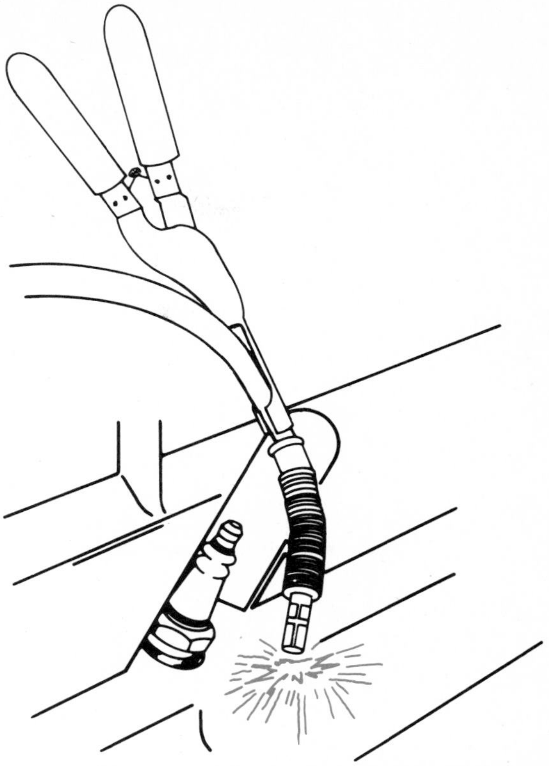

Fig. 42-1 To make a spark test, disconnect a spark-plug wire and hold the clip close to the engine block while cranking the engine.

● 42-6 Engine Runs but Misses

A missing engine is a rough engine. If one or more cylinders fail to fire, the engine is thrown out of balance. The result is roughness and loss of power. It is sometimes hard to track down a miss. The miss might occur at some speeds and not others. Also, a miss may skip around. The modern method of checking out a missing engine is to use an oscilloscope and a dynamometer. The oscilloscope is discussed in ● 31-2 to 31-5. The dynamometer is discussed in ● 41-10. If these testing instruments are not available, then the test can be made as follows.

Use insulated pliers to remove each spark-plug wire in turn, to locate the missing cylinder. Removing the plug wire prevents the spark from reaching the plug, and the plug will not fire. If removing the wire changes the engine rhythm or speed, then the cylinder was delivering power before you removed the wire. But if there is no change in engine speed or rhythm, then that cylinder was missing before you removed the wire.

1. Check a missing cylinder further by removing the spark-plug lead. Use insulated pliers to hold the plug clip close to the engine block while the engine is running. If no spark occurs, there is probably a high-voltage leak due to a bad lead or a cracked or burned distributor cap. If a good spark occurs, install a new spark plug in the cylinder (or swap plugs between two cylinders). Then reconnect the lead, and see whether

the cylinder still misses. If it does, the cause of the trouble is probably defective engine parts, such as valves or rings.

2. If the miss is hard to locate, perform a general tune-up (Chap. 43). This will disclose, and maybe eliminate, various causes of missing. These could include defects in the ignition system or fuel system, loss of engine compression, sticky or damaged engine valves, overheated engine, sticky manifold heat control, and clogged exhaust.
3. With most oscilloscopes, you can make a power balance test that will quickly pinpoint the missing cylinder. When the oscilloscope is connected to the running engine, you turn a knob or push a button. This shorts out the cylinders one by one, in the firing order. The scope shows which cylinder is shorted out. If shorting a cylinder changes the engine rpm as registered on the tester, you know the cylinder was delivering power. But if no change in rpm takes place, then you know that cylinder was not delivering power.

In the diesel engine, a miss is most likely caused by failure of the fuel system to deliver an adequate amount of fuel to the cylinder. This could be due to a clogged nozzle, air in the line. or injector-pump problems.

● 42-7 Engine Vacuum Gauge

This is an important engine tester for tracking down troubles in an engine that runs but does not perform satisfactorily. It measures intake-manifold vacuum. The intake-manifold vacuum varies with different operating conditions, and with different engine defects. The manner in which the vacuum varies from normal indicates the type of engine trouble (● 41-6 explains how to use the vacuum gauge).

● 42-8 Engine Lacks Power

This is a general complaint that is often difficult to analyze. The best procedure is to do a tune-up job (Chap. 43). This will disclose various engine conditions that could cause loss of power. To get some idea of the cause, find out whether the engine lacks power both when hot and cold, only when hot, or only when cold. Also, find out if the problem suddenly developed or if the power fell off over a period of many months or many miles of operation. A chassis dynamometer (● 41-10) or an oscilloscope (● 31-2 to 31-5) can be used to help locate the cause of trouble.

1. Engine lacks power and acceleration either hot or cold The fuel system may not be enriching the mixture as the throttle is opened. This could be due to a faulty accelerator pump or a defective high-speed or full-power circuit in the carburetor. Also, the fuel system could be supplying an excessively lean or rich mixture. This could be due to a defective fuel pump, clogged lines, clogged filter, worn carburetor jets or lines, air leaks at the carburetor or manifold joints, malfunctioning PCV valve, and so on. Carburetor and fuel-system action can be checked with an exhaust-gas analyzer (● 41-7).

Another condition could cause lack of power with the engine hot or cold. This is an improper linkage adjustment that prevents full throttle opening. Also, the ignition system may be causing trouble, owing to incorrect timing, a "weak" coil, reversed polarity (see Fig. 31-20), wrong spark-plug heat range, and so on. The wrong fuel or oil for the engine could reduce performance. In the engine, numerous conditions could cause loss of power: engine deposits (carbon), lack of compression (faulty valves, rings, worn cylinder walls, pistons, and so on), and defective bearings. A clogged exhaust (bent or collapsed exhaust pipe or tail pipe or clogged muffler) could create back pressure that would cause poor engine performance (see Fig. 42-2). Also, any sort of excessive rolling resistance would absorb engine power and hold down engine acceleration and speed. This would include dragging brakes, underinflated tires, misaligned wheels, and excessive friction in the transmission or power train. Also, the automatic transmission may not be downshifting, or the torque converter may be defective.

2. Engine lacks power only when hot The engine may be overheating (● 42-10). Also, the automatic choke may not be opening normally as the engine warms up. The manifold heat-control valve may be stuck. Or there may be a vapor lock in the fuel pump or line.

3. Engine lacks power when cold or reaches operating temperature too slowly The automatic choke may be leaning out the mixture too soon (before the engine warms up). The manifold heat-control valve may not be closed (so that insufficient heat reaches the intake manifold). Or, the cooling-system thermostat may be stuck open. In this case, water circulation goes on between the engine and radiator even with the engine cold, and so warm-up is delayed. Occasionally, engine valves may stick when the engine is cold, but as the engine warms up, the valves become free and work normally.

● 42-9 Exhaust-Gas Analyzer

At one time the major use of the exhaust-gas analyzer was to adjust the carburetor. Today, its major job is to check the emission controls on the car. If the emission controls are not working properly, there will be excess HC and CO in the exhaust. The exhaust-gas analyzer measures the amount of HC and CO in the exhaust gases coming out the tail pipe. Use of the exhaust-gas analyzer is discussed in ● 41-7.

● 42-10 Engine Overheats

Most engine overheating is caused by loss of coolant due to leaks in the cooling system. Other causes include a loose or broken fan belt, a defective water pump, clogged

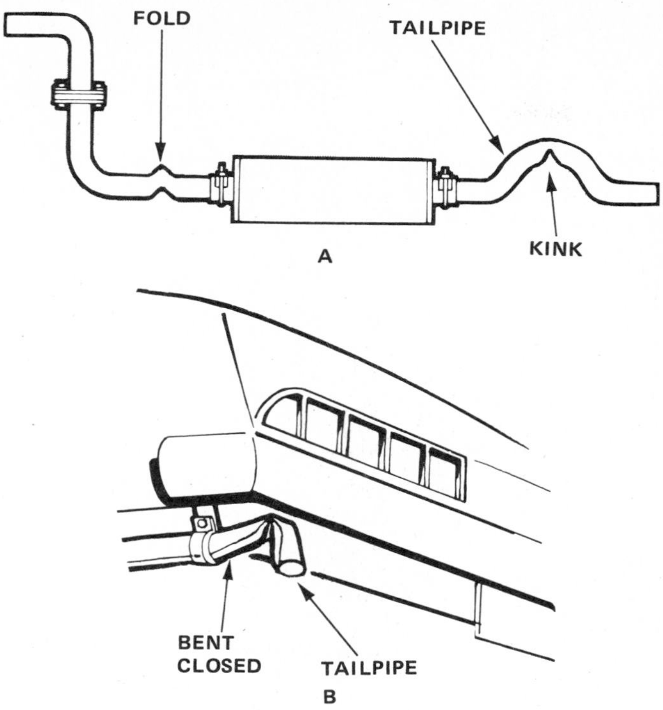

Fig. 42-2 A clogged or restricted exhaust system may cause the engine, cold or hot, to lack power. (*Ford Motor Company*)

water jackets, a defective radiator hose, and a defective thermostat or fan clutch. Also, late ignition or valve timing, lack of engine oil, overloading the engine, or high-speed, high-altitude, or hot-climate operation can cause engine overheating. Also, freezing of the coolant could cause lack of coolant circulation so that local hot spots and boiling develop. In addition, if a faulty TCS system prevents vacuum advance in any gear, or if the distributor vacuum advance is defective, overheating may result. Note: If you are stuck in slow-moving traffic with an overheating engine, try this: Open the car windows and turn off the air conditioner. Turn on the heater to maximum. This lightens the load on the engine and also takes some heat away from the engine. You will be temporarily uncomfortable, but your actions may make the difference between an engine boilover with possible stalling and simply a hot engine.

● 42-11 Rough Idle

If the engine idles roughly but runs normally above idle, chances are the idle speed and idle mixture are incorrectly adjusted. A rough idle could also be due to many other causes. See items 6 to 10 in the Trouble-Diagnosis Chart.

● 42-12 Engine Stalls

If the engine starts and then stalls, note whether the stalling takes place before or after the engine warms up, after idling or slow-speed driving, or after high-speed or full-load driving. Special note should be made of the PCV valve. If this valve becomes clogged or sticks, it will cause poor idling and stalling.

1. Engine stalls before it warms up This could be due to an improperly set fast or slow idle, or to improper adjustment of the idle fuel-mixture needle in the carburetor. Also, it could be due to a low carburetor float setting or to insufficient fuel entering the carburetor. The condition could result from a thermostatic air cleaner not working, dirt or water in the fuel lines or filter, a defective fuel pump, or a plugged fuel-tank vent. Also, the carburetor could be icing. Certain ignition troubles could cause stalling after starting. But, as a rule, if the ignition troubles are bad enough to cause stalling, they would also prevent starting. However, burned contact points might permit starting but could fail to keep the engine going. One other condition might be an open primary resistance wire. When the engine is cranked, this wire is bypassed. Then, when the engine starts and cranking stops, this wire becomes part of the ignition primary circuit. If the wire were open, the engine would then stall.

2. Engine stalls as it warms up This could result if the choke valve were stuck closed. The mixture becomes too rich for a hot engine, and the engine stalls. If the manifold heat-control valve sticks closed, the air-fuel mixture might become overheated and too lean, causing the engine to stall. If the hot-idle speed is too low, the engine may stall as it warms up because the idling speed drops too low. Also, stalling may be caused by overheating of the engine, which could cause vapor lock. In addition, stalling could be caused by the damper in the thermostatic air cleaner sticking closed. This would have the same effect as a stuck manifold heat-control valve. The mixture can become overheated and too lean to support combustion so the engine stalls.

3. Engine stalls after idling or slow-speed driving This could occur if the fuel pump has a cracked diaphragm, weak spring, or defective valve. The pump fails to deliver enough fuel for idling or slow-speed operation (although it could deliver enough for high-speed operation). If the carburetor float level is set too high or the idle adjustment is too rich, the engine may "load up" and stall. A lean idle adjustment may also cause stalling. The engine may overheat during sustained idling or slow-speed driving. With this condition, air movement through the radiator may not be sufficient to keep the engine cool. Overheating, in turn, could cause vapor lock and engine stalling. (See ● 42-10 for causes of overheating.)

One action that may help when you are stuck in slow-moving traffic so your engine begins to overheat is this: Open the car windows and turn off the air conditioner. Turn on the heater to maximum. This may make you temporarily uncomfortable, but the heater will take some of the heat from the engine. Often, this makes the difference between an engine boilover and simply a hot engine.

4. Engine stalls after high-speed driving This could occur if enough heat accumulates to cause a vapor lock.

The remedy here would be to shield the fuel line and fuel pump or use a less volatile fuel. Failure of the venting or idle-compensator valve in the carburetor may also cause stalling after high-speed or full-load operation. Excessive overheating of the engine is also a primary cause of stalling (● 42-10).

● 42-13 Engine Backfires

Most backfiring is caused by a faulty antibackfire valve. It could also be due to late ignition timing, or ignition cross-firing (caused by the spark jumping across the distributor cap or through the cable insulation). In addition, it could be due to spark plugs of the wrong heat range (which overheat and cause preignition), excessively rich mixtures (caused by fuel-pump or carburetor troubles), overheating of the engine (● 42-10), carbon in the engine, hot valves, or intake valves that stick or seat poorly. Carbon in the engine, if excessive, may retain enough heat to cause the air-fuel mixture to preignite as it enters the cylinder, so that backfiring occurs. Carbon also increases the compression ratio and thus the tendency for knocking and preignition. Hot plugs may cause preignition; cooler plugs should be installed. If intake valves hang open, combustion may be carried back into the carburetor. Valves which have been ground excessively so that they have sharp edges, valves which seat poorly, or valves which are carboned so that they overheat often produce backfiring.

● 42-14 Engine Run-On or Dieseling

Modern engines, with their emission controls, require a fairly high hot idle for best operation. This makes run-on, or dieseling, possible. If there are hot spots in the combustion chambers, the engine can continue to run if the throttle is not completely closed. The hot spots take the place of the spark plugs. If the throttle is slightly open, enough air-fuel mixture could get past it to keep the engine running. Ignition in the combustion chambers would be caused by the hot spots. Modern engines have an idle-stop solenoid to close the throttle completely when the ignition switch is turned off.

If an engine runs on, or diesels, check the idle-stop solenoid (if present). Make sure it is releasing when the ignition is turned off. It could require adjustment to permit the throttle to close completely. Be sure the engine speed is not set too high. The trouble could also be due to advanced ignition timing. Correction of engine overheating is covered in ● 42-10. Correcting hot spots may require spark-plug service, or removing the cylinder head for cleaning, plus valve service.

● 42-15 Too Much HC and CO in Exhaust

If the exhaust-gas analyzer (● 41-7) discloses that there is too much HC and CO in the exhaust, correction must be made. Excessive CO can usually be brought within specifications by proper adjustment of the carburetor. Excessive HC usually can be corrected by proper service and adjustment of the ignition system, and by replacement of worn parts in the engine. Some states require exhaust-gas testing of all cars during state inspection. Cars that emit too much CO and HC must be repaired before they can be licensed. This restriction is designed to get the "smoggers" off the highway. Here are the possible causes (the corrections are obvious):

1. Missing due to ignition problems such as faulty plugs, high-voltage wiring, distributor cap, ignition coil, condenser, contact points, or electronic control.
2. Incorrect ignition timing.
3. Carburetor troubles such as the choke sticking closed, worn jets, high float level, and other conditions listed in ● 42-18.
4. Faulty air-injection system which does not inject enough air into the exhaust manifold to completely burn the HC and CO. This could be caused by a faulty air pump or a leaking hose or air manifold.
5. Defective transmission-controlled spark system which permits vacuum advance in all gear positions instead of high and reverse only.
6. Defective catalytic converters which must be replaced or serviced to restore the catalytic action.
7. Too much HC and CO in the exhaust gas can also be caused by excessive carbon deposits in the combustion chamber or by stuck or burned valves.

● 42-16 Excessive Oil Consumption

Oil is lost from the engine in three ways: (1) by burning in the combustion chamber, (2) by leakage in liquid form, and (3) by passing out of the crankcase through the crankcase ventilating system in the form of mist or vapor.

External leakage can often be detected by inspecting the seals around the oil pan, valve cover plate, and timing-gear housing, or at oil-line and filter connections.

Burning of oil in the combustion chamber gives the exhaust gas a bluish tinge. Oil can enter the combustion chamber through the PCV system, through the clearance between intake-valve or exhaust-valve stems and valve guides, and past piston rings.

If intake-valve-stem clearance is excessive, oil is "pulled" through this clearance, and into the combustion chamber, on each intake stroke. The appearance of the intake-valve stem often indicates that this is occurring. Some of the oil remains on the underside of the valve and stem to form carbon. Oil can also seep down past the exhaust-valve stem if the clearance is excessive. The remedy is to install valve seals or a new valve guide and possibly a new valve.

Probably the most common cause of excessive oil consumption is passage of oil into the combustion chamber between the piston rings and the cylinder walls. This is often called "oil pumping." It is due to worn, tapered, or out-of-round cylinder walls or worn or carboned rings.

In addition, when engine bearings are worn, excessive oil is thrown on the cylinder walls. The rings are not able to control all of it. Too much oil works up into the combustion chamber.

High speed must also be considered if there is excessive oil consumption. High speed means high temperatures and thus thin oil. More oil, and thinner oil, is thrown on the cylinder walls at high speed. The piston rings, moving at high speed, cannot function so effectively. So more oil works up into the combustion chamber. In addition, the churning effect of the oil in the crankcase creates more oil vapor and mist at high speed. More oil is thus lost through the crankcase ventilating system. Tests show that an engine uses several times as much oil at 60 miles per hour (mph) [97 km/h] as at 30 mph [48 km/h].

● 42-17 Low Oil Pressure

Low oil pressure is often a warning of worn oil-pump or engine bearings. The bearings can pass so much oil that the oil pump cannot maintain oil pressure. Also, the end bearings will probably be oil-starved and may fail. Other causes of low oil pressure are a weak relief-valve spring, a worn oil pump, a broken or cracked oil line, and a clogged oil line. Oil dilution, or foaming, sludge, insufficient oil, or oil made too thin by engine overheating will cause low oil pressure.

● 42-18 Excessive Fuel Consumption

This condition can be caused by almost anything in the car, from the driver to underinflated tires or a defective choke. A fuel-mileage tester can be used to accurately check fuel consumption (Fig. 42-3). The compression or leakage tester and the vacuum gauge (● 41-3 to 41-6) will help determine whether the trouble is in the engine, fuel system, ignition system, or elsewhere. Also, the exhaust-gas analyzer, dynamometer, and fuel-flow meter are useful in analyzing the problem.

A rough test of mixture richness that does not require any testing instruments is to install a set of new or cleaned spark plugs of the correct heat range for the engine and to operate the car for 15 or 20 minutes. Then stop the car, and remove and examine the plugs. If they are coated with a black carbon deposit, the mixture is too rich. (See a to g under item 4 below.) Black exhaust smoke is another indication of an excessively rich mixture: The mixture is too rich to burn fully, and so the exhaust gas contains "soot," or unburned fuel.

If the trouble seems to be in the fuel system, consider the following:

1. A driver who pumps the accelerator when idling and insists on being the first to get away when the stoplight changes will use excessive amounts of fuel.
2. Operation with the choke partly closed after warm-up will use excessive amounts of fuel.
3. Short-run operation means the engine will be operating on warm-up most of the time. This means fuel consumption will be high.

These three conditions are due to the type of operation. Changing operating conditions is the only cure.

4. If excessive fuel consumption is not due to operating conditions, the trouble is likely to be in the carburetor. It could be any of the following:
   a. If the car is equipped with an automatic choke, the choke may not be opening rapidly enough during warm-up or may not open fully. This can be checked by removing the air cleaner and observing choke operation during warm-up.
   b. A clogged air cleaner that does not admit sufficient air will act somewhat like a partly closed choke valve. The cleaner element should be cleaned or replaced.
   c. If the float level is high in the float bowl, it will cause flooding and delivery of excessive fuel to the carburetor air horn. The needle valve may be stuck open or may not be seating fully. The float level should be checked and adjusted.
   d. If the idle is set too rich or the idle speed too high, excessive fuel consumption will result. These should be checked and adjusted as necessary.
   e. Where the accelerator-pump circuit has a check valve, failure of the check valve to close properly may allow fuel to feed into the carburetor air horn. The carburetor will require disassembly for repair.
   f. The metering rod may be stuck in the high-speed full-throttle position, or the economizer valve may be held open. These permit the high-speed full-power circuit to function, supplying an excessively rich mixture. The carburetor will require disassembly for repair.
   g. Worn jets, permitting the discharge of too much fuel, require replacement during carburetor rebuilding.
5. Faulty ignition can also cause excessive fuel consumption. The ignition system could cause engine miss and thus failure of the engine to use all the fuel. This sort of trouble also is found with loss of power, acceleration, or high-speed performance (● 42-8). Conditions in the ignition system that might add to the trouble include a "weak" coil or condenser, incorrect timing, faulty advance-mechanism action, dirty or worn plugs or contact points, or defective wiring.
6. Inferior engine action can produce excessive fuel consumption. Examples are loss of engine compression from worn or stuck rings, worn or stuck valves, or a loose or burned cylinder-head gasket. Power is lost under these conditions, and more fuel must be burned to achieve the same speed. (Refer to ● 41-3 to 41-5 for compression and leakage checking procedures.)
7. Excessive fuel consumption can also result from conditions that make it hard for the engine to move the car along the road. Such factors as low tires, dragging brakes, defective automatic transmission, and misalignment of wheels increase the rolling resistance of

Fig. 42-3 Fuel mileage tester. A small container holding exactly 0.1 gallon [0.4 L] is mounted on the driver's door. *(Ford Motor Company)*

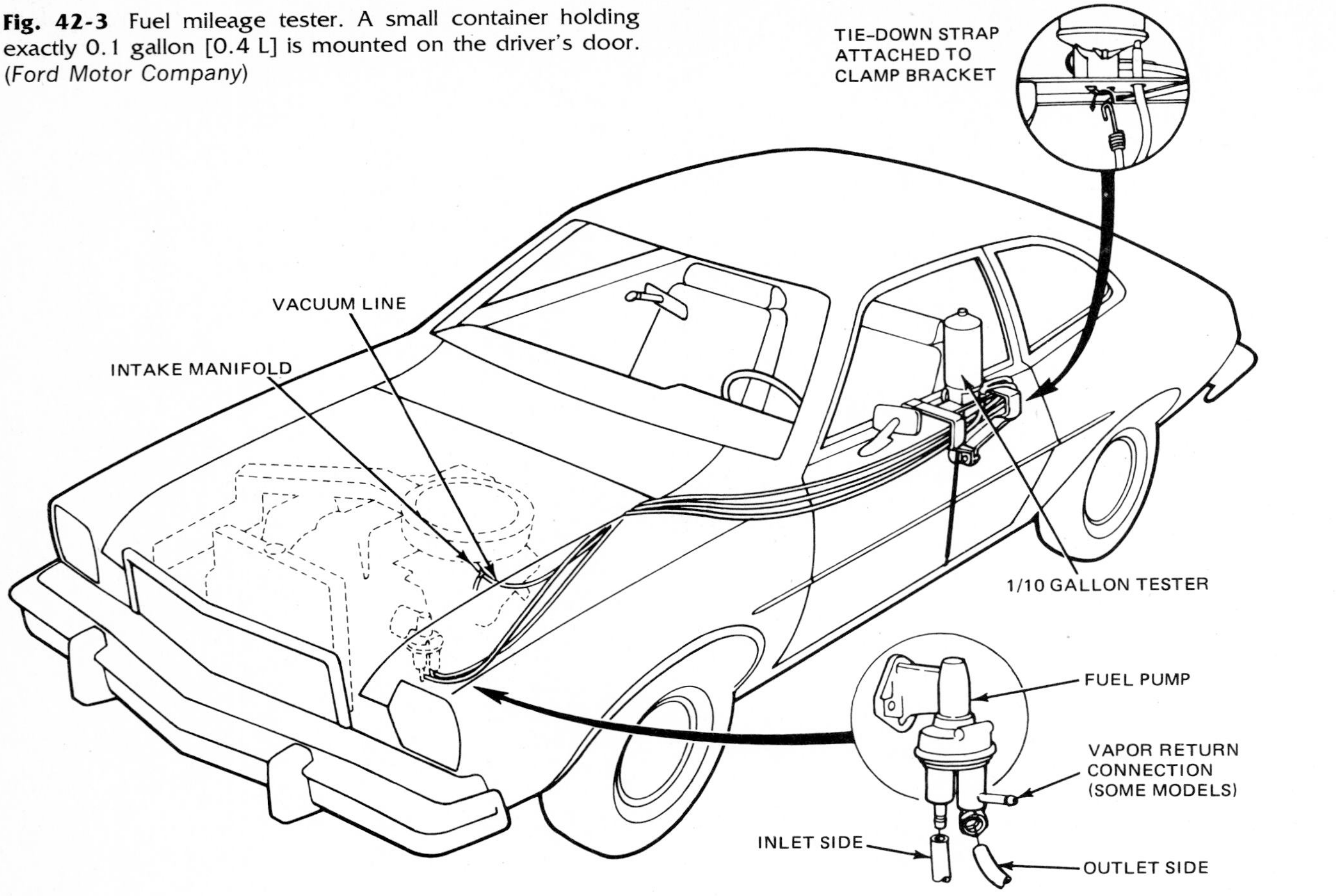

the car. The engine must use up more fuel to overcome this excessive rolling resistance.

● 42-19 Engine Noises

Some engine noises have little meaning. Other noises may indicate serious engine trouble that requires prompt attention to prevent major damage to the engine. Various noises and their causes are described below, along with tests that may be necessary to confirm a diagnosis.

A listening rod or stethoscope is of help in locating the source of a noise. The rod acts like the stethoscope that a doctor uses to listen to a patient's heartbeat or breathing. When one end is placed at the ear and the other end at some part of the engine, noises from that part of the engine are carried along the rod to the ear. A long screwdriver or one of the engine stethoscopes now available can be used. When using the listening rod to locate the source of a noise, put the engine end at various places on the engine until the noise is loudest. You can also use a piece of garden hose (about 4 feet [1.2 m] long) to place engine noises. Hold one end of the hose to your ear, and move the other end of the hose around the engine until the noise is loudest. In this way you can, for example, locate a broken and noisy ring in a particular cylinder, or a main-bearing knock.

CAUTION: Keep away from the moving fan belt and fan when using the listening rod.

1. Valve and tappet noise This is a regular clicking noise that usually gets louder as engine speed increases (sometimes disappears at higher engine speeds). The cause is usually excessive valve clearance or a defective hydraulic valve lifter. A feeler gauge inserted between the valve stem and lifter or rocker arm reduces the clearance. If the noise also is reduced, then the cause is excessive clearance. The clearance should be readjusted. If inserting the feeler gauge does not reduce the noise, it is the result of such conditions in the valve mechanism as weak springs, worn lifter faces, lifters loose in the block, rough adjustment-screw face, or rough cams. Or else the noise is not from the valves at all. (See other conditions listed below.)

2. Detonation Spark knock or detonation is a pinging or chattering sound most noticeable during acceleration or when the car is climbing a hill. Some spark knock is considered normal under such extreme operating conditions. When it becomes excessive, it is due to conditions such as the use of fuel of too low an octane rating for the engine, carbon deposits in the engine which increase compression ratio, advanced ignition timing, or the conditions described in ● 42-13.

3. Connecting-rod noises Connecting-rod noises usually have a light knocking or pounding character. The sound is most noticeable when the engine is "floating" (not accelerating or decelerating). The sound becomes more noticeable as the accelerator is eased off with the car running at medium speed. To locate connecting-rod noise, short out spark plugs one at a time. It is difficult to short some of the plugs on some engines. The easiest way to make the check is to use the oscilloscope (● 31-2 to 31-5) which can short out cylinders with the turn of a knob or push of a button. The noise will be considerably reduced when the cylinder that is responsible is not delivering power. A worn bearing or crankpin, a misaligned connecting rod, inadequate oil, and excessive bearing clearances cause connecting-rod noise.

4. Piston-pin noise Piston-pin noise is similar to valve and tappet noise, but it has a unique metallic double knock. It is usually most audible during idle with the spark advanced. However, on some engines, the noise becomes most audible at car speeds of around 30 mph. A check can be made by running the engine at idle with the spark advanced, and then shorting out spark plugs (or use the oscilloscope. See item 3 above). Piston-pin noise will be reduced somewhat when a plug in a noisy cylinder is shorted out. Causes of this noise are a worn or loose piston pin, a worn bushing, and lack of oil.

5. Piston-ring noise Piston-ring noise is also similar to valve and tappet noise. It is a clicking, snapping, or rattling noise. This noise, however, is most evident on acceleration. Low ring tension, broken rings, worn rings, or worn cylinder walls produce this noise. Since the noise can sometimes be confused with other engine noises, a test can be made as follows: Remove the spark plugs, and add an ounce or two of heavy engine oil to each cylinder. Crank the engine for several revolutions to work the oil down past the rings. Then replace the plugs, and start the engine. If the noise has been reduced, the rings are probably at fault.

6. Piston slap Piston slap is a muffled, hollow, bell-like sound. It is due to the rocking back and forth of the piston in the cylinder. If it occurs only when the engine is cold, it is not serious. When it occurs under all operating conditions, it should be checked further. It is caused by inadequate oil, worn cylinder walls, worn pistons, collapsed piston skirts, excessive piston clearances, or misaligned connecting rods.

7. Crankshaft knock Crankshaft knock is a heavy, dull metallic knock. It is most noticeable when the engine is under a heavy load or accelerating, particularly when cold. When the noise is regular and more of a rumble, it probably results from worn main bearings. Worn rod bearings produce a more distinct knock. When the noise is irregular and sharp, it is probably due to a worn end-thrust bearing. This latter condition, when unusually bad, will cause the noise to be produced each time the clutch is released and engaged, and when first accelerating.

8. Miscellaneous noises Other noises result from loosely mounted accessory parts, such as the alternator, starting motor, horn, water pump, manifolds, flywheel, crankshaft pulley, and oil pan. Other automotive components, such as the clutch, transmission, and differential, may also develop noises.

REVIEW QUESTIONS

1. Name five possible causes of the engine failing to turn over when starting is attempted. How do you use the lights to check for the cause?
2. Name four possible causes of the engine turning over slowly but not starting.
3. Name three possible causes of the engine turning over normally but not starting.
4. Name four possible causes of the engine missing.
5. Name eight causes of the engine lacking power, acceleration, or high-speed performance.
6. Name eight possible causes of the engine overheating.
7. What could cause the engine to stall as it warms up?
8. What could cause stalls after prolonged idling?
9. Name five possible causes of engine backfire.
10. Name three possible causes of engine run-on.
11. Name five possible causes of too much HC and CO in the exhaust.
12. Name three possible causes of excessive oil consumption.
13. What could cause black exhaust smoke?
14. What could cause blue exhaust smoke?
15. What could cause white exhaust smoke?
16. Name ten causes of excessive fuel consumption.
17. Explain the difference between valve-and-tappet noise and connecting-rod noise.
18. Describe piston-pin noise.
19. Describe piston-ring noise.
20. Describe piston slap.

SELF PROJECT

Here is a way to really learn engine troubles and their causes—that is, how to relate causes with troubles. Get yourself a package of 3-by-5-inch cards. As you study each trouble and its causes in the chapter, write the trouble on one side of a card, and the possible causes on the other side. For example, you read the complaint "1. Engine will not turn over" in the chart, and then study ● 42-2, which discusses the complaint in detail. You would write "Engine will not turn over" on one side of a card. You would also write "8 causes" on the same side of the card. This is because there are 8 causes (a to h) listed in the chart. Then, on the other side of the card, you would write the causes listed in the chart. Now, after you go through the entire chart doing this, and have studied the entire chapter, you are ready to play a game of cards. First you pick a card and read the complaint side. Then you try to remember what is written on the other side—that is, what are the possible causes of the complaint. Continue to do this until you know engine troubles and their causes.

CHAPTER 43
Engine Tune-up

After you have studied this chapter you should be able to:

- **Explain what tune-up means to the expert technician.**
- **Describe the tune-up procedure.**
- **List the items to be checked or serviced during a tune-up.**
- **Under the instructor's supervision, perform a tune-up job.**

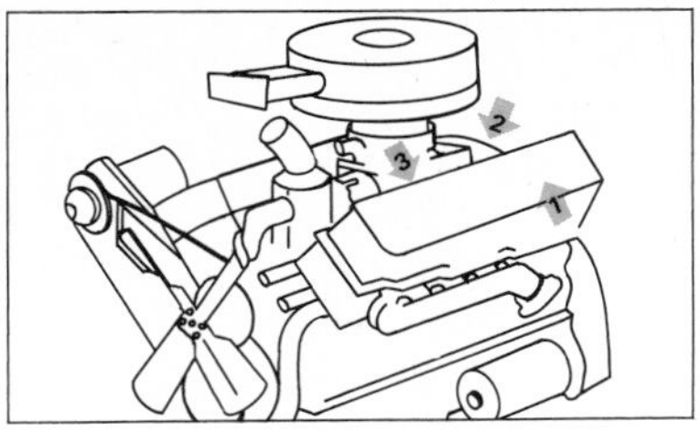

This chapter describes the procedure known as *engine tune-up*. Tune-up includes testing the various components and accessory systems involved in engine operation. Tune-up goes farther than just testing, however. It also includes readjusting or replacing parts as required to restore engine performance. In some cases, during a tune-up, serious problems may be uncovered that will require major repair work. Previous chapters have described the various service jobs that may be performed on engines and accessory systems. The various testing procedures have also been described. None of these will be repeated in the pages that follow. When we refer to a specific test or repair operation, turn back to the chapter in which the procedure is described if you need to refresh your memory.

● 43-1 What Tune-Up Is

Engine tune-up means different things to different people. To some, it means a light once-over check of the engine that takes in only the more obvious trouble spots. To others, it means use of the proper test instruments to do a careful, complete analysis of all engine components. In addition, it means adjusting everything to "specs" and repairing or replacing all worn parts. The latter is the proper meaning of engine tune-up; it is the procedure outlined in this chapter.

In this chapter, we have put together two separate programs, engine tune-up and complete car-care inspection. Engine tune-up refers to the things you check and service on the engine and its systems. Car-care inspection includes checking all other components on the car, such as brakes, steering, tires, and so on. An engine tune-up, plus a car-care inspection, covers everything on and in the car that could cause trouble. Some shops are finding that, when a customer comes in for an engine tune-up, they can also sell a car-care inspection. Likewise, when a customer comes in for a car-care inspection, they can also sell an engine tune-up.

● 43-2 Tune-Up Procedure

An engine tune-up follows a more or less set procedure. Many mechanics use a printed form supplied by automotive or test-equipment manufacturers. By following the form and checking off the items listed, one by one, the mechanic is sure of not overlooking any part of the procedure. However, all tune-up forms are not the same. Different companies have different ideas about what should be done, and the order in which it should be done. In addition, the tune-up procedure depends on the equipment available. If the shop has an oscilloscope or a dynamometer, it is used as part of the tune-up procedure. If these test instruments are not available, then a tune-up is performed differently.

The procedure that follows includes car-care inspection. It lists all essential checks and adjustments, in what authorities believe is the most logical sequence.

43-3 Tune-Up and Car Care

The tune-up procedure restores driveability, power, and performance that have been lost through wear, corrosion, and deterioration of engine parts. These changes take place gradually in many automotive parts during normal car operation. Because of federal laws limiting automotive emissions, the tune-up procedure must include checks of all emission controls. Here is the procedure:

1. If the engine is cold, operate it for 15 to 20 minutes at 1,500 rpm (revolutions per minute), or until it reaches operating temperature.
2. Connect the oscilloscope, if available, and perform an electronic diagnosis. Check for any abnormal ignition-system conditions that appear on the pattern. Make a note of any abnormality and the cylinders in which it appears.
3. Remove all spark plugs. Fully open the throttle and choke valves. Disconnect the distributor primary lead from the coil so the engine will not start.
4. Check the compression of each cylinder. Record the readings. If one or more cylinders read low, squirt about a tablespoon of engine oil through the spark-plug hole. Recheck the compression, and record the new readings.
5. Clean, inspect, file, gap, and test the spark plugs. Discard worn or defective plugs. Gap all plugs, old and new. Install the plugs.
6. Inspect and clean the battery case, terminals, cables, and hold-down brackets. Test the battery. Add water, if necessary. If severe corrosion is present, clean the battery and cables with brushes and a solution of baking soda and water.
7. Test the starting voltage. If the battery is in good condition but cranking speed is low, test the starting system.
8. If the battery is low, or the customer complains that the battery keeps running down, check the charging system (alternator and regulator). If the battery is old, it may have worn out. A new battery is required.
9. Check the drive belts, and replace any in poor condition. If you have to replace one belt of a two-belt drive, replace both belts. Tighten belts to the correct tension, using a tension gauge.
10. Inspect the distributor rotor, cap, and primary and high-voltage (spark-plug) wires (Fig. 30-2).
11. Clean or replace and adjust distributor contact points by setting the point gap. Lubricate the distributor breaker cam if specifications call for this. On distributors with cam lubricators (Fig. 31-21), specifications call for turning the cam lubricator 180 degrees every 12,000 miles [19,311 km], and replacing the cam lubricator every 24,000 miles [38,624 km].
12. Check the centrifugal and vacuum advances. Set the contact dwell, and then adjust ignition timing. Make sure idle speed is not excessive because this could produce centrifugal advance.
13. Use the oscilloscope to recheck the ignition system. Any abnormal conditions that appeared in step 2, above, should now have been eliminated.
14. Check the manifold heat-control valve. Lubricate it with heat-valve lubricant. Free up or replace the valve, if necessary.
15. Check fuel-pump operation with a fuel-pump tester. Replace the fuel filter. Check the fuel-tank cap, fuel lines, and connections for leakage and damage.
16. Clean or replace the air-cleaner filter. If the engine is equipped with a thermostatically controlled air cleaner, check the operation of the control damper.
17. Check the operation of the choke and the fast-idle cam. Check the throttle valve for full opening, and the throttle linkage for free movement.
18. Inspect all engine vacuum fittings, hoses, and connections. Replace any brittle or cracked hose.
19. Clean the engine oil-filler cap, if a filter-type oil-filler cap is used.
20. Check the cooling system (Fig. 21-1). Inspect all water hoses and connections, the radiator, water pump, and fan clutch, if used. Check the strength of the antifreeze, and record the reading. Pressure-check the system and the radiator cap. Replace any defective hose.
21. Check and replace the PCV valve, if necessary (see Fig. 43-1). Clean or replace the PCV filter, if required. Inspect the PCV hoses and connections. Replace any cracked or brittle hose.
22. If the engine is equipped with an air-pump type of exhaust-emission control, replace the pump-inlet air filter, if used. Inspect the system hoses and connections. Replace any brittle or cracked hose.
23. If the vehicle is equipped with a fuel-vapor recovery system, replace the charcoal-canister filter.
24. Check the transmission-controlled vacuum spark-advance system, if the vehicle is so equipped.
25. On engines equipped with an EGR system, inspect and clean the exhaust-gas-recirculation valve. Inspect and clean the EGR discharge port.
26. Tighten the intake-manifold and exhaust-manifold bolts to the proper tension in the proper sequence.
27. Adjust the engine valves, if necessary.
28. Adjust the carburetor idle speed. Use an exhaust-gas analyzer, and adjust the idle-mixture screw. Check the CO and HC in the exhaust gas.

Many mechanics check the CO and HC both before and after the tune-up job to show how much the tune-up has reduced these pollutants.

29. Road test the car on a dynamometer or on the road. Check for driveability, power, and idling. Any abnormal condition should be noted on the repair order before you return the car to the customer.

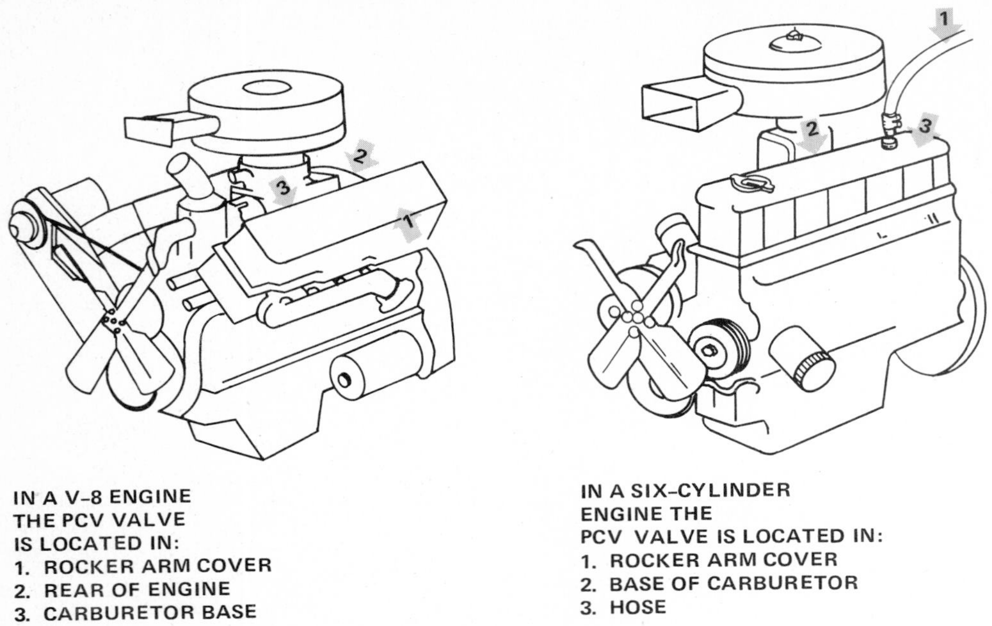

Fig. 43-1 PCV-valve locations.

30. Check the door-jamb sticker to determine if an oil and oil-filter change is due. Also note the schedule for chassis lubrication. Recommend an oil change and a lube job if they are due. Note that car manufacturers recommend changing the oil filter every time—or every other time—the oil is changed.

31. Whenever the car is on the lift, check the exhaust system for leaks which could admit CO into the car. Also check for loose bolts, rust spots, and other under-the-car damage.

Items 32 to 37 that follow are not actually part of the tune-up job. They are included here so you will have the complete car-care program all in one place.

32. Check the brakes for even braking and adequate braking power.

33. Check the steering system for ease and smoothness of operation. Check for excessive play in the system. Record any abnormal conditions.

34. Check the tires for inflation and for abnormal wear. Abnormal wear can mean suspension trouble, and a front-alignment job should be recommended.

35. Check the suspension system for looseness, excessive play, and wear.

36. Check the front wheels and ball joints for excessive wear or loose bearings. Adjust the bearings if necessary.

37. Check the headlights and horns to make sure they are in good working order. Check all other lights. Replace any burned-out lights. Check the headlight alignment, if possible.

As you can see, the comprehensive list above covers about everything on the vehicle that could cause trouble. The complete procedure therefore will uncover any problems that might affect driveability and performance. If all necessary corrective steps are taken, new-car performance will be restored to the vehicle.

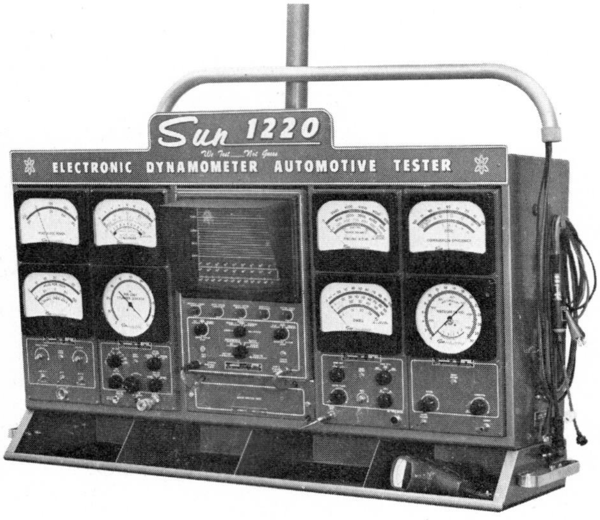

Fig. 43-2 Electronic diagnosis engine tester, or engine analyzer. The tester includes an oscilloscope (top center) and meters to check the condenser, distributor contact-point dwell, engine speed, and other items. (*Sun Electric Corporation*)

● 43-4 Engine Analyzers

Figure 43-2 shows an engine analyzer. It includes an oscilloscope and other instruments for making comprehensive tests of all engine components. Once you learn how to use this equipment, you can make a complete engine analysis in a very short time.

In addition, there are testers that run many of the tests almost automatically. They produce a printed record of the tests and their results.

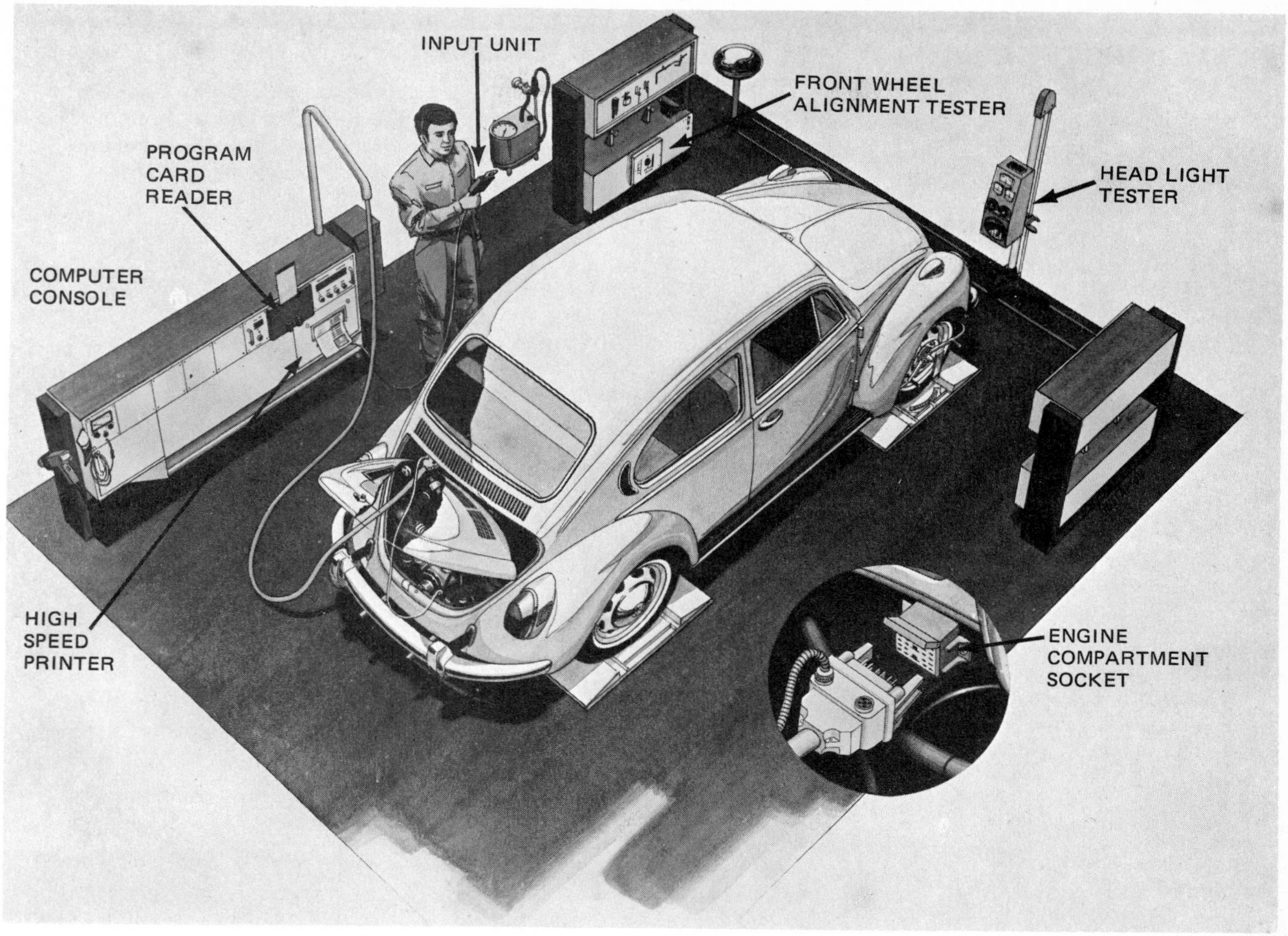

Fig. 43-3 Computerized automobile diagnostic system. (*Volkswagen of America, Inc.*)

For example, Figure 43-3 shows a computerized diagnostic system introduced by Volkswagen. Wiring and sensors built into the car are connected to the computer through a socket in the engine compartment. The system makes more than 70 checks. A special program card contains the specifications for the year and model of car being checked. One type of card, used in another system, is shown in Fig. 43-4 on page 452. The computer compares the operation of components on the car with values it reads from the card. The results are recorded by a high-speed printer.

One further refinement has been suggested. This is to put into the computer information on the costs of parts and repair operations. Then the computer could print out, along with the test information, the cost of fixing any troubles. That is, it would print out the costs of parts and labor. It has also been suggested that the computer could be programmed to schedule the work, depending on the the manpower and space available in the shop.

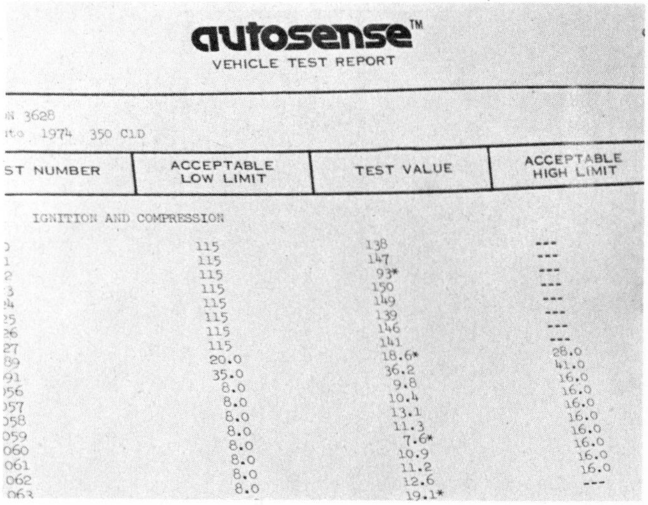

autosense™
VEHICLE TEST REPORT

N 3628
to 1974 350 CID

ST NUMBER	ACCEPTABLE LOW LIMIT	TEST VALUE	ACCEPTABLE HIGH LIMIT
IGNITION AND COMPRESSION			
0	115	138	---
1	115	147	---
2	115	93*	---
3	115	150	---
4	115	149	---
25	115	139	---
26	115	146	---
27	115	141	---
89	20.0	18.6*	28.0
91	35.0	36.2	41.0
56	8.0	9.8	16.0
57	8.0	10.4	16.0
58	8.0	13.1	16.0
059	8.0	11.3	16.0
060	8.0	7.6*	16.0
061	8.0	10.9	16.0
062	8.0	11.2	16.0
063	8.0	12.6	16.0
		19.1*	---

Fig. 43-4 The computer printout tells the mechanic and the customer what work is needed. (*Hamilton Standard Division of United Aircraft Corporation*)

REVIEW QUESTIONS

1. Define the term "engine tune-up."
2. What does "car-care inspection" mean?
3. What is the purpose of the printed tune-up form?
4. What is the purpose of the door-jamb sticker?

SELF PROJECT

Study a standard printed tune-up form. Decide whether or not it is arranged in the most logical manner. Does it include emission-control-system checks? Make up a form that lists everything in the proper testing order.

Automotive Power Trains

8

In this part, we describe the automotive power train, which carries power from the engine to the car wheels. It consists of the clutch (on some cars), transmission, drive line, and differential. The front-engine, rear-wheel drive is by far the most common arrangement. This requires a long drive shaft to connect between the engine at the front and the rear wheels. However, front-wheel drive with the engine up front is becoming more popular, especially on smaller cars. With this arrangement, the long drive shaft to the rear is eliminated. This also eliminates the tunnel in the floor pan of the car required for the drive shaft. There are eight chapters in Part Eight.

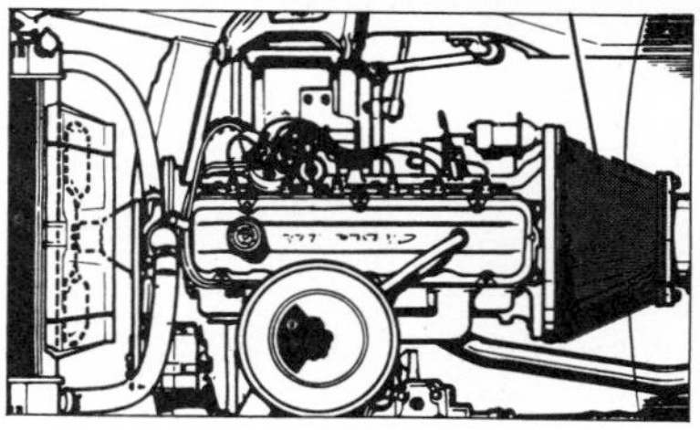

Chapter 44 Clutches

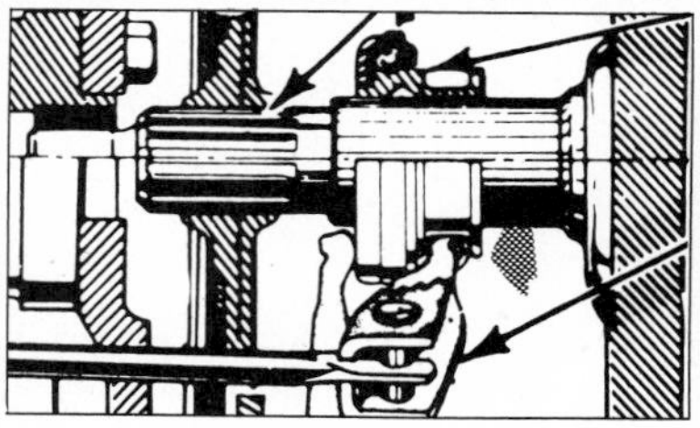

Chapter 45 Clutch Service

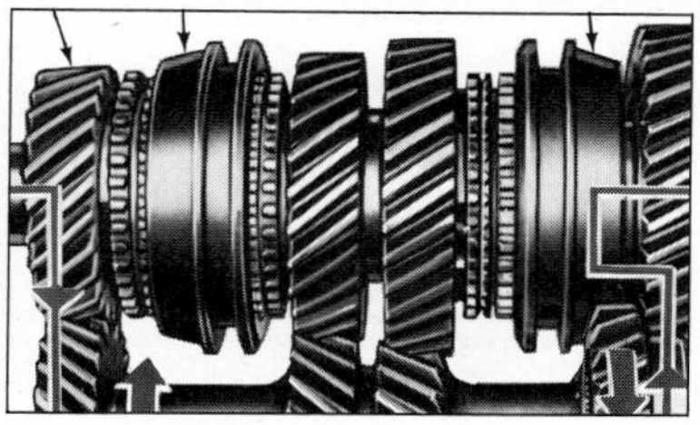

Chapter 46 Manual Transmissions

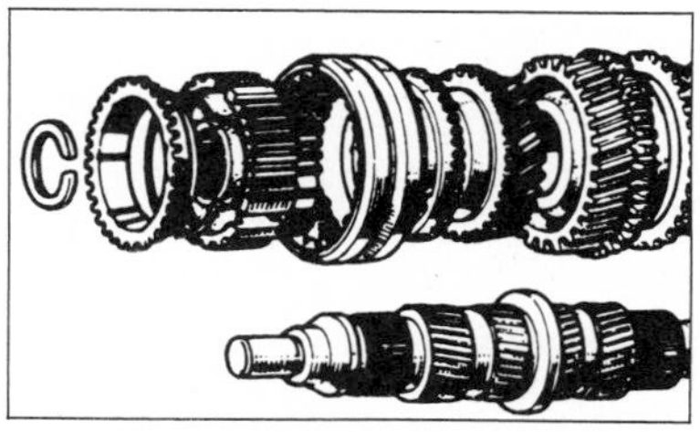

Chapter 47 Manual-Transmission Service

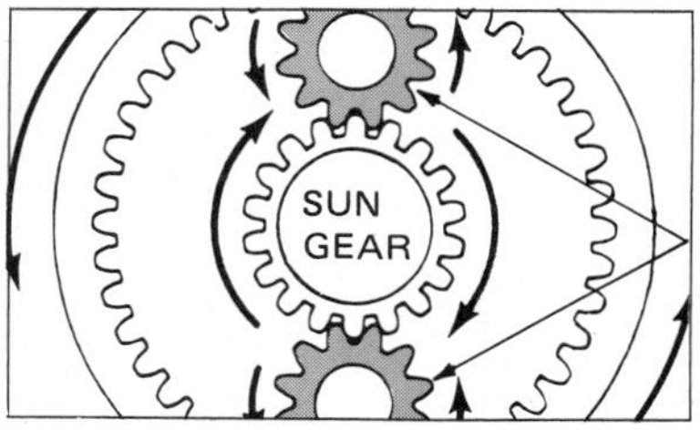

Chapter 48 Automatic Transmissions

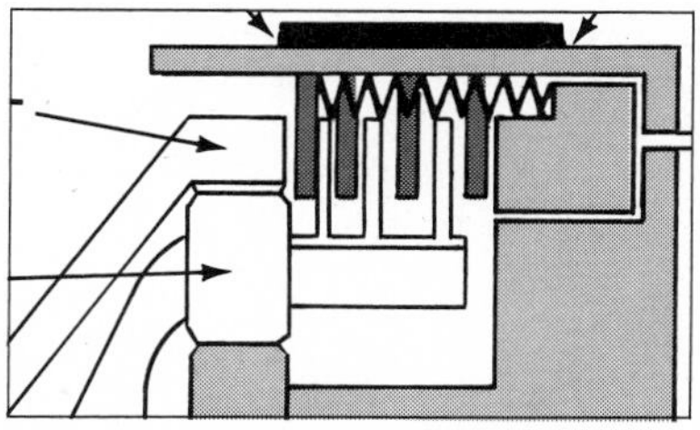

Chapter 49 Automatic-Transmission Service

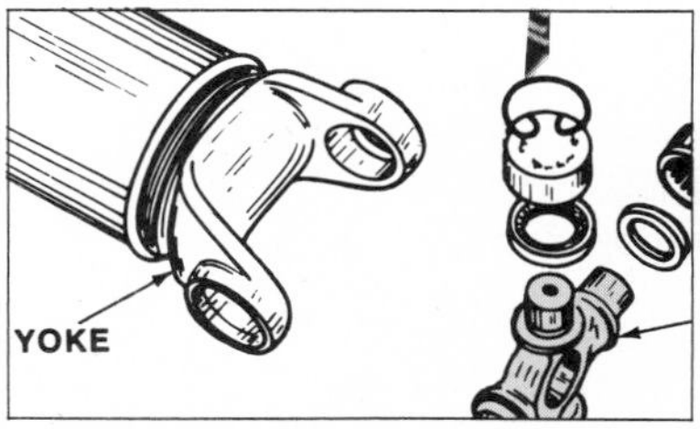

Chapter 50 Drive Lines and Universal Joints

Chapter 51 Rear Axles and Differentials

CHAPTER 44
Clutches

After studying this chapter, you should be able to:

- **Explain why a clutch is needed on cars with manual transmissions.**
- **Describe the construction and operation of a typical clutch.**
- **List and describe the various types of clutches. Explain how they work.**

This chapter discusses the purpose, construction, and operation of automotive clutches. The clutch is located in the power train, between the engine and the transmission (Fig. 44-1). At one time, all cars had clutches. However, most cars now have automatic transmissions which do not require clutches. There are still many cars on the road with manually shifted transmissions, and many new cars are so equipped. So you should know about clutches.

● 44-1 Purpose of the Clutch

The clutch is used on cars with transmissions that are shifted by hand. It allows the driver to couple the engine to, or uncouple the engine from, the transmission. The driver operates the clutch with a foot pedal which is linked to the clutch (Fig. 44-2). When the clutch is applied (in the normal running position), the power from the engine can flow through the clutch and enter the transmission. When the driver pushes down on the clutch pedal, the clutch is released and the engine is uncoupled from the transmission. Then no power can flow through. It is necessary to interrupt the flow of power—to uncouple the engine—in order to shift gears.

● 44-2 Location of the Clutch

The clutch is located just behind the engine, between the engine and the transmission (Fig. 44-1). Figure 44-3 shows the parts of one type of clutch. When the parts are assembled, the flywheel is bolted to the crankshaft. The friction disk (also called the *driven plate*) is mounted on the end of the transmission shaft, which sticks through the clutch housing. The pressure-plate-and-cover assembly is bolted to the flywheel. Then, the other parts are installed (● 44-3). All parts are enclosed by the clutch housing, which is a protective cover. The clutch linkage pivots on the clutch housing; the housing also serves as part of the transmission housing.

● 44-3 Construction of the Clutch

There are different kinds of clutches. We start by looking at one of the most common types, the *coil-spring* clutch. This clutch has a series of coil springs set in a circle. Figures 44-2 and 44-4 show sectional and cutaway views of this clutch. Figure 44-5 shows this type of clutch disassembled, without the flywheel or the transmission shaft. The friction disk (or driven plate) is about a foot in diameter, and it is mounted on the transmission input shaft. The disk has splines in its hub that match the splines on the input shaft (see Fig. 44-4). These splines consist of two sets of teeth. The internal teeth in the hub of the friction disk match the external teeth on the shaft. When the friction disk is driven, it turns the transmission input shaft. The end of the transmission input shaft rides in a pilot bushing in the end of the crankshaft (see Fig. 44-6).

The clutch also has a pressure-plate assembly, which includes a series of coil springs. The pressure-plate-and-cover assembly is attached to the engine flywheel (Fig. 44-3). The springs provide the pressure to hold the friction disk against the flywheel. Then, when the flywheel turns, the pressure plate and the friction disk also turn. However, when the clutch is released, the spring pressure

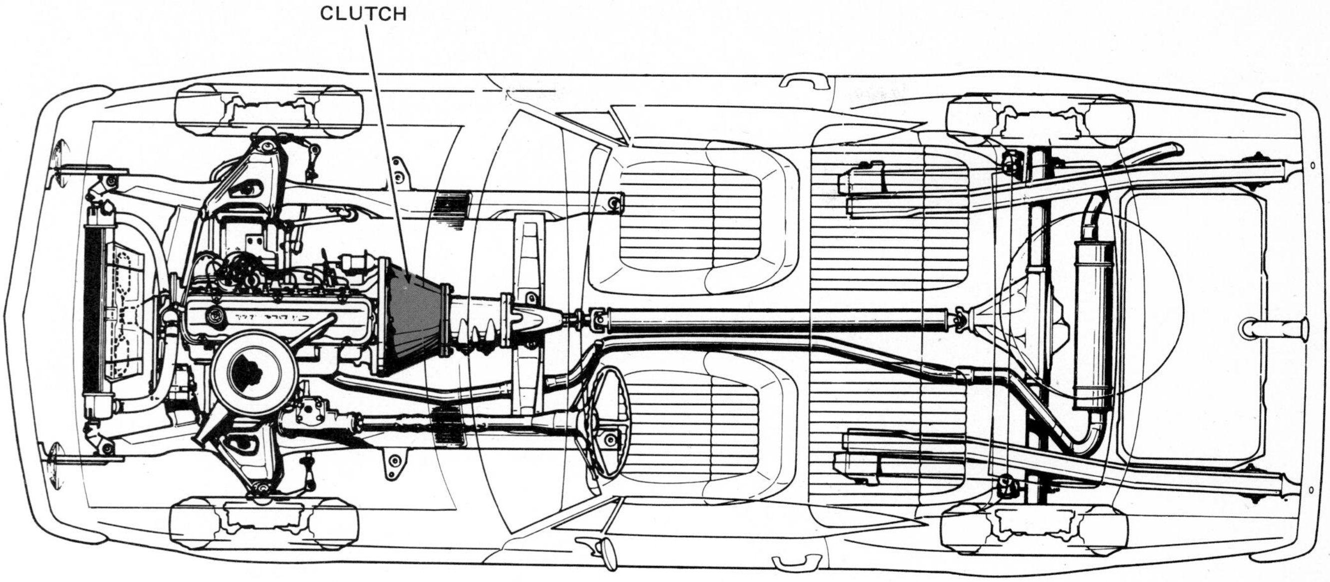

Fig. 44-1 Location of the clutch in the power train.

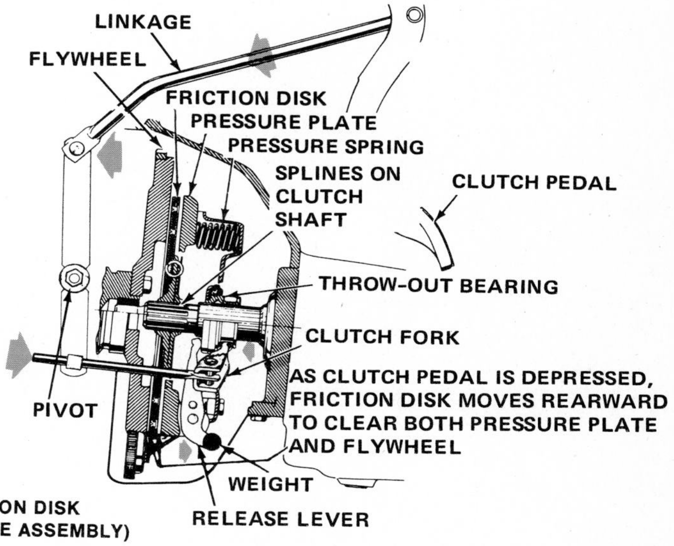

Fig. 44-2 Sectional view of a clutch, with the linkage to the clutch pedal. (*Buick Motor Division of General Motors Corporation*)

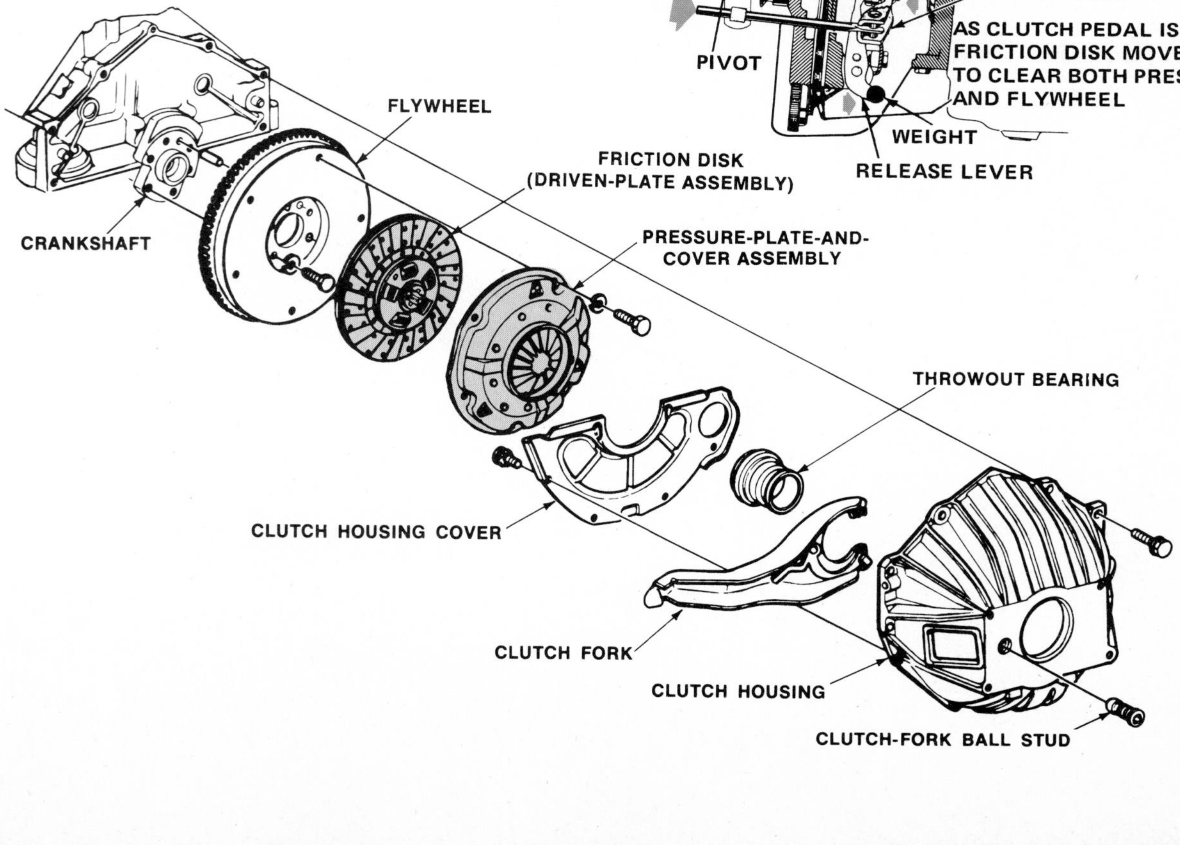

Fig. 44-3 Disassembled clutch and flywheel assembly. (*Chevrolet Motor Division of General Motors Corporation*)

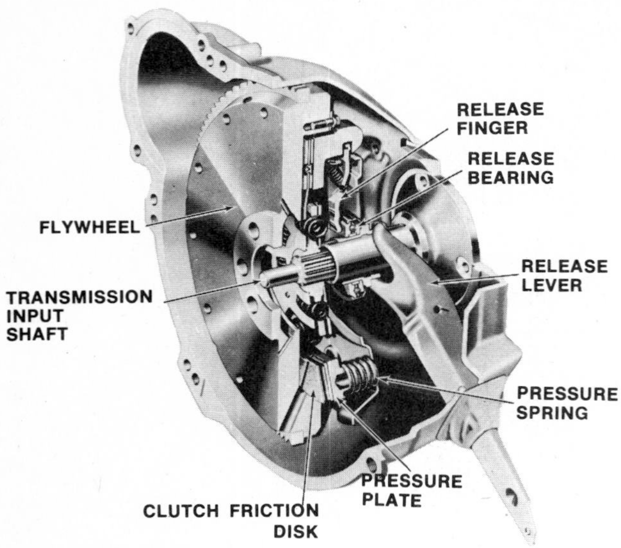

Fig. 44-4 Partial cutaway of a clutch. (*Ford Motor Company*)

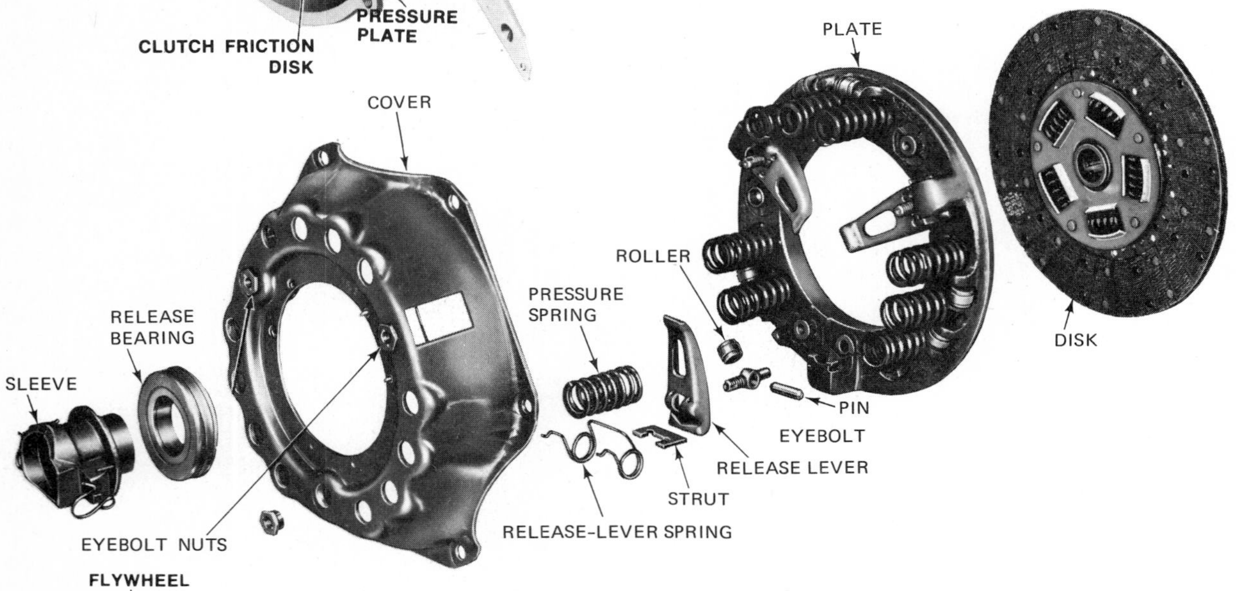

Fig. 44-5 Disassembled nine-spring clutch. Flywheel and transmission shaft are not shown. (*Chrysler Corporation*)

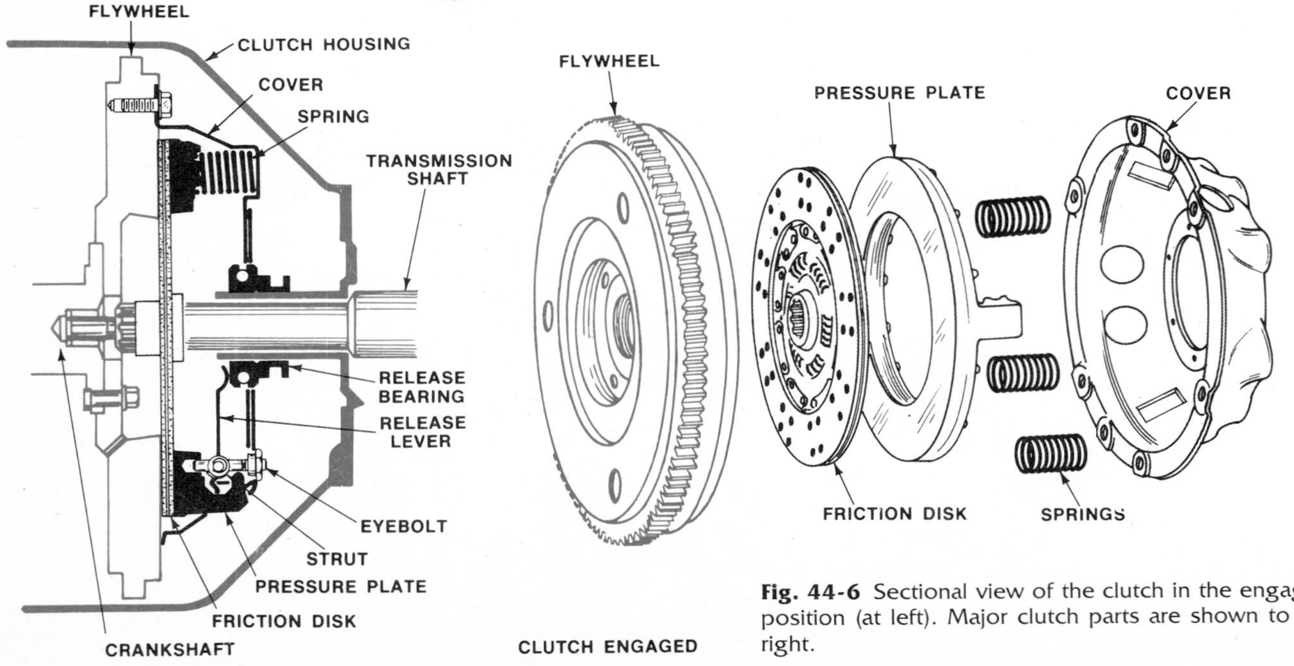

Fig. 44-6 Sectional view of the clutch in the engaged position (at left). Major clutch parts are shown to the right.

is relieved so that the friction disk and the flywheel can rotate separately.

● 44-4 Operation of the Clutch

Let's look at the operation of the coil-spring clutch. Figure 44-6 shows, at the left, a sectional view of the clutch. The major parts are shown disassembled at the right; nine springs are used, although only three are shown. Now look at the assembled view at the left. Note that the springs are held between the clutch cover and the pressure plate. In the condition shown in Fig. 44-6, the springs are clamping the friction disk (driven plate) tightly between the flywheel and the pressure plate (at the right in Fig. 44-7). As we mentioned, this forces the friction disk to rotate with the flywheel. In other words, the clutch is engaged.

Now look at what happens when the driver operates the clutch pedal to disengage, or release, the clutch (Fig. 44-8). When the clutch is released, the linkage from the pedal forces the release bearing inward (to the left in Fig. 44-8). Note that the release bearing is also called the *throwout* bearing. We will get to the linkage in ● 44-5. First, let's see what happens when the release bearing is forced to the left.

As the release bearing moves to the left, it pushes against the inner ends of three release levers (also called *release fingers*). The release levers are pivoted on eyebolts, as shown in Figs. 44-2, 44-4, and 44-8. When the inner ends of the release levers are pushed in by the release bearing, the outer ends are moved to the right. This motion is carried by struts to the pressure plate (Fig. 44-9). The pressure plate is thus moved to the right (in Fig. 44-8), and the springs are compressed. With the spring pressure off the friction disk, space appears between the disk, the flywheel, and the pressure plate (at the left in Fig. 44-7). Now the clutch is released, and the flywheel can rotate without sending power through the friction disk.

Releasing the clutch pedal takes the pressure off the release bearing. The springs push the pressure plate to the left (in Fig. 44-6). The friction disk is again clamped tightly between the flywheel and the pressure plate (at the right in Fig. 44-7). The friction disk must again rotate with the flywheel. In other words, the clutch is engaged.

Note that there must be some free play in the linkage system. That is, the clutch pedal must move an inch or so

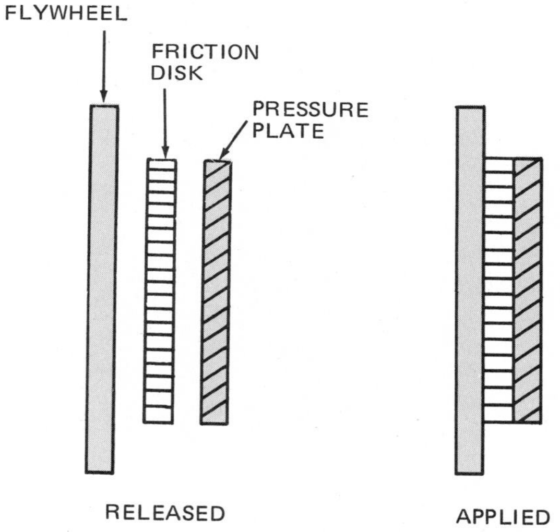

Fig. 44-7 Basic clutch action. Left, clutch released. Pressure plate and friction disk are separated and have moved away from flywheel. Right, clutch applied. Pressure plate is clamping friction disk to flywheel so all have to rotate together.

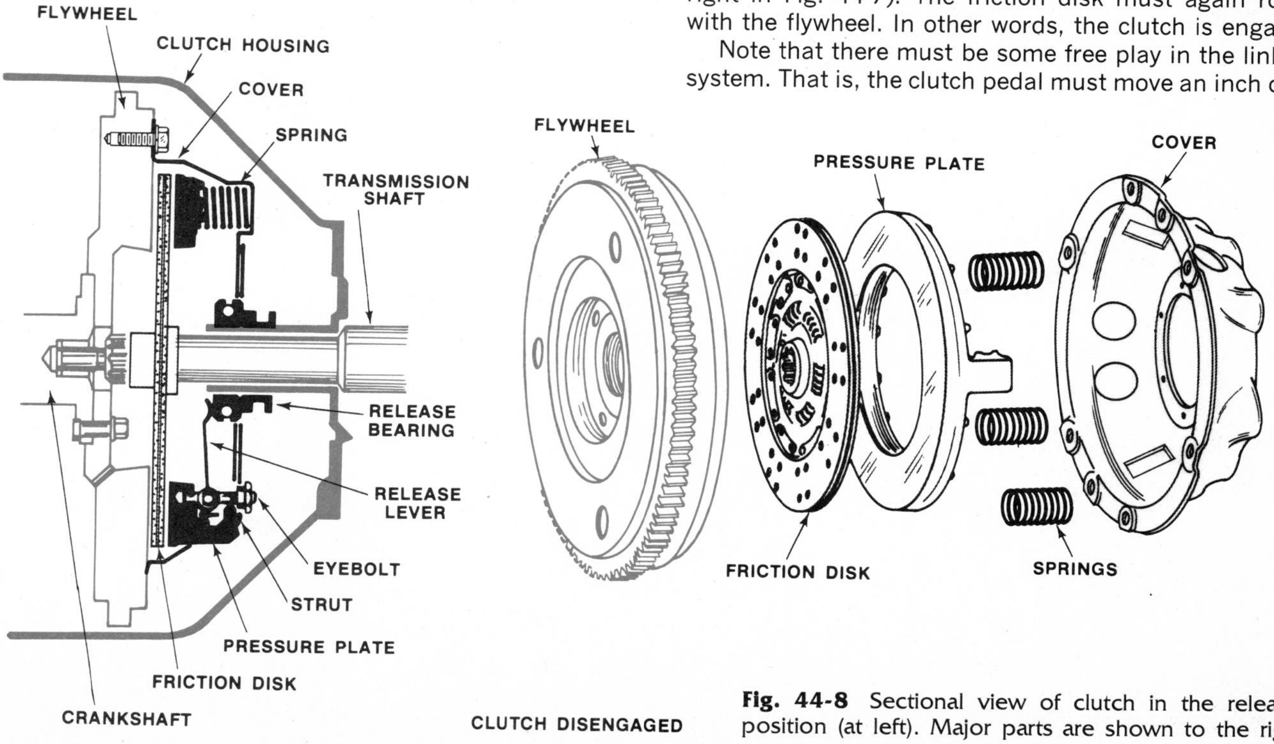

Fig. 44-8 Sectional view of clutch in the released position (at left). Major parts are shown to the right.

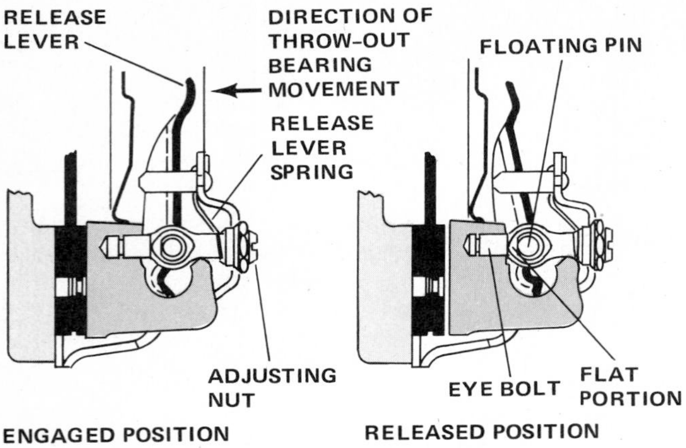

Fig. 44-9 The two limiting positions of the pressure plate and the release lever. (*Oldsmobile Division of General Motors Corporation*)

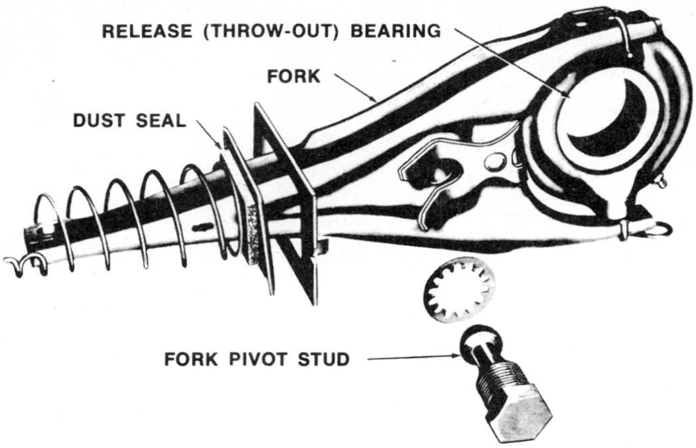

Fig. 44-10 Clutch-fork assembly with release (throwout) bearing. (*Buick Motor Division of General Motors Corporation*)

before all the play in the linkage is taken up and the release bearing comes up against the release levers. Without this free play, the release bearing would be riding on the ends of the release levers. This would cause rapid wear of the bearing and release levers.

● 44-5 Clutch Linkage

The clutch linkage carries the movement of the clutch pedal to the release bearing. A variety of clutch linkages are used. One of the simplest arrangements is shown in Fig. 44-2. Here, pushing down on the clutch pedal pushes the linkage, as shown by the arrows. This action causes the clutch fork to pivot and force the release (throwout) bearing in (at the left in Fig. 44-2). A typical clutch-fork assembly, with the release (throwout) bearing, is shown in Fig. 44-10. Note that the clutch fork has a spring-held dust seal to prevent dust from entering the clutch.

On the linkage for the coil-spring clutch shown in previous illustrations, there is an over-center spring, as shown in Fig. 44-11. The purpose of the over-center spring is to reduce the pedal pressure required in the final stages of declutching. Note that, in the illustrations of the coil-spring clutch, the coil springs have to be compressed to produce declutching. This requires an increasing amount of pressure to reach the fully declutched position. The over-center spring is arranged so that, as the lever to which it is attached pivots upward, the spring tension is applied to the clutch-pedal lever. That is, the over-center spring helps, in the final stages of declutching, to take up the coil-spring pressure.

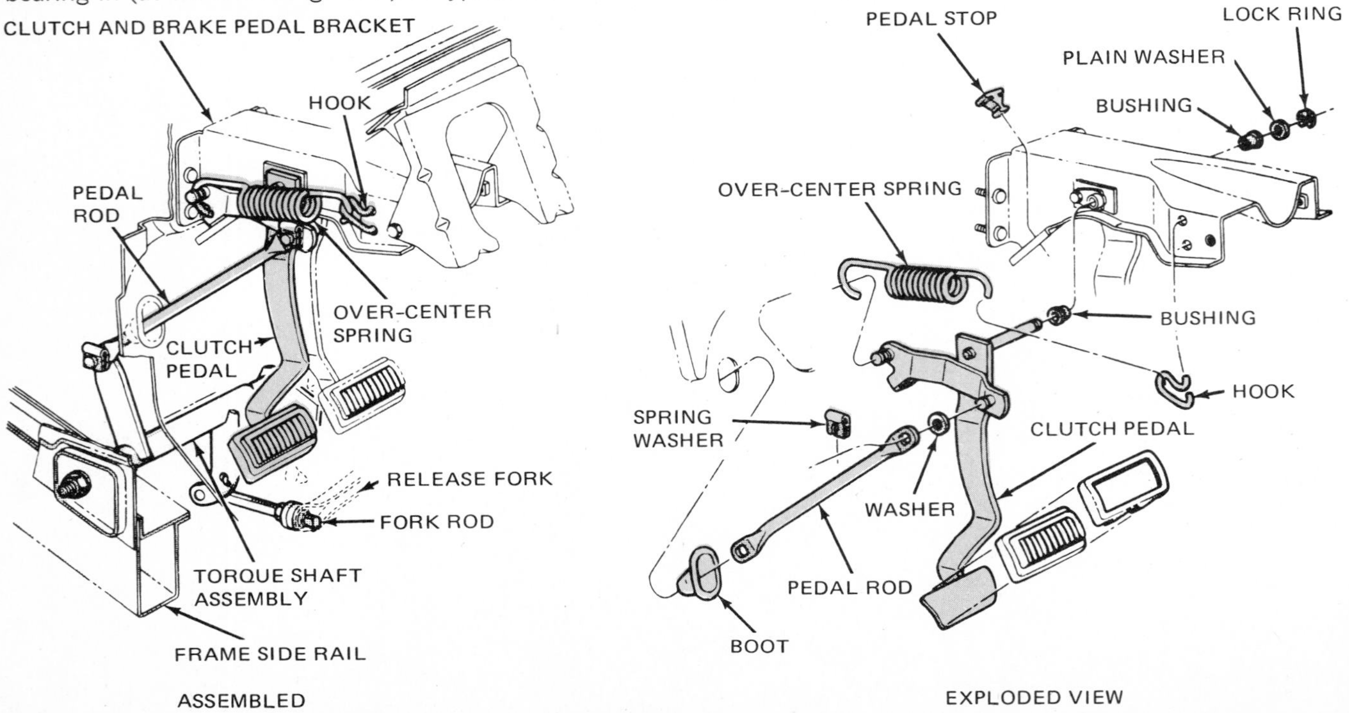

Fig. 44-11 One clutch-pedal-and-linkage arrangement. (*Chrysler Corporation*)

The linkage for the *diaphragm-spring* clutch (● 44-8) does not require an over-center spring. This is because no great increase in pedal pressure is required to complete declutching.

● 44-6 Friction Disk

The friction disk, or driven plate, is shown partly cut away in Fig. 44-12. It consists of a hub and a plate, with facings attached to the plate. The friction disk has cushion springs and dampening springs. The cushion springs are waved, or curled, slightly. The cushion springs are attached to the plate, and the friction facings are attached to the springs. When the clutch is engaged, the springs compress slightly to take up the shock of engagement. The dampening springs are spiral springs set in a circle around the hub. The hub is driven through these springs. They help to smooth out the power pulses from the engine so that the power flow to the transmission is smooth.

Note that there are grooves in both sides of the friction-disk facings. These grooves prevent the facings from sticking to the flywheel face and pressure plate when the clutch is released. The grooves break any vacuum that might form and cause the facings to stick to the flywheel or pressure plate.

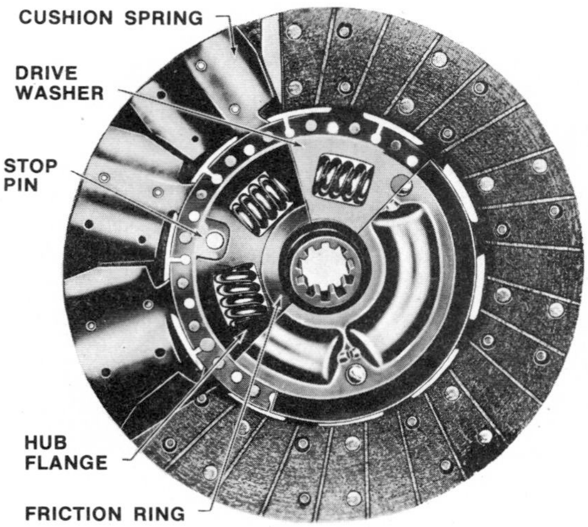

Fig. 44-12 Friction disk, or driven plate. Facings and drive washer have been partly cut away to show springs. (*Buick Motor Division of General Motors Corporation*)

● 44-7 Types of Clutches

We have already described the nine-spring clutch (illustrated in Figs. 44-4 to 44-9). Now let's look at other types of clutches.

A three-coil-spring clutch is shown in Fig. 44-13. It works the same way as the nine-spring clutch. There are also the *diaphragm-spring* clutch, the *centrifugal* clutch, and the *hydraulic* clutch. These are described in the following sections.

● 44-8 Diaphragm-Spring Clutch

The diaphragm-spring clutch is shown disassembled in Fig. 44-3. It is shown in sectional view in Fig. 44-14. The diaphragm spring is a round sheet of steel with a series of fingers pointing toward the center. You can see these fingers in the pressure-plate-and-cover assembly in Fig. 44-3. When the release bearing is forced inward against these fingers, the entire diaphragm dishes inward. This action is somewhat like the flexing action of the bottom of an oil can. The movement lifts the pressure plate away from the friction disk. Figures 44-15 and 44-16 show the two positions of the diaphragm spring and the clutch parts.

There is another model of diaphragm-spring clutch in which the tapering fingers are bent outward. This clutch operates in the same manner, however. The diaphragm-spring clutch also comes in a two-friction-disk design. The purpose of adding a second friction disk is to permit the clutch to carry heavier loads. This clutch is used in trucks and some cars.

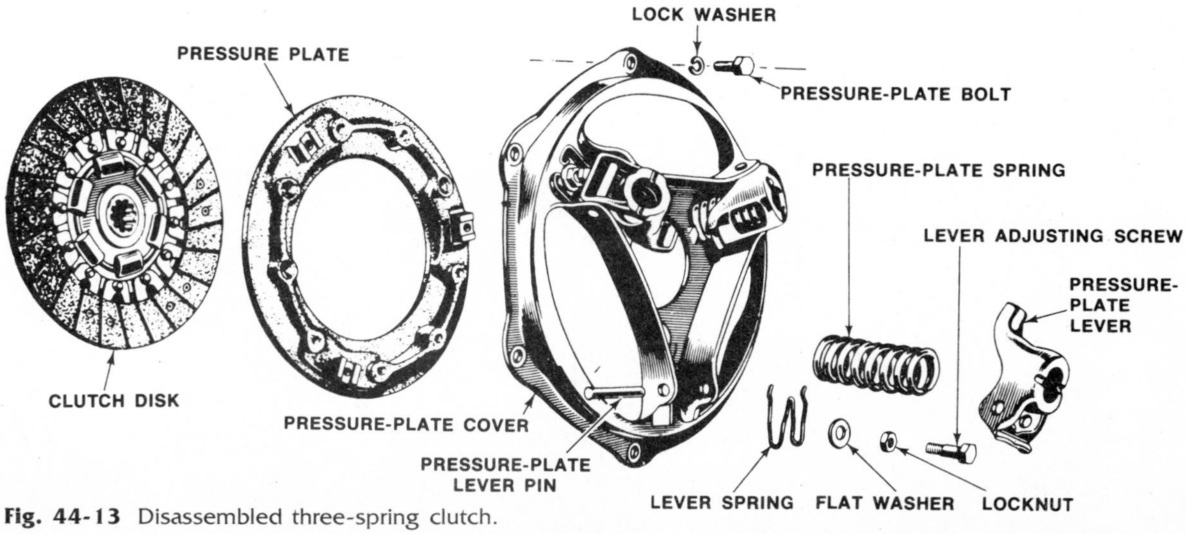

Fig. 44-13 Disassembled three-spring clutch.

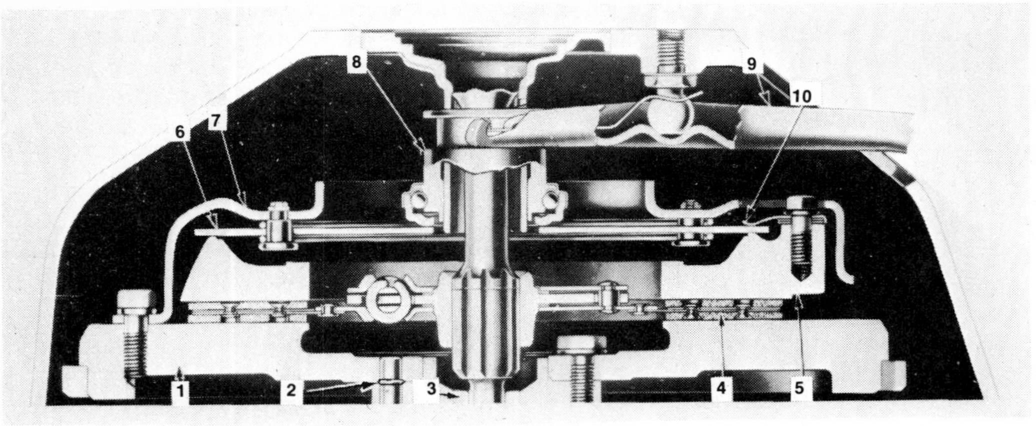

Fig. 44-14 Diaphragm-spring clutch in sectional view, looking down from the top. *(Chevrolet Motor Division of General Motors Corporation)*

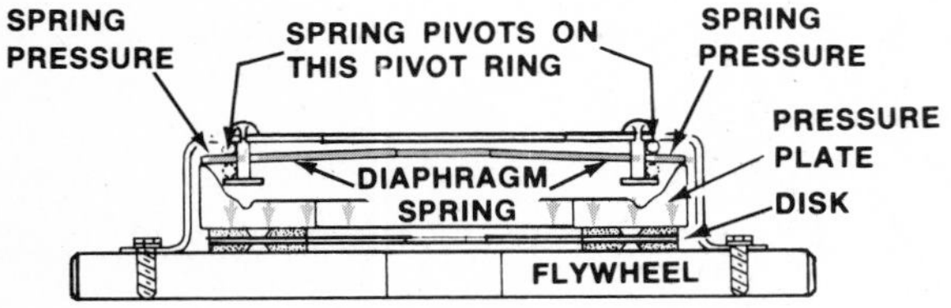

Fig. 44-15 Diaphragm-spring clutch in the engaged position. *(Chevrolet Motor Division of General Motors Corporation)*

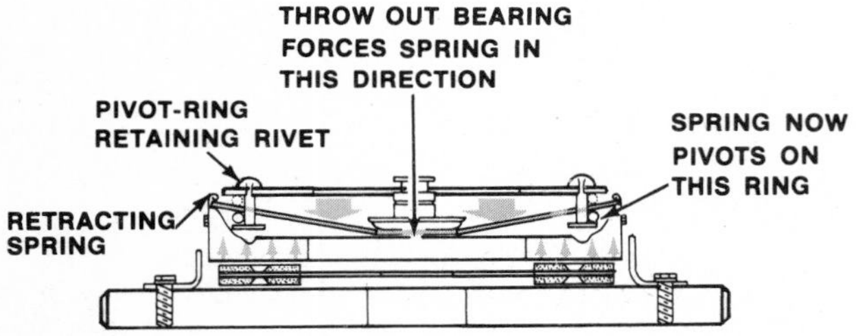

Fig. 44-16 Diaphragm-spring clutch in the released position. *(Chevrolet Motor Division of General Motors Corporation)*

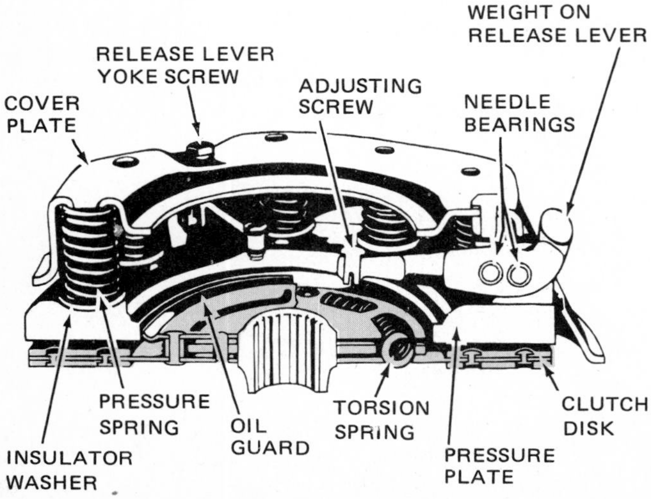

Fig. 44-17 Cutaway view of a centrifugal clutch.

● 44-9 Centrifugal Clutch

Some clutches have a centrifugal mechanism to increase clutch holding power at higher speeds. As clutch speed increases, weights on the clutch-release levers are forced outward by centrifugal force. With this arrangement, fairly light coil pressure springs can be used, so that the clutch operates without too much foot pressure. But at the same time, the centrifugal effect produces a higher pressure against the friction disk with increasing speed. This provides the extra holding power the clutch needs at higher speeds.

Figure 44-17 shows one arrangement. The release levers have weights on their outer ends. As centrifugal force acts on the weights, the levers press down tightly against the pressure plate. This adds to the coil-spring force holding the friction disk locked between the pressure plate and flywheel. Figure 44-5 shows a somewhat different arrangement. Here, the centrifugal effect acts on the rollers that are placed under the release levers. This centrifugal effect wedges the roller between the cover and pressure plate to produce the increased pressure on the pressure plate.

● 44-10 Hydraulic Clutch

The hydraulic clutch is used in vehicles in which the clutch is located far from the foot pedal. Instead of trying to install a complicated linkage between the clutch pedal and the clutch, car manufacturers use a hydraulic device. Figure 44-18 shows the clutch, the hydraulic cylinders, and the clutch pedal. This arrangement is also used on heavy-duty clutches with high clutch-spring pressure. High clutch-spring pressure means high clutch-pedal pressure. The hydraulic clutch can be used to reduce the amount of pedal pressure needed.

Now let's see how the hydraulic clutch works. When the driver pushes down on the clutch pedal, a push rod is forced down into a master cylinder. As the push rod moves down into the master cylinder, the rod forces a

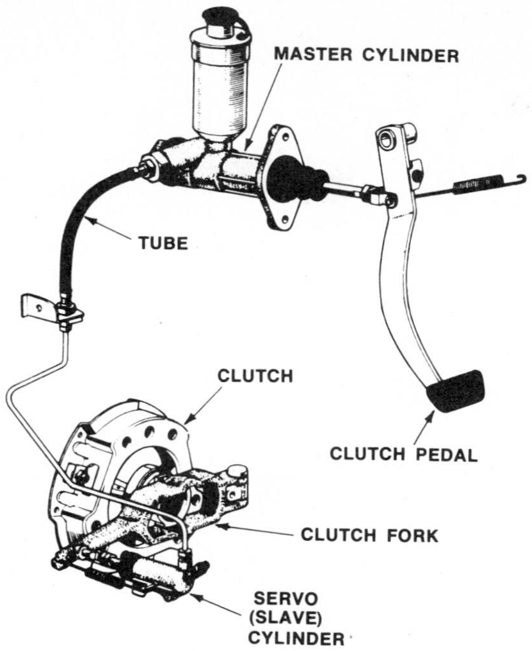

Fig. 44-18 Hydraulically operated clutch. (*Toyota Motor Sales, Limited*)

piston down into the cylinder. This action puts pressure on the fluid in the cylinder, and some of the fluid is forced out. The fluid flows through a tube and into a servo cylinder at the clutch. The servo cylinder also has a piston. The fluid, flowing into the servo cylinder from the master cylinder, forces the piston in the servo cylinder to move. This movement is carried through a push rod to the release lever, thus releasing the clutch.

● 44-11 Clutch Safety Switch

Late-model cars have a clutch safety switch that prevents starting if the clutch is engaged. That is, the clutch pedal must be depressed at the same time that the ignition switch is turned to START. The movement of the clutch pedal closes the safety switch so that the circuit to the starting motor can be completed. The purpose of the switch is to prevent starting with the transmission in gear and the clutch engaged. If this happened, the engine might start and the car might move before the driver was ready for it. And that could lead to accidents.

REVIEW QUESTIONS

1. To what part is the pressure-plate-and-cover assembly attached?
2. When the release levers are pushed in by the release bearing, what is happening at the outer ends of the release levers?
3. When the clutch is engaged, between what parts is the friction disk clamped?
4. Which is simpler, the diaphragm-spring clutch or the coil-spring clutch?
5. In the hydraulic clutch, is there direct mechanical linkage between the clutch pedal and the clutch?
6. When does the release, or throwout, bearing turn?
7. What is free play?
8. What is the over-center spring? Why is it used?
9. Which type of clutch requires the least pedal pressure, the coil-spring or the diaphragm-spring type?
10. What is the clutch safety switch? What is its purpose?

SELF PROJECT

Learn more about clutch operation. Get in the car, observing all safety precautions. Set the parking brake, and put the transmission in neutral. With the engine off, depress the clutch pedal with your hand until all free play is eliminated. As you do this, remember that you are moving the release bearing from the clutched position to where it is in contact with the release levers. Now try it with your foot. Note that your foot is less sensitive; it is harder to feel the point at which all free play is eliminated.

Now start the engine. Depress the clutch pedal with your fingers. Note the vibration you get when the pedal moves down far enough to release the clutch. Then operate the clutch pedal with your foot. Note that you will not normally feel the vibrations with your foot unless you are really looking for them. Work with the pedal carefully until you can distinguish the sound made as the release bearing starts to turn and the clutch is released.

Listening and feeling are two "tools" used constantly in automotive-service work. The expert automotive mechanic can tell a great deal about how components are working by just looking, listening, and feeling.

CHAPTER 45
Clutch Service

After studying this chapter, you should be able to:

- **List and describe the various types of clutch trouble.**
- **Make the various tests discussed to pinpoint the cause of clutch troubles.**
- **Under the instructor's supervision, make clutch-linkage adjustments.**
- **Under the instructor's supervision, remove and replace clutches.**

This chapter covers trouble diagnosis, removal, overhaul, adjustment, reassembly, and installation of clutches used on passenger cars.

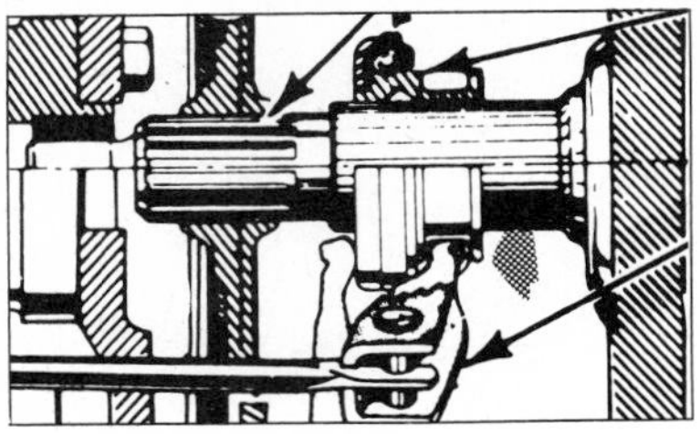

● 45-1 Trouble-Diagnosis Chart

The chart that follows lists various clutch complaints, their possible causes, and checks or corrections to be made. The information in the chart will shorten the time you need to correct a trouble. If you follow a logical procedure, you can locate the cause of the trouble quickly. On the other hand, haphazard guessing wastes time and effort.

Most clutch troubles show up as shifting problems such as gear clash, grinding, high shifting effort, and bearing squeal. See ● 45-2 to 45-8 for detailed explanations of the trouble causes and corrections listed below.

Note: The troubles and possible causes are not listed according to how often they occur. That is, item 1 (or item a under "Possible Cause") does not necessarily occur more often than item 2 (or item b).

CLUTCH TROUBLE-DIAGNOSIS CHART

Complaint	Possible Cause	Check or Correction
1. Clutch slips while engaged (● 45-2)	a. Binding or incorrectly adjusted linkage	Lubricate, adjust; check for broken return spring
	b. Broken engine mount	Replace
	c. Internal damage or out of adjustment	Remove clutch for service
2. Clutch chatters or grabs when engaging (● 45-3)	a. Binding linkage	Lubricate, adjust
	b. Broken engine mount	Replace
	c. Misaligned clutch housing	Realign or replace
	d. Internal problems: disk hub binding on shaft; grease or oil on facing; loose facings	Free disk hub on shaft; replace disk, pressure-plate assembly
3. Clutch spins or drags when disengaged (● 45-4)	a. Linkage out of adjustment	Readjust
	b. Broken engine mount	Replace
	c. Internal problems: worn facings; weak springs; oil or grease on facings; incorrect lever adjustment	Replace disk, pressure-plate assembly; adjust levers
4. Clutch noises (● 45-5)	a. Worn disk hub or shaft	Replace
	b. Misalignment	Align transmission and clutch with engine

CLUTCH TROUBLE-DIAGNOSIS CHART (Continued)

Complaint	Possible Cause	Check or Correction
	c. Worn throwout bearings	Replace
	d. Worn pilot bearing in crankshaft	Replace
	e. Linkage pivots need lubrication	Lubricate
	f. Release levers rubbing	Adjust
	g. Worn or weak retracting springs (diaphragm)	Replace
5. Clutch-pedal pulsation (● 45-6)	a. Misalignment	Align transmission and clutch with engine
	b. Flywheel not seated or warped	Seat flywheel on flange or replace
	c. Release levers out of adjustment	Adjust
	d. Warped disk or pressure plate	Replace
6. Rapid friction-disk-facing wear (● 45-7)	a. Driver abuse	Avoid riding clutch, speed shifts, popping clutch, etc.
	b. Cracks in flywheel or pressure-plate faces	Replace
	c. Internal problems: weak springs; oil on facings; release levers out of adjustment	Replace parts; adjust levers
	d. Linkage binding or out of adjustment	Lubricate, adjust
7. Clutch pedal stiff (● 45-8)	a. Binding in linkage	Lubricate, adjust
	b. Pedal shaft binding on floor-board seal	Free up seal, lubricate
	c. Over-center spring out of adjustment or broken	Adjust; replace
8. Hydraulic-clutch troubles	a. Gear clashing, trouble shifting into and out of mesh	Hydraulic clutches can have any of the troubles listed in the chart. Also, if the hydraulic system is not working, gear clash and shifting trouble will result (● 45-12)

● 45-2 Clutch Slips While Engaged

This is extremely hard on the clutch facings and mating surfaces of the flywheel and pressure plate. The slipping clutch generates considerable heat. The clutch facings wear rapidly and may char and burn. The flywheel face and pressure plate wear. They may groove, crack, and score. The heat in the pressure plate can cause the springs to lose their tension, and this makes the situation worse.

Clutch slippage is very noticeable during acceleration, especially from a standing start or in low gear. A rough test for clutch slippage can be made by starting the engine, setting the hand brake, and shifting into high gear. Then slowly release the clutch while accelerating the engine slowly. If the clutch is in good condition, it should hold so that the engine stalls immediately after clutch engagement is completed. The dynamometer can also be used to detect a slipping clutch. Connect a tachometer to read engine rpm (revolutions per minute). Run the vehicle at intermediate speed at part throttle. Note the engine rpm and speedometer reading. Then push the accelerator all the way down, using the dynamometer to load the engine while opening the throttle. Any increase in engine rpm at the same vehicle speed is clutch slippage.

Several conditions can cause clutch slippage. The linkage may not be correctly adjusted. If the incorrect adjustment reduces pedal lash too much, the throwout bearing may be up against the release fingers even with a fully released pedal. This can take up part of the spring pressure, as the pressure plate is not locking the friction disk to the flywheel. The remedy here is to readjust the linkage.

Binding linkage or a broken return spring may prevent full return of the linkage to the engaged position. Replace the spring if it is broken. Lubricate the linkage. Much of the clutch linkage is pivoted in nylon or neoprene bushings. These should be lubricated with silicone spray, not with oil.

If the linkage is not at fault, the slippage could be caused by a broken engine mount. This could allow the engine to shift enough to prevent good clutch engagement. The remedy here is to replace the mount. See ● 39-23.

If none of the above is causing slipping, then the clutch should be removed for service. Conditions in the clutch that could cause slipping include worn friction-disk facings, weak pressure-plate or diaphragm springs, oil or grease on the facings, or incorrectly adjusted release levers.

The recommendation of most manufacturers is to replace the disk and pressure-plate assembly if there is internal wear or damage, or weak springs. Pressure-plate assemblies can be rebuilt, but this is a job for the clutch-rebuilding shop.

One clue to a slipping clutch is metal and facing material in the clutch housing. This can be detected by removing the inspection cover from under the clutch and flywheel.

Careful: If the clutch disk and pressure-plate assembly are replaced, the flywheel should be carefully inspected for damage—wear, cracks, grooves, and checks. Any of these, if well advanced, will require replacement of the flywheel. Putting a new disk facing against a damaged flywheel will lead to rapid facing wear.

● 45-3 Clutch Chatters or Grabs When Engaging

This trouble is most likely inside the clutch. The clutch will require removal for service or replacement. Before this is done, however, check the clutch linkage to make sure it is not binding. If it binds, it could release suddenly to throw the clutch into quick engagement, with a resulting heavy jerk.

A broken engine mount can also cause the problem. The engine is free to move excessively, and this can cause the clutch to grab or chatter when engaging. The remedy is to replace the mount. See ● 39-23.

Inside the clutch, the trouble could be due to oil or grease on the disk facings or to glazed or loose facings. It could also be due to binding of the friction-disk hub on the clutch shaft. This latter condition requires cleaning of the splines in the hub and on the shaft, and lubrication of the splines.

Clutch chatter after removal and installation of an engine may be caused by a misaligned clutch housing. Some clutch housings have small shims that could be lost during engine or clutch-housing removal. These shims must be replaced in the same positions to assure housing alignment. It is also possible for dirt to get between the clutch housing and cylinder block, or either could be nicked or burred. Any of these could throw off the housing alignment.

Other clutch problems—glazed or loose facings, oil or grease on the facings—require disk and pressure-plate replacement.

● 45-4 Clutch Spins or Drags When Disengaged

The clutch friction disk spins briefly after disengagement, when the transmission is in neutral. This normal spinning should not be confused with a dragging clutch. When the clutch drags, the friction disk is not being fully released from the flywheel or pressure plate as the clutch pedal is depressed. Therefore, the friction disk continues to rotate with or to rub against the flywheel or pressure plate. The common complaint of the driver is that he or she has trouble shifting the automobile into gear without clashing; the dragging disk keeps the transmission rotating.

The first thing to check with this condition is the pedal-linkage adjustment. If there is excessive pedal lash, or free travel, even full movement of the pedal will not release the clutch fully. If making linkage adjustment does not correct the problem, then the trouble is in the clutch.

Internal clutch troubles could be due to a warped friction disk or pressure plate or loose friction-disk facings. One cause of loose friction-disk facings is abuse of the clutch. This includes "popping" the clutch for a quick getaway (letting the clutch out suddenly with the engine turning at high rpm), slipping the clutch for drag-strip starts, and increasing engine power output ("souping up" the engine).

The release levers may not be correctly adjusted so they do not fully disengage the clutch. Also, the friction-disk hub may be binding on the clutch shaft. This condition is corrected by cleaning and lubricating the splines.

● 45-5 Clutch Noises

Clutch noises are usually most noticeable when the engine is idling. To determine the cause, note whether the noise is heard when the clutch is engaged, when it is disengaged, or during pedal movement to engage or disengage the clutch.

Noises while the pedal is in motion are probably due to dry or dirty linkage pivot points. Clean and lubricate them as already noted in ● 45-2.

Noises heard in neutral, but that disappear when the pedal is depressed, are transmission noises. (This could also be due to a dry or worn pilot bushing in the crankshaft.) They are usually rough-bearing sounds. The cause is worn transmission bearings, sometimes caused by clutch-popping and shifting gears too fast. These throw an extra load on the transmission bearings, as well as on the gears.

Noises that come from the clutch when the clutch is engaged could be due to a friction-disk hub that is loose on the clutch shaft. This would require replacement of the disk or clutch shaft, or perhaps both if both are excessively worn. Friction-disk dampener springs that are broken or weak will cause noise. This requires replacement of the complete disk. Misalignment of the engine and transmission will cause a backward-and-forward movement of the friction disk on the clutch shaft. The alignment must be corrected.

Noises that come from the clutch when it is disengaged could be due to a clutch throwout bearing that is worn, is binding, or has lost its lubricant. Such a bearing squeals when the clutch pedal is depressed and the bearing comes into operation. The bearing should be relubricated or replaced. If the release levers are not properly adjusted, they rub against the friction-disk hub when the clutch pedal is depressed. The release levers should be readjusted. If the pilot bearing in the crankshaft is worn or lacks lubricant, it produces a high-pitched whine when the transmission is in gear, the clutch is disengaged, and the car is stationary. Under these conditions, the clutch shaft, which is piloted in the bearing in the crankshaft, is stationary, but the crankshaft and bearing are turning. The bearing should be lubricated or replaced.

In the diaphragm-spring clutch, worn or weak retracting springs will cause a rattling noise when the clutch is disengaged and the engine is idling. Eliminate by replacing the springs without removing the clutch from the engine.

● 45-6 Clutch-Pedal Pulsation

Clutch-pedal pulsation is noticeable when a slight pressure is applied to the clutch pedal with the engine running. The pulsations can be felt by the foot as a series of slight pedal movements. As pedal pressure is increased, the pulsations cease. This condition often indicates trouble that must be corrected before serious damage to the clutch results. One possible cause is misalignment of the engine and transmission. If the two are not in line, the friction disk or other clutch parts move back and forth with every revolution. The result is rapid wear of clutch parts. Correction is to detach the transmission, remove the clutch, and then check the housing alignment with the engine and crankshaft. At the same time, the flywheel can be checked for wobble. A flywheel that is not seated on the crankshaft flange will also produce clutch-pedal pulsations. A flywheel that is not seated on the crankshaft flange should be removed and remounted to make sure that it seats evenly.

If the clutch housing is distorted or shifted so that alignment between the engine and transmission has been lost, it is sometimes possible to restore alignment. This is done by installing shims between the housing and engine block and between the housing and transmission case. Otherwise, a new clutch housing will be required.

Note that these causes of clutch-pedal pulsation—housing misalignment, bent flywheel, flywheel not seated on the crankshaft flange—are not conditions that would usually arise during normal operation. They would most likely result from faulty reassembly after a service job.

Another cause of clutch-pedal pulsations is uneven release-lever adjustment (so that release levers do not meet the throwout bearing and pressure plate together). Still another is a warped friction disk or pressure plate. Release levers of the adjustable type should be readjusted. A warped friction disk must be replaced. If the pressure plate is out of line because of a distorted clutch cover, the cover sometimes can be straightened to restore alignment.

● 45-7 Rapid Wear of Friction-Disk Facings

Rapid wear of the friction-disk facings is caused by slippage between the facings and the flywheel or pressure plate. Thus, if the driver has the habit of "riding" the clutch (that is, if she or he keeps her or his foot resting on the clutch), part of the pressure-plate spring pressure will be taken up so that slipping may take place. Likewise, frequent use of the clutch, incorrect clutching and declutching, overloading the clutch, and slow clutch release increase clutch-facing wear. Speed, or "snap," gear shifting, increasing engine output ("souping up"), and drag-strip starts shorten clutch life. Also, the installation of wide oversize tires increases the clutch load. (Some manufacturers will not warranty the clutch if oversize tires are installed.)

Rapid facing wear after installation of a new friction disk can be caused by heat checks and cracks in the flywheel and pressure-plate faces. The sharp edges act like tiny knives. They shave off a little of the facing during each engagement. This is the reason we mentioned, in ● 45-2, that when a friction disk is replaced, the pressure-plate assembly should also be replaced. In addition, the flywheel face should be inspected and, if it is damaged, the flywheel should be replaced.

Several conditions in the clutch itself can cause rapid friction-disk-facing wear. For example, weak or broken pressure springs will cause slippage and facing wear. In this case, the springs must be replaced. If the pressure plate or friction disk is warped or out of line, it must be replaced or realignment must be reestablished. In addition, improper pedal-linkage adjustment or binding of the linkage may prevent full spring pressure from being applied to the friction disk. With less than full spring pressure, slippage and wear may take place. The linkage must be readjusted and lubricated at all points of friction.

● 45-8 Stiff Clutch Pedal

A stiff clutch pedal, or a pedal that is hard to depress, is likely to result from lack of lubricant in the clutch linkage, from binding of the clutch-pedal shaft in the floorboard seal, or from misaligned linkage parts that are binding. In addition, the over-center spring (on cars so equipped) may be out of adjustment or broken. Also, if the clutch pedal has been bent so that it rubs on the floor board, it may not operate easily. The remedy in each of these cases is obvious: Parts must be realigned, lubricated, or readjusted as necessary.

● 45-9 Clutch Spin-Down Time

Clutch spin-down time is the time it takes for the clutch to stop spinning when it is disengaged. To make the test, run the engine at normal idle with the transmission in neutral and the clutch engaged.

Disengage the clutch, wait nine seconds, and shift the transmission into reverse. No grinding noise should be heard. A grinding noise indicates incorrect clutch-linkage adjustment, clutch misalignment, or internal clutch problems. Clutch problems would require clutch removal for correction.

● 45-10 Clutch-Linkage Adjustment

Clutch-pedal-linkage adjustment may be required from time to time to compensate for friction-disk-facing wear. The linkage also requires periodic lubrication. The adjustment must provide the proper amount of free clutch-pedal travel (also called pedal *lash*). The free travel is the pedal movement before the throwout bearing comes up against the clutch-release levers. After this occurs, there is a definite increase in the amount of pressure required to actuate the release levers and disengage the clutch. If the pedal lash is too great, the clutch may not release fully. This could cause clutch spinning during disengagement (● 45-4). If the pedal lash is too small, the clutch may not be able to engage fully (● 45-2). This could cause rapid friction-disk-facing wear. Methods of adjustment vary in different cars. Refer to the manufacturer's shop manual for details and specifications.

● 45-11 Clutch Removal and Replacement

Variations in construction and design require that different removal and replacement procedures be used on different cars. First, the transmission must be removed (● 47-3). Then, the clutch-housing pan or flywheel lower cover must be removed, and the clutch linkage detached. Finally, the clutch can be detached from the flywheel and removed. Refer to the shop manual.

● 45-12 Clutch Overhaul

No general instructions that would apply to all types of clutches can be given. Whenever a clutch is to be disassembled, serviced, reassembled, and adjusted, refer to the shop manual describing these procedures.

If the trouble is located in the clutch itself, replacement of the complete clutch assembly is recommended by most manufacturers' shop manuals. At one time, these shop manuals carried instructions on the disassembly and repair of clutches. But now, complete replacement is considered the way to go. Some shops specialize in rebuilding clutches.

● 45-13 Hydraulic-Clutch Troubles

The hydraulic clutch (● 44-10) can display any of the troubles previously discussed, plus several in the hydraulic system. These special troubles include gear clashing and difficulty in shifting into or out of gear. The cause is usually loss of fluid from the hydraulic system. This prevents the hydraulic system from completely declutching for gear shifting. The hydraulic system is checked and serviced in the same way as the hydraulic system in hydraulic brakes (Chap. 56). Leaks may be located in the master or servo cylinder, or in the line or connections between the two.

● 45-14 Inspecting and Servicing Clutch Parts

The various clutch parts can be checked as follows after the clutch is removed from the vehicle.

1. Use compressed air to blow dust out of the clutch housing.

CAUTION: Do not breathe the dust you blow out from the clutch housing. It may contain asbestos. Asbestos is used in the clutch facings in many clutches. Authorities say that breathing asbestos dust can cause lung cancer. To be on the safe side, medical authorities say you should use damp cloths to wipe out the clutch housing. After working on a clutch, wash your hands carefully to remove any trace of dust.

2. Check for oil leakage through the engine rear main-bearing oil seal and the transmission drive-pinion seal. If leakage is noted, replace the seal.
3. Check the friction face of the flywheel for uniform appearance and for cracks, grooving, and uneven wear. If there is uneven wear, check the flywheel runout with a dial indicator. A warped or otherwise damaged flywheel should be replaced.
4. Check the bushing in the end of the crankshaft. Replace it if it is worn.
5. Check the journal on the end of the transmission input (clutch) shaft for wear. Replace it if it is worn.
6. Handle the disk with care. Do not touch the facings. Any trace of oil or grease will cause clutch slippage and rapid facing wear. Replace the disk if the facings show evidence of oil or grease, wear to within 0.015 inch [0.381 mm] (Plymouth) of the rivet heads, or if the facings are loose. The disk should also be replaced if there is other damage—worn splines, loose rivets, evidence of heat, and so on. Note the Caution under item 1, above, about the disk facings having asbestos in them.
7. Wipe the pressure-plate face with solvent. Check the face for flatness with a straightedge. Check the face for burned, cracked, grooved, and ridged areas.

If the friction disk is replaced, then as a rule the pressure-plate assembly should also be replaced.

8. Check the condition of the release levers. The inner ends should have a uniform wear pattern.
9. Test the cover for flatness on a surface plate.
10. If any of the pressure-plate parts are not up to specifications, replace the assembly. Also replace the friction disk.
11. When a clutch is being rebuilt, it is normal practice to install a new throwout bearing. However, if you wish to reuse the old bearing, examine it carefully to make sure it is usable. The bearing should turn freely when held in the hand under a light thrust load. There should be no noise. The bearing should turn

smoothly, without roughness. Note the condition of the face where the release levers touch. Replace the bearing if it is not in good condition.

Careful: Never clean the bearing in solvent or degreasing compound. It is prelubricated and sealed, and such cleaning would remove the lubricant and ruin the bearing.

12. Check the fork for wear on throwout-bearing attachments or other damage. On reassembly, be sure that the dust seal or cover is in good condition to prevent dirt from entering.

Steam cleaning can cause clutch trouble. Steam may enter and condense on the facings of the friction disk, pressure plate, and flywheel. The disk facings will absorb moisture. If the car is allowed to stand for a while with the facings wet, they may adhere to either the flywheel or the pressure plate. This means the clutch would not disengage. To prevent this from happening, start the engine immediately after steam cleaning. Slip the clutch in order to heat up and dry off the facings.

REVIEW QUESTIONS

1. Name four causes of clutch slippage.
2. Name four causes of clutch chatter or grabbing when the clutch is engaging.
3. Name four causes of clutch spinning or dragging when the clutch is disengaged.
4. Into what two general groups can clutch noises be divided?
5. What could produce clutch noise when the clutch is engaged?
6. What could produce clutch noise when the clutch is disengaged?
7. Name two possible causes of clutch-pedal pulsations.
8. Name three causes of rapid friction-disk-facing wear.
9. Give some general instructions for making a clutch-linkage adjustment.
10. Give some general instructions for inspecting clutch conditions in the car.

SELF PROJECTS

1. Refer to a shop manual, and make a list of the steps required to remove and replace a clutch. File this list in your notebook.
2. Refer to a shop manual, and make a list of the steps required to adjust a clutch linkage.
3. Make a set of 3-by-5-inch trouble-diagnosis cards, based on the trouble-diagnosis chart at the beginning of the chapter. Complaints go on one side of a card, possible causes on the other.
4. Make a notebook page on every clutch job you do. Include the customer complaint, conditions found, repair performed, and special tools used.

CHAPTER 46
Manual Transmissions

After studying this chapter, you should be able to:

- **Explain the purpose of the transmission.**
- **Explain what gear ratios are and how they are achieved.**
- **Explain how torque is related to gear ratio.**
- **Describe the operation of a manual transmission.**
- **Explain the purpose of overdrive and how it is achieved.**

This chapter discusses the purpose, construction, and operation of manually shifted transmissions. The transmission is located between the clutch and the drive line (Fig. 46-1). Its purpose is to provide a varying gear ratio between the engine and the car wheels, as explained in following sections.

46-1 Purpose of the Transmission

The transmission allows the engine crankshaft to turn fast while the wheels turn slowly. The transmission can then change the ratio of crankshaft speed to car speed as car speed increases. Thus, the engine crankshaft may turn about four, eight, or twelve times for each wheel revolution. In addition, the transmission includes a reverse gear so the car can be backed. Each of these gear ratios is selected manually by the driver.

In automatic transmissions, the varying ratios between the engine crankshaft and the wheels are achieved automatically. That is, the driver does not need to shift gears. The automatic transmission automatically selects the ratio required by engine speed and operating conditions. Automatic transmissions use a torque converter as well as hydraulic controls. All these are covered in later chapters.

46-2 Gear Ratios

Before we look into manual-transmission construction and operation, let us find out more about gears. Many

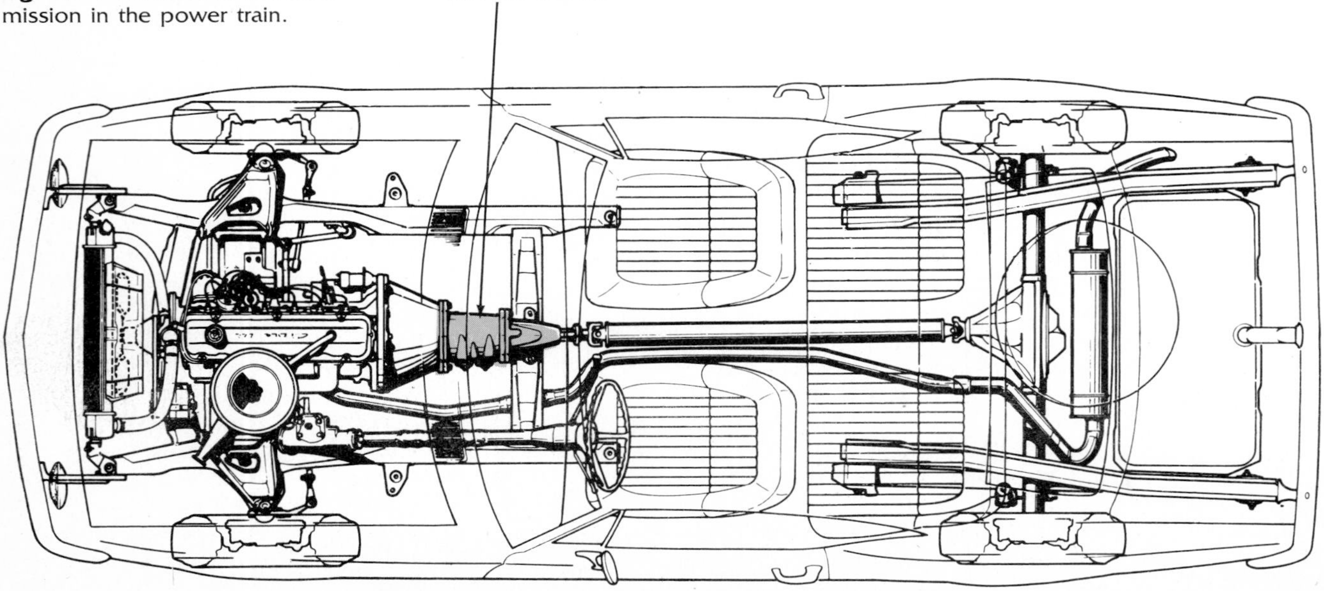

Fig. 46-1 Location of the transmission in the power train.

different gears are used on automobiles. All are basically similar. They all have teeth of one sort or another that mesh to carry motion from one gear to another. The simplest gear is the spur gear (Fig. 46-2). It is like a wheel with teeth. Two spur gears are shown meshed in Fig. 46-3. By "meshed" we mean that the teeth of one gear are fitted between the teeth of the other gear. When one gear rotates, the other gear must also rotate.

The relative speed of the two meshing gears (or gear ratio) is determined by the number of teeth of the two gears. For instance, when two gears have the same number of teeth (Fig. 46-3), they both rotate at the same speed. But if one gear has 12 teeth and the other gear has 24 teeth (Fig. 46-4), the smaller gear will rotate twice as fast as the larger gear. This is a two-to-one gear ratio (written 2:1). If the 12-tooth gear were meshed with a 36-tooth gear, the gear ratio would be 3:1.

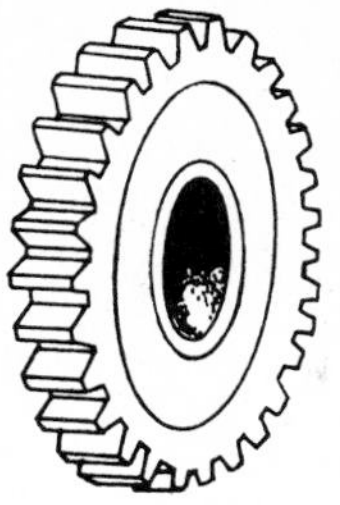

Fig. 46-2 Spur gear.

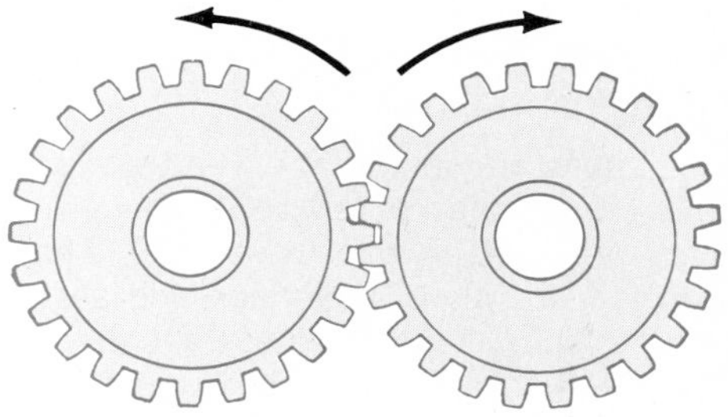

Fig. 46-3 Two meshing gears with the same number of teeth.

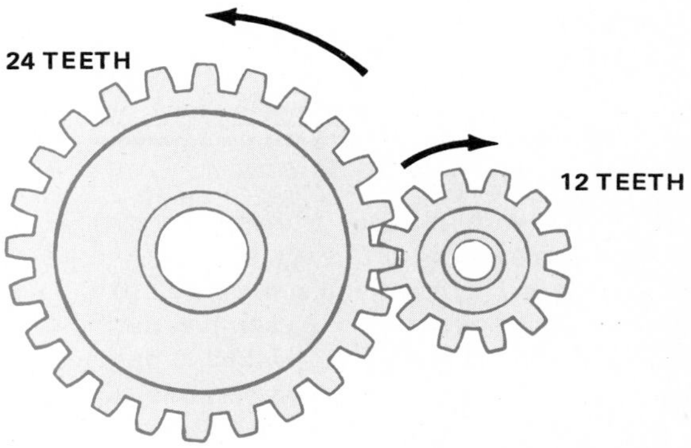

Fig. 46-4 Two meshing gears with different numbers of teeth. The smaller gear with fewer teeth turns more rapidly than the larger gear.

46-3 Torque

The *torque*, as well as the gear ratio, changes with the relative number of teeth in the meshing gears. Torque is twisting or turning effort. It is measured in pound-feet (lb-ft) or kilogram-meters (kg-m).

Any shaft or gear that is being turned has torque applied to it. The engine pistons and connecting rods push on the cranks on the crankshaft. This applies torque to the crankshaft and causes it to turn. The crankshaft applies torque to the gears in the transmission, so that the gears turn. This turning effort, or torque, is carried through the power train to the driving wheels so that they turn.

46-4 Torque in Gears

Torque on shafts or gears is measured as a straight-line force at a distance from the center of the shaft or gear. For instance, suppose that we want to measure the torque in the gears shown in Fig. 46-4. If we could hook a spring scale to the gear teeth and get a measurement of the pull, we could determine the torque. (Actually, a spring scale could not be used, although there are devices to measure the torque of rotating parts.) Suppose, for example, that the tooth of the driving gear is pushing against the tooth of the driven gear with a 25-pound [11.34-kg] force. This force, at a distance of 1 foot [0.305 m] (the radius, or distance from the center of the driving gear), means a torque of 25 pound-feet. That is, the smaller, driving gear is delivering a torque of 25 pound-feet [3.455 kg-m].

The 25-pound [11.34-kg] push from the gear teeth of the smaller gear is applied to the gear teeth of the larger gear. But it is applied at a distance of 2 feet [0.61 m] from the center of the larger gear. Therefore, the torque on the larger gear is 50 pound-feet (25 $\times$ 2). The same *force* is acting on the teeth of the larger gear, but it is acting at twice the *distance* from the shaft center.

46-5 Torque and Gear Ratio

Now, the important point of all this is that, if the smaller gear is driving the larger gear, the gear ratio will be 2:1. But the torque ratio will be 1:2. The larger gear will turn only half as fast as the smaller gear. But the larger gear will have *twice the torque* of the smaller gear. In gear systems, *speed reduction means torque increase.* For example, suppose that, when the transmission is in first gear, there is a speed reduction (or gear reduction) of 12:1 from engine to wheels. That is, the crankshaft turns 12 times to turn the rear wheels once. This means that the torque *increases* 12 times (ignoring losses due to friction). In other words, if the engine produced a torque of 100 pound-feet [13.82 kg-m] then 1,200 pound-feet [165.84 kg-m] torque would be applied to the rear wheels.

Wheel radius is assumed to be 1 foot [0.305 m] (for ease of figuring). With the torque acting on the ground at a distance of 1 foot [0.305 m] (the radius of the wheel), the push of the tire on the ground is 1,200 pounds.

Consequently, the push on the wheel axle, and thus the car, is 1,200 pounds [544.31 kg].

Actually, the torque is split between the two rear wheels. Thus in our example the torque on each rear wheel would be 600 pound-feet [82.92 kg-m]. Each tire thus pushes on the ground with a force of 600 pounds [272.15 kg]. Both tires together push with a force of 1,200 pounds [544.31 kg], giving the car a forward thrust of 1,200 pounds [544.31 kg].

46-6 Other Gears

The gears discussed above are spur gears. The teeth are parallel to, and align with, the centerline of the gear. Many types of gears are used in the automobile. They differ from the spur gear mainly in the shape and alignment of the gear teeth. Thus, helical gears are like spur gears except that the teeth have, in effect, been twisted at an angle to the gear centerline. Bevel gears are shaped like cones with the tops cut off; the teeth point inward toward the *apex* or tip, of the cone. Bevel gears are used to transmit motion through angles. Some gears have their teeth pointing inward; these are internal gears.

46-7 Types of Transmissions

Today, most cars with manual transmissions have a four-speed transmission. That is, it has four forward speeds and reverse. There are still some cars being built with three-speed transmissions. For example, during one recent year about 7 percent of Ford Motor Company cars were equipped with four-speed transmissions. About 1 percent were equipped with three-speed transmissions. The other 92 percent had automatic transmissions.

In some four-speed transmissions, the fourth gear is actually an overdrive. That is, the transmission output shaft turns faster then, or overdrives, the input shaft. Some cars are equipped with five-speed transmissions. In these, fifth gear is overdrive. We describe these transmissions in following sections.

Heavy-duty equipment such as large trucks uses transmissions with additional forward speeds. One heavy-duty transmission has 10 forward speeds. Actually, these heavy-duty transmissions work in the same general way as the simpler three-speed and four-speed transmissions. The essential difference is they have more gears and the shift lever has more operating positions.

46-8 A Simplified Transmission

We now describe how the transmission works, using a simplified version of a three-speed unit. This is shown in Fig. 46-5. It has three shafts and eight spur gears of varying sizes. The transmission housing and bearings are not shown. Four of the gears are rigidly connected to the countershaft. These are the driven gear, second-speed gear, first-speed gear, and reverse gear. When the clutch

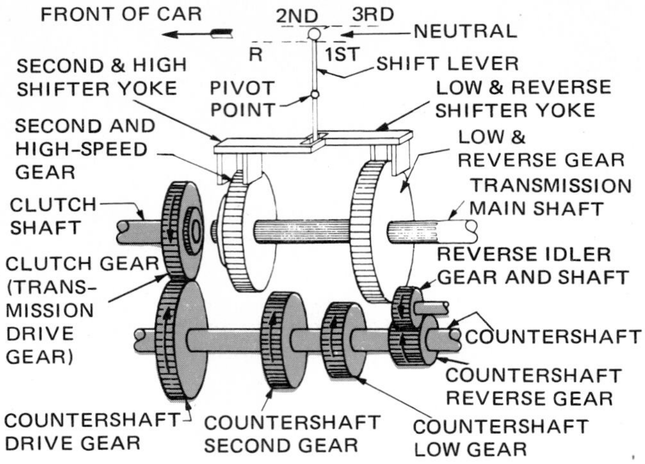

Fig. 46-5 Transmission with gears in neutral.

is engaged and the engine is running, the clutch-shaft gear drives the countershaft driven gear. This turns the countershaft and the other gears on the countershaft. The countershaft rotates in a direction opposite, or counter, to the rotation of the clutch-shaft gear. With the gears in neutral, as shown in Fig. 46-5, and the car stationary, the transmission main shaft is not turning.

The transmission main shaft is mechanically connected by shafts and gears in the rear axle to the car's rear wheels and to the car's front wheel in front drive. The two gears on the transmission main shaft may be shifted back and forth along the splines on the shaft. This is done by operating the gearshift lever in the driving compartment. The splines are matching internal and external teeth that permit endwise (axial) movement of the gears but cause the gears and shaft to rotate together. Note that a floor-type shift lever is shown in the illustrations. This type of lever illustrates more clearly the lever action in shifing gears. The transmission action is the same, regardless of whether a floor-type shift lever or a steering-column shift lever is used.

In reality, there is no transmission made today that has gears which slide along a shaft. But we use this simplified version to get across to you how the transmission produces various gear and torque ratios.

1. Shifting into first Suppose the gearshift lever is operated to place the gears in *first* (Fig. 46-6). The large gear on the transmission main shaft is moved along the shaft until it meshes with the small gear on the countershaft. The clutch is disengaged for this operation, so that the clutch shaft and the countershaft stop rotating. When the clutch is again engaged, the transmission main shaft rotates, as the driving gear on the clutch shaft drives it through the countershaft. The countershaft is turning more slowly than the clutch shaft, and the small countershaft gear is engaged with the large transmission mainshaft gear. A gear reduction of approximately 3:1 is achieved. That is, the clutch shaft turns three times for each revolution of the transmission main shaft. There is further gear reduction in the differential at the rear

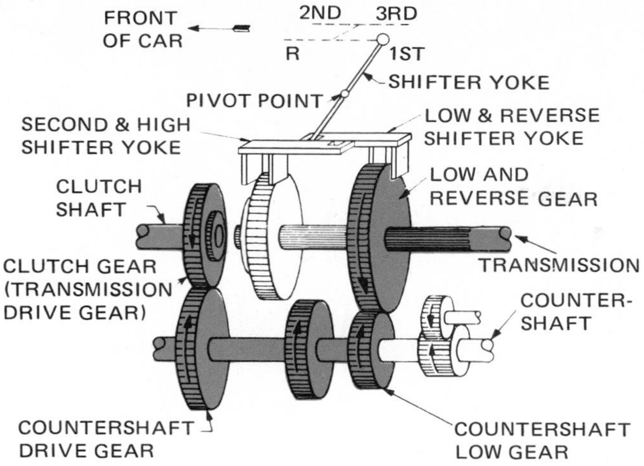

Fig. 46-6 Transmission with gears in first.

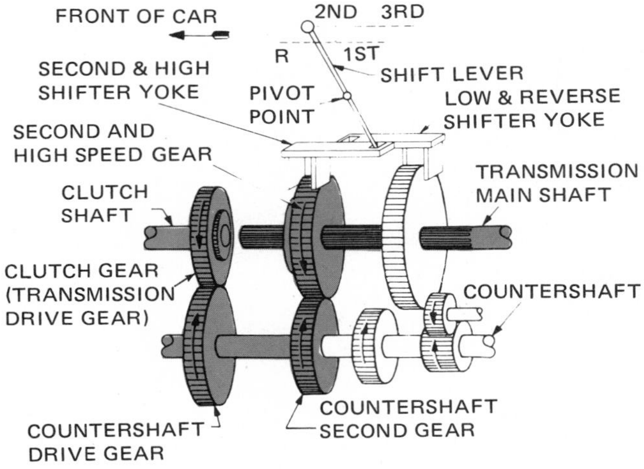

Fig. 46-7 Transmission with gears in second.

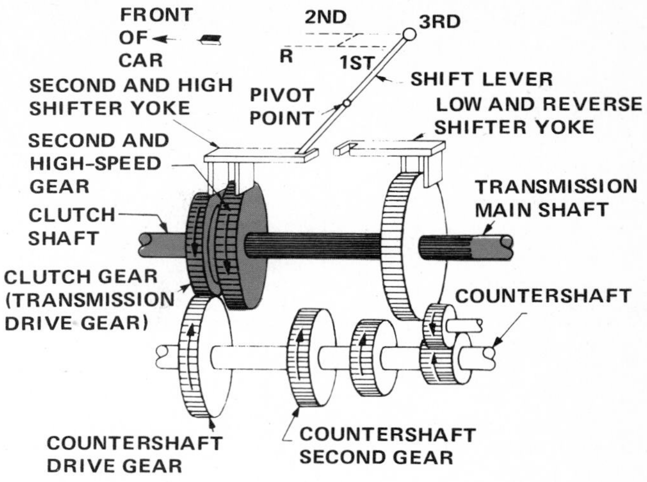

Fig. 46-8 Transmission with gears in third.

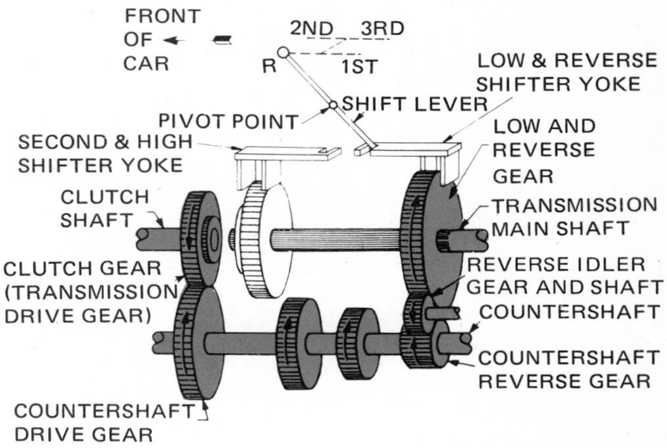

Fig. 46-9 Transmission with gears in reverse.

wheels. This produces a still higher gear ratio (approximately 12:1) between the engine crankshaft and the wheels.

2. Shifting into second Now suppose the clutch is operated and the gearshift lever is moved to *second* (Fig. 46-7). The large gear on the transmission main shaft de-meshes from the small first-speed countershaft gear. The smaller transmission main-shaft gear is slid into mesh with the large second-speed countershaft gear. This provides a somewhat reduced gear ratio, so that the engine crankshaft turns only about twice when the transmission main shaft turns once. The differential gear reduction increases this gear ratio to approximately 8:1.

3. Shifting into third When the gears are shifted into *third* (Fig. 46-8), the two gears on the transmission main shaft are de-meshed from the countershaft gears. Also, the second-and-third-speed gear is forced to slide up against the clutch-shaft gear. External teeth on the clutch-shaft gear mesh with internal teeth in the second-and-third-speed gear. Then the transmission main shaft turns with the clutch shaft, and a ratio of 1:1 is obtained. The differential reduction produces a gear ratio of about 4:1 between the engine crankshaft and the wheels.

4. Shifting into reverse When the gears are placed in *reverse* (Fig. 46-9), the larger of the transmission main-shaft gears is meshed with the reverse idler gear. This reverse idler gear is always in mesh with the small reverse gear on the end of the countershaft. Putting the idler gear between the countershaft reverse gear and the transmission main-shaft gear causes the transmission shaft to rotate in the opposite direction. That is, it rotates in the same direction as the countershaft. This reverses the rotation of the wheels, so that the car moves backward.

The description above outlines the basic operation of all transmissions. However, more complex transmissions are used on modern cars. These include helical gears and synchromesh devices that synchronize the rotation of gears that are about to be meshed. This eliminates clashing of the gears and makes gearshifting easier.

46-9 Gearshifting

Gearshift levers on manual transmissions are located either on the steering column or on the floor board. With either location, the operation of the gearshift lever does two things. First, it selects the gear assembly to be engaged. Second, it engages the gear to provide the desired gear position.

Figure 46-10 shows the shifting patterns for a column and a floor-board shift lever. Let us look at the column shift first. To shift into first or reverse, the driver depresses the clutch pedal to momentarily disconnect the engine from the transmission. Then the shift lever is lifted and moved forward for reverse or back for first gear. When the lever is lifted, it pivots on its mounting pin. This forces a rod on its inner end to be pushed down the steering column. This downward movement pushes a crossover blade at the bottom of the steering column downward. A slot in the blade engages a pin on the first-and-reverse shift lever. Now, when the shift lever is moved, say into first, the first-and-reverse lever is rotated.

This movement is carried by linkage to the transmission (Fig. 46-11). At the transmission, the movement causes the transmission first-and-reverse lever to move. This lever is on a shaft that extends through the transmission side cover and is connected inside the transmission to another lever (Fig. 46-12). A shift fork is mounted on the end of this lever. The arrangement does this: When the shaft is rotated by the movement of the first-and-reverse lever, it causes the shift fork to move forward or backward in the transmission. This pushes a gear or synchronizer assembly forward into the first-gear position, or backward, as we shall explain later.

Figure 46-13 shows a floor-board shift lever and the linkages to the transmission. This arrangement operates in the same manner as the column-mounted shift lever.

46-10 A Three-Speed Transmission

Now let us look at an actual three-speed transmission (Fig. 46-14). The transmission has the same type of countergear assembly as in our simplified version (Fig. 46-5). The clutch shaft and gear are also similar. However, the transmission main shaft and the gears on it are a little different. We show the gears separately in Fig. 46-15, so it is easier to see what happens when gears are shifted.

Note that the second-speed and first-speed gears do not move back and forth along the main shaft. They are in constant mesh with their respective gears on the countergear assembly. Also, these two gears are supported on bearings that allow them to rotate independently on the main shaft.

The hubs of the two synchronizer assemblies are splined to the main shaft and rotate with it. However, the synchronizer sleeves of the two assemblies can slide back and forth along the splines on the synchronizer hub. The forks shown in Fig. 46-12 fit in the grooves of the synchronizers (shown in Fig. 46-15). Now, you can see the connection between the gearshift lever and the synchronizers. When the gearshift lever is moved, the linkage selects one

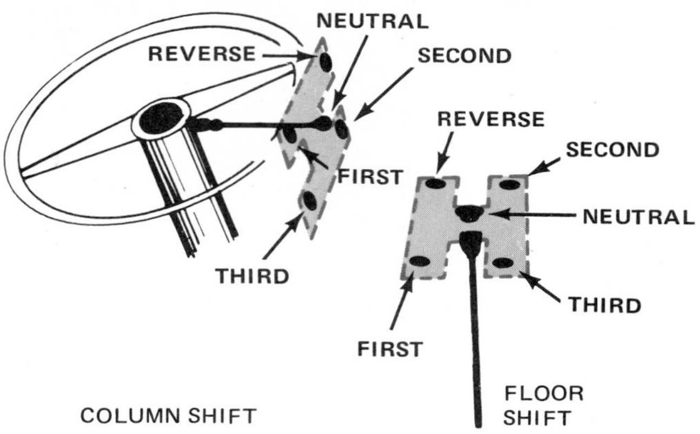

Fig. 46-10 Gearshift patterns for steering-column and floor-board shift levers.

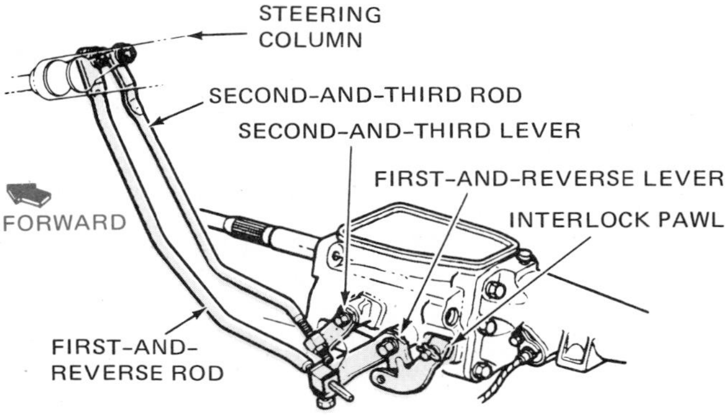

Fig. 46-11 Gearshift linkage between shift levers at the bottom of the steering column and transmission levers on the side of the transmission. *(Chrysler Corporation)*

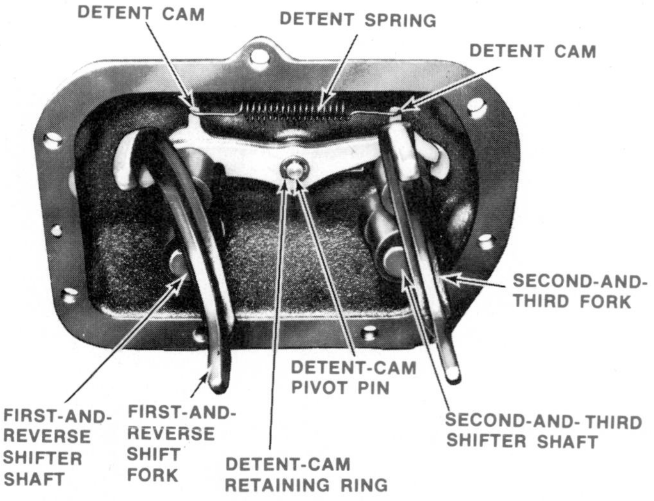

Fig. 46-12 Transmission side cover, viewed from inside the transmission. The shift forks are mounted on the ends of levers attached to shafts. The shafts can rotate in the side cover. The detent cams and springs prevent more than one of the shift forks from moving at any one time. *(Chevrolet Motor Division of General Motors Corporation)*

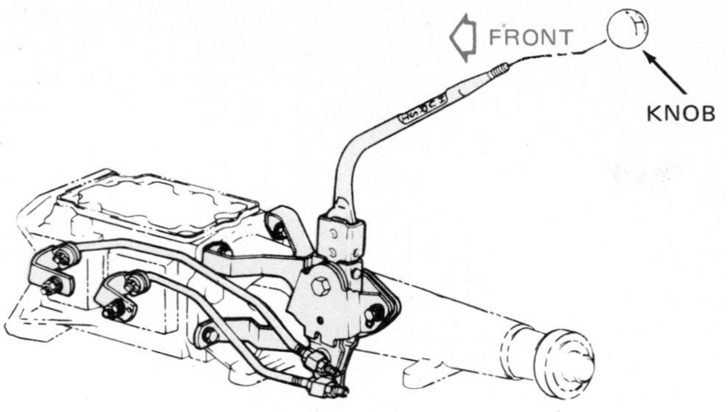

Fig. 46-13 Gearshift-linkage arrangement for transmission with floor-board shift lever. (*Pontiac Motor Division of General Motors Corporation*)

or the other of the two synchronizers. It then moves this synchronizer.

1. First-speed gear Now let us see what happens when the gearshift lever is moved to first. The clutch is disengaged for this operation. Movement of the gearshift lever causes the linkage to select the first-reverse synchronizer sleeve and move it to the left (in Fig. 46-16). As the

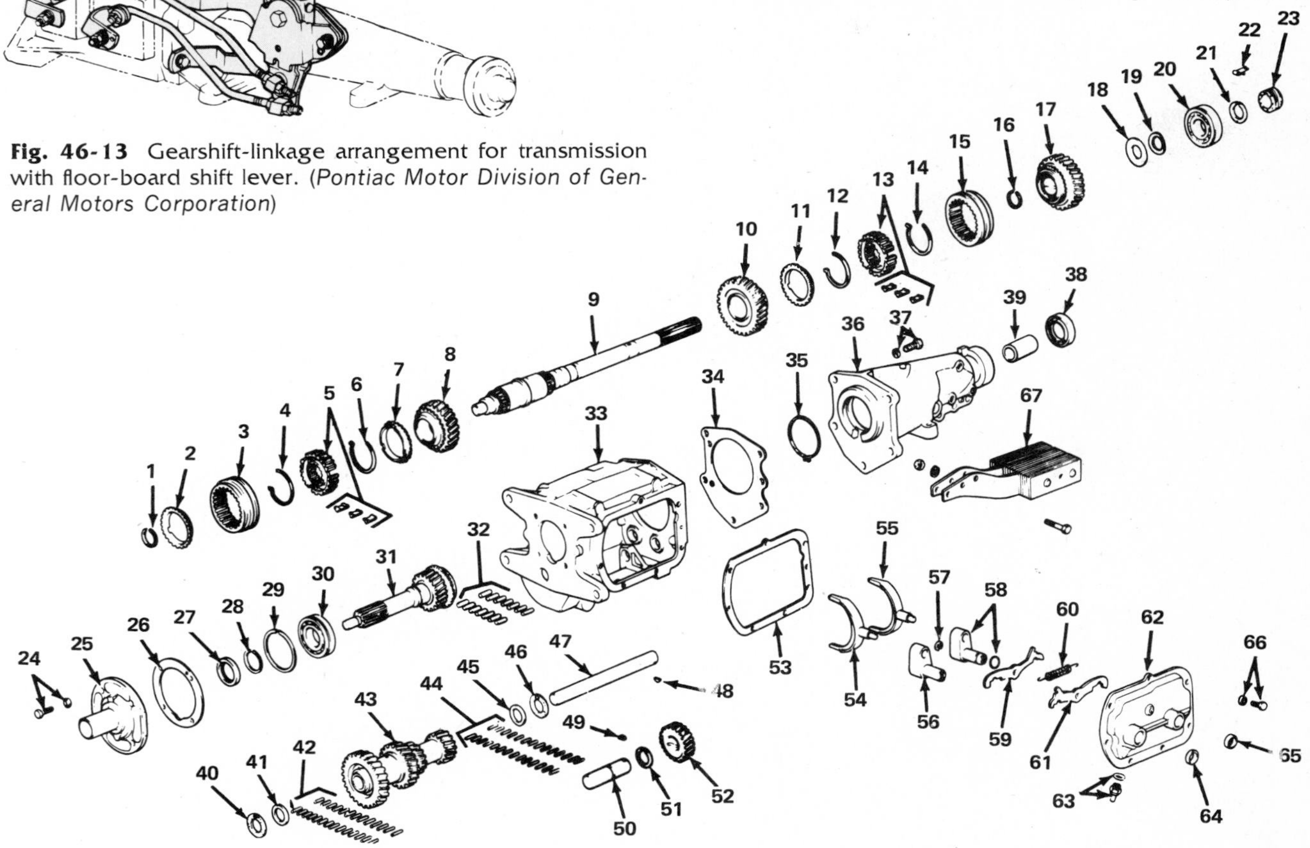

1. Snap ring
2. Synchronizer ring
3. 2–3 Synchronizer sleeve
4. Synchronizer key spring
5. Synchronizer hub and keys
6. Synchronizer key spring
7. Synchronizer ring
8. Second gear
9. Main shaft
10. First gear
11. Synchronizer ring
12. Synchronizer key spring
13. Synchronizer hub and keys
14. Synchronizer key spring
15. 1–2 Synchronizer sleeve
16. Snap ring
17. Reverse gear
18. Thrust washer
19. Waved washer
20. Rear bearing
21. Snap ring
22. Speedometer gear clip
23. Speedometer drive gear
24. Bearing retainer bolts and washers (4)
25. Front bearing retainer
26. Bearing retainer gasket
27. Bearing retainer oil seal
28. Snap ring
29. Bearing snap ring
30. Front bearing
31. Drive gear
32. Pilot bearings
33. Case
34. Extension to case gasket
35. Rear bearing to extension retaining ring
36. Rear extension
37. Extension to case retaining bolts and washers
38. Rear extension bushing
39. Rear seal
40. Thrust washer
41. Spacer
42. Countergear shaft bearings
43. Countergear
44. Countergear shaft bearings
45. Spacer
46. Thrust washer
47. Countergear shaft
48. Countergear shaft key
49. Idler shaft key
50. Reverse idler shaft
51. Snap ring
52. Reverse idler gear
53. Side cover gasket
54. 2–3 Shift fork
55. 1–Rev shift fork
56. 2–3 shifter shaft
57. Retaining "E" ring
58. 1–Rev shifter shaft with "O" ring
59. 2–3 detent cam
60. Detent cam spring
61. 1–Rev detent cam
62. Shift cover
63. TCS switch and gasket
64. Shifter shaft seal
65. Shifter shaft seal
66. Shift cover bolts and washers
67. Damper assembly

Fig. 46-14 Disassembled view of a three-speed transmission. (*Chevrolet Motor Division of General Motors Corporation*)

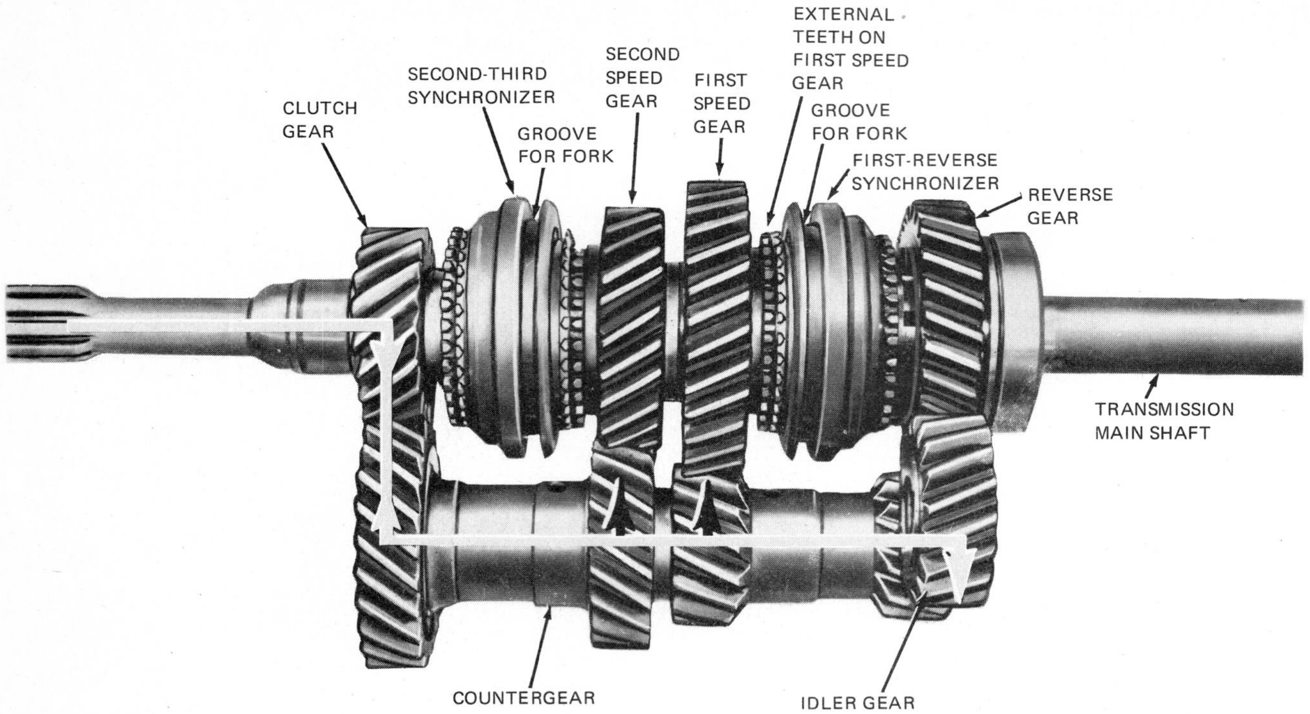

Fig. 46-15 Gear train and shafts in the three-speed transmission. (*Chevrolet Motor Division of General Motors Corporation*)

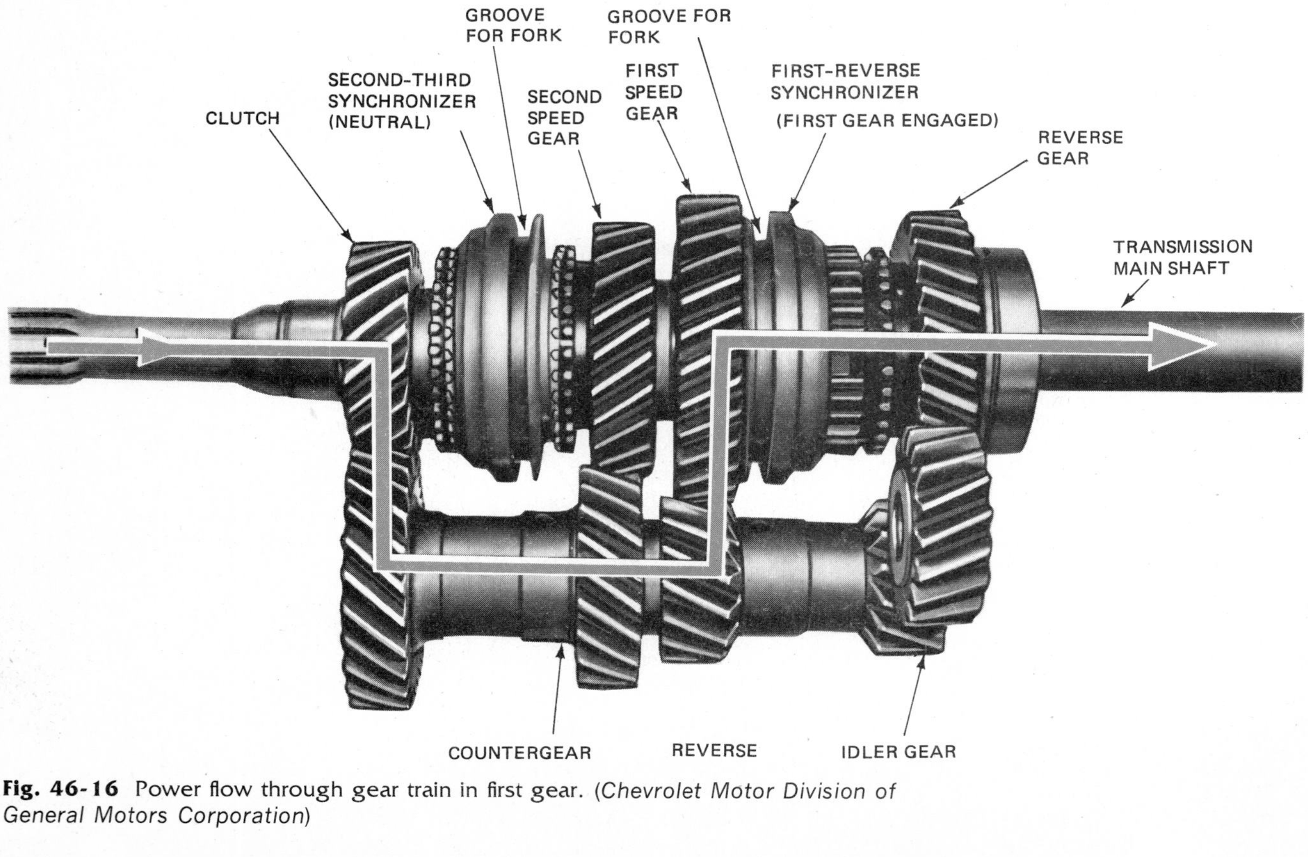

Fig. 46-16 Power flow through gear train in first gear. (*Chevrolet Motor Division of General Motors Corporation*)

synchronizer is moved to the left (in Fig. 46-16), internal teeth in the synchronizer engage with the external teeth on the first-speed gear. This locks the first-speed gear, through the synchronizer, to the main shaft. Figure 46-17 shows a first-speed gear by itself so you can see the external teeth. The synchronizer is shown by itself in Fig. 46-18, so you can see the internal teeth.

Now, when the clutch is engaged, the power flow through the transmission is as shown in Fig. 46-16. Note that there is gear reduction, as the smaller clutch gear drives the larger gear on the countergear assembly. There is also gear reduction as the small gear on the countergear assembly drives the large first-speed gear. The total gear reduction in first-speed gear is a little less than 3:1 in most transmissions (2.636:1 in one model).

2. Second-speed gear Figure 46-19 shows the transmission in second. The first-reverse synchronizer has been moved to its center position and out of mesh with the first-speed gear. The 2–3 synchronizer has been moved to the right so that its internal teeth engage the external teeth on the second-speed gear. The power flow is as shown by the arrow. The second-speed gear is smaller than the first-speed gear. And the countergear that meshes with the second-speed gear is larger than the countergear meshed with the first-speed gear. Therefore, the gear reduction is less. It is a little less than 2:1 in most transmissions (1.605:1 in one model).

3. Third-speed gear In this gear, the power flow is straight through the transmission, as shown in Fig. 46-20. The gear ratio is 1:1. Third gear is achieved by moving the synchronizer to the left, as shown. Its internal teeth engage the external teeth of the clutch gear, so the drive is through these teeth and the synchronizer splines to the main shaft.

4. Reverse To achieve reverse, an extra gear is inserted into the gear train. This extra gear, called the reverse idler gear, is in constant mesh with the fourth gear on the countergear assembly. See Fig. 46-21. When the shift lever is moved to reverse, the linkage moves the first-reverse synchronizer to the right, as shown in Fig. 46-21. The synchronizer internal teeth engage the external teeth of the reverse gear, so the flow of power through the transmission is as shown in Fig. 46-21. Because of the extra gear in the gear train, the main shaft turns in the reverse direction. The wheels therefore turn in the reverse direction, and the car moves backward. The gear ratio is a little less than 3:1 in most transmissions (2.636:1 in one model).

● 46-11 Synchronizers

To avoid the clashing of gears during shifting, and to simplify the shifting action for the driver, synchronizing devices are used in transmissions. These devices assure that gears which are about to mesh will be rotating at the same speed and thus will engage smoothly. One type uses synchronizing cones on the gears and on the synchronizing drums (Fig. 46-22). In the neutral position, the sliding sleeve is held in place by spring-loaded balls resting in detents in the sliding sleeve (or ring gear). When a shift starts, the drum and ring gear, as an assembly, are moved toward the selected gear. The first contact is between the synchronizing cones on the selected gear and the drum. This contact brings the two into synchronization. Both rotate at the same speed. Further movement of the shift fork forces the sliding sleeve on toward the selected gear. The internal teeth on the sliding sleeve match the external teeth on the selected gear. Now, the gears are locked up, or engaged, and the shift is completed. Note that the sliding sleeve moves off center from the drum for engagement. This pushes the balls down against the spring.

The pin-type synchronizer (Fig. 46-23), another type, has a pair of stop rings. Each has three pins which pin it

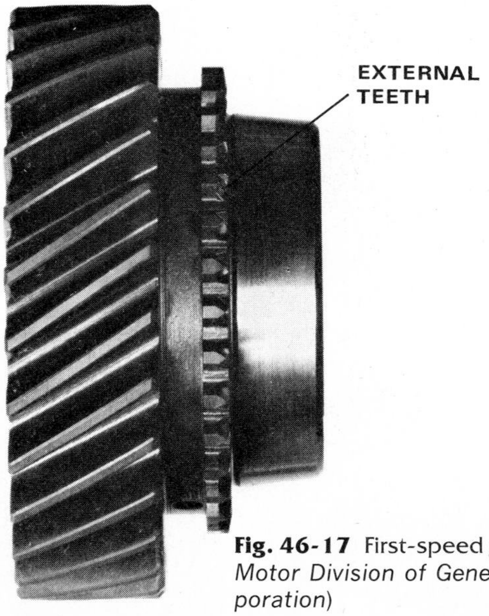

Fig. 46-17 First-speed gear. (*Chevrolet Motor Division of General Motors Corporation*)

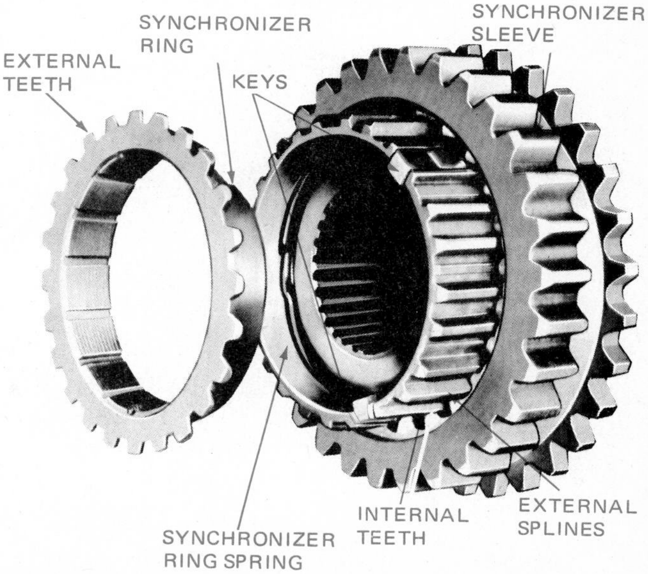

Fig. 46-18 Synchronizing assembly. (*Chevrolet Motor Division of General Motors Corporation*)

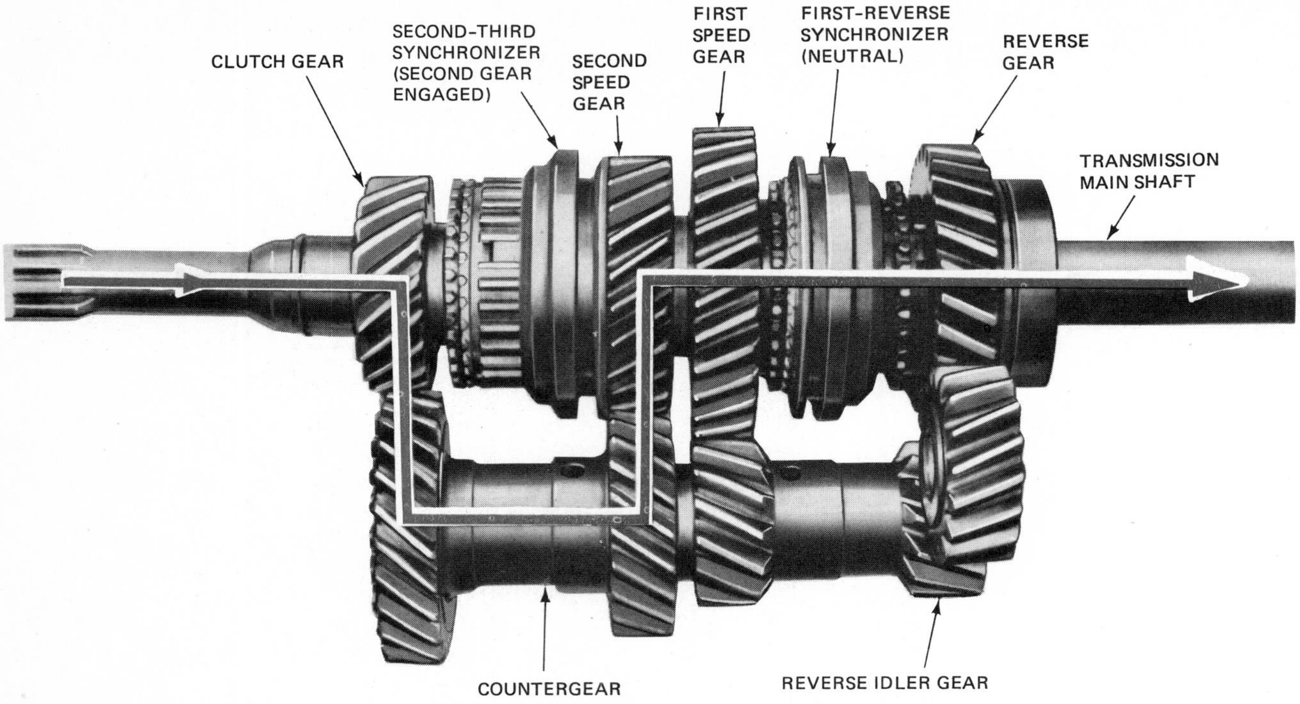

Fig. 46-19 Power flow in second gear. (*Chevrolet Motor Division of General Motors Corporation*)

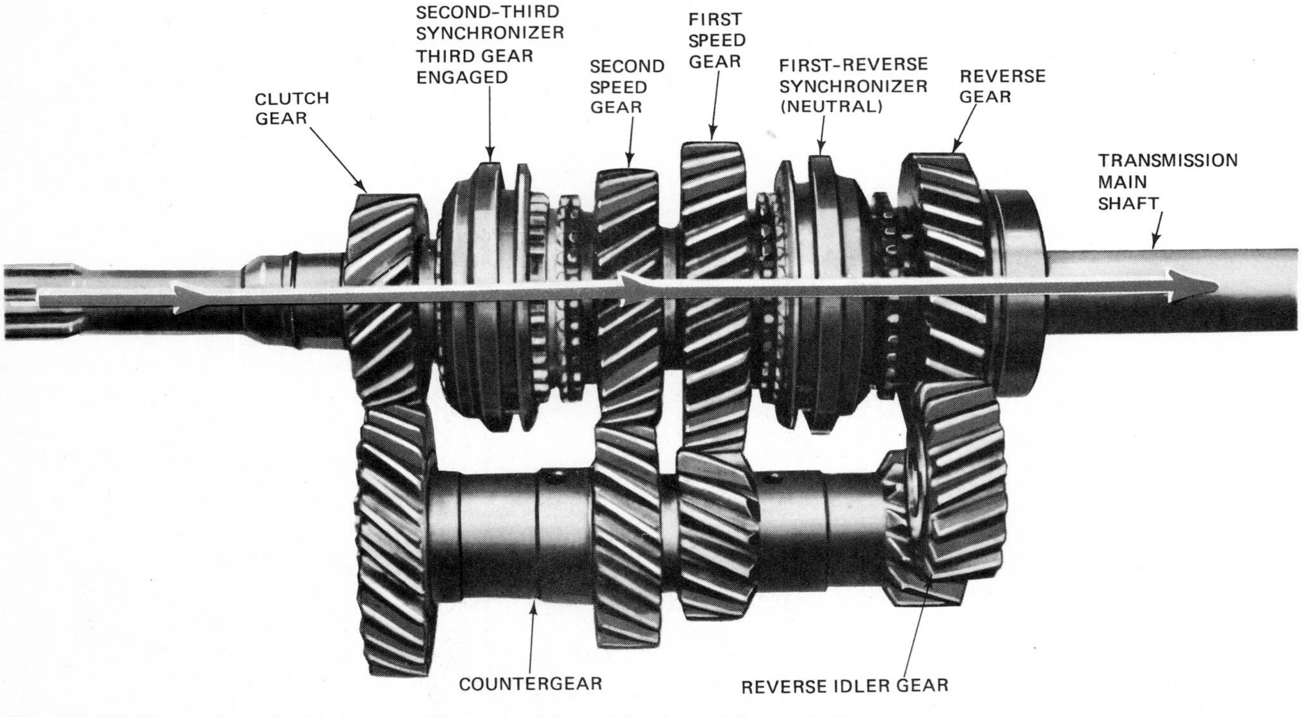

Fig. 46-20 Power flow in third gear. (*Chevrolet Motor Division of General Motors Corporation*)

to the clutch-gear sleeve. The clutch gear is splined to the main shaft. External teeth on the clutch gear mesh with internal teeth in the clutch-gear sleeve. Thus, the clutch gear, clutch-gear sleeve, and two stop rings are always rotating with the main shaft. When the shift is made into second, for example, the main shaft and associated parts may be rotating at a different speed from the second-speed gear. However, as the clutch-gear sleeve is moved toward the second-speed gear, the rear inner stop ring moves against the face of the second-speed gear. This

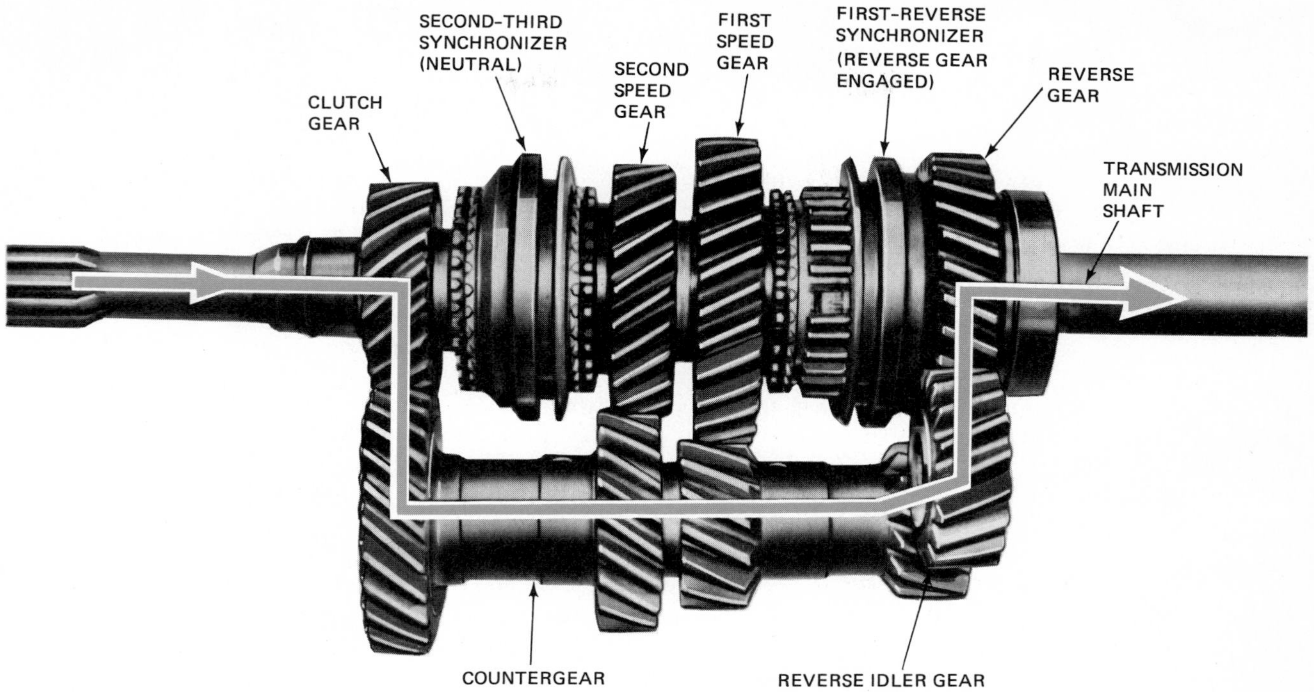

Fig. 46-21 Power flow in reverse. (*Chevrolet Motor Division of General Motors Corporation*)

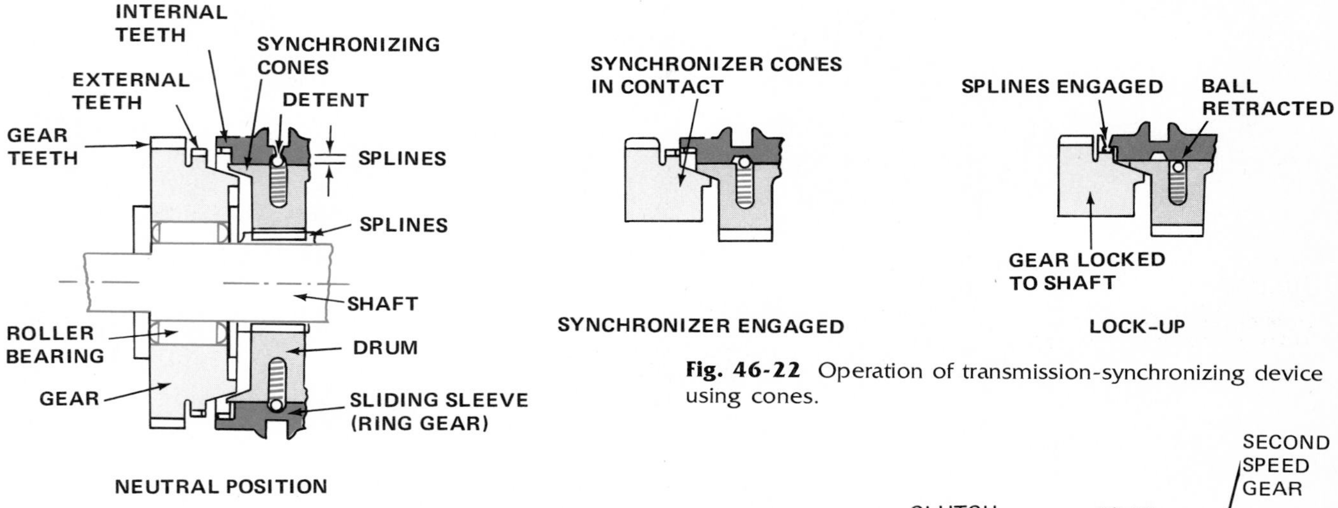

Fig. 46-22 Operation of transmission-synchronizing device using cones.

brings it into synchronous rotation with the clutch-gear sleeve. This permits alignment of the external teeth on the clutch gear and the teeth on the small diameter of the second-speed gear. Now, the clutch-gear sleeve can slip over the teeth of the second-speed gear to couple the second-speed gear and the clutch gear. Then, when the clutch is engaged and the engine again delivers power through it, the second-speed gear drives the main shaft through the clutch gear and the clutch-gear sleeve.

The action in a shift to third is very similar. The third-speed gear is supported on roller bearings.

Another type of synchronizer is shown partly disassembled in Fig. 46-24. Instead of retracting balls, as in Fig. 46-22, this synchronizer has three keys and a pair of

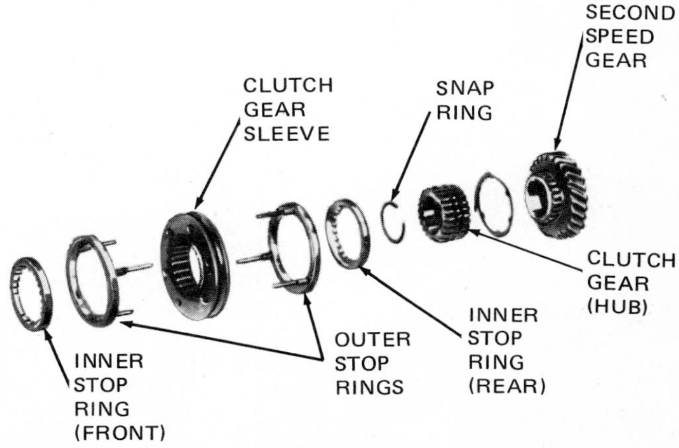

Fig. 46-23 Disassembled view of a pin-type synchronizing device. (*Dodge Division of Chrysler Corporation*)

ring-shaped synchronizing springs. The keys are assembled in slots in the hub. The hub is splined to the main shaft. Assembled outside the hub is the synchronizing sleeve. The hub has external splines that fit the internal splines of the sleeve. The three keys have raised sections that fit in the annular groove of the sleeve.

Synchronizing is a three-stage action. First, the sleeve is moved toward the first-speed gear (when shifting to first). The sleeve slides on the hub splines. It carries the three keys with it. Second, the keys move up against the synchronizer ring and push the ring toward the first-speed gear. The ring presses against the cone of the first-speed gear. Third, further sleeve movement causes the keys to be pressed out of the annular groove in the sleeve. The sleeve continues to move toward the first-speed gear. The friction between the synchronizing ring and the first-speed gear brings the two into synchronous rotation. Now, the final movement of the sleeve allows the internal teeth of the sleeve to engage the external teeth of the first-speed gear. Meshing is completed. The action in the shifts to second and high is similar.

46-12 Four-Speed Transmission

There are two types of four-speed transmissions. One type has four forward speeds with fourth gear being direct drive. The other also has four forward gears but fourth is overdrive. That is, the output shaft turns faster, or overdrives, the input shaft. We look at the transmission in which fourth is direct drive. Figures 46-25 to 46-30 shows the various gear positions.

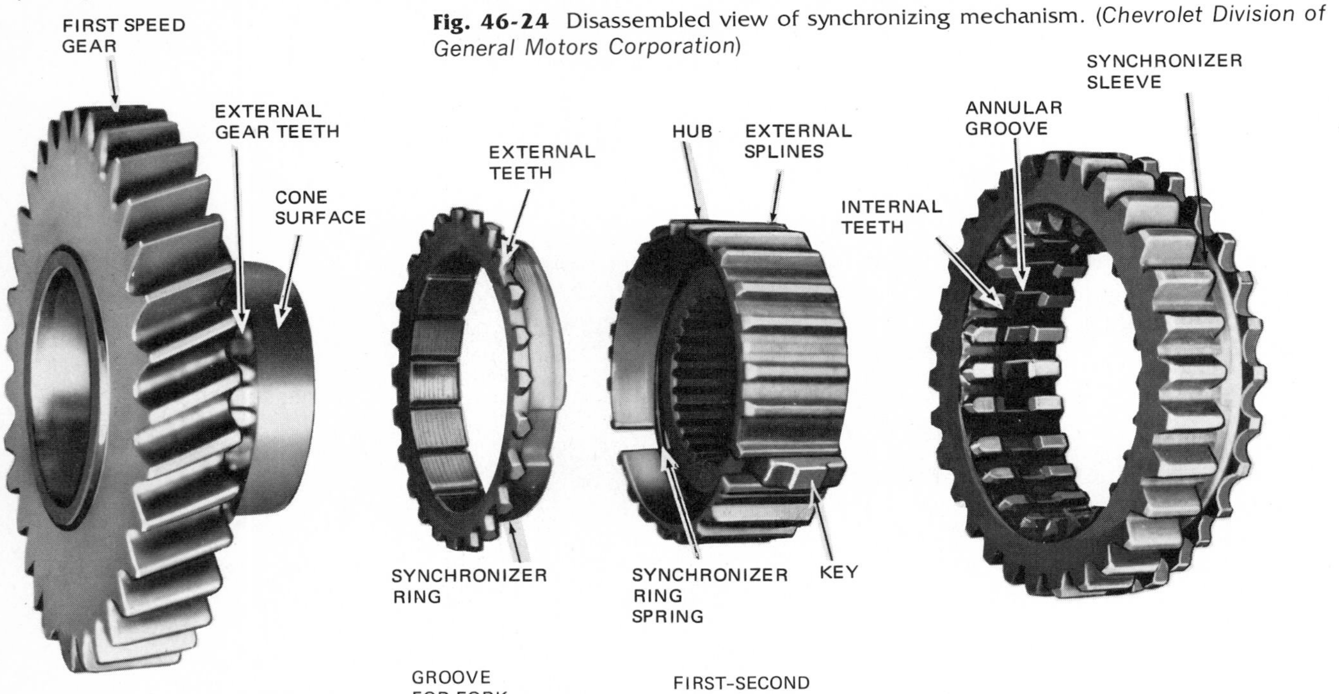

Fig. 46-24 Disassembled view of synchronizing mechanism. (*Chevrolet Division of General Motors Corporation*)

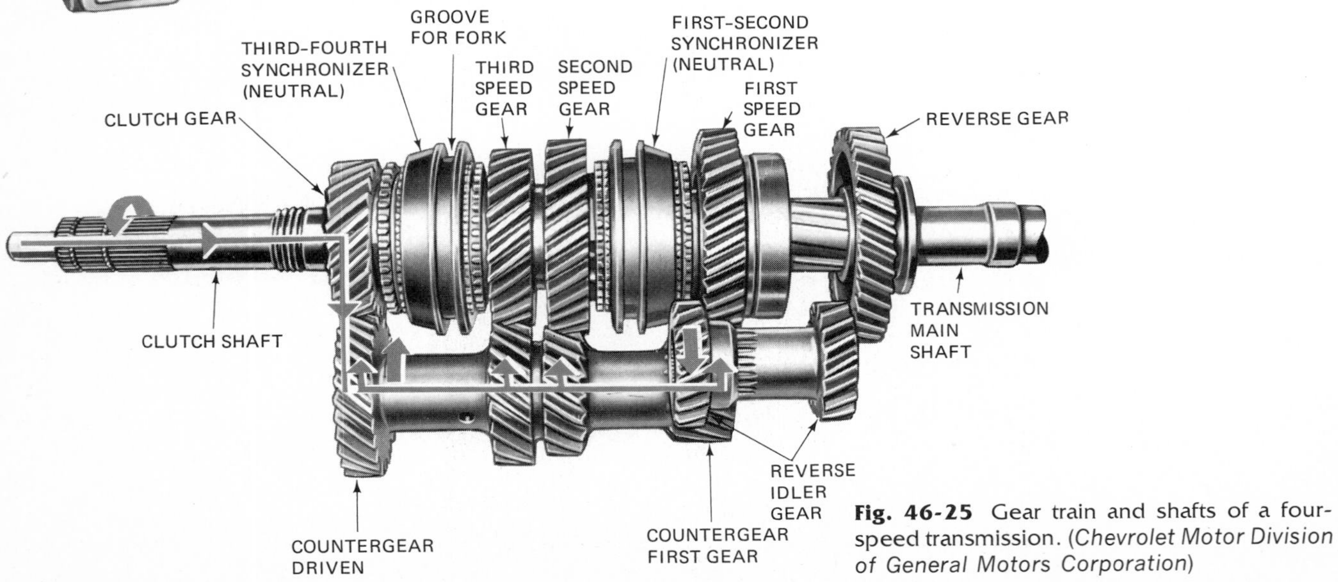

Fig. 46-25 Gear train and shafts of a four-speed transmission. (*Chevrolet Motor Division of General Motors Corporation*)

The purpose of overdrive is to reduce engine wear and improve gasoline mileage. When a shift is made from direct into overdrive, engine speed will drop while still maintaining car speed. For example, in one application, the engine would be turning 3,000 rpm to maintain highway speed in direct drive. By shifting into overdrive, the engine would have to turn only 2,600 rpm to maintain the same highway speed. This means less engine wear and lower fuel consumption.

1. First gear Figure 46-26 shows the power flow through the transmission in first. The 1–2 synchronizer has been moved to the right so its internal teeth engage the external teeth of the first-speed gear.

2. Second gear Figure 46-27 shows the power flow through the transmission in second gear. The 1–2 synchronizer has been moved to the left so its internal teeth engage the external teeth of the second-speed gear.

3. Third gear Figure 46-28 shows the power flow through the transmission in third gear. The 3–4 synchronizer has been moved to the right so its internal teeth engage the external teeth of the third-speed gear.

4. Fourth gear Figure 46-29 shows the power flow through the transmission in fourth gear. The 3–4 synchronizer has been moved to the left so its internal teeth engage the external teeth of the clutch gear.

5. Reverse In reverse, both synchronizers are in the neutral position. The reverse gear has been moved to the left, as shown in Fig. 46-30, so it engages the reverse idler gear. Now, the extra gear in the train causes the main shaft to turn in the reverse direction, so the car moves backward.

6. Gearshift linkage Figure 46-31 shows a typical linkage between the floor-mounted gearshift lever and the trans-

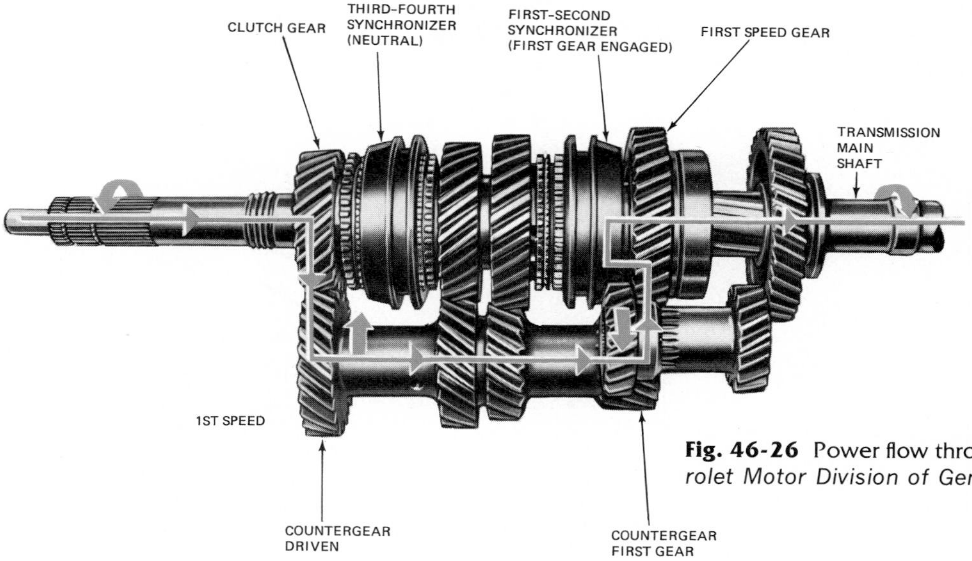

Fig. 46-26 Power flow through gear train in first gear. *(Chevrolet Motor Division of General Motors Corporation)*

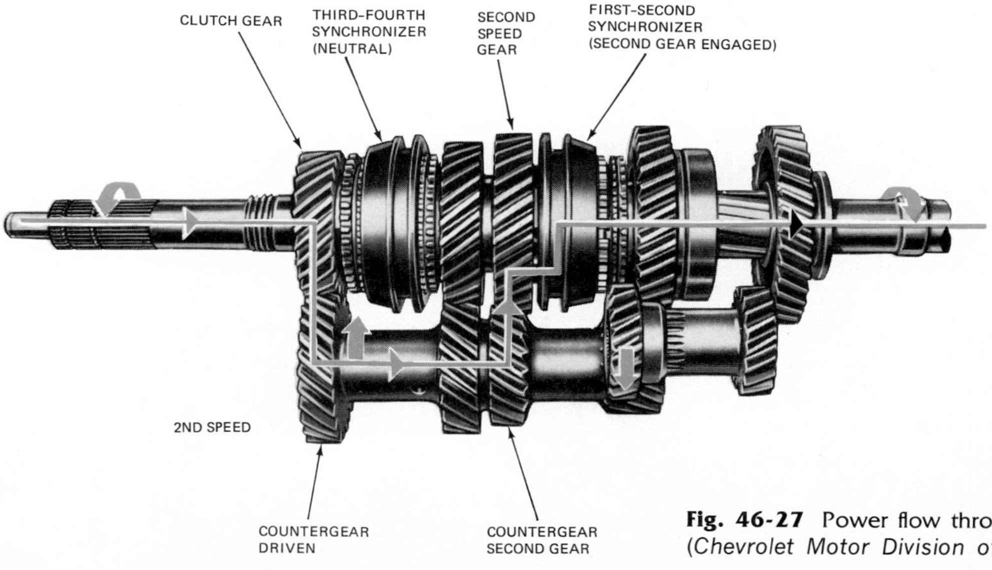

Fig. 46-27 Power flow through gear train in second gear. *(Chevrolet Motor Division of General Motors Corporation)*

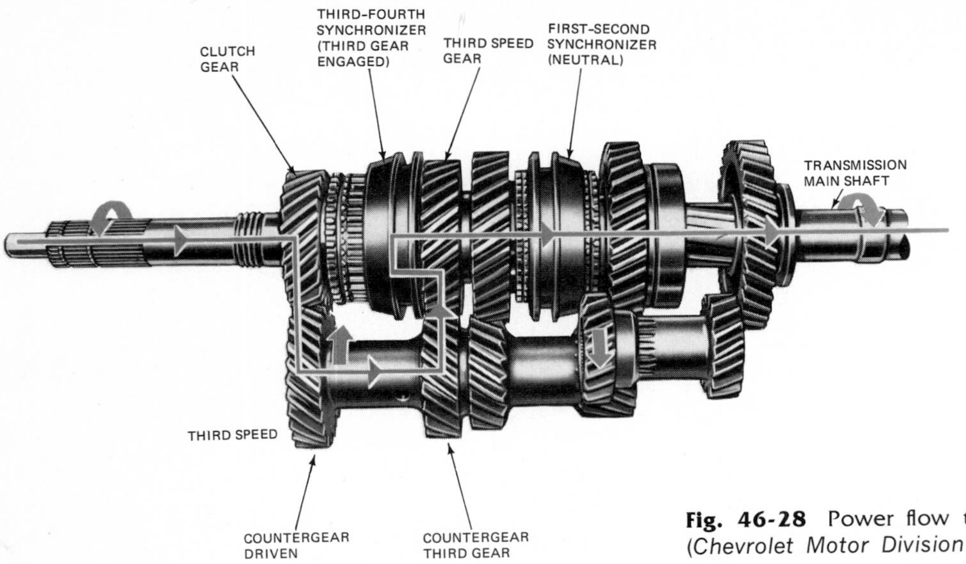

Fig. 46-28 Power flow through gear train in third gear. *(Chevrolet Motor Division of General Motors Corporation)*

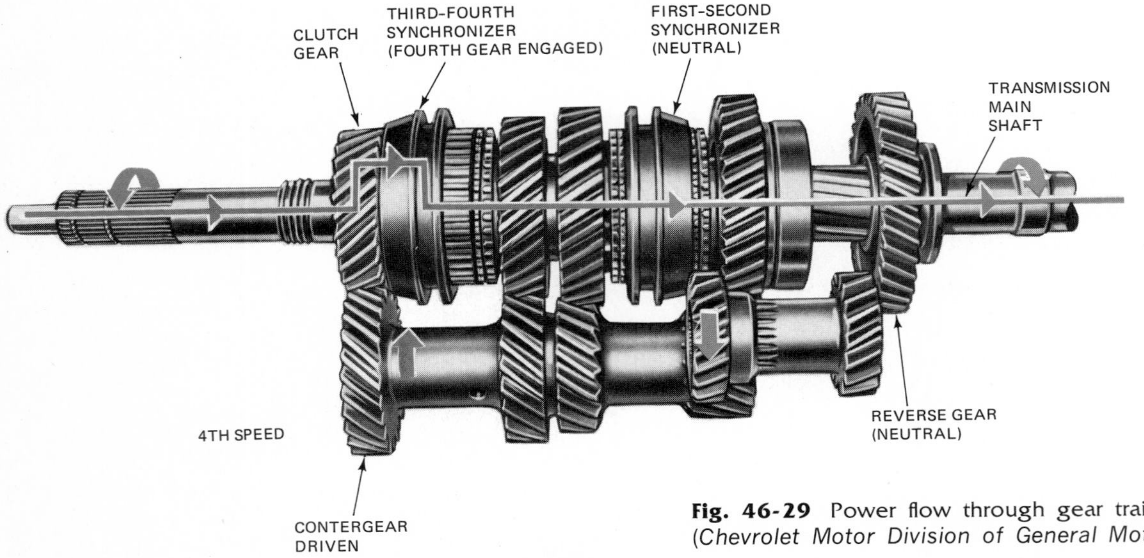

Fig. 46-29 Power flow through gear train in fourth gear. *(Chevrolet Motor Division of General Motors Corporation)*

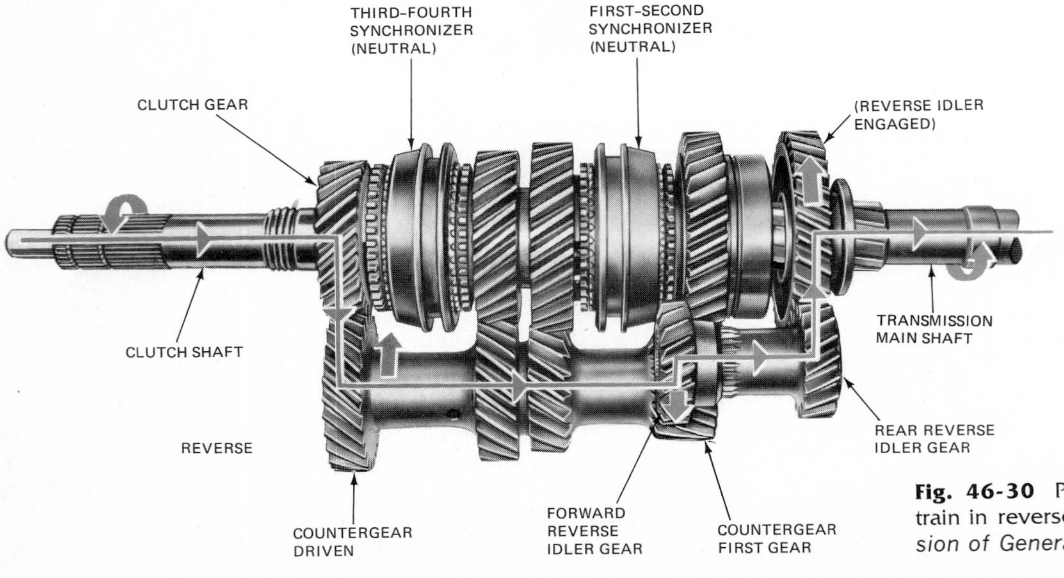

Fig. 46-30 Power flow through gear train in reverse. *(Chevrolet Motor Division of General Motors Corporation)*

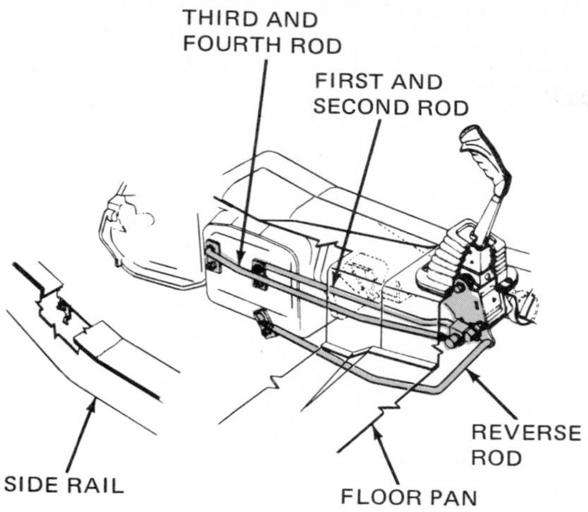

Fig. 46-31 Floor-board gearshift lever and linkage for a four-speed transmission. *(Chrysler Corporation)*

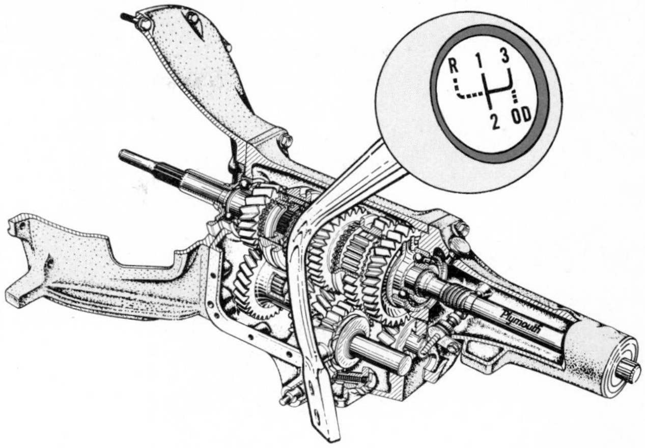

Fig. 46-32 Cutaway view of a four-speed transmission in which fourth speed is an overdrive. In fourth speed, the output shaft turns faster than the input shaft. That is, the output shaft overdrives the input shaft. *(Chrysler Corporation)*

mission. This is the "four on the floor" arrangement. Note that there is an extra rod—the reverse rod—linking the shift lever and transmission. This carries the movement to the reverse fork that moves the reverse gear.

● 46-13 Four-Speed Transmission with Overdrive

This transmission is shown in cutaway view in Fig. 46-32. In third gear, the ratio through the transmission is 1:1. This means the input and output shafts turn at the same speed. However, in fourth gear, the output shaft turns faster than the input shaft. As previously mentioned, in overdrive the engine turns slower to maintain the same car speed. This saves engine wear and gasoline.

● 46-14 Five-Speed Transmission with Overdrive

The five-speed transmission provides five forward speeds. The fifth speed is overdrive. The fifth gear position is achieved by extra gears. Figure 46-33 shows the difference between a four-speed transmission and a five-speed transmission with overdrive. The parts shown in the two boxes are used in the five-speed transmission. The parts not boxed go together to make a four-speed transmission.

● 46-15 Other Transmissions

A different arrangement is required for rear-engine, rear-drive cars and for front-engine, front-drive cars. A simplified view of one arrangement for a rear-engine, rear-drive car is shown in Fig. 46-34. The transmission in Figure 46-35 is for a front-drive car. The engine is mounted crosswise so that the transmission is also crosswise in the car. The transmission and axle are combined and the assembly is known as a *transaxle* (from *trans*mission and *axle*). Note that the clutch is hydraulically operated. The transmission is a four-speed unit.

Figure 46-36 is another transaxle. It uses a standard diaphragm-spring clutch which is operated by a push rod through the main shaft. The clutch is to the right and the clutch lever and release bearing are to the extreme left.

● 46-16 Transmission-Controlled Spark Switch

Several emission-control devices are used on cars. One is the transmission-controlled spark system. This system prevents vacuum advance in any gear but high. The switch that controls the system is screwed into a boss in the side cover of the transmission, as shown in Fig. 46-37. This switch is open in all gears but high, thereby preventing vacuum advance. In high, the switch is closed, allowing vacuum advance.

● 46-17 Back-up Light

When the gearshift lever is moved to reverse, the linkage closes a switch that connects the backup lights to the battery. Thus they come on automatically. This is a warning that the car is about to be backed. The light also gives the driver a chance to see where the car is going.

● 46-18 Speedometer Drive

The speedometer is described in ● 32-17. It is driven by a pair of gears in the transmission extension housing. One of the gears is mounted on the transmission main shaft. The other gear is on the end of the flexible shaft connected to the speedometer.

Fig. 46-33 Gears and shafts in the four- and five-speed transmissions. The boxes enclose the additional parts needed for the five-speed unit. (*Toyota Motor Sales Company, Ltd.*)

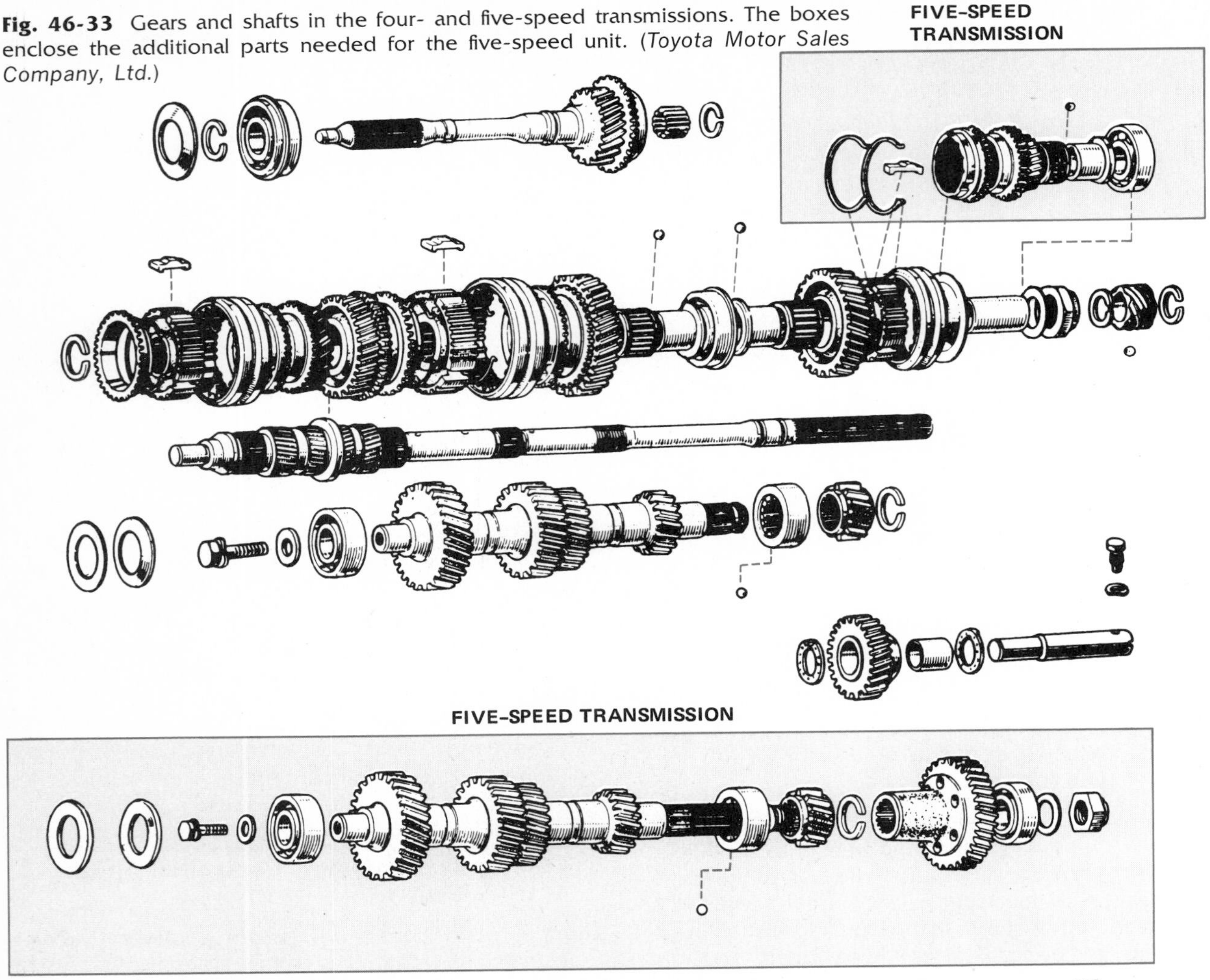

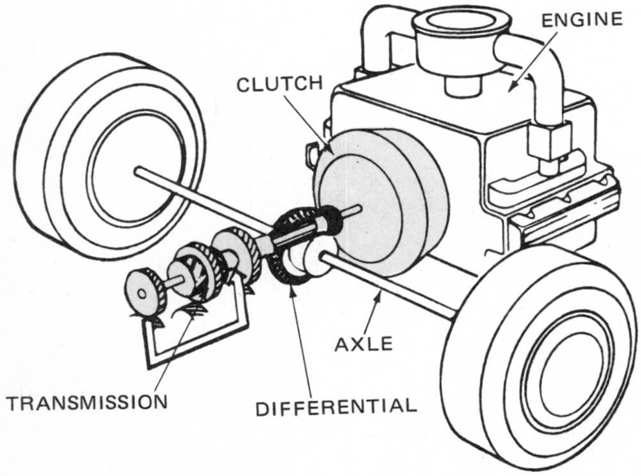

Fig. 46-34 Schematic view of the Corvair power train, showing location of the engine, clutch, and transmission. (*Chevrolet Motor Division of General Motors Corporation*)

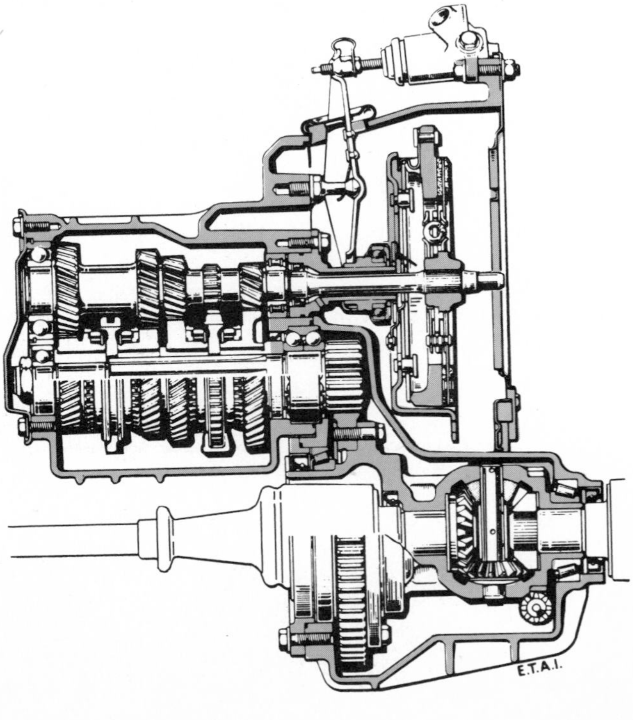

Fig. 46-35 Cutaway view of a four-speed transmission for a cross-mounted engine used in a front-drive car. (*Simca*)

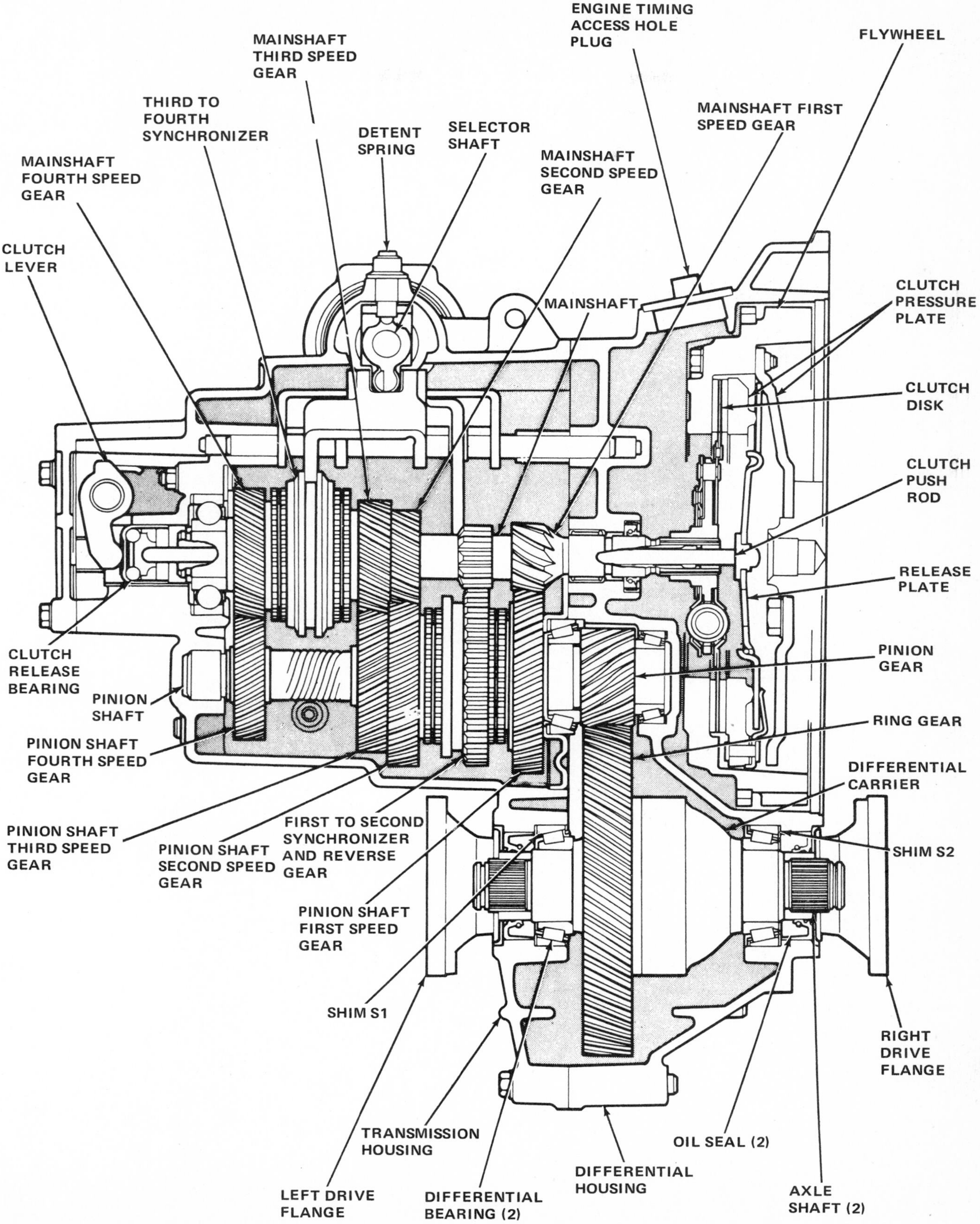

Fig. 46-36 Four-speed transaxle for a front-drive car. (*Chrysler Corporation*)

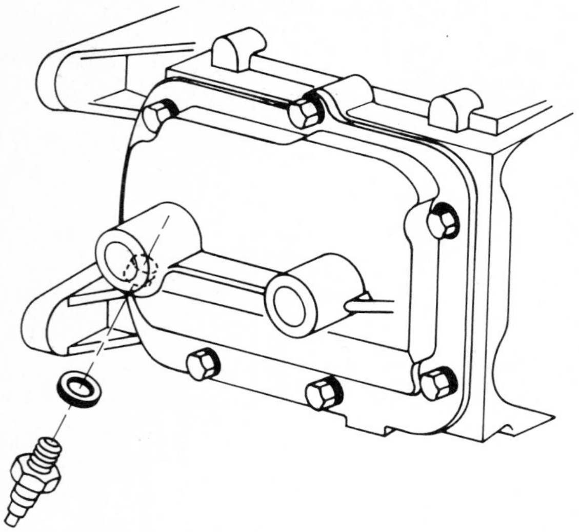

Fig. 46-37 Location of transmission-controlled spark switch on transmission. (*Pontiac Motor Division of General Motors Corporation*)

46-19 Steering-Column Lock

The steering-column lock requires that the transmission be in reverse gear before the ignition key can be removed from the ignition switch. In addition, the clutch safety switch will not permit the engine to be cranked unless the clutch pedal is pushed down far enough to completely disengage the clutch.

REVIEW QUESTIONS

1. What is the basic purpose of the transmission?
2. If a 30-tooth gear is meshed with a 12-tooth gear, what is the gear ratio? If the 30-tooth gear turns at 1,000 rpm, how fast will the 12-tooth gear turn?
3. What is torque?
4. If the 30-tooth gear in question 2 has 100 pound-feet [13.82 kg-m] of torque applied to it, what is the torque on the 12-tooth gear (ignoring friction)?
5. What manual transmission is most commonly used on cars?
6. Through what gears does the power flow, in first, in the transmission gearing pictured in Fig. 46-6?
7. Through what gears does the power flow in second?
8. Through what gears does the power flow in reverse?
9. What is the purpose of the synchronizer? Explain how the synchronizer in Fig. 46-24 works.
10. What is meant by the expression "four on the floor"?
11. What are the two gearshift-lever locations for manual transmissions?
12. Describe the linkage actions when gears are shifted into second from first.

SELF PROJECTS

1. Look at several manufacturers' shop manuals for manual transmissions. Make a list of car models and the types of manual transmissions they use. Late-model Chevrolets, for example, use three-speed Saginaw and four-speed Muncie and Saginaw transmissions.
2. Examine disassembled transmissions and transmission parts. If you can find a transmission that has been removed from a car, take off the side or top cover so you can see the gearing. Now, trace the power flow as you move the gears into the forward speeds and reverse. Make drawings, similar to those in this chapter, showing the gears through which the power flows. With the gears in each of the forward speeds, turn the input shaft (clutch gear). Count the number of times you have to turn this shaft to turn the transmission main shaft (output shaft) once. This tells you the gear ratio in each of the gears.

CHAPTER 47
Manual-Transmission Service

After studying this chapter, you should be able to:

- **List and describe various manual-transmission troubles and complaints.**
- **Explain what might cause the troubles and how they can be corrected.**

This chapter discusses trouble diagnosis, removal, overhaul, and installation of manual transmissions. Following chapters describe automatic transmissions, their operation, and servicing.

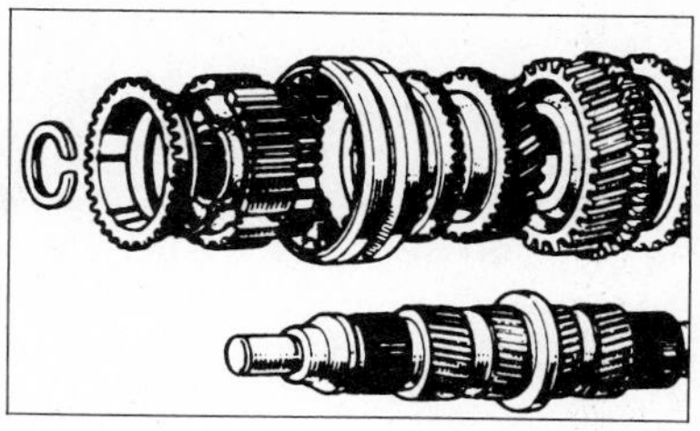

● 47-1 Transmission Troubles

As a first step in any transmission service, the trouble should be diagnosed. This will help to pinpoint the trouble. Sometimes it is not possible to determine the exact location of the trouble. The unit must be removed from the car so that it can be torn down and examined. At other times, diagnosis will lead to the point of trouble, so that it can be eliminated without major overhaul.

The chart that follows lists various transmission complaints, together with possible causes, checks to be made, and corrections needed.

● 47-2 Transmission Trouble-Diagnosis Chart

Most transmission troubles can be grouped under such headings as hard shifting, transmission slips out of gear, transmission noisy, and so on, as listed in the chart.

Note: The complaints and possible causes are not listed according to how often they occur. That is, item 1 (or item a under "Possible Cause") does not necessarily occur more often than item 2 (or item b).

TRANSMISSION TROUBLE-DIAGNOSIS CHART

Complaint	Possible Cause	Check or Correction
1. Hard shifting into gear	a. Clutch not releasing	Adjust (● 45-10)
	b. Excessive clutch-pedal free play	Adjust
	c. Shifter tube binding in steering column	Correct tube alignment
	d. End of transmission input shaft binding in crankshaft pilot bushing	Lubricate; replace bushing
	e. Gearshift linkage out of adjustment	Adjust
	f. Linkage needs lubrication	Lubricate
	g. Shift fork bent	Replace or straighten
	h. Synchronizing unit damaged or springs improperly installed (after a service job)	Replace unit or defective parts; install spring properly
2. Transmission sticks in gear	a. Clutch not releasing	Adjust (● 45-10)
	b. Gearshift linkage out of adjustment or disconnected	Adjust; reconnect
	c. Linkage needs lubrication	Lubricate
	d. Incorrect or insufficient lubricant in transmission	Replace with correct lubricant and correct amount
	e. Synchronizing unit stuck	Free; replace damaged parts
	f. Detent balls (lockouts) stuck	Free

TRANSMISSION TROUBLE-DIAGNOSIS CHART (Continued)

Complaint	Possible Cause	Check or Correction
3. Transmission slips out of gear	a. Gearshift linkage out of adjustment	Adjust
	b. Excessive end play of shaft or gears	Replace worn or loose parts
	c. Bearings worn	Replace
	d. Synchronizer worn or defective	Repair; replace
	e. Transmission loose on clutch housing or misaligned	Tighten mounting bolts; correct alignment
	f. Clutch housing misaligned	Correct alignment
	g. Pilot bearing loose or worn	Replace
	h. Input-shaft retainer loose or broken	Replace
	i. Insufficient lockout-spring tension	Replace
4. No power through transmission	a. Clutch slipping	Adjust (● 45-10)
	b. Gear teeth stripped	Replace gears
	c. Shifter fork or other linkage part broken	Replace
	d. Gear or shaft broken	Replace
	e. Drive key sheared	Replace
5. Transmission noisy in neutral	a. Bearings worn or dry	Replace; lubricate
	b. Defective input-shaft bearing	Replace
	c. Gears worn or teeth broken or chipped	Replace gears
	d. Countershaft worn or bent, or damaged thrust plate or washers	Replace worn or damaged parts
	e. Pilot bushing loose in crankshaft	Replace
6. Transmission noisy in gear	a. Incorrect or insufficient lubricant	Replace with correct lubricant and correct amount
	b. Main rear bearing worn or dry	Replace or lubricate
	c. Gears loose on main shaft	Replace worn parts
	d. Worn or damaged synchronizers	Replace worn or damaged parts
	e. Speedometer gears worn	Replace
	f. Conditions noted in item 5, "Transmission noisy in neutral"	See item 5
7. Gears clash in shifting	a. Clutch not releasing—incorrect pedal lash	Adjust
	b. Synchronizer defective	Repair or replace
	c. Excessive idle speed	Readjust
	d. Pilot bushing binding	Replace
	e. Incorrect gearshift-linkage adjustment	Adjust
	f. Incorrect lubricant	Replace with correct lubricant
8. Oil leaks	a. Foaming due to incorrect lubricant	Replace with correct lubricant
	b. Lubricant level too high	Use proper amount, no more
	c. Gaskets broken or missing	Replace
	d. Side cover loose	Tighten
	e. Oil seals damaged or missing	Replace
	f. Oil slingers damaged, improperly installed, or missing	Replace correctly
	g. Drain plug loose	Tighten
	h. Transmission retainer bolts loose	Tighten
	i. Transmission or extension case cracked	Replace
	j. Speedometer-gear retainer loose	Tighten
	k. Extension-housing seal or drive-line yoke worn	Replace

● 47-3 Transmission Removal and Installation

The transmissions on different automobiles are constructed differently. Thus, there is no single procedure to follow in the removal, disassembly, repair, assembly, and installation of transmissions. The operation requires about 5 to 7 hours; the difference in time is due to variations in the procedures required. The basic procedures are similar. However, before attempting such work, refer to the manufacturer's shop manual. In general, the following steps are required.

1. Drain the lubricant from the transmission.
2. Disconnect the rear axle or the front end of the propeller shaft or universal joint, according to type. Where needle bearings are used, tape the bearing retainers to the shaft to avoid losing needles.
3. Disconnect the shifting linkages from the transmission, the hand-brake linkage or spring, and the speedometer cable. On some floor shifts, you may have to remove the shift controls before you can remove the transmission.
4. Install engine supports, where specified.
5. Remove the attaching bolts or stud nuts. Where recommended, two pilot, or guide, pins should be used. These pins are substituted for transmission bolts. They prevent damage to the clutch friction disk as the transmission is moved back. The transmission is then moved toward the rear until the shaft clears the clutch disk. Then it can be lowered to the floor.

CAUTION: The transmission is heavy. Always use a transmission jack if available. If not, get another person to help you lower the transmission from the car. Do not support the transmission on the hub of the clutch friction disk. This will damage the friction disk. That is the purpose of the guide pins—to carry the weight of the transmission until the shaft splines slip out of the disk hub.

6. With the transmission out, inspect the clutch, flywheel, and flywheel bolts for tightness, and the transmission input-shaft pilot bushing in the crankshaft.
7. In general, installation is the reverse of removal. Just before installation, shift the transmission into each gear, and turn the input shaft to see that the transmission works as it should. Be sure the matching faces of the transmission and the flywheel housing are clean. Place a small amount of lubricant on the splines of the input shaft. Prealign the splines on the input shaft and the friction-disk hub by turning the input shaft so the splines line up. Install guide pins, and lift the transmission. Slide the transmission forward into position. Turn the shaft, if necessary, to secure alignment of the shaft and the friction-disk hub splines. Put the bolts in place, and tighten them. Replace the guide pins with bolts, and tighten them.

Careful: If the transmission does not fit snugly against the flywheel housing, or you cannot move it easily into place, do not force it. It may be that the splines on the shaft and hub are not aligned. Or perhaps roughness or dirt, or a loose retainer ring in the transmission, may be blocking the transmission. If the bolts are tightened under such circumstances, the transmission case may be broken. And there will not be proper alignment.

As a final step in the procedure, fill the transmission with the proper kind and amount of lubricant.

● 47-4 Transmission Overhaul

Overhaul procedures differ for different transmissions. Before disassembling, servicing, and reassembling a transmission, refer to the shop manual that covers the model being repaired.

● 47-5 Gearshift-Linkage Adjustments

The linkage between the gearshift lever and the shifter levers on the transmission must be properly adjusted. This permits proper selection of gears and completion of the shifts. Typically, the adjustment is made with the transmission levers positioned in neutral. Then position the shift lever in neutral. The rods that were disconnected may require some minor adjustment, but in most cases they will slip in and clip in—that is, provided you disconnected the rods at the transmission levers, did not bend the rods, did not unscrew clevis pins, etc.

If the linkage has been tampered with, or the rods do not fit into the transmission levers, then you will need to make a linkage adjustment. Refer to the shop manual.

REVIEW QUESTIONS

1. Name five causes of hard shifting into gear.
2. Name four causes of the transmission sticking in gear.
3. Name five causes of the transmission slipping out of gear.
4. Name four causes of no power through the transmission.
5. Name four causes of noise in the transmission.
6. Name three causes of gear clash when shifting.
7. Name six causes of oil leakage from the transmission.
8. List the steps required to remove a transmission.

SELF PROJECTS

1. Make yourself a set of cause-and-effect cards on manual-transmission troubles. Get some 3-by-5-inch cards. Write the complaint on one side, and the possible causes on the other. For instance, you would write "Hard shifting into gear" on one side, and the eight possible causes listed in the chart on the other side. Carry these cards around with you, and study them. Look at the complaint side of a card, and try to remember the causes written on the other side. Keep doing this, and soon you will know the troubles and their causes backward and forward.
2. Make yourself sets of transmission guide pins. Get a couple of extra bolts of the same size as the transmission attaching bolts, about 4 inches [102 mm] long. Cut off the heads, and cut a groove about $\frac{1}{4}$ inch [6.35 mm] deep through the head end of the bolt with a hack saw. This makes a screwdriver slot. Chamfer the worked end on a bench grinder. Run a die or thread chaser over the threads if they are battered. Coat the threads with a touch of oil.
3. Collect and save transmission input shafts that have been discarded. They can be used to align the clutch friction-disk hub splines when installing the clutch. They can also be used to remove the pilot bushing from the crankshaft. To do this, fill the space in the crankshaft in and back of the bushing. Then push the input shaft into place and drive it in lightly with a plastic hammer. The grease will float the bushing out.

CHAPTER 48
Automatic Transmissions

After studying this chapter, you should be able to:

- **Describe the purpose, construction, and operation of a torque converter.**
- **Discuss planetary-gear construction and operation.**
- **Explain how bands and clutches control the planetary gears.**
- **Describe the hydraulic control system and explain how it controls the clutches and gears.**

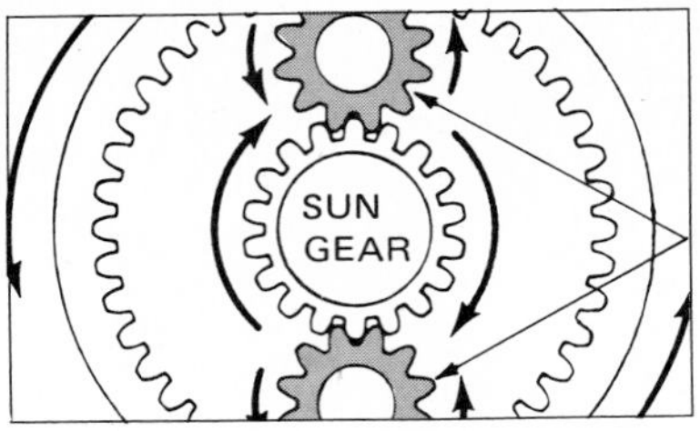

This chapter describes the construction and operation of the modern automatic transmissions used in automobiles. Although they vary in details, all operate in the same way. All have a torque converter and planetary gears controlled by bands and clutches. Almost all provide either two or three forward speeds and reverse. A few models provide four forward speeds. In some of these models, fourth is overdrive.

● 48-1 Function of Automatic Transmissions

Automatic transmissions do the job of shifting gears without assistance from the driver. They start out in low as the car begins to move forward. They shift from low gear into intermediate and then high as the car picks up speed. Automatic transmissions operate hydraulically, that is, by oil pressure.

There are two basic parts to the automatic transmission, the torque converter and the gear system. The torque converter passes the engine power to the gear system. Hydraulic pressure works on the gear system to produce the shifts.

Torque Converter

● 48-2 Fluid Coupling

The torque converter is a special type of fluid coupling. It uses a fluid to transmit rotation from one shaft to another. Two fans can be used to demonstrate a simple fluid coupling (Fig. 48-1). When one fan is turned on and faced toward the other, the stream of air causes the second fan to rotate, even though it is not plugged in. In this case, the "fluid" is air. To improve the efficiency of the fluid coupling, the two members must be closely coupled.

A simple version of an actual fluid coupling is shown in Fig. 48-2. The assembly is like a hollow doughnut, sliced in two. Each hollow half has a series of semicircular plates, called *vanes*. The two halves, or members, are enclosed in an outer cover that is attached to the fly-

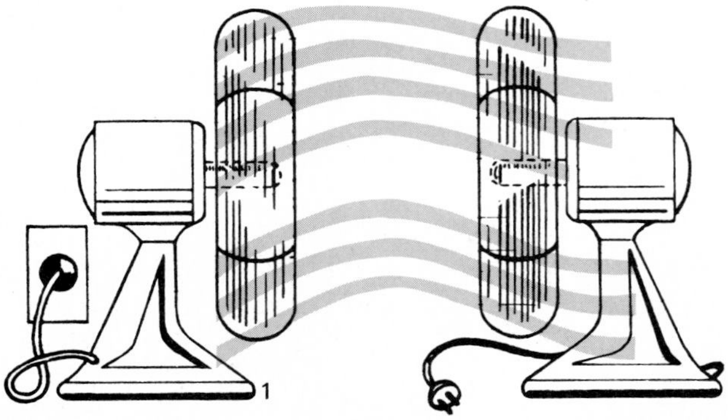

Fig. 48-1 Rotation of fan 1 causes fan 2 to rotate. This is a simple fluid coupling, with air serving as the fluid.

Fig. 48-2 Simple version of two members of a fluid coupling. (*Chevrolet Motor Division of General Motors Corporation*)

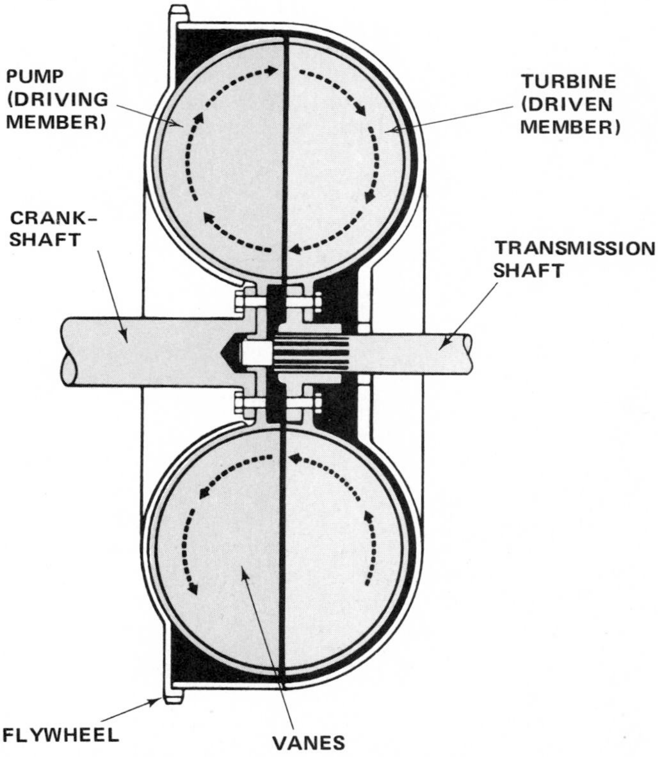

Fig. 48-3 Sectional view of a fluid coupling.

wheel. Figure 48-3 shows the arrangement in sectional view. The driving half of the fluid coupling, called the *pump,* or *impeller,* is attached to the crankshaft. The driven half, called the *turbine,* is attached to the transmission shaft.

There is no direct mechanical connection between the pump (driving member) and turbine (driven member). If there were no oil in the assembly, the two members could rotate independently of each other. However, filling the fluid coupling with oil makes the difference. When the pump rotates, the oil between the vanes of the pump is thrown out by centrifugal force. It is thrown into the turbine.

Centrifugal force is the force that pushes things outward from a center around which they are revolving. The oil caught between the vanes of the pump is thrown out. It has no place to go except into the turbine. The oil is thrown into the turbine with great force, and it hits the vanes of the turbine at an angle. In other words, the moving oil applies pressure to the vanes of the turbine, forcing the turbine to turn. The action is shown in Fig. 48-4.

48-3 Torque Converter

The torque converter is a special sort of fluid coupling. In the torque converter, the vanes are curved and not flat, as shown in Fig. 48-2. Curving the vanes reduces "bounce-back" of the fluid. With flat vanes, the fluid, as it hit the vanes of the driven member, would tend to bounce back into the driving member. This would remove some of the driving torque, and power would be lost.

But with curved vanes (Fig. 48-5), the fluid is unable to bounce back. In the illustration, the split guide ring is a smaller doughnut-shaped ring. It tends to keep the fluid in the outer part of the driving and driven members, where it can do the most good.

Note: From this point on, we will use the term "pump" for the driving member, and "turbine" for the driven member.

48-4 The Third Member

The coupling shown in Fig. 48-5 would not be very efficient. The reason is that, as the fluid leaves the inner part of the turbine, it is thrown back into the pump in the wrong direction. That is, it opposes the rotation of the pump (Fig. 48-6). As you can see by the arrows, the fluid hits the pump vanes in the direction opposite to the direction in which the vanes are moving. This greatly reduces the efficiency of the fluid coupling. That is, the pump has to overcome this opposing push to get the oil moving in the right direction again.

To eliminate this, a third member, called a *stator,* is installed between the inner ends of the pump and the turbine vanes (Figs. 48-7 and 48-8). The stator has curved vanes. They change the direction of the fluid coming out of the turbine to one that helps the rotation.

Figure 48-9 shows, in a simplified form, how the third member changes the direction of the fluid to a helping direction. To the left, a jet of fluid is shown hitting a round bucket attached to a wheel. The fluid pushes on the bucket, but only a little. The fluid leaves the bucket with about the same energy it had when it entered. A single pass through the bucket does not give the bucket much of a push.

However, if a curved vane is added, as shown to the right in Fig. 48-9, the fluid makes more than one pass through the bucket. That is, the curved vane redirects the fluid back into the bucket, giving the bucket an added push. Actually, the fluid could complete the circuit many times, adding a push to the bucket each time. This effect is what we mean by *torque multiplication.*

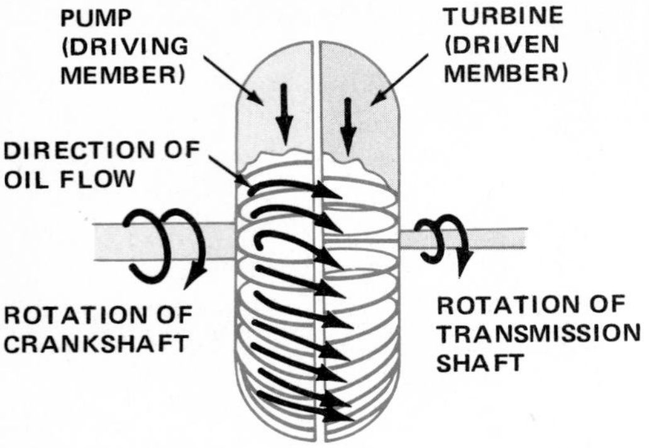

Fig. 48-4 Fluid coupling in action. Oil is thrown from the driving member into the driven member. The outer casings have been cut away so that the vanes can be seen.

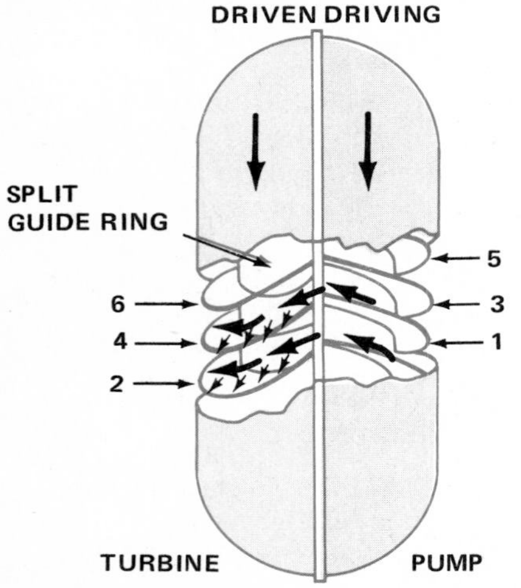

Fig. 48-5 Simplified cutaway view of two members of a torque converter. The heavy arrows show how oil circulates between the driving-member and the driven-member vanes. In operation, the oil is forced by vane 1 downward toward vane 2; thus it pushes downward against vane 2, as shown by the small arrows. Oil then passes around behind the split guide ring and into the driving member again, or between vanes 1 and 3. Then, it is thrown against vane 4 and continues this circulatory pattern, passing continuously from one member to the other.

Now, let us see how torque multiplication applies to the torque converter. As the oil leaves the turbine, it hits the stator vanes. There, it is redirected into the pump in a helping direction. The pump then throws the oil back into the turbine. This is a continuous action. The repeated pushes of the fluid on the turbine vanes increase the torque on the turbine. In many torque converters, the torque is more than doubled. That is, for each pound-foot [0.138 kg-m] of torque entering the pump, the turbine delivers more than 2 pound-feet [0.276 kg-m] of torque to the transmission shaft. This is torque multiplication.

48-5 Stator Action

The stator, as we have seen, causes the torque converter to multiply torque *when the pump is turning faster than*

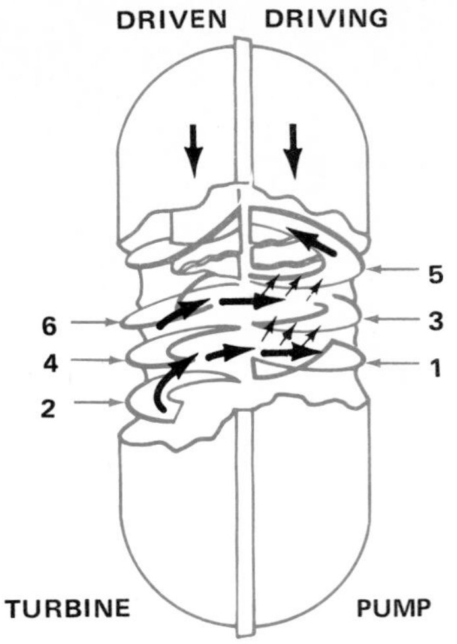

Fig. 48-6 This shows what would happen if the vanes in Fig. 48-5 were continuous. Actually, the inner ends of the vanes are not as shown here, but are as in figures that follow. Here, the split guide ring and outer ends of the vanes have been cut away. If the vanes were as shown here, the oil leaving the trailing edges of the driven member would be thrown upward against the forward faces of the driving-member vanes. They would thus oppose the driving force. This effect, shown by the small arrows, would waste power and torque.

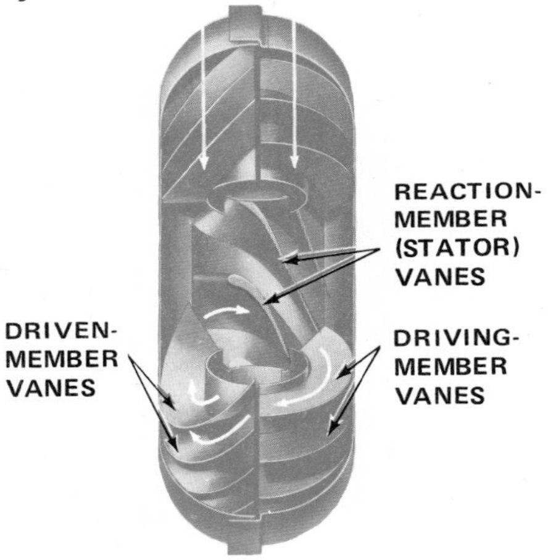

Fig. 48-7 Cutaway view of a torque converter with three members. The third member, the stator, serves as a reaction member. It changes the direction of oil flow, as shown by the curved arrows, under certain operating conditions. (*Chevrolet Motor Division of General Motors Corporation*)

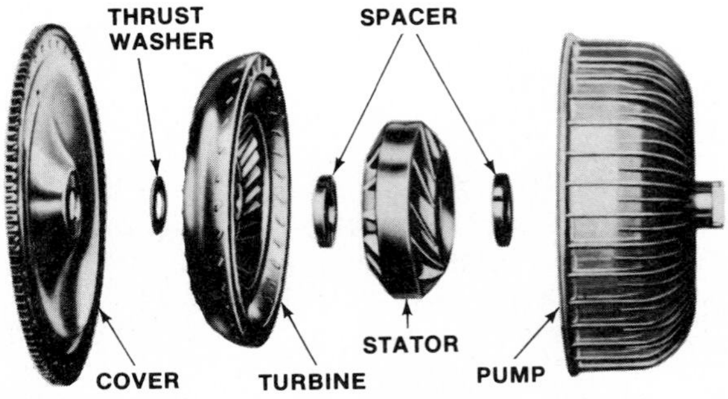

Fig. 48-8 Turbine, stator, and pump (impeller) in a torque converter. (*Ford Motor Company*)

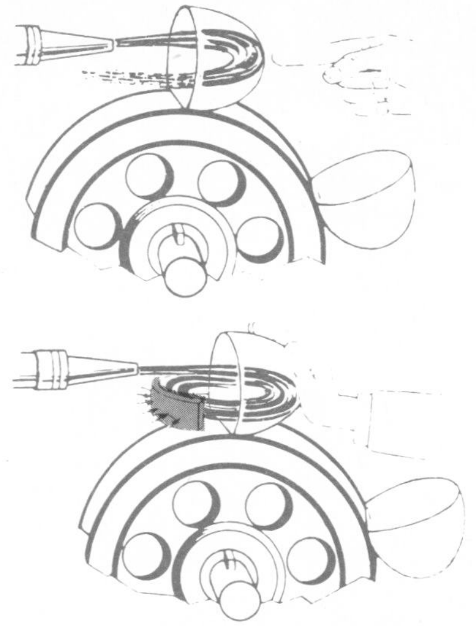

Fig. 48-9 Effect of a jet of oil on a bucket attached to a wheel. If the oil enters and leaves as at the top, the push on the bucket and wheel is small. But, if the oil jet is redirected into the bucket by a curved vane, as at the bottom, the push is increased. (*Chrysler Corporation*)

the turbine. This speed difference and increase in torque have the same effect as a low gear in the manual transmission. They allow the engine to turn fast while the car wheels are turning slowly. Thus, a high torque can be applied, and the car can accelerate.

However, as the car comes up to speed, the turbine begins to "catch up" with the pump. When this happens, the fluid leaving the trailing edges of the turbine vanes is moving at about the same speed as the pump. Therefore, it could pass directly into the pump in a helping direction, without being given an assist by the stator. In fact, under these conditions, the stator vanes are in the way. The fluid begins to hit the back sides of the stator vanes. To allow the stator vanes to move out of the way, the stator is mounted on a *freewheeling* mechanism.

The freewheeling mechanism is a one-way clutch that allows the stator to revolve freely, or "freewheel," in only one direction. The mechanism locks the stator if the stator tries to turn in the other direction. The freewheeling mechanism uses an overrunning clutch. (Recall from our discussion in ● 26-5 that the starting motor uses a very similar overrunning clutch.) The stator overrunning clutch is somewhat more complicated than that in the starting motor. Figure 48-10 shows, in sectional view, the location of the overrunning clutch. Note that it is also called a *one-way clutch*. Figure 48-11 shows the clutch in detail. It includes a hub, an outer ring that is part of the stator, and a series of rollers. The rollers are located in notches in the outer ring. The outer ring is called the overrunning-clutch *cam* in Fig. 48-11. The notches are smaller at one end than at the other. The rollers have springs behind them. When there is a push on the front of the stator vanes from the fluid leaving the turbine, the stator attempts to roll backward. This causes the rollers to roll into the smaller ends of the notches. There, they jam and lock the stator to the hub. Now the stator cannot turn backward. Instead, the stator vanes change the

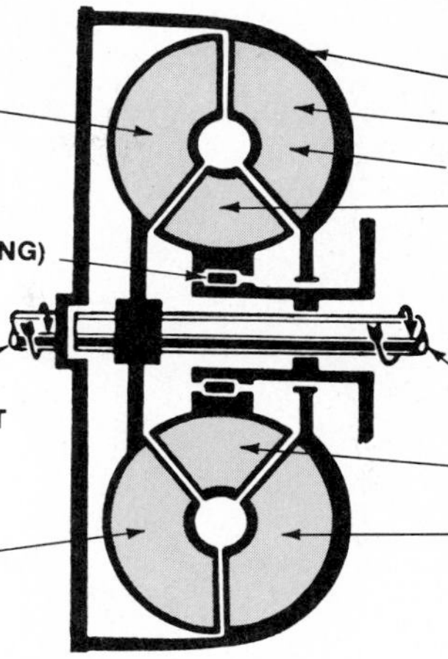

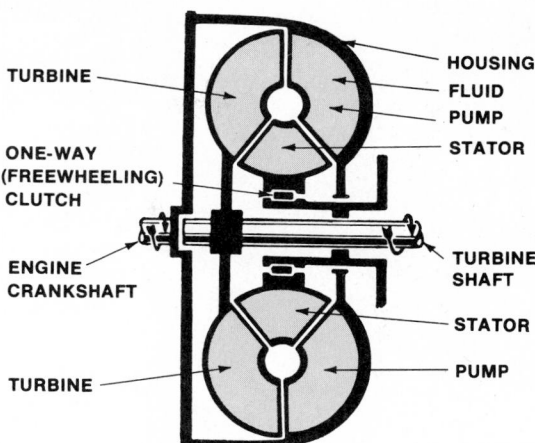

Fig. 48-10 Simplified sectional view of a torque converter showing locations of turbine, stator, pump, and one-way clutch. (*Ford Motor Company*)

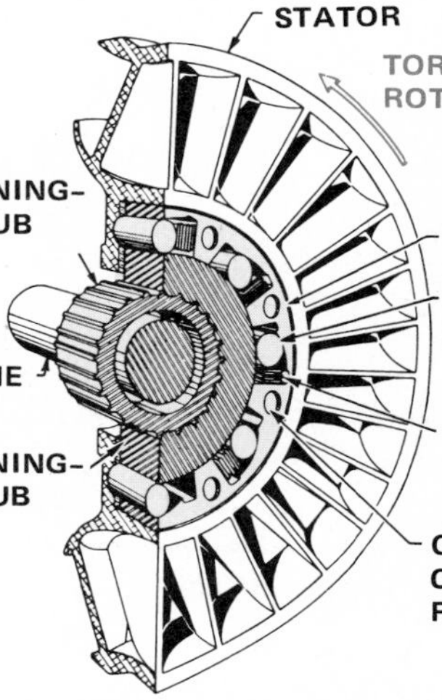

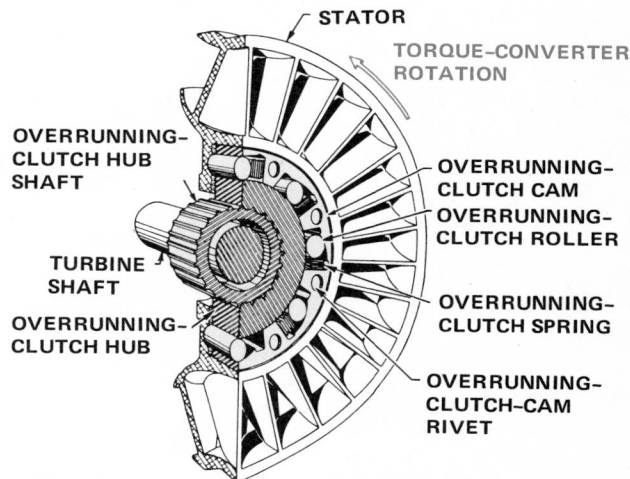

Fig. 48-11 Details of an overrunning, or one-way, clutch used to support a stator in a torque converter. (*Chrysler Corporation*)

direction of the fluid into a helping direction, as previously explained.

However, as the turbine speed approaches the pump speed, the direction of the fluid no longer has to be changed as it leaves the turbine. The fluid now begins to hit the other side of the stator vanes. The stator begins to revolve in a forward direction. The rollers roll out of the smaller ends of the notches, and into the larger ends. There, they cannot jam, and the stator is able to run freely—to freewheel. That is, the vanes simply move forward to get out of the way of the fluid.

Planetary Gears

● 48-6 Planetary Gears

An automatic transmission has two or more planetary-gear sets. The simple-looking planetary-gear assembly, shown in Fig. 48-12, can do many things. It can:

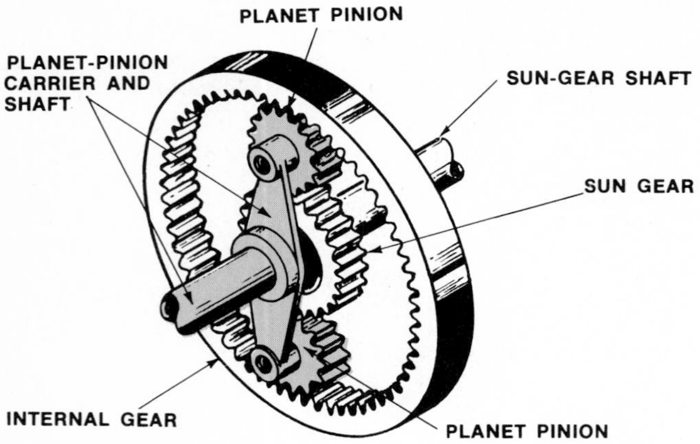

Fig. 48-12 Planetary-gear system. The planet pinions rotate on shafts that are mounted on a planet-pinion carrier. The planet-pinion carrier is attached to a shaft that is exactly aligned with the sun-gear shaft. These shafts are centered in the internal gear.

Increase speed and reduce torque
Reduce speed and increase torque
Reverse the direction of rotation
Act as a solid shaft
Disconnect the driving shaft from the driven shaft

The planetary-gear set, or assembly, includes an *internal* gear (also called a *ring* gear), a *sun* gear, and two or more *planet pinions* on a carrier and a shaft. Before we describe how the planetary-gear set does its job, let us review gears.

48-7 Gear Combinations

When two gears are in mesh, as shown in Fig. 46-3, they turn in opposite directions. But if another gear is put into the gear train, as shown in Fig. 48-13, the two outside gears turn in the same direction. Note that the middle gear is called an *idler* gear. It doesn't work—it is idle.

To get a combination of two gears to rotate in the same direction, you can use one internal gear. The internal gear, or ring gear, has teeth on the inside. The spur gear and the internal gear both rotate in the same direction (Fig. 48-14).

Now, if another spur gear is added in the center and meshed with the small spur gear, the combination is a simple planetary-gear system (Fig. 48-15). The center gear is called the *sun gear* because the other gears revolve around it. This is similar to the way the planets in our solar system revolve around the sun. The spur gear between the sun gear and the internal gear is called the *planet pinion.* This is because it revolves around the sun gear, just as planets revolve around the sun. Now let's look at the planetary-gear system.

48-8 Planetary-Gear Operation

We complete the planetary-gear set by adding another planet pinion, as shown in Fig. 48-16. This gives us the combination shown in Fig. 48-12.

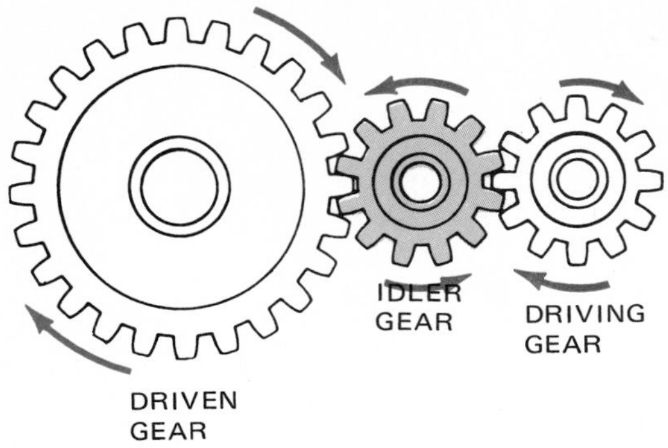

Fig. 48-13 The idler gear causes the driven gear to turn in the same direction as the driving gear.

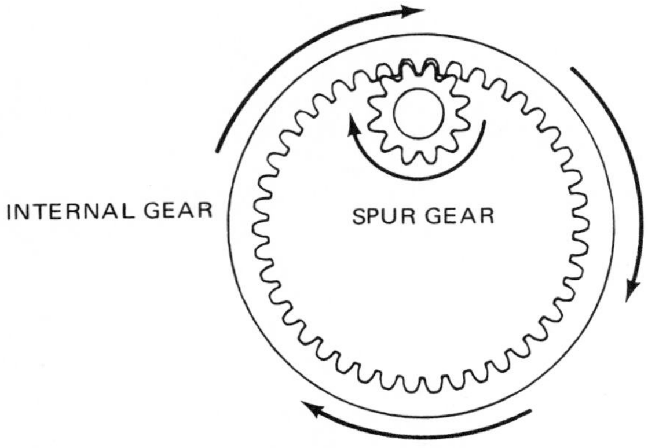

Fig. 48-14 If one internal gear is used with one external gear, both the driven and the driving gears turn in the same direction.

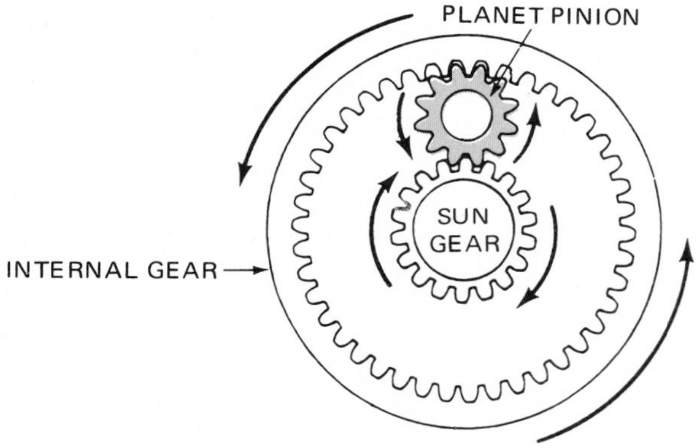

Fig. 48-15 If a sun gear is added to the arrangement shown in Fig. 48-14, the result is a simple planetary-gear system.

The planetary-gear sets used in automatic transmissions usually have three or four planet pinions. But there are various other combinations, as we shall see when we discuss the complete assemblies. In the explanation that follows, we shall use the simple two-planet-pinion gear set shown in Fig. 48-12.

The two planet pinions rotate on shafts that are a part of a planet-pinion carrier (Fig. 48-12). There are thus three members in the planetary-gear set: internal gear, sun gear, and planet-pinion carrier assembly. As mentioned previously, the gear set can increase speed and reduce torque, reduce speed and increase torque, reverse

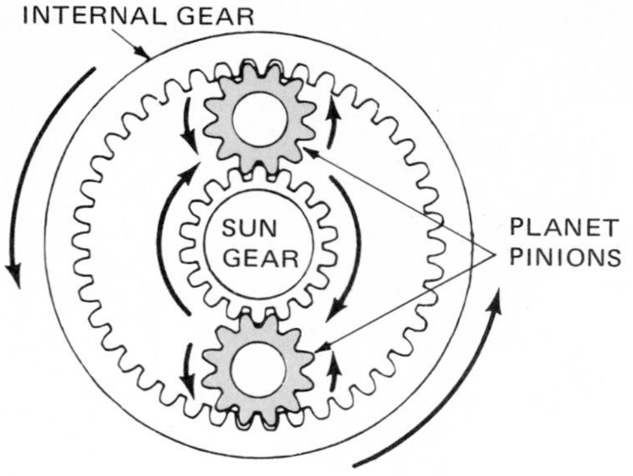

Fig. 48-16 To complete the planetary-gear system, a second planet pinion is added. This pinion balances the forces so that the system runs smoothly. Planetary-gear systems in automatic transmissions have three or four planet pinions.

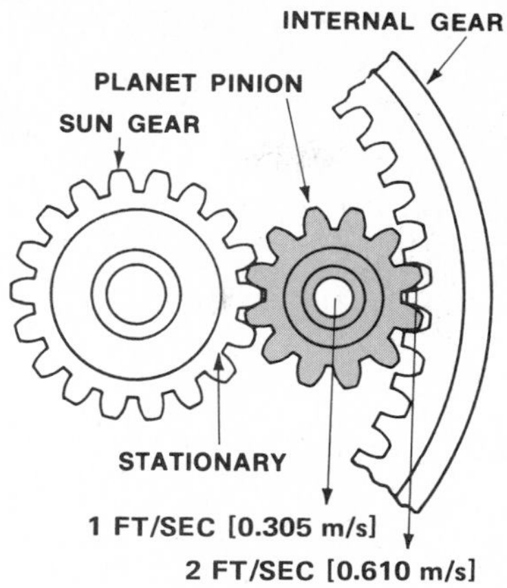

Fig. 48-17 If the sun gear is stationary and the planet-pinion carrier is turned, the ring gear turns faster than the carrier. The planet pinion pivots around the stationary teeth. If the center of the pinion shaft is moving at 1 foot per second, the tooth opposite the stationary tooth must move at 2 feet per second. (It is twice as far away from the stationary tooth as the center of the shaft.)

the direction of rotation, act as a solid shaft, and disconnect the driving and driven shafts.

If one member is held stationary and another is turned, there is either a speed increase, a speed reduction, or reverse. If two members are locked together, the gear set acts like a solid shaft. If no members are locked, no power is transmitted through the gear set. Let's see how all this comes about.

1. Speed increase 1 Suppose the sun gear is stationary, and the planet-pinion carrier turns. There would be a speed increase. Why? When the carrier revolves, it carries the planet pinions around with it. This movement makes the planet pinions rotate on their shafts. As the pinions rotate, they cause the internal gear to rotate also (see Fig. 48-17). Note the conditions. The sun gear is stationary. The planet-pinion carrier is moving, carrying the pinions around with it. The planet pinions "walk around" the sun gear, which means they rotate on their shafts. The inside pinion tooth, meshed with the sun gear, is stationary because the sun gear is stationary. That means the outside pinion tooth, meshed with the internal gear, is moving twice as fast as the shaft on which the planet pinion is turning. If the planet-pinion shaft is moving at 1 foot per second [0.305 meter per second], the outer tooth is moving at 2 feet per second [0.610 meter per second]. In other words, speed increases.

This condition is not used in automatic transmissions.

2. Speed increase 2 Another combination is to hold the internal gear stationary and turn the planet-pinion carrier. In this case, the sun gear is forced to rotate faster than the planet-pinion carrier, and there is a speed increase as in case 1.

3. Speed reduction 1 If the internal gear turns while the sun gear is held stationary, the planet-pinion carrier turns more slowly than the internal gear. This is just the opposite of what we described as speed increase 1. With the internal gear turning the planet-pinion carrier, the planetary-gear set acts as a speed-reducing system.

4. Speed reduction 2 If the internal gear is held stationary and the sun gear turns, there is speed reduction. The planet pinions must rotate on their shafts. They must also walk around the internal gear, since they are in mesh with it. As the pinions rotate, the planet carrier rotates. But it rotates at a slower speed than the speed at which the sun gear is turning.

5. Reverse 1 To get reverse, the planet-pinion carrier can be held stationary, and the internal gear turned. In this case, the planet pinions act as idlers and cause the sun gear to turn in the reverse direction. The system acts as a direction-reversing system, with the sun gear turning faster than the internal gear.

This condition is not used in automatic transmissions.

6. Reverse 2 A second way to get reverse is to hold the planet-pinion carrier stationary and turn the sun gear. The internal gear turns in the reverse direction, but slower than the sun gear.

7. Direct drive If any two members are locked together, then the entire planetary-gear system acts as a solid shaft. Locking two members together locks up the system.

All the conditions discussed above are listed in the chart in Fig. 48-18. The three conditions used in automatic transmissions are listed in columns 3, 4, and 6. The planetary-gear sets used in transmissions are designed to give direct drive, speed reduction, and reverse.

Hydraulic Controls

48-9 Hydraulic Shift Controls

We have seen how the power flows into the transmission through the torque converter, and how planetary gears operate. Figure 48-19 shows, in simplified form, how the power flows through the torque converter and into the planetary gear. Note that it flows to the internal (ring)

Conditions	1	2	3	4	5	6
Ring Gear	D	H	T	H	T	D
Carrier	T	T	D	D	H	H
Sun Gear	H	D	H	T	D	T
Speed	I	I	L	L	IR	LR

D—Driven
H—Hold (Stationary)
I—Increasing of Speed
L—Reduction of Speed
T—Turn or Drive

Fig. 48-18 Various conditions that are possible in the planetary-gear system if one member is held and another is turned.

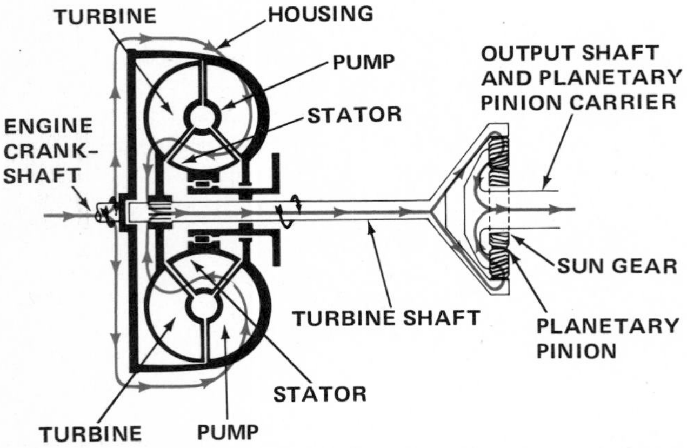

Fig. 48-19 Power flow through torque converter to planetary gears.

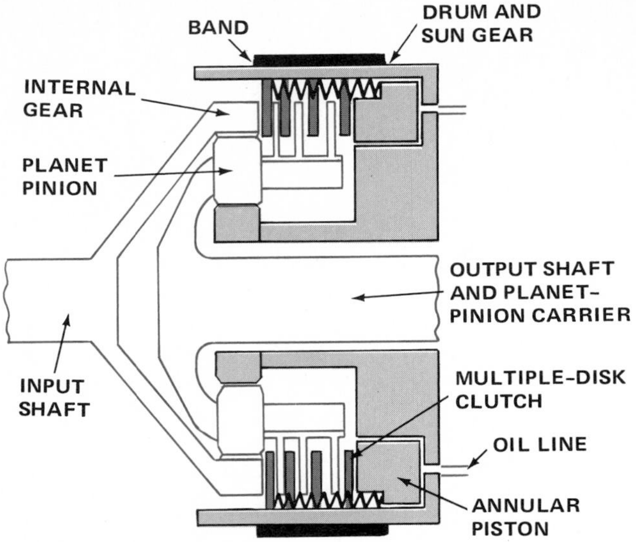

Fig. 48-20 Sectional view showing the two controlling mechanisms used in the front planetary gear set in an automatic transmission. One mechanism consists of a brake drum and brake band. The other is a multiple-disk clutch.

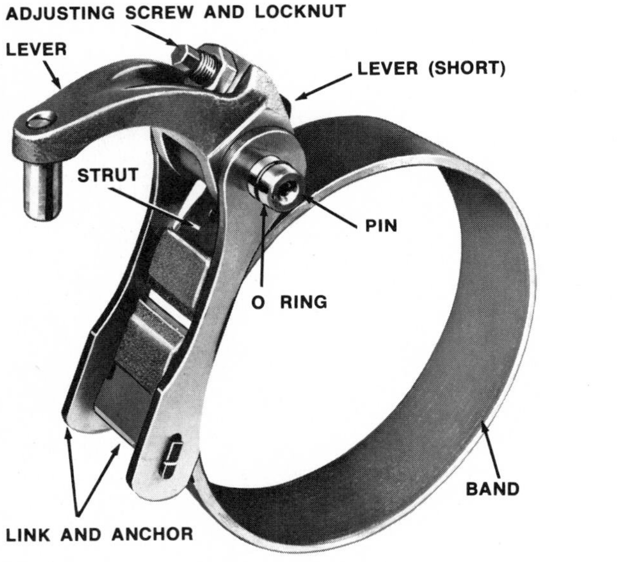

Fig. 48-21 Transmission band. (*Chrysler Corporation*)

gear. Now, let's see how the planetary gears are controlled.

There are two controls, a band and a clutch. The band consists of a brake band that surrounds a metal drum. The drum may be attached to the sun gear, as shown in Fig. 48-20, or it may be the outer surface of the planetary ring (internal) gear. The clutch consists of a series of clutch plates. Half the plates are splined to an outer ring, called the *clutch drum,* and the other half are splined to the clutch hub. The clutch hub is splined to one of the members of the planetary-gear set. When oil pressure forces the two sets of clutch plates together, the clutch is engaged. That means the planetary-gear set is locked up, rotating as a single unit. When the oil pressure is released, the clutch is released. This means that the two sets of clutch plates can rotate independently.

● 48-10 Band and Clutch

Let's put the two controls—the clutch and the band—onto the planetary-gear set. Figure 48-20 is a sectional view of a planetary-gear set with a band and clutch. We shall explain later how the band and clutch are applied. First, we want to see what happens when they are applied.

A band is shown in Fig. 48-21. Note that the band is positioned around the sun-gear drum (Fig. 48-20). When the band is applied, the sun gear is held stationary. This means that the planetary-gear set acts as a speed reducer. The internal gear is turning—it is mounted on the input shaft. This arrangement forces the planet pinions to rotate. They walk around the stationary sun gear and carry the pinion carrier around with them. The carrier rotates at a slower speed than the internal gear.

Now suppose that the clutch is applied instead of the band. Oil pressure that enters through the oil line causes the clutch to apply. The oil pressure forces the piston in the sun-gear drum to the left (in Fig. 48-20). The clutch plates are pushed together so that the clutch is engaged. With this situation, the planet-pinion carrier and the sun gear are locked together. The planetary-gear set is now in direct drive. In other words, the system is locked up. (Figure 48-22 shows the clutch plates. Note that they are alternately splined to the drum and the sun gear.)

The arrangement shown in Fig. 48-20 is only one of several arrangements used in automatic transmissions. In some transmissions, when the band is applied, it holds the internal gear or the planet-pinion carrier stationary. Different transmissions may lock different members together when the clutch is applied. The principle is the same in all transmissions, however. There is gear reduction when the band is applied, and there is direct drive when the clutch is applied.

● 48-11 Hydraulic Circuits

Figure 48-23 is a simplified diagram of a hydraulic control circuit for a single planetary-gear set in an automatic transmission. Later we will look at the circuits for automatic transmissions that use two or more planetary-gear sets. As we have noted, automatic transmissions use more than one planetary-gear set.

The major purpose of the hydraulic circuit is to control the shift from gear reduction to direct drive. The shift must take place at the right time, and this depends on car speed and throttle opening. These two factors produce two varying oil pressures that work against the two ends of the shift valve.

The shift valve is a spool valve inside a bore, or hole, in the valve body. Figure 48-24 shows what the spool valve looks like. Pressure at one end of the spool valve comes from the governor.

A governor is a device that controls, or governs, another device. In the hydraulic circuit, the governor controls pressure on one end of the shift valve.

Pressure at the other end of the spool valve changes as vacuum in the intake manifold changes. Let us explain. First, the governor pressure changes with car speed. The governor is driven by the output shaft from the transmission. As output-shaft speed and car speed go up, the governor pressure increases proportionally. This pressure works against the right end of the shift valve, as shown in Fig. 48-23.

The governor pressure is actually a modified line pressure. That is, a pump in the transmission produces the line pressure. This line pressure passes into the governor. The governor releases part of this pressure to the right end of the shift valve. The higher the car speed, the more pressure the governor releases. It is this modified pressure—the governor pressure—that works on one end of the shift valve.

Working on the left end of the shift valve is a pressure that changes as intake-manifold vacuum changes. Line

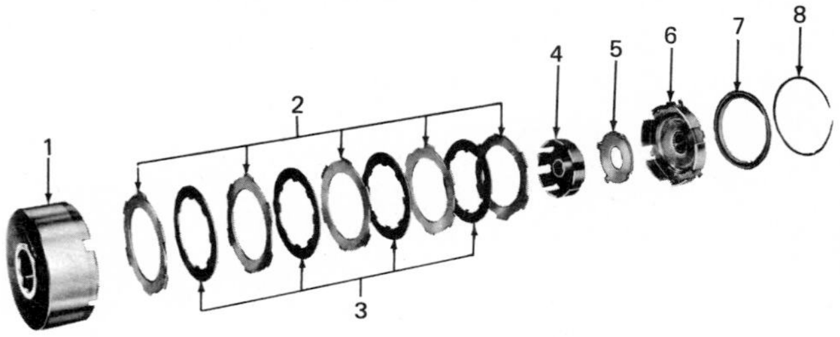

1 CLUTCH-DRUM ASSEMBLY
2 CLUTCH DRIVEN PLATE
3 CLUTCH DRIVE PLATE
4 CLUTCH HUB
5 CLUTCH-HUB THRUST WASHER
6 LOW-SUN-GEAR-AND-CLUTCH-FLANGE ASSEMBLY
7 CLUTCH-FLANGE RETAINER
8 RETAINER SNAP RING

Fig. 48-22 Parts of a clutch. (*Chevrolet Motor Division of General Motors Corporation*)

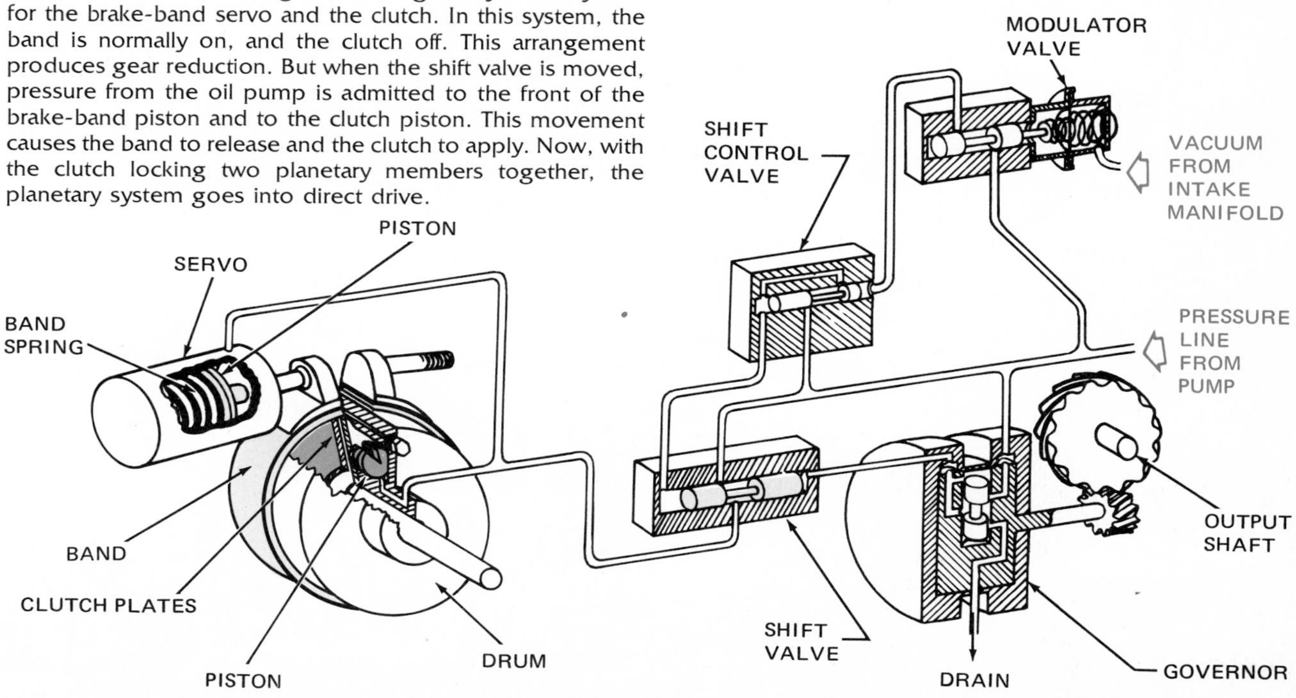

Fig. 48-23 Schematic diagram showing the hydraulic system for the brake-band servo and the clutch. In this system, the band is normally on, and the clutch off. This arrangement produces gear reduction. But when the shift valve is moved, pressure from the oil pump is admitted to the front of the brake-band piston and to the clutch piston. This movement causes the band to release and the clutch to apply. Now, with the clutch locking two planetary members together, the planetary system goes into direct drive.

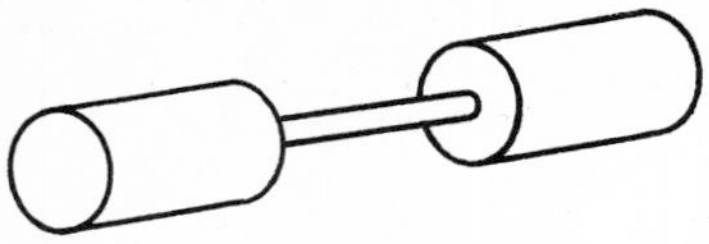

Fig. 48-24 Spool valve for a shift valve.

pressure enters the modulator valve at the upper right in Fig. 48-23. The modulator valve contains a spool valve attached to a spring-loaded diaphragm. Vacuum increases in the intake manifold when the throttle is partly closed. This vacuum pulls the diaphragm in and moves the modulator spool valve to the right. The motion cuts off the line pressure going to the shift control valve. When this happens, the shift control valve moves to the right, cutting off pressure from the left end of the shift valve. This means that the shift valve is pushed to the left by governor pressure. As a result, line pressure can pass through the shift valve. Therefore, line pressure is applied to the clutch and the servo at the planetary-gear set. With this condition, the band is released and the clutch is applied. This puts the planetary-gear set into direct drive.

Now let's put it all together and see how the hydraulic control circuit works. To start with, there is no pressure going to the planetary-gear controls. The clutch is released and the band is applied. The band is applied by the heavy spring in the servo.

Remember that Fig. 48-23 is a *simplified* version of the actual system found in modern automatic transmissions. In modern transmissions oil pressure is used to help the spring hold the band tight. The basic principles, however, are as shown in Fig. 48-23.

With the clutch released and the band applied, the planetary-gear set is in gear reduction, or low. As car speed increases, the governor releases more and more pressure. This pressure is applied to one end of the shift valve, as mentioned. The pressure on the other end of the shift valve depends on intake-manifold vacuum, that is, on engine speed plus throttle opening. As long as the throttle is held open, there is little manifold vacuum. The pressure on the left end of the shift valve is high. This holds the planetary-gear set in low for good acceleration. However, as car speed continues to increase, the governor pressure becomes great enough to push the shift valve to the left. This lets line pressure through to the planetary-gear set. Now the band is released and the clutch is applied. This shifts the planetary-gear set into direct drive.

The upshift will also take place if the throttle is partly closed after the car reaches intermediate speed. Closing the throttle increases the intake-manifold vacuum. The vacuum cuts off line pressure to the modulator valve. This cuts off line pressure to the shift control valve. When this happens, the shift control valve moves to the right, cutting off pressure to the left end of the shift valve. The shift valve then moves to the left, pushed by governor pressure. This applies line pressure to the planetary-gear set. The clutch applies and the band releases. Now the planetary-gear set goes into direct drive.

There is a reason for this roundabout way of getting pressure to the left end of the shift valve. It is to vary the point of upshift according to driving conditions. When the vehicle is accelerating, the driver wants high engine torque. That is, the gears must stay in low. Then, when the car reaches the desired cruising speed, less torque is needed, so the driver eases up on the throttle. This increases intake-manifold vacuum so that the upshift takes place. As you can see, the upshift can take place at any speed from medium to high. Also, for fast acceleration, the driver opens the throttle. This reduces intake-manifold vacuum. As a result, the planetary-gear set drops into gear reduction to increase torque.

● 48-12 Transmission Fluid

We have been saying "oil" and "fluid" in our discussions of how the hydraulic system of the transmission works. Transmission fluid could be considered a form of oil, but it is a very special sort of oil. It has several additives, including viscosity-index improvers, oxidation and corrosion inhibitors, extreme-pressure and antifoam agents, detergents, dispersants, pour-point depressants, and fluidity modifiers. A number of these additives are discussed in the chapter on engine lubricating systems (see ● 19-2).

Careful: It is extremely important to use the transmission fluid recommended by the automobile manuacturer. Use of a transmission fluid that is not on the recommended list can cause serious transmission trouble.

● 48-13 Transmission-Oil Coolers

Transmission oil may become very hot, especially under severe operating conditions. Thus, cars are equipped with transmission-oil-coolers, as described in ● 21-6 and illustrated in Figs. 21-15 and 21-16. The transmission oil circulates through a tube located in the engine cooling-system radiator. Heat from the transmission oil passes to the coolant circulating through the engine cooling system. This cools the transmission oil. In addition, a vehicle used for towing or trailering is frequently equipped with an auxiliary transmission-oil cooler.

● 48-14 Chrysler Torque-Flite Transmission

This transmission has two clutches and two bands to provide three forward speeds. A sectional view of the transmission is shown in Fig. 48-25, and the various gear positions are shown in Figs. 48-26 to 48-31. In the drive position, the transmission starts out in low and upshifts through second to high automatically, as conditions require. There is also a kick-down condition: Depressing the accelerator, as, for instance, when passing, will cause the transmission to drop back from high into second gear.

Note that this transmission has two semiseparate planetary-gear systems. There are two ring gears and two planet-pinion carriers, but only one sun gear. This is another compound planetary-gear system. If you study the various illustrations, you will see how operation of the clutches and bands produces the various gear positions. Figure 48-32, which shows the planetary-gear train in exploded view, may be of help.

● 48-15 Automatic Lock-Up Torque Converter

When cruising at road speed, the impeller is turning only a little faster than the turbine. It must turn faster, how-

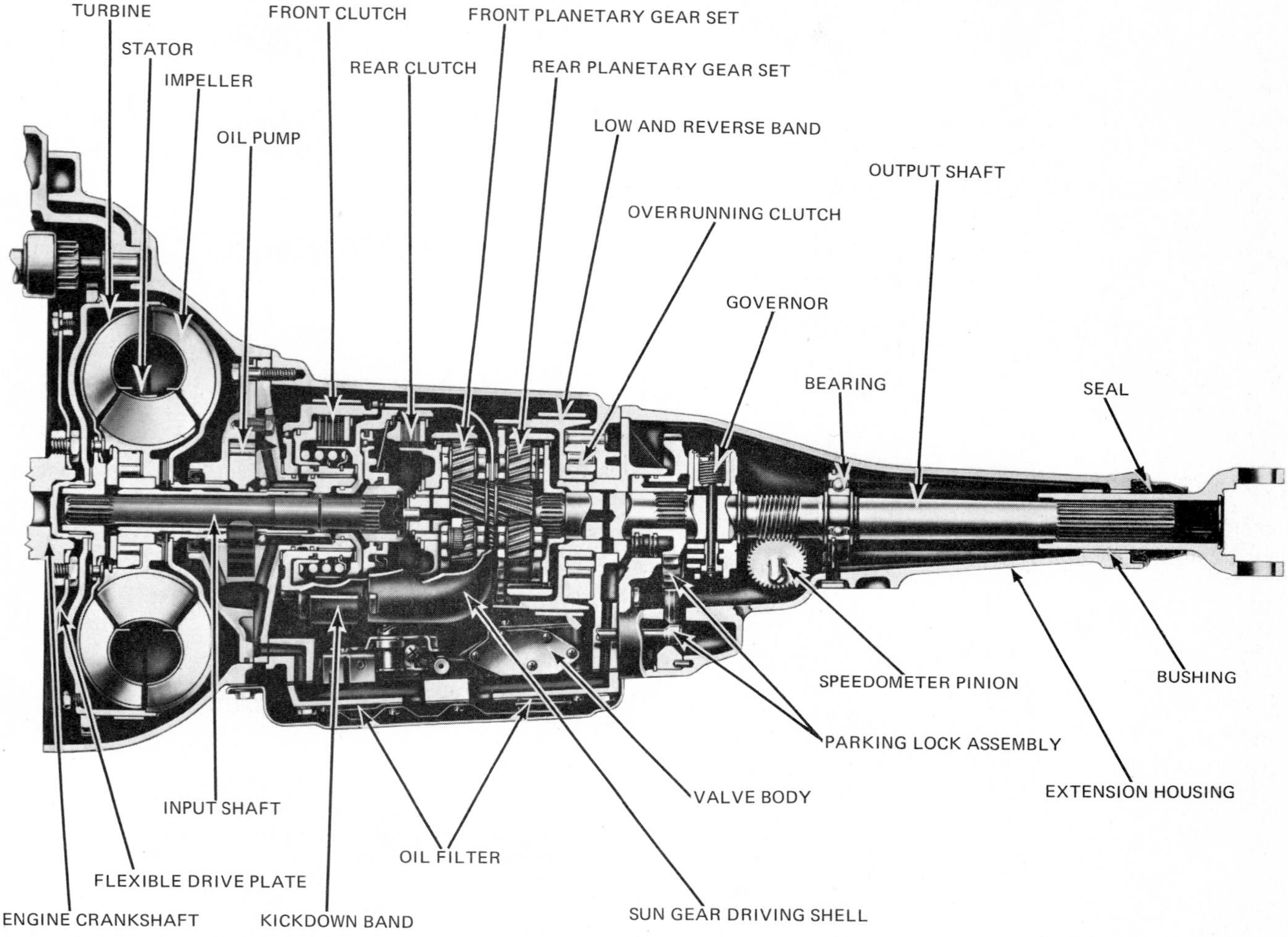

Fig. 48-25 Sectional view of Torque-Flite transmission. (*Chrysler Corporation*)

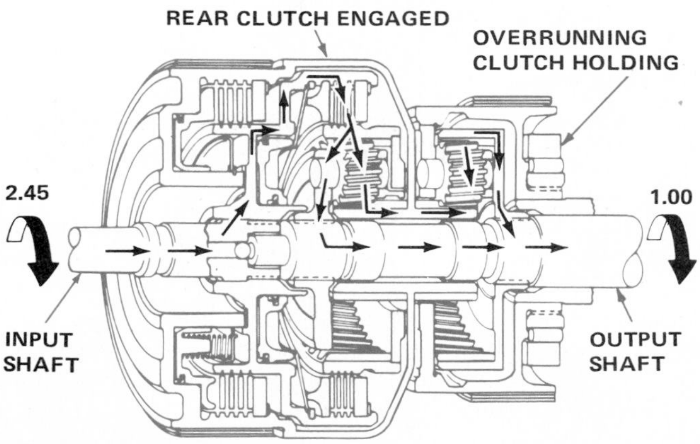

Fig. 48-26 Power flow in D (drive) position when first breaking away from a standing start. (*Chrysler Corporation*)

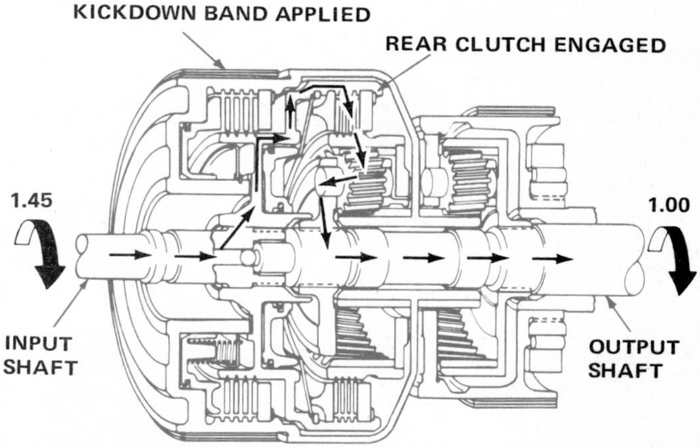

Fig. 48-27 Power flow in D after upshift from first to second, or after kickdown from D to second. (*Chrysler Corporation*)

ever, so that it can continue to discharge oil against and apply torque to the turbine blades. This difference in speed, or rpms, represents a power loss. For this reason, some transmissions are being introduced that have a lock-up torque converter. That is, when the car reaches cruising speed and is neither accelerating nor decelerating, the torque converter locks up. This results in better fuel economy. Also, the transmission oil does not get so hot because it is not working in the lock-up mode.

Figure 48-33 is a cutaway view of a lock-up torque

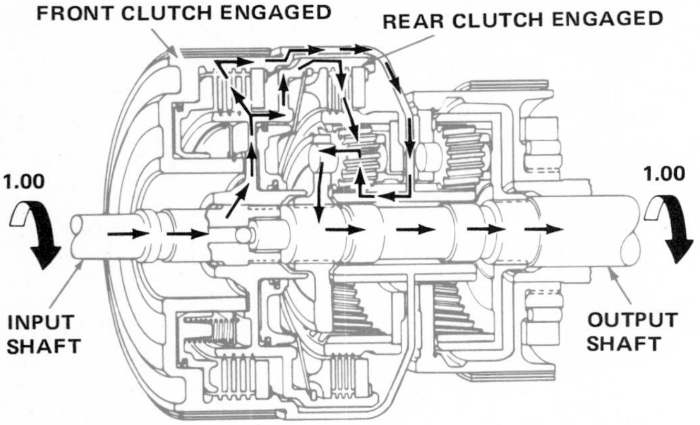

Fig. 48-28 Power flow in D after upshift to direct drive. (*Chrysler Corporation*)

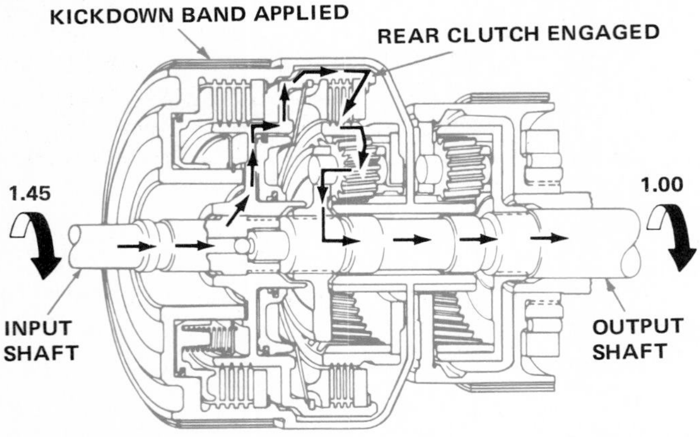

Fig. 48-29 Power flow with shift lever in 2, or second. (*Chrysler Corporation*)

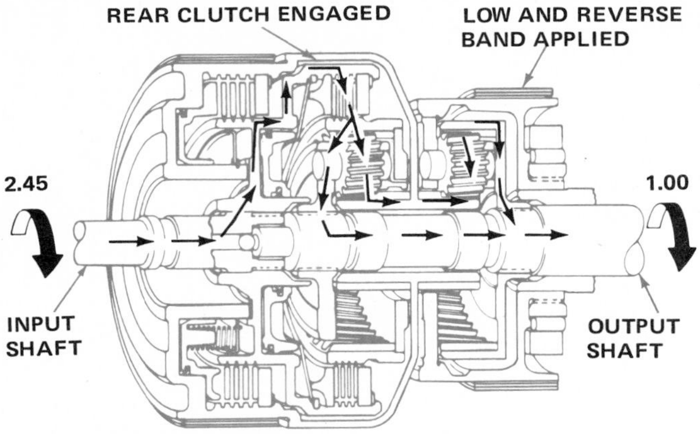

Fig. 48-30 Power flow with shift lever in 1, or low. (*Chrysler Corporation*)

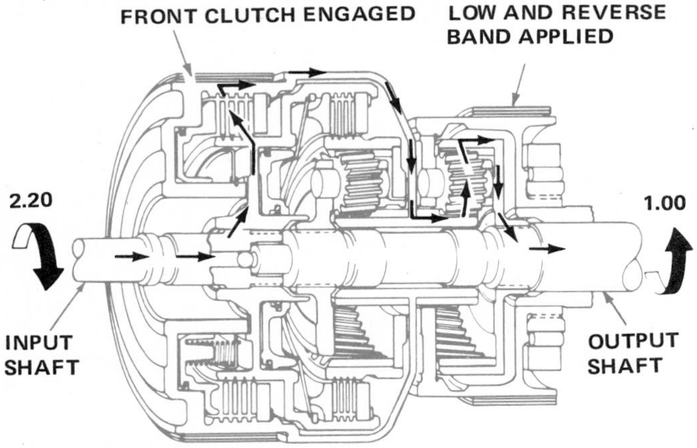

Fig. 48-31 Power flow in R (reverse). (*Chrysler Corporation*)

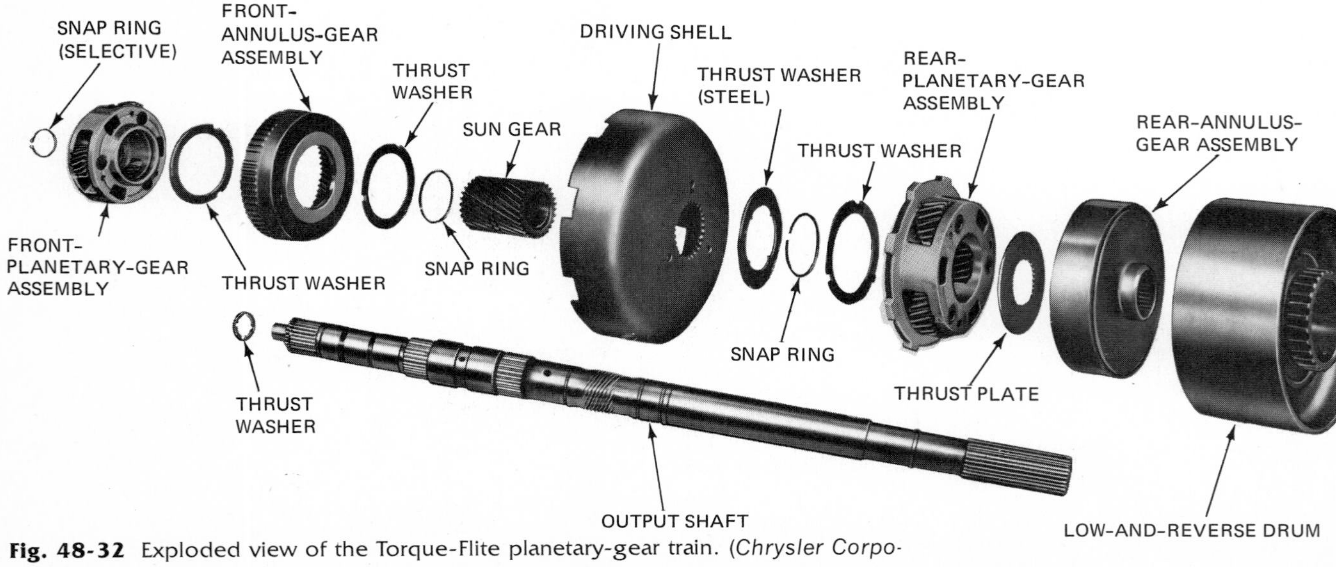

Fig. 48-32 Exploded view of the Torque-Flite planetary-gear train. (*Chrysler Corporation*)

converter. It has a clutch and a clutch apply piston. The isolator springs on the clutch help dampen the clutching action as the torque converter goes into the lock-up mode. These isolator springs also dampen out the power pulses from the engine when the transmission is in direct drive and the converter is locked up. They do the same job as the torsional springs in the standard clutch disk.

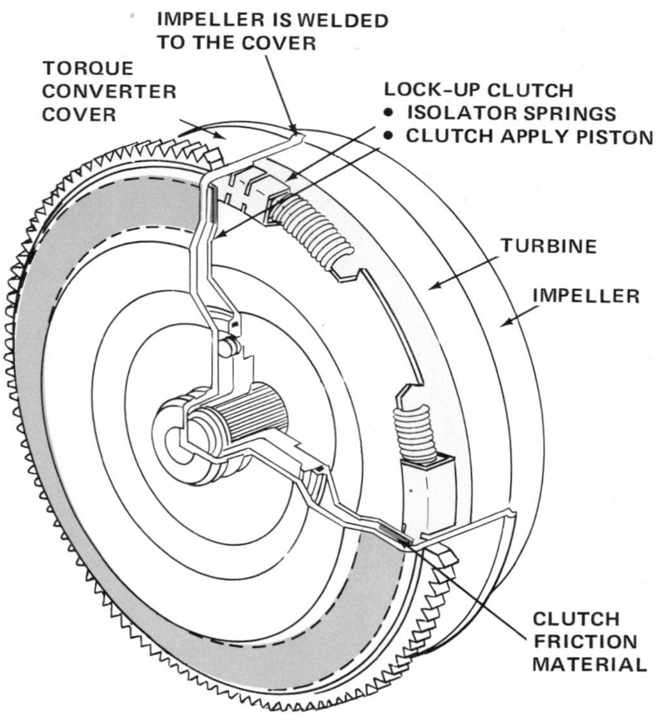

Fig. 48-33 Partial cutaway view of lock-up torque converter. *(Chrysler Corporation)*

Figure 48-34 shows how the lock-up works. In the center picture, the torque converter is not locked up. The piston is released. The picture to the right shows the lock-up position. The torque-converter cover has a ring of friction material bonded to it. When the converter goes into lock-up, oil pressure back of the piston forces it to the left (in Fig. 48-34). Now, the piston, and output shaft, must turn together. Since the input is to the converter cover, the torque converter turns as a unit. The arrows show the power flow through the converter in the two modes, not locked and locked up.

● 48-16 Automatic Transmission for Front Drive

The transmission in Fig. 48-35 is for the front-drive Cadillac automobile. It is essentially a Turbo Hydra-Matic transmission which has been cut in two just back of the torque converter. The planetary-gear system and controls can then be placed alongside the torque converter.

Automatic transmissions for other front-drive cars are of the transaxle type (● 46-15). That is, they are used with a cross-mounted engine and combine the transmission and the axles. They look much like Figs. 46-35 and 46-36 except that the manual clutch and internal gears have been replaced with a torque converter and hydraulically operated planetary gearing. These assemblies are called automatic transaxles.

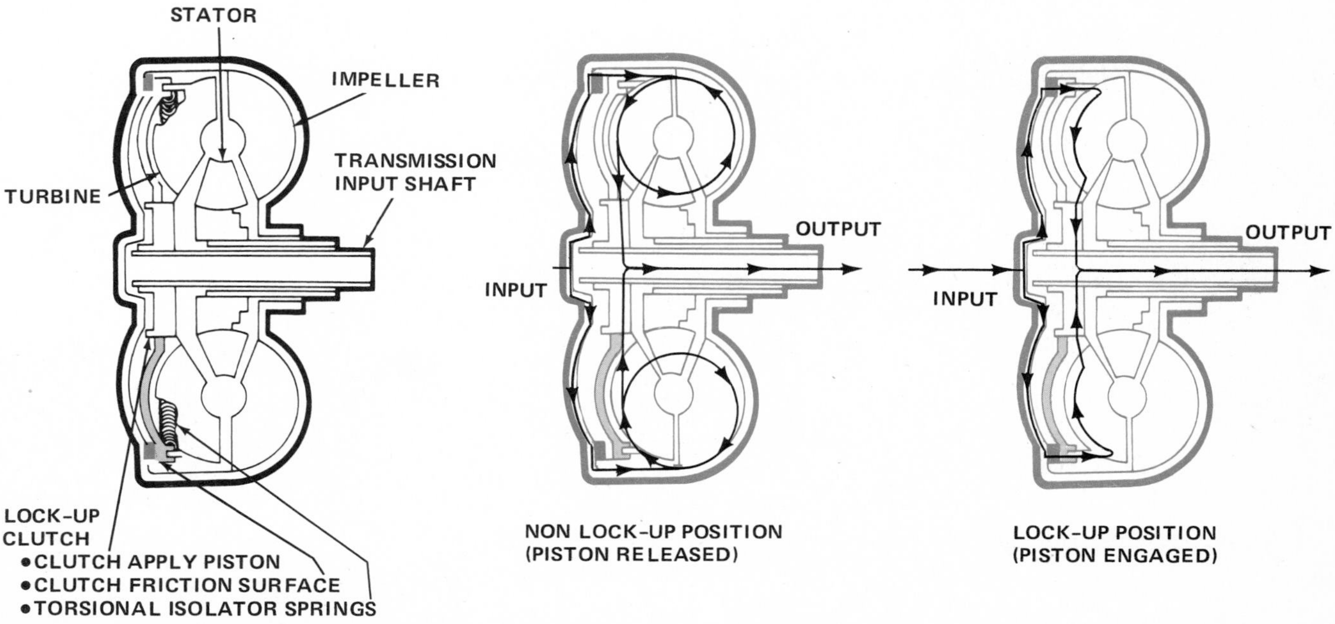

Fig. 48-34 Operation of the lock-up torque converter. To left, parts named. Center, torque converter in the released position. Right, piston locking up the torque converter. *(Chrysler Corporation)*

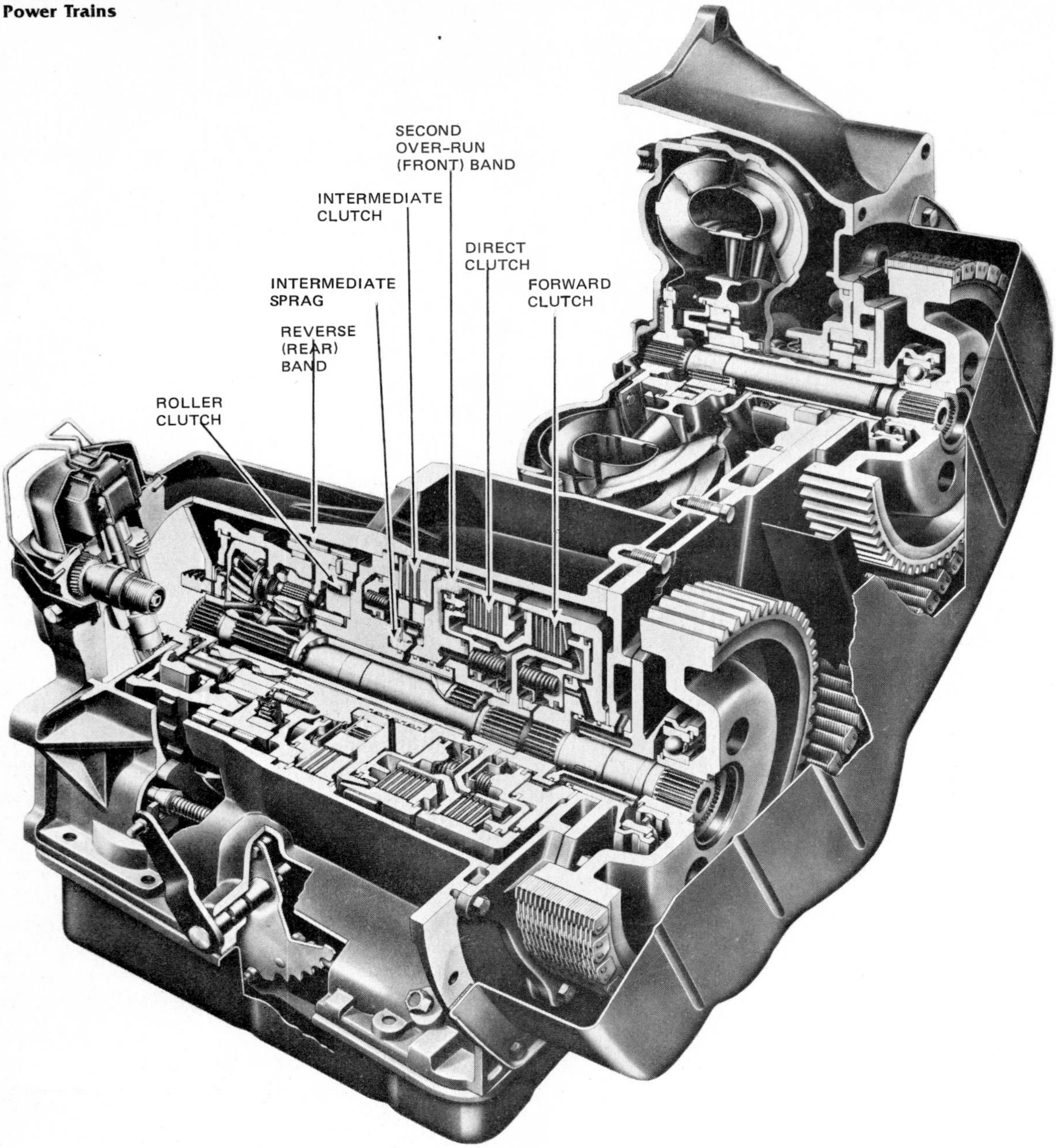

SELECTOR POSITION	PUMP PRESSURE	FORWARD CLUTCH	DIRECT CLUTCH	2ND OVERRUN BAND	INT. CLUTCH	INT. SPRAG	ROLLER CLUTCH	REV. BAND
PARK—NEUT.	60-150	OFF	OFF	OFF	OFF	OFF	OFF	OFF
DRIVE LEFT 1	60-150	ON	OFF	OFF	OFF	OFF	ON	OFF
2	60-150	ON	OFF	OFF	ON	ON	OFF	OFF
3	60-150	ON	ON	OFF	ON	OFF	OFF	OFF
DRIVE RIGHT 1	150	ON	OFF	OFF	OFF	OFF	ON	OFF
2	150	ON	OFF	ON	ON	ON	OFF	OFF
LO 1	150	ON	OFF	OFF	OFF	OFF	ON	ON
2	150	ON	OFF	ON	ON	ON	OFF	OFF
REV.	95 - 230	OFF	ON	OFF	OFF	OFF	OFF	ON

Fig. 48-35 Turbo Hydra-Matic automatic transmission for a front-drive car, partly cut away so that the internal construction can be seen. The table shows the internal conditions for different selector positions. (*Cadillac Motor Car Division of General Motors Corporation*)

REVIEW QUESTIONS

1. Explain the operation of a torque converter. What limits the "bounce-back" effect?
2. Explain how the overrunning clutch works.
3. Explain the purpose of the stator.
4. What are the two devices used in automatic transmissions to control the planetary-gear sets?
5. In the simplified version of the planetary-gear set described in the chapter, which control device puts the gear set into reduction?
6. Which control device puts the planetary-gear set into direct drive?
7. Describe two ways to get gear reduction through a planetary-gear set.
8. Describe two ways to get reverse through a planetary-gear set.
9. What is the purpose of the governor in the automatic transmission?
10. What are the two pressures that work on the two ends of the shift valve?
11. Explain how the modulator valve changes the shift point as throttle opening changes.
12. Will upshifts take place at lower speed or higher speed with the throttle wide open? Why?
13. What is the purpose of the transmission-oil cooler? Where is it located?
14. How many forward speeds does the Chrysler Torque-Flite automatic transmission have?
15. How does the lock-up torque converter work?

SELF PROJECTS

Automatic transmissions are a special study in themselves. You can find out much more about them, however, as follows.

1. Look in manufacturers' service manuals for the sections on automatic transmissions. Turn to the pages that show the hydraulic circuits. These are often printed in color so you can follow the hydraulic pressures more easily. Concentrate on one of these pages —the page showing drive range in low, for example. Find the manual valve. This is the valve that the driver moves when moving the selector lever on the steering column. Now trace the oil pressure from the pump to the manual valve and then from this valve to the clutch that is applied.
2. Now find, in the service manual, the picture showing the power flow through the transmission in low. Make a sketch of this power flow, naming the parts that are involved in carrying the power through the transmission.
3. You are now ready to look at actual transmission parts. If you are able to study the parts as an expert transmission technician takes them off the transmission, you will understand just what each part looks like and what it does. Automatic transmissions look complicated, but they really are very simple in operation. Make notes and sketches, as time allows, so you have a record of your studies of the transmission.

CHAPTER 49
Automatic-Transmission Service

After studying this chapter, you should be able to:

- **List diagnosis and servicing procedures for automatic transmissions.**

This chapter discusses trouble-diagnosis and servicing procedures for automatic transmissions. The instructions apply, in general, to all automatic transmissions. However, you should always refer to the manufacturer's shop manual for the model of transmission you are working on.

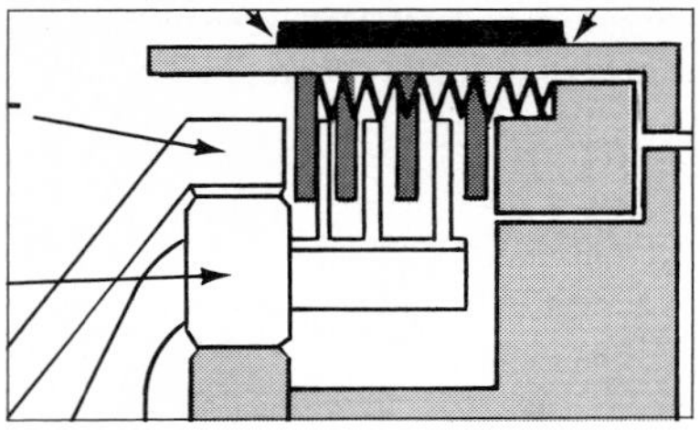

● 49-1 Servicing Automatic Transmissions

Automatic-transmission service can be divided into four parts:

1. Normal maintenance
2. Trouble diagnosis
3. On-the-car repairs
4. Transmission overhaul

● 49-2 Normal Maintenance

Normal maintenance includes:

1. Changing fluid and filter
2. Checking fluid level
3. Adding fluid if necessary
4. Checking throttle and shift linkages
5. Adjusting the neutral starting switch
6. Possibly band adjustment
7. On some transmissions, changing the fluid

Before you undertake any of these procedures, refer to the manufacturer's shop manual for the model of automatic transmission you are about to work on. Then follow the procedures outlined. Because there are many models and designs of automatic transmissions, there are variations in the service procedures. Always make sure you follow the procedure recommended for the model transmission you are working on.

● 49-3 Automatic-Transmission Trouble Diagnosis

Each make and model of automatic transmission has its own specific trouble-diagnosis guides. These are issued by the manufacturers, and they list, in step-by-step fashion, the diagnostic procedure. Locate the trouble-diagnosis section in the service manual covering the transmission you are about to service.

Before making operating tests on an automatic transmission, be sure the engine is in good condition and operating normally. A sluggish or missing engine will not give the transmission a chance to perform normally.

Identify the make and model of transmission in the vehicle you are about to check. Then study the specific reference that applies to that model of transmission. Follow the procedure carefully. Any deviation from the procedure could damage the transmission.

You may be required by the service manual to make stall checks or operating checks. Observe the special cautions outlined in the references. Note that some manufacturers do not recommend stall tests, and in fact forbid them. Other manufacturers caution you to be very quick if you do make a stall test.

The stall test checks the torque converter and the holding ability of the transmission clutches. It is performed by applying the car brakes, blocking the wheels, and measuring engine speed with the transmission in drive and the throttle wide open. The check must be made within five seconds. To take longer risks damage to the transmission.

● 49-4 On-the-Car Repairs of Automatic Transmissions

Each make and model of automatic transmission has certain special on-the-car repairs that can be made. The manufacturer's service manual for the automatic transmission you are about to service will tell you what can be done on the car. If the repair is not one that can be done on the car, the transmission must be removed.

● 49-5 Automatic-Transmission Overhaul

Automatic-transmission overhaul is a specialty in itself. It takes time and study to become an expert automatic-transmission technician. You may not wish to become an automatic-transmission expert, but you should have a good idea of how automatic transmissions are overhauled. If you are given an automatic transmission to overhaul in the shop, first study the service manual that covers that make and model of transmission. Then follow the procedure that is outlined, step by step. If you carefully follow the procedure, and use the special tools required, you will have no trouble doing the job.

Some repair operations require removal of the transmission from the car. But this does not always mean a complete overhaul.

REVIEW QUESTIONS

1. Explain how to change a transmission oil filter. Refer to a shop manual to get the information you need.
2. Explain how to refill an automatic transmission with transmission fluid.
3. Refer to a shop manual, and make a list of the steps required to remove and replace an automatic transmission.
4. Refer to a shop manual, and make a list of the jobs that can be done on the transmission on the car.
5. Refer to a shop manual, and study the detailed overhaul instructions. Make a list of the major jobs that can be performed only with the transmission on the bench.

SELF PROJECTS

1. Refer to a manufacturer's shop manual, and make a list of the normal maintenance operations.
2. Turn to the trouble-diagnosis section of the service manual. Make a list of the troubles in the list, such as "Harsh engagement in D, 1, 2, and R." Note that these lists are extensive. They must be complete, of course, so they include all possible troubles, even those that almost never occur. Expert transmission technicians probably know these troubles and their causes, but even they refer to detailed printed instructions when they work on transmissions they are not especially familiar with. Making a list of troubles will familiarize you with the types of trouble an automatic transmission can have.
3. Make a list of the repairs that can be done on the car—that is, repairs that do not require removing the transmission from the car.

CHAPTER 50
Drive Lines and Universal Joints

After studying this chapter, you should be able to:

- **Explain the purpose of the drive line.**
- **Describe the two types of joint used in drive lines, and explain their purpose.**
- **Describe the two most common drive arrangements (front-engine, rear-wheel drive; and front-engine, front-wheel drive).**

Drive lines, also called drive shafts and propeller shafts, carry the power from the transmission to the car wheels. In most automobiles, the engine is at the front and the rear wheels are driven (Fig. 50-1). A long drive shaft is required. However, when the engine is at the front and the front wheels are driven, two short drive shafts are used. This is the arrangement of the future, according to automotive manufacturers. General Motors, for example, will convert almost all their cars to front drive in the next few years. Cars with engines in the rear driving the rear wheels also use two short drive shafts.

In this chapter, we look at the two most common arrangements. These are front engine with rear wheels driven, and front engine with front wheels driven.

Front Engine with Rear Wheels Driven

● 50-1 Function of Drive Line

The drive line (propeller shaft) is a shaft that connects the transmission main, or output, shaft to the differential at the rear wheels (on rear-drive cars). Thus, the rotary motion of the transmission output shaft is carried to the differential and, from there, to the wheels. The drive-line design must take two facts into account. First, the engine and transmission are more or less rigidly connected to the car frame, and second, the rear-axle housing (with wheels and differential) is attached to the car frame through springs. This means two things:

1. The drive line must change in length as the wheels move up and down.
2. The angle of drive must change as the wheels move up and down.

Figure 50-2 shows how the length of the drive line and the angle of drive change as the wheels move up and down. At the top, the wheels and differential are shown at their up position. The angle of drive is small, and the drive line is at its maximum length. In the lower part of Fig. 50-2, the differential and wheels have moved to their lowest position. This is the position they take when the wheels drop into a depression in the road. In this position, the drive angle is increased. Also, the drive-line length is reduced. The reason for this is that, as the rear wheels and differential swing down, they also move forward. The axle housing is attached to the car frame by springs or control arms, and this makes them move in a shorter arc than the drive line.

To allow for these two variations, two different kinds of joints are necessary: two or more *universal* joints and a

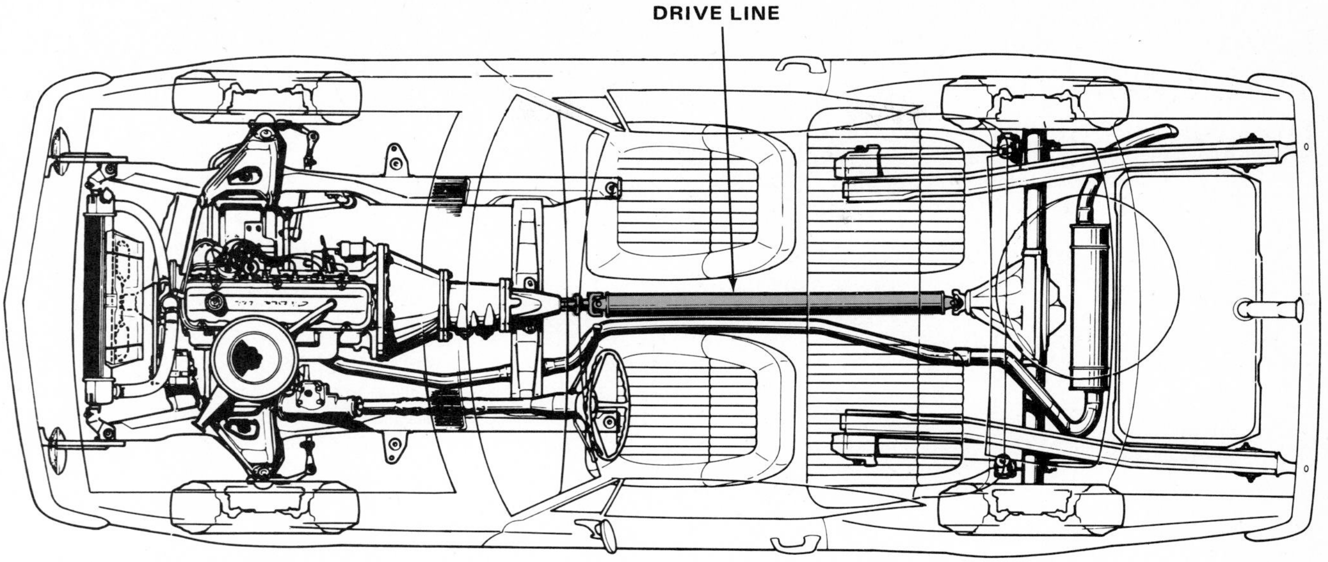

Fig. 50-1 Location of the drive line in the power train.

slip joint (Fig. 50-3). The universal joints take care of the change in drive angle. The slip joint takes care of the change in length. We describe these joints in following sections.

The propeller shaft is a hollow shaft, or tube, on most cars with a front-mounted engine and rear drive. Some drive lines have two propeller shafts, with a third universal joint between them (Fig. 50-4). Note that the center universal joint in Fig. 50-4 is a constant-velocity universal joint. It includes a center bearing and support assembly which supports the center of the drive line.

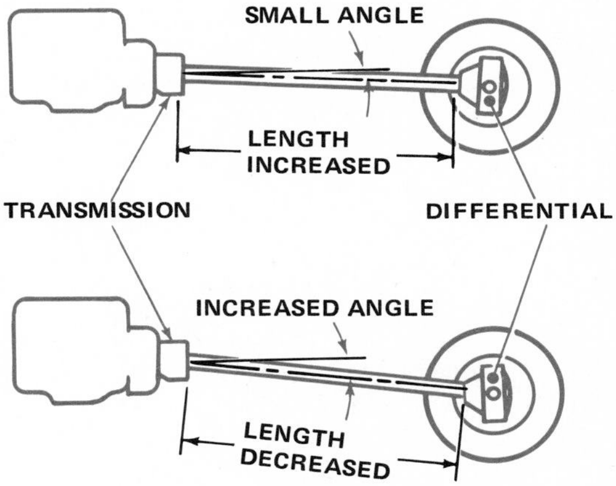

Fig. 50-2 The rear axle housing, with differential and wheels, moves up and down. As it does, the angle between the transmission output shaft and the drive line changes. The length of the propeller shaft also changes. The reason the propeller shaft shortens as the angle increases is that the rear axle and differential move in a shorter arc than the propeller shaft. The center point of the axle-housing arc is the rear spring or control-arm attachments to the frame.

50-2 Universal Joints

The universal joint allows driving power to be carried through two shafts that are at an angle to each other. A simple universal joint is illustrated in Fig. 50-5. It is a double-hinged joint consisting of two Y-shaped yokes and a cross-shaped member called the *spider.* One of the yokes is on the driving shaft, and the other is on the driven shaft. The four arms of the spider, called *trunnions,* are assembled into bearings in the ends of the two shaft yokes. The driving shaft and yoke cause the spider to rotate, and the other two trunnions of the spider cause the driven shaft to rotate. When the two shafts are at an angle to each other, the bearings in the yokes permit the yokes to swing around on the trunnions with each revolution. A variety of universal joints have been used on automobiles, but the types now in most common use are the spider-and-two-yoke, the constant-velocity, and the ball-and-trunnion joints.

The spider-and-two-yoke design is essentially the same as the simple universal joint discussed above, except that the bearings are of the needle type (Fig. 50-6). As will be noted, there are four needle bearings, one for each trunnion of the spider. The bearings are held in place by snap rings that drop into undercuts in the yoke-bearing holes.

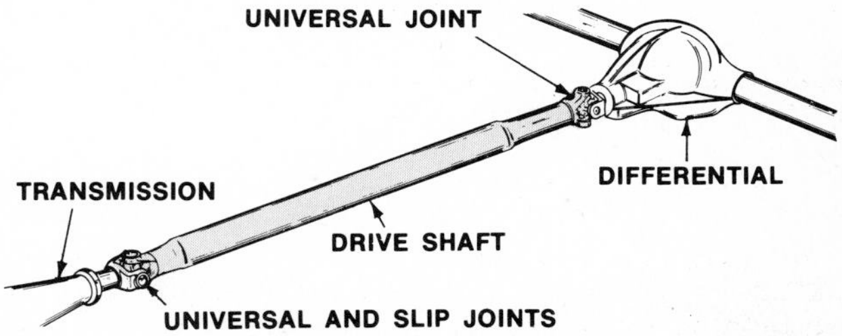

Fig. 50-3 The drive line connects the transmission to the differential. The drive line shown here consists of the drive shaft (propeller shaft) with two universal joints and one slip joint.

The ball-and-trunnion type of universal joint combines both the universal and the slip joint in one assembly.

The universal joint shown in Fig. 50-6 is not a constant-velocity joint. That is, if the two shafts are at an

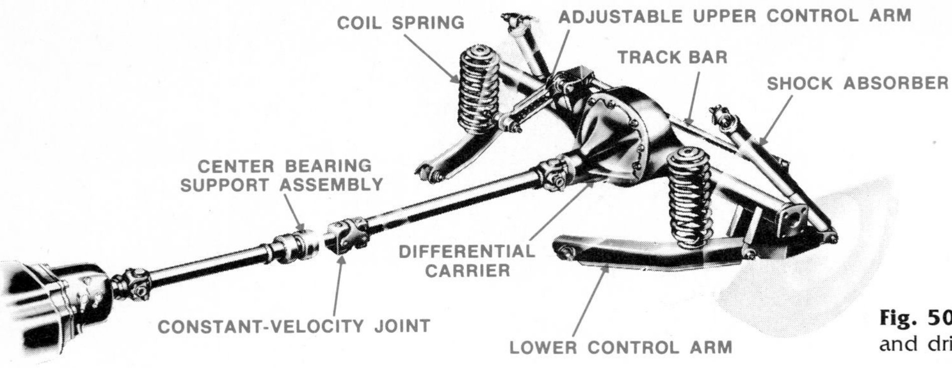

Fig. 50-4 One design of rear suspension and drive line.

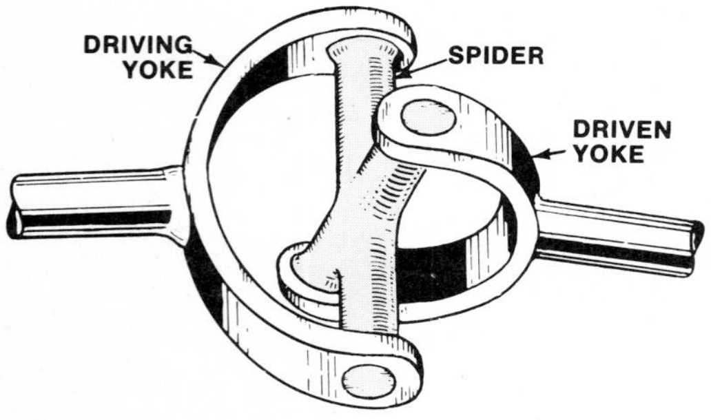

Fig. 50-5 Simple universal joint.

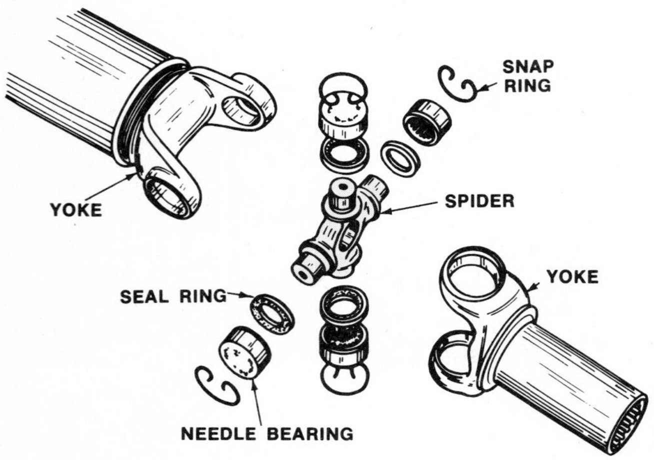

Fig. 50-6 Two-yoke-and-spider universal joint disassembled.

angle, the driven shaft will be given a variable speed, or velocity. With each revolution, the shaft will speed up and slow down slightly two times. The greater the angle, the greater the velocity variation. This can cause wear on the bearings and gears in the differential because the load on them varies, or pulsates. To eliminate this, constant-velocity joints are used. In these, the driven shaft turns at the same speed as the drive shaft, with no variation in the velocity, or speed.

Constant-velocity universal joints are shown on drive lines in Figs. 50-4 and 50-7. The drive line in Fig. 50-7 has three, one at each end and one in the center connecting the two drive shafts. Figure 50-8 is a disassembled view of a constant-velocity joint.

As you will note, the constant-velocity universal joint (also called a *double-Carden* joint) includes two separate universal joints. They are linked by a ball and socket. The ball and socket split the angle of the two drive shafts between the two universal joints. Because the two joints operate at the same angle (half the total), the variations that could result from a single joint are canceled out. That is, the acceleration resulting at any instant from the action of one joint is nullified by the deceleration of the other, and vice versa.

50-3 Slip Joint

A slip joint is illustrated in Fig. 50-9. The slip joint has outside splines on one shaft and matching internal splines on a mating hollow shaft. The splines cause the two shafts to rotate together but permit the two to move

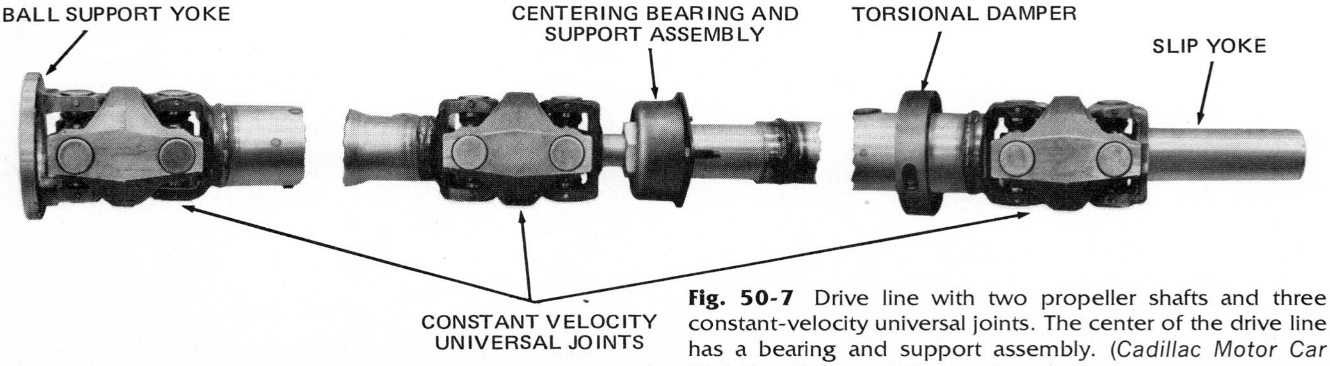

Fig. 50-7 Drive line with two propeller shafts and three constant-velocity universal joints. The center of the drive line has a bearing and support assembly. (*Cadillac Motor Car Division of General Motors Corporation*)

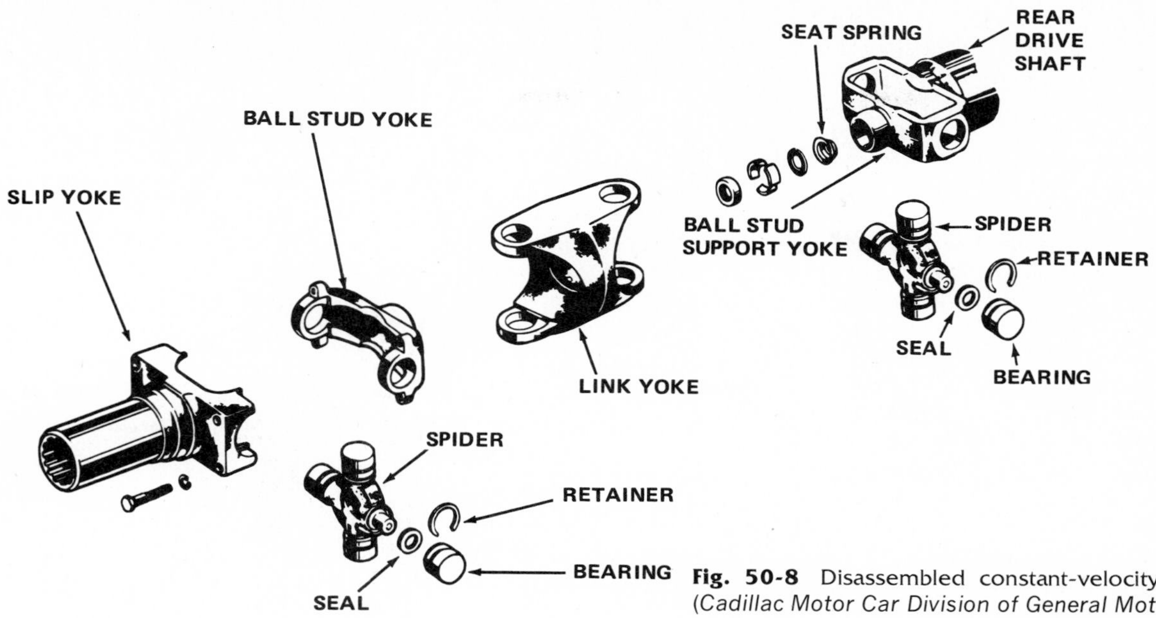

Fig. 50-8 Disassembled constant-velocity universal joint. (*Cadillac Motor Car Division of General Motors Corporation*)

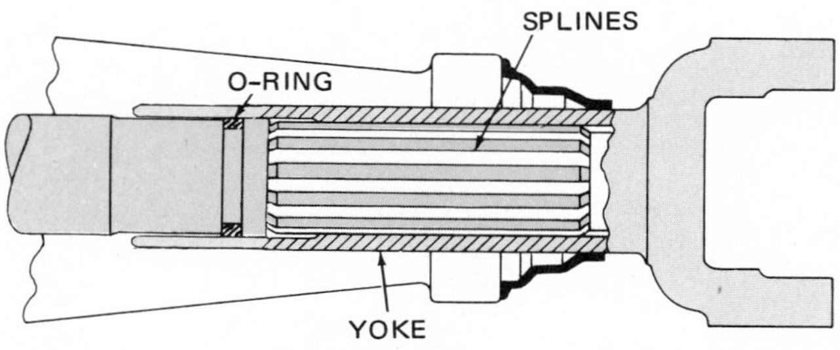

Fig. 50-9 Cutaway view of a slip joint. (*Ford Motor Company*)

endwise in relation to each other. This allows changes in the length of the propeller shaft as the rear axles move toward or away from the car frame.

50-4 Center Support

Many vehicles have drive lines which are supported at the center, as shown in Fig. 50-7. The center support prevents angular movement, or "whipping" of the drive line. The center support has a bearing in which the drive line can rotate. Note that this arrangement requires a universal joint back of the center support.

Front Engine with Front Wheels Driven

50-5 Front Drive

Front-wheel drive, or front drive, is becoming more popular, especially for smaller cars. With front-wheel drive, the

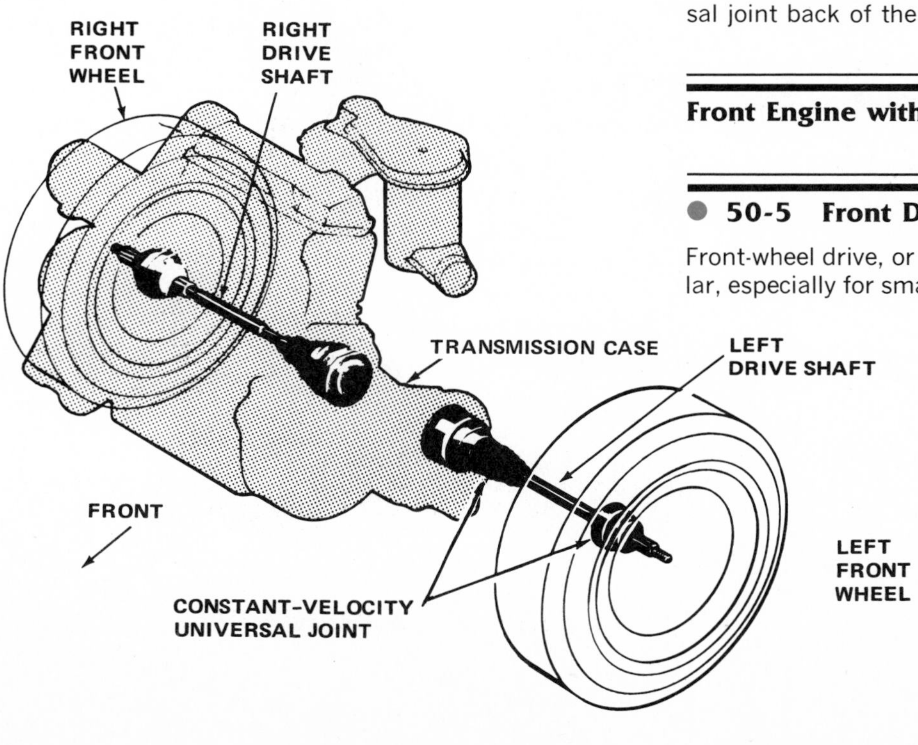

Fig. 50-10 Engine and power-train layout for a front-wheel-drive small car. Two short drive shafts are used, with two constant-velocity universal joints in each shaft. (*Chrysler Corporation*)

long drive shaft to the rear axle is eliminated. This also does away with the long tunnel in the floor pan of the car required for the drive shaft. Figure 50-10 shows the engine and power-train layout used in many small cars today. Figure 50-11 shows the drive train in greater detail. Note, in Fig. 50-11, that the right-hand drive shaft has three universal joints, two of which are the constant-velocity (CV) type.

Driving the front wheels makes the front suspension somewhat more complicated. The front wheels must swing from side to side so that the car can be steered. Also, they must be supported in such a way that they can be driven. This requires two constant-velocity universal joints in each front-wheel drive shaft (Fig. 50-11). Note that these joints are protected by rubber boots. The constant-velocity universal joints can drive the wheels even though the wheels are turned many degrees from straight ahead.

The inner ends of the two drive shafts are connected to a transaxle, which is a combined transmission and differential. Differentials are covered in Chap. 51.

For another view of a front-drive car, see Fig. 50-12, which shows a front-drive car raised on a hoist so that the layout and construction can be seen. See also Fig. 50-13, which shows a flat-four front-mounted engine with the drive train for a front-drive small car.

The constant-velocity universal joints in Fig. 50-11 are the ball-and-race type. A universal joint of this type is shown disassembled in Fig. 50-14. The balls roll in grooves in the inner race and inner housing to take care of changing angles of drive. The joint also acts as a slip joint. Small balls roll in straight grooves in the outside of the inner housing and the outer housing to take care of changing drive-shaft length.

● 50-6 Rear Drive with Rear-Mounted Engine

Some cars have the engine mounted at the rear. They use short stub shafts and universal joints to carry the engine power to the two rear wheels. The rear wheels are independently suspended. Some Volkswagen models use this type of rear drive.

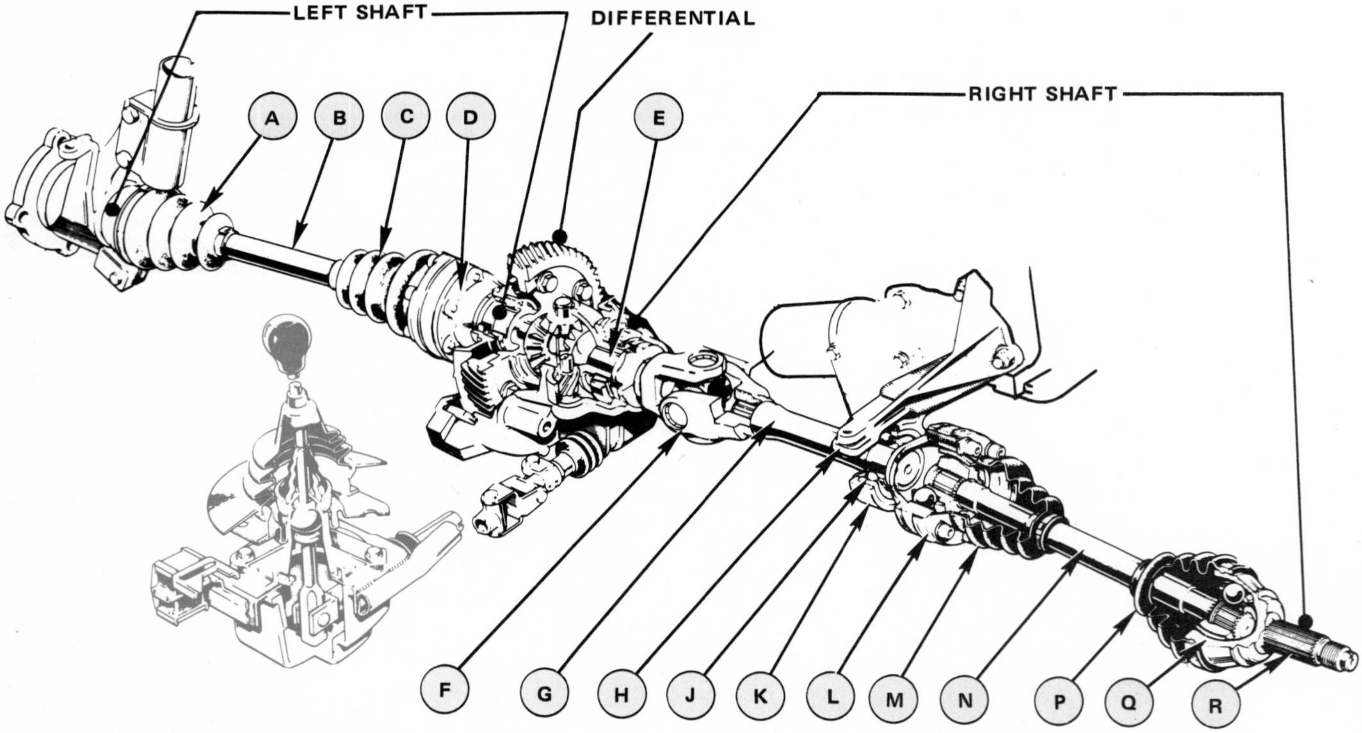

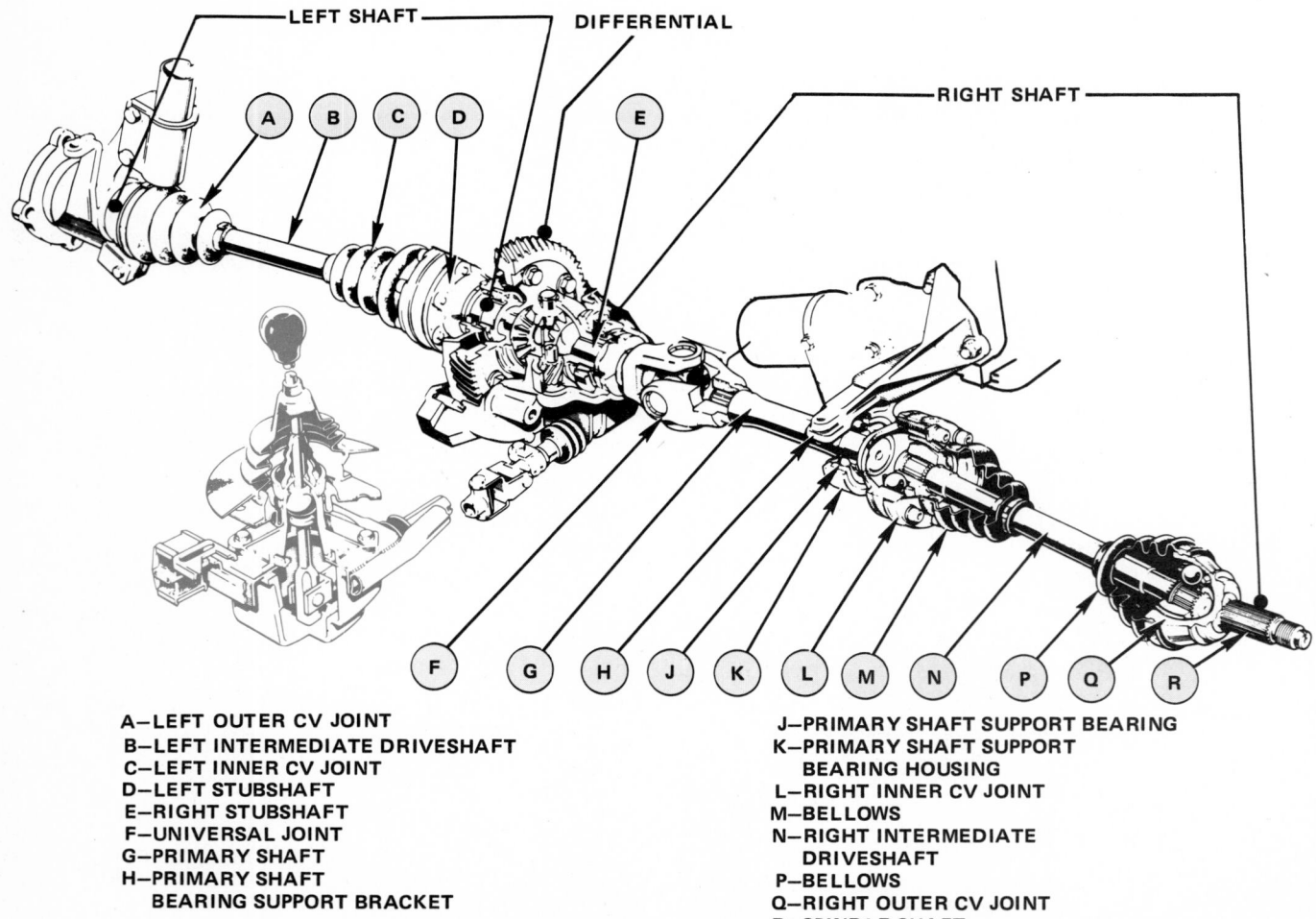

A—LEFT OUTER CV JOINT
B—LEFT INTERMEDIATE DRIVESHAFT
C—LEFT INNER CV JOINT
D—LEFT STUBSHAFT
E—RIGHT STUBSHAFT
F—UNIVERSAL JOINT
G—PRIMARY SHAFT
H—PRIMARY SHAFT BEARING SUPPORT BRACKET
J—PRIMARY SHAFT SUPPORT BEARING
K—PRIMARY SHAFT SUPPORT BEARING HOUSING
L—RIGHT INNER CV JOINT
M—BELLOWS
N—RIGHT INTERMEDIATE DRIVESHAFT
P—BELLOWS
Q—RIGHT OUTER CV JOINT
R—SPINDLE SHAFT

Fig. 50-11 Drive train for front-drive small car. The CV means constant velocity. The drive train includes the transmission and differential. (*Ford Motor Company*)

Fig. 50-12 Front-drive car raised on a hoist so that the layout of the parts can be seen. (*Chrysler Corporation*)

● 50-7 Four-Wheel Drive

Some vehicles, especially those that are used off the road, can drive all four wheels (Fig. 50-15). Each drive shaft to the front and rear axles has universal joints and slip joints.

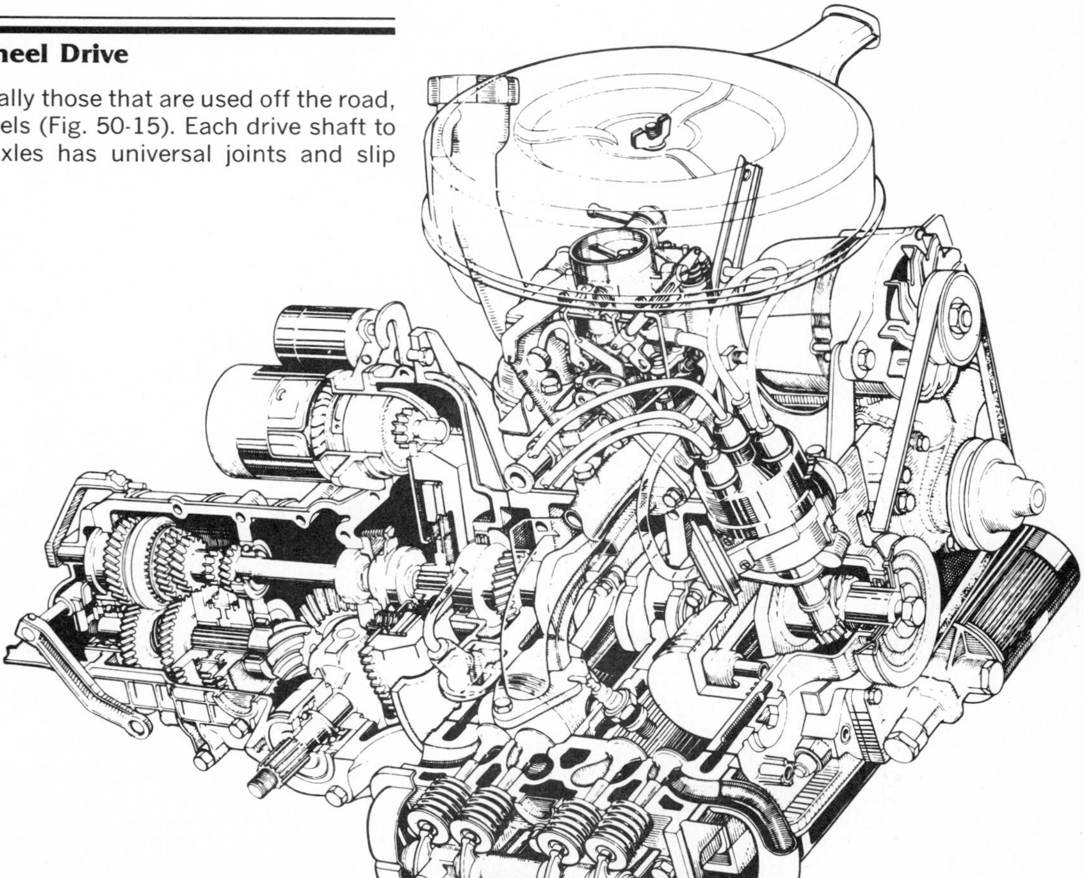

Fig. 50-13 Cutaway view of a flat-four engine with assembled transmission and differential for a front-drive car. (*Subaru*)

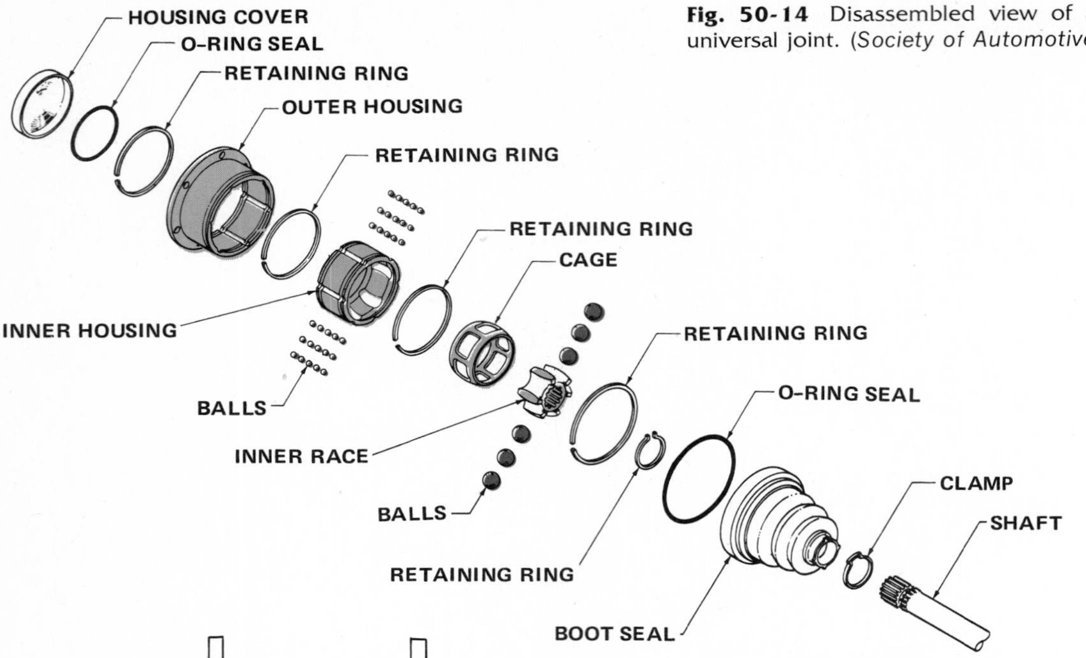

Fig. 50-14 Disassembled view of a ball-spline (Rzeppa) universal joint. (*Society of Automotive Engineers*)

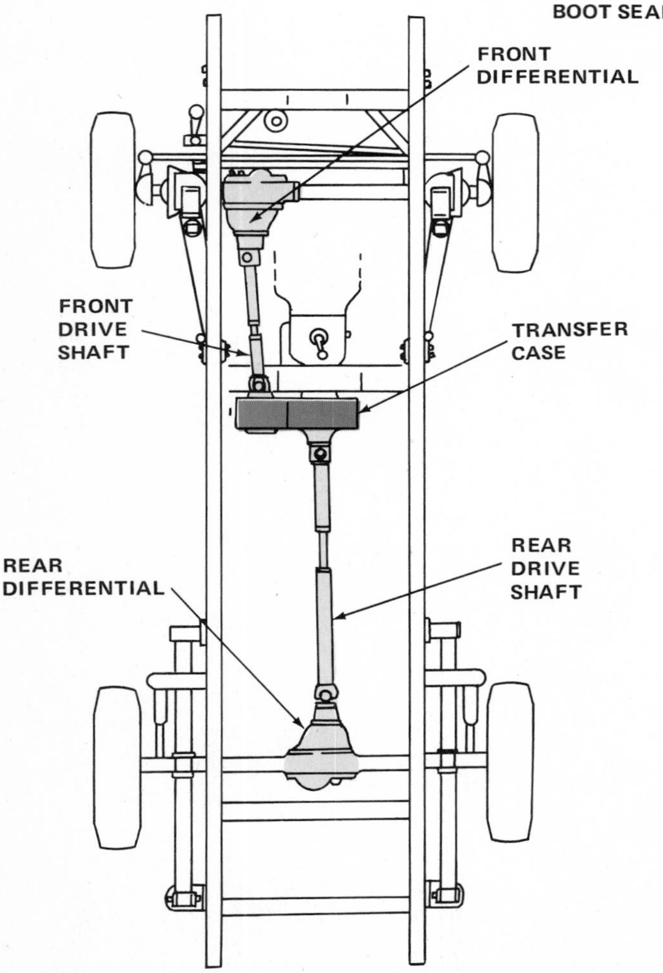

Fig. 50-15 The suspension and drive train for a Ford Bronco. This is a four-wheel drive vehicle. The transfer case allows the driver to use rear-wheel or front-wheel drive whichever is preferred.

● 50-8 Servicing the Universal Joint and Drive Shaft

Universal joints and drive shafts usually require no service. Most universal joints are prelubricated during original assembly and do not need additional lubrication. However, in case of wear or damage, the universal joints should be replaced.

A drive shaft that is out of balance can often be balanced by installation of two worm-type hose clamps.

REVIEW QUESTIONS

1. What is the function of the propeller shaft?
2. What two actions take place in the propeller shaft as the rear springs compress and expand? What two devices are used to allow for these actions?
3. Describe the action of a spider-and-two-yoke universal joint.
4. Describe the action of a constant-velocity universal joint.
5. What is the purpose of the universal joint?
6. What is the purpose of a slip joint?

SELF PROJECTS

1. Refer to several manufacturers' shop manuals, and list the various kinds of drive lines they use. How many and what kinds of universal joints do they have?
2. Examine universal joints, noting how they transmit power even though the input and output shafts are at an angle. Put the two at a considerable angle—say 30 degrees—and turn one shaft at an even speed. Notice how the other shaft speeds up and then slows down. Constant-velocity joints prevent a vibrating effect.

CHAPTER 51
Rear Axles and Differentials

After studying this chapter, you should be able to:

- **Explain the purpose of the differential.**
- **Describe the construction and operation of a differential.**
- **Describe the constuction and operation of a nonslip differential.**
- **Explain what the first sign of differential trouble usually is and how to diagnose the condition.**

This chapter discusses the purpose, construction, operation, and servicing of differentials and axles. Power flows from the engine, through the drive line, to the differential. If the car is moving in a straight line, both drive wheels travel at the same speed. But if the car is making a turn, the outer drive wheel must travel farther and faster than the inner drive wheel. The differential, located between the wheel axles (Fig. 51-1) makes this possible.

51-1 Purpose of the Differential

When the car rounds a turn, the outer wheel must travel farther than the inner wheel. For example, suppose the car makes a sharp turn to the left, as shown in Fig. 51-2 which illustrates the conditions for a rear-drive car. The inner rear wheel, turning on a 20-foot [6.096-m] radius, travels 31 feet [9.449 m] during a 90-degree turn. The outer rear wheel, being nearly 5 feet [1.524 m] from the inner wheel, turns on a $24\frac{2}{3}$-foot [7.519-m] radius (in the car shown), and it travels 39 feet [8.839 m].

If the propeller shaft (drive line) were geared rigidly to both rear wheels, each wheel would have to skid an average of 4 feet [1.219 m] to make the turn. The tires would

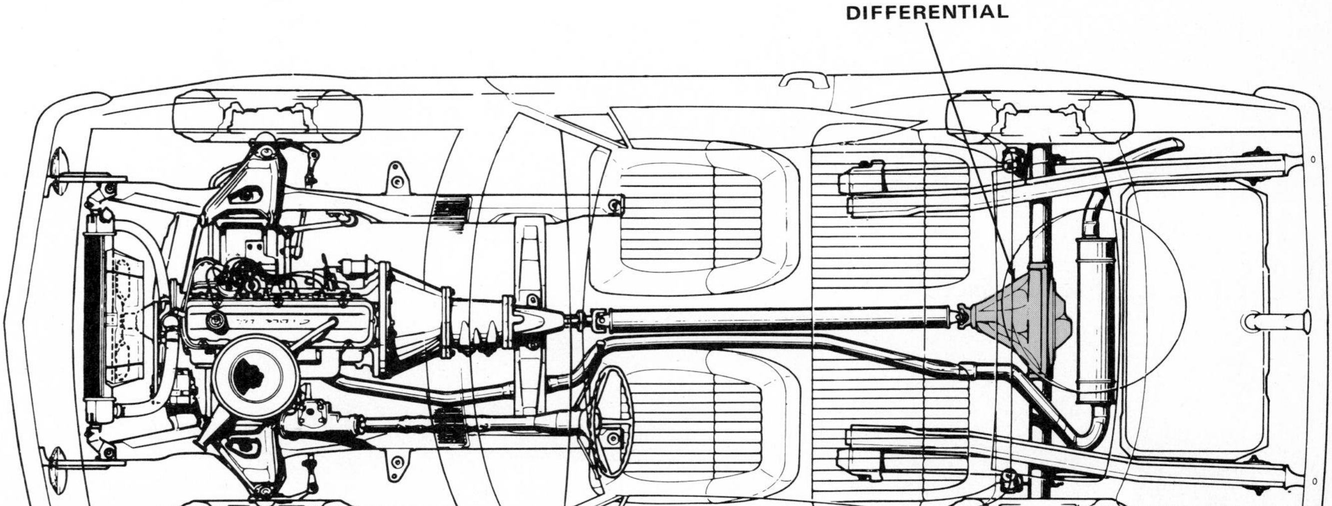

Fig. 51-1 Location of the differential in the power train.

not last very long. But what is worse is that the car could not be controlled during turns. The job of the differential is to avoid these troubles by allowing one rear wheel to turn faster than the other when the car goes around a curve.

The same thing applies to front-drive cars. Front-drive cars also must have differentials. They are usually built into the transaxle as part of the assembly. (See Figs. 46-35 and 46-36.)

● 51-2 Construction of the Differential

To study differential construction and operation, we shall build up, gear by gear, a simple differential of the type used in rear-drive cars. The two rear wheels are mounted on axles. On the inner ends of the axles are bevel gears which are called *differential side gears* (see Fig. 51-3). All the teeth are at an angle. When two bevel gears are put together so that their teeth match, the driving and driven shafts can be at a 90-degree angle (Fig. 51-4).

Now, let's get back to the differential. Figure 51-5 shows all the essential parts of a differential. The parts are separated so that they can be seen clearly. Keep referring to this picture as we put the parts of the differential together.

First, in Fig. 51-6, we show the addition of the differential case to the two wheel axles and bevel (differential side) gears. The differential case has bearings that permit it to rotate on the two axles. Next, we add the two pinion gears and the supporting shaft (Fig. 51-7). The shaft is part of the differential case. The two pinion gears are meshed with the bevel gears.

Actually, the two pinion gears are also bevel gears, but we call them pinion gears so as not to confuse them with the bevel gears (differential side gears) on the ends of the axles.

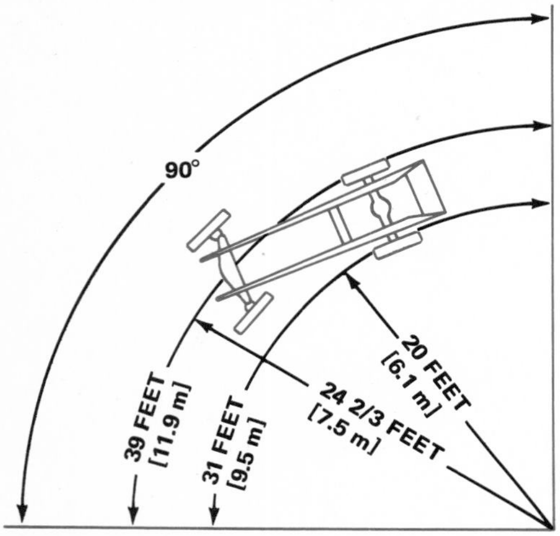

Fig. 51-2 The difference in wheel travel as car makes a 90-degree turn with the inner wheel turning on a 20-foot [6.1-m] radius.

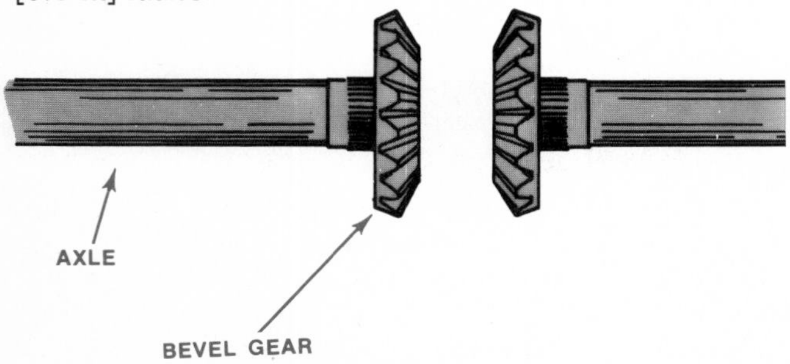

Fig. 51-3 Inner ends of the rear axles with bevel gears (differential side gears) installed on them.

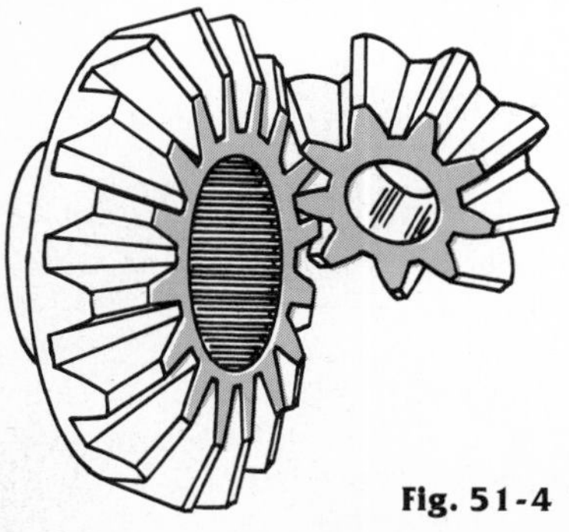

Fig. 51-4 Two meshing bevel gears.

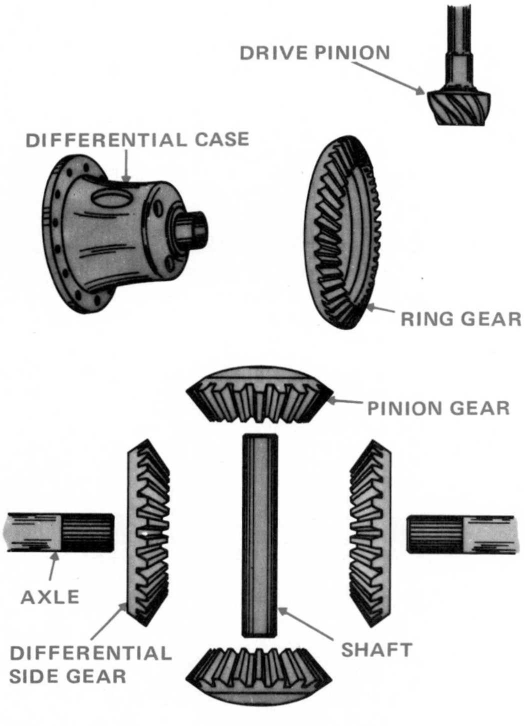

Fig. 51-5 Basic parts of a differential.

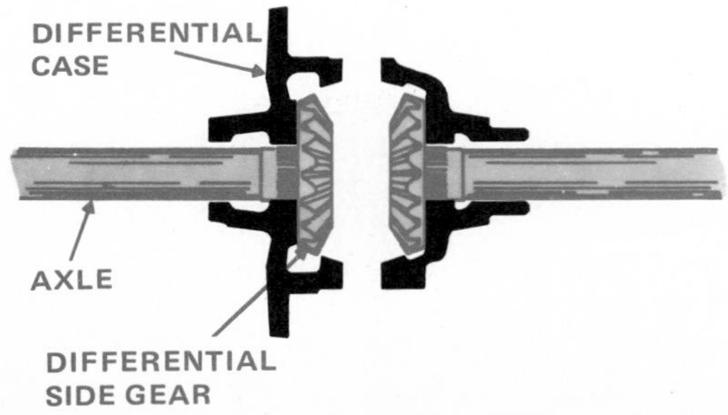

Fig. 51-6 Here, we add the differential case.

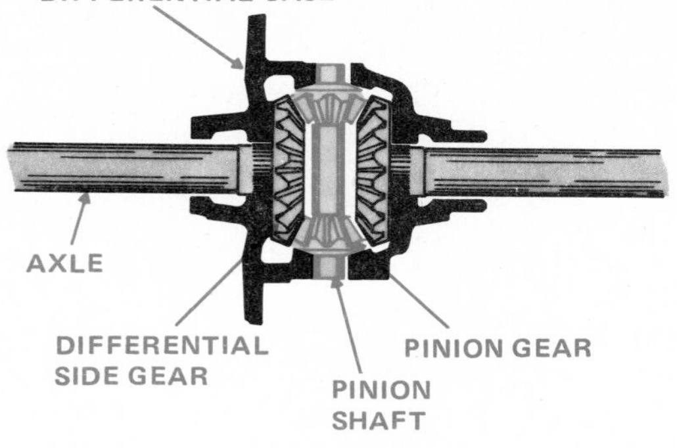

Fig. 51-7 Here, we add the two pinion gears and supporting shaft.

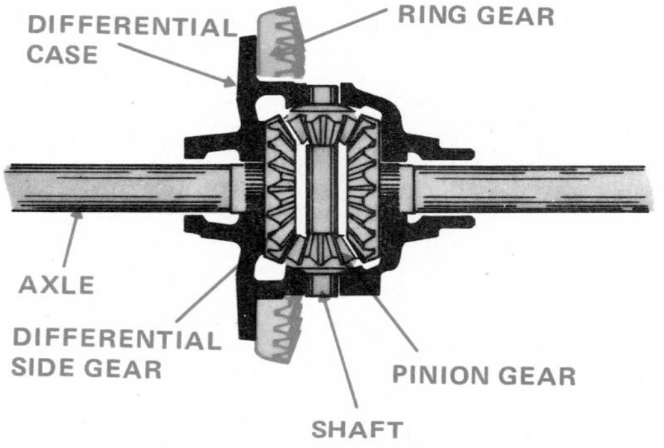

Fig. 51-8 Here, we add the ring gear.

Now we add the ring gear (Fig. 51-8). The ring gear is bolted to the flange on the differential case. Finally, we add the drive pinion (Fig. 51-9). The drive pinion is at the end of the propeller shaft (drive line). When the propeller shaft rotates, the drive pinion rotates, and the drive pinion rotates the ring gear.

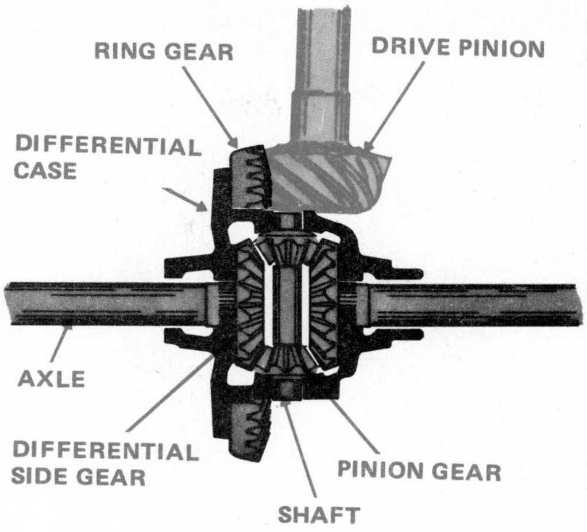

Fig. 51-9 To complete the basic differential, we now add the drive pinion. The drive pinion is on the end of the drive shaft and meshes with the ring gear.

51-3 Operation of the Differential

The drive pinion on the end of the propeller shaft drives the ring gear. The rotation of the ring gear causes the differential case to rotate. When the differential case rotates, the two pinion gears and their shaft move around in a circle with the differential case. Because the two differential side gears are meshed with the pinion gears, the differential side gears must rotate. This causes the rear axles to rotate. The wheels turn, and the car moves.

The outer rear wheel turns faster than the inner rear wheel as the car rounds a curve. As the differential case rotates, the pinion gears have to rotate on their shafts. The reason for this is that the pinion gears must walk around the slower-turning differential side gear. This means that they carry additional rotary motion to the faster-turning outer wheel on the turn. The action is shown in the sample (but typical) condition in Fig. 51-10. Differential-case speed is considered as 100 percent. The rotating action of the pinion gears carries 90 percent of this speed to the slower-rotating inner wheel and 110 percent of the speed to the faster-rotating outer wheel.

You can now see how the differential can allow one rear wheel to turn faster than the other. Whenever the car goes around a turn, the outer rear wheel travels a greater distance than the inner rear wheel. The two pinion gears rotate on their shaft and send more rotary motion to the outer wheel.

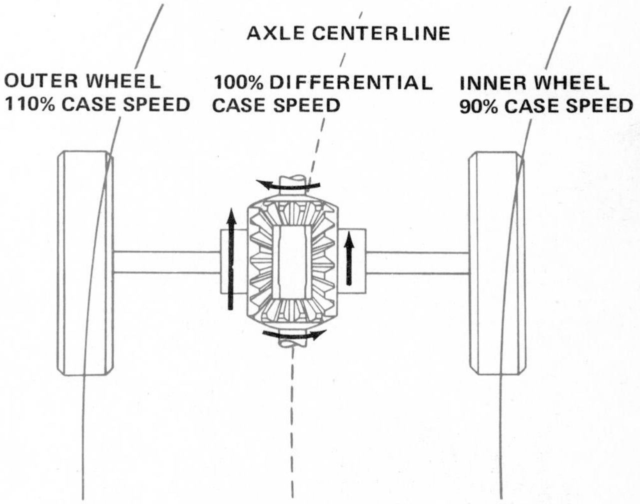

Fig. 51-10 Differential action on turns. *(Chevrolet Motor Division of General Motors Corporation)*

When the car is moving down a straight road, the pinion gears do not rotate on their shaft. They apply equal torque to the bevel gears. Therefore, both rear wheels rotate at the same speed.

51-4 Differential Construction

Figure 51-11 is a partial cutaway view of a differential and rear-axle assembly. As you will note, the pinions and gears are all of heavy construction to carry the power from the drive line. Figure 51-12 is a disassembled view of another differential. Notice that in Fig. 51-12 different names are used for the different gears. For instance, the drive pinion and the ring gear are called "drive gear and

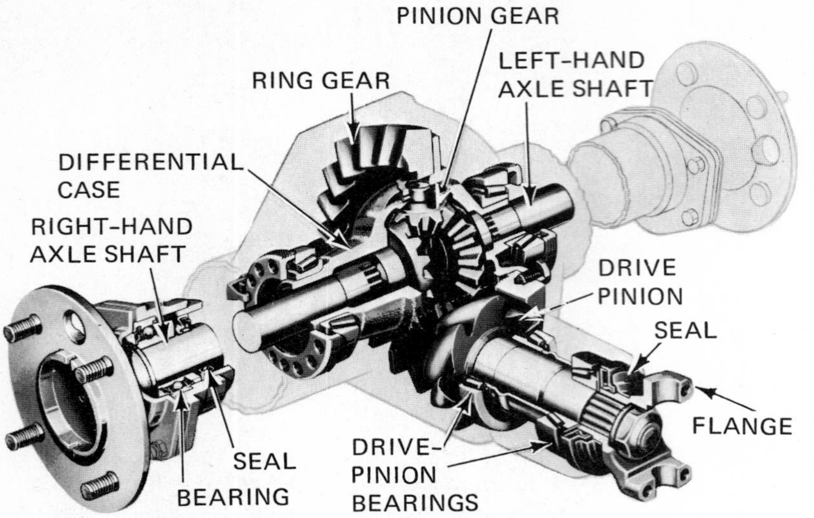

Fig. 51-11 Cutaway of a differential and rear axle. (*Ford Motor Company*)

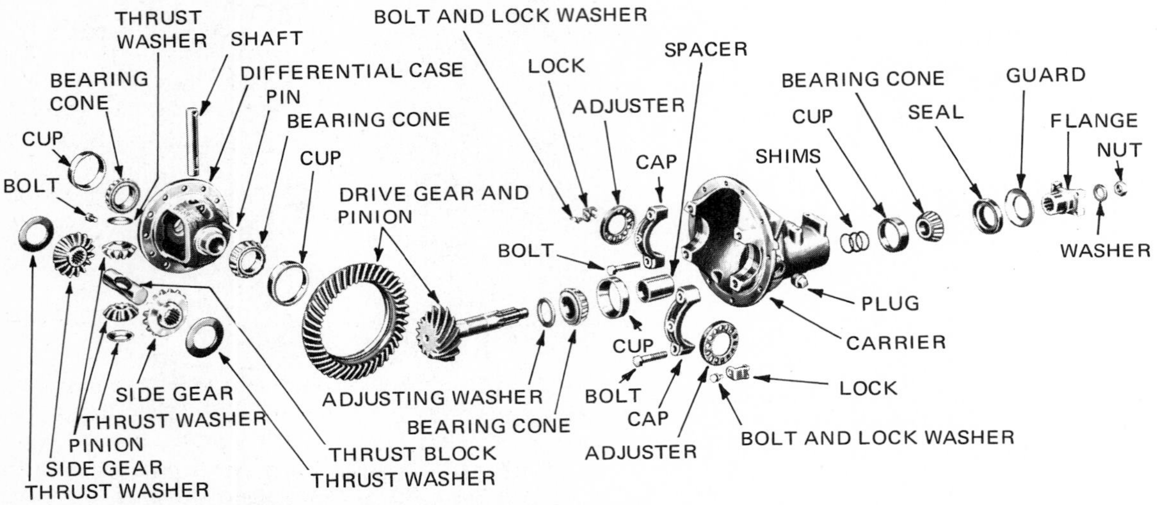

Fig. 51-12 Disassembled differential. (*Chrysler Corporation*)

pinion." The bevel gears are called "side gears." The pinion gears are called simply "pinions." Different manufacturers often use different names for the same parts in their cars. You must watch out for this in automotive work.

51-5 Nonslip Differential

The differential we have just studied delivers the same amount of torque to each rear wheel when both wheels have equal traction. When one wheel has less traction than the other—for example, when one wheel is slipping on ice—the other wheel cannot deliver torque. That is, all the turning effort goes to the slipping wheel. To provide good traction even though one wheel is slipping, a nonslip differential is used in many cars. It is very similar to the standard unit but has some means of preventing wheel spin and loss of traction.

To sum up, the standard differential delivers maximum torque to the wheel with minimum traction. But the nonslip differential delivers maximum torque to the wheel with maximum traction.

One type of nonslip differential is shown in Fig. 51-13. It has two sets of clutch plates. Also, the ends of the pinion-gear shafts lie rather loosely in notches in the two halves of the differential case. Figure 51-14 is a sectional view of the nonslip differential. During normal straight-road driving, the power flow is as shown in Fig. 51-15.

In Figs. 51-14 to 51-16, the ring gear is called the "axle drive gear." The bevel gear is called the "differential side gear." The pinion gears are called the "differential pinions."

Note that the rotating differential case carries the pinion-gear shafts around with it. Since there is considerable side thrust, the pinion shafts tend to slide up the sides of the notches in the two halves of the differential case. As the pinion shafts slide up, they are forced out-

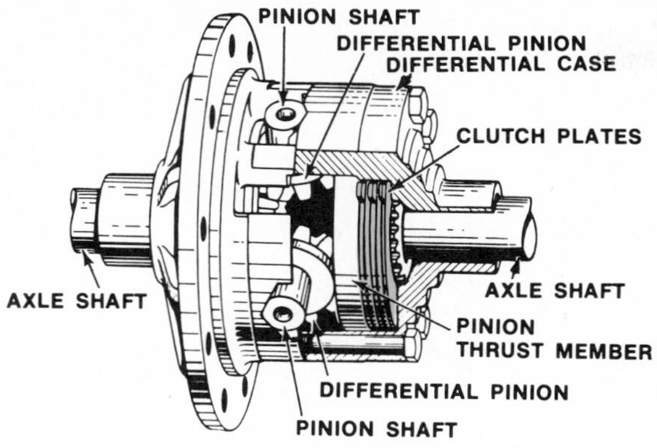

Fig. 51-13 Cutaway of a nonslip differential. (*Chrysler Corporation*)

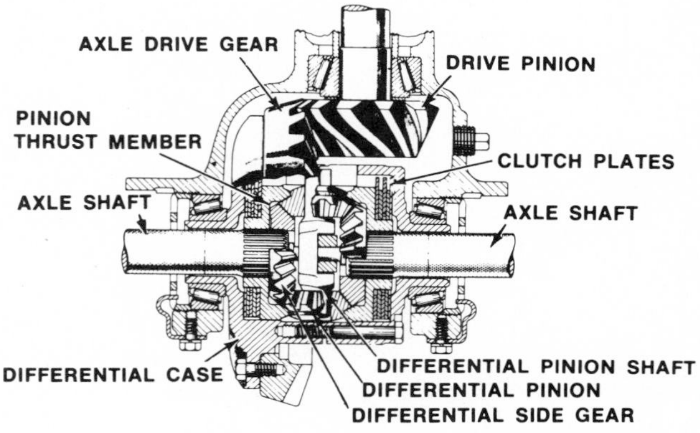

Fig. 51-14 Sectional view of a nonslip differential. (*Chrysler Corporation*)

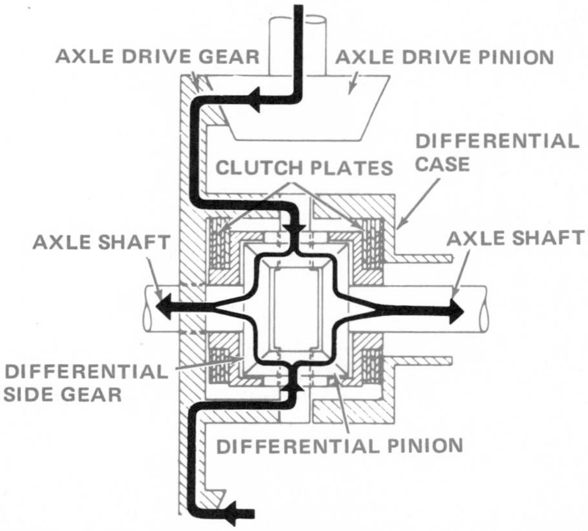

Fig. 51-15 Power flow through nonslip differential on a straightaway. (*Chrysler Corporation*)

ward. This force is carried to the two sets of clutch plates. The clutch plates thus lock the axle shafts to the differential case. Therefore, both wheels turn.

Suppose one wheel encounters a patch of ice or snow and loses traction, or tends to slip. Then the pressure is released on the clutch plates feeding power to that wheel. Thus the torque goes to the other wheel, and the wheel on the ice does not slip.

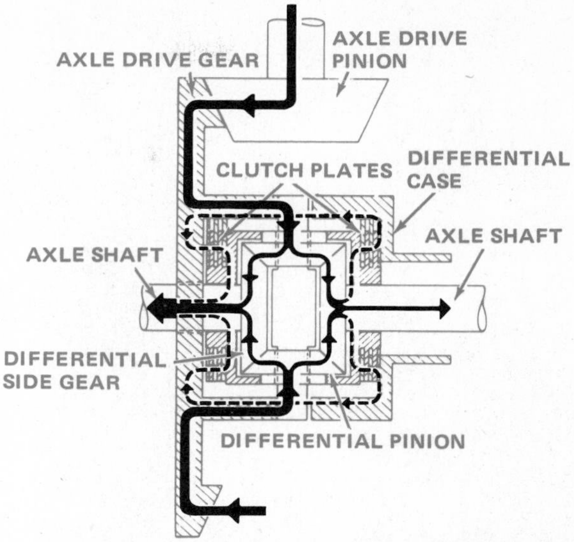

Fig. 51-16 Power flow through nonslip differential when rounding a turn. Heavier arrows show greater torque to the left axle shaft. (*Chrysler Corporation*)

During normal driving, if the car rounds a turn, pressure is released on the clutch feeding the inner wheel. Just enough pressure is released to permit some slipping. Figure 51-16 shows the action. This release permits the outer wheel to turn faster than the inner wheel.

Other types of nonslip differentials have spring-loaded clutch plates, or spring-loaded clutch cones. These are shown in disassembled views in Figs. 51-17 and 51-18. The action is the same as in the type described above. The essential difference is that the plates or cones are preloaded by springs to give a more positive action.

51-6 Differential Gearing

Since the ring gear has many more teeth than the drive pinion, a considerable gear reduction is effected in the differential. The gear ratios vary, depending on car and engine design. Ratios of 2.45:1 upward to about 4:1 are used on passenger cars. This means that the ring gear has 2.45 to 4 times as many teeth as the drive pinion. The drive pinion has to rotate 2.45 to 4 times (according to gear ratio) to cause the ring gear to rotate once.

The gear ratio in the differential is usually referred to as the *axle ratio,* although it would be more accurate to call it the differential ratio.

If you want to figure the axle ratio of a differential, and you have the ring gear and pinion, do this. Count the number of teeth on each. Divide the number of teeth on the ring gear by the number of teeth on the pinion. This gives you the axle, or differential, ratio.

Early differentials used straight bevel gears for the drive pinion and ring gear (Fig. 51-19). A later design used spiral bevel gears (Fig. 51-19). These were quieter running. Modern car bodies are low, and this means the drive line must be dropped as low as possible. To accomplish this, special gears, called *hypoid* gears, are used (Fig. 51-19). These are similar to spiral bevel gears. The difference is that the tooth formation permits the drive

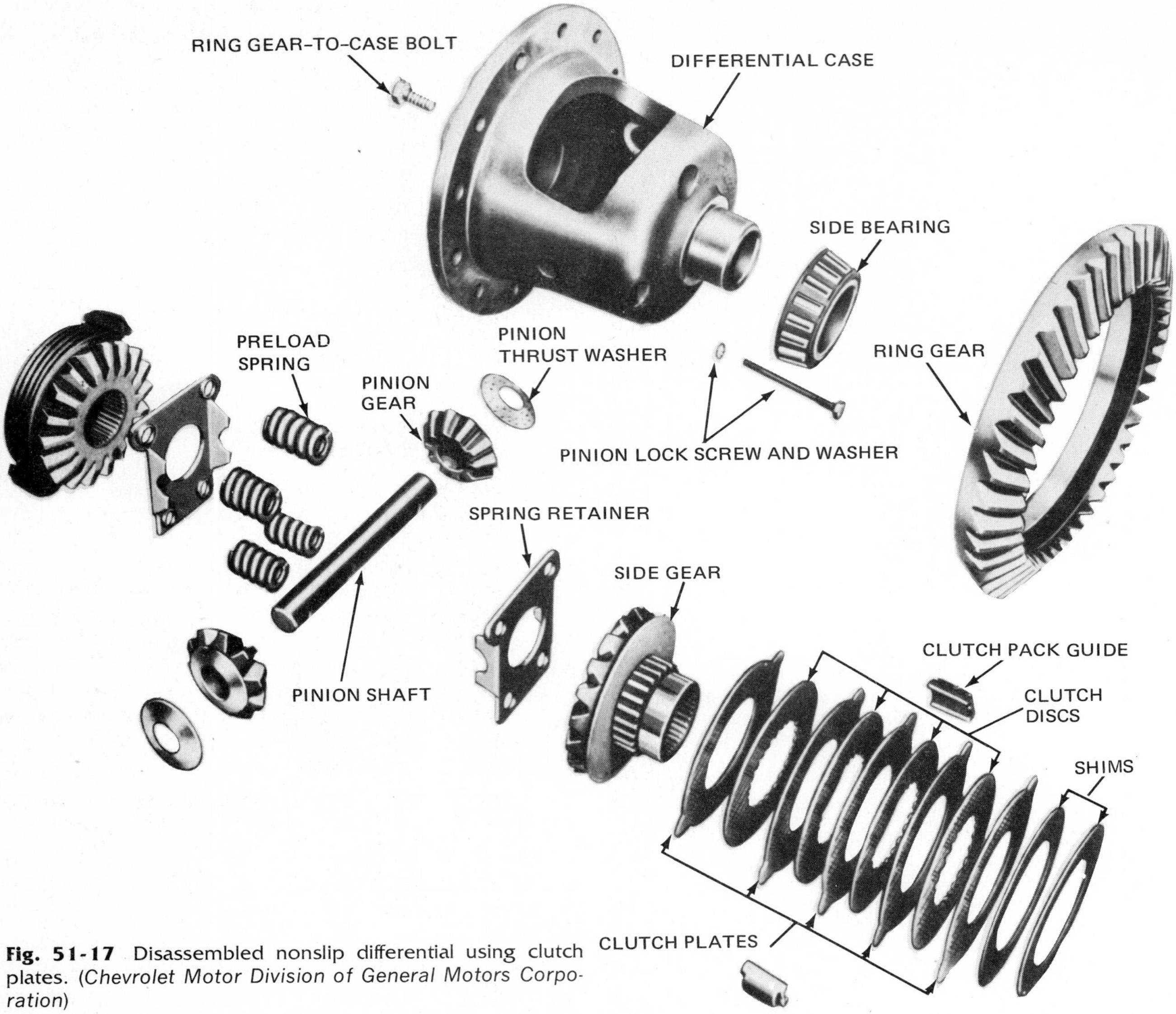

Fig. 51-17 Disassembled nonslip differential using clutch plates. (*Chevrolet Motor Division of General Motors Corporation*)

pinion to be placed well below the centerline of the ring gear, as shown. Also, there is a wiping action between the teeth as they mesh and unmesh. This wiping action is always found in hypoid gears. It requires the use of a special hypoid-gear lubricant.

Figure 51-20 illustrates gear-tooth nomenclature. The mating teeth to the left illustrate clearance and backlash. The tooth to the right has its various parts named. *Clearance* is the distance between the top of the tooth of one gear and the valley between adjacent teeth of the mating gear. *Blacklash* is the distance between adjacent meshing teeth in the driving and driven gears. It is the distance one gear can rotate backward, or backlash, before it will cause the other gear to move. The *toe* is the smaller section of the gear tooth; the *heel* is the larger section.

51-7 Types of Rear Axle

There are two basic types of axle: dead axles and live axles. The dead axle does not rotate; the wheel rotates on it. A common example is the axle on a horse-drawn wagon. Live axles are attached to the wheel so that the wheel and axle rotate together. Live axles are classified according to the manner in which they are supported: semifloating, three-quarter-floating, and full-floating. The difference among the three is the way in which the wheel and axle are supported. The semifloating axle carries some of the vehicle weight. The full-floating axle carries no weight.

A special type of rear axle using two universal joints is illustrated in Fig. 52-8 (in Chap. 52). Each rear wheel is independently suspended by a control arm.

51-8 Rear-End Torque

The rotation of the drive shaft sends torque through the differential to the rear wheels. The wheels rotate and the car moves because torque is applied to them. This torque not only rotates the wheels in one direction, it also tries to rotate the differential housing in the opposite direction.

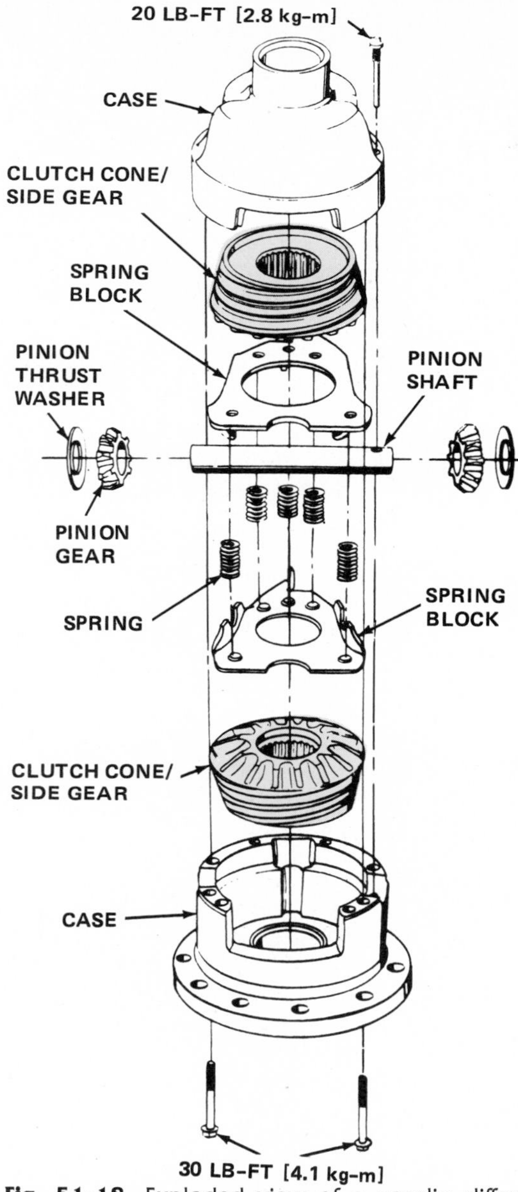

Fig. 51-18 Exploded view of a nonslip differential using cones. (*Pontiac Motor Division of General Motors Corporation*)

To understand this, let us review differential construction. The ring gear is connected through other gears to the rear-wheel axles. The torque applied through the drive pinion forces the ring gear and wheels to rotate. It is the side thrust of the drive-pinion teeth against the ring-gear teeth that makes the ring gear rotate. This side thrust also causes the drive-pinion shaft to push against the shaft bearing. This pushes the differential housing up.

Here's another way to look at it. The ring gear gets a downward push from the drive pinion when in forward drive. The drive pinion therefore pushes up against the differential housing. The differential housing tries to rotate in a direction opposite to wheel rotation. The effect is called *rear-end torque*. The differential and axle housing must be braced to resist this. If the rear suspension uses leaf springs, the leaf springs provide the bracing. If the rear suspension uses coil springs, a set of control arms provide the bracing. This is covered in Chap. 52.

Despite the bracing, however, the differential and axle housing and rear of the car do react to rear-end torque. You can see this any time a driver gives the engine the gas, trying to make a quick getaway. The rear end of the car will drop down momentarily as the torque arrives at the differential. This action is called *rear-end squat*.

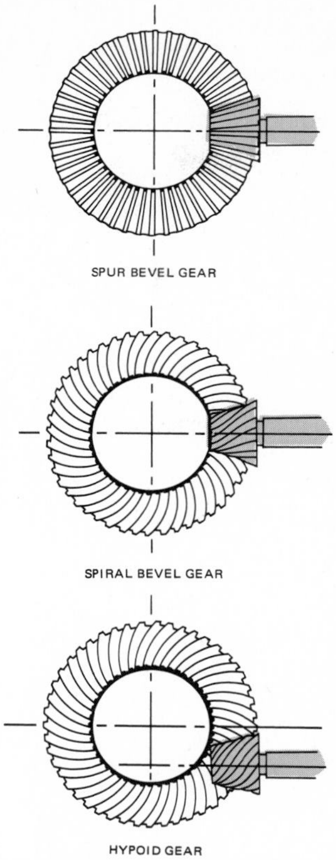

Fig. 51-19 Spur-bevel, spiral-bevel, and hypoid differential drive pinions and ring gears.

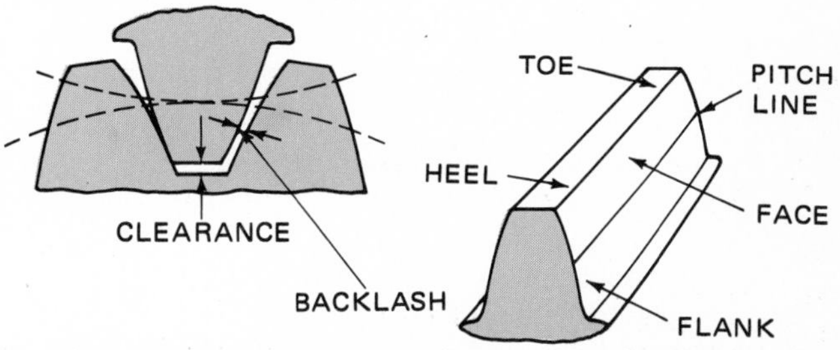

Fig. 51-20 Gear-tooth nomenclature. (*Chrysler Corporation*)

● 51-9 Front Drive

The differentials for front-drive cars are similar to those for rear-drive cars. Both work the same way. Figures 46-35 and 46-36 show differentials built into transaxles used at the front of the car. With the front wheels driving the car, front-end squat may occur when the driver suddenly opens the throttle for a quick start. As noted in ● 51-8, the rear-drive car will have rear-end squat. Front-end squat and rear-end squat both result from the same actions in the differential as explained in ● 51-8.

51-10 Diagnosing Differential Troubles

The first sign of differential trouble is usually noise. The kind of noise you hear can help you determine what is causing the trouble. You have to be sure, however, that the noise is actually coming from the differential. It is sometimes possible to be fooled by universal-joint, wheel-bearing, or tire noise. Note whether the noise is a hum, a growl, or a knock. Note whether the noise is produced when the car is operating on a straight road or only on turns. Note whether the noise is more evident when the engine is driving the car or when the car is coasting. It is difficult to diagnose differential noise by running the car with the wheels jacked up. Power must be flowing through the differential to the wheels, or the wheels must be sending power back (car coasting).

1. Humming A humming noise is often caused by incorrect internal adjustment of the drive pinion or the ring gear. Incorrect adjustment prevents normal tooth contact and can cause rapid tooth wear and early failure of the differential. The humming noise will take on a growling sound as the wear progresses. Check the shop manual covering the car you are servicing when you make differential adjustments.

2. Noise on acceleration Noise that is more evident when the car is accelerating probably means there is heavy contact on the heel ends of the gear teeth. Noise that is more evident when the car is coasting probably means there is heavy toe contact. Both these conditions must be corrected. Refer to the manufacturer's shop manual for servicing procedures.

3. Noise on curves If the noise is present only when the car is going around a curve, the trouble is inside the differential case. Pinion gears tight on the pinion shaft, damaged gears or pinions, too much backlash between gears, or worn differential-case bearings can cause this trouble. When you round a curve, it is these parts inside the differential case that are moving relative to each other.

This could also be due to a defective axle bearing because the outside bearing on a curve takes an increased load.

4. Nonslip differential The nonslip differential requires a special type of lubricant. The wrong lubricant can cause the clutch surfaces to grab, which will produce chattering noise during a turn. The remedy is to drain the old lubricant and put in the specified lubricant designed for nonslip differentials.

Wheel spin can occur under some conditions, even if the differential is in good condition. For instance, if one wheel is on dry pavement and the other is on smooth ice, sudden acceleration can produce wheel spin. The remedy is to open the throttle *slowly* to allow the differential to do its job properly.

Careful: On cars with nonslip differentials, it is important to use the same type of tire on both rear wheels. Both tires should have the same pattern, the same air pressure, and the same amount of wear. If one tire is larger than the other, the differential will be working all the time. This will shorten the life of the differential.

51-11 Servicing the Differential

Repair and overhaul procedures on rear axles and differentials vary among car models. Always make sure you have the manufacturer's shop manual that covers the car you are working on when you repair a differential.

REVIEW QUESTIONS

1. What is the purpose of the differential?
2. Name, in order, the parts of the differential through which power is carried from the drive pinion to the axle shafts.
3. When a car is operated on a straight road, do the differential pinion gears rotate on their shafts?
4. What happens to the differential pinion gears when the car rounds a curve?
5. About what gear reduction is obtained in the differential on passenger cars? Does it vary from car to car?
6. What is the purpose of the nonslip differential?
7. Describe the operation of a nonslip differential.
8. In gears, what is clearance? Backlash? Face? Flank?
9. What are the two basic types of axle? In what way do they differ?
10. What is it that most often draws attention to trouble in the differential?
11. What does a humming noise in the differential often indicate?
12. When differential noise is most evident during car acceleration, what is the probable trouble?
13. If the noise is most evident when the car is coasting in gear, what is the probable trouble?
14. If the noise is present only when the car rounds a turn, what is the probable location of the trouble?

SELF PROJECTS

1. Refer to several manufacturers' shop manuals, and, make a list of the different axle ratios offered by the car companies in their cars.
2. You can figure the axle ratio of almost any car as follows: With the rear end of the car raised, turn both rear wheels the same amount, while someone watches the rotation of the drive shaft. When the shaft rotates exactly one full turn, you will have turned the wheels more than two times. Suppose the wheels have turned $2\frac{3}{4}$ times while the shaft rotated once. The axle ratio is 2.75:1.
3. Refer to a manufacturer's service manual. Make a list of the differential troubles in the trouble-diagnosis section. Note the possible causes of the troubles.
4. Study the servicing section of a manufacturer's shop manual, and note the steps in disassembly, repair, reassembly, and adjustment of the differential.

Automotive Chassis

9

In this part, we describe the automotive chassis, which includes the car frame, springs and shock absorbers, steering system, brakes, tires, and wheels. Also included are automotive heating and air conditioning. There are seven chapters in Part Nine.

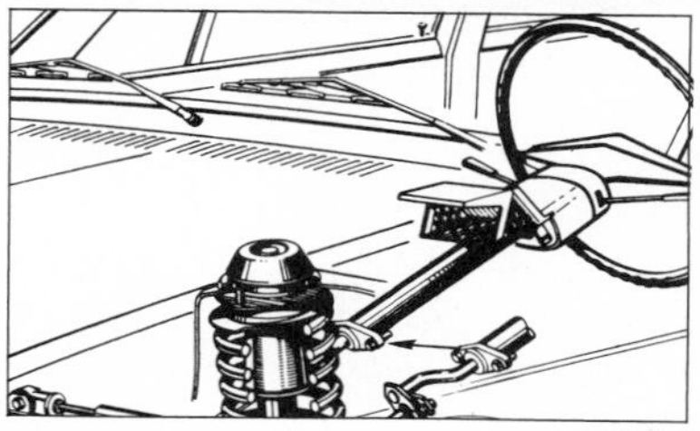

Chapter 52 Springs and Shock Absorbers

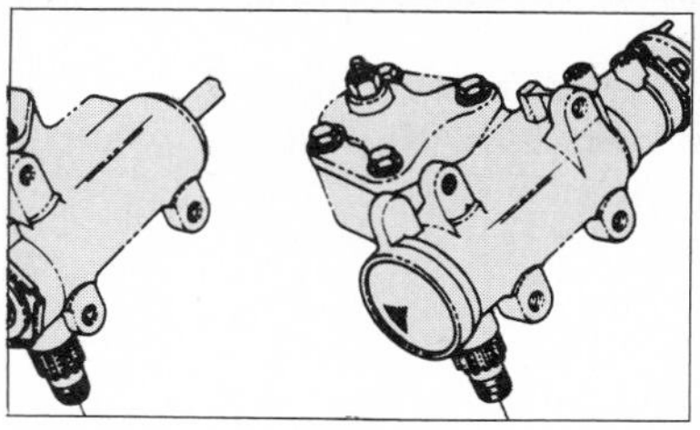

Chapter 53 Automotive Steering Systems

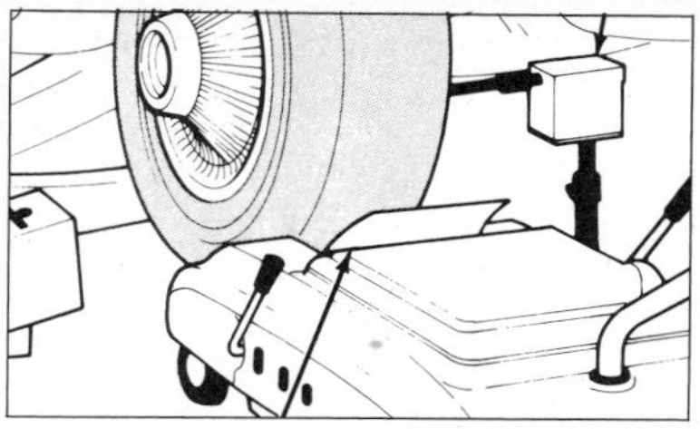

Chapter 54 Steering and Suspension Service

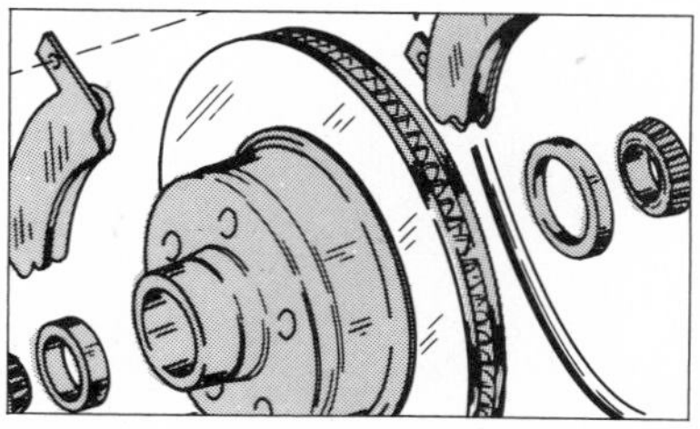

Chapter 55 Automotive Brakes

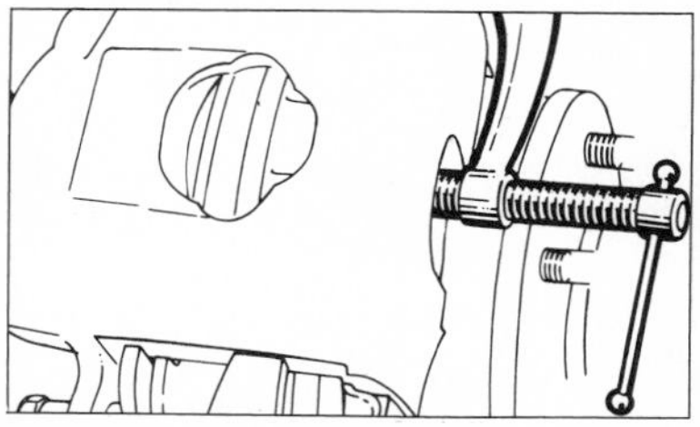

Chapter 56 Brake Service

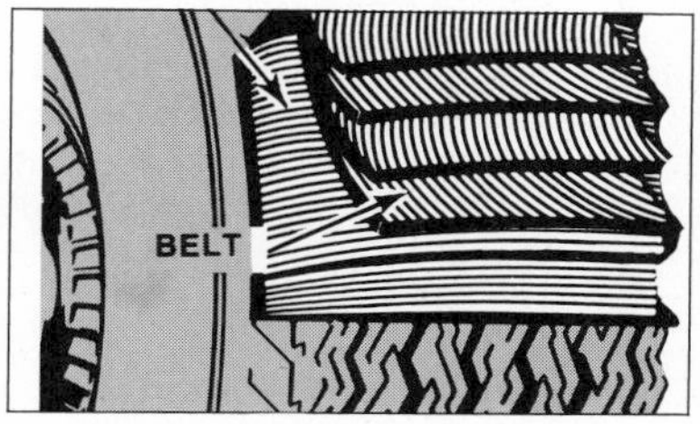

Chapter 57 Tires and Tire Service

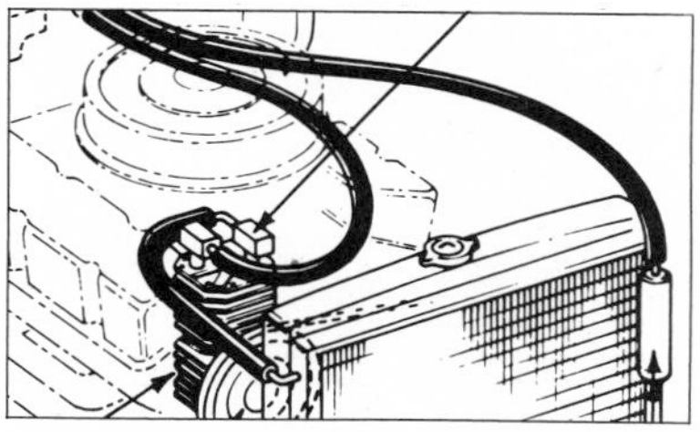

Chapter 58 Automotive Heating and Air Conditioning

CHAPTER 52
Springs and Shock Absorbers

After studying this chapter, you should be able to:

- **Describe the three basic types of springs and explain how they work.**
- **Describe and identify on cars the various types of rear suspension. Explain how they work.**
- **Describe and identify on cars the various types of front suspension. Explain how they work.**
- **Describe the operation and construction of shock absorbers.**
- **Describe the construction and operation of automatic level controls. Identify the components of the system on various cars.**

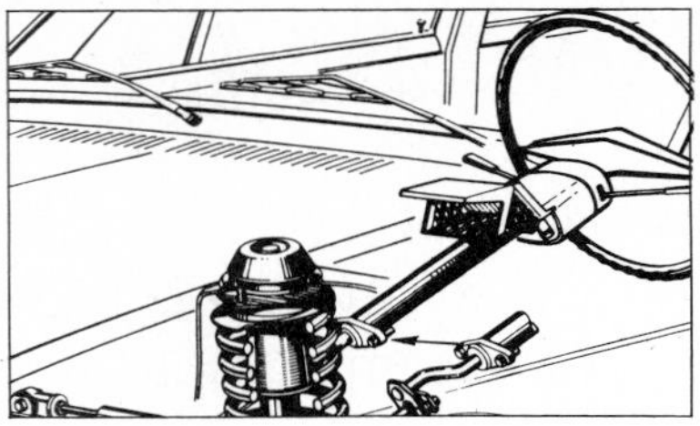

The springs are part of the suspension system of the automobile. Their purpose is to absorb any road shocks that result from the wheels hitting holes or bumps in the road. In this chapter, we discuss the various kinds of springs and shock absorbers and learn how they work.

● 52-1 Function of Springs

The car frame supports the weight of the engine, the power train, the car body, and the passengers. The frame is supported by the springs. There is a spring at each wheel. The weight of the car frame, body, and so on, applies an initial compression to the springs. The springs compress further or expand as the car wheels hit bumps or holes in the road. The springs cannot do the complete job of absorbing road shocks. The tires absorb some of the irregularities in the road. The springs in the car seats also help to absorb shock. As a rule, very little shock from road bumps and holes gets to the passenger.

● 52-2 Types of Springs

There are three basic types of automotive springs: leaf, coil, and torsion bar. In addition, air suspension is used in some trucks and buses. Air suspension was offered for passenger cars some years ago, but few people wanted it so the option was dropped.

Most cars use either coil springs or torsion-bar springs at the front wheels (Figs. 52-1 and 52-2). Some cars use coil springs at the rear wheels. Others use leaf springs. We describe all these types of springs in detail in the following sections.

● 52-3 Characteristics of Springs

The ideal spring for automotive suspension would be one which would absorb road shock rapidly and then return to its normal position slowly. Such an ideal is not possible, however. An extremely flexible, or soft, spring would allow too much movement. A stiff, or hard, spring would give too rough a ride. However, satisfactory riding qualities are attained by using a fairly soft spring with a shock absorber (● 52-18).

The softness or hardness of a spring is referred to as its *rate*. The rate of a spring is the weight required to deflect it 1 inch [25.4 mm]. The rate of automotive springs is about constant through their operating range, or deflection, in the car. This is stated by Hooke's law, as applied to coil springs: The spring will compress in direct

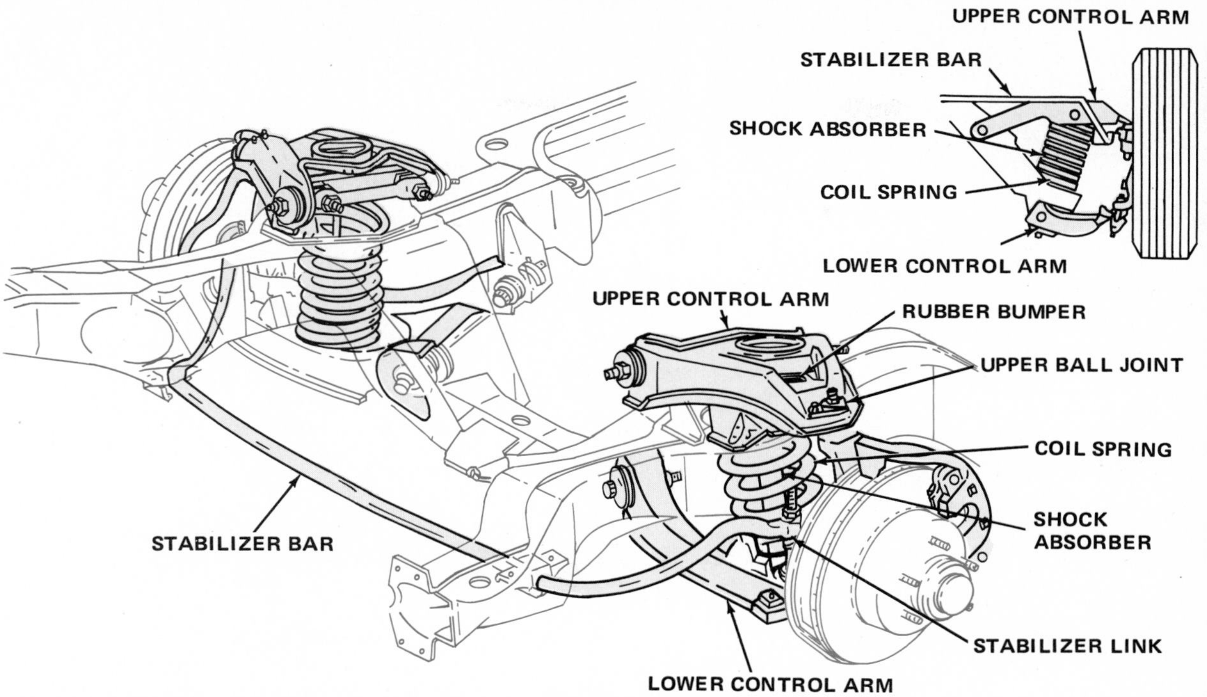

Fig. 52-1 Front-wheel suspension. The wheels mount on bearings on the tapered spindles of the steering knuckles. The support of one wheel is shown to the right in simplified line drawing. (*Buick Motor Division of General Motors Corporation*)

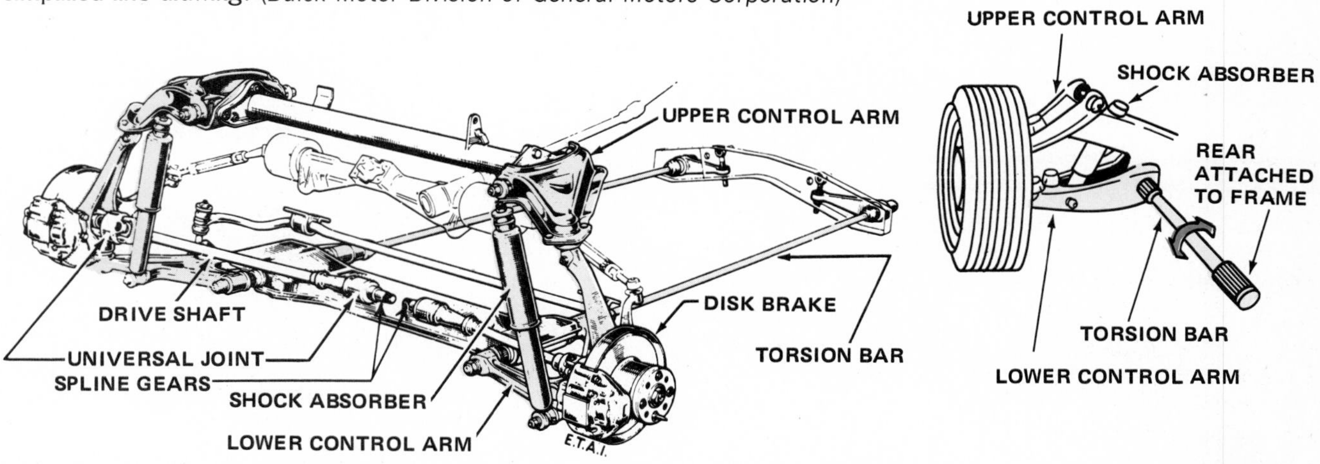

Fig. 52-2 Front-suspension system using torsion bars. To the right, a simplified drawing showing how the torsion bar is attached to the lower control arm and to the frame. (*Simca*)

proportion to the weight applied. Thus, if 600 pounds [272.154 kg] will compress the spring 3 inches [76.2 mmn] 1,200 pounds [544.328 kg] will compress the spring twice as far, or 6 inches [152.4 mm].

52-4 Leaf Springs

The leaf spring most commonly used in automobile rear suspensions is made up of several long plates, or leaves. Figure 52-3 shows a typical leaf-spring installation at a rear wheel. Figure 52-4 shows how the spring at each rear wheel is mounted on the frame. Because the leaf spring consists of a series of thin leaves, one on top of another, it does not break when bent.

The plates, or leaves, are held together at the center by a center bolt which passes through holes in the leaves. Clips are placed at intervals along the spring, as shown in Fig. 52-3. They keep the leaves in alignment and prevent excessive separation during rebound. Also, many leaf springs have special inserts between the leaves to permit the leaves to slip over one another when the spring bends.

Some lightweight cars use single-leaf springs. Single-leaf springs are tapered from the center to the ends so that they work in the same way as multileaf springs.

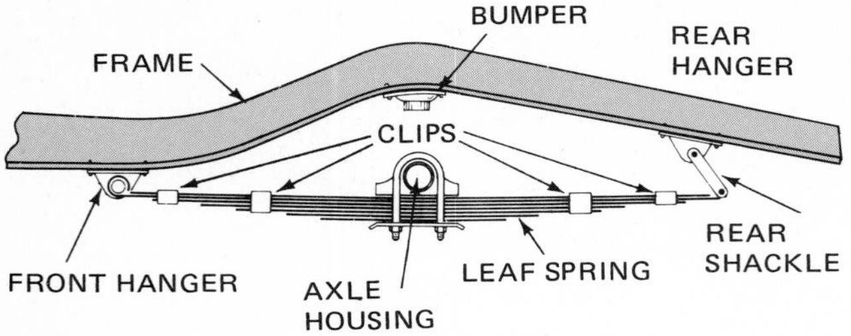

Fig. 52-3 Rear suspension at one wheel. This is a leaf-spring system. The leaf spring is anchored at three places: the front hanger, the rear shackle, and the axle housing.

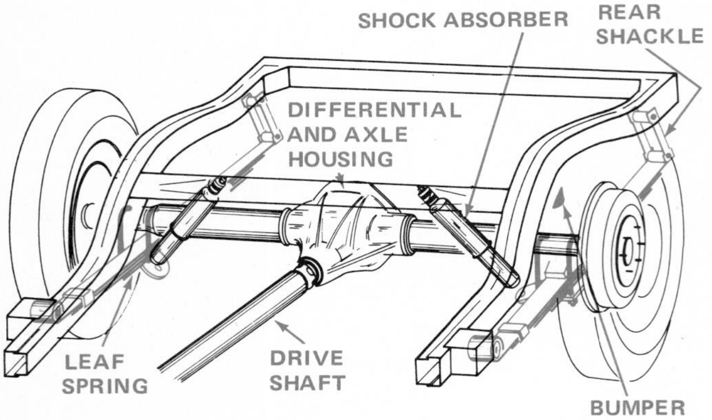

Fig. 52-4 Rear suspension using leaf springs.

Figure 52-5 shows a rear-suspension system using two single-leaf springs. They are attached in the same way as multileaf springs.

● 52-5 Leaf-Spring Installation

Now look again at Figs. 52-3 and 52-4. The spring leaves are of graduated length, as shown. The front end of the longest leaf is bent into a circle to form a spring eye. The spring eye is attached to the spring hanger by a bolt. Rubber bushings insulate the bolt from the spring hanger (see Fig. 52-6). The rubber bushings serve two purposes: They absorb vibration and thus prevent it from getting up to the car frame. The bushings also allow the spring eye to twist back and forth as the leaf spring bends.

The rear end of the spring is also bent to form a spring eye. This spring eye is attached to the car frame through a spring shackle. The shackle allows for changes in the length of the leaf spring as it bends. As the spring is pushed upward or downward by bumps or holes in the road, the distance between the two spring eyes changes. The shackle forms a swinging support that permits this change in length. Figure 52-7 shows a disassembled spring shackle. The shackle is always located at the rear of the spring, as shown in Fig. 52-4. Note that the shackle includes rubber bushings. They absorb vibration and prevent it from getting up to the car frame.

The center of the spring is hung from the rear-axle housing by a pair of U bolts. The rear of the car is, in effect, hung from the axle housing by two pairs of U bolts. Note that there are rubber bumpers on the car

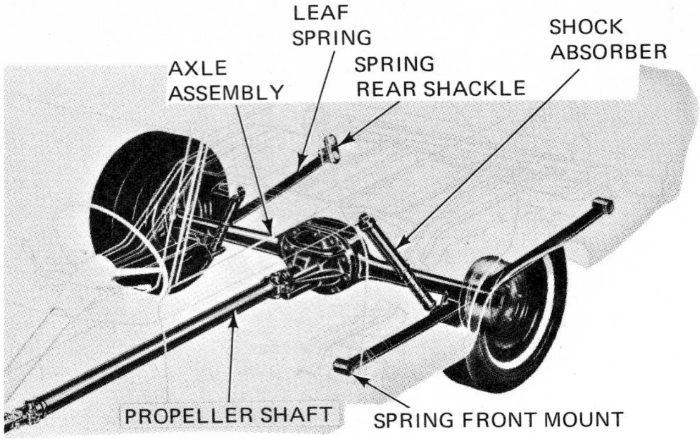

Fig. 52-5 Rear suspension using tapered-plate, or single-leaf, springs. (*Chrysler Corporation*)

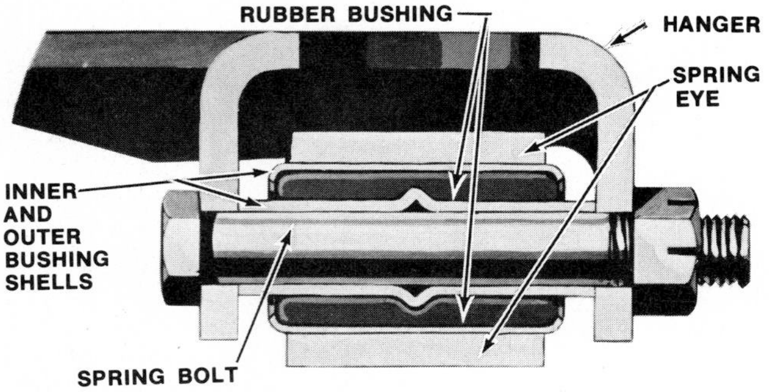

Fig. 52-6 Details of the bushing in a spring eye through which the spring eye of the leaf spring is attached to the hanger on the car.

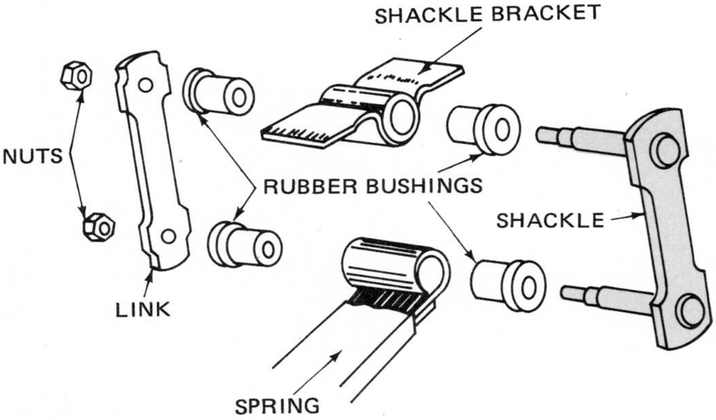

Fig. 52-7 Disassembled spring shackle for a leaf spring.

frame above the axles. The purpose of these bumpers is to absorb the shock that would result if the axle housing actually moved up far enough to hit the frame. The axle housing would move up this far only if the wheels hit a large bump, or if the rear were carrying a heavy load.

Two shock absorbers, one for each spring, are shown in Fig. 52-4. We describe shock absorbers in ● 52-18.

● 52-6 Transverse Leaf-Spring Mounting

Figures 52-8 shows a rear suspension which uses a single leaf spring, mounted transversely; that is, from side to

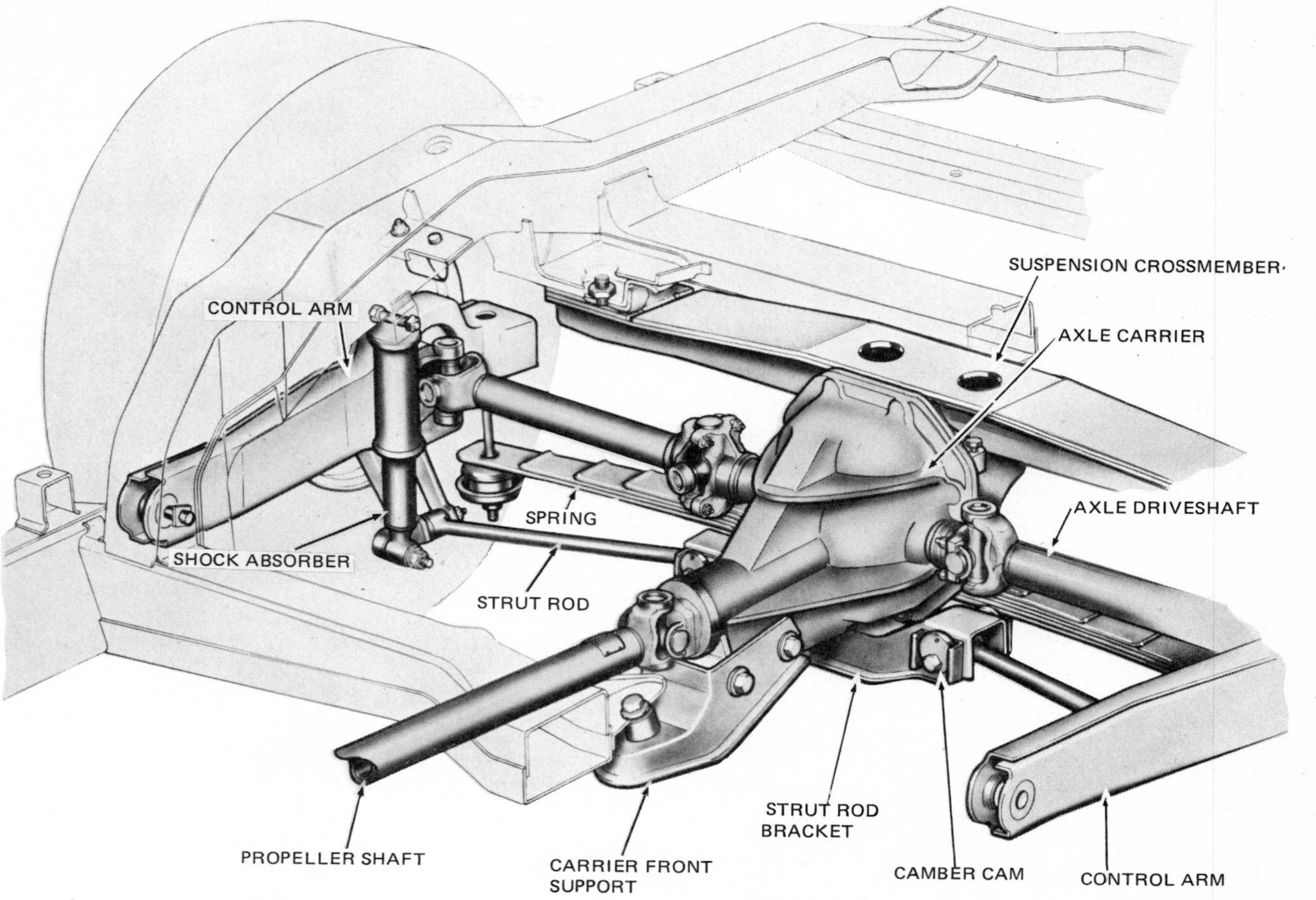

Fig. 52-8 Rear-suspension and drive-line components in the Corvette. Note the transverse leaf spring and the axle drive shafts with their two universal joints each. (*Chevrolet Motor Division of General Motors Corporation*)

side. Each rear wheel is thus independently suspended by one end of the leaf spring. Note that each drive shaft has two universal joints.

52-7 Coil-Spring Rear Suspension

In the rear-suspension systems of many cars, coil springs are used instead of leaf springs. The coil spring is made from a length of steel rod wound into a coil (Fig. 52-9). The coil spring is very elastic and will compress when a weight is put on it. The heavier the weight, the more the spring will compress. Figure 52-10 shows the chassis of a car using a coil spring at each wheel.

A rear suspension using coil springs is shown in Fig. 52-11. Each spring is assembled between spring seats in the car frame and lower control arms, or pads on the axle housing. When the rear wheels hit a hole or a bump in the road, the springs expand or compress to absorb the shock. On expansion, the control arms or shock absorbers prevent excessive movement. This keeps the coil springs from moving out of their seats.

Notice that the coil-spring rear-suspension system shown in Fig. 52-11 has four control arms. Two of the arms are upper control arms, and two are lower control arms. The purpose of the arms is to keep the rear-axle housing in alignment with the frame. The two upper control arms are pivoted on the frame and the differential housing. The upper control arms prevent sideward movement of the axle housing. The two lower control arms are pivoted on the frame and the axle housing. The lower

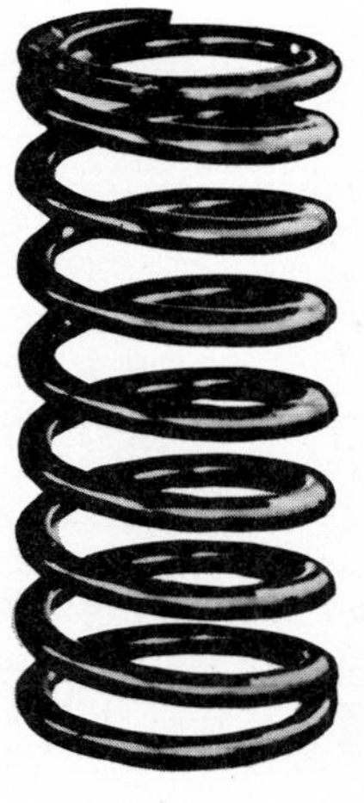

Fig. 52-9 Coil spring used in a front-suspension system.

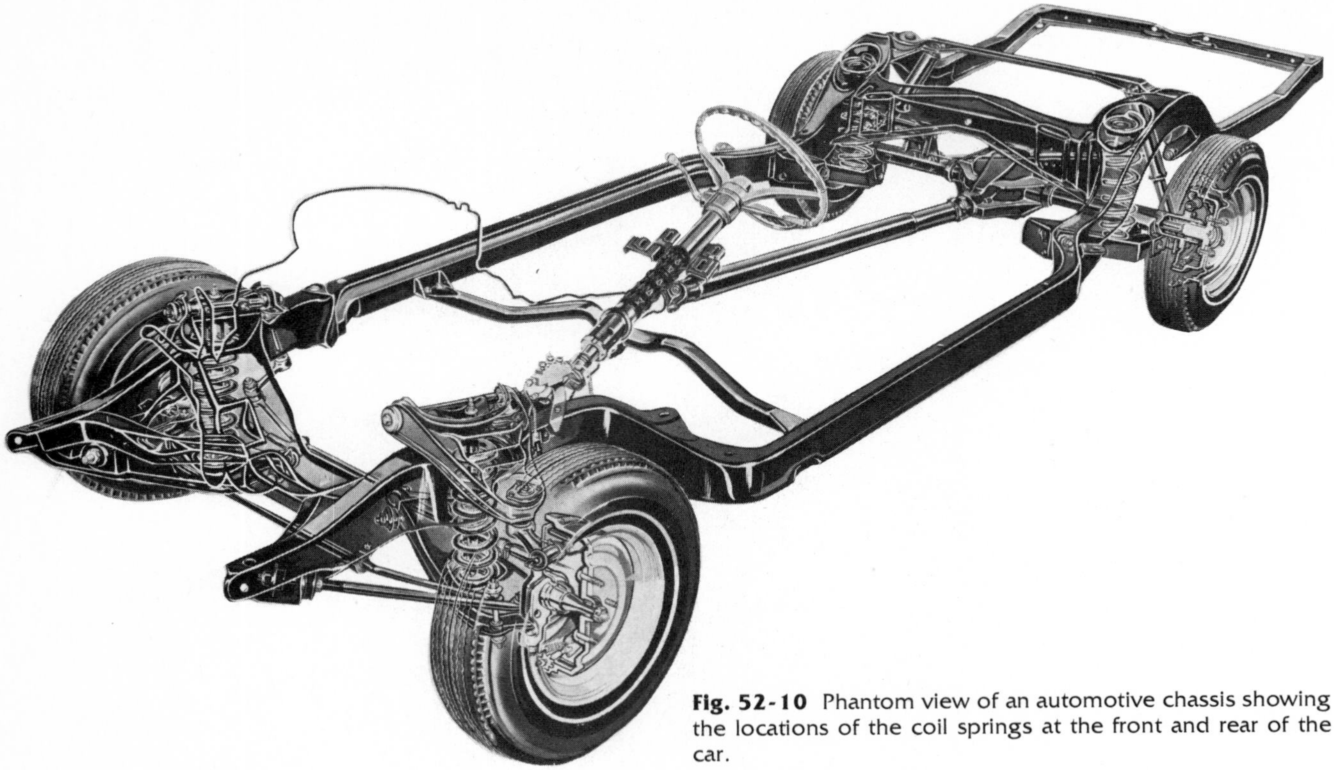

Fig. 52-10 Phantom view of an automotive chassis showing the locations of the coil springs at the front and rear of the car.

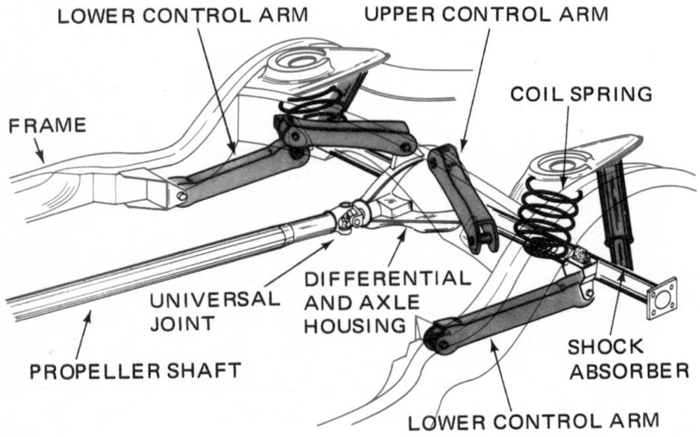

Fig. 52-11 Rear suspension system using coil springs. Note the four control arms.

control arms prevent forward-and-backward movement of the housing. In other words, the arms permit the rear-axle housing to move up and down. But they prevent sideward or forward-and-backward movement. Note that this system also uses shock absorbers at each wheel.

A rear suspension for a front-drive car is shown in Fig. 52-12. This type of suspension is relatively simple. There is no drive through the wheels. Coil springs are used with one pair of control arms.

● 52-8 Rear-End Torque

Whenever the rear wheels are being driven by the power train, they rotate, as shown in Fig. 52-13. At the same time, the rear-axle housing tries to rotate in the opposite direction (Fig. 52-13). The twisting motion thus applied to the axle housing is called *rear-end torque*. This rear-end torque is absorbed by the rear springs and control arms. On the leaf-spring rear suspension, the leaf spring absorbs the rear-end torque. On the coil-spring rear suspension, the control arms absorb the rear-end torque.

One effect of rear-end torque is rear-end "squat" on acceleration (Fig. 52-14). When a car is accelerated from a standing start, the drive pinion tries to climb the teeth of the ring gear. Thus, the drive pinion and the differential carrier move upward. The result is that the rear springs are pulled downward, or compressed, so the rear end of the car moves down, or squats. On braking, the rear of the car moves up, owing to the inertia of the car (Fig. 52-14).

● 52-9 Coil-Spring Front Suspension

The front-suspension system is more complicated than the rear-suspension system. The front-suspension system must allow the wheels to move up and down. It must also allow the wheels to pivot from side to side so that the car can be steered. Figure 52-1 shows a front-suspension system using coil springs. In these *short-arm, long-arm* suspension systems, the upper and lower control arms are of different lengths. The upper control arm is the shorter. Note that the spring is held between the frame at one end and a control arm at the other.

In the system shown in Fig. 52-1, the coil spring is held between a spring seat in the car frame and a lower con-

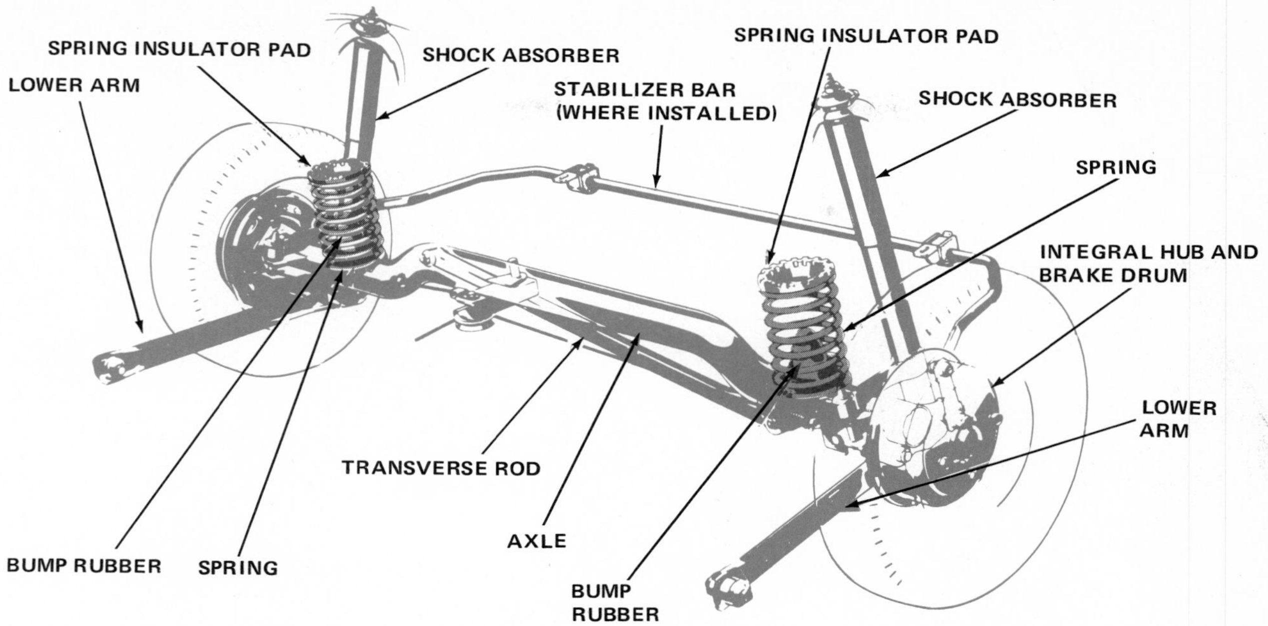

Fig. 52-12 Rear suspension system using coil springs. This is for a front-drive car. (*Ford Motor Company*)

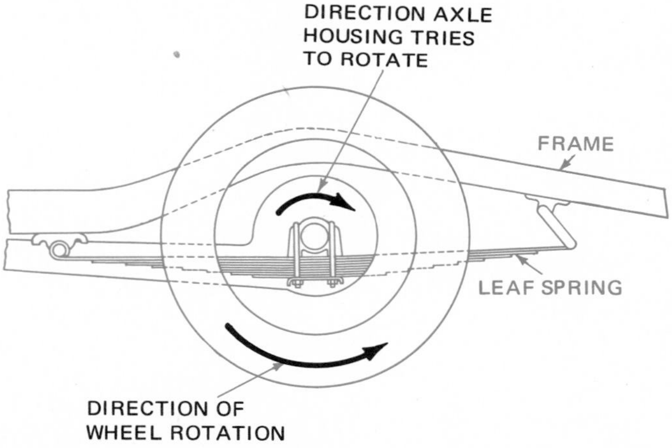

Fig. 52-13 Axle housing tries to rotate in a direction opposite to wheel rotation.

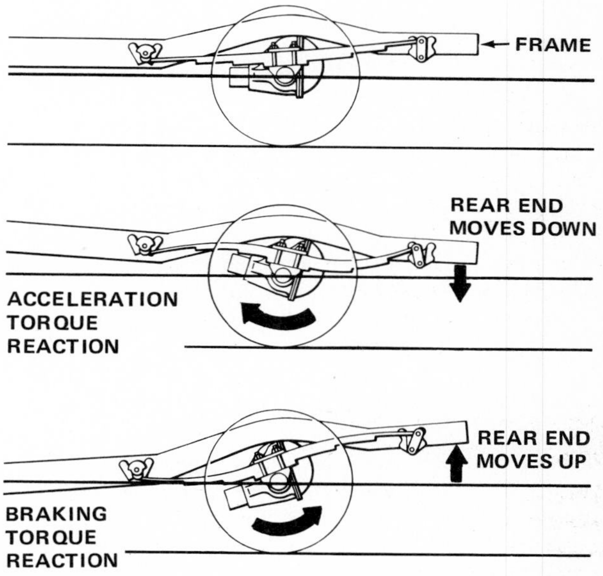

Fig. 52-14 Actions of the spring and rear end when the car is accelerated or braked. (*Ford Motor Company*)

trol arm. The inner ends of both the lower and upper control arms are pivoted on the car frame. The outer ends of the control arms are connected to the steering knuckle. The steering knuckle is attached to the control arms through ball joints (Fig. 52-15). These ball joints allow the steering knuckles to swing to the left or right for steering. In the assembled car, the wheels are mounted on the steering knuckles. Thus, swinging the knuckles from left to right pivots the front wheels so that the car can be steered. We cover steering systems in detail in Chap. 53.

Now let's see how the coil-spring front-suspension system works. Figure 52-16 shows what happens when a front wheel hits a bump in the road. Figure 52-17 shows what happens when a front wheel drops into a hole in the road. When a front wheel hits a bump (Fig. 52-16), the wheel moves up, as shown by the dashed lines. As the wheel moves up, the two control arms pivot upward. This action compresses the spring between the lower control arm and the car frame. When a front wheel hits a hole (Fig. 52-17), the control arms pivot downward. This allows the spring to expand. Notice, on both expansion and compression, how far the spring can allow the wheel and tire to travel, but how little movement is carried to the frame. The passengers enjoy a relatively smooth ride.

In the coil-suspension system shown in Figs. 52-16 and 52-17, the shock absorbers are centered in the springs. We cover shock absorbers in ● 52-18.

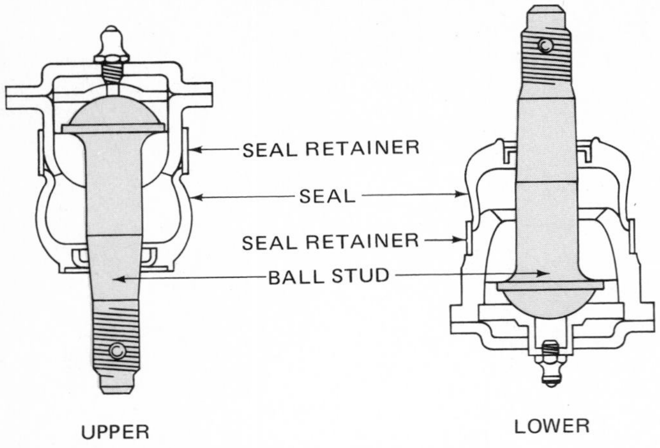

Fig. 52-15 Sectional views of ball joints used in a front-suspension system. (*Chevrolet Motor Division of General Motors Corporation*)

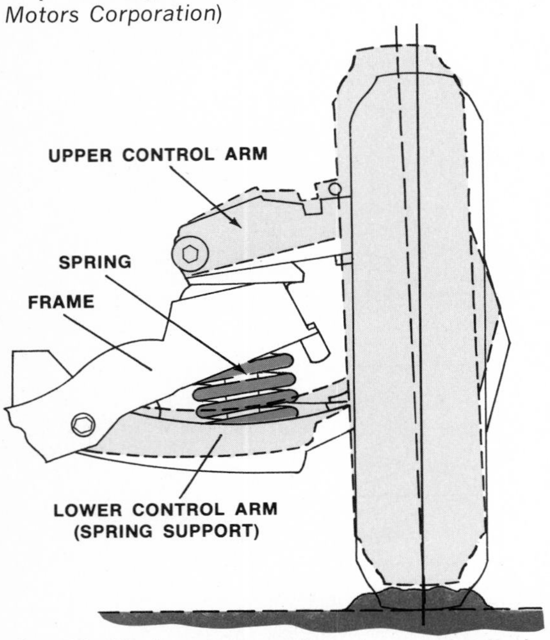

Fig. 52-16 Front suspension at one wheel, showing the actions as the wheel meets a bump in the road. Note how the upward movement of the wheel, shown dashed, raises the lower control arm, causing the spring to compress.

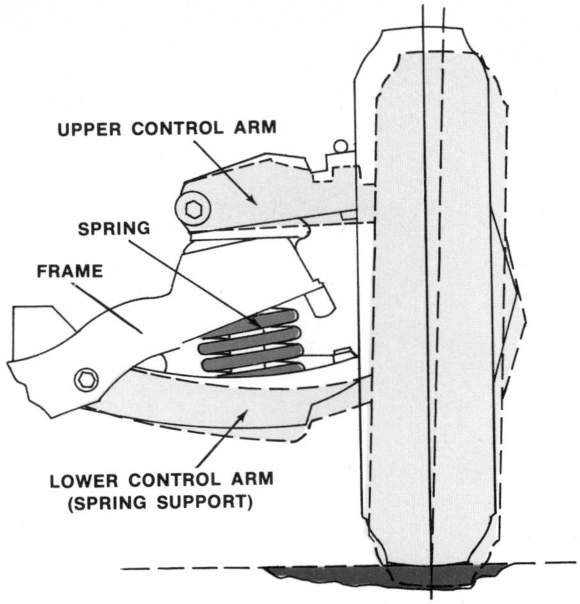

Fig. 52-17 Front suspension at one wheel, showing the actions as the wheel drops into a hole in the road. Note how the downward movement of the wheel, shown dashed, lowers the lower control arm, permitting the spring to expand.

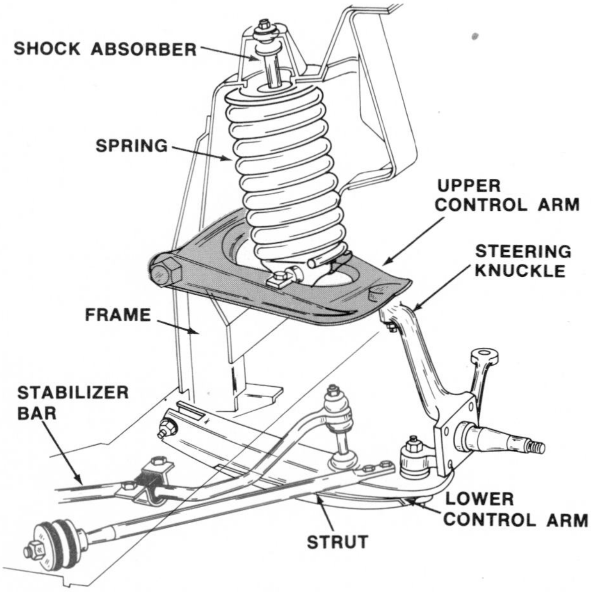

Fig. 52-18 Coil-spring front-suspension system of the type having the spring above the upper control arm.

● 52-10 Coil-Spring Front Suspension—Spring Above Control Arm

Another type of front-suspension system has the coil springs mounted between the upper control arm and a spring tower. The tower is part of the front-end sheet-metal work (see Fig. 52-18). The action of this type of coil-spring front-suspension system is the same as the action of the system explained in ● 52-9. When the wheel hits a bump or a hole, the control arms pivot and compress or expand the spring. Note that the system shown in Fig. 52-18 has a lower control arm with only one point of attachment to the frame. This system uses a strut to prevent the outer end of the lower control arm from swinging forward or backward.

● 52-11 Strut

In many front-suspension systems, the lower control arm has only one point of attachment to the frame. Figure 52-1 shows this type of system. With this system an extra part is required. It is called a *strut* or a *brake-reaction rod*. The strut is fastened between the outer end of the lower control arm and the car frame. The strut prevents the outer end of the lower control arm from swinging forward or backward during braking or when the wheel hits holes or bumps in the road.

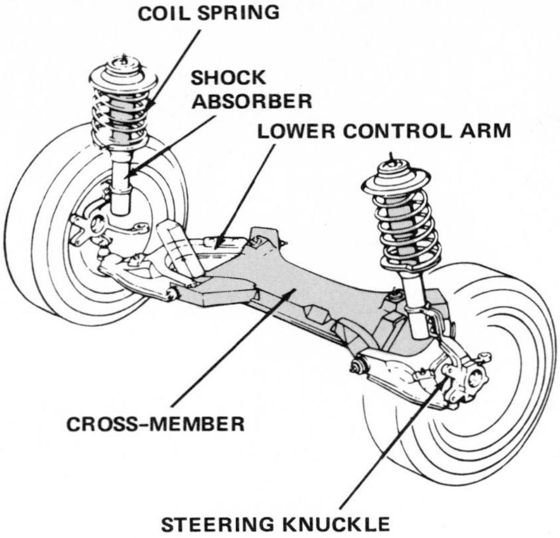

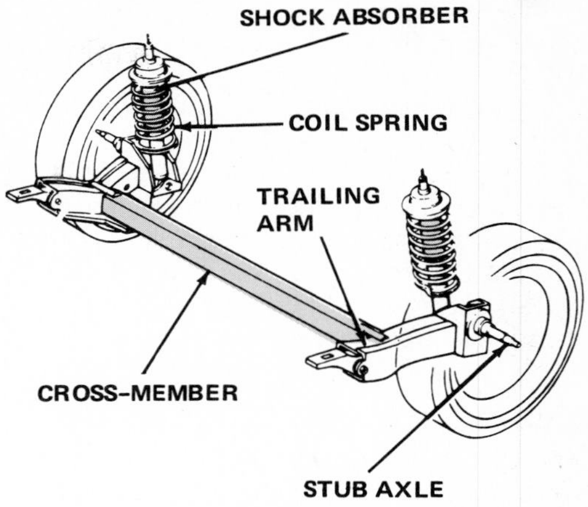

Fig. 52-19 Left, front-suspension system with MacPherson struts. Right, modified MacPherson rear-suspension system. These suspension systems are used on a small front-drive car. (*Chrysler Corporation*)

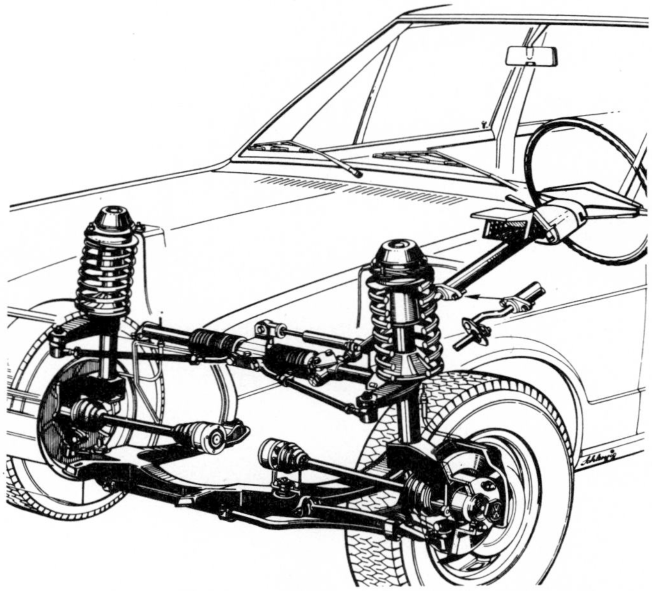

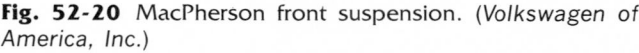

Fig. 52-20 MacPherson front suspension. (*Volkswagen of America, Inc.*)

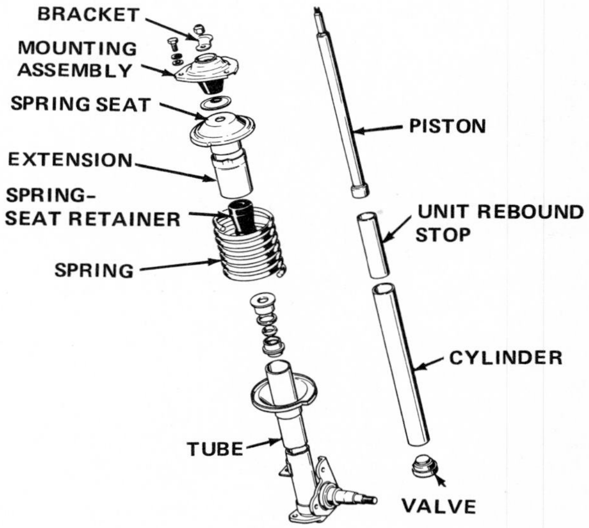

Fig. 52-21 Disassembled MacPherson front suspension. (*Volkswagen of America, Inc.*)

● 52-12 MacPherson Front Suspension

The MacPherson front-suspension system (Fig 52-19) is similar to the coil-spring suspension system shown in Fig. 52-18. The top of the coil spring fits into a tower that is part of the body sheet metal. No upper control arm is needed. The shock absorber is built into a strut that connects the lower control arm to the mounting assembly in the tower. Figures 52-19 and 52-20 show the MacPherson front suspension in a car. Figure 52-21 shows the strut partly disassembled. The MacPherson front-suspension system is widely used on small cars. A variation is also used for the rear suspension in some cars.

● 52-13 Stabilizer Bar

In Fig. 52-1, you can see a part labeled "stabilizer bar." This bar is a long steel rod, fastened at each end to the two lower control arms. The stabilizer bar is sometimes called a *sway bar*.

When the car goes around a curve, centrifugal force tends to keep the car moving in a straight line. The car therefore "leans out" on the turn. This *lean out* is also called *body roll*. With lean out, or body roll, additional weight is thrown on the outer spring. This puts additional compression on the outer spring, and the control arm pivots upward. As the control arm pivots upward, it carries its end of the stabilizer bar up with it. At the inner wheel on the turn, there is less weight on the spring. Weight has shifted to the outer spring because of centrifugal force. Therefore, the inner spring tends to expand. The expansion of the inner spring tends to pivot the lower control arm downward. As this happens, the lower control arm carries its end of the stabilizer bar downward.

The outer end of the stabilizer bar is carried upward by the outer control arm. The inner end is carried downward. This combined action twists the stabilizer bar. The resistance of the bar to twisting combats the tendency of the car to lean out on turns. In other words, there is less body roll than there would be without the stabilizer bar.

52-14 Torsion-Bar Front Suspension

In the torsion-bar front-suspension system, two long steel bars serve as the springs (see Figs. 52-2 and 52-22). The rear ends of the bars are locked to a cross member of the frame. The front ends of the bars are attached to the lower control arms. In operation, the lower control arms pivot up and down, twisting the torsion bars. The effect is very similar to the actions of the coil and leaf springs. The car weight places an initial twist on the torsion bars, just as it places an initial compression on the coil springs of cars with coil-spring suspension. Figure 52-23 shows the car-leveling device at the rear end of one torsion bar. Turning the height-adjustment bolt causes the hub-and-anchor assembly to turn. This rotates the rear end of the torsion bar so that the front end of the car is raised or lowered.

Torsion bars that are mounted from front to rear, or the long way of the car (Fig. 52-22), are called longitudinal torsion bars. Some torsion bars on smaller cars are mounted sideways, or transversely, in the car (Figs. 52-24 and 52-25). They work the same way in either location. They twist more or less as the wheels meet bumps and holes in the road.

52-15 Geometry of Front Suspension

"Geometry" means the relationship among angles and plane surfaces. So far as the front suspension is concerned, it refers to the angles among the front-suspension parts as the wheels move up and down. This is shown in Fig. 52-26. Note that the lower arm is considerably longer than the upper arm. The points of attachment are arranged to allow the steering-knuckle support to move up and down almost vertically (due to the shorter arc of the upper arm) as the wheels meet road holes and bumps. At the same time, the steering knuckle and wheel can pivot on the ball joints to steer the car (see Chap. 53).

Note that the steering knuckle is in nearly vertical alignment as it moves up and down. This keeps the wheel in almost vertical alignment also. This is desirable from the standpoint of steering control and tire wear. The geometry of the front suspension produces this vertical alignment.

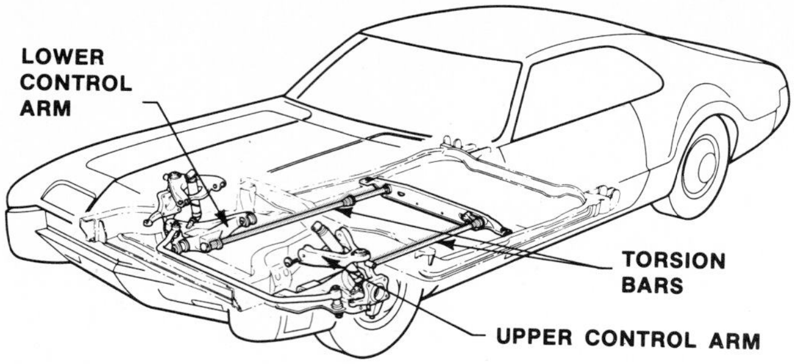

Fig. 52-22 Front suspension of a front-drive car using torsion bars. The bars are locked at the rear to the frame. They are attached at the front to the inner ends of the lower control arms. They twist varying amounts as varying loads are applied. The front wheels can move up and down, as with other suspension systems. (*Oldsmobile Division of General Motors Corporation*)

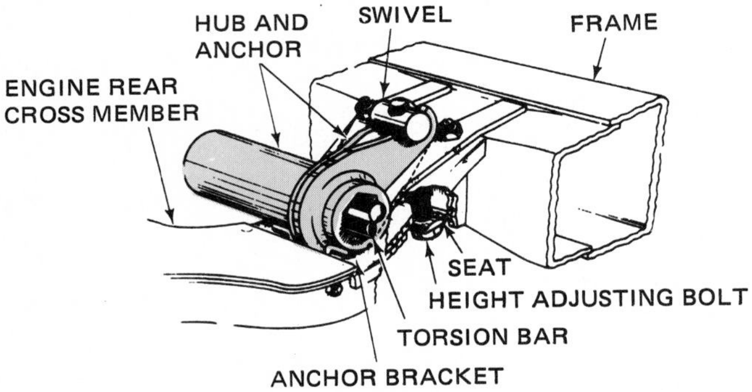

Fig. 52-23 Method of attaching the rear of a torsion bar to the frame. The hub and anchor, swivel, and adjusting bolt are for adjusting the height of the car. (*Chrysler Corporation*)

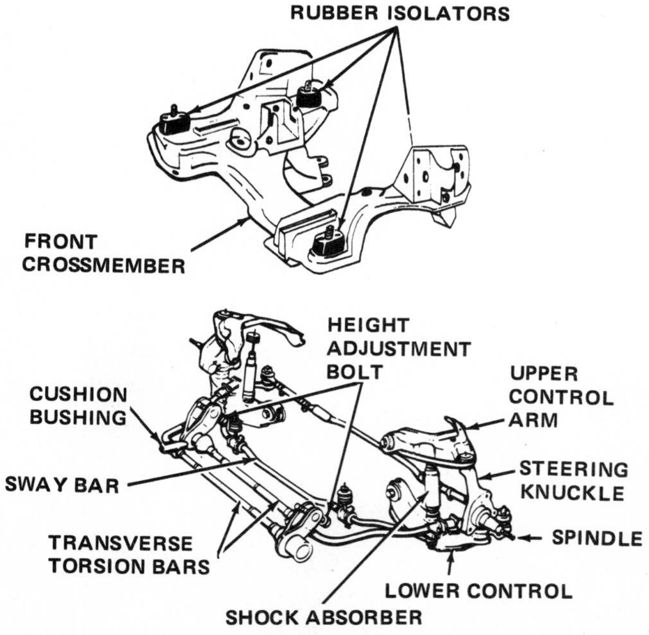

Fig. 52-24 Front-suspension system for a small car using transverse torsion bars. (*Chrysler Corporation*)

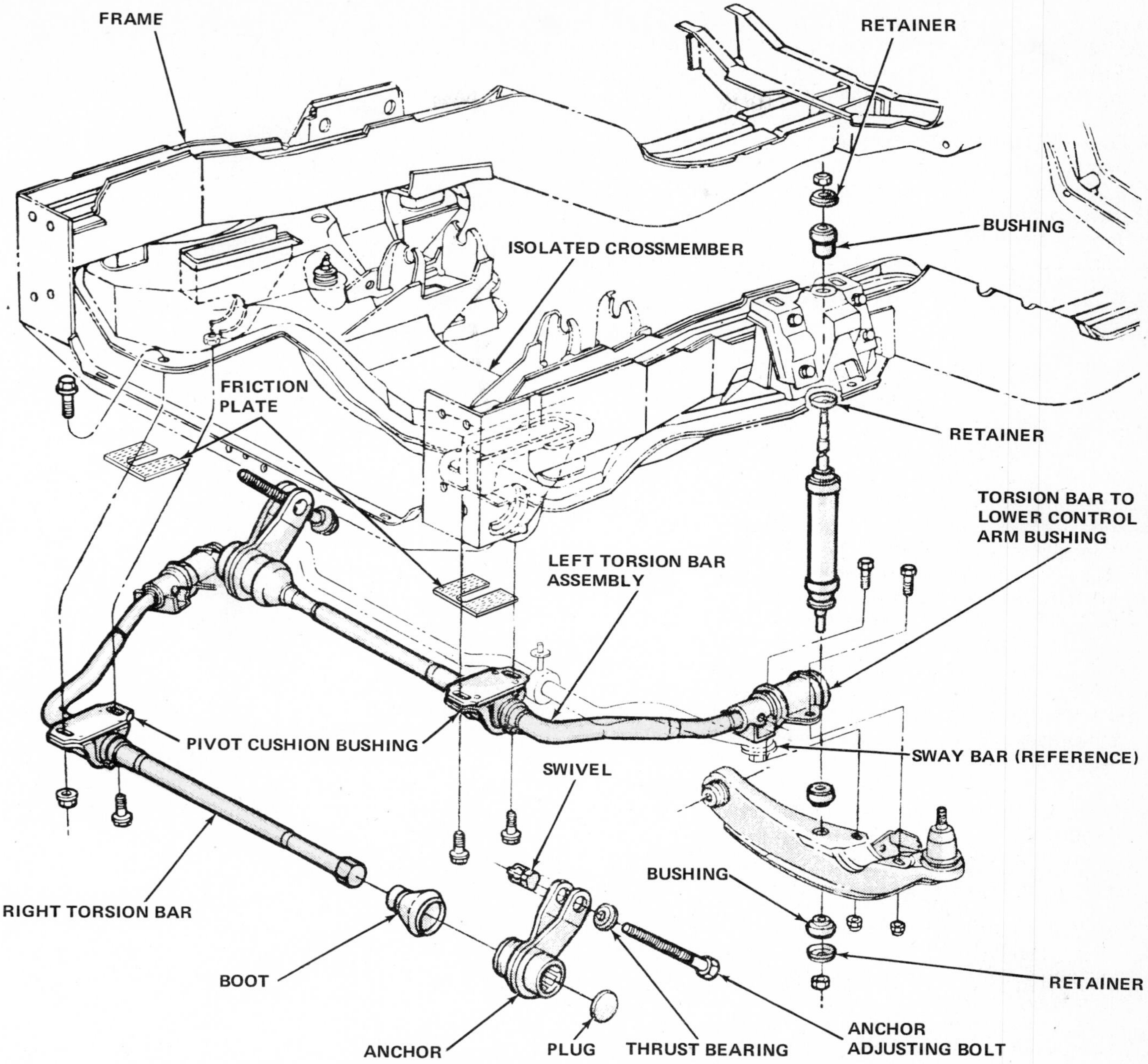

Fig. 52-25 Disassembled transverse torsion-bar front-suspension system. (*Chrysler Corporation*)

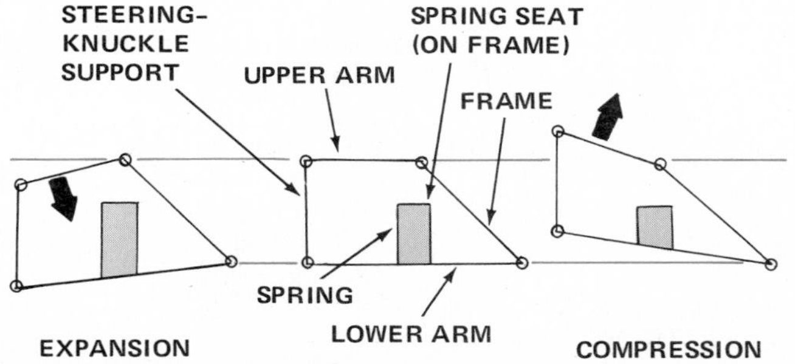

Fig. 52-26 Geometry of a front-suspension system. As the steering-knuckle support and wheel move up and down, the upper and lower arms pivot on the frame, causing the spring to expand or compress.

● 52-16 Air Suspension

In air suspension, the four conventional springs are replaced by four air bags, or air-spring assemblies. Essentially, the air-spring assembly is a flexible bag enclosed in a metal dome or girdle. The bag is filled with compressed air which supports the weight of the car. When the wheel meets a bump in the road, the air is further compressed to absorb the shock. An air compressor, driven by the engine, maintains pressure in the air bags.

A variation of air suspension uses four gas-filled spring units (called *Hydragas* springs), one at each wheel (Fig. 52-27). Each unit has a sealed chamber containing a quantity of nitrogen gas at high pressure. Below this chamber is a displacement chamber filled with a water-based fluid. When the wheel meets a bump, the fluid is pushed upward, compressing the gas. In addition, the two units on each side are interconnected front to back. Therefore, when the left front wheel meets a bump, for example, part of the fluid from the left front unit is forced through a pipe to the left rear unit. This raises the left rear wheel also. The shock of the bump is thus distributed between the left front and left rear wheels. This is said to improve the ride.

52-17 Sprung and Unsprung Weight

The term "sprung weight" refers to the part of the car that is supported on springs. The term "unsprung weight" refers to the part that is not. The frame and the parts attached to the frame are sprung. That is, their weight is supported on the car springs. However, the wheels and wheel axles (and rear-axle housing and differential) are not supported on springs. They represent unsprung weight. Generally speaking, unsprung weight should be kept as low as possible. The reason is that unsprung weight increases the roughness of the ride. For example, consider a single wheel. If it is light, it can move up and down without causing much reaction of the car frame. But if the weight of the wheel is increased, then its movement becomes more noticeable to the car occupants. To take a ridiculous example, suppose the unsprung weight at the wheel is equal to the sprung weight above the wheel. In such a case, the sprung weight tends to move as much as the unsprung weight. The unsprung weight, which must move up and down over road irregularities, tends to cause a like motion of the sprung weight. This is the reason for keeping the unsprung weight as low as possible.

52-18 Shock Absorbers

Shock absorbers are necessary because springs do not "settle down" fast enough. In other words, after a spring has been compressed and released, it continues to shorten and lengthen, or *oscillate*, for a time.

Let's see what would happen if the car springs were not controlled. A wheel hits a bump. The spring compresses. Then the spring expands after the wheel passes the bump. The expansion of the spring causes the car to be thrown upward. Now, having over-expanded, the spring shortens again. This action causes the wheel to momentarily leave the road, and the car drops down. The action is repeated until the oscillations gradually die out.

Such spring action on a car would produce a very bumpy and uncomfortable ride. It could also be dangerous, because a bouncing wheel makes the car impossible to control. This would be very dangerous on a curve. Therefore, a device is needed to control the oscillating action of the spring. This device is the shock absorber.

52-19 Operation of the Shock Absorber

The shock absorber absorbs the shock of the wheel meeting a bump or hole. As soon as the wheel passes the hole

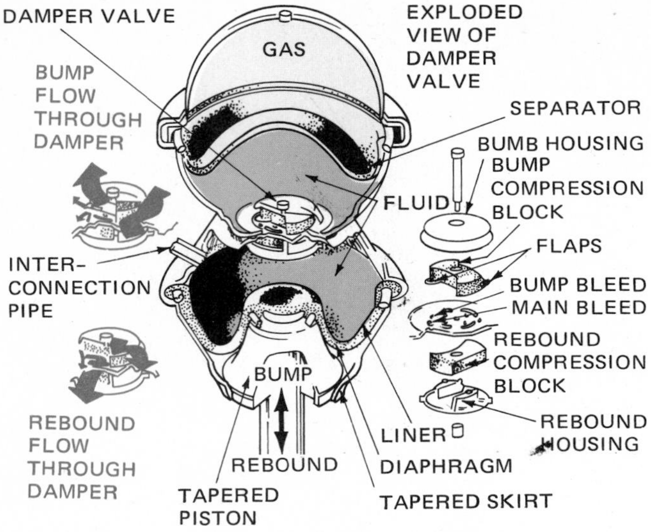

Fig. 52-27 Partial cutaway view of a Hydragas spring. (*British Motor Corporation*)

or bump, the shock absorber returns the wheel to contact with the road and keeps the wheel from bouncing.

The most commonly used shock absorber is the direct-acting, or telescope, type (Fig. 52-28). Several illustrations in this chapter show the locations of the shock absorbers at the front and back of the car. In operation, the shock absorbers lengthen and shorten as the wheels meet irregularities in the road. As they do this, a piston inside the shock absorber moves in a chamber filled with fluid. The fluid is therefore put under high pressure and is forced to flow through small openings. The fluid can only pass through the openings slowly. It thus slows the piston motion and places a restraint on spring action.

Figure 52-28 shows a shock absorber in cutaway view. Figure 52-29 shows two sectional views of a shock absorber. Actions during compression and rebound are shown. With either action, the piston is moving. The fluid in the shock absorber is being forced through small openings and is restraining spring movement. Actually, there are small valves in the shock absorber that open when the internal pressure becomes excessive. When the valves are open, a slightly faster spring movement occurs, but restraint is still imposed on the spring.

52-20 Automatic Level Control

The automatic-level-control system takes care of changes in the load at the rear of the car. In a car without automatic level control, adding weight at the rear will make the rear end of the car squat. This changes the handling characteristics of the car. It also causes the headlights to point upward. The automatic level control prevents this by automatically raising the rear end of the car to level when a load is added. The system also automatically lowers the rear end to level when the load is removed.

The automatic-level-control system includes a compressor, an air-reverse tank, a height-control valve, and two special shock absorbers with built-in air chambers

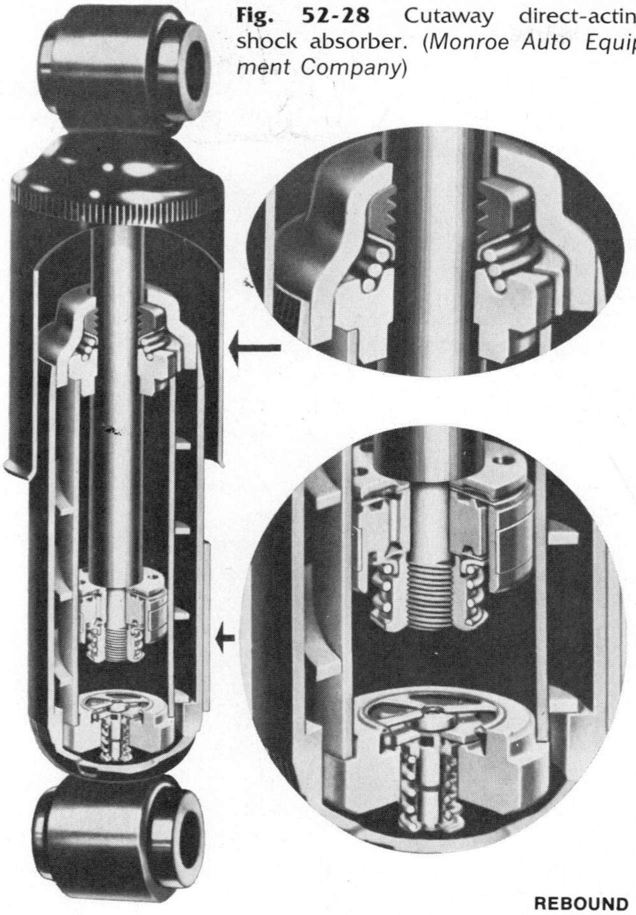

Fig. 52-28 Cutaway direct-acting shock absorber. (*Monroe Auto Equipment Company*)

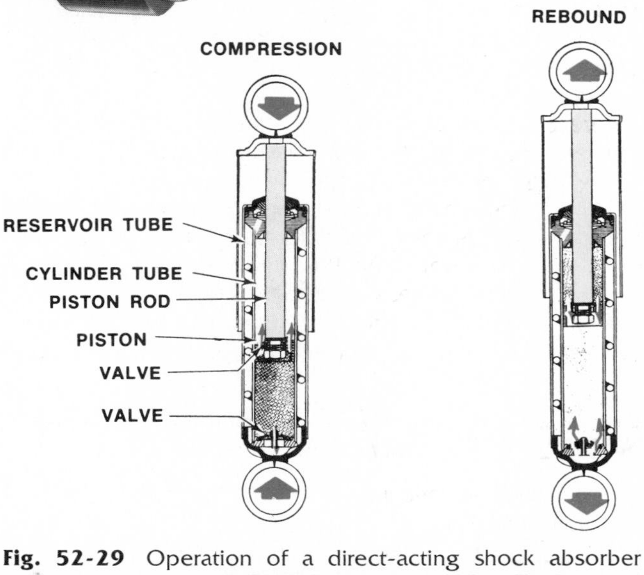

Fig. 52-29 Operation of a direct-acting shock absorber during compression (left) and extension (right). Fluid movement is shown by the arrows. (*Chrysler Corporation*)

(Figs. 52-30 and 52-31). The compressor is operated by engine intake-manifold vacuum. The vacuum operates a pump that builds up air pressure in the reserve tank. When a load is added to the rear of the car, additional air passes through the height-control valve to the two rear shock absorbers. This raises the air domes (Fig. 52-31) The movement brings the rear up to normal level.

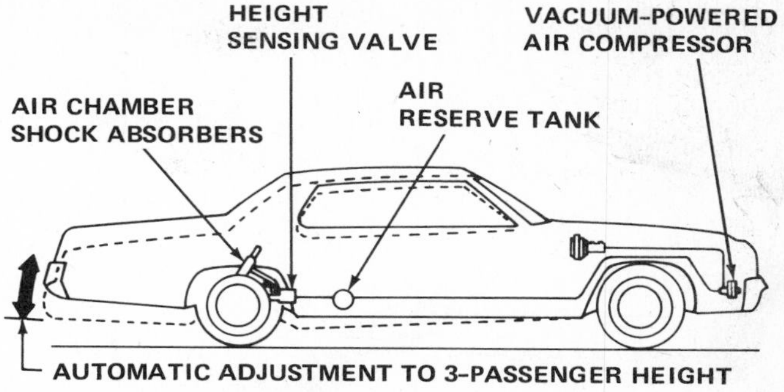

Fig. 52-30 An automatic-level-control system. The dotted lines show the lower height of the car before the automatic level control restores the correct height.

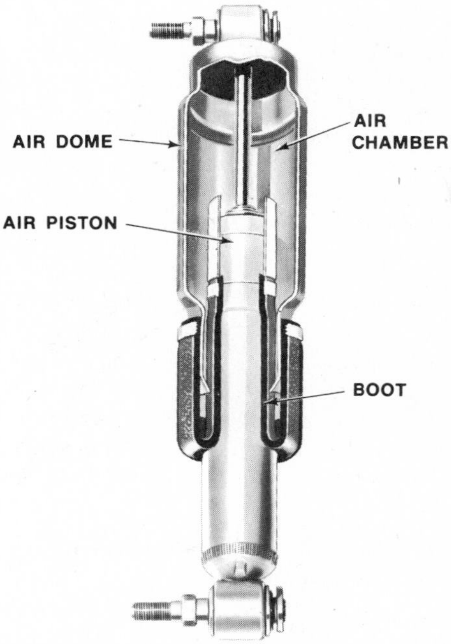

Fig. 52-31 Cutaway view of a special shock absorber (called a *Superlift* by the manufacturer) used in the automatic level control.

The height-control valve (Fig. 52-32) has a linkage to the rear-suspension system. When this linkage is operated by the addition of a load, it opens the intake valve, thus admitting air to the shock absorber. Note that the height-control arm is moved upward by the addition of a load. When the load is removed, the rear of the car moves up, and the height-control valve operates. The exhaust valve opens to allow air to exit from the shock absorber. The rear of the car then settles down to normal height.

The height-control valve has a time-delay mechanism that allows the valve to operate only after several seconds. This mechanism prevents fast valve action, which could operate the system after each bump or hole in the road. In other words, the automatic-level-control system works only when loads are added or removed from the rear of the car.

Figure 52-33 is the wiring circuit of an automatic electronic control introduced in 1978 by General Motors. This system has a height sensor that operates electronically rather than mechanically. Also, the compressor is different and is assembled with an air dryer (Fig. 52-34). The electronic height sensor has a shutter connected to the control arm. The control arm is connected to a suspension arm so that it moves as the height of the car rear end changes. The shutter can interrupt a beam of light inside the height sensor if the height is not correct. If it is either too low or too high, the sensor triggers either the compressor relay or the solenoid exhaust valve. If the height is low, it triggers the compressor relay. This causes the compressor relay to close its points so the compressor is connected to the battery. It runs and supplies air to the rear shock absorbers. This raises the rear of the car to level.

If the height is too high, as it would be after a load is removed from the rear seat or trunk, the sensor triggers the solenoid exhaust valve. The solenoid then opens the valve to allow some of the air in the shock absorbers to escape. This lowers the car rear end to level.

The air dryer (Fig. 52-34) has a supply of chemical which absorbs any moisture in the air being pumped in by the compressor. This assures a supply of dry air for the shock absorbers. When air is released from the shock absorbers by the solenoid-exhaust-valve action, this air passes back through the air dryer where it absorbs the moisture from the chemical as it exits. The air dryer also has a valve that maintains some air pressure in the shock absorbers to improve ride characteristics.

Figure 52-35 shows an electronic automatic-level-control system that is similar to the one shown in Fig. 52-33. The major difference is that the electronic control is installed inside one of the rear shock absorbers. A photo-optic sensor (an electric eye) is built into the shock absorber. The sensor "tells" the electronic control module when any change in height has occurred. Then the electronic module either sends air to the shock absorbers or releases air from them to adjust the height to the

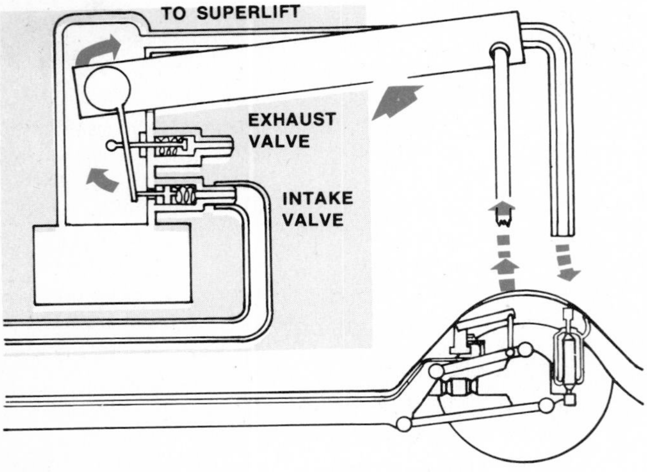

Fig. 52-32 Schematic view of a height-control valve and the actions that take place when a load is added to the car. The components are not in proportion. (*Buick Motor Division of General Motors Corporation*)

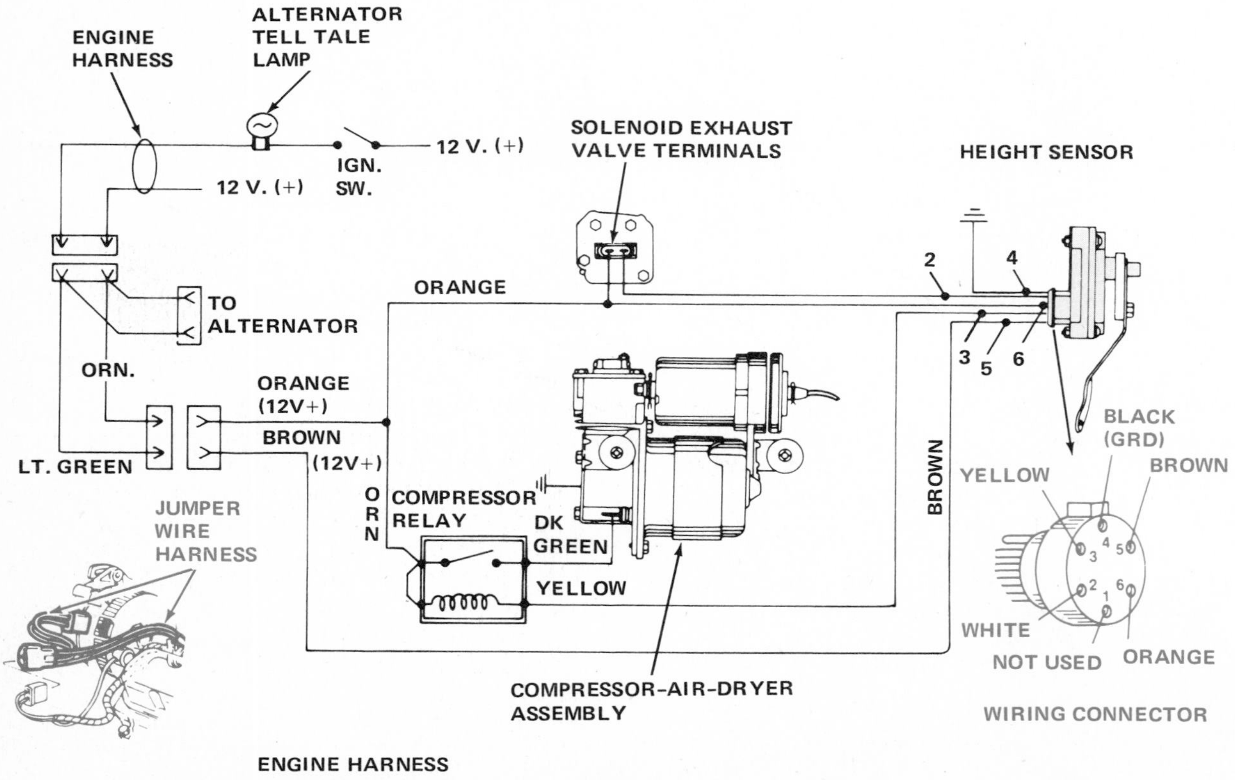

Fig. 52-33 Schematic wiring diagram of an electronic automatic-level-control system using an electronic height sensor. (*General Motors Corporation*)

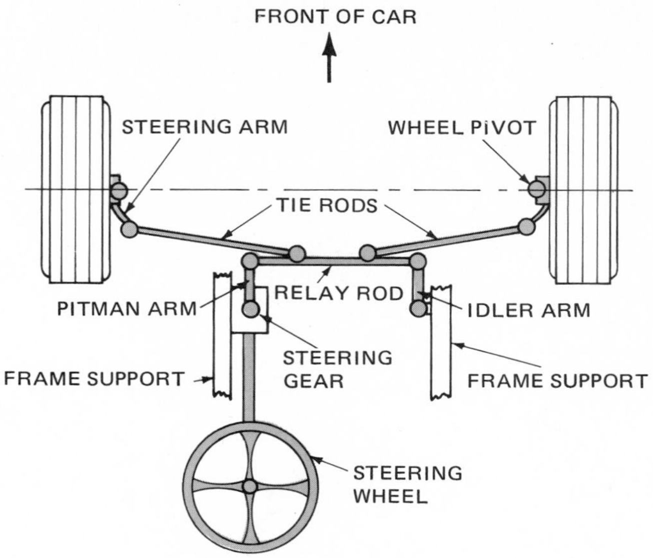

Fig. 53-1 Simplified drawing of a steering system.

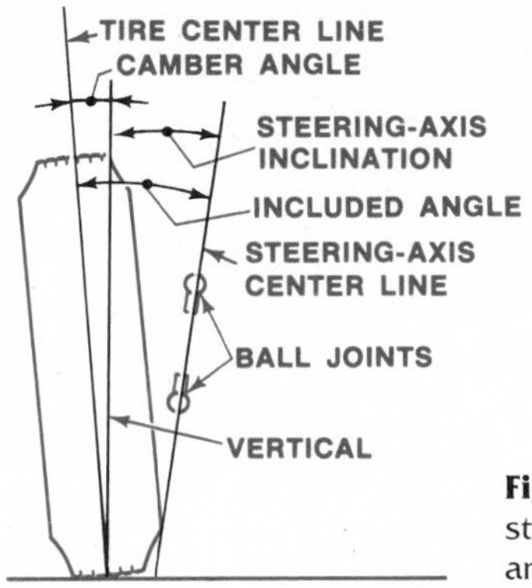

Fig. 53-2 Camber angle and steering-axis inclination. The angles are exaggerated.

angles during an alignment job. Each angle is important. If the angles are off, the car is harder to control, and the tires wear rapidly.

● 53-3 Camber

Camber is the tilting of the front wheels from the vertical (straight up and down). When the wheel tilts outward at the top, the camber is positive (see Fig. 53-2). When the tilt is inward, the camber is negative. The amount of tilt is measured as a number of degrees from the vertical. This measurement is called the *camber angle*. On a moving car, an average running camber of zero provides the longest tire life.

The purpose of camber is to give the wheels a slight outward tilt to start with. Then, when the car is loaded and moving, the load just about brings the wheels to vertical again. If the wheels started with no camber, then loading the car could give the front wheels a negative camber. That is, the wheels would tilt inward.

Any amount of camber—positive or negative—will cause uneven tire wear. The tilt puts more of the load on one side of the tread, rather than centering it. When the car is rolling forward, the car pulls toward the wheel with the most positive camber.

● 53-4 Steering-Axis Inclination

Steering-axis inclination is the amount the ball joints are tilted inward from the vertical. You can see this angle in the simplified view in Fig. 53-2. An actual cutaway front suspension at one wheel is shown in Fig. 53-3. You measure the steering-axis inclination by drawing a line through the centers of the two ball joints. Then you measure how many degrees this line is off from the vertical. There are three reasons for having steering-axis inclination:

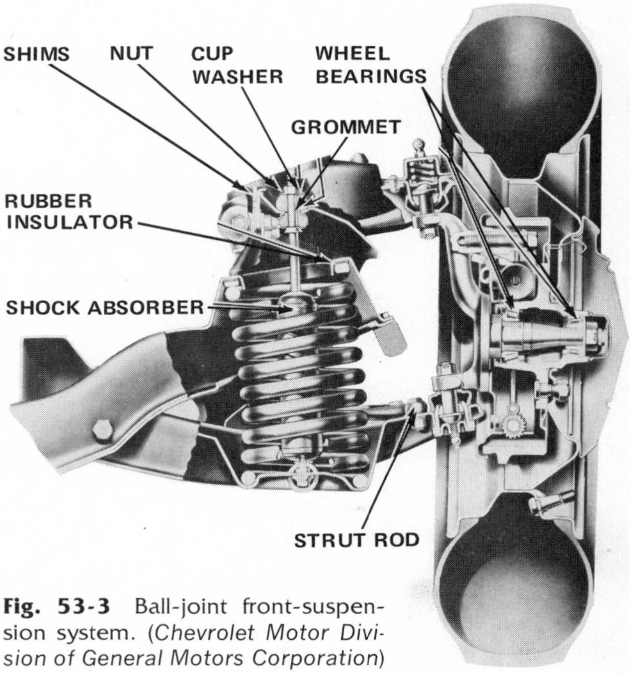

Fig. 53-3 Ball-joint front-suspension system. (*Chevrolet Motor Division of General Motors Corporation*)

1. It helps provide steering stability.
2. It reduces steering effort.
3. It reduces tire wear.

The inward tilt, or inclination, of the steering axis tends to keep the wheels pointed staight ahead. It also helps recovery, or the return of the wheels to the straight-ahead position, after a turn. To demonstrate this, cut out a cardboard disk and a cardboard brace, as shown in Fig. 53-4. Tape them together, and attach them to a pencil with a rubber band, as shown. The cardboard disk represents the wheel. The pencil represents the ball-joint center line. The brace at the top holds the disk and the pencil apart so as to get steering-axis inclination. Of course, the angle is greatly exaggerated.

Now, hold the pencil at an angle to the table top so that the wheel is vertical, as shown in Fig. 53-5. Then rotate the pencil, but do not change its angle. Notice that, as you do this, the wheel is carried around and down toward the table top. If the wheel could not move down, the pencil would be moved up.

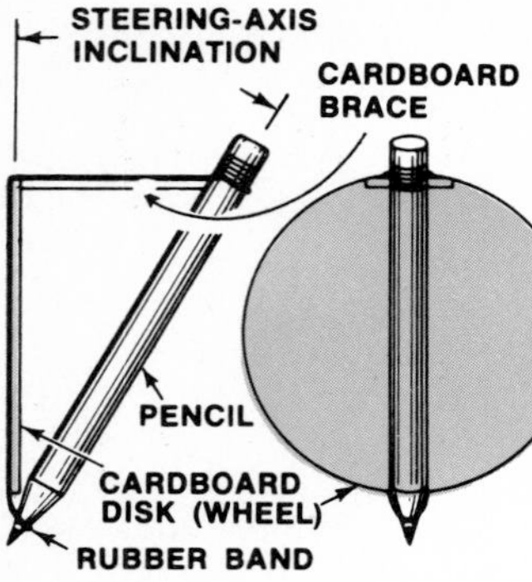

Fig. 53-4 A cardboard disk to serve as the wheel, a rubber band, and a cardboard brace demonstrate the effects of steering-axis inclination.

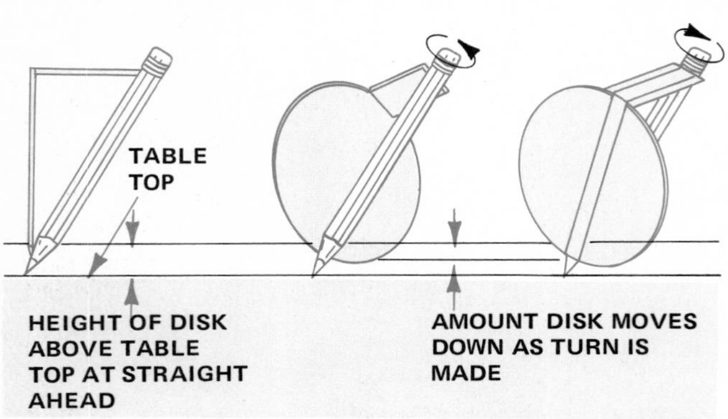

Fig. 53-5 The cardboard disk represents the left front wheel as viewed from the driver's seat.

This actually is what happens in the car. The wheel is in contact with the ground. It cannot move down. Therefore, when the wheel is swung away from straight ahead, the end of the wheel spindle moves down. This forces the ball joints to move up. The result is that the car body is actually lifted. The lift is not very much—an inch [25.4 mm] or less. But it is enough to help bring the wheels back to the straight-ahead position when the car completes a turn. That is one purpose of steering-axis inclination. When the front wheels are turned away from straight ahead, the front end of the car is raised slightly. Then the weight of the car helps to bring the wheels back to the straight-ahead position when the turn is completed. Steering-axis inclination provides steering stability and reduces the effort required to return the steering wheel to staight ahead. Steering-axis inclination is not adjustable. It can be measured; if incorrect, the steering knuckle is bent.

53-5 Included Angle

The included, or combined, angle is the camber angle plus the steering-axis angle (Fig. 53-2). The included angle is important because it determines the point of intersection of the wheel and the steering-axis center lines (Fig. 53-6). This, in turn, determines whether the wheel will tend to toe out or toe in. "Toe-out" is a term used to describe the tendency of the wheel to point outward. A soldier standing at attention has his or her feet "toed out." "Toe-in" is just the opposite; a pigeon-toed person turns the toes of his or her feet inward. Likewise, a wheel that toes in tends to point inward as it rolls. Figure 53-11 shows what toe-in is on a vehicle. The tire on a wheel that is toed in or toed out will wear more rapidly. The tire has to go in the direction in which the car is moving. But since it is not pointed in that direction (it is toed out or toed in), it is dragged sideways as it rolls forward. The more toe-out or toe-in, the more it is dragged sideways, and the faster the tire wears.

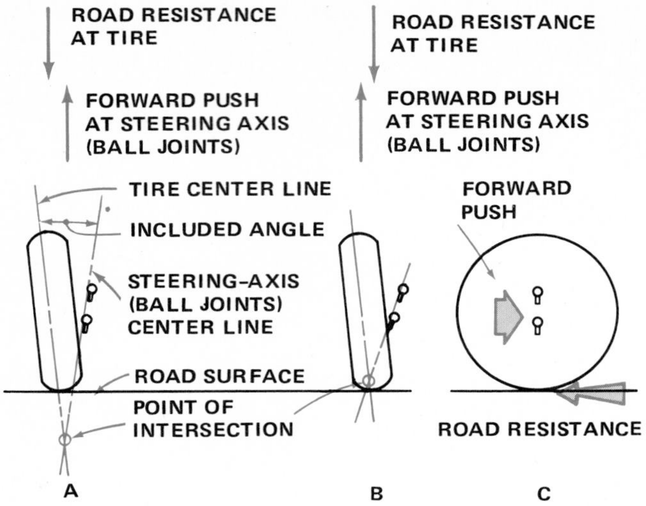

Fig. 53-6 Effect when the point of intersection is below the road surface (*A*), and above the road surface (*B*). The left front wheel as viewed from the driver's seat is shown in *A* and *B*. *C* is a side view of the wheel to show two forces acting on the wheel and ball joints.

When the point of intersection (Fig. 53-6) is below the road surface, then the wheel tends to toe out. This is because the forward push (which is through the steering axis or centerlines of the ball joints) is inside the tire centerline at the road surface. In the right-hand picture in Fig. 53-6, the two opposing forces working on the wheel are shown. One is the forward push through the ball joints. The other is the road resistance to the tire. If these two forces are exactly in line, then the wheel has no tendency to toe out or toe in. The two forces are in line with each other only when the point of intersection is at the road surface. When it is below the road level, as shown at *A* in Fig. 53-6, then the wheel attempts to swing outward, or toe out. When the point of intersection is above the road level (*B* in Fig. 53-6), then the wheel attempts to swing inward, or toe in.

53-6 Caster

In addition to being tilted inward toward the center of the car, the steering axis may also be tilted forward or backward from the vertical (Fig. 53-7). Backward tilt from the vertical is called *positive caster*. Positive caster aids directional stability, since the centerline of the ball joints passes through the road surface ahead of the centerline of the wheel. Thus, the push on the ball joints is ahead of

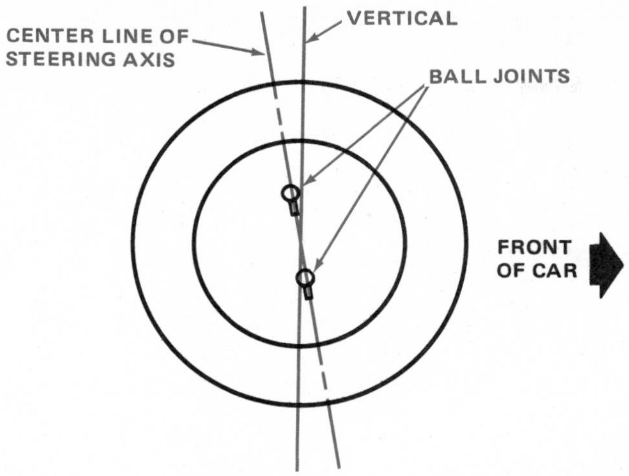

Fig. 53-7 Left front wheel (as viewed from driver's seat). The view is from inside, so that the backward tilt of the steering axis from the vertical can be seen. This backward tilt is called *positive caster.*

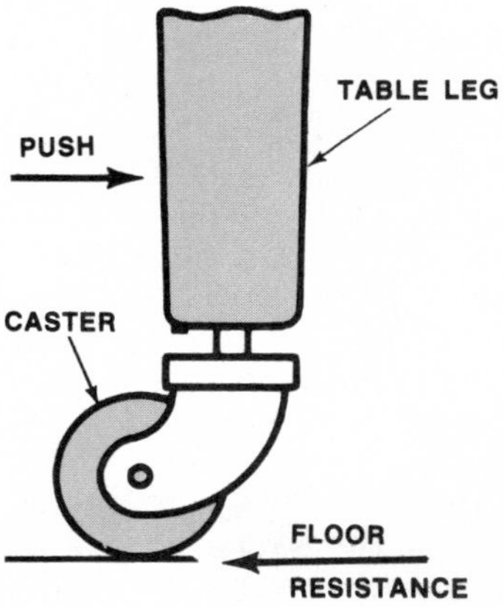

Fig. 53-8 The wheel of the caster trails behind and follows in the direction of the push when the table leg is moved.

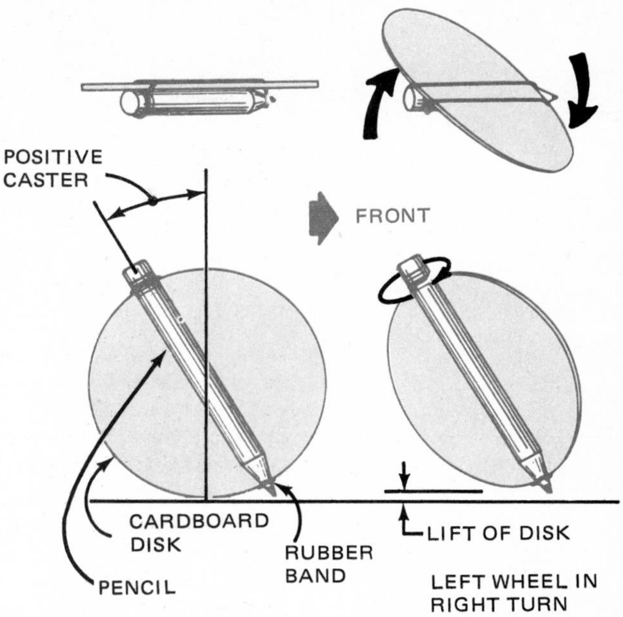

Fig. 53-9 A cardboard disk, a pencil, and a rubber band show the effects of positive caster in a right turn. The disk represents the left front wheel.

the road resistance to the tire. The tire is trailing behind, just as the caster on a table leg "tails behind" when the table is pushed (Fig. 53-8).

Positive caster helps to keep the wheels pointed straight ahead. This helps steering stability. It helps to overcome any tendency for the car to wander away from straight ahead.

Positive caster also increases steering-wheel returnability. Because of this, cars with power steering often have slightly more positive caster than manual-steering cars. The additional positive caster helps overcome the tendency of power steering to hold the wheels in a turn. But because of the power steering, the driver does not notice that additional positive caster requires greater steering effort. Cars with manual steering often have a small negative caster.

Caster has another effect that is important. When both front wheels have positive caster, the car tends to roll out or lean out on turns. But if the front wheels have negative caster, then the car tends to bank, or lean in, on turns. Use a pencil, rubber band, and cardboard disk to demonstrate why this is so (Fig. 53-9). Fasten the cardboard disk and the pencil together as shown. The disk represents the left front wheel. Note that we do not include any steering-axis inclination here; we want to show only the effect of positive caster. Hold the disk vertical with the pencil at an angle, so that both the pencil point and the edge of the disk rest on the table top. Now, rotate the pencil as shown. Note that the disk is lifted from the table top. Actually, in the car, the wheel (disk) could not be lifted. Instead, the ball joints (pencil) would move down. In other words, on a right turn, the left side would drop.

Now, see what happens at the right front wheel (Fig. 53-10). As the right turn is made, the wheel pivots on the road surface. This causes the ball joints (pencil) to be lifted. The right side of the car is lifted.

The left side of the car is lowered and the right side of the car is lifted as a right turn is made (as described above). This makes the car roll, or lean out, on the turn. This is just the opposite of what is most desirable, since it adds to the effect of centrifugal force on the turn. By using negative caster (tilting the steering axis forward), the car can be made to lean in on a turn (and thus decrease the effect of centrifugal force). For instance, with negative caster the left side of the car would lift during a right turn. The right side of the car would drop. This would combat the roll-out effect of centrifugal force.

There is another important effect that caster has. Positive caster tends to make the front wheels toe in. With positive caster, the car is lowered as the wheel pivots inward. Thus, the weight of the car is always trying to make the wheel toe in. With negative caster, the wheels would tend to toe out.

As we mentioned, positive caster increases the effort required to steer. Positive caster tends to keep the wheels straight ahead. To make a turn, this tendency must be overcome. Note, too, that steering-axis inclination also

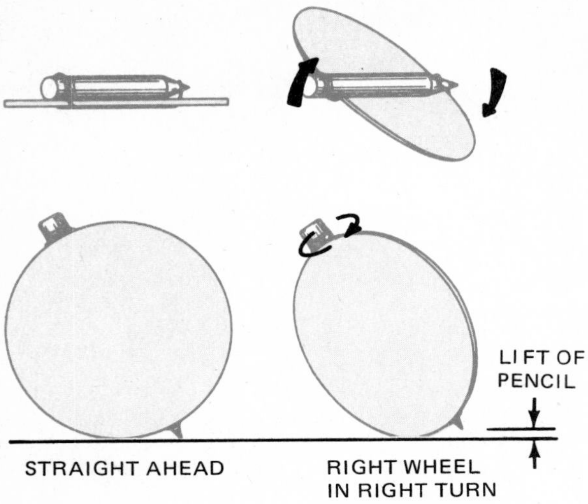

Fig. 53-10 Effect of positive caster on the right front wheel during a right turn.

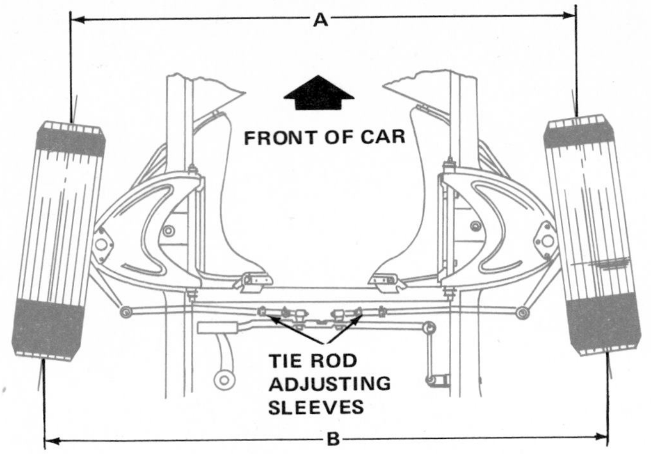

Fig. 53-11 Front-wheel toe-in, shown greatly exaggerated. The wheels are viewed from above, and the front of the car is at the top. (*Oldsmobile Division of General Motors Corporation*)

tends to keep the wheels straight ahead. Thus, to make a turn, the effects of both caster (when positive) and steering-axis inclination must be overcome. Late-model vehicles, and particularly heavy-duty trucks, tend toward a negative caster. This makes steering easier, and there is still sufficient tendency toward recovery, or the return of the wheels to straight ahead. This is provided by steering-axis inclination.

53-7 Toe-In

Toe-in is the pointing inward of the front wheels. On a car with toe-in (Fig. 53-11), the distance between the front wheels is less at the front (A) than at the rear (B). The actual amount of toe-in is normally only a fraction of an inch (a few millimeters). The purpose of toe-in is to ensure parallel rolling of the front wheels, to stabilize steering, and to prevent sideslipping and excessive wear of the tires. Toe-in is set with the car standing still. Then, when the car starts to move forward, the resistance of the road against the tires causes the wheels to swing into a parallel position. In other words, even though the wheels are set to toe in slightly when the car is standing still, they tend to roll parallel when the car is moving forward.

53-8 Toe-Out on Turns

Toe-out on turns is also called *steering geometry*. It refers to the difference in the angles between the two front wheels and the car frame during turns. The inner wheel rotates on or follows a smaller radius than the outer wheel when the car is rounding a curve. Thus, its axle must be at a sharper angle with the car frame. That is, it must toe out more. This condition is shown in Fig. 53-12. When the front wheels are steered to make the turn illustrated, the inner wheel turns to an angle of 23 degrees with the car frame. But the outer wheel turns to only 20 degrees with the car frame. This permits the inner wheel

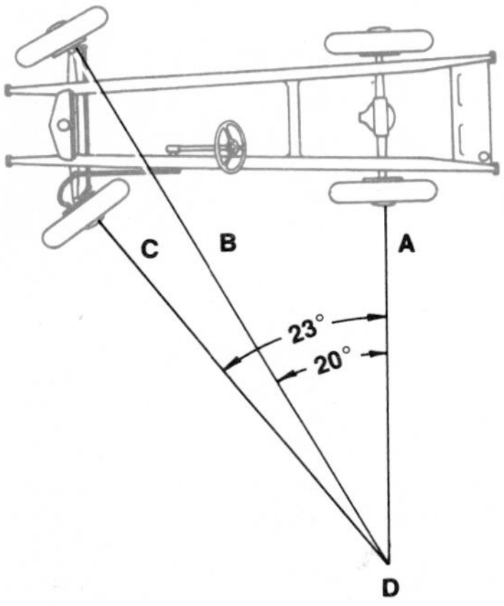

Fig. 53-12 Toe-out on turns.

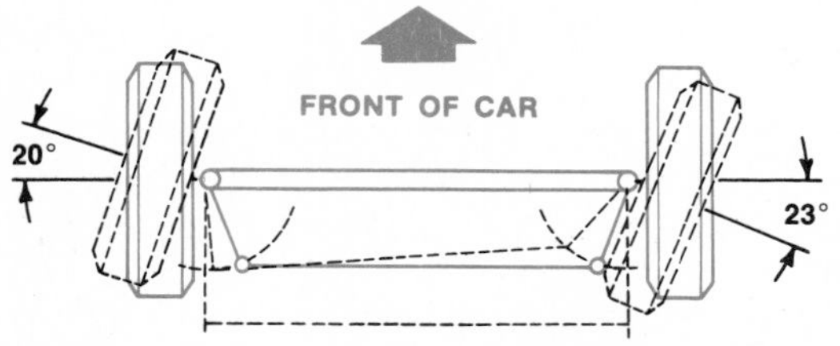

Fig. 53-13 Manner in which toe-out on turns is achieved. (*Chevrolet Motor Division of General Motors Corporation*)

to follow a shorter radius than the outer wheel. The circles on which the two front wheels turn are then concentric; that is, their centers are at the same place, D. Toe-out is secured by providing the proper relationship among the steering-knuckle arms, tie rods, and pitman arm. This relationship is such that the inner wheel on a curve always toes out more than the outer wheel. Figure 53-13 illustrates how it is done. When the tie rod is moved to the left during a right turn, it pushes at almost a right angle against the left steering-knuckle arm. The right end of the tie rod, however, not only moves to the left, but also swings forward, as shown by the dotted line.

Fig. 53-14 Steering linkage. (*Pontiac Motor Division of General Motors Corporation*)

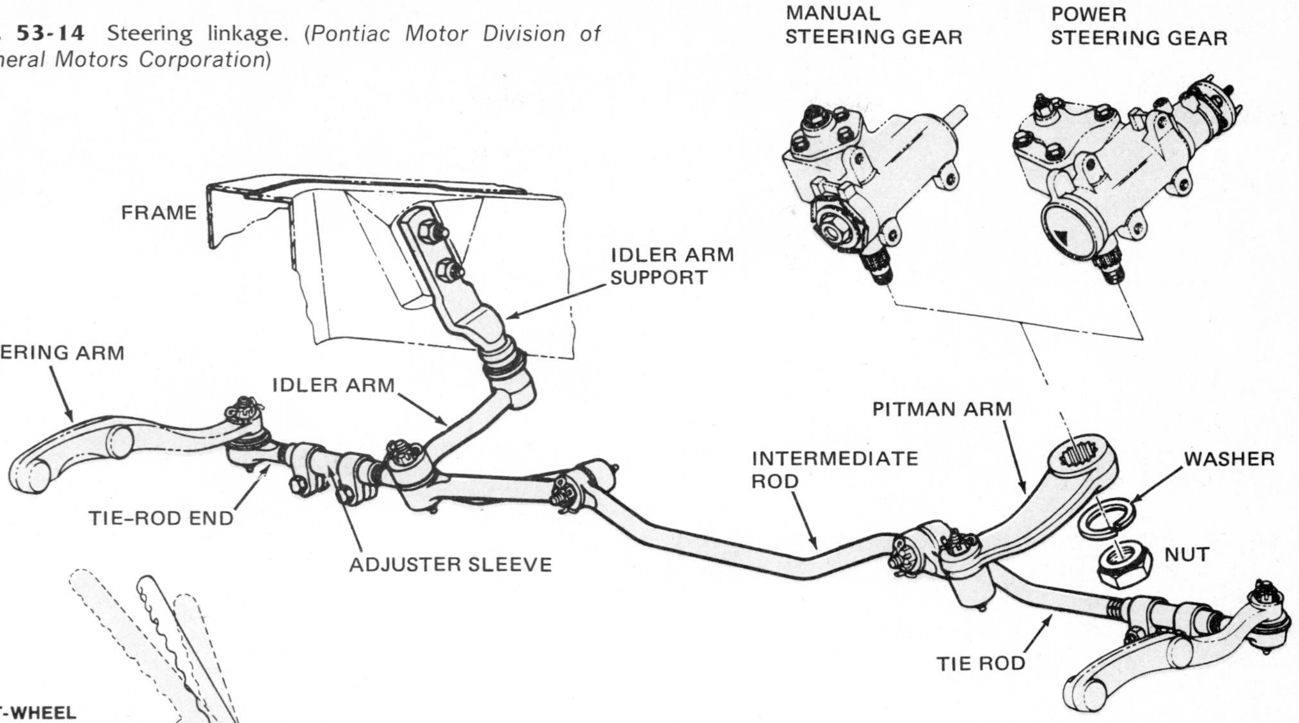

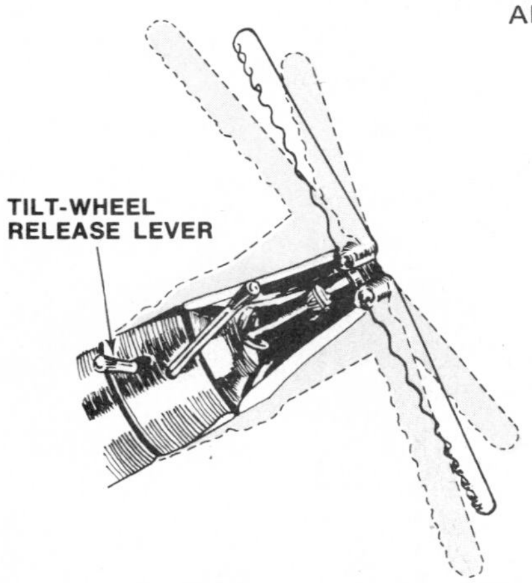

Fig. 53-15 Tilt steering wheel. Lifting the release lever permits the steering wheel to be tilted to various positions, as shown. (*Buick Motor Division of General Motors Corporation*)

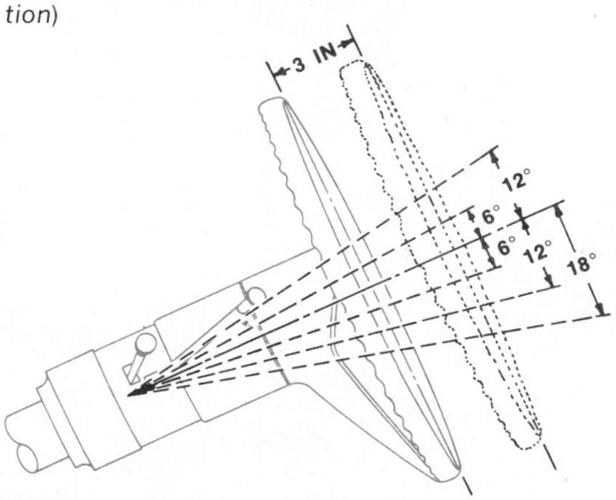

Fig. 53-16 Tilt-and-telescoping steering wheel. (*Cadillac Motor Car Division of General Motors Corporation*)

Thus, the right wheel is turned an additional amount. When a left turn is made, the left wheel is turned more than the right wheel. Figure 53-13 shows a parallelogram type of linkage (see ● 53-9). Other types of linkage give a similar effect and provide a like toe-out on turns.

● 53-9 Steering Linkages

The steering linkage connects the steering knuckles of the front wheels and the pitman arm of the steering gear (except on rack-and-pinion units, which have no pitman arms. See ● 53-15). Recall that the pitman arm swings from one side to the other (or forward and backward on some cars) as the steering wheel is turned. This movement must be carried to the steering knuckles at the wheels by some form of linkage. And there must be some means of adjusting the lengths of the tie rods or links for proper alignment between the front wheels. This alignment gives the front wheels a slight toe-in when the car is at rest. Then, when the car begins to move forward, this toe-in practically disappears as all looseness, or "sloppage," in the steering system is taken up.

Many types of steering linkages have been used. Figure 53-14 shows one form of parallelogram linkage. The intermediate rod assembly is connected at one end to the pitman arm, and at the other end to the steering idler arm. The steering-knuckle (spindle) arms are connected to the connecting-rod assembly by two secondary connecting-rod, or tie-rod, assemblies.

● 53-10 Tilt Steering Wheel and Column

Many cars have steering wheels that can be tilted at various angles. The purpose of the tilt steering wheel is to allow drivers to change the angle to suit their builds. They can also change the position of the steering wheel during a long drive. Figure 53-15 shows a tilt wheel.

Some cars also have a telescoping steering column. The wheel (Fig. 53-16) can be moved out or in to suit the driver's convenience. Another innovation is a steering wheel and column that can be swung to the right when the driver gets into or out of the car. All these devices have locking mechanisms which lock the wheel into the position selected. Also, the tilt steering column has an interlock to the transmission steering lever. This mechanism locks out the transmission until the steering column is returned to the driving position. The interlock is a safety feature which prevents the steering column from being accidentally moved while the car is in operation.

53-11 Collapsible Steering Column

The collapsible steering column is a safety device. It will collapse on impact. For example, in a front-end crash, the driver will be thrown forward and into the steering wheel. This will cause the steering column to collapse and thereby cushion the driver's impact.

Of course, if drivers wear safety belts, they will not be thrown into the steering wheel.

There are several types of collapsible steering column. These include the "Japanese-lantern" type, the tube-and-ball type, and the shear-capsule type. Figure 53-17 shows the Japanese-lantern type. It gets its name from the fact that on impact the column collapses like a Japanese lantern. The tube-and-ball type, shown in Fig. 53-18, has two tubes with balls between them. On impact, the balls must make grooves in the tubes to permit the column to shorten. This absorbs the shock. The shear-capsule type (Fig. 53-19) absorbs shock by the cutting of the capsule, which permits the column to collapse.

53-12 Steering Lock

The combination ignition switch and steering-wheel lock (Fig. 53-20) does several jobs. It locks the steering wheel when the ignition switch is turned off. When the ignition key is inserted and the ignition switch is turned to ON, the gear rotates. This pulls the rack and plunger out of the locking position. Then, when the switch is turned to OFF, the rotation of the gear moves the rack and plunger toward the locked position. If the plunger is lined up with a notch in the disk, the plunger moves in to lock the steering wheel. If it is not lined up, the plunger is spring-loaded against the disk. Then a slight turn of the steering wheel brings a notch into line. The plunger drops into the notch to lock the steering wheel.

In Chap. 30 we mention that the ignition switch also serves as a starting switch. In other words, when the switch is turned past ON to START, it connects the starting motor to the battery. The engine is then cranked for starting.

53-13 Steering Gears

There are two types of steering gears: manual and power. We discuss the manual type first. The steering gear converts the rotary motion of the steering wheel into

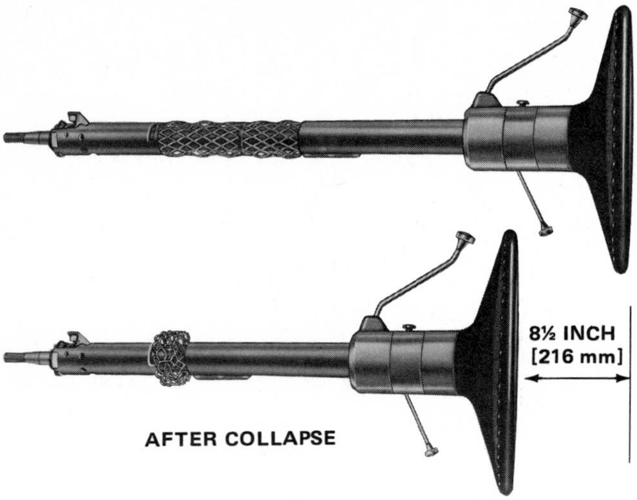

Fig. 53-17 Energy-absorbing, or collapsible, steering column of the "Japanese-lantern" design. The column collapses on impact, as shown at the bottom. (*General Motors Corporation*)

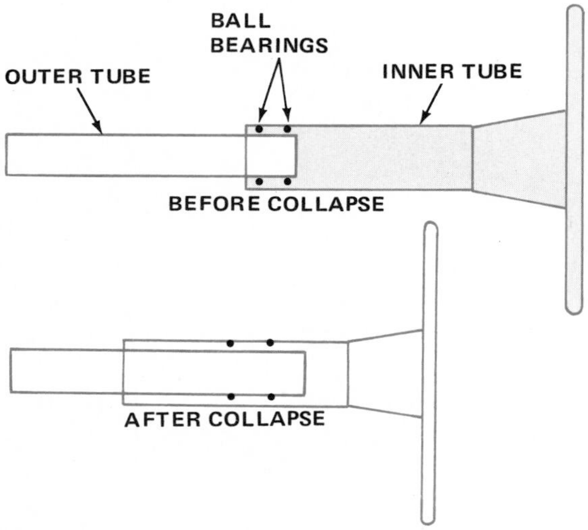

Fig. 53-18 Energy-absorbing, or collapsible, steering column of the tube-and-ball type. (*General Motors Corporation*)

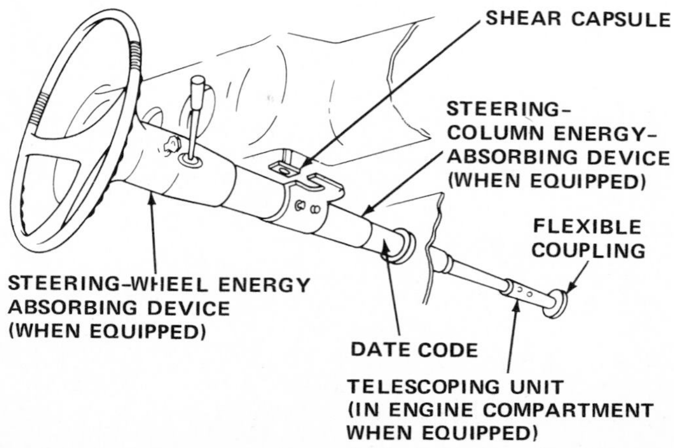

Fig. 53-19 Shear-capsule type of collapsible steering column.

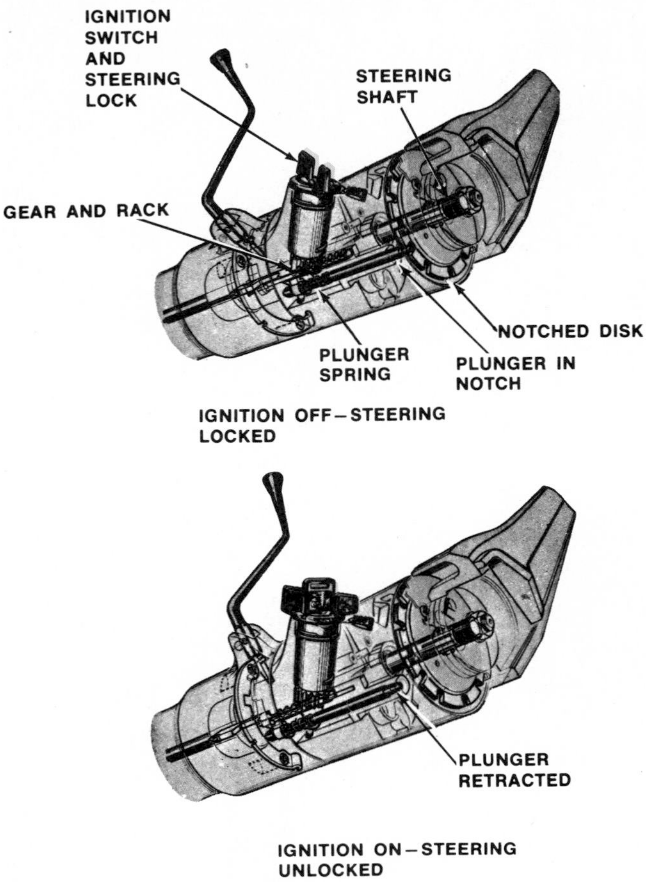

Fig. 53-20 Combination ignition and steering lock in phantom views, showing the two positions of the lock. (*General Motors Corporation*)

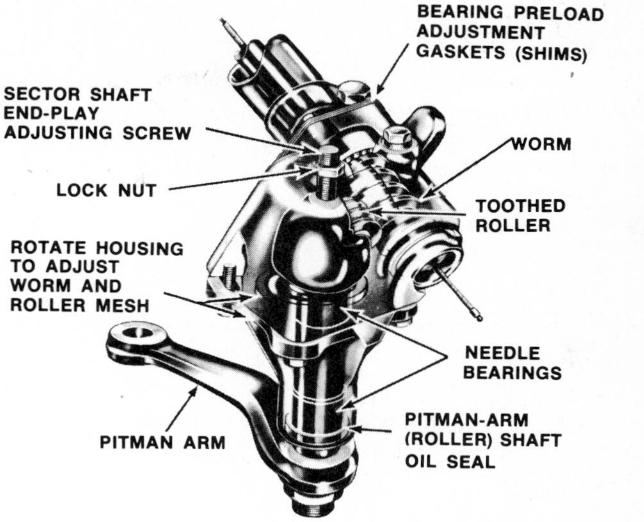

Fig. 53-21 Phantom view of a steering gear using a toothed roller attached to the pitman shaft. The worm and roller teeth mesh. (*Ford Motor Company*)

straight-line motion. This straight-line motion then moves the linkage to the steering arms on the steering knuckles. The front wheels are thus swung left or right for steering.

The steering gear has two essential parts: a worm on the end of the steering shaft and a matching sector gear or toothed roller attached to the pitman arm. (The rack-and-pinion steering gear is differently constructed as explained in ● 53-15.)

Figure 53-21 shows one type of steering gear. As the steering shaft and worm rotate, the toothed roller must follow the worm threads. The roller is moved toward one end of the worm or the other. The toothed-roller movement causes the pitman arm to swing one way or the other.

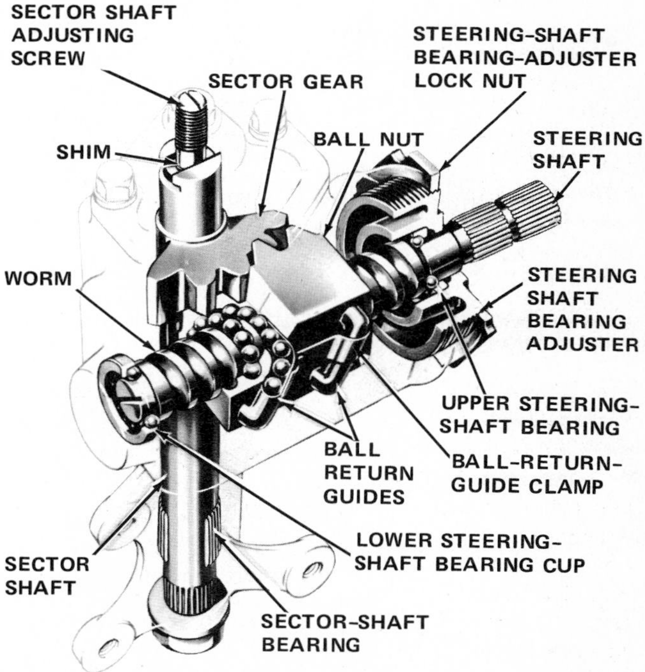

Fig. 53-22 Phantom view of a recirculating-ball steering gear. (*Ford Motor Company*)

● 53-14 Recirculating-Ball Steering Gear

Figure 53-22 shows another type of steering gear. In this unit, friction is kept low by using balls between the major moving parts. The balls roll between the worm teeth and the grooves cut in the hole in the ball nut. As the worm turns, the balls roll in the worm teeth. The balls must also roll in the grooves inside the ball nut. Thus, as the worm rotates, the balls cause the nut to move up or down along the worm. This motion is carried by the teeth on the outside of the ball nut to the teeth on the sector gear. Thus the sector gear must move, and this movement swings the pitman arm.

The steering gear in Fig. 53-22 is a *recirculating-ball steering gear*. The balls move from one end of the worm to the other. When they reach the end, they enter ball-return guides, which take them back into the ball nut.

● 53-15 Rack-and-Pinion Steering Gear

The rack-and-pinion steering gear, used on some imported and domestic small cars, is shown in Fig. 53-23. This system has a pinion on the end of the steering shaft (2 in Fig. 52-23). The pinion is meshed with a rack (3 in

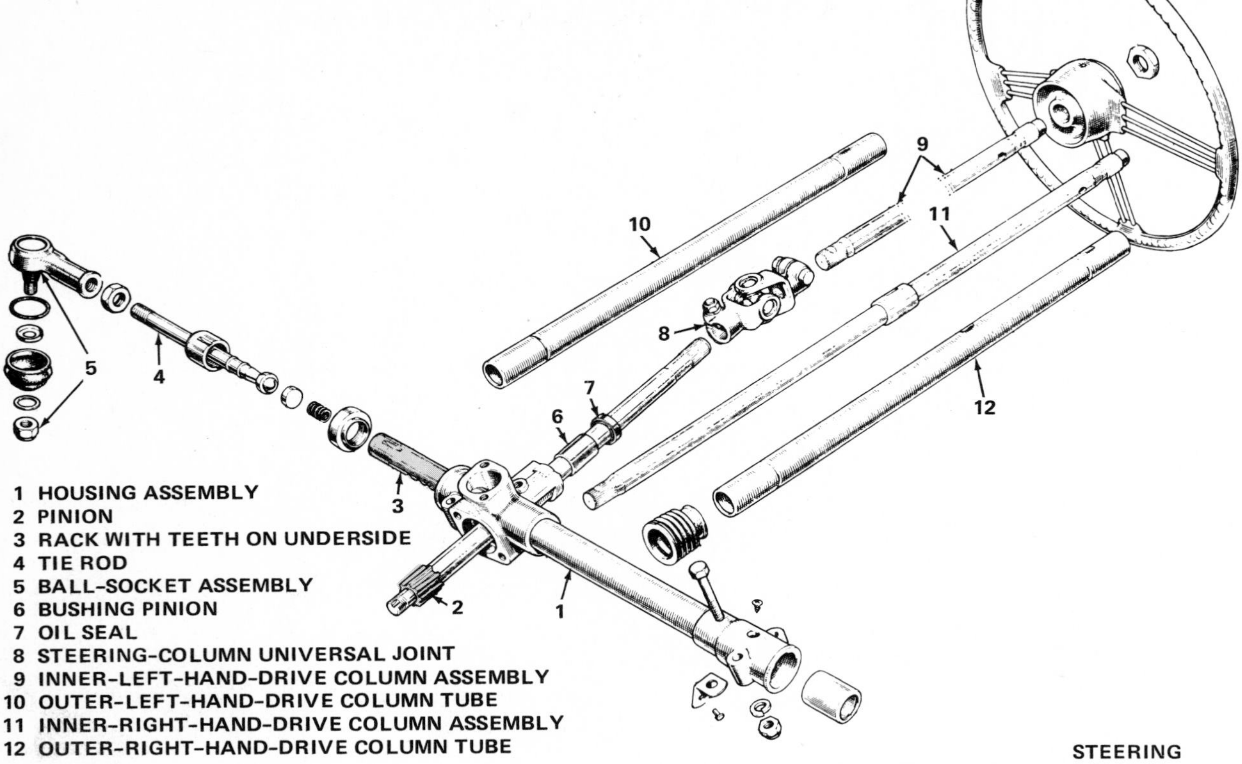

Fig. 53-23 Disassembled steering system using a rack-and-pinion steering gear. (*British Motor Corporation, Limited*)

Fig. 53-23). When the wheel is turned, the pinion turns. This moves the rack to left or right. Movement is carried through tie rods to the steering arms at the front wheels.

● 53-16 Power Steering

In the power-steering system, a pump sends a special fluid—oil—under pressure into the steering gear. The high-pressure fluid does about 80 percent of the work of steering. Figure 53-24 shows the power-steering system for a car with a six-cylinder engine. The steering gear looks almost like manual gear except that it is larger (also Fig. 53-14) because it contains a power piston that is moved by the high-pressure fluid from the pump. The pump is driven by a belt from the crankshaft pulley.

Figure 53-25 shows the working parts of a typical power-steering pump. The rotor rotates, and the vanes move in and out of slots in the pump rotor. As the vanes move out, the spaces between the vanes increase. Fluid is drawn into these spaces. Then, as the rotor turns further, the vanes are pushed back into the rotor. This decreases the spaces between the vanes. The fluid is forced out under high pressure. It goes through hoses to the steering gear. The pump has a pressure-relief valve which opens if the pressure goes too high.

There are two types of power steering, the integral type, shown in Figs. 53-14 and 53-24, and the linkage type, shown in Fig. 53-29. The linkage type puts the power piston in the steering linkage.

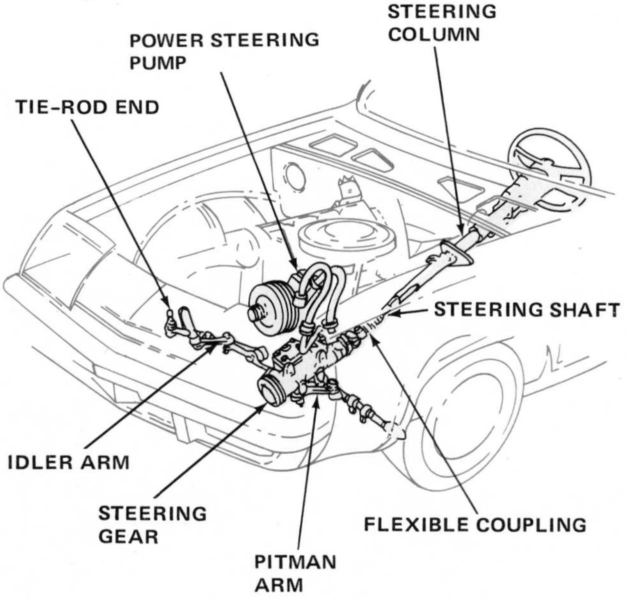

Fig. 53-24 Power-steering system for a car with a six-cylinder engine.

● 53-17 General Motors Power Steering

We now describe two General Motors power-steering units. Both are integral and operate in the same manner.

They are a little different in construction. Both Chrysler and Ford have similar integral power-steering units.

The two General Motors models are the in-line and rotary-valve (also called *torsion-bar*). Both include a recirculating-ball-and-nut steering gear (Fig. 53-22) to which has been added the power system.

1. Rotary-valve power steering This unit (Figs. 53-26 and 53-27) uses a small torsion bar and a rotary valve. The bar twists as the steering wheel is turned. In the straight-ahead position, the rotary valve is centered. Oil can flow freely through the valve and no pressure builds up. When a turn is made, the turning effort applied to the steering wheel carries through the torsion bar. The bar twists slightly. This turns the rotary valve. Now, the rotary valve cuts off the return line and opens the line to one side of the piston. The action during a right turn is shown in Fig. 53-28. Fluid under pressure acts on the piston, as shown, and assists in making the turn. During a left turn, the fluid pressure is applied to the opposite end of the

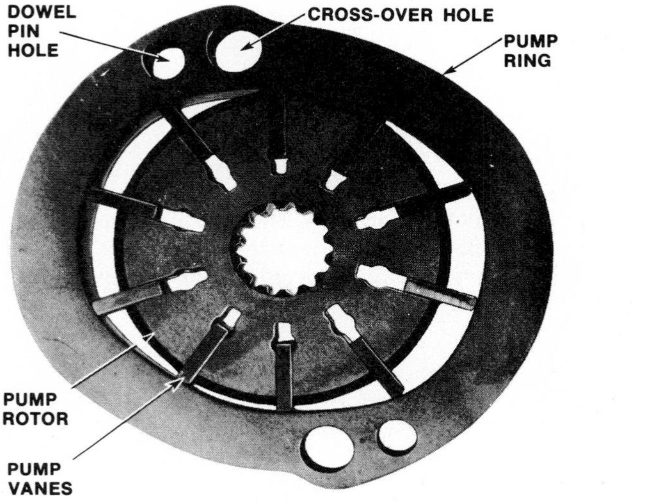

Fig. 53-25 Internal working parts of a vane-type power-steering pump. The vanes follow the oval shape of the housing and move in and out of the slots in the rotor. (*Chevrolet Motor Division of General Motors Corporation*)

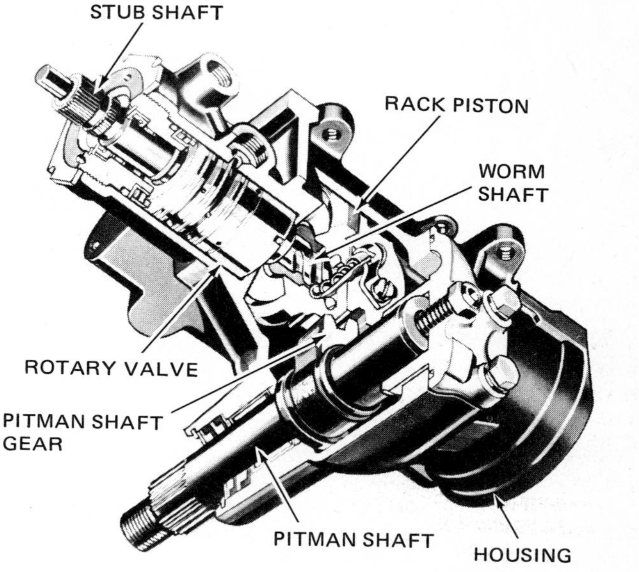

Fig. 53-26 Cutaway power-steering gear using a rotary valve. (*Cadillac Motor Car Division of General Motors Corporation*)

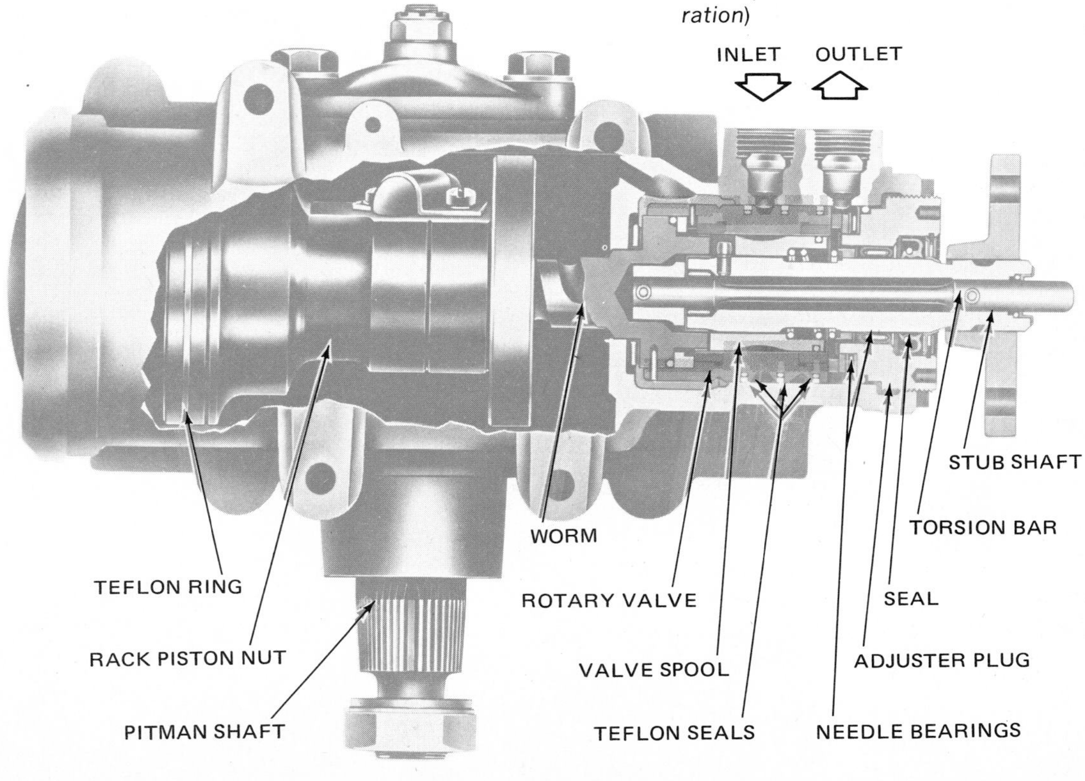

Fig. 53-27 Rotary-valve power-steering gear, partly cut away. (*Pontiac Motor Division of General Motors Corporation*)

piston. The amount of pressure applied depends on how much resistance to turning the wheel the driver encounters. If the resistance is high, as it might be in making a parking maneuver, the torsion bar and rotor are turned more. This sends more pressure to the piston.

2. In-line power steering The in-line power-steering unit is slightly different in construction. It depends on shifting of the worm endwise as torque is applied to the steering wheel. The worm is kept centered during straight-ahead driving, by springs. When a turn is made, the worm moves endwise slightly because of the resistance to turning at the car wheels. This slight movement opens a valve. The valve sends high-pressure fluid to the end of the piston. Which end depends on which direction—right or left—the car is turning. The amount of pressure sent to the piston depends on how much resistance to turning the driver encounters at the steering wheel.

● 53-18 Linkage-Type Power Steering

In the linkage-type power-steering system, the power cylinder is not part of the steering gear. Instead, the power cylinder (or *booster* cylinder, as it is also called) is connected into the steering linkage. In addition, the valve assembly is included in the steering linkage. It is either a separate assembly or integral with the power cylinder. Figure 53-29 shows one linkage-type power-steering system in which the booster cylinder and valve assembly are separate units. In operation, the steering gear works in exactly the same way as the mechanical types described in ● 53-14. However, the swinging end of the pitman arm is not directly connected to the steering linkage. Instead, it is connected to a valve assembly. As the end of the pitman arm swings when a turn is made, it actuates the valve assembly. Then, the valve assembly directs hydraulic oil pressure from the oil pump to the booster cylinder. Inside the booster cylinder, pressure is applied to one or the other side of a piston. Movement then takes place (actually, in this unit, the cylinder moves instead of the piston). This movement is transferred to the connecting rod in the steering linkage. Thus, most of the effort required to move the connecting rod and steer the car is furnished by the booster cylinder.

1. Valve-assembly operation In many ways, the valve assembly is similar to that used in the in-line power-steering unit. That is, when turning effort passes through the steering gear and into the steering linkage, it must pass through the valve assembly. This causes the valve to open and send high-pressure fluid to the power cylinder.

2. Booster cylinder Figure 53-30 is a cutaway view of the booster cylinder. The cylinder is made up of two concentric shells. Oil flows between the two shells to enter the piston-rod end of the cylinder. The end of the piston rod is attached to the car frame by a flexible connection. This allows some movement of the rod, to permit alignment of the rod with the cylinder as the cylinder moves back and forth during steering.

3. Integral valve and power cylinder In this design, the linkage-type power-steering unit contains both the valve assembly and the power cylinder. The piston rod of the power cylinder is attached to the car frame. The cylinder is linked to the steering linkage, forming a part of the linkage. This assembly is called the *power link,* since it is a part of the linkage and, at the same time, supplies steering power.

● 53-19 Power Rack-and-Pinion Steering Gear

This unit is similar to the manual type (Fig. 53-23) except that a rotary valve and power cylinder have been incorporated into the assembly (Fig. 53-31). The valve works in the same way as the rotary valve for other power-steering

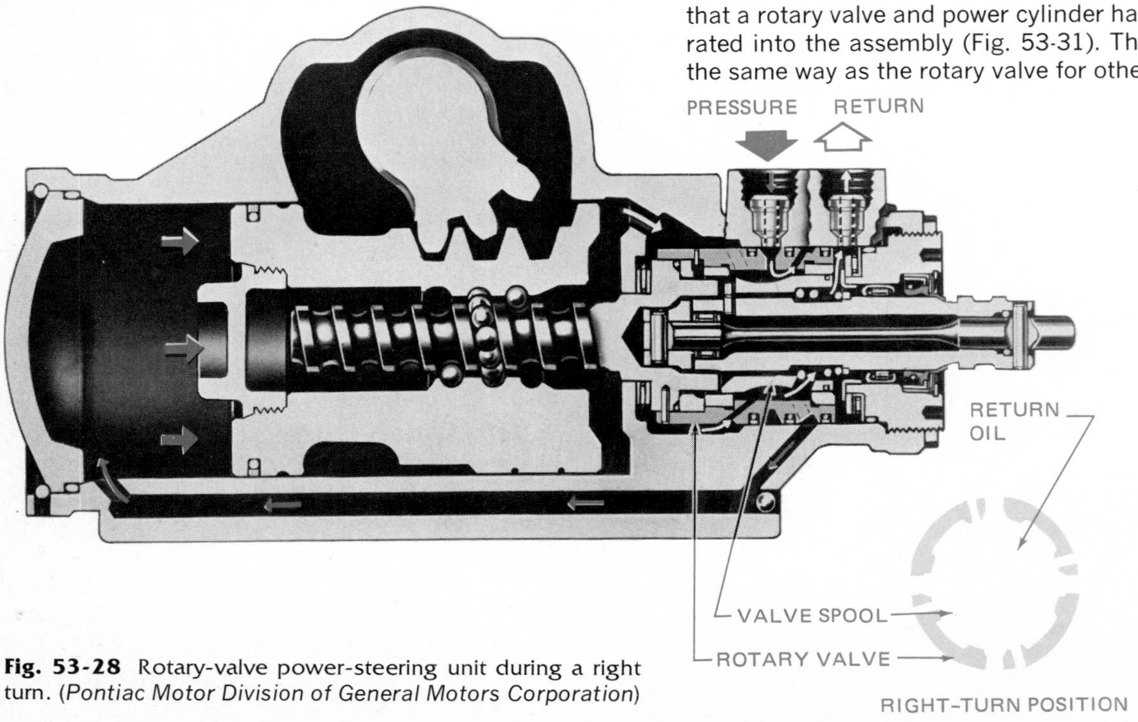

Fig. 53-28 Rotary-valve power-steering unit during a right turn. (*Pontiac Motor Division of General Motors Corporation*)

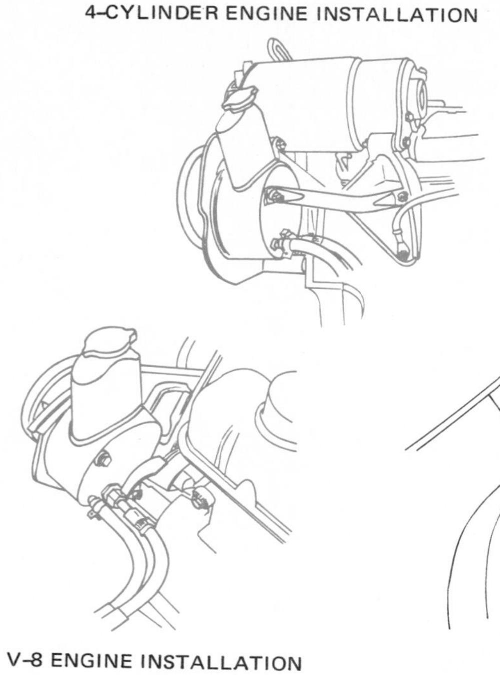

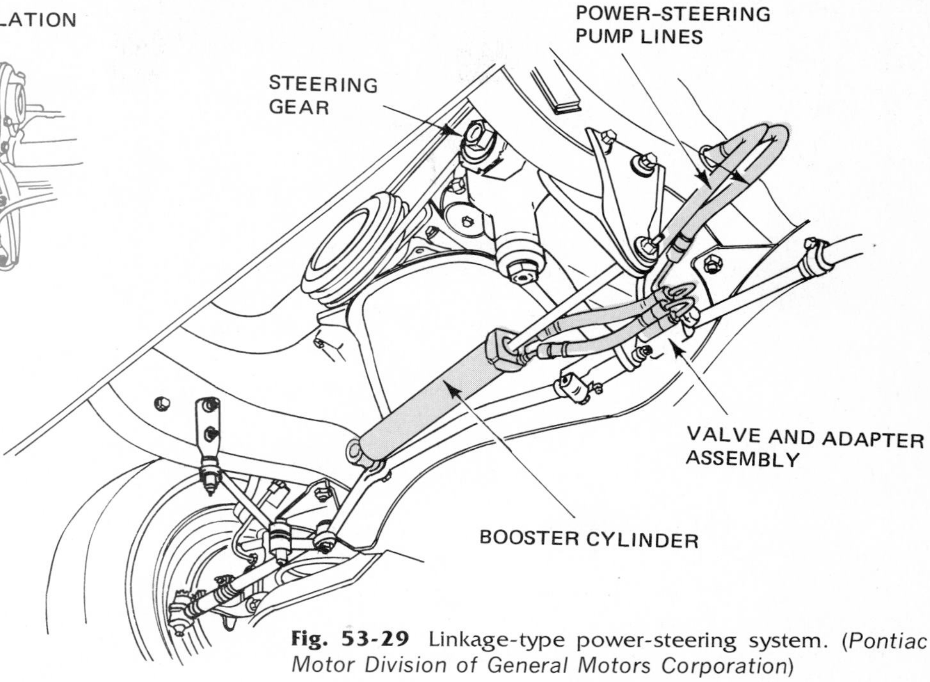

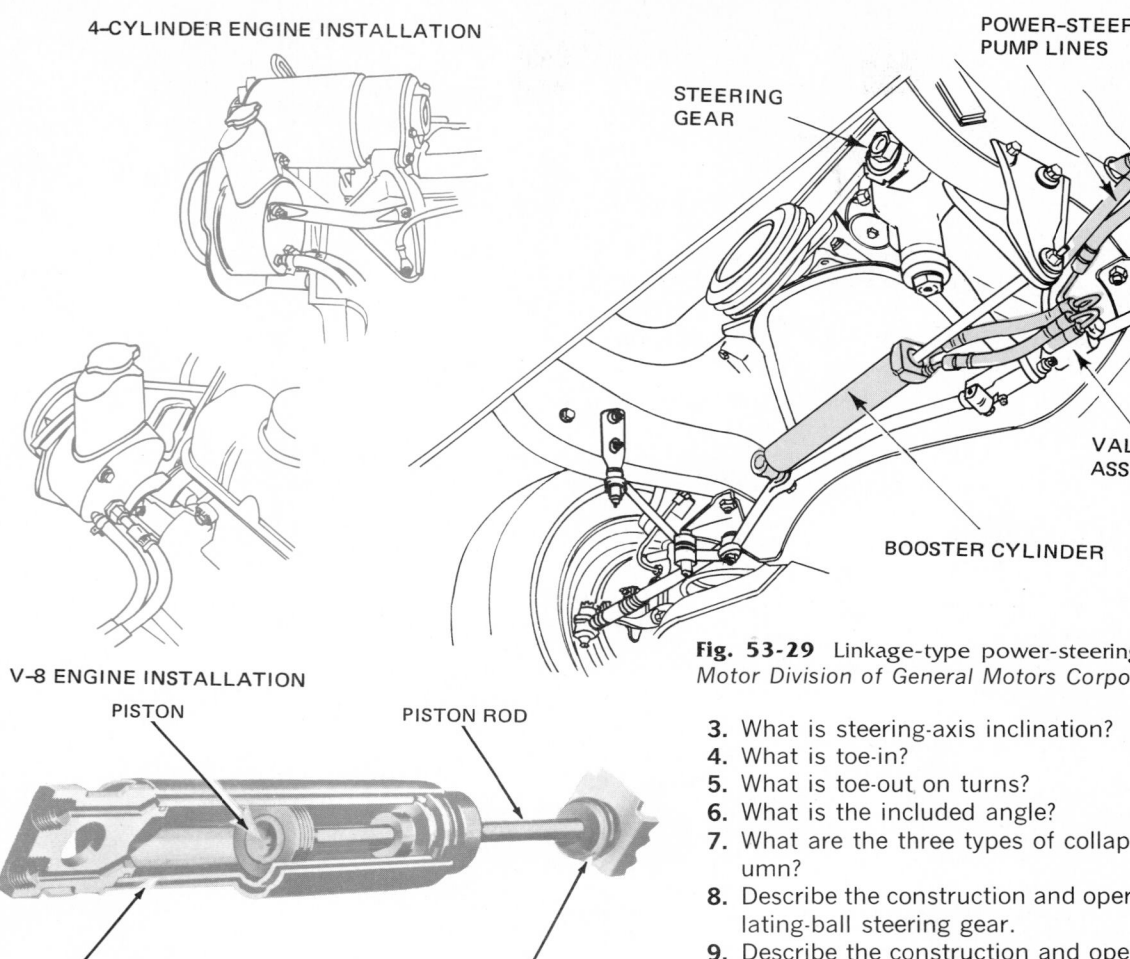

Fig. 53-29 Linkage-type power-steering system. (*Pontiac Motor Division of General Motors Corporation*)

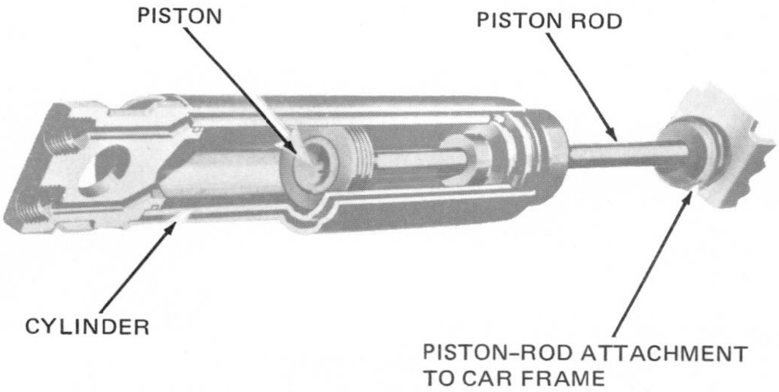

Fig. 53-31 Two views of power rack-and-pinion steering gears. (*Ford Motor Company*)

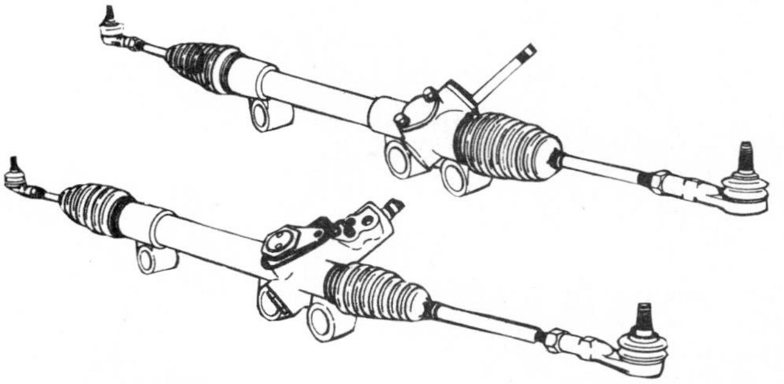

Fig. 53-31 Two views of power rack-and-pinion steering gears. (*Ford Motor Company*)

units previously described. It sends fluid to one or the other side of the piston when a turn is made. The fluid, under pressure, then provides most of the steering effort.

REVIEW QUESTIONS

1. What is camber?
2. What is caster?
3. What is steering-axis inclination?
4. What is toe-in?
5. What is toe-out on turns?
6. What is the included angle?
7. What are the three types of collapsible steering column?
8. Describe the construction and operation of a recirculating-ball steering gear.
9. Describe the construction and operation of a rotary-valve power-steering gear.
10. Describe the construction and operation of the linkage-type power-steering system.

SELF PROJECTS

1. One of the self projects for the previous chapter was to get under cars on lifts to study suspension systems. This is a good time to study steering systems, too. Keep a safe distance when the car is being raised or lowered. When you are under a car, locate the steering gear and all the linkages that carry the steering motion to the wheels. Make sketches of the systems in several cars, and identify the various parts, the pitman arm, idler arm, tie rods, knuckle arm, and so on. Note particularly the location of the tie rod and clamps, because this is where toe-in is adjusted.
2. You can make the same study of steering systems by referring to manufacturers' service manuals.
3. On cars with power steering, note the location of the power-steering pump and the hoses that connect it to the steering gear.
4. Examine steering gears and steering-gear parts. Notice that the manual types are simple in construction. The power-steering gears are more complicated and harder to disassemble.

CHAPTER 54
Steering and Suspension Service

After studying this chapter, you should be able to:

- **List the 15 troubles or complaints in steering and suspension systems.**
- **For each trouble or complaint, explain what could cause the condition.**
- **Explain how to make the checks and corrections for each condition.**
- **Relate types of abnormal tire wear with causes.**
- **Explain how to do a front-end alignment job. Under the instructor's supervision, perform this job.**

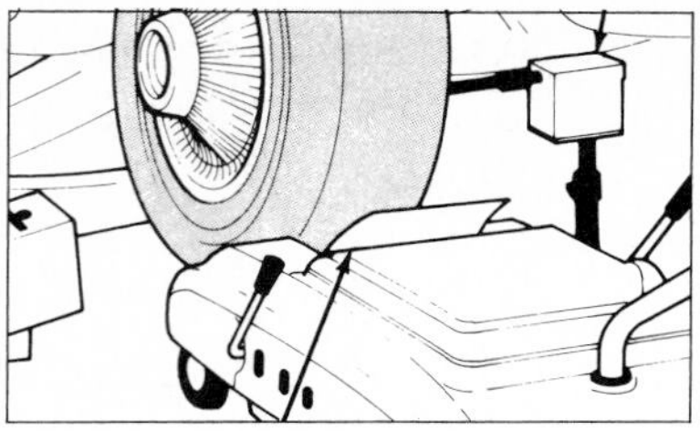

This chapter discusses various steering and suspension troubles and their possible causes. It also outlines the servicing and overhaul procedures that might be required on steering gears and suspension systems. In addition, it describes the different front-alignment checks and adjustments.

● 54-1 Need for Logical Procedure

If you are able to relate steering and suspension complaints to the conditions that cause them, you are much better off than the person who blindly tries to find the cause of the trouble. You will know what to check and correct to eliminate the trouble. You can save a great deal of time and effort when you know where to look.

The following sections tell you where to look when handling complaints about the steering and suspension.

● 54-2 Trouble-Diagnosis Chart

A variety of steering and suspension troubles will bring the driver to the mechanic. But it is a rare driver who will have a clear idea of what is causing the trouble. Drivers can detect steering difficulty, hard steering, or excessive play in the steering system, but they probably will not have a very good idea of what is causing the condition. The chart that follows lists possible troubles in the steering and suspension, their possible causes, and checks or corrections to be made. Following sections then describe the trouble causes and corrections in detail.

Note: The troubles and possible causes are not listed according to how often they occur. That is, item 1 (or item a under "Possible Cause") does not necessarily occur more often than item 2 (or item b).

STEERING AND SUSPENSION TROUBLE-DIAGNOSIS CHART

See ● 54-3 to 54-17 for detailed explanations of trouble causes and corrections listed below.

Complaint	Possible Cause	Check or Correction
1. Excessive play in system (● 54-3)	a. Improper linkage or steering-gear adjustment	Adjust
	b. Worn linkage parts	Replace
	c. Worn ball joints	Replace
	d. Loose wheel bearings	Adjust

STEERING AND SUSPENSION TROUBLE-DIAGNOSIS CHART (Continued)

Complaint	Possible Cause	Check or Correction
2. Hard steering (● 54-4)	a. Tighten steering gear or linkage	Readjust
	b. Low or uneven tire pressure	Inflate properly
	c. Excessive friction in steering gear or linkage	Lubricate, readjust
	d. Excessive friction in ball joints	Lubricate or replace
	e. Improper front alignment	Align
	f. Frame bent	Straighten; replace damaged parts
	g. Sagging springs	Replace
	h. Power steering not working	Service
3. Car wander (● 54-5)	a. Low or uneven tire pressure	Inflate properly
	b. Binding or excessive play in linkage or steering gear	Adjust
	c. Improper front end alignment	Align
4. Car pulls to one side during normal driving (● 54-6)	a. Uneven tire pressure	Inflate properly
	b. Uneven caster or camber	Adjust
	c. Tight wheel bearing	Adjust
	d. Uneven springs	Replace
	e. Uneven torsion-bar adjustment	Adjust
	f. Improper wheel tracking	Straighten bent frame; replace damaged parts
5. Car pulls to one side when braking (● 54-7)	a. Brakes not applying evenly	Adjust; replace parts causing trouble
	b. Causes listed in item 4 above	
6. Front-wheel shimmy—low speed (● 54-8)	a. Low or uneven tire pressure	Inflate properly
	b. Loose linkage	Adjust
	c. Soft springs	Replace; adjust
	d. Incorrect or uneven camber	Adjust
	e. Irregularities in tire tread	Replace; match tires
	f. Dynamic imbalance	Balance wheel
7. Front-wheel tramp (● 54-9)	a. Unbalanced wheels	Balance
	b. Wheel run-out	Adjust bearings; replace bent wheel
	c. Defective shock absorbers	Replace
	d. Causes listed in item 6 above	
8. Steering kickback (● 54-10)	a. Uneven or incorrect tire inflation	Inflate correctly
	b. Sagging springs	Replace; adjust
	c. Defective shock absorbers	Replace
	d. Looseness in linkage or steering gear	Adjust
9. Tires squeal on turns (● 54-11)	a. Excessive speed	Slow down
	b. Front-end misalignment	Align
	c. Low or uneven tire inflation	Inflate correctly
	d. Worn tires	Replace
10. Abnormal tire wear (● 54-12)	a. The type of wear indicates the possible causes. See ● 54-12 for details.	
11. Hard or rough ride (● 54-13)	a. Excessive tire pressure	Inflate properly
	b. Defective shock absorbers	Replace
	c. Excessive friction in suspension	Lubricate; adjust
12. Sway on turns (● 54-14)	a. Loose stabilizer bar	Tighten
	b. Sagging or weak springs	Replace; adjust
	c. Defective shock absorbers	Replace
	d. Excessive positive caster	Adjust

STEERING AND SUSPENSION TROUBLE-DIAGNOSIS CHART (Continued)

Complaint	Possible Cause	Check or Correction
13. Leaf-spring breakage (● 54-15)	a. Loose U bolt or center bolt	Tighten properly when replacing spring
	b. Excessive overloading	Avoid overloads
	c. Defective shock absorbers	Replace
	d. Tight spring shackle	Lubricate or replace
14. Sagging springs (● 54-16)	a. Overloads	Avoid overloading
	b. Loss of shim on coil-spring seat	Always replace when servicing springs
	c. Defective shock absorbers	Replace
15. Noises (● 54-17)	a. Rattles	Tighten loose parts
	b. Squeaks	Lubricate springs, suspension parts
	c. Squeaks when steering	Lubricate linkage

● 54-3 Excessive Play in System

Excessive play, or looseness in the steering system, means there is excessive free movement of the steering wheel without corresponding movement of the front wheels (Fig. 54-1). A small amount of steering-wheel play is desirable to provide easier steering. But when the play becomes excessive, it can make steering harder. The steering-wheel rim should move less than 2 inches [51 mm] before the front wheels begin to move. Excessive play can be due to:

Wear or improper adjustments in the steering linkage
Worn ball joints
Loose wheel bearings

To check for loose tie rods and linkage, jack up the front end of the car. Then grasp both front wheels. Push out on both at the same time, and then pull in on both at the same time (Fig. 54-2). Excessive relative movement between the two wheels means that the linkage connections are worn or out of adjustment.

To check for worn steering-knuckle parts and loose wheel bearings, jack up the front end of the car. Then grasp the wheel top and bottom, and check it for side play (Fig. 54-3). Try to see how much you can wobble the wheel. Excessive looseness indicates worn or loose parts either in the steering knuckle or in the wheel bearing. Have someone apply the brakes as you again try to rock the wheel. If applying the brakes eliminates the wobble, the wheel bearing is loose. It must be adjusted. If applying the brakes does not eliminate the wobble, the looseness is in the steering knuckle and the ball joints should be checked. Figures 54-4 and 54-5 show how to support the wheel for different designs. Axial play is checked by moving the wheel up and down. Radial play is measured by rocking the wheel back and forth. Many cars have wear-indicating ball joints. On these, a visual check may be all that is necessary (Fig. 54-6). When the grease-fitting nipple has receded into the ball-joint socket, the ball joint is worn and should be replaced.

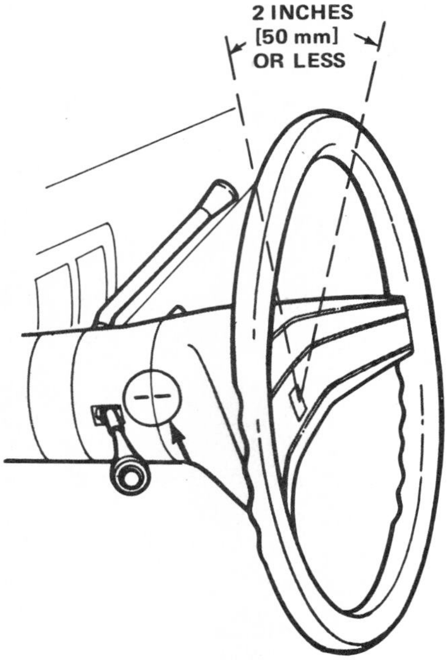

Fig. 54-1 Checking free play in the steering wheel. It should move less than 2 inches [51 mm]. (*Ford Motor Company*)

Fig. 54-2 Checking tie rods and linkage for looseness. (*Bear Manufacturing Company*)

Fig. 54-3 Checking for wear at the steering knuckle and wheel bearings. *(Bear Manufacturing Company)*

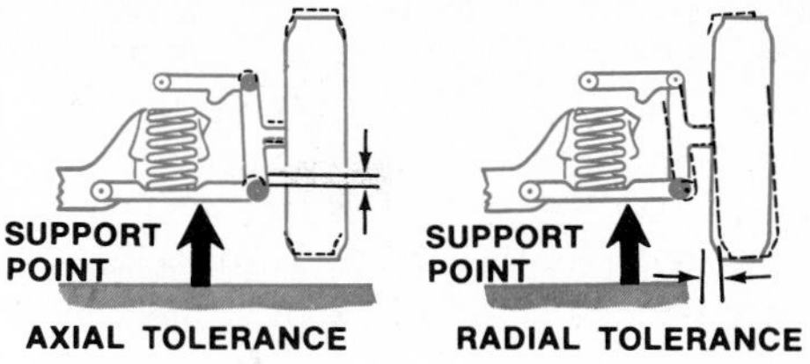

Fig. 54-4 To check ball joints for wear on a suspension system with the spring on the lower arm, support the wheel under the arm, as shown.

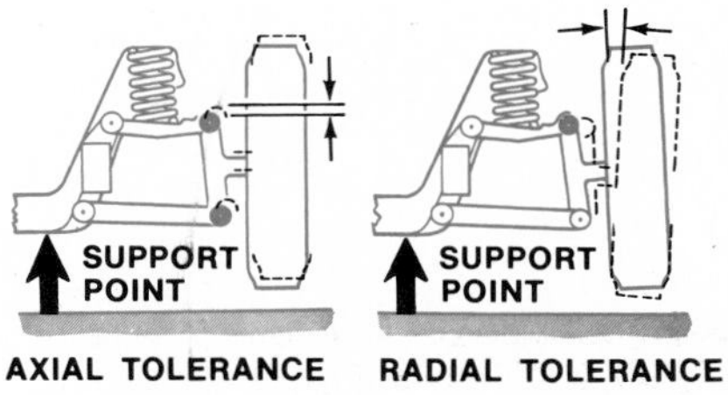

Fig. 54-5 To check ball joints for wear on a suspension system with the spring on the upper arm, support the front end on the frame, as shown.

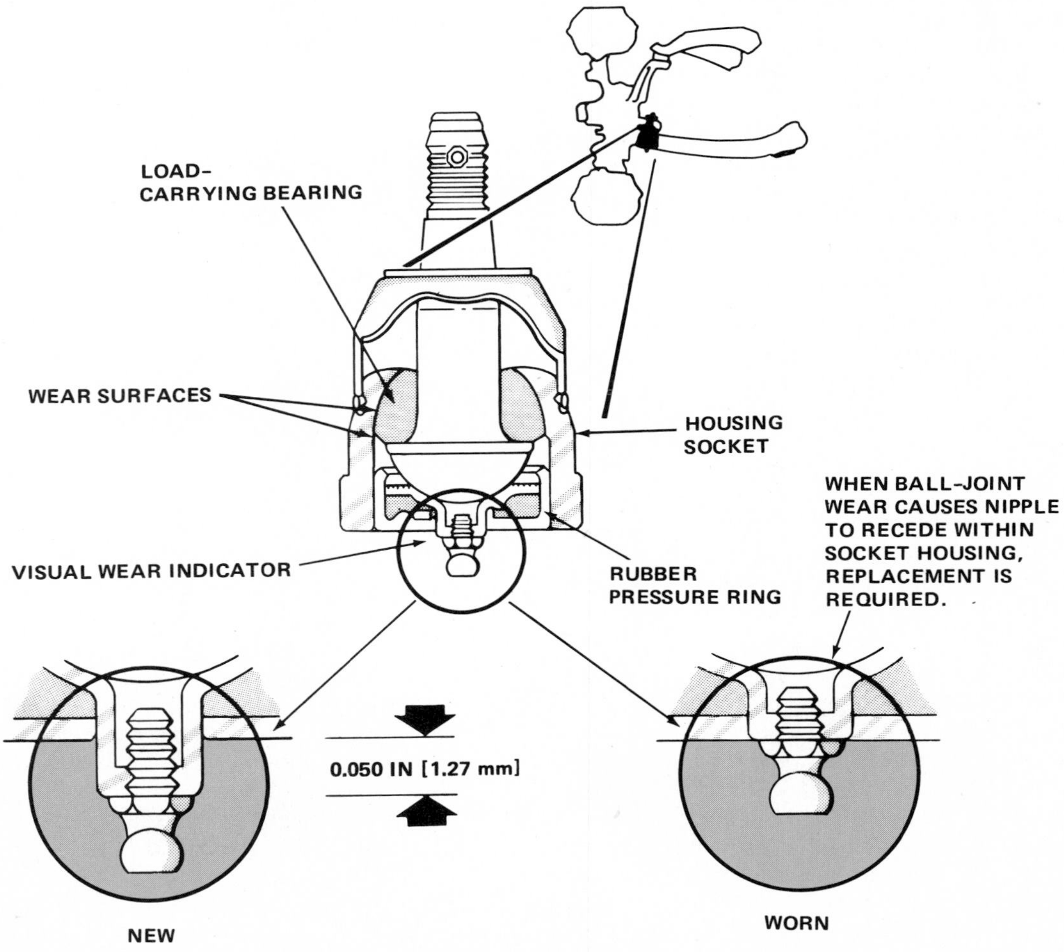

Fig. 54-6 How wear-indicating ball joints show when ball-joint replacement is necessary. In a worn ball joint, the grease-fitting nipple has receded into the socket, as shown to the right.

A quick check for looseness in the steering gear can be made as follows: Watch the pitman arm while an assistant turns the steering wheel one way and then the other, with the front wheels on the floor. Watch especially after reversal of steering-wheel rotation. If a lot of steering-wheel movement is required to set the pitman arm in motion, the steering gear is worn or needs adjustment.

● 54-4 Hard Steering

Hard steering that occurs just after steering-system work is probably due to excessively tight adjustments in the steering gear or linkages. If hard steering develops at other times, it could be due to low or uneven tire pressure, to abnormal friction in the steering gear or linkage or at the ball joints (or kingpin), to improper front-wheel or frame alignment, or to sagging springs.

On a car with power steering, the power-steering mechanism may have failed. This would require straight mechanical operation and, thus, considerably greater steering effort. When this happens, the power-steering system and pump should be checked. First check the fluid level in the pump fluid reservoir. If the problem is not low fluid level, install a pressure gauge in the system to check the pressures.

To check the steering system for excessive friction, jack up the front end of the car. Then turn the steering wheel, and watch the steering-system components to locate the source of excessive friction. Disconnect the linkage at the pitman arm. If this eliminates the frictional drag that makes it hard to turn the steering wheel, then the friction is either in the linkage itself or at the ball joints. If the friction is not eliminated when the linkage is disconnected at the pitman arm, then the steering gear is at fault.

If the trouble is not excessive friction in the steering system, it may be incorrect front-wheel alignment or a misaligned frame or sagging springs. Excessive caster, especially, will cause hard steering.

● 54-5 Car Wander

Car wander is experienced as difficulty in keeping the car moving straight ahead. Frequent steering-wheel movements are necessary to prevent the car from wandering from one side of the road to the other. Inexperienced drivers may sometimes complain of car wander. This is because they tend to oversteer. They have to keep moving the wheel back and forth to stay on their side of the road.

Several conditions can cause car wander. Low or uneven tire pressure, binding or excessive play in the linkage or steering gear, or improper front-wheel alignment causes car wander. Any condition that causes tightness in the steering system keeps the wheels from automatically seeking the straight-ahead position. The driver therefore has to correct the wheels constantly. This condition would probably also cause hard steering (● 54-4). Looseness, or excessive play, in the steering system might also cause car wander. This would tend to allow the wheels to waver somewhat and permit the car to wander.

Excessively low caster, uneven caster, or a point of intersection too far above or below the road surface (from wrong camber angle) tends to cause the wheels to swing away from straight ahead. The driver must steer continuously. Excessive toe-in causes the same condition.

● 54-6 Car Pulls to One Side during Normal Driving

If the car keeps pulling to one side, pressure must be applied to the steering wheel more or less constantly to maintain forward movement. The trouble could be due to any of the following: uneven tire pressure, uneven caster or camber, a tight wheel bearing, uneven springs, uneven torsion-bar adjustment, or improper wheel tracking (rear wheels not following in the tracks of the front wheels because of a bent frame). Anything that makes one wheel drag or toe in or toe out more than the other will make the car pull to that side.

● 54-7 Car Pulls to One Side during Braking

The most likely cause of this condition is grabbing brakes. This could be due to the brake linings becoming soaked with oil or brake fluid, to brake shoes unevenly or improperly adjusted, to a loose strut rod, or to a brake backing plate loose or out of line. Anything that would cause the brake at one wheel to apply harder than the brake at the opposite wheel could be at fault. The conditions listed in ● 54-6 could also cause pulling to one side during braking. The condition, from whatever cause, tends to become more noticeable when the car is braked.

● 54-8 Front-Wheel Shimmy (Low Speed)

Front-wheel shimmy and front-wheel tramp (● 54-9) are sometimes confused. Low-speed shimmy is the rapid oscillation of the wheel on the ball joints. The wheel tries to turn in and out alternately. The action causes the front end of the car to shake from side to side. On the other hand, front-wheel tramp, or high-speed shimmy, is a tendency for the wheel-and-tire assembly to hop up and down. When the condition is severe, the tire may actually leave the pavement. Even when the tire does not leave the pavement, tramp can be observed as a rapid flexing-unflexing action of the part of the tire in contact with the pavement. That is, the bottom of the tire first appears deflated (as the wheel moves down) and then inflated (as the wheel moves up).

Low-speed shimmy can result from low or uneven tire pressure, loose linkage, excessively soft springs, incorrect or uneven wheel camber, irregularities in the tire treads, or dynamic imbalance.

54-9 Front-Wheel Tramp

As explained in the previous section, front-wheel tramp is often called *high-speed shimmy.* This condition causes the front wheels to hop up and down alternately. Three of the most common causes of front-wheel tramp are unbalanced wheels, wheels that have too much run-out, and defective shock absorbers. An unbalanced wheel is heavy at one part. As it rotates, the heavy part sets up a circulating outward thrust that makes the wheel hop up and down. A similar action occurs if the wheel has too much run-out. Run-out is the amount the wheel is out of line with the axle—one part of the wheel "runs out," or moves to the side, more than other parts of the wheel. Defective shock absorbers, which fail to control natural spring oscillations, also cause wheel tramp. Any of the causes described in the previous sections may also cause wheel tramp. Following sections describe the servicing of the wheel and tire so that they can be restored to proper balance and alignment.

54-10 Steering Kickback

Steering shock, or kickback, is felt as sharp and rapid movements of the steering wheel. It becomes evident when the front wheels encounter obstructions in the road. Normally, some kickback to the steering wheel always occurs. But when it becomes excessive, an investigation should be made. This condition could result from incorrect or uneven tire inflation, sagging springs, defective shock absorbers, or looseness in the linkage or steering gear. These defects permit road shock to carry back excessively to the steering wheel.

54-11 Tires Squeal on Turns

Excessive speed causes tires to skid or squeal on turns. If this is not the cause, then it is probably low or uneven tire pressure, worn tires, or misalignment of the front wheels (particularly camber and toe-in).

54-12 Abnormal Tire Wear

Various types of abnormal tire wear can be experienced. The type of wear is often a good indication of the defect. For example, if the tire is operated with insufficient air pressure (underinflated), the sides will bulge over. the center of the tread will be lifted clear of the road. The sides of the tread will take all the wear, the center being barely worn (Fig. 54-7). Uneven tread wear shortens tire life.

Overinflation causes the tire to ride on the center of its tread. Thus, only the center of the tread wears (Fig. 54-7). This uneven tread wear shortens tire life. But more damaging is the fact that the overinflated tire does not have normal "give" when it meets a rut or bump in the road.

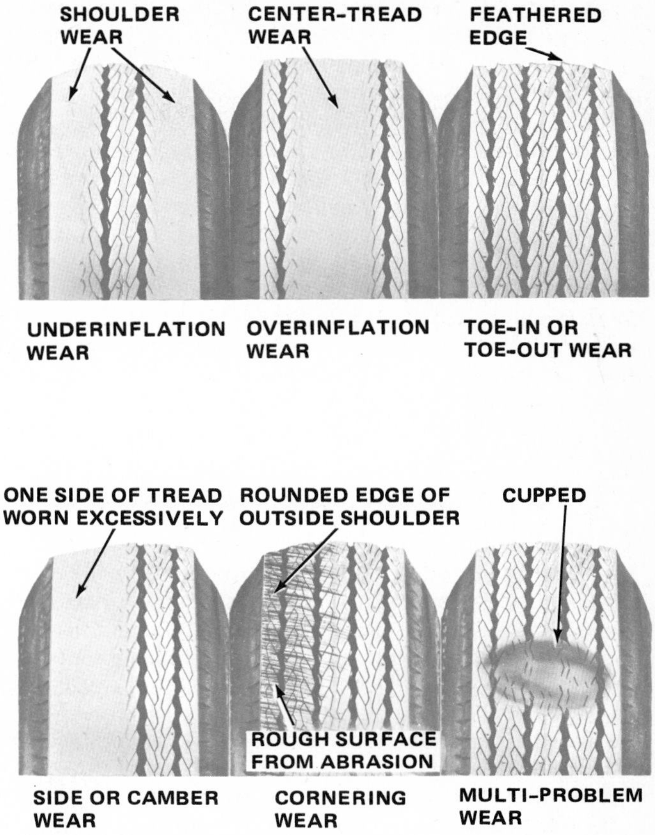

Fig. 54-7 Patterns of tire wear tell what caused the abnormal wear. (*Buick Motor Division of General Motors Corporation*)

Instead, the tire fabric takes the major shock. This may cause the fabric to crack or break and the tire to fail.

Excessive camber of the wheel causes one side of the tire tread to wear more rapidly than the other, as shown in Fig. 54-7.

Excessive toe-in or toe-out on turns causes the tire to be dragged sideways while it is moving forward. The tire on a front wheel that toes in 1 inch from straight ahead will be dragged sideways about 150 feet every mile. This sideward drag scrapes off rubber, as shown in Fig. 54-7. One characteristic of this type of wear is the feather edges of rubber that appear on one side of the tread design. If both front tires show this type of wear, the front suspension is misaligned. But if only one tire shows this type of wear (and if both front tires have been on the car for some time), then it indicates a bent steering arm. This causes one wheel to toe in more than the other.

Cornering wear (Fig. 54-7) caused by taking curves at excessively high speeds, can be mistaken for camber wear or toe-in or toe-out wear. Cornering wear is due to centrifugal force acting on the car and causing the tires to roll as well as skid on the road. This produces a diagonal type of wear which rounds the outside shoulder. In severe cornering wear, fins or sharp edges will be found along the inner edges of the tire treads. There is no adjustment that can be made to correct the steering system for this type of wear. The only preventive is for the driver to slow down on curves.

Uneven tire wear such as shown in Fig. 54-7, where the tread is unevenly or spottily worn, can result from a number of mechanical conditions. These include misaligned wheels, unequally or improperly adjusted brakes, unbalanced wheels, and incorrect linkage adjustments.

High-speed operation causes much more rapid tire wear because of the high temperature, greater amount of scuffing, and rapid flexing to which the tires are subjected. The chart (Fig. 54-8) shows how tire wear increases with car speed. According to the chart, tires wear more than three times faster at 70 mph (miles per hour) [112.65 kmph] than they do at 30 mph [48.280 kmph]. More careful, slower driving and correct tire inflation increase tire life greatly.

54-13 Hard or Rough Ride

A hard or rough ride could be due to excessive tire pressure, improperly operating shock absorbers, or excessive friction in the spring suspension. Correct it by lubricating springs, shackles, and bushings (on types where lubrication is specified), and by loosening the shock-absorber arm linkages, shackle bolts, and U bolts. Then retighten the U bolts, shackle bolts, and shock-absorber linkages in that order. This procedure permits realignment of parts that have slipped to cause excessive friction.

Shock-absorber action on cars giving a hard or rough ride may be first checked by bouncing each corner of the car in turn. Seize the bumper, pull up and push down on it several times so that the car bounces, and then release the bumper. (Some direct-acting shock absorbers, as, for instance, those used on some Plymouths, cannot be tested in this way. They are valved to permit slow spring oscillations in the interest of smoother riding.) If the shock absorber is operating normally, the car will come to rest immediately. If the car continues to bounce after the bumper is released, the shock absorber is probably defective. A more accurate check can be made by disconnecting the shock-absorber linkage. Then operate the shock absorber, and note the resistance to movement. If the resistance is small or is not uniform through the full stroke, or if the movement is very stiff, the shock absorber requires service. Shock absorbers are serviced by complete replacement.

54-14 Sway on Turns

Sway of the car body on turns or on rough roads may be due to a loose stabilizer bar. The attachments of the bars to the frame, axle housing, and suspension arms should be checked. Weak or sagging springs could also cause excessive sway. If the shock absorbers are ineffective, they may permit excessive spring movement. This could cause strong body pitching and sway, particularly on rough roads. If the caster is excessively positive, it will cause the car to roll out, or lean out, on turns. This requires front-wheel realignment.

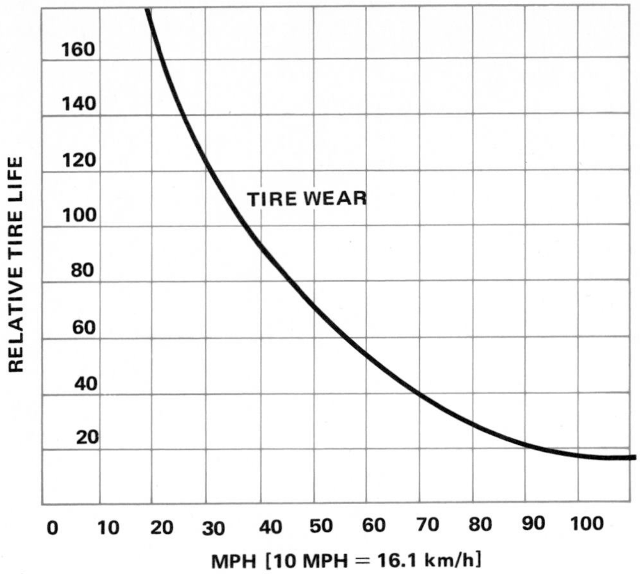

Fig. 54-8 How tire wear increases with speed.

54-15 Leaf-Spring Breakage

Breakage of leaf springs can result from excessive overloading; loose U bolts, which cause breakage near the center bolt; loose center bolt, which causes breakage at the center-bolt holes; an improperly operating shock absorber, which causes breakage of the master leaf; or tight spring shackle, which causes breakage of the master leaf near or at the spring eye. Determining the point at which breakage has occurred will indicate the cause.

54-16 Sagging Springs

Springs will sag from overloading or if they have become weak (as, for example, from habitual overloading). Loss of the shim from the coil-spring seat on the coil-spring suspension (from failure to return it during overhaul) will cause the spring to seem shorter and to sag. Not all coil springs require or use shims. Defective shock absorbers may tend to restrict spring action and thus make them appear to sag more than normal.

54-17 Noises

Noises produced by spring or shock-absorber difficulties are usually either rattles or squeaks. Rattling noises can be produced by looseness of such parts as spring U bolts, metal spring covers, rebound clips, spring shackles, and shock-absorber linkages or springs. These can generally be located by a careful examination of the suspension parts. Spring squeaks can result from lack of lubrication in the spring shackles or at spring bushings (on the type requiring lubrication), or in the spring itself (on leaf springs requiring lubrication). Shock-absorber squeak could result from tight or dry bushings. Steering-linkage rattles may develop if linkage components become loose. Under exceptional circumstances, squeaks during turns could develop owing to lack of lubrication in steering-linkage joints or bearings. This would, of course, also produce hard steering.

● 54-18 Servicing Steering Linkages and Suspensions

Steering and suspension service includes removal, replacement, and adjustment of tie rods; removal and replacement of other linkage parts, such as the steering idler and the upper and lower control arms; removal and replacement of springs; and removal and replacement of the wheel hub and drum. In addition, the steering gear may require adjustment or removal, overhaul, and reinstallation. Also, the front wheels may require alignment (adjustment of caster, camber, and toe-in).

For service on any of these components, refer to the shop manual for the model and make of car being worked on. Discussions of front-end alignment, wheel balance, and steering-gear adjustments and service follow.

● 54-19 Wheel Alignment

There are many different types of wheel aligners. Some are mechanical types that attach to the wheel spindles (Fig. 54-9). Others have lights that display the measurements on a screen in front of the car (Fig. 54-10). In the front-wheel alignment job, you check camber, caster, and toe-in. These are the three elements you adjust. You also measure steering-axis inclination and toe-out on turns. These are not adjustable. If they are out of specifications, it means parts are bent or damaged and must be replaced. Before you make the alignment checks, however, here are the preliminary things you must do.

Check and correct tire pressure.
Check and adjust wheel bearings.
Check and adjust wheel run-out.
Check ball joints and, if they are too loose, replace them.
Check steering linkages, and make any corrections necessary.
Check wheel balance, and correct it if necessary.
Check rear leaf springs for cracks, broken leaves, and loose U bolts. Make any corrections necessary.
Check front-suspension height on Chrysler-built cars.
Check shock absorbers, and replace them if they are defective.
Check wheel tracking. That is, check whether the rear wheels follow the front wheels or are off the track. If the wheels are off the track, it usually means a bent frame. The frame must be straightened before you can do an alignment job.

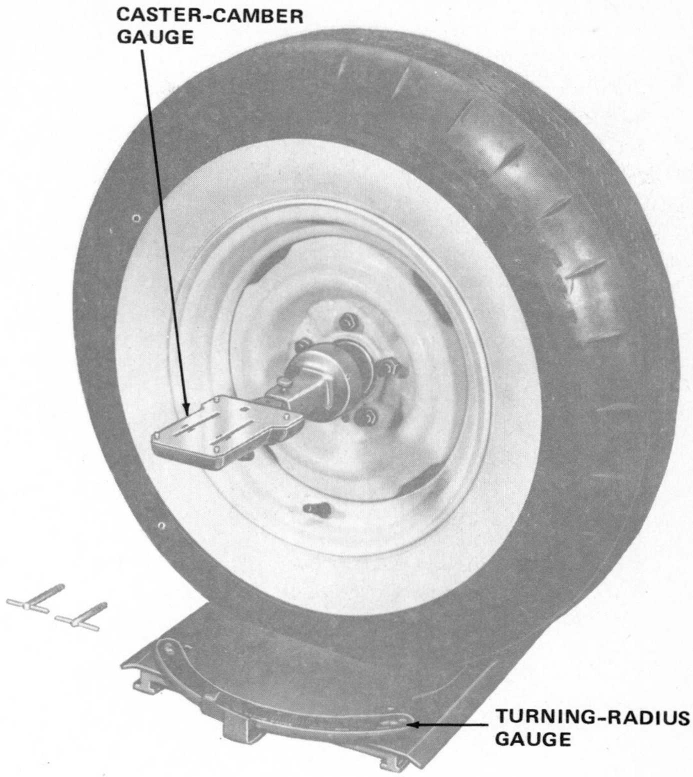

Fig. 54-9 Front wheel on a turning-radius gauge with the caster-camber tester attached to the wheel hub. (*Bear Manufacturing Company*)

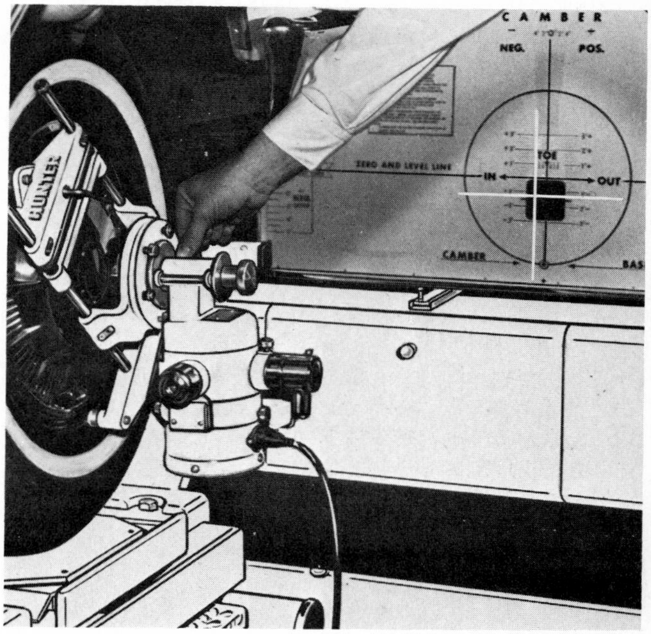

Fig. 54-10 Adjusting the wheel gauge to make a wheel-alignment check. Note the horizontal and vertical lines of light shining on the screen. (*Hunter Engineering Company*)

● 54-20 Wheel Balance

The wheel may be checked for balance on or off the car. This job is done by either of two methods: *static* or *dynamic*. In static balancing, the wheel is taken off the car and put on a machine that detects any imbalance (Fig. 54-11). A wheel that is out of balance is heavier in one section. This will cause the bubble in the center of the

Fig. 54-11 Bubble-type static balancer. (*John Bean Division of FMC Corporation*)

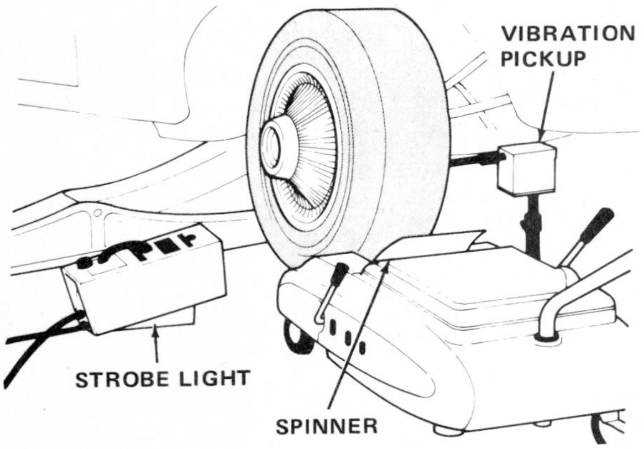

Fig. 54-12 An electronic wheel balancer. A magnet is attached to the brake backing plate. Through a short arm, any movement of the magnet is sensed by a vibration pickup. This causes the strobe light to flash, indicating where to attach a wheel weight. (*Ford Motor Company*)

balancer to move off center. To balance the wheel, weights are added until the bubble returns to center.

To dynamic-balance (or "spin-balance") a wheel, the wheel is run at high speed either on or off the car. Figure 54-12 shows an elctronic wheel balancer being used to balance a wheel on a car. Lack of balance shows up as a tendency for the wheel to move off center or out of line as it spins.

If the wheel is out of balance, one or more weights are installed on the wheel rim (Fig. 54-13). In the shop, you will learn how to balance wheels.

CAUTION: To prevent injury from stones thrown out of the spinning tire, off-the-car wheel spinners should have a safety hood (Fig. 54-14). The hood fits around or over the tire while it is spinning, and will catch any stones that are thrown out.

54-21 Adjusting Camber and Caster

Several different ways to adjust camber and caster have been used. Some of the methods include removing and installing shims, turning a cam, shifting the inner control-arm shaft, and changing the length of the strut rod. Let us look at some of these methods.

1. Adjustment by installing or removing shims The shims are located at the upper control-arm shafts. They

Fig. 54-13 Balancing a wheel by placing a weight on the wheel rim. (*Bear Manufacturing Company*)

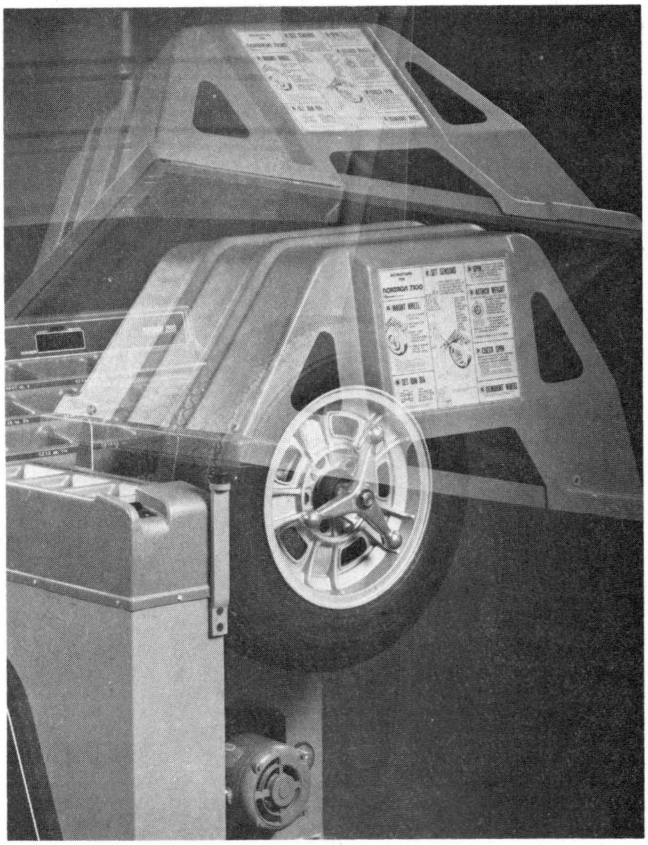

Fig. 54-14 Wheel balancer with a safety hood installed over the tire. (*Hennessy Industries, Inc.*)

are placed either inside or outside the frame bracket. Figure 54-15 shows the location of the shims in many General Motors cars. The shims are inside the frame bracket. Figure 54-16 shows the location of the shims in many Ford cars. The shims are outside the frame bracket. When the shims are inside the frame bracket (Fig. 54-15), adding shims moves the upper control arm inward. This

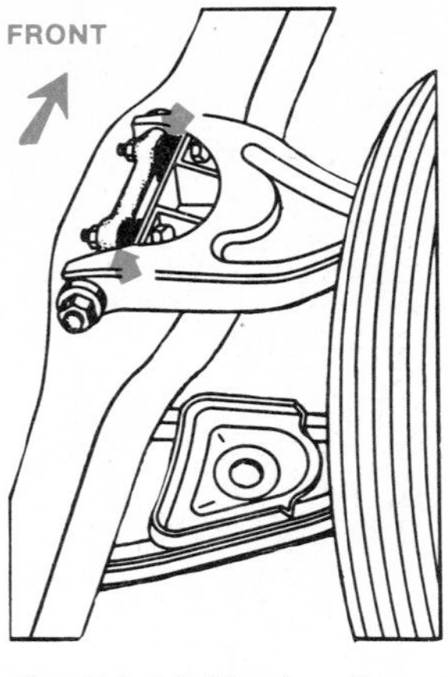

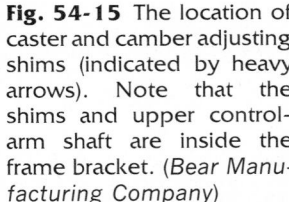

Fig. 54-15 The location of caster and camber adjusting shims (indicated by heavy arrows). Note that the shims and upper control-arm shaft are inside the frame bracket. (*Bear Manufacturing Company*)

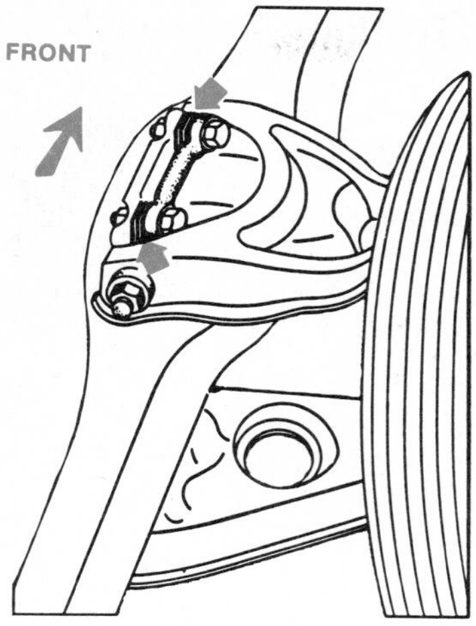

Fig. 54-16 The location of caster and camber adjusting shims (indicated by heavy arrows). Note that the shims and upper control-arm shaft are outside the frame bracket. (*Bear Manufacturing Company*)

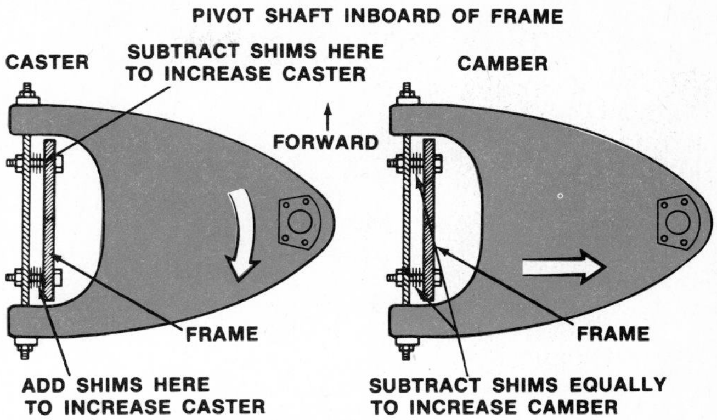

Fig. 54-17 Caster and camber adjustments on some cars using shims. (*Chevrolet Motor Division of General Motors Corporation*)

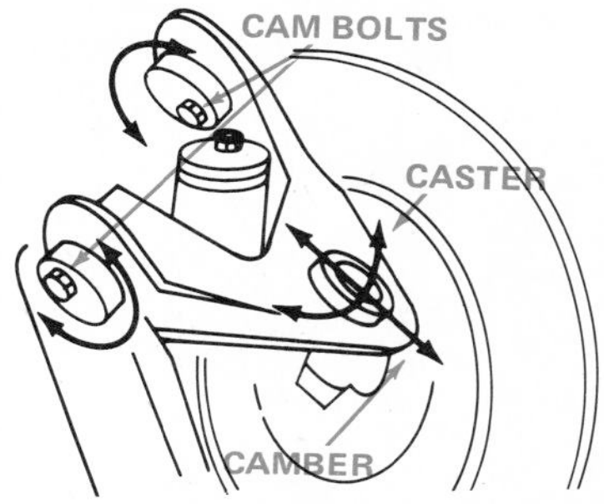

Fig. 54-18 Turning the cam bolts moves the upper control arm toward or away from the frame to adjust caster and camber. (*Ammco Tools, Inc.*)

reduces positive camber. When the shims and shaft are outside the frame bracket (Fig. 54-16), adding shims moves the upper control arm outward. This increases positive camber. If shims are added at one attachment bolt and removed from the other, the outer end of the upper control arm shifts one way or the other. This increases or decreases caster. Figure 54-17 shows these adjustments.

2. Adjustment by turning a cam There have been several variations of this method. Figure 54-18 shows the arrangement used on some Chrysler-built cars. The two bushings at the inner end of the upper control arm are attached to the frame brackets by two attachment bolts and cam assemblies. When the cam bolts are turned, the camber and caster are changed. If both are turned the same amount and in the same direction, the camber is changed. If only one cam bolt is turned, or if the two are turned in opposite directions, the caster is changed.

3. Adjustment by shifting inner shaft This system uses slots in the frame at the two points where the inner shaft is attached (Fig. 54-19). When the attaching bolts are loosened, the inner shaft can be shifted in or out to change camber. Only one end is shifted to change caster.

4. Adjustment by changing length of strut rod This type of adjustment is shown in Fig. 54-20.

54-22 Adjusting Toe-In

After correcting caster and camber, toe-in is adjusted (Fig. 54-21). Place the front wheels in the straight-ahead position. Then check the positions of the spokes in the steering wheel. If they are not centered, they can be

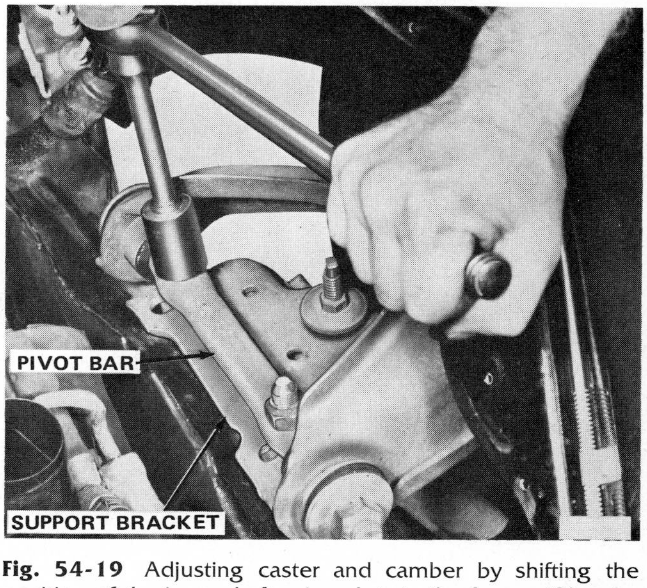

Fig. 54-19 Adjusting caster and camber by shifting the position of the inner shaft using slots in the frame. (*Chrysler Corporation*)

properly positioned when toe-in is set. Toe-in is adjusted by turning the adjuster sleeves in the linkage. If the adjuster sleeves are turned to lengthen the tie rods, the toe-in is increased.

54-23 Servicing the Steering Gear

Manual steering gears have two basic adjustments. One of these takes up the worm-gear and steering-shaft end

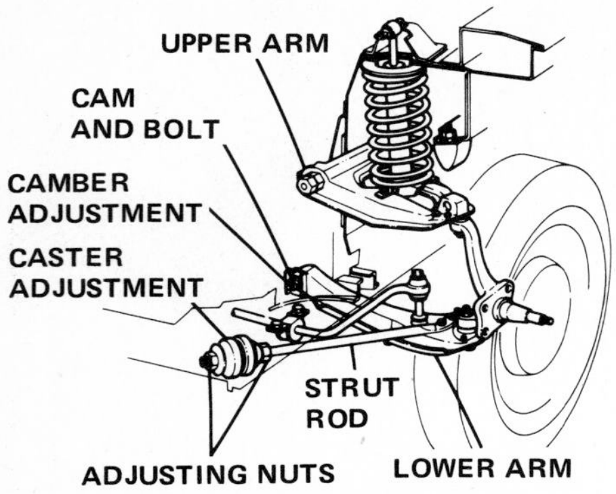

Fig. 54-20 Adjusting caster by changing the length of the strut rod. (*Ford Motor Company*)

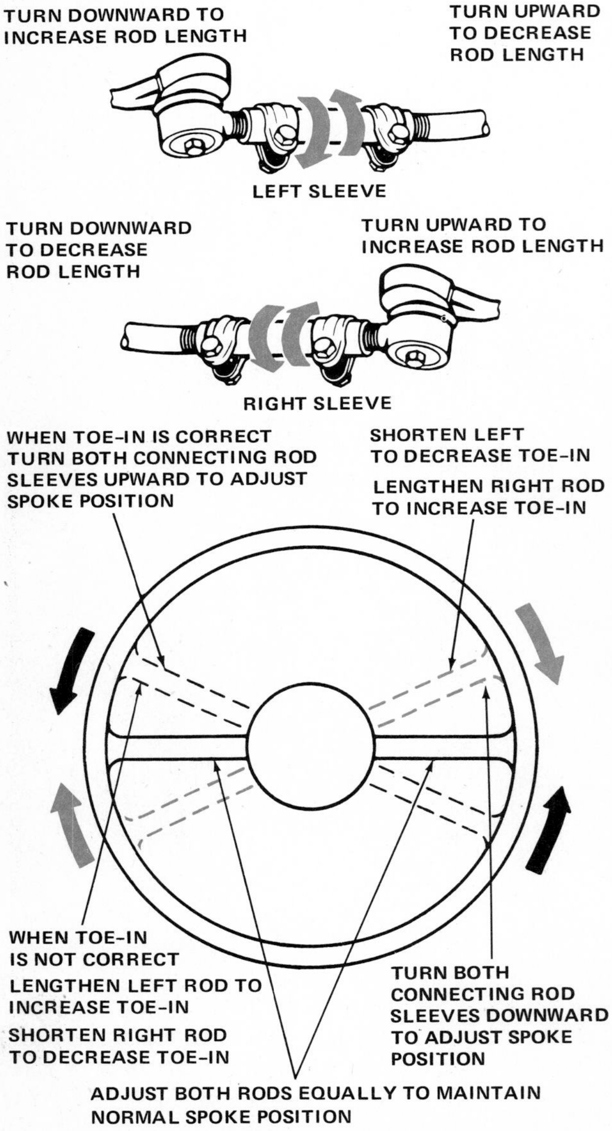

Fig. 54-21 Adjustments for toe-in and steering-wheel alignment. Top, spindle connecting-rod (tie-rod) adjustments. Bottom, adjustments to align steering wheel. (*Ford Motor Company*)

play. The other adjusts the backlash, or free play, between the worm and sector.

Other adjustments are required on power-steering gears. Refer to the manufacturer's shop manual covering the unit being serviced before attempting to adjust or repair a power-steering gear.

REVIEW QUESTIONS

1. Name four causes of excessive play in the steering system.
2. Name five causes of hard steering.
3. Name three causes of car wander.
4. Name five causes of the car pulling to one side during normal driving.
5. What could cause the car to pull to one side during braking?
6. Explain the difference between front-wheel shimmy and front-wheel tramp.
7. Name four causes of front-wheel shimmy.
8. Name three causes of front-wheel tramp.
9. Name four causes of steering kickback.
10. Name four causes of tire squeal on turns.
11. What would cause the tire tread to wear in the center? On the sides, or shoulders?
12. What would cause the tire tread to wear on one side or shoulder only?
13. Name seven things that should be done before caster, camber, and toe-in are checked and adjusted.
14. Describe the two locations of caster and camber adjusting shims.
15. Explain how to adjust caster by installing or removing shims.
16. Explain how to adjust camber by installing or removing shims.
17. Explain how to adjust camber by turning a cam.
18. Explain how to adjust toe-in.

SELF PROJECTS

1. Make a list of possible troubles in the steering and suspension systems. Making the list will help you remember them. Notice the various causes of the troubles.
2. If you really want to learn the troubles and their causes backward and forward, make yourself a set of 3-by-5-inch cards. On one side, write the complaints and the number of possible causes. These are listed in the trouble-diagnosis chart in the first part of the chapter. Then, on the other side, list the possible causes. Study these cards whenever you have a chance. Look at a complaint, and then try to remember what is on the other side of the card.
3. Refer to the instruction booklet covering the front-alignment equipment in the shop. Read the procedures outlined for checking and adjusting caster, camber, and toe-in. Read the procedure for balancing the wheels. Make notes of the key points in these procedures.

CHAPTER 55
Automotive Brakes

After studying this chapter, you should be able to:

- **Describe the construction and operation of a dual-brake system.**
- **Describe the operation of drum brakes.**
- **Describe the operation of a disk-brake system.**
- **Explain the purpose of the brake warning light and how it works.**
- **Describe the construction and operation of an antilock brake system.**
- **Explain how a power brake works.**

This chapter describes the construction and operation of the brakes used on automotive vehicles. Most automotive brakes used today are hydraulically operated. Therefore, we review hydraulic principles and explain their application to brakes. Also, because brakes operate by friction, the principles of friction are discussed. There are two types of wheel brake mechanisms, *drum* and *disk*. The drum type uses curved shoes that fit the curvature of the drums. The disk type uses flat shoes or pads that fit the flat surfaces of the disks.

55-1 FRICTION

Friction is the resistance to motion between two objects in contact with each other. Three types of friction are dry, greasy, and viscous. Generally, we are concerned only with dry friction in connection with brakes. (But sometimes we have greasy friction if the brake linings are greasy or oil-soaked.) Friction varies according to the pressure applied between the sliding surfaces, the roughness of the surfaces, and the material of which the surfaces are made. Suppose, for example, that a platform and its load weigh 110 pounds (lb) or 50 kilograms (kg). Suppose it takes 55 pounds or 25 kilograms of pull to move the platform along the floor (Fig. 55-1). Now we reduce the load, so that the platform and load weigh only 11 pounds or 5 kilograms. We would find that it requires only 5.5 pounds or 2.5 kilograms of pull to move it along the floor. *Friction varies with the load.*

Now suppose we smoothed the floor and the sliding part of the platform with sandpaper. Then it would require less pull to move the platform on the floor. *Friction varies with the roughness of the surfaces.*

Friction also varies with the *type of material*. For example, dragging a 110-pound (50-kilogram) bale of rubber across a concrete floor might require a pull of 66 pounds or 30 kilograms (Fig. 55-2). But to drag a 110-pound or 50-kilogram cake of ice across the same floor might require a pull of only 2.2 pounds or 1 kilogram.

55-2 Friction of Rest and Motion

It requires more force to put an object into motion than it does to keep it in motion (Fig. 55-3). In the example shown, it takes two men to get the object started. Once it is started, one man can keep it moving. Thus, the friction of an object at rest is greater than the friction of an object in motion.

Engineers refer to these two kinds of friction as *static friction* and *kinetic friction*. The word "static" means at rest. The word "kinetic" means in motion, or moving.

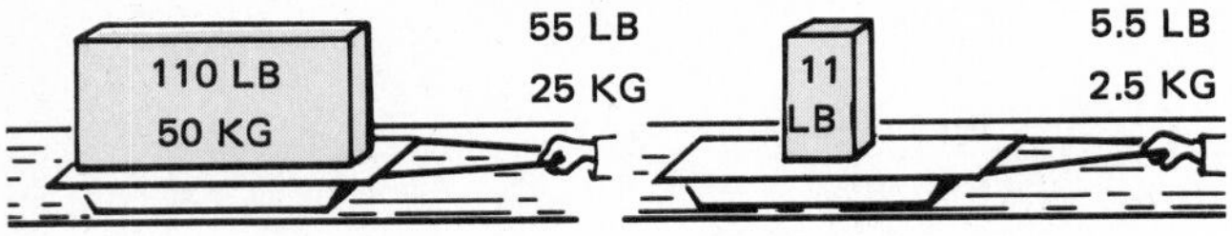

Fig. 55-1 Friction varies with the load applied between the sliding surfaces.

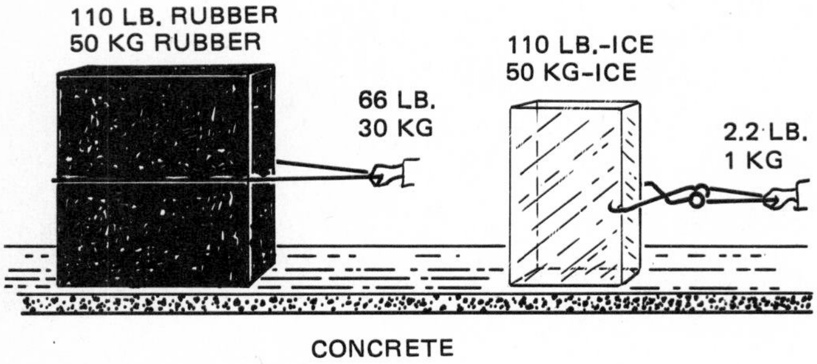

Fig. 55-2 Friction varies with the type of material.

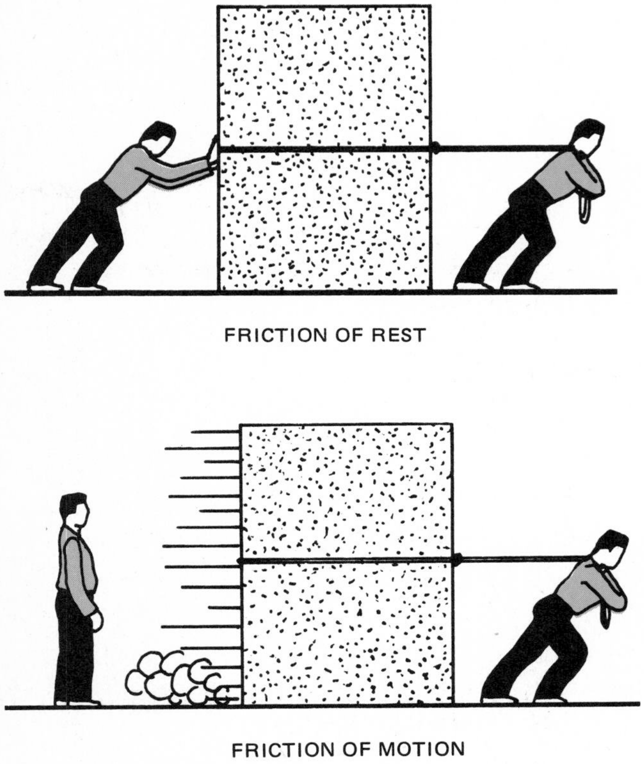

Fig. 55-3 Friction of rest is greater than friction of motion. In the example shown, it takes two men to overcome the friction of rest. But one man can keep the object moving by overcoming the friction of motion.

Thus, static friction is friction of rest, and kinetic friction is friction of motion.

● 55-3 Friction in the Car Brakes

Friction is used in the car braking system. The friction between the brake drums or disks and brake shoes slows or stops the car. This friction slows the rotation of the wheels. Then friction between the tires and the road slows the motion of the car. Note that it is the friction between the tires and the road that stops the car. That being the case, would the car stop more quickly if the wheels were locked (so that the tires skidded on the road)? The answer is that the car would not. If the brakes are applied so hard that the wheels lock, then the friction between the tires and road is kinetic friction (friction of motion as the tires skid on the road). When the brakes are applied a little less hard, the wheels are permitted to continue rotating. Then it is static friction that works between the tires and road. The tire surface is not skidding on the road but is rolling on it. Since this produces static friction between the road and tires, there is greater braking effect. The car stops more quickly if the brakes are applied just hard enough to get maximum static friction between the tires and road. If the brakes are applied harder than this, then the wheels will lock. The tires will skid, or slide, and the lesser kinetic friction will result. See ● 55-24 on anti-wheel-lock devices.

● 55-4 Brake Action

A typical hydraulic braking system includes two essentials, the master cylinder with brake pedal and the wheel brake mechanism. The other parts are the connecting tubing, or brake lines, and the supporting arrangements.

Braking action starts at the brake pedal. When the pedal is pushed down, brake fluid is sent from the master cylinder to the wheels. At the wheels, the fluid pushes brake shoes, or pads, against revolving drums or disks. The friction between the stationary shoes or pads and the revolving drums or disks slows and stops them. This slows or stops the revolving wheels which, in turn, slow or stop the car.

Figure 55-4 shows the brake lines, or tubes, through which the fluid flows. Note that there are two chambers and two pistons in the master cylinder. One chamber is connected to the front-wheel brakes. The other chamber is connected to the rear-wheel brakes. This braking arrangement is called a *dual braking system*. Dual braking systems are used on modern cars. The purpose of splitting the system into two parts is that, if one part fails, the other can still work and stop the car.

In earlier braking systems, there was only one chamber in the master cylinder. It was connected to all four wheel brakes. If one part failed, the whole system failed. The dual braking system provides extra protection, because the rear and front sections seldom fail at once.

● 55-5 Dual Braking System

In older-model cars, the master cylinder contained only one piston, as shown in Fig. 55-5. The modern dual braking system uses a two-piston master cylinder. A schematic layout of a dual braking system is shown in Fig. 55-6. The system shown uses a power-brake unit, which we describe later. As you can see, one brake line

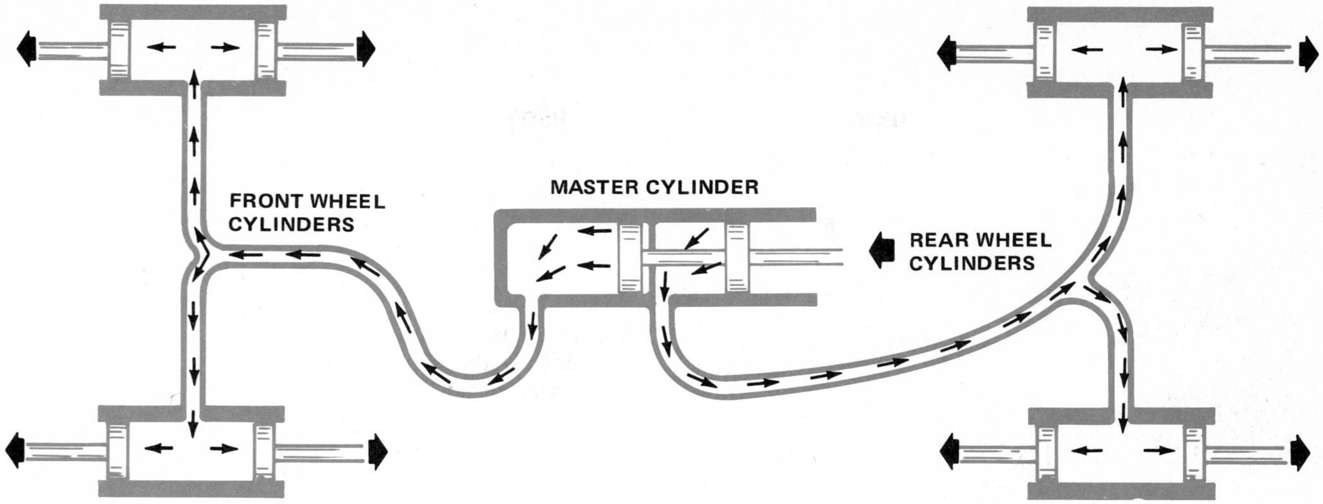

Fig. 55-4 Flow of brake fluid to the four wheel cylinders when the pistons are pushed into the master cylinder.

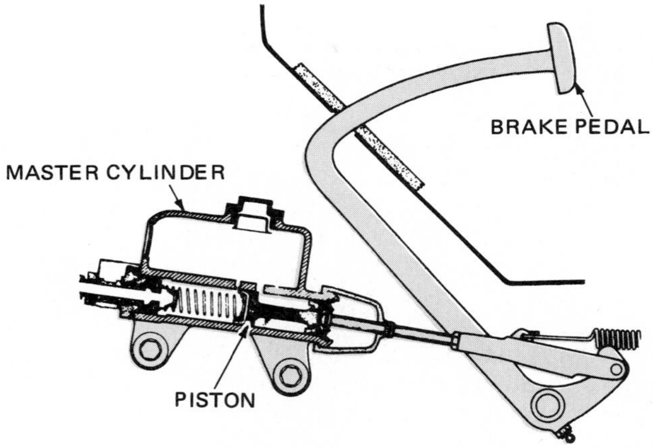

Fig. 55-5 Linkage between brake pedal and master cylinder. (*Pontiac Motor Division of General Motors Corporation*)

from the master cylinder goes to the rear wheels. The other brake line goes to the front wheels. The brake-fluid pressure must first pass through a pressure-differential valve. This valve turns on a red light on the instrument panel if either the rear or the front brakes fail. We cover the pressure-differential valve in ● 55-21.

Two different types of braking devices are used at the wheels: a drum brake and a disk brake. Although both the drum brake and the disk brake do the same job, they are constructed differently. We cover drum brakes and disk brakes in later sections.

Most cars have dual braking systems today, with one brake line working the front-wheel brakes and the other brake line working the rear-wheel brakes. However, some cars, particularly the smaller ones, have a diagonally split braking system. That is, one brake line takes care of the right front and left rear brakes. The other takes care of the left front and right rear brakes. With this system, if one part fails, there will always be braking at the front and at the rear. This provides a more balanced control during an emergency. This system uses the same type of

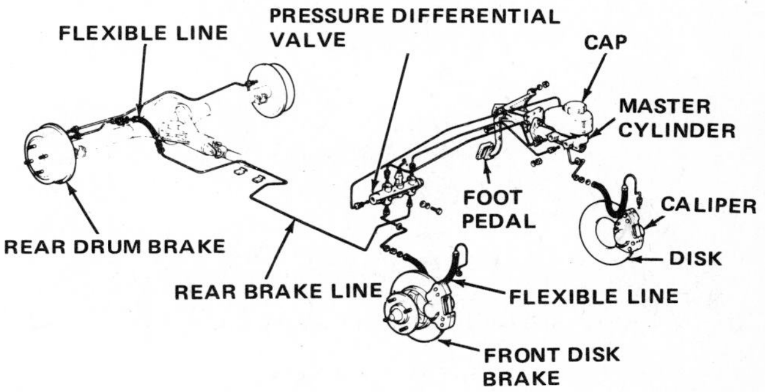

Fig. 55-6 Layout of a dual braking system. (*Ford Motor Company*)

warning light to tell the driver if one of the two systems has failed.

CAUTION: If the red light comes on, warning of failure of one of the two circuits, service should be sought at once. Although the second circuit can give emergency braking, it is not safe to operate the car depending on only half the brakes.

● 55-6 Master Cylinder

Figure 55-7 is a cutaway view of a dual master cylinder. Figure 55-8 shows a similar dual master cylinder cut away and, above it, the component parts disassembled. When the push rod is pushed into the cylinder, the two pistons push brake fluid out of the master cylinder and through the brake lines to the brakes. One piston sends brake fluid to the front-wheel brakes. The other piston sends brake fluid to the rear-wheel brakes.

● 55-7 Brake Lines

The brake fluid is carried by steel pipes, called *brake lines,* from the master cylinder to connecting points.

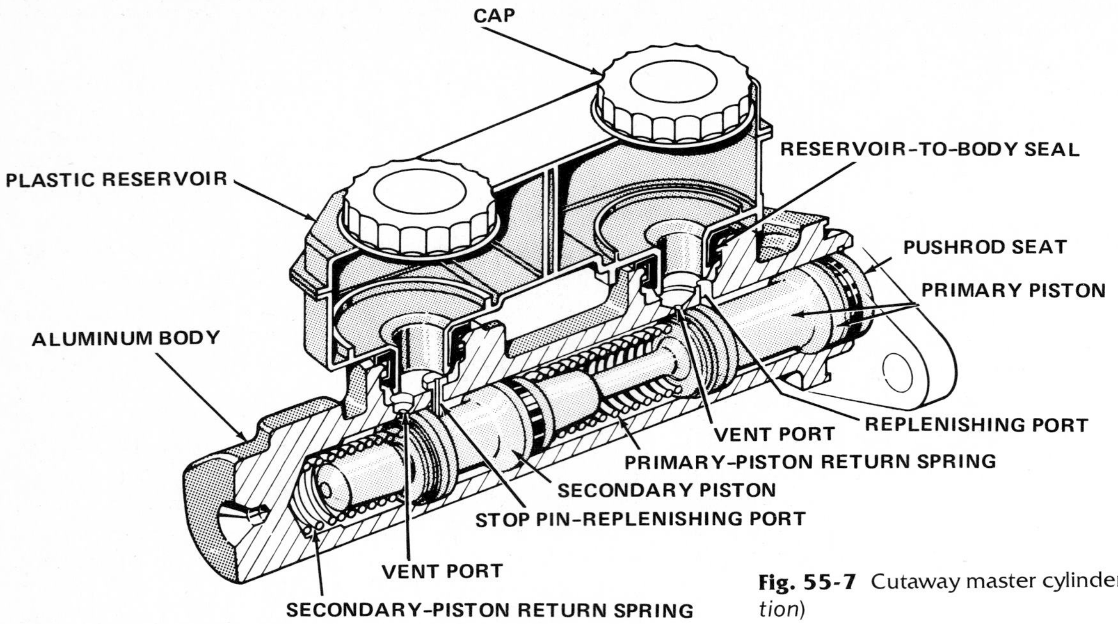

Fig. 55-7 Cutaway master cylinder. (*Chrysler Corporation*)

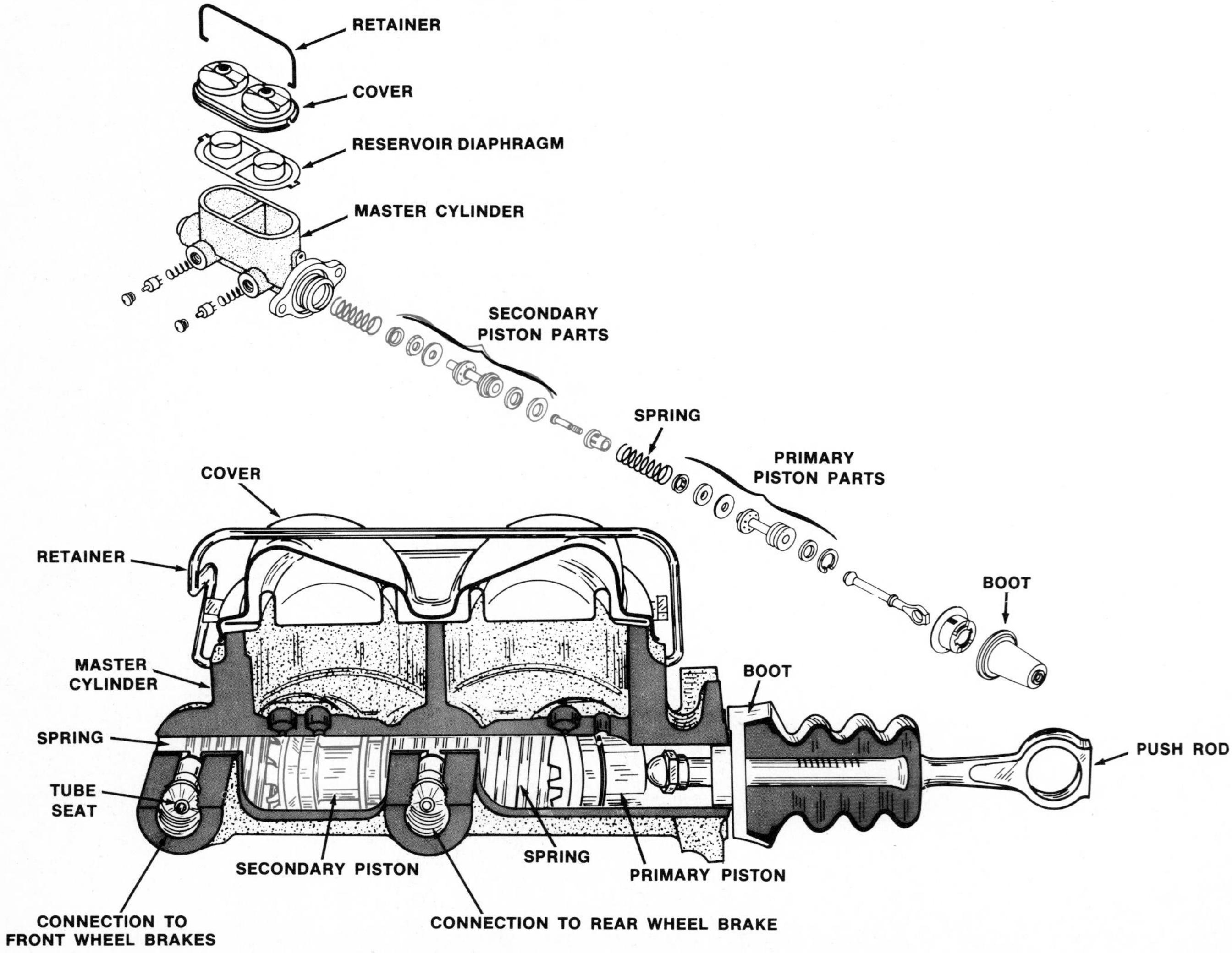

Fig. 55-8 Cutaway of a master cylinder with all parts shown above it in exploded view.

These points are located on the car frame near the four wheels. From these connecting points, a special flexible hose is used to connect the brake lines to the brake mechanisms at the wheels.

Copper or aluminum brake lines must not be used. They may fail from the high pressure.

● 55-8 Brake Fluid

The brake fluid is *not oil!* Brake fluid is a very special fluid that is little affected by high or low temperatures. Also, brake fluid does not damage the metal and rubber parts in the braking system. Ordinary oil will damage these parts. For this reason, only the brake fluid recommended by the manufacturer should be put into the brake system.

CAUTION: Never put ordinary oil in a brake system. Ordinary oil will cause rubber parts in the system, such as the piston cups, to swell and go to pieces. This could cause complete brake failure and lead to a fatal accident. Never use anything but the brake fluid recommended by the car manufacturer!

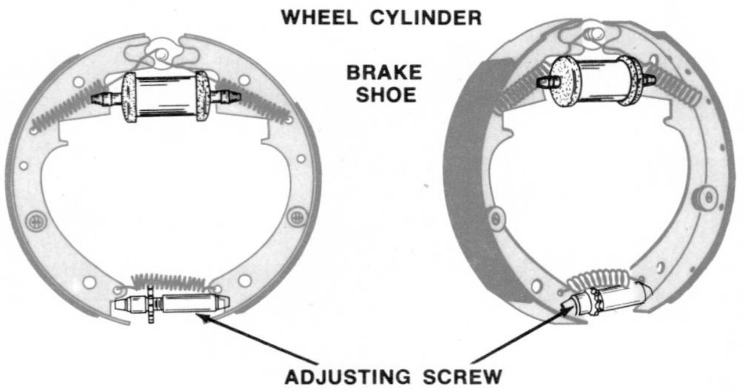

Fig. 55-9 Operating components of drum brakes for rear (left) and front (right) wheels.

● 55-9 Drum Brakes

The drum brake has a steel or iron drum to which the wheel is bolted. The drum and wheel rotate together. Inside the drum, attached to the steering knuckle or axle housing, is the brake mechanism. The brake mechanism at the front wheels is attached to the steering knuckle. In Fig. 52-1 you can see the boltholes by which the front-wheel brake mechanism is attached. The brake mechanism at the rear wheels is attached to the axle housing. The boltholes in the axle housing are shown in Fig. 52-11.

Figure 55-9 shows the moving parts of the drum brake. There are two brake shoes at each wheel. The bottoms of the shoes are held apart by an adjustment screw—also known as a *star wheel*. The tops of the shoes are held apart by a wheel cylinder. The shoes are made of metal. Riveted to each shoe is a facing of friction material. The facing is called the *brake lining*. These brake linings are made of tough asbestos material that can hold up under the rubbing and heat produced by braking. (See ● 55-10.)

Figure 55-10 shows the addition of the brake backing plate and the wheel spindle. The brake backing plate is bolted to the axle housing at the back wheels, or to the steering knuckle at the front wheels. At the right in Fig. 55-10, you can see the brake drum in place over the brake shoes. The drum has been partly cut away so you can see one of the shoes.

The shoes are attached at one point to the brake backing plate. The attachment is loose so that the shoes can move around a little.

● 55-10 Brake Shoes

The shoes are made of metal to which a tough asbestos lining has been added (Fig. 55-11). The lining has to be tough. During hard braking, the shoe may be pressed against the drum with a force as great as 1,000 pounds

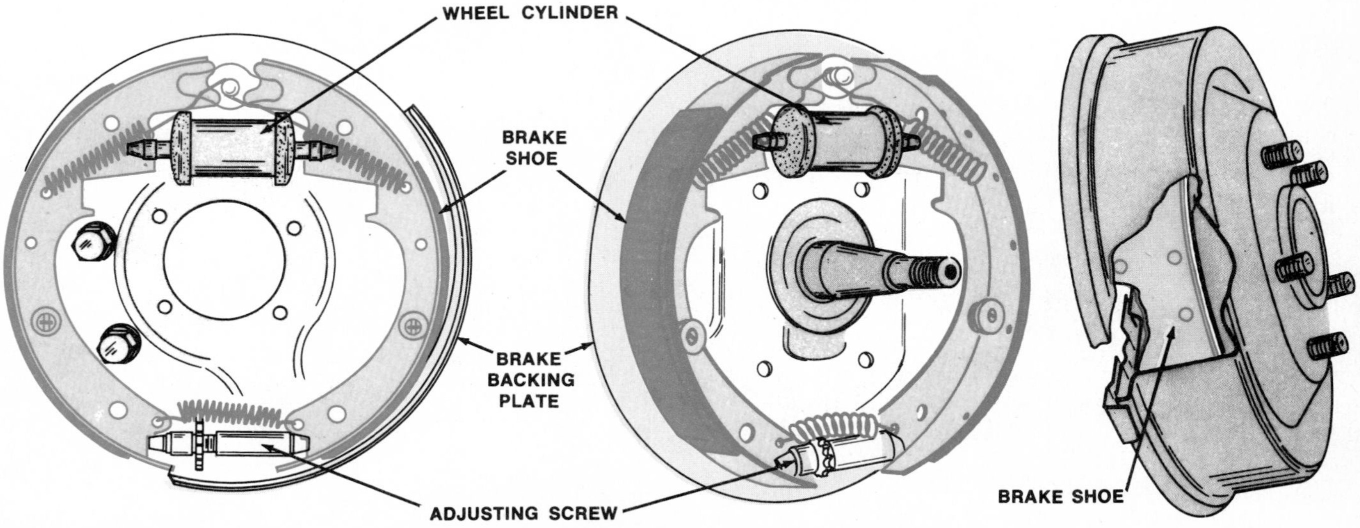

Fig. 55-10 Here, we have added the brake backing plates to both brakes, and the wheel spindle to the front brake. (right) The brake drum has been partly cut away to show the shoe inside.

[453 kg]. Since friction increases with load (force), this produces a strong frictional drag on the brake drum. It has a strong braking effect on the wheel.

A great deal of heat is also produced by the friction between the brake shoes and drum. When you rub your hands together vigorously, they become warm. Similarly, when the drum rubs against the shoe, the drum and shoe get warm. In fact, under extreme braking conditions, temperatures may reach 500°F (260°C). Some of this heat goes through the brake linings to the shoes and backing plate. From there, it is radiated to the surrounding air. But most of it is absorbed by the brake disk or drum. Some brake drums have cooling fins to get rid of the heat more quickly. Some disks are ventilated. That is, they have hollow spaces between the two sides. There are fins in these areas to help get rid of the heat. Excessive temperatures are not good for brakes, since they may char the brake linings. Also, with the linings and drums hot, braking action is less effective. This is the reason that brakes "fade" or lose braking ability when they are used continuously for long periods, as in driving down a mountain.

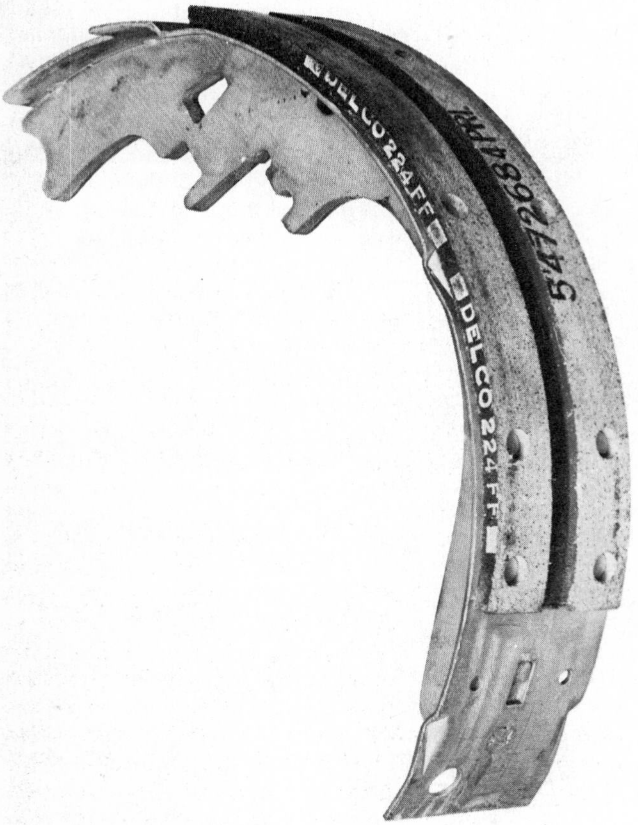

Fig. 55-11 Brake shoe of the type for drum brakes.

The pistons in the wheel cylinders (● 55-11) are usually larger at the front wheels. This is because, when the brakes are applied, the forward momentum of the car throws more of the weight on the front wheels. A stronger braking effort at the front wheels is therefore necessary to achieve balanced braking.

● 55-11 Wheel Cylinders

Figures 55-12 and 55-13 show the construction of a wheel cylinder for a drum-type brake. Hydraulic pressure from the master cylinder, applied between the two piston cups, forces the pistons out (Fig. 55-13). Then the brake-shoe actuating pins force the brake shoes into contact with the brake drums. The piston cups are so formed that the hydraulic pressure forces them tightly against the cylinder wall of the wheel cylinder. This produces a good sealing action that holds the fluid in the cylinder.

● 55-12 Return Stroke

On the return stroke, when the driver's foot is lifted from the pedal, spring tension on the brake linkage and spring pressure against the master-cylinder piston force the piston to move back in its cylinder. Fluid now flows from the wheel cylinders to the master cylinder, as shown in Fig. 55-14. The tension of the brake-shoe springs forces the brake shoes away from the brake drums and thus

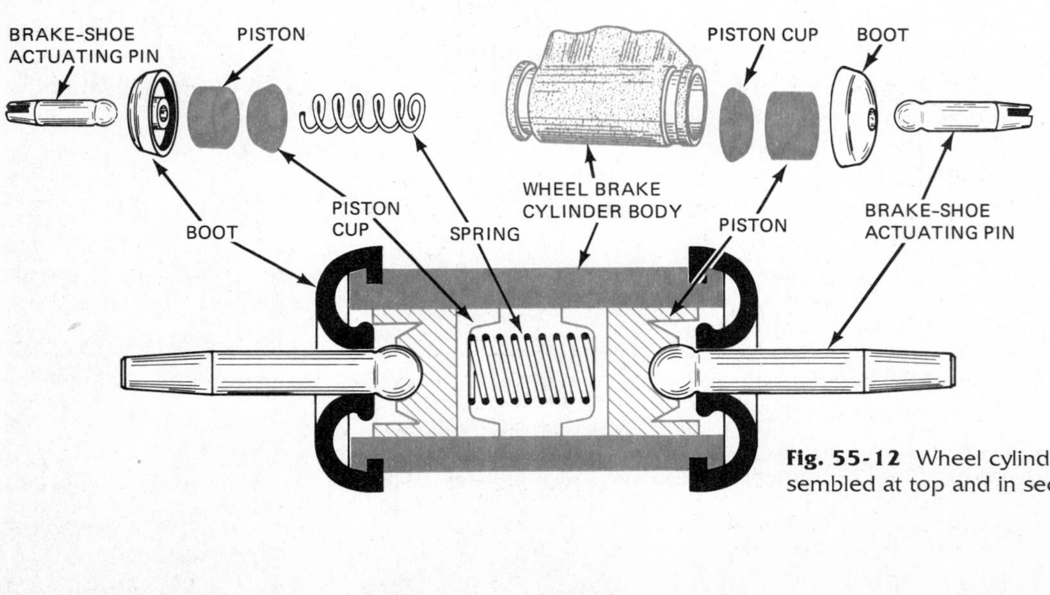

Fig. 55-12 Wheel cylinder for a drum brake, disassembled at top and in sectional view at the bottom.

pushes the wheel-cylinder pistons inward. Fluid is thus returned from the wheel cylinders to the master cylinder, as shown by the arrows. However, in the system shown, some pressure is trapped in the lines by the check valve at the end of the master cylinder (see Fig. 55-13). As the pressure drops, the check valve closes, trapping a few pounds per square inch of pressure in the lines and wheel cylinders. This pressure keeps the wheel cylinders from leaking and reduces the chances of air leaking into the system.

55-13 Self-Adjusting Brakes (Drum Type)

Most modern automotive brakes have a self-adjusting feature. It automatically adjusts the brakes when they need it as a result of brake-lining wear. Figures 55-15 and 55-16 illustrate typical arrangements. The adjustment takes place only when the car is moving backward and the brakes are applied. When this happens, an adjustment is made, if it is necessary.

Suppose the brakes are applied when the car is moving backward. Friction between the primary shoe and brake drum forces the primary shoe against the anchor pin. Then hydraulic pressure from the wheel cylinder forces the upper end of the secondary shoe away from the anchor pin and downward. This causes the adjustment lever to pivot on the secondary shoe. The lower end of the lever is forced against the sprocket on the adjustment screw. If the brake linings have worn enough, the adjustment screw is turned a full tooth. This spreads the lower ends of the brake shoes a few thousandths of an inch, or enough to compensate for the lining wear. On some cars, the adjustment feature operates with the car moving forward when the brakes are applied.

55-14 Disk Brakes

The disk (also spelled disc) brake has a metal disk instead of a drum. Figure 55-17 shows a disk brake. It has a flat shoe, or pad, located on each side of the disk. In

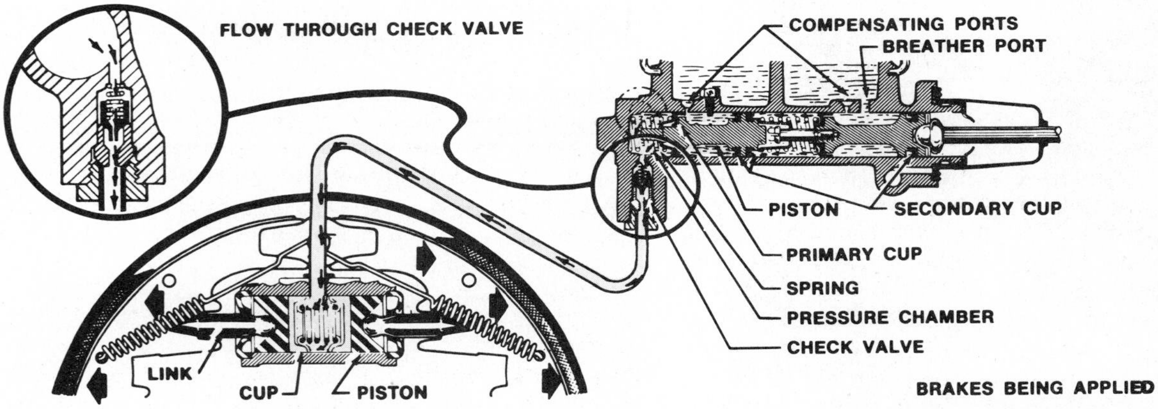

Fig. 55-13 Conditions in a drum-brake system with the brakes applied. Brake fluid flows from the master cylinder to the wheel cylinder, as shown. This causes the wheel-cylinder pistons to move outward and apply the brakes.

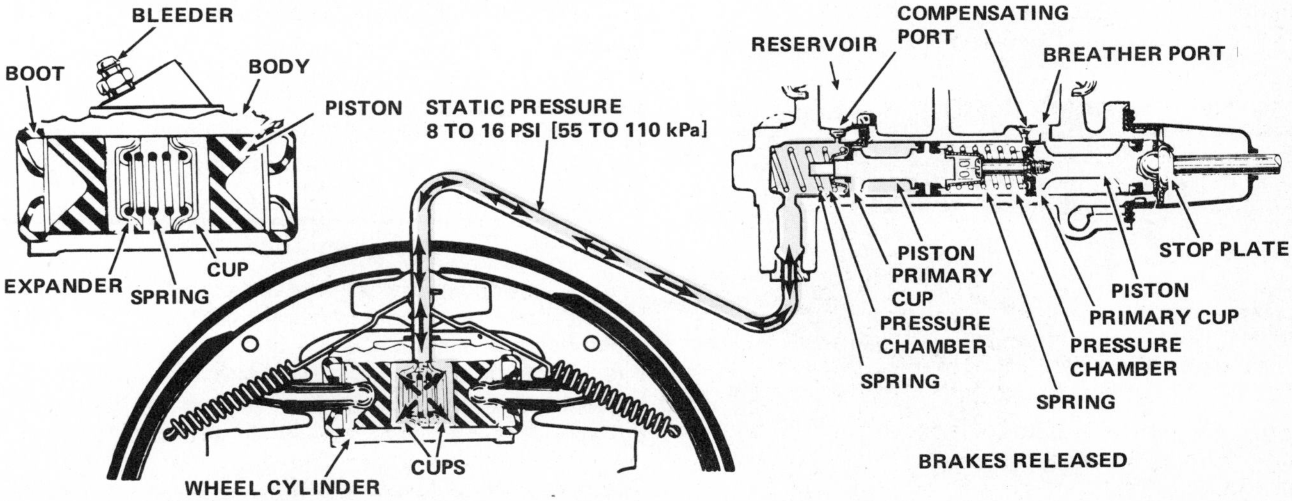

Fig. 55-14 Conditions in a drum-brake system when the brakes are released. Brake fluid flows to the master cylinder, as shown. (*Buick Motor Division of General Motors Corporation*)

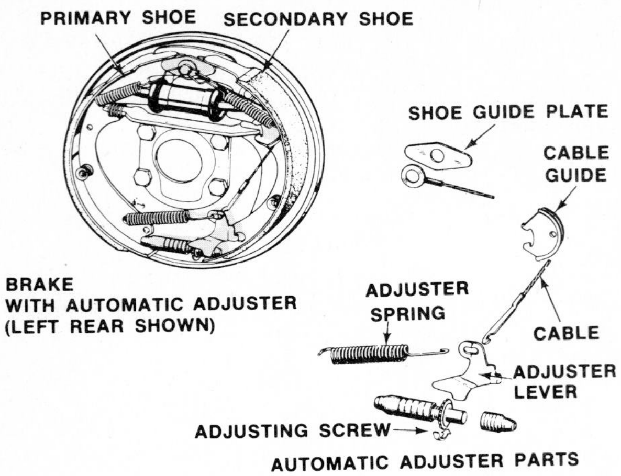

Fig. 55-15 Drum brake of the self-adjusting type. The automatic self-adjuster parts are shown to the right. *(Bendix-Westinghouse Automotive Brake Company)*

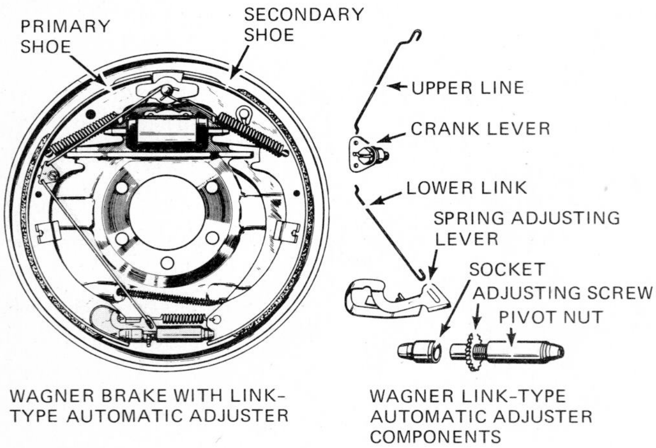

Fig. 55-16 Wagner self-adjusting drum brake with link-type adjuster parts to the right. *(Bendix-Westinghouse Automotive Brake Company)*

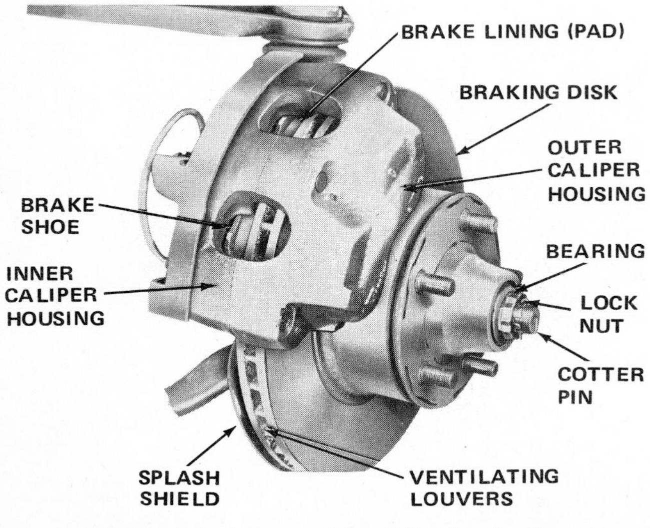

Fig. 55-17 Disk-brake assembly of the fixed-caliper type. *(Chrysler Corporation)*

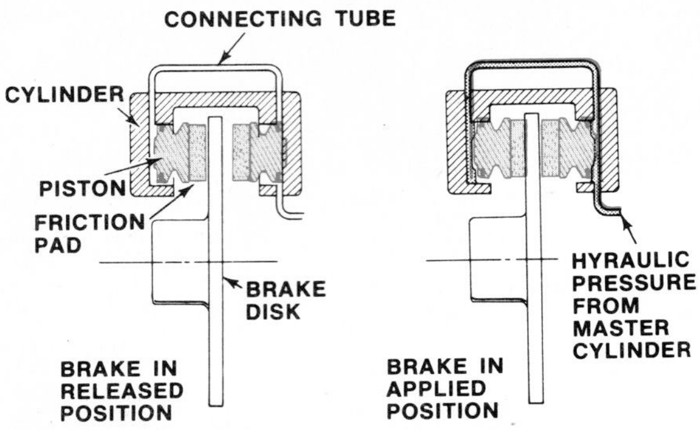

Fig. 55-18 Sectional views showing how hydraulic pressure forces friction pads inward against the brake disk to produce braking action.

operation these two flat shoes are forced tightly against the rotating disk. The shoes grip the disk, just as you would squeeze a piece of paper between your finger and thumb as you picked it up. Figure 55-18 shows how the disk brake works. Fluid pressure from the master cylinder forces the pistons to move in. This action pushes the friction pads of the brake shoes tightly against the disk. The friction between the shoes and the disk slows and stops the disk, thus providing the braking action.

● 55-15 Floating-Caliper Disk Brake

There are three general types of disk brakes: floating-caliper, fixed-caliper, and sliding-caliper. Figure 55-19 is a sectional view of a floating-caliper disk brake. Figure 55-20 is a disassembled view of the brake. The caliper is the part that holds the brake shoes on each side of the disk. In the floating-caliper brake, two steel guide pins are threaded into the steering-knuckle adapter. The caliper floats on four rubber bushings which fit on the inner and outer ends of the guide pins. The bushings allow the caliper to swing in or out slightly when the brakes are applied.

When the brakes are applied, the brake fluid flows to the cylinder in the caliper and pushes the piston out. The piston then forces the shoe against the disk. At the same time, the pressure in the cylinder causes the caliper to pivot inward. This movement brings the other shoe into tight contact with the disk. As a result, the two shoes "pinch" the disk tightly to produce the braking action.

● 55-16 Fixed-Caliper Disk Brake

This brake works in the same way as the floating-caliper brake. The only real difference is that the fixed-caliper brake has four pistons, two on each side of the disk. In this brake the caliper is firmly attached and cannot swing. In operation, the pistons on both sides of the disk push the brake shoes in against the disk.

● 55-17 Sliding-Caliper Disk Brake

The sliding-caliper brake is similar to the floating-caliper brake. The difference is that the sliding caliper is sus-

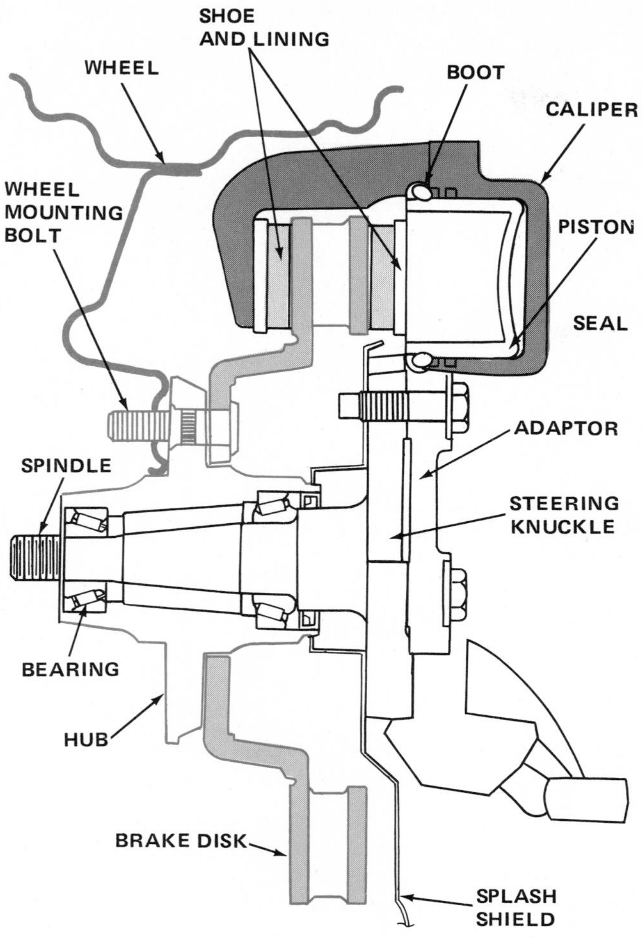

Fig. 55-19 Sectional view of a floating-caliper disk brake.

pended from rubber bushings on bolts. The arrangement permits the caliper to slide on the bolts when the brakes are applied.

● 55-18 Self-Adjustment of Disk Brakes

Disk brakes are self-adjusting. In one type of disk brake there are piston seals at the back of each piston. When the brakes are applied, the piston moves toward the disk. This distorts the piston seal, as shown at the top in Fig. 55-21. Then, when the brakes are released, the seal relaxes, or returns to its original position. This movement draws the piston away from the disk. As the brake linings wear, the piston takes a new position in relation to the seal. The brakes remain in good adjustment.

● 55-19 Telltale Tabs

Many brake shoes have "telltale" tabs, as shown in Fig. 55-22. The purpose of these tabs is to warn the driver that the brake linings are worn down and that new linings are required. When the brake linings are worn down, the tabs touch the disk when the brakes are applied. This gives off a scraping noise that warns the driver to get brake service.

● 55-20 Metering Valve

Disk-brake systems have a metering valve. This valve keeps the front brakes from applying until the rear brakes apply. If the front brakes were applied first, the car could be thrown into a rear-end skid.

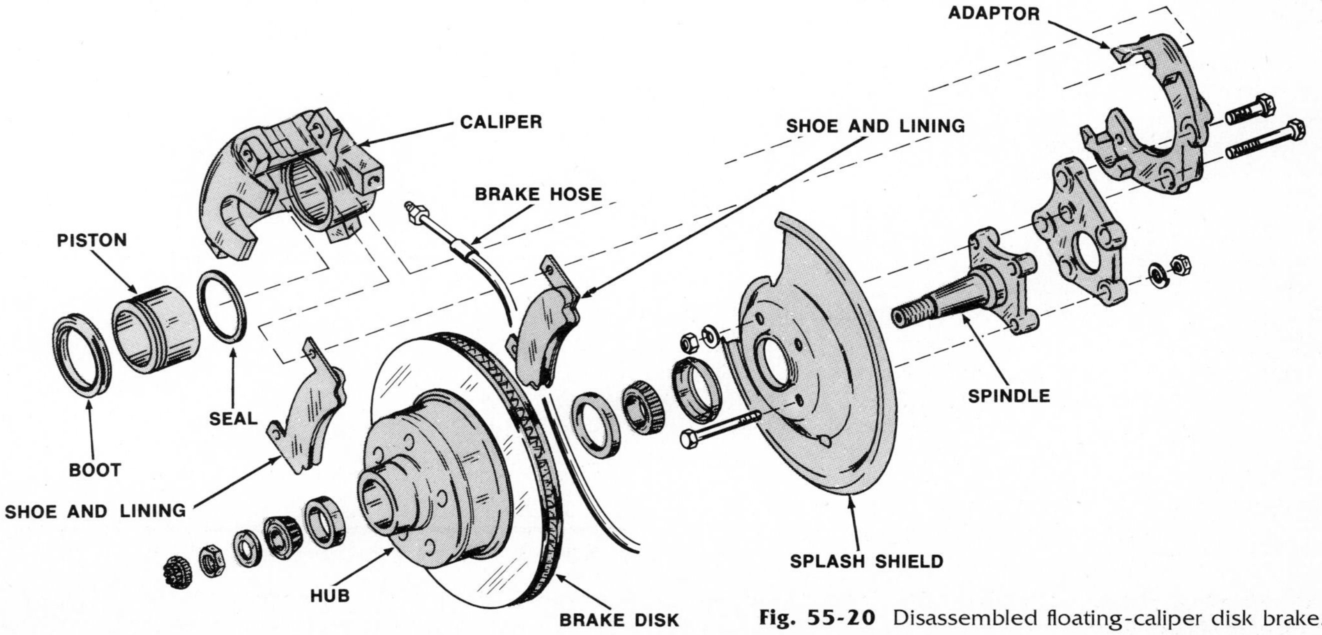

Fig. 55-20 Disassembled floating-caliper disk brake.

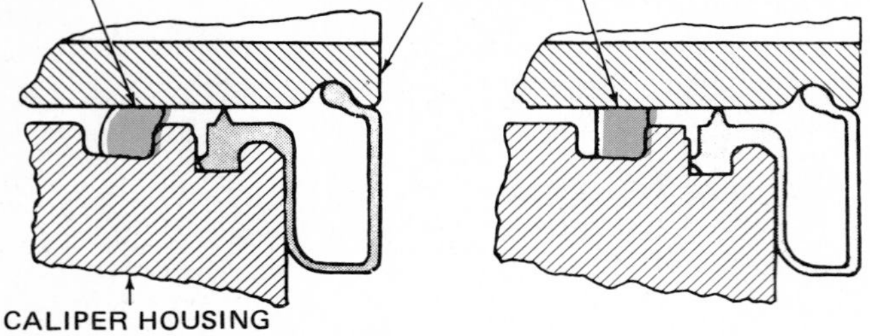

Fig. 55-21 Self-adjusting action of the piston seal when the brakes are applied and when they are released. (*Ford Motor Company*)

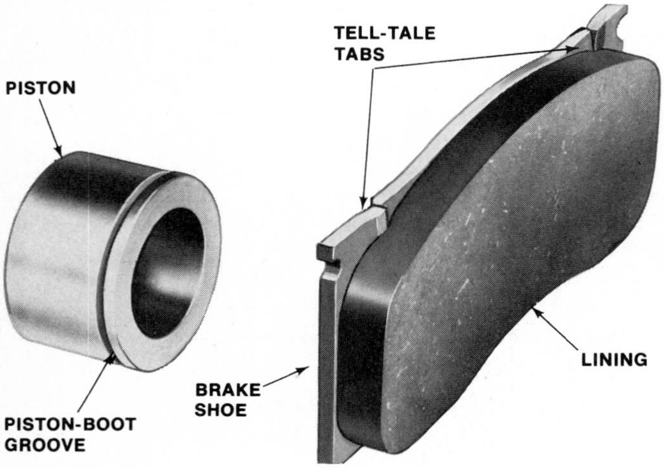

Fig. 55-22 Disk-brake piston and shoe. Note the telltale tabs. (*Chrysler Corporation*)

55-21 Warning Light

The dual braking system uses a pressure-differential valve to operate a warning light. The purpose of this light is to warn the driver if one or the other half of the braking system has failed. Figure 55-23 is a sectional view of the valve. The valve has a piston that is centered when both front and rear brakes are operating normally, as shown in Fig. 55-23. However, if one section fails, then there is low pressure on one side of the piston. The high pressure from the normally operating side pushes the piston to one side, as shown in Fig. 55-24. This movement pushes up the plunger of the brake-warning-light switch. Contacts inside the switch close. This connects a warning light on the instrument panel to the battery. The light then comes on to warn the driver that there is trouble.

55-22 Proportioning Valve

The proportioning valve improves braking action during hard braking. During hard braking, more of the car weight is transferred to the front wheels. As a result, more braking is needed at the front wheels and less at the rear wheels. If normal braking continued, the rear wheels could lock up and skid. The proportioning valve reduces maximum pressure to the rear brakes when hard braking and high fluid pressures develop. This transfers more effort to the front brakes where it is needed.

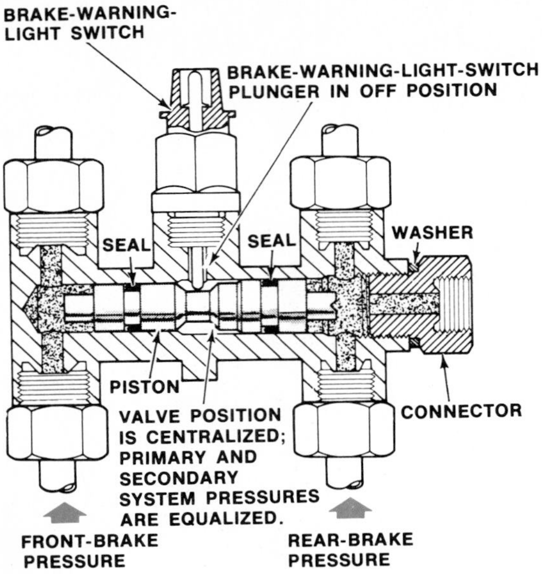

Fig. 55-23 Sectional view of pressure-differential valve with both rear and front brake systems operating normally. (*Ford Motor Company*)

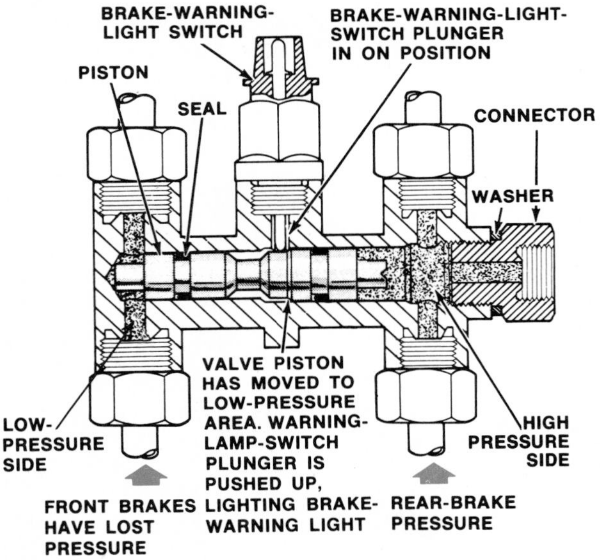

Fig. 55-24 Sectional view of pressure-differential valve showing conditions when front brake system has failed. (*Ford Motor Company*)

55-23 Combination Valve

In many cars the warning-light valve, the metering valve, and the proportioning valve are combined in a single unit. Figure 55-25 is a sectional view of this combination valve.

55-24 Antilock Devices

The most efficient braking takes place when the wheels are revolving. Once the brakes lock the wheels and the wheels begin to skid, braking is much less effective. Antilock systems relieve hydraulic pressure at wheels that are about to skid. This action reduces braking effort that would cause a skid. Here is the way one system works.

Figure 55-26 shows the antilock mechanism at a front wheel. Figure 55-27 shows the mechanism at a rear wheel. The action is the same at either wheel. At the front

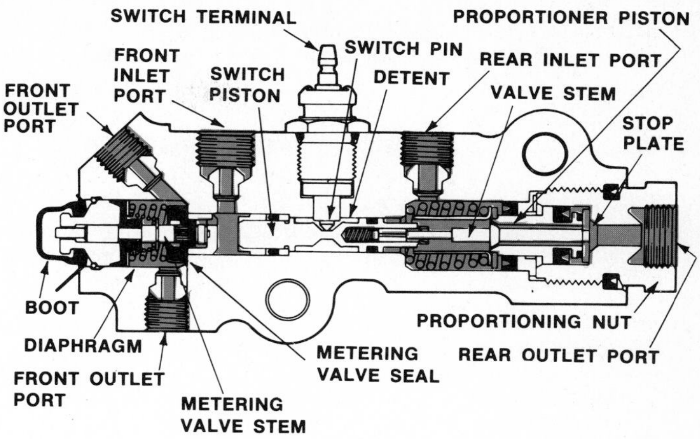

Fig. 55-25 Combination valve with warning-light, metering, and proportioning valves all in the same assembly. (*Chevrolet Motor Division of General Motors Corporation*)

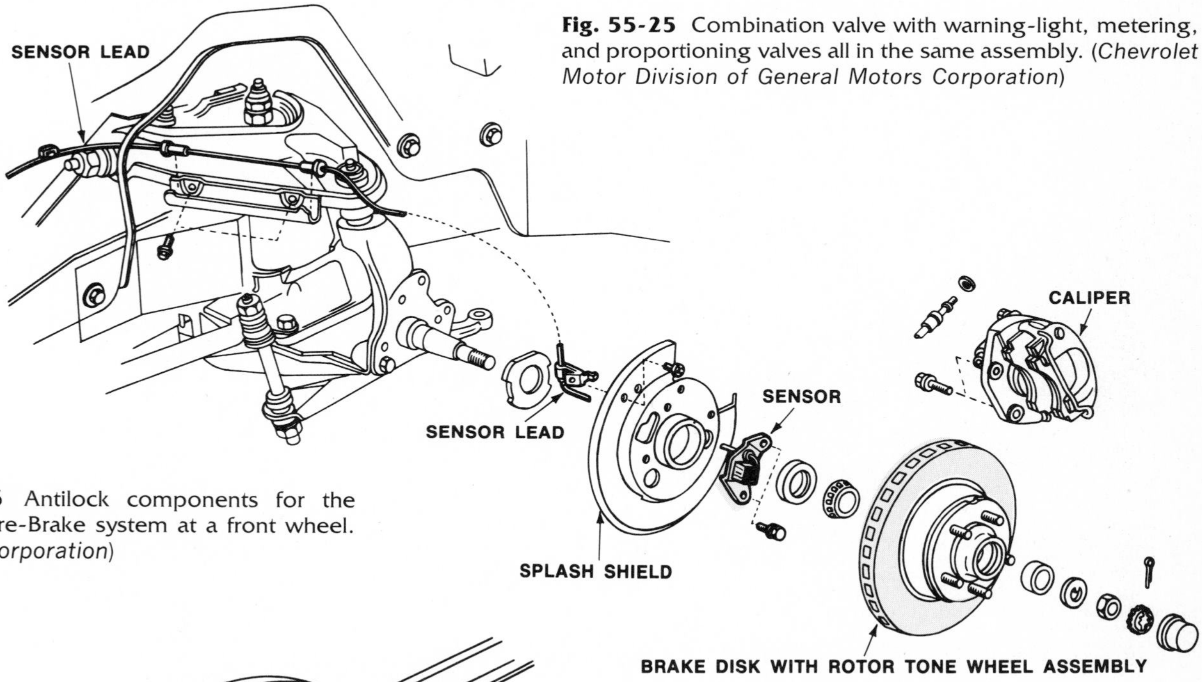

Fig. 55-26 Antilock components for the Chrysler Sure-Brake system at a front wheel. (*Chrysler Corporation*)

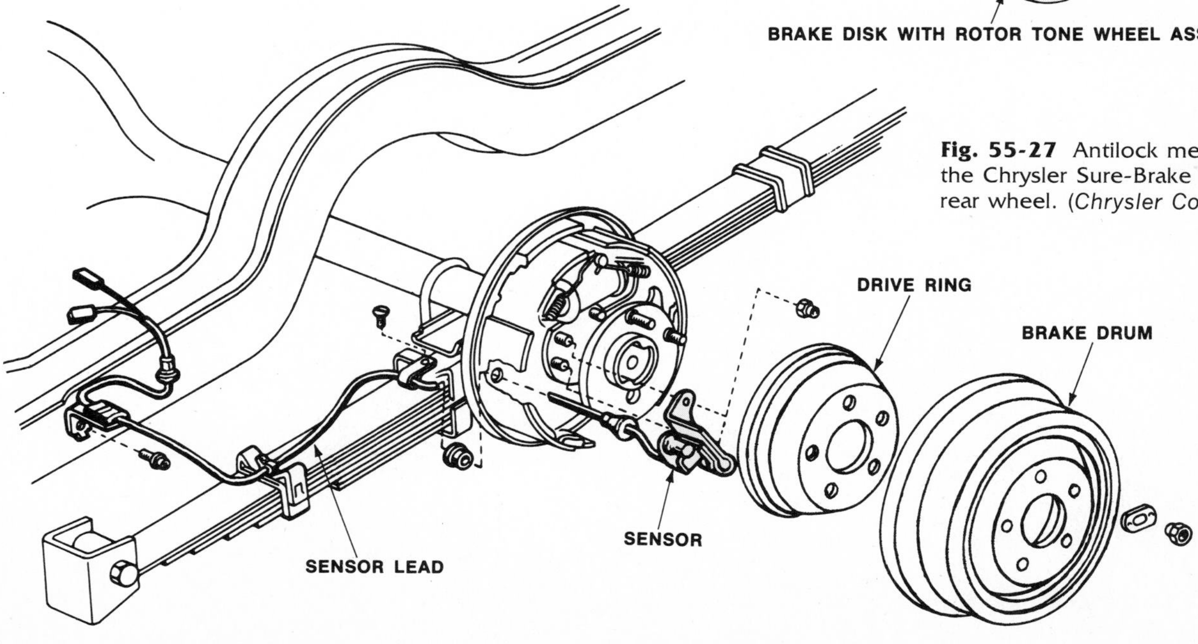

Fig. 55-27 Antilock mechanism for the Chrysler Sure-Brake system at a rear wheel. (*Chrysler Corporation*)

wheel there is a magnetic wheel attached to the brake disk. As the wheel and disk revolve, the magnetic wheel produces an alternating current in the sensor. The sensor is a coil of wire, or a winding. Chapter 28, which covers the alternator in the electrical system, explains that the rotor carries a magnetic field through the stator windings. This produces an alternating current (ac) in the stator windings. In the same way, the magnetic wheel produces ac in the sensor. A similar action takes place at the other wheels. These ac signals from the car wheels are fed into a logic control unit, located in the trunk, as shown in Fig. 55-28. When the brakes are applied, the logic control compares the ac signals from the wheels. The frequency of the ac increases with speed. As long as the frequency of the ac from all wheels is about the same, normal braking is indicated. However, if the ac from any wheel shows a rapid decrease in frequency, it means that the wheel is slowing down too fast. It is beginning to lock and skid.

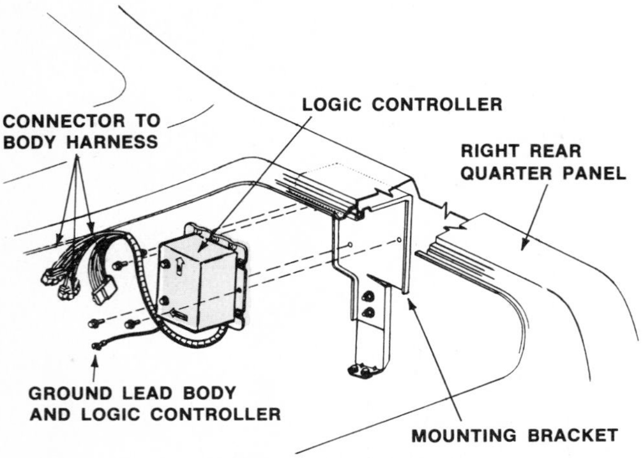

Fig. 55-28 Location of the logic control unit. (*Chrysler Corporation*)

When the logic control unit senses a rapid drop in the frequency of the ac, it signals modulators at the front of the car. Figure 55-29 shows the locations of the front-wheel and rear-wheel modulators. The hydraulic pressure from the master cylinder to the wheel cylinders or calipers passes through these modulators. When the logic control unit senses that a wheel is about to skid, it "tells" the modulator for that wheel to "ease up." In other words, the logic control signals the modulator to reduce the hydraulic pressure to the brake for that wheel. When the pressure is reduced, the braking effect at that wheel is reduced, so that lockup is prevented.

55-25 Stoplight Switch

Figure 55-30 shows the stoplight switch used in many cars. When the brake pedal is pushed down for braking, it causes the switch contacts in the stoplight switch to close. This action connects the stoplights to the battery so that the stoplights come on. Figure 55-30 shows the action during braking.

55-26 Power Brakes

About 90 percent of all American-built cars have power brakes. With power brakes, only a relatively light pressure is required to brake the car. When the brake pedal is pushed down, a vacuum-operated device takes over and does most of the job of pushing the pistons into the master cylinder. The vacuum comes from the intake manifold of the engine. Figure 55-31 is a simplified drawing that shows how the system works. The system includes a cylinder in which a tight-fitting piston can move. When vacuum is applied to one side of the piston, atmospheric pressure causes the piston to be pushed to the right, as shown. This movement pushes the piston rod

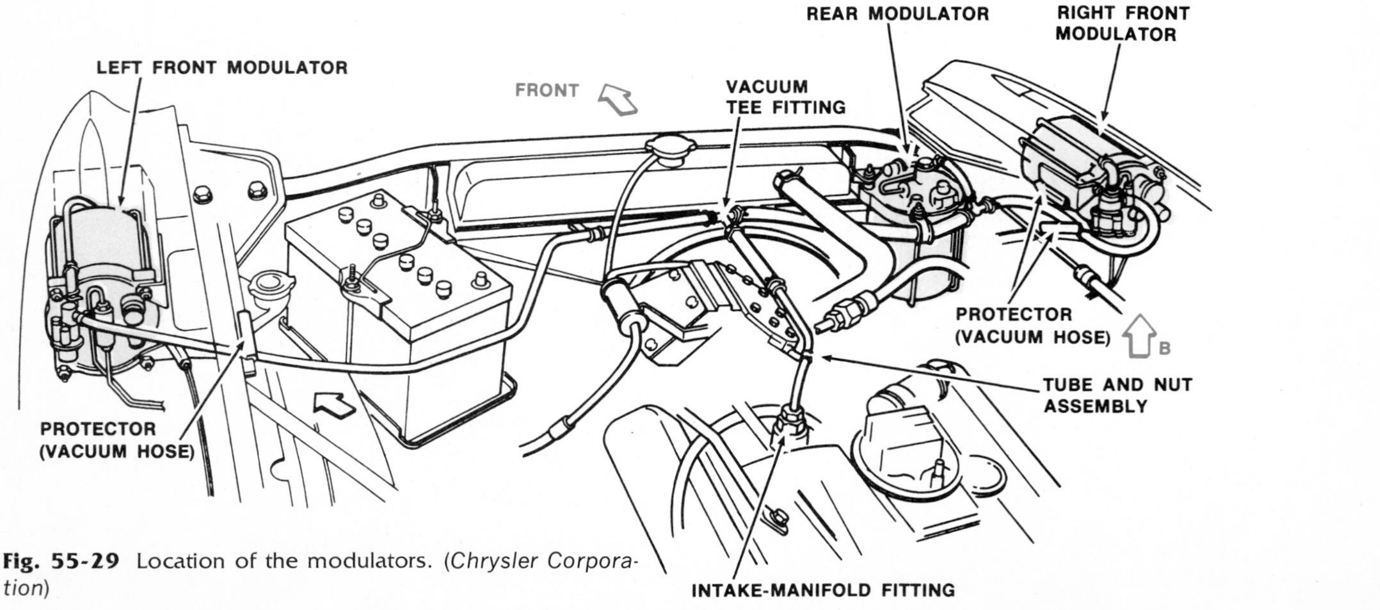

Fig. 55-29 Location of the modulators. (*Chrysler Corporation*)

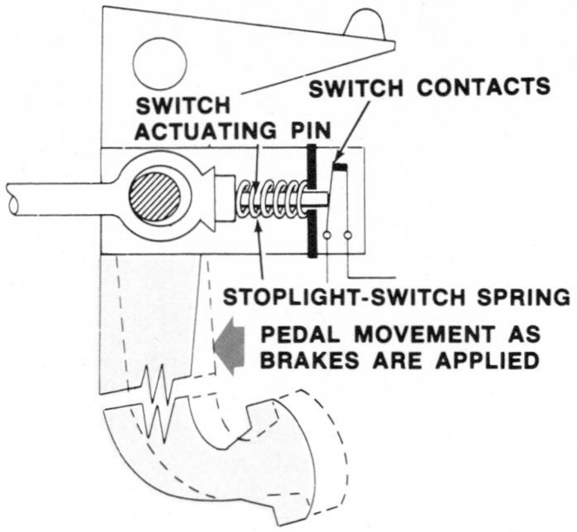

Fig. 55-30 Mechanical stoplight switch shown closed, with brakes applied. (*Ford Motor Company*)

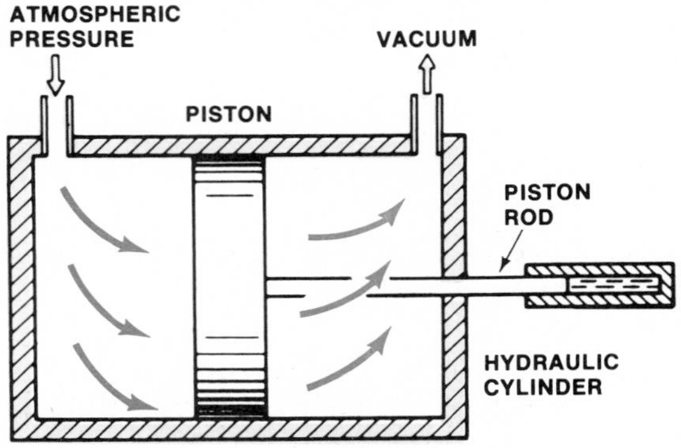

Fig. 55-31 If there is vacuum on one side of the piston and atmospheric pressure on the other side, the piston must move toward the vacuum side.

into the hydraulic cylinder (called the master cylinder in the actual power-brake system).

In the actual power-brake system, the brake pedal does not directly work on the master cylinder. Instead, the brake pedal works a vacuum valve, which then admits vacuum to the power cylinder. Figure 55-32 shows the layout of a power-brake system. Figure 55-33 is a sectional view of one widely used power-brake unit.

Figure 55-34 shows what happens when the brake pedal is pushed down to apply the brakes. First, the push rod moves the air valve away from the floating control valve. Now, atmospheric pressure can flow past the valves and into the space to the right of the plate and valve body, or the piston. The atmospheric pressure forces the piston to the left, so that the master-cylinder pistons are forced into the master cylinder. This action causes braking to take place.

The reaction disk, next to the air valve, gives the driver some braking "feel." In other words, a small proportion

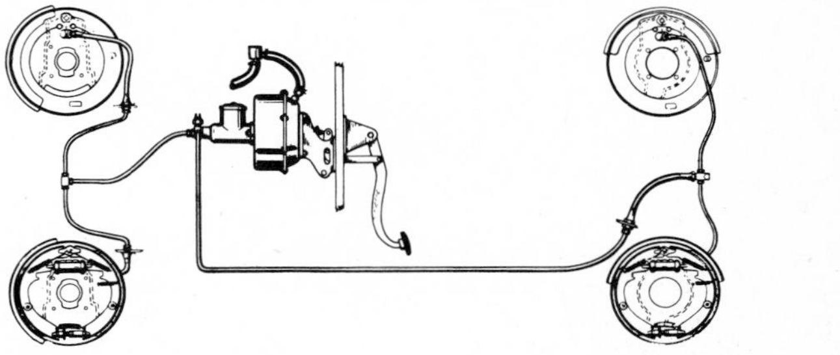

Fig. 55-32 Typical power-brake system. (*Bendix-Westinghouse Automotive Brake Company*)

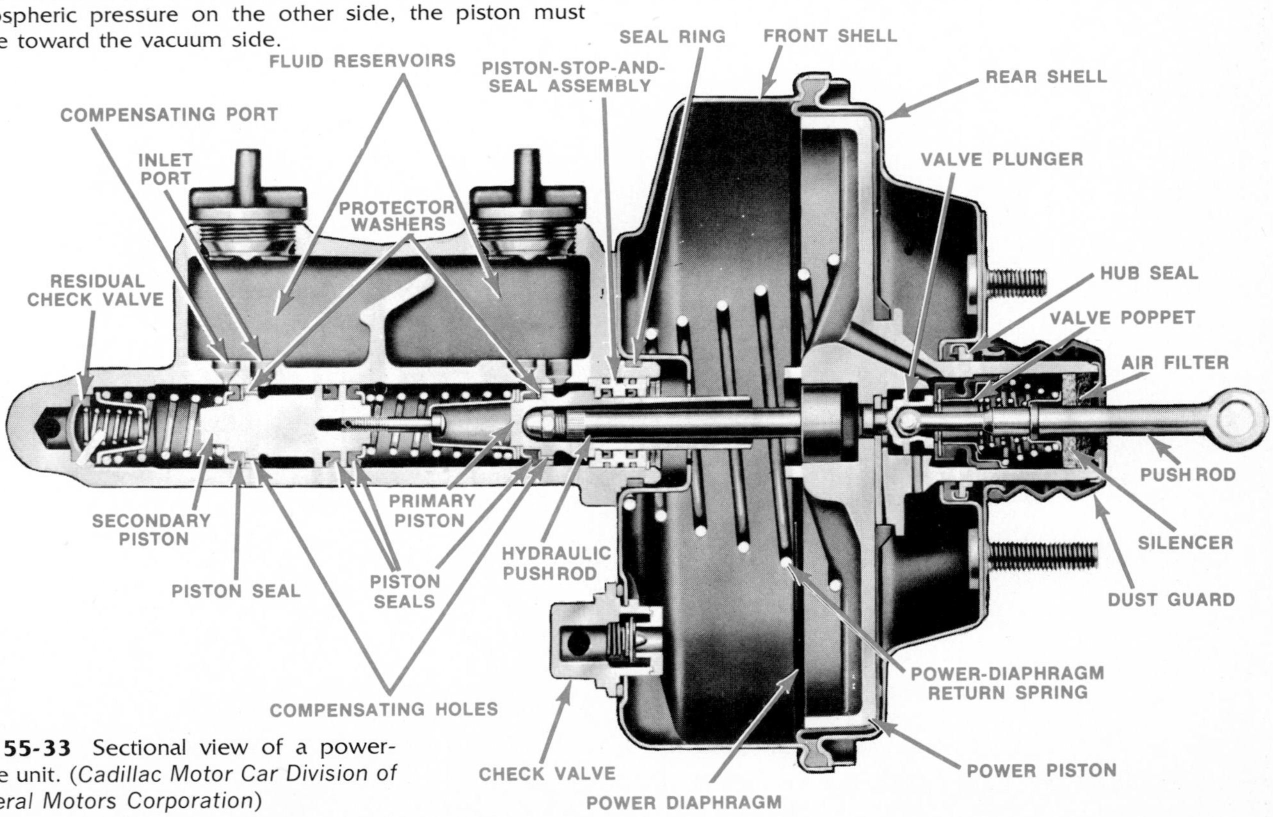

Fig. 55-33 Sectional view of a power-brake unit. (*Cadillac Motor Car Division of General Motors Corporation*)

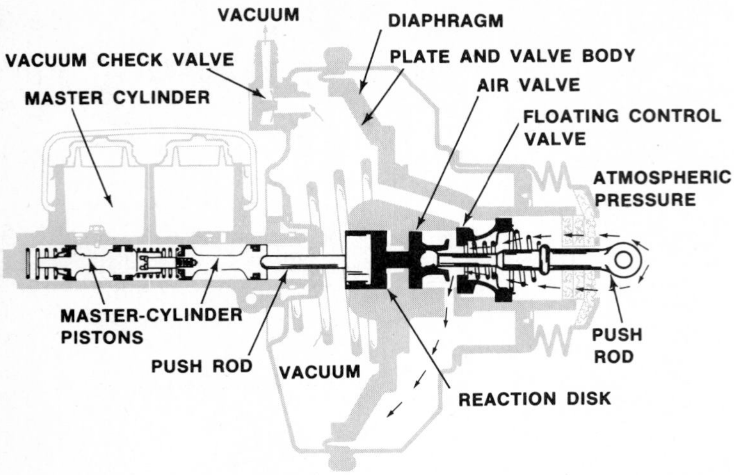

Fig. 55-34 Positions of the internal parts of the power brake when the brakes are applied.

of the braking effort being applied by the power-brake unit feeds back through the reaction disk. This feedback is felt by the driver through the push rod and linkage to the brake pedal.

When the brake pedal is released, the air valve moves back to contact the floating control valve. This contact reseals the power-brake unit from atmospheric pressure so that the brakes are released.

Some cars have a hydraulic brake booster (Fig. 55-35). It uses hydraulic pressure supplied by the power-steering pump to assist in applying the brakes. The ports, shown in Fig. 55-35, are connected to the power-steering system. The hydraulic booster is smaller than the vacuum-type power-brake unit. Also, the hydraulic booster can supply about twice the assist power of a vacuum-type brake.

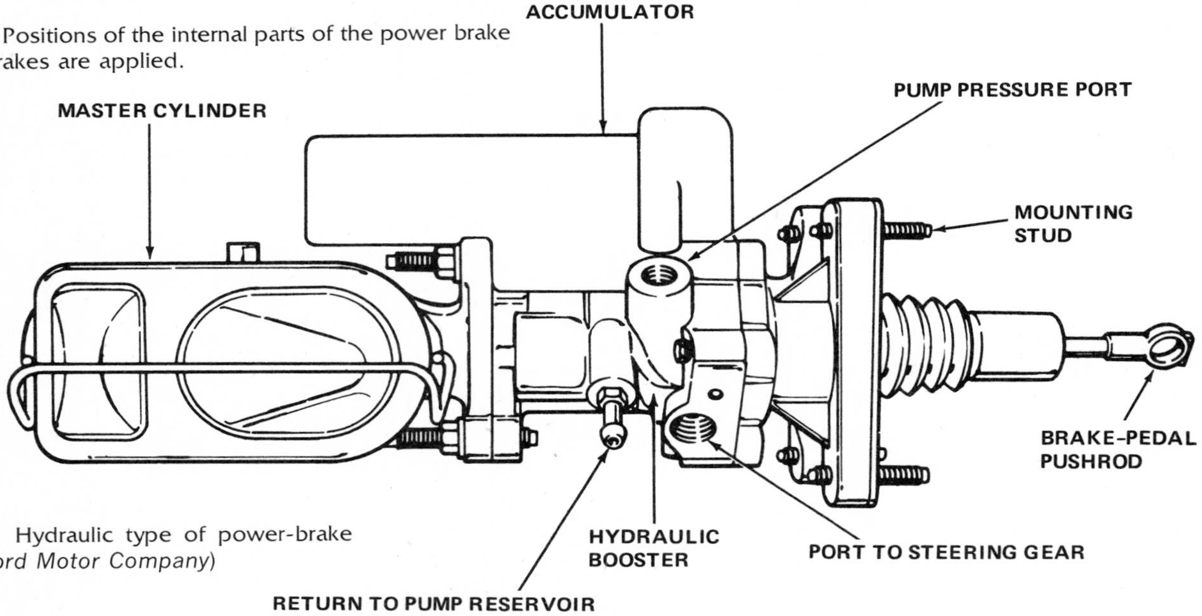

Fig. 55-35 Hydraulic type of power-brake booster. (*Ford Motor Company*)

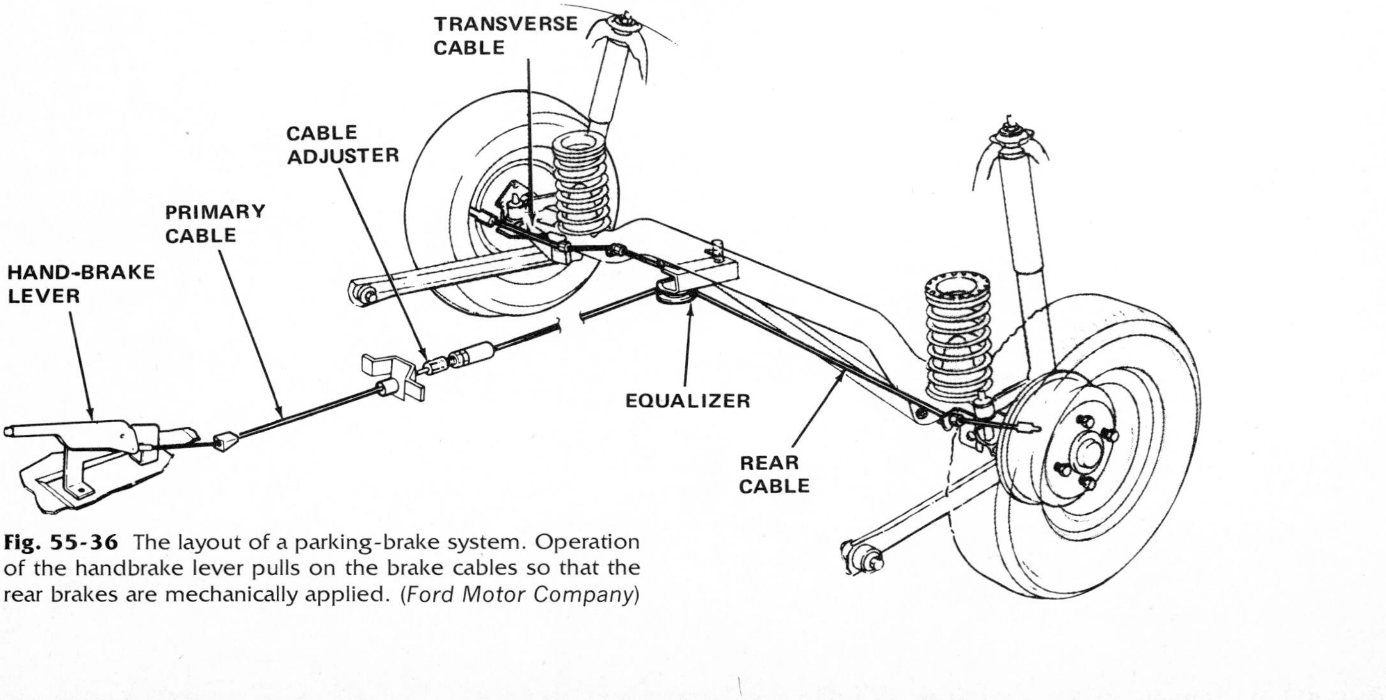

Fig. 55-36 The layout of a parking-brake system. Operation of the handbrake lever pulls on the brake cables so that the rear brakes are mechanically applied. (*Ford Motor Company*)

55-27 Parking Brakes

The parking brake is operated by a separate foot pedal or handbrake lever. Operation of the parking brake causes the rear brakes to be applied mechanically for parking. The pedal or lever is connected to the rear-wheel brake shoes by cables or a linkage. One system is shown in Fig. 55-36. When the pedal or lever is operated, the linkage forces the brake shoes to move into contact with the brake drums.

In some cars the parking brake is released by a vacuum cylinder. This happens when the engine is started and the transmission selector lever is moved out of PARK.

REVIEW QUESTIONS

1. What is static friction? What is kinetic friction?
2. Name the two conditions that cause friction to vary.
3. Generally speaking, will the car stop more quickly if the wheels are locked or if the wheels continue to turn when the brakes are applied?
4. What are the two types of wheel brake mechanisms?
5. Describe a dual braking system. Contrast it with the system that was in use earlier.
6. Describe the actions that take place in a braking system when the pistons are pushed into the master cylinder.
7. Explain how self-adjusting drum brakes work.
8. What are three types of disk brakes?
9. Explain how a disk brake works.
10. Why must mineral oil never be put into a braking system?
11. What is the purpose of the proportioning valve?
12. Describe the operation of an antilock mechanism.
13. Describe the operation of a power brake.
14. What is meant by brake feel? How does the power brake produce this?

SELF PROJECTS

1. When you are under a car on a lift, trace the brake lines to the wheel brakes. Then, with the car on the ground, find the master cylinder. Trace the brake lines to where they go out of sight under the engine and body. Find the proportioning valve.
2. Get your hands on master and wheel cylinders, as well as disk-brake assemblies. If possible, disassemble these units, noting the size, shape, and location of each part. This "hands-on" experience will help you understand just how these components work. Make notes of any important facts you run across for filing in your notebook.

CHAPTER 56
Brake Service

After studying this chapter, you should be able to:

- **List and describe the 12 complaints, or troubles, that might occur in a drum-brake system.**
- **List and describe the 9 complaints, or troubles, that might occur in a disk-brake system.**
- **Under each complaint or trouble, list possible causes and the checks or corrections to make.**
- **Describe typical service procedures that a drum-brake and a disk-brake system might require.**
- **Under the supervision of the instructor, perform these services.**

This chapter discusses trouble diagnosis, adjustments, and servicing of the various components in automotive hydraulic brake systems.

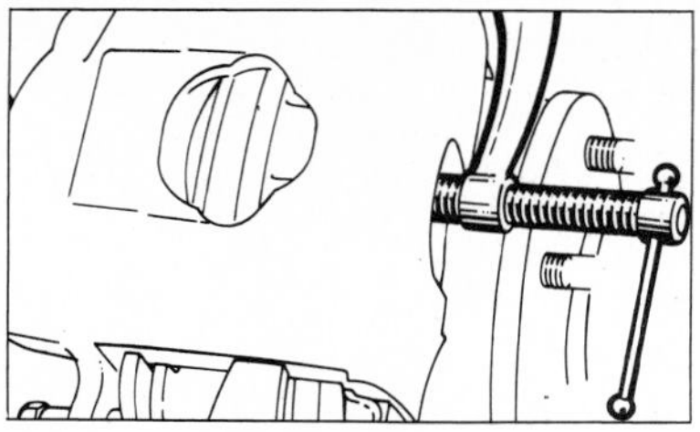

● 56-1 Brake Trouble Diagnosis

The charts and the sections that follow them give you a means of tracing troubles in the brakes to their causes. This permits quick location of causes and, thus, quick correction of troubles. If the cause is known, the trouble is usually easy to correct. Following the trouble-diagnosis sections are sections on the adjustment and repair procedures for automotive bakes.

Trouble diagnosis is divided into two parts. The first chart with the explanatory sections following it covers drum brakes. The second chart, covering disk brakes, then follows. Note, however, that some of the troubles can occur in either system, often from the same causes. A careful study of the two charts, and the explanatory sections that follow, will clarify all of this for you.

CAUTION: Brake linings are made of asbestos and other materials. Breathing the asbestos dust from drums, disks, linings, or other brake parts could cause lung cancer. For this reason, do not use the air hose to blow away dust from the brakes. Instead, wipe it away with a damp cloth. Avoid breathing any brake dust. Wash your hands after handling dusty brake parts or linings.

● 56-2 Drum-Brake Trouble-Diagnosis Chart

A variety of braking problems bring the driver to the mechanic. It is a rare driver who will know exactly what is causing a trouble. The chart that follows lists possible troubles in drum-brake systems, their possible causes, and checks or corrections to be made. Following sections describe the troubles and causes or corrections in detail. The chart in ● 56-15 covers possible disk-brake troubles.

Note: The troubles and possible causes are not listed according to how often they occur. That is, item 1 (or item a under "Possible Cause") does not necessarily occur more often than item 2 (or item b).

DRUM-BRAKE TROUBLE-DIAGNOSIS CHART

See ● 56-3 to 56-14 for detailed explanations of the trouble causes and corrections listed in the chart.

Complaint	Possible Cause	Check or Correction
1. Brake pedal goes to floorboard (● 56-3)	a. Linkage or shoes out of adjustment	Adjust
	b. Brake linings worn	Replace
	c. Lack of brake fluid	Add fluid; bleed system (see item 10 below)
	d. Ruptured brake line	
	e. Air in system	Add fluid; bleed system (see item 9 below)
	f. Worn master cylinder	Repair
2. One brake drags (● 56-4)	a. Shoes out of adjustment	Adjust
	b. Clogged brake line	Clear or replace line
	c. Wheel cylinder defective	Repair or replace
	d. Weak or broken return spring	Replace
	e. Loose wheel bearing	Adjust bearing
3. All brakes drag (● 56-5)	a. Incorrect linkage adjustment	Adjust
	b. Trouble in master cylinder	Repair or replace
	c. Mineral oil in system	Replace damaged rubber parts; use only recommended brake fluid
4. Car pulls to one side when braking (● 56-6)	a. Brake linings soaked with oil	Replace linings and oil seals; avoid over-lubrication
	b. Brake linings soaked with brake fluid	Replace linings; repair or replace wheel cylinder
	c. Brake shoes out of adjustment	Adjust
	d. Tires not uniformly inflated	Inflate correctly
	e. Brake line clogged	Clear or replace line
	f. Defective wheel cylinder	Repair or replace
	g. Brake backing plate loose	Tighten
	h. Mismatched linings	Use same linings all around
5. Soft or spongy pedal (● 56-7)	a. Air in system	Add brake fluid; bleed system (see item 9 below)
	b. Brake shoes out of adjustment	Adjust
6. Poor braking action requiring excessive pedal pressure (● 56-8)	a. Brake linings soaked with water	Will be all right when dried out
	b. Shoes out of adjustment	Adjust
	c. Brake linings hot	Allow to cool
	d. Brake linings burned	Replace
	e. Brake drum glazed	Turn or grind drum
	f. Power-brake assembly not operating	Overhaul or replace
7. Brakes too sensitive or grab (● 56-9)	a. Shoes out of adjustment	Adjust
	b. Wrong linings	Install correct linings
	c. Brake linings greasy	Replace; check oil seals; avoid over-lubrication
	d. Drums scored	Turn or grind drums
	e. Backing plates loose	Tighten
	f. Power-brake assembly malfunctioning	Overhaul or replace
	g. Brake linings soaked with oil	Replace linings and oil seals; avoid over-lubrication
	h. Brake linings soaked with brake fluid	Replace linings; repair or replace wheel cylinders

DRUM-BRAKE TROUBLE-DIAGNOSIS CHART (Continued)

Complaint	Possible Cause	Check or Correction
8. Noisy brakes (● 56-10)	a. Linings worn b. Shoes warped c. Shoe rivets loose d. Drums worn or rough e. Loose parts	Replace Replace Replace shoe or lining Turn or grind drums Tighten
9. Air in system (● 56-11)	a. Defective master cylinder b. Loose connections, damaged tube c. Brake fluid lost	Repair or replace Tighten connections; replace tube See item 10 below
10. Loss of brake fluid (● 56-12)	a. Master cylinder leaks b. Wheel cylinder leaks c. Loose connections, damaged tube NOTE After repair, add brake fluid and bleed system	Repair or replace Repair or replace Tighten connections; replace tube
11. Brakes do not self-adjust (● 56-13)	a. Adjustment screw stuck b. Adjustment lever does not engage star wheel c. Adjuster incorrectly installed	Free and clean up Repair; free up or replace adjuster Install correctly
12. Warning light comes on when braking (dual system) (● 56-14)	a. One section (front or back) has failed b. Pressure-differential valve defective	Check both sections for braking action; repair defective section Replace

● 56-3 Brake Pedal Goes to Floor Board

When this happens, there is no pedal reserve. Full pedal movement does not produce adequate braking. This would be a very unlikely situation with a dual brake system. One section (front or rear) might fail, but it would be rare for both to fail at the same time. It is possible that the driver has continued to operate the car with one section out. (Either the driver ignored the warning light, or the light or pressure-differential valve has failed.) Causes of failure could be linkage or brake shoes out of adjustment, linings worn, ruptured brake line, air in the system or lack of brake fluid, or a worn master cylinder.

● 56-4 One Brake Drags

This means the brake shoes are not moving away from the brake drum when the brakes are released. This could be due to incorrect shoe adjustment, to a clogged brake line which does not release pressure from the wheel cylinder, to sticking pistons in the wheel cylinder, to weak or broken brake-shoe return springs, or to a loose wheel bearing which permits the wheel to wobble so that the brake drum comes in contact with the brake shoes even though they are retracted.

● 56-5 All Brakes Drag

When all brakes drag, it may be that the brake pedal does not have sufficient play. In that case, the pistons in the master cylinder do not fully retract. This would prevent the lip of the piston cup from clearing the compensating port. Hydraulic pressure would not be relieved as it should be. As a result, the wheel cylinders would not release the brake shoes. A similar condition could result if mineral oil had been added to the system. This would be likely to cause the piston cup to swell. If it swelled enough, it would not clear the compensating port even with the piston in the "fully retracted" position. A clogged compensating port would have the same result. Do not use a wire or drill to clear the port. This might produce a burr that would cut the piston cup. Instead, clear it with alcohol and compressed air. Clogging of the reservoir vent might cause dragging brakes. This could trap pressure in the reservoir which would prevent release of pressure. But this would be just as likely to cause leakage of air into the system (see ● 56-11).

● 56-6 Car Pulls to One Side

If the car pulls to one side when the brakes are applied, more braking pressure is being applied to one side than to the other. This happens if some of the brake linings have become soaked in oil or brake fluid (so that they lose braking effectiveness), if brake shoes are unevenly or improperly adjusted, if tires are not evenly inflated, or if defective wheel cylinders or clogged brake lines are preventing uniform braking action at all wheels. A loose brake backing plate or the use of two different types of brake lining will cause the car to pull to one side when the brakes are applied. A misaligned front end or a broken spring could also cause this.

Rear brake linings will become soaked with oil if the lubricant level in the differential and rear axle is too high.

This usually causes leakage past the oil seal (Fig. 56-1) and onto the brake linings. A defective seal might also allow leakage. At the front wheel, brake linings may become oil soaked if the front-wheel bearings are improperly lubricated or if the oil seal is defective or not properly installed. Wheel cylinders will leak brake fluid onto the brake linings if they are defective or if an actuating pin has been improperly installed (see ● 56-12). If the linings at a right wheel become soaked with brake fluid or oil, for example, the car pulls to the left. This is because the brakes are more effective on the left side.

● 56-7 Soft, or Spongy, Pedal

If the pedal action is soft, or spongy, the chances are that their is air in the system. Out-of-adjustment brake shoes could also cause this. Refer to ● 56-11 for conditions that could allow air to get into the system.

● 56-8 Braking Requires Excessive Pedal Pressure

Excessive pedal pressure could be caused by improper brake-shoe adjustment. The use of the wrong brake lining could cause the same trouble. Sometimes brake linings that have become wet after a hard rain or after driving through puddles will not hold well. Normal braking action is usually restored after the brake linings have dried out. But if the linings are soaked with oil or brake fluid, they must be replaced. It is not feasible to cleanse the linings of these fluids. Another possible cause of poor braking action is excessive temperature. After the brakes have been applied for long periods, as in coming down a long hill, they begin to overheat. This overheating reduces braking effectiveness so that the brakes "fade." Often, if brakes are allowed to cool, braking efficiency is restored. However, excessively long periods of braking at high temperature may char the brake linings so that they must be replaced. Also, this may glaze the brake drum so that it becomes too smooth for effective braking action. In this case, the drum must be ground or turned to remove the glaze. Glazing can also take place even though the brakes are not overheated. Failure of the power-brake assembly will considerably increase the amount of pedal pressure required to produce braking.

● 56-9 Brakes Too Sensitive or Grab

If linings are greasy, soaked with oil or brake fluid, the bakes tend to grab with slight pedal pressure. In any case, the linings must be replaced. If the brake shoes are out of adjustment, if the wrong linings are being used, or if drums are scored or rough (Fig. 56-2), grabbing may result. A loose backing plate may cause the same condition. As the linings come into contact with the drum, the backing plate shifts to give hard braking. A defective power-brake assembly can also cause grabbing.

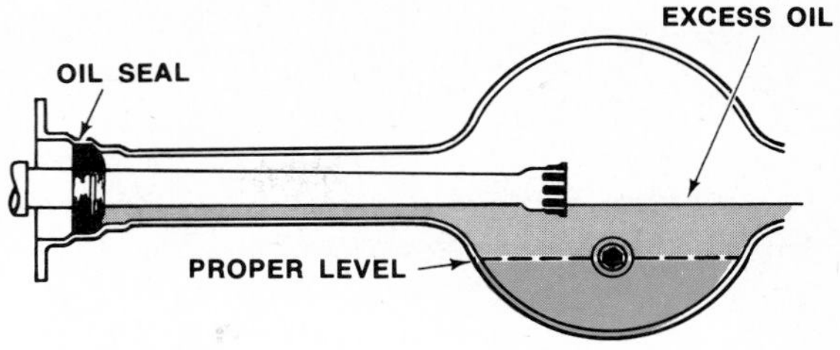

Fig. 56-1 A high lubricant level in the differential and rear-axle housing may cause leakage past the oil seal, resulting in soaked brake linings. (*Pontiac Motor Division of General Motors Corporation*)

Fig. 56-2 Various types of brake-drum defects that require drum service. (*Bear Manufacturing Company*)

● 56-10 Noisy Brakes

Brakes become noisy if the brake linings wear so much that the rivets come into contact with the brake drum (see Fig. 56-2); if the shoes become warped so that pressure on the drum is not uniform; if shoe rivets become loose so that they contact the drum; or if the drum becomes rough or worn. Any of these conditions is likely to cause a squeak or squeal when the brakes are applied. Also, loose parts, such as the brake backing plate, may rattle.

● 56-11 Air in System

If air gets into the hydraulic system, poor braking and a spongy pedal will result. Air can get into the system if the filler vent becomes plugged (Fig. 56-3), since this tends to create a partial vacuum in the system on the return stroke of the piston. Air could then bypass the rear piston cup, as shown by the arrows, and enter the system. It is possible accidentally to plug the vent when the filler plug or cover is removed. Always check the vent and clean it when the plug or cover is removed. Air can also get into the system if the master-cylinder valve is leaky and does not hold pressure in the system. This could allow air to seep in around the wheel-cylinder piston cups, since there would be no pressure holding the cups tight against the cylinder walls. Probably the most common cause of

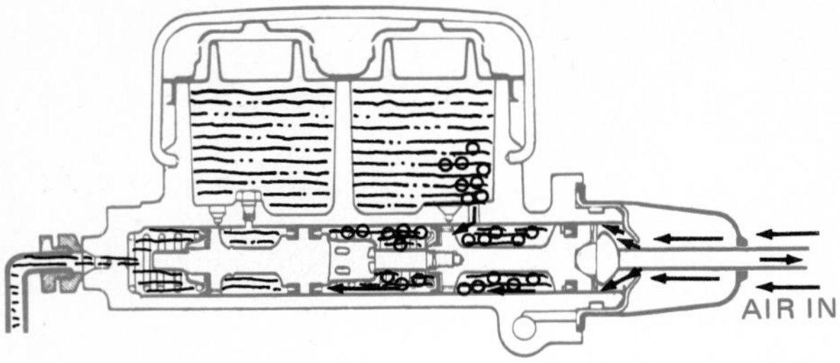

Fig. 56-3 If the filler vent becomes plugged, air may be drawn into the system on the return stroke of the piston, past the rear piston cup, as shown by the small arrows and bubbles. (*Pontiac Motor Division of General Motors Corporation*)

air in the braking system is insufficient brake fluid in the master cylinder. If the brake fluid drops below the compensating port, then the hydraulic system will draw air in as the piston moves forward on the braking stroke. Air in the system must be removed by adding brake fluid and bleeding the system, as described in ● 56-33.

● 56-12 Loss of Brake Fluid

Brake fluid can be lost if the master cylinder leaks, if the wheel cylinder leaks, if the line connections are loose, or if the line is damaged. One possible cause of wheel-cylinder leakage is incorrect installation of the actuating pin. If the pin is cocked, then the side thrust on the piston may permit leakage past the piston. Leakage from other causes at the master cylinder or wheel cylinder requires removal and repair, or replacement, of the defective parts.

● 56-13 Brakes Do Not Self-Adjust

Either the adjustment screw is stuck, the adjustment lever does not engage the star wheel, or else the adjuster was incorrectly installed. It is necessary to inspect the brakes to find and correct the trouble.

● 56-14 Warning Light Comes on When Braking (Dual System)

This is a signal that one of the two braking sections has failed. Both sections (front and rear) should be checked so that the trouble can be found and eliminated. It is dangerous to drive with this condition, even though braking takes place, because only half the wheels are being braked. It is possible that the pressure-differential valve or the light itself has failed so that the driver was not warned when one of the two sections went out.

● 56-15 Disk-Brake Trouble-Diagnosis Chart

The chart that follows lists troubles in disk-brake systems, their possible causes, and checks or corrections to be made. Following sections describe the troubles and causes or corrections in detail.
Note: The troubles and possible causes are not listed according to how often they occur. That is, item 1 (or item a under "Possible Cause") does not necessarily occur more often than item 2 (or item b).

DISK-BRAKE TROUBLE-DIAGNOSIS CHART

See ● 56-16 to 56-24 for detailed explanations of the trouble causes and corrections listed in the chart.

Complaint	Possible Cause	Check or Correction
1. Excessive pedal travel (● 56-16)	a. Excessive disk run-out	Check run-out; if excessive, install new disk
	b. Air leak, or insufficient fluid	Check system for leaks
	c. Improper brake fluid (boil)	Drain and install correct fluid
	d. Warped or tapered shoe	Install new shoe
	e. Loose wheel-bearing adjustment	Readjust
	f. Damaged piston seal	Install new seal
	g. Power-brake malfunction	Check power unit
2. Brake roughness or chatter (pedal pulsating) (● 56-17)	a. Excessive disk run-out	Check run-out; if excessive, install new disk
	b. Disk out of parallel	Check run-out; if excessive, install new disk
	c. Loose wheel bearing	Readjust
3. Excessive pedal effort (● 56-18)	a. Power-brake malfunction	Check power unit
	b. Brake fluid or grease on linings	Install new linings
	c. Lining worn	Install new shoe and linings
	d. Incorrect lining	Install correct lining
	e. Frozen or seized pistons	Disassemble caliper, and free up pistons

DISK-BRAKE TROUBLE-DIAGNOSIS CHART (Continued)

Complaint	Possible Cause	Check or Correction
4. Car pulls to one side (● 56-19)	a. Brake fluid or grease on linings	Install new linings
	b. Frozen or seized pistons	Disassemble caliper, and free up pistons
	c. Incorrect tire pressure	Inflate tires to recommended pressures
	d. Distorted brake shoes	Install new brake shoes
	e. Front end out of alignment	Check and align front end
	f. Broken rear spring	Install new rear spring
	g. Restricted hose or line	Check hoses and lines and correct as necessary
	h. Unmatched linings	Install correct lining
5. Noise:		
Groan	Brake noise when slowly releasing brakes (creep-groan). Not detrimental to function of disk brakes—no corrective action required. This noise may be eliminated by slightly increasing or decreasing brake-pedal effort.	
Rattle	Brake noise or rattle at low speeds on rough roads may be due to excessive clearance between the shoe and the caliper. Install new shoe and lining assemblies to correct.	
Scraping	a. Mounting bolts too long	Install mounting bolts of correct length
	b. Disk rubbing housing	Check for rust or mud buildup on caliper housing; check caliper mounting and bridge bolt tightness
	c. Loose wheel bearings	Readjust
	d. Linings worn, allowing telltale tabs to scrape on disk	Replace linings
6. Brakes heat up during driving and fail to release (● 56-21)	a. Power-brake malfunction	Check and correct power unit
	b. Sticking pedal linkage	Free up sticking pedal linkage
	c. Operator riding brake pedal	Instruct owner how to drive with disk brakes
	d. Frozen or seized piston	Disassemble caliper, hone cylinder bore, clean seal groove, and install new pistons, seals, and boots
7. Leaky wheel cylinder (● 56-22)	a. Damaged or worn piston seal	Install new seal
	b. Scores or corrosion on surface of piston	Disassemble caliper, hone cylinder bore; if necessary, install new pistons
8. Grabbing or uneven braking action (● 56-23)	a. Power-brake malfunction	Check and correct unit
	b. Causes listed under Item 4	Corrections listed under item 4
9. Brake pedal can be depressed without braking effect (● 56-24)	a. Piston pushed back in cylinder bores during servicing of caliper (shoe and lining not properly positioned)	Reposition brake shoe and lining assemblies. Depress pedal a second time and if condition persists, check the following causes
	b. Leak in system or caliper	Check for leak, repair as required
	c. Damaged piston seal in one or more cylinders	Disassemble caliper and replace piston seals as required
	d. Air in hydraulic system, or improper bleeding procedure	Bleed system
	e. Bleeder screw open	Close bleeder screw and bleed entire system
	f. Leak past primary cup in master cylinder	Recondition master cylinder

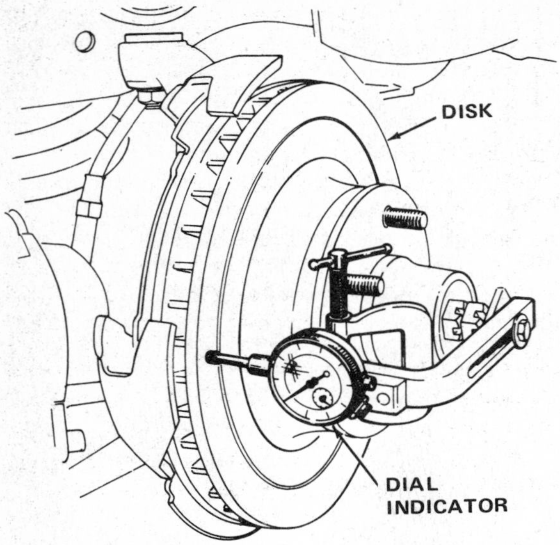

Fig. 56-4 Checking a disk for run-out with a dial indicator. *(Chevrolet Motor Division of General Motors Corporation)*

● 56-16 Excessive Pedal Travel

Anything that requires excessive movement of the caliper pistons requires excessive pedal travel. For example, if the disk has excessive run-out, it will force the pistons farther back in their bores when the brakes are released. This will require additional pedal travel when the brakes are applied. The runout of the disk can be checked with a dial indicator (Fig. 56-4).

Warped or tapered shoes, a damaged piston seal, or a loose wheel bearing could cause the same thing. In addition, air in the lines, insufficient fluid in the system, or incorrect fluid, which boils, will cause a spongy pedal and excessive pedal travel. Also, if the power brake is malfunctioning, it could cause excessive pedal travel.

● 56-17 Brake Roughness–Brake-Pedal Pulsation

This is probably due to a disk with excessive run-out or to a loose wheel bearing.

● 56-18 Excessive Pedal Effort

The power brake may not be operating properly. In addition, if the linings are worn or have brake fluid on them, they will not produce normal braking. High pedal pressure will be required. Also, if the pistons are jammed in the calipers, high pedal effort will be required.

● 56-19 Car Pulls to One Side

This is due to uneven braking action. It could be caused by incorrect front-end alignment, uneven tire inflation, or a broken or weak suspension spring. Within the braking system itself, such things as brake fluid on the linings, unmatched linings, warped brake shoes, jammed pistons, or restrictions in the brake lines could cause the car to pull to one side when braking.

● 56-20 Noise

The chart covers the various noises and their causes. Refer to item 5 in the chart for details.

● 56-21 Brakes Fail to Release

This could mean anything from a sticking pedal linkage or malfunctioning power brake to pistons stuck in the calipers. It could also be due to the driver's riding the brake pedal, or to failure of the master cylinder to release the pressure when the brakes are released.

● 56-22 Leaky Wheel Cylinder

This could be due to a damaged or worn piston seal or to roughness on the surface of the piston, as a result of scores, scratches, or corrosion.

● 56-23 Grabbing or Uneven Braking Action

This could be due to any of the causes listed in ● 56-19. It could also be due to a malfunctioning power brake which hangs up and then applies suddenly.

● 56-24 Pedal Can Be Depressed Without Braking Effect

If the brake calipers have been serviced, the pistons may be pushed back so far in their bores that a single full movement of the brake pedal will not produce braking. Thus, after any service on disk brakes, the brake pedal should be pumped many times, and the master-cylinder reservoir properly filled, before the car is moved. Pumping the pedal several times gradually moves the pistons in to normal position so that normal brake-pedal application causes braking.

Of course, other conditions can prevent braking action when the pedal is depressed. Leaks or air in the system can also cause this. Leaks can occur at the piston seals, bleeder screws, brake-line connections, or in the master cylinder.

Consider also the possibility that the pressure-differential valve is stuck. Or, the warning light may be burned out, if both the front and rear sections have failed. The driver might have been driving for some time with one section defective, but the warning system did not work to warn of the trouble.

● 56-25 Brake Service

Any complaint of faulty braking action should be analyzed to determine its cause, as noted in previous sections. Sometimes, all that is necessary (on earlier drum-type brakes) is a minor brake adjustment to compensate for lining wear. On later brakes with the self-adjuster (Figs. 55-15 and 55-16), the brakes automatically adjust themselves to compensate for lining wear. Other brake services include addition of brake fluid, bleeding the hydraulic system to remove air, repair or replacement of master cylinder and wheel cylinders, replacement of brake linings, and refinishing of brake drums.

There are several types of automatic brake testers which can check braking action and efficiency in a few moments. To use one type, you drive the car onto four tread plates at about 5 mph (miles per hour) [8.047 km] and apply the brakes. The braking effect at each wheel is registered on a separate meter or glass column. A second type is similar to a dynamometer. It has two sets of rollers set in the floor. Either the front or rear wheels of the car are placed on the rollers. Now, electric motors drive the rollers and wheels. The brakes are then applied, and their efficiency, or braking effect, is registered on a meter.

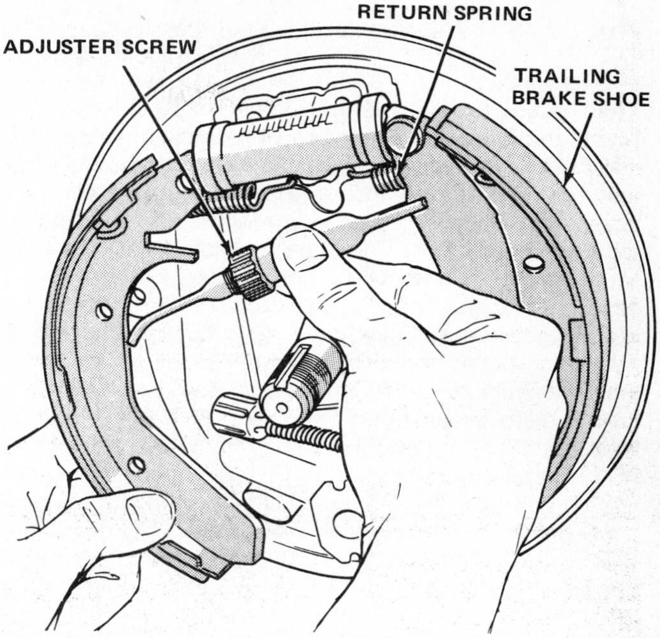

Fig. 56-5 Replacing the brake shoes on a rear drum brake of a front-wheel-drive car. (*Chrysler Corporation*)

● 56-26 Adjustment of Brakes

On earlier drum-type brakes, periodic adjustments were required to compensate for brake-lining wear. On the self-adjusting type, no adjustment is required except perhaps just after new linings are installed. Procedures vary from car to car. Always consult the manufacturer's shop manual before attempting any adjustment of a brake.

● 56-27 Brake Linings

When linings wear, the shoes must be replaced. To replace shoes in a drum brake, the wheel and brake drum must be removed (Fig. 56-5). Brake linings on drum brakes are either riveted or cement-bonded to the shoe. Both types are replaced as assemblies. That is, shops now do not try to put new linings on brake shoes even though this was common practice years ago.

At one time new linings were given a preliminary grinding to improve their fit. This is no longer recommended because of the hazards resulting from the asbestos dust raised by the grinding job. See the Cautions in ● 56-1.

Replacement of disk-brake shoes is described in ● 56-30.

● 56-28 Brake Disks and Drums

Brake disks require replacement only if they become deeply scored or are warped out of line. Light scores and grooving are normal and will not affect braking. Some manufacturers say never to grind down or reface a scored brake disk. This will adversely affect performance. Instead, they say to replace it. Other manufacturers (Ford, for example) supply a special disk-refinishing tool. Check the manufacturer's shop manual for details.

Many disks have a dimension (a number) cast into them. This dimension is the minimum to which the disk can be refinished. If it is necessary to refinish to a smaller dimension, discard it. It will be too thin to work safely.

On drum-type brakes, the drums should be inspected for distortion, cracks, scores, roughness, or excessive glaze or smoothness. Glaze lowers friction and braking efficiency. Drums that are distorted or cracked should be discarded, and new drums installed. Light score marks can be removed with fine emery cloth. All traces of emery must be removed after smoothing the drum. Deeper scores, roughness, and glaze can be removed by turning or grinding the drum.

Many brake drums have a dimension (a number) cast into them. This dimension is the maximum allowable diameter. If it is necessary to turn or grind the drum to a larger diameter, discard it. The drum would be too thin to work safely. It would probably warp or crack when it heated up and cause faulty braking.

Note that the diameters of the left and right drums on the same axle should be within 0.010 inch [0.254 mm] of each other. When the drum diameters on the same axle vary more than this, replace both drums.

● 56-29 Wheel and Master Cylinders

Some shops do not attempt to overhaul wheel and master cylinders. Instead, they install new or rebuilt cylinders obtained from a parts dealer specializing in this work.

Wheel and master cylinders must be disassembled and assembled with extreme care. The mechanic must avoid getting the slightest trace of grease or dirt in them. Hands must be clean—washed with soap and water, not gasoline. Any trace of oil or gasoline on the cylinder parts may ruin them. The bench and the tools must be clean.

To remove a wheel cylinder from the car, the wheel and the drum must be off. The brake pedal should be blocked up to prevent its operation. Then the tube or hose should be disconnected from the cylinder. The cylinder is removed by taking out the attaching bolts. The tube end at the wheel should be taped closed to keep dirt from entering. The cylinder can be disassembled by rolling off the rubber boots or taking off the covers. All parts should be washed in brake-system cleaning fluid. Old boots and piston cups should be discarded if they are not in excellent condition. Some manufacturers recommend replacement of these parts every time the cylinder is disassembled. If the cylinder is scored, it should be polished with crocus cloth (not sandpaper or emery cloth). Some manufacturers permit the use of a hone if the diameter of the cylinder is not increased more than a few thousandths of an inch. If scores do not come out, the cylinder should be replaced. Also, the cylinder and the pistons should be replaced if the clearance between them is excessive. When reassembling the cylinder, lubricate all parts with brake fluid.

CAUTION: Never allow any grease or oil to come in contact with rubber parts of the brake system. This would cause them to swell so that braking action might be destroyed. This could lead to a serious accident.

To remove a master cylinder, detach the brake pedal and the brake line. Take out the bolts holding the cylinder to the frame. Then, drain out the brake fluid. Disassemble the cylinder by taking off the boot and removing the push rod, snap ring, or stop plate so that piston, cup, spring, valve, and other parts will come out. Use extreme care to keep all parts clean and free of grease or oil. All rubber parts that appear at all deteriorated should be discarded. Some manufacturers supply master-cylinder repair kits. They recommend that the parts in such a kit be used to replace the old parts whenever a master cylinder is disassembled.

All parts should be washed in brake-system cleaning fluid. If the cylinder is scored, it should be polished with crocus cloth (not sandpaper or emery cloth). Some manufacturers permit the use of a hone, provided the diameter of the cylinder is not increased more than a few thousandths of an inch. However, if light polishing or honing does not remove scores, the cylinder should be replaced.

Check the fit of the piston to the cylinder. If they do not fit within specifications, replace them.

On reassembly, lubricate the parts with brake fluid. Never allow grease or oil to come in contact with any rubber parts of the brake system.

Before installing the master cylinder, bleed it. This is done by installing plastic plugs at the two outlet ports and filling both reservoirs with clean brake fluid. Slowly push the pistons in with a wood dowel several times or until no more bubbles appear.

After installing the cylinder, fill and bleed the brake system (● 56-33).

● 56-30 Disk-Brake Service

On fixed-caliper brakes, the brake shoes can be replaced without removing the caliper assembly (Fig. 56-4). With the car on a hoist or jack stand, remove the wheel. Then use two pairs of pliers and pull on the tabs to pull each shoe out. Before installing new shoes, push the pistons in with slip-joint pliers (right in Fig. 56-4). However, before you do this, you should remove some fluid from the master cylinder. Otherwise, pushing the pistons in will force enough fluid back into the master cylinder so it will overflow. Discard this fluid; it may be dirty.

On the floating-caliper brake, you must first remove both the wheel and caliper to replace the brake shoes. First, remove two-thirds of the fluid from the master-cylinder section feeding the disk brakes. Discard the fluid. Raise the car and remove the wheel cover and wheel. Use a C clamp as shown in Fig. 56-6 and tighten it to force the piston back into its cylinder. Remove the two mounting bolts and lift the caliper off. Support it with a wire hook so it does not hang from the brake hose. Remove the old shoes. Remove the sleeves and bushings from the four caliper ears.

On reinstallation, first install new sleeves and bushings and the shoes. Put each shoe back on the same side of the caliper from which it was removed. Make sure the piston is pushed back into its cylinder. Position the caliper over the disk and attach it with the bolts. Clinch the upper ears of the outboard shoe to hold it in place. The ears should be flat against the caliper. Add fresh brake fluid to the master cylinder and pump the brake pedal several times to seat the linings against the disk and to get a firm pedal.

CAUTION: Do not attempt to move the car before pumping the brake pedal. You will not have brake action until you do.

Check and refill the master-cylinder reservoir as necessary. Drive the car, and make several heavy 40-mph [64.37 kmph] stops to seat the brake shoes.

If pistons or seals require replacement, the caliper assembly must be removed. A special tool should be used to pull the pistons from the caliper. They must come out straight, without cocking. If they cock, the bore is apt to be damaged. When installing new seals in the piston bores, be sure that they go in straight and are not twisted. Dipping them in clean brake fluid will help.

Slight roughness or corrosion in the piston bores can be cleaned up with a special hone. If the bore diameter must be increased more than 0.002 inch [0.051 mm] discard the old caliper and use a new one. New pistons are required, also, if the old ones show signs of wear sufficient to remove the chrome plating.

Wheel bearings (front wheels) must be adjusted to specifications. Excessive play will have a bad effect on disk-brake action.

56-31 Installing Brake Tubing

Special steel tubing must be used for hydraulic brakes. It is made to withstand the high pressures developed in the system. Tubing must be cut off square with a special tube cutter. Tubing must not be cut with a jaw-type cutter or with a hacksaw. Either of these may distort the tube and leave heavy burrs that prevent flaring of the tube. After the tube has been cut, a special flaring tool must be used to lap and flare it (Fig. 56-7).

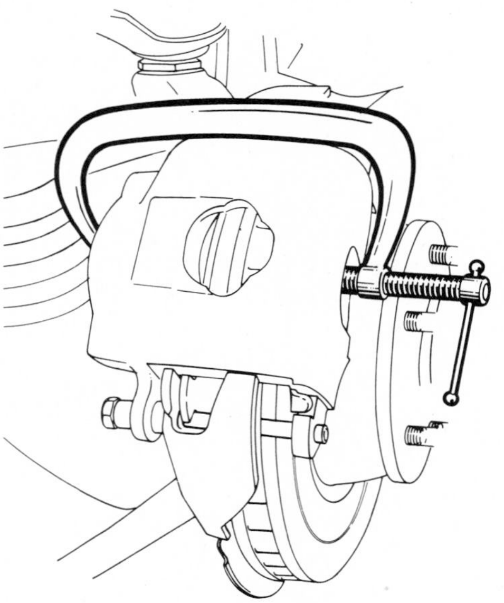

Fig. 56-6 Using a C clamp to force the piston into the bore *(Buick Motor Division of General Motors Corporation)*

56-32 Flushing Hydraulic System

If dirt or damaging liquid has been introduced into the hydraulic system, it is necessary to flush out the system. We repeat that mineral oil should never be put into the system. It will cause the rubber parts to swell and deteriorate so that braking action may be completely lost. In flushing the system, use only the special flushing compound recommended by the manufacturer. Anything else is likely to cause damage in the system.

To flush the system, remove the bleeder-valve screw at one wheel cylinder, and attach bleeder drain tube.

CAUTION: Clean dirt and grease from around the valve, to avoid getting any dirt into the cylinders. Any dirt at a valve or in a drain tube may get sucked into the cylinder on the brake-pedal return stroke. This could cause failure of the wheel cylinder and brakes.

Put the lower end of the drain tube into a clean glass jar. Unscrew the bleeder valve about three-quarters of a turn. Then operate the brake pedal full strokes, to force all fluid from the system. When all fluid is out, fill the master cylinder with brake-system cleaning fluid. (Use only recommended brake-system cleaning fluid.) Use a master-cylinder filler so that the reservoir will be replenished as the cleaning fluid passes through the system. Operate the brake pedal full strokes until all the cleaning fluid in the reservoir and in the filler has passed through the system. Then use dry, clean air, applied through the master cylinder, to blow out all the liquid from the system. Do not apply too much air pressure. Repeat the procedure at the other wheels. Finally, add new brake fluid, and bleed the system.

56-33 Filling and Bleeding Brake System

When a hydraulic brake system has been flushed, when the fluid has become low, or when air has leaked into the system, the system must be bled to eliminate the air. Air in the system will cause soft, or spongy, brake-pedal

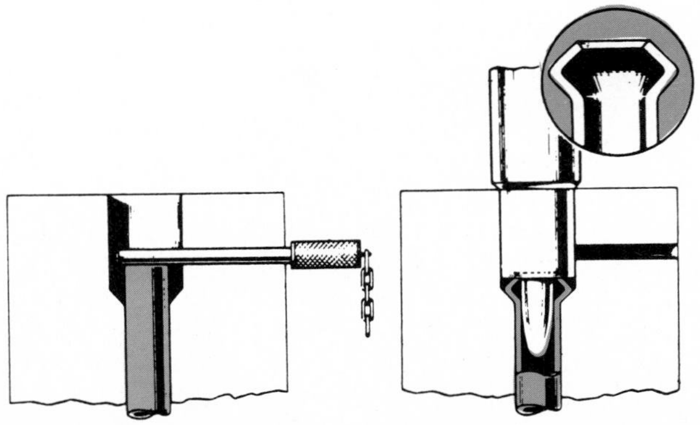
SECTIONAL VIEW OF DIE SHOWING TUBING LOCATED AGAINST STOP PIN.
FIRST OPERATION TOOL SHOWN FORMING OUTSIDE FLARE.
SECOND OPERATION TOOL SHOWN FORMING INSIDE FLARE AND SEAT. COMPLETED DOUBLE-LAP FLARE SHOWN IN INSERT

Fig. 56-7 The three steps required to double-lap flare hydraulic-brake tubing. *(Ford Motor Company)*

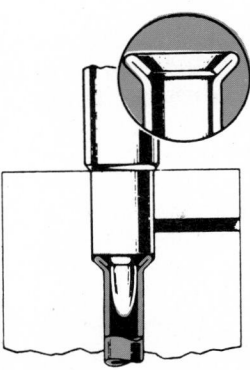

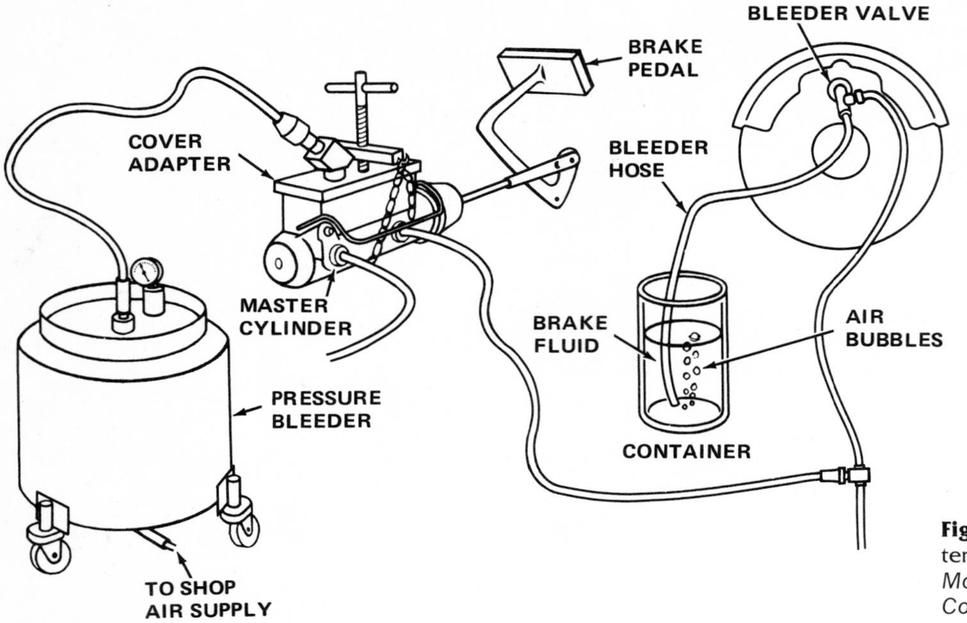

Fig. 56-8 Bleeding a hydraulic system with a pressure tank. (*Pontiac Motor Division of General Motors Corporation*)

action. The air compresses when the brakes are applied, and poor braking action results. Air is eliminated by adding brake fluid and bleeding off a little of the fluid from each wheel cylinder. To add brake fluid, first make sure that the bleeder valves are closed at all cylinders. Then, use either a master-cylinder reservoir filler or a pressure tank (Fig. 56-8). In either case, the reservoir filler or pressure tank should contain approved brake fluid.

When the reservoir is filled and the filler is in place (or pressure tank connected), install a bleeder drain and jar at one wheel cylinder. (Make sure that dirt is cleaned from around the connection so that dirt will not get into the wheel cylinder.) Open the bleeder drain. When using the reservoir, have someone get into the car and pump the brake pedal full strokes, allowing it to return slowly. Continue until the fluid flows from the drain tube into the glass jar in a solid stream that is free of air bubbles. Make sure the end of the tube is below the liquid level in the jar. This prevents air from being sucked into the system on the brake-pedal return strokes. Tighten the bleeder valve. Remove the drain tube, and replace the screw in the valve. Repeat the operation at the other wheel cylinders. Be sure to maintain proper fluid level in the master-cylinder reservoir. When the bleeding operation is complete, remove the master-cylinder filler. Make sure that the fluid level in the reservoir is correct. Then install the filler plug and gasket. Be sure the vent is open.

When the pressure tank is used (Fig. 56-8), no assistant is needed. The pressure tank is partly filled with brake fluid. Then, air is compressed in the tank by use of the tire-inflating equipment. The brake fluid is therefore under pressure in the tank. The tank is connected to the master cylinder, as shown in Fig. 56-8. When the valve is turned on, brake fluid flows from the tank, under pressure, to the master-cylinder reservoir. Brake fluid is forced through the brake line and the wheel cylinder to which the drain tube has been connected, as shown. With the pressure tank, the valve in the line from the tank to the reservoir is turned on. Brake fluid is allowed to flow from the tank into the brake system until the brake fluid runs from the drain tube in a solid stream without air bubbles. Then, the valve is tightened, the drain tube removed, and the screw replaced. The operation is repeated at each wheel cylinder.

Do not attempt to reuse the brake fluid in the glass jar. It is likely to be contaminated or dirty.

REVIEW QUESTIONS

1. Name five conditions that would allow the brake pedal to go to the floor board without producing braking (drum brakes).
2. Name five conditions in the drum-brake system that would cause one or more brakes to drag.
3. Name six conditions that would cause the car to pull to one side when braking (drum brakes).
4. Name six conditions that would cause poor braking action requiring excessive pedal pressure (drum brakes).
5. Name three conditions that would cause loss of brake fluid or air in the system (drum brakes).
6. Name six conditions in a disk-brake system that would require excessive pedal travel to produce braking.
7. Name two conditions in a disk-brake system that would cause pedal pulsation during braking.
8. Name five conditions in a disk-brake system that would require excessive pedal effort to produce braking.
9. Name six conditions in a disk-brake system that would cause the car to pull to one side when braking.

10. Name three types of noise from a disk-brake system. Explain what could cause each.
11. Name five possible causes for the brakes to fail to release (disk-brake system).
12. Name five reasons why the brake pedal could be depressed all the way without producing any braking action (disk-brake system).
13. Describe the procedure for flushing the hydraulic system.
14. Describe the procedure for filling and bleeding a brake system.
15. Why is it necessary to bleed the brake system when fluid is added?

SELF PROJECTS

1. Make a list of all the possible complaints of trouble in a drum-brake system. Do the same for disk brakes.
2. If you really want to learn the various brake complaints and their possible causes, do this: Make two sets of 3-by-5-inch cards, one for drum brakes and the other for disk brakes. Copy from the trouble-diagnosis charts in the chapter. On one side of each card, write a complaint, the number of possible causes, and whether the complaint applies to drum or disk brakes. Then, on the other side of the card, write the possible causes of the complaint. When you have finished your two sets of cards, study them every chance you get. Read a complaint, and then try to remember what possible causes there could be. By the time you have memorized all the complaints and their possible causes, you will be well on your way to being a brake expert.
3. If you can get your hands on junked master cylinders, wheel cylinders, and disk brakes, disassemble them to see if you can find out what is wrong with them. When you find a cause, write it out on a tag. Attach the tag to the defective part. This is the procedure automotive mechanics use when they return defective parts to the manufacturer on warranty. If you find enough defective parts, make up a display board for the school shop. This will help other students.

CHAPTER 57
Tires and Tire Service

After studying this chapter, you should be able to:

- **Describe tire construction and explain the difference between bias and radial tires.**
- **Interpret the markings on the side of the tire and explain what each means.**
- **Describe various kinds of tire wear and explain what causes each.**
- **Explain how to remove a wheel from a car, take the tire off the wheel and repair it. Then explain how to replace the tire on the wheel and install the wheel on the car.**
- **Perform this job under the supervision of the instructor.**

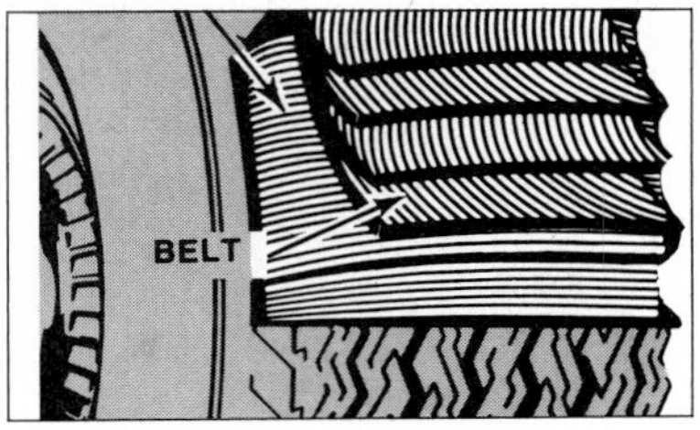

This chapter describes tires and tire service. Improved tire design and improved highways have combined to make tire trouble much less common today than it was years ago. Yet, there is still a need for tire service. The automotive technician should know how tires are made and how they do their job. Furthermore, the technician should know what various patterns of abnormal wear mean. Abnormal tire wear usually is a sign of trouble in the steering, suspension, or brake system. Such trouble should be fixed before more serious trouble develops.

● 57-1 Purpose of Tires

Tires have two functions. First, they are air-filled cushions that absorb most of the shocks caused by road irregularities. The tires flex, or give, as they meet these irregularities. Thus they reduce the effect of the shocks on the passengers in the car. Second, the tires grip the road to provide good traction. Good traction enables the car to accelerate, brake, and take turns without skidding.

● 57-2 Tire Construction

There are two general types of tires: those with inner tubes and those without tubes, called *tubeless* tires. On the inner-tube type, both the tube and the tire casing are mounted on the wheel rim. The tube is a hollow rubber doughnut. It is inflated with air after it is installed inside the tire and the tire is put on the wheel rim (Fig. 57-1). This inflation causes the tire to resist any change of shape.

Tubes are used in some truck tires and in motorcycle tires. Tubes are seldom used in passenger-car tires today. Cars use tubeless tires. The tubeless tire does not use an inner tube. Instead, the tubeless tire is mounted on the wheel rim so that the air is retained between the rim and the tire (Fig. 57-2).

The amount of air pressure used in the tire depends on the type of tire and the operation. Passenger tires are inflated to about 22 to 30 pounds per square inch, or psi [155 to 205 kPa]. Heavy-duty tires on trucks or buses may be inflated to 100 psi [690 kPa].

The tire casings, and tubeless tires, are made in about the same way. Layers of cord, called *plies*, are shaped on a form and impregnated with rubber. The rubber sidewalls and treads are then applied, as shown in Fig. 57-3. They are vulcanized into place to form the completed tire. The term "vulcanizing" means heating the rubber under pressure. This process molds the rubber into the desired

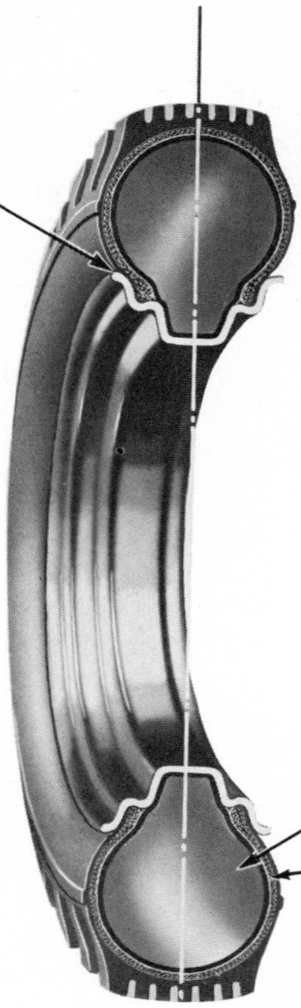

Fig. 57-1 Tire and tire rim cut away so the tube can be seen.

form and gives it the proper wear characteristics and flexibility. The number of layers of cord, or plies, varies according to the intended use of the tire. Passenger-car tires have 2, 4, or 6 plies. Heavy-duty truck and bus tires may have up to 14 plies. Tires for heavy-duty service, such as earthmoving machinery, may have up to 32 plies.

57-3 Bias vs. Radial Plies

There are two ways to apply the plies, on the bias and radially. For many years most tires were of the bias type, as shown to the left in Fig. 57-4. These tires had the plies criss-crossed. That is, one layer runs diagonally one way, and the other layer runs diagonally the other way. This arrangement makes a carcass that is strong in all directions because of the overlapping plies. However, the plies tend to move against each other in bias tires. This movement generates heat, especially at high speed. Also, the tread tends to "squirm," or close up, as it meets the road. This increases tire wear.

Tires with radial plies, as shown to the right in Fig. 57-4 and in Fig. 57-5, were introduced to remedy these problems. In a radial tire, all plies run parallel to each other and vertical to the tire bead. Belts are then applied on top of the plies to provide strength parallel to the bead. The tread is then vulcanized on top of the belts which are made of rayon, nylon, glass fiber, or steel mesh.

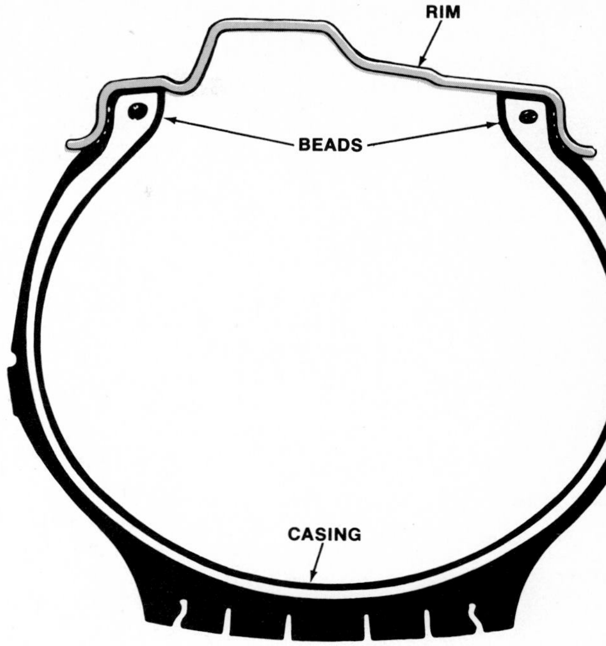

Fig. 57-2 Sectional view of a tubeless tire, showing how the bead rests between the ledge and flanges of the rim to produce a good seal.

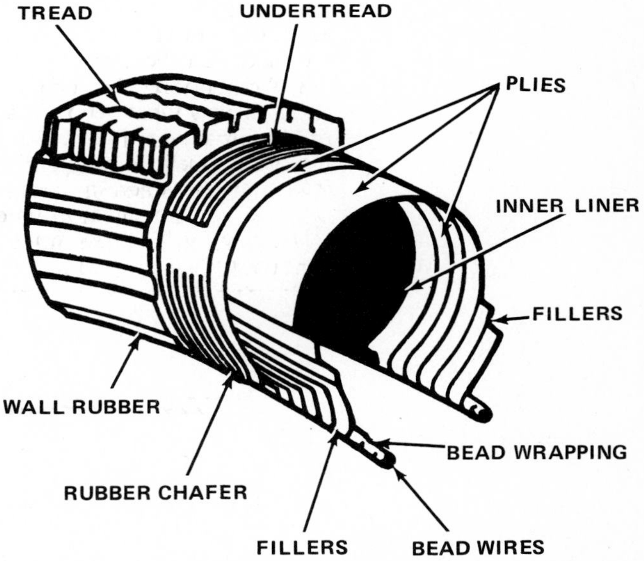

Fig. 57-3 A cutaway view of a tubeless tire, showing tire construction. (*Chevrolet Motor Division of General Motors Corporation*)

Radial tires are installed on about 80 percent of all cars built in the United States. All radial tires work in the same way, regardless of the belt material. The belt provides added strength. Radial tires put more rubber on the road than a bias-ply tire. The radial is more flexible, so more of the tread stays on the pavement, as shown in Fig. 57-6. Also, the tread has less tendency to heel up when the car goes around a curve, as shown in Fig. 57-7. This keeps more rubber on the road and reduces the tendency of the tire to skid.

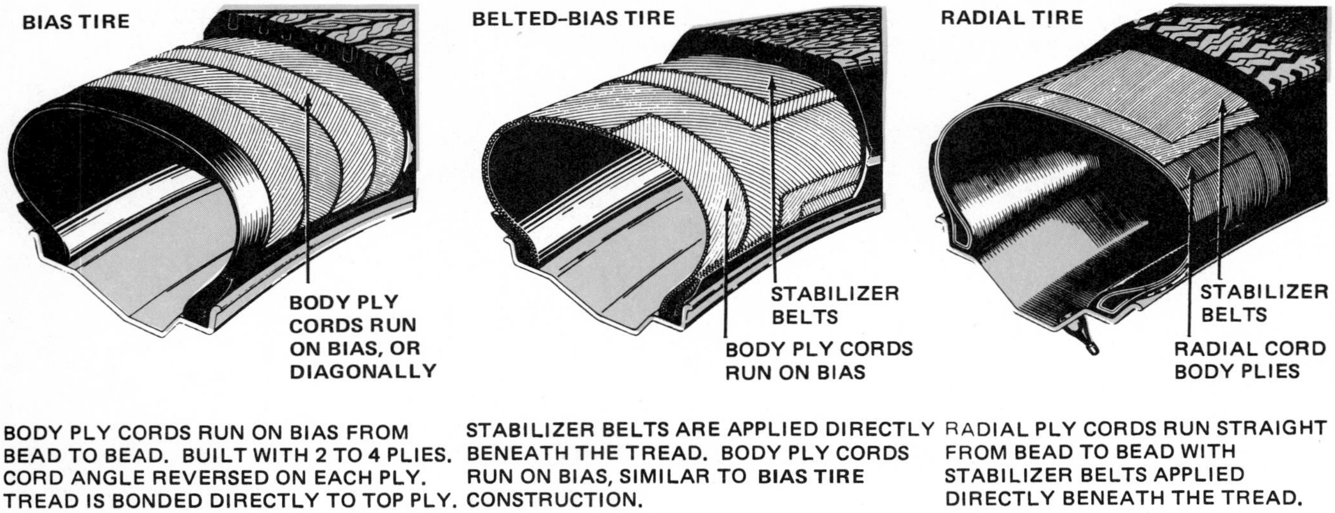
BIAS TIRE
BODY PLY CORDS RUN ON BIAS, OR DIAGONALLY
BELTED-BIAS TIRE
STABILIZER BELTS
BODY PLY CORDS RUN ON BIAS
RADIAL TIRE
STABILIZER BELTS
RADIAL CORD BODY PLIES
BODY PLY CORDS RUN ON BIAS FROM BEAD TO BEAD. BUILT WITH 2 TO 4 PLIES. CORD ANGLE REVERSED ON EACH PLY. TREAD IS BONDED DIRECTLY TO TOP PLY.
STABILIZER BELTS ARE APPLIED DIRECTLY BENEATH THE TREAD. BODY PLY CORDS RUN ON BIAS, SIMILAR TO BIAS TIRE CONSTRUCTION.
RADIAL PLY CORDS RUN STRAIGHT FROM BEAD TO BEAD WITH STABILIZER BELTS APPLIED DIRECTLY BENEATH THE TREAD.

Fig. 57-4 Cutaway views of the three basic tire constructions. (*Firestone Rubber Company*)

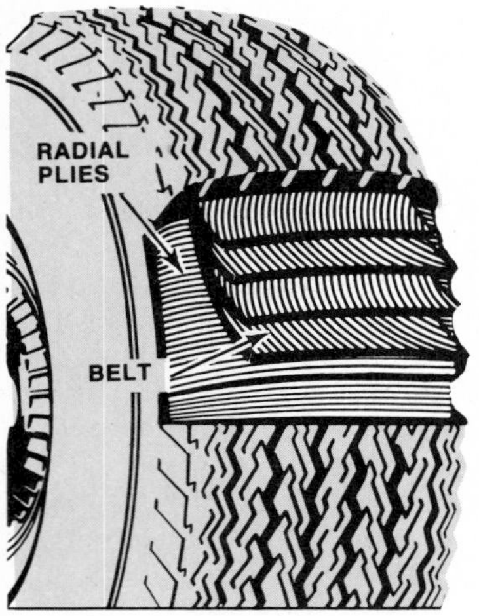
RADIAL PLIES
BELT

Fig. 57-5 Belted radial tire, partly cut away to show radial plies and belt. (*B. F. Goodrich Company*)

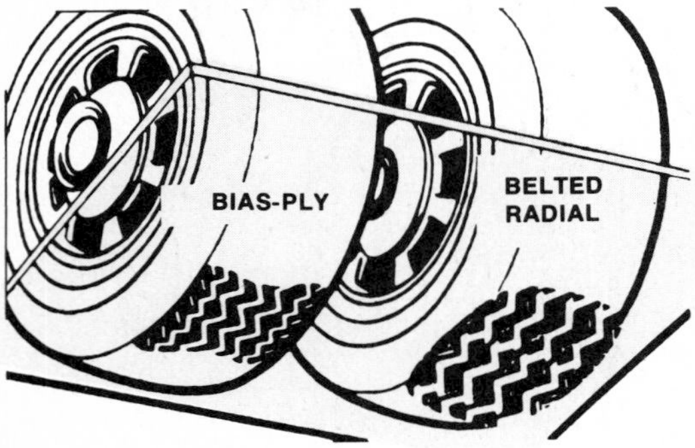
BIAS-PLY
BELTED RADIAL

Fig. 57-6 Footprints of a nonbelted bias-ply tire and a belted-radial tire on a flat surface. Note that the belted-radial tire puts more rubber on the road.

Radial tires wear more slowly than bias-ply tires. This is because the radial-tire tread does not squirm as the tire meets the pavement. The bias-ply tire tread tends to squirm, as shown in Fig. 57-8. As the treads pinch together, they slide sideways. This causes tread wear. There is less heat buildup on the highway in the radial tire. This also slows radial-tire wear.

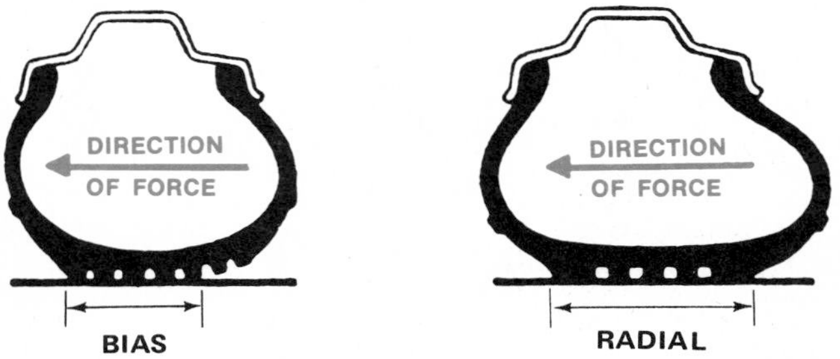
DIRECTION OF FORCE
BIAS
DIRECTION OF FORCE
RADIAL

Fig. 57-7 Difference in the amount of tread a nonbelted bias-ply tire and a belted-radial tire apply to the pavement during a turn.

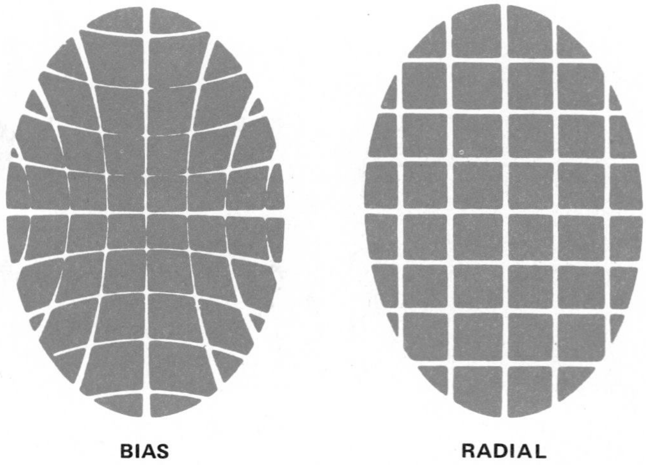
BIAS
RADIAL

Fig. 57-8 Bias-ply tire tread tends to "squirm" as it meets the road. Belted-radial tire tread tends to remain apart.

Bias-ply tires may also be belted, as shown in Fig. 57-9, but even some manufacturers who make the belted-bias tire openly recommend the belted radial as the superior tire.

CAUTION: Never mix belted-radial and bias-ply tires, either belted or unbelted, on a car. Mixing the two types can cause poor car handling and increase the possibility of skidding. This is especially important with snow tires.

Regular snow tires on the rear and belted radials on the front can result in oversteer and spin-out on wet or icy roads.

● 57-4 Tire Tread

The tread is the part of the tire that rests on the road. There are many different tread designs. Figure 57-10 shows a few. Snow tires have large rubber cleats molded into the tread. The cleats cut through snow to improve traction.

Some tires have steel studs that stick out through the tread. Studs help the tire get better traction in ice and snow. However, some people claim studded tires shorten the life of the road surface. For this reason, studded tires are banned in some localities.

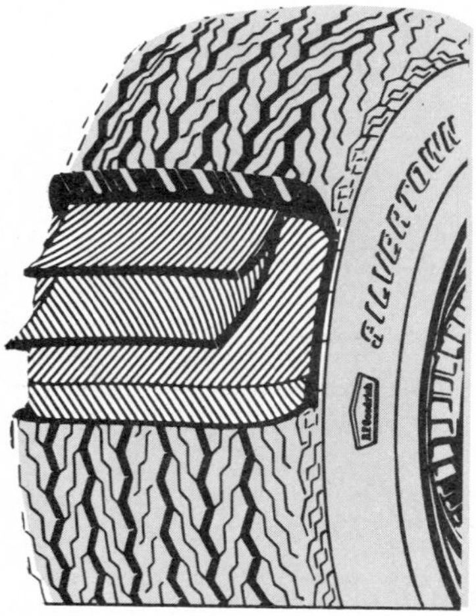

Fig. 57-9 Belted-bias tire, partly cut away to show bias plies and belt. (*B. F. Goodrich*)

● 57-5 Tire Valve

Air is put into the tire, or into the inner tube, through a valve that opens when an air hose is applied to it. The valve is sometimes called a Schrader valve. On a tubed tire, the valve is mounted in the inner tube and sticks out through a hole in the wheel rim. On the tubeless tire, the valve is mounted in a hole in the wheel rim. When the valve is closed, spring pressure and air pressure inside the tire or tube hold the valve on its seat. Most valves carry a valve cap. The cap is screwed down over the end of the valve. It protects the valve from dirt and acts as an added safeguard against air leaks.

● 57-6 Tire Size and Markings

Tire size is marked on the sidewall of the tire. An older tire might be marked 7.75-14. This means that the tire fits on a wheel that is 14 inches [356 mm] in diameter at the rim where the tire bead rests. The 7.75 means the tire itself is about 7.75 inches [197 mm] wide when it is properly inflated.

Figure 57-11 shows a tire with an explanation of each mark on it. Tires carry several markings on the sidewall. The markings include a letter code to designate the type of car the tire is designed for. D means a lightweight car. F means intermediate. G means a standard car. H, J, and L are for large luxury cars and high-performance vehicles. For example, some cars use a G78-14 tire. The 14 means a rim diameter of 14 inches [356 mm]. The 78 indicates the ratio between the tire height and width, as shown in Fig. 57-12. This tire is 78 percent as high as it is wide. The ratio of the height to the width is called the *aspect ratio* or the *profile ratio*. There are four aspect ratios at present: 83, 78, 70, and 60. The lower the number, the

Fig. 57-10 Types of tire tread. (*B. F. Goodrich Company, Goodyear Tire and Rubber Company*)

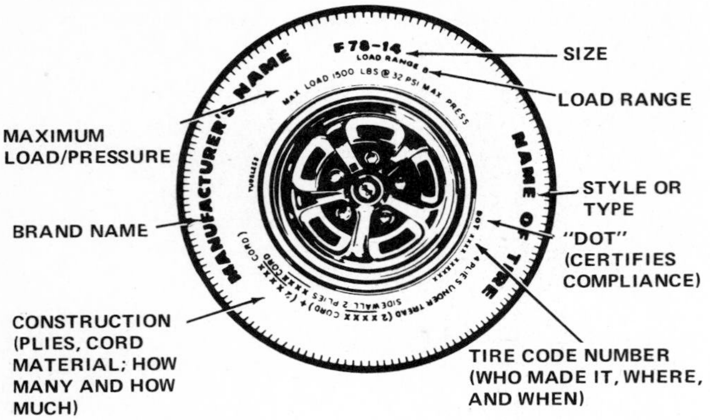

Fig. 57-11 Every marking on a tire is important information. Note location of tire size, and the pressure and load limit.

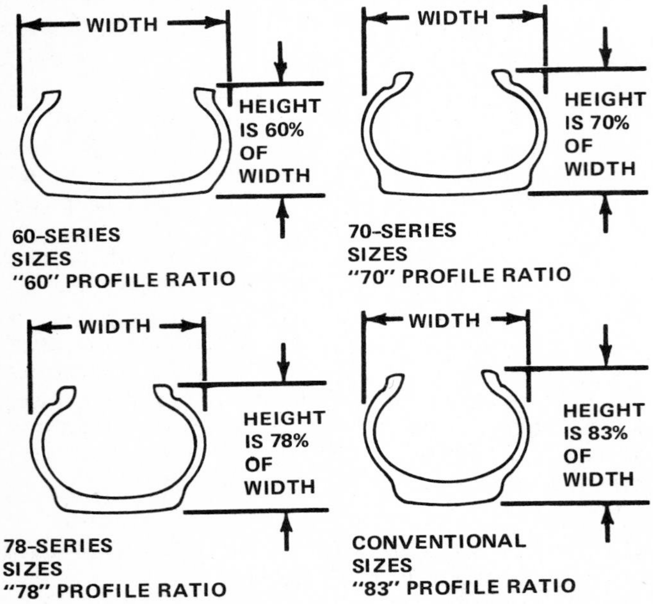

Fig. 57-12 Four aspect ratios of car tires. *(American Motors Corporation)*

wider the tire looks. A 60 tire is only 60 percent as high as it is wide.

The addition of an R to the sidewall marking, such as GR78-14, indicates that the tire is a radial. Also, if a tire is a radial, the word "radial" must be molded into the sidewall. Some radials are marked in the metric system. For example, a tire marked 175R13 is a radial tire which measures 175 mm (6.9 inches) wide. It mounts on a wheel with a rim diameter of 13 inches [330 mm].

Some cars use metric size tires. The meaning of each letter and number of a metric tire is shown in Fig. 57-13. This is the latest size designation for tires. Comparing the two tire-size labels, a tire formerly marked as an ER78-14 is now marked P195/75R14.

To identify the load that a tire can safely carry, each tire is classified into a load range. The load range indicates the allowable load for the tire as inflation pressure is increased. In Fig. 57-11, you can see that the tire is marked "Load Range B." Most passenger car tires are in

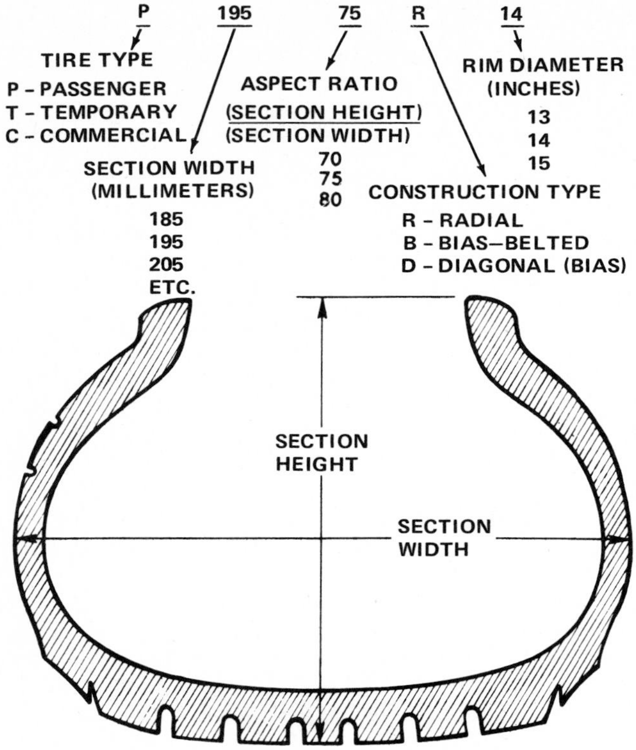

Fig. 57-13 Meanings of the size designations for a metric tire. *(Chevrolet Motor Division of General Motors Corporation)*

Fig. 57-14 A collapsible spare tire.

load range B. There are three load ranges for passenger-car tires, B, C, and D. Under the old system, "ply rating" was used to indicate load range. The load range B tire has the same load-carrying capacity as a tire with a 4-ply rating. Load range C equals a 5-ply rating tire. Load range D equals an 8-ply rating tire.

57-7 Collapsible Spare Tire

This tire (Fig. 57-14) saves space in the luggage compartment. It is installed on the wheel in a deflated condition. It barely protrudes beyond the rim. A pressure can of inflation propellant, called the inflator, is stored in the luggage compartment. Instructions on how to install and safely inflate collapsible spare tires are printed on the inflator can. We describe the procedure in ● 57-29.

CAUTION: This tire must not be driven more than 150 miles [240 km] and at a speed of only 50 mph [80 km/h] or less. To exceed these limits is to risk a blowout of the collapsible spare tire. Also, the tire must not be inflated from the usual air hose at the shop or in a service station. This can cause the tire to explode.

● 57-8 Compact Spare Tire

Another tire that saves space in the luggage compartment is the compact spare tire (Fig. 57-15). This spare is lighter and considerably smaller than the standard tire. The compact spare tire can be driven for the 1,000- to 3,000-mile [1,609- to 4,800-km] life of the tread. It is mounted on a narrow 51 x 4 wheel. The tire must not be mounted on any other wheel. No other tire, wheel cover, or trim ring should be installed on the special wheel. Also, the compact spare tire should not be used on the rear of a car equipped with a nonslip differential. The collapsible spare tire is smaller in diameter than the tire on the other side of the rear axle. Differential action would have to take place continuously. This may cause damage and failure of the differential.

The compact spare tire is for emergency use only. As soon as the standard tire has been repaired, reinstall it on the car. The compact spare carries an inflation pressure of 60 psi [415 kPa]. It gives a rough and noisy ride.

● 57-9 Wheels

Most cars use a pressed steel or disk wheel. This type of wheel also is called a *safety-rim* wheel. Figure 57-16 shows how a disk wheel is made. The outer part, called the rim, is in one piece and it is welded to the disk. This forms the seamless and airtight wheel that is needed to mount a tubeless tire. The center of the rim is smaller in diameter than the rest. This gives the rim the name "drop center." The center well is necessary to permit removal and installation of the tire. The bead of the tire must be pushed off the bead seat and into the smaller diameter. Only then can the tire beads be worked up over the rim flange. We discuss tire service later in the chapter.

The 14-inch [356-mm] wheel is used on most cars today. However, some smaller cars have 12-inch [305-mm] or 13-inch [330-mm] wheels. Fourteen-inch [356-mm] wheels are available in three widths. The width used on a car depends on the tire specified for the car. The rim widths are 4.5 inches [114 mm], 5 inches [127 mm], and 6 inches [152 mm]. Optional larger tires usually require wider rims.

Most manufacturers recommend that a wheel be replaced if it is bent or leaks air. The new wheel must be exactly the same as the old wheel. Installation of the wrong wheel could cause the wheel bearing to fail, the brakes to overheat, the speedometer to read inaccurately, and the tire to rub the body and frame.

Fig. 57-15 A compact spare tire. (*Pontiac Motor Division of General Motors Corporation*)

● 57-10 Special Wheels

Plain steel wheels, decorated with hub caps or wheel covers, are used on cars today. A variety of special wheels are available. These special wheels can be classified as styled steel or styled aluminum wheels. The "mag" wheel is very popular. It looks like the magnesium wheels used on some race cars (Fig. 57-17). Magnesium metal is very light. However, for passenger cars, "mag" wheels are made of aluminum. Actually, the term "mag wheel" can mean almost any chromed, aluminum offset, or wide-rim wheel of spoke design.

Some aluminum wheels are lighter than the steel wheels they replace. Lighter wheels reduce unsprung weight. This improves handling and performance. Also, some aluminum wheels can improve brake and tire performance by allowing them to run cooler. Aluminum transmits and radiates heat faster than steel.

● 57-11 Split-Rim Wheels

Split-rim wheels are used on heavy trucks, trailers, earthmovers, and so on. They are heavily made and require a different method of tire installation. The drop-rim wheel is not satisfactory for a heavy-duty tire. One split-rim wheel has two pieces, the wheel-assembly piece and the flange. The flange locks into the wheel assembly. The other kind also has these two pieces but has, in addition, a rim lock ring. With either type, the whole wheel and rim, with tire, are removed as an assembly for tire service. Then the lock ring and flange can be removed so that the tire can be taken off the rim. These tires require tubes because the rim cannot be made airtight.

CAUTION: If you ever work on tires mounted on split-rim wheels, make sure all air pressure has been released from the tube before beginning to remove the lock ring or flange. If air pressure is still in the tube, it could blow the tire off the rim when the lock ring or flange is removed and seriously injure or kill anyone nearby. Heavy-duty tires carry high air pressures. Make sure the lock ring or flange is securely in place before attempting to inflate the tire. *Never* stand over the tire while inflating it! It could explode.

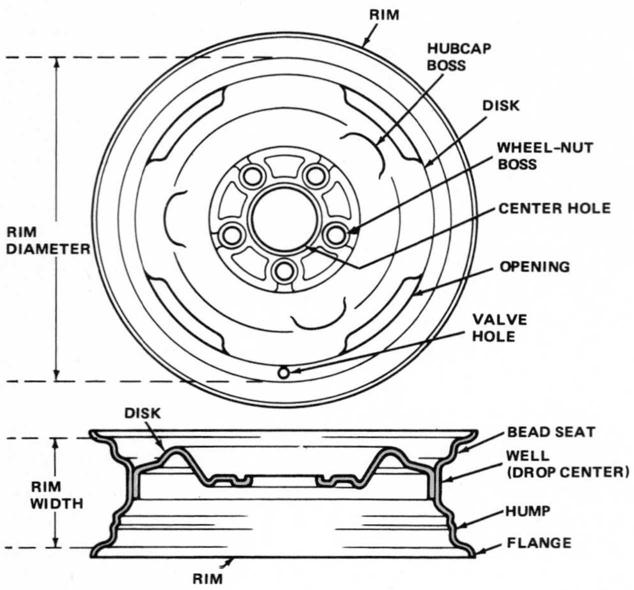

Fig. 57-16 Construction of a car wheel. (*American Motors Corporation*)

Fig. 57-17 A "mag" wheel. (*Shelby International, Inc.*)

Tire Service

● 57-12 Tire Service

Tire service includes periodic checking of the air pressure and addition of air as needed. Failure to maintain correct air pressure can cause rapid tire wear and early tire failure. Incorrect air pressure can also cause handling problems, as we explain in ● 57-13. Tire service also includes periodic inspection of the tire for abnormal wear, cuts, bruises, and other damage. In addition, tire service includes removal, repair, and replacement of tires.

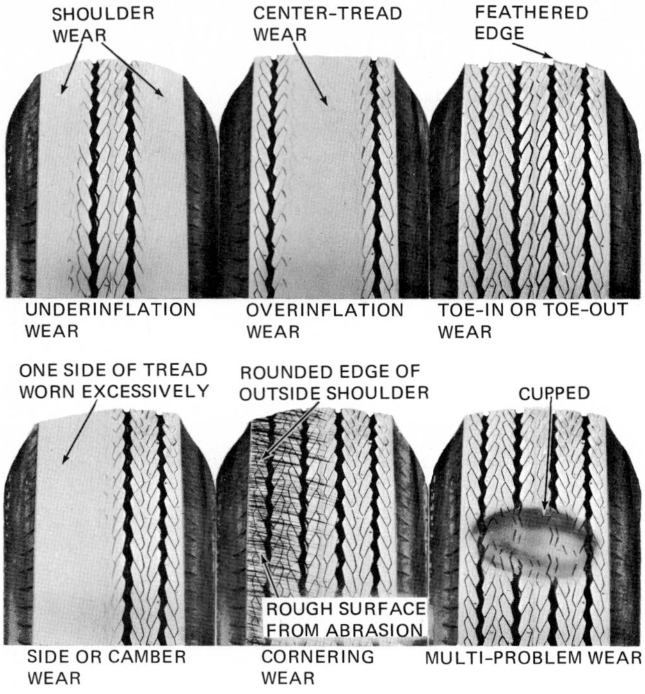

Fig. 57-18 Patterns of abnormal tire-tread wear. (*Buick Motor Division of General Motors Corporation*)

● 57-13 Tire Inflation and Tire Wear

The driver has more effect on tire life than anything else. Good drivers usually get longer tire life than careless drivers. Rapid tire wear results from quick starts and stops, heavy braking, high speed, taking corners too fast, and striking or rubbing curbs. Too little air in the tire can cause hard steering, front-wheel shimmy, steering kickback, and tire squeal on turns. A tire with too little air will wear on the shoulders and not in the center of the tread, as shown at the upper left in Fig. 57-18.

Also, the underinflated tire is subject to rim bruises. That is, if the tire strikes a rut or stone, or bumps a curb too hard, it flexes so much that it is pinched against the rim. Any of these kinds of damage can lead to early tire failure.

With overinflation, the tire rides on the center of the tread. Then, only the center wears, as shown at the upper center in Fig. 57-18. In addition, the overinflated tire will not flex normally. As a result, the tire fabric can be weakened or even broken.

● 57-14 Toe-In or Toe-Out Tire Wear

Excessive toe-in or toe-out on turns causes the tire to be dragged sideways as it moves forward. For example, a tire on a front wheel that toes in 1 inch [25.4 mm] from straight ahead will be dragged sideways about 150 feet

[45.72 m] every mile [1.60 km]. This sideward drag scrapes off rubber, as shown at the upper right in Fig. 57-18. Note the feather edges of rubber that appear on one side of the tread. If both sides show this type of wear, the front end is misaligned. If only one tire shows this type of wear, chances are that the steering arm is bent. This causes one wheel to toe in more than the other.

57-15 Camber Wear

If a wheel has excessive camber, the tire runs more on one shoulder than the other. The tread wears excessively on that side, as shown at the lower left in Fig. 57-18.

57-16 Cornering Wear

Cornering wear, shown at the lower center in Fig. 57-18, is caused by taking curves at excessive speeds. The tire not only skids, but it tends to roll, producing the diagonal type of wear shown. This is one of the more common causes of tire wear. The only remedy is to have the driver slow down around curves.

57-17 Uneven Tire Wear

Uneven tire wear, with the tread unevenly or spottily worn, is shown at the lower right in Fig. 57-18. It can result from several mechanical problems. These include misaligned wheels, unbalanced wheels, uneven or "grabby" brakes, overinflated tires, and out-of-round brake drums.

57-18 High-Speed Wear

Tires wear more rapidly at high speed than at low speed. Tires driven consistently at 70 to 80 miles per hour [112.65 to 128.75 kmph] will give less than half the miles of tires driven at, say, 30 miles per hour [48.28 kmph].

57-19 Radial-Tire Waddle

Waddle is side-to-side movement of the front or rear of the car as it moves forward. It is caused by the steel belt not being straight inside the tire. It is most noticeable at low speeds, 5 to 30 mph [8.047 to 48.280 kmph]. It may also be felt as ride roughness at 50 to 70 mph [80.467 to 112.67 kmph].

To determine where the faulty tire is on the car (front or rear), make a road test. If the faulty tire is on the rear, the rear end of the car will shake from side to side. From the driver's seat it feels as though someone were pushing on the side of the car.

If the faulty tire is on the front, the waddle is more visual. The front sheet metal will seem to move back and forth or from side to side.

57-20 Checking Tire Pressure and Inflating Tires

To check tire pressure and inflate the tires, you must know the correct tire pressure for the tire you are servicing. You find this spec on the tire sidewall (on many tires), in the shop manual, in the driver's operating manual, and on one of the door jambs (or at some similar place) on the car. Specifications are for cold tires. Tires that are hot from being driven or from sitting in the sun will have an increased air pressure. Air expands when hot. Tires that have just come off an interstate highway may show as much as a 5- to 7-psi [35- to 48-kPa] increase.

As a hot tire cools, it loses pressure. So, never bleed a hot tire to reduce the pressure. If you do this, then when the tire cools, its pressure could drop below the specified minimum.

There are times when the tire pressure should be on the high side. For instance, one tire manufacturer recommends adding 4 psi [28 kPa] for turnpike speed, trailer pulling, or extra-heavy loads. But never exceed the maximum pressure specified on the tire sidewall.

CAUTION: Never stand over a tire while inflating it. It could explode and you would be injured.

If the tire valve has a cap, always replace the cap after checking pressure or adding air.

Inflation specifications are often given in kiloPascals (kPa) instead of pounds per square inch (psi). The conversion table (Table 57-1) will help you convert from one to the other.

57-21 Tire Rotation

The amount of wear a tire gets depends on its location on the car. For example, the right rear tire wears about twice as fast as the left front tire. This is because roads are

TABLE 57-1.

INFLATION PRESSURE CONVERSION CHART (KILOPASCALS TO PSI)

kPa	psi	kPa	psi
140	20	215	31
145	21	220	32
155	22	230	33
160	23	235	34
165	24	240	35
170	25	250	36
180	26	275	40
185	27	310	45
190	28	345	50
200	29	380	55
205	30	415	60

Conversion: 6.9 kPa = 1 psi

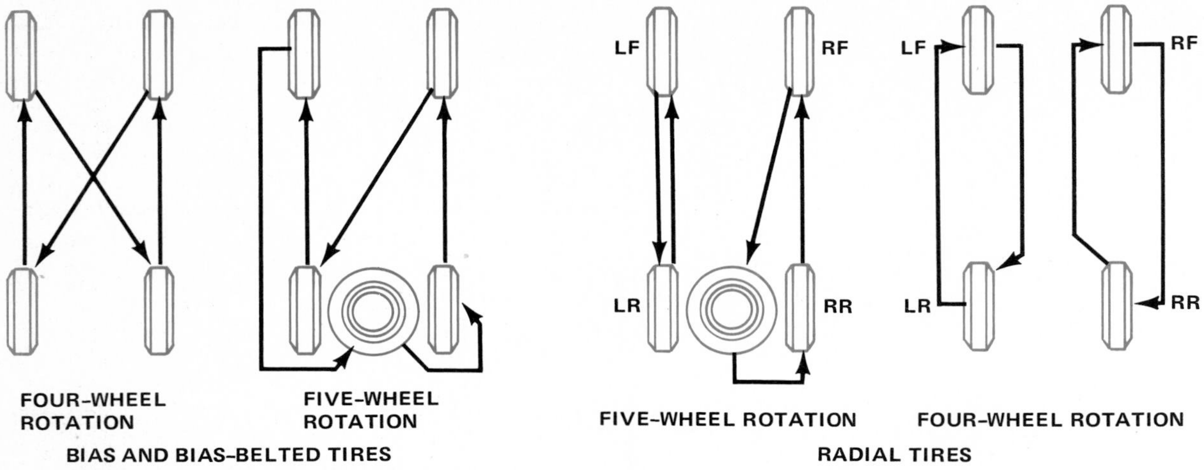

Fig. 57-19 Tire-rotation patterns for cars with and without a rotatable spare tire. *(Chevrolet Motor Division of General Motors Corporation)*

slightly crowned (higher in the center) and also because the right rear tire is driving. The crown causes the car to lean out a little, so the right tires carry more weight. The combination of this and carrying power through the right rear tire causes it to wear faster. To equalize wear as much as possible, tires should be rotated any time uneven wear is noticed and at the milage specified by the car manufacturer. One manufacturer recommends rotating radial tires after the first 7,500 miles [12,000 km] and then every 15,000 miles [24,000 km]. Bias tires should be rotated every 7,500 miles [12,000 km]. Figure 57-19 shows the recommended rotation pattern for bias, bias-belted, and radial tires. Bias and bias-belted can be switched from one side of the car to the other. However, radial tires must not be switched from one side to the other. This would reverse their direction of rotation and cause handling and wear problems.

On cars with a collapsible or a compact spare tire, use the four-wheel rotation pattern shown in Fig. 57-19.

Always check tire pressure after switching tires. Many cars require that the front tires carry different pressures from the rear tires. Thus, when you switch tires from front to back, the pressure will require adjustment to meet the specifications.

CAUTION: Studded tires should never be rotated. A studded tire should be put back on the wheel from which it was removed. Before you remove a studded tire, mark its location (LR or RR) on the sidewall. A studded tire should be put back on the same wheel from which it was removed, mounted so that it rolls in the same direction. Reversing its rotation can result in serious handling problems.

57-22 Tire Inspection

The purpose of inspecting the tires is to determine whether they are safe for further use. When an improper wear pattern is found, the technician must know the cause of the abnormal tread wear (Fig. 57-18). The technician must correct the cause or tell the driver what is wrong. When the tires are found to be in good condition, they can be rotated (Fig. 57-19). After the tires are cool, check and adjust inflation pressures.

When inspecting a tire, check for bulges in the sidewalls. A bulge is a danger signal. It can mean that the plies are separated or broken and that the tire is likely to go flat. A tire with a bulge should be removed. If the plies are broken or separated, the tire should be thrown away. To make a complete tire inspection, remove all stones from the tread. This is to make sure that no tire damage is hidden by the stones. Also, any time the tire is to be spin-balanced, remove all stones from the tread. This will ensure that no person is struck and injured by stones thrown from the tread as the tire rotates.

Many tires have tread-wear indicators, which are filled-in sections of the tread grooves. When the tread has worn down enough to show the indicators (Fig. 57-20), the tire should be replaced. There also are special tire-tread gauges that can be inserted into the tread grooves to measure how much tread remains. A quick way to check tread wear is with a Lincoln penny, as shown in Fig. 57-21. If at any point you can see all of Lincoln's head, the tread is excessively worn. Some state laws require a tread depth of at least $\frac{1}{32}$ inch [0.79 mm] in any two adjacent grooves at any location on the tire. A tire with little or no tread has poor traction on the road and will produce poor braking.

A tire can look okay from the outside and still have internal damage. To completely inspect a tire, remove it from the rim. Then examine it closely, inside and out.

57-23 Tube Inspection

Tubes usually give little trouble if correctly installed. However, careless installation can cause trouble. For example, if the wheel rim is rough or rusty or if the tire bead is rough, the tube may wear through. Dirt in the casing can cause the same trouble. Another condition that can

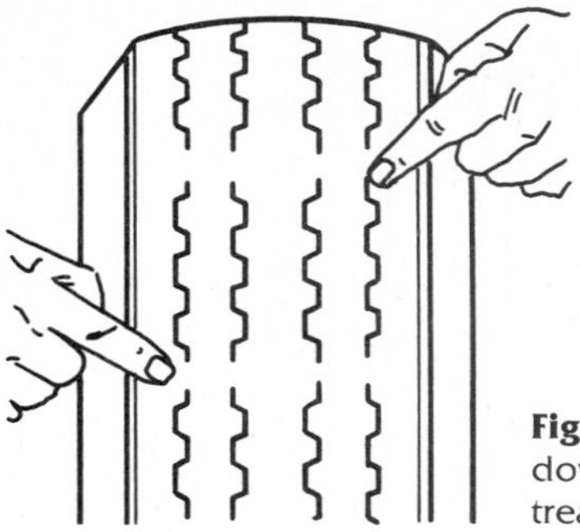

Fig. 57-20 A tire tread worn down so much that the tread-wear indicator shows up.

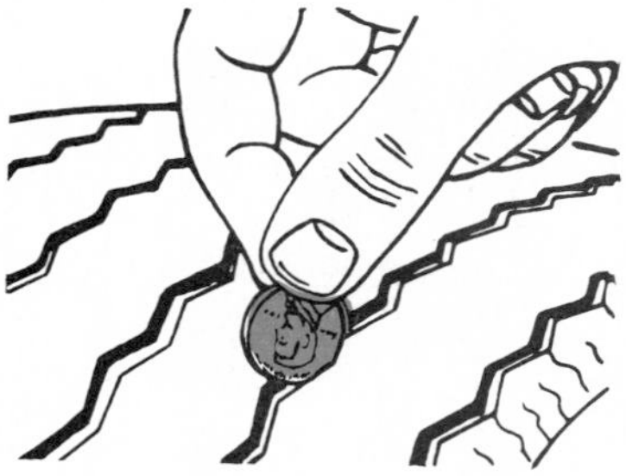

Fig. 57-21 Using a penny to check tread wear.

Fig. 57-22 Using an impact wrench to remove the lug nuts. *(Mobil Oil Company)*

cause trouble is installing a tube that is too large in the tire. Sometimes an old tube (which may have stretched) is put in a new tire. When a tube that is too large is put into a casing, the tube can overlap at some point. The overlap will rub and wear and possibly cause early tube failure.

A radial tire that is used with a tube must have a special radial-tire inner tube in it. If regular tubes are used in radial tires, the tube splice may come apart.

Always check carefully around the valve stem when inspecting a tube. If the tube has been run flat or at low pressure, the valve stem may be broken or tearing away from the tube. Valve-stem trouble requires installation of a new tube.

57-24 Removing a Wheel from the Car

Radial tires must be removed from the car to be repaired. Also, if the tire has been run flat, remove the tire for inspection. To repair a tire, first take off the hub cap or wheel cover. Then remove the wheel from the car. If you are using a lug wrench, loosen the lug nuts before raising the car. It is easier to loosen the lug nuts first, because the wheel will not turn if the car weight is on it. On some cars the lug nuts on the right side of the car have right-hand threads. The lug nuts on the left side have left-hand threads. The reason is that the forward rotation of the wheels tends to tighten the nuts, not loosen them. Figure 57-22 shows the technician using an air-powered impact wrench to remove the lug nuts.

When using an impact wrench, use an impact socket with it. A regular socket may crack when used on an impact wrench.

57-25 Demounting a Tire from a Drop-Center Rim

With the wheel off the car, make a chalk mark across the tire and rim so you can reinstall the tire in the same position. This preserves the balance of the wheel and tire. Next, release the air from the tire. This can be done by holding the tire valve open or removing the valve core.

CAUTION: The air coming out could shoot dirt particles into your eyes, so protect them with glasses or a shield.

The tire should then be removed from the rim, using a shop tire changer. At one time, tire irons—flat strips of steel—were used to remove and install tires. They can damage the tire bead so it will not seal and this ruins the tire. Do not use tire irons!

57-26 Using Shop Tire Changers

Today many shops have an air-powered tire changer, such as shown in Fig. 57-23. After the bead is pushed off the rim (this is called "breaking the bead"), a tool is used with the tire changer to lift the bead up over the rim (Fig. 57-24). The powered tire changer also has a tool to re-mount the tire on the rim.

57-27 Remounting the Tire on the Rim

To mount the tire on the rim, use the tire changer. Coat the rim and beads with rubber lubricant or a soap-and-water mixture. This will make the mounting procedure

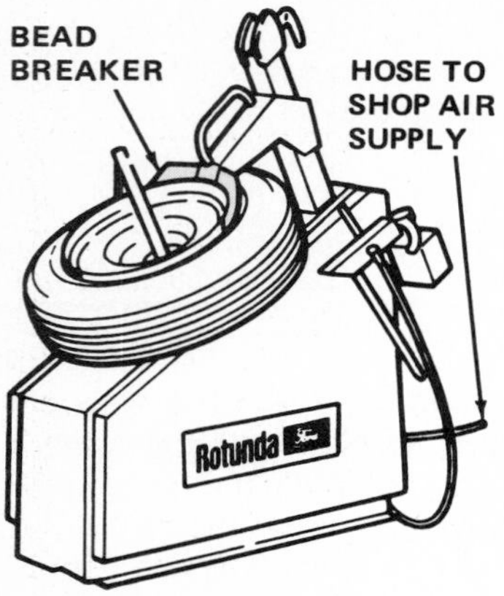

Fig. 57-23 Using an air-powered tire changer to "break the bead" so the tire can be removed from the wheel.

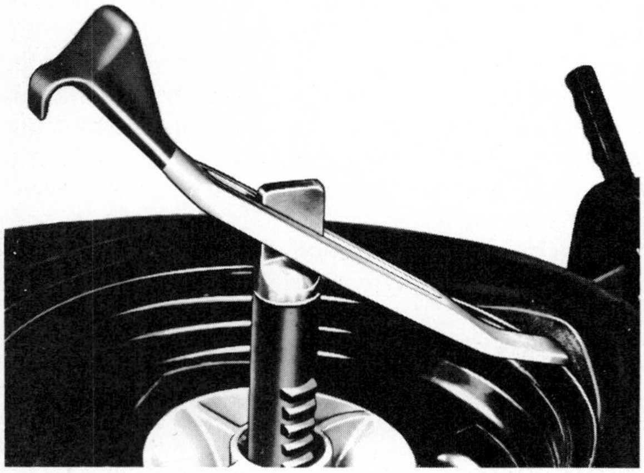

Fig. 57-24 Using the powered tire changer to lift the upper bead above the rim. The center post rotates, carrying the bead-lifting tool around with it.

easier. Do not use a nondrying lubricant, such as antifreeze, silicone, grease, or oil. They will allow the tire to "walk around" the rim so that the tire balance is lost. Oil or grease will damage the rubber. When you are remounting the same tire that was removed from the rim, make sure the chalk marks on the tire and rim align. After the tire is on the rim, reposition the beads against the bead seat. Slowly inflate the tire (Fig. 57-25). If the beads do not hold air, use a tire-mounting band to spread the beads. You usually will hear a "pop" as the beads seat on the rim. Then install the valve core and inflate the tire to the recommended pressure.

CAUTION: Do not stand over the tire while inflating it. If the tire should explode, you could be injured.

● 57-28 Checking the Wheel

When the tire is off the wheel, check the rim for dents and roughness. Steel wool can be used to clean rust spots from standard steel wheels. Aluminum wheels can be cleaned only with mild soap and water. File off nicks or burrs. Then clean the rim to remove all filings and dirt. A wheel that has been bent should be discarded. A bent wheel may be weakened by heating, welding, or straightening so that it could fail on the highway.

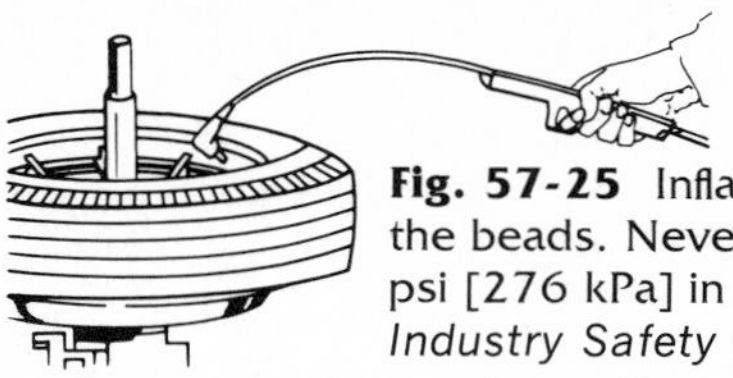

Fig. 57-25 Inflating the tire while seating the beads. Never exceed a pressure of 40 psi [276 kPa] in a passenger-car tire. (*Tire Industry Safety Council*)

Some wheels now have decorative plastic inserts. The plastic can be cleaned by using a sponge and soap and water.

● 57-29 Tire Valve

If the valve in the wheel requires replacement, remove the old valve and install a new one. There are two types: the snap-in type, and the type that is secured with a nut. To remove the snap-in type, cut off the base of the old valve. Lubricate the new valve with rubber lubricant. Then attach a tire-valve installing tool to the valve and pull the new valve into place.

On the clamp-in type that is secured with a nut, remove the nut to take the old valve out. Be sure to tighten the nut sufficiently when installing the new valve.

● 57-30 Servicing the Collapsible Spare Tire

We described the space-saving collapsible tire in Sec. 57-7. It is installed on the wheel deflated (Fig. 57-14). The wheel must be installed on the car before the tire is inflated.

CAUTION: Do not inflate the tire before the wheel is mounted on the car. Follow the safety cautions listed on the inflator can.

The inflator can has detailed instructions on how to install and inflate the tire. Briefly, here is the inflation procedure.

1. If the temperature is 10°F [-12°C] or below, the inflator must be heated. Put the inflator over the defroster outlet of the car. Set the heater at "Defrost" at the highest temperature. Then run the blower at the fastest speed for 10 minutes.
2. Do not inflate the tire off the car. First bolt the wheel to the car with the air valve at the bottom. Remove the plastic cap from the inflator and the cap from the tire valve.
3. Push the inflator onto the valve stem until you hear the sound of gas entering the tire.

CAUTION: Keep your hands off the metal parts of the inflator. They become extremely cold during discharge. You could freeze your fingers!

4. When the sound stops, wait 1 minute. Then remove the inflator and install the valve cap. The gas in the inflator, when completely used up, will properly fill the tire.

After you have inflated the tire and have removed the jack so that the tire rests on the pavement, the tire may look underinflated. This can happen especially in cold weather. Drive slowly for the first mile [16 km]. This will warm up the tire and increase the pressure.

5. If the inflator is the nonrefillable type, dispose of it in a safe waste receptacle. Do not burn or puncture the inflator. If it is the refillable type, it can be recharged with the proper equipment.
6. The collapsible spare tire must not be driven farther than necessary. The maximum distance is 150 miles [240 km]. As soon as the regular tire has been repaired and installed in place of the collapsible spare tire, remove the valve core from it. This will allow the gas to escape so that the tire collapses. It can then be stored, as before, in the luggage compartment.

Instead of a can of inflator, some manufacturers supply an electric air compressor to inflate the tire. Porsche, for example, supplies an electric air compressor. It is plugged into the cigarette-lighter socket and gets power from the car battery. The manufacturer warns that you should not use any other equipment to inflate the tire.

CAUTION: See the Caution in ● 57-7 about speed and mileage limits for the collapsible tire. Also, the tire must not be inflated with an ordinary shop or service-station air hose.

Tube and Tire Repair

● 57-31 Tube Repair

If a tire has been punctured but has no other damage, it can be repaired with a patch. Remove the tube from the tire to find the leak. Inflate the tube and then submerge it in water. Bubbles will appear where there is a leak. Mark the spot. Then deflate the tube and dry it.

There are two ways to patch a tube leak. They are the cold-patch method and the hot-patch method. With the cold-patch method (also known as "chemical vulcanizing"), first make sure the rubber is clean, dry, and free of oil or grease. Buff, or roughen, the area around the leak. Then cover the area with vulcanizing cement. Let the cement dry until it is tacky. Press the patch into place. Roll it from the center out with a "stitching tool" or with the edge of a patch-kit can.

With the hot-patch method, prepare the tube in the same way as for the cold patch. Put the hot patch into place and clamp it. Then, with a match, light the fuel on the back of the patch. As the fuel burns, the heat vulcanizes the patch to the tube. After the patch has cooled, recheck the tube for leaks by submerging the tube in water.

Another kind of hot patch uses a vulcanizing hot plate. The hot plate supplies the heat required to bond the patch to the tube.

The hot-patch method is preferred to the cold-patch method by many technicians.

● 57-32 Tire Repair

No attempt should be made to repair a tire that has been badly damaged. If the plies are torn or have holes in them, the tire should be thrown away. A puncture bigger than $\frac{1}{4}$ inch [6.36 mm] should not be patched. Instead, the tire should be replaced. Even though you might be able to patch the tire, it would be dangerous to use. The tire might blow out on the highway.

To repair small holes in a tubeless tire, first make sure that the object that caused the hole has been removed. Check the tire for other puncturing objects. Sometimes a tubeless tire can carry a nail for a long distance without losing air.

A radial tire should be removed from the wheel for repair. The plug should be of the head type and applied from inside the tire (Fig. 57-26). Figure 57-27 shows the area of a tire in which a puncture can be repaired. Punctures outside this area require replacement of the tire.

Leaks from a tubeless tire are located in the same way as leaks from a tube. With the tire on the wheel and inflated, submerge the tire and wheel in water. Bubbles will show the location of any leaks. If a water tank is not available, coat the tire with soapy water. Soap bubbles will show the location of leaks.

If air leaks from around the spoke welds of the wheel, you can repair the leaks. Clean the area and apply two coats of cold-patch vulcanizing cement on the inside of the rim. Allow the first coat to dry before applying the second coat. Then cement a strip of rubber patching material over the area.

● 57-33 Repairing a Puncture with a Rubber Plug (Tire on Rim)

A temporary repair of a small puncture can be made with the tire still mounted on the rim. However, this repair is only a temporary fix. At the first opportunity, the tire must be removed from the rim and repaired from the inside, as explained in ● 57-32.

Remove the puncturing object, and clean the hole with a rasp. Apply the special vulcanizing fluid, supplied with the repair kit, to the outside of the hole. Push the snout of the vulcanizing-fluid can into the hole to get fluid inside the tire. There are different kinds of rubber plugs. One kind is installed with a plug needle. To use this plug, first cover the hole with vulcanizing fluid. Then select a plug of the right size for the hole. The plug should be at least twice the diameter of the hole. Roll the small end of the plug into the eye of the needle. Dip the plug into vulcanizing fluid. Push the needle and plug through the hole (Fig. 57-28). Then pull the needle out. Trim off the plug one-eighth inch (or 3.2 mm) above the tire surface. Check for leakage. If there is no leakage, the tire is ready for service after it is inflated.

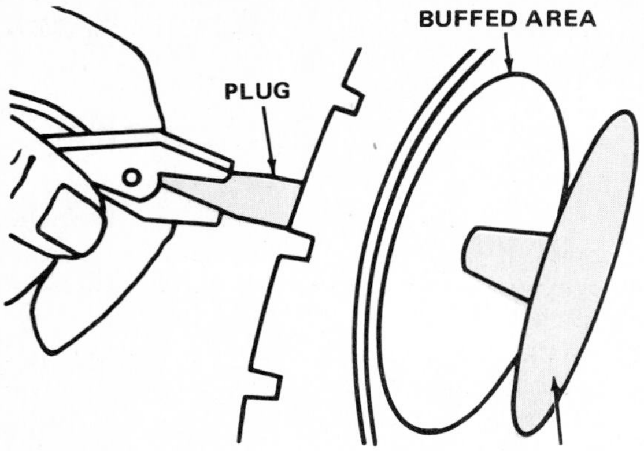

Fig. 57-26 Installing a head-type plug in a radial tubeless tire. (*Rubber Manufacturer's Association*)

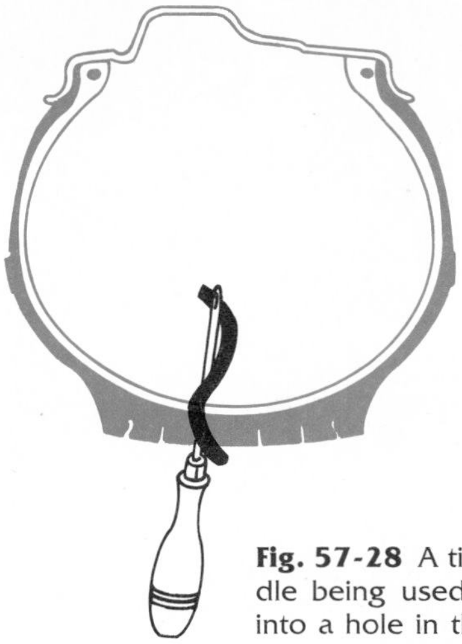

Fig. 57-28 A tire cut away to show a needle being used to insert the rubber plug into a hole in the tire.

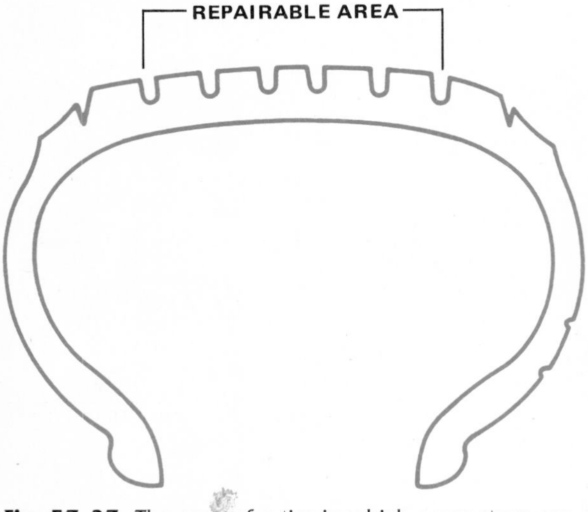

Fig. 57-27 The area of a tire in which a puncture can safely be patched. (*Chrysler Corporation*)

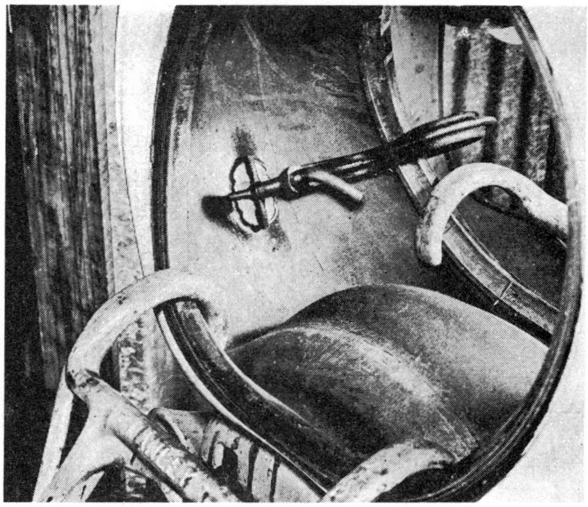

Fig. 57-29 Repairing a tubeless tire by the hot-patch method.

● 57-34 Repairing a Tire Removed from the Rim

There are three methods of repairing holes in tires: the rubber-plug method, the cold-patch method, and the hot-patch method. Permanent repairs are made from inside the tire—with the tire off the rim.

1. Rubber-plug method Rubber plugs can be used in the same way as explained in ● 57-33. The basic difference is that the repair is made from inside the tire, and the inside area around the puncture is buffed and cleaned. Then the plug is installed from inside the tire. See ● 57-32.

2. Cold-patch method In the cold-patch method, first clean and buff the inside area around the puncture. Then pour a small amount of self-vulcanizing fluid around the injury. Allow it to dry for 5 minutes. Next, remove the backing on the patch, and place the patch over the puncture. Stitch it down with the stitching tool. Start stitching at the center and work out, making sure to stitch down the edges.

Careful: Make sure no dirt gets on the fluid or patch during the repair job. Dirt could allow leakage.

3. Hot-patch method The hot-patch method is very similar to the cold-patch method. The difference is that heat is applied after the patch has been put into place over the area. This is done by lighting the patch with a match, or with an electric hot plate, according to the type of patch being used (Fig. 57-29).

After the repair job is done, mount the tire on the rim. Inflate it and test it for leakage, as explained in ● 57-32.

● 57-35 Recapping

Recapping is a specialized trade. The process involves applying new tread material to the old casing and vulcanizing it into place. Only casings that are in good condition should be recapped. Recapping cannot repair a casing with broken or separated plies or other damage. Recapping requires a special machine (Fig. 57-30).

Fig. 57-30 One type of recapping machine. (*Akron Equipment Company*)

The tire is cleaned, and the tread area is roughened by rasping it or buffing it on a wire wheel. Then a strip of new rubber tread, called "camelback," is placed around the tread. The casing with the camelback then goes into the recapping machine. The machine is clamped shut, and heat is applied for the specified time. Thus a new tread is vulcanized onto the old casing.

● 57-36 Repairing a Tire that Uses a Tube

If a tire that uses a tube has a small hole, clean out the hole. Repair the tube. The tube will hold the air. No repair to the tire is necessary. However, if the hole is of any size, $\frac{1}{4}$ inch [6.3 mm] or larger, it should be repaired with a patch on the inside. This prevents dirt or water from working in between the tire tube and causing tube failure.

● 57-37 Balancing the Wheel

After a tire change or repair, the tire and wheel assembly should be checked for balance (see ● 54-20).

REVIEW QUESTIONS

1. If a tire wears on the shoulders and not in the center of the tread, what is the probable cause?
2. If a tire wears in the center of the tread and not on the shoulders, what is the probable cause?
3. If a tire wears more on one side of the tread than the other, what is the probable cause?
4. Should you inflate the tire to spec with the tire hot or cold?
5. What is a tire-wear indicator?
6. Can you completely inspect a tire without removing it from the wheel?
7. Is it better to loosen the wheel nuts before or after you jack up the car? Why?
8. What are the two ways to patch a hole in a tube? Which way is preferred?
9. What are three methods of repairing holes in a tubeless tire?
10. What is camelback?

SELF PROJECTS

1. Get acquainted with the markings on tire sidewalls. Make lists of the markings you find, along with what they mean. If you are not sure about a marking, ask your instructor to explain it.
2. Make a list of different kinds and causes of abnormal tire wear. As you become aware of these different kinds of wear and what they mean, you will be able to offer a diagnosis when you see a tire with abnormal wear. You could save some trouble and expense if you spotted a tire wearing abnormally because of incorrect toe-in or tire inflation.
3. If you are able to go to a tire-repair shop, watch the tire experts as they remove and replace tires and repair them. Make notes of any interesting facts you learn.

CHAPTER 58
Automotive Heating and Air Conditioning

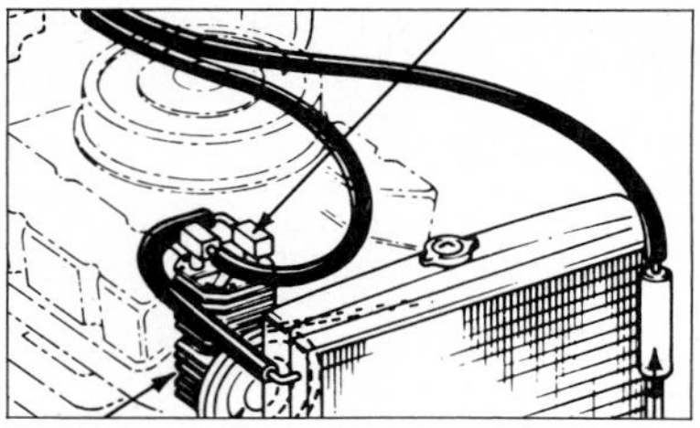

After reading this chapter, you should be able to:

- **Explain how a car heater works.**
- **Explain how a refrigerator works.**
- **Explain how a car air conditioner works.**
- **Point out on cars the various components of air conditioners, such as the condenser, evaporator, compressor, and valves, and explain how each works.**
- **Under the supervision of the instructor, connect a gauge set and vacuum pump to check the operation of the air-conditioner system, and also evacuate and charge the system.**

All modern cars have car heaters, and more than half have air conditioners. In this chapter we cover the various kinds of heating and air-conditioning systems. Some systems are adjusted manually. Other systems automatically maintain a selected temperature. If heat is needed inside the car, the heating system goes to work. Or, if the interior of the car needs cooling, the air-conditioning system goes to work. All automatically! We begin with the simple systems and work up to the most complex, fully automatic systems.

58-1 Car Heater

Early car heaters consisted of a small radiator, or heater core, mounted under the car dash. The radiator or heater core was connected by two hoses to the engine cooling system. When heat was needed, a small electric motor behind the heater radiator was turned on. The motor turned a fan that blew air through the heater radiator. Thus the interior of the car was warmed by heated air. The heat came from the engine cooling system. Modern car heaters work the same way, but they have additional controls to provide better regulation of the heat.

Figure 58-1 shows the hose connections between the engine cooling system and the car heater. In operation, the water pump in the engine cooling system keeps hot water flowing through the heater radiator core. The heater core is therefore hot. The amount of heat that gets into the car interior is determined by the amount of air that is allowed to flow through the heater core. In a modern car-heating system, three doors are provided to adjust the air flow.

Figure 58-2 shows a schematic diagram of the system. The amount of air that enters is determined by the blower action. The blower motor is connected to the battery through a switch. The switch is turned to operate the blower slow or fast (and at an intermediate speed in many systems).

After the air enters the system, its direction of flow is determined by the position of the temperature door. Note in Fig. 58-2 that three positions of this door are indicated (labeled *A*, *B*, and *C*). If the temperature door is open wide (position *C*), then most of the air entering has to pass through the heater core. Thus maximum heating is obtained. On the other hand, if the door is closed (position *A*), then no air can pass through the heater core. The air temperature is unchanged.

The air door determines the amount of air, either hot or cold, flowing through the system. It can be adjusted for full air flow (position 1) or for no air flow at all (position 2), or any place in between.

The defroster door can be adjusted so that heated air is directed up through the defroster outlets to the inside of the windshield. Or the defroster door can be adjusted so that most, or all, of the heated air is directed into the car.

The doors shown in Fig. 58-2 are operated by cables attached to heater controls on the instrument panel.

● 58-2 Vacuum-Operated Heater Controls

Figure 58-3 shows a heater-control system that uses vacuum motors. The unit is mounted under the instrument

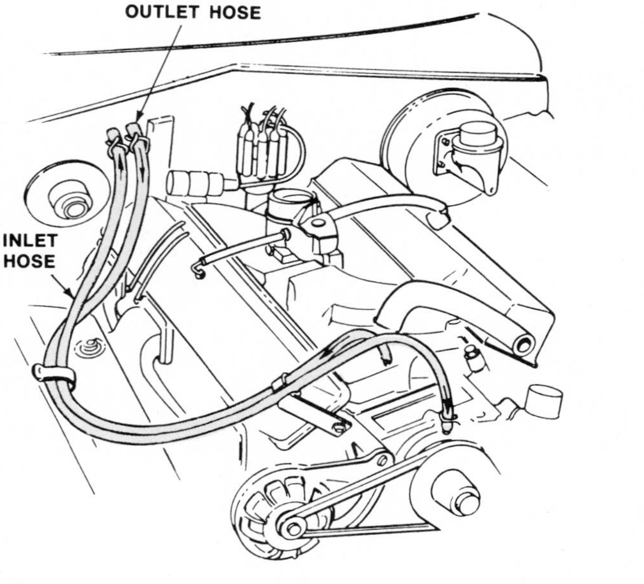

Fig. 58-1 Heater-hose routing in a car with a V-8 engine. (*Chrysler Corporation*)

panel. The control assembly is mounted in the instrument panel. It is attached to the main unit by various cables and hoses. Vacuum to operate the vacuum motors comes from the engine intake manifold when the engine is running.

Figure 58-4 is a schematic drawing of the heater system. Outside air, drawn into the system through the cool-air intake, flows into the blower housing. From the blower housing, the air flows through the heater case to the temperature-blend door (5 in Fig. 58-4). This door directs air through or around the heater core, depending on the position of the door. The door position is controlled by the temperature-control lever, which is a manual control on the instrument panel (at the top in Fig. 58-4). Setting the control lever at the COOL position causes the door to block off passage of the air through the heater core. The air passes through the system unchanged in temperature. Moving the lever to WARM causes the door to direct all the air through the heater core so that the air is heated. At various positions between COOL and WARM, part of the air goes through the heater core and part bypasses the heater core. The heated and unheated air blend and enter the car. So, as you can see, moving the temperature-control lever moves the temperature-blend door to give the desired amount of heat in the car.

A feature of this system is that the air can be directed in the following three ways:

1. To the floor
2. Through registers into the car
3. To the defroster vents

The registers are the two louvered outlets mounted at the two extreme ends of the instrument panel. The louvers can be adjusted to aim the flow of air. Also, the heat can be shut off entirely so that untreated outside air flows into the car. When the functional-control lever (see top of Fig. 58-4) is set at VENT, a vacuum motor opens the vent-heat door (6 in Fig. 58-4). At the same time, another vacuum motor closes the water-heater valve (4 in Fig. 58-4) so that the flow of hot water to the heater core is cut off.

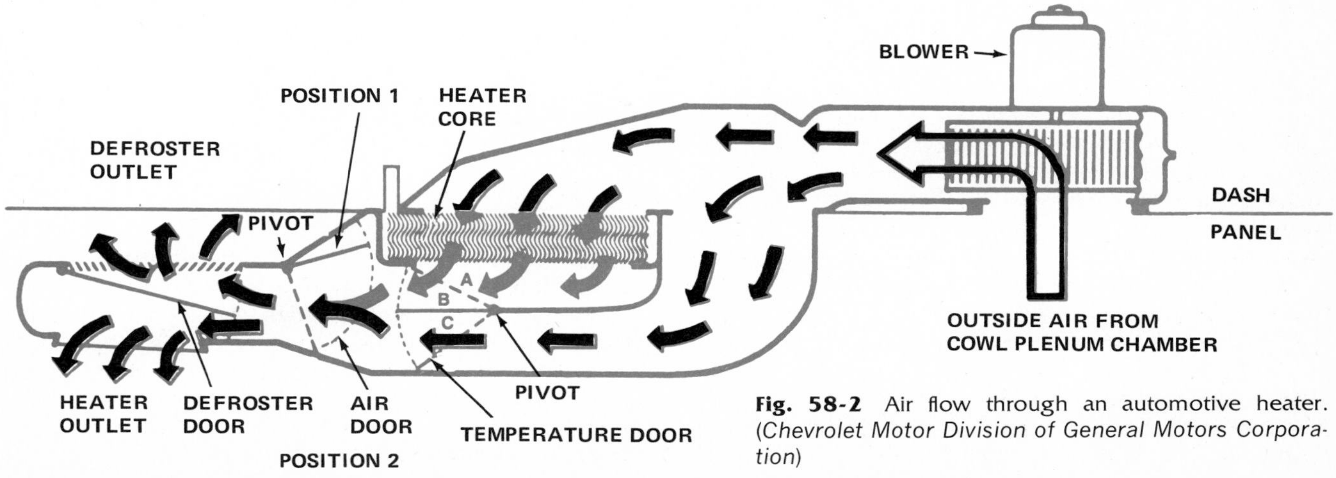

Fig. 58-2 Air flow through an automotive heater. (*Chevrolet Motor Division of General Motors Corporation*)

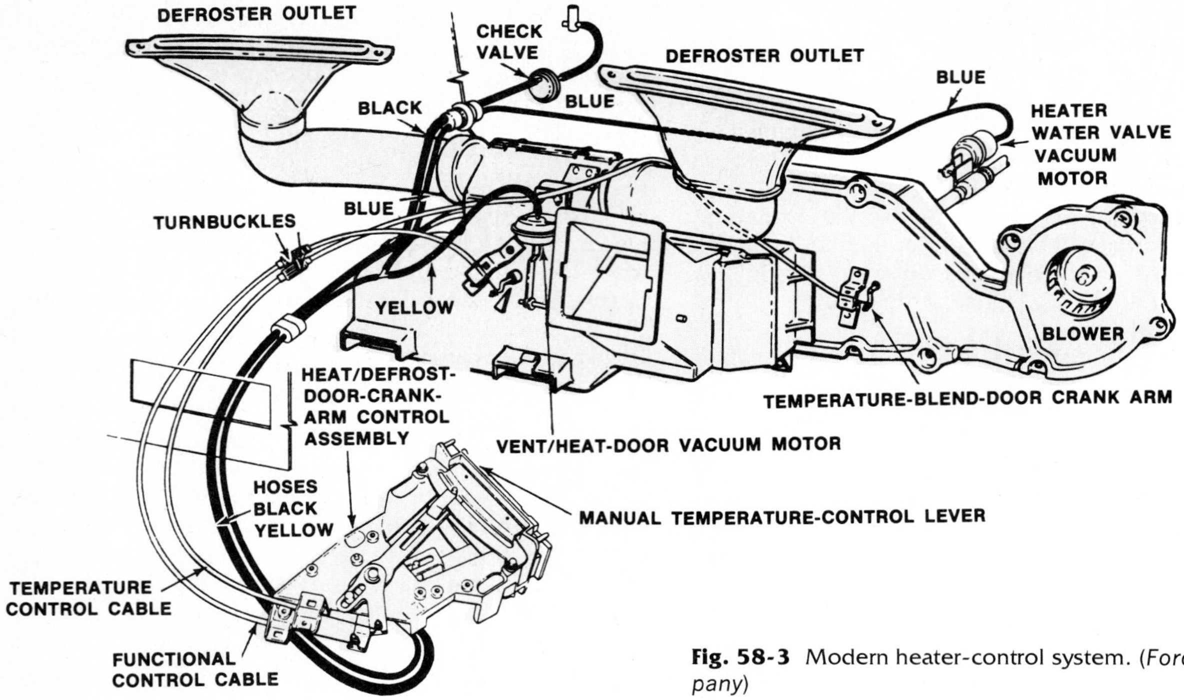

Fig. 58-3 Modern heater-control system. (*Ford Motor Company*)

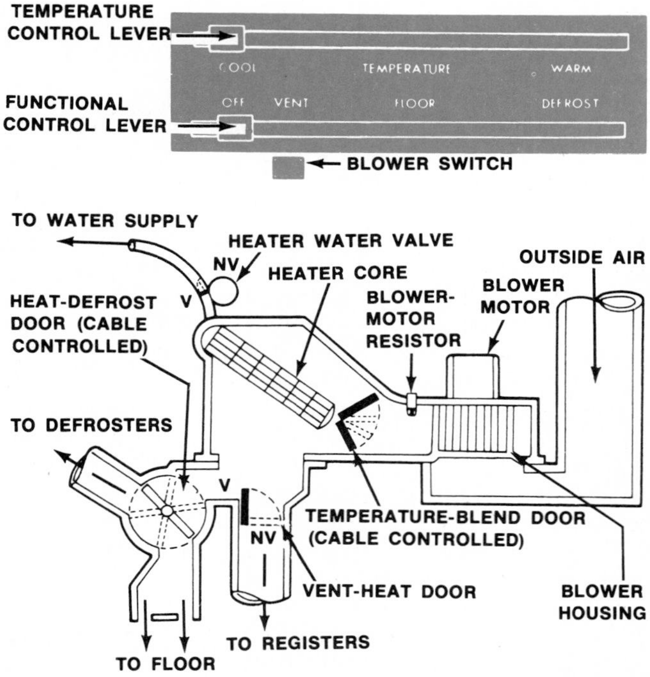

Fig. 58-4 Instrument-panel heater controls and a schematic layout of the car heater system. (*Ford Motor Company*)

Refrigeration

● 58-3 Refrigeration

The air conditioner uses a refrigerator to cool and dry the air. It works the same way as refrigerators used in homes. So before we discuss air conditioning, we should understand how refrigerators work.

Refrigeration works by evaporation. Evaporation is what happens when a liquid turns into a gas, or vapor. Put a little water on your hand. Blow on your hand, or wave it vigorously in the air. Your hand feels cold because, as the water evaporates, it takes heat from your hand.

The removal of heat by evaporation is the basic principle of refrigeration. For example, heat is removed from the refrigerator by the evaporation of a liquid in the refrigeration system. Figure 58-5 is a simple example of what we mean.

The illustration shows a refrigerator containing food, plus a jug of a liquid called Freon-12 (sometimes referred to as R-12—the R means refrigerant). Freon-12 is a liquid at temperatures below −22°F [−30°C] (that is, 54°F below freezing). At temperatures above −22°F [−30°C] Freon-12 will boil, or turn to vapor. Since the temperature in the refrigerator shown in Fig. 58-5 is well above −22°F [−30°C] the Freon-12 boils. That is, it evaporates. As Freon-12 evaporates, it takes heat from the surrounding objects and therefore cools the refrigerator. The refrigerating action continues as long as there is any Freon-12 left in the jug to evaporate. Because the Freon-12 refrigerates, it is called a *refrigerant*.

There are a couple of drawbacks to the system shown in Fig. 58-5. In the first place, it is wasteful to allow all the Freon-12 to escape. Besides, having Freon-12 vapor floating around the house could be dangerous. Also, since the refrigerant boils, or evaporates, at −22°F [−30°C], it tends to bring everything in the refrigerator down to this temperature. Thus everything would be frozen.

Fig. 58-5 Simple refrigerator. Evaporation of Freon-12 (R-12) removes heat from the refrigerator.

Obviously, two things must be done. First, the Freon-12 must be recaptured and turned back into a liquid. Second, refrigerant action must be controlled so the proper temperature can be maintained in the refrigerator.

58-4 Freon-12

Freon-12 was developed by chemists seeking the ideal refrigerant. They found it in Freon-12, commonly referred to as R-12. This substance is stable in the refrigerating system (does not change to other chemicals), is nontoxic (not poisonous to breathe), nonflammable, nonexplosive, and will not corrode or rust the air-conditioner parts.

In fact, it was found to be so safe to use that the spray can became practical. Almost every kind of liquid that could be sprayed was put in cans, along with a quantity of R-12. This meant the liquid was under pressure. When the button was pressed, the liquid sprayed out, forced out by the pressure of the R-12. Liquids in spray cans include paints, antiperspirants, whipped cream, shaving cream, hair sprays, cheese spreads, household cleaners and polishes, and several automotive service items such as lubricants and carburetor cleaners.

Then scientists sounded a warning. All this R-12 being released into the atmosphere from the billions of spray cans used each year was causing trouble in the upper atmosphere. At very high altitudes (several hundred thousand feet or meters) there is a protective layer of ozone. This is a special form of oxygen and it is protective because it shields the earth from potentially harmful light from the sun. This is light in the ultraviolet range—the range that gives you a sunburn if you stay out in the sun too long. The ozone layer allows only a little of this ultraviolet light to get through.

However, the R-12 rises to this ozone layer and damages it. If this ozone layer were destroyed, life on earth as

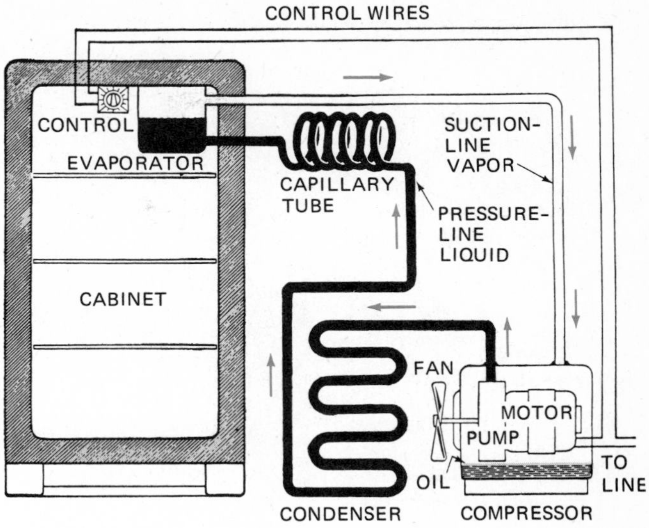

Fig. 58-6 Simplified schematic diagram of an electric refrigerator.

we know it would become impossible. A few minutes in the sun would give you a fatal sunburn. Many growing things would die. For this reason, the use of R-12 in spray cans has been banned.

The reason we discuss the damage that escaping R-12 can cause is this. The servicing procedures for air conditioners and refrigerators may change in future years. That is, new rules may be made that would prohibit the dicharge of R-12 into the air as is now being done during certain stages of air-conditioner service. The new procedures would require special containers into which R-12 could be discharged. These containers would then go back to the factory for processing. We mention this again when we discuss evacuation and charging of the air-conditioner system.

58-5 A Refrigerator

Look at Fig. 58-6, which is a simplified diagram of an electric refrigerator. The evaporator serves the same purpose as the Freon-12 in the jug in Fig. 58-5. As the refrigerant evaporates, it is carried through a suction line to the compressor pump. The compressor works something like an engine oil pump but is not constructed in the same way. The compressor pump takes in the vaporized refrigerant and applies a high pressure to it (up to 200 pounds per square inch [1,378.90 kPa]). The high pressure causes the temperature of the vapor to be increased to well above 100°F. The hot vapor, under high pressure, is then sent into the condenser. The condenser is a long tube, usually equipped with radiating fins. The hot vapor, as it passes through the condenser, loses its heat and therefore condenses into liquid again.

Do you see what has happened? The refrigerant has carried heat out of the refrigerator. The refrigerant then gets rid of this heat in the condenser. Now, the liquid flows through the capillary tube and back into the evaporator. The capillary tube has a very small diameter. The

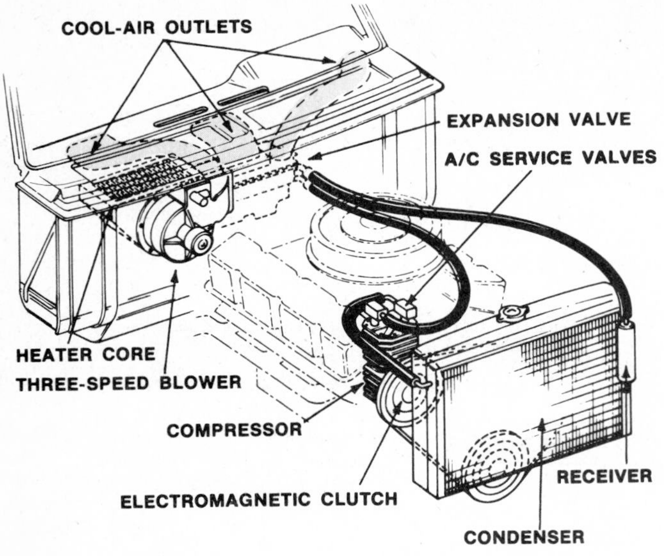

Fig. 58-7 Locations of major components in a car air conditioner. (*American Motors Corporation*)

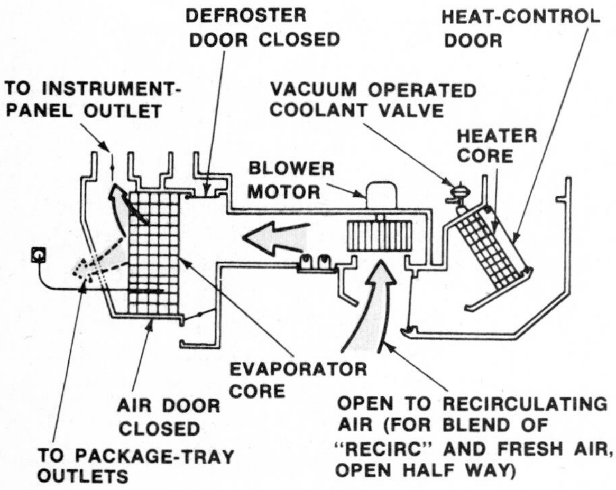

Fig. 58-8 Schematic diagram of an air conditioner for a car. The package tray is the shelf back of the back seat. Some cars have outlets there. (*American Motors Corporation*)

small diameter restricts the flow of the liquid. The flow of refrigerant must be restricted because there must be a pressure difference between the condenser and the evaporator. In other words, the pressure must be high on the pump side, but low on the evaporator side. You can see why. High pressure is necessary to condense the refrigerant. Low pressure is necessary to allow the refrigerant to evaporate.

58-6 Refrigerator Control

The refrigerator control bellows and bulb are connected by a tube. The bulb contains a liquid that has a low-temperature boiling point. The bulb is placed in the refrigerator cabinet so that it is subject to refrigerator temperatures. As the refrigerator temperature goes up, the liquid in the bulb starts to boil. This creates pressure in the bellows. The pressure expands the bellows, causing the contact points to close. When the contact points close, the pump motor is turned on, and refrigerating action is started.

The bellows must expand against a spring. The spring tension is controlled by an adjusting knob. When the knob is turned in, more tension is put on the spring. Then the bellows must exert a higher pressure to close the contacts. This means that the pump will not come on until a higher temperature is reached. If the adjusting knob is backed off to relieve spring tension, the bellows pressure needed to close the contacts is lower. Thus, a lower temperature is maintained in the refrigerator.

Air Conditioning

58-7 Air-Conditioner Action

In previous sections we explained how the car heater puts heat in the car interior. The air conditioner does just the opposite. It takes heat out of the car interior. The blower that is used for heating is also used for air conditioning. For heating, the blower blows air through the heater core. For cooling, it blows air through the cooler coil, called the *evaporator*.

In automotive shop manuals you will often see the term "air conditioner" shortened to "A/C."

In addition to cooling the air, the evaporator also takes moisture out of the air. This action is the same as the action of moisture condensing on a cold glass. What happens is that as the air flows through the cold evaporator, moisture condenses on the evaporator core. The moisture runs off the evaporator core and drops outside the car. Thus the air conditioner not only cools the air, but also dries it. Dry air feels cooler than moist air, so the air conditioner helps keep the car interior cool in two ways: by cooling the air and by drying it.

58-8 Automotive Air Conditioner

Figure 58-7 shows the installation of an air conditioner in a car. The assembly includes the same basic parts as the refrigerator we discussed previously. There are a compressor, a condenser, and an evaporator. You cannot see the evaporator in Fig. 58-7; it is in the assembly under the instrument panel. Figure 58-8 is a schematic layout of the assembly. You can see the location of the heater core and of the evaporator core in Fig. 58-8.

Notice that the assembly has several doors. These doors can be opened or closed to direct the flow of air either through the heater core or through the evaporator core. Also, the doors can be operated so that the air flowing through the evaporator core can be taken from outside or picked up from inside the car (that is, recirculated). The heated air can also be directed to the defroster outlets or into the car by operation of other doors.

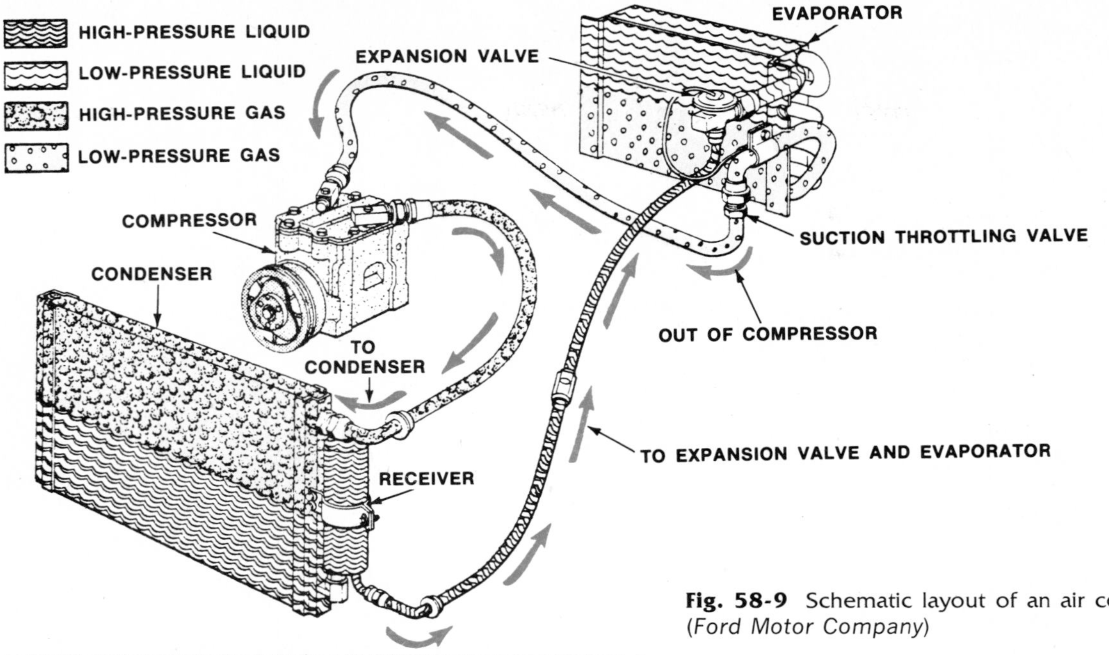

Fig. 58-9 Schematic layout of an air conditioner for a car. (*Ford Motor Company*)

58-9 Air-Conditioner Operation

Figure 58-9 shows the basic air-conditioning system. The compressor takes in the vapor from the evaporator and puts the vapor under pressure. The high-pressure vapor, or gas, is sent to the condenser. In the condenser, the vapor loses heat and returns to liquid form. The liquid then flows to the evaporator, where it evaporates. As the liquid turns to vapor, it takes in heat. The vapor then passes back into the compressor. This flow continues as long as the air conditioner is operating.

The components in the automotive air conditioner are: the receiver, the expansion valve, the suction throttling valve, and the magnetic clutch on the compressor.

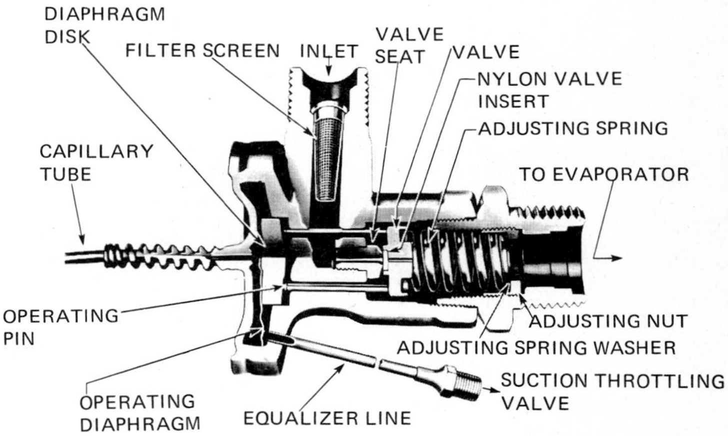

Fig. 58-10 Cutaway view of a thermostatic expansion valve.

58-10 Receiver

The purpose of the receiver is to ensure a supply of liquid refrigerant to the evaporator. The liquid refrigerant flows through an outlet at the bottom of the receiver, so no vapor can mix with the liquid refrigerant. In some systems the outlet is at the top of the receiver, but it is connected to a pipe that goes to the bottom. The liquid refrigerant always settles at the bottom of the receiver, so the outlet always picks up liquid rather than vapor.

The receiver also contains a substance that absorbs water. Thus any moisture in the system is absorbed so that it can do no harm. If moisture were not absorbed, it might freeze on a valve and prevent normal air-conditioner action.

58-11 Expansion Valve

The expansion valve is located in the circuit between the condenser and the evaporator, as shown in Figs. 58-7 and 58-9. The expansion valve regulates the flow of liquid refrigerant to the evaporator. Figure 58-10 is a cutaway view of an expansion valve. The capillary tube is connected to a bulb attached to the evaporator. The bulb and tube are filled with a gas that expands or contracts with changing temperatures. As the temperature of the evaporator goes up, the gas expands. The expansion of the gas exerts pressure on the diaphragm. Operating pins from the diaphragm disk then force the valve off its seat. With the valve off its seat, liquid refrigerant is able to flow into the evaporator. Cooling takes place. As the temperature in the car goes down, the gas in the bulb and tube contracts. The pressure on the diaphragm drops, and the spring pushes the valve closed. This cuts off the flow of liquid refrigerant to the evaporator. Cooling stops. In operation, the valve takes the position needed to provide the right amount of cooling in the car. If cooling needs are low, the valve is almost closed, and only a little cooling is provided. If cooling needs are high, the valve is

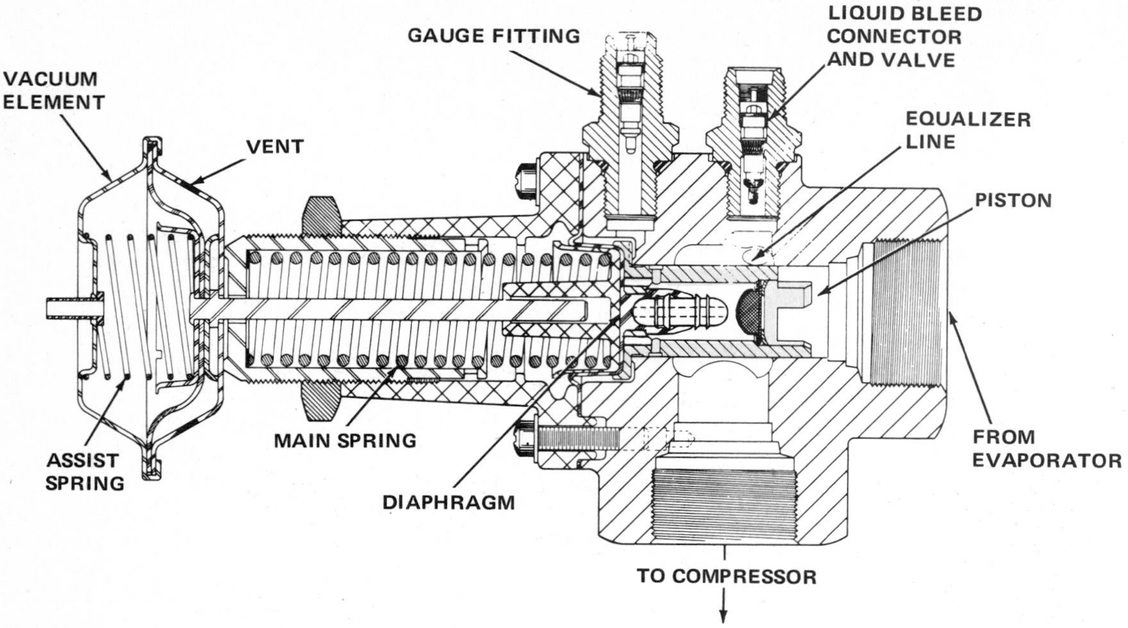

Fig. 58-11 Sectional view of a suction throttling valve. *(Pontiac Motor Division of General Motors Corporation)*

opened wider. More liquid refrigerant flows, and cooling is increased.

● 58-12 Suction Throttling Valve

The suction throttling valve is located between the evaporator and the compressor. Its main purpose is to prevent freezing of moisture on the evaporator. If the evaporator temperature goes below 32°F [0°C], any moisture in the air going through the evaporator will freeze. If the evaporator froze, cooling efficiency would be reduced. Figure 58-11 is a sectional view of a suction throttling valve. Spring pressure and atmospheric pressure on one side of the valve piston, and evaporator pressure on the other side, control the valve operation. When evaporator pressure exceeds the specified maximum, it pushes the piston back. This action opens the valve so that refrigerant can flow from the evaporator to the compressor.

The vacuum element comes into operation when the driver turns the control to full cooling. Then, vacuum from the engine is admitted to the end of the vacuum element. Now, the assist spring is compressed. The compression of the assist spring allows the valve to move farther out so that more refrigerant can flow.

● 58-13 Controls

The air-conditioning system is controlled by levers or buttons on the instrument panel. On some models the controls are vacuum-operated, just as in vacuum-operated heater systems (● 58-2). In one system the levers are set manually and adjusted by the driver for different conditions. In a fully automatic system, the driver simply sets the temperature as desired. The system takes over from there. The system will cool when cooling is needed to maintain the temperature set. The system will also heat when heating is needed to maintain the preset temperature.

● 58-14 Magnetic Clutch

As part of the control system, the compressor has a magnetic clutch. This clutch is located in the pulley. The pulley is driven by a belt from the crankshaft pulley. When the system calls for cooling, the magnetic clutch is engaged. The compressor is then driven through the magnetic clutch so that cooling is obtained. The purpose of the magnetic clutch, therefore, is to engage the compressor for cooling and to disengage it when cooling is not needed.

● 58-15 Manual Temperature-Control System

Figure 58-12 is a schematic drawing of the manual temperature-control system. Air enters either from outside or from inside the car, depending on the positions of doors 1 and 2. These are the outside-air door and the recirculating door. The position of these doors is controlled by the movement of the functional-control lever (see Fig. 58-12). When the lever is moved to the left—to MAX A/C—the outside-air door is closed, and the recirculating door is opened. If the lever is set at FRESH A/C, fresh air from outside is brought into the system.

The speed with which the air moves is controlled by the speed of the blower. The setting of the FAN knob (Fig.

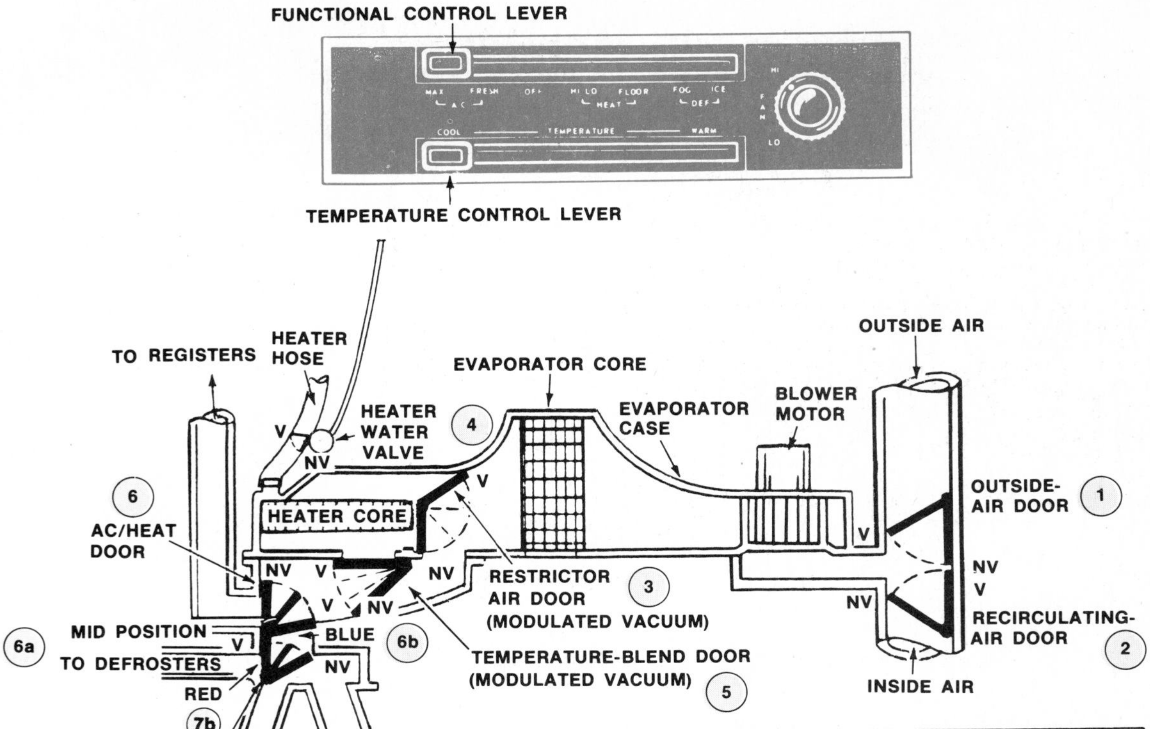

LINCOLN MANUAL A/C-HEATER SYSTEM-VACUUM MOTOR TEST CHART

Functional Control Lever Position	Temperature Control Lever Position	Vacuum Motors Applied With Vacuum								
MAX A/C	COOL	1	2	3	4	5	6a	6b	7a	7b
	WARM	1	2	3	–	5	6a	6b	7a	7b
FRESH A/C	COOL	–	–	3	4	5	6a	6b	7a	7b
	WARM	–	–	–	–	–	6a	6b	7a	7b
OFF	WARM	1	2	–	–	–	6a	6b	7a	7b
HI/LO HEAT	WARM	–	–	–	–	–	6a	–	7a	7b
FLOOR HEAT	WARM	–	–	–	–	–	–	–	7a	7b
FOG DEF	WARM	–	–	–	–	–	–	–	7a	–
ICE DEF	WARM	–	–	–	–	–	–	–	–	–

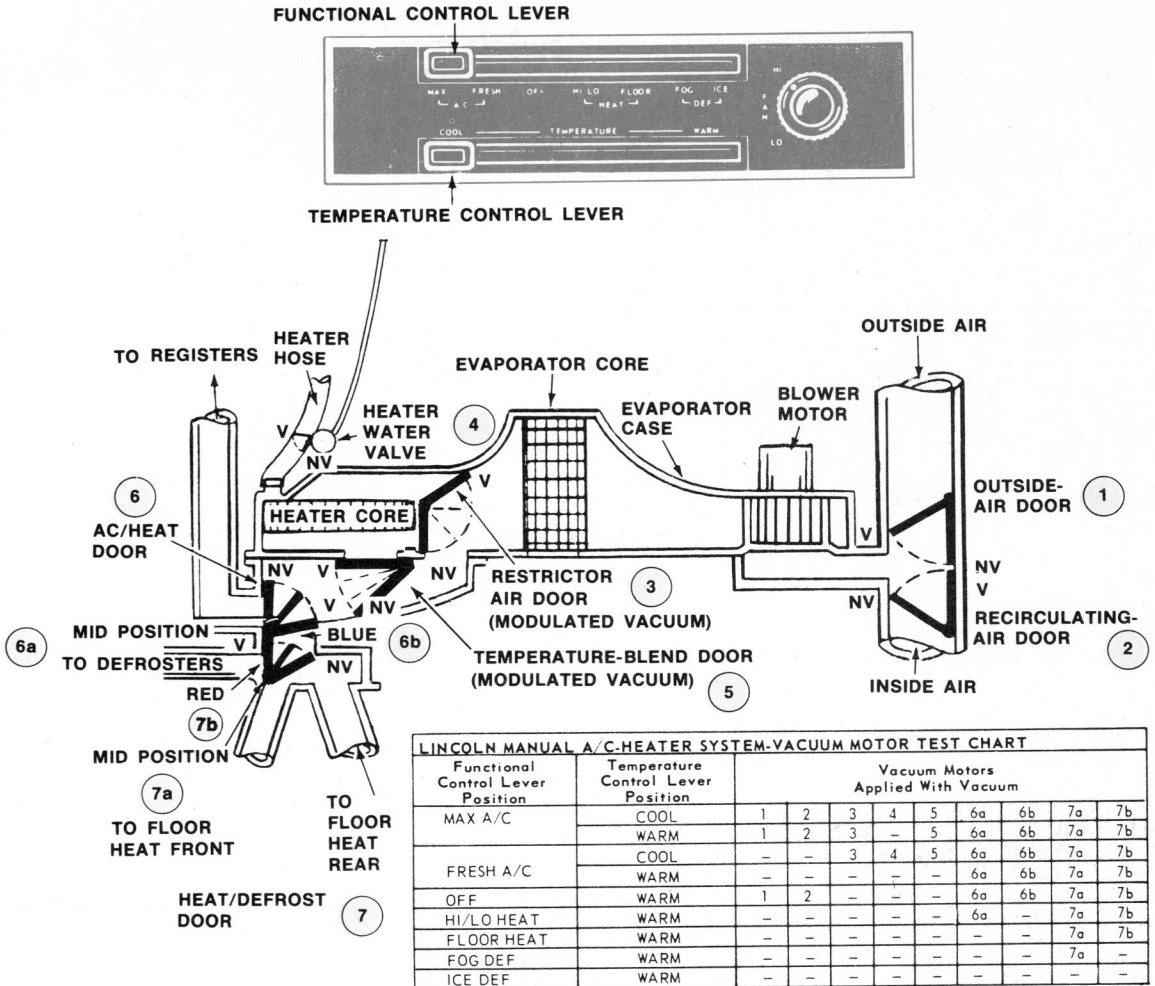

Fig. 58-12 Manually controlled air-conditioning and heating system. (*Ford Motor Company*)

58-12) determines the speed of the blower and the speed with which air is brought into the system.

The air first goes through the evaporator core. It may or may not be cooled, depending on the position of the temperature-control lever. If the lever is set at A/C, then the magnetic clutch on the compressor is actuated. The compressor operates, and cooling results. If the lever is set at HEAT or DEF (for defrost), the magnetic clutch does not operate, and no compressor action takes place. Thus there is no cooling.

After passing through the evaporator core, the air meets the air-restrictor door (3 in Fig. 58-12). This door can be swung up to admit air to the heater core, or swung down to prevent air from entering the heater core. The temperature-blend door (5) can be moved varying amounts to control the percentage of heater-core air and evaporator-core air that mix at this point. Next, the air meets the air-conditioner-heat door (6), which can be moved one way or the other. In the up position, the passage to the registers is blocked, and the air moves down into the defroster-floor-heat position. Now the air, which has been warmed, moves to the defrosters or to the floor-heat outlets, depending on the position of the heat-defrost door (7).

The system also includes a water-heater valve (4). This valve shuts off the flow of hot water from the engine cooling system when the air conditioner is running. Both the water valve and the air door are operated by vacuum motors.

58-16 Automatic Temperature Control

The automatic temperature-control system is simple from the standpoint of the driver. But it is complicated from the standpoint of the automotive technician. Figure 58-13 is a schematic layout of the system. Note that the control panel has two levers. The upper lever can be moved to select the desired temperature (from 65°F to 85°F) [18.3°C to 29.4°C]. The lower lever can be moved all the way to the left to get automatic action in high. That is, the blower will operate at the maximum speed until the desired temperature is reached. Then the blower will slow down and operate just fast enough to maintain the de-

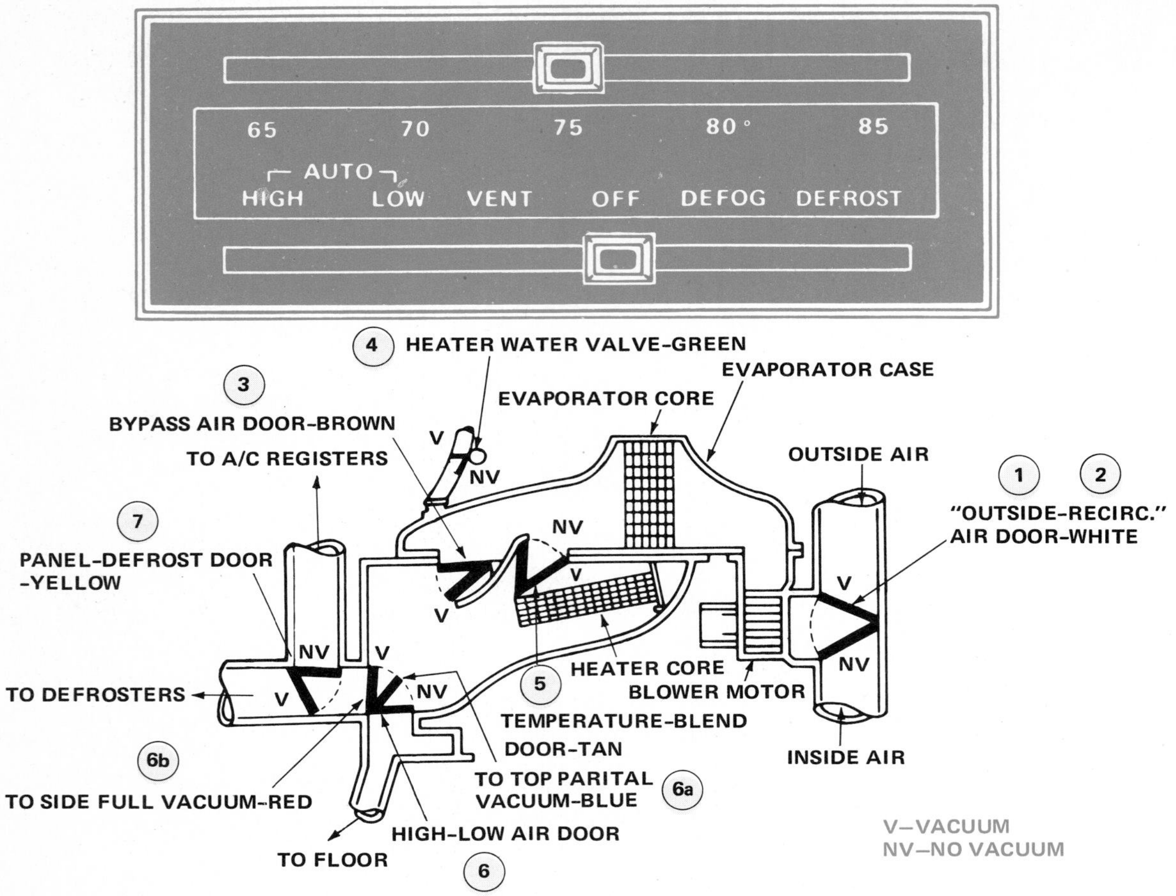

Fig. 58-13 Instrument control panel and schematic diagram for a fully automatic heater–air-conditioner system. (*Ford Motor Company*)

sired temperature. If slower action is desired, the driver can set the lever at LOW. If the driver only wants untreated outside air to enter, the lever is set at VENT. At this setting the outside-recirculating door (1 and 2 in Fig. 58-13) is moved down so that outside air enters. To defog or defrost the windshield, the driver sets the lever at either of these positions.

The air passes through the evaporator core after it leaves the blower. Whether the air is cooled depends on whether the air conditioner is working. Next, the air either bypasses the heater core or passes through the heater core. If the system calls for heat, then the air conditioner does not work, but the heater core is hot. That is, hot water from the engine cooling system is circulating through the heater core. With this operating condition, the water heater valve (4) is open. Also, the bypass air door (3) is closed, and the temperature-blend door (5) is open. The direction the heated air then takes depends on the positions of the high-low door (6) and the panel-defrost door (7).

If the system calls for cooling, the evaporator core is cold. That is, the air-conditioning system is working. The air passing through the evaporator core is therefore cooled. The air then goes through the open bypass door and to the air-conditioning registers.

In this system there is a delay circuit in the heating section. This delay circuit prevents the blower from coming on until the water circulating in the heater core is warm. This prevents the circulation of cold air, which would be uncomfortable for people in the car.

Vacuum motors operate the water-heater valve and the air doors, just as in the systems previously covered in the chapter.

Servicing Heaters and Air Conditioners

The heater is part of the engine cooling system. Manually operated heaters may require occasional service. Air-conditioning systems, however, require more work. In this part of the chapter we see how to troubleshoot and service both car heaters and air conditioners.

58-17 Servicing Heating Systems

Heater problems usually result in one of three complaints: leaks, no heat, or failure of the blower to work.

Leaks are obvious and easily found. They could be caused by leaky hoses or connections, leaks in the radiator core, or leaks in the control valve. Lack of heat could be due to a bad control valve, to clogged hoses or heater core, or to a faulty engine cooling system thermostat. Failure of the blower to work is probably an electrical problem. It usually requires replacement of the blower, switch, or wiring.

If the system automatically adjusts heating according to the temperature-control setting, failure to heat could be due to several things. A vacuum motor may not be working, the water-heater valve may be stuck, the thermostatic control may not be working, or any of the conditions listed in the previous paragraph may be causing the trouble.

In any case, be sure you have the shop manual that covers the model car you are working on. It will tell you how to check and service the system.

58-18 Servicing Air Conditioners

You must never work on air conditioners—that is, attempt to troubleshoot or service them—unless you have the proper equipment and know exactly how to use it. If you use the proper equipment in the right way, there is no special danger in servicing air conditioners. However, if you go at the job the wrong way, you can get yourself badly hurt! Remember that the air-conditioning system contains high-pressure liquid and vapor. The liquid, if released, can turn to vapor almost instantly, freezing anything it touches—and this includes skin and eyes! Here are essential cautions you *must* observe:

1. *Undercoating*. Never apply undercoating to any connections in the refrigerator lines or to the air-conditioning parts.
2. *Steam-cleaning and welding*. Never apply any form of heat to any refrigerant line or to any component of the system. The refrigerant system is under pressure. Heating the refrigerant could increase the pressure excessively and cause a terrible explosion.
3. *Handling refrigerant*. Freon-12 is about the safest refrigerant available, but you can be seriously hurt if you handle it carelessly.

Be sure to use only the Feon-12 specially prepared for air-conditioning systems. The kind used in equipment such as boat air horns and fire-alarm signals is not pure and can ruin an air conditioner.

Freon-12 that escapes into the air can evaporate so quickly that it will freeze anything it touches, including skin and eyes. You could be blinded if you get Freon-12 in your eyes!

CAUTION: Always wear safety goggles when servicing any part of an air-conditioning system!

Keep a bottle of sterile mineral oil and a bottle of weak boric-acid solution handy. If you get Freon-12 in your eyes, instantly wash it out with a few drops of mineral oil, followed by the boric-acid solution. See a doctor right away! Don't wait, even if the irritation seems to have gone away.

When discharging refrigerant from an air conditioner, discharge it into the garage exhaust system or into the open air. The refrigerant evaporates so quickly that it will displace all the air around the car. This could cause you to suffocate if the immediate area in which you are working is enclosed and without ventilation!

Do not discharge Freon-12 in a room where there is an open flame. Freon-12 turns into a poisonous gas when it comes in contact with a flame. The gas is very dangerous if inhaled. This poisonous gas is also produced when you use a flame-type leak detector. But only a small amount of gas is produced, so you don't need to worry as long as you don't inhale it.

Never heat a container of Freon-12 when charging an air-conditioning system. This could cause an explosion.

Remember, also, to keep heat away from the air conditioner in the car when steam-cleaning or welding.

58-19 Troubleshooting Air Conditioners

A quick check of the system operation can be made by examining the sight glass with the system on. The sight glass is a glass-covered peephole in the receiver. Watch for bubbles with the engine running at about 1,500 rpm and the system set for maximum cooling. A continuous stream of bubbles means that the system needs more refrigerant. If this is the case, the system must be checked for leaks. More refrigerant must be added after the leak has been fixed.

No bubbles means that there is either too much refrigerant or no refrigerant at all. To determine which is the case, cycle the magnetic clutch on the compressor. This is done by moving the air-conditioner controls with the engine running. If there is no refrigerant, no bubbles will appear. This condition requires a leak test, leak repair, and recharging of the system with refrigerant.

Some late-model cars do not have a sight glass.

58-20 Air-Conditioner Performance Test

To test the air conditioner for performance, you need a manifold gauge set (Fig. 58-14). This gauge set is attached to the compressor to measure the pressures in the system. These pressures tell you where there is trouble. Refer to the car shop manual for details on how to use the manifold gauge set and what the various results mean.

58-21 Evacuating and Charging the System

"Evacuating" means removing any air or moisture from the system after the R-12 has been discharged from the system. If the system has been opened, air and moisture have entered. All this air and moisture must be removed by applying a vacuum to the system. The vacuum is produced by a special vacuum pump required for air-conditioning work.

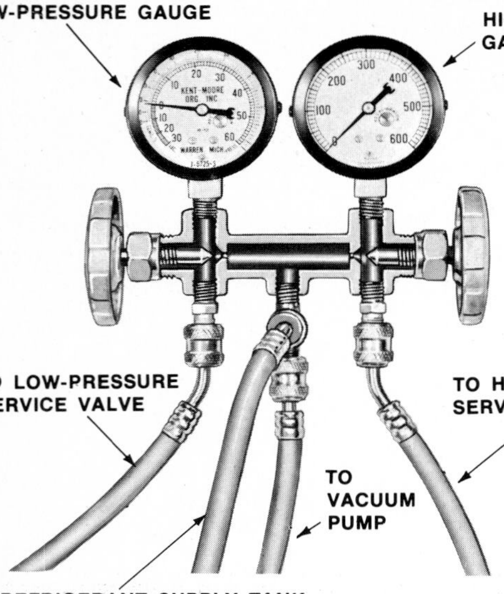

Fig. 58-14 Manifold gauge set for checking pressures in the air-conditioning system. *(Ford Motor Company)*

As we mentioned in ● 58-4, you may be required at some future time to discharge the R-12 into a container instead of into the air.

To charge the system, connect a refrigerant tank to the system and allow fresh refrigerant to flow in. This job is done with the engine running so that the compressor will be in operation.

You should have a thorough understanding of air-conditioning service before attempting to work on an air-conditioning system. Also, you must have the proper equipment and the car shop manual covering the system you are working on.

REVIEW QUESTIONS

1. Where does the heat for the car heater come from?
2. What are the names of the three doors in the vacuum-motor-operated heating system in Fig. 58-4?
3. Explain the basic principle of refrigeration.
4. Refrigerant vaporizes in one component of the refrigerator. What is the name of this component?
5. The refrigerant changes from a vapor to a liquid in one component of the refrigerator. What is the name of this component?
6. What are the three things that the car air conditioner does?
7. What is the basic purpose of the compressor?
8. What is the purpose of the expansion valve?
9. What is the main purpose of the suction throttle valve?
10. What is the purpose of the magnetic clutch on the compressor?
11. What are the three most common complaints about car heater systems?
12. Why should you wear goggles when working with Freon-12?
13. Why shouldn't you discharge Freon-12 into a room where there is an open flame?
14. What is the purpose of the sight glass?
15. What is the purpose of the manifold gauge set?
16. Why has R-12 been banned for use in spray cans?

SELF PROJECTS

1. Look at the heating and air-conditioning sections of several manufacturers' service manuals to see what kinds of systems they offer on their cars. Some of the more complex systems are optional. That is, the buyer has to order it when ordering the car. In more expensive cars, the fully automatic system (Fig. 58-13) is standard equipment. The buyer gets it without paying anything extra for it. Make a list of several of the more popular cars. Under each model, write down the types of systems the buyer can get.
2. Refer to a service manual covering one of the more complex heating and air-conditioning systems. Make a list of the troubles and possible causes as outlined in the manual. Complete the list with the steps to be taken to analyze trouble with the system.

Automotive Safety Devices and Safety Inspection

10

In this part, we describe the various safety devices used on automobiles and their operation and servicing. There is one chapter in Part Ten.

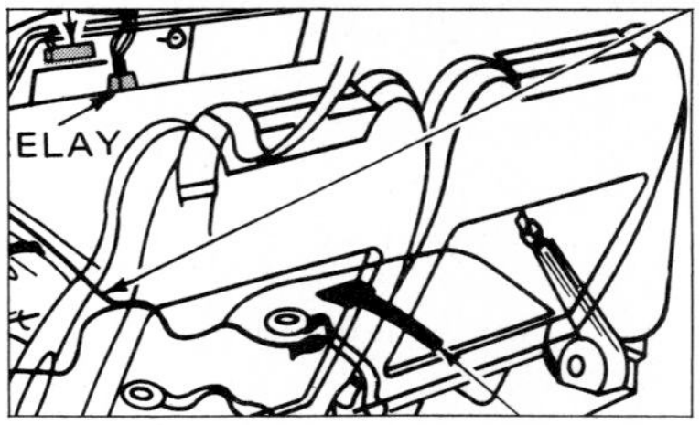

CHAPTER 59
Automotive Safety Devices

After you have studied this chapter, you should be able to:

- **Explain the purpose of seat belts and describe the operation of seat-belt warning and interlock systems.**
- **Explain how the open-door and headlights-on warning-buzzer systems work.**
- **Explain how security alarm systems work.**
- **Explain the purpose of air bags and describe their operation.**
- **Describe three types of energy-absorbing bumpers and explain how they work. Point them out on cars.**
- **Explain how to service the shock-absorber type of energy absorber.**

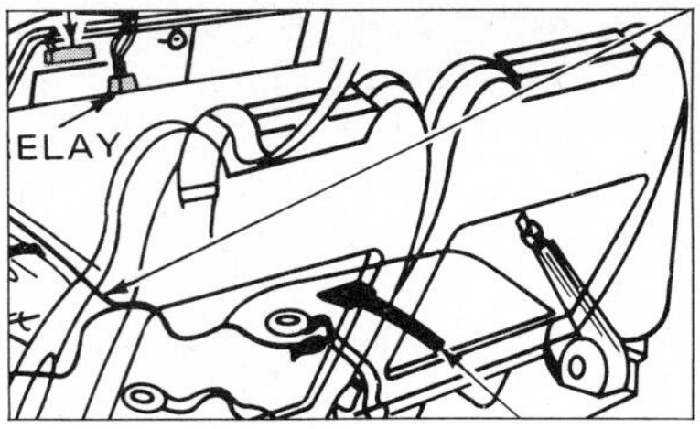

There are more safety devices on the automobile than the average person realizes. We immediately think of seat belts or air bags. But there are structural safety features: steel beams in the doors, safety glass, steel support beams for the roof, collapsible steering columns, and so on. In addition, the headlights, stop lights, and turn signals are all safety devices. Indicating devices such as the oil-pressure and engine-temperature indicators are safety devices that warn the driver if something is wrong. All of these, if not previously described in the book, are reviewed in this chapter.

● 59-1 Automotive Safety Devices

Let us first list the devices that protect the people in the car, the car itself, and the environment from possible injury or damage. We have divided the list into two parts—those items that are covered elsewhere in the book, and those covered in this chapter.

Devices covered elsewhere in the book

1. Blinker warning system (● 32-10), which causes all exterior lights to flash on and off. This signals trouble and the need for help.
2. Brakes (Chaps. 55 and 56).
3. Circuit breakers and fuses (● 32-3 and 32-4) which protect the electric circuits in the car from overloads.
4. Collapsible steering column (● 53-11), which collapses if the driver is thrown forward into the steering wheel during a crash.
5. Automotive emission controls (Chaps. 33 to 36), which reduce emissions from the automobile and thus protect the atmosphere from excessive pollution.
6. Headlights (● 32-6).
7. Indicating devices which permit the driver to monitor engine performance. The driver can then do what is necessary if something goes wrong. Engine temperature (● 21-11), fuel level in the tank (● 12-6), engine oil pressure (● 19-14), and charging rate to the battery (● 23-4) are the four items monitored in most cars. In addition, some cars have a tachometer which indicates engine rpm (revolutions per minute).

8. Neutral safety switch (● 26-6), which prevents starting if the transmission is in any gear.
9. Padded instrument panel. Padding is used on the instrument panel, steering wheel, and other items against which a person might be thrown in case of a front-end crash.
10. Side-guard beams in the doors to protect against side collisions.
11. Speedometer (● 32-17).
12. Steering lock (● 53-12), which locks the steering wheel when the ignition switch is turned to OFF.
13. Stop light (● 32-11), which signals any car following that the brakes have been applied.
14. Tires (Chap. 57).
15. Turn signals (● 32-10).
16. Windshield wipers (● 32-18).

Devices covered in this chapter

1. Optional seat belts (● 59-2)
2. Front-seat-belt warning system (● 59-3)
3. Seat-belt–starter interlock system (● 59-4)
4. Open-door warning-buzzer system (● 59-6)
5. Headlights-on warning buzzer (● 59-7)
6. Security alarm system (● 59-8)
7. Air bags (● 59-9)
8. Energy-absorbing bumpers (● 59-10 and 59-11)

● 59-2 Optional Seat Belts

The purpose of seat belts is to restrain the driver and passengers if there is an accident. During a front-end crash, for example, the car is brought to a sudden stop. But everything inside the car continues to move forward until it hits some solid object. An unrestrained passenger would continue to move forward until he or she hit the windshield or the instrument panel. It is these so-called second collisions that hurt and kill people. However, if the passengers and driver are wearing seat belts, they will be restrained. They will not continue to move forward and will not hit some solid object in the car.

There are two different kinds of seat belts, lap belts and shoulder belts. The lap belt has been credited with saving many lives and preventing injuries. The lap belt plus the shoulder belt is even more effective. The lap belt prevents the passenger or driver from being thrown forward. The shoulder belt keeps the passenger or driver from jackknifing. That is, the upper body is kept from bending at the waist and moving forward. The driver who jackknifed would be thrown into the steering shaft. A passenger who jackknifed would strike the instrument panel with his or her head.

The seat belts installed on earlier-model cars were optional. That is, the driver and passengers did not have to use them. Then, on later-model cars, a warning buzzer was incorporated which buzzed if the seat belts were not fastened. The latest arrangement is called the *seat-belt–starter interlock system.* This system requires the driver and any front-seat passengers to buckle their seat belts before the engine can be started. The purpose of these devices is to urge, or require, the driver and front-seat passengers to use their seat belts, and thus protect themselves from injury or death in case the car is involved in an accident.

● 59-3 Front-Seat-Belt Warning System

This system includes a buzzer and a red warning light which remind the driver and passengers to buckle their seat belts. Figure 59-1 is a schematic drawing of the system. Figure 59-2 is a wiring diagram of the system. The reminder signals (light and buzzer) come on if the seat belts are not buckled and the driver starts the engine and then:

1. Releases the parking brake (manual transmission)
2. Shifts into gear (automatic transmission)

The outboard seat-belt retractors have switches which are closed when the seat belts are retracted. When the seat belts are pulled out and buckled, the retractor switches open, thus preventing the buzzer and light from coming on.

There is a sensing switch under the passenger side of the front seat. This switch is interconnected with the right-hand seat-belt-retractor switch. The sensing switch closes when a weight of more than a few pounds (a passenger) is placed on the seat. Then, if the passenger fails to buckle up, the retractor switch, also closed, completes the circuit to the buzzer and light. Now, when the drive releases the parking brake (manual transmission), or shifts into gear (automatic transmission), the reminders come on.

Some manufacturers supply adjusting information for the passenger sensing switch. It goes like this: If the switch is too sensitive, apply your full weight on one knee

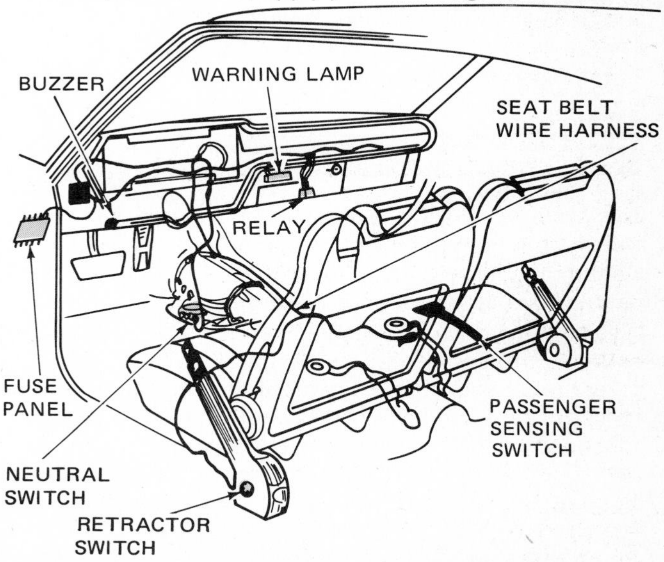

Fig. 59-1 Seat-belt warning system. (*American Motors Corporation*)

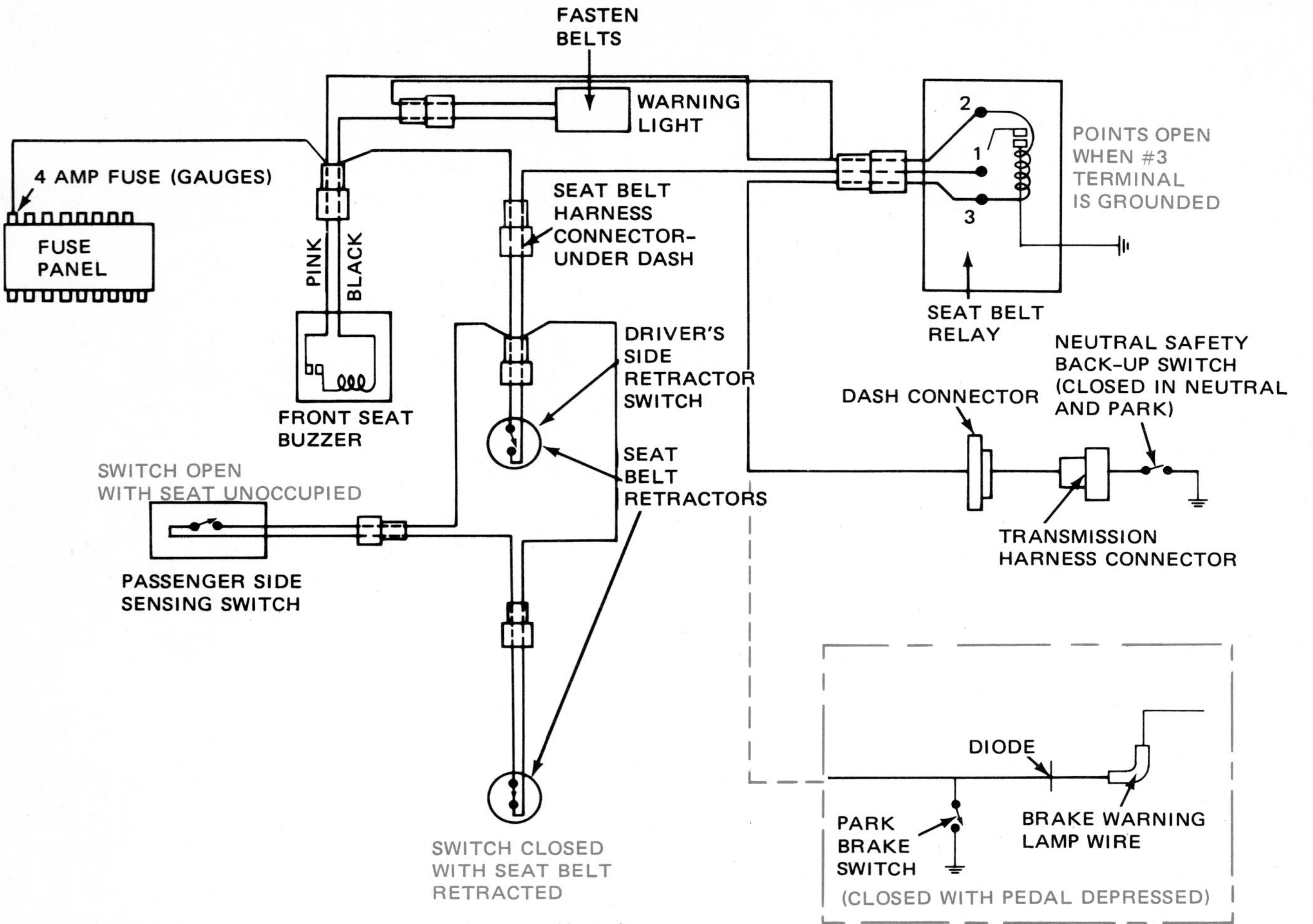

Fig. 59-2 Seat-belt warning-system wiring diagram. (*American Motors Corporation*)

directly above the sensing switch. When the cushion bottoms, the switch contacts will be bent. This should increase the amount of weight necessary to actuate the switch. Other manufacturers state that the switch should be replaced if it does not work satisfactorily.

59-4 Seat-Belt—Starter Interlock System

This system makes it necessary to buckle the front seat belts before the engine can be started. It was required by federal law starting in the mid 1970s. That is, all automotive manufacturers were required by law to install the system in all cars they built. However, after the public became acquainted with the complexities of the system, so many people objected to it that the United States Congress rescinded the law. Manufacturers stopped installing the system on their cars. It also became legal to deactivate the system on cars having it. Thus today you may not find very many cars with the system still working. However, in case you do, you should have the following information.

The system works like this. A specific sequence of actions was required to start the car. Here is how it went:

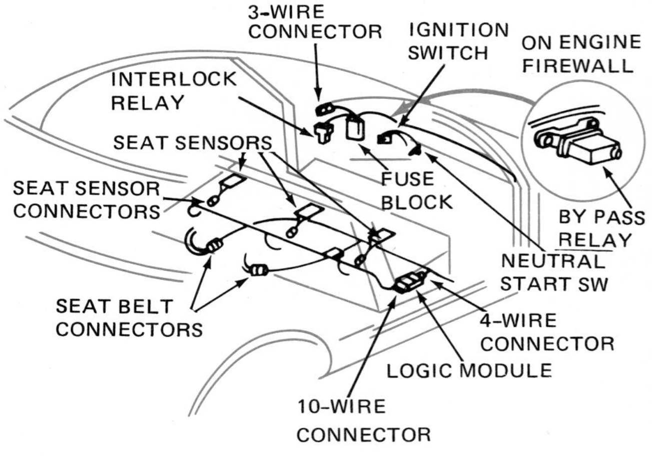

Fig. 59-3 Location of components in the seat-belt-starter interlock system. (*Chevrolet Motor Division of General Motors Corporation*)

1. Get in the car, and sit down.
2. Buckle the seat belt. If a passenger gets in, the passenger must also buckle up.
3. Insert the ignition key, and turn the switch to START.

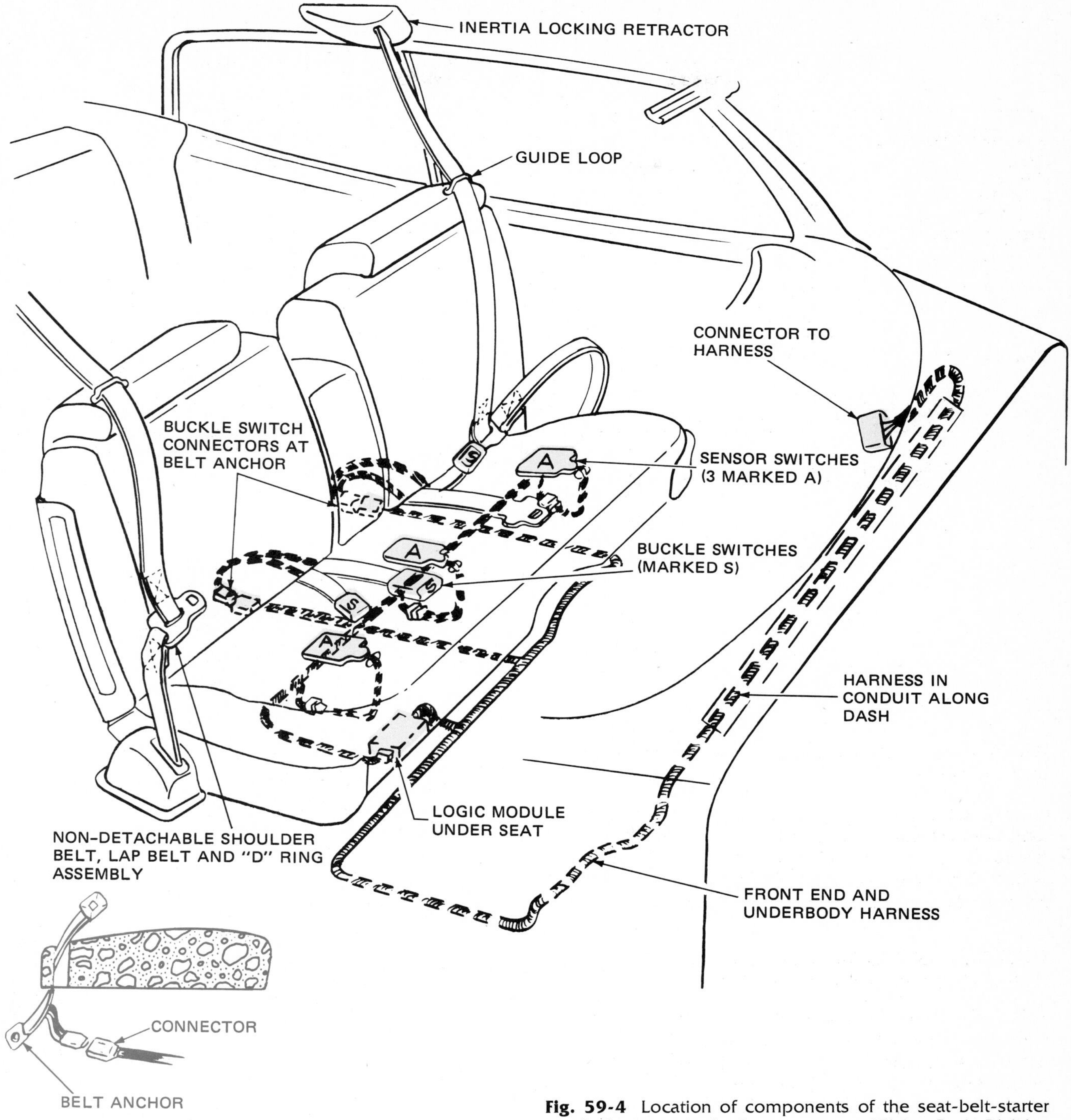

Fig. 59-4 Location of components of the seat-belt-starter interlock system in the front-seat area. (*Fisher Body Division of General Motors Corporation*)

This is the *only* sequence that will get the engine started. Figure 59-3 shows the parts in the system. Figure 59-4 shows how the shoulder belt, lap belt, and other parts are arranged.

If the seat belts are not buckled, or are buckled before the occupants are seated, a relay operates to activate the warning light and buzzer. Also, the light and buzzer will come on if any front-seat occupant unbuckles after the transmission has been shifted to a forward drive position.

The light and buzzer will not come on if a front seat belt is unbuckled and the engine is running with the:

Transmission in PARK or NEUTRAL (automatic and column-shift)

Parking brake engaged (floor-shift transmission)

1. Bounce The logic module is under the front seat. It has a time delay that deactivates the interlock system for a few seconds if a buckled-in occupant momentarily lifts off the seat. Without this feature, the logic module would sense that the occupant had left the car and then buckled the seat belt before being reseated. And that would prevent starting. However, the time delay allows the system to ignore the momentary lifting of the occupant off the seat. But if the occupant is off the seat for more than 5 to 10 seconds, the seat belts must be unbuckled and then rebuckled before the engine will start.

2. Restarting Once the car has been started, using the correct sequence noted above, it can be restarted with the seat belts unbuckled as long as the driver remains seated. If the driver leaves the seat, the three-step sequence must be repeated to start the engine.

3. Mechanic's start The engine can be started with the seat belts in any position when the front seats are unoccupied. Simply reach inside the car and turn the ignition key to START without sitting on the front seat. The light and buzzer will come on if the front seat is then occupied and the transmission is shifted to DRIVE. But they will turn off as soon as the seat belts at the occupied positions are buckled up.

4. Override relay An override, or bypass, relay is located on the firewell under the hood (Fig. 59-5). This relay can be used to start the engine when the interlock system has failed and is preventing starting. To use the relay, turn the ignition switch to ON. Then open the hood, and *press and release* the button on the relay. *Holding the button in will damage the override mechanism*. The engine can now be started by turning the ignition key to START. The relay will remain engaged until the ignition key is turned to OFF or LOCK.

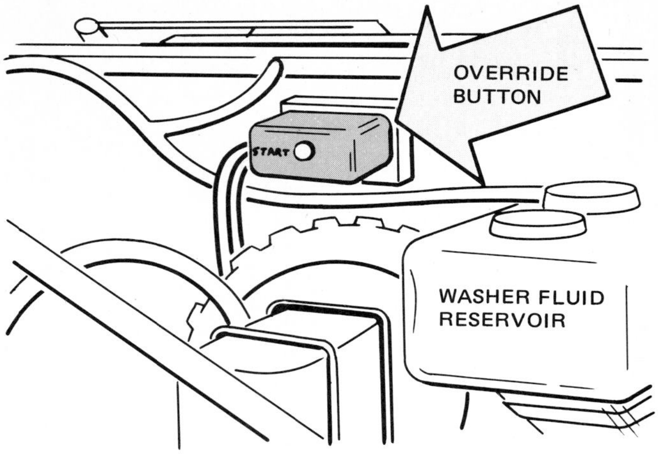

Fig. 59-5 Location of the override relay and button in the engine compartment.

● 59-5 Seat-Belt—Starter Interlock System Trouble Diagnosis

There are four possible troubles:

1. Starting motor will not crank the engine.
2. Starting motor cranks with the seat belts unbuckled.
3. Buzzer and light will not operate.
4. Buzzer and light remain on.

The service manuals of the automotive manufacturers supply step-by-step procedures to follow with any of these four conditions. Chrysler has developed an interlock tester which quickly checks out the system. It has a plug connector which is connected into the system. Then the tester switch is turned on, and the tester goes through an automatic testing sequence.

● 59-6 Open-Door Warning Buzzer System

This system operates a buzzer if the door in the driver's side is opened with the ignition key in the ignition switch. It makes no difference what position the ignition switch is in. Figure 32-22 shows the system. When the ignition key is left in the ignition switch, a warning switch is closed. The warning switch is in the ignition switch and is connected to the door switch (see Fig. 32-22). Then, if the door is opened, the door switch is closed. This completes the circuit to the horn relay. This relay now buzzes, as explained in ● 32-12. The system warns the driver that the ignition key is still in the ignition switch and should be removed. An ignition key in the ignition switch of an unoccupied car is an invitation to thieves.

● 59-7 Headlights-On Warning Buzzer

This system is usually combined with the open-door warning-buzzer system (Fig. 59-6). When the headlights are on and the driver's door is open, a warning buzzer sounds. The system warns the driver to turn off the headlights before leaving the car.

● 59-8 Security Alarm System

This system, shown in one version in Figs. 59-7 and 59-8, sounds an alarm if a thief tries to force entry to any door, the hood, the trunk, or the tailgate. The alarm consists of the horns sounding at about 90 cycles per minute for 3 to 5 minutes. In addition, the headlights, taillights, and side marker lights flash on and off at the same rate. Some systems have a siren which sounds off continuously.

When either front door is locked with the key, the system shown is armed. That is, it is ready to sound the alarm if any attempt is made to force entry into the car. The system can be turned off by using the door key to unlock the door.

● 59-9 Air Bags

Air bags are a passive safety feature that protects the driver and passengers in a car accident. "Passive" means that the driver and passengers do not have to do anything to be protected by the air bags. This is in contrast to the seat belts, described above, that require an action—buckling up. Many people do not bother to buckle up because it is "too much trouble." As a result, there are far more highway injuries and deaths than there need be. Air-bag advocates believe that the air bag will save many lives and prevent injuries.

The principle is simple. At the instant that a crash occurs, the air bags are blown up. They then give the driver and passengers a cushion into which to move. The air bags absorb the forward motion of the occupants and

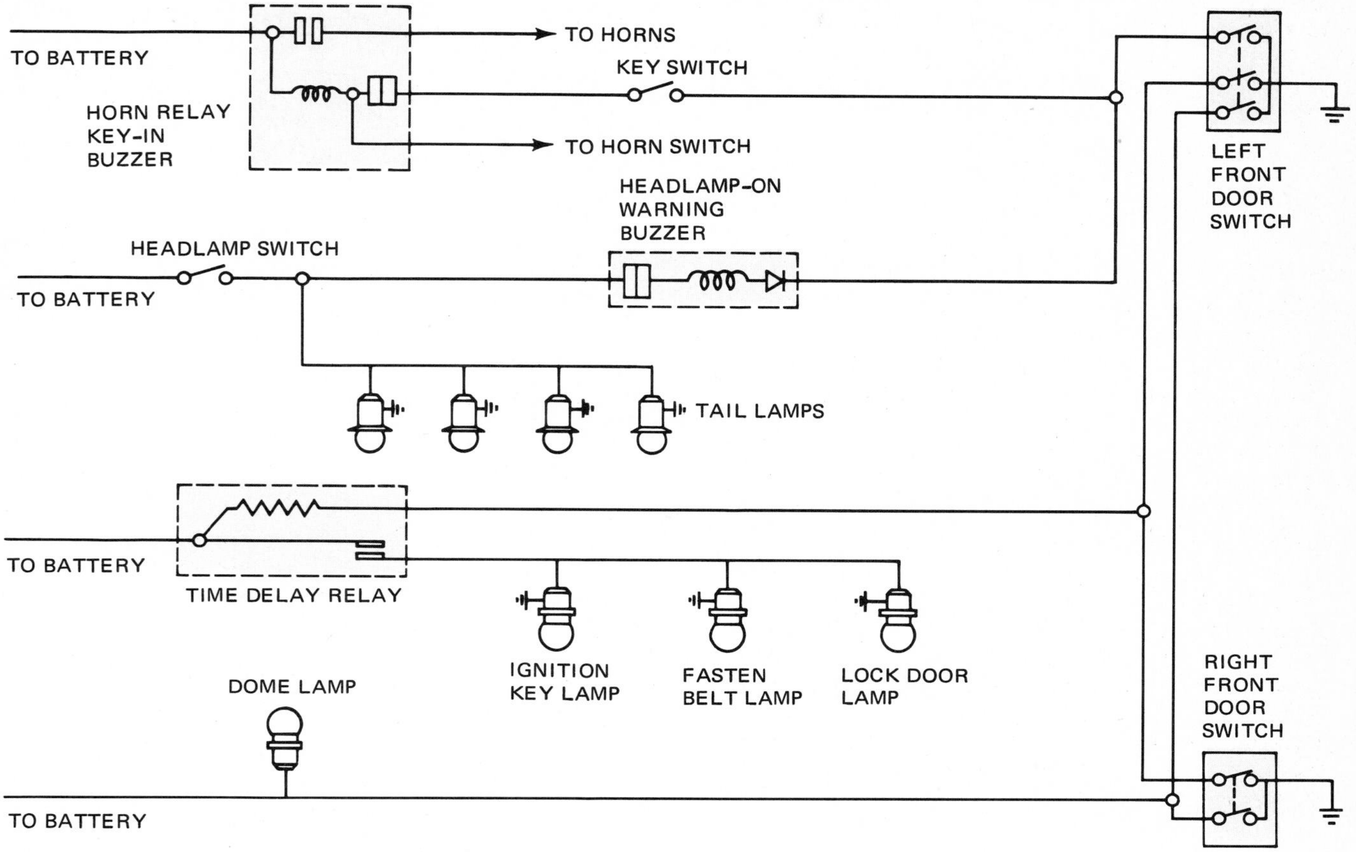

Fig. 59-6 Open-door headlights-on warning-buzzer system. (*Chrysler Corporation*)

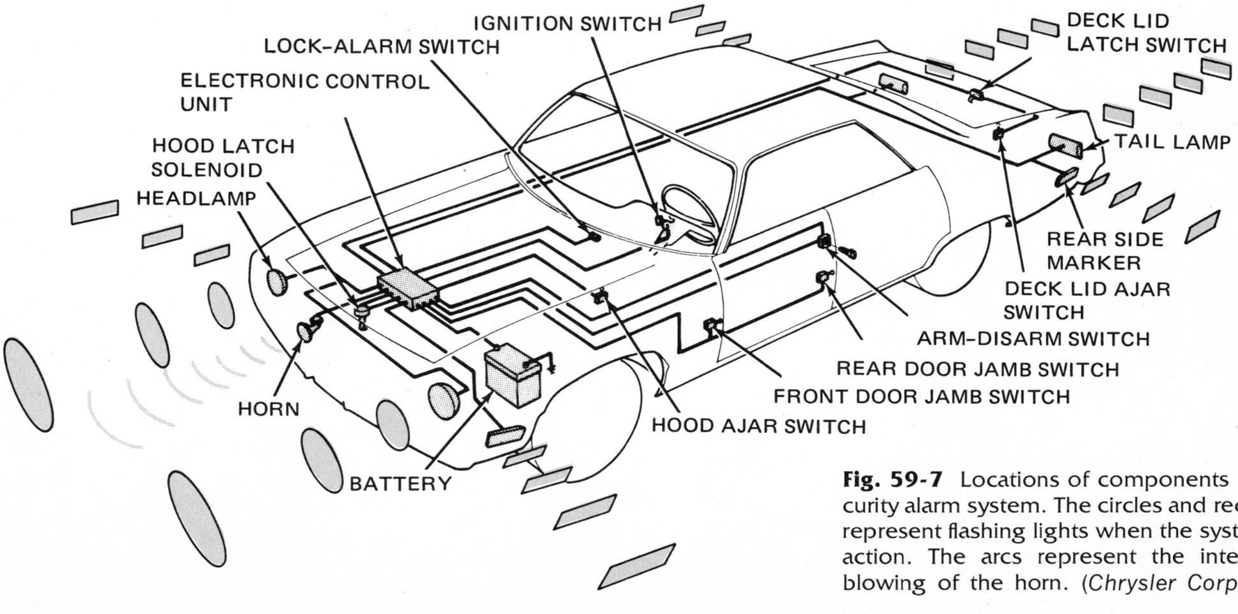

Fig. 59-7 Locations of components of a security alarm system. The circles and rectangles represent flashing lights when the system is in action. The arcs represent the intermittent blowing of the horn. (*Chrysler Corporation*)

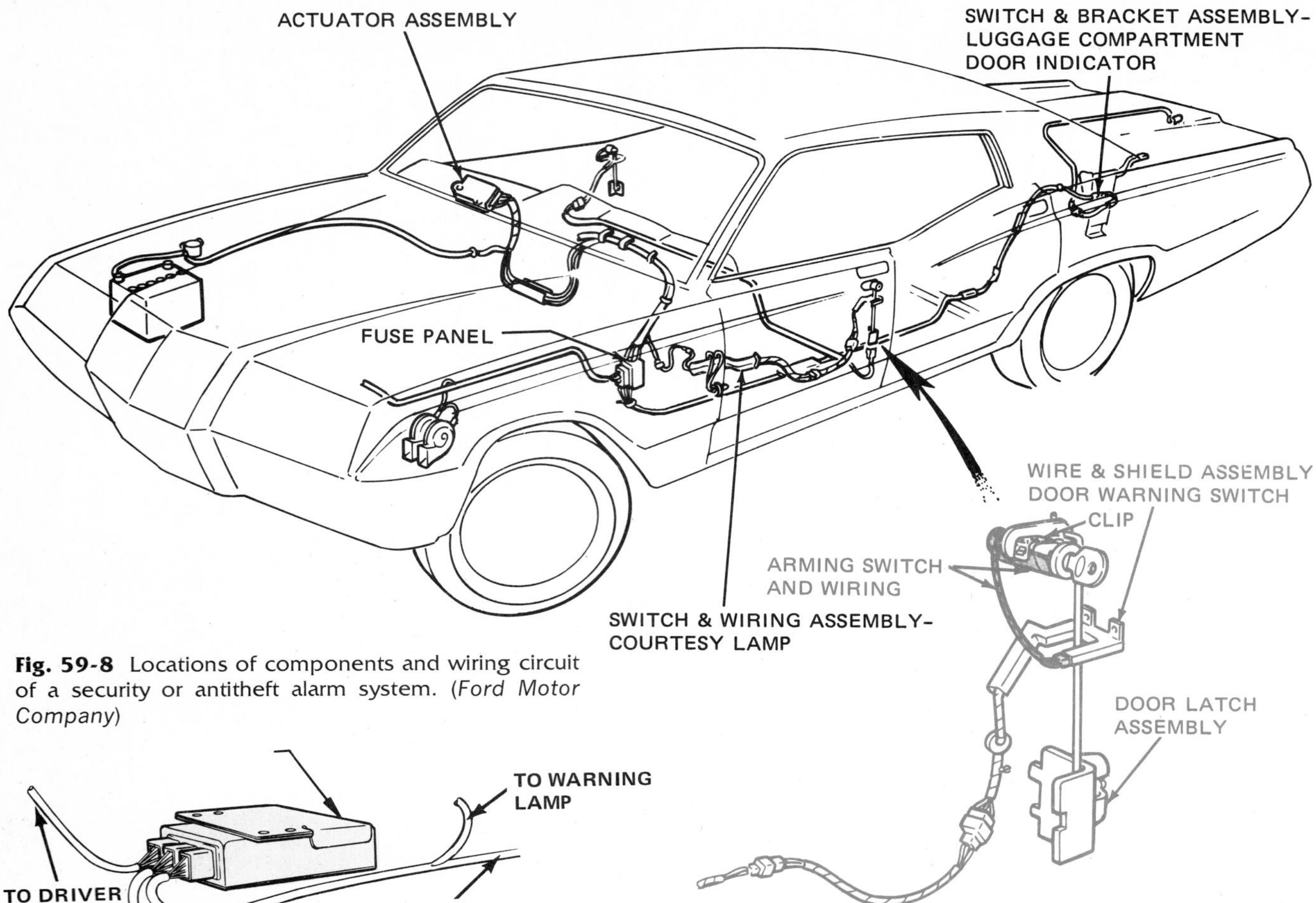

Fig. 59-8 Locations of components and wiring circuit of a security or antitheft alarm system. (*Ford Motor Company*)

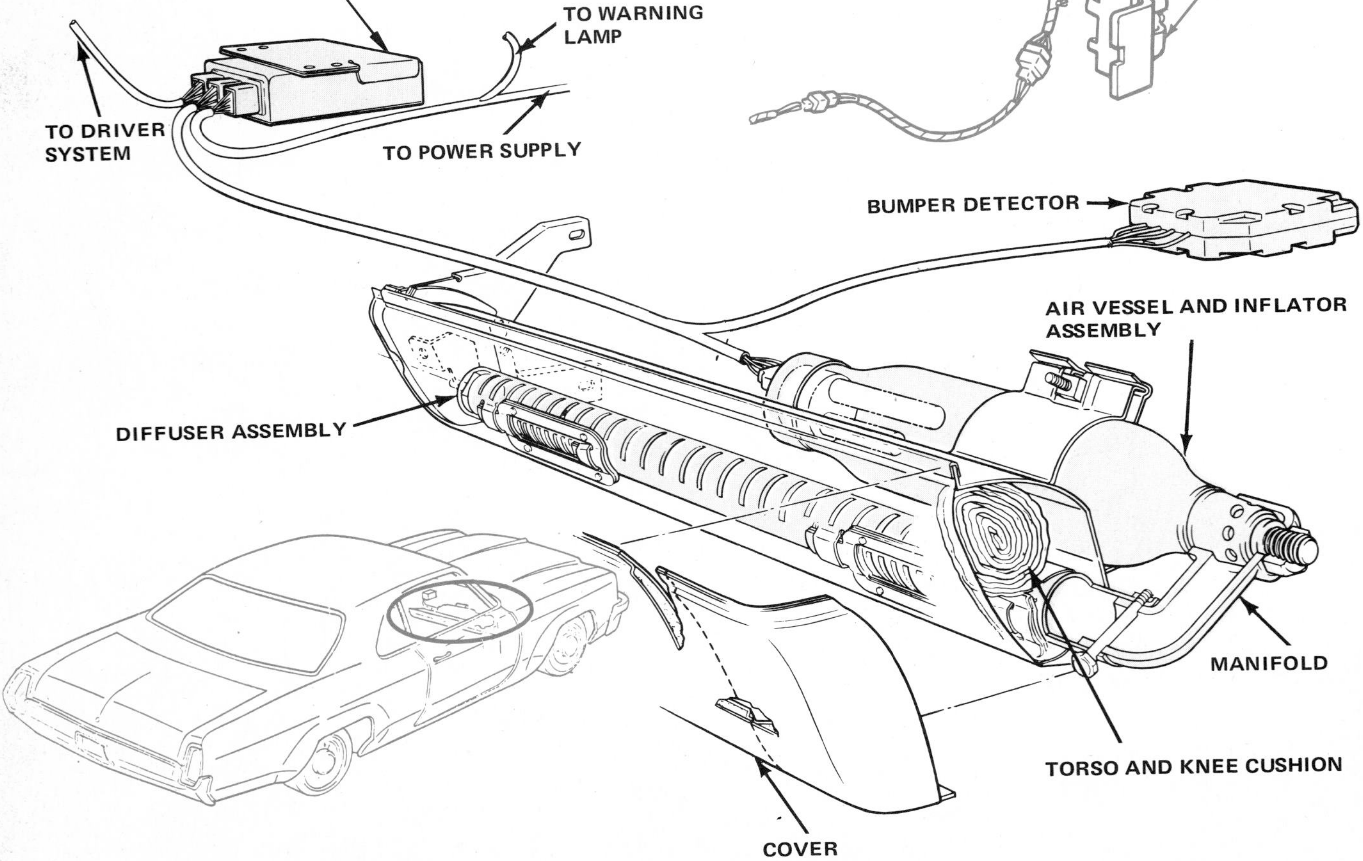

Fig. 59-9 Air-bag restraint system for front-seat passengers. (*General Motors Corporation*)

save them from hitting anything hard that could injure them.

Figure 59-9 shows one arrangement for the passengers in the front seat. Figure 59-10 shows how the system works. A dummy was used in this crash to show how fast the system works. At impact, the dummy is sitting back in the seat in a normal position. At $\frac{1}{30}$ second after the crash starts (upper right in Fig. 59-10), the air-bag system has actuated. The bag is already pushing out from its position in the instrument panel.

The system is actuated by a bumper detector. This detector contains a switch that is closed when the car is suddenly decelerated—that is, when it is brought to a quick stop in an accident. When the switch closes, the air-vessel-and-inflater assembly is actuated (see Fig. 59-9). The air vessel is filled with compressed gas. When it is actuated, this gas is released, and it flows into the air bag. The action is almost instantaneous. Now see 3 at the lower left in Fig. 59-10. This is the condition $\frac{2}{30}$ second after impact. Note that the air bag is almost fully inflated, and that the dummy has moved forward into the air bag. In 4 (lower right in Fig. 59-10), the dummy is all the way forward into the air bag. Now, a fraction of a second later, when the force of the dummy's forward motion has been completely absorbed, the air bag begins to deflate. The air bag has relief holes for this purpose. This quick deflation, after the air bag has done its job, permits the passenger to get out of the car.

The air bag on the driver's side is located in the steering wheel. Figure 59-11 shows the complete system for the driver and passengers, and Fig. 59-12 shows the air bag inflated. Figure 59-13 shows the action with a dummy occupying the driver's seat. The times between the four parts of the picture, showing the inflation of the air bag and the forward motion of the dummy, are approximately the same as in Fig. 59-11.

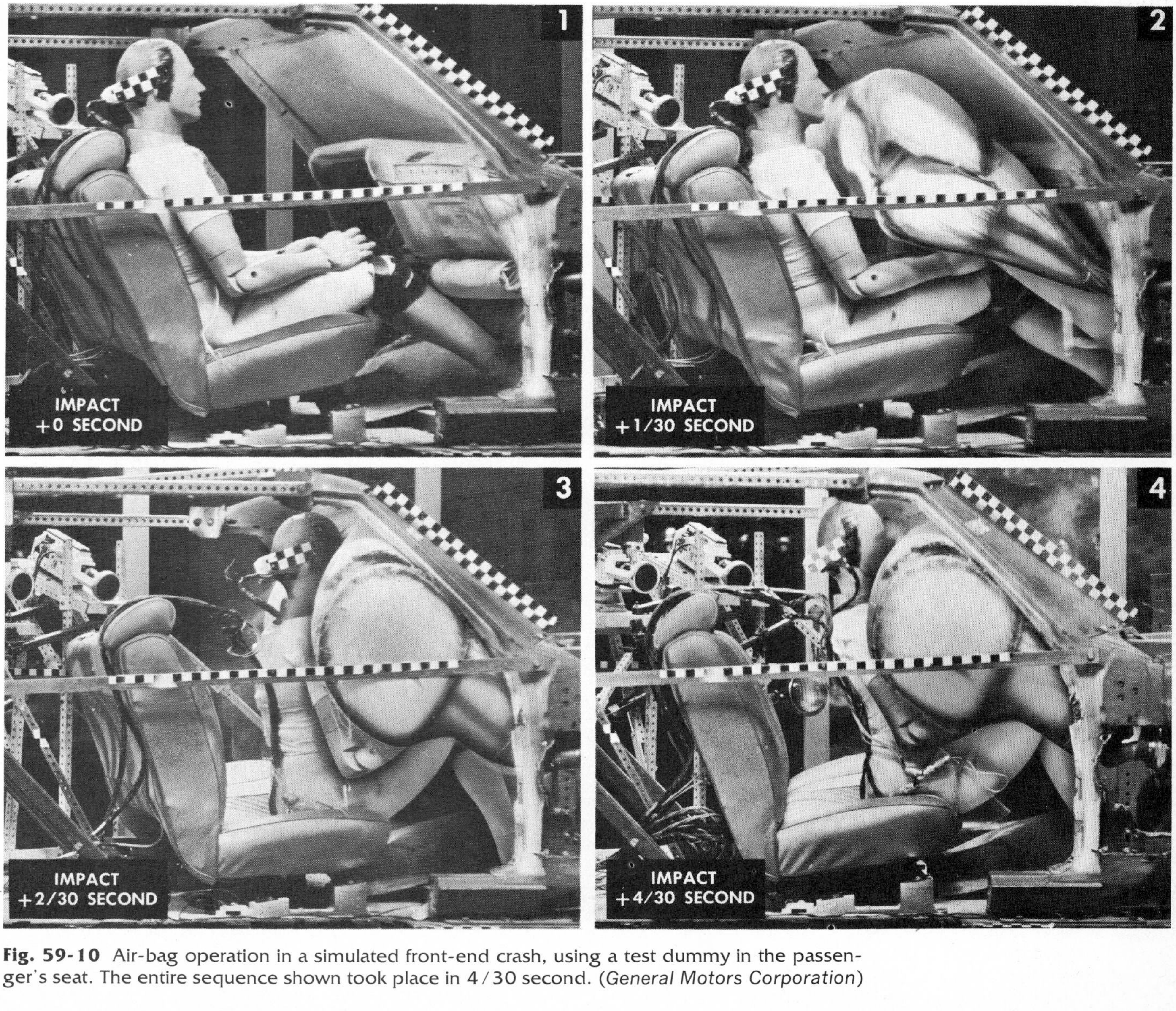

Fig. 59-10 Air-bag operation in a simulated front-end crash, using a test dummy in the passenger's seat. The entire sequence shown took place in 4/30 second. (*General Motors Corporation*)

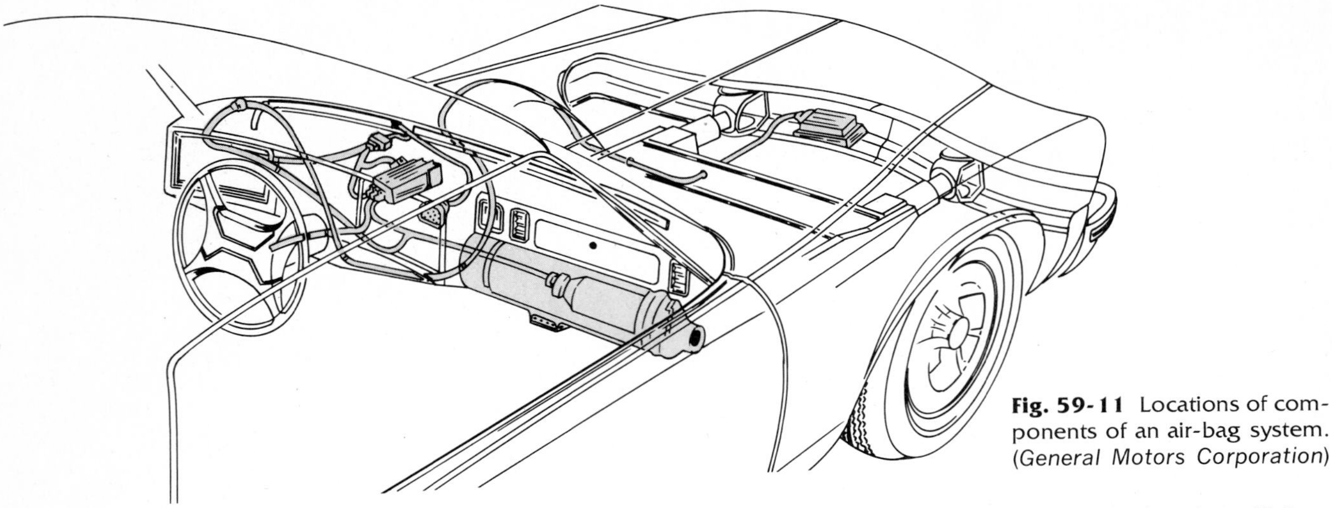

Fig. 59-11 Locations of components of an air-bag system. *(General Motors Corporation)*

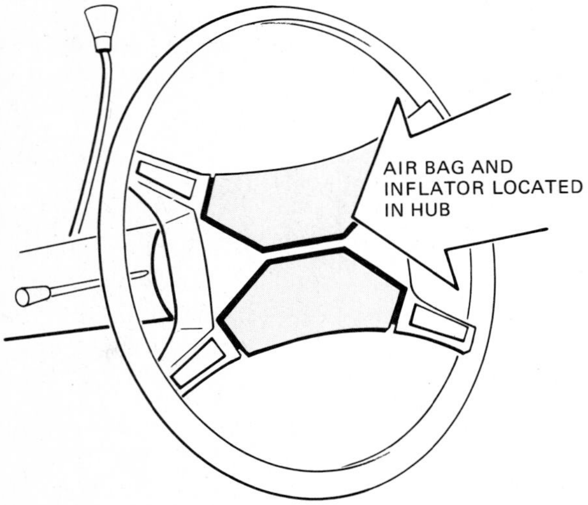

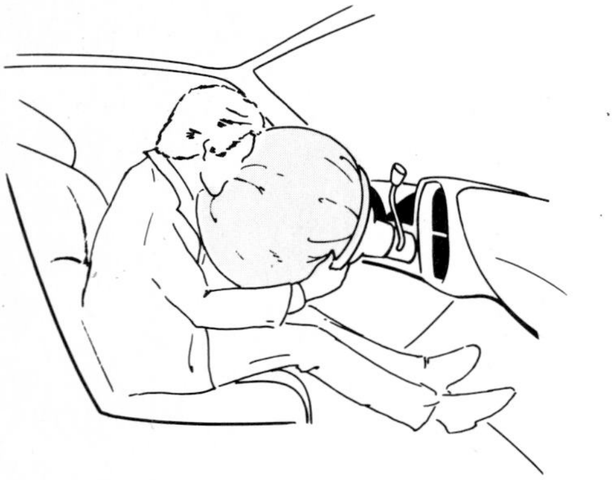

Fig. 59-12 Location of the air bag in the steering wheel. The lower picture shows the action when the air bag is inflated and the driver is thrown forward into it. *(General Motors Corporation)*

● 59-10 Energy-Absorbing Bumpers

The energy-absorbing bumper is required by law on late-model cars. It will withstand collisions at low speed without damage to the bumper or car. Further, most of these bumpers resume their original position after the collision. There are several types, used at both the front and the rear of late-model cars. One uses a leaf-spring-assembly which supports the bumper as shown in Fig. 59-14. On impact, the spring gives and absorbs the blow. It then returns to its original position, if the impact was within the designed limits. If the impact was greater than the designed limits, damage may have occurred and repair may be required.

A second type uses a pair of special bolts and two bolt dies (Fig. 59-15). The dies are steel rings having an inner diameter smaller than the energy-absorbing bolts. During impact, the dies are pushed along the bolts. This action reduces the diameter of the special energy-absorbing bolts. In the system shown, the bolt diameters are reduced from 0.33 inch (8.38 mm) to 0.31 inch (7.87 mm). The bolts are elongated about 1 inch (25.4 mm) during an impact that moves the bumper $3\frac{1}{2}$ inches (88.9 mm). If the impact is severe enough to elongate the bolts, then the energy-absorbing-bolt assembly, dies, and related parts must be replaced. If the impact was greater than the designed limit of the system, other damage may have occurred which would require repair.

A third type of energy-absorbing bumper uses a pair of energy absorbers that are like shock absorbers. Figure 59-16 shows how the two absorbers are located between the frame and the bumper reinforcement. During a front-end impact, the energy absorbers shorten, just like telescope-type shock absorbers. Following the impact, if it is not beyond the designed limits, the energy absorbers return to their original length.

Figure 59-17 is an external view of an energy absorber. Figure 59-18 is a sectional view of the energy absorber in its normal extended position. Figure 59-19 shows the absorber action. At the top, the absorber is shown in the extended position at the start of impact. The impact forces the piston tube to the right (in Fig. 59-19). This action forces the hydraulic fluid to flow around the metering pin and through the orifice in the end of the piston tube. As the piston tube continues to move, the flow of hydraulic fluid into the piston tube pushes the floating piston to the left in Fig. 59-19. This compresses the gas in the piston tube, as shown in the bottom picture.

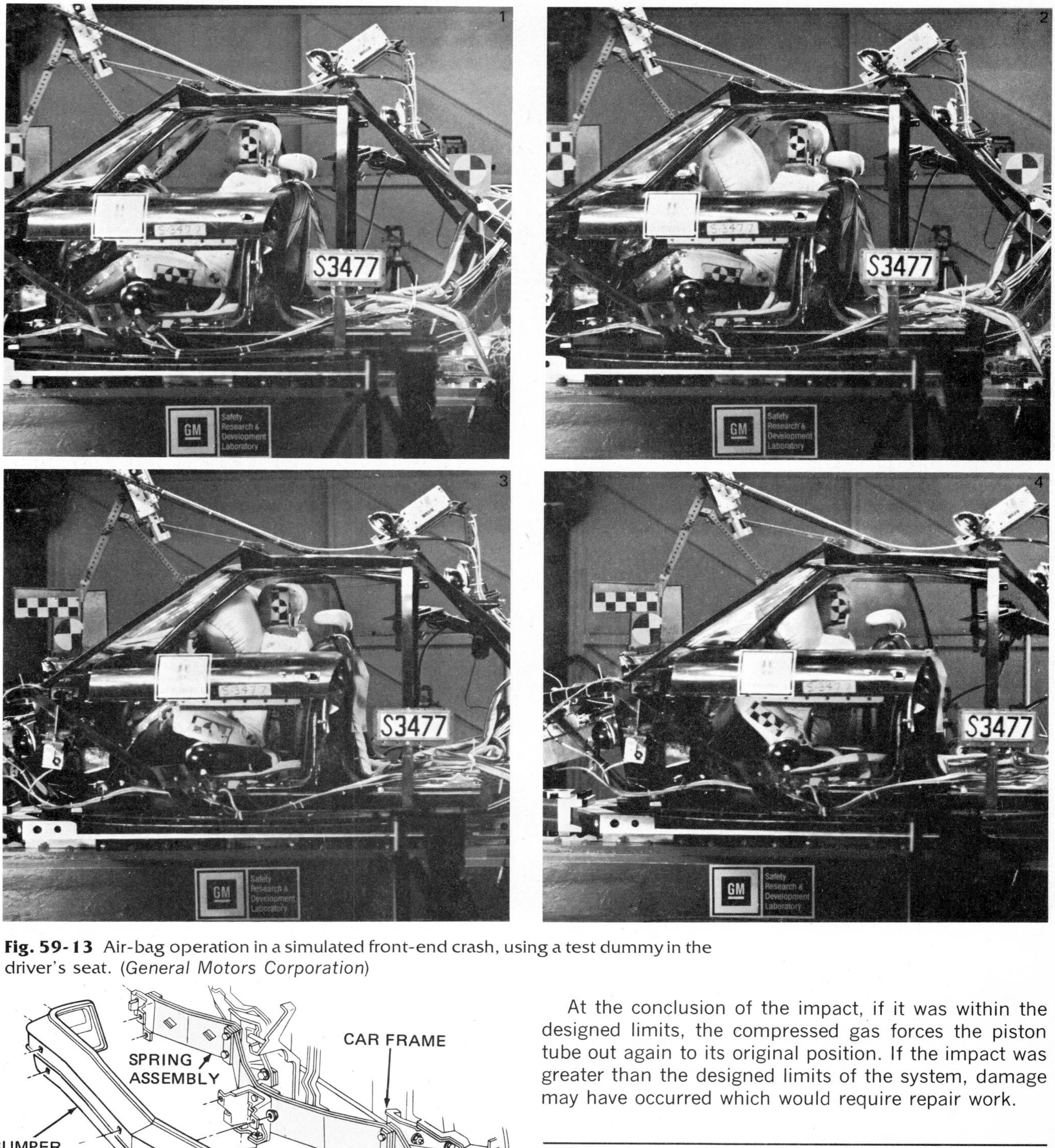

Fig. 59-13 Air-bag operation in a simulated front-end crash, using a test dummy in the driver's seat. (*General Motors Corporation*)

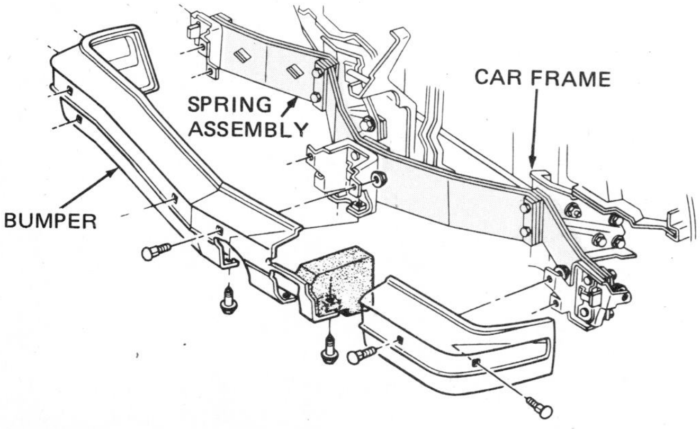

Fig. 59-14 Front bumper system using a leaf-spring assembly to absorb the energy of a front-end impact. (*Chevrolet Motor Division of General Motors Corporation*)

At the conclusion of the impact, if it was within the designed limits, the compressed gas forces the piston tube out again to its original position. If the impact was greater than the designed limits of the system, damage may have occurred which would require repair work.

● 59-11 Servicing the Energy Absorber

This absorber is shown in Figs. 59-16–59-19. It is serviced by unit replacement. If a unit is to be scrapped, internal pressure must be released as explained below.

CAUTION: The energy absorber contains gas at high pressure. Under no conditions should you attempt to repair, weld, or apply heat to the unit. This could cause the unit to explode.

To relieve the gas pressure, put the energy absorber in a vise. Drill a small hole in the piston tube, as shown in Fig. 59-20. Use the caution label as a locator for drilling. Drill either in front of or through the label.

CAUTION: Wear safety glasses when drilling the hole. When the gas is released, it can drive metal chips at high speed. One of these could enter an eye and *put it out.*

1. Handling a bound-up energy absorber When an energy absorber is bound up as a result of a collision, extra care must be used to remove it from the vehicle.

a. Stand clear of the bumper.

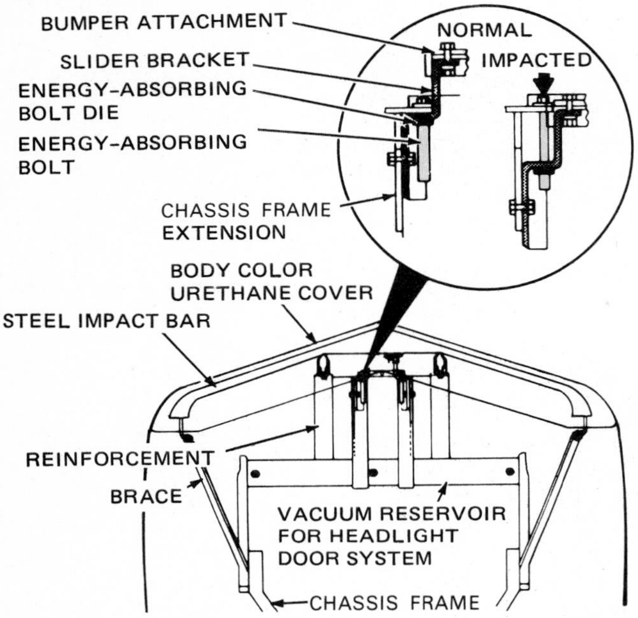

Fig. 59-15 Front bumper system using a pair of special bolts and dies. The dies draw the bolts to a smaller diameter to absorb the energy of a front-end impact. (*Chevrolet Motor Division of General Motors Corporation*)

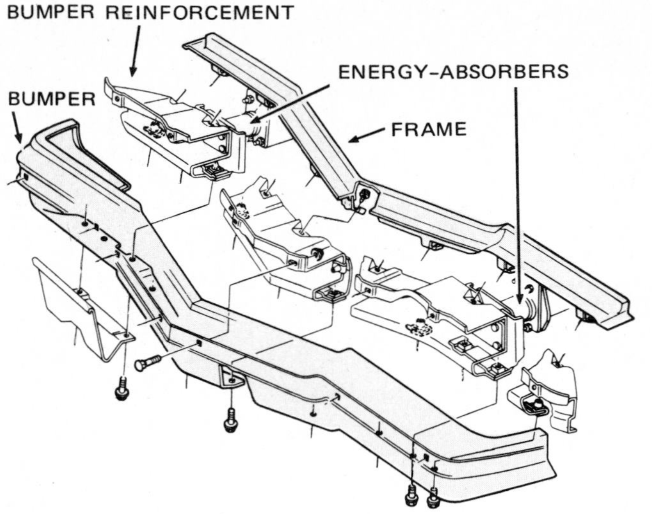

Fig. 59-16 Locations of components in a front bumper system using two energy absorbers of the shock-absorber type. (*Chevrolet Motor Division of General Motors Corporation*)

Fig. 59-17 Energy absorber. (*Chevrolet Motor Division of General Motors Corporation*)

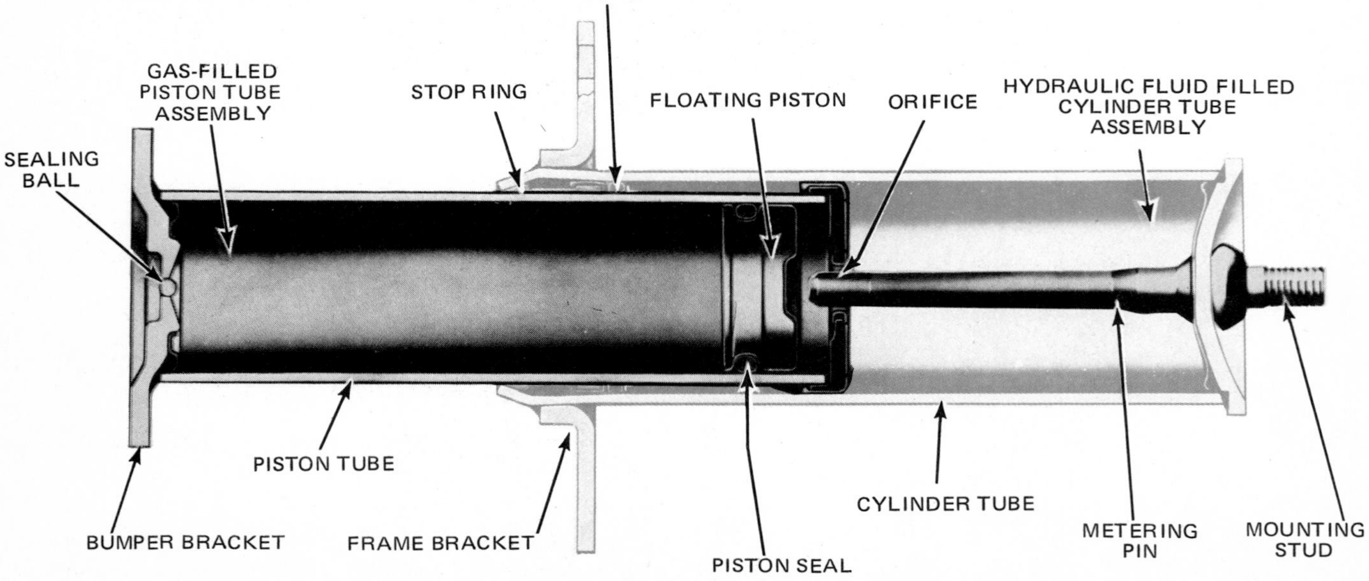

Fig. 59-18 Sectional view of the energy absorber in the extended position. (*Chevrolet Motor Division of General Motors Corporation*)

b. Use a chain or cable to apply positive restraint to the energy absorber, so it will not suddenly return to its original length.
c. Drill a small hole in the piston tube near the bumper bracket to relieve the gas pressure.
d. After the gas has escaped, remove the absorber from the vehicle.

2. Checking the energy absorber on the car Examine it for leakage around the seal between the cylinder tube and the piston tube. A stain or trace of oil on the piston tube near the seal is normal. But if oil is dripping continuously from the seal or stud end, it should be replaced.

Examine the bumper bracket, piston tube, frame bracket, and cylinder tube for visible distortion. Scuffing of the piston tube, if the unit has been stroked, is considered normal. If there is obvious damage, the unit and associated damaged parts should be replaced.

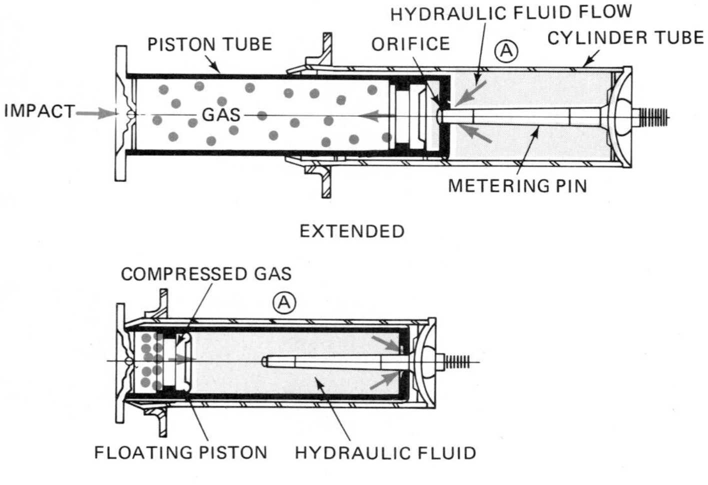

Fig. 59-19 Operation of the energy absorber during a front-end impact. Top: Action at the start of impact; the piston tube starts to enter the cylinder tube, and hydraulic fluid flows through the orifice. Bottom: The piston tube has reached the inner limit of motion; hydraulic fluid has flowed through the orifice, forcing the floating piston to compress the gas. *(General Motors Corporation)*

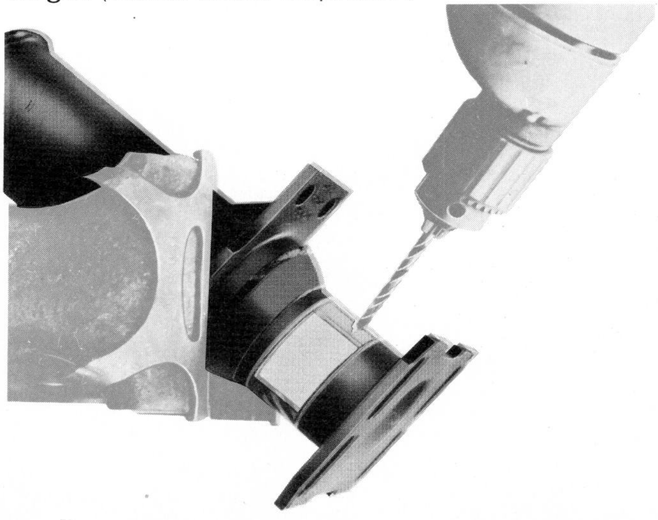

Fig. 59-20 Drilling a hole in the piston tube to relieve the gas pressure before discarding a defective energy absorber. *(General Motors Corporation)*

Energy absorbers can be checked on the car. Each should be checked. The test is made with the engine not running, the transmission in PARK, the parking brake set, and a brake-applying tool holding the service brakes. Any suitable barrier can be used, such as a wall, or post. Install a device that can apply pressure. Pressure is applied to see if the energy absorber will move in $\frac{3}{8}$ inch [9.5 mm] or more. Then, when pressure is released, the bumper should return to its original position. The pressure device can be a hydraulic or mechanical jack. Pressure must be applied squarely to avoid slippage.

CAUTION: Driving into a post, wall, or other barrier to perform the test is not recommended!

3. Checking the energy absorber on the bench This can be done in a suitable arbor press. The unit should compress at least $\frac{3}{8}$ inch [9.5 mm], and return to normal length when pressure is released. If not, discard the unit.

REVIEW QUESTIONS

1. What is the purpose of seat belts?
2. What does the shoulder belt do that the lap belt cannot?
3. Describe the front-seat-belt warning system.
4. In cars with the seat-belt–starter interlock system, what is the sequence required to start the engine?
5. What is the purpose of the bounce system in the starter interlock system? How does it work?
6. Explain how the engine can be started with the seat belts in any position and the front seats unoccupied.
7. What is the override relay? How does it work?
8. Explain how the open-door warning-buzzer system works.
9. Explain how the security alarm system works.
10. Describe the operation of the air-bag system. About how long does it take for the air bag to fill during a front-end crash?
11. What is the purpose of energy-absorbing bumpers?
12. Describe the operation of the shock-absorber type of energy absorber.
13. Describe the operation of the energy absorber shown in Fig. 59-15.
14. What is the purpose of drilling a hole in the energy-absorber piston tube if the unit is to be discarded?
15. Why must you never apply heat to the energy absorber?
16. Explain how to check the operation of the energy absorber with the unit on the car. On the bench.

SELF PROJECTS

Build up a file of stories about accidents in which lives were saved or injuries minimized by seat belts and air bags. Seat belts prevent injuries and save lives. Examine late-model cars, and note the types of bumpers and the locations of the components of the air-bag system.

APPENDIX
Conversion Tables

MILLIMETERS TO INCHES

mm	1	2	3	4	5	6	7	8	9	10	11	12	13
inches	0.0394	0.0787	0.1181	0.1575	0.1968	0.2362	0.2756	0.3150	0.3543	0.3937	0.4331	0.4724	0.5118
mm	14	15	16	17	18	19	20	21	22	23	24	25	26
inches	0.5512	0.5905	0.6299	0.6693	0.7087	0.7480	0.7874	0.8268	0.8661	0.9055	0.9449	0.9842	1.0236
mm	27	28	29	30	31	32	33	34	35	36	37	38	39
inches	1.0630	1.1024	1.1417	1.1811	1.2205	1.2598	1.2992	1.3386	1.3779	1.4173	1.4567	1.4961	1.5354
mm	40	41	42	43	44	45	46	47	48	49	50	51	52
inches	1.5748	1.6142	1.6535	1.6929	1.7323	1.7716	1.8110	1.8504	1.8898	1.9291	1.9685	2.0079	2.0472
mm	53	54	55	56	57	58	59	60	61	62	63	64	65
inches	2.0866	2.1260	2.1653	2.2047	2.2441	2.2835	2.3228	2.3622	2.4016	2.4409	2.4803	2.5197	2.5590
mm	66	67	68	69	70	71	72	73	74	75	76	77	78
inches	2.6984	2.6378	2.6772	2.7165	2.7559	2.7953	2.8346	2.8740	2.9134	2.9527	2.9921	3.0315	3.0709
mm	79	80	81	82	83	84	85	86	87	88	89	90	91
inches	3.1102	3.1496	3.1890	3.2283	3.2677	3.3071	3.3464	3.3858	3.4252	3.4646	3.5039	3.5433	3.5827
mm	92	93	94	95	96	97	98	99	100				
inches	3.6220	3.6614	3.7008	3.7401	3.7795	3.8189	3.8583	3.8976	3.9370				

INCHES TO MILLIMETERS

inches	1/64	1/32	3/64	1/16	5/64	3/32	7/64	1/8	9/64	5/32	11/64	3/16	13/64
mm	0.3969	0.7937	1.1906	1.5875	1.9844	2.3812	2.7781	3.1750	3.5719	3.9687	4.3656	4.7625	5.1594
inches	7/32	15/64	1/4	17/64	9/32	19/64	5/16	21/64	11/32	23/64	3/8	25/64	13/32
mm	5.5562	5.9531	6.3500	6.7469	7.1437	7.5406	7.9375	8.3344	8.7312	9.1281	9.5250	9.9219	10.3187
inches	27/64	7/16	29/64	15/32	31/64	1/2	33/64	17/32	35/64	9/16	37/64	19/32	39/64
mm	10.7156	11.1125	11.5094	11.9062	12.3031	12.7000	13.0969	13.4937	13.8906	14.2875	14.6844	15.0812	15.4781
inches	5/8	41/64	21/32	43/64	11/16	45/64	23/32	47/64	3/4	49/64	25/32	51/64	13/16
mm	15.8750	16.2719	16.6687	17.0656	17.4625	17.8594	18.2562	18.6531	19.0500	19.4469	19.8437	20.2406	20.6375
inches	53/64	27/32	55/64	7/8	57/64	29/32	59/64	15/16	61/64	31/32	63/64		
mm	21.0344	21.4312	21.8281	22.2250	22.6219	23.0187	23.4156	23.8125	24.2094	24.6062	25.0031		

FAHRENHEIT TO CELSIUS (Centigrade)

°F	−20	−15	−10	−5	0	1	2	3	4	5	10	15	20
°C	−28.9	−26.1	−23.3	−20.6	−17.8	−17.2	−16.7	−16.1	−15.6	−15.0	−12.2	−9.4	−6.7
°F	25	30	35	40	45	50	55	60	65	70	75	80	85
°C	−3.9	−1.1	1.7	4.4	7.2	10.0	12.8	15.6	18.3	21.1	23.9	26.7	29.4
°F	90	95	100	105	110	115	120	125	130	135	140	145	150
°C	32.2	35.0	37.8	40.6	43.3	46.1	48.9	51.7	54.4	57.2	60.0	62.8	65.6
°F	155	160	165	170	175	180	185	190	195	200	205	210	212
°C	68.3	71.1	73.9	76.7	79.4	82.2	85.0	87.8	90.6	93.8	96.1	98.9	100.0
°F	215	220	225	230	235	240	245	250	255	260	265		
°C	101.7	104.4	107.2	110.0	112.8	115.6	118.3	121.1	123.9	126.6	129.4		

FEET TO METERS

ft	0	1	2	3	4	5	6	7	8	9	ft
	m	m	m	m	m	m	m	m	m	m	
—		0.305	0.610	0.914	1.219	1.524	1.829	2.134	2.438	2.743	—
10	3.048	3.353	3.658	3.962	4.267	4.572	4.877	5.182	5.486	5.791	10
20	6.096	6.401	6.706	7.010	7.315	7.620	7.925	8.230	8.534	8.839	20
30	9.144	9.449	9.754	10.058	10.363	10.668	10.973	11.278	11.582	11.887	30
40	12.192	12.497	12.802	13.106	13.411	13.716	14.021	14.326	14.630	14.935	40
50	15.240	15.545	15.850	16.154	16.459	16.764	17.069	17.374	17.678	17.983	50
60	18.288	18.593	18.898	19.202	19.507	19.812	20.117	20.422	20.726	21.031	60
70	21.336	21.641	21.946	22.250	22.555	22.860	23.165	23.470	23.774	24.079	70
80	24.384	24.689	24.994	25.298	25.603	25.908	26.213	26.518	26.822	27.127	80
90	27.432	27.737	28.042	28.346	28.651	28.956	29.261	29.566	29.870	30.175	90
100	30.480	30.785	31.090	31.394	31.699	32.004	32.309	32.614	32.918	33.223	100

Source: Buick Motor Division of General Motors Corporation

MILES TO KILOMETERS

mile	0	1	2	3	4	5	6	7	8	9	mile
	km	km	km	km	km	km	km	km	km	km	
—		1.609	3.219	4.828	6.437	8.047	9.656	11.265	12.875	14.484	—
10	16.093	17.703	19.312	20.921	22.531	24.140	25.750	27.359	28.968	30.578	10
20	32.187	33.796	35.406	37.015	38.624	40.234	41.843	43.452	45.062	46.671	20
30	48.280	49.890	51.499	53.108	54.718	56.327	57.936	59.546	61.155	62.764	30
40	64.374	65.983	67.593	69.202	70.811	72.421	74.030	75.639	77.249	78.858	40
50	80.467	82.077	83.686	85.295	86.905	88.514	90.123	91.733	93.342	94.951	50
60	96.561	98.170	99.779	101.39	103.00	104.61	106.22	107.83	109.44	111.04	60
70	112.65	114.26	115.87	117.48	119.09	120.70	122.31	123.92	125.53	127.14	70
80	128.75	130.36	131.97	133.58	135.19	136.79	138.40	140.01	141.62	143.23	80
90	144.84	146.45	148.06	149.67	151.28	152.89	154.50	156.11	157.72	159.33	90
100	160.93	162.54	164.15	165.76	167.37	168.98	170.59	172.20	173.81	175.42	100

SQUARE INCHES TO SQUARE CENTIMETERS

in^2	0	1	2	3	4	5	6	7	8	9	in^2
	cm^2	cm^2	cm^2	cm^2	cm^2	cm^2	cm^2	cm^2	cm^2	cm^2	
—		6.452	12.903	19.355	25.806	32.258	38.710	45.161	51.613	58.064	—
10	64.516	70.968	77.419	83.871	90.322	96.774	103.226	109.677	116.129	122.580	10
20	129.032	135.484	141.935	148.387	154.838	161.290	167.742	174.193	180.645	187.096	20
30	193.548	200.000	206.451	212.903	219.354	225.806	232.258	238.709	245.161	251.612	30
40	258.064	264.516	270.967	277.419	283.870	290.322	296.774	303.225	309.677	316.128	40
50	322.580	329.032	335.483	341.935	348.386	354.838	361.290	367.741	374.193	380.644	50
60	387.096	393.548	399.999	406.451	412.902	419.354	425.806	432.257	438.709	445.160	60
70	451.612	458.064	464.515	470.967	477.418	483.870	490.322	496.773	503.225	509.676	70
80	516.128	522.580	529.031	535.483	541.934	548.386	554.838	561.289	567.741	574.192	80
90	580.644	587.096	593.547	599.999	606.450	612.902	619.354	625.805	632.257	638.708	90
100	645.160	651.612	658.063	664.515	670.966	677.418	683.870	690.321	696.773	703.224	100

CUBIC INCHES TO CUBIC CENTIMETERS

in^3	0	1	2	3	4	5	6	7	8	9	in^3
	cm^3	cm^3	cm^3	cm^3	cm^3	cm^3	cm^3	cm^3	cm^3	cm^3	
—		16.387	32.774	49.161	65.548	81.935	98.322	114.709	131.097	147.484	—
10	163.871	180.258	196.645	213.032	229.419	245.806	262.193	278.580	294.967	311.354	10
20	327.741	344.128	360.515	376.902	393.290	409.677	426.064	442.451	458.838	475.225	20
30	491.612	507.999	524.386	540.773	557.160	573.547	589.934	606.321	622.708	639.095	30
40	655.483	671.870	688.257	704.644	721.031	737.418	753.805	770.192	786.579	802.966	40
50	819.353	835.740	852.127	868.514	884.901	901.289	917.676	934.063	950.450	966.837	50
60	983.224	999.611	1015.998	1032.385	1048.772	1065.159	1081.546	1097.933	1114.320	1130.707	60
70	1147.094	1163.482	1179.869	1196.256	1212.643	1229.030	1245.417	1261.804	1278.191	1294.578	70
80	1310.965	1327.352	1343.739	1360.126	1376.513	1392.200	1409.288	1425.675	1442.062	1458.449	80
90	1474.836	1491.223	1507.610	1523.997	1540.384	1556.771	1573.158	1589.545	1605.932	1622.319	90
100	1638.706	1655.093	1671.481	1687.863	1704.255	1720.642	1737.029	1753.416	1769.803	1786.190	100

CUBIC FEET TO CUBIC METERS

ft^3	0	1	2	3	4	5	6	7	8	9	ft^3
	m^3	m^3	m^3	m^3	m^3	m^3	m^3	m^3	m^3	m^3	
—		0.0283	0.0566	0.0850	0.1133	0.1416	0.1699	0.1982	0.2265	0.2549	—
10	0.2832	0.3115	0.3398	0.3681	0.3964	0.4248	0.4531	0.4814	0.5097	0.5380	10
20	0.5663	0.5947	0.6230	0.6513	0.6796	0.7079	0.7362	0.7646	0.7929	0.8212	20
30	0.8495	0.8778	0.9061	0.9345	0.9628	0.9911	1.0194	1.0477	1.0760	1.1044	30
40	1.1327	1.1610	1.1893	1.2176	1.2459	1.2743	1.3026	1.3309	1.3592	1.3875	40
50	1.4159	1.4442	1.4725	1.5008	1.5291	1.5574	1.5858	1.6141	1.6424	1.6707	50
60	1.6990	1.7273	1.7557	1.7840	1.8123	1.8406	1.8689	1.8972	1.9256	1.9539	60
70	1.9822	2.0105	2.0388	2.0671	2.0955	2.1238	2.1521	2.1804	2.2087	2.2370	70
80	2.2654	2.2937	2.3220	2.3503	2.3786	2.4069	2.4353	2.4636	2.4919	2.5202	80
90	2.5485	2.5768	2.6052	2.6335	2.6618	2.6901	2.7184	2.7468	2.7751	2.8034	90
100	2.8317	2.6800	2.8884	2.9167	2.9450	2.9733	3.0016	3.0300	3.0583	3.0866	100

GALLONS (U.S.) TO LITERS

U.S. gal	0	1	2	3	4	5	6	7	8	9	U.S. gal
	l	l	l	l	l	l	l	l	l	l	
—		3.7854	7.5709	11.3563	15.1417	18.9271	22.7126	26.4980	30.2834	34.0638	—
10	37.8543	41.6397	45.4251	49.2105	52.9960	56.7814	60.5668	64.3523	68.1377	71.9231	10
20	75.7085	79.4940	83.2794	87.0648	90.8502	94.6357	98.4211	102.2065	105.9920	109.7774	20
30	113.5528	117.3482	121.1337	124.9191	128.7045	132.4899	136.2754	140.0608	143.8462	147.6316	30
40	151.4171	155.2025	158.9879	162.7734	166.5588	170.3442	174.1296	177.9151	181.7005	185.4859	40
50	189.2713	193.0568	196.8422	200.6276	204.4131	208.1985	211.9839	215.7693	219.5548	223.3402	50
60	227.1256	230.9110	234.6965	238.4819	242.2673	246.0527	249.8382	253.6236	257.4090	261.1945	60
70	264.9799	268.7653	272.5507	276.3362	280.1216	283.9070	287.6924	291.4779	295.2633	299.0487	70
80	302.8342	306.6196	310.4050	314.1904	317.9759	321.7613	325.5467	329.3321	333.1176	336.9030	80
90	340.6884	344.4738	348.2593	352.0447	355.8301	359.6156	363.4010	367.1864	370.9718	374.7573	90
100	378.5427	382.3281	386.1135	389.8990	393.6844	397.4698	401.2553	405.0407	408.8261	412.6115	100

POUNDS TO KILOGRAMS

lb	0	1	2	3	4	5	6	7	8	9	lb
	kg	kg	kg	kg	kg	kg	kg	kg	kg	kg	
—		0.454	0.907	1.361	1.814	2.268	2.722	3.175	3.629	4.082	—
10	4.536	4.990	5.443	5.897	6.350	6.804	7.257	7.711	8.165	8.618	10
20	9.072	9.525	9.979	10.433	10.886	11.340	11.793	12.247	12.701	13.154	20
30	13.608	14.061	14.515	14.969	15.422	15.876	16.329	16.783	17.237	17.690	30
40	18.144	18.597	19.051	19.504	19.958	20.412	20.865	21.319	21.772	22.226	40
50	22.680	23.133	23.587	24.040	24.494	24.948	25.401	25.855	26.308	26.762	50
60	27.216	27.669	28.123	28.576	29.030	29.484	29.937	30.391	30.844	31.298	60
70	31.751	32.205	32.659	33.112	33.566	34.019	34.473	34.927	35.380	35.834	70
80	36.287	36.741	37.195	37.648	38.102	38.555	39.009	39.463	39.916	40.370	80
90	40.823	41.277	41.730	42.184	42.638	43.092	43.545	43.998	44.453	44.906	90
100	45.359	45.813	46.266	46.720	47.174	47.627	48.081	48.534	48.988	49.442	100

POUNDS PER SQUARE INCH TO KILOPASCALS

psi	0	1	2	3	4	5	6	7	8	9	psi
	kPa	kPa	kPa	kPa	kPa	kPa	kPa	kPa	kPa	kPa	
—	0.0000	6.8948	13.7895	20.6843	27.5790	34.4738	41.3685	48.2663	55.1581	62.0528	—
10	68.9476	75.8423	82.7371	89.6318	96.5266	103.4214	110.3161	117.2109	124.1056	131.0004	10
20	137.8951	144.7899	151.6847	158.5794	165.4742	172.3689	179.2637	186.1584	193.0532	199.9480	20
30	206.8427	213.7375	220.6322	227.5270	234.4217	241.3165	248.2113	255.1060	262.0008	268.8955	30
40	275.7903	282.6850	289.5798	296.4746	303.3693	310.2641	317.1588	324.0536	330.9483	337.8431	40
50	344.7379	351.6326	358.5274	365.4221	372.3169	379.2116	386.1064	393.0012	399.8959	406.7907	50
60	412.6854	420.5802	427.4749	434.3697	441.2645	448.1592	455.0540	461.9487	468.8435	475.7382	60
70	482.6330	489.5278	496.4225	503.3173	510.2120	517.1068	524.0015	530.8963	537.7911	544.6858	70
80	551.5806	558.4753	565.3701	572.2648	579.1596	586.0544	592.9491	599.8439	606.7386	613.6334	80
90	620.5281	627.4229	634.3177	641.2124	648.1072	655.0019	661.8967	668.7914	675.6862	682.5810	90
100	689.4757	696.3705	703.2653	710.1601	717.0549	723.9497	730.8445	737.7393	744.6341	751.5289	100

POUND-FEET TO KILOGRAM-METERS

lb-ft	0	1	2	3	4	5	6	7	8	9	lb-ft
	kg-m	kg-m	kg-m	kg-m	kg-m	kg-m	kg-m	kg-m	kg-m	kg-m	
—		0.138	0.276	0.415	0.553	0.691	0.829	0.967	1.106	1.244	—
10	1.382	1.520	1.658	1.796	1.934	2.073	2.211	2.349	2.487	2.625	10
20	2.764	2.902	3.040	3.178	3.316	3.455	3.593	3.731	3.869	4.007	20
30	4.146	4.284	4.422	4.560	4.698	4.837	4.975	5.113	5.251	5.389	30
40	5.528	5.666	5.804	5.942	6.080	6.219	6.357	6.495	6.633	6.771	40
50	6.910	7.048	7.186	7.324	7.462	7.601	7.739	7.877	8.015	8.153	50
60	8.292	8.430	8.568	8.706	8.844	8.983	9.121	9.259	9.397	9.535	60
70	9.674	9.812	9.950	10.088	10.227	10.365	10.503	10.641	10.779	10.918	70
80	11.056	11.194	11.332	11.470	11.609	11.747	11.885	12.023	12.161	12.300	80
90	12.438	12.576	12.714	12.855	12.991	13.129	13.267	13.405	13.544	13.682	90
100	13.820	13.958	14.096	14.235	14.373	14.511	14.649	14.787	14.925	14.064	100

GLOSSARY

Accelerator A foot-operated pedal, linked to the throttle valve in the carburetor; used to control the flow of gasoline to the engine.

Accelerator pump In the carburetor, a pump (linked to the accelerator) which momentarily enriches the air-fuel mixture when the accelerator is depressed at low speed.

Air cleaner A device, mounted on or connected to the carburetor, for filtering dirt and dust out of the air being drawn into the engine.

Air-cooled engine An engine that is cooled by the passage of air around the cylinders, not by the passage of a liquid through water jackets.

Air filter A filter that removes dirt and dust particles from air passing through it.

Air-injection system A type of exhaust-emission control system. Air is injected at low pressure into the exhaust manifold or thermal reactor to complete the combustion of unburned hydrocarbons and carbon monoxide in the exhaust gas.

Alternator In the vehicle electric system, a device that converts mechanical energy into electric energy for charging the battery and operating electrical accessories. Also known as an ac generator.

Antidieseling solenoid See **idle-stop solenoid.**

Antifreeze A chemical, usually ethylene glycol, that is added to the engine coolant to raise its boiling point and lower its freezing point.

Antilock system A system installed with the brakes to prevent wheel lockup during braking.

Automatic transmission A transmission in which gear ratios are changed automatically, eliminating the necessity of hand-shifting gears.

Backfiring Preexplosion of the air-fuel mixture so that the explosion passes back around the opened intake valve and through the intake manifold and carburetor; also applied to the loud explosion of overly rich exhaust gas in the exhaust manifold, which exits through the muffler and tail pipe with a loud popping or banging noise.

Ball joint A flexible joint consisting of a ball within a socket, used in front-suspension systems and valve-train rocker arms.

Battery An electrochemical device for storing energy in chemical form so that it can be released as electricity; a group of electric cells connected together.

BDC Abbreviation for bottom dead center.

Bearing A part that transmits a load to a support and, in so doing, absorbs the friction of moving parts.

Bearing caps In the engine, caps held in place by bolts or nuts which, in turn, hold bearing valves in place.

Belt In a tire, a flat strip of material—glass fiber, rayon, or woven steel—which underlies the tread, all around the circumference of the tire.

Belted-bias tire A tire in which the plies are laid diagonally, crisscrossing one another, with a circumferential belt on top of them. The rubber tread is vulcanized on top of the belt and plies.

Belted-radial tire A tire in which the plies run parallel to one another and perpendicular to the tire bead. Belts running parallel to the tire tread are applied over this radial section.

Bias-ply tire A conventionally constructed tire in which the plies are laid diagonally, crisscrossing one another at an angle of about 30 to 40 degrees.

Blow-by Leakage of compressed air-fuel mixture and burned gases (from combustion) past the piston rings into the crankcase.

Brake An energy-conversion device used to slow, stop, or hold a vehicle or mechanism; a device which changes the kinetic energy of motion into useless and wasted heat energy.

Brake drum A metal drum mounted on a car wheel to form the outer shell of the brake; the brake shoes press against the drum to slow or stop drum-and-wheel rotation for braking.

Brake fluid A special non–mineral-oil fluid used in the hydraulic braking system to transmit pressure through a closed system of tubing known as the brake lines.

Brake shoes In drum brakes, arc-shaped metal pieces lined with a high-friction material (the brake lining) which are forced against the revolving drums to produce braking action. In disk brakes, flat metal pieces lined with brake lining which are forced against the rotor face.

Brake system A combination of one or more brakes and their operating and control mechanism.

Cam A rotating lobe or eccentric which can be used with a cam follower to change rotary motion to reciprocating motion.

Camber The tilt of the top of the wheels from the vertical; when the tilt is outward, the camber is positive. Also, the angle which a front-wheel spindle makes with the horizontal.

Camshaft The shaft in the engine which has a series of cams for operating the valve mechanisms. It is driven by gears or sprockets and a toothed belt or chain from the crankshaft.

Carbon (C) A black deposit which is left on engine parts such as pistons, rings, and valves by the combustion of fuel and which inhibits their action.

Carbon dioxide (CO_2) A colorless, odorless gas which results from complete combustion; usually considered harmless. The gas absorbed from air by plants in photosynthesis; also used to carbonate beverages.

Carbon monoxide (CO) A colorless, odorless, tasteless, poisonous gas which results from incomplete combustion. A pollutant contained in engine exhaust gas.

Carburetion The actions that take place in the carburetor; converting liquid fuel to vapor and mixing it with air to form a combustible mixture.

Carburetor The device in an engine fuel system which mixes fuel with air and supplies the combustible mixture to the intake manifold.

Caster Tilting of the steering axis forward or backward to provide directional steering stability. Also, the angle which a front-wheel kingpin makes with the vertical.

Catalytic converter A mufflerlike device for use in an exhaust system. It converts harmful exhaust gases into harmless gases by promoting a chemical reaction between a catalyst and the pollutants.

Centrifugal advance A rotating-weight mechanism in the distributor that advances and retards ignition timing through the centrifugal force resulting from changes in the rotational speed of the engine distributor.

Charcoal canister A container filled with activated charcoal; used to trap gasoline vapor from the fuel tank and carburetor while the engine is off.

Choke In the carburetor, a device used when starting a cold engine. It "chokes off" the airflow through the air horn, producing a partial vacuum in the air horn for greater fuel delivery and a richer mixture. It operates automatically on many newer cars.

Clutch A coupling which connects and disconnects a shaft from its drive while the drive mechanism is running. In an automobile power train, the device which engages and disengages the compressor shaft from its continuously rotating drive-belt pulley.

Coil In an automobile ignition system, a transformer used to step up the battery voltage (by induction) to the high voltage required to fire the spark plugs.

Coil spring A spring made of an elastic metal such as steel, formed into a wire and wound into a coil.

Combustion chamber The space between the top of the piston and the cylinder head, in which the air-fuel mixture is burned.

Compression Reduction in the volume of a gas by squeezing it into a smaller space. Increasing the pressure reduces the volume and increases the density and temperature of the gas.

Compression ignition The ignition of fuel solely by the heat generated when air is compressed in the cylinder; the method of ignition in a diesel engine.

Compression ratio The volume of the cylinder and combustion chamber when the piston is at BDC, divided by the volume when the piston is at TDC.

Compression stroke The piston movement from BDC to TDC immediately following the intake stroke, during which both the intake and exhaust valves are closed while the air-fuel mixture in the cylinder is compressed.

Condenser In the ignition system, a device that is also called a capacitor; connected across the contact points to reduce arcing by providing a storage place for electricity (electrons) as the contact points open. In an air-conditioning system, the radiatorlike heat exchanger in which refrigerant vapor loses heat and returns to the liquid state.

Connecting rod In the engine, the rod that connects the crank on the crankshaft to the piston. Sometimes called a conrod.

Contact points In the conventional ignition system, the stationary and the movable points in the distributor which open and close the ignition primary circuit.

Coolant The liquid mixture of about 50 percent antifreeze and 50 percent water used to carry heat out of the engine.

Cooling system The system that removes heat from the engine by the forced circulation of coolant, and thereby prevents engine overheating. It includes the water jackets, water pump, radiator, and thermostat.

Crankcase The lower part of the engine in which the crankshaft rotates; includes the lower section of the cylinder block and the oil pan.

Crankcase emissions Pollutants emitted into the atmosphere from any portion of the engine-crankcase ventilation or lubrication system.

Crankcase ventilation The circulation of air through the crankcase of a running engine to remove water, blow-by, and other vapors; prevents oil dilution, contamination, sludge formation, and pressure buildup.

Crankshaft The main rotating member or shaft of the engine, with cranks to which the connecting rods are attached; converts up and down (reciprocating) motion into circular (rotary) motion.

Cylinder block The basic framework of the engine, in and on which the other engine parts are attached. It includes the engine cylinders and the upper part of the crankcase.

Cylinder head The part of the engine that covers and encloses the cylinders. It contains cooling fins or water jackets and, on I-head engines, the valves.

Cylinder hone An expandable rotating tool with abrasive stones turned by an electric motor; used to clean and smooth the inside surface of a cylinder.

Cylinder sleeve A replaceable sleeve, or liner, set into the cylinder block to form the cylinder bore.

Detonation Commonly referred to as spark knock or ping. In the combustion chamber, an uncontrolled second explosion (after the spark occurs at the spark plug) with spontaneous combustion of the remaining compressed air-fuel mixture, resulting in a pinging noise.

Diesel cycle An engine operating cycle in which air is compressed and fuel oil is injected into the compressed air at the end of the compression stroke. The heat produced by the compression ignites the fuel oil, eliminating the need for spark plugs or a separate ignition system.

Diesel engine An engine operating on the diesel cycle and burning oil instead of gasoline.

Dieseling A condition in which an automobile engine continues to run after the ignition is off; caused by carbon deposits or hot spots in the combustion chamber glowing sufficiently to furnish heat for combustion.

Differential A gear assembly between axles that permits one wheel to turn at a different speed from the other, while transmitting power from the drive shaft to the wheel axles.

Dipstick See **oil-level indicator.**

Direct current Electric current that flows in one direction only.

Disk brake A brake in which brake shoes, on a viselike caliper, grip a revolving disk to stop it.

Distributor Any device that distributes. In the ignition system, the rotary switch that directs high-voltage surges to the engine cylinders in the proper sequence. See **ignition distributor.**

Distributor advance See **centrifugal advance, Ignition advance,** and **vacuum advance.**

DOHC A double-overhead-camshaft engine with two camshafts per cylinder head.

Drive line The driving connection between the transmission and the differential; made up of one or more drive shafts.

Drive shaft An assembly of one or two universal joints and slip joints connected to a heavy metal tube; used to transmit power from the transmission to the differential. Also called the propeller shaft.

Drum brake A brake in which curved brake shoes press against the inner circumference of a metal drum to produce the braking action.

Dual-brake system A brake system consisting of two separate hydraulic systems; usually, one operates the front brakes, and the other operates the rear brakes.

Dwell meter A precision electrical instrument used to measure the cam angle, or dwell, or number of degrees the distributor points are closed while the engine is running.

Dynamic balance The balance of an object when it is in motion (for example, the dynamic balance of a rotating wheel).

Dynamometer A device for measuring the power output, or brake horsepower, of an engine. An engine dynamometer measures the power output at the flywheel; a chassis dynamometer measures the power output at the drive wheels.

ECU See **electronic control unit.**

EGR system Abbreviation for exhaust-gas recirculation system.

Electric current The movement of electrons through a conductor such as a copper wire; measured in amperes.

Electric system In the automobile, the system that electrically cranks the engine for starting; furnishes high-voltage sparks to the engine cylinders to fire the compressed air-fuel charges; lights the lights; and powers the heater motor, radio, and other accessories. Consists, in part, of the starting motor, wiring, battery, alternator, regulator, ignition distributor, and ignition coil.

Electrolyte The mixture of sulfuric acid and water used in lead-acid storage batteries. The acid enters into chemical reaction with active material in the plates to produce voltage and current.

Electronic control unit A solid-state device that receives information from sensors and is programmed to operate various circuits and systems based on that information.

Electronic fuel-injection system A system that injects gasoline to a spark-ignition engine and includes an electronic control unit to time and meter the fuel flow.

Electronic ignition system A transistorized ignition system which does not have mechanical contact points in the distributor, but uses the distributor for distributing the secondary voltage to the spark plugs. Also called a solid-state ignition system.

Electronics Electrical assemblies, circuits, and systems that use electron devices such as transistors and diodes.

Emission control Any device or modification added onto or designed into a motor vehicle for the purpose of reducing air-polluting emissions.

Engine tuneup A procedure for inspecting, testing, and adjusting an engine, and replacing any worn parts, to restore the engine to its best performance.

Evaporative emission-control system A system which prevents the escape of gasoline vapors from the fuel tank or carburetor float bowl to the atmosphere while the engine is off. The vapors are stored in a canister or in the crankcase until the engine is started.

Evaporation control system A system which prevents the escape of gasoline vapors from the fuel tank or carburetor to the atmosphere while the engine is off. The vapors are stored in a charcoal canister or in the engine crankcase until the engine is started.

Exhaust emissions Pollutants emitted into the atmosphere through any opening downstream of the exhaust ports of an engine.

Exhaust-gas analyzer A device for sensing the amounts of air pollutants in the exhaust gas of a motor vehicle. The analyzers used in automotive shops check HC and CO; those used in testing laboratories can also check NO_x.

Exhaust-gas recirculation system An NO_x control system that recycles a small part of the inert exhaust gas back through the intake manifold at all throttle positions except idle and wide open, to lower the combustion temperature.

Exhaust manifold A device with several passages through which exhaust gases leave the engine combustion chambers and enter the exhaust piping system.

Exhaust stroke The piston stroke (from BDC to TDC) immediately following the power stroke, during which the exhaust valve opens so that the exhaust gases can escape from the cylinder to the exhaust manifold.

Exhaust valve The valve that opens during the exhaust stroke to allow burned gases to flow from the cylinder to the exhaust manifold.

Fan The bladed device on the front of the engine that rotates to draw cooling air through the radiator or around the engine cylinders; an air blower, such as the heater fan and the a/c blower.

Fast-idle cam A mechanism on the carburetor, connected to the automatic choke, that holds the throttle valve slightly open when the engine is cold; causes the engine to idle at a higher rpm as long as the choke is applied.

Filter A device through which air, gases, or liquids are passed to remove impurities.

Firing line The high-voltage vertical spike, or line, that appears on the oscilloscope pattern of the ignition-system secondary circuit. The firing line shows when the spark plug begins to fire as well as the voltage required to fire it.

Firing order The order in which the engine cylinders fire, or deliver their power strokes, beginning with No. 1 cylinder.

Fixed-caliper disk brake Disk brake using a caliper which is fixed in position and cannot move; the caliper usually has four pistons, two on each side of the disk.

Floating-caliper disk brake Disk brake using a caliper mounted through rubber bushings which permit the caliper to float, or move, when the brakes are applied; there is one large piston in the caliper.

Float level The float position at which the needle valve closes the fuel inlet to the float bowl in the carburetor to prevent further delivery of fuel.

Flooded Term used to indicate that the engine cylinders received "raw" or liquid gasoline, or an air-fuel mixture too rich to burn.

Flywheel A heavy metal wheel that is attached to the crankshaft and rotates with it; helps smooth out the power surges from the engine power strokes; also serves as part of the clutch and engine cranking system.

Flywheel ring gear A gear, fitted around the flywheel, that is engaged by teeth on the starting-motor drive to crank the engine.

Four-barrel carburetor A carburetor with four throttle valves. In effect, two two-barrel carburetors in a single assembly.

Four-stroke cycle The four piston strokes—intake compression, power, and exhaust—that make up the complete cycle of events in the four-stroke-cycle engine. Also called four-cycle and four-stroke.

Front geometry The angular relationship between the front wheels, wheel-attaching parts, and car frame. Includes camber, caster, steering-axis inclination, toe-in, and toe-out on turns.

Fuel Any combustible substance. In an automobile engine, the fuel (gasoline) is burned, and the heat of combustion expands the resulting gases, which force the pistons downward, causing the crankshaft to rotate.

Fuel filter A device located in the fuel line, ahead of the float bowl; removes dirt and other contaminants from fuel passing through.

Fuel-injection system A system which delivers fuel under pressure into the combustion chamber, or into the airflow just as it enters each individual cylinder. Replaces the conventional carburetor.

Fuel pump The electrical or mechanical device in the fuel system which forces fuel from the fuel tank to the carburetor.

Fuel system In an automobile, the system that delivers the combustible mixture of vaporized fuel and air to the engine cylinders. Consists of the fuel tank and lines, gauge, fuel pump, carburetor, and intake manifold.

Fuel tank The storage tank for fuel on the vehicle.

Fuse A device designed to open an electric circuit when the current is excessive, to protect equipment in the circuit. An open, or "blown," fuse must be replaced after the circuit problem is corrected.

Fusible link A type of fuse in which a special wire melts to open the circuit when the current is excessive. An open, or "blown," fusible link must be replaced after the circuit problem is corrected.

Gasket A layer of material, usually made of cork or metal or both, that is placed between two machined surfaces to provide a tight seal.

Gasket cement A liquid adhesive material, or sealer, used to install gaskets; in some applications, a layer of gasket cement is used as the gasket.

Gasoline A liquid blend of hydrocarbons, obtained from crude oil; used as the fuel in most automobile engines.

Gear ratio The number of revolutions of a driving gear required to turn a driven gear through one complete revolution. For a pair of gears, the ratio is found by dividing the number of teeth on the driven gear by the number of teeth on the driving gear.

Gears Mechanical devices that transmit power or turning effort from one shaft to another; gears contain teeth that mesh as the gears turn.

Gearshift A linkage-type mechanism by which the gears in an automobile transmission are engaged.

Glaze breaker A tool, rotated by an electric motor, used to remove the glaze from engine-cylinder walls.

HC Abbreviation for hydrocarbon.

Headlights Lights at the front of a vehicle; designed to illuminate the road ahead of the vehicle.

Heat-control valve In the engine, a thermostatically operated valve in the exhaust manifold; diverts heat to the intake manifold to warm it before the engine reaches normal operating temperature.

Heated-air system A system in which a thermostatically controlled air cleaner supplies hot air from a stove around the exhaust manifold to the carburetor during warmup; improves cold-engine operation.

High-Energy Ignition (HEI) system A General Motors electronic ignition system without contact points and with all ignition-system components contained in the distributor. Capable of producing 35,000 volts.

High-speed system In the carburetor, the system that supplies fuel to the engine at speeds above about 25 mph [40 km/h]. Also called the main-metering system.

High-voltage cables The secondary (or spark-plug) cables or wires that carry high voltage from the ignition coil to the spark plugs.

Horn An electrical noise-making device on a vehicle, used for signaling.

Horn relay A relay connected between the battery and the horns. When the horn button is pressed, the relay is energized; it then connects the horns to the battery.

Horsepower A measure of mechanical power, or the rate at which work is done. One horsepower equals 33,000 ft-lb (foot-pounds) [4554 kg-m (kilogram-meters)] of work per minute; it is the power necessary to raise 33,000 pounds a distance of one foot in 1 minute.

Hydraulic brakes A braking system that uses hydraulic pressure to force the brake shoes against the brake drums, or rotors, as the brake pedal is depressed.

Hydraulic clutch A clutch that is actuated by hydraulic pressure; used in heavy-duty equipment, and where the engine is some distance from the driver's compartment so it would be difficult to use mechanical linkages.

Hydraulic valve lifter A valve lifter that, by means of oil pressure, maintains zero valve clearance so that valve noise is reduced.

Hydrocarbon (HC) An organic compound containing only carbon and hydrogen, usually derived from fossil fuels such as petroleum, natural gas, and coal; an agent in the formation of photochemical smog. Gasoline is a blend of liquid hydrocarbons refined from crude oil.

Hydrogen (H) A colorless, odorless, highly flammable gas whose combustion produces water; the simplest and lightest element.

Hydrometer A device used to measure specific gravity. In automotive servicing, a device used to measure the specific gravity of battery electrolyte to determine the state of the battery charge; also a device used to measure the specific gravity of coolant to determine its freezing temperature.

IC See **internal combustion engine.**

Idle limiter A device that controls the maximum richness of the idle air-fuel mixture in the carburetor; also aids in preventing overly rich idle adjustments. Limiters are of two types: the external plastic-cap type, installed on the head of the idle-mixture adjustment screw, and the internal needle type, located in the idle passages of the carburetor.

Idle mixture The air-fuel mixture supplied to the engine during idling.

Idle-mixture screw The adjustment screw (on some carburetors) that can be turned in or out to lean out or enrich the idle mixture.

Idle-stop solenoid An electrically operated two-position plunger used to provide a predetermined throttle setting at idle.

Idle system In the carburetor, the passages through which fuel is fed when the engine is idling.

Ignition The action of the spark in starting the burning of the compressed air-fuel mixture in the combustion chamber.

Ignition advance The moving forward, in time, of the ignition spark relative to the piston position. TDC or one degree ATDC is considered advanced as compared with 2 degrees ATDC.

Ignition coil The ignition-system component that acts as a transformer to step up (increase) the battery voltage to many thousands of volts; the high-voltage surge from the coil is transmitted to the spark plug to ignite the compressed air-fuel mixture.

Ignition distributor The ignition-system component that closes and opens the primary circuit to the ignition coil at the proper times and distributes the resulting high-voltage surges from the ignition coil to the proper spark plugs.

Ignition switch The switch in the ignition system (usually operated with a key) that opens and closes the ignition-coil primary circuit. May also be used to open and close other vehicle electric circuits.

Ignition system In the automobile, the system that furnishes high-voltage sparks to the engine cylinders to fire the compressed air-fuel mixture. Consists of the battery, ignition coil, ignition distributor, ignition switch, wiring, and spark plugs.

Ignition timing The delivery of the spark from the coil to the spark plug at the proper time for the power stroke, relative to the piston position.

I-head engine An overhead-valve (OHV) engine; an engine with the valves in the cylinder head.

Inner tube See **tire tube.**

Intake manifold A device with several passages through which the air-fuel mixture flows from the carburetor to the ports in the cylinder head or cylinder block.

Intake stroke The piston stroke from TDC to BDC immediately following the exhaust stroke, during which the intake valve opens and the cylinder fills with air-fuel mixture from the intake manifold.

Internal-combustion (IC) engine An engine in which the fuel is burned inside the engine itself, rather than in a separate device (as in a steam engine).

Kingpin In older cars and trucks, the steel pin on which the steering knuckle pivots; attaches the steering knuckle to the knuckle support or axle.

Kingpin inclination Inward tilt of the kingpin from the vertical. See **steering-axis inclination.**

Knock A heavy metallic engine sound which varies with engine speed; usually caused by a loose or worn bearing; name also used for detonation, pinging, and spark knock. See **detonation.**

Leaded gasoline Gasoline to which small amounts of tetraethyl lead (pronounced *led*) are added to improve engine performance and reduce detonation.

Leaf spring A spring made up of a single flat steel plate, or several plates of graduated lengths assembled one on top of another; used on vehicles to absorb road shocks by bending, or flexing in the middle.

Lean mixture An air-fuel mixture that has a relatively high proportion of air and a relatively low proportion of fuel. An air-fuel ratio of 16:1 is a lean mixture, compared with an air-fuel ratio of 13:1.

Limited-slip differential A differential designed so that when one wheel is slipping, a major portion of the drive torque is supplied to the wheel with the better traction; also called a nonslip differential.

Linkage-type power steering A type of power steering in which the power-steering units (power cylinder and valve) are part of the steering linkage; frequently a bolt-on type of system.

Lubricating system The system in the engine that supplies engine parts with lubricating oil to prevent contact between any two moving metal surfaces.

Magnetic switch A switch with a winding (a coil of wire); when the winding is energized, the switch is moved to open or close a circuit.

Main bearings In the engine, the bearings that support the crankshaft.

Manifold A device with several inlet or outlet passageways through which a gas or liquid is gathered or distributed. See **exhaust manifold** and **intake manifold.**

Manifold vacuum The vacuum in the intake manifold that develops as a result of the vacuum in the cylinders on their intake strokes.

Master cylinder The liquid-filled cylinder in the hydraulic braking system or clutch where hydraulic pressure is developed when the driver depresses a foot pedal.

MISAR Microprocessed Sensing and Automatic Regulation; a high-energy or ignition system in which the centrifugal and vacuum advance units are replaced with sensors and an electronic control unit.

Misfire In the engine, a failure to ignite the air-fuel mixture in one or more cylinders. This condition may be intermittent or continuous in one or more cylinders.

Monolithic timing Making accurate spark-timing adjustment with an electronic timing device which can be used with the engine running. Also called magnetic timing.

mph Abbreviation for miles per hour, a measure of speed.

Muffler In the engine-exhaust system, a device through which gases must pass and which reduces the exhaust noise. In an air-conditioning system, a device to minimize pumping sounds from the compressor.

Multiple-viscosity oil An engine oil which has a low viscosity when cold (for easier cranking) and a higher viscosity when hot (to provide adequate engine lubrication).

Neutral-start switch A switch wired into the ignition switch to prevent engine cranking unless the transmission shift lever is in neutral.

NO_x control system Any device or system used to reduce the amount of NO_x produced by an engine.

Octane rating A measure of the antiknock properties of a gasoline. The higher the octane rating, the more resistant the gasoline is to spark knock or detonation.

Oil A liquid lubricant; made from crude oil (or other base material) and used to produce lubrication between moving parts. In a diesel engine, oil is used for fuel.

Oil cooler A small radiator which lowers the temperature of oil flowing through it.

Oil dilution Thinning of oil in the crankcase; caused by liquid gasoline leaking past the piston rings from the combustion chamber.

Oil filter A filter which removes impurities from crankcase oil passing through it.

Oil-level indicator The indicator that is removed and inspected to check the level of oil in the crankcase of an engine or compressor. Usually called the dipstick.

Oil pan The detachable lower part of the engine, made of sheet metal, which encloses the crankcase and acts as an oil reservoir.

Oil-pressure indicator A gauge that indicates (to the driver) the oil pressure in the engine lubricating system.

Oil pump In the lubricating system, the device that forces oil from the oil pan to the moving engine parts.

Oil pumping Leakage of oil past the piston rings and into the combustion chamber, usually as a result of defective rings or worn cylinder walls.

Oil ring The lower ring or rings on a piston; designed to prevent excessive amounts of oil from working up the cylinder walls and into the combustion chamber. Also called an oil-control ring.

Oil seal A seal placed around a rotating shaft or other moving part to prevent leakage of oil.

Oscilloscope A high-speed voltmeter which visually displays voltage variations on a television-type picture tube. Used to check engine ignition systems; also used to check charging systems and electronic fuel-injection systems.

Overhead-camshaft (OHC) engine An engine in which the camshaft is mounted over the cylinder head, instead of inside the cylinder block.

Overhead-valve (OHV) engine An engine in which the valves are mounted in the cylinder head above the combustion chamber, instead of in the cylinder block. In this type of engine, the camshaft is usually mounted in the cylinder block, and the valves are actuated by pushrods.

Parade pattern An oscilloscope pattern showing the ignition voltages on one line, from left to right across the scope screen in engine firing order.

PCV Abbreviation for positive crankcase ventilation.

PCV valve The valve that controls the flow of crankcase vapors in accordance with ventilation requirements for different engine speeds and loads.

Ping Engine spark knock that occurs during acceleration. Usually associated with medium to heavy throttle, acceleration, or lugging at relatively low speeds, especially with a manual transmission. However, it may occur in higher-speed ranges under heavy-load conditions. Caused by too much advance of ignition timing or low-octane fuel.

Piston A movable part, fitted to a cylinder, which can receive or transmit motion as a result of pressure changes in a fluid. In the engine, the cylindrical part that moves up and down within a cylinder as the crankshaft rotates.

Piston rings Rings fitted into grooves in the piston. There are two types: compression rings for sealing the compression pressure in the combustion chamber and oil rings to scrape excessive oil off the cylinder wall. See **oil ring.**

Pollution Any gas or substance, in the air, which makes the air less fit to breathe. Also, noise pollution is the name applied to excessive noise from machinery or vehicles.

Positive crankcase ventilation (PCV) A crankcase-ventilation system; uses intake-manifold vacuum to return the crankcase vapors and blow-by gases from the crankcase to the intake manifold to be burned, thereby preventing their escape into the atmosphere.

Power steering A steering system that uses hydraulic pressure (from a pump) to multiply the driver's steering effort.

Power stroke The piston stroke from TDC to BDC immediately following the compression stroke, during which both valves are closed and the air-fuel mixture burns, expands, and forces the piston down to transmit power to the crankshaft.

Power train The mechanisms that carry the rotary motion developed in the engine to the car wheels; includes the clutch, transmission, drive shaft, differential, and axles.

Precombustion chamber In some diesel engines, a separate small combustion chamber into which the fuel is injected and where combustion begins.

Preignition Ignition of the air-fuel mixture in the combustion chamber by some unwanted means, before the ignition spark occurs at the spark plug.

Pressure-relief valve A valve in the oil line that opens to relieve excessive pressure.

Pump A device that transfers gas or liquid from one place to another.

Pushrod In the I-head engine, the rod between the valve lifter and the rocker arm; transmits cam-lobe lift.

Quick charger A battery charger that produces a high charging current which charges, or boosts, a battery in a short time.

Rack-and-pinion steering gear A steering gear in which a pinion on the end of the steering shaft meshes with a rack on the major cross member of the steering linkage.

Radial-ply tire A tire in which the plies are placed radially, or perpendicular to the rim, with a circumferential belt on top of them. The rubber tread is vulcanized on top of the belt and plies.

Radiator In the cooling system, the device that removes heat from coolant passing through it; takes hot coolant from the engine and returns the coolant to the engine at a lower temperature.

Rectifier A device which changes alternating current to direct current; in the alternator, a diode.

Refrigerant A substance used to transfer heat in an air conditioner, through a cycle of evaporation and condensation.

Relay An electrical device that opens or closes a circuit or circuits in response to a voltage signal.

Reluctor In an electronic ignition system, the metal rotor (with a series of tips) which replaces the conventional distributor cam. Also called "trigger wheel" and "armature."

Rich mixture An air-fuel mixture that has a relatively high proportion of fuel and a relatively low proportion of air. An air-fuel ratio of 13:1 indicates a rich mixture, compared with an air-fuel ratio of 16:1.

Ring gear A large gear carried by the differential case; meshes with and is driven by the drive pinion.

Ring ridge The ridge formed at the top of a cylinder as the cylinder wall below is worn away by piston-ring movement.

Rocker arm In an I-head engine, a device that rocks on a shaft (or pivots on a stud) as the cam moves the pushrod, causing a valve to open.

Rod bearing In an engine, the bearing in the connecting rod in which a crankpin of the crankshaft rotates. Also called a connecting-rod bearing.

Rotor A revolving part of a machine, such as an alternator rotor, disk-brake rotor, distributor rotor, or Wankel-engine rotor.

rpm Abbreviation for revolutions per minute, a measure of rotational speed.

Run-on See **dieseling.**

SAE Abbreviation for Society of Automotive Engineers. Used to indicate a grade or weight of oil measured according to Society of Automotive Engineers standards.

Schrader valve A spring-loaded valve through which a connection can be made to a refrigeration system; also used in tires.

Scuffing A type of wear in which there is a transfer of material between parts moving against each other; shows up as pits or grooves in the mating surfaces.

Secondary circuit The high-voltage circuit of the ignition system; consists of the coil, rotor, distributor cap, spark-plug cables, and spark plugs.

Self-adjuster A mechanism used on drum brakes; compensates for shoe wear by automatically keeping the shoe adjusted close to the drum.

Servo A device in a hydraulic system that converts hydraulic pressure to mechanical movement. Consists of a piston which moves in a cylinder as hydraulic pressure acts on it.

Shim A slotted strip of metal used as a spacer to adjust the front-end alignment on many cars; also used to make small corrections in the position of body sheet metal and other parts.

Shimmy Rapid oscillation. In wheel shimmy, for example, the front wheel turns in and out alternately and rapidly; this causes the front end of the car to oscillate, or shimmy.

Shock absorber A device placed at each vehicle wheel to regulate spring rebound and compression.

Single-overhead-camshaft (SOHC) engine An engine in which a single camshaft is mounted over each cylinder head, instead of inside the cylinder block.

Slip joint In the power train, a variable-length connection that permits the drive shaft to change its effective length.

Sludge An accumulation of water, dirt, and oil in the oil pan; sludge is very viscous and tends to reduce lubrication.

Smog A term coined from the words "smoke" and "fog." First applied to the foglike layer that hangs in the air under the certain atmospheric conditions; now generally used to describe any condition of dirty air and/or fumes or smoke. Smog is compounded from smoke, moisture, and numerous chemicals which are produced by combustion.

Smoke Small gasborne or airborne particles, exclusive of water vapor, that result from combustion; such particles emitted by an engine into the atmosphere in sufficient quantity to be observable.

SOHC See **single-overhead-camshaft engine.**

Solenoid An electromechanical device which, when connected to an electrical source such as a battery, produces a mechanical movement. This movement can be used to control a valve or to produce other movements.

Solenoid relay A relay that connects a solenoid to a current source when its contacts close; specifically, the starting-motor solenoid relay.

Solenoid switch A switch that is opened and closed electromagnetically, by the movement of a solenoid core. Usually, the core also causes a mechanical action, such as the movement of a drive pinion into mesh with flywheel teeth for cranking.

Spark knock See **detonation.**

Spark plug A device that screws into the cylinder head of an engine; provides a spark to ignite the compressed air-fuel mixture in the combustion chamber.

Spark-plug heat range The distance heat must travel from the center electrode to reach the outer shell of the spark plug and enter the cylinder head.

Specific gravity The weight per unit volume of a substance as compared with the weight per unit volume of water.

Speedometer An instrument that indicates vehicle speed; usually driven from the transmission.

Stabilizer bar An interconnecting shaft between the two lower suspension arms; reduces body roll on turns.

Starting motor The electric motor that cranks the engine, or turns the crankshaft, for starting.

Starting-motor drive The drive mechanism and gear on the end of the starting-motor armature shaft; used to couple the starting motor to, and disengage it from, the flywheel ring-gear teeth.

Steering gear That part of the steering system that is located at the lower end of the steering shaft; carries the rotary motion of the steering wheel to the car wheels for steering.

Steering system The mechanism that enables the driver to turn the wheels for changing the direction of vehicle movement.

Stoplights Lights at the rear of a vehicle which indicate that the driver is applying the brakes.

Storage battery A device that changes chemical energy into electric energy; that part of the electric system which acts as a reservoir for electric energy, storing it in chemical form.

Supercharger In the intake system of the engine, a pump that pressurizes the ingoing air-fuel mixture. This increases the amount of mixture delivered to the cylinders, which increases the engine output. If the supercharger is driven by the engine-exhaust gas, it is called a turbocharger.

Synchronizer A device in the transmission that synchronizes gears about to be meshed, so that no gear clash will occur.

Tachometer A device for measuring engine speed, or rpm.

Taillights Steady-burning low-intensity lights used on the rear of a vehicle.

Tetraethyl lead A chemical which, when added to engine fuel, increases its octane rating, or reduces its knocking tendency. Also called ethyl.

Thermostat A device for the automatic regulation of temperature; usually contains a temperature-sensitive element that expands or contracts to open or close off the flow of air, a gas, or a liquid.

Thermostatically controlled air cleaner An air cleaner in which a thermostat controls the preheating of intake air.

Throttle A disk valve in the carburetor base that pivots in response to accelerator-pedal position; allows the driver to regulate the volume of air-fuel mixture entering the intake manifold, thereby controlling the engine speed. Also called the throttle plate or throttle valve.

Timing In an engine, delivery of the ignition spark or operation of the valves (in relation to the piston position) for the power stroke. See **ignition timing** and **valve timing.**

Timing light A light that can be connected to the ignition system to flash each time the No. 1 spark plug fires; used for adjusting the timing of the ignition spark.

Tire The casing-and-tread assembly (with or without a tube) that is mounted on a car wheel to provide pneumatically cushioned contact and traction with the road.

Torque converter In an automatic transmission, a fluid coupling which incorporates a stator to permit a torque increase.

Torsion-bar spring A long, straight bar that is fastened to the vehicle frame at one end and to a suspension part at the other. Spring action is produced by a twisting of the bar.

Transistor An electronic device that can be used as an electric switch; used to replace the contact points in electronic ignition systems.

Transmission An assembly of gears that provides the different gear ratios, as well as neutral and reverse, through which engine power is transmitted to the differential to rotate the drive wheels.

Transmission-oil cooler A small radiator, either mounted separately or as part of the engine radiator, which cools the transmission fluid.

Tread The part of the tire that contacts the road. It is the thickest part of the tire and is cut with grooves to provide traction for driving and stopping.

Turbocharger A supercharger driven by the engine-exhaust gas.

Universal joint In the power train, a jointed connection in the drive shaft that permits the driving angle to change.

Unleaded gasoline Gasoline to which no lead compounds have been intentionally added. Gasoline that contains 0.05 g (gram) or less of lead per gallon; required by law to be used in 1975 and later vehicles equipped with catalytic converters.

Upper beam A headlight beam intended primarily for distant illumination; not for use when other vehicles are being met or followed.

Vacuum advance The advancing (or retarding) of ignition timing by changes in intake-manifold vacuum, reflecting throttle opening and engine load. Also, a mechanism on the ignition distributor that uses intake-manifold vacuum to advance the timing of the spark to the spark plugs.

Vacuum switch A switch that closes or opens its contacts in response to changing vacuum conditions.

Valve A device that can be opened or closed to allow or stop the flow of a liquid or gas.

Valve guide A cylindrical part in the cylinder block or head in which a valve is assembled and in which it moves up and down.

Valve lifter A cylindrical part of the engine which rests on a cam of the camshaft and is lifted, by cam action, so that the valve is opened. Also called a lifter, tappet, valve tappet, or cam follower.

Valve seat The surface against which a valve comes to rest to provide a seal against leaking.

Valve-seat inserts Metal rings inserted in cylinder heads to act as valve seats (usually for exhaust valves). They are made of special metals able to withstand very high temperatures.

Valve-stem seal A device placed on or around the valve stem to reduce the amount of oil that can get on the stem and then work its way down into the combustion chamber.

Valve timing The timing of the opening and closing of the valves in relation to the piston position.

Valve train The valve-operating mechanism of an engine; includes all components from the camshaft to the valve.

Vapor lock A condition in the fuel system in which gasoline vaporizes in the fuel line or fuel pump; bubbles of gasoline vapor restrict or prevent fuel delivery to the carburetor.

Vapor-recovery system An evaporative emission-control system that recovers gasoline vapor escaping from the fuel tank and carburetor float bowl. See **evaporation control system.**

Variable-venturi carburetor A carburetor in which the size of the venturi changes according to engine speed and load.

Vibration damper A device attached to the crankshaft of an engine to oppose crankshaft torsional vibration (the twist-untwist actions of the crankshaft caused by the cylinder firing impulses). Also called a harmonic balancer.

Viscosity rating An indicator of the viscosity of engine oil. There are separate ratings for winter driving and for summer driving. The winter grades are SAE5W, SAE10W, and SAE20W. The summer grades are SAE20, SAE30, SAE40, and SAE50. Many oils have multiple viscosity ratings, as, for example, SAE20W–30.

Voltage regulator A device that prevents excessive alternator or generator voltage by alternately inserting and removing a resistance in the field circuit.

Voltmeter A device for measuring the

potential difference (voltage) between two points, such as the terminals of a battery or alternator or two points in an electric circuit.

VV carburetor See **variable-venturi carburetor.**

Wankel engine A rotary engine in which a three-lobe rotor turns eccentrically in an oval chamber to produce power.

Water jackets The spaces between the inner and outer shells of the cylinder block or head, through which coolant circulates.

Water pump In the cooling system, the device that circulates coolant between the engine water jackets and the radiator.

Wheel tramp Tendency for a wheel to move up and down so it repeatedly bears down hard, or "tramps," on the road. Sometimes called high-speed shimmy.

Wiring harness A group of individually insulated wires, wrapped together to form a neat, easily installed bundle.

INDEX